Marine Ecological Processes

Editor-in-Chief

John H. Steele
Marine Policy Center, Woods Hole Oceanographic Institution, Woods Hole, Massachusetts, USA

Editors

Steve A. Thorpe

National Oceanography Centre, University of Southampton, Southampton, UK
and
School of Ocean Sciences, Bangor University, Menai Bridge, Anglesey, UK

Karl K. Turekian
Yale University, Department of Geology and Geophysics, New Haven, Connecticut, USA

Subject Area Volumes from the Second Edition

Climate & Oceans edited by Karl K. Turekian
Elements of Physical Oceanography edited by Steve A. Thorpe
Marine Biology edited by John H. Steele
Marine Chemistry & Geochemistry edited by Karl K. Turekian
Marine Ecological Processes edited by John H. Steele
Marine Geology & Geophysics edited by Karl K. Turekian
Marine Policy & Economics guest edited by Porter Hoagland, Marine Policy Center, Woods Hole Oceanographic Institution, Woods Hole, Massachusetts
Measurement Techniques, Sensors & Platforms edited by Steve A. Thorpe
Ocean Currents edited by Steve A. Thorpe
The Coastal Ocean edited by Karl K. Turekian
The Upper Ocean edited by Steve A. Thorpe

MARINE ECOLOGICAL PROCESSES

A DERIVATIVE OF ENCYCLOPEDIA OF OCEAN SCIENCES, 2ND EDITION

Editor

JOHN H. STEELE

AMSTERDAM • BOSTON • HEIDELBERG • LONDON • NEW YORK • OXFORD
PARIS • SAN DIEGO • SAN FRANCISCO • SINGAPORE • SYDNEY • TOKYO
Academic Press is an imprint of Elsevier

Academic Press is an imprint of Elsevier
32 Jamestown Road, London NW1 7BY, UK
30 Corporate Drive, Suite 400, Burlington, MA 01803, USA
525 B Street, Suite 1900, San Diego, CA 92101-4495, USA

Material in the work originally appeared in *Encyclopedia of Ocean Sciences* (Elsevier Ltd., 2001) and *Encyclopedia of Ocean Sciences*, 2nd Edition (Elsevier Ltd., 2009), edited by John H. Steele, Steve A. Thorpe and Karl K. Turekian.

British Library Cataloguing in Publication Data
A catalogue record for this book is available from the British Library

Library of Congress Cataloging-in-Publication Data
A catalog record for this book is available from the Library of Congress

ISBN: 978-0-08-096488-1

For information on all Academic Press publications
visit our website at www.elsevierdirect.com

CONTENTS

CLIMATE IMPACTS

FISHERIES

MARICULTURE

METHODS OF ANALYSIS

ECOLOGICAL MODELS

APPENDIX

METHODS OF ANALYSIS

ECOLOGICAL MODELS

APPENDIX

INDEX

MARINE ECOLOGICAL PROCESSES: INTRODUCTION

This is the second of two volumes in this series covering general marine biological and ecological topics. The separation between "biology" and "ecology" is arbitrary since taxonomical and physiological features are determined by environmental patterns and, correspondingly, ecological processes depend on the species mix within communities. In this volume the focus is on ecosystem structure, ecological processes, and the forcing of these systems by external stresses, principally climate and fisheries related.

Given their fluid nature, marine ecosystems are difficult to define in terms of geographical boundaries (see Longhurst, 1998). Thus they are often described by physical processes such as upwelling or ocean gyres. Larger scale systems on the continental shelf may be associated with their fisheries. Many of the smaller ecosystems are determined by special habitats such as coral reefs or hydrothermal vents.

These different kinds of functions can prescribe the distinguishing features of such systems, but their dynamics are a result of their internal food web structure that involves a complicated set of prey-predator and competitive interactions. Study of these food web processes is the dominant impetus for marine ecological research.

Understanding the complex responses of food webs to environmental changes provides a major theme in this research. Some of the most perplexing yet important features are termed regime shifts. These relatively rapid switches in the whole structure of a web have been observed in different oceans in response to physical changes in the environment. These shifts are one example of the more general concern with ecosystem reactions to the increasing stresses imposed by society. The role of the oceans in determining longer period climatic trends is described in other volumes. Outlined here are the responses of major food web components — plankton, marine mammals, seabirds and fish stocks — to climatic change.

The other dominant concern is over-fishing, especially of fish communities on the world's continental shelves. It is now recognized that such fisheries not only affect the targeted fish stocks but also alter the rest of their ecosystem. The articles collected here describe the main types of fisheries and their impacts. Questions relating to management are considered in another volume (*Marine Policy and Economics*).

Mariculture is now the one method of harvesting food from the sea that is still increasing. Salmonid farming demonstrates the economic benefits but also the ecological hazards in terms not only of diseases but also in relation to those ecosystems that must be fished to provide food for the salmonids. Articles in this volume describe the great variety of products that are now farmed. A separate volume (*Marine Policy and Economics*) deals with the economic aspects.

A major limitation on marine research is caused by our inability to sample the oceans adequately. Marine ecosystems, especially, are seriously under-sampled compared with those on land. This is the greatest barrier to improving our understanding of life in the sea. Five articles describe methods for investigating phytoplankton, zooplankton, benthos, fish and marine mammals. Another problem in the open ocean is our inability to conduct controlled experiments at the scale of pelagic ecosystems. As a compromise large containers, called mesocosms, are used to capture parts of planktonic and benthic communities. Two articles discuss the benefits and limitations of this approach.

These problems with sampling and experimenting in the ocean have led to major initiatives to construct numerical models of all aspects of ocean processes. Articles on physical and chemical modeling are in other volumes. Gathered here are a set of articles that describe numerical simulations of ecological processes.

John H. Steele
Editor

REFERENCES

Longhurst, A.R. 1998. *Ecological Geography of the Sea*, 398pp. Academic Press.

ECOSYSTEM STRUCTURE

LARGE MARINE ECOSYSTEMS

K. Sherman, Narragansett Laboratory, Narragansett, RI, USA

Introduction

Coastal waters around the margins of the ocean basins are in a degraded condition. With the exception of Antarctica, they are being degraded from habitat alteration, eutrophication, toxic pollution, aerosol contaminants, emerging diseases, and overfishing. It has also been recently argued by Pauly and his colleagues that the average levels of global primary productivity are limiting the carrying capacity of coastal ocean waters for supporting traditional fish and fisheries and that any further large-scale increases in yields from unmanaged fisheries are likely to be at the lower trophic levels in the marine food web and likely to disrupt marine ecosystem structure.

Large Marine Ecosystems

Approximately 95% of the world's annual fish catches are produced within the geographic boundaries of 50 large marine ecosystems (LMEs) (**Figure 1A**). The LMEs are regions of ocean space encompassing coastal areas from river basins and estuaries out to the seaward boundary of continental shelves, and the outer margins of coastal currents. They are relatively large regions, on the order of 200 000 km^2 or greater, characterized by distinct bathymetry, hydrography, productivity, and trophically dependent populations. The close linkage between global ocean areas of highest primary productivity and the locations of the large marine ecosystems is shown in **Figure 1B**. Primary productivity at the base of marine food webs is a critical factor in the determination of fishery yields. Since the 1960s through the 1990s, significant changes have occurred within the LMEs, attributed in part to the affects of excessive fishing effort on the structure of food webs in LMEs.

Food Webs and Large Marine Ecosystems

Since 1984, a series of LME conferences, workshops, and symposia have been held during the annual meeting of the American Association for the Advancement of Science (AAAS). In the subsequent intervening 15 years, 33 case studies of LMEs were prepared, peer-reviewed, and published (see Further Reading). From the perspective of actual and potential fish yields of the LMEs an 'ECOPATH'-type trophic model, based on the use of a static system of linear equations for different species in the food web, has been developed by Polovina, Pauly and Christensen (eqn [1]).

$$P_i = \mathrm{Ex}_i + \sum B_i(Q/B_i)(\mathrm{DC}_{ji}) + B_i(P/B) - (IEE_i) \quad [1]$$

P_i is the production during any normal period (usually one year) of group i; Ex_i represents the exports (fishery catches and emigration) of i; $\sum_i$ represents the summation over all predators of I; B_j and B_i are the biomasses of the predator J and group I, respectively; Q/B_j is the relative food consumption of j; DC_{ji} the fraction that i constitutes of the diet of B_i is the biomass of i and $(\mathrm{I} - \mathrm{EE_i})$ is the other mortality of I, that is the fraction of i's production that is not consumed within or exported from the system under consideration. A practical consideration of food web dynamics in LMEs is the effect that changes in the structure of marine food webs could have on the long-term sustainability of fish species biomass yields.

Biomass Yields and Food Webs

South China Sea Large Marine Ecosystems

An example of the use of fisheries yield data in constructing estimates of combined prey consumption by trophic levels is depicted in **Figure 2** for shallow waters of the South China Sea (SCS) LME. The trophic transfers up the food web from phytoplankton to apex predators is shown in **Figure 3** for open-ocean areas of the SCS. The differences in fish/fish predation is approximately 50% of the fish production in the shallow-water subsystem and increases to 95% in the open-ocean subsystem.

Application of the ECOPATH model to the SCS LME by Pauly and Christensen produced an initial outcome of an additional 5.8 Mt annually. This is a rate that is nearly double the average annual catch reported for the SCS up through 1993, indicating some flexibility for increasing catches from the ecosystem, but not fully realizing its potential because of technical difficulties in fishing methodologies.

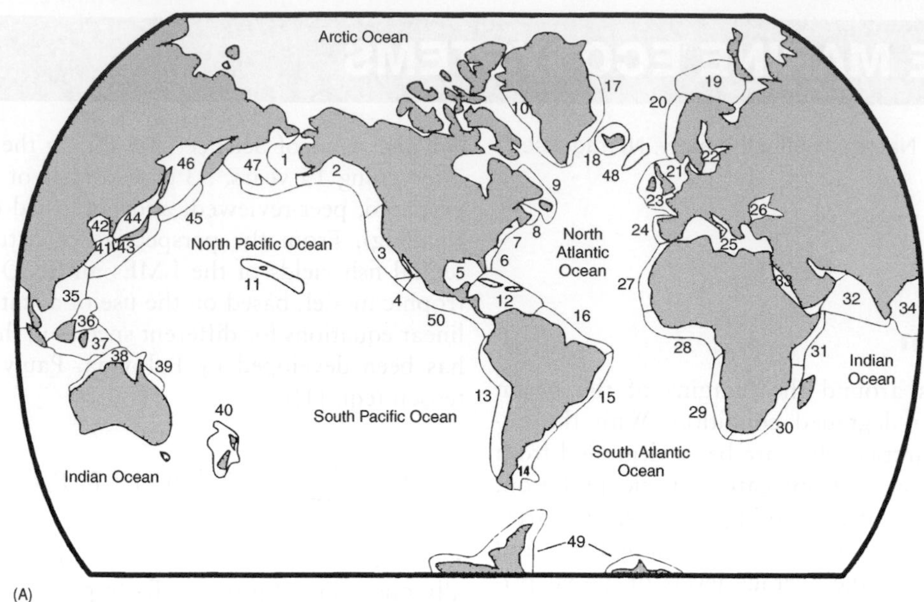

1. Eastern Bering Sea
2. Gulf of Alaska
3. California Current
4. Gulf of California
5. Gulf of Mexico
6. South-east US Continental Shelf
7. North-east US Continental Shelf
8. Scotian Shelf
9. Newfoundland Shelf
10. West Greenland Shelf
11. Insular Pacific-Hawaiian
12. Caribbean Sea
13. Humboldt Current
14. Patagonian Shelf
15. Brazil Current
16. North-east Brazil Shelf
17. East Greenland Shelf
18. Iceland Shelf
19. Barents Sea
20. Norwegian Shelf
21. North Sea
22. Baltic Sea
23. Celtic-Biscay Shelf
24. Iberian Coastal
25. Mediterranean Sea
26. Black Sea
27. Canary Current
28. Gulf of Guinea
29. Benguela Current
30. Agulhas Current
31. Somali Coastal Current
32. Arabian Sea
33. Red Sea
34. Bay of Bengal
35. South China Sea
36. Sulu-Celebes Seas
37. Indonesian Seas
38. Northern Australian Shelf
39. Great Barrier Reef
40. New Zealand Shelf
41. East China Sea
42. Yellow Sea
43. Kuroshio Current
44. Sea of Japan
45. Oyashio Current
46. Sea of Okhotsk
47. West Bering Sea
48. Faroe Plateau
49. Antarctic
50. Pacific Central American Coastal

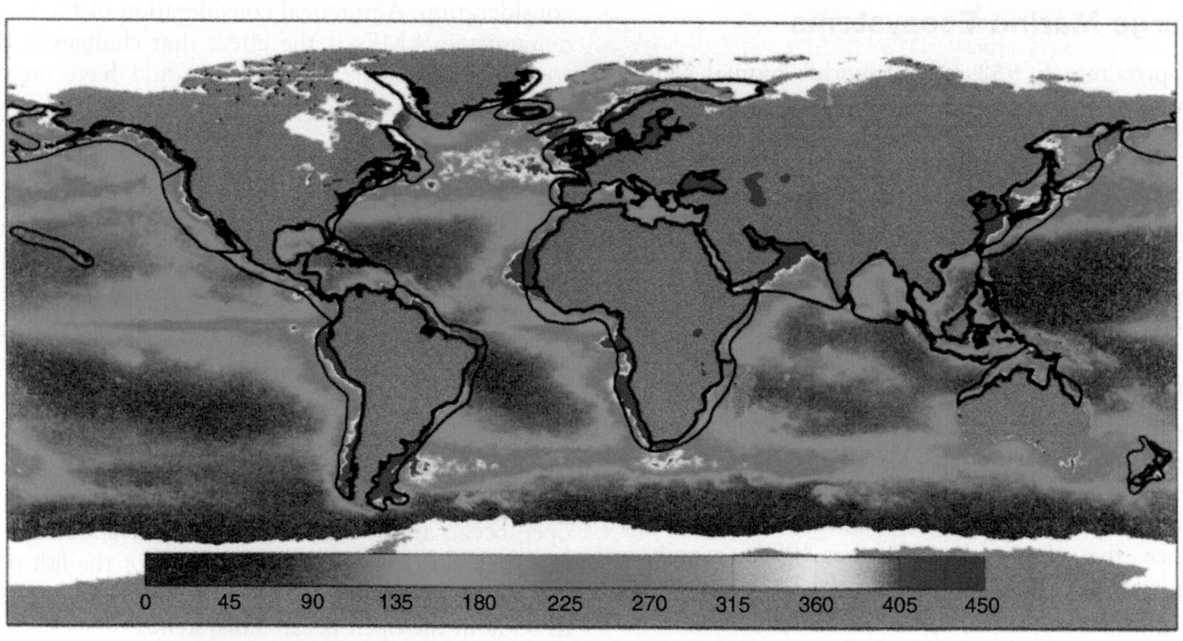

(B)

Figure 1 Boundaries of 50 large marine ecosystems (and) (B) SeaWiFS chlorophyll and outlines of LME boundaries.

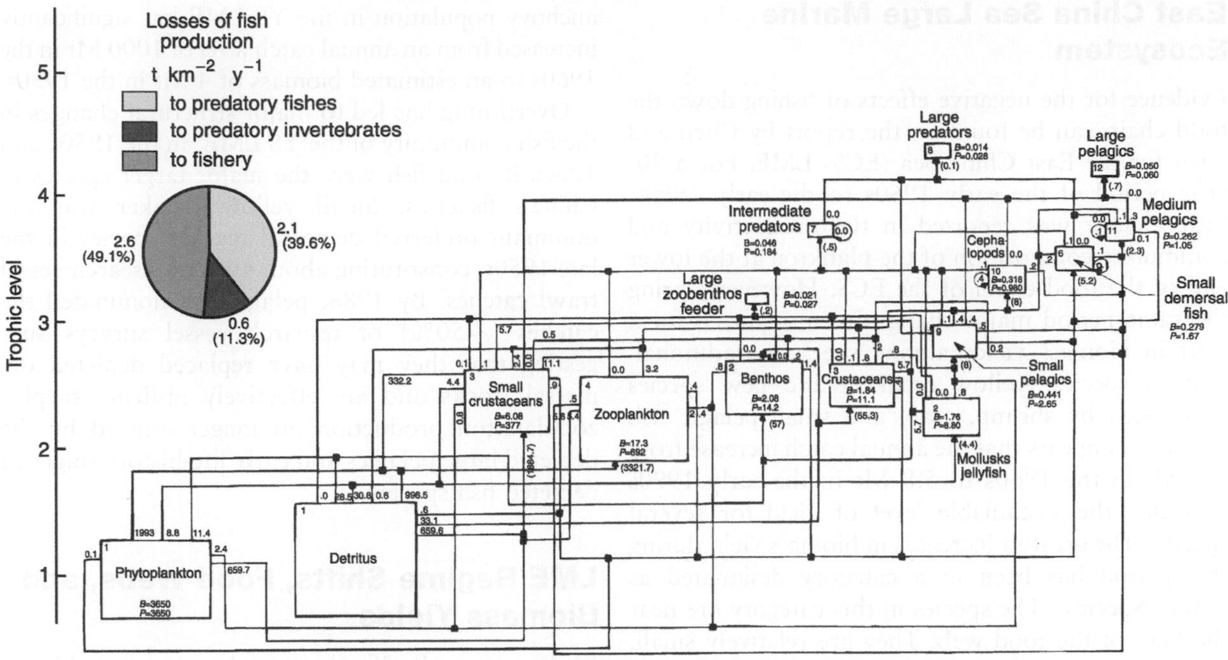

Figure 2 South China Sea shallow-water food web based on the ECOPATH model. (From Pauly and Christensen (1993).)

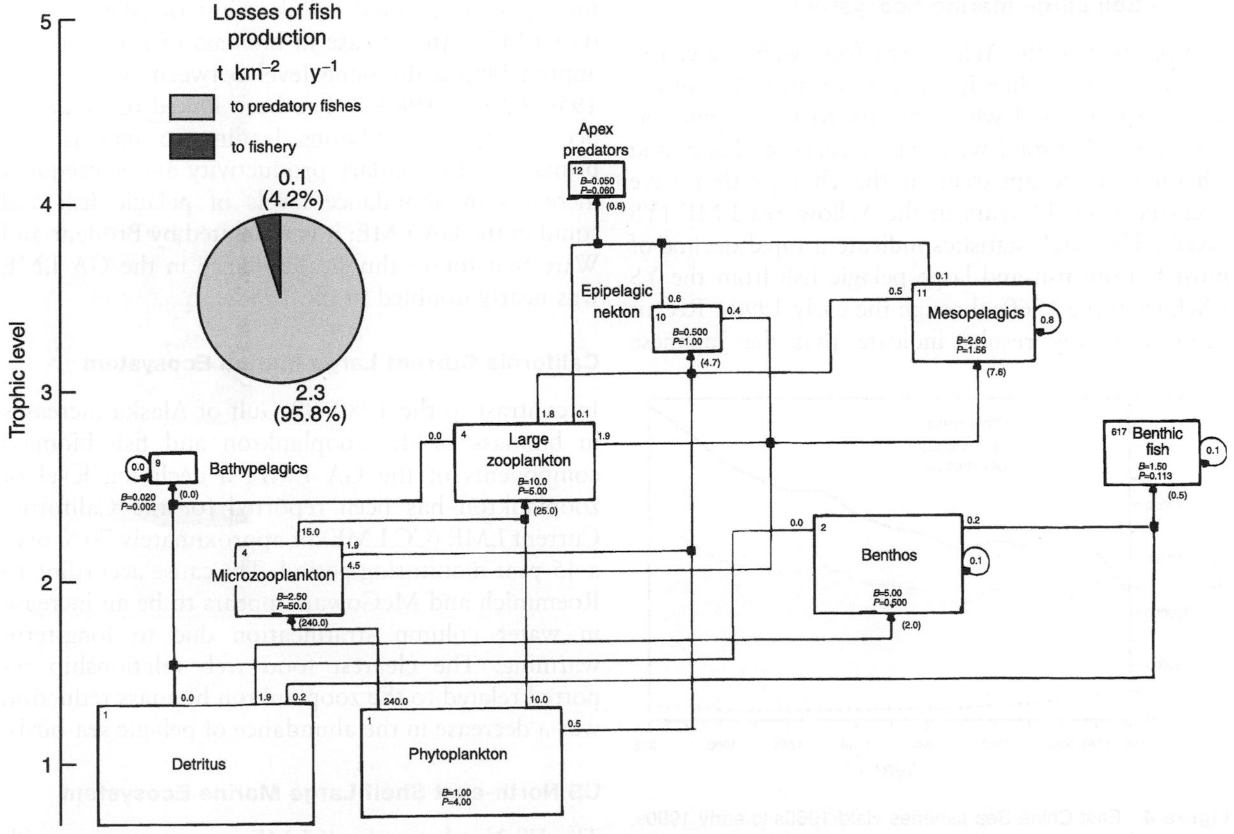

Figure 3 South China Sea open-ocean food web. (From Pauly and Christensen (1993).)

East China Sea Large Marine Ecosystem

Evidence for the negative effects of fishing down the food chain can be found in the report by Chen and Shen for the East China Sea (ECS) LME. For a 30-year period of the early 1960s to the early 1990s, little change was reported in the productivity and community composition of the plankton at the lower end of the food chain of the ECS. However, during the same period major changes were reported for a shift in biomass yields among the 'old traditional' bottom species (yellow croaker) and new species dominated by shrimp, crab, and small pelagic fish species. It appears that the annual catch increase from 0.9 Mt in the 1960s to 5.8 Mt in the early 1990s exceeded the sustainable level of yield for several species. The greatest increases in biomass yield during this period has been in a category designated as 'Other Species.' The species in this category are near the base of the food web. They are relatively small, pelagic, and fast growing, and are not used for human consumption but are used for feeding 'cultured fish or poultry' (**Figure 4**). Collectively, the catches of 'Other Species' provide additional evidence of the effects of 'fishing down the food web.'

Yellow Sea Large Marine Ecosystem

A projection of the Yellow Sea food web is given in **Figure 5**. The decline in the east Asian LMEs of demersal species and what appears to be 'trophic-forcing' down the food web hypothesized by Pauly and Christensen are apparent in the changes that have occurred over 30 years in the Yellow Sea LME (YS LME). The catch statistics indicate a rapid decline of most bottom fish and large pelagic fish from the YS LME from the 1960s through the early 1990s. Recent acoustic survey results indicate that the Japanese anchovy population in the YS LME has significantly increased from an annual catch level of 1000 Mt in the 1960s to an estimated biomass of 4 Mt in the 1990s.

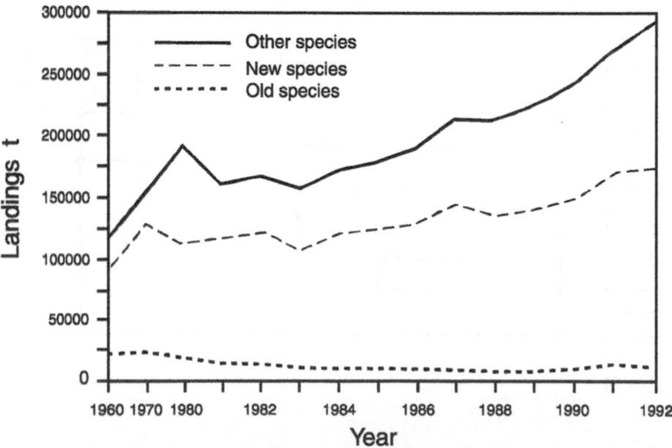

Figure 4 East China Sea fisheries yield 1960s to early 1990s, showing increased annual catches of 'Other Species' used mostly for fish and poultry food. (From Chen and Shen (1999).)

Overfishing has led to major structural changes in the fish community of the YS LME. In the 1950s and 1960s bottom fish were the major target species in China's fisheries. Small yellow croaker was the dominant preferred demersal market species in the late 1950s, constituting about 40% of research vessel trawl catches. By 1986, pelagic fish dominated the catches (~50%) of research vessel surveys suggesting that they may have replaced depleted demersal stocks and are effectively utilizing surplus zooplankton production no longer utilized by the depleted large pelagics and early life-history stages of depleted fish species.

LME Regime Shifts, Food Webs, and Biomass Yields

In the eastern Pacific, large-scale oceanographic regime shifts have been a major cause of changes in food web structure and biomass yields of LMEs.

Gulf of Alaska Large Marine Ecosystem

Evidence of the food web effects from oceanographic forcing was reported for the Gulf of Alaska LME (GA LME). An increase in biomass of zooplankton, approaching a doubling level between two periods 1956–62 and 1980–89 has been linked to favorable oceanographic conditions leading to increases in primary and secondary productivity and subsequent increases in abundance levels of pelagic fish and squid in the GA LME; it is estimated by Brodeur and Ware that total salmon abundance in the GA LME was nearly doubled in the 1980s.

California Current Large Marine Ecosystem

In contrast to the 1980–89 Gulf of Alaska increases in biomass of the zooplankton and fish biomass components of the GA LME, a declining level of zooplankton has been reported for the California Current LME (CC LME) of approximately 70% over a 45-year monitoring period. The cause according to Roemmich and McGowan appears to be an increase in water column stratification due to long-term warming. The clearest food web relationship reported related to the zooplankton biomass reduction was a decrease in the abundance of pelagic sea-birds.

US North-east Shelf Large Marine Ecosystem

The US North-east Shelf LME is an ecosystem with more structured coherence in the lower food web

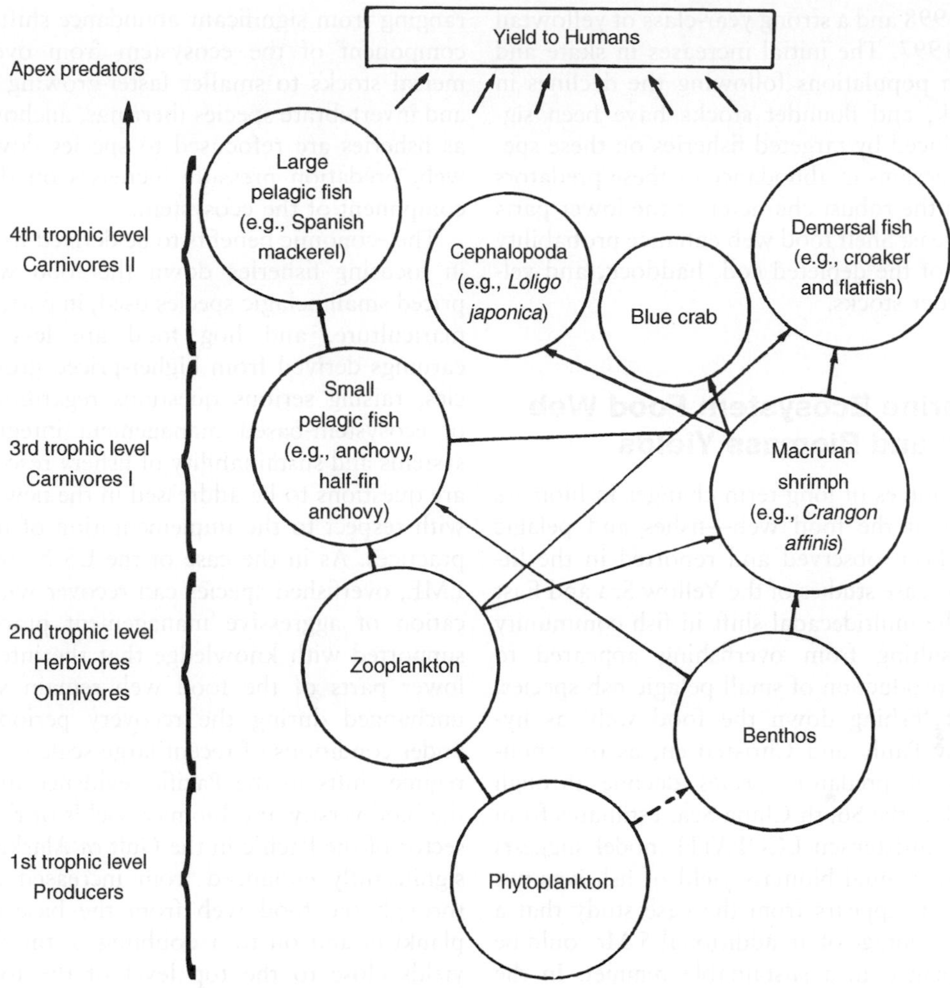

Figure 5 A simplified version of the Yellow Sea food web and trophic structure based on the main resources populations in 1985–1986. (From Tang 1993).)

than in the Gulf of Alaska or California Current systems. Following a decade of overfishing beginning in the mid-1960s, the demersal fish stocks, principally haddock, cod, and yellowtail flounder, declined to historic low levels of spawning biomass. In addition, the herring and mackerel spawning stock levels were reduced in the mid-1970s. By the mid-1980s, the demersal fish biomass had declined to less than 50% of levels in the early 1960s.

Following the 1975 extension of jurisdiction by the United States to 200 miles of the continental shelf, the rebuilding of the spawning stock biomass (SSB) of herring and mackerel commenced. Beginning in 1982 there was a sharp reduction in fishing effort from foreign vessels excluded from the newly designated US Exclusive Economic Zone (EEZ). Within four years the mackerel population recovered from just under 0.5 Mt to 1 Mt in 1986 and an estimated 2 Mt by 1994. Herring recovery was also initiated in the absence of any significant fishing effort from 1982 to 1990 when increases in SSB went from less than 0.2 Mt to 1 Mt. An unprecedented 3.5 Mt level of herring SSB was reached by 1994.

The NOAA-NMFS time-series of zooplankton collected from across the entire North-east Shelf ecosystem from 1977 to 1999 is indicative of an internally coherent structure of the zooplankton component of the North-east Shelf ecosystem. During the mid to late 1990s and the unprecedented abundance levels of SSB of herring and mackerel, the zooplankton component of the ecosystem showed no evidence of significant changes in biomass levels with annual values close to the long-term annual median of 30 ml/100 m^3 for the North-east Shelf ecosystem. In keeping with the robust character of the zooplankton component is the initiation of spawning stock recovery subsequent to reductions in fishing effort for cod, haddock, and yellowtail flounder. Accompanying the recovery of spawning stock biomass is the production of a strong year-class of haddock in 1998 and a strong year-class of yellowtail flounder in 1997. The initial increases in skate and spiny dogfish populations following the declines in cod, haddock, and flounder stocks have been significantly reduced by targeted fisheries on these species. The reductions in abundance of these predators coupled with the robust character at the lower parts of the North-east Shelf food web enhance probability for recovery of the depleted cod, haddock, and yellowtail flounder stocks.

Large Marine Ecosystem Food Web Dynamics and Biomass Yields

Two major sources of long-term changes in biomass near the top of the food web—fishes and pelagic birds—have been observed and reported in the literature. In the case studies of the Yellow Sea and East China Sea, the multidecadal shift in fish community structure resulting from overfishing appeared to promote the production of small pelagic fish species, indicative of 'fishing down the food web' as hypothesized by Pauly and Christensen, as the abundance levels of predator species decline through overfishing. For the South China Sea, estimates from a Pauly and Christensen ECOPATH model suggests that the mean annual biomass yield of fish was not fully utilized. It appears from the case study that a significant percentage of an additional 5 Mt could be fished if managed in a sustainable manner. In the eastern Pacific the results of oceanographic regime shifts had direct impact in increasing zooplankton and fish biomass in the Gulf of Alaska LME, whereas a multidecadal warming trend in the California Current LME lowered productivity at the base of the food web and resulted in a decrease in pelagic bird biomass. The importance of fish and fisheries to the structure of marine food webs is also an important cause of variability in biomass yields. A clear demonstration of this relationship is found in the application of the ECOPATH model to four continental shelf ecosystems, where it was shown that fish preying on other fish was a principal source of fish biomass loss. The level of predation ranged from 3 to 35 times the loss to commercial fisheries.

Fish are keystone components of food webs in marine ecosystems. The worldwide effort to catch fish using highly effective advanced electronics to locate them, and efficient trawling, gill-netting, and longline capture methodologies, has had an impact on the structure of marine food webs. From case studies examined, evidence indicates that the fishing effort of countries bordering on LMEs has resulted in changes in the structure of marine food webs, ranging from significant abundance shifts in the fish component of the ecosystem from overfished demersal stocks to smaller faster-growing pelagic fish and invertebrate species (herrings, anchovies, squids) as fisheries are refocused to species down the food web, predation pressure increases on the plankton component of the ecosystem.

The economic benefits to be derived from the trend in focusing fisheries down the food web to low-priced small pelagic species used, in part, for poultry, mariculture, and hog food are less than from earnings derived from higher-priced groundfish species, raising serious questions regarding objectives of ecosystem-based management integrity of ecosystems and sustainability of fishery resources. These are questions to be addressed in the new millennium with respect to the implementation of management practices. As in the case of the US North-east Shelf LME, overfished species can recover with the application of aggressive management practices, when supported with knowledge that the integrity of the lower parts of the food web remain substantially unchanged during the recovery period. However, under conditions of recent large-scale oceanographic regime shifts in the Pacific, evidence indicates that the biodiversity and biomass yields of the north-east sector of the Pacific in the Gulf of Alaska LME were significantly enhanced from increased productivity through the food web from the base to the zooplankton and on to a doubling of the fish biomass yields close to the top level of the food web. In contrast, in the California Current ecosystem the apparent heating and deepening of the thermocline effectively reduced phytoplankton and zooplankton production over a 40-year period, suggesting that in upwelling regions prediction of oceanographic events effecting food web dynamics require increased commitment to long-term monitoring and assessment practices if forecasts on effects of regime shifts on biomass yields are to be improved.

If ecosystem-based management is to be effective, it will be desirable to refine ECOPATH-type models for estimating the carrying capacity of LMEs in relation to sustainability levels for fishing selected species. It was assumed in the early 1980s by Skud, based on the historic record, that herring and mackerel stocks inhabiting the US North-east Shelf ecosystem could not be supported at high biomass levels simultaneously by the carrying capacity of the ecosystem. However, subsequent events have demonstrated the carrying capacity of the ecosystem is now of sufficient robustness to support an unprecedented almost 5.5 Mt of spawning biomass of both species combined. In addition, the ecosystem in its present state apparently has the carrying capacity to support the growing spawning biomass of recovering haddock and flounder stocks. Evidence of the production of strong year-classes for both species supported by high average levels of primary production of $350\,\mathrm{g\,C\,m^2\,y^{-1}}$, a robust level of zooplankton biomass, relatively high levels of epibenthic macrofauna, and apparent absence of any large-scale oceanographic regime shift suggests that integrity of the ecosystem food web will enhance the return of the fish component of the ecosystem to the more balanced demersal–pelagic community structure inhabiting the shelf prior to the massive overfishing perturbation of the 1960s to the 1980s.

Prospectus: Food Webs and Large Marine Ecosystem Management

It is clear from the LME studies examined that time-series measurements of physical oceanographic conditions that are coupled with appropriate indicators of food web integrity (e.g., phytoplankton, chlorophyll primary productivity, zooplankton, fish demography) are essential components of a marine science program designed to support the newly emergent concept of ecosystem-based management.

It is important to consider the dynamic state of LMEs and their food webs in considering management protocols, recognizing that they will need to be considered from an adaptive perspective. To assist economically developing countries in taking positive steps toward achieving improved understanding of food web dynamics and their role in contributing to longer-term sustainability of fish biomass yields, reducing and controlling coastal pollution and habitat degradation, and improving oceanographic and resource forecasting systems, the Global Environment Facility (GEF) and its $2 billion trust fund has been opened to universal participation that builds on partnerships with several UN agencies (e.g., World Bank, UNDP, UNEP, UNIDO). The GEF, located within the World Bank, is an organization established to provide financial support to post-Rio Conference actions by developing nations for improving global environmental conditions in accordance with GEF operational guidelines.

See also

Ecosystem Effects of Fishing. Fisheries and Climate. Fisheries Overview. Marine Fishery Resources, Global State of. Network Analysis of Food Webs. Population Dynamics Models. Upwelling Ecosystems.

Further Reading

Chen Y and Shen X (1999) Changes in the biomass of the East China Sea ecosystem. In: Sherman K and Tang Q (eds.) *Large Marine Ecosystems of the Pacific Rim: Assessment, Sustainability, and Management*, pp. 221–239. Malden MA: Blackwell Science.

Kumpf H, Stiedinger K, and Sherman K (1999) *The Gulf of Mexico Large Marine Ecosystem: Assessment, Sustainability, and Management*. Malden, MA: Blackwell Science.

Pauly D and Christensen V (1993) Stratified models of large marine ecosystems: a general approach and an application to the South China Sea. In: Sherman K, Alexander LM, and Gold BD (eds.) *Large Marine Ecosystems: Stress, Mitigation, and Sustainability*, pp. 148–174. Washington DC: AAAS.

Sherman K and Alexander LM (eds.) (1986) *Variability and Management of Large Marine Ecosystems, AAAS Selected Symposium 99*. Boulder, CO: Westview Press.

Sherman K and Alexander LM (eds.) (1989) *Biomass Yields and Geography of Large Marine Ecosystems, AAAS Selected Symposium 111*. Boulder, CO: Westview Press.

Sherman K, Alexander LM, and Gold BD (eds.) (1990) *Large Marine Ecosystems: Patterns, Processes, and Yields, AAAS Symposium*. Washington, DC: AAAS.

Sherman K, Alexander LM, and Gold BD (eds.) (1991) *Food Chains, Yields, Models, and Management of Large Marine Ecosystems*. AAAS Symposium Boulder, CO: Westview Press.

Sherman K, Alexander LM, and Gold BD (eds.) (1992) *Large Marine Ecosystems: Stress, Mitigation, and Sustainability*. Washington, DC: AAAS.

Sherman K, Jaworski NA, and Smayda TJ (eds.) (1996) *The Northeast Shelf Ecosystem: Assessment, Sustainability, and Management*. Cambridge, MA: Blackwell Science.

Sherman K, Okemwa EN, and Ntiba MJ (eds.) (1998) *Large Marine Ecosystems of the Indian Ocean: Assessment, Sustainability, and Management*. Malden, MA: Blackwell Science.

Sherman K and Tang Q (eds.) (1999) *Large Marine Ecosystems of the Pacific Rim: Assessment, Sustainability, and Management*. Malden, MA: Blackwell Science.

Tang Q (1993) Effects of long-term physical and biological perturbations on the contemporary biomass yields of the Yellow Sea ecosystem. In: Sherman K, Alexander LM, and Gold BD (eds.) *Large Marine Ecosystems: Stress, Mitigation, and Sustainability*, Washington DC, AAAS Press, pp. 79–93.

OCEAN GYRE ECOSYSTEMS

M. P. Seki and J. J. Polovina, National Marine Fisheries Service, Honolulu, HI, USA

The Generalized Open-ocean Food Web

Food webs, simply put, describe all of the trophic relationships and energy flow between and among the component species of a community or ecosystem. A food chain depicts a single pathway up the food web. The first trophic level of a simple food chain in the open ocean begins with the phytoplankton, the autotrophic primary producers, which build organic materials from inorganic elements. Herbivorous zooplankton that feed directly on the phytoplankton are the primary consumers and make up the second trophic level. Subsequent trophic levels are formed by the carnivorous species of zooplankton that feed on herbivorous species and by the carnivores that feed on smaller carnivores, and so on up to the highest trophic level occupied by those adult animals that have no predators of their own other than humans; these top-level predators may include sharks, fish, squid, and mammals.

In ocean gyres, the dominant phytoplankton, especially in oligotrophic waters, are composed of very small forms, marine protozoans such as zooflagellates and ciliates become important intermediary links, and the food chain is lengthened. There are thus typically about five or six trophic levels in these ecosystems. In contrast, large diatoms dominate in nutrient-rich upwelling regions, resulting in shorter food chains of three or four trophic levels since large zooplankton or fish can feed directly on the larger primary producers. Production of larger flagellates/diatoms in specialized habitats of the open ocean may lead to shortened energy paths. As there are seldom simple linear food chains in the sea, a food web with multiple and shifting interactions between the organisms involved portrays more accurately the trophic dynamics of a given ecosystem. Examples of food webs are presented for the North Pacific Subarctic and Subtropical gyres in **Figures 1** and **2**, respectively. The place a particular species occupies in the ecosystem food web is not necessarily constant, since feeding requirements change as organisms grow. Some species change diets (and trophic levels) as they grow or as the relative abundance and availability of different food items change. In some species, cannibalism of their own young may be important.

Before proceeding, we need to recognize that in ocean gyre ecosystems, very large percentages of organic matter are cycled through microbes before entering the linear arrangement of the classic food web. The role of viruses, bacteria, heterotrophic nanoflagellates, nano- and microplanktonic protozoans, and the microbial loop in the open ocean ecosystem is discussed in detail elsewhere in the Encyclopedia and we only note here that if several trophic steps are involved in a microbial food web, there must be a significant loss of carbon at each transfer and therefore little transfer of carbon from microbial to classic planktonic food webs.

The Ocean Gyre Ecosystem

In each of the major ocean basins, surface winds drive currents that form the large anticyclonic subtropical gyres as well as the smaller, cyclonic subarctic gyres. The details of both vertical structure and trophic links in the ocean gyre food web differ regionally and seasonally; opportunism is generally the rule of prey selection and is driven by regional and seasonal differences in vertical physical structure of the upper water column.

The surface waters in the warm subtropical gyres away from seasonal meridional frontal systems tend to be highly stratified, with a permanent thermocline that limits vertical enrichment of the euphotic zone throughout the year. In this nutrient depleted environment, recycled nitrogen (primarily in the form of ammonia and urea) excreted by the zooplankton or released by the microbial community provides the primary nitrogen for a continued low level of phytoplankton 'regenerated' production that remains in approximate balance by zooplankton grazing. These oligotrophic waters are characteristically recognized as some of the least productive waters in the world's oceans. Alternatively, 'new' primary production based on the input of new nitrogen (primarily nitrate) into the euphotic zone occurs at much lower levels through oceanographic physical forcing, atmospheric input, or nitrogen fixation. As in most other ocean systems, a deep chlorophyll maximum (DCM) is present, but this will vary spatially and temporally with consequent

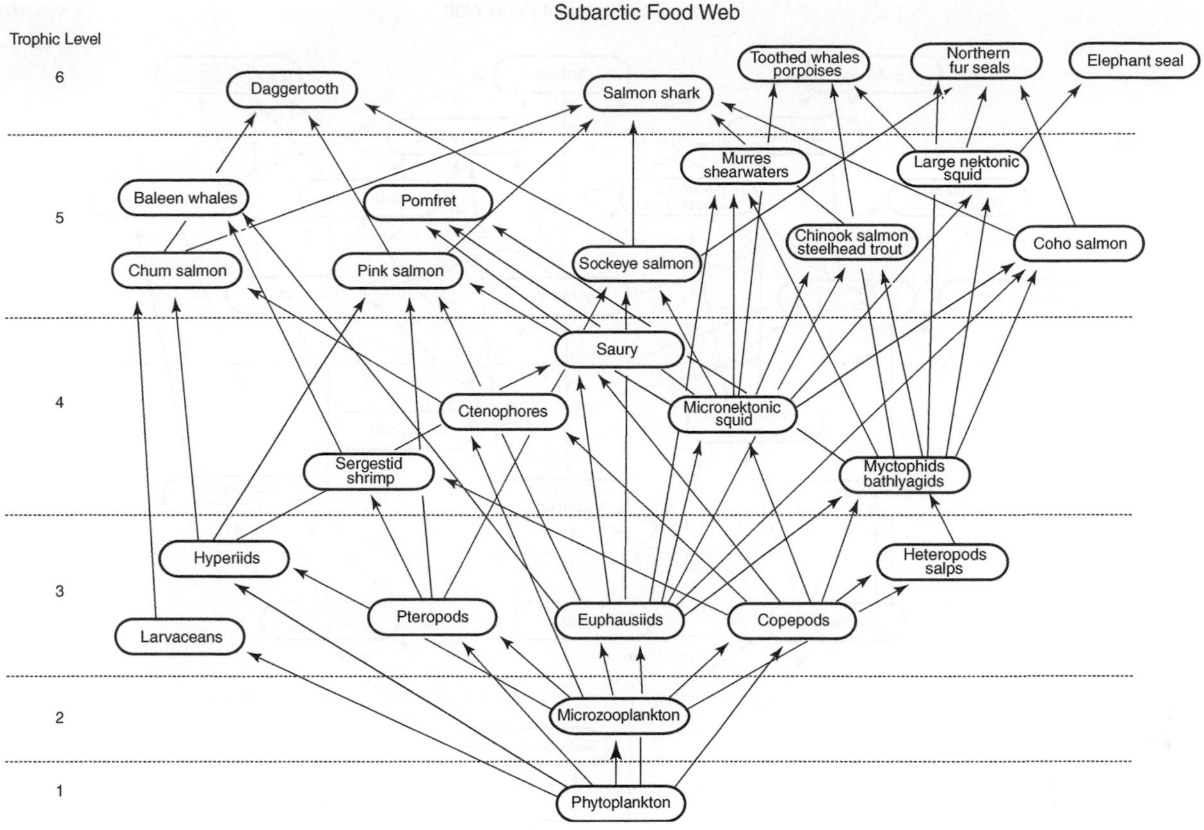

Figure 1 Food web of the Subarctic North Pacific oceanic ecosystem; arrows point in the direction of energy flow. (Adapted from Brodeur *et al.* (1999).)

changes in the depth of the thermocline, nutrient flux, and the maximum primary production rate.

In contrast, subarctic oceans benefit from alternating periods of summer stratification and winter mixing, so that there is a seasonal injection of nutrients into surface waters. In the North Atlantic the trophic framework is set up by the classic temperate bloom cycle; i.e., a 'spring bloom' followed by a 'crash', a 'stable state summer equilibrium', a smaller 'fall bloom', and a 'winter decline'. Winter storms create deeper well-mixed conditions including the presence of high nitrate levels. With the onset of spring comes an increase in solar irradiation and water column stability. Phytoplankton (normally diatoms) 'blooms' with nutrient injection. Slightly lagging behind the phytoplankton bloom is a bloom of zooplankton that feeds on the increased phytoplankton. As the increased phytoplankton population exhausts the nitrate supply and zooplankton grazing overshoots phytoplankton growth, a 'crash' occurs, returning nitrate, phytoplankton, and zooplankton to low, stable levels. During the summer months, water is well stratified with balanced processes (i.e., phytoplankton growth (limited by lower levels of available nitrate) is balanced by zooplankton grazing). During fall, early episodic storms induce some mixing and hence phytoplankton blooms followed by periods of stability. The onset of increased winter storm activity mixes nutrients in surface waters; however, available light energy decreases, limiting phytoplankton production. Thus the low phytoplankton production periods are light limited during the winter and nutrient-limited during the summer.

In the subarctic North Pacific there is virtually no bloom, diatoms populations are low, and high abundances of microzooplankton graze the small phytoplankton, keeping the abundance down. The subarctic North Pacific waters are thus high in nutrients (nitrogen) but low in chlorophyll (HNLC), similar to those ecosystems found near the equator and in the Antarctic. Modest winter mixing in the subarctic Pacific results in a more favorable light regime and a continued modest level of biological production throughout the year and reduced biomass accumulation that is less strongly pulsed than that occurring in the Atlantic. The constant planktonic biomass in the Pacific favors pelagic fish production in contrast to the Atlantic, where much of the plankton biomass is lost to sinking without being grazed, thereby supporting more benthic resources.

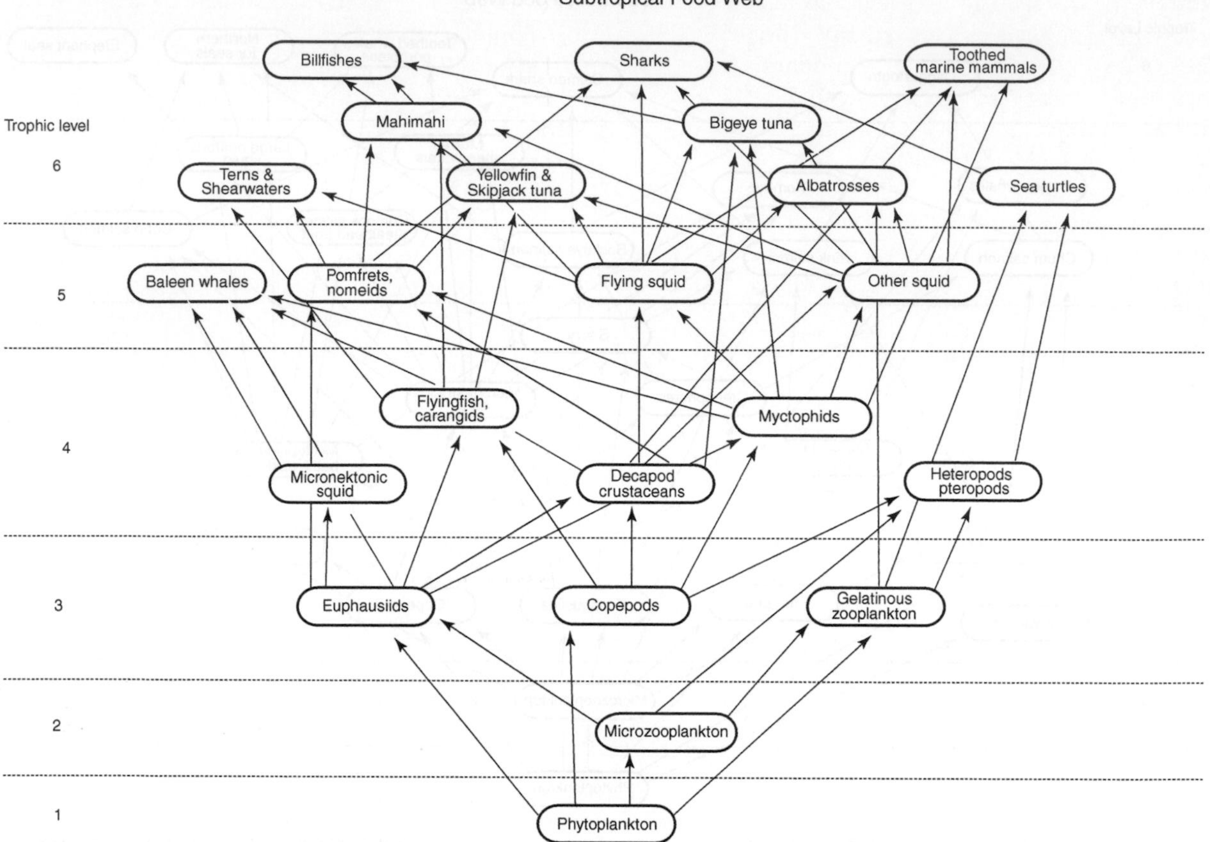

Figure 2 Food web of the Subtropical North Pacific oceanic ecosystem; arrows point in the direction of energy flow.

The Intermediate Trophic Levels

The animals occupying the intermediate trophic levels that link the primary producers and consumers with the top of the food web include all of the zooplankton and micronekton (smaller organisms 10–100 cm in length) large enough to swim in inertial conditions. Two groups of animals in particular play a key role in the ocean gyre food web: those that compose the vertically migrating deep scattering layer (DSL) and the small pelagic 'forage fishes'.

As night approaches, myriads of animals make an ascent from various depths to grazing or hunting grounds near the surface. Before daybreak, the DSL organisms descend back to their deeper, daytime residence. These migrations range 200–700 m in vertical extent and are confined to the epipelagic and mesopelagic zones. With the diel DSL migration, concentrations of food near the surface are much greater (a hundredfold increase in population numbers) at night than during day, establishing a forage base for larger, aggressive, nocturnally active carnivores; e.g., dolphins and pelagic squid. In the tropics, large filter-feeding organisms such as whale sharks, baleen whales, and megamouth sharks probably also utilize these high-density layers of plankton and micronekton. Diel vertical movements of bigeye tuna (*Thunnus obesus*) exhibit high correlation with the migration of the DSL and their associated prey species. The DSL micronekton, which includes fishes (e.g., myctophids, gonostomids), squid, and crustaceans (especially euphausiids and sergestid shrimps) are likely the primary zooplanktivores in the open ocean.

A large component of the DSL faunal composition is the gelatinous zooplankton (combjellies, pelagic tunicates, medusae, ctenophores, siphonophores, etc.). Despite their ubiquitous and abundant nature, the significance of these zooplankton in the pelagic food web has been and still is somewhat equivocal. These gelatinous animals had been thought to represent an ecological dead end. Because of their low food value, energy taken up through their phytoplankton grazing was not passed up to higher trophic levels. From the growing list of fish, sea turtle, and seabird species identified as habitually feeding on gelatinous prey, there is strong evidence that this resource plays more of a key role in the gyre food web than previously believed.

In the open ocean, small pelagic 'forage fishes' are the keystone prey species among apex predators. By definition, forage fishes are '... abundant, schooling fishes (and squid) preyed upon by many species of seabirds, marine mammals, and other fish species' according to A. M. Springer and S. G. Speckman. These animals are generally short-lived, highly fecund, and heavily preyed on as juveniles as well as adults. In the subtropical gyre, loose aggregations of flying squid, pomfret, saury, or mackerel, and in temperate waters tighter schools of anchovies, sardines, or herring, are representative of this guild. From a trophic standpoint, the importance of pelagic ommastrephid squids in the gyre ecosystems cannot be understated. These forage animals have rapid population turnover rates and can dominate the epipelagic nektonic biomass, as is the case with the red flying squid *Ommastrephes bartramii* at the summer Subarctic Boundary. Throughout the gyre food webs, ommastrephids often dominate the diets of toothed cetaceans, pinnipeds, fishes, and seabirds, as well as of tropical seabirds and marine mammals.

The Higher Trophic Levels

The faunal compositions of epipelagic nekton communities tend to be more diverse and more complex in the subtropics than those found in the temperate subarctic. At the top levels of the epipelagic (0–400 m) food web (and particularly within the 0–100 m euphotic zone) are the fast-swimming predators such as sharks, billfishes, and large tunas as well as some of the marine mammals and surface-feeding seabirds that feed at the fourth and fifth trophic levels. Generally, most of these breed in warmer tropical fall/winter waters, some seabirds notwithstanding, but make major excursions to the northern subtropics–subarctic during spring/summer to feed; all are opportunistic predators within their respective niches.

Large pelagic fishes are the most abundant and diversified of the apex predators but are nevertheless still highly dispersed, with typical population densities of only one fish per square kilometer integrated over the upper mixed layer. Blue sharks (*Prionace glauca*) are by far the most abundant shark species in the ocean gyres and in subtropical waters; frequent encounters with fishing gear make them seem ubiquitous. Dominant prey items for blue shark include the abundant ommastrephid squids and forage fishes. Billfishes, including the marlins (*Makaira* spp. and *Tetrapturus* spp.) and swordfish (*Xiphias gladius*), are capable of extensive vertical excursions of several hundred meters but concentrate feeding at the surface on smaller fishes and, in the case of swordfish, exhibit a particular predilection for pelagic squid. Tunas (Scombridae) are diurnal predators and feed from the surface down to about 400 m. With the exception of bigeye tuna, tunas feed almost solely on small surface dwelling fauna – i.e., they rarely feed on the vertically migrating DSL. Juvenile and post-larval crustaceans, cephalopods, and oceanic- (Bramidae, Nomeidae, etc.) and shore-fishes (Mullidae, Chaetodontidae, etc.) dominate the tuna diet. Insular-based prey are likely artifacts of the inherent bias introduced from land-based fishing operations that underlie our present perception of oceanic predatory fish diets; these diminish in importance to the food web with distance from land.

Among marine mammals, baleen whales are filter feeders of large zooplankton and hence feed at the third trophic level. Toothed whales and delphinids all feed at the fifth or sixth trophic level on pelagic cephalopods and, depending on the respective species' distributional range, on epipelagic and mesopelagic fish. In the North Pacific, only two species of pinnipeds, the northern fur seal and northern elephant seal use the open ocean regions of the gyres to any extent. For these mammals, stomach contents suggest that oceanic squid are the key food web link.

Most seabirds foraging in the high seas of the ocean gyres are Procellariiformes, namely the albatrosses, shearwaters, petrels, and fulmars. These seabirds are opportunistic surface feeders on small 'forage fishes'. In the oligotrophic subtropics close to nesting colonies, a feeding guild of shearwaters and terns represents numerically the most abundant among breeding seabird species, relying on predatory fish such as tuna to drive schools of small pelagics to the ocean surface, thereby facilitating feeding.

Enhanced Feeding Habitats

The pelagic zone of the open ocean gyre is often perceived as the most monotonous living space of our planet, with few visual cues to maintain spatial orientation, and spatial heterogeneity effectively restricted to vertical gradients of light, temperature, and abundance of organisms. However, regions of horizontal oceanographic variability abound throughout the gyre ecosystem in the form of large-scale frontal systems and mesoscale dynamic features where productivity is enhanced and trophic transfer facilitated. A discussion of the gyre food web would not be complete without highlighting these dynamic features where so much of the ecosystem energy is mobilized.

Large-scale fronts and associated frontal zones in the open ocean have profound effects on the distribution and movement patterns of pelagic animals. Generally, the distribution of pelagic animals tends to be coupled with preferred thermal habitats, a response to forage accumulation (enhanced forage opportunity), migration cues, and/or energetic gains from riding currents. On one hand, these basin-scale features form the boundaries that divide some of the large, core pelagic biogeographic provinces; on the other, the fronts are recognized as regions of convergence where life-forms on all trophic levels are concentrated and so support feeding and spawning aggregations. In the North Pacific, these thermohaline fronts mark the bounds for many of the Transition Zone keystone species, including the Pacific pomfret (*Brama japonica*), the red flying squid, and blue shark, which undergo extensive seasonal migrations northward during summer to feed along the Subarctic Boundary convergence and southward during the winter and spring to spawn in the subtropics. The aggregation of all trophic levels at the Subarctic Boundary results in a complex but trophically rich community where considerable energy is transferred. During the winter–spring, shoaling of the thermocline at the North Pacific Subtropical Front is believed to elevate the nutricline into the euphotic zone, resulting in enhanced local productivity, particularly at the depth of the DCM, as well as physically concentrating loose aggregations of prey closer to the ocean surface for predation. A basin-wide surface chlorophyll front, the Transition Zone Chlorophyll Front, is also manifested in the surface waters of the North Pacific. This biological front lies at the boundary between the low-chlorophyll subtropical and high-chlorophyll subarctic gyres. Recently, neuston-feeding loggerhead sea turtles, *Caretta caretta*, have been found to exhibit strong affinities for this specialized habitat.

Mesoscale variability on spatial scales of 10–100 km – the 'weather' of the ocean – includes high-gradient dynamic features such as frontal meanders, eddies, and jets. These features often give rise to localized regions of higher productivity, leading to aggregation and development of a forage base while physical gradients provide cues for predators to locate prey or more directly aggregate or concentrate food items. In stratified, oligotrophic waters of the large ocean gyres, recycling of nutrients between the grazers and the phytoplankton typically maintains primary production at uniformly low levels. Transient episodes of upwelled nutrient-rich water from mesoscale events, particularly strong cyclonic eddies and meanders, have been shown to induce 'new' production, thus providing a mechanism to shorten the trophic pathway and facilitate energy transfer. The injection of nutrients at these features increases the biomass of phytoplankton through the contribution of larger eukaryotic phytoplankton, notably diatoms and dinoflagellates, over the recycled-nutrient-based picophytoplankton (e.g., photosynthesizing cyanobacteria species) that normally typify the system.

Seamounts, also known as submarine rises or table mounts, can have a strong influence on adjacent open-ocean food webs in a variety of ways, particularly those that rise within the upper few hundred meters of the surface. Waters overlying seamounts are often characterized by high standing stocks of plankton and, at some locations, concentrate and transfer energy not only among the pelagic community but also to the demersal resident ecosystems. Several phenomena contribute to the maintenance and enhancement of the unique seamount communities. For one, some seamount ecosystems possess midwater micronekton communities that include both a unique resident assemblage as well as a DSL component advected from adjacent waters. At shallower seamounts, the members of the DSL community rise in the water column at night, are advected over the seamount, and are subsequently trapped on the summit as they descend at dawn, creating an enhanced feeding regime during the morning hours. The resident faunal assemblages are analogous to the land-associated mesopelagic boundary communities identified on larger island and shelf slopes. Seamounts are also hypothesized to benefit from the development of Taylor columns, or semistationary eddies located above the seamount, that would help retain planktonic populations and enhance productivity.

Food Web Dynamics

Food webs of ocean gyres are altered by a range of factors, including fishing and variations in physical forcing resulting from climate fluctuations. The relationship between climate variation and ecosystem dynamics has been a focus of considerable research. Changes in abundance of organisms at many trophic levels have been documented to vary coherently with various atmospheric indices, including the Aleutian Low in the Pacific and the North Atlantic Oscillation in the Atlantic. For example, in the North Pacific subarctic gyre, changes in the fishery catches of all species of salmon vary on timescales of decades, coherent with the intensity of the Aleutian Low Pressure System. While the link between the atmosphere and salmon productivity has not been resolved, a number of hypotheses suggest that changes in salmon carrying capacity result from changes in physical forcing that impact productivity at the bottom of the food web and propagate up the food web to salmon. A change in the energy flow through the food web from the bottom up in response to climate variation is typical of many relationships described between climate and food web dynamics in ocean gyres.

Fisheries in ocean gyres typically impact food webs through removals at top or mid-trophic levels. For example, longline or troll fisheries target apex predators including tunas, swordfish, marlins, and salmon, although some mid-trophic level species including squid can also be the target species. By-catch and incidental catches in ocean gyre fisheries also include apex species including sharks and protected species such as seabirds, marine mammals, and sea turtles. Data on the responses of oceanic gyre food webs to fishing are generally limited to harvested species, so the food web impacts at lower trophic levels are not documented. However, a dynamic ecosystem model has been used to investigate possible impacts and has found no evidence that the removal of any single high trophic level species significantly altered the food web. The lack of a keystone species appears to be due to a high degree of diet overlap among the high trophic level species. Fisheries in oceanic gyres alter the food web by reducing biomass at the top of the food web. When this reduction becomes substantial, it may result in some increase in biomass at mid-trophic levels.

See also

Ecosystem Effects of Fishing. Large Marine Ecosystems.

Further Reading

Boehlert GW (1986) Productivity and population maintenance of seamount resources and future research directions. In: Uchida RN, Hayasi S, and Boehlert GW (eds) *Environment and Resources of Seamounts in the North Pacific*, pp. 95–101 Washington, DC: US Department of Commerce. NOAA Technical Report NMFS 43.

Brodeur R, McKinnell S, Nagasawa K, *et al.* (1999) Epipelagic nekton of the North Pacific Subarctic and Transition Zones. *Progress in Oceanography* 43: 365–397.

Francis RC, Hare CR, Hollowed AB, and Wooster WS (1998) Effects of interdecadal climate variability on the oceanic ecosystems of the NE Pacific. *Fisheries Oceanography* 7: 1–21.

Josse E, Bach P, and Dagorn L (1988) Simultaneous observations of tuna movements and their prey by sonic tracking and acoustic surveys. *Hydrobiologia* 371/372: 61–69.

Kajimura H and Loughlin TR (1988) Marine mammals in the oceanic food web of the eastern subarctic Pacific. In: Nemoto T and Pearcy WG (eds.) *The Biology of the Subarctic Pacific. Bulletins of the Ocean Research Institute*, 187–223.

Kitchell JF, Boggs CH, He X, and Walters CJ (1999) Keystone predators in the central Pacific. *Ecosystem Approaches for Fisheries Management*, pp. 665–683. Fairbanks, AK: Alaska Sea Grant College.

Longhurst AR and Pauly D (1987) *Ecology of Tropical Oceans*. San Diego: Academic Press.

Mann KH and Lazier JRN (1991) *Dynamics of Marine Ecosystems. Biological–Physical Interactions in the Oceans*. Cambridge, MA: Blackwell Scientific.

Olson DB, Hitchcock GL, Mariano AJ, *et al.* (1984) Life on the edge: marine life and fronts. *Oceanography* 7: 52–60.

Polovina JJ (1994) The case of the missing lobsters. *Natural History* 103(2): 50–59.

Polovina JJ, Howell E, Kobayashi DR and Seki MP (2001) The Transition Zone Chlorophyll Front, a dynamic, global feature defining migration and forage habitat for marine resources. *Progress in Oceanography* in press.

Springer AM and Speckman SG (1997) A forage fish is what? Summary of the symposium. In: *Proceedings. Forage Fishes in Marine Ecosystems*, pp. 773–805. Fairbanks, AK: Alaska Sea Grant College.

Steele JH (1974) *The Structure of Marine Ecosystems*. Cambridge, MA: Harvard University Press.

Valiela I (1995) *Marine Ecological Processes*. New York: Springer-Verlag.

POLAR ECOSYSTEMS

A. Clarke, British Antarctic Survey, Cambridge, UK

Introduction

The Arctic Ocean and the Southern Ocean together comprise a little under one-fifth of the world's oceans (the precise fraction depending on how these oceans are defined). The two polar oceans are similar in being cold, seasonal, productive, and heavily influenced by ice, and both have long been fished by man. They differ markedly, however, in geography, age, and many aspects of their biology.

In both polar oceans sea water temperatures are typically low, and in many areas are close to freezing (-1.86°C) for long periods. In areas of seasonal ice cover the surface waters undergo a summer warming, but even at the lowest latitudes this rarely amounts to more than a few degrees. Typically summer sea water temperatures in the seasonal sea ice zone of Antarctica are between $+0.5$ and $+1.5$°C.

The inclination of the earth's rotational axis means that the seasonality of received radiation increases towards the poles. This seasonality, exacerbated throughout much of the polar oceans by the accompanying seasonal ice cover, means that primary production (the conversion of inorganic carbon and other essential nutrients into organic matter by plants) is also intensely seasonal.

The combination of a low and fairly seasonal temperature with a highly seasonal primary production (**Figure 1**) distinguishes polar marine ecosystems from all others on earth. Temperate systems are typically also highly seasonal, but here both temperature and primary production tend to co-vary. Tropical marine ecosystems experience high temperatures that are fairly aseasonal, but there are often strong seasonal variations in other important environmental variables (for example, rainfall and fresh water run-off in monsoon areas).

These features of the polar regions set two particular challenges for organisms living there. The first is that the low temperatures will tend to slow the rates of many biological processes. This does not apply to mammals and birds, whose internal body temperatures are maintained by metabolic processes; for these endothermic ('warm-blooded') organisms the challenge is to avoid losing heat, which they do by a variety of mechanisms but principally by enhanced insulation. The subtle molecular mechanisms used by ectothermic ('cold-blooded') fish, invertebrates, plants, and bacteria are the subject of extensive physiological investigation. Despite these adaptations, however, it is a widespread observation that many important biological processes proceed slowly in polar systems. The second challenge for polar organisms is that the marked seasonality of primary production means that many of these processes are constrained to the summer months. These two features of generally slow physiological rates and intense seasonality together have a profound influence on the structure and dynamics of polar marine ecosystems.

General Features of Polar Marine Food Webs

Food webs can be constructed to emphasize different aspects of their structure or dynamics (for example, to highlight trophic interactions, energy flux, or biomass). When comprehensive, such food webs can be extremely complex. There is, however, usually

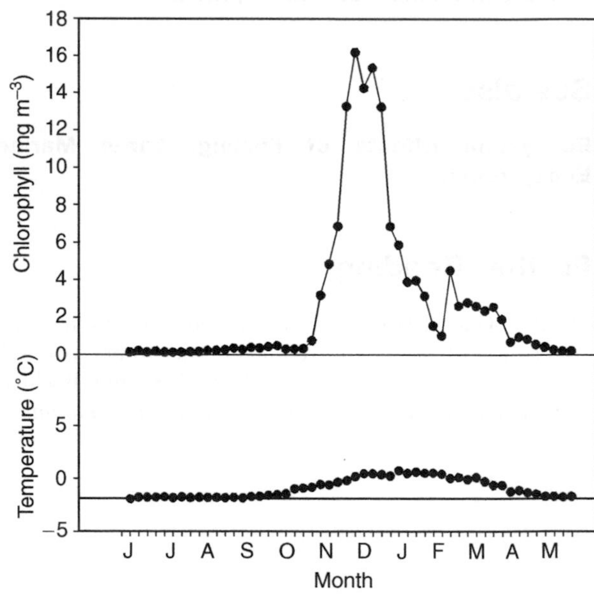

Figure 1 Seasonal variation in chlorophyll biomass (>0.2 μm) and temperature at a typical polar locality (Signy Island, maritime Antarctic, 60°S). Both plots show weekly means over the period 1988–1994. The horizontal lines show zero chlorophyll biomass and -1.86°C (the equilibrium freezing point of sea water). The marked seasonality of chlorophyll biomass coupled with the small annual variation in water temperature is typical of polar oceans.

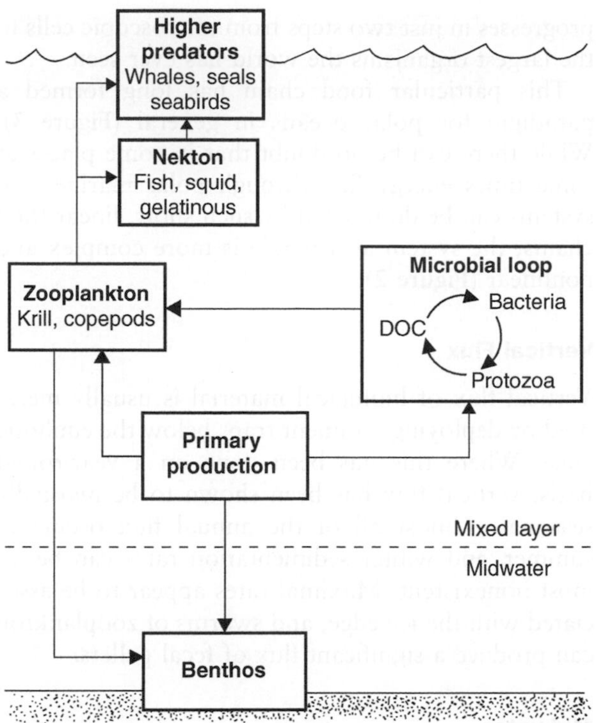

Figure 2 The basic structure of all marine food webs. Carbon fixed in primary production flows in three directions: to zooplankton herbivores (and thence to zooplanktonic and nektonic carnivores, and vertebrate higher predators), to the microbial loop, and via sedimentation to the benthos. In oceanic areas, the vertical flux of phytodetritus (from primary production) and particulate matter from zooplankton (primarily fecal pellets) passes out of the mixed layer and through the midwater to the benthos. In coastal areas, the benthos is within the mixed layer. Although specific food webs for individual oceanic locations will vary in complexity and the relative importance of different pathways, all have the same underlying structure. DOC, dissolved organic carbon.

a basic underlying structure, and a simplified food web for all marine systems can be produced (**Figure 2**).

The key feature of this basic structure is that it is nonlinear; energy fixed by marine plants (principally phytoplankton but also macroalgae in shallow water areas) passes in three important directions. These are (1) to planktonic herbivores and through a sequence of higher trophic levels eventually to top predators (including man); (2) to the microbial loop; and (3) through vertical flux (sedimentation) to the benthos. In deeper water the flux passes through the vast range of the midwater before reaching the abyssal plain. This basic structure applies to all marine ecosystems, but there are a number of extra features that are peculiar to polar regions. Particularly important are the influence of sea ice, the dominance at times of short linear food chains, and the importance of top predators.

Sea Ice

Sea ice has profound and pervasive effect on polar marine ecosystems. It forms a barrier to the exchange of energy, momentum, and gases, thereby acting as a major regulator of primary production in the water column beneath. In particular where snow lies on the ice, surface incident radiation can be reduced to a sufficiently low level to prevent photosynthesis. The isolation of the water column from wind also reduces turbulence and many biological particles (both living and inanimate) sediment out of the photic zone.

Sea ice also has a positive effect on primary production in polar oceans. In areas where sea ice is melting (the marginal ice zone) the release of fresh water can lead to an area of increased water column stability that encourages rapid primary production. It is not at present clear to what extent the enhanced primary productivity is caused by the increased stability, by seeding by phytoplankton cells released from the ice, or by other oceanographic features; indeed, it is likely that each of these factors may be important in different places at different times.

A second important influence by sea ice is its role as a habitat, supporting a diverse microbial community in the interstices between ice crystals. This distinctive assemblage (sometimes called the sympagic or epontic community) can grow throughout annual sea ice but is almost always most highly developed at the interface between the ice and the underlying sea water. The growth of phytoplankton and other microbial primary producers can be so intense as to color the undersides of the ice deep green or brown, colors that are easily seen as floes are overturned by ships passing through ice in late winter or spring. The contribution of ice-associated primary production to production in the Arctic or Southern Oceans as a whole is not known with any certainty. Current estimates for the Southern Ocean are of the order of 25–30%.

In studies of sea ice communities much attention has been directed to diatoms, but the microbial community growing within ice can be rich and diverse. It includes bacteria and flagellates as primary producers and a variety of flagellates, ciliates, and other protozoans as consumers. Associated with these are a range of meiofauna, and the whole assemblage is grazed by consumers. Particularly noteworthy among these are amphipods (in both Arctic and Antarctic systems) and euphausiids (principally in the Antarctic, where sea ice plays an important role in the biology of the Antarctic krill, *Euphausia superba*). These herbivores are themselves subject to predation by specialist consumers associated with the undersides of ice, most notably gelatinous zooplankton such as ctenophores and in the Antarctic the cryopelagic fish *Pagothenia borchgrevinki*.

Primary Production and the Microbial Loop

The microbial loop (**Figure 2**) is known to be a feature of all ocean systems, though its importance varies from place to place. It is undoubtedly present in polar regions, and can be important at times. Summer chlorophyll biomass is, however, dominated by diatoms rather than the small cells that contribute directly to the microbial loop. Nevertheless, all the components of the microbial loop are well documented from polar oceans, and there is considerable release of dissolved organic matter from rapidly growing phytoplankton cells in summer. It is likely that the microbial loop is present and active in all polar seas, but that production is dominated by larger cells for much of the summer open water season.

One group of primary producers that are important in many oceanic areas, but that are absent from truly polar seas, are coccolithophorids. This may be because the low temperatures are unfavorable for calcification.

Short Linear Food Chains

The polar regions have long been famous for the presence of short, linear food chains. The classic example of this is the trophic relationship that characterizes the lower latitude regions of the Southern Oceans: diatoms–krill–whales. This is a three trophic level food chain through which energy progresses in just two steps from microscopic cells to the largest organisms the world has ever seen.

This particular food chain has long formed a paradigm for polar oceans in general (**Figure 3**). While there can be no doubt that in some places at some times energy flux through polar marine ecosystems can be dominated by such short, linear food chains, the system as a whole is more complex and nonlinear (**Figure 2**).

Vertical Flux

Vertical flux of biological material is usually measured by deploying sediment traps below the euphotic zone. Where this has been done on a year-round basis, vertical flux has been shown to be markedly seasonal. Almost all of the annual flux occurs in summer, and winter sedimentation rates can be almost nonexistent. Maximal rates appear to be associated with the ice edge, and swarms of zooplankton can produce a significant flux of fecal pellets.

Midwater

The midwater zone is one of the least explored in polar regions. It is known that many species characteristic of lower latitudes extend into polar regions, and this is probably because the oceanographic features that define the polar regions are typically surface features and their influence may not extend to the mesopelagic realm.

As with the midwater elsewhere, the food web in polar mesopelagic waters is characterized by

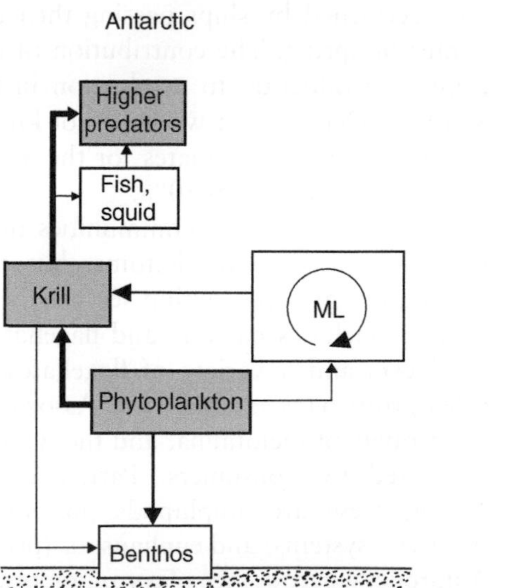

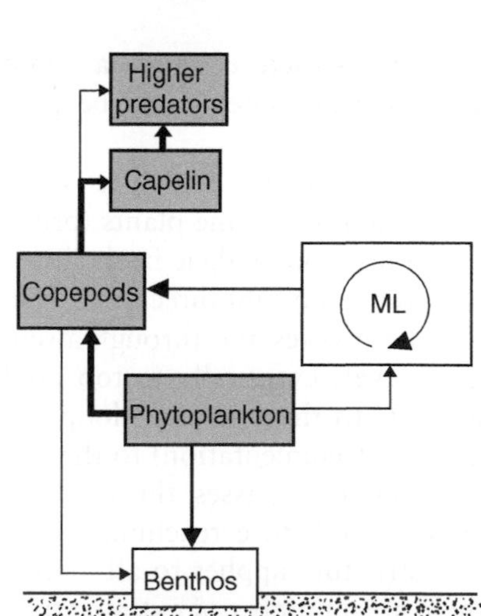

Figure 3 Representative polar food webs for the Antarctic (South Georgia) and the Arctic (Barents Sea), showing how energy flow can be dominated in some places at some times by a short, effectively linear, food chain. The major routes of energy flow are highlighted by the shaded boxes and the thicker arrows. The South Georgia example has become something of a paradigm for polar oceans in general. ML, microbial loop.

crustaceans, fish, and gelatinous zooplankton. The latter are little known from polar regions because they are badly damaged by traditional net sampling techniques, and to date there have been no submersible expeditions to polar waters to complement those from temperate midwaters where gelatinous zooplankton have been shown to be very important.

Myctophid fish are a major component of the midwater fauna and they form an important link to the surface food web through their vertical migration. They rise into surface waters by night, and here they are both an important consumer of crustacean zooplankton and an important prey item for sea birds and other top predators.

Benthos

The benthos is an important component of the marine food web in all oceans. In shallow waters benthic organisms can feed directly within the zone of surface production, whereas in deeper waters they rely primarily on vertical flux from the euphotic zone.

In polar regions suspension feeders are particularly important, with sponges, ascidians (sea squirts), bryozoans, brachiopods, holothurians (sea cucumbers) and fan-worms all well represented.

Top Predators

Polar marine ecosystems have long been recognized as being rich in top predators (seabirds, seals, and whales). In the Arctic the important top predators are seabirds (especially alcids and gulls) and the smaller toothed wales, although in the past baleen whales were also important. In the Southern Ocean the important top predators are penguins, albatrosses, and petrels among the sea birds, and both toothed and baleen whales.

These top predators achieve greater biomass in polar regions than anywhere else in the world's oceans, and their role within the marine food web is very significant. Many feed directly on crustacean zooplankton, thereby creating the short linear food chains that can characterize the system at some times. Others feed on fish and squid, and in some polar regions the status of these top predators is being used as a simple measure expressing the overall integrated well-being of the polar marine ecosystem.

Seasonality

Two features contribute to the intense seasonality of the polar marine ecosystem. The first is that primary production, and hence food availability for herbivores, is constrained almost completely to the summer months. In consequence, the growth and reproduction of organisms at all higher levels in the food web (both plankton and benthic) is itself often constrained to summer. Many species, however, have life cycles longer than one year and must overcome the long winter period of food shortage. This they do either through the utilization of reserves, typically fat, laid down the previous summer (many zooplankton), or by utilizing general body tissue and 'de-growing' over winter (some zooplankton and many benthos).

This seasonality leads to very large differences in energy flux through the polar food web between summer and winter. There is also an important seasonal variation in food web structure in that many top predators migrate to low latitudes in winter, thereby decreasing predation pressure on lower levels of the food web.

Man and the Polar Food Web

Although the polar regions lie far from many home ports and can be extremely inhospitable, they have been subject to intense disturbance by man. The Arctic has long been a traditional fishing ground for both demersal and midwater fish, and at both poles a variety of top predators have been subject to intense exploitation. In neither the Arctic nor the Southern Ocean can the structure or dynamics of the marine ecosystem be regarded as pristine (in the sense of reflecting the position before the intervention of man).

In the Arctic three species of large whale have been either effectively eradicated or reduced to very low populations; several species of fish are now at levels that render them economically unviable; and a major benthic crustacean predator (Alaskan king crab) has been reduced to very low population levels. In the Antarctic the large baleen whales have been reduced to very low populations (though the smaller Minke whale may have increased as a result) and the Southern fur seal is recovering from the brink of extinction. Many demersal fish stocks have been reduced to very low levels, and intermediate levels of the food web (Antarctic krill) are now being fished.

We can only speculate as to the structure of these polar food webs prior to the intervention of man. The complexity of the system and the nonlinearity of many of the dynamical processes involved make both prediction and hindcasting almost impossible. It is distinctly possible, however, that following the disturbance of these systems, they have settled (or will eventually do so) to a new and different stable structure. We may never again have a Southern Ocean dominated by the great whales, no matter how long we wait. Such complexities also make it very difficult to predict the effect of further disturbances, such as the development of a new fishery, or climate change.

Although there are many general features common to both polar marine food webs, there are also features peculiar to each.

The Arctic Food Web

The Arctic is a deep-water basin surrounded by wide shallow continental shelves and almost completely enclosed by large land masses. Extensive river systems drain into the Arctic basin, delivering large volumes of fresh water and sediment. Exchange of water with lower latitudes is highly constrained, and most of the surface is covered with multiyear ice. The Arctic basin food web is probably relatively young, as it is likely that very little planktonic or benthic fauna was present at the height of the last glaciation.

Relatively little is known of the food web of the high Arctic basin as it is so difficult to work there. The benthos is low in diversity, though it is not clear to what extent this is a function of the relative youth of the system, the heavy input of fresh water and sediment in some areas, or the intense disturbance from bottom-feeding marine mammals. At lower latitudes there are important stocks of shoaling plankton-feeding fish, which have long been exploited by man.

The Southern Ocean Food Web

The Antarctic is in many ways a mirror image of the Arctic. It is a large continental land mass entirely surrounded by a deep ocean. The marine food web is likely to have been in existence for many millions of years, and while the zooplankton community appears to be relatively low in species richness, the benthos exhibits a diversity fully comparable with all but the richest habitats elsewhere.

The summer phytoplankton bloom is formed predominantly of the larger diatoms, and the haptophyte *Phaeocystis* appears not to be as important here as in the Arctic. The zooplankton is dominated by copepods and euphausiids, with *Euphausia superba* at lower latitudes and *E. crystallarophias* closer to the ice. Midwater planktivorous fish are almost absent, and the fish fauna is dominated by the radiation of two predominantly benthic/demersal groups: notothenioids on the continental shelf and lipariids in the deeper water of the continental slope. As with the Arctic, relatively little is known of the deep sea.

See also

Fisheries and Climate. Microbial Loops. Southern Ocean Fisheries.

Further Reading

Knox GA (1994) *The Biology of the Southern Ocean.* Cambridge: Cambridge University Press.

Fogg GE (1998) *The Biology of Polar Habitats.* Oxford: Oxford University Press.

El-Sayed SZ (ed.) (1994) *Southern Ocean Ecology.* Cambridge: Cambridge University Press.

Smith WO (ed.) (1990) *Polar Oceanography,* vols 1 and 2. San Diego: Academic Press.

FIORDIC ECOSYSTEMS

K. S. Tande, Norwegian College of Fishery Science, Tromsø, Norway

Introduction

Fiords and semienclosed marine systems are characterized by distinct vertical gradients in environmental factors such as salinity, temperature, nutrients, and oxygen; sampling of biological variables is considerably less adversely influenced by currents and weather conditions than in offshore systems. Therefore, the fiords are particularly well-suited for detailed studies of the pelagic habitat and have traditionally attracted marine scientists, mainly as a site for curiosity-driven research. They are easily accessible and have therefore been used as experimental laboratories for testing new methodologies, exploring trophic relations, and building new theories.

Compared with the fisheries on the shelf outside or in open oceanic waters, the catch from the fiords does not seem proportionately great. Nevertheless, fiords play an important role in many local communities in productive coastal areas in Norway, Scotland, British Columbia, and Greenland. There are fisheries locally renowned for local herring stocks – Loch Fyne for example – and for early year classes of Atlantic–Scandian herring. The salmon fisheries of British Columbia, Scotland, and Norway, highly seasonal affairs catching the fish as they return from the open sea to spawn in fresh water, are also a prominent resource. In more recent time, fiords have become refuse pits of cities and highly populated areas. A growing awareness of the importance of these sites for the bordering societies has led to a more cautionary use of the fiords. This is in particular true for the development of salmon and mariculture production facilities, where future demand for area is being met in coastal developmental plans developed from our knowledge of the physical and the biological setting of the particular fiord habitats.

The trophic relationships play an essential role in shaping the pattern of species interactions. Delineating their existence in the form of food chains and webs provides important knowledge of ecosystem functioning and dynamics. This basic knowledge is essential for the understanding that has to form the basis for ecosystem and management models for the marine biota; this article presents some of this generic knowledge for fiordic ecosystems.

General Features

Globally, most of the fiords are found in high latitudes: in the Northern Hemisphere north of 50°N in Canada, the United States, Greenland, Scandinavia, Scotland and Svalbard; and south of 40°S in Peru, Chile, and New Zealand. They are subjected to temperate, boreal, or arctic climates, and have many oceanographic processes that occur over a wide range of space and timescales. At the high-frequency end of the spectrum, tidal flows can generate turbulence, internal waves, and eddies, while processes such as renewal of oxygen in deep enclosed basins may in extreme cases have timescales of tens of years. Another property of fiords is the presence of strong gradients in a number of abiotic factors, which may affect organisms. The gradients may exist in the horizontal or vertical plane and may be permanent or variable.

Topography and Estuarine Circulation

The fiord systems of the world are all quite young in evolutionary time, yet they offer opportunities that are quite unique among coastal systems. Their common origin as glaciated coastal U-shaped valleys and the common hydrographic conditions give them a number of characteristic features. They are generally long, narrow, and often deeper than the continental shelf outside. A deep basin connects directly or indirectly with the open sea at one end over a relatively shallow sill, typically one-half to one-tenth the basin depth (**Figure 1**). The sill, formed by moraine material as it accumulated at the ice edge, limits the exchange of water between the fiord and the coastal region outside. Most fiords have large rivers that often enter at the head of the fiord, but a substantial amount of fresh water is drained into the fiord from the bordering mainland.

The water masses can be separated into three vertical layers: brackish surface, intermediate, and basin water. The brackish surface layer varies in time and space, dictated by the fresh water runoff, wind, and tide. At the river outlet, a local elevation of the surface generates a density-driven fresh water current out of the fiord, which is gradually being mixed with saline water from below. This vertical mixing process is called entrainment and leads to an overall increase in

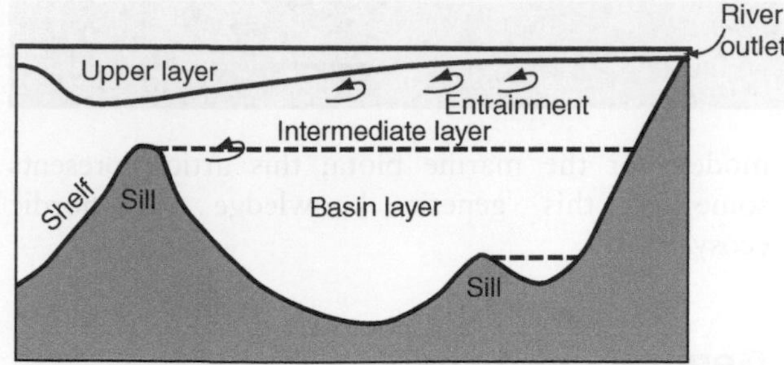

Figure 1 Basic topography and water layers of a fiord with estuarine circulation.

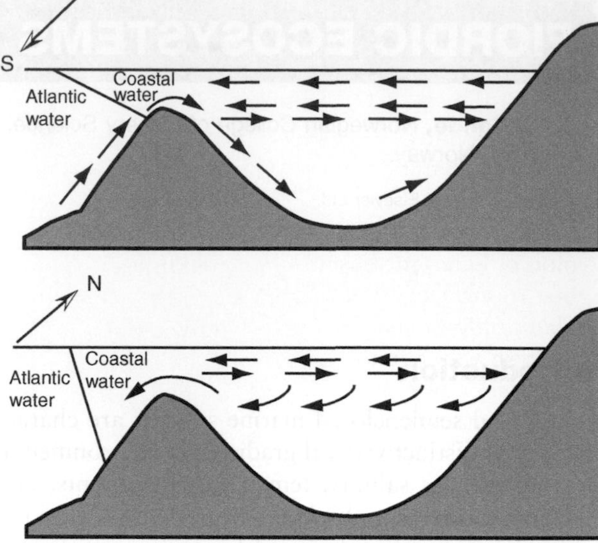

Figure 2 General pattern of wind-induced water exchanges between coastal and fiords on the western European continental margins, with prevailing northerly wind (upper panel) with exchange of basin water, and southerly wind (lower panel) with exchange of water in the intermediate layer.

salinity in the surface brackish water. Entrainment leads to a compensation current below that is going in the opposite direction. This pattern of water transport is called estuarine circulation and is a prominent feature in most fiords. Owing to the Coriolis force there is a tendency for water currents to be deflected to the right (in the Northern Hemisphere), which will generate a larger outflow of surface water on the right side of the fiords compared to the left side.

The tides vary locally, but will in general add only little to the net water transport. Therefore the wind-generated and estuarine circulations, are important for the exchange of water between fiords and the outside shelf. The wind-driven circulation is dictated by the wind stress on the surface and the vertical density gradient. The effect of the wind decreases vertically with increasing density gradient. Since the wind blows in or out the fiord, the wind-driven circulation sets up the estuarine circulation, as either an inward or an outward flowing surface layer depending on the magnitude, duration, and direction of the wind.

The main reason for water exchange in the intermediate layer is the existence of density gradients of water between the fiord and the shelf outside, which follow wind driven upwelling and downwelling at the shelf (**Figure 2**). Characteristically, this exchange occurs mainly as a two-layer circulation with inflow in the surface layer and a compensating outflow at the lower intermediate layer. This happens when the prevailing wind is northward along the western side of the continent (downwelling), and an oppositely directed circulation occurs when the wind is coming from the north (upwelling).

The sill creates a natural barrier that prevents water exchange between the basin and the corresponding depth outside. Partial or total renewal of the basin water takes place when water of higher density than the deep basin water is found above the sill. The major periods of deep-water renewals are found in late winter during upwelling conditions on the shelf outside. Total renewal of the deep water in the fiords happens occasionally, with a frequency of 5–10 years. The basin water in most fiords is being renewed at a rate that prevents anoxia. Nevertheless, in some fiords with very shallow sills, the renewal of the basin water is low and takes place through vertical diffusion. Here one can find anoxic conditions with high levels of hydrogen sulfide, due to the organic drain-off from land and sedimentation from the surface plankton production.

The above outline of the basic physical functioning of fiords underlines the open-ended nature of fiord communities, which nevertheless seem to maintain individual and group characteristics in terms of their biology. There may be a useful distinction between fiords *sensu stricto* and smaller, more shallow enclosed systems (i.e., 'polls'). The basic distinction between these two systems is that fiords have a sill depth greater than the depth of the pycnocline, while shallow enclosed systems normally have a sill depth less than the depth of the pycnocline. In biological terms, the latter are in general an ultraplankton–microzooplankton community, whereas the fiord community is chiefly a net phytoplankton–macro/megazooplankton community.

Fresh Water Runoff and Nutrient Cycling

The fresh water runoff plays a major modifying role in the dynamics of the lower trophic organisms in a fiord. The magnitude of this fresh water runoff varies depending on the extent of the surrounding landmass. Fiords can be classified as high-runoff and low-runoff, arbitrarily separated at an annual mean river runoff of $150\,m^3\,s^{-1}$. On the other hand, peak values of a few weeks' duration are found in the range from 500 to $20\,000\,m^3\,s^{-1}$. There is a strong seasonal variation in fresh water runoff, with annual maximum around April in southern fiords and around June in subarctic regions. Low runoff in the beginning of the season will enhance the stabilization of the water masses, facilitating the onset of the spring bloom. Later, increased runoff enhances the washout of the surface water layer, and modifies the phytoplankton biomass and species composition.

Fresh water runoff plays a minor direct role in the nutrient cycle of fiords, where the nitrate supplied has been found to account for a maximum of 10–20% of the potential uptake by phytoplankton in June in western Norwegian fiords. Indirect factors influencing the transport of new nutrients into the photic zone from below are therefore of major importance for the new production within the fiord. Although most of the nitrate loss out of fiords takes place below the euphotic zone, such losses may affect the future vertical transport of nitrate from the nonphotic to the photic zone. Hence, advective nutrient loss in the layer close to the photic zone may lead to reduced new production within the fiord. Since loss rates of nutrient appears to exceed loss rates from phytoplankton, an advective nutrient exchange, even below the euphotic zone, may therefore be of greater importance to the primary production than the exchange of the phytoplankton itself. The above picture prevails under specific wind conditions on the coast, and in case of reversed winds a nutrient loss may be turned into net nutrient supply to the fiord.

Seasonality in Energy Input to the System

Primary Production

Year-round measurements of primary production exist for a number of fiords and demonstrate that they are in the range of the general level of production in coastal waters of $100\text{–}150\,g\,C\,m^{-2}\,y^{-1}$ (**Table 1**). Some of the data indicate that remarkably high levels of primary production can exist where boundary conditions or land runoff enhance the nutrient load to the fiord. Vertical stability can also generate off-seasonal blooms, even in November and December in fiords at lower latitudes, but it is unlikely that these blooms effectively stimulate secondary production.

A significant proportion of the annual phytoplankton production occurs before the macrozooplankton grazing population becomes established. The estimated annual flux of phytoplankton based carbon to the bottom of fiords is ~10% of the total particulate material sedimentation. This means the majority of the particulate material reaching the bottoms of the fiord is inorganic. Although there is a strong component of copepods in fiord communities, krill, when present, are the major pellets producers contributing to the recognizable organic matter reaching the fiord bottom. Surface sediments show negligible seasonal variation in total organic matter, organic carbon and nitrogen, amino acids, and lipids. This may be due to a rapid conversion of sediment material into a pool of sediment microorganisms. The macrobenthos in the deep basin can be dominated by specialized deposit-eaters, such as the echinoderm (*Ctenodiscus crispatus*), which accumulates fatty acids indicative of an extensive microbial input.

Table 1 Estimates of primary production in representative fiords on the Northern Hemisphere

Region	*Year*	*Primary production* ($g\,Cm^{-2}\,y^{-1}$)
Norway		
Balsfiorden	1977	110
Lindspolls	1976	90–100
Sweden		
Kungsbackafjord	1970	100
Greenland		
Godthaabfjord	1953–56	98
Canada		
How Sound		
entrance	1973	300
entrance	1974	516
inner stations	1973	118
inner stations	1974	163
Port Moody Arm	1975–76	532
Indian Arm and the Narrows regions	1975–76	260
Strait of Georgia	1965–68	120
River Plume	1975–77	149
USA		
Puget Sound	1966–67	465
Port Valdez	1971–72	150
Valdez Arm	1971–72	200

Variability in Time and Space

Scaling of Exchange Processes

The pelagic community in fiords is often found to have a higher variability in time and space than that in oceanic water. This is mostly related to differences in advection between the habitats, where the strongest flushing is usually found in fiords.

As the physical scale of fiords varies, the balance between internal and external forcing is also likely to vary. The cross-sectional area above the sill is therefore an important boundary property, and the ratio between the cross-sectional area and the total fiord volume may indicate the impact of the sill boundary conditions on the fiord. This ratio varies considerably from one fiord to another (**Table 2**). In a sill fiord, the cross-sectional area of the fiord mouth (A) is believed to constrain the water exchange over the sill. The advective impact has been found to be larger in fiords with a high ratio of cross-sectional area to total fiord volume (see A/V in **Table 2**).

The extent to which the planktonic part of a fiord system is controlled by internal biological processes rather than advective processes depends on the physical scale of the fiord versus the timescale of these processes. As a general simplification we may write eqns [1] and [2].

$$\frac{\delta B}{\delta t} = rB + 0.5vR(B_{\mathrm{B}} - B) \quad [1]$$

$$R = \frac{A}{V} \quad [2]$$

Here B = biomass concentration within the system (mg m^{-3}); t = time (s); r = local instantaneous growth rage of B (s^{-1}); v = mean absolute current above the sill (m s^{-1}); B_B = biomass concentration in incoming current (mg m^{-1}); A = cross-sectional area above the sill (m^2); and V = fiord volume (m^3).

As B approaches B_{B}, the net advective effect becomes zero. This does not mean, however, that the advective effect has ceased, since the biomass renewal within the system may still be dominated by advection rather than local growth. The growth rate (r) and the advective rate ($\beta = 0.5\ vR$) have the same dimension (s^{-1}) and the ratio $r/\beta > 1$ decides which of the two processes dominates the biomass formation within the system. If $r/\beta > 1$, growth is the dominating process, while $r/\beta < 1$ indicates advective dominance.

The importance of advection relative to the growth of phytoplankton and zooplankton is given in **Figure 3**. The much lower growth rate of zooplankton compared with that of phytoplankton implies that the transport influences primarily the zooplankton biomass. Phytoplankton is, however, constrained to the upper, photic, zone where transport processes are most prominent. Zooplankton may utilize the entire water column and the advective influence may thereby be diminished. The zooplankton confined to the advective layer (dotted line) in **Figure 3** is influenced three times more strongly by advection than is zooplankton distributed in the entire fiord volume (solid line). Similarly, vertical migration (diel and seasonal) may also reduce the influence of advection for populations depending on the food availability in the advective layer.

From eqns [1] and [2] we see that the value of R (ratio between the cross-sectional area above the sill and the fiord volume) gives the order of magnitude of the advective influence in a particular system. This ratio varies considerably from one fiord to another (**Table 2**), and it may serve as an index indicating the potential advective influence on a system.

Table 2 Examples of ratios of cross-sectional area (A) to total fiord volume (V) in Norwegian fiords

Fiord	*A/V*	*Advective influence*
Lindåspolls	10^{-7}	Advective influence of *Calanus* population < internal processes
Ryfylke fjord	2×10^{-6}	Advective exchange of zooplankton < internal production
Masfjord	8×10^{-6}	Advective exchange of zooplankton = internal production
Malangen	4×10^{-5}	Advective exchange of zooplankton biomass > internal production
Korsfjorden	10^{-4}	*Calanus* heavily influenced by advective processes

Impact of Tidal Currents

The tidal amplitude varies by approximately a factor of 2 within the regions where fiords are found. Tidal currents result in no net water exchange when averaged over a tidal period (12.42 h). On the other hand, there is a significant vertical difference in tidal currents, where the highest flow rates are found in surface layer, with concurrent flow in the opposite direction close to the bottom. Therefore, total exchange of organisms is not necessarily proportional to the net exchange of water, and the vertical position of the organisms plays a major role for the advection rate.

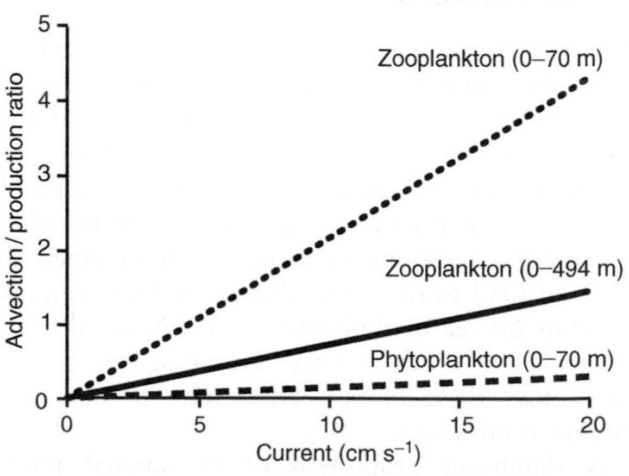

Figure 3 The relation between the ratio advection/production and current velocity for phytoplankton and zooplankton in a fiord. Modified from Aksnes *et al.* (1989).

The renewal rate of water and zooplankton can be calculated as a percentage of the total water and zooplankton biomass volume in a fiord with moderate to high tidal amplitudes (**Table 3**). Comparing contrasting periods in spring and autumn, the total renewal rate of the water is ~6% and 14% per day in spring and autumn, respectively. The numbers for the two given size categories of zooplankton are 6% in spring, 3.5% for zooplankton >500 μm and 12% for zooplankton between 180 and 500 μm in size. The importance of the tidal current for water and zooplankton transport can be estimated by comparing the residual current (i.e., the current velocity after the removal of the tidal component) and the total current. For example, the tidal water is responsible for 23% and 66% of the renewal of zooplankton >500 μm during spring and autumn, respectively. Equivalent numbers for the smallest size fraction of zooplankton are 19% and 39%. This demonstrates that the tide can play an important role in the transport of zooplankton in fiords.

Impact of Vertical Behavior

The zooplankton community structure in fiords appears to be determined primarily by the species–environment relationship, rather than species–species relationships as in the central oceanic gyres. Different spatial distribution patterns are often observed along the length of the fiord, where populations decrease as they are moved away from their population centers. This along-fiord difference in abundance usually evolves from homogenous low stocks of overwintering populations early the same year.

Neritic species prevail at the inner regions, with a downstream decline in the abundance. Most oceanic species have been unable to establish viable populations even in the deepest part of the fiords. However, some, mesopelagic oceanic species have succeeded and the species-specific along-fiord differences in abundance vary and are linked to their differences in vertical behavior.

Table 3 Advective daily transport of water and zooplankton biomass in Malangen, expressed as percentage of the entire volume inside the fjord

	Spring (%)		*Autumn* (%)	
	Total	*Residual*	*Total*	*Residual*
Water	6.1	3.6	14.6	8.1
Zooplankton >500 μm	6.6	5.1	3.5	1.2
Zooplankton 180–500 μm	6.4	5.2	12.0	7.4

Most zooplankton migrates on different timescales, and by diel vertical migration (DVM) the population tends to move from deep waters during daytime to surface during night. The residence time during day and night is normally much longer than the transition time between upper and deep waters. Combining the residence time of the animals with the total current in the same depth strata of the water column, a simple prediction of the displacement over time can be estimated. For a species with wide and regular vertical migration, for instance, *Chiridius armatus*, there is a strong tendency to reside in the same geographical region over time (**Figure 4**). This species does best at deeper and more oceanic sites, and has been found to reduce its population size by a factor of 4 over a distance of 3–4 km, with an overall decline in abundance toward the head of the fiord.

Other species have very low migration amplitude, where ontogenetic migration tends to keep the recruits in surface water for a long time, with an induction of a slow downward migration at older life stages. This is the case for the oceanic species *Calanus finmarchicus*, which has its population center in the North Atlantic and the Nordic Seas but is a very widely distributed and quantitative species in the fiords of the Northern Hemisphere. *C. finmarchicus* spends as much as 8–9 months in the deep waters

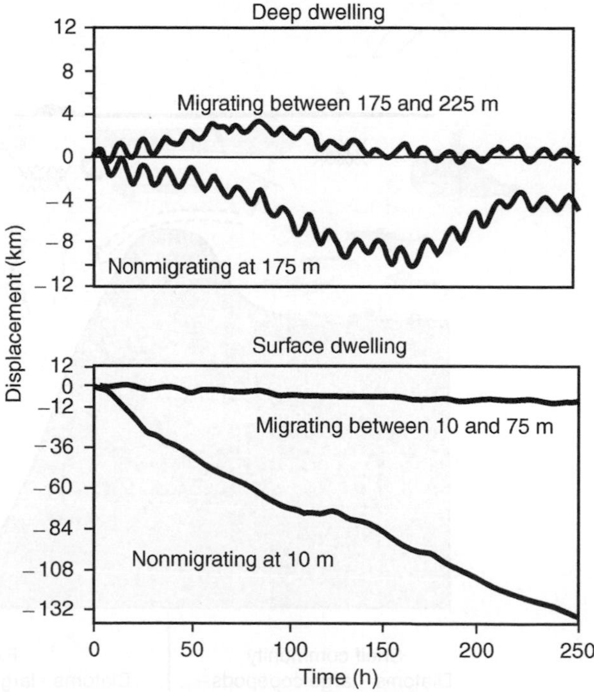

Figure 4 The travel distance for copepods migrating between 175 and 225 m and residing at 175 m during a 24 h cycle (upper panel), and for copepods migrating between 10 and 75 m and residing at 10 m during a 24 h cycle (lower panel).

>500 m while reproduction and growth takes place at the surface from April to July.

C. finmarchicus is a very prominent species on the shelves and in the fiords during the productive season but descends during summer and early autumn, being drained off from the shallower areas toward overwintering depths in deep basins in the fiords and the continental shelves. In some fiords the biomass build-up can reach 10–20% per day, and the high biomass, in particular in the basins during the autumn, clearly indicates that physical/biological aggregation mechanisms overrule the local production. The mechanisms can be explained by the same simple model given above (**Figure 4**), where a declining flush rate with depth tends to retain older stages as they initiate the downward migration during the summer. The biomass aggregation will thus continue for several months, draining older stages from the upper and intermediate water layers in the fiord.

Food Web Structure and Functioning

The Zooplankton Community

The community structure varies extensively between fiords but reflects mostly the shelf habitats found at similar latitudes (**Figure 5**). In subarctic waters, the zooplankton is composed of few species, but with high biomass. Small copepods may be abundant, especially during summer and autumn, and are not major pathways to the juvenile and adult fish. The larger copepods forms (i.e., *Calanus finmarchicus* and *Metridia longa*), the chaetognath *Sagitta elegans*, and the two krill species (*Thysanoessa* spp.) form easily identifiable trophic links in the transfer of materials to higher trophic levels. They all spawn during spring, matching the spring bloom to variable degrees, and each has a restricted growth period within the time-window from April to October. During the long overwintering period a marked decrease in organic lipid-based reserves takes place in both copepods and krill, accounting for 40–70% of that present at the end of the primary production season. Copepods and krill are often found as sound-scattering layers (SSLs) in the basin water of the fiord, and are heavily preyed upon both by demersal and pelagic fish.

The Higher Trophic Animals

In fiords there may be around 30 species that have a commercial potential, although only half of these are exploited regularly by man. Some pelagic and demersal fishes are separated into ocean and coastal

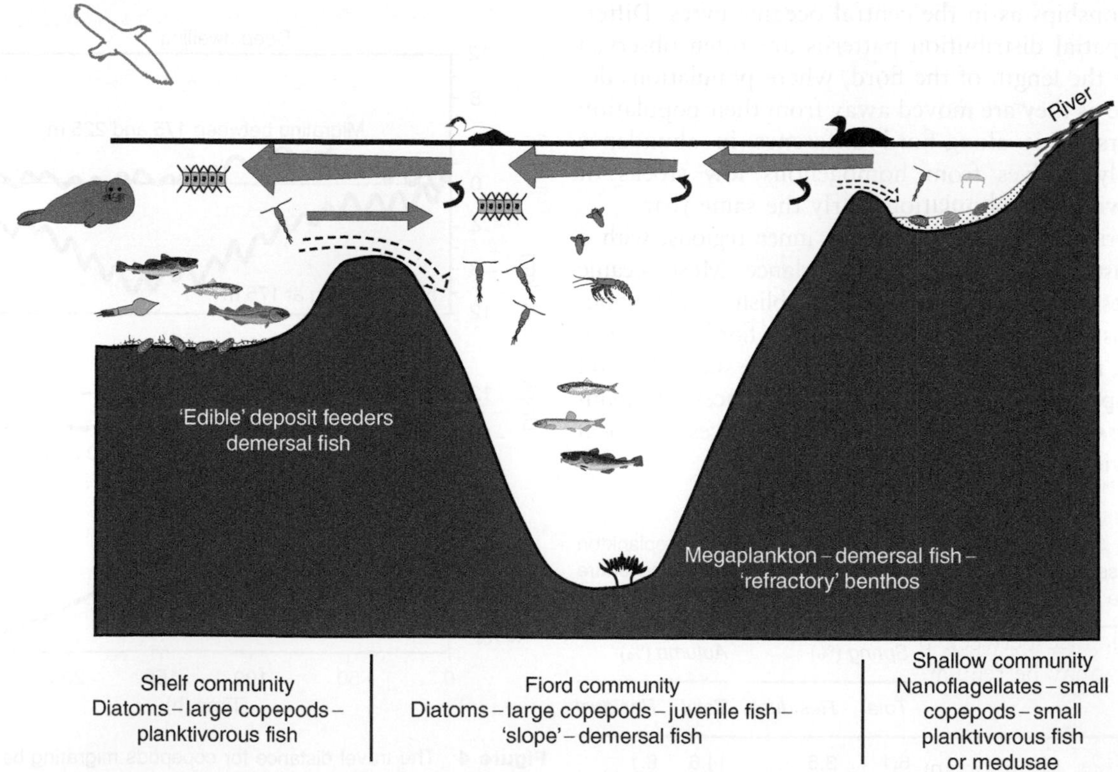

Figure 5 A generalized structure of the biological community from the shelf to the inner part of the fiord. Modified from Matthews and Heimdal (1980).

stocks, where the former spends their time mostly in the ocean and visit the coast during spawning or overwintering. The latter are more confined to the fiords and coastal zones during their entire lifetime.

The communities in fiords may have from one to several apex predators. In some fiords cod plays this role, and as many as 70 taxa have been found in cod stomachs from Balsfiorden, northern Norway. Although as many as 15 fish species were identified, krill, capelin, and herring are considered its dominant prey. This emphasizes that cod is weakly linked with the benethic community. Prawns, also, may be more trophically dependent on the pelagial, a feature contrary to the general tendency in the literature to classify them as part of the benthic food chain. The cod stocks in fiords are mostly coastal cod and undertake very little migration. Tagging experiments have documented that individuals are confined to an area from 5 to 8 km in extent during their entire lifetime. Artificially released cod tend to migrate over longer distance and are more easily trapped by fishing gears and other predators, and are thereby subjected to a higher risk of predation. Cod stocks in fiords may have lower growth rate, length, and age at maturity than oceanic stocks, which indicates that they may be subjected to different management strategies in future.

Consumption of cod by mammals other than man is substantial, although it varies extensively between fiords. Species such as sea otter, harbor seals, and porpoises do visit fiords for short time-windows, having a more limited affect as predators on the local fish fauna. Cormorants have a less varied diet and are known to roost and feed in fiords from late September to early April, within 8 km of their night roost. Although their toll on local cod stocks may be low, around 1/20 in terms of numbers, compared to cannibalism within the cod population, their local impact is still important since they prefer juvenile cod in the length range 4–50 cm.

Other Structural Forces of the Pelagic Community in Fiordic Ecosystems

Light is an important limiting factor for the visual foraging process in fishes, and the light regime may potentially affect the competition between visual and tactile predators. Food demand and risk of mortality are regulated by balance between catching and avoidance between predator and prey, which ultimately may be regulated by visibility in the water column. The seasonal variation in the light may therefore be an important structural force for vertical distributions of important components in the community in fiords. Zooplankton size and density increase with depth; the most visible forms are found only in deep waters. At night, macroplankton and mesopelagic fishes are dispersed in the water column, with a tendency for dispersion of the SSL. All components of the SSL respond to changes in light intensity during day, in order to balance vision versus visibility. The pelagic juvenile fishes in the upper part of the SSL migrate to the surface to maximize their feeding period. The more visible fishes stay in the deeper part of the SSL. At dawn, euphausiids (*Meganyctiphanes norvegica*) descend from surface waters to midwater depths, and the larger mesopelagic fishes (*Benthosema glaciale*) and the pelagic prawns (*Sergestes arcticus* and *Pasiphaea multidentata*) migrate to even darker water. Large pelagic fishes are found in the entire water column, feeding with the highest densities mainly below and at the deepest end of the SSL.

A strong faunal difference is often found between adjacent fiords. This has been linked to the differences in the light climate, fiords with low visibility tending to have a higher component of jellyfish, with the jellyfish being replaced by fish in fiords with higher visibility. This has prompted the hypothesis that the visibility regime may affect the distribution of tactile and visual predators such as jellyfish and fish. The implication is that light is a forcing factor on the marine ecosystem dynamics through the visual feeding process, with potential secondary links to eutrophication. The mechanistic relationship between differences in sea water absorbance and the biological components is not established, but fiords with different visibility regimes will still play an important basis for research needed for ecologically based future management of these important marine biotopes.

See also

Patch Dynamics. Phytoplankton Blooms.

Further Reading

Aksnes DL, Aure J, Kaartved S, Magnesen T, and Richard J (1989) Significance of advection for the carrying capacities of fiord populations. *Marine Ecology Progress Series* 50: 263–274.

Falkenhaug T (1997) *Studies of Spatio-temporal Variations in a Zooplankton Community. Interactions between Vertical Behaviour and Physical Processes*. Dr Scient thesis, University of Troms; ISBN 82-91086-12-5.

Howell B, Moksness E, and Svsand T (eds.) (1999) *Stock Enhancement and Sea Ranching*. Oxford: Fishing News Books.

Matthews JBL and Heimdal BR (1980) Pelagic productivity and food chains in fiord systems. In: Freeland HJ, Farmer DM, and Levings CD (eds.) *Fiord Oceanography*. New York: Plenum.

UPWELLING ECOSYSTEMS

R. T. Barber, Duke University Marine Laboratory, Beaufort, NC, USA

Introduction

An ecosystem is a natural unit in which physical and biological processes interact to organize the flow of energy, mass, and information. The result of this self-organizing activity is that each kind of ecosystem has a characteristic trophic structure and material cycle, some degree of internal homogeneity, objectively definable boundaries, and predictable patterns of seasonality. Oceanic ecosystems are those ecosystems that exist in the open ocean independently of solid substrates; for example, oceanic ecosystems are fundamentally distinct from coral or intertidal ecosystems.

Upwelling ecosystems are those that occupy regions of the ocean where there is a persistent upward motion of sea water that transports subsurface water with increased inorganic plant nutrients into the sunlit surface layer. The upwelling water is not only rich in nutrients, but also frequently cooler than the surface water it replaces; this results in a variety of atmospheric changes, such as coastal deserts or arid zones. The increased nutrient supply and favorable light regime of upwelling ecosystems, however, distinguish them from other oceanic ecosystems and generate characteristic food webs that are both quantitatively and qualitatively different from those of other oceanic ecosystems.

For persistent upwelling to take place it is necessary for the surface layer to be displaced laterally in a process physical oceanographers call divergence and then for subsurface water to flow upward to replace the displaced water. The physical concept of upwelling is simple in principle but, as with many ocean processes, it becomes surprisingly complex when real examples are studied. To begin with, there are two fundamental kinds of upwelling ecosystems: coastal and oceanic. They differ in the nature of their divergence. In coastal upwelling, the surface layer diverges from the coastline and flows offshore in a shallow layer; subsurface water flows inshore toward the coast, up to the surface layer, then offshore in the surface divergence. In contrast, oceanic upwelling, which occurs in many regions of the ocean, depends on the divergence of one surface layer of water from another. One such oceanic divergence is created when an increasing gradient in wind strength forces one surface layer to move faster, thereby leaving behind, or diverging from, another surface layer. Major regions of this kind of oceanic upwelling are found in high latitudes in the Subpolar gyres of the Northern Hemisphere and the Antarctic divergence in the Southern Ocean. The food webs of polar upwelling ecosystems are described elsewhere in the Encyclopedia and this article will focus on coastal and equatorial upwelling ecosystems that occur in low and mid-latitude regions of the world's oceans.

The physical boundary organizing oceanic divergence in equatorial upwelling is the Coriolis force, which changes sign at the equator, causing the easterly Trade Winds to force a northward divergence north of the Equator and a southward divergence south of the Equator. Both coastal and equatorial upwelling ecosystems have been well studied in recent years, so they are among the best known of oceanic ecosystems. The physical processes of equatorial upwelling are described elsewhere in the Encyclopedia. This article describes the quantitative and qualitative character of the food webs of coastal upwelling ecosystems, focusing especially on how their physical forces and chemical conditions affect the way food webs pass organic material to higher trophic level organisms such as fish, birds, marine mammals, and humans.

Why are Upwelling Ecosystem Food Webs different?

In low- and mid-latitude oceanic systems where there is annual net positive heat flux, warming of the surface layer produces a density barrier, the pycnocline, that prevents subsurface nutrient-rich water from mixing into the sunlit surface layer. The nutrient-depleted condition of these surface waters severely limits their annual quantity of new primary productivity, and the food webs of these stratified oceanic regions have low phytoplankton biomass, as shown especially clearly in satellite images of ocean color.

In the high-latitude polar regions where there is annual net negative heat flux, the surface waters cool, become unstable, and mix with the underlying nutrient-rich subsurface waters. The concentration of inorganic nutrients in well-mixed high-latitude waters is high during polar fall, winter, and spring during periods of strong winds, heat loss to the atmosphere, short day length, and low sun angle. But during periods of deep convective mixing the phytoplankton population may spend so much time below the euphotic zone that there is no net positive primary production. The primary producers under these conditions are said to be light-limited.

Upwelling is a circulation pattern that overrides both the nutrient limitation of stratified low- and mid-latitude waters and the light limitation of high-latitude polar waters. Upwelling ecosystem food webs are different from those of other oceanic ecosystems because (1) optimal conditions of nutrient supply are provided by the upward flow of cool, nutrient-rich subsurface water into the sunlit surface layer and (2) optimal light conditions are provided for maximal photosynthetic production of new organic matter in the divergent horizontal flow of upwelled water as it gains heat from the sun, producing a well-stabilized, stratified surface flow.

Optimal nutrient conditions are formally defined as having nutrient [NO_3^-, PO_4^{3-}, $Si(OH)_4$] concentrations well above those that saturate the phytoplankton cell's nutrient uptake mechanism; i.e., $[N] \gg K_s$, where [N] is nutrient concentration in mole units and K_s in one-half the concentration required for nutrient uptake saturation. Optimal light conditions are formally defined as having a level of irradiance, or photon flux density, in the upper waters that exceeds considerably the irradiance required to saturate the photosynthetic capacity of the phytoplankton assemblage; i.e., $[E] \gg K_E$ where [E] is irradiance in mol photons $m^{-2} s^{-1}$ in surface waters and K_E is irradiance at saturation.

In coastal and equatorial upwelling ecosystems, optimal nutrient and light conditions for high primary production are maintained for several months or longer each year, and in low-latitude Trade Wind regions they persist for the entire year; therefore, the annual quantity of new organic matter generated by primary productivity is much higher in upwelling regions than in other oceanic ecosystems that are nutrient- or light-limited or dependent on one or two seasonal pulses of convective mixing.

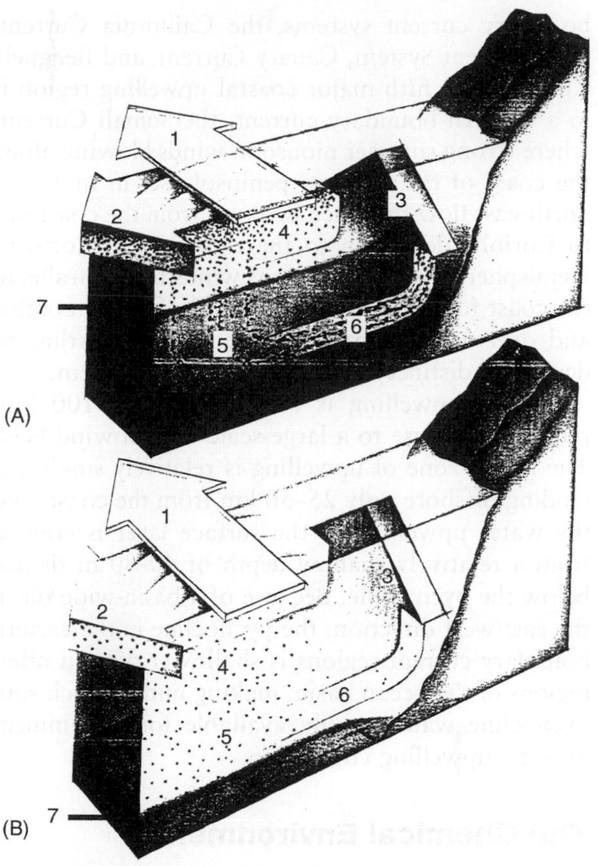

Figure 1 Conceptual diagram of the coastal upwelling ecosystem during (A) normal (cool) conditions and (B) El Niño (warm) conditions. (1) is the alongshore wind blowing toward the Equator; (2) is the wind-driven net offshore surface layer, called the Ekman layer, whose direction of flow is 90° to the left of the wind direction in the Southern Hemisphere because of the Coriolis force; (3) is the upwelling that replaces the water moved offshore in (2); (4) is the euphotic zone where productivity is high relative to other oceanic ecosystems and where high-density blooms of large diatoms accumulate; (5) is the downward flux of ungrazed diatoms and other components of the food web, such as macrozooplankton and fish eggs and larvae; (6) is the subsurface (40–80 m) onshore flow of nutrient-rich water (shown in darker shading) that feeds into the upwelling and recycles material and organisms that sink out of the Ekman layer; (7) is the thermocline and nutricline that separate cool, nutrient-rich subsurface water from the surface layer of warm and nutrient-depleted water. This is an original figure designed by RT Barber in 1983.

The Physical Setting

Upwelling is a response of the ocean to wind-driven divergence of the surface layer. As the wind begins to blow across the surface of the ocean, a thin surface slab of water (25–50 m thick) is set in motion by friction of the wind (**Figure 1**). This wind-driven layer or Ekman layer (named for the Swedish oceanographer who in 1905 worked out how wind drives ocean currents), as a result of the Coriolis force, has a net movement 90° to the right (left) of the wind in the Northern (Southern) Hemisphere. Four of the major coastal upwelling systems are located in the eastern boundary of the ocean basins along the west coasts of the continents where equatorward winds are part of stationary or seasonal mid-ocean high-pressure systems. These four coastal upwelling regions off the west coasts of North America, South America, north-west Africa, and south-west Africa are in the four great eastern boundary current systems, the California Current, Peru Current System, Canary Current, and Benguela Current. The fifth major coastal upwelling region is in a western boundary current, the Somali Current, where strong summer monsoon winds blowing along the coast of the Arabian peninsula set in motion a north-east flow that then diverges from the coast due to Coriolis deflection to the right in the Northern Hemisphere. In all five regions winds blow parallel to the coast for a long enough period of time (months) and over a sufficiently large length of coastline to develop a distinct coastal upwelling ecosystem.

Coastal upwelling is a mesoscale (10–100 km) physical response to a large-scale coastal wind field. The major zone of upwelling is relatively small, extending offshore only 25–50 km from the coast, and the water upwelling to the surface layer is coming from a relatively shallow depth of 40–80 m or just below the pycnocline. Because of a basin-wide tilt in the east/west direction, the pycnocline in the eastern boundary current regions is shallower than in other regions of the ocean basin, making nutrient-rich subpycnocline water readily available for entrainment into the upwelling circulation.

The Chemical Environment

The *sine qua non* of coastal upwelling is high concentrations of the new inorganic plant nutrients nitrate (NO_3^-), phosphate (PO_4^{3-}), and silicate or silicic acid ($Si[OH]_4$) that are well in excess of the half-saturation concentrations for nutrient uptake. Typical concentrations are as high as 15–20 μmol l^{-1} of NO_3, with the other macronutrients occurring in appropriate proportional concentrations according to the Redfield ratio. The highest nutrient concentrations and lowest water temperatures are inshore in the most recently upwelled water; there is frequently a strong offshore spatial gradient in nutrient concentration, but the spatial domains of the five great coastal upwelling ecosystems vary remarkably. In the Peruvian upwelling near 15°S latitude the onshore/offshore gradients are steep, with nitrate concentrations decreasing from 20 to 2 μmol l^{-1} in an offshore distance $\leqslant$50 km; in the Somali Current off the coast of Oman the initial inshore concentrations are lower, about 10 μmol l^{-1}, but remain elevated for 500–700 km offshore.

The supply of new nutrients advected into the euphotic zone sets up the highly productive character of upwelling ecosystems, but nutrients regenerated or recycled in the euphotic zone are also unusually abundant in coastal upwelling. High-productivity fuels increased heterotrophic consumption by protozoans, crustaceans, and vertebrates, and these consumers, along with heterotrophic bacteria, bring about increased regeneration of nutrients. Regeneration of nutrients from particulate organic matter that sinks out of the offshore surface flow and into the subsurface inshore flow results in nutrient 'trapping' that maintains elevated nutrient concentrations in bottom waters of the continental shelf. These regenerated nutrients, together with short-term storage of regenerated nutrients in surficial sediments beneath the upwelling circulation, provide a flywheel to the nutrient supply process that dampens variations in the wind-driven vertical transport of new nutrients from deep water.

An additional important chemical consequence of trapping by the two-layered partitioning of organic particles in coastal upwelling is the generation of zones of intense oxygen depletion. The great oxygen minimum zones of the four eastern boundary currents and the Somali Current are fueled by enhanced productivity in the narrow coastal upwelling zone. In addition to water column oxygen depletion, shelf and slope sediments under coastal upwelling are frequently anoxic and colonized by large anaerobic bacterial mats. These benthic hypoxic and anoxic zones are two sites of intense denitrification, a microbial process by which nitrate is converted to nitrogen gas. Occasionally, oceanographers have found complete denitrification in a midwater anoxic layer beneath upwelling systems; these processes of benthic and water column denitrification may be a major global feedback mechanism involved in the regulation of fixed, or biologically available, nitrogen.

Another important chemical consequence of the reducing conditions generated in anoxic and hypoxic sediments beneath coastal upwelling involves the cycling of iron. Iron is an essential micronutrient for the maintenance of high rates of primary productivity. Studies in the coastal upwelling ecosystem of the California Current System showed that resuspension and dissolution of iron from sediments generated enhanced concentrations of iron in the bottom boundary currents. Subsequent upwelling of this subsurface water during episodes of strong upwelling resulted in elevated iron concentrations in the euphotic layer. Particle sedimentation to anoxic or hypoxic sediments followed by resuspension and dissolution is a positive feedback that enhances the productive potential of coastal upwelling, especially compared to open ocean equatorial upwelling.

A Milestone in Quantifying Food Web Function

The basic food webs of upwelling ecosystems differ in both quantity and quality from those of other oceanic ecosystems. A milestone in understanding these differences was made by John Ryther, who in 1969 provided a quantitative explanation of why fish yields vary by about 200-fold from the richest coastal upwelling ecosystems to the poorest ocean gyres. Variations in productivity are, of course, well known from terrestrial ecosystems, but on land a lack of water from either aridity in deserts or freezing in polar regions is responsible for the productive poverty of the poorest regions. Understanding why the food web of the benign low-latitude gyre ecosystem was so poor in fish production was much more difficult. Part of the explanation was proposed in 1955 by Sverdrup who stated simply that reduced physical supply of nutrients to the euphotic zone is the reason for the low productivity, biomass, and fish yields of stratified oceanic gyre ecosystems. Ryther amplified this simple physical explanation by considering, along with the physics and chemistry, the biological properties of the food web that lead to fish production.

First, Ryther estimated that about half the fish caught in the world are caught in coastal upwelling ecosystems, the smallest of the ocean ecosystems. Why? To begin, Sverdrup was correct: the physical processes of upwelling and subsequent stratification provide optimal nutrients and light to support high primary productivity. However, more is involved. The phytoplankton, especially diatoms, that thrive in coastal upwelling are large – so large that some portion of the diatoms can be eaten directly by fish or other large grazers such as euphausids. This means that in coastal upwelling the food web leading to fish is often very short, involving only one, or at most two, trophic transfers. Ryther estimated that in the Peru upwelling ecosystem half of the diet of the small pelagic clupeid fish such as anchovies is phytoplankton and the other half is composed of crustacean zooplankton such as euphausids. On average, then, the length of the food web from primary producers to fish had 1.5 transfers: large diatoms to anchovies, or large diatoms to euphausids to larger fish such as mackerel. At each ecological transfer, a large portion (80–90%) of the energy of the food is used to support the organism and that portion cannot be passed up the food web. Ryther also noted that in the phytoplankton-rich waters of the spatially small coastal upwelling regions, grazers do not have to work so hard to get food; therefore, the efficiency of transfer through the food web is increased relative to that of a poor environment such as the low-latitude gyre, where grazers have to cover larger distances and filter large volumes of water to get adequate food. Ryther proposed that fish yields are high in the coastal upwelling ecosystem because of (1) high initial primary productivity, (2) large phytoplankton that can be grazed directly, (3) short food webs with few transfers, and finally, (4) increased efficiency at each transfer. These effects multiply and lead to high yields of fish that are 200 times the yield of gyre ecosystems. These high yields are exploited by seabirds, marine mammals and, of course, humans.

Food Web Structure and Function

Coastal upwelling ecosystems are typically dominated by chain-forming and colonial diatoms with individual cell diameters of 5–30 μm. The growth rates of these large cells are surprisingly as fast as those of the much smaller autotrophic pico- and nanoplankton that are the basis of the microbial loop. The larger diatoms are more effective than pico- or nanoplankton at taking up high concentrations of new nutrients; this property, together with their more favorable photosynthesis/respiration ratio, makes diatoms considerably more efficient at new production.

New production uses nutrients carried into the system by upwelling, while regenerated production is based on nutrients recycled in the euphotic zone. The *f*-ratio measures the proportion of new production; *f*-ratios of coastal upwelling are as high as any in the oceans, with values ranging from 0.3 to 0.8 and 0.5 being a representative value. Primary productivity values in the most productive portion of the upwelling ecosystem range from 1.0 to 6.0 mg C m^{-2} d^{-1}. Representative inshore values for the California Current System are 1.0–3.0 mg C m^{-2} d^{-1}; for the Peru Current System 2.0–6.0 mg C m^{-2} d^{-1}; for the Canary Current 1.0 – 3.0 mg C m^{-2} d^{-1}; and for the Somali Current 1.0–2.0 mg C m^{-2} d^{-1}. High *f*-ratios and high primary productivity indicate that more organic material can be exported via the food web to higher trophic levels such as fish, birds, and marine mammals or exported vertically as particle flux to deep water or sediments.

A second element in Ryther's hypothesis was that large diatoms could be grazed directly by clupeid fishes. Why are the phytoplankton in coastal upwelling large? One explanation comes from a model study of diatom sinking and circulation in the Peruvian upwelling region. Small phytoplankton that sank slowly or maintained themselves in the euphotic zone were consistently carried in the surface Ekman layer to the oligotrophic offshore waters; large diatoms that sank rapidly fell into the subsurface onshore circulation and were carried back into the upwelling cycle (**Figure 1A**). Large size that confers fast sinking is an adaptation that keeps diatoms in the highly productive upwelling habitat for several growth cycles. In addition, newly upwelled water contains large numbers of diatom resting spores, indicating that diatoms sink to sediment, remain there in a resting stage, then become resuspended and transported into the euphotic zone by episodes of strong upwelling. Large size confers rapid sinking, which enhances both recirculation and resuspension, but it also makes the large diatoms efficient prey for fish and large zooplankton like euphausids.

The biomass of larger phytoplankton such as diatoms is more variable in time and space than the biomass of pico- and nanophytoplankton. The abundance of small phytoplankton is efficiently controlled by their fast-growing protozoan microzooplankton grazers. The micrograzers can grow as rapidly as their prey, so there is no opportunity for uncoupling of prey and predator abundance; pico- and nanoplankton, therefore, rarely form blooms. In contrast, diatoms are grazed by larger organisms with longer reproductive cycles, such as clupeid fish with a 1-year cycle or copepods and euphausids with a cycle of 10–40 days or longer. Clearly, the zooplankton or fish cannot reproduce fast enough to keep up their abundance in pace with a diatom bloom; at times, therefore, large diatoms can accumulate in dense blooms with low initial grazing losses. While fish and zooplankton cannot match growth rates with diatoms, they do have mobility and behavior that enable them to find and move into patches of abundant food. In practice, however, coupling of the growth rates of diatoms and their animal grazers frequently breaks down, and when this happens high biomass blooms become evident in the ocean color satellite images in upwelling regions.

Phytoplankton cells, especially large cells that are not grazed or consumed by heterotrophic microorganisms, rapidly sink out of the water column when ungrazed biomass accumulates in a dense bloom (**Figure 1A**). If nutrients are depleted by the high-biomass bloom, phytoplankton lose the ability to regulate their buoyancy and sink rapidly at rates as high as $100\,\mathrm{m\,d^{-1}}$. Sediments under coastal upwelling ecosystems are characterized by the highest rate of organic deposition found in the ocean. These high deposition rates indicate that the large diatom/large grazer food path is relatively more important to the throughput of material than the microbial or picophytoplankton/nanophytoplankton/protozoan grazer path. The microbial path is always present in the two-path upwelling food web and it does increase in absolute productivity during increased upwelling; however, the huge increase in biomass and productivity of the large diatom/large grazer food path dominates export of new organic material. The large diatom food path does not replace the picophytoplankton/nanophytoplankton path, but it becomes so numerically overwhelming that it appears as though there is a shift in the character of the food web.

In coastal upwelling ecosystems there is enough time and space constancy in the physical response that macrozooplankton and shoaling pelagic fish have been able to evolve adaptations that enable them to exploit this rich but small habitat, and these adaptations affect the efficiency of transfer of primary production to higher trophic levels. Zooplankton such as copepods and euphausids have limited ability to swim against onshore–offshore currents, but they have considerable ability to migrate up and down rapidly. Some upwelling zooplankton species have evolved behavior that causes them, when saturated with food in the offshore flow, to migrate down into the onshore flow, which then carries them back into the upwelling circulation for another cycle. Other species remain in the food-rich habitat by having eggs or juvenile life stages that sink into the subsurface onshore flow. The adaptations of macrozooplankton to the physics of upwelling are remarkable examples of how the evolution of upwelling organisms differs from the evolution of organisms of other oceanic ecosystems. Parallel adaptations are present in the shoaling pelagic fish that dominate the fish biomass of coastal upwelling ecosystems. These behavioral adaptations have optimized feeding, reproduction, and growth for the sardines, anchovies, and mackerel that make up the bulk of the fish harvested from coastal upwelling ecosystems.

Climatic Forcing and Food Web Responses

Adaptations to the specific upwelling circulation pattern confer great fitness advantage to phytoplankton, zooplankton, fish, birds, and marine mammals when the upwelling pattern is prevalent, but the coastal upwelling ecosystems are buffeted by strong interannual and interdecadal climate variability. The El Niño–Southern Oscillation (ENSO) phenomenon is the best-known example of large-scale, climate-driven biological variability. El Niño is defined by the appearance and persistence, for 6–18 months, of anomalously warm water in the coastal and equatorial ocean off Peru and Ecuador. The anomalous ocean conditions of El Niño are accompanied by large reductions of plankton, fish, and sea birds in the normally rich upwelling region. To understand how this climate variability causes these large decreases in abundance, consider how El Niño temporarily alters the physical pattern of the upwelling circulation. One discovery of recent decades is that during El Niño events the coastal winds that drive coastal upwelling do not stop entirely (**Figure 1B**). In fact, coastal winds sometimes intensify during El Niño because of increased thermal differences between land and sea. Therefore, coastal upwelling as a physical process continues, but because the ENSO process has depressed the thermocline and nutricline to a depth below the depth at which water is entrained into the upwelling circulation (40–80 m), the water upwelled is warm and low in nutrients. As a result, during El Niño the upwelling circulation transports only warm, nutrient-depleted water to the surface layer. The physics of upwelling continues, but the chemistry of upwelling stops very dramatically.

This conceptual model of El Niño forcing and food web response, shown in **Figure 1B**, indicates that El Niño affects the upwelling ecosystem by decreasing the nutrients supplied to the euphotic layer, which causes primary production to decrease proportionally. In this manner the supply of nutrients is reduced as El Niño strengthens in intensity, and the decrease in new primary production available to fuel the food web causes proportional reductions in the growth and reproductive success of fish, birds, and marine mammals. Obviously, temperature, nutrients, primary productivity and higher trophic level productivity are tightly linked in coastal upwelling ecosystems, but by far the most dramatic link is the climate variability/fish variability link. That is, the most impressive biological consequence of El Niño is its effect on the abundance and catch of Peruvian anchovy (*Engraulis ringens*), the basis of the world's largest single-species fishery. **Figure 2** shows the covariation of thermal conditions and anchovy harvest from the 1950s to the present. This relationship is causal in the sense that temperature is a proxy for nutrients, and nutrient decreases (temperature increases) are always accompanied by reduction in the productivity of the food web including the catch of anchovies. Note that the temperature/nutrient variability works in both directions. Each local minimum in catch is associated with a warm anomaly and each local maximum is associated with cool, nutrient-rich conditions. The period of very low catch from 1976 to 1985 is often cited as an example of the destruction of a fishery by overfishing, but **Figure 2** indicates that the anchovy stock failed to recover from 1972 and 1976 El Niño events because there was little upwelling of cool, nutrient-rich waters during that decade. The coastal winds were normal

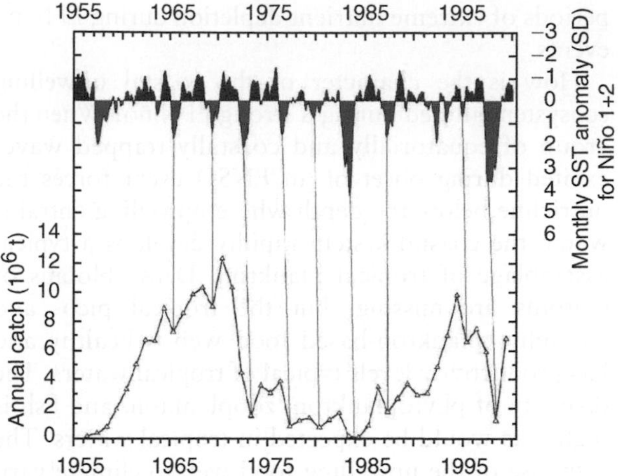

Figure 2 The association of sea surface temperature (SST) anomaly along the coast of Peru and Ecuador with the annual catch of Peruvian anchovy, showing that each minimum of catch is associated with a period of anomalously warm water. Note that the SST anomaly scale is inverted, with red showing the warm anomalies. The anomaly is calculated from SST in the Peru coastal area (Niño 1) and the eastern equatorial Pacific (Niño 2). Warmer water at the sea surface means that warmer water is being entrained into the upwelling cell because the thermocline has deepened owing to large-scale, basin-wide responses to changes in Trade Winds. The nutricline also deepens, so that the warmer water is also lower in nutrient concentration. Temperature is a proxy in this figure for nutrient concentration. The close association of SST anomaly and anchovy catch suggests that natural thermal and nutrient variability, not overfishing, is the process controlling the interannual variability of this particular fish stock.

or even stronger than normal during this decade, but the increased heat storage in the upper ocean apparently kept the thermocline and nutricline anomalously deep from 1976 to 1985.

The extreme variability of anchovy abundance sends shock waves into the global economy, because fishmeal from upwelling ecosystems is a commodity that is necessary for a variety of animal production processes. The social hardship of this climate-driven variability affects many people, but the upwelling ecosystem is not in the least damaged by ENSO variability. The food web has evolved to exploit the productive phase of the ENSO cycle and persist through the unproductive phase. **Figure 2** shows that as long as a period of cool, high-nutrient conditions follows the warm event, the coastal upwelling system recovers to its previous high productivity. The climate process that appears to have the potential to alter or disrupt this ecosystem is the lower-frequency, decadal anomaly that prevailed from 1976 to 1985 and again in the mid-1990s. A decadal anomaly that causes relative nutrient poverty appears to have greater long-term food web consequences than short periods of extreme nutrient depletion during El Niño events.

How is the character of the coastal upwelling ecosystem altered during a strong El Niño? When the group of equatorially and coastally trapped waves excited during onset of an ENSO event forces the nutricline below the depth where upwelling entrains water, the coastal system rapidly develops a typical assemblage of tropical plankton. Dense blooms of diatoms are missing, but the tropical pico- and nanophytoplankton-based food web is healthy and has productivity levels typical of tropical waters. The diversity of phytoplankton, zooplankton, and fish is high – as would be expected in tropical waters. The response of the upwelling food web to climate variability emphasizes the resilience of oceanic ecosystems to strong transient perturbations; their resilience to the effects of persistent change, however, is unknown.

Glossary

Antarctic divergence The zone of upwelling driven by the Antarctic Circumpolar Current (ACC).

Convective mixing Vertical mixing produced by the increasing density of a fluid in the upper layer, especially during winter in temperate and polar regions.

Denitrification A microbial process that takes place under anoxic conditions, converting nitrate to N_2 gas.

Diatom A taxonomic group of phytoplankton that are nonmotile, have silicon frustules, and are capable of rapid growth.

Ecosystem A natural unit in which physical and biological processes interact to organize the flow of energy, mass, and information.

Ekman layer The surface layer of the ocean that responds directly to the wind.

Euphotic zone The surface layer of the ocean where there is adequate sunlight for net positive photosynthesis.

Nutrients Dissolved mineral salts necessary for primary productivity and phytoplankton growth; macronutrients are phosphate, nitrate, and silicate; micronutrients are iron, zinc, manganese, and other trace metals.

Oxygen minimum zone A mid-water layer along the eastern boundary regions of the oceans in which oxygen concentrations are significantly reduced relative to the layers above and below it.

Phytoplankton Photosynthetic single-called plants or bacteria that drift with ocean currents and are the major primary producers for oceanic food webs; very small phytoplankton are called picoplankton and small phytoplankton are called nanoplankton; all of these are $<2\,\mu m$ in diameter.

Primary productivity The use of chemical or radiant energy to synthesize new organic matter from inorganic precursors.

Pycnocline The layer where density changes most rapidly with depth and separates the surface mixed layer from deeper ocean waters.

Southern Ocean The circumpolar ocean in the Southern Hemisphere between the Subtropical Front and the continent of Antarctica.

Stratification The formation of distinct layers with different densities (see 'pycnocline' above); stratification inhibits mixing or exchange between the nutrient-rich deeper water and the sunlit surface layer.

Subpolar gyres Large cyclonic water masses in the Northern Hemisphere between the subtropical front and the polar front.

Tropical Pertaining to the regions that, under the influence of the Trade Winds, are permanently stratified.

Upwelling Upward vertical movement of water into the surface mixed layer produced by divergence of the surface waters.

Zooplankton Animals that float or drift with ocean currents; microzooplankton are protozoan plankton that graze on small phytoplankton; mesozooplankton are crustaceans that graze on larger phytoplankton such as diatoms.

See also

Fisheries and Climate. Microbial Loops. Network Analysis of Food Webs. Ocean Gyre Ecosystems. Plankton and Climate. Polar Ecosystems. Small Pelagic Species Fisheries.

Further Reading

Bakun A (1990) Global climate change and intensification of coastal ocean upwelling. *Science* 247: 198–201.

Barber RT and Chavez FP (1983) Biological consequences of El Niño. *Science* 222: 1203–1210.

Barber RT and Smith RL (1981) Coastal upwelling ecosystems. In: Longhurst A (ed.) *Analysis of Marine Ecosystems*, pp. 31–68. New York: Academic Press.

Longhurst A (1998) *Ecological Geography of the Sea*. San Diego: Academic Press.

Pauly D and Christensen V (1995) Primary production required to sustain global fisheries. *Nature* 374: 255–257.

Richards FA (ed.) (1981) *Coastal Upwelling*. Washington, DC: American Geophysical Union.

Ryther JH (1969) Photosynthesis and fish production in the sea. *Science* 166: 72–76.

Smith RL (1992) Coastal upwelling in the modern ocean. In: Summerhayes CP, Prell WL and Emeis K-C (eds) *Upwelling Systems: Evolution Since the Early Miocene*, Geological Society Special Publication 64, pp. 9–28. London: The Geological Society.

Summerhayes CP, Emeis K-C, Angel MV, Smith RL and Zeitschel B (eds.) *Upwelling in the Ocean Modern Processes and Ancient Records*. Chichester: Wiley

Sverdrup HU (1955) The place of physical oceanography in oceanographic research. *Journal of Marine Research* 14: 287.

CORAL REEFS

J. W. McManus, University of Miami, Miami, FL, USA

Introduction

Coral reefs are highly diverse ecosystems that provide food, income, and coastal protection for hundreds of millions of coastal dwellers. They are found in a diverse range of geomorphologies, from small coral communities of little or no relief, to calcareous structures hundreds of kilometers across. The most diverse coral reefs occur in the waters around south-east Asia. This is primarily because of extinctions that have occurred in other regions due to the gradual restriction of global oceanic circulation associated with continental drift. However, rates of speciation of coral reef organisms in this region may also have been high within the last 50 million years.

Human activities have caused the degradation of coral reefs to varying degrees in all areas of the world. A major focus of present research is on the resilience of coral reefs to disturbances such as storms, diseases of reef species, bleaching (the expulsion of endosymbiotic photosynthetic zooxanthellae) and harvesting to local extinction. Along many coastlines, a combination of increased eutrophication due to coastal runoff and the extraction of herbivorous fish and invertebrates appears to favor the replacement of corals with macroalgae following disturbances. Another important aspect of resilience is the degree to which a depleted population of a given species on one reef can be replenished from other reefs. Most coral reef organisms undergo periods of free-swimming (pelagic) life ranging from a few hours to a few months. Genetic studies show evidence of broad dispersal of the progeny of some species among reefs, but most of the replenishment of a given population from year to year is believed to be from the same or nearby coral reefs.

Coral reefs are complex biophysical systems that are generally linked to similarly complex socioeconomic systems. Their proper management calls for system-level approaches such as integrated coastal management.

General

Types of Coral Reefs

The term 'coral reef' commonly refers to a marine ecosystem in which a prominent ecological functional role is played by scleractinian corals. A 'structural coral reef' differs from a 'nonstructural coral community' in being associated with a geomorphologically significant calcium carbonate (limestone) structure of meters to hundreds of meters height above the surrounding substrate, deposited by components of a coral reef ecosystem. The term 'coral reef' is often applied to both structural and nonstructural coral ecosystems or their fossil remains, although many scientists, especially geomorphologists, reserve the term for structural coral reefs and their underlying limestone. Both types of ecosystem occur within a wide range of tropical and subtropical marine environments, although structural development tends to be greater in waters of lower silt or mud concentration and oceanic salinity. Many reefs survive well amid open ocean waters with low nutrients, aided by efficient 'combing' of waters for plankton, high levels of nitrogen fixation and fast and thorough nutrient cycling. However, extensive coral reefs also occur in coastal waters of much higher nutrient concentrations.

Scleractinian (stony) corals grow as colonies or solitary polyps on a wide variety of substrates, including fallen trees, metal wreckage, rubber tires and rocks. Rates of settlement are often enhanced by the presence of calcareous encrusting algae. Soft sand, silt and mud tend to inhibit the settlement of stony corals, and so few coral ecosystems occur in modern or ancient deltaic deposits. However, coral can grow very near the mouths of small rivers and steams, and fresh or brackish groundwater often percolates through reef structures or emerges periodically through tunnels and caves. Vertical caves are often called 'blue holes'.

Nonstructural coral communities are common on rocky outcrops in shallow seas in many tropical and subtropical regions. They can range from a few clumps of coral to very substantial communities covering many square kilometers of wave-cut shelves near deeper areas.

Structural coral reefs come in many shapes and sizes, from less than a kilometer to many tens of kilometers in linear dimension. It is helpful to differentiate individual coral reefs from systems of coral reefs, such as the misnamed 'Great Barrier Reef' of

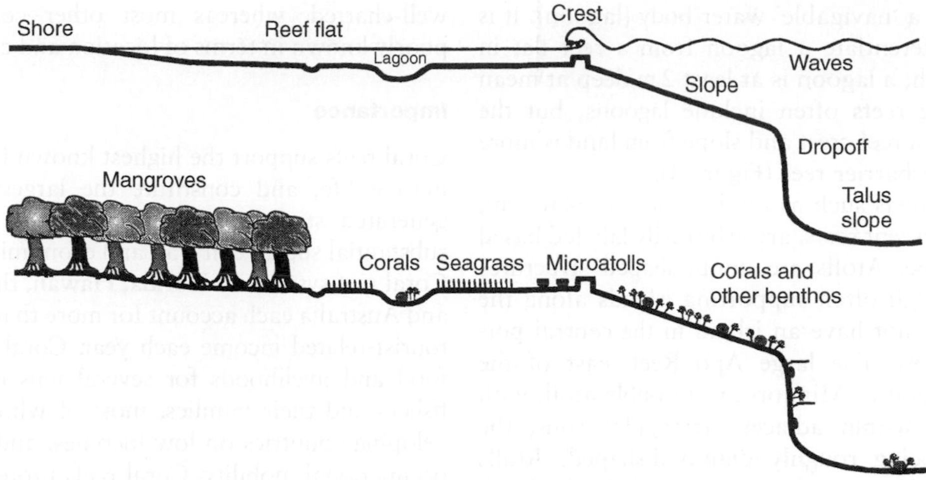

Figure 1 Profiles of a hypothetical fringing reef showing geomorphological and ecological zonation relative to wave action.

Australia, which actually consists of thousands of densely packed coral reefs. Common types of structural coral reefs include fringing reefs, barrier reefs, knoll reefs, pinnacle reefs, platform reefs, ribbon reefs, crescent reefs, and atolls. The term 'patch reef' may refer either to a patch of coral and limestone a few meters across within a structural coral reef (typically in a lagoon or on a reef flat), or to a platform or knoll reef. Thus, the term is best avoided. There is a wide range of structures intermediate between crescent reefs, platform reefs, and atolls in areas such as the Great Barrier Reef System.

A fringing reef is, by definition, always found adjacent to a land mass. Most fringing reefs include a wave-breaking reef crest, one or more meters above the rest of the reef, forming a thin strip offshore (**Figure 1**). Between the crest and the land, there is usually a relatively level area, broken by channels, called a 'reef flat' (**Figure 2**). Fringing reefs differ from barrier reefs, in that the latter are separated

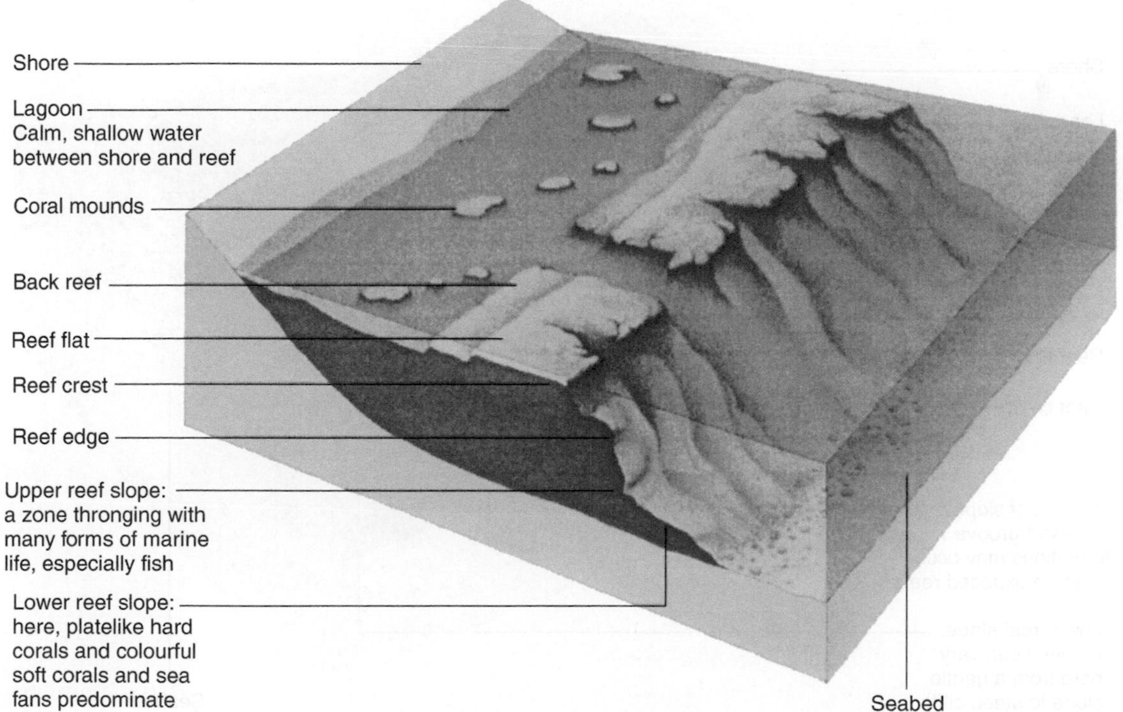

Figure 2 Geomorphology of a typical fringing reef. (Adapted from Holliday, 1989.)

from land by a 'navigable' water body (lagoon). It is useful to differentiate a lagoon from a reef flat in terms of depth; a lagoon is at least 2 m deep at mean tide. Fringing reefs often include lagoons, but the separation of a reef crest and slope from land is more complete in a barrier reef (**Figure 3**).

Structural reefs such as knoll, pinnacle, platform, ribbon, or crescent reefs, are arbitrarily labeled based on their shape. Atolls are donut-shaped structures which, although often supporting islands along the outer rim, do not have an island in the central portion (**Figure 4**). The large Apo Reef, east of the Philippine island of Mindoro, is a double atoll, with two lagoons within adjacent triangular rims, the whole reef being roughly diamond-shaped. Atolls can be quite large, such as the North Male Atoll, which houses the capital of the Maldives (**Figure 5**).

Although one commonly thinks of structural coral reefs as reaching to the sea surface, most of the coral reefs of the world do not. For example, there is a system of atolls and other reefs to approximately 50 km off the north-west of Palawan Island (Philippines) that closely resembles portions of the Australian Great Barrier Reef. However, very few reefs of the Palawan 'barrier system' come within 10 m of the sea surface. Some estimates of coral reef area are based on reefs at or near the sea surface, partly because the larger examples of such reefs tend to be well-charted, whereas most other coral reefs are poorly known in terms of location and characteristics.

Importance

Coral reefs support the highest known biodiversity of marine life, and constitute the largest biologically generated structures on Earth. Coral reefs are of substantial social, cultural, and economic importance. Coral reef systems in Florida, Hawaii, the Philippines, and Australia each account for more than $1 billion in tourist-related income each year. Coral reefs provide food and livelihoods for several tens of millions of fishers and their families, most of whom live in developing countries on low incomes, and have limited occupational mobility. Coral reefs protect coastal developments and farm lands from erosion. In many countries, particularly among Pacific islands, coral reefs are culturally very important, as they are involved in social structuring and interaction, and in religion. Despite poor taxonomic understanding and increasingly strict controls on bioprospecting, coral reef species are yielding substantial numbers of important drugs and other products.

Zonation

The ecological zonation of most coral reefs depends on physical factors, including depth, exposure to

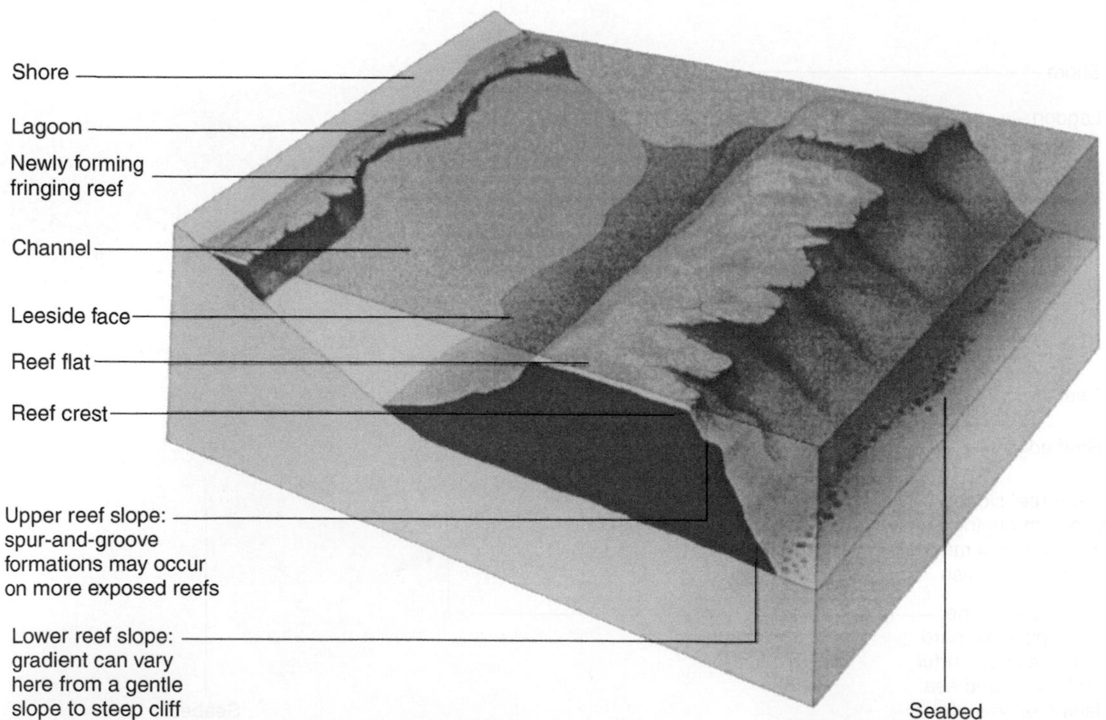

Figure 3 Geomorphology of a typical barrier reef. (Adapted from Holliday, 1989.)

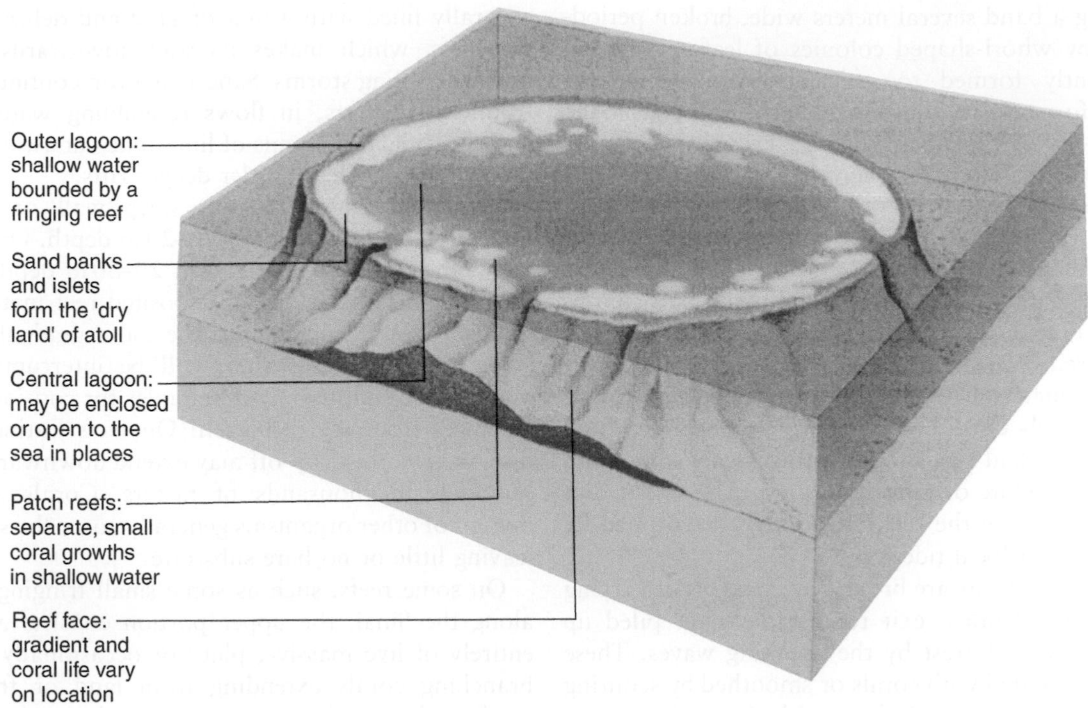

Figure 4 Geomorphology of a typical atoll. (Adapted from Holliday, 1989.)

waves and currents, and oxygen limitation. A fringing reef is basically a large slab of limestone jutting out from land. The landward margin may support extensive mangrove forests. A sandy channel may

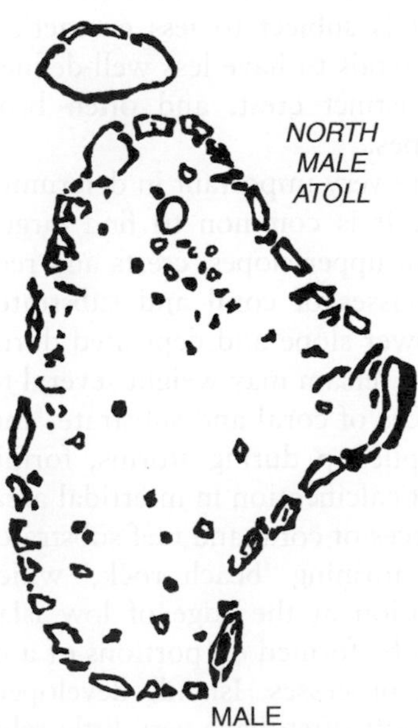

Figure 5 North Male Atoll of the Maldives. Large atolls often support substantial human population that will be threatened as sealevels rise. (Adapted from Holliday, 1989.)

separate these from extensive seagrass beds on a reef flat. Other channels, depressions, and basins may be predominated by sandy bottoms, studded with clumps of coral. Clumps in deeper waters, rising 2 m or more from the bottom and consisting of several species of coral are known as 'bommies.' Some lagoons hold large numbers of 'pillars,' tall shafts of limestone, similarly supporting corals. Throughout the reef flat, there are, typically, low clumps of coral. Macroalgae and seagrass are often kept away from these clumps by the feeding action of the herbivorous fish and sea urchins, particularly active at night and resting during the day in the clumps of coral. Low massive coral colonies may form 'microatolls,' in which central portions of the colony are dead, while the raised outer edge and sides continue to flourish. On some reef flats, branching corals form patches which extend over hectares. Other reef flats and lagoons may be packed with high densities of diverse coral colonies, and have little or no seagrass.

The seaward edge of the reef flat often leads into low branching or massive corals, including microatolls, which become increasingly dense to seaward. A thin band of macroalgae, such as *Turbinaria*, may be present, just before the clumps of coral and hard substrate rise to form a reef crest or 'pavement.' In some areas, the crest may be made of living coral. In areas protected from heavy wave action, the crest may consist of fingerlike *Montipora* or *Porites* forming a band several meters wide, broken periodically by whorl-shaped colonies of leafy corals, all apparently formed to efficiently comb breaking waves for zooplankton. On higher-energy coasts, the corals may be primarily dense growths of wave-adapted *Pocillopora*. Other crests may be covered with various species and growth forms of the brown alga *Sargassum*. The edible and commercially important alga *Caulerpa* may also be abundant. Other reef crests may be densely packed with small clumps of articulated calcareous algae such as *Halimeda* and *Amphiroa*. Still others consist of a pavement of calcareous material or sheared-off ancient corals, and are relatively devoid of all macrobenthos except tiny clumps of algal turf or encrusting algae, sometimes forming a white or pink algal crest. The height of a reef crest above the reef flat is often determined by the height of local tides.

Most reef crests are broken by channels of varying width and depth – exit routes for water piled up behind the reef crest by the breaking waves. These may be studded with corals or smoothed by scouring sand and rock carried along with the exiting water. They are particularly important as breeding grounds for a variety of reef fish.

Beyond the reef crest, on the upper reaches of the 'reef slope,' one often finds small thick clumps of branching *Pocillopora* coral or various species of *Acropora* colonies in similar, wave-resistant growth forms. Encrusting and low lump-like (submassive) colonies may be common. More of the brown algae *Turbinaria* and *Sargassum* may be present. Along a rounded or gentle upper reef slope in the Indo-Pacific, there may be table-shaped *Acropora* colonies of gradually larger size as one proceeds to deeper waters. On many reefs, the reef slope is the most active area of coral growth, because of the oxygen, nutrients, and plankton brought in on the waves and currents. Coral cover may exceed 100%, as colonies overgrow colonies, all competing for light. Soft corals dominate some reef slopes, *Sargassum* others, and on some, the profusion of corals, sponges, algae, and other benthic organisms may prevent the identification of a dominant group. The mean and median stony coral cover (the percentage of the substrate covered with coral) on a reef slope globally are both approximately 40%, with a broad variance.

On most reefs, small channels on the upper slope consolidate into increasingly wider and deeper channels along lower reaches of the slope. Some may be steepened or converted into tunnels by coral growth. The channels may occur fairly regularly at distances of tens of meters, resulting in a 'ridge and rift' or 'spur (or buttress) and groove' structure resembling the toes of giant feet. The bottoms are generally filled with a mix of sand and debris from the reefs, which makes its way downwards, particularly during storms. Sand may drop continuously from escarpments, in flows resembling waterfalls, and spectacular columns of limestone cut away from the reef proper may border deeper rifts.

Many reef slopes lead to a steep 'wall' or 'drop-off,' often beginning about 10–20 m depth. On shelf areas, the drop-off may end at 20–30 m depth, followed by a 'talus' slope of deposited reef materials, often 30°–60°, leading into the more gradual shelf slope. Typically, this shelf will be interrupted by outcrops of limestone and bommies for considerable distances from the reef itself. On reefs jutting into deep waters, the drop-off may extend downwards for hundreds of thousands of meters. Corals and a myriad of other organisms generally cover the slopes, leaving little or no bare substrate.

On some reefs, such as some small fringing reefs along the Sinai, the upper portion consists almost entirely of live massive, platy or occasionally thick branching corals extending from land or from a small reef flat. The corals are joined together tangentially, leaving large spaces of water between. There may be no identifiable reef crest, and the mass simply juts out over a steep dropoff to hundreds of meters depth.

The zonation of atolls and other surface reefs is generally similar. However, as one proceeds across a lagoon basin from the windward to the leeward side of an atoll, one encounters a 'backreef' area (note that the term is also applied by some to the coral-dominated area behind a reef crest). The leeward side of the atoll is subject to less energetic waves and currents. It tends to have less well-defined zonation, less of a distinct crest, and often broader, more gradual slopes.

Storms are very important in determining features of the reef. It is common to find large chunks of limestone on upper slopes, crests and reef flats, representing masses of coral and substrate pulled up from the lower slope and deposited during a storm. Pieces of this jetsam may weight several tonnes each. Smaller pieces of coral and substrate, and masses of sand may pile up during storms, forming islands. Processes of calcification in intertidal areas may glue together pieces of coral and reef substrate thrown up by storms, forming 'beach rock,' which helps to prevent erosion at the edge of low islands. Other islands may be formed on portions of a reef uplifted by tectonic processes. Islands developed by storm action or uplift often have very little relief, but may support human communities. Entire nations, as with the Maldives or some Pacific island nations, may consist of such low islands on atoll reefs.

Geology

Calcification

Carbon dioxide and water combine to form carbonic acid, which can then dissociate into hydrogen ions (H^+) and (HCO_3^-) bicarbonate, or carbonate (CO_3^{2-}) ions as follows:

$$CO_2 + H_2O \Leftrightarrow H^+ + HO_3{}^- \Leftrightarrow 2H^+ + CO_3{}^{2-}$$

Colder water can hold more CO_2 in solution than warmer water. Calcium carbonate reacts with CO_2 and water as follows:

$$CO_2 + H_2O + CaCO_3 \Leftrightarrow Ca^{2+} + 2HCO_3{}^-$$

The dissolution of $CaCO_3$ depends directly on the amount of dissolved CO_2 present in the water. Thus, in colder waters the higher levels of CO_2 inhibit calcification. Structural coral reefs are primarily limited to a circumglobal band stretching from the Line Islands in Hawaii to Perth, Australia, with nonstructural coral communities being increasingly dominant along the fringes. The distribution band varies, such that cold waters (e.g., Peru) narrow the range of coral reefs closer to the equator, and warm currents facilitate higher latitude development (e.g., Bermuda).

Calcification in stony corals is greatly enhanced by the presence of zooxanthellae in the tissues. Zooxanthellae are nonmotile stages of a dinoflagellate algae. They are contained within specialized coral cells, growing on excess nutrients and trace metals from the carnivorous corals, producing sugars for utilization by the host, and using up excess CO_2 to enhance calcification by the coral. Zooxanthellae are found in other reef organisms, including giant clams, and the tiny foraminifera, amoeba-like organisms (Order Sarcodina) with calcareous skeletons that create substantial amounts of the sand on coral reefs and nearby white-sand beaches. Much of the biology and ecology of zooxanthellae is poorly known.

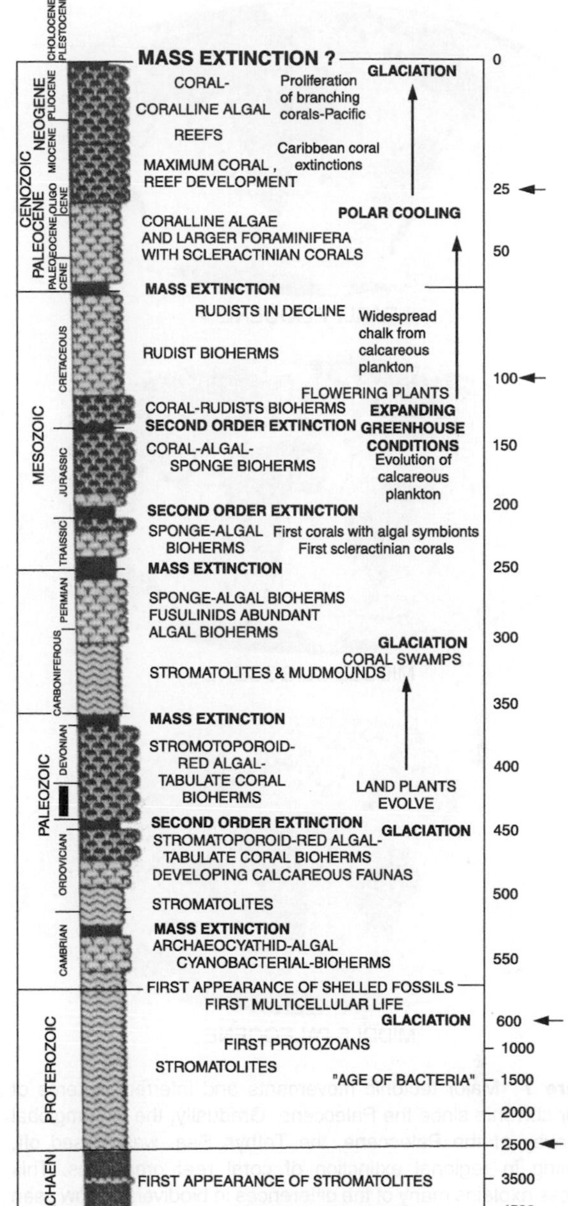

Figure 6 Geological time-scale (in thousand years) highlighting biogenic reef development. Arrows denote changes in scale. (Adapted from Hallock, 1989.)

Paleoecology

Structural coral reefs are forms of 'biogenic reefs,' distinct geomorphological structures constructed by living organisms. Biogenic reefs have existed in various forms since approximately 3.5 billion years ago, at which time cyanobacteria began building stromatolites (**Figure 6**). Bioherms, biogenic reefs constructed from limestone produced by shelled animals, became prominent by 570 million years ago. During the mid-Triassic, the Jurassic and early Cretaceous periods (roughly 200–100 million years ago), scleractinian corals had become significant components of coral–algal–sponge bioherms. Some of these corals may have included zooxanthellae. However, by the mid-Cretaceous, rudist bivalves dominated shallow-water reefs, and scleractinian corals were mainly restricted to deeper shelf-slope environments. This may have been due to seawater chemistry or competition with rudists or both. The massive extinction at the end of the Cretaceous, which ended the reign of dinosaurs on land, also led to the extinction of rudists. Many scleractinian corals survived in their deeper habitats, and evolved into

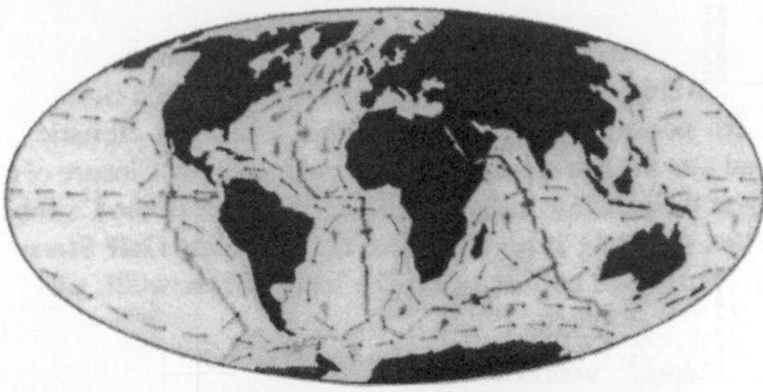
EARLY PLIOCENE

MIDDLE OLIGOCENE

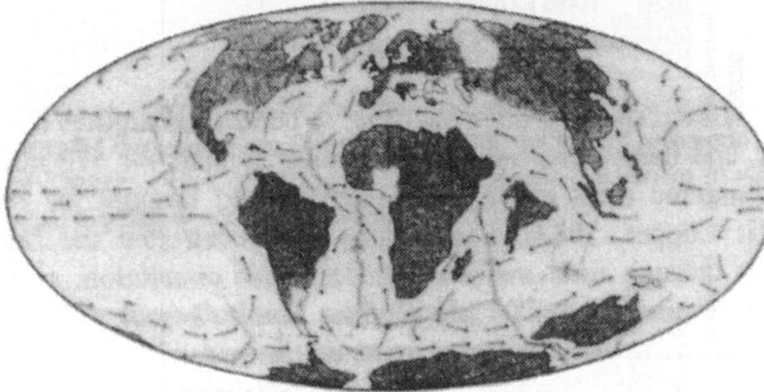
MIDDLE PALEOCENE

Figure 7 Major tectonic movements and inferred patterns of major currents since the Paleocene. Gradually, the circumglobal waterway of the Paleocene, the Tethys Sea, was closed off, resulting in regional extinction of coral reef organisms. This process explains many of the differences in biodiversity now seen among the world's coral reefs. (Adapted from Hallock, 1997.)

most of the modern genera by the Eocene. However, CO_2 concentrations in the atmosphere were high at the time, apparently inhibiting reef development by aragonite-producing corals. Instead, major limestone deposits of the time were formed by calcite-producing organisms such as larger foraminifera (including the limestones from which the Egyptian pyramids were later to be built) and red coralline algae. Calcite is less stable over time than aragonite, but can form in waters of higher CO_2 concentration. Major reef-building by stony corals and their associates did not occur until the middle to late Oligocene. By then, CO_2 concentrations were falling and sea water was warming in tropical areas, whereas at high-latitudes it became cooler. By the late Oligocene, Caribbean coral reefs achieved their greatest development, and by the early Miocene, coral reefs globally had extended to beyond 10° north and south.

During the late Eocene and early Oligocene, tropical oceans were openly connected, in a system of equatorial waterways known as the Tethys Sea. During this period, most scleractinian corals were cosmopolitan. The upward movements of Africa and India restricted circulation, particularly with the closure of the Qatar arch in the Middle East around the time of the Oligocene–Miocene boundary (**Figure** 7). The Central American passageway became restricted, but did not close until the middle Pliocene. However, nearly half of the Caribbean coral genera became extinct at the Oligocene–Miocene boundary, and many more disappeared during the Miocene. Larger foraminifera suffered similar extinctions. Far less dramatic extinctions occurred in the Indo-Pacific.

Reef Geomorphology

Charles Darwin proposed that atolls were formed by a process involving the sinking of islands (**Figure 8**). He noted that most high islands in the Pacific were

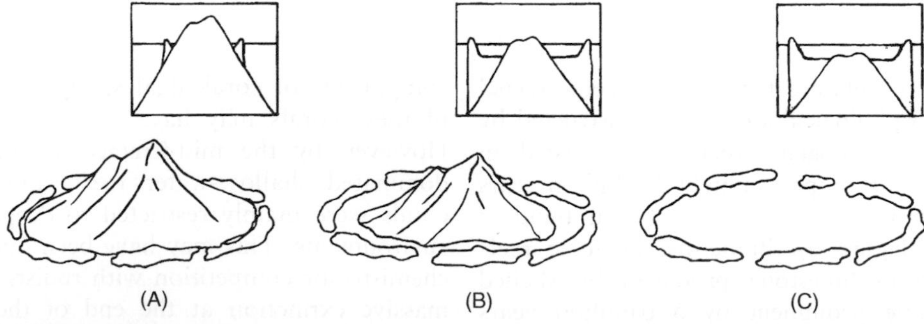

Figure 8 Stages in the development of an atoll as envisioned by Darwin. Fringing reefs surrounding a volcano (A) become barrier reefs as the volcano sinks (B) and finally form a donut-shaped atoll (C) Sinking islands and/or rising sea levels do help to explain the structures of many reefs, but other factors have been more important in the case of other reefs. (Adapted from Slafford-Deitsch, 1991.)

the tops of volcanic cones. A rocky, volcanic island would tend to form fringing reefs around its shores. Should the island begin to sink, the highly oxygenated outer edges would be able to maintain themselves at the sea surface, while the less actively growing reef flats would tend to sink, forming lagoons defining barrier reefs. Eventually, a sinking cone would disappear altogether, leaving a lagoon in the midst of an atoll. The feasibility of this explanation was confirmed by the mid-twentieth century when drilling on Enewetak and Bikini reefs yielded signs that the atolls were perched over islands that had gradually subsided over thousands of years.

Modern researchers have identified a variety of explanations for the morphology of various reefs. Variations in relative sea level have had profound influences on most structural coral reefs. Few coral reefs are believed to have exhibited continuous growth prior to 8000 years ago, and most are considerably younger. Coral reefs located on extensive continental shelf areas, such as the Florida reef tract and the vast shelf areas off the Yucatan and peninsular south-east Asia, grow on substrates that were generally far inland during the previous ice age.

The three-dimensional shapes of many coral reefs have been highly influenced by underlying topography, often an ancient reef. Wind, storms, waves, and currents have helped to shape many reefs, resulting in tear-drop shapes with broad, raised, steep edges facing predominant winds and currents. The shapes of some reefs are believed to have been particularly dependent on the erosion of much larger blocks of reef limestone during low stands of sea level. It has been demonstrated in the laboratory that small blocks of limestone subjected to rain-like erosion can form many features now identifiable on coral reefs, including lagoons, rifts and ridges, and channels. Evidence has also been found that the ridge and rift structures and some other features of coral reefs can result from collective ecological growth processes as the biota accommodates effects of waves, channeled water, and scouring by sand and other reef matter as calcifying benthic organisms compete for exposure to sunlight. It has been suggested that the double lagoon structure of the Atoll reef off Mindoro, Philippines, resulted from coral reef growth around the subsurface rim of a volcanic crater. Drilling on some reefs has demonstrated continuous growth for 30 m or more. On other reefs, little or no substantial calcification has occurred, the extreme case being nonstructural coral communities.

The Hawaiian and Line Islands are part of a vast chain of islands formed as a tectonic plate has drifted north and west across the Pacific. Volcanic lava has tended to erupt in a fixed 'hot-spot', forming volcanic islands one after the other. As the islands have drifted north, they have subsided gradually into deeper waters until coral growth has ceased. This process has had a strong influence over the structure of coral reefs along these islands and seamounts. However, at smaller scales, coral reefs along the coasts of the Hawaiian islands also show the effects of heavy storm action, exhibiting mound-like features at the sea surface often dominated by calcareous algae, and sometimes broad platforms which produce the famous surfing waves of the islands.

Role in Oil Exploration

Coral reefs are generally highly porous, filled with holes, tunnels, and caves of all sizes. The most famous 'blue-holes' of the Bahamas are the vertical seaward entrances of huge cave and tunnel systems formed by erosion at low stands of sea level, through which sea water flows beneath the islands as water through a sponge. Much of the porosity of coral reefs results from incomplete filling of spaces between coral heads during the active growth of the reef.

Much of the world's crude oil comes from ancient coral reefs that have been subsequently overlain with terrigenous (land-based) sediments. Dense, horizontally layered sediments have often trapped oil within the mounds of porous coral reef limestone. In some areas, subsequent tectonic activity has produced faults, and much of the oil has been lost. In other areas, however, the ancient reefs have yielded vast amounts of oil.

In some parts of the world, modern coral reefs have grown in areas where ancient reefs once occurred and were subsequently covered by dense sediments that trapped oil. Thus, it is common to find oil drilling platforms on modern reefs. This has frequently raised environmental concerns, centering on the damage done to modern corals by the construction and associated activities, the occasional spillages of drilling muds of various compositions and the leakage of oil.

Biology

Biogeography

The highest species diversity on coral reefs occurs in the seas of south-east Asia, often referred to as the Central Indo-West Pacific region. In the Caribbean or Hawaii, there are 70–80 species of reef-building corals. In south-east Asia, there are more than 70 genera and 400 or more species of reef-building corals (**Figure 9**). The diversities of most other groups of reef-associated species, including fish, tend to be higher in south-east Asia than in other regions.

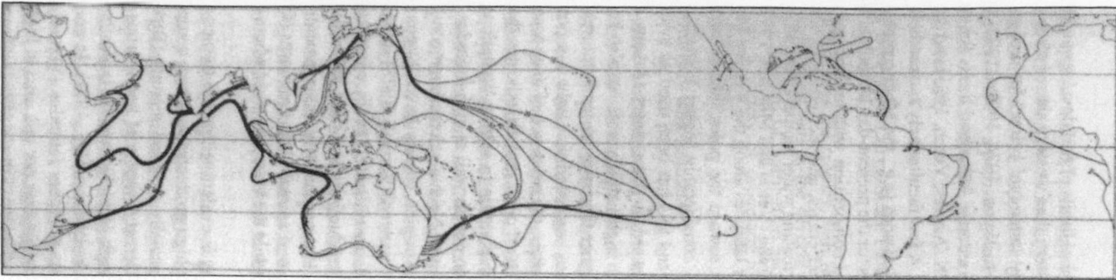

Figure 9 Patterns of generic diversity is scleractinian corals. Contours were estimated from generic patterns and existing data on species distributions. (From Veron 1993; reprinted with permission of the author.)

Diversity begins to drop off gradually to the south below Indonesia, and north above the Philippines. The decline eastward is very dramatic, such that only a few species of corals live in Tahiti and the Galapagos. Westward, the decline is only slight; at least 50 genera of corals live along the coasts of East Africa and the Red Sea. There are only a few species of reef-building corals off the coast of West Africa. Species in south-east Asia tend to have very broad ranges, and endemism is higher in peripheral areas such as the central Pacific.

Much of the global pattern of diversity, particularly at the level of genera and families, can be explained in terms of selective regional extinctions following the gradual breakup of the circumglobal Tethys Sea during the Cenozoic. This is well supported in the fossil records of corals, seagrasses and mangrove trees. Although fish fossils are uncommon, there is a highly diverse assemblage of fossil coral reef fish in Italy, a remnant from the Tethys Sea.

However, there is some reason to believe that rates of speciation have been higher in south-east Asia than elsewhere. The evidence is best known for mollusks, and include unusually high ratios of species to genera and particularly highly evolved armament (such as the spines on many murex shells) in this region. There is a myriad of explanations for this potentially higher speciation rate. Two popular explanations, the Pacific island vicariance hypothesis and the basin isolation hypothesis, both involve changes in sea level. In the former hypothesis, high levels of the sea tend to isolate groups of Pacific islands, facilitating speciation. At low sea levels, the species spread to the heterogeneous refugia habitat of south-east Asia and are gradually lost on the less-heterogeneous Pacific islands due to competition from other species. In the latter hypothesis, low stands of sea level, and, more importantly, periods of mountain building, have isolated particular marine basins within south-east Asia in the past, possibly permitting rapid speciation among whole biotas.

Further understanding of these processes must await increased efforts in taxonomy and systematics. The vast majority of coral reef organisms, particularly in the Indo-West Pacific region, have never been formally identified. The ranges of known organisms are poorly known. Furthermore, recent decades have seen a rapid decline in interest in and support for taxonomy and systematics. It is likely that human impacts will result in vast changes to the distributions and abundances of coral reef species long before existing biogeography and ecology have been well understood.

Coral Reef Ecosystem Health

Resilience and phase shifts Most coral reefs are subjected to periodic disturbances, such as storms. The ability of an ecosystem to maintain constancy in terms of ecological functions and in the abundances and distributions of organisms is ecological resistance. The capacity of an ecosystem to revert to a previous state (or near to it) in terms of these characteristics is ecological resilience. Terms such as ecosystem health, integrity, and stability usually infer degrees of resistance and resilience. Because profound changes often occur to a coral reef following a perturbation such as a storm, there is increasing focus on the resilience of a coral reef.

Many coral reefs have been known to undergo losses of coral cover from greater than 50% to less than 10%, and then to recover to the former level within 4–10 years. Naturally, in reefs in which colonies may have been decades or centuries old, the age structure of the corals present have often been disrupted. This, in turn, may affect resilience to future perturbations, as certain corals are believed to exhibit higher fecundities in older colonies.

Some coral reefs do not seem to return to high levels of coral cover after a perturbation, especially if they have been subsequently overgrown by fleshy macroalgae. Human intervention, in the form of increased eutrophication and the removal of herbivorous fish and invertebrates, are suspected to favor the growth of macroalgae following perturbations.

Reefs around Jamaica have shown little recovery in more than a decade following coral damage by a hurricane, and both forms of intervention are suspected of being causes of this loss of resilience. Widespread losses of resilience are a concern throughout most coral reef areas.

Reef interconnectivity Most coral reef organisms undergo pelagic life stages before settling into a reef community. Most corals live as planulae for a few hours to a few days before setting, although longer periods have been recorded. The average benthic invertebrate spends roughly two weeks in waterborne stages, but some survive for months. The average coral reef fish appears to require nearly a month before settling, and many require two months.

The pelagic stages are by no means passive, and although the sizes are very small, the organisms may be adapted to swim into currents and eddies that facilitate their retention or return to particular reefs or groups of reefs. Their success in doing so is believed to depend on factors such as reef geomorphology and the nature and predictability of local oceanography. Analyses indicate that some fish populations regularly exchange genetic material over thousands of kilometers. Although most coral-reef populations are believed to be replenished each year from local progeny, a severely depleted reef may be restocked to some degree from other reefs. This process is crucial to the problem of resilience, and is an active area of controversy and research. The results of this work may have profound implications for the design of marine protected areas, for international agreements on the coordination of management schemes, and for the regulation of harvesting on coral reefs. Furthermore, climate change is likely to alter local oceanographic processes, and has important implications for reef management as such disruptions occur.

Coral Reef Management

Disturbances

Various types of perturbations have affected coral reefs with increasing frequency within recent decades. Some of these are clearly directly related to human activities, whereas others are suspected to be the indirect results of human interventions.

The majority of coral reefs are located in developing countries. In many of these, crowded low-income human populations increasingly overfish, often reducing herbivorous fish and invertebrates, thereby decreasing reef resilience. In a process known as Malthusian overfishing, social norms break down and fishers turn to destructive fishing methods such as the use of poisons and explosives to capture fish.

Organic pollution from coastal habitations and agricultural fertilizing activities are common along coastlines. Runoff from deforested hillsides, mining operations, and coastal construction often contains materials that favor macroalgal growth, and silt and mud that restrict light to zooxanthellae, abrade reef benthos or bury portions of coral reefs entirely.

Increasingly common outbreaks of the coral-eating crown-of-thorns starfish, *Acanthaster planci*, may be related to reductions in predators, such as lethrinid fishes. During 1997–98, a worldwide epidemic of coral bleaching occurred, in which the zooxanthellae of many colonies were expelled and high rates of coral mortality resulted. The cause was unusually warm patches of sea water associated with a strong El Niño event, which some believe to be related to increasing levels of atmospheric CO_2 and global warming. A more controversial suggestion is that the increasing CO_2 levels will cause acidification of the oceans sufficient to result in the net erosion of some coral reefs, especially at high latitudes. Although there have recently been major epidemics of diseases killing corals and associated organisms in western Atlantic reefs, only certain coral diseases appear to be directly linked to stress from human activities.

Assessments

Efforts are underway to gather empirical information on coral reefs via the quantification of benthic organisms and fish by divers. These efforts are being supplemented by the use of remotely sensed data from satellites, space shuttles, aircraft, ships, manned and unmanned underwater vehicles. The usefulness of these approaches ranges from identifying or mapping reefs, to quantifying bleaching and disease. A global database, ReefBase, has been developed by the International Center for Living Aquatic Resources Management (ICLARM) to gather together existing information about the world's reefs and to make it widely available via CD-ROM and the Internet.

Integrated Coastal Management

Management is a process of modifying human behavior. Biophysical scientists can provide advice and predictions concerning factors such as levels of fishing pressure, siltation, and pollution. However, the management decisions must account for social, cultural, political, and economic considerations. Furthermore, almost all management interventions will have both positive and negative effects on various aspects of the ecosystem and the societies that impact it and depend upon it. For example, diverting fishers into forestry may lead to increased deforestation, siltation, and further reef degradation.

It is increasingly recognized that effective management is achieved only through approaches that integrate biophysical considerations with socioeconomic and related factors. Balanced stakeholder involvement is generally a prerequisite for compliance with management decisions. The field of integrated coastal management is rapidly evolving, as is the set of scientific paradigms on which it is based. To a large degree, the future of the world's coral reefs is directly linked to this evolution.

See Also

Diversity of Marine Species. Lagoons. Mangroves. Rocky Shores. Sandy Beaches, Biology of.

Further Reading

Birkeland C (ed.) (1997) *Life and Death of Coral Reefs.* New York: Chapman and Hall.

Bryant D, Burke L, McManus J, and Spalding M (1998) *Reefs at Risk: a Map-based Indicator of the Threats to the World's Coral Reefs.* Washington, DC: World Resources Institute.

Davidson OG (1998) *The Enchanted Braid: Coming to Terms with Nature on the Coral Reef.* New York: John Wiley.

Holliday L (1989) *Coral Reefs: a Global View by a Diver and Aquarist.* London: Salamander Press.

McManus JW, Ablan MCA, Vergara SG, *et al.* (1997) *ReefBase Aquanaut Survey Manual.* Manila, Philippines: International Center for Living Aquatic Resources Management.

McManus JW and Vergara SG (eds.) (1998) *ReefBase: a Global Database of Coral Reefs and Their Resources. Version 3.0 (Manual and CD-ROM).* Manila, Philippines: International Center for Living Aquatic Resource Management.

Polunin NVC and Roberts C (eds.) (1996) *Reef Fisheries.* New York: Chapman and Hall.

Sale PF (ed.) (1991) *The Ecology of Fishes on Coral Reefs.* New York: Academic Press.

Stafford-Deitsch J (1993) *Reef: a Safari Through the Coral World.* San Francisco: Sierra Club Books.

Veron JEN (1993) *A Biogeographic Database of Hermatypic Corals. Species of the Central Indo-Pacific, Genera of the World. Monograph Series 10.* Townsville, Australia: Australian Institute of Marine Science.

COLD-WATER CORAL REEFS

J. M. Roberts, Scottish Association for Marine Science, Oban, UK

Introduction and Historical Background

Corals and coral reefs are not restricted to shallow, tropical waters. Deep-ocean exploration around the world is now revealing coral ecosystems at great depths in the cooler waters of the continental shelf, slope, and seamounts. Here, in permanent darkness and without the algal symbionts (zooxanthellae) of many tropical species, cold-water corals grow to form true deep-water scleractinian reefs or 'forests' of flexible gorgonian, black, gold, and bamboo corals.

Such corals have been known since the eighteenth century; the Reverend Pontoppidan, Bishop of Bergen, discussed corals as 'sea-vegetables' in his 1755 book *The Natural History of Norway* and Linnaeus subsequently described several cold-water coral species. The following century, the British naturalist Philip Henry Gosse summarized British corals in his 1860 book *Actinologica Britannica. A History of the British Sea-Anemones and Corals* (**Figure 1**). Further records and samples were obtained during the pioneering nineteenth century expeditions of HMS *Porcupine* (1869, 1870) and HMS *Challenger* (1872–76), and in the first half of the twentieth century scientific dredging studies by Dons and Le Danois identified sizeable coral banks off the Norwegian and Celtic margins, respectively. However, until relatively recently, cold-water coral banks remained best known to fishermen, especially trawlermen, who risked damaging their nets and marked coral areas on fishing charts. In the latter half of the twentieth century, advances in acoustic survey techniques and research submersibles allowed the first mapping and direct observations of coral colonies in deep water. Using the early research submersible *Pisces*, Wilson described cold-water coral patch development on the Rockall Bank west of the UK and was among the first to show the value of video surveys to document and understand these structurally complex habitats (**Figure 2**). In the last 20 years, there have been further advances in deep-ocean exploration, notably the development of multibeam echo sounders, remotely operated vehicles, and, most recently, autonomous underwater vehicles that are now beginning to reveal the true extent of cold-water coral ecosystems (**Figure 3**).

Cold-Water Corals

Cold-water corals are all members of the phylum Cnidaria and include species belonging to a number of lower taxonomic groups. Among these the colonial scleractinian or stony corals can develop sizable reef frameworks and are the focus of this article. *Lophelia pertusa* dominates reef frameworks in the Northeast Atlantic where *Madrepora oculata* is an important secondary species. *L. pertusa* is also abundant on the other side of the Atlantic Ocean in the US South Atlantic Bight where other framework corals

Figure 1 Color plate from Gosse's 1860 book *Actinologica Britannica*, showing a colony of *Lophohelia prolifera* (*Lophelia pertusa*) together with cup corals and zoanthids.

Figure 2 *Lophelia pertusa* reef patches on Rockall Bank in the Northeast Atlantic were among the first cold-water coral habitats observed from a manned submersible. (a) The *Pisces III* submersible in 1973. (b) Live coral framework and surrounding rubble. (c) Large antipatharian coral colony, possibly *Parantipathes hirondelle*. Images courtesy of Dr. John Wilson. (b) Reproduced from Wilson JB (1979) 'Patch' development of the deep-water coral *Lophelia pertusa* (L.) on Rockall Bank. *Journal of the Marine Biological Association of the UK* 59: 165–177, with permission of the Marine Biological Association of the UK.

include *Enallopsammia profunda*, *M. oculata*, and *Solenosmilia variabilis*. Along the eastern Florida shelf, the facultatively zooxanthellate coral *Oculina varicosa* forms banks up to 35 m in height at depths of 70–100 m. *S. variabilis* is also reported forming tightly branched frameworks on the Little Bahama Bank and Reykjanes Ridge in the Atlantic and on Tasmanian seamounts in the South Pacific. *Goniocorella dumosa* is only reported from the Southern Hemisphere, where it forms reef frameworks around New Zealand on the Chatham Plateau. Thus, there are just six cold-water scleractinian coral species currently known to form significant reef frameworks in deep water, compared to more than 800 species of shallow reef-building tropical corals. While the scleractinian reef framework-forming species form the basis of this article, other cold-water corals, notably gorgonians, antipatharians, and hydrocorals, can develop dense assemblages that also provide significant structural habitat for other species.

Reef Distribution and Development

The robust anastomosing skeletons of colonial cold-water scleractinians produce dense frameworks that over time can develop structures with significant topographic expression from the seafloor that alter local sedimentary conditions, are subject to the dynamic process of (bio)erosion, and provide habitat to many other species (**Figure 4**). By these criteria, cold-water scleractinian corals form reefs and these reefs can persist for tens of thousands of years.

Cold-water coral reef distribution is controlled by a suite of environmental factors. They are largely restricted to water masses with temperatures of 4–12 °C and salinities of 35 psu. Although they are often generically referred to as deep-sea corals, their wide depth distribution, from shallow fiordic sills at just 40 m to shelf and slope depths of 200–1000 m where the majority of cold-water coral reefs are found, reflects bathyal environments (200–2000 m) and not the abyssal depths (2000–6000 m) more often associated with the term 'deep sea'. On a global scale, the importance of seawater carbonate chemistry as a control on cold-water coral occurrence is now becoming apparent. Scleractinian corals secrete calcium carbonate skeletons in the form of aragonite. The boundary between saturated and undersaturated seawater, the aragonite saturation horizon (ASH), is relatively deep (>2000 m) in the Northeast Atlantic and relatively shallow (50–500 m) in the North Pacific. As shown in **Figure 5**, there are many records of reef framework-forming cold-water corals from the Northeast Atlantic and very few from the North Pacific where coral assemblages are dominated by octocorals and

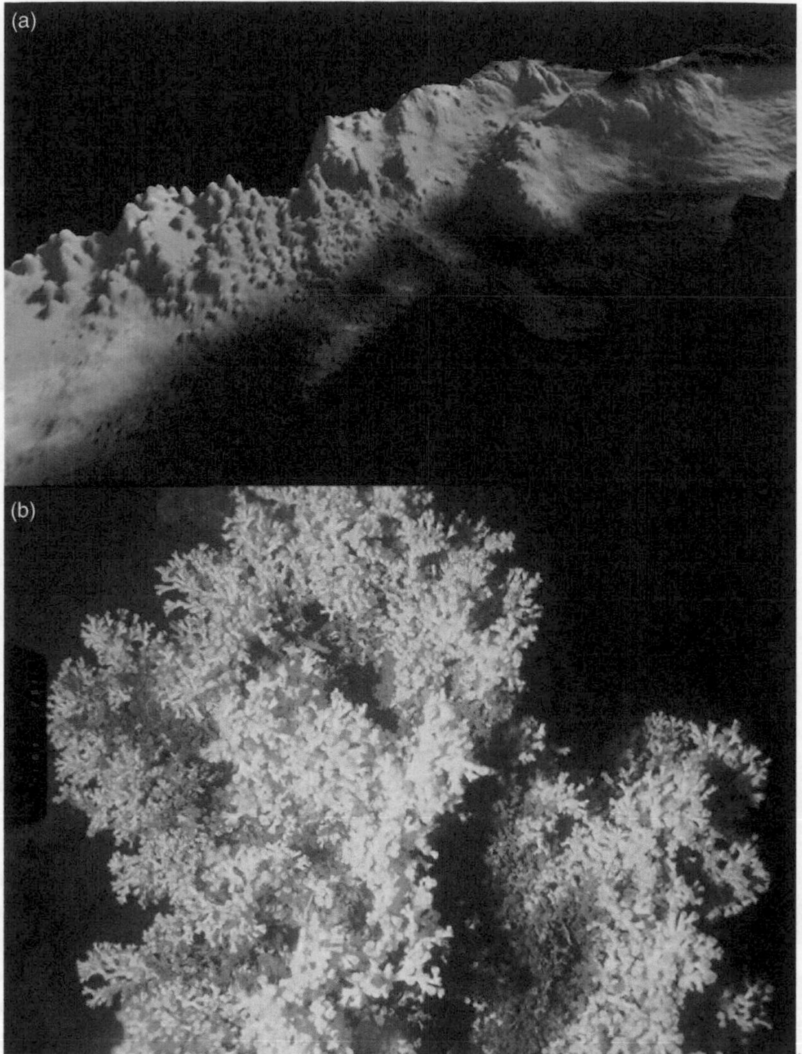

Figure 3 Multibeam echosounders are valuable tools for wide area mapping in deep waters. (a) Three-dimensional bathymetry, exaggerated sixfold in the vertical, showing the many seabed mounds formed by *Lophelia pertusa* reefs of the Mingulay Reef Complex, Northeast Atlantic. (b) Photograph of a *L. pertusa* colony from these mounds. Reproduced from Roberts JM, Brown CJ, Long D, and Bates CR (2005) Acoustic mapping using a multibeam echosounder reveals cold-water coral reefs and sourrounding habitats. *Coral Reefs* 24: 654–669.

hydrocorals that do not form robust aragonitic skeletons. Indeed, the scleractinians beneath the ASH from the Aleutian Islands in the North Pacific are dominated by species adapted to low-calcium-carbonate environments and the majority of the hydrocorals found there form calcitic rather than aragonitic coralla. Compellingly, calcite is *c.* 50% less soluble in seawater than aragonite.

In the 1990s, the hydraulic theory was advanced proposing that cold-water coral reefs, notably the well-developed *Lophelia* reefs of the Norwegian continental margin, were coupled to the geosphere via the seepage of light hydrocarbons. This intriguing idea led to a large research effort searching for evidence that the corals and associated fauna were reliant on local seepage. Despite proximity to seeps in certain areas, stable isotope analyses of coral tissue reflect material derived from surface productivity and recent ocean drilling through a coral carbonate mound in the Porcupine Seabight failed to find any evidence of gas accumulation or that mound growth had been initiated by local hydrocarbon seepage.

Thus while broad trends in the factors controlling cold-water coral reef distribution are becoming apparent, our understanding of their global distribution remains biased to those parts of the world, notably the Northeast and Northwest Atlantic, where most surveys have been carried out. Predictive modeling approaches suggest that suitable conditions for cold-water coral reef development may be found in areas of the continental slope that have not yet been surveyed and mapped in sufficient detail to test for their

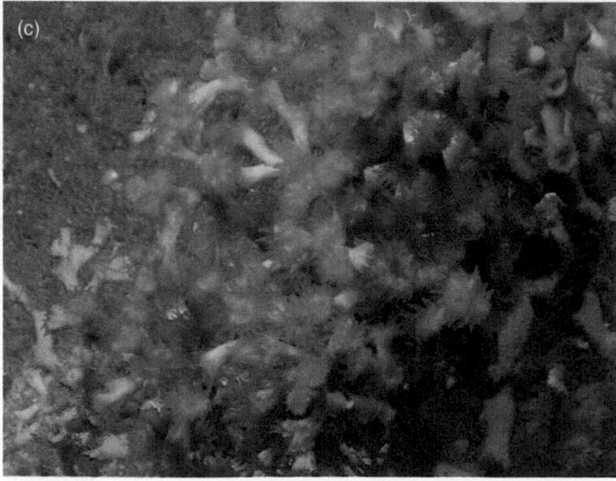

Figure 4 Cold-water coral reefs form highly complex three-dimensional structural habitat. (a) The sloping flank of a giant coral carbonate mound in the Porcupine Seabight, Northeast Atlantic. (b) Dense scleractinian coral framework (*Lophelia pertusa* and *Madrepora oculata*) with purple octocorals (probably *Anthothelia grandiflora*) and glass sponges (probably *Aphrocallistes*). (c) Live polyps of *L. pertusa*. Images courtesy of *VICTOR-Polarstern* cruise ARKXIX/3a, Alfred-Wegener-Institut für Polar- und Meeresforschung and the Institut Français de Recherche pour l'Exploitation de la Mer.

occurrence. Similarly, only a small proportion of the tens to hundreds of thousands of seamounts that have been estimated to exist have ever been surveyed. Many of the seamounts that have been examined reveal abundant cold-water corals like the *S. variabilis* reefs on Tasmanian seamounts in the South Pacific or the assemblages of gorgonian, black, and bamboo corals on the Davidson Seamount off Monterey in the East Pacific.

Feeding, Growth, and Reproduction

Cold-water corals are typically reported from areas with locally accelerated currents or from regions offshore where internal tidal waves impinge on the slope and break, thus enhancing mixing and flux of food material from the surface to the seabed. Both productive surface waters and hydrographic conditions that transport this material to the benthos are needed to support cold-water coral growth. The reef framework-forming corals *L. pertusa* and *M. oculata* seem able to use both zoo- and phytoplankton and, as with other cnidarians, are likely to feed from a cosmopolitan mix of zooplankton prey, detritus, and dissolved organic matter.

Understanding cold-water coral growth has largely been limited to observations derived in two ways from dead coral skeletons. First, coral colonies found on man-made structures can be measured and used to estimate approximate extension rates based on the age of the man-made structure. Second, cycles in the stable isotopes of carbon and oxygen in the coral's skeleton can be used to infer annual extension rates, although recent work has shown this method is complicated by poor understanding of skeletal banding patterns. Using these approaches to study *L. pertusa*, annual extension rates of between 0.5 and 3 cm have been derived, with the faster growth rates associated with shallower colonies growing on North Sea oil platforms where enhanced food availability and competition for space may combine to accelerate linear extension rates. To date there have been no detailed reports on cold-water coral calcification rate or mode – a worrying deficiency in a time of predicted climate change and ocean acidification, discussed below.

While we lack detailed understanding of cold-water coral growth and calcification, some promising initial progress in unraveling the reproductive ecology of some species has been made. Histological studies show that *L. pertusa* polyps have separate sexes (gonochoristic) and that in the Northeast Atlantic they produce gametes over winter following phyto-detrital flux to the seafloor (**Figure 6**). Spawning has

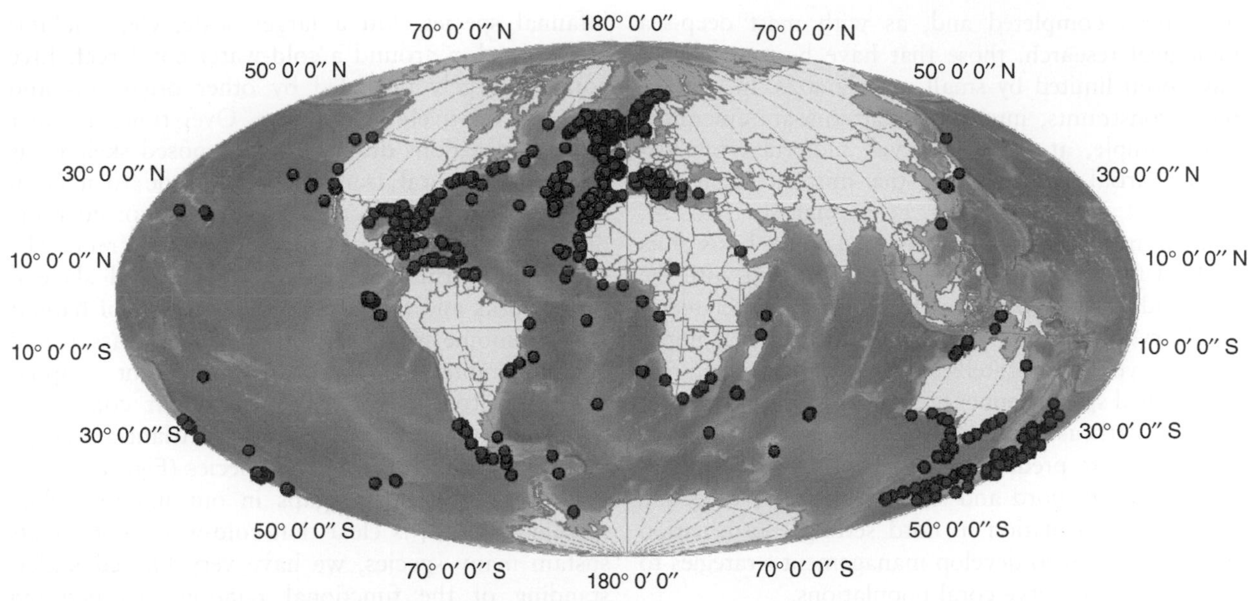

Figure 5 Global distribution of reef framework-forming cold-water corals. Map courtesy of Dr. Max Wisshak, University of Erlangen, and Dr. Andrew Davies, Scottish Association for Marine Science.

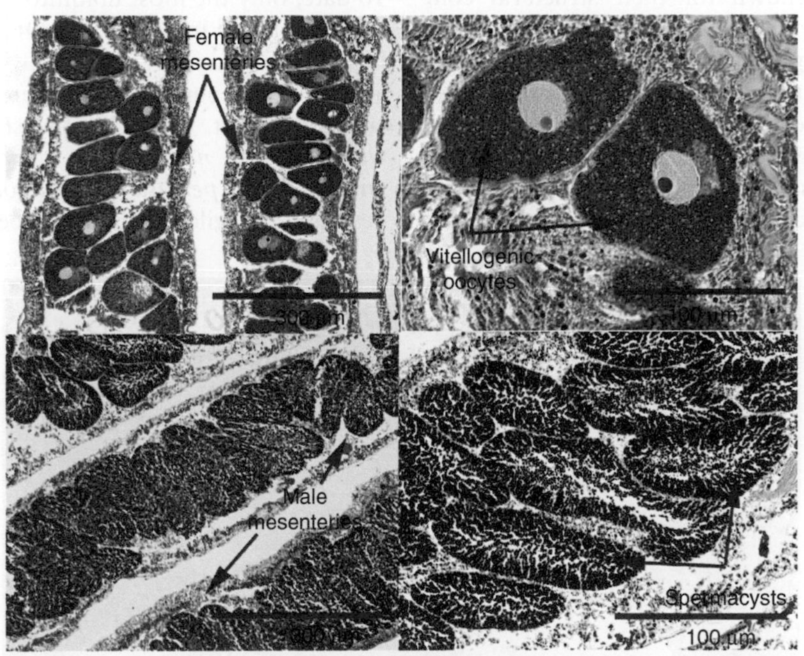

Figure 6 Histological sections showing female and male reproductive tissues of *Lophelia pertusa* sampled from a North Sea oil platform. Image courtesy of Dr. Rhian Waller, University of Hawaii.

not been observed yet and while no coral larvae have been sampled, the widespread occurrence and rapid colonization of man-made structures both point to a dispersive planula larva capable of remaining competent in the water column for several weeks.

Hidden Diversity and Molecular Genetics

Molecular genetic analysis has tremendous potential to reveal both species- and population-level information. Few molecular studies of cold-water corals have been completed and, as with most deep-sea biological research, those that have been attempted have been limited by small sample sizes. In spite of these constraints, interesting patterns are emerging. For example, at a species level, molecular analysis using partial sequences of the mitochondrial 16S ribosomal RNA encoding gene suggests that *M. oculata* may have been misclassified on the basis of its skeletal morphology. Rather than grouping with the Oculinidae, *M. oculata* may be more closely related to the Caryophyliidae and Pocilloporidae. On a population level, microsatellite and ribosomal internal transcribed spacer sequence analyses have shown that in the Northeast Atlantic some slope populations of *L. pertusa* are predominantly clonal and this species forms discrete fiord and shelf populations reflecting geographical isolation in fiord settings. Information like this is vital to develop management strategies to protect and conserve coral populations.

Habitats and Biodiversity

Coral reefs are renown for their structural complexity and cold-water coral reefs are no exception. Corals are ecological engineers, their skeletons forming complex three-dimensional structures that provide a multitude of surfaces for attached epifauna and shelter for mobile fauna. Coral frameworks trap resuspended sand and mud, forming sediment-clogged frameworks and providing further niches for infaunal species. On a larger scale, clear habitat zones develop around a cold-water coral reef. Live coral is largely unfouled by other organisms and supports relatively few species. Over time, as older polyp generations die back and exposed skeleton is (bio)eroded, coral frameworks degrade to a coral rubble apron that can extend for considerable distances downslope at the foot of the reef. The small-scale structural complexity of coral skeletal frameworks and the larger-scale diversity of habitat types combine to support highly diverse associated animal communities. For example, a recent compilation of European studies showed that cold-water coral reefs along the Northeast Atlantic margin supported 1317 other animal species (**Figure 7**).

However, significant gaps in our understanding remain. While it is clear that cold-water coral reefs sustain many species, we have very limited understanding of the functional relationships between these species. Largely because of the great expense and technical difficulties of working on structurally complex habitats in deep water, few examples of the natural history of these systems have been described. To date, only the most ubiquitous relationships have been examined in any detail. Of these the symbiosis formed between reef framework-forming scleractinians and eunicid polychaetes appears particularly significant. In the Northeast Atlantic, the large worm *Eunice norvegica* is very frequently found with both *L. pertusa* and *M. oculata*. The worm develops a fragile parchment tube through the coral

Figure 7 Examples of diverse fauna sampled from a giant coral carbonate mound in the Porcupine Seabight, Northeast Atlantic. (a) Isopod *Natatolana borealis*. (b) Gastropod *Boreotrophon clavatus*. (c) Brachiopod *Macandrevia cranium*. (d) Hydrocoral *Pliobothrus symmeticus*. Images courtesy of Dr. Lea-Anne Henry, Scottish Association for Marine Science.

framework that becomes calcified by the coral. As well as apparently strengthening the overall framework, recent aquarium observations have shown that the worms repeatedly aggregate small coral colonies. *In situ*, this behavior could be a significant factor in enhancing patch formation and accelerating reef growth.

We also know little of what organisms prey upon cold-water corals with only a few direct submersible observations showing asteroids apparently grazing on live coral colonies. However, some parasitic relationships have now been described from samples recovered from cold-water coral reefs. *L. pertusa* is parasitized by the foraminiferan *Hyrrokkin sarcophaga* and gall-forming copepods are sometimes associated with large gorgonian corals such as *Paragorgia arborea*. It is intriguing to think what other interactions and behaviors might remain to be described and *in situ* video and photographic records from unobtrusive benthic landers have great potential to provide these observations (**Figure** 8).

There has been great interest in whether cold-water coral reefs form essential fish habitat, for example, providing areas for spawning or nursery areas for juvenile fish. Investigations are again at an early stage but some broad trends are becoming apparent. The degree to which cold-water corals provide habitat to fish seems to depend on the coral habitat in question. Coral assemblages dominated by gorgonian 'forests' seem to sustain fish communities similar to those found near seabeds with other structural habitat such as large rocks. In contrast, there is reasonably compelling evidence that the large, structurally complex reefs formed by scleractinians can provide important fish habitat. For example, higher numbers of gravid female redfish (*Sebastes marinus*) were found associated with large *Lophelia* reefs than neighboring off-reef areas. However, the studies to date rely on sparse data collected in different ways (e.g., long line catches vs. submersible observations) in different regions, making it hard to draw out clear patterns.

Thus although cold-water coral reefs form biodiversity hotspots on the continental slope, it is proving hard to understand their true significance in biogeographic and speciation terms. Studies are typically biased by the methodology used to collect samples and the taxonomic expertise applied to the fauna. Cold-water coral reef biodiversity studies rely on material gathered by trawl, dredge, box core, grab, or on megafaunal descriptions from photographs or lower-resolution video records. Each technique differs in the type of sample it can recover and further biases may be introduced by different sample-processing methods. Despite these frustrations, when the animal communities recovered from cold-water coral reefs are examined by taxonomic specialists, they frequently reveal undescribed species and often new records or range extensions of species known from other areas. For example, one study examining just 11 box core samples from coral carbonate mounds in the Porcupine Seabight reported 10 undescribed species and that coral-rich cores on-mound were 3 times more species-rich than off-mound cores.

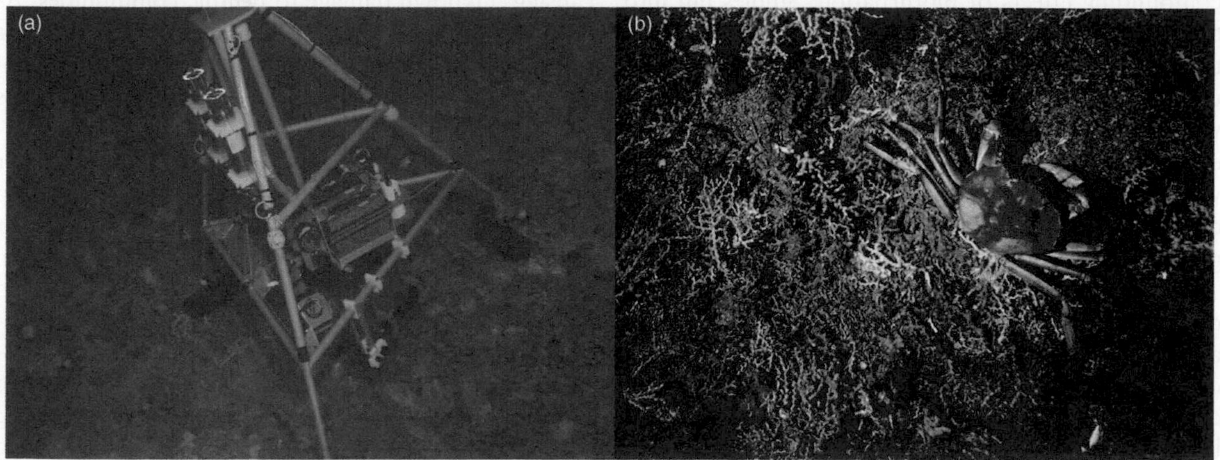

Figure 8 Developments in benthic landers and seafloor observatories will allow long-term environmental data recording and unobtrusive observations of cold-water coral reef fauna. (a) A 'photolander' deployed at 800 m water depth on a giant coral carbonate mound. (b) Lander photograph showing coral framework, glass sponges (probably *Aphrocallistes*), and deep-sea red crab (*Chaceon affinis*). (a) Courtesy of *VICTOR-Polarstern* cruise ARKXIX/3a, Alfred-Wegener-Institut für Polar- und Meeresforschung and the Institut Français de Recherche pour l'Exploitation de la Mer.

Timescales and Archives

As individual animals, corals can be extremely long-lived. ^{14}C dating has shown that one colony of the gold coral *Gerardia* (Zoanthidea) was approximately 1800 years old when collected, making it possibly the oldest marine animal known. Although scleractinian polyps are unlikely to live for more than 10–20 years, reef frameworks can persist for thousands to tens of thousands of years. In the Northeast Atlantic, the ages of scleractinian reefs clearly correspond to glacial history with relatively young reefs (8000 yr BP) found at latitudes affected by Pleistocene glaciation and far older reef frameworks (50 000 yr BP) further south well beyond glacial influence. Initial interpretation of the cores obtained by drilling through a large carbonate mound in the Porcupine Seabight suggest that the entire mound structure has developed over 1.5–2 My with periods of interglacial coral framework growth interspersed with periods of glacial sediments when conditions were unsuitable for corals.

An extensive literature now exists, showing the value of shallow, tropical corals as paleoenvironmental archives, and interest in studying historical patterns in ocean temperature and circulation has increased as evidence of anthropogenic climate change grows. Cores extracted from glacial ice sheets provide invaluable high-resolution climate archives but the potential of deep-ocean sediments to reveal similar high-resolution temporal patterns is limited by the mixing action of bioturbating infauna. Recent research has shown that cold-water coral skeletons not only provide a long-lasting archive but one that can be analyzed at high temporal resolution without confounding effects of bioturbation. Studies fall into two categories: those that use coral skeletal chemistry to estimate past seawater temperature and those that use combinations of dating techniques to trace ocean ventilation. The most convincing paleotemperature estimates come from gorgonian and bamboo corals, which, unlike reef framework-forming scleractinians, contain clear skeletal banding patterns that allow a good chronology to be developed (**Figure 9**).

We continue to learn more about the importance of deep-ocean circulation in regulating global climate and once again recent research demonstrates that cold-water corals provide a unique archive of ocean circulation patterns. By dating coral skeletons with both ^{14}C and U/Th techniques, it is possible to discriminate the coral's actual age from the age of the seawater in which it grew. This is possible because as corals calcify they use dissolved carbon from the ambient seawater, thus providing material that can later be ^{14}C-dated. This means that cold-water coral skeletons can record ocean ventilation history as water masses of differing ^{14}C age exchange. This approach has so far successfully followed ventilation patterns in the Southern Ocean and North Atlantic and offers great potential to study past ocean circulation at key oceanographic gateways.

Threats

As shallow-water fish stocks have diminished on continental shelves around the world, the fishing industry has expanded into deeper slope and even seamount waters. This move, made possible with larger, powerful refrigerated vessels and improved navigational technology (Global Positioning System), has seen the development of a series of boom-and-bust deep-water fisheries for species such as the orange roughy (*Hoplostethus atlanticus*) around New Zealand, the roundnose grenadier (*Coryphaenoides rupestris*) in the Northwest Atlantic, marbled rockcod (*Notothenia rossii*) around Antarctica, and pelagic armorheads (*Pseudopentoceros pectoralis*) on Pacific seamounts. Trawling for fish in deep waters requires heavier ground gear and large trawl doors that plough across the seafloor and can easily damage epifaunal communities dominated by corals and sponges (**Figure 10**). Visual and acoustic evidence of damage to cold-water coral habitats has now been recorded from the territorial seas of many nations including the Canada, Ireland, New Zealand, Norway, UK, and USA, and in each case measures to limit or ban trawling in some coral-rich areas have been instigated. However, on the High Seas beyond the jurisdiction of any one nation, deep-water trawling continues without regulation causing unknown damage to benthic communities. As well as direct physical impacts, deep-water trawling disturbs sediment producing a plume that could smother epifauna over a wider area. To date, few studies have examined this wider area impact.

While evidence for damage from deep-water trawling is visually clear and now known to be a concern in several regions, the impacts of other human activities are harder to pin down. As with the fishing industry, technological developments and reduced shallow-water reserves have made it economically viable for the oil industry to explore progressively deeper and deeper waters. At the time of writing, there are producing oil fields in the deep waters of the Atlantic Frontier (UK), Campos Basin (Brazil), and Gulf of Mexico (USA), where exploratory drilling has now taken place to depths over 3000 m. For example, in late 2003, the Toledo well was drilled at 3051 m water depth by Chevron Texaco in the Gulf of Mexico.

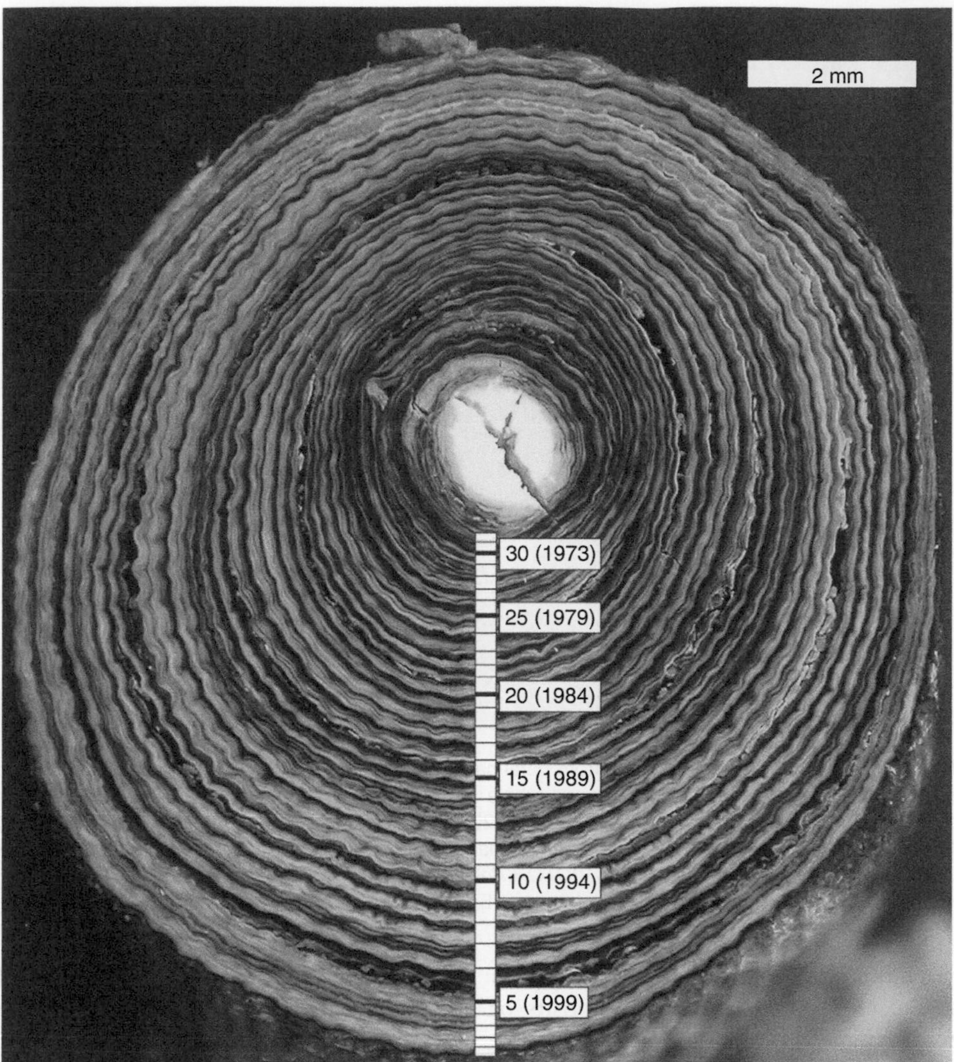

Figure 9 Cross section of the gorgonian *Primnoa resedaeformis* showing clear growth banding. Courtesy of Dr. Owen Sherwood, Memorial University of Newfoundland, reproduced from Sherwood OA, Scott DB, Risk MJ, and Guilderson TP (2005) Radiocarbon evidence for annual growth rings in the deep-sea octocoral *Primnoa resedaeformis*. *Marine Ecology Progress Series* 301: 129–134, with permission of Inter-Research.

As with all suspension-feeding invertebrate animals, cold-water corals are vulnerable to increased sediment loads that could smother polyps or even bury whole colonies. Tropical studies have shown corals are vulnerable to discharges, notably the muds and cuttings released during drilling operations, but it is frequently hard to disentangle physical effects of increased particle exposure from any toxic effects of exposure to drill muds or other additives. Intriguingly, both tropical and cold-water corals will settle and grow on producing oil platforms, sometimes close to drilling discharge points. Some of the older platforms in the northern North Sea support large colonies of *L. pertusa* that must have grown there continuously for over 20 years to have reached their present size and now form a reproductive population (**Figure 6**). Visual surveys have shown that coral polyps directly exposed to drill cuttings may be smothered but other polyps, even on the same colonies, that are not directly exposed can survive. This and the relatively small geographical extent of drilling suggest that if these activities are restricted in areas supporting cold-water coral reefs, their impacts will be considerably less than those of deep-water trawling now known to abrade large areas of the slope and to target seamount fish stocks. However, as tropical studies have shown, coral responses to drill discharges are hard to interpret and to date no detailed studies of reef framework-forming cold-water corals exposed to muds and cuttings have been carried out.

Figure 10 Fisheries damage to cold-water corals from coral carbonate mounds in the Porcupine Seabight and Porcupine Bank, Northeast Atlantic. (a) Nets and ropes with crushed coral rubble. (b) Closer view showing scavenging crabs. (c) Lost trawl net. (d) Abandoned trawl rope. (a), (b) Reproduced from Grehan *et al.* (2004) *Proceedings, ICES Annual Science Conference*, 22–25 Sep., Vigo, Spain, ICES CM 2004/AA07, with permission of International Council for the Exploration of the Sea (ICES). (c), (d) Courtesy of Dr. Anthony Grehan, National University of Ireland; *VICTOR-Polarstern* cruise ARKXIX/3a, Alfred-Wegener-Institut für Polar- und Meeresforschung and the Institut Français de Recherche pour l'Exploitation de la Mer.

Deep seabed mining for valuable materials such as manganese nodules or the rich mineral deposits of ridge and vent systems has yet to develop as a commercial proposition, although some forecasts suggest this could happen within the next 20 years. As with trawling and drilling, mining activities would have localized impacts on the area mined but could also disturb a sediment plume that would disperse affecting a wider area. Again, great care is needed to limit the impact of any developments in deep seabed mining on cold-water coral ecosystems.

Recent analysis suggests that rising atmospheric carbon dioxide levels will not only cause global warming, but also could cause the most rapid 'acidification' of the oceans seen in the last 300 My. Once again, no studies of cold-water coral response to ocean acidification have been carried out but tropical coral calcification could be reduced by over 50% if atmospheric carbon dioxide concentrations doubled. Perhaps of greatest concern are modeled predictions that the depth of the ASH could shallow by several hundred meters in as little as next 50–100 years, leading to concerns that regions currently suitable for cold-water coral reef development will become inhospitable. Corals living in lowered aragonite saturation states produce weaker skeletons more vulnerable to (bio)erosion; consequently, entire reef systems may shift from a phase of overall growth to erosion with severe implications for habitat integrity.

Conclusions

The last decade has seen an explosion of interest in cold-water coral reefs. This article attempts to summarize the many and exciting advances in our understanding of reef development, longevity, and diversity in deep waters while realizing that this work has only just begun. Work to map and characterize cold-water coral reefs is still needed, as shown by the geographical bias in studies so far, and efforts to unify sampling methodologies and species identification are vital. However, in the coming years, we will enter a phase in research on cold-water coral ecosystems that requires a shift away from baseline mapping to one centered on process-oriented questions that will examine the factors driving reef development in the deep ocean and allow us to really understand the significance and vulnerability of cold-water coral reefs and develop protected areas for their conservation.

See also

Coral Reefs.

Further Reading

Cairns SD (2007) Deep-water corals: An overview with special reference to diversity and distribution of deep-water Scleractinia. *Bulletin of Marine Science* 81: 311–322.

Expedition Scientists (2005) Modern carbonate mounds: Porcupine drilling. *Integrated Ocean Drilling Program Report Number 307*. Washington, DC: Integrated Ocean Drilling Program.

Freiwald A, Fosså JH, Grehan A, Koslow T, and Roberts JM (2004) *Cold-Water Coral Reefs*. Cambridge, UK: United Nations Environment Programme – World Conservation Monitoring Centre.

Freiwald A and Roberts JM (eds.) (2005) *Cold-Water Corals and Ecosystems*. Berlin: Springer.

Gage JD and Tyler PA (1991) *Deep-Sea Biology: A Natural History of Organisms at the Deep-Sea Floor.* Cambridge, UK: Cambridge University Press.

Grehan A, Unnithan V, Wheeler A, *et al.* (2004) *Proceedings, ICES Annual Science Conference*, 22–25 Sep., Vigo, Spain, ICES CM 2004/AA07.

Roberts JM, Brown CJ, Long D, and Bates CR (2005) Acoustic mapping using multibean echosounder reveals cold-water coral reefs and surrounding habitats. *Coral Reefs* 24: 654–669.

Roberts JM, Wheeler AJ, and Freiwald A (2006) Reefs of the deep: The biology and geology of cold-water coral ecosystems. *Science* 312: 543–547.

Rogers AD (1999) The biology of *Lophelia pertusa* (LINNAEUS 1758) and other deep-water reef-forming corals and impacts from human activities. *International Review of Hydrobiology* 4: 315–406.

Sherwood OA, Scott DB, Risk MJ, and Guilderson TP (2005) Radiocarbon evidence for annual growth rings in the deep-sea octocoral *Primnoa resedaeformis*. *Marine Ecology Progress Series* 301: 129–134.

Wilson JB (1979) 'Patch' development of the deep-water coral *Lophelia pertusa* (L.) on Rockall Bank. *Journal of the Marine Biological Association of the UK* 59: 165–177.

Relevant Website

http://www.lophelia.org

– Lophelia.org, an information resource on the cold-water coral ecosystems of the deep ocean.

ARTIFICIAL REEFS

W. Seaman and W. J. Lindberg, University of Florida, Gainesville, FL, USA

Introduction

Artificial reefs are intentionally placed benthic structures built of natural or man-made materials, which are designed to protect, enhance, or restore components of marine ecosystems. Their ecological structure and function, vertical relief, and irregular surfaces vary according to location, construction, and degree to which they mimic natural habitats, such as coral reefs. For ages, humans have taken advantage of the behavior of some organisms to seek shelter at submerged objects by introducing structures into shallow waters, where biological communities could form and fishes could be harvested. Contemporary purposes for artificial reefs include increasing the efficiency of artisanal, commercial, and recreational fisheries, producing new biomass in fisheries and aquaculture, boosting underwater recreation and ecotourism opportunities, preserving and renewing coastal habitats and biodiversity, and advancing research. This article reviews the evolution, spread, and increased scale of such practices globally, discusses their scientific basis, and describes trends concerning artificial reef planning, evaluation, and their appropriate application in natural resource management.

Utilization of Reefs in the World's Oceans

The near-shore ocean ecosystems of all inhabited continents contain artificial reefs. Their scales and purposes vary widely, ranging, for example, from placement in local waters of small structures by individuals in artisanal fishing communities, to deploying complex systems of heavy modules in distant areas by organizations concerned with commercial seafood production, to pilot studies using experimentally based structures for restoration of seagrass, kelp, and coral ecosystems. The origins of artificial reef technology are traced to places as diverse as Japan and Greece. Modern deployment of reefs has the longest history and has been most widespread in Eastern Asia, Australia, Southern Europe, and North America. In recent years, the technology has been introduced or more extensively established in scores of nations worldwide, including in Central and South America and the Indo-Pacific basin. No organization maintains a database for artificial reef development globally, so that assessments of reef-related activity must be derived from the records of economic development, fishery, and environment organizations, as well as from scientific journal articles which serve as a proxy for gauging national efforts.

For centuries structures of natural materials have been used to support artisanal and subsistence fishing in coastal communities, particularly in tropical areas (**Figure 1**). In India, traditionally tree branches have been weighted and sunken. Brush parks made of

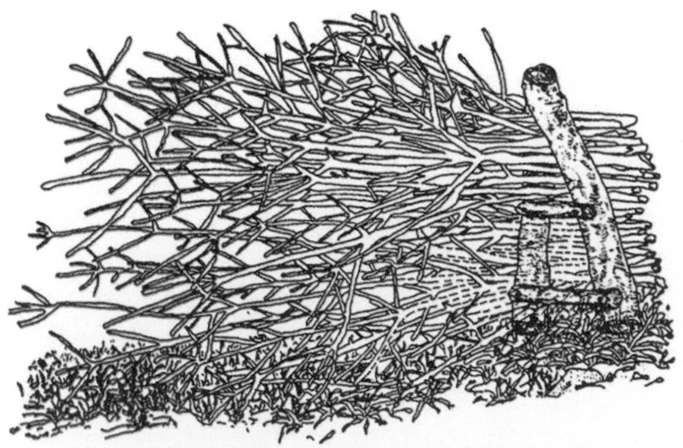

Figure 1 In some areas of long-standing artisanal fisheries, structures such as the *pesquero* (left), a bundle of mangrove tree branches, are used for benthic structure, as in Cuba. More recently, small fabricated modules, such as this example from India (right), are used to complement or replace natural materials.

branches stuck into the substrates have been used in estuaries of several African countries, Sri Lanka, and Mexico. In the Caribbean basin, casitas of wood logs provide benthic shelters from which spiny lobsters (*Panulirus argus*) are harvested. Indigenous knowledge of local fisherfolk is important in sustainably managing these reef systems, even when in their initial stages they may function principally to aggregate larger mobile fishes, such as around bundles of mangrove branches, *pesqueros*, in Cuba.

In recent decades, some artisanal fishing communities have deployed newer designs of artificial reefs, and at larger scales. In part, this has been in response to damage of habitat and fisheries by coastal land-use practices and more intense fishing practices, such as trawling. In India, for example, a national effort has deployed reef modules of steel plates, to increase the time fishing cooperatives actually devote to harvest as a means of promoting social and economic well-being. The coastal waters of Thailand received concrete modules as part of a plan to balance small- and large-scale fisheries, whereby areas of 14–22 km^2 received between 2400 and 3300 units.

The most extensive deployment of artificial reefs, for any purpose, is in Japan. There, fishing enhancement is the goal of a national plan begun in 1952 and since greatly expanded. By 1989, 9% of the coastal shelf (<200 m depth) had been affected by reef development. Early emphasis was on engineering and construction of reefs to withstand rigorous high seas environmental conditions. Industrial manufacturers have used materials such as steel, fiberglass, and concrete in structures that are among the largest units in the world, attaining heights of 9 m, widths of 27 m, and volumes of 3600 m^3 (**Figure 2**). Research at sea has been augmented by laboratory studies in which scale models of reefs have been analyzed for effects including deflection of bottom water currents to create nutrient upwellings for enhancing primary production. Observations of fish behavior led to a classification of the affinity of fishes to associate with physical structure. The Japanese aims have been to develop nursery, reproduction, and fishing grounds to supply seafood including abalone, clams, sea urchins, and finfishes. This partially was in response to the closure of distant waters to fishing by other nations.

Figure 2 The largest reef modules have been fabricated and deployed in Japan.

In contrast to Japan (and many other nations more recently involved), the United States, while also an early entrant in this field, built thousands of artificial reefs to enhance recreational fishing through numerous, independent small-scale efforts organized by local interest groups or governmental organizations at the state level. Until recently, materials of opportunity predominated, including heavier and more durable materials such as concrete rubble from bridge and building demolition sites. Designed structures are becoming more common, especially as reefs gain acceptance in habitat restoration. One state, Florida, has about half of the nation's total number of reefs, and in one four-county area annual expenditures by nonresidents and visitors on fishing and diving at dozens of artificial reefs were US $1.7 billion (2001 values). Recreational fishing has also been the focus of widespread reef building in Australia, and lesser efforts in areas such as Northern Ireland and Brazil.

Other regions more recently deploying artificial reefs to enhance or simply maintain fisheries to supply human foodstuffs include southern Europe and Southeast Asia. From inception, their approach has been to design, experimentally test, and document reefs. As early as the 1970s, 8 m^3, 13 t concrete blocks (and other designs) were deployed in the Adriatic Sea of Italy, forming reefs of volumes up to 13 000 m^3 and coverage up to 2.4 ha. They serve multiple, fishery-related objectives that are sought in many nations: shelter for juvenile and adult organisms; reproduction sites; capture nutrients in shellfish biomass; protect fish spawning and nursery areas from illegal trawling; and protect artisanal set fishing gear from illegal trawling damage. Higher catches and profits for small-scale fisheries were the result. This is especially true in Spain, where heavy concrete and steel rod reef designs have been used effectively to protect seagrass beds from illegal trawling (**Figure 3**).

Artificial reefs are being used as part of a larger strategy to implement marine ranching programs. In Korea, for example, the decline of distant water fisheries led to a transition from capture-based to culture-based fisheries. From 1971 to 1999, artificial reef placements have affected 143 000 ha of submerged

Figure 3 Concrete reef modules with projecting steel beams such as railroad tracks are deployed in protected marine areas of Spain to discourage illegal trawling, so as to promote restoration of seagrass habitats and fish populations. Photograph Courtesy of Juan Goutayer.

Figure 4 This larval enclosure tent is for settlement of the coral, *Montastraea faveolata*, at the site of a ship grounding on Molasses Reef, Florida Keys National Marine Sanctuary. Coral are mass-spawned, larvae cultured through the planktonic phase, and then introduced into the tent. In the laboratory, larvae may settle onto rubble pieces which subsequently are cemented to the reef. Photograph by D. Paul Brown.

coastal areas, out of the total 307 000 ha planned. At certain reef sites, hatchery-reared juvenile invertebrates and fishes have been released, and enhanced fishery yield reported, as part of long-term experiments. Artificial reefs are also used exclusively for aquaculture, such as in settings where excess nutrients stimulate primary production that is transferred into biomass of harvestable species, such as mussels (*Mytilus galloprovincialis*) and oysters (*Ostrea edulis* and *Crassostrea gigas*) in Italy.

Restoration of marine ecosystems using artificial reefs has focused on plant and coral communities. On the Pacific coast of the United States, for example, a project to mitigate for electric powerplant impacts is developing a 61 ha artificial reef for kelp (*Macrocystis pyrifera*) colonization, survival, and growth. Along the coast of the Northwest Mediterranean Sea, artificial reefs are deployed in protected areas to preserve and promote colonization of seagrasses (*Posidonia*). In numerous tropical areas, artificial reefs have been used in repairs to damaged coral reefs (**Figure 4**), or to hasten replacement of dead or removed coral. In the Philippines, for example, hollow concrete cubes are deployed as sites for coral colonization, sometimes being located in marine reserves.

Increasing popularity of artificial reefs to promote tourism is seen in the development of diving opportunities, both for people in submersible vehicles and for scuba divers. In places such as Mexico, the Bahamas, Monaco, and Hawaii, submarine operators provide trips to artificial reefs. On a larger scale, obsolete ships have been sunk to create recreational dive destinations in areas as diverse as British Columbia, Canada, Oman, New Zealand, Mauritius, and Israel. Purposes include generation of economic revenues in local communities and diversion of diver pressure from natural reefs.

As strictly works of engineering, structures such as rock jetties for shore protection also serve as *de facto* reefs with biological communities. Submerged breakwaters have been used to create waves for recreational surfboarding.

Progress toward Scientific Understanding

Advances in research on artificial reefs include methodologies, subjects and rigor of approaches. Methodological advances include improved underwater biological census techniques, and the application of remote sensing. Investigations have become more ecological process-oriented in explaining biological phenomena such as sheltering, reproduction, recruitment, and feeding, better able to explain physical and hydrodynamic behavior of reefs, and have led to specifying reef designs consistent with the ecology of organisms. Newer approaches include an expansion of hypothesis-driven and experimental research, scaling up of pilot designs into larger reefs, ecological modeling and forecasting, and interdisciplinary studies with various combinations of physical, social, life, and mathematical sciences. Comparative studies of natural and artificial habitats have advanced understanding of both systems. The earlier history of research on artificial reefs may be observed in the subjects, methods, and findings reported at eight international conferences held from

1974 to 2005. Subsequent to a peak number of peer-reviewed articles contained in a proceedings from the 1991 conference (84), fewer resulted from the latest conferences (1999 (56) and 2005 (20)). Scientists studying artificial reefs increasingly are reporting in journals devoted to fish biology, fisheries, and marine ecology.

Colonization of artificial reefs has the oldest and most extensive record of scientific inquiry. The appearance of large fishes at some reef sites almost immediately after placement of structure is the most visible form of colonization (and also a factor in one of the most controversial artificial reef issues, discussed below). However, microbes, plants, and invertebrates also colonize reefs, and patterns of succession occur that converge toward naturally occurring assemblages of species (**Figure** 5). Augmenting earlier studies describing diversity and abundance of artificial reef flora and fauna, later research addressed the influence of environmental variables upon reef ecology, interactions of individual species with the artificial reef structure and biota, and long-term life history studies of selected species. Comparison of patch reefs of stacked concrete blocks with translocated coral reefs in the Bahamas (**Figure** 6) determined that the reefs had similar fish species composition, while the natural reefs supported more individuals and species, owing to their greater structural complexity (i.e., variety of hole sizes) and associated forage base. Meanwhile, long-term studies of the spiny lobster experimentally established that artificial reef structures, with dark recessed crevices approximately one body-length deep and multiple entrances, indeed augmented recruitment in a situation of habitat limitation in the Florida Keys, USA. While these situations allowed observation by scuba diving, in deeper waters hydroacoustic and video-recording instruments have been used to document fish species, such as at petroleum production platforms in the North Sea and Gulf of Mexico.

Prominent among the scientific issues regarding the efficacy of artificial reefs for fisheries production and management is the 'attraction-production question': Do artificial reefs merely attract fishes thereby improving fishing efficiency, or do they contribute to biological production so as to enhance fisheries stocks? While this popularized dichotomy is an oversimplification, and an expectation of one all-encompassing answer is unrealistic, this question has had great heuristic value for the advance of reef science and responsible reef development. The origins of this issue derive from early observations and assumptions about artificial reefs. High catch rates and densities of fish at artificial reefs were popularly taken as proof that artificial reefs benefited fisheries stocks, based on assumptions that hard substrata were limiting and reef fish standing stocks were dependent on food webs that have hard substrata as their foundations. However, a few fisheries scientists challenged this reasoning in the 1980s. They noted that before heavy exploitation the existing natural habitat supported abundant reef fishes, presumably at or near carrying capacity. Fishing then reduced populations to some lower level, yet the amount of natural habitat remained the same, still capable of supporting higher population numbers. The fisheries scientists reasoned that with fish stocks substantially below carrying capacity, the amount of hard-bottom habitat could not be the factor limiting population size, so the addition of artificial reefs would not benefit fish stocks. Thus, observed high densities of

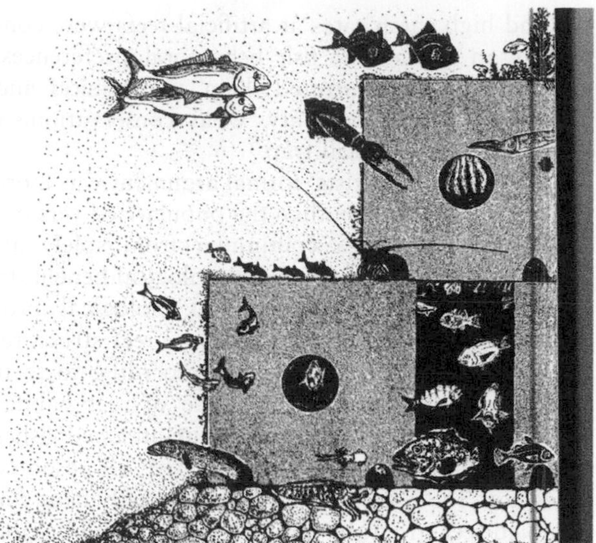

Figure 5 Aspects of attachment surfaces for marine plants and shelter and feeding sites for invertebrates and fishes provided by artificial reefs. Drawing courtesy of S. Riggio.

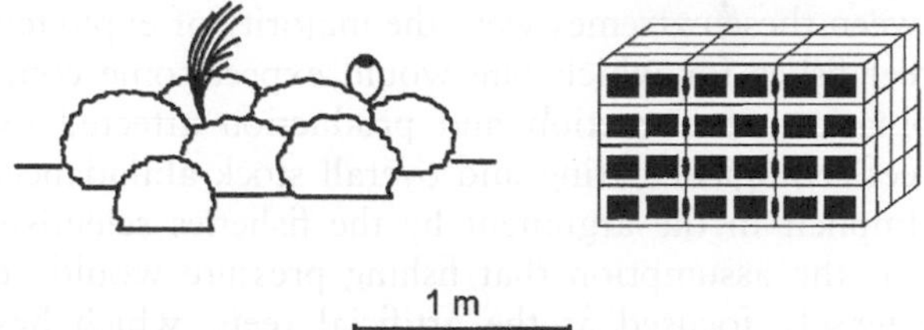

Figure 6 Experimental modules such as the one pictured on the right from the Bahamas exemplify a global trend toward greater manipulation of artificial reefs as tools for testing of hypotheses concerning ecological processes in the ocean, and comparison with systems such as coral reefs, pictured left. Drawing courtesy of American Fisheries Society.

fish and high catch rates at artificial reefs were considered an artifact of fish behavioral preferences, which simply concentrated them at reef sites and intensified fishing mortality, what conservationists today call an ecological trap.

The fisheries scientists' formal argument was more thorough and complex than the popularized version, as they recognized a continuum from solely attraction, with little if any direct biological benefit, to potentially high levels of added biological production from each unit of additional reef structure. Species fit along that continuum according to their life history characteristics and use of reef habitat ecologically. Thus small, highly site-attached fishes (e.g., blennies and gobies) that derive all resource needs (i.e., food, shelter, and mates) directly from the occupied reef structure, and that complete all but the planktonic phase of their life cycle in one place, would have the greatest net production potential from artificial reef development. Conversely, large, transient reef-visiting fishes (e.g., jacks and mackerels) were expected to have the least production potential from artificial reef development. In between these extremes were the majority of exploited reef fishes for which one would expect some combination of attraction and production affected by reef ecological setting and overall stock abundance. (Implicit in the argument by the fisheries scientists was the assumption that fishing pressure would be intensely focused at the artificial reefs, which has more to do with the management of human activities than the intrinsic ecological characteristics of the reefs themselves.)

Differing professional positions concerning attraction-production still remain, and the debate continues to stimulate research questions and artificial reef management issues. In part, this is because the complexity of reef ecological functions and the effects of spatially explicit fishing mortality are beyond simple answers to the seemingly dichotomous question that has been popularized among lay audiences. And in part, this is because researchers necessarily address this issue from their own disciplinary perspectives and study systems, generally not integrating the multiple levels of biological organization or spatial-temporal domains over which this question can be legitimately addressed. Pertinent studies can focus on life history stages of populations, communities, or ecosystems and encompass individual reefs, broad landscapes, or geographic ranges of fisheries stocks, while covering seasonal, interannual, or long-term time frames. Practical solutions to the fisheries management implications will require more specific attention to the desired species and assemblages, and the processes that sustain them.

Now a pivotal question is how the manipulation of habitat affects demographic rates and exploitation rates across spatial and temporal scales. An example of a way to integrate large sets of ecological data to analyze different scenarios of artificial reef development is through ecosystem simulations such as for Hong Kong where fishing levels in and out of protected areas have been predicted as part of a fishery recovery program.

Advances in Planning, Design, and Construction

The increase of artificial reefs in the world's coastal seas, as measured by the growing number, cost, size, intricacy, and footprint of structures deployed, has prompted an emphasis on their planning. This is to promote more efficient and cost-effective structures, enable work at larger scales and with more precision, satisfy regulatory requirements, reduce conflicts with other natural and human aspects of ecosystems, and minimize the prospects of unintended consequences from improperly constructed reefs. The scientific basis for reefs has been strengthened by research efforts in dozens of countries, coupled with the practical experiences of numerous individuals and groups that have worked independently to build, manage, and, increasingly, evaluate reefs.

Independently in different geographic areas, common approaches and some published guidelines for developing artificial reefs have emerged. The earliest handbooks focused on physical and oceanographic conditions as first encountered by the Coastal Fishing Ground Enhancement and Development Project of Japan. (Translations of certain documents into English in the 1980s by the United States National Marine Fisheries Service disseminated information.) The United States produced a national plan in 1984 (in revision in 2006) which spurred promulgation of plans among the coastal states. In some nations, planning is done through a coordinated reef program run by a federal or provincial marine or fishery resources organization, as in Italy or China. Finally, where reef deployment is directed to selected localities or areas, but not nationwide, master plans or other guidelines for siting, materials, and other aspects have been produced, such as for the Aegean Sea of Turkey and northern Taiwan.

With a legitimate intent assured, the initial component of reef planning is to frame measurable objectives, define expectations for success, and forecast interactions of that reef with the ecosystem. Subsequently, a valid design and site plan for the reef and procedures for its construction must be developed and meet regulatory requirements. The context for management of the reef must be defined. For example, an artificial reef program in Hong Kong was based on an extensive consultation with community interests, leading to stakeholder support for regulatory practices in the local fisheries and protection of marine reserve areas containing some of the reefs. Finally, protocols for evaluation of reef performance and management of assessment data must be established, with communication of results to all interested parties.

Construction is a three-phase process: fabrication, transportation, and placement. Representative costs of larger reefs include US $16 000 000 for a 56 ha reef in California, USA, and between US $38 and $57 million annually during 1995–99 for a national program in Korea. The ecological and economic impacts of poorly designed artificial reefs are exemplified by an effort in the United States – projected to cost US $5 000 000 – to remove 2 000 000 automobile tires that had been deployed in the 1970s, but which were drifting onto adjacent live coral reefs and beaches.

The design phase of reef development has changed dramatically due to growing emphasis on making the structural attributes of intentional, fabricated reef materials conform and contribute to the biological life history requirements of organisms that are particularly desired as part of the reef assemblage. Reef design should be dictated by: (1) a set of measurable and justified objectives for the reef, (2) the ecology of species of concern associated with the reef, and (3) predicted and understandable environmental and socioeconomic consequences of introduction of the reef into the aquatic ecosystem. This requires a multidisciplinary approach using expertise of biologists, engineers, economists, planners, sociologists, and others.

From a physical standpoint, key aspects of design and construction include stability of a reef site, durability of the reef configuration, and potential adverse impacts in the environment. Also, the use of physical processes to enhance reef performance is a desirable aspect. To build a quantitative understanding of the environment into which a reef is to be placed, large-scale oceanographic processes and local conditions must be determined by site surveys early in reef planning, including water circulation driven by tides, wind and baroclinic/density fields, locally generated wind waves, swells propagated from distance, sediment/substrate composition, distribution and transport, and depth. These factors are then coupled with the attributes of the reef material, such as weight, density, dimensions, and strength, in order to forecast reef physical performance.

From an ecological standpoint, abiotic and biotic influences considered in design include geographic location, type and quality of substrate surrounding the reef site, isolation, depth, currents, seasonality, temperature regime, salinity, turbidity, nutrients, and productivity. Substrate attributes affecting reef ecology include its composition and surface texture, shape, height, profile, surface area, volume and hole size, which taken together contribute to the structural complexity of the reef. Spacing of individual and groups of reef structure is important.

A plethora of designs for artificial reefs exist (**Figure 7**). For a mitigation reef aiming to create new kelp beds off California, USA, biologists and engineers concluded that the most effective design should place boulders and concrete rubble in low-relief piles (<1 m height), at depths of 12–14.5 m on a sand layer of 30–50 cm overlaying hard substrate; success criteria include (1) support of four adult plants per 100 m^2 and (2) invertebrate and fish populations similar to natural reefs. As part of a system of artificial reefs in Portugal, experiments quantified production of sessile invertebrates (upon which fishes could feed) according to location on different facets of cubic settlement structures placed on reefs. For fishes, meanwhile, one of five designs used in Korea is the box reef, a $3 \times 3 \times 3$ m concrete cube (**Figure 8**), targeted to two species: small, dark spaces in the lower two-thirds of the reef are provided for rockfish (*Sebastes schlegeli*), while the upper third is more open to satisfy behavioral preferences of porgy (*Pagrus major*). Repeatedly, authorities cite complexity of structure as a primary factor in design.

As large individual and sets of reefs are planned, more organizations are using pilot projects to determine physical, biological, economic, and even

Figure 7 Artificial reefs of concrete modules are used worldwide, with designs intended to meet ecological requirements of designated species and habitats.

Figure 8 The box reef design used in Korea resulted from a collaboration of engineers and biologists. Modules have been placed in research plots within Tongyong Marine Reserve for use in experiments with stocking juvenile fishes. Photograph by Kim Chang-gil.

political feasibility. The 42 000 t Loch Linne Artificial Reef in Scotland was started in 2001 as a platform for scientific investigation of the performance of different structures, ultimately to establish fisheries for target species, specifically lobster (*Homarus gammarus*). This effort also is representative of a trend toward increased predeployment research – in this case a 4-year study of seabed, water-column, and biological parameters – intended to enable better forecasting of reef impacts and measurement of results.

A special case of working with reef materials concerns the deployment of obsolete ships and petroleum/gas production platforms, due to the need to handle, prepare, and place them in environmentally compatible ways. In Canada, for example, decommissioned naval vessels require extensive removal of electrical wiring and other components to eliminate release of pollutants into the sea, in conformance with strict federal rules. In the northern Gulf of Mexico (where over 4000 platforms provide a considerable area of hard surface for sessile organism attachment), eastern Pacific, North Sea, and Adriatic Sea offshore platforms either act as *de facto* reefs or in a limited number of cases are being toppled in place or transported to new locations to serve as dedicated reefs, which costs less than removal to land.

Integration of Reefs in Ecological and Human Systems

Increasingly, artificial reef technology is being applied globally in fisheries and ecosystem management. Early disappointments and healthy skepticism have led to more realistic expectations bolstered by two decades of scientific advances. National plans (e.g., Japan, Korea, and United States), regional programs (e.g., Hong Kong, Singapore, and Turkey), and large-scale pilot projects (e.g., Loch Linne, Scotland; San Onofre, USA) concerning artificial reefs are better defining their role in ecosystem and fishery management. International scientific bodies such as the North Pacific Marine Sciences Organization (PICES) have addressed the relevance of artificial reefs to core fisheries issues, including stock enhancement, fishing regulations, and conservation. Integrated coastal management in the Philippines, India, Spain, and elsewhere now includes artificial reefs in multifaceted responses to issues of habitat destruction, fishery decline, and socioeconomic development. Finally, private consultants and businesses have found markets for their services and products, with one patented design being deployed in over 40 countries, and ecological engineers have recognized reefs as constructed ecosystems for use in restoration ecology.

The evaluation of reef performance is fostering an increased acceptance and utilization of artificial reef technology by resource management organizations. Quantitative evaluation is increasingly driven by agency concerns for demonstrating positive returns on their investments, and the increasing scale of artificial reef projects. Evaluations vary in intensity and complexity, ranging from descriptive studies of short duration (e.g., pre- and postdeployment) to extensive studies of ecological processes and dynamics, to the synthesis of complex databases through quantitative modeling. One response of the scientific community was through formation of the European Artificial Reef Research Network, which promulgated priorities and protocols for research across its membership.

A significant trend is for artificial reef projects to be planned and evaluated in an adaptive management framework, in which expectations are more explicitly stated, the projects implemented and rigorously evaluated, and then adjustments made to the management practices based on findings from the evaluations. When applied consistently, this cycle will continue to evolve the application of artificial reef technologies toward an ever-increasing standard of practical effectiveness.

See also

Cold-Water Coral Reefs. Coral Reefs. Fiordic Ecosystems. Fisheries Overview. Fishery Manipulation through Stock Enhancement or Restoration. Large Marine Ecosystems. Mariculture Diseases

and Health. Mariculture, Economic and Social Impacts. Mariculture of Aquarium Fishes. Mariculture of Mediterranean Species. Mariculture Overview. Ocean Gyre Ecosystems. Upwelling Ecosystems.

Further Reading

American Fisheries Society (1997) *Special Issue on Artificial Reef Management. Fisheries* 22: 4–36.

International Council for the Exploration of the Sea (2002) *ICES Journal of Marine Science* 59(supplement S1-5363).

Jensen AC, Collins KJ, and Lockwood APM (eds.) (2000) *Artificial Reefs in European Seas*. Dordrecht: Kluwer.

Lindberg WJ, Frazer TK, Portier KM, *et al.* (2006) Density-dependent habitat selection and performance by a large mobile reef fish. *Ecological Applications* 16(2): 731–746.

Love MS, Schroeder DM, and Nishimoto MM (2003) The ecological role of oil and gas production platforms and natural outcrops on fishes in southern and central California: A synthesis of information. OCS Study MMS 2003-032. Seattle. WA: US Department of the Interior, US Geological Survey, Biological Resources Division.

Seaman W (ed.) (2000) *Artificial Reef Evaluation.* Boca Raton, FL: CRC Press.

Svane I and Petersen JK (2001) On the problems of epibioses, fouling and artificial reefs: A review. *Marine Ecology* 22: 169–188.

CORALS AND HUMAN DISTURBANCE

N. J. Pilcher, Universiti Malaysia Sarawak, Sarawak, Malaysia

Introduction

Coral reefs are the centers of marine biodiversity on the planet. Reefs are constructed by a host of hermatypic (reef-building) coral species, but also are home to ahermatypic (non-calcium-carbonate depositing) corals, such as soft corals, black corals and gorgonians. The major structural components of reefs are the scleractinian corals. Much like their terrestrial rivals the tropical rainforests, reefs combine a host of microhabitats and a diverse array of life forms that is still being discovered and described. Coral reefs are mostly distributed throughout the tropical belt, and a large fraction are located in developing countries.

To understand how human activities affect coral reefs, it is necessary to briefly review their basic life history. Coral reefs are mostly made up of numerous smaller coral colonies; these colonies are in turn made up of thousands of minute polyps, which secrete a calcium carbonate skeleton. The deposition rate for individual coral species varies, but generally ranges between 0.1 mm and 10.0 cm per year. The accumulation of these skeletons over a long period of time results in massive, three-dimensional geological structures. The actual living tissue, however, is only the thin layer of living coral polyps on the surface. Corals are particularly susceptible to contaminants in sea water because the layer of tissue covering the coral skeleton is thin ($\sim 100\,\mu m$) and rich in lipids, facilitating direct uptake of chemicals. Coral polyps feed by filtering plankton using nematocyst (stinging cell)-tipped tentacles, and also receive organic matter through their symbiotic relationship with minute dinoflagellates called zooxanthellae. Zooxanthellae live within the gastrodermal tissues, and chemical communication (exchange) occurs via the translocation of metabolites. These small algal cells use sunlight to photosynthesize carbonates and water into organic matter and oxygen, both of which are used by the polyp.

Coral reefs support complex food and energy webs that are interlinked with nutrient inputs from outside sources (such as those brought with ocean currents and runoff from nearby rivers) and from the reef itself (where natural predation and die-off recirculate organic matter). These complex webs mean that any effect on one group of individuals will ultimately impact another, and single disturbances can have multiple effects on reef inhabitants. For example, the complete eradication of the giant Triton *Charonia trinis* through overfishing can result in outbreaks of the crown-of-thorns starfish *Acanthaster planci*. This can lead to massive coral mortality as the starfish reproduce and feed on the coral polyps. The mortality in turn may reduce habitats and food sources for reef fishes, which again, in turn, could lead to declines of larger predatory fishes. Similarly, the introduction of an invasive species either by accident or through ignorance (e.g., dumping of a personal aquarium contents into local habitats) might disrupt feeding processes and kill resident fishes. The death of key organisms on the reef (which then shifts from an autotrophic to a heterotrophic, suspension/detritus-feeding community) changes the dominant ecological process from calcium carbonate deposition to erosion, and ultimate loss of coral reef. Reef ecosystems may respond to environmental change by altering their physical and ecological structure, and through changes in rates of accretion and biogeochemical cycling. However, the potential for adaptation in reef organisms may be overwhelmed by today's anthropogenic stresses. The following sections provide a review of human disturbances and their general effects on coral reefs.

Collection of Corals

Corals have been mined for construction purposes in numerous Pacific Ocean islands and in South-East Asia. Usually the large massive life forms such as *Porites*, *Platygyra*, *Favia*, and *Favites* are collected and broken into manageable sizes or crushed for cement and lime manufacture. Similarly, the shells of the giant clam *Tridacna* and the conch shell *Strombus* are collected. Coral blocks and shells are used for construction of houses, roads, and numerous other projects. Corals are also collected for use in the ornamental trade, either as curios and souvenirs or as jewelry. Entire, small colonies of branching species such as *Acropora* and *Seriatophora* are used in the souvenir trade and for decoration, while black corals *Antipathes* and blue coral *Heliopora* are used for jewelry. The aquarium industry is also responsible for coral collection either for direct sale as live colonies or through the process of fish collecting. In many cases entire colonies full of fish are brought to the surface and are then smashed and discarded.

The removal of coral colonies decreases the shelter and niche areas available to numerous other reef inhabitants. Juvenile stages of fish that seek shelter among the branching species of corals, and worms and ascidians that take up residence on massive life forms, are deprived of protection and may become prey to other reef organisms. Removal of adult colonies also results in a reduction of overall reproductive output, as the corals no longer serve as a source of replenishment larvae. Further, removal of entire colonies reduces the overall structural stability of the reef, and increases rates of erosion through wave and surge damage.

Destructive Fishing

Destructive fishing pressures are taking their toll on coral reefs, particularly in developing countries in South-East Asia. The use of military explosives and dynamite was common shortly after the Second World War, but today this has shifted to the use of home-made explosives of fertilizer, fuel, and fuse caps inserted into empty beer bottles. Bombs weigh approximately 1 kg and have a destructive diameter of 4–5 m. Blast fishers hunt for schooling fish such as sweetlips and fusiliers, which aggregate in groups in the open or hide under large coral heads. Parrot fish and surgeon fish schools grazing on the reef crest are also actively sought. The bombs are usually set on five-second fuses and are dropped into the center of an area judged to have many fish. After the bomb has exploded, the fishers use dip nets, either from the boats or from underwater, to collect the stunned and dying fish. Many larger boats collect the fish using 'hookah' compressors and long air hoses to support divers working underwater.

The pressure wave from the explosion kills or stuns fish, but also damages corals. Natural disturbances may also fragment stony reef corals, and there are few quantitative data on the impacts of skeletal fragmentation on the biology of these corals. Lightly bombed reefs are usually pockmarked with blast craters, while many reefs in developing countries comprise a continuous band of coral rubble instead of a reef crest and upper reef slope. The lower reef slope is a mix of rubble, sand, and overturned coral heads. Typically at the base of the reef slope is a mound of coral boulders that have been dislodged by a blast and then rolled down the slope in an underwater avalanche. The reef slopes are mostly dead coral, loose sand, rubble, or rock and occasionally have overturned clams or coral heads with small patches of living tissue protruding from the rubble. The blasts also change the three-dimensional structure of reefs, and blasted areas no longer provide food or shelter to reef inhabitants. Further, once the reef structure has been weakened or destroyed by blast fishing, it is much more susceptible to wave action and the reef is unable to maintain its role in coastline protection. Larvae do not settle on rubble and thus replenishment and rehabilitation is minimal. Additionally, the destruction of adult colonies also results in a reduction of overall reproductive output, and reefs no longer serve as a source of replenishment larvae. Experimental findings, for instance, indicate that fragmentation reduces sexual reproductive output in the reef-building coral *Pocillopora damicornis*. The recovery of such areas has been measured in decades, and only then with complete protection and cessation of fishery pressure of any kind.

Another type of destructive fishing is 'Muro Ami', in which a large semicircular net is placed around a reef. Fish are driven into the net by a long line of fishermen armed with weighted lines. The weights are repeatedly crashed onto the reef to scare fish in the direction of the net, reducing coral colonies to rubble. The resulting effects are similar to the effects of blast fishing, spread over a larger area.

Cyanide fishing is also among the most destructive fishing methods, in which an aqueous solution of sodium cyanide is squirted at fish to stun them, after which they are collected and sold to the live-fish trade. Other chemicals are also used, including rotenone, plant extracts, fertilizers, and quinaldine. These chemicals all narcotize fish, rendering them inactive enough for collection. The fish are then held in clean water for a short period to allow them to recover, before being hauled aboard boats with live fish holds. In the process of stunning fish, the cyanide affects corals and small fish and invertebrates. The narcotizing solution for large fish is often lethal to smaller ones. Cyanide has been shown to limit coral growth and cause diseases and bleaching, and ultimately death in many coral species.

Among other destructive aspects of fishing are lost fishing gear, and normal trawl and purse fishing operations, when these take place near and over reefs. Trawlers operate close to reefs to take advantage of the higher levels of fish aggregated around them, only to have the trawls caught on the reefs. Many of these have to be cut away and discarded, becoming further entangled on the reefs, breaking corals and smothering others. Similarly, fishing with fine-mesh nets that get entangled in coral structures also results in coral breakage and loss. In South-East Asia, fishing companies have been reported to pull a chain across the bottom using two boats to clear off corals, making it accessible to trawlers.

Spearfishing also damages corals as fishermen trample and break coral to get at fish that disappear into crevices, and crowbars are frequently used to break coral. The collection of reef invertebrates along the reef crest results in breakage of corals that have particular erosion control functions, reducing the reef's potential to act as a coastal barrier.

Discharges

Mankind also effects corals through the uncontrolled and often unregulated discharge of a number of industrial and domestic effluents. Many of these are 'point-source' discharges that affect local reef areas, rather than causing broad-scale reef mortality. Sources of chemical contamination include terrestrial runoff from rivers and streams, urban and agricultural areas, sewage outfalls near coral reefs, desalination plants, and chemical inputs from recreational uses and industries (boat manufacturing, boating, fueling, etc.). Landfills can also leach directly or indirectly into shallow water tables. Industrial inputs from coastal mining and smelting operations are sources of heavy metals. Untreated and partially treated sewage is discharged over reefs in areas where fringing reefs are located close to shore, such as the reefs that fringe the entire length of the Red Sea. Raw sewage can result in tumors on fish, and erosion of fins as a result of high concentrations of bacteria. The resulting smothering sludge produces anaerobic conditions under which all benthic organisms perish, including corals. In enclosed and semi-enclosed areas the sewage causes eutrophication of the coral habitats. For instance, in Kaneohe bay in Hawaii, which had luxuriant reefs, sewage was dumped straight into the bay and green algae grew in plague quantities, smothering and killing the reefs. Evidence indicates that branching species might be more susceptible to some chemical contaminants than are massive corals.

Abattoir refuse is another localized source of excessive nutrients and other wastes that can lead to large grease mats smothering the seabed, local eutrophication, red tides, jellyfish outbreaks, an increase in biological oxygen demand (BOD), and algal blooms. Similarly, pumping/dumping of organic compounds such as sugar cane wastes also results in oxygen depletion.

The oil industry is a major source of polluting discharges. Petroleum hydrocarbons and their derivatives and associated compounds have caused widespread damage to coastal ecosystems, many of which include coral reefs. The effects of these discharges are often more noticeable onshore than offshore where reefs are generally located, but nonetheless have resulted in the loss of reef areas, particularly near major exploration and drilling areas, and along major shipping routes. Although buoyant eggs and developing larvae are sometimes affected, reef flats are more vulnerable to direct contamination by oil. Oiling can lead to the increased incidence of mortality of coral colonies.

In the narrow Red Sea, where many millions of tonnes per annum pass through the region, there have been more than 20 oil spills along the Egyptian coast since 1982, which have smothered and poisoned corals and other organisms. Medium spills from ballast and bilgewater discharges, and leakages from terminals, cause localized damage and smothering of intertidal habitats. Oil leakage is a regular occurrence from the oil terminal and tankers in Port Sudan harbor. Seismic blasts during oil exploration are also a threat to coral reefs. Refineries discharge oil and petroleum-related compounds, resulting in an increase in diatoms and a decrease in marine fauna closer to the refineries. Throughout many parts of the world there is inadequate control and monitoring of procedures, equipment, and training of personnel at refineries and shipping operations.

Drilling activities frequently take place near reef areas, such as the Saudi Arabian shoreline in the Arabian Gulf. Drilling muds smother reefs and contain compounds that disrupt growth and cause diseases in coral colonies. Field assessment of a reef several years after drilling indicated a 70–90% reduction in abundance of foliose, branching, and platelike corals within 85–115 m of a drilling site. Research indicates that exposure to ferrochrome lignosulfate (FCLS) can decrease growth rates in *Montastrea annularis*, and growth rates and extension of calices (skeleton supporting the polyps) decrease in response to exposure to $100\,\mathrm{mg\,l^{-1}}$ of drilling mud.

Oil spills affect coral reefs through smothering, resulting in a lack of further colonization, such as occurred in the Gulf of Aqaba in 1970 when the coral *Stylophora pistillata* did not recolonize oil-contaminated areas after a large spill. Effects of oil on individual coral colonies range from tissue death to impaired reproduction to loss of symbiotic algae (bleaching). Larvae of many broadcast spawners pass through sensitive early stages of development at the sea surface, where they can be exposed to contaminants and surface slicks. Oiling affects not only coral growth and tissue maintenance but also reproduction. Other effects from oil pollution include degeneration of tissues, impairment of growth and reproduction (there can be impaired gonadal development in both brooding and broadcasting species, decreased egg size and decreased fecundity), and decreased photosynthetic rates in zooxanthellae.

In developing countries, virtually no ports have reception facilities to collect these wastes and the problem will continue mostly through a lack of enforcement of existing regulations. The potential exists for large oil spills and disasters from oil tank ruptures and collisions at sea, and there are no mechanisms to contain and clean such spills. The levels of oil and its derivatives (persistent carcinogens) were correlated with coral disease in the Red Sea, where there were significant levels of diseases, especially Black Band Disease. In addition to the impacts of oils themselves are the impacts of dispersants used to combat spills. These chemicals are also toxic and promote the breakup of heavier molecules, allowing toxic fractions of the oil to reach the benthos. They also promote erosion through limiting adhesion among sand particles. The full effect of oil on corals is not fully understood or studied, and much more work is needed to understand the full impact. Although natural degradation by bacteria occurs, it is slow and, by the time bacteria consume the heavy, sinking components, these have already smothered coral colonies.

Industrial effluents, from a variety of sources, also impact coral reefs and their associated fauna and habitats. Heavy metal discharges lead to elevated levels of lead, mercury, and copper in bivalves and fish, and to elevated levels of cadmium, vanadium, and zinc in sediments. Larval stages of crustaceans and fish are particularly affected, and effluents often inhibit growth in phytoplankton, resulting in a lack of zooplankton, a major food source for corals. Industrial discharges can increase the susceptibility of fish to diseases, and many coral colonies end up with swollen tissues, excessive production of mucus, or areas without tissue. Reproduction and feeding in surviving polyps is affected, and such coral colonies rarely contribute to recolonization of reef areas.

Organisms in low-nutrient tropical waters are particularly sensitive to pollutants that can be metabolically substituted for essential elements (such as manganese). Metals enter coral tissues or skeleton by several pathways. Exposed skeletal spines (in response to environmental stress), can take up metals directly from the surrounding sea water. In Thailand, massive species such as *Porites* tended to be smaller in areas exposed to copper, zinc, and tin, there was a reduced growth rate in branching corals, and calcium carbonate accretion was significantly reduced.

Symbiotic algae have been shown to accumulate higher concentrations of metals than do host tissues in corals. Such sequestering in the algae might diminish possible toxic effects to the host. In addition, the symbiotic algae of corals can influence the skeletal concentrations of metals through enhancement of calcification rates. There is evidence, however, that corals might be able to regulate the concentrations of metals in their own tissues. For example, elevated iron in Thai waters resulted in loss of symbiotic algae in corals from pristine areas, but this response was lower in corals that has been exposed to daily runoff from an enriched iron effluent, suggesting that the corals could develop a tolerance to the metal.

Cooling brine is another industrial effluent that affects shoreline-fringing reefs, often originating from industrial installations or as the outflow from desalination plants. These effluents are typically up to 5–10°C higher in temperature and up to 3–10 ppt higher in salinity. Discharges into the marine environment from desalination plants in Jeddah include chlorine and antiscalant chemicals and 1.73 billion $m^3\,d^{-1}$ of brine at a salinity of 51 ppt and 41°C. The higher temperatures decrease the water's ability to dissolve oxygen, slowing reef processes. Increases in temperature are particularly threatening to coral reefs distributed throughout the tropics, where reef-building species generally survive just below their natural thermal thresholds. Higher-temperature effluents usually result in localized bleaching of coral colonies. The higher-salinity discharges increase coral mucus production and result in the expulsion of zooxanthellae and eventual bleaching and algal overgrowth in coral colonies. Often these waters are chlorinated to limit growth of fouling organisms, which increases the effects of the effluents on reef areas. The chlorinated effluents contain compounds that are not biodegradable and can circulate in the environment for years, bringing about a reduction in photosynthesis, with blooms of blue/green and red algae. Chlorinated hydrocarbon compounds include aldrin, lindrane, dieldrin, and even the banned DDT. These oxidating compounds are absorbed by phytoplankton and in turn by filter-feeding corals. Through the complex reef food webs these compounds concentrate in carnivorous fishes, which are often poisonous to humans.

Many airborne particles are also deposited over coral reefs, such as fertilizer dust, dust from construction activities and cement dust. At Ras Baridi, on the Red Sea coast of Saudi Arabia, a cement plant that operates without filtered chimneys discharges over 100 $t\,d^{-1}$ of partially processed cement over the nearby coral reefs, which are now smothered by over 10 cm of fine silt.

Solid Waste Dumping

The widespread dumping of waste into the seas has continued for decades, if not centuries. Plastics, metal, wood, rubber, and glass can all be found littering coral reefs. These wastes are often not biodegradable, and those that are can persist over long periods. Damage to reefs through solid waste dumping is primarily physical. Solid wastes damage coral colonies at the time of dumping, and thereafter through natural tidal and surge action. Sometimes the well-intentioned practice of developing artificial reefs backfires and the artificial materials are thrown around by violent storms, wrecking nearby reefs in the process.

Construction

Construction activities have had a major effect on reef habitats. Such activity includes coastal reclamation works, port development, dredging, and urban and industrial development. A causeway across Abu Ali bay in the northern Arabian Gulf was developed right over coral reefs, which today no longer exist. Commercial and residential property developments in Jeddah, on the Red Sea, have filled in reef lagoon areas out to reef crest and bulldozed rocks over reef crest for protection against erosion and wave action. Activities of this type result in increased levels of sedimentation as soils are nearly always dumped without the benefit of screens or silt barriers.

Siltation is invariably the consequence of poorly planned and poorly implemented construction and coastal development, which can result in removal of shoreline vegetation and sedimentation. Coral polyps, although able to withstand moderate sediment loading, cannot displace the heavier loads and perish through suffocation. Partial smothering also limits photosynthesis by zooxanthellae in corals, reducing feeding, growth, and reproductive rates.

The development of ports and marinas involves dredging deep channels through reef areas for safe navigation and berthing. Damage to reefs comes through the direct removal of coral colonies, sediment fallout, churning of water by dredger propellers, which increases sediment loads, and disruption of normal current patterns on which reefs depend for nutrients.

Landfilling is one of the most disruptive activities for coastal and marine resources, and has caused severe and permanent destruction of coastal habitats and changed sedimentation patterns that damage adjacent coral resources. Changes in water circulation caused by landfilling can alter the distribution of coral communities through redistribution of nutrients or increased sediment loads.

Recreation

The recreation industry can cause significant damage to coral reefs. Flipper damage by scuba divers is widespread. Some will argue that today's divers are more environmentally conscious and avoid damaging reefs, but certain activities, such as irresponsible underwater photography finds divers breaking corals to get at subjects and trampling reef habitats in order to get the 'perfect shot'. In areas where divers walk over a reef lagoon and crest to reach the deeper waters, there is a degree of reef trampling, heightened in cases where entry and exit points are limited.

Anchor damage from boats is a common problem at tourist destinations. In South-East Asia many diving operations are switching to nonanchored boat operations, but many others continue the practice unabated. Large tracts of reef can be found in Malaysia that have been scoured by dragging anchors, breaking corals and reducing reef crests to rubble. Experiments have shown that repeated break-age of corals, such as is caused by intensive diving tourism, may lead to substantially reduced sexual reproduction in corals, and eventually to lower rates of recolonization. In the northern Red Sea, another popular diving destination, and in the Caribbean, efforts are underway to install permanent moorings to minimize the damage to reefs from anchors.

Shipping and Port Activities

Congested and high-use maritime areas such as narrow straits, ports, and anchorage zones often lead to physical damage and/or pollution of coral reef areas. Ship groundings and collisions with reefs occur in areas where major shipping routes traverse coral reef areas, such as the Spratley Island complex in the South China Sea, the Red Sea, the Straits of Bab al Mandab and Hormuz, and the Gulf of Suez, to name only a few. Major groundings have occurred off the coast of Florida in the United States, such as the one off Key Largo in State park waters in the 1980s, causing extensive damage to coral reefs. Often these physical blows are severe and destroy decades, if not centuries, of growth. Fish and other invertebrates lose their refuges and foraging habitats, while settlement of new colonies is restricted by the broken-up nature of the substrate. Seismic cables towed during seabed surveys and exploration activities may damage the seabed. Cable damage from towing of vessels (e.g., a tug and barge) has been reported snagging on shallow reefs in the Gulf of Mexico, causing acute damage to sensitive reefs.

Discharges from vessels include untreated sewage, solid wastes, oily bilge, and ballast water. On the high seas these do not have a major noticeable effect on marine ecosystems, but close to shore, particularly at anchorages and near ports, the effects become more obvious. At low tides, oily residues may coat exposed coral colonies, and sewage may cause localized eutrophication and algal blooms. Algal blooms in turn deplete dissolved oxygen levels and prevent penetration of sunlight.

Port activities can have adverse effects on nearby reefs through spills of bulk cargoes and petrochemicals. Fertilizers, phosphates, manganese, and bauxite, for instance, are often shipped in bulk, granular form. These are loaded and offloaded using massive mechanical grabs that spill a little of their contents on each haul. In Jordan, the death of corals was up to four times higher near a port that suffered frequent phosphate spills when compared to control sites. The input of these nutrients often reduces light penetration, inhibits calcification, and increases sedimentation, resulting in slower feeding and growth rates, and limited settlement of new larva.

War-related Activities

The effects of war-related activities on coral reef health and development are often overlooked. Nuclear testing by the United States in Bimini in the early 1960s obliterated complete atolls, which only in recent years have returned to anything like their original form. This redevelopment is nothing like the original geologic structure that had been built by the reefs over millenia. The effects of the nuclear fallout at such sites is poorly understood, and possibly has long-term effects that are not appreciable on a human timescale. The slow growth rate of coral reefs means that those blast areas are still on the path to recovery.

Target practice is another destructive impact on reefs, such as occurred in 1999 in Puerto Rico, where reefs were threatened by aerial bombing practice operations. In Saudi Arabia, offshore islands were used for target practice prior to the Gulf war in 1991. The bombs do not always impact reefs, but those that do cause acute damage that takes long periods to recover.

In the Spratley islands, the development of military structures to support and defend overlapping claims to reefs and islands has brought about the destruction of large tracts of coral reefs. Man-made islands, aircraft landing strips, military bases, and housing units have all used landfilling to one extent or another, smothering complete reefs and resulting in high sediment loads over nearby reefs. Dredging to create channels into reef atolls has also wiped out extensive reef areas.

Indirect Effects

Most anthropogenic effects and disturbances to coral reefs are easily identifiable. Blast fishing debris and discarded fishing nets can be seen. Pollutant levels and sediment loads can be measured. However, many other man-made changes can have indirect impacts on coral reefs that are more difficult to link directly to coral mortality. Global warming is generally accepted as an ongoing phenomenon, resulting from the greenhouse effect and the buildup of carbon dioxide in the atmosphere. Temperatures generally have risen by 1–2°C across the planet, bringing about secondary effects that have had noticeable consequences for coral reefs. The extensive coral beaching event that took place in 1998, which was particularly severe in the Indian Ocean region, is accepted as having been the result of surface sea temperature rise. Bleaching of coral colonies occurs through the expulsion of zooxanthellae, or reductions in chlorophyll content of the zooxanthellae, as coral polyps become stressed by adverse thermal gradients. Some corals are able to survive the bleaching event if nutrients are still available, or if the period of warm water is short.

Coupled with global warming is change of sea level, which is predicted to rise by 25 cm by the year 2050. This sea level rise, if not matched by coral growth, will mean corals will be submerged deeper and will not receive the levels of sunlight required for zooxanthellae photosynthesis. Additionally, the present control of erosion by coral reefs will be lost if waves are able to wash over submerged reefs.

Coral reef calcification depends on the saturation state of carbonate minerals in surface waters, and this rate of calcification may decrease significantly in the future as a result of the decrease in the saturation level due to anthropogenic release of CO_2 into the atmosphere. The concentration of CO_2 in the atmosphere is projected to reach twice the preindustrial level by the middle of the twenty-first century, which will reduce the calcium carbonate saturation state of the surface ocean by 30%. Carbonate saturation, through changes in calcium concentration, has a highly significant short-term effect on coral calcification. Coral reef organisms do not seem to be able to acclimate to the changing saturation state, and, as calcification rates drop, coral reefs will be less able to cope with rising sea level and other anthropogenic stresses.

The Future

Mankind has contributed to the widespread destruction of corals, reef areas, and their associated fauna through a number of acute and chronic pollutant discharges, through destructive processes, and through uncontrolled and unregulated development. These effects are more noticeable in developing countries, where social and traditional practices have changed without development of infrastructure, finances, and educational resources. Destructive fishing pressures are destroying large tracts of reefs in South-East Asia, while the development of industry affects reefs throughout their range. If mankind is to be the keeper of coral reefs into the coming millennium, there is going to have to be a shift in fishing practices, and adherence to development and shipping guidelines and regulations, along with integrated coastal management programs that take into account the socioeconomic status of people, the environment, and developmental needs.

Glossary

Ahermatypic Non-reef-building corals that do not secrete a calcium carbonate skeletal structure.

DDT Dichlorodiphenyltrichloroethane.

Dinoflagellates One of the most important groups of unicellular plankton organisms, characterized by the possession of two unequal flagella and a set of brownish photosynthetic pigments.

Eutrophication Pollution by excessive nutrient enrichment.

Gastrodemal The epithelial (skin) lining of the gastric cavity.

Hermatypic Reef-building corals that secrete a calcium carbonate skeletal structure.

Quinaldine A registered trademark fish narcotizing agent.

Scleractinians Anthozoa that secrete a calcareous skeleton and are true or stony corals (Order Scleractinia).

Zooxanthellae Symbiotic algae living within coral polyps.

Further Reading

Birkland C (1997) *Life and Death of Coral Reefs*. New York: Chapman and Hall.

Connel DW and Hawker DW (1991) *Pollution in Tropical Aquatic Systems*. Boca Raton, FL: CRC Press.

Ginsburg RN (ed.) (1994) *Global Aspects of Coral Reefs: Health, Hazards and History*, 7–11 June 1993, p. 420. Miami: University of Miami.

Hatziolos ME, Hooten AJ, and Fodor F (1998) Coral reefs: challenges and opportunities for sustainable management. In: *Proceedings of an Associated Event of the Fifth Annual World Bank Conference on Environmentally and Socially Sustainable Development*. Washington, DC: World Bank.

Peters EC, Glassman NJ, Firman JC, Richmonds RH, and Power EA (1997) Ecotoxicology of tropical marine ecosystems. *Environmental Toxicology and Chemistry* 16(1): 12–40.

Salvat B (ed.) (1987) *Human Impacts on Coral Reefs: Facts and Recommendations*, p. 253. French Polynesia: Antenne Museum E.P.H.E.

Wachenfeld D, Oliver J, and Morrisey JI (1998) *State of the Great Barrier Reef World Heritage Area 1998*. Townsville: Great Barrier Reef Marine Park Authority.

Wilkinson CR (1993) Coral reefs of the world are facing widespread devastation: can we prevent this through sustainable management practices. In: *Proceedings of the Seventh International Coral Reef Symposium* Guam, Micronesia. Mangilao: University of Guam Marine Laboratory.

Wilkinson CR and Buddemeier RW (1994) *Global Climate Change and Coral Reefs: Implications for People and Reefs*. Report of the UNEO-IOC-ASPEI-IUCN Task Team on Coral Reefs. Gland: IUCN.

Wilkinson CR, Sudara S, and Chou LM (1994) Living coastal resources of Southeast Asia: Status and review. In: *Proceedings of the Third ASEAN-Australia Symposium on Living Coastal Resources*, vol. 1. Townsville: ASEAN-Australia Marine Science Project, Living Coastal Resources.

HYDROTHERMAL VENT BIOTA

R. A. Lutz, Rutgers University, New Brunswick, NJ, USA

On 17 February 1977, the deep-submergence vehicle *Alvin* descended 2500 m to the crest of the Galapagos Rift spreading center to first visit an ecosystem that would forever change our view of life in the deep sea. Cracks and crevices in the ocean floor were emanating fluids with temperatures up to 17°C. None of the bizarre organisms clustering around these 'hydrothermal vents' had ever been encountered; they comprised new species, genera, families, superfamilies, and bizarre 'tubeworms,' up to 2 m long, which were subsequently placed in a new phylum (Vestimentifera) (**Figure 1**). Since the Galapagos Rift discovery, numerous hydrothermal vent sites have been found throughout the world's oceans and over 500 new species have been described from these regions. **Figure 2** depicts many of the major hydrothermal systems from which organisms have been collected to date. Fluids with temperatures as high as 403°C exit from polymetallic sulfide chimneys in many of these regions (**Figure 3**).

Most ecosystems on earth ultimately rely on photosynthesis, with the energy source being solar. In marked contrast, deep-sea hydrothermal ecosystems are based predominantly on chemosynthesis, with the energy source being geothermal. Many of the chemosynthetic microbes are fueled by hydrogen sulfide, which is present at low-temperature vents in concentrations up to several hundred micromoles per liter and at high-temperature vents in concentrations up to 100 milimoles per liter. These microbial organisms can be either 'free-living' (in the water or on the surface of various substrates) or symbiotic in association with certain vent organisms. The vestimentiferan tubeworms *Riftia pachyptila* and *Tevnia jerichonana* (**Figures 1, 4, 5,** and **6**), for example, each have a specialized 'tissue,' known as the trophosome, which is comprised entirely of chemosynthetic bacteria. The tubeworms have no mouth, no digestive system, and no anus; in short, no opening to the external environment. Hydrogen sulfide diffuses across cell membranes and is transported via the hemoglobin-containing circulatory system to the trophosome, where it is utilized by the associated symbionts. Mussels (*Bathymodiolus thermophilus*) (**Figures** 7 and 8) and vesicomyid clams (*Calyptogena magnifica*) (**Figure** 9), common along both the Galapagos Rift and East Pacific Rise (EPR), represent two of the other dominant members of the vent megafauna that house chemosynthetic symbionts. In the case of each of these bivalves, the symbionts are associated with the gills and both species have modified feeding apparatuses relative to those of shallow-water related species (likely a result of their

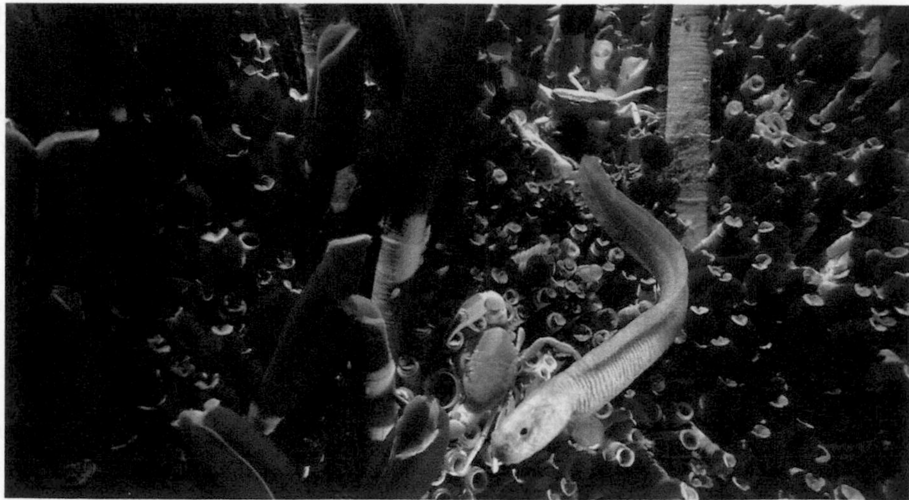

Figure 1 A cluster of vestimentiferan tubeworms (*Riftia pachyptila* and *Tevnia jerichonana*), together with a zoarcid fish (*Thermarces andersoni*) and brachyuran crabs (*Bythograea thermydron*) inhabiting low-temperature hydrothermal vents at 9°50′N along the East Pacific Rise (depth 2500 m).

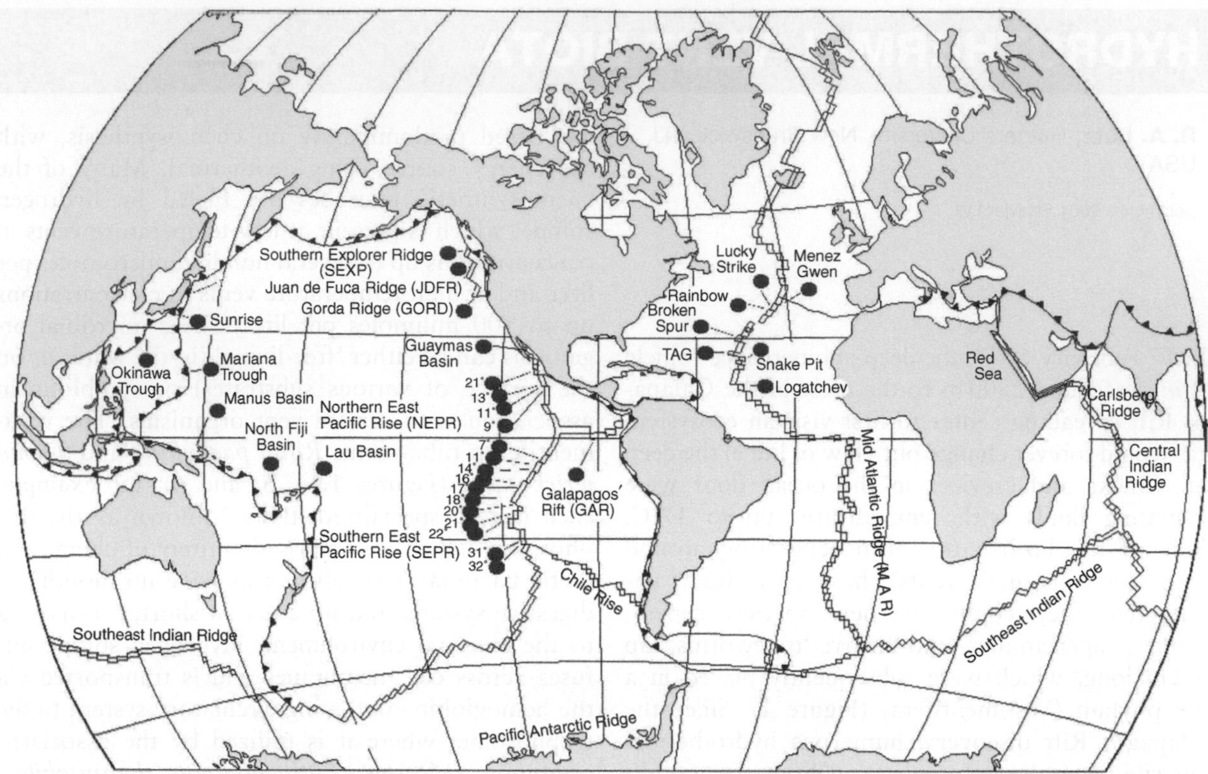

Figure 2 Deep-sea hydrothermal vent sites along mid-oceanic and back-arc ridge systems from which vent organisms have been collected to date. Numbers indicate approximate latitude of site along the East Pacific Rise.

predominant reliance on the associated symbionts for nutrition). Closely related mussels and clams within the families Mytilidae and Vesicomyidae are common constituents of the fauna associated with vents along mid-oceanic ridge and back-arc spreading centers (as well as at many cold-water hydrocarbon seeps) throughout the world's oceans. All of these bivalve mollusks retrieved to date appear to contain thiotrophic ('sulfur-feeding') or methanotrophic ('methane-feeding') chemosynthetic symbionts.

It should be mentioned that, while chemosynthetic bacteria play a critical role in food chains associated with many vent systems, there are numerous other microbes at the vents belonging to two other recognized kingdoms. Both eukaryotes and Archaea occupy specialized niches within various vent ecosystems and certain Archaea, which have been isolated from environments associated with high-temperature sulfide chimneys, have been reported to occupy the most primitive 'node' on the phylogenetic tree of life. Such reports have led to considerable speculation as to whether or not life itself may have originated in hydrothermal vent environments.

At various high-temperature vent sites, numerous organisms colonize the sides of active sulfide edifices. **Figure 4** depicts a sulfide edifice (named 'Tubeworm Pillar') that is 11 m high, the sides of which are covered with tubeworms (both *R. pachyptila* and *T. jerichonana*), crabs, zoarcid fish, and a spectrum of smaller, associated vent fauna. At other sites, the sides of 'black smoker' chimneys are frequently covered with the tubes of polychaetes, such as *Alvinella pompejana*, a 'Pompeii' worm named after the

Figure 3 A 'black smoker' polymetallic sulfide chimney at 20°50′N along the East Pacific Rise (depth 2615 m). Temperatures as high as 403°C have been recorded at the orifice of such edifices from which mineral-rich fluids emanate violently in many deep-sea hydrothermal systems.

Figure 4 A portion of an 11 m high polymetallic sulfide edifice known as Tubeworm Pillar at 9°50′N along the East Pacific Rise (depth 2500 m). The top two-thirds of the edifice is covered with vestimentiferan tubeworms, both *Riftia pachyptila* (larger organisms) and *Tevnia jerichonana* (smaller organisms), as well as numerous brachyuran crabs (*Bythograea thermydron*) and zoarcid fish (*Thermarces andersoni*).

submersible *Alvin* (**Figure 10**). This organism has been reported to routinely withstand long-term exposure to temperatures ranging from 2°C to 35°C and short-term exposure to temperatures in excess of 100°C, rendering this annelid perhaps the most eurythermal organism on the planet.

In addition to the numerous species of bivalves, such as mussels and clams mentioned earlier, there are myriad other common mollusks at vents, including numerous gastropods, particularly archeogastropod limpets. Thirteen different gastropod species have been reported from vents along the narrow range of the East Pacific Rise from 9°17′N to 9°54′N. These organisms can achieve high population densities and are found on a wide variety of substrates including the tubes of *Riftia pachyptila* and the shells of *Bathymodiolus thermophilus* (**Figures 8 and 11**). Limpets graze on free-living microbes that coat the majority of surfaces associated with both low-temperature and high-temperature (e.g., sulfide chimney) vents. Over 100 species of mollusks have been collected to date from the mid-oceanic and back-arc spreading centers visited to date.

Virtually all of the invertebrates inhabiting deep-sea hydrothermal vents have planktonic larval stages. These free-swimming stages serve as the primary

Figure 5 Higher magnification of the side of Tubeworm Pillar (depicted in **Figure 4**), showing tubeworms (*Riftia pachyptila* and *Tevnia jerichonana*) and scavenging brachyuran crabs (*Bythograea thermydron*).

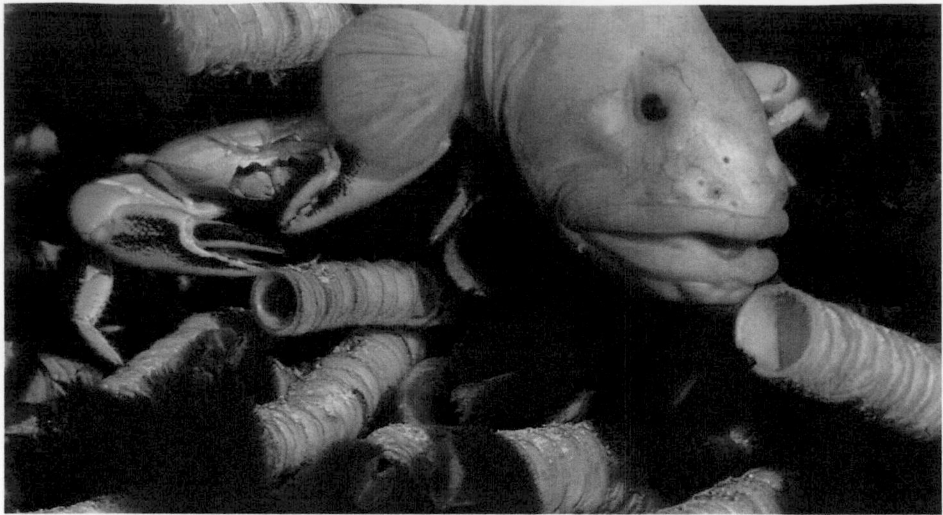

Figure 6 Close-up image of a cluster of *Tevnia jerichonana*, together with a brachyuran crab (*Bythograea thermydron*) and a zoarcid fish (*Thermarces andersoni*). Note the 'accordion-like' morphology of the tube of this species of vestimentiferan tubeworm.

Figure 7 A dense population of mussels (*Bathymodiolus thermophilus*) inhabiting a low-temperature hydrothermal vent field along the East Pacific Rise. Associated fauna in the field of view include tubeworms (*Riftia pachyptila*), brachyuran crabs (*Bythograea thermydron*), zoarcid fish (*Thermarces andersoni*), and a galatheid crab (*Munidopsis subsquamosa*) (lower left).

means of dispersal between isolated vent systems, which can be separated by hundreds of kilometers. The larval stages can be either planktotrophic (feeding within the water column) or lecithotrophic (utilizing yolk reserves for nutrition). In either case, it appears that the early life history stages of the vast majority of vent organisms are capable of staying in the water column for considerable lengths of time (likely months in the case of many vent species). Such extended planktonic durations facilitate passive transport over vast distances via ocean currents (velocities of ocean currents between 15 and 30 cm s^{-1} are commonly encountered along the crest of ridge systems).

Numerous crustaceans inhabit the vent environment and represent perhaps the dominant scavengers

Figure 8 Close-up of mussels (*Bathymodiolus thermophilus*) attached to the tubes of the tubeworm *Riftia pachyptila*. Limpets (*Lepetodrilus elevatus*) are seen attached to the external surfaces of both the mussel shells and tubeworm tubes.

Figure 9 Vesicomyid clams (*Calyptogena magnifica*) line cracks and crevices from which low-temperature fluids are venting in an area known as Clam Acres at 20°50′N along the East Pacific Rise (depth 2615 m). This species is common at many vent sites that are believed to be in relatively late stages of succession along the Galapagos Rift and the entire stretch of the East Pacific Rise from 21°N to 18°S.

of the ecosystem. Brachyuran crabs (e.g., *Bythograea thermydron* on the Galapagos Rift and the East Pacific Rise) (**Figures 1, 5, 6,** and 7) appear to be one of the earliest colonizers of hydrothermal vent environments. When new vents were formed during the April, 1991 volcanic eruption along the crest of the East Pacific Rise at 9°50′N, several regions were referred to as 'crab nurseries' owing to the relatively large abundance of crab larvae (megalopae) in areas of low-temperature discharge. Eleven months after the eruptive event, the region was populated by tremendous numbers of large crabs. These were frequently observed to be holding with their claws various substrates, such as empty tubeworm tubes or pieces of basalt covered with microbial mats, and voraciously scraping the surfaces with their mouth parts, suggesting that microbes and/or their products may represent an important source of nutrition for

Figure 10 The polychaete *Alvinella pompejana* emerging from its tube on the side of a black smoker chimney at 9°50′N along the East Pacific Rise. It has been suggested that the organism may be the most eurythermal invertebrate on the planet, capable of withstanding short-term exposures to temperatures ranging from less than 2°C to in excess of 100°C.

Figure 11 Limpets (*Lepetodrilus elevatus*) coating the surfaces of the tubes of *Riftia pachyptila* and the shell of *Bathymodiolus thermophilus* (left). A zoarcid fish (*Thermarces andersoni*) is seen emerging from the tubeworm tubes.

these organisms. They are also frequently seen extending their claws into the tubes of *Riftia pachyptila* and will readily devour the tissues of any tubeworm or mussel damaged during routine submersible activities. They are easily captured in 'crab traps' baited with a wide variety of dead fish. Intraspecific attacks appear to be relatively common occurrences and crabs with only one claw or a missing leg are frequently seen on the bottom, presumably reflecting an encounter with another crab.

Galatheid crabs (e.g., *Munidopsis subsquamosa*) (**Figure 12**) inhabit peripheral areas of the vents during earlier stages of succession and, as hydrogen sulfide levels gradually decrease over time, they are commonly encountered in central areas as well, often among tubeworms and mussels. While these 'squat lobsters' may also be scavengers, they are seldom, if ever, caught in traps. One chance encounter with a relatively large dead octopod in the peripheral area of a vent revealed a large quantity of galatheid crabs on the carcass and a noticeable absence of brachyuran crabs or other vent organisms (**Figure 13**). Close-up imagery revealed that the crabs were actively feeding on the dead tissues of the carcass. Galatheids that have been collected to date have generally been 'transported' back to the surface in random regions of the submersible itself. The carapaces of this species are frequently covered with fine, filamentous bacteria (**Figure 12**).

Figure 12 A galatheid crab (*Munidopsis subsquamosa*) with filamentous bacteria on its carapace. This species is a common inhabitant of peripheral vent environments and is also frequently observed in ambient deep-sea environments.

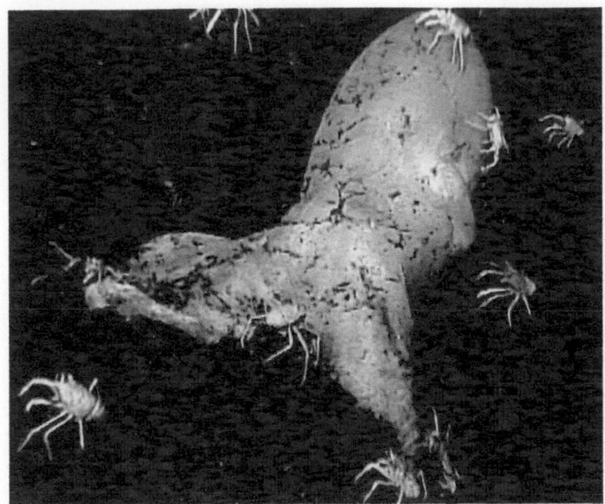

Figure 13 Numerous galatheid crabs (*Munidopsis subsquamosa*) on and in the immediate vicinity of the carcass of a cirrate octopod approximately 50 m from an active hydrothermal vent at 9°50′N along the East Pacific Rise. Close-up imagery revealed that the crabs were actively feeding on the dead tissues of the carcass.

Numerous species of shrimp (**Figure 14**) have been encountered at vent sites visited to date along ridge axes in all ocean basins. At several sites (e.g., the TAG hydrothermal vent field) along the Mid-Atlantic Ridge, thousands of shrimp are frequently seen 'swarming' on top of one another, completely carpeting the sides of large sulfide chimneys. It has been suggested that a large 'eye spot' on the back of the carapace of the shrimp is capable of sensing long-wavelength light emitted by high-temperature smokers. Shrimp are also relatively common inhabitants of vents visited to date in the western Pacific (e.g., at the Alice Springs vent field along the back-arc spreading center in the Mariana Trough).

Other smaller, common vent-endemic crustaceans include numerous species of amphipods, copepods, and leptostracans. *Halice hesmonectes*, an amphipod common at various vent sites along the East Pacific Rise is frequently seen 'swarming' above mussel and tubeworm colonies in regions of active low-temperature venting. It has been reported that these dense swarms represent the highest concentration of planktonic invertebrates in the ocean.

Serpulid polychaetes (**Figure 15**) are common inhabitants of the peripheral area of many vent fields. Commonly referred to as 'feather dusters,' they were seen extending over large expanses of lava during the early expeditions to the Galapagos Rift in 1977 and 1979 and were subsequently reported at numerous vents along the East Pacific Rise. Their small tubes, which generally reach lengths of about 5 cm, consist of calcite. When the tentacular plumes are withdrawn into the tube, a small 'plug' seals the tube from the external environment.

Numerous species of apparently vent-endemic fish have been reported from hydrothermal systems along ridge systems throughout the world's oceans. *Bythites hollisi* is a bythitid that was encountered on the first dive to the Galapagos Rift vent field in 1977. Bythitids have been observed in large numbers at several other vent sites, such as the hydrothermal field at 9°50′N along the East Pacific Rise (**Figure 16**). One large 'pit,' with a diameter of several meters, from which cloudy, shimmering water was emanating, had a concentration of over 20 bythitids,

Figure 14 The shrimp *Alvinocaris lusca* perched atop a tube of *Riftia pachyptila* at a low temperature vent along the East Pacific Rise.

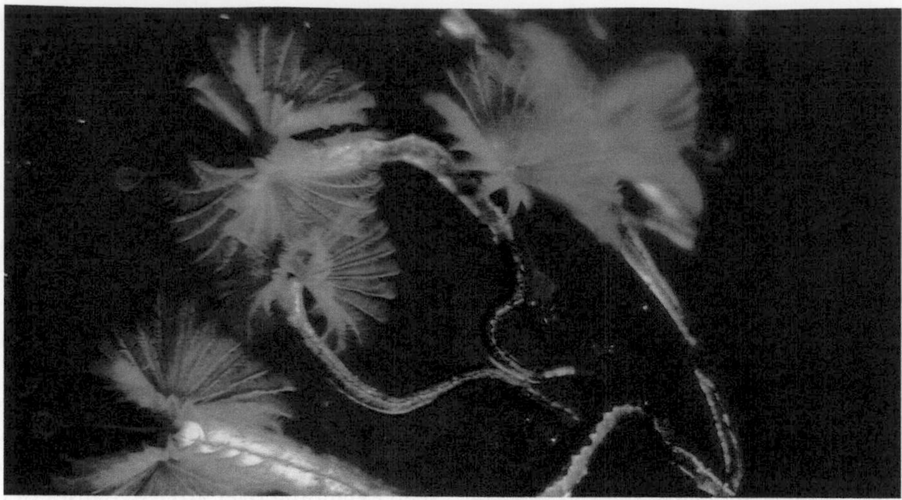

Figure 15 Serpulid polychaetes in the peripheral area of a vent field at 9°50′N along the East Pacific Rise.

which were frequently observed with their heads projecting downward into the cloudiest portions of the water at the base of the pit.

Zoarcids (eel pouts) are common members of the vent fauna at various sites along the East Pacific Rise and the Mid-Atlantic Ridge, with several different species having been encountered at the various vent fields visited to date. The common species that inhabits many East Pacific Rise vents is *Thermarces andersoni* (**Figures 1, 4, 6,** and **11**). From analyses of extensive video footage and stomach contents of retrieved specimens of this species, it appears to commonly feed on a wide range of organisms, including bacteria, amphipods, leptostracans, and shrimp. It has also been observed scavenging on the plumes of specimens of vestimentiferan tubeworms that have been damaged in the process of sampling or maneuvering with the submersible.

Enteroptneusts, commonly referred to as 'spaghetti worms' were first observed in 1977 draped over pillow lava in peripheral regions of the Galapagos Rift vent field. Few individuals were observed at any of the other many vent fields visited over the next 20 years throughout the world. In 1997, six years after the volcanic eruption at 9°50′N along the East Pacific Rise, numerous enteroptneusts were observed at distances ranging from a few meters to a few hundred meters from active vent sites within the region. No individual organisms were observed living in direct association with venting fluids or at distances in excess of a kilometer from active hydrothermal systems. This unusual, soft-bodied

Figure 16 The vent fish *Bythites hollisi* emerging from a high-temperature vent region known as Hole-to-Hell at 9°50′N along the East Pacific Rise. The species is common at various vent fields along the Galapagos Rift and East Pacific Rise.

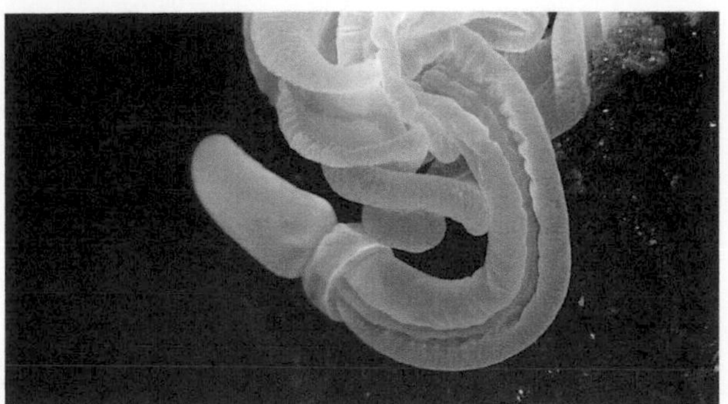

Figure 17 The enteroptneust *Saxipendium coronatum* (commonly known as a 'spaghetti worm') on the surface of basalt about 30 m from an active low-temperature vent field at 9°50′N along the East Pacific Rise. The image was taken with a prototype high-resolution video camera system equipped with a macro-lens.

invertebrate may represent an organism that is uniquely adapted to an ecotone between active vents and the ambient deep sea. Analyses of video images of numerous individuals taken with a high-resolution video camera system equipped with a macro-lens (**Figure 17**) have revealed behavioral patterns suggesting that the organism may be 'grazing' directly on basaltic surfaces, potentially consuming microbes or organic substances ultimately originating in the vent environment.

One of the most unusual and spectacular organisms inhabiting vent ecosystems is the vestimentiferan tubeworm *Riftia pachyptila* which thrives at numerous vent fields along the Galapagos Rift and East Pacific Rise (**Figures 1, 4,** and **8**). As mentioned earlier, it lives in a symbiotic relationship with chemosynthetic bacteria concentrated within its body. Such a relationship provides an internal, hydrogen sulfide-nourished 'garden' that, in turn, nourishes the tubeworm. Although the mechanism by which the host obtains energy from the bacteria is unclear, the energy transfer appears remarkably efficient. An unique opportunity to determine the growth rate potential of *R. pachyptila* arose as a result of the April, 1991 volcanic eruptive event at 9°50′N along the East Pacific Rise mentioned above. In March,

Figure 18 (Right)Temporal sequence of vent community development at a low-temperature vent in a region known as Hole-to-Hell at 9°50′N along the East Pacific Rise. This was the site of a volcanic eruption in April, 1991 that entirely decimated previously existing communities within the region. The field of view of each image and the heading of the camera system utilized are approximately the same for each image taken at the following times: (A) April, 1991; (B) March, 1992; (C) December, 1993; (D) October, 1994; and (E) November, 1995. (Shank *et al.*, 1998.)

1992, no *R. pachyptila* were present within the region, which had been devastated by the eruption. By December, 1993, less than 2 years later, huge colonies of this tubeworm had colonized the active low-temperature vents; the tube lengths of many individuals were in excess of 1.5 m (**Figure 18** and **19**). Such growth rates of more than 85 cm y^{-1} increase in tube length represent the fastest rates of growth documented for any marine invertebrate. It is interesting to note that one of the other fastest-growing marine invertebrates is the giant clam *Tridacna squamosa*, a bivalve that has a symbiotic association with photosynthetic algae (zooxanthellae). The efficiency with which energy is transferred from the symbiont to the host in certain invertebrates may well be a contributing factor to the remarkable growth rates of these organisms. The rapid succession of a tubeworm-dominated vent community over the 5-year period following the April, 1991 eruption is dramatically illustrated in **Figures 18** and **19**.

Inhabitants of deep-sea hydrothermal vents are among the most spectacular and unusual organisms on the planet. Given the relatively small number of vent ecosystems found to date, many questions have been raised concerning whether conservation measures need to be taken to protect vent communities from anthropogenic disturbances. Any assessment of the potential consequences of anthropogenic impacts needs to consider that the organisms inhabiting deep-sea hydrothermal vents thrive in an environment that is constantly being altered radically by geological process. Periodic devastation of entire biological communities is a relatively common occurrence as volcanic and tectonic processes proceed along active ridge axes. Over the past two decades we have learned that many vent communities are remarkably resilient. Populations of essentially all vent organisms indigenous to regions decimated by massive volcanic eruptions have, like a Phoenix rising from the ashes, reestablished themselves in less than a decade. This remarkable resilience in the face of huge natural disasters has profound implications as one considers the potential impacts of exploitation of

Figure 19 (Left)Temporal sequence of vent community development at a low-temperature vent located approximately 500 m from the community depicted in **Figure 18**. This was also a region that was buried with fresh lava by the April, 1991 volcanic eruptive event, decimating previously existing communities within the area. The field of view of each image and the heading of the camera system utilized are approximately the same for each image taken at the following times: (A) April, 1991; (B) March, 1992; (C) December, 1993; (D) October, 1994; and (E) November, 1995. (Shank *et al.*, 1998.)

precious mineral and biological resources associated with active hydrothermal systems throughout the world's oceans.

Acknowledgments

I thank the pilots and crew of the DSV *Alvin* and the R/V *Atlantis* for their expertise, assistance, and patience over the years; W. Lange and the Woods Hole Oceanographic Institution for technical expertise and the provision of the camera and recording systems critical to the generation of the majority of images presented in this article; Emory Kristof, Stephen Low, and Michael V. DeGruy for inspiration and assistance with the procurement of video images using a variety of camera systems; and Matt Tieger for assistance in the generation of video prints. Supported by National Science Foundation Grants OCE-95-29819 and OCE-96-33131.

See also

Hydrothermal Vent Ecology. Hydrothermal Vent Fauna, Physiology of.

Further Reading

Childress JJ and Fisher CR (1992) The biology of hydrothermal vent animals: physiology, biochemistry, and autotrophic symbioses. *Oceanography and Marine Biology Annual Review* 30: 337–441.

Gage JD and Tyler PA (1991) *Deep-Sea Biology: A Natural History of Organisms at the Deep-Sea Floor.* Cambridge: Cambridge University Press.

Lutz RA (2000) Deep sea vents. *National Geographic* 198(4): 116–127.

Shank TM, Fornari DJ, and Von Damm KL (1998) *Temporal and spatial patterns of biological community development at nascent deep-sea hydrothermal vents (9°N, East Pacific Rise). Deep-Sea Research.* 45: 465–515.

Jones ML (ed.) (1985) *The Hydrothermal Vents of the Eastern Pacific: An Overview. Bulletin of the Biological Society of Washington*, vol. 6. Washington, DC: Biological Society of Washington.

Rona PA, Bostrom K, Laubier L, and Smith KL Jr (eds.) (1983) *Hydrothermal Processes at Seafloor Spreading Centers.* New York: Plenum Press.

Tunnicliffe V (1991) The biology of hydrothermal vents: ecology and evolution. *Oceanography and Marine Biology Annual Review* 29: 319–417.

Van Dover CL (2000) *The Ecology of Deep-Sea Hydrothermal Vents.* Princeton, NJ: Princeton University Press, Princeton.

HYDROTHERMAL VENT ECOLOGY

C. L. Van Dover, The College of William and Mary, Williamsburg, VA, USA

Introduction

Most of the ocean floor is covered with a thick layer of sediment and is populated by sparse and minute, mud-dwelling and mud-consuming invertebrates. In striking contrast, the volcanic basalt pavement of mid-ocean ridges hosts hydrothermal vents and their attendant lush communities of large invertebrates that ultimately rely on inorganic chemicals for their nutrition. Vents themselves are sustained by tectonic forces that fracture the basalt and allow sea water to penetrate deep within the ocean crust, and by volcanism, which generates the hot rock at depth that strips sea water of oxygen and magnesium. The hot rock gives up to the nascent vent fluid a variety of metals, especially copper, iron, and zinc, as well as reduced compounds such as hydrogen sulfide and methane. The vent fluid, thermally buoyant, rises to exit as hot springs on the seafloor.

Discovered first by geologists in 1977 along a stretch of mountain range known as the Galapagos Rift, near the equator in the eastern Pacific Ocean, hydrothermal vents are now known to occur along every major ridge system on the planet. Several of these ridge systems – in the Arctic and Antarctic, in the Indian Ocean, in the southern Atlantic – are only just beginning to be explored. Regional species composition of the vent fauna differs between ocean basins and, sometimes, even within a basin. Most of the species that occur at vents have never been found in the adjacent, non-vent deep sea and are considered to be endemic, adapted to the chemical milieu of the vent environment. Reduced compounds carried in hydrothermal fluids, together with oxygen from the surrounding sea water, fuel the microbial fixation of inorganic carbon into organic carbon that forms the chemosynthetic base of the vent food web.

Hydrothermal vents on midocean ridges are thus globally distributed, insular ecosystems that support endemic faunas through chemosynthetic processes rather than through photosynthesis. They are effectively decoupled by depth (typically $>1000\,m$) from climatic variations and anthropogenic activities, but are tightly coupled to geophysical processes of tectonism and volcanism. Vents thus offer unique opportunities for biologists to study adaptations that allow life to persist in this extreme environment and to explore planetary controls on biodiversity and biogeography along submarine, hydrothermal 'archipelagoes', where propagules are water-borne and subject to dispersal in an open system. Further, because vents are thought to have been a primordial component of the oceans and because there is increasing speculation that early life on this planet may have thrived in hot environments and on chemicals rather than on an organic soup for nourishment, extant hydrothermal systems are thought by many to be analogues for sites where early life may have evolved on this and other planets or planetary bodies in our solar system.

Microorganisms and the Chemosynthetic Basis for Life at Vents

The terms 'chemosynthesis' and 'photosynthesis' are imprecise. While a voluminous nomenclature is available to differentiate among variations in these processes, for simplicity, chemosynthesis and photosynthesis are used here.

In photosynthesis, sunlight captured by proteins provides energy for the conversion of inorganic carbon (carbon dioxide, CO_2) and water (H_2O) into organic carbon (carbohydrates, $[CH_2O]$ and oxygen (O_2) (eqn [1]).

$$CO_2 + H_2O \overset{\text{light}}{\rightarrow} [CH_2O] + O_2 \qquad [1]$$

Photosynthesis by plants is the basis for consumer and degradative food webs both on land and, as a rain of organic detritus derived from surface phytoplankton productivity, on the seabed. In the deep sea, detrital inputs of organic carbon are exceedingly small, accounting for the paucity of consumer biomass in abyssal muds. At hydrothermal vents, the supply of surface-derived organic material is overwhelmed by the supply of new organic carbon generated through chemical oxidation of hydrogen sulfide (H_2S) (eqn [2]).

$$CO_2 + H_2O + H_2S + O_2 \rightarrow [CH_2O] + H_2SO_4 \qquad [2]$$

Metabolic fixation pathways for carbon can be identical in photosynthetic plants and chemosynthetic microorganisms, namely the Calvin–Benson cycle, but the energy-yielding processes that fuel the Calvin–Benson cycle (photon capture versus chemical oxidation) are distinctive. High biomass at hydrothermal vents is in part a consequence of the aerobic nature of the process described in eqn [2]. Oxygen is used to oxidize the hydrogen sulfide, generating a large energy yield that in turn can fuel the production of large amounts of organic carbon (**Figure 1**). Nonaerobic chemical reactions, such as oxidation of vent-supplied hydrogen (H_2) by carbon dioxide (CO_2), can also support chemosynthesis at vents, but energy yields under such anaerobic conditions are much lower than from aerobic oxidation. Microorganisms using these anaerobic reactions cannot by themselves support complex food webs and large invertebrates.

Figure 1 Photosynthetic and chemosynthetic processes in the ocean. Sunlight fuels the generation of organic material (CH_2O) from inorganic carbon dioxide (CO_2) and water (H_2O) by phytoplankton in surface, illuminated waters. At depths where hydrothermal vents exist (typically>2000 m), no sunlight penetrates. In place of sunlight, the chemical oxidation of sulfide (H_2S) by oxygen (O_2) fuels the conversion of carbon dioxide to organic carbon by chemosynthetic bacteria.

Symbiosis and the Host–Symbiont Relationship

One of the hallmarks of many hydrothermal vent communities is the dominance of the biomass by invertebrate species that host chemosynthetic microorganisms within their tissues. Giant, red-plumed, vestimentiferan tubeworms (*Riftia pachyptila*; **Figure 2**) so far provide the ultimate in host accommodation of endosymbiotic bacteria. These worms live in white, chitinous tubes, with their plumes extended into the zone of turbulent mixing of warm (~20°C), sulfide-rich, hydrothermal fluid and cold (2°C), oxygenated sea water. When discovered

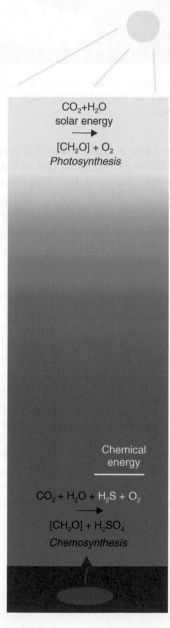

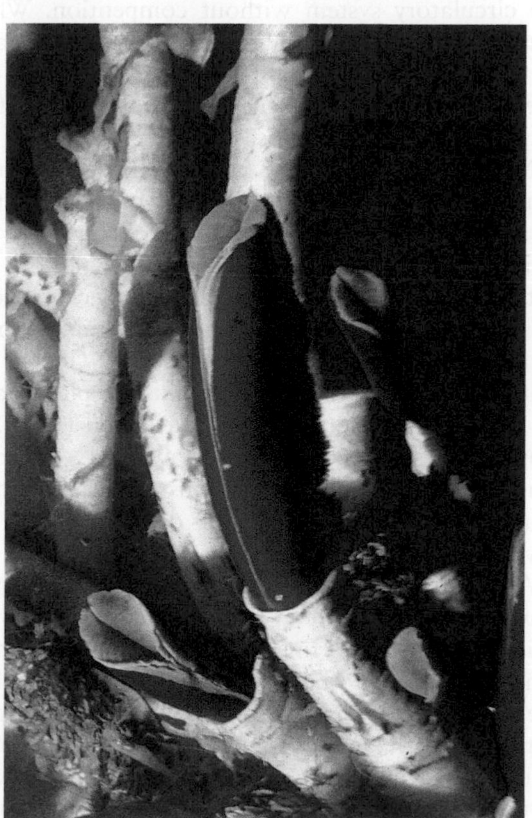

Figure 2 The giant tubeworm *Riftia pachyptila*. The red plume of the tubeworm acts as a gill for uptake of dissolved gases. The trunk of the worm is found inside the white, chitinous tube. (Photograph by Dudley Foster, Woods Hole Oceanographic Institution.)

in 1977, vestimentiferan tubeworms were remarkable for their size (up to several meters in length) and the complete absence of a digestive system in adults. In fact, the digestive system has been replaced by the trophosome, which is a specialized, paired organ derived from the larval gut. The trophosome is richly infiltrated with blood capillaries and each of its lobes lies within a blood-filled body cavity. The blood itself is rich in hemoglobin. Host bacteriocyte cells in the trophosome house chemosynthetic bacteria that use hydrogen sulfide and oxygen to fuel the production of organic carbon, as described above.

The metabolic requirements of the tubeworm endosymbiotic bacteria place some remarkable burdens on the host. First, there is a novel requirement for delivery of sulfide to the bacteria, which reside at a location remote from the site of gas exchange (the plume). Sulfide is normally a potent toxin to animals, poisoning the cellular enzyme system that generates ATP (adenosine triphosphate), the currency of metabolism. Sulfide also competes with oxygen for binding sites on hemoglobin. Tubeworm hemoglobin has separate binding sites for oxygen and sulfide, so that both can be transported throughout the worm in the circulatory system without competition. When bound to hemoglobin, the sulfide is not reactive and so enzyme systems remain unchallenged. Once delivered to the bacteria in the trophosome, the sulfide is quickly oxidized and loses its toxic potential.

Novel requirements for carbon dioxide are also found in tubeworms. The usual flow of CO_2 is out of an animal, as the end-product of metabolism, but the resident bacteria of the trophosome require a net uptake of CO_2. Maintenance of high concentrations of inorganic carbon in the blood of the tubeworm is facilitated by the high partial pressure of CO_2 in the water surrounding the site of uptake (the plume) and by the alkaline internal pH of the blood (7.3–7.4), which favors the bicarbonate form (HCO_3^-) of carbon dioxide and thus maintains a steep concentration gradient for diffusion of CO_2 from the environment into the blood.

As described above, the anatomy of the tubeworm is well adapted for life in sulfide- and CO_2-rich vent fluids and for supporting its endosymbiotic, chemosynthetic bacteria. The bacteria provide nearly all of the nutrition for the host, with the exception, perhaps, of small amounts of dissolved organic materials taken up across the tissues of the plume. In turn, the bacteria are provided with a chemically rich and stable environment for growth.

Other large invertebrates at vents also derive much of their nutrition from endosymbiotic, chemosynthetic bacteria, including 20 to 30-cm long to vesicomyid clams and bathymodiolid mussels. While vent mussels have a fairly normal digestive system and are capable of filter-feeding just as shallow-water mussels do, vent clams lack a functional digestive system. Both types of bivalves have enlarged gills and it is within these gills that the endosymbiotic bacteria are found. Clams are thought to take up hydrogen sulfide via their highly vascularized foot, with which they probe cracks in the basalt where vent fluids emanate.

Figure 3 Swarming shrimp (*Rimicaris exoculata*) at a hydrothermal vent on the Mid-Atlantic Ridge. (Photograph by C.L. Van Dover.)

Not all chemosynthetic bacteria that nourish vent invertebrates are endosymbiotic. Shrimp that dominate vents in the Atlantic (**Figure 3**) host chemosynthetic bacteria on their carapace (i.e., the bacteria are episymbiotic) and seem to depend on these bacteria for a significant portion of their diet. Other chemosynthetic bacteria are free-living, suspended in the water column, providing nourishment to suspension-feeding invertebrates such as barnacles, or grow as mats or films on surfaces, where grazers such as limpets and polychaetes forage. Heterotrophic bacteria (using organic rather than inorganic compounds) may also be important for consumers within the vent invertebrate food web, but this has yet to be examined carefully.

Thermal Adaptations

While hydrothermal vent communities live at temperatures slightly elevated above ambient sea water temperature, the existence of highly productive communities at vents is a consequence of fluid chemistry rather than thermal input. Nevertheless, there are some invertebrates, notably the large, thumb-sized polychaetes in the family Alvinellidae (**Figure 4**), that are especially tolerant of high temperatures and that compete with desert ants for the

Guinness Book of World Records in the category of multicellular animal living at a thermal extreme. Alvinellids live on the sides of black smoker sulfide chimneys, and are reported to survive brief exposures to temperatures as high as 105°C. They routinely experience a thermal gradient of 50–60°C over the length of their bodies. Temperature tolerance in these worms is not completely understood, but thermal stability of enzymes has been shown to increase in alvinellid species that occupy increasingly warmer habitats. Membranes may also be adapted for thermal stability through an increase in the degree of double bonding in fatty acids in species living in warmer portions of the vent habitat.

Shrimp (*Rimicaris exoculata*) that swarm on black smoker chimneys in close proximity to extremely high temperature (350°C) fluids are not bathed in excessively hot water, but they do have paired photoreceptive organs that may function to detect the glow emitted by the hot water. The organs are derived from ordinary shrimp eyes, but the photoreceptive surfaces are hypertrophied and rich in visual pigment (rhodopsin). The eyestalks are lost and the derived eyes extend back along the dorsal surface of the shrimp, beneath the transparent carapace. These 'eyes' have also lost their optic (lens) systems, so they cannot form an image, but they are optimized to detect gradients of dim light. While light from black smokers has now been well documented, the behavioral response of the shrimp to this light has not been studied.

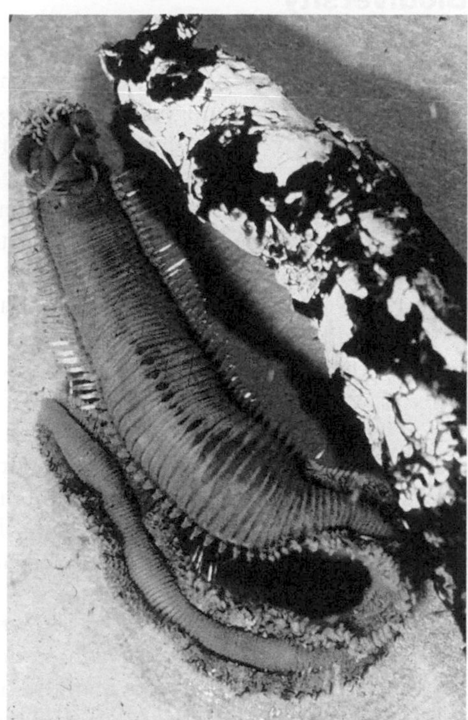

Figure 4 The alvinellid polychaete *Alvinella caudata*, beside its fragile tube. (Photograph by J. Porteous, Woods Hole Oceanographic Institution.)

Community Dynamics

There could hardly be a greater dynamic contrast than that found between the cold, food-limited, relentlessly stable and vast deep sea environment, and the thermally complex, trophically rich, ephemeral, and insular deep-sea hydrothermal vent fields. Since the discovery of vents, ecologists have attempted to predict the cycle of community development over the life span of a vent field by interpolation and extrapolation from snapshot observations. The 1991 seafloor volcanic eruption at the Venture Hydrothermal Field on the East Pacific Rise was witnessed within days to weeks of the event, providing biologists with the first submarine equivalent of Krakatau. At the time of the Venture eruption, existing vent communities were obliterated by fresh lava flows, pervasive warm-water venting was observed along ~1.5 km of ridge axis, and a dense 'bloom' of flocculent material poured from cracks between lobes of lava, obscuring visual navigation (**Figure** 5). The flocs were determined to be the mineral by-product of microbial production. Within one year, venting became focused at numerous sites along the ridge axis and the first colonists had arrived, including a small tubeworm species (*Tevnia jerichonana*) and dense aggregations of several species of limpets. Mobile vent organisms, including vent crabs and squat lobsters, zoarcid fish, and swarming amphipods were also well represented. After 2.5 years, some vents had shut off but, where venting persisted, mature colonies of the giant tubeworm (*Riftia*

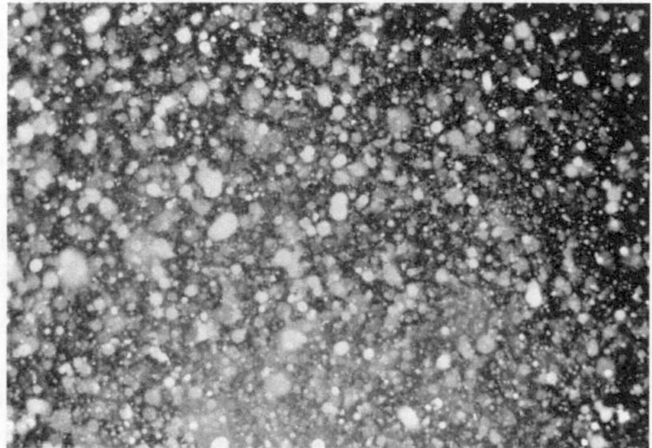

Figure 5 Flocculent material suspended in the water column 1 m above new ocean crust. The floc is a mineral by-product of microbial production and emanated from cracks in the seafloor. (Photograph by Alvin, Woods Hole Oceanographic Institution.)

pachyptila) were established, along with a variety of smaller invertebrates that live among the tubeworm tubes, including shrimp, limpets, amphipods, and polychaetes. Growth rates of *R. pachyptila* were measured to be among the most rapid of any aquatic invertebrate. Within 5 years, 75% of the regional species pool could be found at the new vents and mussel beds were well-established and beginning to overwhelm the *R. pachyptila* thickets. In this example, vesicoymid clams were the last of the big megafaunal species to arrive at the site, despite the presence of adult populations within several kilometers of the area overrun by fresh lava. Mussels appear to have a competitive edge among the larger taxa at vents, in part because they are mobile and can relocate as necessary to cope with changing flow patterns of vent fluids, while tubeworms are stationary and have few options for tracking vent flow. In addition, because mussels can filter-feed as well as derive nutrition from their endosymbiotic bacteria, they are among the last species to disappear as a vent shuts down. Ultimately, it is the mobile scavengers – the crabs and fish and octopus – that witness the final demise of a vent field.

The eruption at the Venture Hydrothermal Field was not entirely unexpected – geologists were studying this region of the ridge axis because it was so inflated and appeared to be ripe for an eruptive event. Localization of seafloor eruptions on the Juan de Fuca Ridge in the northeast Pacific (off the coast of Vancouver Island) now takes place in real-time, facilitated by a legacy of the cold war era, namely through acoustic signals received by underwater sound-surveillance systems originally designed to track enemy submarines. As eruptions take place, T-waves (low-frequency, tertiary waves characteristic of lava in motion) are transmitted into the water column. Navy hydrophone networks allow the T-waves to be placed in a geographical context and traced from start to finish. One migration of lava in June 1993 was traced for about 40 km below the ocean crust over a two-day period before it finally erupted onto the seafloor. As at the Venture Field, fresh lava on the seafloor was observed, along with venting of flocculent material derived from microbial production. Sites of persistent venting were colonized by populations of vent invertebrates within one year, some of which were reproductively mature.

The two examples of community dynamics cited here suggest that the species composition of all vent communities is constantly changing. But some vent sites are long-lived, and repeat visits to these long-lived sites over a 15-year period document essentially no change in the nature of the fauna. The best example of this to date is the TAG site on the Mid-Atlantic Ridge. TAG is a large sulfide mound (100 m diameter) that has occupied the junction of cross-cutting faults and fissures for more than 100 000 years. Swarming shrimp (*Rimicaris exoculata*) and anemones (*Maractis rimicarivora*) dominated the site when it was first discovered in 1985 and continued to dominate through the most recent set of observations in 1998, despite massive disruptions of hydrothermal flow caused by drilling in 1994. There have been shifts in the precise location of the primary masses of shrimp as they track local natural and anthropogenic changes in vent flow on the mound, and we assume that there has been replacement of generations by recruitment. But there has been no succession observed, no invasion by other taxa. The absence of change does little to attract ecologists, who thrive on dynamic systems, but the stability of the species composition at TAG and other sites has profound implications regarding the selective pressures encountered by the species that inhabit these sites compared to sites that are constantly threatened by lava overruns or tectonic shifts in plumbing.

Origins of Vent Faunas, Biogeography, and Biodiversity

Evolutionary paths that brought invertebrate taxa to vents are varied. Several major taxa, including the galatheid squat lobsters, pycnogonid sea spiders, and echinoderms are likely to be immigrants from the surrounding deep sea. Some species are closely related to (and presumably derived from) shallow-water genera. Many of the most familiar vent taxa – the vesicomyid clams, bathymodiolid mussels, vestimentiferan tubeworms, alvinocarid shrimp – are allied to genera and families found in a variety of deep-sea chemosynthetic ecosystems (i.e., seeps and whale-falls as well as vents). The direction of invasion (seep-to-vent or vice versa) can be inferred using molecular techniques. For example, molecular phylogenetics suggests that the bathymodiolid mussel group invaded vents from seeps, with several seep species resulting from reinvasion of the seep habitat by vent ancestors.

There are also specialized taxa so far known only from hydrothermal vents. The most conspicuous of these is the alvinellid polychaete family, whose members often occupy the warmest habitable waters of a vent site. Still other taxa appear to be relicts of ancient lineages that have found refuge in the vent environment. These relict taxa include the stalked

Figure 6 The ancient barnacle *Neolepas zevinae*. (Photograph by C.L. Van Dover.)

barnacle, *Neolepas zevinae* (Figure 6) and the archaeogastropod limpet, *Neomphalus fretterae*.

Fossil vent communities found on land provide a glimpse of Silurian vent assemblages. Vestimentiferans are reported from some of the oldest vent deposits known, accompanied by brachiopods and monoplacophorans. Brachiopods and monoplacophorans are so far poorly represented in modern vent communities, if at all.

When vents were first discovered, some biologists hypothesized that the vent fauna would be globally cosmopolitan. Subsequent explorations have shown this hypothesis to be false, and mechanisms that allow isolation and differentiation of faunas have been postulated. One of the best examples comes from a comparison of faunas from the Juan de Fuca Ridge and East Pacific Rise. At one time (56 Ma), these ridges were one continuous ridge system, but around 37 Ma the North American Plate began to

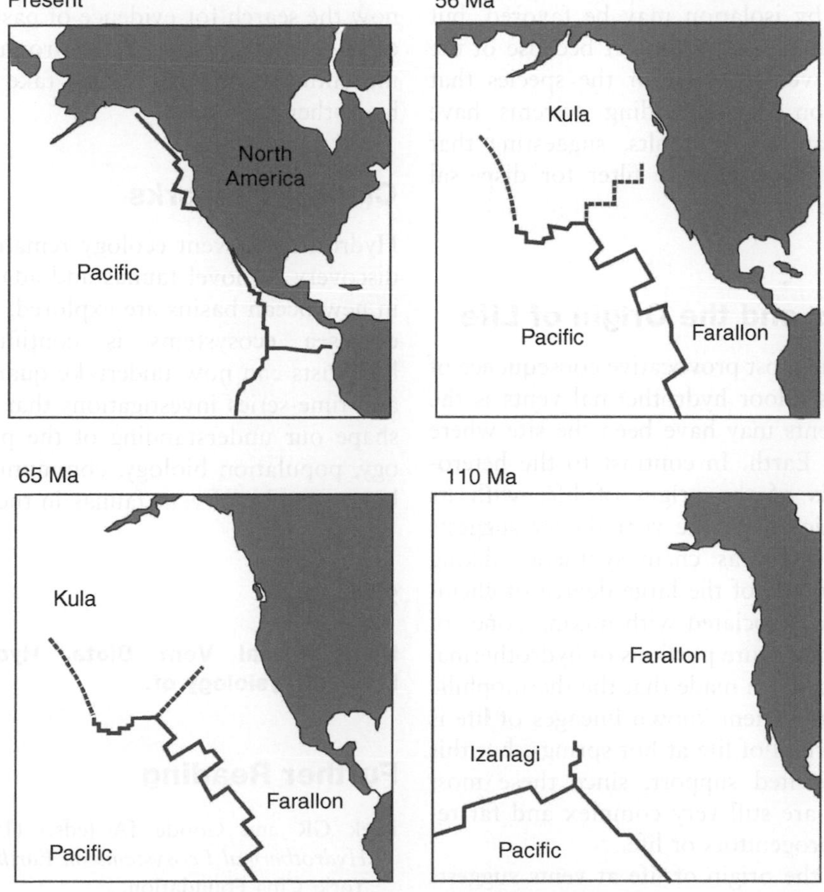

Figure 7 Bisection of a mid-ocean ridge by the North American Plate. At one time (56 Ma), there was a continuous mid-ocean ridge in the eastern Pacific basin but, as the North American Plate overrode the ridge system, it was bisected to form the northeast Pacific ridge system (Juan de Fuca, Explorer, Endeavour, and Gorda Ridges) and the East Pacific Rise (emerging in the Gulf of California and running south toward Antarctica). (Reproduced from Tunnicliffe *et al.* 1996.)

override this ridge (**Figure** 7), splitting the parent faunal assemblage into two daughter assemblages that are distinct yet closely allied at the generic level.

Similarities among faunas at the species level may be reduced to nearly zero at vent sites that are in separate ocean basins. This observation suggests that major additions to the global vent faunal inventory await us in the unexplored ocean basins; the potential for discovery of novel adaptations to the vent ecosystem is extremely high.

Vent biologists are just beginning to examine global patterns in biodiversity. Preliminary measures show a strong correlation between spreading rate and species richness, raising the hypothesis that patterns of volcanism may be an ultimate control on species diversity in hydrothermal vent ecosystems. Where ridges are fast-spreading, the magma budget is high and the temporal and spatial frequency of vents along the ridge axis is high. In contrast, on slow-spreading ridges, the magma budget is low and vents are far apart. Because distances between vents are short on fast-spreading ridges, species are less likely to go extinct. At slow-spreading ridges, allopatric speciation by isolation may be favored, but species are more likely to go extinct because of the distance between vents. Many of the species that dominate vents on slow-spreading systems have mobile rather than sessile adults, suggesting that distance may act as a selective filter for dispersal capability in these systems.

Vent Systems and the Origin of Life

Without doubt, the most provocative consequence of the discovery of seafloor hydrothermal vents is the suggestion that vents may have been the site where life originated on Earth. In contrast to the heterotrophic hypothesis of the origin of life, with its nourishing 'organic soup', the vent theory suggests that the earliest life was chemosynthetic, taking biochemical advantage of the large degree of chemical disequilibrium associated with mixing zones of low- and high-temperature portions of hydrothermal systems. A case has been made that the thermophilic nature of the most ancient known lineages of life is indicative of an origin of life at hot springs, but this argument finds limited support, since these most ancient cell types are still very complex and far removed from the progenitors of life.

One model for the origin of life at vents suggests the following acellular precursor: a monomolecular, negatively-charged organic layer bonded to positively charged mineral surfaces at the interface of hot water. In this model, pyrite, which forms exergonically from iron monosulfide and hydrogen sulfide (both components of vent fluids), serves as the mineral surface. The pyrite-forming reaction yields two free electrons that can be used for building biochemical constituents; simple organic molecules interacting with the pyrite could be reduced to more complex organic molecules. The simple organics also derive from the vent fluids, through abiogenic synthesis. In a secondary stage of development, the precursor evolves to being semicellular, still supported by minerals, but with a lipid membrane and internal broth, with increasing metabolic capabilities. In the final stage of origin, the pyrite support is abandoned and true cellular organisms arise. The elegance of this model is that it uses an energetically realistic inorganic chemical reaction to create a cationic substrate that can bind with organic compounds, all within a single setting.

Discovery of vent ecosystems and the appreciation of their chemosynthetic basis has influenced the search for life elsewhere in the solar system. When the Viking Mission to Mars took place, the emphasis was on a search for photoautotrophic processes, but now the search for evidence of past or extant life on other planets highlights environments where chemosynthetic processes may take place, including hydrothermal areas.

Closing Remarks

Hydrothermal vent ecology remains a field ripe for discovery of novel faunas and adaptations, as vents in new ocean basins are explored. Because access to deep-sea ecosystems is continually improving, biologists can now undertake quantitative sampling and time-series investigations that are certain to reshape our understanding of the physiological ecology, population biology, community dynamics, and biogeography of vent faunas in the near future.

See also

Hydrothermal Vent Biota. Hydrothermal Vent Fauna, Physiology of.

Further Reading

Bock GR and Goode JA (eds.) (1996) *Evolution of Hydrothermal Ecosystems on Earth (and Mars?)*. New York: Ciba Foundation.

Childress JJ and Fisher CR (1992) The biology of hydrothermal vent animals: physiology, biochemistry, and autotrophic symbioses. *Oceanography and Marine Biology Annual Review* 30: 337–441.

Desbruyères D and Segonzac M (eds.) (1997) *Handbook of Deep-Sea Hydrothermal Vent Fauna*. Brest: IFREMER.

Fisher CR (1990) Chemoautotrophic and methanotrophic symbioses in marine invertebrates. *Critical Reviews in Aquatic Science* 2: 399–436.

Humphris SE, Zierenberg RA, Mullineaux LS, and Thomson RE (eds.) (1995) *Seafloor Hydrothermal Systems: Physical, Chemical, Biological and Geological Interactions*. Washington, DC: American Geophysical Union.

Karl DM (ed.) (1995) *The Microbiology of Deep-Sea Hydrothermal Vents*. New York: CRC Press.

Shank TM, Fornari DJ, Von Damm KL *et al.* (1998) Temporal and spatial patterns of biological community development at nascent deep-sea hydrothermal vents (9°N, East Pacific Rise). *Deep-Sea Research* 45: 465–516.

Tunnicliffe V (1991) The biology of hydrothermal vents: ecology and evolution. *Ocenography and Marine Biology Annual Review* 29: 319–407.

Tunnicliffe V, Fowler CMR, and McArthur AG (1996) Plate tectonic history and hot vent biogeography. In: MacLeod CJ, Tyler PA, and Walker CL (eds.) *Tectonic, Magmatic, Hydrothermal and Biological Segmentation of Mid-Ocean Ridges*, Geological Society Special Publication 118, pp. 225–238.

Tyler PA and Young CM (1999) Reproduction and dispersal at vents and cold seeps: a review. *Journal of the Marine Biology Association of the UK* 79: 193–208.

Van Dover CL (2000) *The Ecology of Deep-Sea Hydrothermal Vents*. Princeton: Princeton University Press.

HYDROTHERMAL VENT FAUNA, PHYSIOLOGY OF

A. J. Arp, Romberg Tiburon Center for Environment Studies, Tiburon, CA, USA

Hydrothermal Vent Environments are Dynamic, Hot, and Toxic

The hydrothermal vent environments, lying at the bottom of the ocean at depths of 2.5 km or more, were discovered in 1977 by a group of geologists exploring spreading centers at midocean ridges on the sea floor. As fissures open up in the earth's surface, lava is extruded onto the ocean floor and sea water is pulled towards the center of the earth so deeply that it comes into contact with hot, molten magma. The sea water is superheated and then discharged back into the environment through fissures in the ocean floor. As sea water moves from the center of the earth and into the vent habitat on the seafloor, it becomes laden with inorganic chemicals. In particular, hydrogen sulfide, an essential chemical in this unique environment, leaches into the water at depth, and water discharging into the environment can be highly enriched in this toxic but energy-rich chemical.

There are two types of hydrothermal vents. In those characterized by diffuse venting, sea water percolates out at a moderate rate and is approximately 10–20°C in temperature. As the bottom water temperature of the majority of the earth's deep ocean is about 2°C, these hydrothermal fluids are elevated in temperature, but rapidly mix with the surrounding sea water. One of the first hydrothermal vent environments discovered, Rose Garden at the hydrothermal vent environment near the Galapagos Islands, is an example of a diffuse vent habitat. In the black smoker environment of the hydrothermal vents, things are a lot hotter, such as at those on the Juan de Fuca Ridge off the coast from the state of Washington. This is a very dynamic, high-temperature environment where water issuing forth from large chimneylike structures can be as hot as 400°C in temperature.

In spite of the hydrogen sulfide-enriched environment, elevated temperatures, and the dynamic volcanic activity, numerous and varied animals cluster around these sites, taking advantage of the hard substrate provided by the extruded pillow lava. A typical vent environment begins with a stretch of pillow lava in the foreground, with clams wedged into the numerous fissures. At the heart of the vent environment are diffuse venting hydrothermal fluids or actively spewing white smoker chimneys and black smoker chimneys, teeming with life. Typical inhabitants include dense clusters of tubeworms and many free-ranging animals roaming in and out of the vent environment such as brachyuran crabs, galatheid crabs, numerous amphipods, a few species of fish, and a host of other smaller animals.

Chemosynthesis — The Basis of All Life in the Vent Environment

Possibly the most revolutionary outcome of the discovery of hydrothermal vents is the story of how life exists in this challenging habitat and the unique nature of the food chain and the source of basic energy in this remote location. Prior to the discovery of the hydrothermal vents, most biologists believed that all life depended upon the energy of sunlight and that the basis of all food chains was photosynthesis. When the hydrothermal vents were discovered, it was immediately clear that they represent a very enriched biological environment very remote from the surface sunlight. It was difficult to imagine that organic material could drift down in large enough quantities to provide the energy to fuel this environment. Other interesting data materializing rapidly after the discovery of the vents indicated that most of the animals, especially the large invertebrates, have no digestive systems. For example, the large tubeworm, *Riftia pachyptila*, has no mouth and no intestine; however, it is a large animal approximately 1.5 m in length and up to 2 cm in diameter, and large colonies of these animals flourish in the remote vent habitat.

The question of how animal life is supported in the deep sea communities of the hydrothermal vents became, and remains to this day, a focus of intense research. Several lines of evidence lead to the realization that some of the major invertebrates endemic to the vent environment harbor bacteria within their body cavities. Free-living bacteria in this environment and in our own backyard have been known for years to be able to use chemical energy as a basis of their metabolism. In the case of the free-living bacteria, there are many hydrogen sulfide-oxidizing bacteria that can use this chemical as the basis of their metabolic pathways and produce organic compounds that form the basis of their nutrition. Hydrothermal vent animals harboring these chemical-utilizing bacteria within their body tissues as symbionts include the large tubeworm *Riftia pachyptila*, which has dense aggregations of bacteria in a residual gutlike organ, and the clam *Calyptogena magnifica*, which harbors bacteria in the gills. These symbiotic bacteria are able to utilize the inorganic chemical hydrogen sulfide, so plentiful in this environment, in a manner analogous to what plants do with energy from the sun. This process is therefore termed chemosynthesis rather than photosynthesis.

The symbiotic bacteria living within the bodies of the larger invertebrate animals have been demonstrated to oxidize hydrogen sulfide, and the energy released from this biochemical process is used to power the fixation of carbon dioxide into small organic compounds – just as in free-living bacterial sulfide oxidation. The Calvin–Benson cycle employed in both cases is the same metabolic pathway that is utilized by plants in photosynthesis to transform inorganic carbon dioxide into organic compounds that are then utilized as food higher up in the food chain. The critical difference with chemosynthetic metabolism is that, rather than using sunlight, these animals and bacteria utilize chemical energy to power that reaction and are completely independent of sunlight (**Figure 1**). The net result is that free-living bacteria in the environment and symbiotic bacteria living within animal tissues are able to live independently of sunlight by utilizing chemicals from the core of the earth, thus forming a very different basis for the food chain in the hydrothermal vent environment. The discovery of the hydrothermal vent environment was a fundamental discovery of a well-defined ecosystem that is completely independent of sunlight at any level of the food chain.

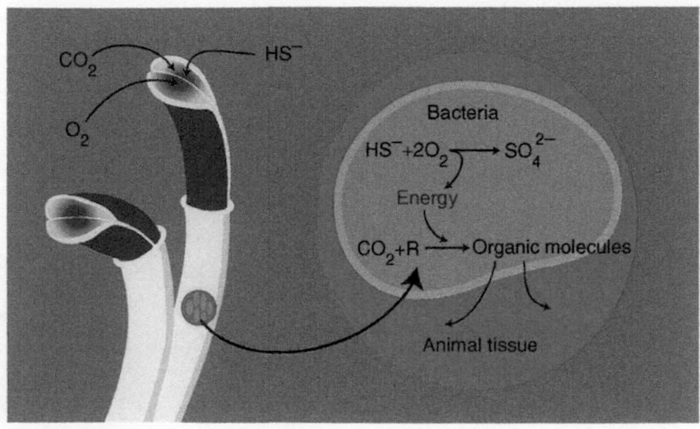

Figure 1 Chemosynthetic pathways in *Riftia pachyptila*.

The Ecophysiology of the Giant Tubeworm *Riftia pachyptila*

One of the most dramatic and best-known of the animals endemic to the hydrothermal vent environment is the giant tubeworm *Riftia pachyptila*. Colonies of these worms are clumped together around effluent points in the hydrothermal vent habitat, growing toward and into the water that is percolating out from the seafloor. An individual animal lives inside a single, unbranched chitinous tube and the red structure protruding out of the end of the tube is the respiratory plume. The animal can retract the plume back into the tube if disturbed by a roaming predator. There is a collarlike vestimentum organ that positions the animal within the tube, and a large trunk region of the animal is filled with an organ termed the trophosome. This organ is believed to be the vestigial gut of the worm and is composed, literally, of masses of bacteria. A pool of coelomic fluid bathes the trophosome and contains a large-molecular-weight extracellular respiratory hemoglobin (**Figure 2**). *Riftia pachyptila* also has a separate pool of blood that contains high concentrations of an extracellular hemoglobin that circulates in an elaborate closed circulatory system powered by a heartlike structure in the vestimentum region. The blood is pumped in a complete circuit

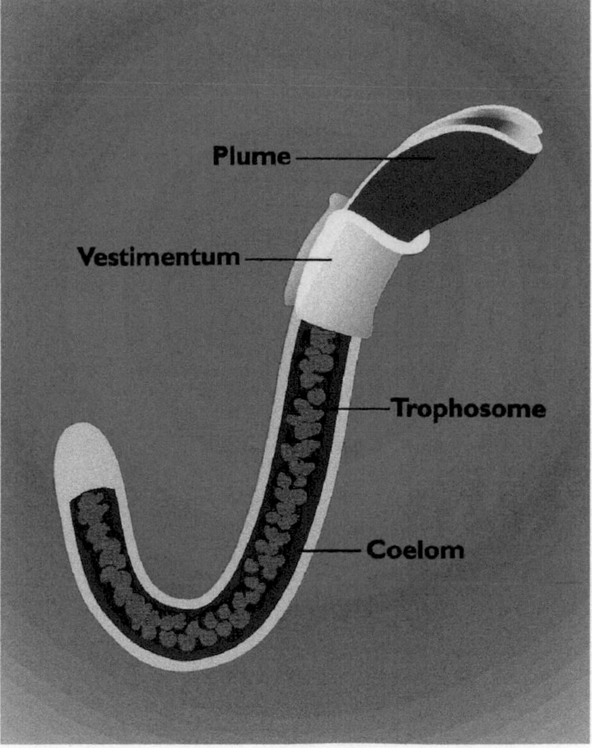

Figure 2 The external anatomy of *Riftia pachyptila*.

from the respiratory plume to body tissues, and on to elaborate capillary beds in the region of the trophosome and bacteria (**Figure 3**). The red color of the blood is due to the high concentration of hemoglobin and gives the characteristic red color to the plumes.

The respiratory plume and the circulating hemoglobin are essential for the transport of the key metabolites oxygen, hydrogen sulfide, and carbon dioxide, which are the principal components of the metabolism of the symbiotic bacteria. The respiratory hemoglobins present in the plume and the coelomic fluid of the animal bind oxygen with a very high affinity. The binding is reversible and cooperative, such that oxygen uptake is enhanced at the respiratory plume, and oxygen delivery is augmented at the tissues and trophosome organ.

Hydrogen sulfide is a highly toxic molecule that typically acts in a similar manner to cyanide by binding at the iron center of cytochrome molecules and hemoglobin molecules, thus arresting aerobic metabolism. Although *Riftia pachyptila* and other hydrothermal vent animals utilize hydrogen sulfide for their metabolism, they also have tissues that are highly sensitive to sulfide poisoning. Detoxification of hydrogen sulfide is essential for aerobic life in this dynamic, chemically enriched environment. The key to the simultaneous needs for transportation and detoxification is the respiratory hemoglobin present in the plume and coelomic fluid. *Riftia pachyptila* hemoglobin binds hydrogen sulfide with a very high affinity. The toxic hydrogen sulfide is transported to the trophosome region in the center of the worm's body as a tightly bound molecule that cannot chemically interact with sulfide-sensitive tissues. Oxygen and sulfide are simultaneously bound to the hemoglobin at separate binding sites and are transported to the trophosome, where they are believed to be delivered to the symbiotic bacteria for metabolism. In this way hydrogen sulfide is taken up from the surrounding sea water and transported to the site of bacterial metabolism while interaction is prevented with other tissues, such as the body wall, that are highly aerobic and sensitive to the toxic effects of hydrogen sulfide. These unusual adaptations function for respiratory gas transport and metabolism as well as for detoxification and tolerance of toxic chemicals in what would be a very inhospitable environment for most animals.

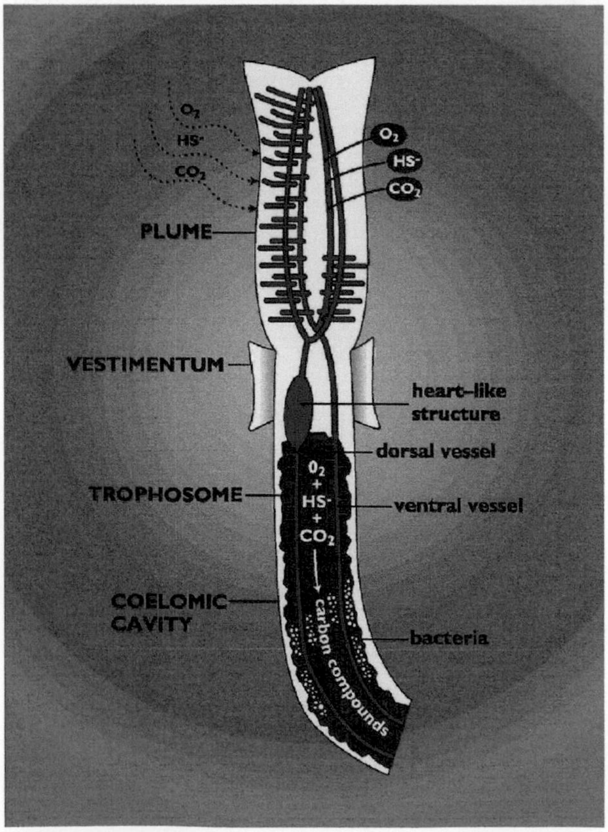

Dense colonies of *Riftia pachyptila* flourish in a specialized microhabitat within the vent environment. The worms anchor themselves on the rocks where the hydrothermal vent fluid is issuing out into the seafloor. The base of the tube is bathed in hydrothermal fluid enriched in hydrogen sulfide and carbon dioxide, but devoid of oxygen. Temperatures are relatively elevated here, and a gradient develops along the length of the tube. The respiratory plume extends into ocean-bottom sea water that is 2°C in temperature, devoid of hydrogen sulfide and enriched in oxygen (**Figure 4**). By occupying the interface between the hydrothermal fluids and the surrounding bottom water, animals are exposed to both of these essential metabolites which are then taken up by the circulating hemoglobin and

Figure 3 The internal anatomy of *Riftia pachyptila*.

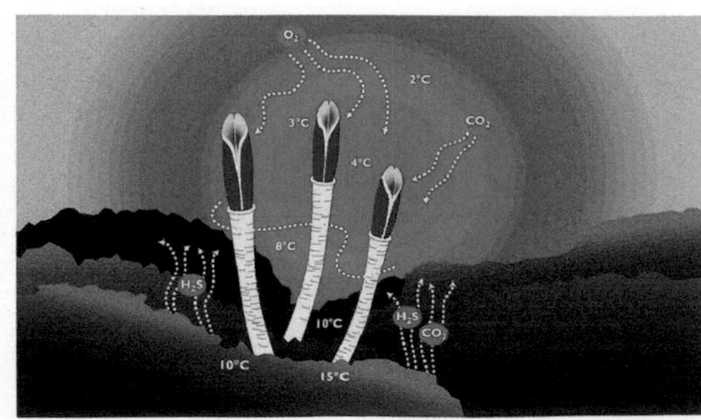

Figure 4 The microhabitat of *Riftia pachyptila*.

transported to internal tissues and symbiotic chemosynthetic bacteria.

The Ecophysiology of the Giant Clam *Calyptogena magnifica*

The vesicomyid clam *Calyptogena magnifica* is a common inhabitant of hydrothermal vents that orients in the fissures of the pillow lava near the periphery of the vent environment. Individual clams can reach up to 20 cm in length. The clams position themselves in cracks in the pillow lava and wedge their muscular foot down into the region where the hydrothermal plume is percolating out (**Figure** 5). This water is enriched in hydrogen sulfide and is elevated in temperature. This position orients the siphon end-up into surrounding sea water and enables the uptake of oxygen and carbon dioxide. *Calyptogena magnifica* possess a high concentration of an intracellular, circulating hemoglobin that functions for oxygen binding and transport. The clam's blood also contains a separate component that binds hydrogen sulfide and transports this essential metabolite to internal symbionts in the gill region. In this manner, the clams are able to accumulate hydrogen sulfide from the hydrothermal fluids bathing the foot, and oxygen via binding to the circulating hemoglobin through gill ventilation at the siphon region. These essential metabolites are transported as bound substances to symbiotic chemosynthetic bacteria in the gill via the circulatory system of the clam, providing both essential respiratory gas transport and detoxification of toxic hydrogen sulfide. The vent clams, like the tubeworms, seek out, exploit, and flourish in a unique microhabitat in the hydrothermal vent community.

Other hydrothermal vent animals include the mytilid mussel *Bathymodiolus thermophilus* that also harbors chemosynthetic bacteria in its gill tissues. There are many free-ranging animals that are not fueled by chemosynthesis but may feed on the larger invertebrates that benefit from that symbiotic, chemoautotrophic metabolism. Animals such as the brachyuran crabs, *Bythograea thermydron*, that wander through the environment scavenging dead or dying material, numerous swarming amphipods, as well as slow-moving fishes, are plentiful. Along with the giant tubeworms, clams, and mussels, these animals benefit directly or indirectly from the chemical-based metabolism that supports this dynamic and robust deep-sea community. It is the specialized physiological adaptations for transport and detoxification of hydrogen sulfide and other processes essential for life that provide the underlying mechanisms that make this possible.

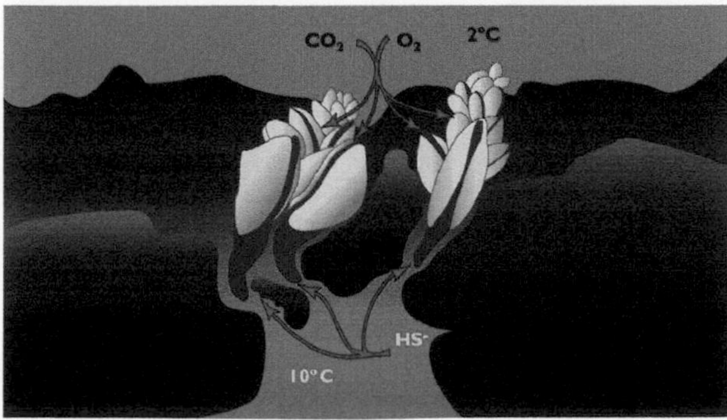

Figure 5 The microhabitat of *Calyptogena magnifica*.

Summary

A fascinating variety of marine invertebrates occur in dense assemblages in organically enriched deep-sea hydrothermal vent environments. Intensive studies on the hydrothermal vent fauna have been conducted since their discovery in the late 1970s. Many investigations have focused on the fact that these organisms, including vestimentiferan tubeworms and vesicomyid clams emphasized here, are nutritionally dependent upon the chemical-based metabolism of large populations of symbiotic bacteria that they harbor internally in dense concentrations. These bacteria utilize hydrogen sulfide as an energy source to fix inorganic carbon into nutrients. Hydrogen sulfide is extremely toxic to aerobic organisms in nanomolar to micromolar concentrations. However, uptake and transport of sulfide to internal symbionts is essential for the host animal's metabolism and survival.

Chemical-based metabolism, or chemoautotrophy, and detoxification of sulfide through binding to blood-borne components occur in vent tubeworms and clams, and are particularly well-characterized for the tubeworm *Riftia pachyptila*. These chemosynthetic endosymbiont-harboring worms simultaneously transport sulfide bound to the respiratory hemoglobin, providing an electron donor for the bacterial symbiont metabolism and protection against sulfide toxicity at the tissues.

How animals living in sulfide-rich environments like hydrothermal vent communities transport, metabolize, and detoxify hydrogen sulfide has been one of the major questions posed by hydrothermal vent researchers. The clarification of both the phylogenetic importance and the ecological status of these animals requires knowledge of their physiological adaptations to low oxygen conditions and high concentrations of sulfide, and provides answers to the puzzle of how these animals flourish in such seemly hostile environments.

See also

Hydrothermal Vent Biota. Hydrothermal Vent Ecology.

Further Reading

Arp AJ, Childress JJ, and Fisher CR (1985) Blood gas transport in Riftia pachyptila. In: Jones ML (ed.) *The Hydrothermal Vents of the Eastern Pacific: An Overview, Bulletin of the Biological Society of Washington, No. 6, p. 289*. Washington DC: Biological Society of Washington.

Childress JJ and Fisher CR (1992) The biology of hydrothermal vent animals: physiology, biochemistry and autotrophic symbioses. *Oceanography and Marine Biology Annual Review* 30: 337.

Delaney JR (1998) Life on the seafloor and elsewhere in the solar system. *Oceanus* 41(2): 10.

Feldman RA, Shank TB, Black MB, *et al.* (1998) Vestimentiferan on a whale fall. *Biology Bulletin* 194: 116.

Fisher CR (1990) Chemoautotrophic and methanotrophic symbioses in marine invertebrates. *Review of Aquatic Science* 2: 399.

Lutz RA (2000) Deep sea vents: science at the extreme. *National Geographic* (October): 116.

MacDonald IR and Fisher CR (1996) Life without light. *National Geographic* (October): 313.

Macdonald KC (1998) Exploring the global mid-ocean ridge. *Oceanus* 41(1): 2–9.

Mullineaux L and Manahan D (1998) The LARVE Project explores how species migrate from vent to vent. *Oceanus* 41(2)

Nelson K and Fisher CR (2000) Speciation of the bacterial symbionts of deep-sea vestimentiferan tube worms. *Symbiosis* 28: 1–15.

Somero GN, Childress JJ, and Anderson AE (1989) Transport, metabolism, and detoxification of hydrogen sulfide in animals from sulfide-rich environments. *Review of Aquatic Science* 1: 591–614.

Shank T, Fornari DJ, Von Damm KL *et al.* (1998) Temporal and spatial patterns of biological community development at nascent deep-sea hydrothermal vents (9°50′N, East Pacific Rise). *Deep-Sea Research II* 45: 465.

Tivey MK (1998) How to build a smoker chimney. *Oceanus* 41(2): 22.

Van Dover CL (2000) *The Ecology of Deep-Sea Hydrothermal Vents*. Princeton, NJ: Princeton University Press.

COASTAL REGIMES

MANGROVES

M. D. Spalding, UNEP World Conservation Monitoring Centre and Cambridge Coastal Research Unit, Cambridge, UK

Definition

The term mangrove is used to define both a group of plants and also a community or habitat type in the coastal zone. Mangrove plants live in or adjacent to the intertidal zone. Mangrove communities are those in which these plants predominate. Other terms for these communities include coastal woodland, intertidal forest, tidal forest, mangrove forest, mangrove swamp and mangal. The word mangrove can be clearly traced to the Portugese word 'mangue' and the Spanish word 'mangle', both of which are actually used in the description of the habitats, rather than the plants themselves, but still have been joined to the English word 'grove' to give the word 'mangrove.' It has been suggested that the original Portugese word has been adapted from a similar word used locally by the people of Senegal, however an alternative derivation may be the word 'manggi-manggi', which is still used in parts of eastern Indonesia to describe one genus (*Avicennia*).

Mangrove Species

Mangrove plants are not a simple taxonomic group, but are largely defined by the ecological niche where they live. The simplest definition describes a shrub or tree which normally grows in or adjacent to the intertidal zone and which has developed special adaptations in order to survive in this environment. Using such a definition a broad range of species can be identified, coming from a number of different families. Although there is no consensus as to which species are, or are not, true mangroves, there is a core group of some 30–40 species which are agreed by most authors. Furthermore, these 'core' species are the most important, both numerically and structurally, in almost all mangrove communities. **Table 1** lists a large range of mangrove species (of tree, shrub, fern, and palm), and highlights those which might be regarded as core species.

All of these plants have adapted to a harsh environment, with regular inundation of the soil and highly varied salinities, often approaching hypersaline conditions. Soils may be shallow, but even where they are deep they are usually anaerobic within a few millimeters of the soil surface. Many mangrove species show one or more of a range of physiological, morphological or life-history adaptations in order to cope with these conditions.

Coping with Salt

All mangroves are able to exclude most of the salt in sea water from their xylem. The exact mechanisms for this remain unclear, but it would appear to be an ultrafiltration process operating at the endodermis of the roots. One group, which includes *Bruguiera, Lumnitzera, Rhizophora* and *Sonneratia*, is highly efficient in this initial salt exclusion and shows only minor further mechanisms for salt secretion. A second group, which includes *Aegialitis, Aegiceras* and *Avicennia* appear to be less efficient at this initial salt exclusion and hence also need to actively secrete salt from their leaves. This is done metabolically, using special salt glands on the leaf surface. The salt evaporates, leaving crystals which may be washed or blown off the leaf surface. In these latter species such exuded salt is often visible on the leaf surface (**Figure 1**).

Anaerobic Soils

The morphological feature for which mangroves are best known is the development of aerial roots. These have developed in most mangrove species in order to cope with the need for atmospheric oxygen at the absorbing surfaces and the impossibility of obtaining such oxygen in an anaerobic and regularly inundated environment. Various types of roots are illustrated in **Figure 1**.

The stilt root, exemplified by *Rhizophora* (**Figure 1B**) consists of long branching structures which arch out away from the tree and may loop down to the soil and up again. Such stilt roots also occur in *Bruguiera* and *Ceriops* although in older specimens they fuse to the trunk as buttresses. They also occur sporadically in other species, including *Avicennia*.

A number of unrelated groups have developed structures known as pneumatophores which are simple upward extensions from the horizontal root into the air above. These are best developed in *Avicennia* and *Sonneratia* (**Figure 1C**), the former typically having narrow, pencil-like pneumatophores, the

Table 1 List of mangrove species: species in bold typeface are those which are considered 'core' species

Family	*Species*	*Family*	*Species*
Pteridaceae	*Acrostichum aureum*		***Bruguiera parviflora***
	Acrostichum danaeifolium		***Bruguiera sexangula***
	Acrostichum speciosum		***Ceriops australis***
Plumbaginaceae	*Aegialitis annulata*		***Ceriops decandra***
	Aegialitis rotundifolia		***Ceriops tagal***
Pellicieraceae	***Pelliciera rhizophorae***		***Kandelia candel***
Bombacaceae	*Camptostemon philippensis*		***Rhizophora apiculata***
	Camptostemon schultzii		***Rhizophora harrisonii***
Sterculiaceae	*Heritiera fomes*		***Rhizophora mangle***
	Heritiera globosa		***Rhizophora mucronata***
	Heritiera littoralis		***Rhizophora racemosa***
Ebenaceae	*Diospyros ferrea*		***Rhizophora samoensis***
Myrsinaceae	*Aegiceras corniculatum*		***Rhizophora stylosa***
	Aegiceras floridum		*Rhizophora* × *lamarckii*
Caesalpiniaceae	*Cynometra iripa*		*Rhizophora* × *selala*
	Mora oleifera	Euphorbiaceae	*Excoecaria agallocha*
Combretaceae	***Conocarpus erectus***		*Excoecaria indica*
	Laguncularia racemosa	Meliaceae	*Aglaia cucullata*
	Lumnitzera littorea		***Xylocarpus granatum***
	Lumnitzera racemosa		***Xylocarpus mekongensis***
	Lumnitzera* × *rosea	Avicenniaceae	***Avicennia alba***
Lythraceae	*Pemphis acidula*		***Avicennia bicolor***
Myrtaceae	*Osbornia octodonta*		***Avicennia germinans***
Sonneratiaceae	***Sonneratia alba***		***Avicennia integra***
	Sonneratia apetala		***Avicennia marina***
	Sonneratia caseolaris		***Avicennia officinalis***
	Sonneratia griffithii		***Avicennia rumphiana***
	Sonneratia lanceolata		***Avicennia schaueriana***
	Sonneratia ovata	Acanthaceae	*Acanthus ebracteatus*
	Sonneratia × *gulngai*		*Acanthus ilicifolius*
	Sonneratia × *urama*	Bignoniaceae	*Dolichandrone spathacea*
Rhizophoraceae	***Bruguiera cylindrica***		*Tabebuia palustris*
	Bruguiera exaristata	Rubiaceae	*Scyphiphora hydrophyllacea*
	Bruguiera gymnorrhiza	Arecaceae	***Nypa fruticans***
	Bruguiera hainesii		

latter with secondary thickening so that they can become quite tall and conical.

One adaptation on the theme of pneumatophores is that of root knees where more rounded knobs are observed to extend upwards from the roots. In *Xylocarpus mekongensis* these are simply the result of localized secondary cambial growth, but in *Bruguiera* (**Figure 1D**) and *Ceriops* they are the result of a primary looping growth. In these species branching may also occur on these root knees.

Buttress roots are a common adaptation of many tropical trees, but in *Xylocarpus granatum* (**Figure 1E**) and to some degree in *Heritiera* such flange-like extensions of the trunk continue into plank roots which are vertically extended roots with a sinuous plank-like form extending above the soil.

The surfaces of the aerial roots are amply covered with porous lenticels to enable gaseous exchange, and the internal structure of the roots is highly adapted, with large internal gas spaces, making up around 40% of the total root volume in some species. It is further widely accepted that there must be some form of ventilatory mechanism to aid gaseous exchange. A system of tidal suction is the probable mechanism in most species: during high tides, oxygen is used by the plant, while carbon dioxide is readily absorbed in the sea water, leading to reduced pressure within the roots. As the tide recedes and the lenticels open, water is then sucked into the roots.

Seeds and Seedlings

Establishment of new mangrove plants in the unstable substrates and regular tidal washing of the mangrove environment presents a particular evolutionary challenge. All mangroves are dispersed by water and particular structures in the seed or the fruit are adapted to support flotation. In a number of groups a degree of vivipary is observed which is unusual in most nonmangroves. The Rhizophoraceae have developed this to its fullest extent and here the embryo grows out of the seed coat and then out of the fruit while still attached to the parent plant, so that the propagule which is eventually released is actually a seedling rather than a seed (**Figure 1F**). In a number of other groups, including *Aegiceras, Avicennia, Nypa* and *Pelliciera* cryptovivipary exists in which the embryo emerges from the seed coat, but not the fruit, prior to abcission.

Longevity of seedlings is clearly important for many species. Most species are able to survive (float and remain viable) for over a month, whereas some *Avicennia* propagules have been shown to remain viable for over a year while in salt water.

Distribution and Biogeography

As a result of their restriction to intertidal areas, mangroves are limited in global extent (**Figure 2**), and are, in fact, one of the most globally restricted of all forest types. **Figure 2** clearly shows the absolute limits to mangrove distribution. Mangroves are largely confined to the regions between 30° north and south of the equator, with notable extensions beyond this to the north in Bermuda (32°20′N) and Japan (31°22′N), and to the south in Australia (38°45′S), New Zealand (38°03′S) and South Africa (32°59′S). Within these confines they are widely distributed, although their latitudinal development is restricted along the western coasts of the Americas and Africa. In the Pacific Ocean natural mangrove

(A)

(B)

(C)

(D)

Figure 1 Mangrove adaptations: (A) salt crystals secreted onto the surface of a leaf, *Avicennia*; (B) stilt roots of *Rhizophora*; (C) pneumatophores in *Sonneratia*; (D) root knees in *Bruguiera*; (E) plank roots in *Xylocarpus*; (F) *Rhizophora* propagule.

(E)

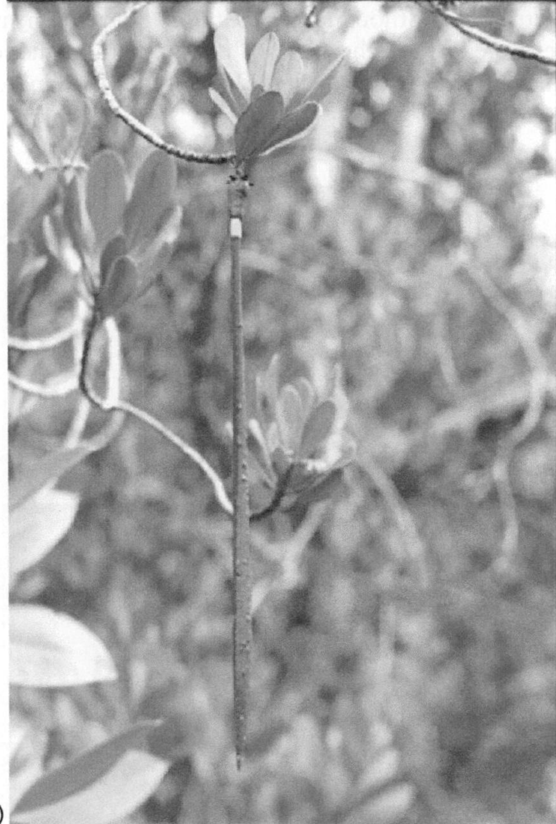
(F)

Figure 1 *Continued*

communities are limited to western areas, and they are absent from many Pacific islands.

In all, an estimated 114 countries and territories have mangroves, however for many nations the total area is very small indeed, and the total global area of these forests is only 181 000 km^2. **Table 2** provides a summary of total mangrove areas by region.

Although these statistics suggest a relatively wide distribution, the distribution of individual species within these areas is clearly far more restricted, and **Figure 3** provides a plot of mangrove biodiversity patterns. A number of points of particular interest are clearly illustrated.

1. There is a distinct region of very high mangrove diversity, sometimes referred to as the 'diversity anomaly', centered over south-east Asia.
2. Away from this high diversity region mangroves generally show relatively even levels of low diversity, although there is smaller peak of diversity around southern Central America.
3. There is a wide area of the central and western Pacific Ocean from 120 to 160°W where mangroves do not occur.
4. Even in the area of highest mangrove diversity there is a very rapid latitudinal decline in species numbers away from the tropics.

One further observation, which is not fully illustrated in the figures, concerns the division of the global mangrove flora into two highly distinct subregions. An eastern group (sometimes known as the Indo-West Pacific) forms one vast and contiguous block stretching from the Red Sea and East Africa to the central Pacific. This group has a totally different species composition from the western group (the Atlantic–East Pacific or Atlantic–Caribbean–East Pacific), which includes both Pacific and Atlantic shores of the Americas, the Caribbean and the shores of West Africa. A number of these patterns are explored more fully, below.

Latitudinal Patterns

Mangroves limits are closely correlated to minimum temperature requirements. There is only one genus (*Avicennia*) which survives in environments where frosts may occur, but many species appear to have their latitudinal limits set by less extreme cold temperatures; air temperatures of 5°C appear inimical to most mangrove species. Sea-surface temperatures may be more important than air temperatures for some species. The 24°C mean annual isotherm appears to be the minimum water temperature tolerated by mangroves in most areas, although this

Table 2 Total mangrove area by region

Region	*Area (km^2)*	*Proportion of global total*
South and south-east Asia	75 173	41.5%
Australasia	18 789	10.4%
The Americas	49 096	27.1%
West Africa	27 995	15.5%
East Africa and the Middle East	10 024	5.5%
Total area	181 077	

Data calculated from best available national sources in Spalding *et al.* (1997).

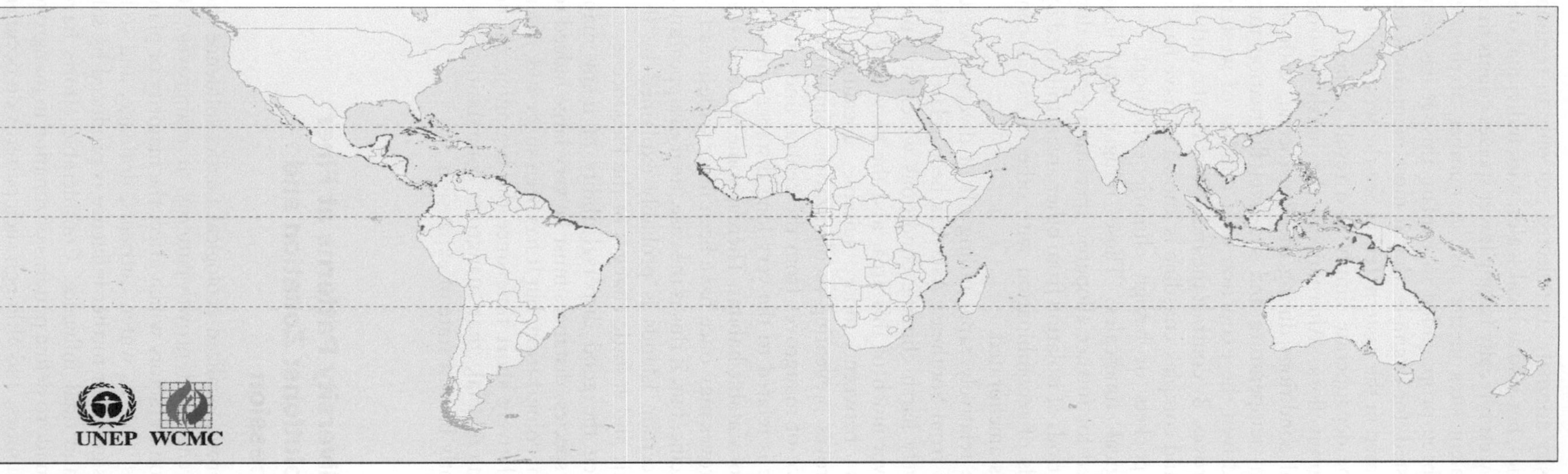

Figure 2 The global distribution of mangrove forests (data kindly provided by the UNEP World Conservation Monitoring Centre).

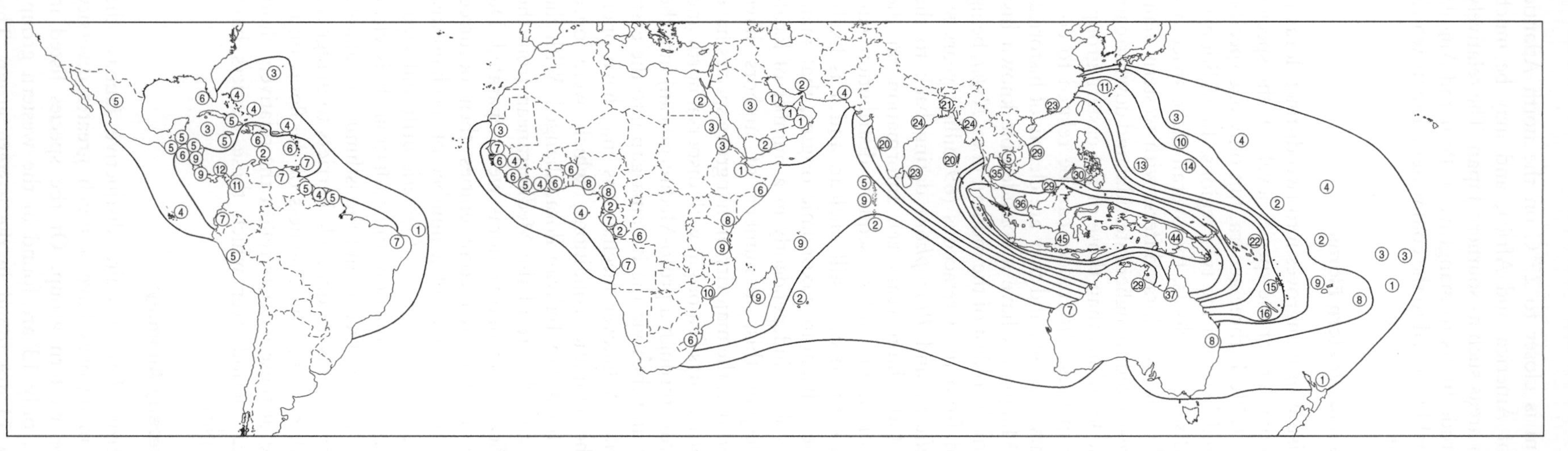

Figure 3 A global map of mangrove diversity plotting contours of equal diversity (1–5 species, 6–10, 11–15, 16–20, 21–25, 26–30, 31–35, 36–40, and 41–45). (Reproduced with permission from Spalding (1998).)

minimum is closer to 27°C on the north Atlantic coasts of America and Africa, and may be much lower in areas such as southern Japan. The relatively low-latitude limits to mangrove in Peru and Angola are probably related to the cold water currents which affect these coastlines.

Eastern and Western Floras

The division of mangroves into two distinct floras is almost complete at the species level. Of the species listed in **Table 1**, three genera, but only one species, are shared between the two regions. In addition to having distinctive floras, the overall niche-space occupied by the two floras differs, with western mangroves restricted to higher intertidal and downstream estuarine locations than those of the eastern group.

None of these differences can be related to contemporary ecology, and they are clearly of historical origin. Mangroves have a considerable known history, with the oldest of the modern taxa, *Nypa*, being recorded from the Cretaceous (69 million years BP) and *Pellicera* and *Rhizophora* dating back to the Eocene (30 million years BP). Information on the centers of origin and subsequent distribution patterns of mangroves is still unclear, and it is likely, given their disparate taxonomic origins, that mangroves evolved independently in a number of localities. Despite this, a number of authors have suggested that the majority of mangrove species have an eastern Tethys Sea origin with dispersal north and westwards (through a proto-Mediterranean) into the Atlantic and then via the Panama gap into the eastern Pacific. Whatever mechanisms may have operated, the climatic conditions, which were once suitable for a pan-Tethyan flora, changed. With the cooling and closure of the Mediterranean from the Tethys Sea the mangrove floras were separated. Divergence of the two communities then occurred through one or more of a number of mechanisms, including natural process of genetic drift and separation, possible extinction and radiation. It is clear that the Atlantic Ocean and the isthmus of Panama now represent insurmountable barriers to mangrove dispersal, however the closeness of the floras on either side of these barriers reflect the relatively short geological period over which these barriers have been in place.

The Diversity Anomaly

Apart from having quite distinctive faunas, the eastern mangroves have a much greater diversity than the western group. Of the species listed in **Table 1**, only 13 are found in the western group, whereas 59 are found in the eastern group. This 'diversity anomaly' is reflected not only in regional statistics, but also at local scales, with individual sites in the west typically having lower species counts than equivalent sites in the east. A number of theories have been propounded to explain this. It has been suggested, for example, that if most mangroves had originated in the eastern Tethys Sea the western flora may be depauperate simply as a result of being an immigrant flora. Alternatively the harsh environmental conditions during the Pleistocene, with significant temperature and sea-level fluctuation may have driven the extinction of a number of western mangroves. By contrast the Indo-West Pacific with its long and complex coastline is known to have had at least pockets of benign climatic conditions over geological timescales. These refugia may have allowed for further allopatric speciation events during periods of isolation from other areas, followed by periods of recombination with other areas as conditions ameliorated.

The relatively rapid tailing off of diversity westwards from Southeast Asia has been related to the relatively harsh climatic conditions which still prevail over much of this area, and the very large distances between more suitable localities for mangroves preventing recolonization. Similarly the absence of mangroves from the central and eastern Pacific is related to the very long distances between areas of suitable habitat. There is some evidence that mangroves may once have been more widespread in the Pacific, but, if this is the case, their disappearance from certain islands is probably explained by the climatic and eustatic changes of the Pleistocene.

Given the good dispersal ability of many mangrove species, distances must be very large indeed to prevent colonization, but it has been suggested, given the relatively short time since the beginning of the last interglacial, that mangrove communities may currently be in a state of expansion.

Biodiversity Patterns at Finer Resolutions: Zonation and Succession

Numerous localized ecological factors influence the occurrence and growth-patterns of mangroves. In addition to factors which affect the majority of plant species, such as water, nutrients, drainage, and soil-type, significant further influence is produced by salinity and tidal influence. Considerable efforts have been made to define patterns of zonation in mangrove communities, and although such patterns do occur in many communities, the enormous variation in local conditions makes the preparation of simple summaries of 'typical' zonation patterns very difficult.

In many tidal areas the regular drying out of the soil, often coupled with patterns of restricted water circulation, or high rates of evaporation, serves to increase salinities to considerably higher levels than the surrounding sea water. This is particularly the case in areas of back mangrove where tidal flushing is less frequent and water circulation may be more restricted. It is further exacerbated in arid regions. In many areas this leads to the development of wide areas of stunted mangroves, or even bare salt-pans where mangroves cannot grow. This situation is diminished, or even reversed in areas where the freshwater input is more considerable, either from high rainfall or terrestrial runoff, or in some estuarine environments.

The tides also exert influences in other ways, most notably through inundation, but also through their influence on soils. Different mangrove species show quite different tolerances to inundation. Species such as *Avicennia* and *Rhizophora*, which are relatively tolerant of frequent and quite high tidal waters, typically form the most seaward zone of the mangrove system. Tides also influence the soil through the delivery or removal of nutrients, and also the re-sorting of sediments. Typically finer sediments are found at higher locations in the tidal frame, whereas coarser sediments tend to be deposited or re-distributed lower down. Once again, the complexity of interactions is highly varied between localities.

In many cases mangrove communities may follow a succession and this has been linked to the process of terrestrial advancement (coastal progradation). The patterns shown in zonation often provide a spatial model for such a temporal succession, starting with the more inundation and salt-tolerant species. These are able to bind nutrients and sediments, gradually raising their position in the tidal frame such that they are then replaced by those species requiring slightly less saline and inundated conditions, and then by mangrove associates and then nonmangrove species. Such successional processes occur in many areas; in parts of Southeast Asia where there is a high input of allochthonous material, rates of coastal advancement have been recorded at 120–200 m year^{-1}. In other areas, however, the notion of mangroves 'creating land' is clearly not valid and mangroves show a range of responses to differing impacts of waves, climate, and sediments. In the Florida Everglades there is considerable evidence for the movement of mangrove communities both landwards and seawards, depending on sea-level changes and it may be more accurate to regard mangroves in these areas as opportunistic followers of sedimentation and substrate or elevation changes.

Humans and Mangroves

Humans have lived in close contact with mangrove communities for millennia and in many cases have made considerable use of this association. Archaeological sites have been located which demonstrate human presence in mangrove areas in Venezuela dating back 5000–6000 years, and there is an Egyptian inscription dating back to the time of King Assa (3580–3536 BC) which mentions mangroves. Countries of the Middle East began a vigorous trade in mangrove timber from about the ninth century, largely for boat-building, exporting from outposts along the shores of East Africa. The European nations became involved in the utilization of mangrove bark as a source of tannins, particularly from the Americas from the sixteenth century.

The earliest record of mangrove protection dates to an edict from the King of Portugal in 1760 who restricted the cutting of mangroves for timber in Brazil unless their bark was also used for tannins. Despite such early concerns, the overexploitation of mangroves began in earnest towards the middle of the twentieth century and is continuing, and in many areas accelerating at the present time.

In many areas mangroves are highly productive and their location on the coastline places them in a zone where many other human activities have, until recently, been somewhat restricted. At the same time they often exist in close proximity to centers of human population, and can be relatively easily approached by sea or land. This makes their utilization inevitable in many areas, although the degree of sustainability of such use is highly variable.

Utilization of Mangroves

Timber and wood products One of the commonest uses of mangroves is as a source of wood. Mangrove wood is often used for fuel either directly or after conversion into charcoal. The former is widespread among artisanal communities worldwide, the latter often for commercial purposes. Mangrove wood is also used for timber; the relatively small size of mangrove trees in many areas has meant that the primary usage of timber is the preparation of timber poles for fencing, housing construction, making of fish-traps and other activities. Larger trees can be utilized for preparation of planking, and indeed some species have a very high value associated with their dense wood and resistance to rot, which is important for construction of houses and boats (both for local and commercial use). Further industrial use of mangrove wood is in the production of wood-pulp for the paper industry, and chipboard.

Fisheries Mangroves and the associated channels which run between them, are important areas of fish productivity. Numerous species inhabit mangrove areas and form the basis of artisanal and commercial fisheries, including crab, prawn, and mollusk fisheries. Mangrove areas are also widely used by a number of offshore fish species which are of commercial importance. These species, which include some highly profitable shrimp species, use mangrove areas for spawning or as a nursery ground and loss of mangrove areas has severe negative impacts on fishery productivity. Cage-based fisheries have been established in many of the wider channels, and mangrove areas are widely used for the capture of juvenile prawns for transfer to aquaculture ponds. In recent years, wide areas of mangrove forest have been cut down in the development of intertidal aquaculture ponds, particularly in south-east Asia. Although this is a highly profitable industry, poor planning has led to the rapid and virtually irrevocable degradation of many of these ponds after only a few years. Rehabilitation of these lands is rarely undertaken with the result that local communities lose a source of valuable natural resources, and the shrimp pond developers move on to new areas.

Coastal protection The important role which mangroves play in the stabilization of coastal sediments and the reduction of coastal erosion has already been mentioned. This role is frequently overlooked until such time as the mangroves are removed and major storm events hit coastlines. The massive and devastating cyclones which regularly impact the coastline of the Bay of Bengal have drawn particular attention to these issues and in a number of localities around the globe there are now efforts to establish mangrove plantations precisely to stabilize sediments and reduce the impact of storm surges.

Alongside these three key areas of human importance, mangroves are regularly utilized for other purposes, a number of which are outlined in **Table 3**. It is highly difficult to place values on many of these uses and functions of mangroves. Apart from direct utilization of wood products, the link between particular products or functions and the mangrove communities which provide them is rarely made. Furthermore, for numerous communities the value in economic terms is greatly enhanced by the social value, providing a source of employment, protein and protection for some of the world's poorest communities.

Table 3 Minor or regionally restriced uses of mangroves

Honey production	An important economic activity in some countries
Fodder	For cattle, camels and goats, notably in India and Pakistan
Recreation	Walkways, boat-based tours and other visiting facilities have been established for tourists and local communities in some areas, notably Trinidad, Bangladesh, and Australia
Thatch and matting	Primarily from the leaves of the mangrove palm *Nypa fruticans* in south-east Asia and from introduced populations in West Africa
Tannin extraction	Formerly widespread, this activity has become less significant as synthetic products have become available
Traditional medicine	Still widespread in many traditional communities
Food	*Nypa fruticians* is widely used for the production of sugar, alcohol and vinegar. Fruits of *Avicennia*, *Kandelia* and *Bruguiera* are used as a source of food in some countries

Overexploitation and Loss

Mention has already been made of the widespread loss of mangrove communities worldwide. Apart from conversion into aquaculture ponds, much of this is related to land reclamation activities for agriculture and for urban and industrial development, and large areas have also been severely degraded or removed by commercial timber companies or through overexploitation by local communities. Some further degradation or loss has been related to human-induced changes to the water regime (including upstream dams leading to reductions in sedimentation at river mouths), pollution (mangroves are particularly sensitive to oil spills), and conversion into salt pans for industrial salt production. To date there is no globally available figure for total mangrove loss, however national loss statistics are available for a number of countries. In south-east Asia, for example, the loss figures for four countries are: Malaysia, 12% from 1980 to 1990; the Philippines, a 60% loss from 4000 km^2 originally to 1600 km^2 today; Thailand, a 55% loss from 5500 km^2 in 1961 to 2470 km^2 in 1986; and Vietnam, a 37% loss from 4000 km^2 originally to 2525 today. These figures alone suggest a total of some 7445 km^2 of mangrove loss, representing over 4% of the current global total. The four countries concerned have certainly suffered significant mangrove loss, but they are not alone.

Sea-level rise associated with global climate change must also be considered as a significant threat to mangrove ecosystems. It is important to note that the impacts of proposed changes (most models predict rises in sea-level of 30–100 cm by 2100) are relatively insignificant in some areas where high levels of sediment movement and deposition will counter such rises, or where other eustatic changes, such as those associated with tectonic movements, will remove or further enhance changing sea-level effects. Furthermore, mangrove species and communities are highly opportunistic and will colonize new areas with some rapidity. Sea-level rise remains a problem, however, as mangrove communities in many areas may become squeezed out as sea-level rise forces mangrove communities landwards, but human use prevents landward migration.

Protection and Plantation

Despite the massive losses which mangrove communities have gone through in the past decades there have also been concerted efforts to protect them in some areas, and the growing realization of their value has led to widespread efforts to utilize mangroves in a more sustainable manner, and in some places large areas of mangrove plantations have now been established.

Worldwide, there are currently an estimated 850 protected areas with mangroves spread between 75 countries, which are managed for conservation purposes. These cover over 16 000 km^2 of mangrove, or 9% of the global total. Although this is a far higher proportion than for many other forest types, active protection is absent from many of these areas, and the remaining unprotected sites are probably more threatened than many other forest types because of their vulnerability to human exploitation.

Increasing recognition of the various values of mangrove forests is leading to widescale mangrove plantation in some areas, for coastal defence, as a source of fuel, or for fisheries enhancement. Plantations in Bangladesh, Vietnam, and Pakistan now cover over 1700 km^2, and Cuba is reported to have planted some 257 km^2 of mangroves. Overall, however, when weighed against the statistics of mangrove loss, the area of such plantations remains insignificant.

Active management of these and other existing mangrove areas for economic production is increasing. The Matang Mangrove Reserve in Malaysia is perhaps the best-known example. Studies have shown combined benefits arising from timber and fuelwood products (notably charcoal), but even more importantly from a large nearshore fishery (directly or indirectly providing employment for over 4000 people), from aquaculture on the mud flats below the mangroves, and from tourism. It is rare that such holistic studies have been carried out. Often the human benefits provided by mangrove fall between several sectors of the economy, fisheries, forestry, tourism, and coastal protection, and their combined benefits are not realized. A better perception of these benefits would undoubtedly lead to much wider-scale protection for mangroves globally.

See also

Crustacean Fisheries. Fisheries Overview.

Further Reading

Chapman VJ (1976) *Mangrove Vegetation*. Vaduz, Germany: J Cramer.

Field CD (1995) *Journey Amongst Mangroves*. Okinawa, Japan: International Society for Mangrove Ecosystems.

Field CD (ed.) (1996) *Restoration of Mangrove Ecosystems*. Okinawa, Japan: International Society for Mangrove Ecosytems.

Robertson AI and Alongi DM (eds.) (1992) *Coastal and Estuarine Studies, 41: Tropical Mangrove Ecosystems*. Washington, USA: American Geophysical Union.

Saenger P, Hegerl EJ, and Davie JDS (eds.) (1983) *IUCN Commission on Ecology Papers, 3: Global Status of Mangrove Ecosystems*. Gland, Switzerland: IUCN (The World Conservation Union).

Spalding MD (1998) *Biodiversity Patterns in Coral Reefs and Mangrove Forests: Global and Local Scales*. PhD Dissertation, University of Cambridge.

Spalding MD, Blasco F, and Field CD (eds.) (1997) *World Mangrove Atlas*. Okinawa, Japan: International Society for Mangrove Ecosystems.

Tomlinson PB (1986) *The Botany of Mangroves*. Cambridge, UK: Cambridge University Press.

ROCKY SHORES

G. M. Branch, University of Cape Town, Cape Town, Republic of South Africa

Introduction

Intertidal rocky shores have been described as 'superb natural laboratories' and a 'cauldron of scientific ferment' because a rich array of concepts has arisen from their study. Because intertidal shores form a narrow band fringing the coast, the gradient between marine and terrestrial conditions is sharp, with abrupt changes in physical conditions. This intensifies patterns of distribution and abundance, making them readily observable. Most of the organisms are easily visible, occur at high densities, and are relatively small and sessile or sedentary. Because of these features, experimental tests of concepts have become a feature of rocky-shore studies, and the critical approach encouraged by scientists such as Tony Underwood has fostered rigor in marine research as a whole.

Rocky shores are a strong contrast with sandy beaches. On sandy shores, the substrate is shifting and unstable. Organisms can burrow to escape physical stresses and predation, but experience continual turnover of the substrate by waves. Most of the fauna relies on imported food because macroalgae cannot attach in the shifting sands, and primary production is low. Physical conditions are relatively uniform because waves shape the substrate. On rocky shores, by contrast, the physical substrate is by definition hard and stable. Escape by burrowing may not be impossible, but is limited to a small suite of creatures capable of drilling into rock. Macroalgae are prominent and *in situ* primary production is high. Rocks alter the impacts of wave action, leading to small-scale variability in physical conditions.

Research on rocky shores began with a phase describing patterns of distribution and abundance. Later work attempted to explain these patterns – initially focusing on physical factors before shifting to biological interactions. Integration of these focuses is relatively recent, and has concentrated on three issues: the relative roles of larval recruitment versus adult survival; the impact of productivity; and the effects of stress or disturbance on the structure and function of rocky shores.

Zonation

The most obvious pattern on rocky shores is an upshore change in plant and animal life. This often creates distinctive bands of organisms. The species making up these bands vary, but the high-shore zone is frequently dominated by littorinid gastropods, the upper midshore prevalently occupied by barnacles, and the lower section by a mix of limpets, barnacles, and seaweeds. The low-shore zone commonly supports mussel beds or mats of algae. Such patterns of zonation were of central interest to Jack Lewis in Britain, and to Stephenson, who pioneered descriptive research on zonation, first in South Africa and then worldwide.

In general, physical stresses ameliorate progressively down the shore. In parallel, biomass and species richness increase downshore. Three factors powerfully influence zonation: the initial settlement of larvae and spores; the effects of physical factors on the survival or movement of subsequent stages; and biotic interactions between species.

Settlement of Larvae or Spores

Many rocky-shore species have adults that are sessile, including barnacles, zoanthids, tubicolous polychaete worms, ascidians, and macroalgae. Many others, such as starfish, anemones, mussels, and territorial limpets, are extremely sedentary, moving less than a few meters as adults. For such species, settlement of the dispersive stages in their life cycles sets initial limits to their zonation (often further restricted by later physical stresses or biological interactions).

Some larvae selectively settle where adults are already present. Barnacles are a classic example. This gregarious behavior, which concentrates individuals in particular zones, has several possible advantages. The presence of adults must indicate a habitat suitable for survival. Furthermore, individuals of sessile species that practice internal fertilization (e.g., barnacles) are obliged to live in close proximity. Even species that broadcast their eggs and sperm will enhance fertilization if they are closely spaced, because sperm becomes diluted away from the point of release. Finally, adults may themselves shelter new recruits. As examples, larvae of the sabellariid reef-worm *Gunnarea capensis* that settle on adult colonies suffer less desiccation, and sporelings of kelps that settle among the holdfasts of adults experience less intense grazing than those that are isolated.

Cues used by larvae to select settlement sites are diverse. Barnacle larvae differ in their preferences for light intensity, water movement, substrate texture, and water depth. All of these responses influence the type of habitat or zone in which the larvae will settle. Most barnacle larvae are attracted to species-specific chemicals in the exoskeletons of their own adults, which persists on the substratum even after adults are eliminated. (Incidentally, this behavior is not just of academic interest: gregarious settlement of barnacles on the hulls of ships costs billions of dollars each year due to increased fuel costs caused by the additional drag of 'fouling' organisms.)

Cues influencing larval settlement can also be negative. Rick Grosberg elegantly demonstrated that the larvae of a wide range of sessile species that are vulnerable to overgrowth avoid settling in the presence of *Botryllus*, a compound ascidian known to be an aggressive competitor for space.

A different aspect of 'supply-side ecology' is the rate at which dispersive larvae or spores settle. The relative importance of recruit supply versus subsequent survival is a topic of intense research. In situations of low recruitment, rate of supply critically influences population and community dynamics. At high levels of recruitment, however, supply rates become less important than subsequent biological interactions such as competition.

Control of Zonation by Physical Factors

Early research on the causes of zonation focused strongly on physical factors such as desiccation, temperature, and salinity, all of which increase in severity upshore. Measurements showed a correlation between the zonation of species and their tolerance of extremes of these factors. In some cases – particularly for sessile species living at the top of the shore – physical conditions become so severe that they kill sections of the population, thus imposing an upper zonation limit by mortality.

For those species blessed with mobility, zonation is more often set by behavior than death, and most individuals live within the 'zone of comfort' that they can tolerate. One example illustrates the point. The trochid gastropod *Oxystele variegata* increases in size from the low- to the high-shore. This gradient is maintained by active migration, and animals transplanted to the 'wrong' zone re-establish themselves in their original zones within 24 hours. The underlying causes of this size gradient seem to be twofold: desiccation is too high in the upper shore for small individuals to survive there; and predation on adults is greatest in the lower shore.

Adaptations to minimize the effects of physical stresses are varied. Physiological adaptation and tolerance are one avenue of escape. Avoidance by concealment in microhabitats is another. Morphological adaptations are a third route. For instance, desiccation and heat stress can be reduced by large size (reducing the ratio of surface area to size), and by differences in shape, color, and texture (**Figure 1A**).

Physical factors can clearly limit the upper zonation of species. It is often difficult, however, to imagine them setting lower limits. For this, we turn to biological interactions.

Biological Interactions

Interactions between species – particularly competition and predation – only began to influence the thinking of intertidal rocky-shore ecologists in the mid 1950s. Extremely influential was Connell's work, exploring whether the zonation of barnacles in Scotland is influenced by competition. He noted that a high-shore species, *Chthamalus stellatus*, seldom penetrates down into the midshore, where another species, *Semibalanus balanoides*, prevails. Was competition from *Semibalanus* excluding *Chthamalus* from the midshore? In field experiments in which *Semibalanus* was eliminated, *Chthamalus* not only occupied the midshore, but survived and grew there better than in its normal high-shore zone. *Chthamalus* is more tolerant of physical stresses than *Semibalanus*, and can therefore survive in the high-shore, where it has a 'spatial refuge' beyond the limits of *Semibalanus*. In the midshore, however, *Semibalanus* thrives and competitively excludes *Chthamalus* by undercutting or overgrowing it.

Other forms of competition have since been discovered. For instance, territorial limpets defend patches of algae by aggressively pushing against other grazers. In areas where they occur densely, they profoundly influence the zonation of other species and the nature of their associated communities.

The role of predators came to the fore following the work of Bob Paine who showed that experimental removal of the starfish *Pisaster ochraceus* from open-coast shores in Washington State led to encroachment of the low-shore by mussels, which are usually restricted to the midshore. Thus, predation sets lower limits to the zonation of the

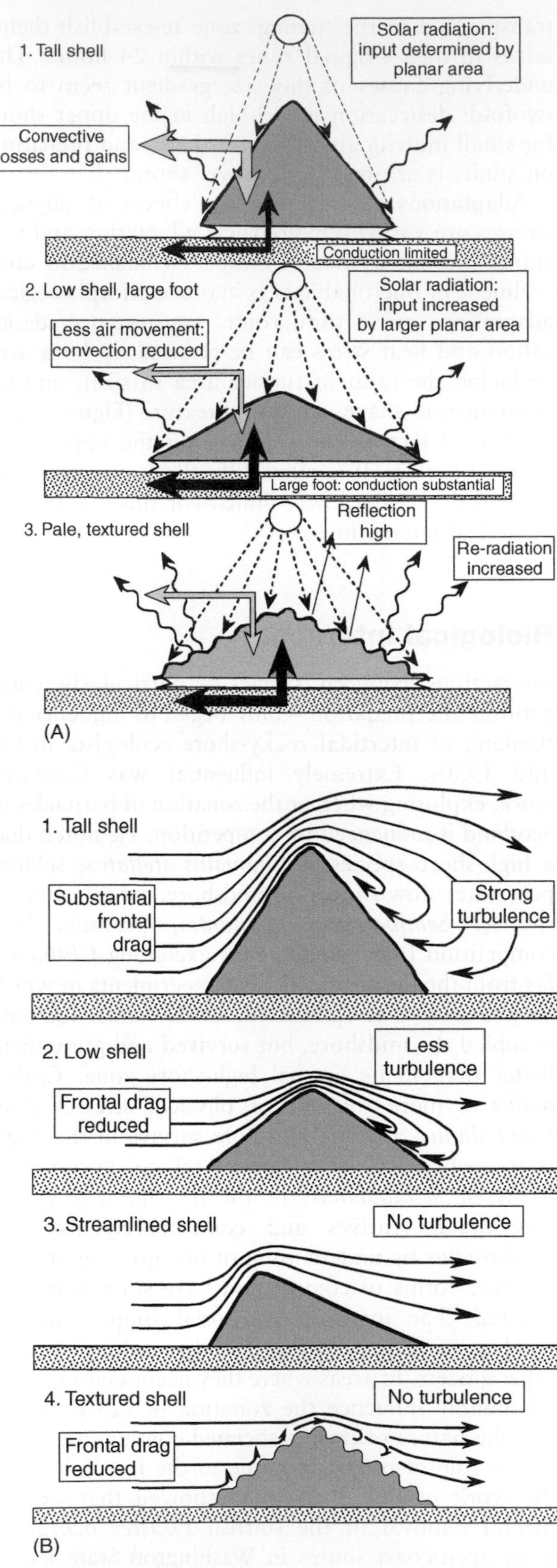

Figure 1 (A) Effects of shell shape, color, and texture on heat uptake and loss. (B) Influence of shell proportions and texture on hydrodynamic drag.

mussels. More importantly, it was shown that the downshore advance of mussel beds reduced the number of space-occupying species there. In other words, predation normally prevents the competitively superior mussels from ousting subordinate species, thus maintaining a higher diversity of species. This concept – the 'predation hypothesis' – has since been broadened to include all forms of biological or physical disturbance. If disturbance is too great, few species survive and diversity is low. On the other hand, if it is absent or has little effect, a few species may competitively monopolize the system, reducing diversity. At intermediate levels of disturbance, diversity is highest – the 'intermediate disturbance hypothesis'. (Incidentally, the idea is not new. Darwin gives an accurate description of this effect in sheep-grazed meadows.)

Paine's work led to the idea of 'keystone predators': those that have a strong effect on community structure and function. Unfortunately the term has been debased by general application to any species that an author feels is somehow 'important'. Consequently, not all scientists are enamoured of the concept. More recently, it has been redefined to mean those species whose effects are disproportionately large relative to their abundance. This more usefully allows recognition of species that can be regarded as 'strong interactors', and which powerfully influence community dynamics.

Many researchers have shown that grazers (particularly limpets) profoundly influence algal zonation, excluding them from much of the shore by eliminating sporelings before they develop. Hawkins and Hartnoll suggested that interactions between grazers and algae depend on the upshore gradient in physical stress, and argued that low on the shore, algae are likely to be sufficiently productive to escape grazing and proliferate, forming large, adult growths that are relatively immune to grazing. In the mid- to high-shore, grazers become dominant and algae seldom develop beyond the sporeling stage. A more subtle reverse effect, that of algae-limiting grazers, has also been described. Low on the shore where productivity is high, algae form dense mats. This not only deprives grazers of a firm substrate for attachment, but also of their primary food source, namely microalgae. Grazers experimentally transplanted into low-shore algal beds starve in the midst of apparent plentitude.

The influences of grazers and predators extend beyond their direct impacts on prey. A bird consuming limpets has positive effects on algae that are grazed by the limpets. Such indirect effects occupy ecologists because their consequences are often difficult to predict. For example, experimental removal of a large, grazing chiton, *Katharina tunicata*, from the shores of Washington State might logically have been expected to improve the lot of intertidal limpets, on the grounds that elimination of a competitor must be good for the remaining grazers. In fact, elimination of the chiton led to starvation of the limpets because macroalgae proliferated, excluding microalgae on which the limpets depend. Indirectly, *Katharina* facilitates microalgae, thus benefiting limpets.

The mix of physical and biological controls affecting zonation was investigated by Bruce Menge in 1976 by using cages to exclude predators from plots in the high-, mid-, and low-shore. In the high-shore, only barnacles became established, irrespective of whether predators were present or absent. In the midshore, mussels became dominant and outcompeted barnacles, again independently of the presence or absence of predators. Competition ruled. In the low-shore, however, mussels only dominated where cages excluded the predators. Elsewhere, predators eliminated mussels, thus allowing barnacles to persist. These results elucidated the interplay between physical stress and biological control and were instrumental in formalizing 'environmental stress models'. These suggest that predation will only be important (exerting a 'top-down' control on community structure) when physical conditions are mild. As stress rises, first predation, and then competition, diminish in importance.

Wave Action

It has been shown that wave action is probably the most important factor affecting distribution patterns along the shore. In a negative sense, waves physically remove organisms, damage them by throwing up logs and boulders, reduce their foraging excursions, and increase the amount of energy devoted to clinging on. One manifestation is a reduction of grazer biomass (**Figure 2F**). Adaptations can, however, counter these adverse effects. Tenacity can be increased by cementing the shell to the rock face (e.g., oysters), developing temporary attachments (e.g., the byssus threads of mussels), or employing

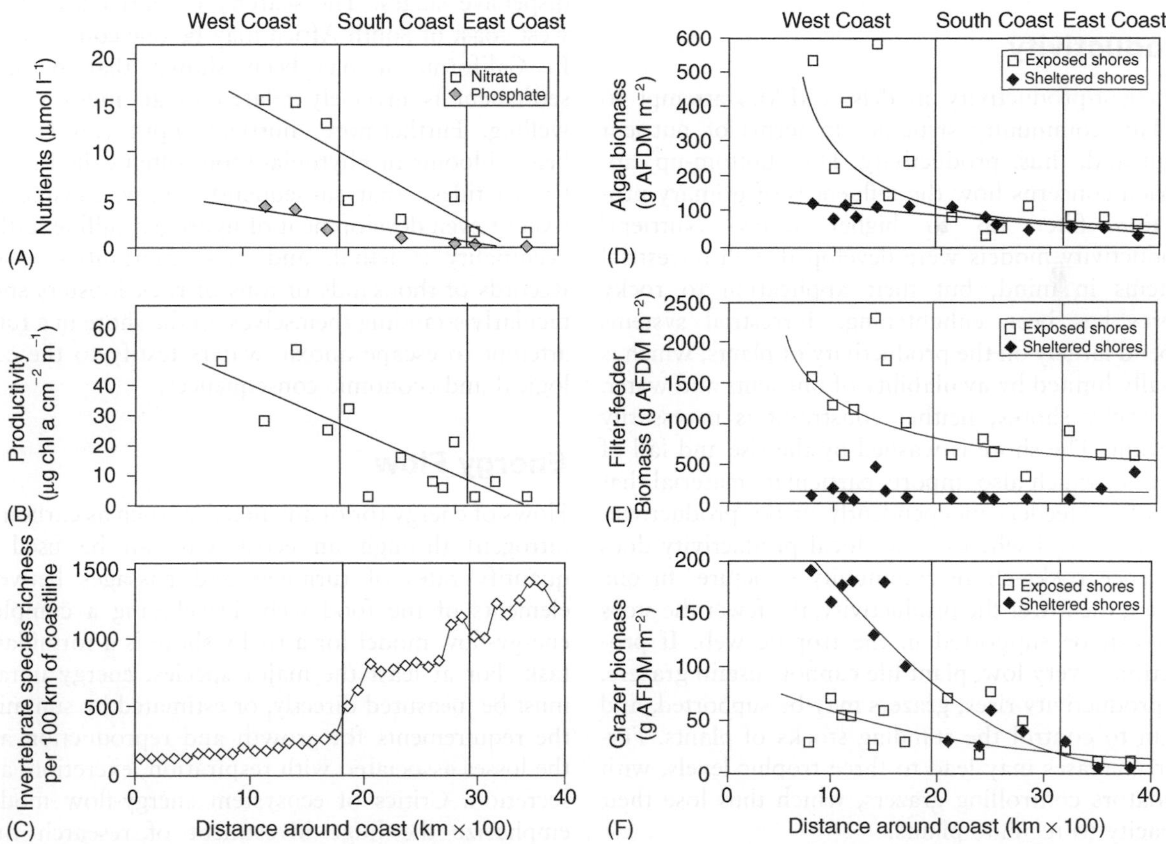

Figure 2 Moving west to east around the coastline of southern Africa (from northern Namibia to southern Mozambique), (A) nutrient input and (B) intertidal primary productivity decrease, (C) invertebrate species richness rises, and there are declines in the biomasses (ash-free dry mass) of (D) algae, (E) filter-feeders, and (F) grazers. Note that biomass is strongly influenced by wave action, either positively (algae and filter-feeders) or negatively (grazers). AFDM.

adhesion (e.g., the feet of limpets and chitons). Shape can be modified to reduce drag, turbulence, and lift (**Figure 1B**). Each organism is, however, a compromise between conflicting stresses. For instance, desiccation is reduced if a limpet has a small aperture; but this implies a smaller foot and thus less ability to cling to the rocks.

Wave action also brings benefits. It enhances nutrient supply, reduces predation and grazing, and increases food supply for filter-feeders. The biomass on South African rocky shores has been shown to rise steeply as wave action increases, mostly due to increases in filter-feeders (**Figure 2E**). Possible explanations include enhanced larval supply and reduced predation, but measurements of food abundance and turnover showed that wave action vitally enhances particulate food.

Wave action varies over short distances. Headlands and bays influence the magnitude of waves at a scale of kilometers, but even a single large boulder will alter wave impacts at a scale of centimeters to meters. As a result, community structure can be extremely patchy on rocky shores.

Productivity

'Nutrient/productivity models' (NPMs) attempt to explain community structure in terms of nutrient input and, thus, productivity. This 'bottom-up' approach concerns how the influences of primary production filter up to higher levels. Nutrient/productivity models were developed with terrestrial systems in mind, but their application to rocky shores has been enlightening. Terrestrial systems depend largely on the productivity of plants, which is usually limited by availability of nutrients and water. On rocky shores, neither constraint is necessarily relevant. The shore is washed by the rise and fall of the tide, which also imports particulate material that fuels filter-feeders independently of the productivity on the shore itself. Even so, local productivity does influence rocky-shore community structure. In one theory, the lower the productivity, the fewer the steps that can be supported in the trophic web. If production is very low, plant life cannot sustain grazers. As productivity rises, grazers may be supported, and begin to control the standing stocks of plants. Further increases may lead to three trophic levels, with predators controlling grazers, which thus lose their capacity to regulate plants.

Nowhere else in the world is the coastline better configured to test ideas about NPMs than in South Africa. The cold, nutrient-rich, upwelled Benguela Current bathes the west coast. The east coast receives fast-moving, warm, nutrient-poor waters from the southward-flowing Agulhas Current. Between the two, the Agulhas swings away from the south coast, creating conditions that are intermediate. From west to east, the coast has a strong gradient of nutrients and productivity (**Figure 2A, B**). As productivity drops, so do the average biomasses of algae, filter-feeders, and grazers (**Figure 2D–F**), and the total biomass. On the other hand, species richness rises (**Figure 2C**). At a more local scale, guano input on islands achieves the same effects (**Figure 3**).

Productivity also has more subtle effects on the functioning of rocky shores. For instance, the frequency of territoriality in limpets is inversely correlated with productivity. It seems that the need to defend patches of food diminishes as the ratio of productivity : consumption rises. Indirectly, this has profound effects on community dynamics, because these territorial algal species reduce species richness and biomass but greatly increase local productivity.

Increased productivity is, however, not an unmixed blessing. The upwelling that fuels coastal productivity also results in a net offshore movement of water. This may export the recruits of species with dispersive stages. The scarcity of barnacles on the west coast of South Africa may be one consequence. In California it has been shown that barnacle settlement is inversely related to an index of upwelling. Furthermore, nutrient input can lead to heavy blooms of phytoplankton (often called red or brown tides), that subsequently decay, causing anoxia or even development of hydrogen sulfide. Either eventuality is lethal, and mass mortalities ensue. Records of thousands of tons of rock lobsters spectacularly stranding themselves on the shore in a futile attempt to escape anoxic waters testify to the ecological and economic consequences.

Energy Flow

Flows of energy (or of any material such as carbon or nitrogen) through an ecosystem can be used to quantify rates of turnover and passages between elements of the food web. Developing a complete energy-flow model for a rocky shore is a formidable task. For at least the major species, energy uptake must be measured directly, or estimated by summing the requirements for growth and reproduction and the losses associated with respiration, excretion, and secretion. Critics of ecosystem energy-flow models emphasize the huge investment of research time needed to complete the task, and argue that rocky shores differ so much from place to place that a model describing one shore will often be inapplicable elsewhere.

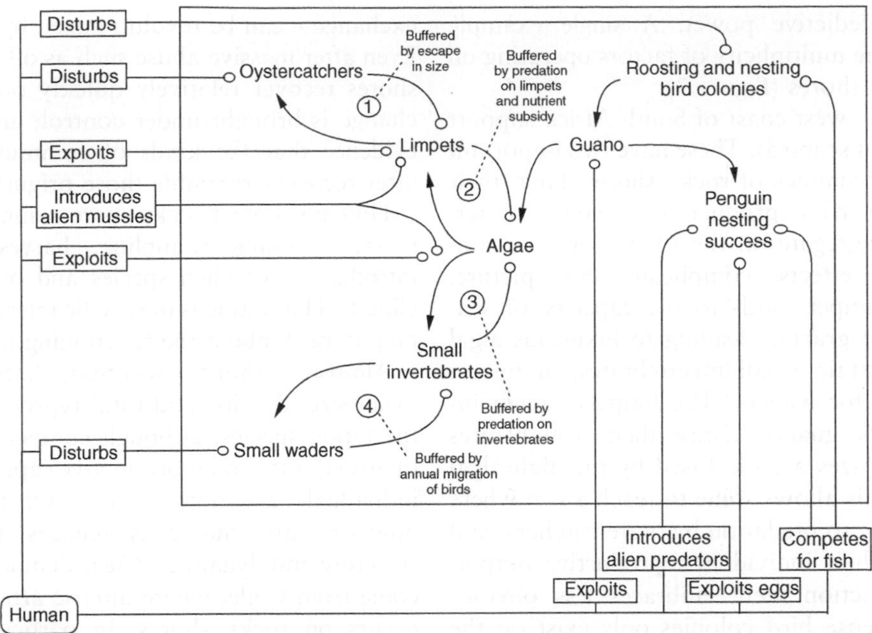

Figure 3 Interactions between organisms on nearshore islands on the west coast of South Africa. Natural interactions, and the processes buffering them (numbered 1–4), are shown inside the box; human impacts influencing them lie outside the box. Lines terminating in arrows and circles indicate positive and negative effects, respectively.

Nevertheless, energy-flow models of rocky shores have revealed features difficult to grasp by other approaches, as the following examples reveal. First, measurements show that most macroalgal productivity is not consumed by grazers, but adds to a detrital pool that fuels filter-feeders. On sheltered shores, production exceeds demands, and they are net exporters of material. On wave-beaten shores, however, the needs of filter-feeders far exceed intertidal production, and these shores depend on a subsidy of materials from the subtidal zone or from offshore. Tides, waves and currents play a vital role in turning over this supply of particulate materials. Intertidal standing stocks of particulate matter are in the order of $0.25\,g\,C\,m^{-3}$, far below the annual requirements of filter-feeders (about $500\,g\,C\,m^{-2}\,y^{-1}$). But if a hypothetical flow of $20\,m^3$ passes over each square meter of shore per day, it will supply $1825\,g\,C\,m^{-2}\,y^{-1}$. This stresses the need for small-scale hydrographic research to predict flows that are meaningful at the level of individual filter-feeders.

Second, some elements of the ecosystem have consistently been overlooked because of their small contribution to biomass. An obvious case is the almost invisible 'skin' of microbiota (sporelings, diatoms, bacteria, and fungi) coating rocks. On most shores, their share of the biomass is an apparently insignificant 0.1%. However, in terms of productivity, they contribute 12%, and most grazers depend on this food source.

Finally, there has always been a tacit assumption that sedentary intertidal grazers must depend on *in situ* algal productivity. However, on the west coast of South Africa, grazers reach extraordinarily high biomasses (up to $1000\,g$ wet flesh m^{-2}). Modeling shows that their needs greatly exceed *in situ* primary production. Instead, they survive by trapping drift and attached kelp. This subtidal material effectively subsidizes the intertidal system. Thus sustained, dense beds of limpets dominate sections of the shore, eliminating virtually all macroalgae and most other grazers. Both 'bottom-up' and 'top-down' processes are at play.

Energy-flow models are seldom absolute measures of how a system operates. Changes in time and space preclude this. Rather, their power lies in identifying bottlenecks, limitations, and overlooked processes, which can then be investigated by complementary approaches.

Integration of Approaches

Ecology as a whole, and that addressing rocky shores in particular, has suffered from polarized viewpoints. Classic examples are arguments over competition versus predation, or the merits of 'top-down' versus 'bottom-up' approaches. In reality, all of these are valid. What is needed is an integration that identifies the circumstances under which one or other model has greatest predictive power. A single example demonstrates the multiplicity of factors operating on intertidal rocky shores (**Figure 3**).

Islands off the west coast of South Africa support dense colonies of seabirds. These have two important effects on the dynamics of rocky shores. First, their guano fuels primary production. Second, oystercatchers that aggregate on the islands prey on limpets. Indirect effects complicate this picture. Reduction of limpets adds to the capacity of seaweeds to escape grazing, leading to luxurious algal mats. These sustain small invertebrates, in turn a source of food for waders. The limpets benefit indirectly from the guano because their growth rates and maximum sizes are increased by the high algal productivity. This allows some to reach a size where they are immune to predation by oystercatchers, and also increases their individual reproductive output. But this 'interaction web' embraces less obvious connections. Dense bird colonies only exist on the islands for two reasons: absence of predators, and food in the form of abundant fish, sustained ultimately by upwelling. Comparison with adjacent mainland rocky shores reveals a contrast: roosting and nesting birds are scarce or absent, oystercatchers are much less numerous, limpets are abundant but small, and algal beds are absent.

This case emphasizes the complexity of driving forces and the difficulty of making predictions about their consequences. Top-down and bottom-up effects, direct and indirect impacts, productivity, grazing, predation, competition, and physical stresses all play their role.

Human Impacts

In one sense intertidal rocky-shore communities are vulnerable to human effects because they are accessible and many of the species have no refuge. In another sense, they are relatively resilient to change. One reason is that humans seldom change the structure and physical factors influencing rocky shores. Tidal excursions and wave action, the two most important determinants of rocky-shore community structure, are seldom altered. The physical rock itself is also rarely modified by human actions. These circumstances are an important contrast with systems such as mangroves, estuaries, and coral reefs, where the structure is determined by the biota. Remove or damage mangrove forests, salt marshes, or corals, and the fundamental nature of these systems is changed and slow to recover. Estuaries are especially vulnerable because their two most important physical attributes – input of riverine water and tidal exchange – can be revolutionized by human actions. Even after massive abuse such as oil pollution, rocky shores recover relatively quickly once the agent of change is brought under control; and there is good evidence that the kinds of communities appearing after recovery resemble those originally present.

Humans impact rocky-shore communities in many ways, including trampling, harvesting, pollution, introduction of alien species and by altering global climate. Harvesting is of specific interest because it has taught much about the functioning of rocky shores.

Almost without exception, harvesting reduces mean size, density, and total reproductive output of the target species, although compensatory increases of growth rate and reproductive capacity of surviving individuals are not unusual. Of greater interest, however, are the consequences for community structure and dynamics. Clear demonstrations of this come from Chile, where intense artisanal harvesting occurs on rocky shores. In particular, a lucrative trade has developed for giant keyhole limpets, *Fissurella* spp., and for a predatory muricid gastropod, *Concholepas concholepas*, colloquially known as 'loco.' Decimated along most of the coast, the natural roles of these species are only evident inside marine protected areas. There, locos consume a small mussel, *Perumytilus purpuratus*, which otherwise outcompetes barnacles. Macroalgae and barnacles compete for space, but only on a seasonal basis. Keyhole limpets control macroalgae and outcompete smaller acmaeid limpets. *Perumytilus* acts as a settlement site for conspecifics and for recruits of keyhole limpets and locos. With a combination of predation, grazing, competition and facilitation, and both direct and indirect effects, the consequences of harvesting locos or keyhole limpets would have been impossible to resolve without the existence of marine protected areas. Even then, careful manipulative experiments inside and outside these areas were required to disentangle these interacting effects.

A second issue of general interest is whether human impacts are qualitatively different from those of other species. The short answer is 'yes', and is best illustrated by a return to an earlier example – interactions between species on rocky shores of islands on the west coast of South Africa (**Figure 3**). In its undisturbed state, each of the key interactions in this ecosystem is buffered in some way. Limpets are consumed by oystercatchers, but some escape by growing too large to be eaten, aided by the high primary production. Limpets and other invertebrates graze on algae, but their effects are muted by predation on them and by the enhancement of algal growth by guano. Waders eat small seaweed-associated invertebrates, but emigrate in winter.

For several reasons, human impacts are not constrained in these subtle ways. First, human populations do not depend on rocky shores in any manner limiting their own numbers. They can harvest these resources to extinction with impunity. Second, modern human effects are too recent a phenomenon for the impacted species to have evolved defenses. Thirdly, humans are supreme generalists. Simultaneously, they can act as predators, competitors, amensal disturbers of the environment, and 'commensal' introducers of alien species. Fourthly, money, not returns of energy, determines profitability. Fifthly, long-range transport means that local needs no longer limit supply and demand. Sixthly, technology denies resources any refuge.

Thus, humans supersede the ecological and evolutionary rules under which natural systems operate; and only human-imposed rules and constraints can replace them in meeting our self-proclaimed goals of sustainable use and maintenance of biodiversity.

See also

Upwelling Ecosystems.

Further Reading

Branch GM and Griffiths CL (1988) The Benguela ecosystem Part V. The coastal zone. *Oceanography and Marine Biology Annual Review*26: 395–486.

Castilla JC (1999) Coastal marine communities: trends and perspectives from human-exclusion experiments. *Trends in Ecology and Evolution* 14: 280–283.

Connell JH (1975) Some mechanisms producing structure in natural communities: a model and evidence from field experiments. In: Cody ML and Diamond JM (eds.) *Ecology and Evolution of Communities*, pp. 460–490. Cambridge, MA: Belknap Press.

Denny MW (1988) *Biology and the Mechanics of the Wave-swept Environment*. Princeton, NJ: Princeton University Press.

Hawkins SJ and Hartnoll RG (1983) Grazing of intertidal algae by marine invertebrates. *Oceanography and Marine Biology Annual Review* 21: 195–282.

Menge BA and Branch GM (2001) Rocky intertidal communities. In: Bertness MD, Gaines SL, and Hay ME (eds.) *Marine Community Ecology*. Sunderland: Sinauer Associates.

Moore PG and Seed R (1985) *The Ecology of Rocky Coasts*. London: Hodder and Stoughton.

Newell RC (1979) *Biology of Intertidal Animals*. Faversham, Kent: Marine Ecological Surveys.

Paine RT (1994) *Marine Rocky Shores and Community Ecology: An Experimentalist's Perspective*. Oldendorf: Ecology Institute.

Siegfried WR (ed.) (1994) *Rocky Shores: Exploitation in Chile and South Africa*. Berlin: Springer-Verlag.

Underwood AJ (1997) *Experiments in Ecology: their Logical Design and Interpretation Using Analysis of Variance*. Cambridge: Cambridge University Press.

LAGOONS

R. S. K. Barnes, University of Cambridge, Cambridge, UK

Introduction

A 'lagoon' is any shallow body of water that is semi-isolated from a larger one by some form of natural linear barrier. In ocean science, it is used to denote two rather different types of environment: 'coastal lagoons' (the main subject of this article) and lagoons impounded by coral reefs. Coastal lagoons are a product of rising sea levels and hence are geologically transient. Whilst they exist, they are highly productive environments that support abundant crustaceans and mollusks, and their fish and bird predators. Harvests from lagoonal aquaculture of up to 2500 kg $ha^{-1}y^{-1}$ of penaeid shrimp, of 400 tonnes $ha^{-1}y^{-1}$ of bivalve molluscs, and in several cases of >200 kg ha^{-1} y^{-1} of fish are taken.

What is a Lagoon?

Coastal lagoons are bodies of water that are partially isolated from an adjacent sea by a sedimentary barrier, but which nevertheless receive an influx of water from that sea. Some 13% of the world's coastline is faced by sedimentary barriers, with only Canada, the western coast of South America, the China Sea coast from Korea to south-east Asia, and the Scandinavian peninsula lacking them and therefore also being without significant lagoons (**Table 1**, **Figure 1**). The lagoons behind such barrier coastlines range in size from small ponds of <1 ha through to large bays exceeding 10 000 km^2. The median size has been suggested to be about 8000 ha. Although several do bear the word 'lagoon' in their name, many do not. Usage of the same titles as for fresh water habitats is widespread (e.g. *Étang* de Vaccarès, France; Swan-*Pool*, UK; Oyster *Pond*, USA; *Lake* Menzalah, Egypt; Benacre *Broad*, UK; *Ozero* Sasyk, Ukraine; Kiziltashskiy *Liman*, Russia, etc.), as is those for coastal marine regions (Peel-Harvey *Estuary*, Australia; Great South *Bay*, USA; Ringkbing *Fjord*, Denmark; *Zaliv* Chayvo, Russia; Pamlico *Sound*, USA; Charlotte *Harbor*, USA; and even Gniloye-*More*, Ukraine; *Mer* des Bibans, Tunisia; *Mar* Menor, Spain, etc.), whilst lagoons liable to hypersalinity are often termed Sebkhas in the Arabic-speaking world (e.g. Sebkha el Melah, Tunisia). Conversely, a few systems with 'lagoon' in their name fall out with the definition; the Knysna Lagoon in South Africa, for example, is an estuarine mouth dilated behind rocky headlands between which only a narrow channel occurs.

Coastal lagoons are most characteristic of regions with a tidal range of <2 m, since large tidal ranges (those >4 m) generate powerful water movements usually capable of breaching if not destroying incipient sedimentary barriers. Furthermore, the usual meaning of 'lagoon' requires the permanent presence of at least some water and large tidal ranges are likely to result in ebb of water from open systems during periods of low tide in the adjacent sea. Thus in Europe, for example, lagoons are abundant only around the shores of the microtidal Baltic, Mediterranean, and Black Seas. However, they are also present along some macrotidal coasts – such as the Atlantic north of about 47°N (in the east) and 40°N (in the west) (**Figure 2**) – where off shore deposits of pebbles or cobbles ('shingle') are to be found as a result of past glacial action. Here, shingle can replace the more characteristic sand of microtidal seas as the barrier material, as indeed it partially does in the lagoon-rich microtidal East Siberian, Chukchi, and Beaufort Seas in the Arctic, because it is less easily redistributed by tidal water movements. Nevertheless, sandy sedimentary barriers have developed (and persisted) in a few relatively macrotidal areas (e.g. in southern Iceland and Portugal), although the regions impounded to landwards retain water only during high tide for the reasons outlined above. Those environments that are true lagoons only during high tide are often referred to as 'tidal-flat lagoons'. They are therefore the relatively rare macrotidal coast equivalent of the typical lagoons of microtidal seas.

In many cases, the salinity of a lagoonal water mass is exactly the same as that of the adjacent sea, although where fresh water discharges into a lagoon its water may be brackish and a (usually relatively stable) salinity gradient can occur between river-mouth and lagoonal entrance channel. In regions where evaporation exceeds precipitation for all or part of a year, lagoons are often hypersaline.

The second environment to which the word lagoon is applied is associated with coral reefs; coral here replaces the unconsolidated sedimentary barrier of the coastal lagoon. Circular atoll reefs enclose the 'lagoon' within their perimeter, whilst barrier reefs are separated from the mainland by an equivalent although less isolated body of water. Indeed barrier reef lagoons are virtually sheltered stretches of coastal sea. Atoll lagoons, however, are distinctive in being floored by coral sand which supports submerged beds of seagrasses and fringing mangrove swamps justas do the coastal lagoons of similar latitudes. The similarity between the two types of lagoon is nevertheless purely physiographic since the atoll lagoon fauna is that typical of coral reefs in general and not related in any way to those of coastal lagoons. Coral lagoons are covered in greater detail in the article Coral Reefs.

Table 1 The contribution of different continents to the world total barrier/lagoonal coastline of 3200 km[a]

Continent	*Percent of coastline barrier/lagoonal*	*Percent of world's lagoonal resource*
N. America	17.6	33.6
Asia	13.8	22.2
Africa	17.9	18.7
S. America	12.2	10.3
Europe	5.3	8.4
Australia	11.4	6.8

[a](Reproduced with permission from estimates by Cromwell JE (1971) *Barrier Coast Distribution: A World-wide Survey*, p. 50. Abstracts Volume, 2nd National Coastal Shallow Water Research Conference, Baton Rouge, Louisiana.)

The Formation of Lagoons

Lagoons have existence only by virtue of the barriers that enclose them. They are therefore characteristic only of periods of rising sea level, when wave action can move sediment on and along shore, and – for a limited period of time – of constant sea level. At times of marine regression, they either drain or their basins may fill with fresh water. Many of the enclosing barriers have today been greatly augmented by wind-blown sand; for example, most of the larger systems along the South African coast (Wilderness, St Lucia, Kosi, etc.), and such enclosed lagoons have a particularly lake-like appearance (**Figure 3**).

The precise physiographic nature of a lagoon then depends on the relationship of the barrier to the adjacent coastline. Starting at a point at which the barriers are some distance offshore, we can erect a sequence of situations in which the barriers are moved ever shorewards. This starting point can be exemplified by Pamlico and Albemarle Sounds in North Carolina, USA (**Figure 4**). There a chain of long narrow barrier islands located some 30 km off the coast enclosesan area of shallow bay of around 5000 km^2.

Further landwards movement of the barriers, often to such an extent that some of the larger barriers become attached to the mainland at one end to produce spits, leads to the situation currently characterizing most lagoonal coastlines and to the typical

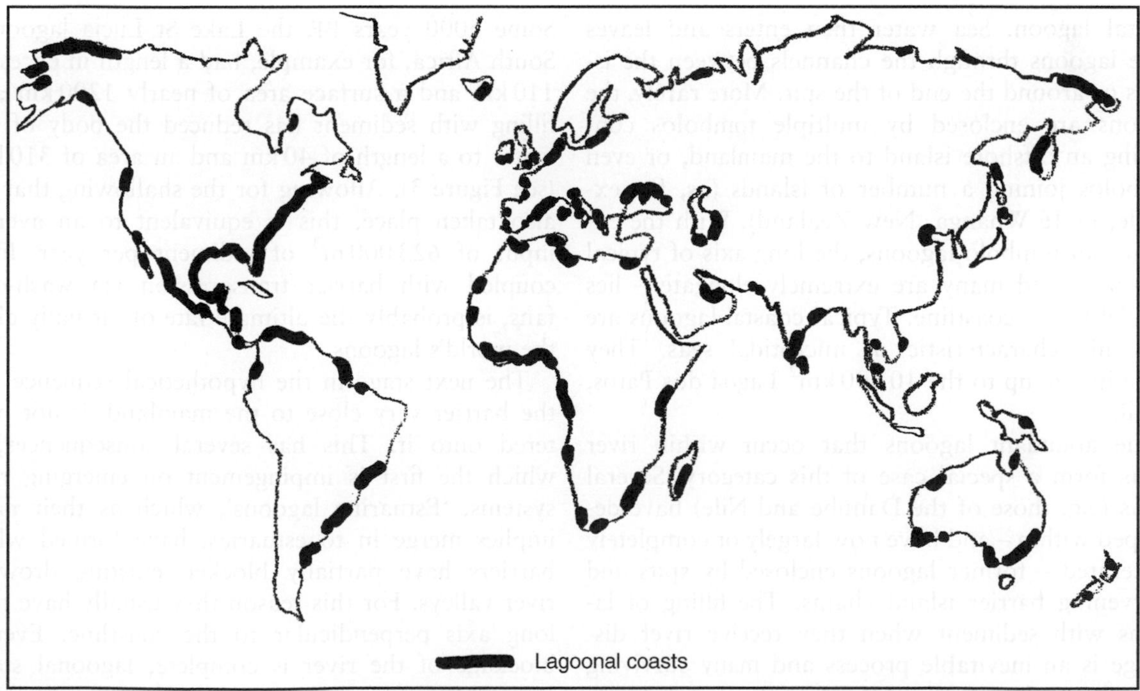

Figure 1 Major barrier/lagoonal coastlines.

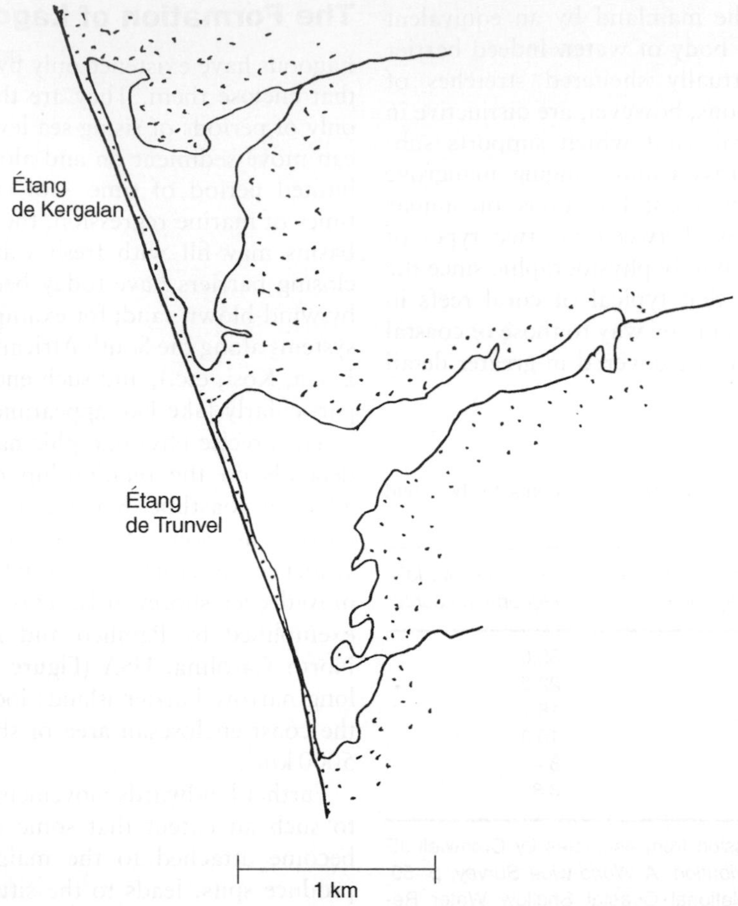

Figure 2 Shallow lagoons along the coast of the Baie d'Audierne (Brittany, France) formed behind a shingle barrier beach. Sea water enters these systems only via overtopping of the barrier; lagoonal water leaves by percolation through the barrier.

coastal lagoon. Sea water then enters and leaves these lagoons through the channels between the islands or around the end of the spit. More rarely, the lagoons are enclosed by multiple tombolos connecting an offshore island to the mainland, or even tombolos joining a number of islands (as, for example, in Te Whanga, New Zealand). With the exception of tombolo lagoons, the long axis of typical lagoons – and many are extremely elongate – lies parallel to the coastline. Typical coastal lagoons are especially characteristic of microtidal seas. They range in size up to the 10 000 km^2 Lagoa dos Patos, Brazil.

The abundant lagoons that occur within river deltas form a special case of this category. Several deltas (e.g. those of the Danube and Nile) have developed within – and have now largely or completely obliterated – former lagoons enclosed by spits and intervening barrier island chains. The filling of lagoons with sediment when they receive river discharge is an inevitable process and many surviving examples are but small remnants of their past extents (**Figure 5**) – islands of water in a sea of marshland. Some 5000 years BP, the Lake St Lucia lagoon in South Africa, for example, had a length in excess of 110 km and a surface area of nearly 1200 km^2; infilling with sediment has reduced the body of free water to a length of 40 km and an area of 310 km^2 (see **Figure 3**). Allowing for the shallowing that has also taken place, this is equivalent to an average input of 623 000 m^3 of sediment per year. Infill, coupled with barrier transgression via wash-over fans, is probably the ultimate fate of virtually all of the world's lagoons.

The next stage in the hypothetical sequence sees the barrier very close to the mainland, if not plastered onto it. This has several consequences, of which the first is impingement on emerging river systems. 'Estuarine lagoons', which as their name implies merge in to estuaries, have formed where barriers have partially blocked existing drowned river valleys. For this reason they usually have their long axis perpendicular to the coastline. Even if blockage of the river is complete, lagoonal status may remain if sea water can enter by overtopping of the barrier during high water of spring tides.

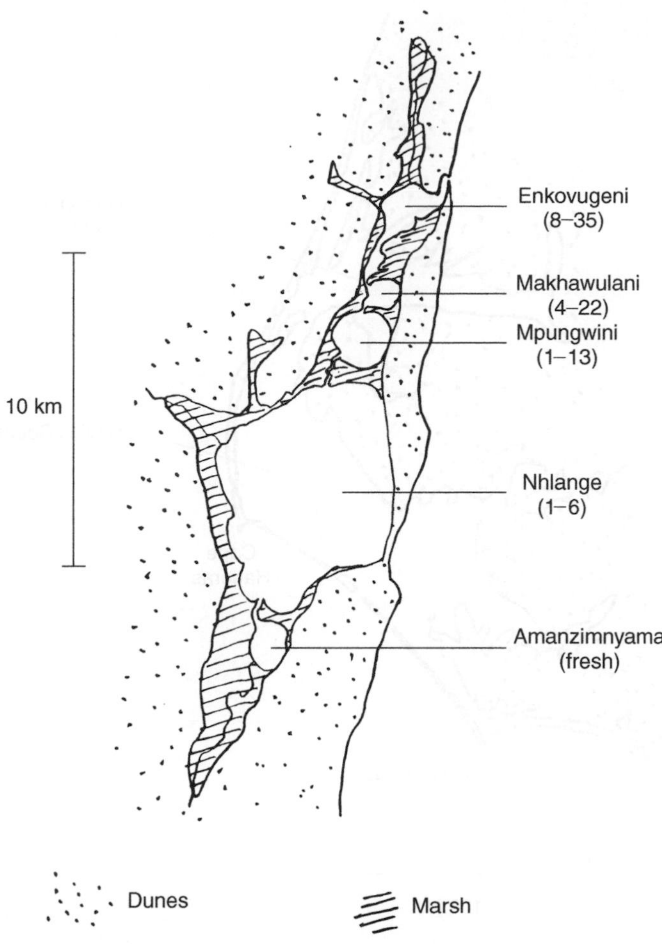

Figure 3 The Kosi Lakes (KwaZulu/Natal, South Africa), with approximate salinity ranges (in practical salinity units) in parentheses. They are enclosed within Pleistocene sand dune fields, and form a classic case of 'segmentation' of an original linear lagoon into aseries of rounded basins. (Salinity data reproduced with permission from Begg G (1978) *The Estuaries of Natal.* Natal Town and RegionalPlanning Report.).

Nevertheless, many former estuarine lagoons are now completely fresh water habitats with seawater entry being prevented by the barrier. In many tropical and subtropical regions, closure of the estuarine lagoons is seasonal. During the wet season, river flow is sufficient to maintain breaches through the barriers. In the dry season, however, flow is reduced and drift can seal the barrier. If fresh water still flows, the whole isolated basin then becomes fresh for several months. The mouths of many natural lagoonal barriers are periodically breached by man for a variety of reasons ranging from ensuring the entry of juveniles of commercially important mollusks, crustaceans, and fish to temporarily lowering water levels to avoid damage to property.

Longshore movement of barrier sediment may also divert the mouths of discharging estuaries several kilometers along the coast, as for example the 30 km deflection of that of the Senegal River by the Langue de Barbarie in West Africa, and the creation of the Indian River Lagoon in Florida, USA. Not infrequently a new mouth is broken through the barrier, naturally or more usually as a result of human intervention, and the diverted stretch can become a dead-end backwater lagoon fed by backflow.

Other consequences of on shore barrier migration also involve human intervention. Many small scale lagoons have (unwittingly) been created by land reclamation schemes. Regions to landwards of longshore ridges that were once, for example, low-lying salt marsh but which are now reclaimed usually receive an influx of water from out of the barrier water-table and this collects in depressions that were once part of the creeks draining the marshes. Equivalently, gravel pits or borrow pits in coastal shingle masses, from which building materials have been extracted, similarly receive an influx of water from out of the shingle. In both cases, the shingle water-table is derived from sea water soaking into the barrier during high tide in the adjacent sea

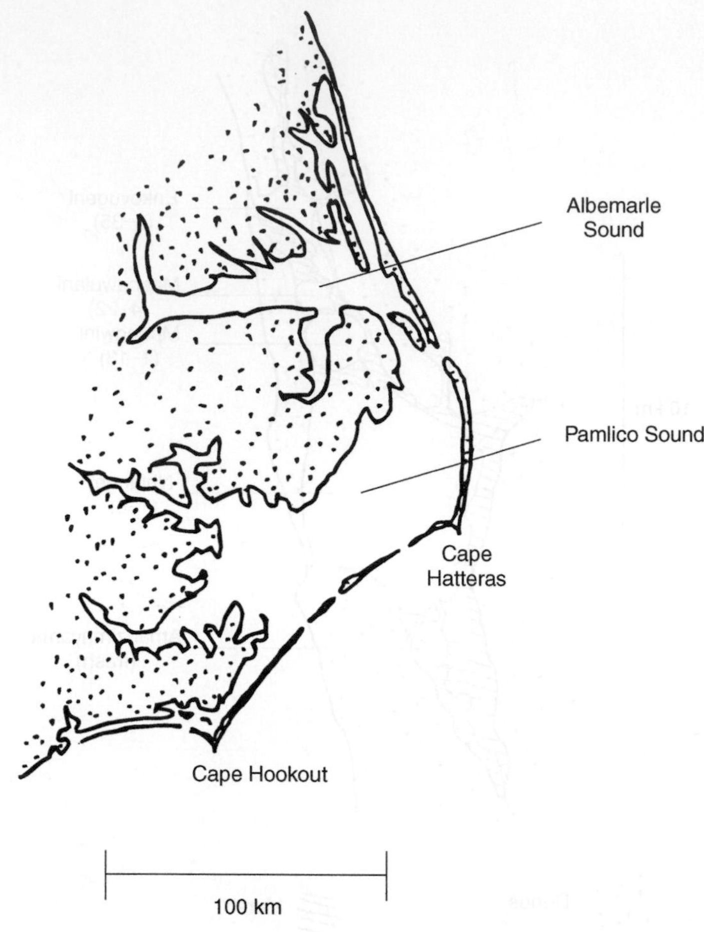

Figure 4 Pamlico and Albemarle Sounds (North Carolina, USA).

together with such rainfall as has also soaked in. Some of the natural limans of the Black Seacoast also receive their salt input via equivalent seepage.

Finally in this category, barriers may come onshore in such a fashion as to straddle the mouth of a pre-existing bay. 'Bahira lagoons' (from the Arabic for 'littlesea') are pre-existing partially land-locked coastal embayments, drowned by the post glacial rise in sea level, that have later had their mouths almost completely blocked by the development of sedimentary barriers. Also included here are systems in which the sea has broken through a pre-existing sedimentary barrier to flood part of the hinterland, but inwhich the entrance/exit channel remains narrow. Bahira lagoons clearly merge into semi-isolated marine bays.

Although lagoons may come and go during geological time, some of the larger lagoon systems are not solely features of the present interglacial period. Some, including the Lagoa dos Patos, the Gippsland Lakes of Victoria, Australia, and the Lake St Lucia lagoon, may incorporate elements of barriers and basins formed during previous marine transgressions. The history of the Gippsland lakes over the last 70 000 years has been reconstructed in some detail (**Figure 6**). Fossil assemblages of foraminiferans suggest that lagoons may have been a feature of the Gulf Coast of the USA and Mexico at intervals ever since the Jurassic.

Lagoonal Environments

It can be argued that the main force structuring lagoons is the extent to which they are connected to the adjacent sea, and they have been divided into three or four types on this basis(**Figure** 7) which largely reflect points along the hypothetical evolutionary sequence described above.

'Leaky lagoons' function virtually as sheltered marine bays (see also **Figure 4**). As a result of large tidal ranges in the adjacent sea or the existence of many and/or wide connecting channels, interactions with the ocean are dominant and the lagoons have tidally fluctuating water levels, short flushing times and a salinity equal to that ofthe local sea water.

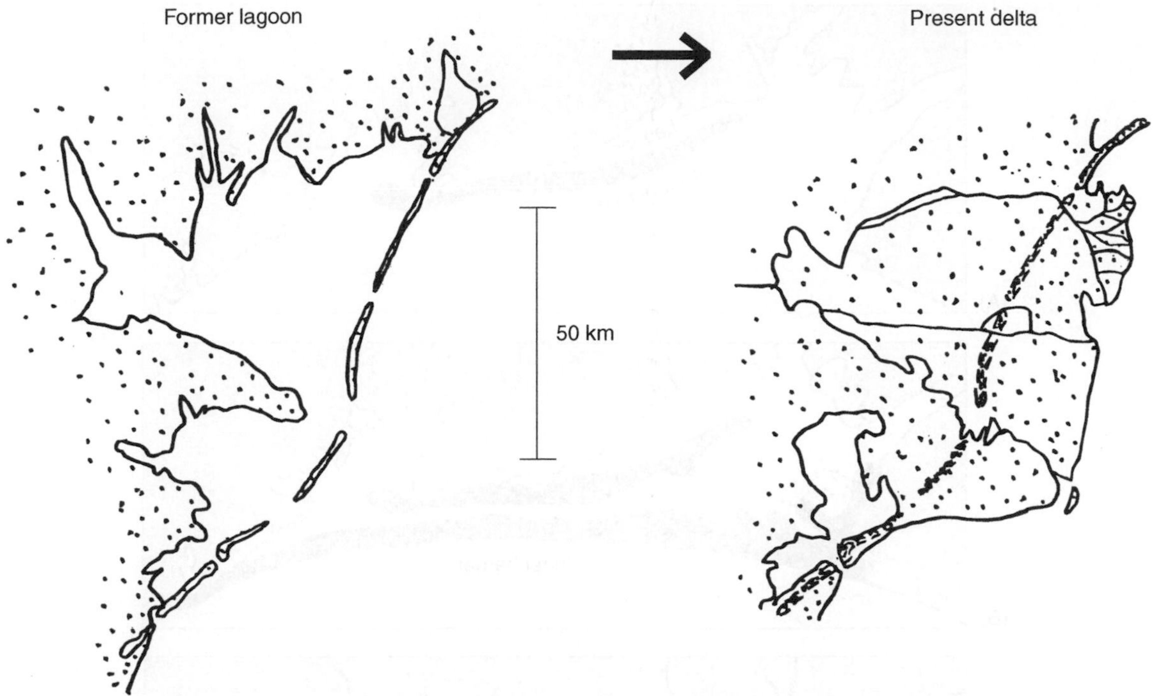

Figure 5 Infill of a former lagoon (established around 6000 years ago) on the Romanian/Ukrainian coast of the Black Sea with sediment carried by the River Danube. The current Danube delta occupies virtually all the former lagoon.

'Restricted lagoons' have asmall number of narrow entrance channels through elements of a barrier is land chain or enclosing spits and are therefore more isolated from the ocean(see also **Figure 7**). Characteristics include: longer flushing times, a vertically well-mixed water column, both wind and tidal water movements as forcing agents, and brackish to oceanic salinity.

'Choked lagoons' are connected to the ocean by a single, often long and narrow, entrance channel that serves as a filter largely eliminating tidal currents and fluctuations in water level(see also **Figure 4**). The 1000 km^2 Chilka Lake lagoon, for example, communicates with the Bay of Bengal via a channel 8 km long and 130 m across at its widest. Choked lagoons are typical of coasts with high wave energy and significant longshore drift. Characteristics include: very long flushing times, intermittent stratification of the water column by thermoclines and/or haloclines, wind action as thed ominant forcing agent, purely freshwater zones near inflowing rivers, and otherwise brackish, and in some climatic zones periodic hypersaline, waters. Wet season rainfall in the 44 km^2 Laguna Unare, Venezuela (at the end of the dry season), for example, can change the salinity from 60–92 to 18–25, at the same time increasing lagoonal depth by 1 m, area by 20 km^2 and volume by 76×10^6m^3.

The 'closed lagoons' shown in **Figure 2** form the limiting condition of effective isolation from direct inputs from the ocean except via occasional overtopping or after percolation through the barrier system. Most are not only 'former lagoons', but also current coastal freshwater lakes; for example, the much studied Slapton Ley in England.

Characteristically restricted and choked lagoons are very shallow (usually about 1 m and almost always <5 m deep), floored by soft sediments, fringed by reedbeds(*Phragmites* and/or *Scirpus*), mangroves (especially *Rhizophora*) or salt marsh vegetation, and support dense beds of submerged macrophytes such as seagrasses (e.g. *Zostera*), pondweeds (*Potamogeton*) or *Ruppia*, together with green algae such as *Chaetomorpha*. The action of the dense beds of submerged vegetation may be to raise levels of pH and to contribute considerable quantities of organic matter. In stratified choked lagoons, decomposition of this organic load below the thermocline or halocline can then lead to anoxic conditions, not withstanding surface waters super saturated with oxygen (**Figure 8**).

Lagoonal Biotas and Ecology

Insofar as is known, although the species may be different, the ecology of coastal lagoons shows the

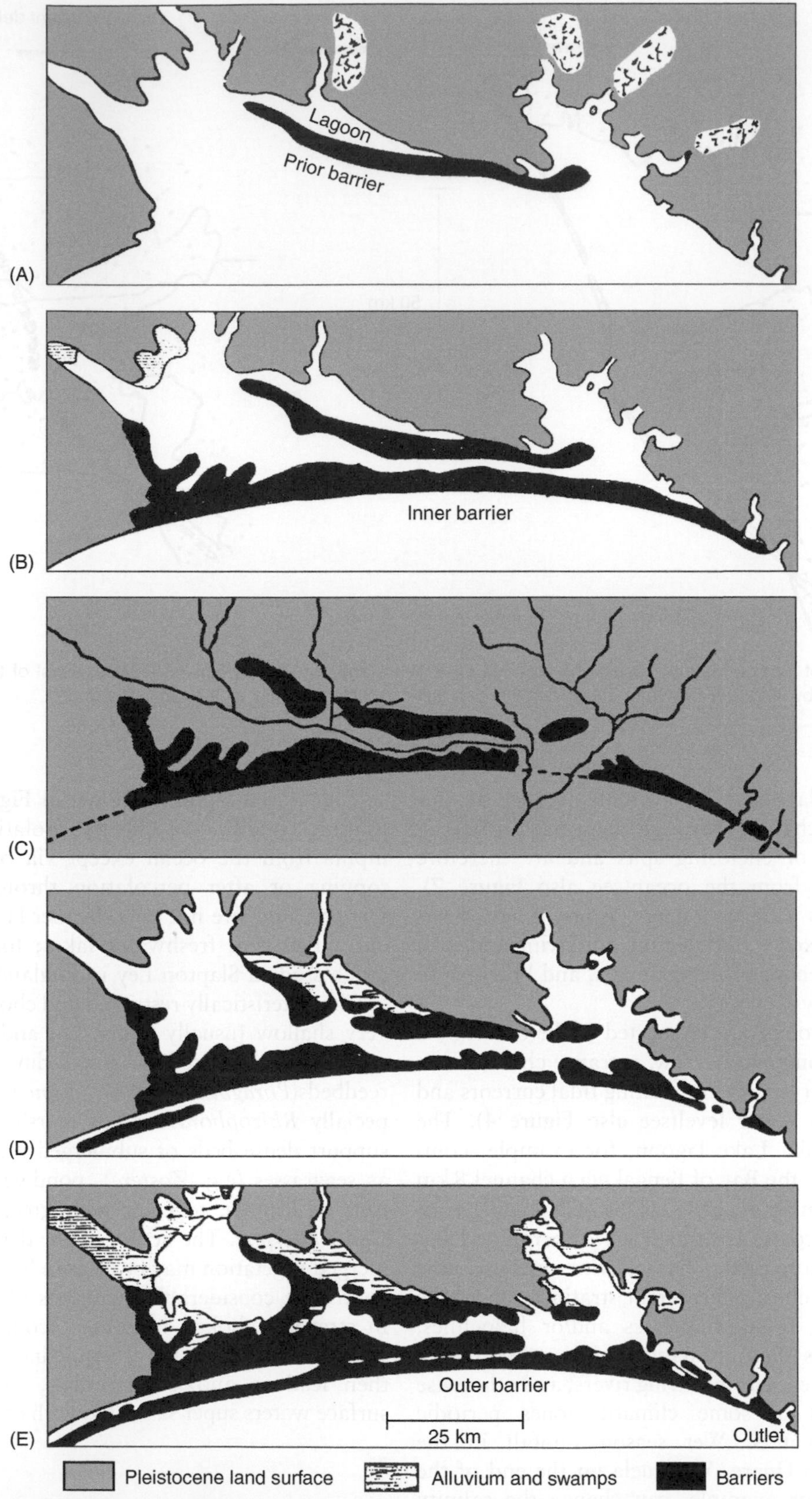

Figure 6 Evolution of the Gippsland Lagoon system (Victoria, Australia). (Reproduced with permission from Barnes 1980; after Bird ECF (1966) The evolution of sandy barrier formations on the East Gippsland coast. *Proceedings of the Royal Society of Victoria* 79:75–88.) The prior barrier (A) can be dated to about 70 000 years BP; the inner barrier (B) was formed during the late Pleistocene; the situation in (C) represents a low sea level phase of the late glacial, and (D) and (E) further evolution during the current marine transgression.

same general pattern as seen in estuaries and other regions of coastal soft sediment, although the relative importance of submerged macrophytes may be greater in lagoons (**Figure 9**). The action of predators as important forces structuring the communities in lagoons as in similar areas, for example, is reflected by the young stages of many species occurring within the dense beds of submerged macrophytes, as the hunting success of predatory species is lower there.

The primary productivities of lagoons appear to vary with their general nature, their latitude, and their depth. Choked lagoons tend to be the most productive, with recorded maxima approaching $2000\,\mathrm{g\,C\,m^{-2}\,y^{-1}}$, and with productivities (allowing for the effect of latitude) some 50% more than inrestricted lagoons. Temperate zone lagoons achieve only about 50–70% of the productivity of low latitude lagoons of comparable nature; and whilst shallow phytoplankton-dominated lagoons are more productive than deeper ones (depth $>2\,\mathrm{m}$), the converse may be true of macrophyte-dominated systems. An order of magnitude calculation (assuming that the total area of the world's lagoons is around $320\,000\,\mathrm{km}^2$ and that average lagoonal productivity is $300\,\mathrm{g\,Cm^{-2}\,y^{-1}}$) yields a total annual lagoonal production of 10^{11} kg fixed C. The contribution of lagoons to total oceanic carbon fixation is therefore minor, although they are as productive per unit area as are estuaries, and are only less productive than some regions of upwelling, coral reefs, and kelp forests.

What happens to this primary production? Data are still scarce. There has been much argument as to whether estuaries are sources or sinks for fixed

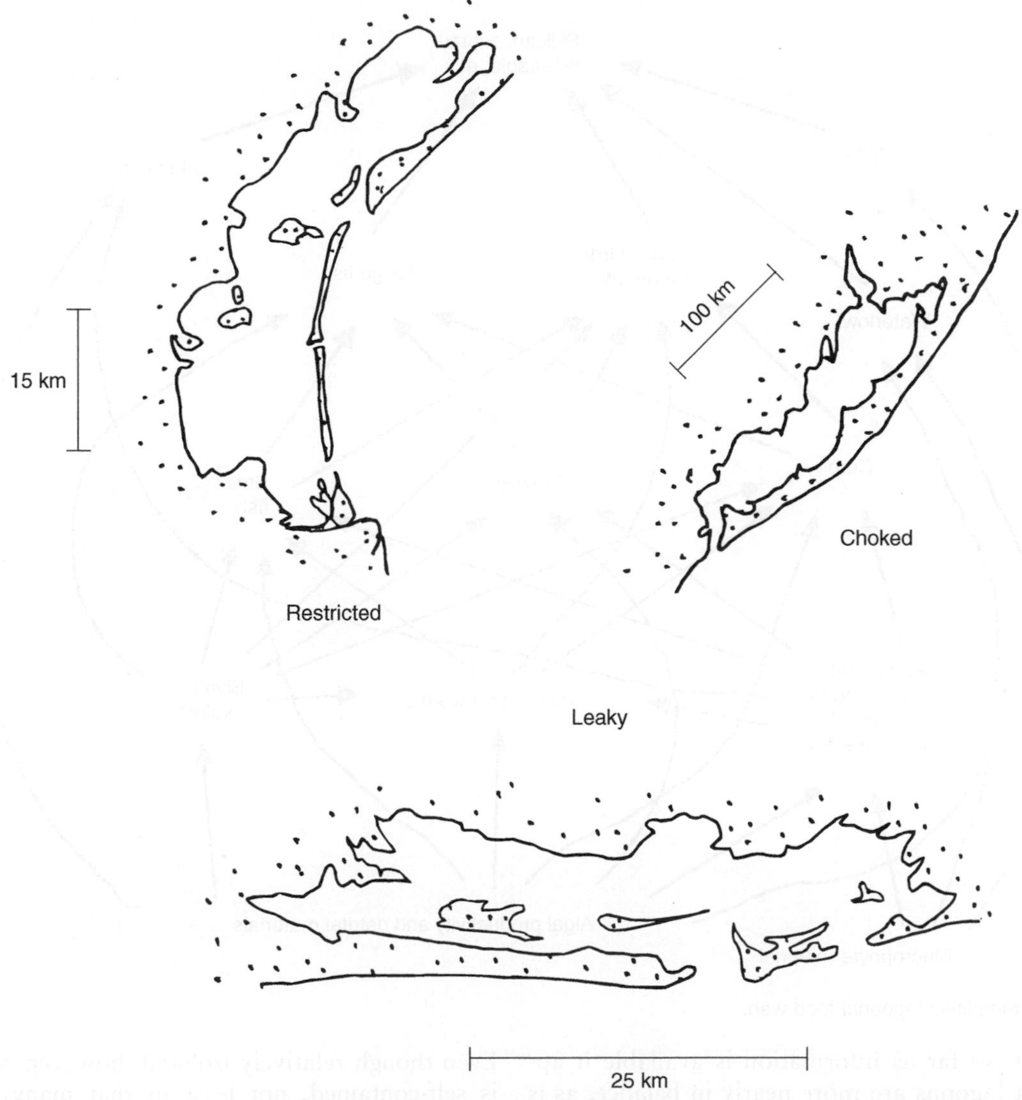

Figure 7 'Choked' (Lagoa dos Patos, Brazil), 'restricted' (Laguna di Venezia, Italy) and 'leaky' (Laguna Jiquilisco, El Salvador) lagoons.

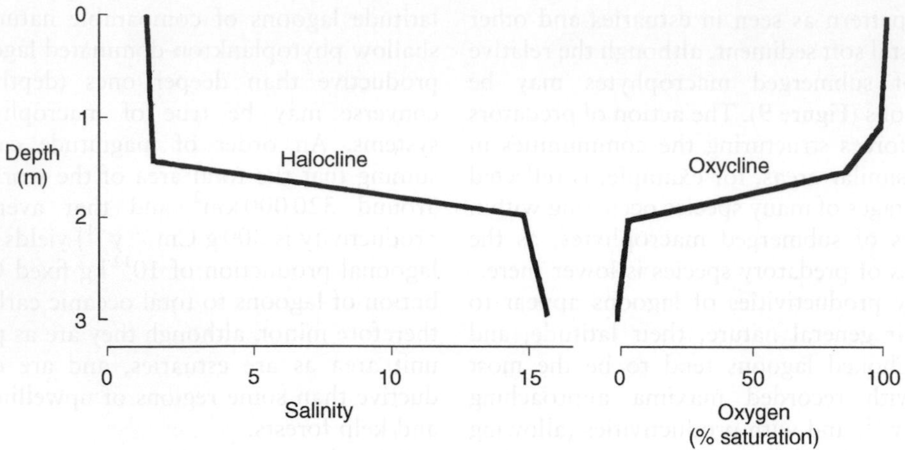

Figure 8 A halocline and consequent oxycline in the choked Swanpool Lagoon, England, after a prolonged period of high freshwater input (salinity inpractical salinity units). (Reproduced with permission from data in Dorey AE *et al.* (1973) An ecological study of the Swanpool, Falmouth II. Hydrography and its relation to animal distributions. *Estuarine Coastla and Marine Science* 1: 153–176.)

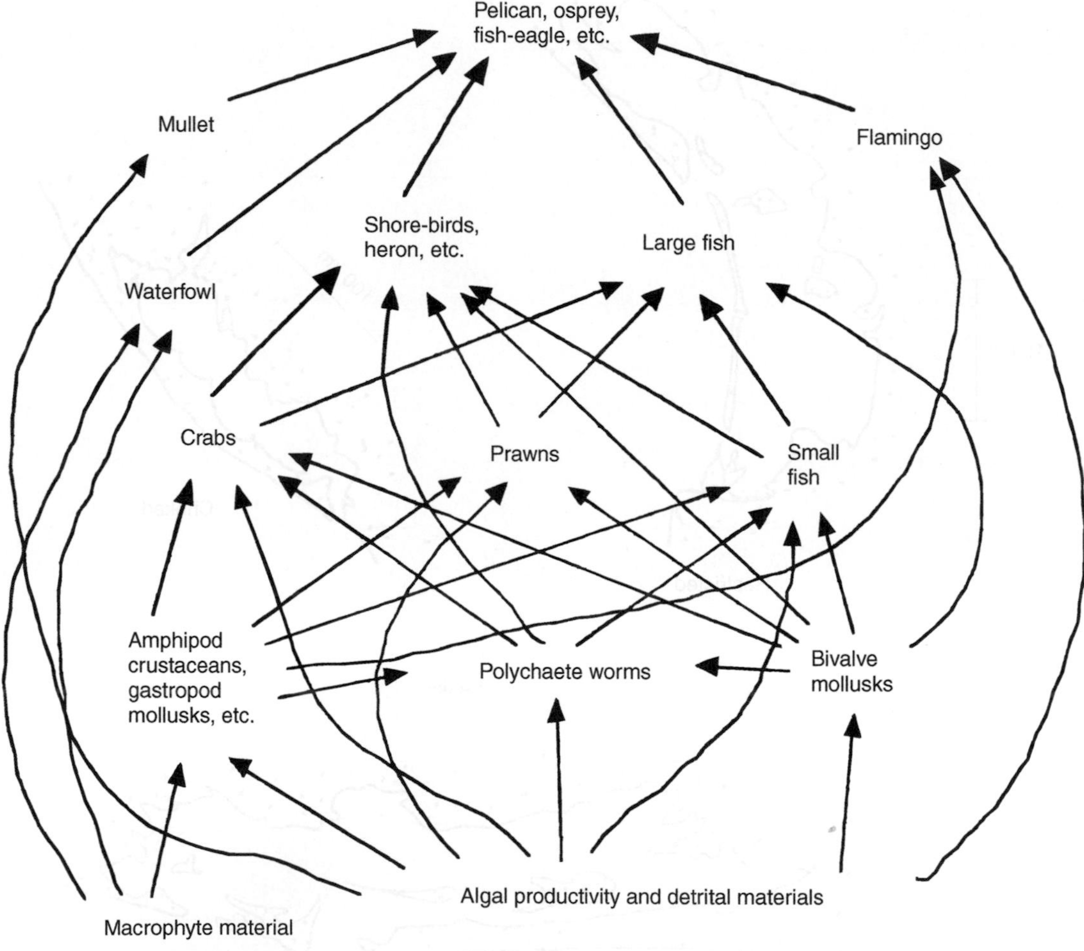

Figure 9 Simplified lagoonal food web.

carbon. In so far as information is available it appears that lagoons are more nearly in balance, as is appropriate for their more isolated nature, although all those so far studied do function as slight sinks. Even though relatively isolated, however, no lagoon is self-contained, not least in that many elements in the faunas migrate between lagoons and other habitats.

Coastal lagoons are heavily used by wetland and shore birds as feeding areas, most noticeably by grebes, pelicans, ibis, egrets, herons, spoonbills, avocets, stilts, storks, flamingoes, cormorants, kingfishers, and fish eagles. Such birds prey on the abundant lagoonal invertebrate and fish faunas. These comprise mixtures of three different elements: both (a) essentially freshwater and (b) essentially marine and/or estuarine species capable of with standing adegree of brackishness, as well as (c) specialist lagoonal or 'paralic' species that are not in fact restricted to lagoons but also (or closely related species) occur in habitats like the Eurasian inland seas (e.g. the Caspian and Aral), as well as in (the usually man-made) tideless brackish ponds and drainage ditches that are abundant in reclaimed coastal regions. The main feature of the lagoonal species category is that they are species of marine ancestry that seem to be restricted to shallow, relatively tideless maritime habitats; over their ranges as a whole they are also clearly capable of inhabiting a wide range of salinity, including full-strength and concentrated sea water, but not fresh water. Their relatedness to marine species is evidenced by the fact that they often form species-pairs with marine or estuarine species. The freshwater component of lagoonal faunas is often greater than would be expected in comparable salinities in estuaries, with the common occurrence of adult water beetles and hemipteran bugs besides the omnipresent dipteran larvae. The reasons for this are not understood, but may relate to decreased water movements.

Because oceanic and freshwater inputs into alagoon occur at different points around its perimeter, these three elements of the fauna tend to be broadly zoned in relation to these inputs. Indeed, an influential French school, the 'Group d'Etude du Domaine Paralique', identifies six characteristic biotic zones in lagoons based not upon salinity but upon 'a complex and abstract value which cannot be measured in the present state of knowledge' which the group terms 'confinement'. This is a function, at least, of the extent to which the water mass at any given point is isolated from oceanic influence. Beginning with the Mediterranean Sea coast, disciples of this school have now published maps of the location of the 'six degrees of confinement' in many lagoons throughout the world.

Lagoons also form the nursery grounds and adult feeding areas for a large number of commercially important fish and crustaceans that migrate between this habitat and the sea, most of which spawn outside the lagoons and only later move in actively or in some cases probably passively. In respect of the fish, temperate regions are amongst others used byeels (Anguillidae), sea bass (Moronidae), drum (Sciaenidae), sea bream (Sparidae), grey mullet (Mugilidae), flounder (Pleuronectidae), and various clupeids, and in warmer waters these are joined by many others, including cichlids, milkfish (Chanidae), silver gars (Belonidae), grouper (Serranidae), puffer fish (Tetraodontidae), grunts(Pomadasyidae), rabbit fish (Siganidae), various flatfish (Soleidae, Cynoglossidae), and rays (Dasyatidae).

Therefore, throughout the world lagoons support (often artisanal) fisheries, as well as being the location of bivalve shellfish culture (mussels, oysters, and clams). For obvious reasons, most data on secondary productivity have been collected in relation to these fisheries. The fish yield from lagoons is greater than from all other similar aquatic systems, averaging (n = 107) 113 kg ha^{-1} y^{-1} and with a median value of 51 kg (**Table 2**) and with 13% exceeding 200 kg ha^{-1} y^{-1}. Yields vary with the intensity of human involvement in the catching and stocking processes. In the Mediterranean Sea, the region for which most data are available, the yield without intensive intervention is some 82 kg $ha^{-1}y^{-1}$; with the installation of permanent fish traps this increases to 185 kg; and with the addition of artificial stocking with juvenile fish it increases to 377 kg. For a 10 year period, the fishermen's cooperative based on the Stagno di Santa Giusta in Sardinia harvested nearly 700 kg ha^{-1} y^{-1}. West African lagoons with 'fish parks' – areas that attract fish because of the provision of artificial refuges – are considerably more productive, with average fish yields of 775 kg ha^{-1} y^{-1}.

Amongst the invertebrates, penaeid shrimp, mudcrabs (*Scylla*), oysters, mussels, and the arcid 'cockle' *Anadara* are the major harvested organisms. Yields of *Penaeus* may attain 2500 kg ha^{-1} y^{-1}in lagoonal systems of aquaculture, whilst those of the oyster *Crassostrea* and the mussel *Perna* can both be 400 tonnesha^{-1} y^{-1}. In the Indian subcontinent wholecommunities can be economically entirely dependent

Table 2 Mean fisheries yield from coastal lagoons in relation to those from other aquatic habitats[a]

Habitat	*Fish yield (kg $ha^{-1}y^{-1}$)*	*Percent of sites exceeding yield of 200 kg $ha^{-1}y^{-1}$*
Coastal lagoons	113	13
Continental shelf	59	5
Coral reef	49	0
Fresh waters	34	2

[a](Data reproduced with permission from Chauvet, 1988.)

Table 3 Discharges in the early 1990s into two lagoons on the southern (Polish/Russian/Lithuanian) coast of the Baltic Sea(inputs in tonnes yr^{-1})[a]

Parameter	*Curonian Lagoon*	*Vistula Lagoon*
Surface area	1584 km^2	838 km^2
Mean depth	3.8 m	2.6 m
Maximum depth	5.8 m	5.2 m
Volume	6.0 km^3	2.3 km^3
Annual water exchange	27.3 km^3	?
BOD input	150 000	47 000
Inorganic N	20 000	25 000
Inorganic P	5000	2100
Chlorinated substances	1	'High concentrations'
Cu	30	?
Zn	50	?
Pb	600	?

[a](Reproduced with permission from data in *Coastal Lagoons and Wetlands in the Baltic.* WWF Baltic Bulletin 1994, no 1.)

on these shrimp, crab, and fish harvests; for example, the 60 000 people living around Chilka Lake, India.

However, many of the world's lagoons are threatened habitats, suffering deterioration as a result of pollution and destruction, through reclamation and from the natural losses resulting from succession to freshwater and swamp habitats and from landwards barrier migration. The discharge of materials into lagoons and its consequences are essentially similar to those in any other semi-enclosed coastal embayment, but there is one specifically lagoonal feature. As they are often used for aquaculture and yields are generally related to primary productivity, lagoonal waters are frequently deliberately enriched to boost catches. Such enrichment varies from domestic organic wastes from the surrounding communities to commercial processed fish foods. Probably in the majority of such cases, however, the result of this nutrient injection has been eutrophication, loss of macrophytes, deoxygenation, and in several areas a change in the primary producers in the direction of a bacteria-dominated plankton and, across wide areas, benthos as well. In Mediterranean France, this all too frequent state of affairs is known as 'malaïgue'. Culture of mussels in the Thau Lagoon in France produces an input to the benthos of some 45 000 tonnes (dry weight) of pseudofecal material. Not surprisingly, at times of minimum throughput of water, malaïgues can cause mass mortality of the cultured animals and degradation of the whole habitat. Thus in Europe, malaïgues in the south, pollution in the Baltic lagoons(**Table 3**), and reclamation ofthose on the Atlantic seaboard have rendered the habitat especially threatened even at a continental level. For this reason they are now a 'priority habitat' under the European Union's Habitats Directive. Intensiv elagoonal aquaculture also injects not only nutrients, but antibiotics, hormones, vitamins, and a variety of other compounds, and the wider effects of these are giving cause for concern.

See also

Crustacean Fisheries. Mangroves. Molluskan Fisheries. Salt Marshes and Mud Flats.

Further Reading

Ayala Castañares and Phleger F.B. (eds.) (1969) *Lagunas Costeras. Un Simposio.* Universidad Nacional Autónoma de México.

Barnes RSK (1980) *Coastal Lagoons.* Cambridge: Cambridge University Press.

Bird ECF (1984) *Coasts,* 3rd edn. London: Blackwell.

Chauvet C (1988) *Manuel sur l'Aménagement des Pêches dans les Lagunes Côtières: La Bordigue Méditerranéenne.* FAO Document Technique sur les Pches No. 290. FAO:

Cooper JAG (1994) Lagoons and microtidal coasts. In: Carter RWG and Woodroffe CD (eds.) *Coastal Evolution,* pp. 219–265. Cambridge: Cambridge University Press.

Emery KO (1969) *A Coastal Pond Studied by Oceanographic Methods.* Elsevier.

Guélorget O, Frisoni GF, and Perthuisot J-P (1983) La zonation biologique des milieux lagunaires: définition d'une échelle de confinement dans le domaine paralique mediterranéen. *Journal de Recherche Océanographique, Paris* 8: 15–36.

Kapetsky JM and Lasserre G (eds.) (1984) *Management of Coastal Lagoon Fisheries*, vol. 2 vols.. Rome: FAO.

Kjerfve B (ed.) (1994) *Coastal Lagoon Processes*. Elsevier.

Lasserre P and Postma H (eds) (1982) Les Lagunes Ctières. *Oceanologica Acta* (Volume Spécial.)

Rosecchi E and Charpentier B (1995) *Aquaculture in Lagoon and Marine Environments*. Tour du Valat.

Sorensen J, Gable F, and Bandarin F (1993) *The Management of Coastal Lagoons and Enclosed Bays*. American Society of Civil Engineers.

Special issue (1992) *Vie et Milieu* 42(2): 59–251.

Yáñez-Arancibia A (ed.) (1985) *Fish Community Ecology in Estuaries and Coastal Lagoons*. Universidad Nacional Autónoma de México.

SALT MARSH VEGETATION

C. T. Roman, University of Rhode Island,
Narragansett, RI, USA

Introduction

Coastal salt marshes are intertidal features that occur as narrow fringes bordering the upland or as extensive meadows, often several kilometers wide. They occur throughout the world's middle and high latitudes, and in tropical/subtropical areas they are mostly, but not entirely, replaced by mangrove ecosystems. Salt marshes develop along the shallow, protected shores of estuaries, lagoons, and behind barrier spits. Here, low energy intertidal mud and sand flats are colonized by halophytes, plants that are tolerant of saline conditions. The initial colonizers serve to enhance sediment accumulation and over time the marsh expands vertically and spreads horizontally, encroaching the upland or growing seaward. As salt marshes mature they become geomorphically and floristically more complex with establishment of creeks, pools, and distinct patterns or zones of vegetation.

Several interacting factors influence salt marsh vegetation patterns, including frequency and duration of tidal flooding, salinity, substrate, surface elevation, oxygen and nutrient availability, disturbance by wrack deposition, and competition among plant species. Moreover, the ability of individual flowering plant species to adapt to an environment with saline and waterlogged soils plays an important role in defining salt marsh vegetation patterns. Morphological and physiological adaptations that halophytes may possess to manage salt stress include a succulent growth form, salt-excreting glands, mechanisms to reduce water loss, such as few stomates and low surface area, and a C4 photosynthetic pathway to promote high water use efficiency. To deal with anaerobic soil conditions, many salt marsh plants have well-developed aerenchymal tissue that delivers oxygen to below-ground roots.

Salt Marsh Vegetation Patterns

Vegetation patterns often reflect the stage of maturation of a salt marsh. Early in the development, halophytes, such as *Spartina alterniflora* along the east coast of the United States, colonize intertidal flats. These initial colonizers are tolerant of frequent flooding. Once established, the plants spread vegetatively by rhizome growth, the plants trap sediments, and the marsh begins to grow vertically. As noted from A. C. Redfield's classic study of salt marsh development along a coast with a rising sea level, salt marshes often extend seaward over tidal flats, while also accreting vertically and encroaching the upland or freshwater tidal wetlands. With this vertical growth and maturity of the marsh, discrete patterns of vegetation develop: frequently flooded low marsh vegetation borders the seaward portion of the marsh and along creekbanks, while high marsh areas support less flood-tolerant species, such as *Spartina patens*. There is some concern that vertical growth or accretion of salt marshes may not be able to keep pace with accelerated rates of sea level rise, resulting in submergence or drowning of marshes. This has been observed in some areas of the world.

Plant species and patterns of vegetation that dominate the salt marsh vary from region to region of the world and it is beyond the scope of this article to detail this variability; however, the general pattern of low marsh and high marsh remains throughout. There is often variation in vegetation patterns from marshes within a region and even between marshes within a single estuary, but in general, the low marsh is dominated by a limited number of species, often just one. On the Atlantic coast of North America, *Spartina alterniflora* is the early colonizer and almost exclusively dominates the low marsh. In European marshes, *Puccinellia maritima* often dominates the intertidal low marsh. *Spartina anglica*, a hybrid of *Spartina alterniflora* and *Spartina maritima*, has invaded many of the muddier low marsh sites of European marshes over the past century. With increasing elevation of the high marsh, species richness tends to increase. *Spartina patens*, *Distichlis spicata*, *Juncus gerardi*, *Juncus roemerianus*, and short-form *Spartina alterniflora* occupy the US east coast high marsh, with each species dominating in patches or zones to form a mosaic vegetation pattern. In European marshes, the low marsh may give way to a diverse high marsh of *Halimione portulacoides*, *Limonium* sp., *Suaeda maritima*, and *Festuca* sp., among others.

Factors Controlling Vegetation Patterns

The frequency and duration of tidal flooding are mostly responsible for the low and high marsh delineation, but many other factors contribute to the wide variation found in salt marsh vegetation patterns (**Figure 1**). Soil salinity is relatively constant within the low marsh because of frequent tidal flooding, but extremes in soil salinity can occur on the less-flooded high marsh contributing to the vegetation mosaic. Concentrations in excess of 100 parts per thousand can occur, resulting in hypersaline pannes that remain unvegetated or are colonized by only the most salt-tolerant halophytes (e.g., *Salicornia*). Salt marshes of southern temperate and subtropical/tropical latitudes tend to have higher soil salinity because of more intense solar radiation and higher evaporation rates. At the other extreme, soil salinity of the high marsh can be dramatically depressed by rainfall or by discharge of groundwater near the marsh upland border. On salt marshes of the New England coast (USA), *Juncus gerardi* and the shrub, *Iva fructescens*, are less tolerant of salt, and thus, grow at higher elevations where tidal flooding is only occasional or near upland freshwater sources.

For successful growth in environments of high soil salinity, halophytes must be able to maintain a flow of water from the soil into the roots. Osmotic pressure in the roots must be higher than the surrounding soil for water uptake. To maintain this osmotic difference, halophytes have high concentrations of solutes in their tissues (i.e., high osmotic pressure), concentrations greater than the surrounding environment. This osmotic adjustment can be achieved by an accumulation of sodium and chloride ions or organic solutes. To effectively tolerate high internal salt concentrations, many halophytes have a succulent growth form (e.g., *Salicornia*, *Suaeda*). This high tissue water content serves to dilute potentially toxic salt concentrations. Other halophytes, such as *Spartina alterniflora*, have salt glands to actively excrete salt from leaves.

Waterlogged or anaerobic soil conditions also strongly influence the pattern of salt marsh vegetation. Plants growing in waterlogged soils must deal with a lack of oxygen at the rhizosphere and the

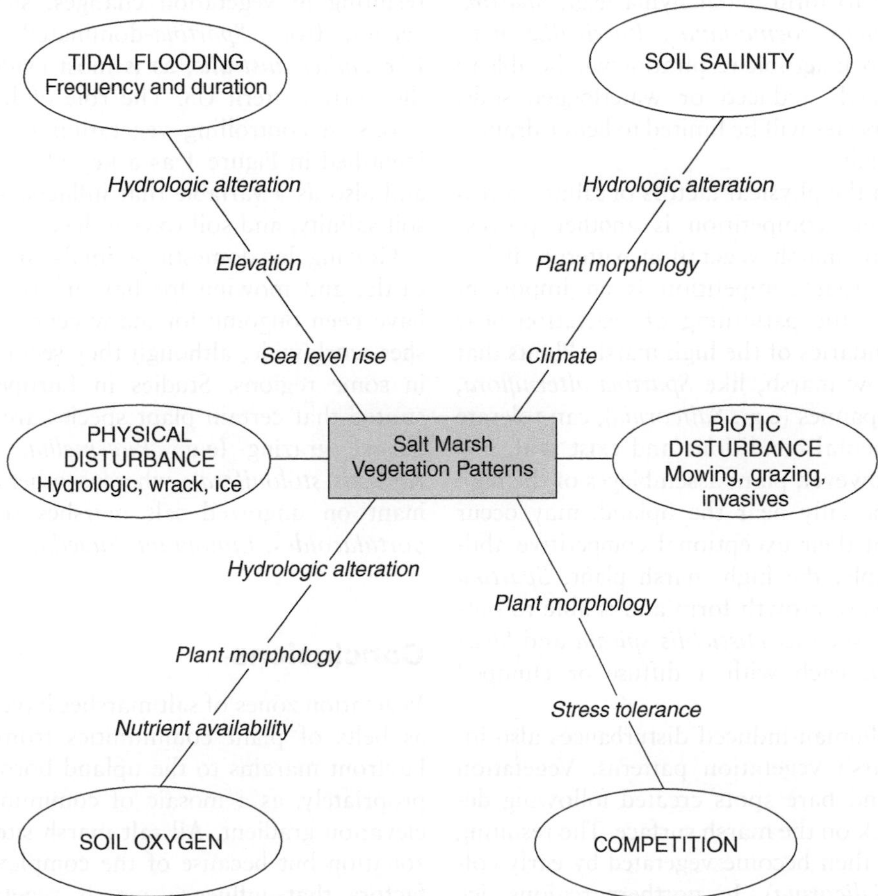

Figure 1 Conceptual model of factors controlling vegetation patterns in salt marshes. Ovals denote major factors and the key interacting parameters are shown in italics. For example, soil salinity responds to hydrologic alterations, climate (e.g., temperature/evaporation and precipitation) influences soil salinity, and a species response to soil salinity is dependent on morphological considerations (e.g., succulence, salt glands, etc.).

accumulation of toxins resulting from biogeochemical soil processes (i.e., sulfate reduction). High concentrations of hydrogen sulfide are toxic to root metabolism, inhibiting nitrogen uptake and resulting in decreased plant growth. Many salt marsh plants are able to survive these soil conditions because they have well-developed aerenchyma tissue, or a network of intercellular spaces that serve to deliver oxygen from above-ground plant parts to below-ground roots. Extensive research has been conducted on the relationships between sediment oxygen levels and growth of *Spartina alterniflora*. Aerenchyma transports oxygen to roots to support aerobic respiration, facilitating nitrogen uptake, and the aerated rhizosphere serves to oxidize reduced soil compounds, such as sulfide, thereby promoting growth. Under severely reducing soil conditions, the aerenchyma may not supply sufficient oxygen for aerobic respiration. *Spartina alterniflora* then has the ability to respire anaerobically, but growth is reduced. It is clear that the degree of soil aeration can dramatically influence salt marsh vegetation patterns. Plants that have the ability to form aerenchyma (e.g., *Spartina alterniflora*, *Juncus roemerianus*, *Puccinellia maritima*) or develop anaerobic respiration will be able to tolerate moderately reduced or waterlogged soils, whereas other species will be limited to better-drained areas of the marsh.

Coupled with the physical factors of salinity stress and waterlogging, competition is another process that controls salt marsh vegetation patterns. It has been suggested that competition is an important factor governing the patterning of vegetation near the upland boundaries of the high marsh. Plants that dominate the low marsh, like *Spartina alterniflora*, or forbs in salt pannes (e.g., *Salicornia*), can tolerate harsh environmental conditions and exist with few competitors. However, plant assemblages of the high marsh, and especially near the upland, may occur there because of their exceptional competitive abilities. For example, the high marsh plant, *Spartina patens*, has a dense growth form and is able to outcompete species such as *Distichlis spicata* and *Spartina alterniflora*, each with a diffuse or clumped morphology.

Natural and human-induced disturbances also influence salt marsh vegetation patterns. Vegetation can be killed and bare spots created following deposition of wrack on the marsh surface. The resulting bare areas may then become vegetated by early colonizers (e.g., *Salicornia*). In northern regions, ice scouring can dramatically alter the creek or bayfront of marshes. Also, large blocks of ice, laden with sediment, are often deposited on the high marsh creating a new microrelief habitat, or these blocks may transport plant rhizomes to mud or sand flats and initiate the process of salt marsh development.

Regarding human-induced disturbances, hydrologic alterations have dramatic effects on salt marsh vegetation patterns throughout the world. Some salt marshes have been diked and drained for agriculture. In others, extensive ditching has drained salt marshes for mosquito-control purposes. In yet another hydrologic alteration, water has been retained or impounded within salt marshes to alter wildlife habitat functions or to control mosquitoes. Impoundments generally restrict tidal inflow and retain fresh water, resulting in conversion from salt-tolerant vegetation to brackish or freshwater vegetation. Practices that drain the marsh, such as ditching, tend to lower the water-table level and aerate the soil, and the resulting vegetation may shift toward that typical of a high marsh. Another type of hydrologic alteration is the restriction of tidal flow by bridges, culverts, roads, and causeways. This is particularly common along urbanized shorelines, where soil salinity can be reduced and water-table levels altered resulting in vegetation changes, such as the conversion from *Spartina*-dominated salt marsh to *Phragmites australis*, as is most evident throughout the north-eastern US. The role of hydrologic alterations in controlling vegetation patterns is clearly identified in **Figure 1** as a key physical disturbance, and also as a variable that influences tidal flooding, soil salinity, and soil oxygen levels.

Grazing by domestic animals, mostly sheep and cattle, and mowing for hay are two practices that have been ongoing for many centuries on salt marshes worldwide, although they seem to be declining in some regions. Studies in Europe have demonstrated that certain plant species are favored by intensive grazing (e.g., *Puccinellia*, *Festuca rubra*, *Agrostis stolonifera*), whereas others become dominant on ungrazed salt marshes (e.g., *Halimione portulacoides*, *Limonium*, *Suaeda*).

Conclusions

Vegetation zones of salt marshes have been described as belts of plant communities from creekbank or bayfront margins to the upland border, or most appropriately, as a mosaic of communities along this elevation gradient. All salt marsh sites display some zonation but because of the complex of interacting factors that influence marsh vegetation patterns, there is extraordinary variability in zonation among individual marshes. Moreover, salt marsh vegetation patterns are constantly changing on seasonal to decadal timescales. Experimental research and long-term monitoring efforts are needed to further evaluate vegetation pattern responses to the myriad of interacting environmental factors that influence salt marsh vegetation. An ultimate goal is to model and predict the response of marsh vegetation as a result of natural or human-induced disturbance, accelerated rates of sea level rise, or marsh-restoration strategies.

See also

Mangroves. Salt Marshes and Mud Flats.

Further Reading

Adam P (1990) *Saltmarsh Ecology*. Cambridge: Cambridge University Press.

Bertness MD (1999) *The Ecology of Atlantic Shorelines*. Sunderland, MA: Sinauer.

Chapman VJ (1960) *Salt Marshes and Salt Deserts of the World*. New York: Interscience.

Daiber FC (1986) *Conservation of Tidal Marshes*. New York: Van Nostrand Reinhold.

Flowers TJ, Hajibagheri MA, and Clipson NJW (1986) Halophytes. *Quarterly Review of Biology* 61: 313–337.

Niering WA and Warren RS (1980) Vegetation patterns and processes in New England salt marshes. *BioScience* 30: 301–307.

Redfield AC (1972) Development of a New England salt marsh. *Ecological Monographs* 42: 201–237.

Reimold RJ and Queen WH (eds.) (1974) *Ecology of Halophytes*. New York: Academic Press.

SALT MARSHES AND MUD FLATS

J. M. Teal, Woods Hole Oceanographic Institution, Rochester, MA, USA

Structure

Salt marshes are vegetated mud flats. They are above mean sea level in the intertidal area where higher plants (angiosperms) grow. Sea grasses are an exception to the generalization about higher plants because they live below low tide levels. Mud flats are vegetated by algae.

Geomorphology

Salt marshes and mud flats are made of soft sediments deposited along the coast in areas protected from ocean surf or strong currents. These are long-term depositional areas intermittently subject to erosion and export of particles. Salt marsh sediments are held in place by plant roots and rhizomes (underground stems). Consequently, marshes are resistant to erosion by all but the strongest storms. Algal mats and animal burrows bind mud flat sediments, although, even when protected along tidal creeks within a salt marsh, mud flats are more easily eroded than the adjacent salt marsh plain.

Salinity in a marsh or mud flat, reported in parts per thousand (ppt), can range from about 40 ppt down to 5 ppt. The interaction of the tides and weather, the salinity of the coastal ocean, and the elevation of the marsh plain control salinity on a marsh or mud flat. Parts of the marsh with strong, regular tides (1 m or more) are flooded twice a day, and salinity is close to that of the coastal ocean. Heavy rain at low tide can temporarily make the surface of the sediment almost fresh. Salinity may vary seasonally if a marsh is located in an estuary where the river volume changes over the year. Salinity varies within a marsh with subtle changes in surface elevation. Higher marshes at sites with regular tides have variation between spring and neap tides that result in some areas being flooded every day while other, higher, areas are flooded less frequently. At higher elevations flooding may occur on only a few days each spring tide, while at the highest elevations flooding may occur only a few times a year.

Some marshes, on coasts with little elevation change, have their highest parts flooded only seasonally by the equinoctial tides. Other marshes occur in areas with small lunar tides where flooding is predominantly wind-driven, such as the marshes in the lagoons along the Texas coast of the United States. They are flooded irregularly and, between flooding, the salinity is greatly raised by evaporation in the hot, dry climate. The salinity in some of the higher areas becomes so high that no rooted plants survive. These are salt flats, high enough in the tidal regime for higher plants to grow, but so salty that only salt-resistant algae can grow there. The weather further affects salinity within marshes and mud flats. Weather that changes the temperature of coastal waters or varying atmospheric pressure can change sea level by 10 cm over periods of weeks to months, and therefore affect the areas of the marsh that are subjected to tidal inundation.

Sea level changes gradually. It has been rising since the retreat of the continental glaciers. The rate of rise may be increasing with global warming. For the last 10 000 years or so, marshes have been able to keep up with sea level rise by accumulating sediment, both through deposition of mud and sand and through accumulation of peat. The peat comes from the underground parts of marsh plants that decay slowly in the anoxic marsh sediments. The result of these processes is illustrated in **Figure 1**, in which the basement sediment is overlain by the accumulated marsh sediment. Keeping up with sea level rise creates a marsh plain that is relatively flat; the elevation determined by water level rather than by the geological processes that determined the original, basement sediment surface on which the marsh developed. Tidal creeks, which carry the tidal waters on and off the marsh, dissect the flat marsh plain.

Organisms

The duration of flooding and the salinities of the sediments and tidal waters control the mix of higher vegetation. Competitive interactions between plants and interactions between plants and animals further determine plant distributions. Duration of flooding duration controls how saturated the sediments will be, which in turn controls how oxygenated or reduced the sediments are. The roots of higher plants must have oxygen to survive, although many can survive short periods of anoxia. Air penetrates into the creekbank sediments as they drain at low tide.

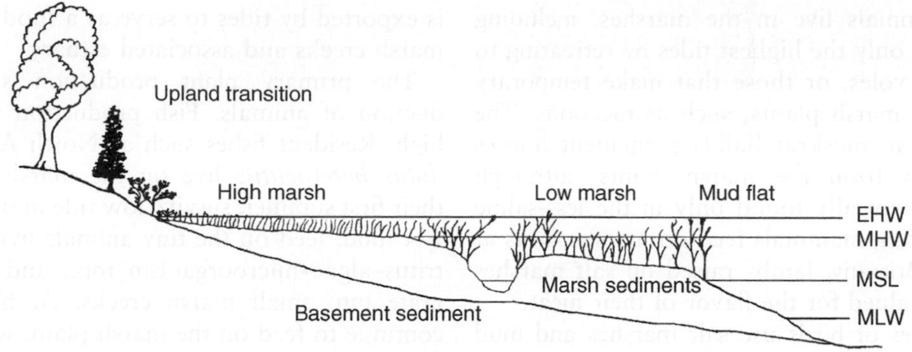

Figure 1 Cartoon of a typical salt marsh of eastern North America. The plants shown are mostly grasses and may differ in other parts of the world. MLW, mean low water; MSL, mean sea level; MHW, mean high water; EHW, extreme high water. The mud flat is shown as a part of the marsh but mud flats also exist independently of marshes.

Evapotranspiration from plants at low tide also removes water from the sediments and facilitates entry of air. Most salt marsh higher plants have aerenchyma (internal air passages) through which oxygen reaches the roots and rhizomes by diffusion or active transport from the above-ground parts. However, they also benefit from availability of oxygen outside the roots.

The species of higher plants that dominate salt marshes vary with latitude, salinity, region of the world, and tidal amplitude. They are composed of relatively few species of plants that have invested in the ability to supply oxygen to roots and rhizomes in reduced sediments and to deal with various levels of salt. Grasses are important, with *Spartina alterniflora* the dominant species from mid-tide to high-tide levels in temperate Eastern North America. *Puccinellia* is a dominant grass in boreal and arctic marshes. The less regularly flooded marshes of East Anglia (UK) support a more diverse vegetation community in which grasses are not dominant. The salty marshes of the Texas coast are covered by salt-tolerant *Salicornia* species. Adjacent to the upper, landward edge of the marsh lie areas flooded only at times when storms drive ocean waters to unusual heights. Some land plants can survive occasional salt baths, but most cannot. An extreme high-water even usually results in the death of plants at the marsh border.

Algae on the marsh and mud flat are less specialized. Depending upon the turbulence of the tidal water, macroalgae (seaweeds) may be present, but a diverse microalgal community is common. Algae live on or near the surface of the sediments and obtain oxygen directly from the air or water and from the oxygen produced by photosynthesis. Their presence on surface sediment is controlled by light. In highly turbid waters they are almost entirely limited to the intertidal flats. In clearer waters, they can grow below low-tide levels. Algae growing on the vegetated marsh plain and on the stems of marsh plants get less light as the plants mature. Production of these algae is greatest in early spring, before the developing vegetation intercepts the light.

Photosynthetic bacteria also contribute to marsh and mud flat production. Blue-green bacteria can be abundant enough to forms mats. Photosynthetic sulfur bacteria occupy a thin stratum in the sediment where they get light from above and sulfide from deeper reduced levels for their hydrogen source but are below the level of oxygen penetration that would kill them. These strange organisms are relicts from the primitive earth before the atmosphere contained oxygen.

Salt marsh animals are from terrestrial and marine sources; mud flat inhabitants are limited to marine sources. Insects, spiders, and mites live in marsh sediments and on marsh plants. Crabs, amphipods, isopods and shrimps, polychaete and oligochaete annelids, snails, and bivalves live in and on the sediments. Most of these marine animals have planktonic larval stages that facilitate movement between marshes and mud flats. Although burrowing animals, such as crabs that live at the water edge of the marsh, may be fairly large (2–15 cm), in general burrowers in marshes are smaller than those in mud flats, presumably because the root mats of the higher plants interfere with burrowing.

Fish are important faunal elements in regularly flooded salt marshes and mud flats. They can be characterized as permanent marsh residents; seasonal residents (species that come into the marsh at the beginning of summer as new post-larvae and live in the marsh until cold weather sets in); species that are primarily residents of coastal waters but enter the marshes at high tide; and predatory fish that come into marshes on the ebb tide to feed on the smaller fishes forced off the marsh plain and out of the smaller creeks by falling water levels.

A few mammals live in the marshes, including those that flee only the highest tides by retreating to land, such as voles, or those that make temporary refuges in tall marsh plants, such as raccoons. The North American muskrat builds permanent houses on the marsh from the marsh plants, although muskrats are typically found only in the less-saline marshes. Grazing mammals feed on marsh plants at low tide. In Brittany, lambs raised on salt marshes are specially valued for the flavor of their meat.

Many species of birds use salt marshes and mud flats. Shore bird species live in the marshes and/or use associated mud flats for feeding during migration. Northern harriers nest on higher portions of salt marshes and feed on their resident voles. Several species of rails dwell in marshes as do bitterns, ducks, and some wrens and sparrows. The nesting species must keep their eggs and young from drowning, which they achieve by building their nests in high vegetation, by building floating nests, and by nesting and raising their young between periods of highest tides.

Functions

Marshes, and to some extent mud flats, produce animals and plants, provide nursery areas for marine fishes, modify nutrient cycles, degrade organic chemicals, immobilize elements within their sediments, and modify wave action on adjacent uplands.

Production and Nursery

Plant production from salt marshes is as high as or higher than that of most other systems because of the ability of muddy sediments to serve as nutrient reservoirs, because of their exposure to full sun, and because of nutrients supplied by sea water. Although the plant production is food for insects, mites and voles, large mammalian herbivores that venture onto the marsh, a few crustacea, and other marine animals, most of the higher plant production is not eaten directly but enters the food web as detritus. As the plants die, they are attacked by fungi and bacteria that reduce them to small particles on the surface of the marsh. Since the labile organic matter in the plants is quickly used as food by the bacteria and fungi, most of the nutrient value of the detritus reaches the next link in the food chain through these microorganisms. These are digested from the plant particles by detritivores, but the cellulose and lignin from the original plants passes through them and is deposited as feces that are recolonized by bacteria and fungi. Besides serving as a food source in the marsh itself, a portion of the detritus–algae mixture is exported by tides to serve as a food source in the marsh creeks and associated estuary.

The primary plant production supports production of animals. Fish production in marshes is high. Resident fishes such as North American *Fundulus heteroclitus* live on the marsh plain during their first summer, survive low tide in tiny pools or in wet mud, feed on the tiny animals living on the detritus–algae–microorganism mix, and grow to migrate into small marsh creeks. At high tide they continue to feed on the marsh plain, where they are joined by the young-of-the-year of those species that use the marsh principally as a nursery area. The warm, shallow waters promote rapid growth and are refuge areas where they are protected from predatory fishes, but not from fish-eating herons and egrets.

The fishes are the most valuable export from the marshes to estuaries and coastal oceans. Some of the fishes are exported in the bodies of predatory fishes that enter the marsh on the ebb tide to feed. Many young fishes, raised in marshes, migrate offshore in the autumn after having spent the summer growing in the marsh.

Nutrient and Element Cycling

Nitrogen is the critical nutrient controlling plant productivity in marshes. Phosphorus is readily available in muddy salt marsh sediments and potassium is sufficiently abundant in sea water. Micronutrients, such as silica or iron, that may be limiting for primary production in deeper waters are abundant in marsh sediments. Thus nitrogen is the nutrient of interest for marsh production and nutrient cycling.

In marshes where nitrogen is in short supply, blue-green bacteria serve as nitrogen fixers, building nitrogen gas from the air into their organic matter. Nitrogen-fixing bacteria associated with the roots of higher plants serve the same function. Nitrogen fixation is an energy-demanding process that is absent where the supply is sufficient to support plant growth.

Two other stages of the nitrogen cycle occur in marshes. Organic nitrogen released by decomposition is in the form of ammonium ion. This can be oxidized to nitrate by certain bacteria that derive energy from the process if oxygen is present. Both nitrate and ammonium can satisfy the nitrogen needs of plants, but nitrate can also serve in place of oxygen for the respiration of another group of bacteria that release nitrogen gas as a by-product in the process called denitrification. Denitrification in salt marshes and mud flats is significant in reducing eutrophication in estuarine and coastal waters.

Phosphorus, present as phosphate, is the other plant nutrient that can be limiting in marshes, especially in regions where nitrogen is in abundant supply. It can also contribute to eutrophication of coastal waters, but phosphate is readily bound to sediments and so tends to be retained in marshes and mud flats rather than released to the estuary.

Sulfur cycling in salt marshes, while of minor importance as a nutrient, contributes to completing the production cycle. Sulfate is the second most abundant anion in sea water. In anoxic sediments, a specialized group of decomposing organisms living on the dead, underground portions of marsh plants can use sulfate as an electron acceptor – an oxygen substitute – in their respiration. The by-product is hydrogen sulfide rather than water. Sulfate reduction yields much less energy than respiration with oxygen or nitrate reduction, so these latter processes occur within the sediment surface, leaving sulfate reduction as the remaining process in deeper parts of the sediments. The sulfide carries much of the free energy not captured by the bacteria in sulfate reduction. As it diffuses to surface layers, most of the sulfide is oxidized by bacteria that grow using it as an energy source. A small amount is used by the photosynthetic sulfur bacteria mentioned above.

Pollution

Marshes, like the estuaries with which they tend to be associated, are depositional areas. They tend to accumulate whatever pollutants are dumped into coastal waters, especially those bound to particles. Much of the pollution load enters the coast transported by rivers and may originate far from the affected marshes. For example, much of the nitrogen and pesticide loading of marshes and coastal waters of the Mississippi Delta region of the United States comes from farming regions hundreds or thousands of kilometers upstream.

Many pollutants, both organic and inorganic, bind to sediments and are retained by salt marshes and mud flats. Organic compounds are often degraded in these biologically active systems, especially since many of them are only metabolized when the microorganisms responsible are actively growing on other, more easily degraded compounds. There are, unfortunately, some organics, the structures of which are protected by constituents such as chlorine, that are highly resistant to microbial attack. Some polychlorinated biphenyls (PCBs) have such structures, with the result that a PCB mixture will gradually lose the degradable compounds while the resistant components will become relatively more concentrated.

Metals are also bound to sediments and so may be removed and retained by marshes and mud flats. Mercury is sequestered in the sediment, while cadmium forms soluble complexes with chloride in sea water and is, at most, temporarily retained.

Since marshes and mud flats tend, in the long term, to be depositional systems, they remove pollutants and bury them as long as the sediments are not remobilized by erosion. Since mud flats are more easily eroded than marsh sediments held in place by plant roots and rhizomes, they are less secure long-term storage sites.

Storm Damage Prevention

While marshes and mud flats exist only in relatively protected situations, they are still subject to storm damage as are the uplands behind them. During storms, the shallow waters and the vegetation on the marshes offer resistance to water flow, making them places where wave forces are dissipated, reducing the water and wave damage to the adjoining upland.

Human Modifications

Direct Effects

Many marshes and mud flats in urban areas have been highly altered or destroyed by filling or by dredging for harbor, channel or marina development. Less intrusive actions can have large impacts. Since salt marshes and mud flats typically lie in indentations along the coast, the openings where tides enter and leave them are often sites of human modification for roads and railroads. Both culverts and bridges restrict flow if they are not large enough. Flow is especially restricted at high water unless the bridge spans the entire marsh opening, a rare situation because it is expensive. The result of restriction is a reduction in the amount of water that floods the marsh. The plants are submerged for a shorter period and to a lesser depth, and the floodwaters do not extend as far onto the marsh surface. The ebb flow is also restricted and the marsh may not drain as efficiently as in the unimpeded case. Poor drainage could freshen the marsh after a heavy rain and runoff. Less commonly, it could increase salinity after an exceptionally prolonged storm-driven high tide.

The result of the disturbance will be a change in the oxygen and salinity relations between roots and sediments. Plants may become oxygen-stressed and drown. Tidal restrictions in moist temperate regions usually result in a freshening of the sediment salinity. This favors species that have not invested in salt control mechanisms of the typical salt marsh plants. A widespread result in North America has been the spread of the common reed, *Phragmites australis*, a brackish-water and freshwater species. Common reed is a tall (3 m) and vigorous plant that can spread horizontally by rhizomes at 10 meters per year. Its robust stem decomposes more slowly than that of the salt marsh cord grass, *Spartina alterniflora* and as a result, it takes over a marsh freshened by tidal restrictions. Since its stems accumulate above ground and rhizomes below ground, it tends to raise the marsh level, fill in the small drainage channels, and reduce the value of the marsh for fish and wildlife. Although *Phragmites australis* is a valuable plant for many purposes (it is the preferred plant for thatching roofs in Europe), its takeover of salt marshes is considered undesirable.

The ultimate modification of tidal flow is restriction by diking. Some temperate marshes have been diked to allow the harvest of salt hay, valued as mulch because it lacks weed seeds. Since some diked marshes are periodically flooded in an attempt to maintain the desired vegetation, they are not completely changed and can be restored. Other marshes, such as those in Holland, have been diked and removed from tidal flow so that the land may be used for upland agriculture. Many marshes and mud flats have been modified to create salt pans for production of sea salt and for aquaculture. The latter is a greater problem in the tropics, where the impact is on mangroves rather than on the salt marshes of more temperate regions.

Indirect Effects

Upland diking The upper borders of coastal marshes were often diked to prevent upland flooding. People built close to the marshes to take advantage of the view. With experience they found that storms could raise the sea level enough to flood upland. The natural response was to construct a barrier to prevent flooding. Roads and railroads along the landward edges of marshes are also barriers that restrict upland flooding. They are built high enough to protect the roadbed from most flooding and usually have only enough drainage to allow rain runoff to pass to the sea. In both cases, the result is a barrier to landward migration of the marsh. As the relative sea level rises, sandy barriers that protect coastal marshes are flooded and, during storms, the sand is washed onto the marsh. As long as the marsh can also move back by occupying the adjacent upland it may be able to persist without loss in area, but if a barrier prevents landward transgression the marsh will be squeezed between the barrier and the rising sea and will eventually disappear. During this process, the drainage structures under the barrier gradually become flow restrictors. The sea will flood the land behind the barrier through the culverts, but these are inevitably too small to permit unrestricted marsh development. When flooding begins, the culverts are typically fitted with tide gates to prevent whatever flooding and marsh development could be accommodated by the capacity of the culverts.

Changes in sediment loading Increases in sediment supplies can allow the marshes to spread as the shallow waters bordering them are filled in. The plant stems further impede water movement and enhance spread of the marsh. This assumes that storms do not carry the additional sediment onto the marsh plain and raise it above normal tidal level, which would damage or destroy rather than extend the marsh.

Reduced sediment supply can destroy a marsh. In a river delta where sediments gradually de-water and consolidate, sinking continually, a continuous supply of new sediment combined with vegetation remains, accumulating as peat, and maintains the marsh level. When sediment supplies are cut off, the peat accumulation may be insufficient to maintain the marsh at sea level. Dams, such as the Aswan Dam on the Nile, can trap sediments. Sediments can be channeled by levées so that they flow into deep water at the mouth of the river rather than spreading over the delta marshes, as is happening in the Mississippi River delta. In the latter case, the coastline of Louisiana is retreating by kilometers a year as a result of the loss of delta marshlands.

Introduction of foreign species Dramatic changes in the marshes and flats of England and Europe occurred after *Spartina alterniflora* was introduced from the east coast of North America and probably hybridized with the native *S. maritima* to produce *S. anglica*. The new species was more tolerant of submergence than the native forms and turned many mud flats into salt marshes. This change reduced populations of mud flat animals, many with commercial value, and reduced the foraging area for shore birds that feed on mud flats. A similar situation has developed in the last decades on the north-west coast of the United States, where introduced *Spartina alterniflora* is invading mud flats and reducing the available area for shellfish.

Marsh restoration Salt marshes and mud flats may be the most readily restored of all wetlands. The source and level of water is known. The vascular plants that will thrive are known and can be planted if a local seed source is not available. Many of the marsh animals have planktonic larvae that can invade the restored marsh on their own. Although many of the properties of a mature salt marsh take time to develop, such as the nutrient-retaining capacity of the sediments, these will develop if the marsh is allowed to survive.

See also

Mangroves. Salt Marsh Vegetation.

Further Reading

Adam P (1990) *Salt Marsh Ecology.* Cambridge: Cambridge University Press.

Mitsch WJ and Gosselink JG (1993) *Wetlands*. New York: Wiley.

Peterson CH and Peterson NM (1972) *The Ecology of Intertidal Flats of North Carolina: A Community Profile*. Washington, DC: US Fish and Wildlife Service, Office of Biological Services, FWS/OBS-79/39.

Streever W (1999) *An International Perspective on Wetland Rehabilitation*. Dordrecht: Kluwer.

Teal JM and Teal M (1969) *Life and Death of the Salt Marsh*. New York: Ballentine.

Weinstein MP and Kreeger DA (2001) *Concepts and Controversies in Tidal Marsh Ecology.* Dordrecht: Kluwer.

Whitlatch RB (1982) *The Ecology of New England Tidal Flats: A Community Profile*. Washington, DC: US Fish and Wildlife Service, Biological Services Program, FES/OBS-81/01.

SANDY BEACHES, BIOLOGY OF

A. C. Brown, University of Cape Town, Cape Town, Republic of South Africa

Introduction

Some 75% of the world's ice-free shores consists of sand. Nevertheless, biological studies on these beaches lagged behind those of other coastal marine habitats for many decades. This was probably because there are very few organisms to be seen on a beach at low tide during the day, when biological activity is at a minimum. A very different impression may be gained at night, but the beach fauna is then much more difficult to study. Even at night few species may be observed but these can be present in vast numbers. Most sandy-beach animals are not seen at all, as they are very tiny, virtually invisible to the naked eye, and they live between the sand grains (i.e., they are interstitial). A large number of these species, comprising the meiofauna, may be present. A further striking characteristic of ocean beaches is the total absence of attached plants intertidally and in the shallow subtidal. The animals are thus deprived of a resident primary food source, such as exists in most other habitats. We may, therefore, suspect that ecosystem functioning differs considerably from that found elsewhere and research has shown that this is, indeed, the case.

The Dominating Factor

In harsh, unpredictably varying environments, physical factors are far more important in shaping the ecosystem than are biological interactions; ocean beaches are no exception. Here the 'superparameter' controlling community structure, species diversity, and the modes of life of the organisms is water movement – waves, tides, and currents. Water movements determine the type of shore present and then interact with the sand particles so that a highly dynamic, unstable environment results. Instability reaches a peak during storms, when many tons of sand may be washed out to sea and physical conditions on the beach become chaotic. To exist in the face of such instability, beach animals must be very mobile and all must be able to burrow. Indeed the ability to burrow before being swept away by the next swash is critical for those animals that habitually emerge from the sand onto the beach face. The more exposed the beach to wave action, the more rapid does burrowing have to be. Furthermore, on most ocean beaches the sand is too unstable not only to support plant growth but also to allow permanent burrows to exist intertidally. Consequently animals such as the bloodworm, *Arenicola*, and the burrowing prawn, *Callianassa*, are found only on sheltered beaches. On exposed beaches, there is virtually nothing the aquatic fauna can do to modify their environment; they are entirely at the mercy of whatever the physical regime may dictate. During storms, behavior patterns must often change if the animals are to survive. Some animals simply dig themselves deeper into the sand, others may move offshore, and some semiterrestrial species (e.g., sand hoppers) move up into the dunes until the storm has passed. Notwithstanding these behavior patterns, mortality due to physical stresses often outweighs mortality due to predation.

Tidal Migrations and Zonation

Despite this extremely harsh environment, the essential mobility of the beach fauna permits exploitation of tidal rise and fall to an extent not available to more-sedentary animals. The meiofauna migrate up and down through the sand column in time with the tides, achieving the most optimal conditions available, and maximizing food resources and, in some cases, promoting photosynthesis. Many of the larger species (i.e., macrofauna) emerge from the sand and follow the tide up and down the slope of the beach face. The aquatic species tend to keep within the swash zone as it rises and falls; this is the area in which their food is likely to be most plentiful. As a bonus, this behavior also reduces predation, as the swash is too shallow for predatory fish to get to them, and most shore birds do not enter the water. Many semiterrestrial crustaceans, above the waterline, follow the falling tide down the slope, feed intertidally, and then migrate back to their positions above high water-mark as the tide rises. They do so only at night, remaining buried during the day, so that predation by birds is at a minimum.

If the macrofauna is sampled at low tide, a zonation of species is often apparent (**Figure 1**) and attempts have been made to relate these zones to those found on rocky shores or with changing physical conditions up the beach. However, the number of zones differs on different beaches and the zones tend

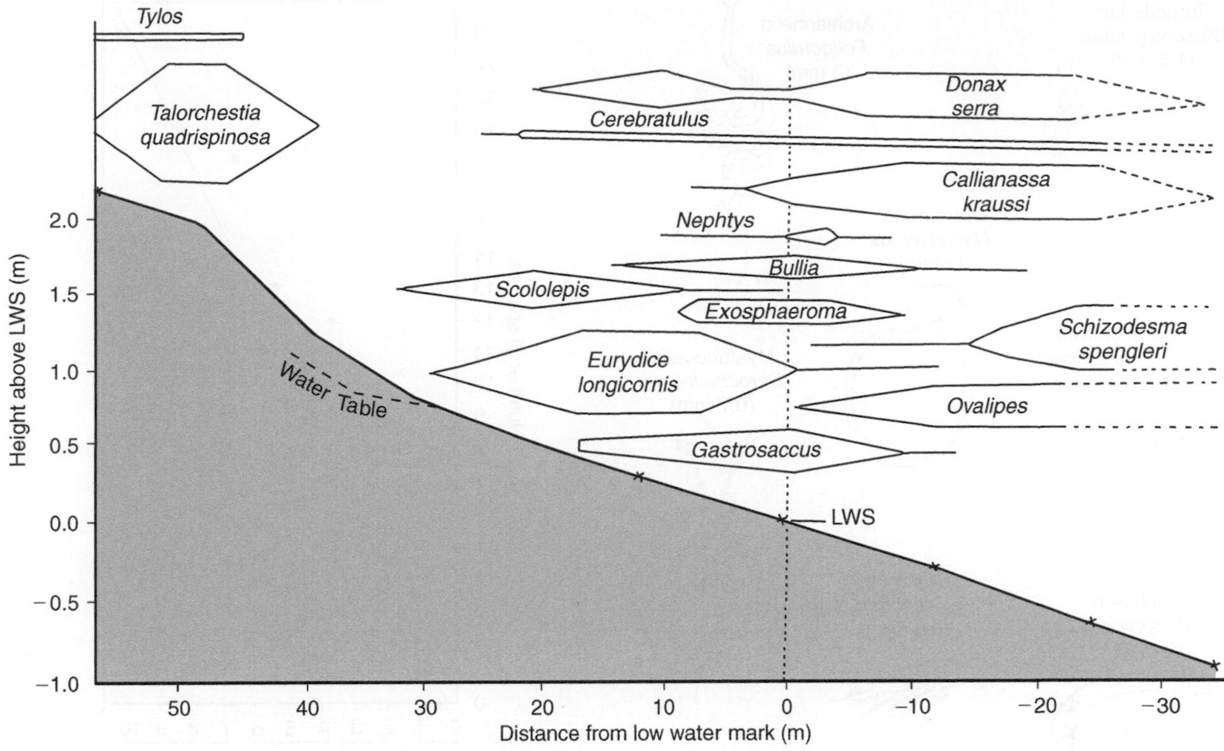

Figure 1 Low-tide distribution of the macrofauna up the beach slope at Muizenberg, near Cape Town. Some zonation is apparent. *Nephtys* and *Scololepis* are annelid worms; *Schizodesma* is a bivalve mollusk, larger than *Donax*; *Cerebratulus* is a nemertean worm; *Eurydice* and *Exosphaeroma* are isopods (pill bugs), *Ovalipes* is a swimming crab. The remaining genera are mentioned in the text. LWS, the level of Low Water of Spring tide.

to be blurred, with considerable overlap. Moreover, due to tidal migrations, the zones change as the tide rises. It has, therefore, been suggested that only two zones are apparent on all beaches at all states of the tide – a zone of aquatic animals and, higher up the shore, a zone of semiterrestrial air breathers; these are known as Brown's Zones.

Longshore Distribution

Not only is the distribution of the macrofauna up the beach slope not uniform but their longshore distribution is also typically discontinuous, or patchy. There is no one reason for this patchy distribution in all species. Variations in penetrability of the sand account for discontinuity in some species, breeding behavior or food maximization in others. In the semiterrestrial pill bug, *Tylos*, aggregations are apparently an incidental effect of their tendency to use existing burrows.

Diversity and Abundance: Meiofauna

The meiofauna consists of interstitial animals that will pass through a 1.0 mm sieve. They tend to be slender and worm-shaped, a necessary adaptation for gliding or wriggling between the grains. Nematoda (round worms) and harpacticoid copepods (small Crustacea, **Figure 2**) are usually dominant. However, most animal phyla are represented, including Platyhelminthes (flat worms), Nemertea (ribbon worms), Rotifera (wheel animalcules), Gastrotricha (**Figure 2**), Kinorhyncha, Annelida (segmented worms) and various Arthropoda such as mites and Collembola (spring tails). Rarer meiofaunal animals include an occasional cnidarian (hydroid), a few species of nudibranchs (sea slugs), tardigrades, a bryozoan, and even a primitive chordate. One whole phylum of animals, the Loricifera, is found only in beach meiofauna, as is the crustacean subclass Mystacocarida (**Figure 2**).

Both the diversity and abundance of the meiofauna are commonly highest on beaches intermediate between those with very gentle slopes (i.e., dissipative) and those with abrupt slopes (reflective). This is largely because these intermediate beaches also present an intermediate particle size range; the occurrence of most meiofaunal species is governed not by particle size itself but by the sizes of the pores between the grains. Pore size (porosity) also governs the depth to which adequate oxygen tensions

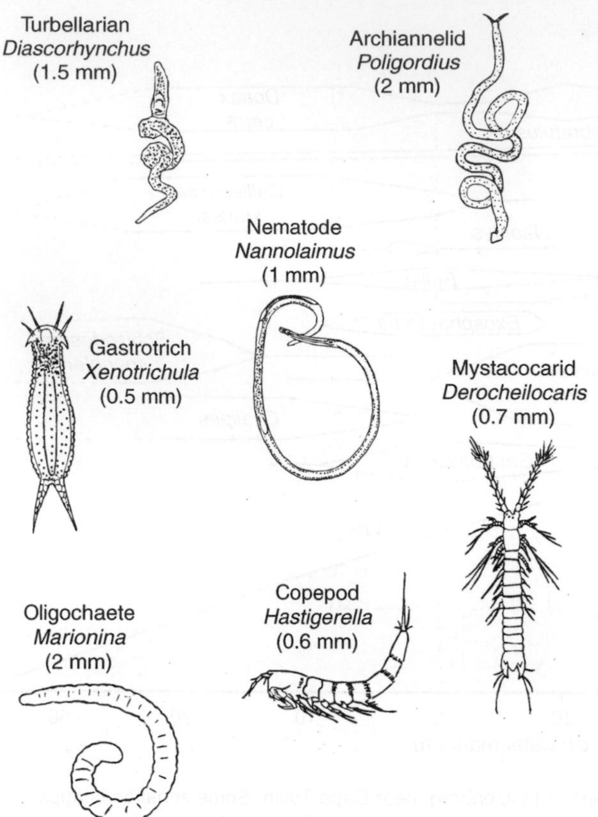

Figure 2 Some characteristic meiofaunal genera found in sandy beaches. (Reproduced from Brown and McLachlan, 1990 with permission from Elsevier Science.)

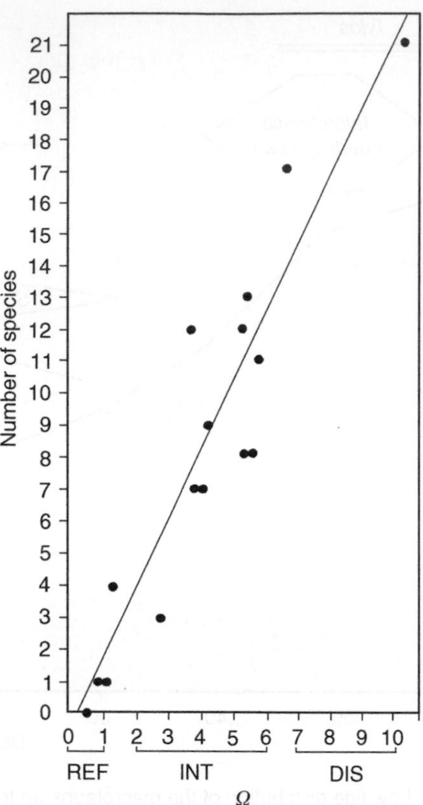

Figure 3 Relationship betwseen macrofaunal diversity and beach morphodynamic state (Dean's parameter, Ω) for 23 beaches in Australia, South Africa, and the USA. REF, reflective; INT, intermediate; DIS, dissipative. (Reproduced from Brown and McLachlan, 1990 with permission from Elsevier Science.)

penetrate, this depth being least in very fine sand. This limits the vertical distribution of aerobic members of the meiofauna. Groups such as the Gastrotricha are completely absent from fine sands and harpacticoid copepods are relatively uncommon. As with the macrofauna, a zonation of the meiofauna is often apparent; maximum diversity and abundance are usually attained in the sand of the mid to upper part of the intertidal beach. Meiofaunal numbers typically average $10^6\,m^{-2}$ but they may be as low as $0.05 \times 10^6\,m^{-2}$ or as high as $3 \times 10^6\,m^{-2}$. One of the few beaches whose meiofauna has been studied intensively is a beach on the island of Sylt, in the North Sea. No less than 652 meiofaunal species have been identified here.

Diversity and Abundance: Macrofauna

As with the meiofauna, the diversity and abundance of the macrofauna vary greatly with beach morphodynamics. Maximum numbers, species, and biomass are found on gently sloping beaches with wide surf zones (i.e., dissipative beaches). Important criteria here, apart from slope, are the relatively fine sand, the extensive beach width and the long, essentially laminar flow of the swash, which results in minimum disturbance of sand, facilitates the transport of some species (see below) and renders burrowing easier. At the other extreme are abruptly sloping, reflective beaches, with waves breaking on the beach face itself, leading to turbulent swash, considerable sand movement and large particle size. Such beaches are inhospitable to virtually all species of aquatic macrofauna. There is, in fact, some correlation between diversity/abundance/biomass and beach slope, breaker height or the particle size of the sand, as these are all related. The mean individual size of the macrofaunal animals correlates best with particle size, whereas biomass correlates best with breaker height. Diversity and abundance are usually best correlated with Dean's Parameter (Ω) (**Figure 3**), which gives a good indication of the overall morphodynamic state of the beach:

$$\Omega = \frac{Hb}{WsT}$$

where Hb is average breaker height, Ws the mean fall velocity of the sand and T the wave period.

Beaches with extensive surf zones and gentle slopes may harbour over 20 intertidal species of aquatic macrofauna but this number decreases with increasing slope until fully reflective beaches may have no aquatic macrofauna at all.

Biomass varies greatly but on dissipative beaches averages about 7000 g dry mass per meter stretch of beach. One beach in Peru was found to have a biomass of no less than 25 700 $g\,m^{-1}$. Although diversity and abundance generally decrease with increasing wave exposure, some very exposed beaches display a surprisingly high biomass due to the larger individual sizes of the macrofauna. Species diversity and biomass tend to decrease from low to high tide marks but often increase again immediately above the high water level, especially if algal debris (e.g., kelp or wrack) is present.

On oceanic beaches, there is a gap in size between the meiofaunal animals and the macrofauna. This is probably because, although the meiofauna move between the grains, macrofaunal animals burrow by displacing the sand. They therefore have to be relatively robust and much larger than the sand grains. Common aquatic macrofauna include filter-feeding bivalve mollusks (clams) of several genera, of which the most wide-spread is *Donax*. Scavenging gastropod mollusks (whelks) may also occur; most is known about the plough-shell, *Bullia* (**Figure 4**). On relatively sheltered shores, a variety of annelid worms occurs, but these decrease in numbers and diversity with increasing wave action, small crustaceans becoming more dominant. On tropical and subtropical beaches, the mole-crabs *Hippa* and *Emerita* (**Figure 5**) may achieve dense populations, whereas on temperate beaches mysid shrimps (**Figure 6**) and aquatic isopods (pill bugs) may be much in evidence.

Figure 4 The 'plough shell' *Bullia*, about to surf.

Above the water-line, semiterrestrial isopods, such as *Tylos* (**Figure 7**), are often abundant, and sand hoppers (or 'beach fleas') (*Talitrus*, **Figure 8**, *Orchestia* and *Talorchestia*) may occur in their millions on temperate beaches. On tropical and subtropical beaches, semiterrestrial crabs, such as the ghost crab, *Ocypode* (**Figure 9**), are important. In addition, insects may be present in some numbers; these include kelp flies and their relatives, staphylinids (rove beetles) and other beetles, and mole crickets.

Although the above forms are those most commonly encountered, some beaches have a quite different fauna. One beach has been found to be dominated by small Tanaidacea (related to pill bugs), another by picno-mole crickets (tiny insects). Seasonal visitors may also make considerable impact; people who know the beaches of the Atlantic seaboard of the United States or the beaches of Japan, Korea, or Malaysia will be familiar with the horseshoe crab, *Limulus* (actually not a crab at all), which may come ashore in vast numbers, and on some tropical beaches female turtles invade the beach to breed, their eggs and hatchlings providing a valuable food supply for a variety of marauding animals.

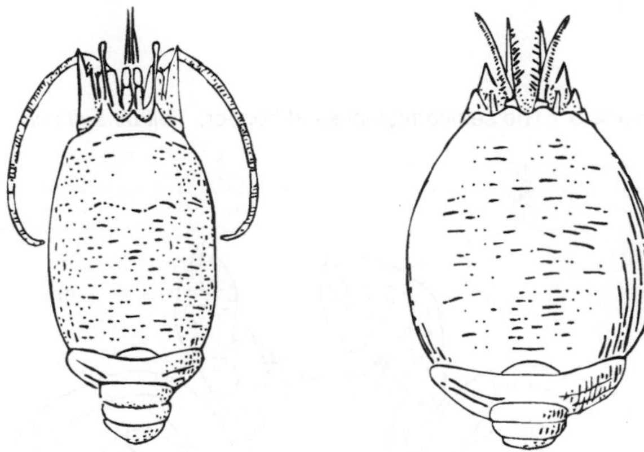

Figure 5 The mole-crabs *Emerita* and *Hippa* in dorsal view. (Reproduced from Brown and McLachlan, 1990 with permission from Elsevier Science.)

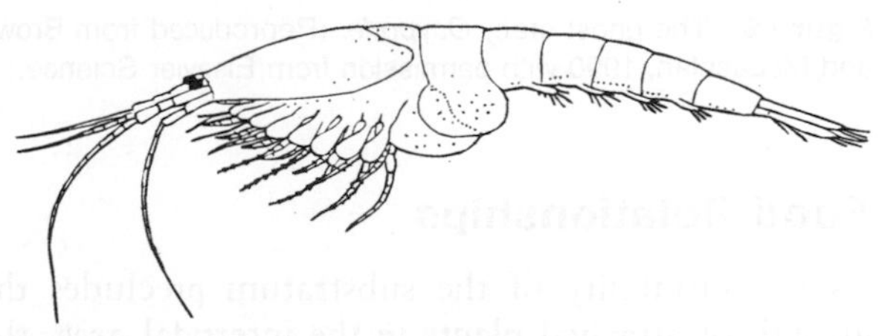

Figure 6 The mysid shrimp, *Gastrosaccus*.

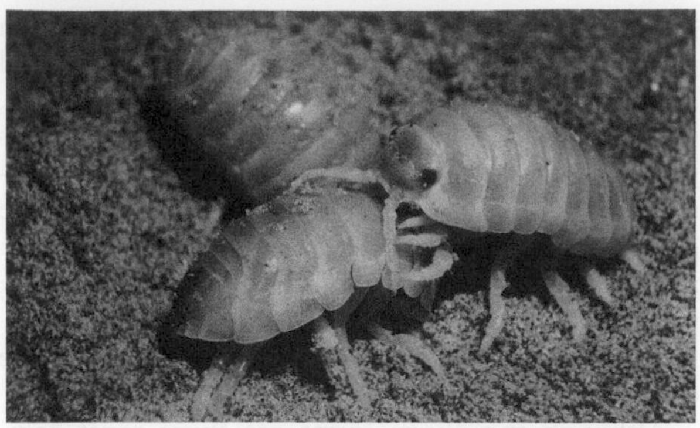

Figure 7 The pill bug, *Tylos granulatus*, showing postcopulatory grooming. The male is mounted back-to-front on the female and is stroking her with his antennae. (Photo: Claudio Velasquez.) (Reproduced with permission from Brown and Odendaal, 1994.)

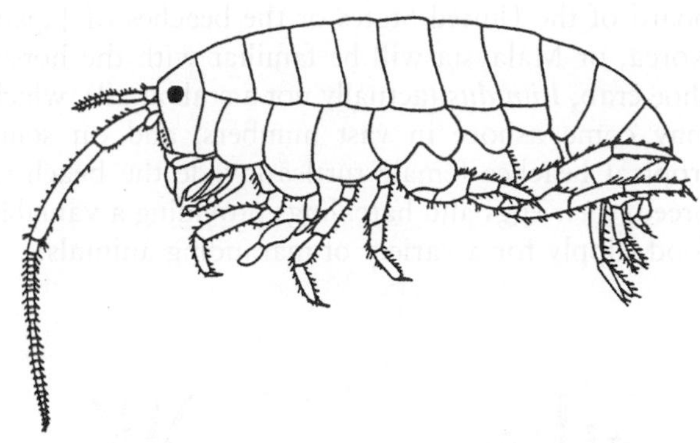

Figure 8 The semiterrestrial sand hopper, *Talitrus saltator*.

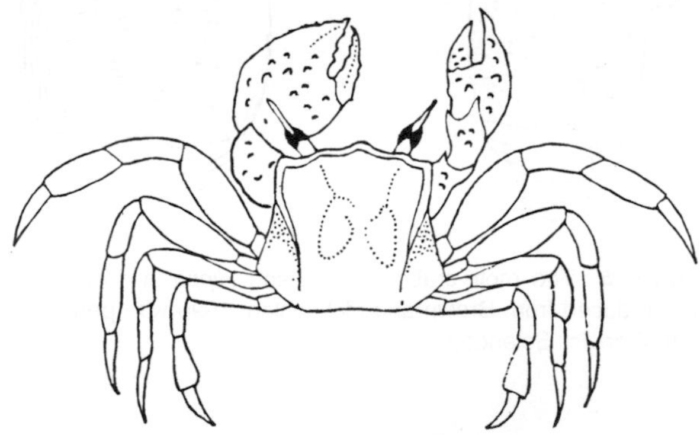

Figure 9 The ghost crab, *Ocypode*. (Reproduced from Brown and McLachlan, 1990 with permission from Elsevier Science.)

Food Relationships

As the instability of the substratum precludes the growth of attached plants in the intertidal area, the sandy-beach community is almost entirely dependent on imported food. A few diatoms may be present in the surface layers of sheltered beaches, but they are in general too sparse to be of much nutritional significance. Some potential food may be blown in from the land (plant debris, insects, etc.) and birds may defecate on the beach or die there; however, food washed up by the sea is of overwhelming importance. Three types of economy based on water-transported food may be distinguished, although they are not mutually exclusive.

One of these is based on detached algae (such as kelp), washed up onto the beach. In most cases, the rising tide pushes this material towards the top of the intertidal slope, where the semiterrestrial fauna make short work of it. However, not only are these animals messy feeders but their assimilation of the ingested material is poor, so that their feces are nutrient-rich. Algal fragments and feces find their way into the sand, where they form a food supply for bacteria and the meiofauna. From this community, mineralized nutrients leach back to sea, where they may support further algal growth – and so the cycle is completed. Some nutrients also pass to the land, because shore birds, and even birds such as passerines (e.g., swallows), feed on the semiterrestrial crustaceans and insects, as do some mammals and reptiles. On reflective beaches with a negligible aquatic macrofauna, this kelp-based economy may be the only nutrient cycle of significance.

A second type of food web depends on carrion, such as stranded jellyfish and animals detached from nearby rocky shores. Even dead seals, penguins, sea snakes, and fish may add to the food supply. Macrofaunal scavengers (swimming crabs, ghost crabs, and whelks such as *Bullia*) come into their own where carrion is plentiful. They will eat any animal matter and their assimilation is good, so they pass very little on to the bacteria and meiofauna. The acquatic scavengers are preyed on by the fishes and crabs of the surf zone, as the tide rises, and ghost crabs and other semiterrestrial crabs are taken by birds during the day and more particularly by small mammals and snakes which invade the beach at night.

The third type of economy is found on beaches with extensive surf zones displaying circulating cells of water. A number of phytoplankton (diatom) species are adapted to live in such cells (**Figure 10**), where they may achieve vast numbers, giving primary production rates of up to $10\,\mathrm{g\,C\,m^{-3}\,h^{-1}}$. Some of this production is inevitably exported to sea but the remaining dissolved and particulate organic material supports three types of community (**Figure 11**). Firstly, it drives a 'microbial loop' in the surf, consisting of bacteria, which are eaten by flagellated protozoans, which in turn are consumed by microzooplankton of various kinds. Secondly, the production supports the interstitial meiofauna of both the surf zone and the intertidal beach, again largely through bacteria. Thirdly, the surf phytoplankton is eaten by a number of species of zooplankton (notably swimming prawns), which in turn are eaten by fish; the phytoplankton also supports the filter-feeders (*Donax*, *Emerita*, etc.) of the beach and these too are consumed by fish as the tide rises. The surf zone of such a beach is thus highly productive, displays considerable diversity and biomass, and forms an important nursery area for fish.

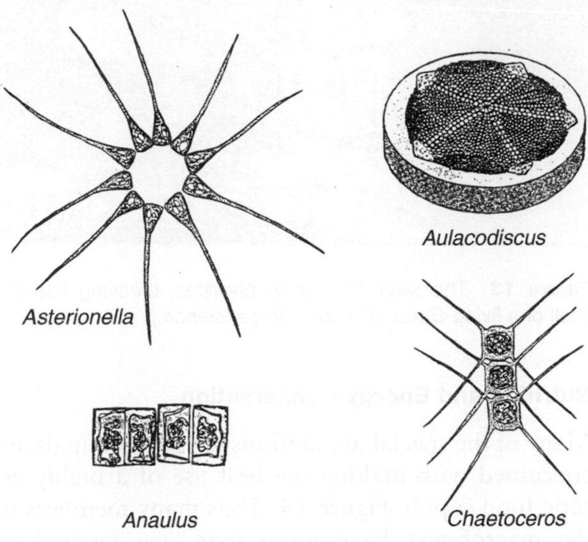

Figure 10 Common surf-zone diatoms (Reproduced from Brown and McLachlan, 1990 with permission from Elsevier Science.)

Adaptations of the Macrofauna

Locomotion

It has already been stressed that beach animals must be highly mobile and able to burrow into the sand. Two distinct types of burrowing are in evidence. The Crustacea, having hard exoskeletons, typically use their jointed walking legs as spades to dig themselves in, often with amazing rapidity; *Emerita* takes less than 1.5 s to completely bury itself. Soft-bodied invertebrates (annelid worms and mollusks) have to employ totally different methods (**Figure 12**). The sand surface is probed repeatedly, to liquefy it and so increase its permeability, and the head (of a worm) or the foot (of a clam or whelk) is inserted deep enough

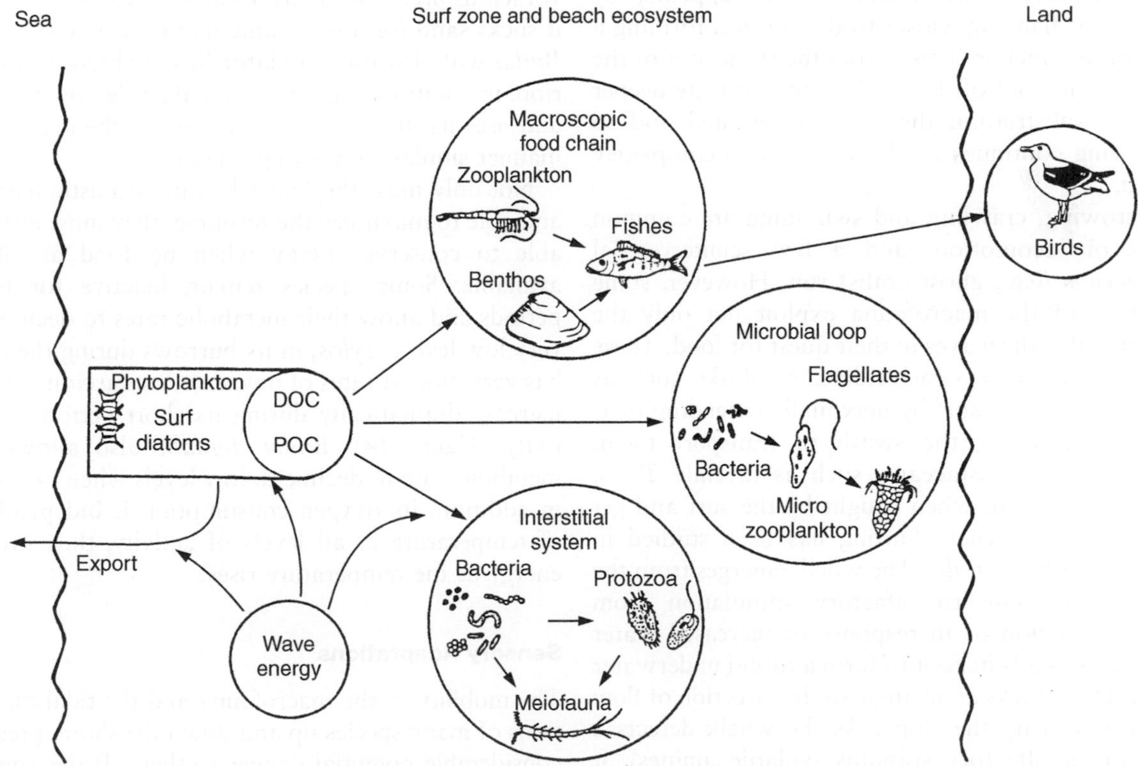

Figure 11 Food relationships on a typical dissipative beach with a rich surf-zone diatom flora, based on carbon flux (greatly simplified). DOC, dissolved organic carbon; POC, particulate organic carbon, including the diatoms themselves. (Reproduced from Brown and McLachlan, 1990 with permission from Elsevier Science.)

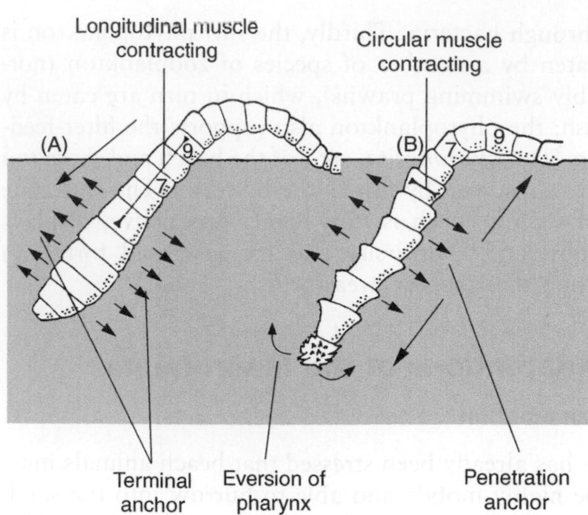

Figure 12 Principal stages of burrowing in the annelid worm *Arenicola*. (A) Anterior segments dilated to form a terminal anchor. (B) Penetration anchor formed by flanging. (Based on Trueman (1966). The mechanism of burrowing in the polychaete worm *Arenicola marina* (L.). *Biological Bulletin* 131: 369–377.

for it to swell to form a terminal anchor. With such an anchor in place, the rest of the body can be drawn down towards it. The swelling of the terminal anchor now subsides, while a new anchor appears, by swelling or flanging, closer to the surface, forming a penetration anchor. This allows the front part of the worm or the foot of the mollusk to penetrate deeper into the substratum; the cycle is repeated and so burrowing continues until the animal is completely buried.

Burrowing, crawling and swimming are common forms of locomotion, and a few semiterrestrial crustaceans (e.g., ghost crabs) run. However, some members of the macrofauna exploit not only the tides but also the waves in their quest for food. These are the surfers or swash-riders. Mollusks such as *Donax* and *Bullia* surf by maximally extending their feet and allowing the swash to transport them, whereas some crustaceans, such as juvenile *Tylos*, roll up into a ball when caught by the surf and get carried up the beach. Surfing has been studied in detail in *Bullia digitalis*. The whelk emerges from the sand in response to olfactory stimulation from stranded carrion or in response to increased water currents, spreads its foot to form a turgid underwater sail and then tacks at an angle to the direction of flow of the swash up the slope. As the whelk detects a decrease in olfactory stimulus (volatile amines), it flips over onto its other side and tacks in the opposite direction. It does this repeatedly and so reaches the carrion quickly and efficiently.

Figure 13 The swimming crab, *Ovalipes*, breaking into the shell of a living *Bullia*. (Photo: George Branch.)

Nutrition and Energy Conservation

Many of the special adaptations of beach animals are concerned with making the best use of a highly erratic food supply **Figure 13**. Thus many members of the macrofauna have more than one method of feeding. The mysid shrimp *Gastrosaccus* filter-feeds while swimming, scoops up detritus deposited on the sand, and will also tear off pieces of carrion. The ghost crab, *Ocypode*, is both a scavenger and a voracious predator **Figure 13** and, when all else fails, it sucks sand for any organic matter it may contain. *Bullia* will also turn predator in the absence of carrion; in addition, it grows an algal garden on its shell and uses its long proboscis to harvest the algae in a manner similar to mowing a lawn.

Not only must the fauna be opportunistic feeders and able to maximize the resource, they must also be able to conserve energy when no food at all is available. Some species remain inactive for long periods and allow their metabolic rates to decline to very low levels. *Tylos*, in its burrows during the day, has very modest rates of oxygen consumption, which increase dramatically during its short period of activity (**Figure 14**). *Bullia digitalis* also allows its metabolic rate to decline to low levels when inactive; in addition, its oxygen consumption is independent of temperature at all levels of activity, thus saving energy as the temperature rises.

Sensory Adaptations

The mobility of the macrofauna and the tidal migrations of many species up and down the shore, present considerable potential danger to them. If the animal moves too far down the slope it may be washed out to sea, and too far up the slope it will die of exposure (heat and desiccation). Sensory adaptations to ensure

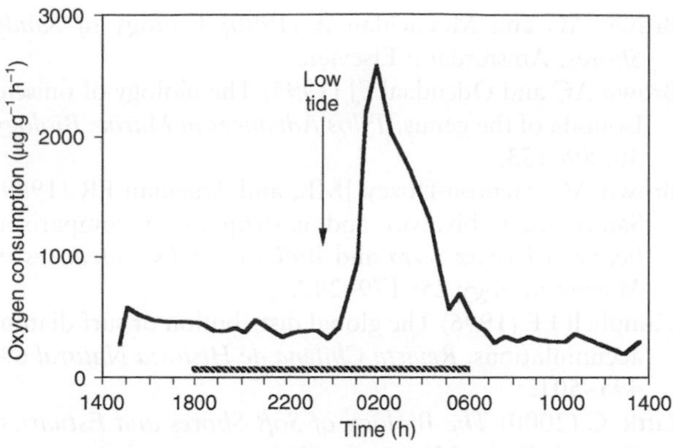

Figure 14 Respiratory rhythm of an individual of *Tylos granulatus*, showing a very low rate of oxygen consumption during the day but with a marked peak immediately after the time of low water at night. The horizontal bar denotes hours of darkness.

maintenance of position in the optimal area are thus essential. Some of these adaptations are relatively simple. For example, *Gastrosaccus* always swims against the current, facing the sea on an incoming swash and turning to face the land as the swash retreats; and the faster the current, the faster it swims. *Donax* and *Bullia* have sensory cells in the foot, which detect the degree of saturation of the sand; as the animal migrates up the beach, it eventually finds itself in unsaturated sand and migrates no further upshore but allows itself to be carried down the slope.

In contrast, the sensory adaptations of semi-terrestrial crustaceans are extremely complex and sophisticated. Most is known about the sand hopper, *Talitrus*. This little amphipod, with a 'brain' that can only be seen under the microscope, nevertheless has a whole suite of responses in its repertoire. It can sense where the shoreline of its particular beach lies relative to the sun and the moon, an orientation learnt while it was carried in the maternal brood pouch. Moreover, it has an internal clock which constantly adjusts this orientation as the sun or moon moves across the sky. It also senses beach slope and will move up the slope if it finds itself on wet sand (i.e., near the sea), or down the slope if the sand is very dry. It will react to broken horizons (e.g., dunes), which reinforces its sense of direction. Some have a magnetic compass and the animal may also orientate to polarized light patterns if the sun or moon is obscured. It also has a second internal clock geared to tidal rise and fall, which tells it when to emerge from the sand (on a falling tide) and when to retreat up the slope (on a rising tide). Some hoppers can detect an approaching storm (how is not known) and then reverse their normal response to slope, migrating as far away from the intertidal zone as possible. It is not surprising that these animals are so successful.

Plasticity of Behavior

In relatively constant environments, animals can often survive with a set of routine behavioral responses but in harsh, unpredictably changing habitats a much greater degree of flexibility (or plasticity) of behavior is called for. The sandy-beach macrofauna provides an outstanding opportunity to study this plasticity among invertebrates. Variations in behavior may in part be determined by genetic selection but the ability to learn plays a most important role in most cases. Both genetically determined plasticity and learning ability are necessary for populations of a species inhabiting different beaches, as no two beaches present exactly the same environment.

For example, *Tylos* typically moves down the intertidal slope to feed but on Mediterranean beaches it commonly moves up the slope after emergence, as food is more plentiful there. In the Eastern Cape Province of South Africa, some populations of *Tylos* have moved permanently into the dunes and have abandoned their tidal rhythm of emergence and reburial; this is also food-related. Another example, *Bullia digitalis* is typically an intertidal, surfing whelk, but in some localities where the beach is inhospitable to it, it occurs offshore, although it may return to the beach and surf if conditions become more favorable.

Perhaps the most remarkable example of the ability to adapt to circumstances concerns observations of *Tylos* on a beach in Japan. Detecting an approaching storm, the animals moved *en masse* towards the sea (i.e., towards the coming danger) and onto an artificial jetty, to which they clung until the storm had passed. This phenomenon has also been observed on a Mediterranean beach. Such behavioral flexibility plays a major role in the survival of the macrofauna and must have been rigorously selected for during the course of evolution.

Conclusion

All of the above phenomena can be linked directly or indirectly to the movements of the water. Waves, tides, and currents determine the type of shore, patterns of erosion, and deposition, the particle size distribution of the substratum, and the slope of the beach. The interaction of waves and sand results in an instability, which precludes attached plants, leading to a unique series of food webs and a series of faunal adaptations to deal with an erratic and unpredictable supply of imported food. Tides and waves are exploited to take maximum advantage of this food. Instability requires that the fauna be mobile and able to burrow, but the three-dimensional nature of the substratum also allows the development of a diverse meiofauna living between the grains. Being of necessity extremely mobile, the macrofauna require sophisticated sensory responses to maintain an optimum position on the beach and the ability to react appropriately to the conditions and circumstances they encounter on their particular beach. Finally, energy conservation is essential, so that metabolic, biochemical adaptations are ultimately dictated by an erratic food supply, again dependent on water movements.

See also

Diversity of Marine Species. Large Marine Ecosystems. Microbial Loops. Network Analysis of Food Webs. Ocean Gyre Ecosystems. Polar Ecosystems. Upwelling Ecosystems.

Further Reading

Brown AC (1996) Behavioural plasticity as a key factor in the survival and evolution of the macrofauna on exposed sandy beaches. *Revista Chilena de Historia Natural* 69: 469–474.

Brown AC and McLachlan A (1990) *Ecology of Sandy Shores*. Amsterdam: Elsevier.

Brown AC and Odendaal FJ (1994) The biology of oniscid Isopoda of the genus. *Tylos Advances in Marine Biology* 30: 89–153.

Brown AC, Stenton-Dozey JME, and Trueman ER (1989) Sandy-beach bivalves and gastropods: a comparison between *Donax serra* and *Bullia digitalis*. *Advances in Marine Biology* 25: 179–247.

Campbell EE (1996) The global distribution of surf diatom accumulations. *Revista Chilena de Historia Natural* 69: 495–501.

Little C (2000) *The Biology of Soft Shores and Estuaries*. Oxford: Oxford University Press.

McLachlan A and Erasmus T (eds.) (1983) *Sandy Beaches as Ecosystems*. The Hague: W. Junk.

McLachlan A, De Ruyck A, and Hacking N (1996) Community structure on sandy beaches: patterns of richness and zonation in relation to tide range and latitude. *Revista Chilena de Historia Natural* 69: 451–467.

FOOD WEB PROCESSES

FOOD WEBS

A. Belgrano, Institute of Marine Research, Lysekil, Sweden
J. A. Dunne, Santa Fe Institute, Santa Fe, NM, USA and Pacific Ecoinformatics and Computational Ecology Lab, Berkely, CA, USA
J. Bascompte, CSIC, Seville, Spain

Introduction

Food webs describe the network of trophic (feeding) interactions among species or populations that co-occur within communities or ecosystems. Food webs are influenced by biotic factors including ecological and evolutionary processes, such as changes in species growth rates, reproduction, mortality, and predation; and by abiotic factors, including changes in the ocean sea surface temperature, nutrient supply, and climate. These different processes are acting at different temporal and spatial scales ranging from local to regional to global. Food webs can be regarded as an emergent property of ecosystems and have been used to explore a variety of different types of questions, for example: what are the impacts of direct versus indirect ecological interactions among species; what is the relationship of network structure of species' interactions to their population dynamics; what is the relationship among ecosystem structure, function, and stability; and what is the response of ecosystems to external forcing such as climate variability and change. Food webs can be seen as assemblages of species of potentially wide-ranging body sizes that occur at different trophic levels (TLs; i.e., autroph vs. herbivore vs. predator vs. parasite) that are connected in a network or web of feeding interactions, represented by links between species. Food webs can be thought of as many interconnected food chains. Thus, the simplest possible depiction of a marine food web is phytoplankton–zooplankton–fish, where fish prey on zooplankton and zooplankton feed on phytoplankton, the autotrophs at the base of the food webs. Food webs are one approach that has been used to investigate the following ecosystem properties:

- predator–prey relationships,
- energy flow and balance,
- biodiversity and genetic diversity (co-evolutionary processes),
- allometry and scaling laws,
- marine macroecology, and
- ecological network structure and dynamics.

By studying interactions among diverse assemblages of species, theoretical and experimental food web research provides a useful way to investigate a wide range of ecosystem properties. For example, biological patterns of species body sizes combined with ecological patterns of predator–prey links among species can be used to provide estimates of ecosystem properties, such as production.

A key, but relatively unexplored issue in marine sciences is to understand the theory that describes the relationships between structure, diversity, and function in marine food webs, in conjunction with abundance/body-size relationships for marine communities that share a common energy source, and to link these ecological properties to ocean biogeochemical cycles. By linking the biology, chemistry, and physics of marine systems, we can improve our ability to manage fisheries and natural marine systems.

Basic Theory

As early as the 1920s, Alfred Lotka and Vito Volterra independently developed a mathematical model, now known as the Lotka–Volterra model, to describe predator–prey dynamics. The model can be described with two differential equations of the following form:

$$\frac{dN}{dt} = N(\alpha - \beta P) \quad [1]$$

$$\frac{dP}{dt} = P(\gamma - \delta N) \quad [2]$$

where N is the prey, P is the predator, and α, β, γ, and δ are parameters that represent the interaction between N and P. **Figure 1(a)** displays population cycles from a Lotka–Volterra model where the prey population is limited by food supply. **Figure 1(b)** shows predator–prey dynamics generated by experimental data where both the prey and the predator are not randomly distributed. While the model produces a cyclicity similar to what is seen in the observational data, the model does not take into account many of the complex dynamics, such as behavioral and evolutionary mechanisms, spatial structure, or interactions with other species underlying patterns observed in natural populations and food web interactions.

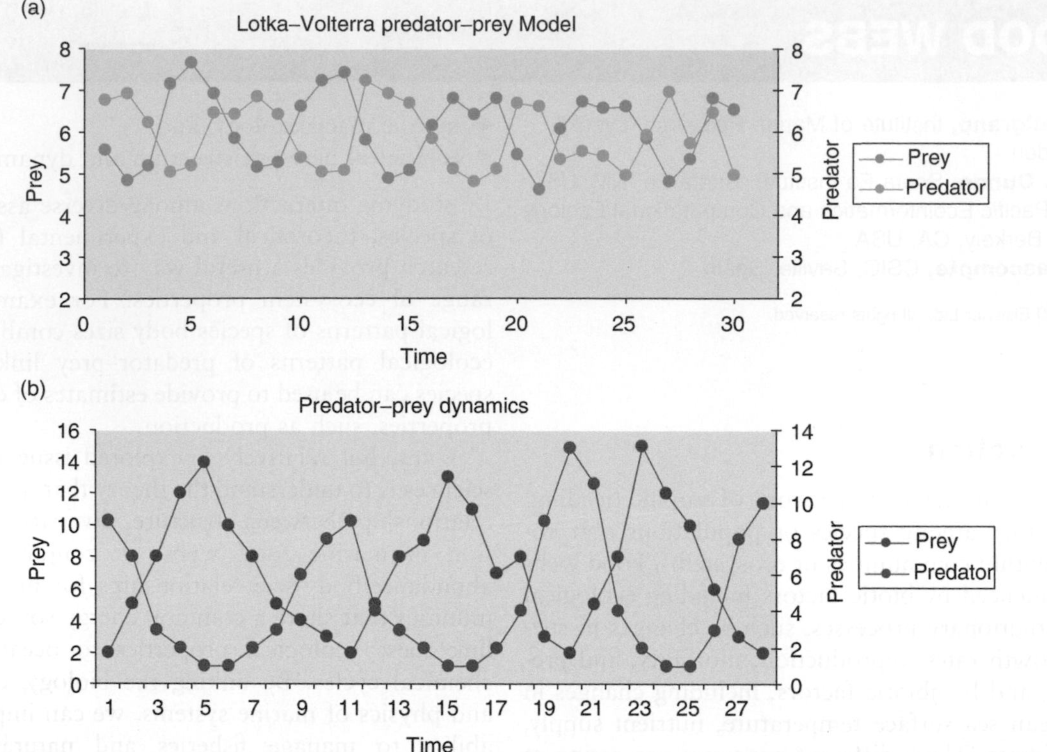

Figure 1 (a) A deterministic Lotka–Volterra simulated predator–prey dynamics. (b) An example of predator–prey dynamics in empirical data.

Much of Lotka and Volterra's theoretical work was later applied by Riley to describe a simple ecosystem with a single nutrient, phytoplankton (i.e., prey) and a grazer zooplankton (i.e., predator). These types of '*NPZ*' models are used to describe species interactions and food web dynamics and are the basis of the representation of ecosystems in today's ocean biogeochemistry models. The model is described with three differential equations, where N, P, and Z are concentrations of the nutrient (i.e., nitrogen), phytoplankton, and zooplankton, and other parameters represent the growth rate, μ, the grazing rate, g, a 'transfer efficiency' coefficient, γ, and a respiration coefficient, K:

$$\frac{dN}{dt} = -\mu P + R \qquad [3]$$

$$\frac{dP}{dt} = \mu P - gPZ \qquad [4]$$

$$\frac{dZ}{dt} = \gamma gPZ - KZ \qquad [5]$$

This type of theoretical approach has been used to study food web and ecosystem properties in marine ecosystems, and especially to understand the carbon cycle in pelagic ecosystem in relation to other macronutrients, such as nitrogen and phosphorus, which follow a 'Redfield ratio' of N:P = 16 for planktonic organisms. In an idealized marine pelagic food web, phytoplankton, such as diatoms, fix carbon via photosynthesis, and zooplankton, such as copepods, feed on the available phytoplankton biomass. These organisms, given their roles in the food web, thus regulate the carbon cycle via respiration (phytoplankton and zooplankton) and excretion (zooplankton). Changes in pelagic food webs due to different constraints and trade-offs can result in shifts of body-size composition of predator and prey species in the food web. These changes, combined with the effect of different external drivers such as climate variability and changes in the physical processes of the upper ocean (e.g., sea surface temperature or SST, mixing), can affect and change carbon pathways in the food webs and affect the way the ocean acts as a sink for CO_2.

Robert May developed food web modeling further in the early 1970s when he used a theoretical framework to relate aspects of local community stability to three community properties: the number of species, S; the connectance, C (i.e., a basic measure of complexity in food webs (the probability that any two species have a trophic (feeding) relationship); and the strength of species' interactions, I. In recent years, simple food-web-structure models (e.g., the cascade and niche models) have been developed that successfully predict the network structure of complex food webs, and have been integrated with nonlinear bioenergetic dynamics to explore various aspects of ecosystem stability within a more realistic model framework. These simple models arrange species over a niche dimension, such as body size, and allow species to probabilistically feed on species over a defined range of sizes below its own.

In many cases, food web models have been developed within a more theoretical framework often because of the lack of adequate data representing trophic interactions. Field and experimental data available for describing the degree of species resolution in food webs are moderate as Jason Link and colleagues showed in assembling published marine food webs covering the past 25 years. Therefore, there is a real need to develop research programs that can provide better data with a more adequate coverage and higher resolution of species and trophic interactions in marine ecosystems. For example, recent advances in marine microbes and marine genomic research may provide useful data and extend the resolution of marine food webs to include molecular networks, thus opening new grounds for testing new hypothesis and validating food web models using both empirical and theoretical frameworks.

However, Michael Follows and colleagues recently presented a marine ecosystem model of microbial communities where the theoretical framework has been based on both field and experimental data. The results from the model simulations corresponded well with the observed phytoplankton distributions, showing the importance of using available field and experimental data to validate theoretical models. Further, this ecosystem model allowed community structure and species diversity to be emergent properties of the system where the species physiological characteristics and resource partitioning were stochastically determined rather then assigned *a priori.* This modeling approach can be seen as an extension of the pioneering work by Tilman, and the more classical *NPZ* representation of marine food webs, in the sense that includes and recognized stochasticity and self-organization as properties of the adaptive nature of ecosystems, contrary to a more deterministic and mechanistic representation of the 'real world'.

Function and Stability

Models of food web structure, such as the niche model, have successfully described and predicted the network structure of feeding interactions, but do not yet provide a strong mechanistic basis for understanding those emergent properties. This limits their utility for understanding how food webs are likely to respond to external perturbations arising from climate change and harvesting. However, by building on earlier work in the 1940s by R. L. Lindeman, researchers are increasingly focused on the way food web dynamics is linked to ecosystem function and stability, for example, in terms of the energy used and transferred within and between different species ensembles in an ecosystem. For example, **Figure 2** depicts the eastern Bering Sea food web, and shows the coupling between the benthos and the pelagic ensemble and how a common energy source is used throughout the food web from the base prey as a food source (e.g., phytoplankton), moving up to higher TLs such as the top predators (e.g., large marine mammals).

This type of food web representation, in which taxa are aggregated into several trophic groups and species, and links are quantified based on energy transfer and biomass consumption rates, have been used extensively to study ongoing changes in the structure, function, stability, and diversity of marine ecosystems. The information provided by this way of aggregating species can be used to study the adaptability of the whole ensemble to respond to different types of pressures, such as a natural increase of predation, changes in mortality rate, and anthropogenic changes via commercial fishing. This approach can also be used to study bottom-up effects in food webs (e.g., variability in nutrients supply related to climate variability), versus top-down effects (e.g., changes in predation pressure). Understanding the consequences and mechanisms that regulate food web dynamics is an important step toward predicting community-wide and ecosystem changes at a different level of organization in relation to an example of overfishing practices.

Food Webs and Fisheries

Daniel Pauly and colleagues analyzed global trends of the mean TL of fisheries landings representing the major types of marine ecosystems worldwide. They were able to show a shift in the fisheries landings from higher-trophic-level catches represented, for example, by long-lived bottom species like cod, to lower-trophic-level species, typically short-lived pelagic fish such as sardines, anchovies, and herrings. Their findings pointed out that 'fishing-down-the-food-web' will result in a first phase characterized by an increase in the landings, but then this pattern will lead to a

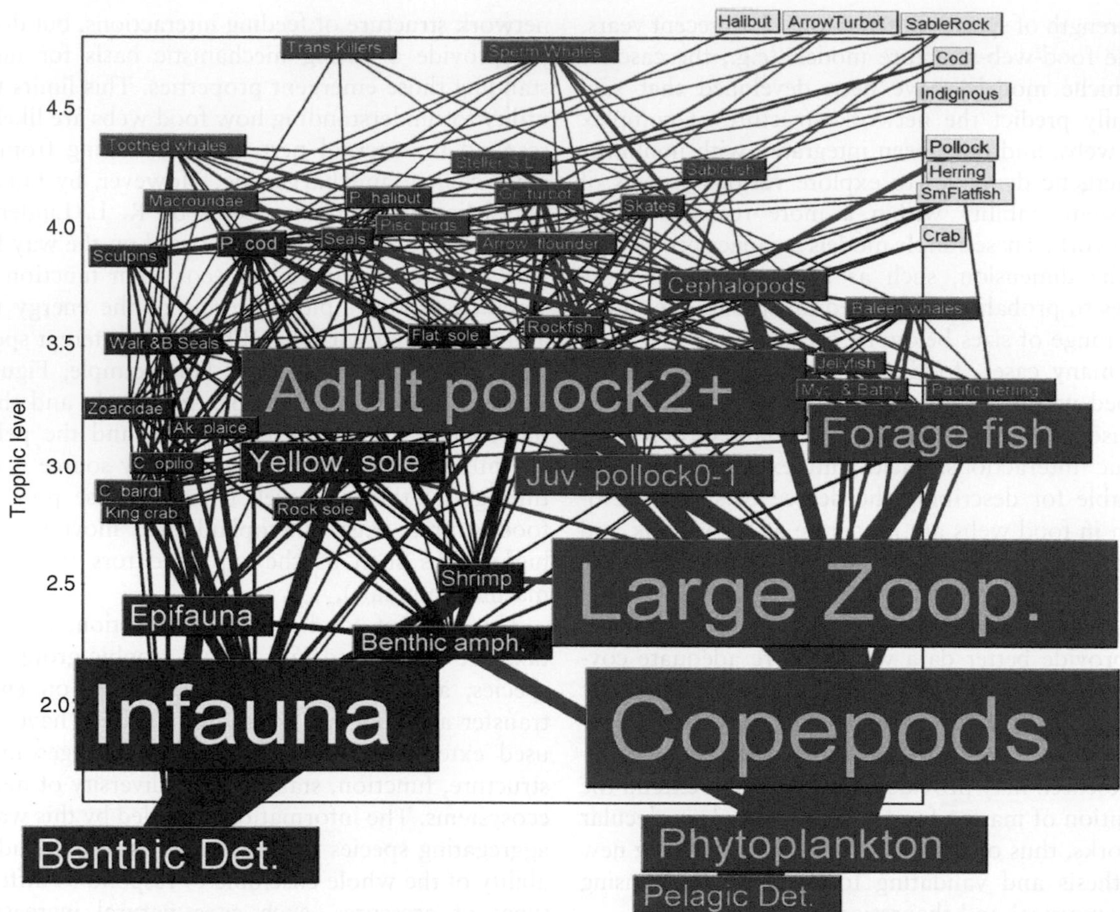

Figure 2 The eastern Bering Sea food web as described in: Aydin KY, Lapko VV, Radchenko VI, and Livingston PA (2002) *A Comparison of the Eastern and Western Bering Sea Shelf and Slope Ecosystems through the Use of Mass-Balance Food Web Models*. US Department of Commerce, NOAA Technical Memoirs NMFS-AFSC-130, 78pp, http://www.afsc.noaa.gov/Publications/AFSC-TM/NOAA-TM-AFSC-130.pdf. Coloration indicates benthic energy (blue) and pelagic energy (red); the *y*-axis represents the food web trophic levels.

shift toward a phase characterized by lower catches, thus suggesting that the current trends in fisheries are not sustainable. Therefore, monitoring changes in the TL of the catch in multispecies fisheries can provide useful information for developing and implementing management strategies toward sustainable fisheries practices and policies worldwide.

One of the first depictions of a marine fishery within a complex food web context was a 29-taxa food web of the Benguela system off the coast of South Africa (**Figure 3**). Canadian ecologist Peter Yodzis used this food web to explore whether culling fur seals is likely to have a positive or negative impact on available biomass of commercially harvested fish species, particularly hake, anchovy, and mackerel. A simplistic view of fish–seal–human dynamics is that if seals and humans both eat particular fish species, when humans remove seals from the system there will be more fish for humans to harvest without a net decrease in the fish stock. However, this is based on a very oversimplified view of a more complex ecosystem, and does not take into account the feeding interactions of seals and commercial fish species with other species in the food web. For example, the 29-taxa version of the Benguela ecosystem includes phytoplankton, bacteria, several types of zooplankton, noncommercial fish species including sharks, whales, birds, and other taxa. Using dynamical modeling of changes in population biomasses of all 29 food web taxa, given reductions in seals, it was shown that culling fur seals is more likely to lead to an overall loss of commercial fish biomass than a gain. This apparently nonintuitive result can occur because the interactions between commercial fish species and fur seals within the Benguela food web are influenced by other species in the web. For example, there are over 28 million different food chains that connect seals and hake.

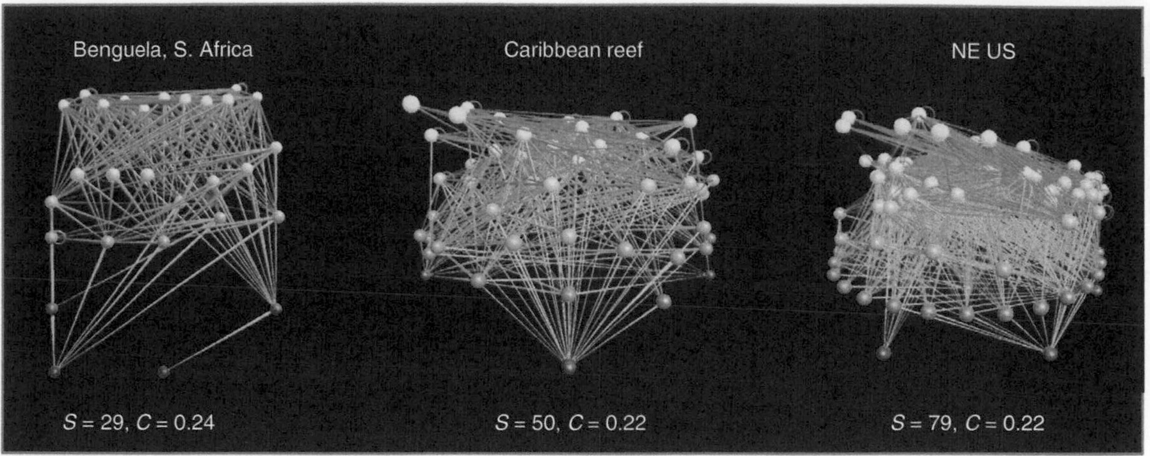

Figure 3 The network structure of three marine food webs. Spheres represent 'trophic species', which are taxa from an originally published food web that share the same set of predators and prey and are grouped together into a single node. Elongated cones represent feeding links, with the narrower part of the cone pointing to the prey taxon. Basal taxa (e.g., phytoplankton and detritus) are shown in red at the bottom of each food web, with highest-trophic-level taxa shown in yellow at the top. S = number of trophic species, C = connectance = L/S^2, where L = feeding links. Images produced with FoodWeb3D software written by R. J. Williams and available through http://www.foodwebs.org.

This type of analysis underscores the dangers of basing marine fisheries policy on overly simplistic models that ignore the broader food web context.

General Properties of Marine Food Web Structure

Because detailed food web data are difficult to compile, only a few complex webs (i.e., with more than 25 trophically distinct taxa represented) have been reported for marine systems. In addition to Benguela, detailed food webs for the Caribbean reef system (**Figure** 3) and for the northeast US shelf (**Figure** 3) have been used to explore marine food web structure. One of the questions that scientists have considered is whether the network structure of marine food webs is fundamentally similar to or different from food webs from other kinds of ecosystems, such as terrestrial, freshwater aquatic, and estuarine habitats. Connectance, or C, is generally measured as the number of total links (L) divided by the number of species (S) squared, or L/S^2. C generally varies from about 0.03 to 0.3 in food webs, which means that *c.* 3–30% of possible feeding links (S^2) actually occur (L). Marine food webs tend to have high C (>0.2, **Figure** 3) compared to other kinds of food webs. This may result from there being a large number of generalist omnivores in marine food webs (i.e., species that eat many taxa at multiple trophic levels). However, once differences in the numbers of species (S) and levels of connectance (C) among food webs are taken into account, using the empirically corroborated 'niche model' of food web network structure, marine food webs appear to have the same fundamental network structure as food webs from other habitats. This suggests that there are strong, universal constraints on how feeding interactions among species are organized in ecosystems.

Overfishing and Biodiversity

The previously described dynamical modeling of the 29-species Benguela food web is only one of many approaches for using food webs to explore their potential response to perturbations, such as overfishing and biodiversity loss. A second approach uses the network structure of complex food webs to look at potential cascading secondary extinctions when focal species or groups of species go extinct. This type of analysis has been done for food webs from a variety of ecosystems including marine habitats, and shows that the structural robustness of marine food webs is also consistent with trends from other types of food webs. As expected, given their relatively high connectance, marine food webs appear fairly robust to the loss of least-connected taxa as well as random taxa. Still, the short, average path length (i.e., the minimum number of steps to connect two randomly picked species through predator–prey interactions) between marine taxa (1.6 feeding links) suggests that effects from perturbations, such as overfishing, can be transmitted more widely throughout marine ecosystems than previously appreciated. In general, understanding the patterns of complex food webs is important for understanding how these communities will respond to human-induced perturbations. The problem of overfishing is normally considered as if target populations are isolated from one another. But, as already discussed, they are not. Depleting one population may originate changes in population abundance that can propagate through the food web. When these snowball effects are strong, they are called trophic cascades. In the Caribbean Sea, for example, overfishing of sharks may increase the population of their prey (groupers and other big fishes), that in turn may reduce the abundance of parrotfish and other grazer fishes they consume. This may ultimately contribute to the observed shift from coral to algae, with profound implications for hundreds of species that depend on coral reefs for food and shelter.

We have discussed two approaches (i.e., focus on structure only; integrate structure with dynamics) for using food webs to think about how scenarios, such as depleting or removing species, might affect marine ecosystems. There is a third approach that has been

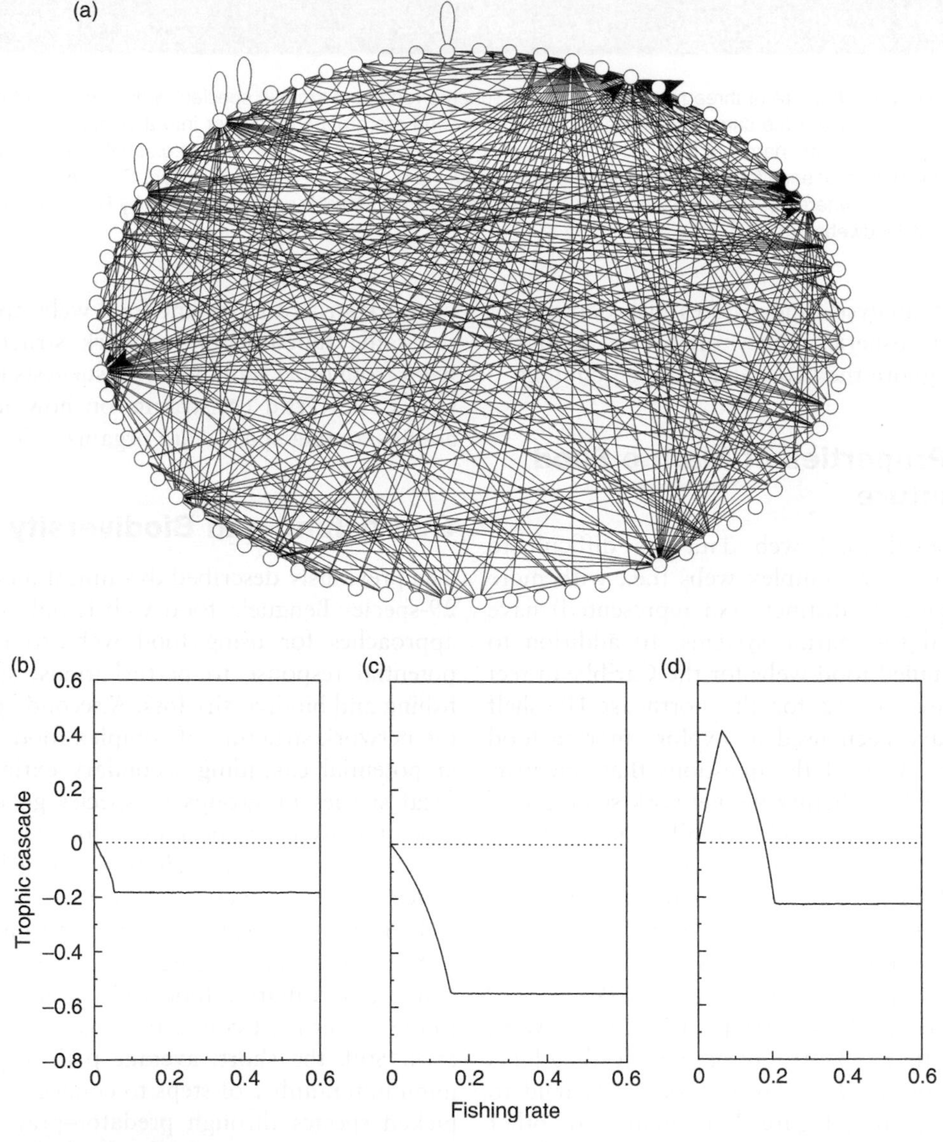

Figure 4 Marine food webs can be extremely complex and their structure greatly affects the community-wide consequences of overfishing. (a) A random subset of the expanded Caribbean food web, where circles represent species and arrows describe predator–prey interactions. The thickness of the arrow is proportional to the strength of the interaction. (b)–(d) The magnitude of trophic cascades measured as changes in the long-term density of the basal species following the fishing of the top predator when the two trophic links are weak (b), strong (c), and strong with similarly strong omnivory (d). The way these interaction strengths are combined to form the basic building blocks of food webs determines the magnitude of trophic cascades after the overfishing of the top predator. Modified from Bascompte J, Melián CJ, and Sala E (2005) Interaction strength combinations and the overfishing of a marine food web. *Proceedings of the National Academy of Sciences of the United States of America* 102: 5443–5447.

demonstrated using a detailed food web from the Caribbean, which not only depicts the network structure, or who eats whom, but also gives the strength of each interaction (**Figure 4**).

Interaction strengths are a sort of per capita effect of predators on their prey populations, and as such they have a large influence on the stability of communities. Previous work has shown that the bulk of interaction strengths are weak, which ensures community persistence by buffering perturbations from being transmitted through the entire food web. But almost nothing has been known about how these interaction strengths combine to form the basic building blocks of marine food webs. Two strong interactions are avoided in two consecutive levels of a food chain, and in the few cases in which they are present, they are accompanied by strong omnivory (top predators eating the prey of their prey) more often than expected by chance. By using a simple food web model, it is possible to explore the dynamic implications of these structural findings. The reported combinations of interaction strengths promote the stability of the community in the face of random overfishing by reducing the likelihood of trophic cascades. However, fishing is not random but targets big top predators such as sharks. Because these sharks are involved in a large fraction of the strongly interacting food chains, their overfishing may have profound community-wide consequences. This aspect has also been recently presented for another coastal location in North Carolina by Meyers and colleagues, in relation to scallop fishery, showing the importance of top-down and indirect effect in oceanic food web dynamics.

A fourth approach for using food webs to think about the stability of ecosystems in the face of perturbations is analyzing how food webs are structured into subwebs. How these subwebs are interrelated determines the compartmentalization of food webs, which is important for their robustness to perturbations. One way to detect compartments is by defining a *k*-subweb, a subset of the food web containing all the species that have at least *k* interactions with species within that group. The way these *k*-subwebs are organized in real food webs is very heterogeneous. The bulk of subwebs have a low *k*-value (a small number of interactions within the subweb), but their abundance distribution is very skewed with a single, most dense subweb. Other subwebs are highly connected to this core, which brings cohesion to the whole food web, and so makes it more robust to the random loss of species, although probably more susceptible to the propagation of contaminants.

Outlook

Food web models can be seen both as an idealized representation of ecosystem complexity and as a source of information for the patterns we observe in natural systems. Food web models can be used as tools to investigate the properties of a system and its dynamics. In the future, combining experiments with theoretical approaches will provide even more useful information on the underlying mechanisms driving marine ecosystem dynamics and linking food web dynamics with ocean's biogeochemical cycles and climate change. Food web models can be used to address a community-wide approach to a growing set of problems observed in marine systems for which management solutions have not been clearly identified. In the marine environment these include depleted stocks of fish, declining marine mammal populations, receding ice cover, and changing pH (oceanic acidification). Taking responsibility for the multiple human contributions in each case is crucial to dealing with the complexity of problems through management. We can use empirical information and patterns that emerge from food web analysis as guidance for estimates of what is sustainable – exemplified by what is called in fisheries policy as the 'ecologically allowable take' (EAT). Finally, the analysis of ecological food webs in marine systems needs to be linked to management issues that require social and economic involvement.

See also

Large Marine Ecosystems. Microbial Loops. Network Analysis of Food Webs. Ocean Gyre Ecosystems. Polar Ecosystems. Upwelling Ecosystems.

Further Reading

Aydin KY, Lapko VV, Radchenko VI, and Livingston PA (2002) *A Comparison of the Eastern and Western Bering Sea Shelf and Slope Ecosystems through the Use of Mass-Balance Food Web Models*. US Department of Commerce, NOAA Technical Memoirs NMFS-AFSC-130, 78pp. http://www.afsc.noaa.gov/Publications/AFSC-TM/NOAA-TM-AFSC-130.pdf (accessed Apr. 2008).

Bascompte J, Melián CJ, and Sala E (2005) Interaction strength combinations and the overfishing of a marine food web. *Proceedings of the National Academy of Sciences of the United States of America* 102: 5443–5447.

Belgrano A, Scharler U, Dunne JA, and Ulanowicz RE (eds.) *Aquatic Food Webs: An Ecosystem Approach*. Oxford, UK: Oxford University Press.

Cohen JE and Newman C (1985) A stochastic theory of community food webs. Part I: Models and aggregated data. *Proceedings of the Royal Society B* 224: 421–448.

Dunne JA, Brose U, Williams RJ, and Martinez ND (2005) Modeling food-web dynamics: Complexity-stability implications. In: Belgrano A, Scharler U, Dunne JA, and Ulanowicz RE (eds.) *Aquatic Food Webs: An Ecosystem Approach*, pp. 117–129. Oxford, UK: Oxford University Press.

Dunne JA, Williams RJ, and Martinez ND (2004) Network structure and robustness of marine food webs. *Marine Ecology Progress Series* 273: 291–302.

Follows MJ, Dutkiewicz S, Grant S, and Chisholm SW (2007) Emergent biogeography of microbial communites in a model ocean. *Science* 315: 1843–1846.

Fowler CW (1999) Management of multi-species fisheries: From overfishing to sustainability. *ICES Journal of Marine Science* 56(6): 927–932.

Li WKW (2002) Macroecological patterns of phytoplankton in the northwestern Atlantic Ocean. *Nature* 419: 154–157.

Lindeman RL (1942) The trophic-dynamic aspect of ecology. *Ecology* 23: 399–418.

Link J (2002) Does food web theory work for marine ecosystems? *Marine Ecology Progress Series* 230: 1–9.

May RM (1973) *Stability and Complexity in Model Ecosystems*. Princeton, NJ: Princeton University Press.

Melián CL and Bascompte J (2004) Food web cohesion. *Ecology* 85: 352–358.

Miller CB (2003) *Biological Oceanography*. Oxford, UK: Blackwell.

Myers RA, Baum JK, Sheperd TD, Powers SP, and Peterson CH (2007) Cascading effects of the loss of apex predatory sharks from a coastal ocean. *Science* 315: 1846–1850.

Myers RA and Worm B (2003) Rapid worldwide depletion of predatory fish communities. *Nature* 423: 280–283.

Pauly D, Christensen V, Dalsgaard J, Froese R, and Torres F, Jr. (1998) Fishing down marine food webs. *Science* 279: 860–863.

Pimm SL (2002) *Food Webs*. Chicago, IL: The University of Chicago Press.

Solé RV and Bascompte J (2006) *Self-Organization in Complex Ecosystems*. Princeton, NJ: Princeton University Press.

Solow AR (1998) On the goodness of fit of the cascade model. *Ecology* 79: 1294–1297.

Steele JH (1976) Application of theoretical models in ecology. *Journal of Theoretical Biology* 63(2): 443–451.

Tilman D (1977) Resource competition between planktonic algae: An experimental and theoretical approach. *Ecology* 58: 338–348.

Yodzis P (1998) Local trophodynamics and the interaction of marine mammals and fisheries in the Benguela ecosystem. *Journal of Animal Ecology* 67: 635–658.

Yodzis P (2000) Diffuse effects in food webs. *Ecology* 81: 261–266.

Relevant Websites

http://www.ecopath.org
– Ecopath with Ecosim (EwS).

http://www.foodwebs.org
– Pacific Ecoinformatics and Computational Ecology Lab.

http://www.sahfos.org
– Sir Alister Hardy Foundation for Ocean Science (SAHFOS).

NETWORK ANALYSIS OF FOOD WEBS

J. H. Steele, Woods Hole Oceanographic Institution, Massachusetts, USA

Introduction

Photosynthesis transforms energy from sunlight into calories within marine plants, predominantly phytoplankton and seaweeds. The plants use this energy to take up carbon and essential nutrients, such as nitrogen and phosphorus, from sea water to produce organic materials.This organic matter forms the base for the food web composed of herbivores thateat those plants and the carnivores that prey on the herbivores. There can beseveral trophic levels of carnivores, including all the fish species that weharvest.

In the open sea, away from the coast and theseabed, microscopic single-celled phytoplankton dominate the plant life, so the organisms tend to get bigger as each trophic level feeds on the onebelow: from small herbivorous crustaceans to larger invertebrates, to small andlarge fish, and finally to human beings and marine mammals.

When any animal consumes food, most of the energy in that food is used for metabolism; some of the remainder is excretedas waste products and only a small fraction goes to growth. In young, cold-blooded animals in the sea, growth can be relatively efficient: 20–30% of energy intake. In older animals growth is replaced by reproduction, which, after all, is the whole point of the life cycle. As an approximate overall figure we usually assume that the energy converted into growth and reproduction is about 10% of the total energy intake. Thus, in asimple trophic pyramid, the energy in successive trophic levels would be 100:10:1:0.1. From this one can see why we are encouraged to eat plants on land, and why fish from the sea are energetically expensive in terms of plant calories.

In practice it is very difficult to measure directly the energy content of marine organisms and, especially the energy transfers between trophic levels. However, carbon is the essential building block for organic matter, being taken up from inorganic form at photosynthesis and returned to sea water during respiration. The carbon content of organisms and the rates of uptake of inorganic carbon can be measured using radioactive carbon, carbon-14, as a tracer; transfers through the food can then bemeasured. Carbon is therefore frequently used as a proxy for the more elusive concept of energy flow.

All organisms also require many essential elements, and many of these are in short supply in sea water. In particular, inorganic nitrogen and phosphorus, as nitrate and phosphate, are regarded as limiting factors in photosynthesis and thus in the rate at which energy and carbon are supplied to the food web. Since organic carbon, nitrogen and phosphorus have a roughly constant ratio in marine organisms (the Redfield ratio), nitrogen can also be used as a proxy for energy flow; it will not, however be considered here.

Carbon and nitrogen fluxes are also important in relation to other issues concerning marine food webs. The biologically mediated flux of carbon to deeper water is important for the calculation of global carbon budgets and climate change. Eutrophication in coastal waters produces imbalance in the food webs of these regions.

Units

Ideally, all presentations would be in units of energy but, as discussed, the actual measurements are often in carbon or, for fisheries, in biomass (wet weight). Conversion from one unit to another will vary with the organism but, as an approximation:

$$10\,\text{kcal} = 1\text{g carbon} = 10\,\text{g biomass.} \qquad [1]$$

History

The first attempt to produce an energy budget for an aquatic food web was made by Lindeman in 1942, for a freshwater lake. In 1965 a budget for the North Sea (**Figure 1**), showed possible pathways from primary production to commercial fisheries. The main conclusion was that the ecological efficiencies needed to be quite high for the budget to balance. This conclusion depends on the number of pathways introduced; for example, the category ‘other carnivores’ is put in to account for the presence of gelatinous invertebrate predators, such as ctenophores, that are ubiquitous but the biomass of which is difficult to estimate. Without this box, the herbivore efficiency could be less than 20%.

In 1969 Ryther wrote a seminal paper that outlined the pathways at the global scale, by dividing the marine ecosystems into three types, upwelling,

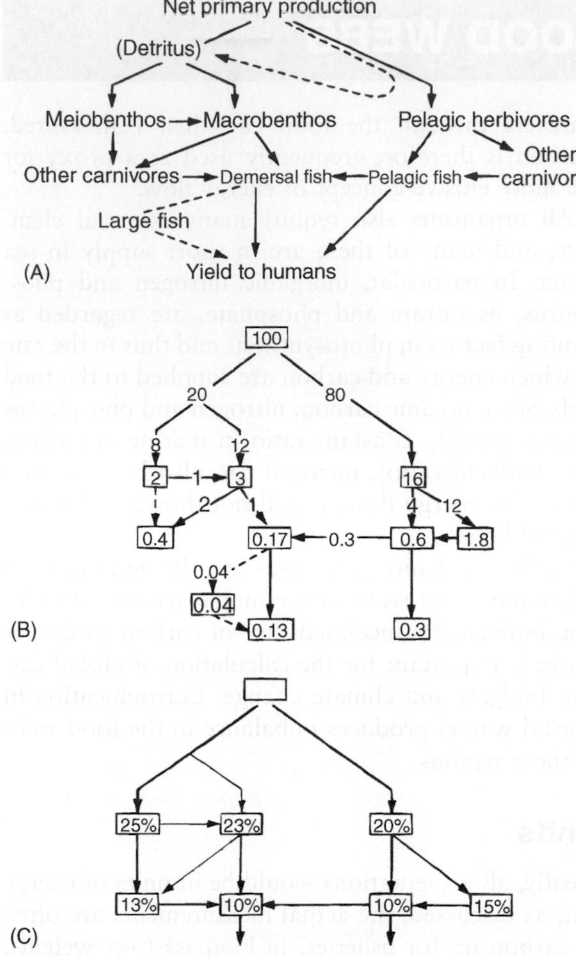

Figure 1 Energy or carbon fluxes in the food web for the North Sea.(A) A simplified food web. (B) Transfers of particulate organic carbon assuming a primary production of 100 gC m^{-2} peryear. Each box indicates the production by that component available to its predators, including humans. (C) The transfer or ecological efficiencies calculated from the output as percentage of input.

continental shelf, and open ocean(**Table 1**). The major implication of Ryther's calculations was that there were no great untapped fish resources in the open ocean. He estimated that the potential sustained yield offish to humans was unlikely to be greater than 100 million tons. In 1998, the yield was about 90 million tons.

The reason for the low open-oceanyield of fish – and the controversial part of Ryther's calculation– is his estimate of five trophic levels from primary production to fish. This choice was based on the very small size of open-ocean phytoplankton. In the intervening years, research on the base of open-ocean food webs has shown that much of the primary production is recycled through the smallest sized categories in the food web – the microbial loop. For the intermediate trophic levels, such as the

Table 1 Primary production, number of trophic levels to commercial fish species, efficiency of transfer of energy (or carbon) between trophic levels and resultant ratios of fish to primary production for three major marine categories

	Ocean	*Coastal*	*Upwelling*
Primary production (gC m^{-2} year^{-1})	50	100	300
Trophic levels	5	3	1.5
Ecological efficiency (%)	10	15	20
Fish production: primary production (%)	0.01	0.3	12

gelatinous predators, there is still relatively little quantitative information on biomass and production; this has led to an extensive development of mathematical treatments to infer energy or carbon flows.

Quantitative Methods

Early calculations, such as those illustrated in **Figure 1**, were put together by a very informal series of iterations until all the flows balanced and the ecological efficiencies were not unreasonable. The major advance has been the development of numerical methods to make more objective estimates of best fit. Essentially, the ecosystem is considered to be at steady state, so that, for any box, such as those in **Figure 1**, there has to be a balance between rates of energy flow or carbon flux entering and leaving:

$$\text{consumption} = \text{predation} + \text{metabolism} + \text{export(input)} \quad [2]$$

The information for each of these terms for each box canbe qualitatively different. In particular, for animals:

$$\text{consumption} = \text{growth} + \text{reproduction} + \text{metabolism} \quad [3]$$

These complexities are eliminated if it is assumed that, for each box, there is a constant ecological efficiency, e_i:

$$e_i = \frac{(\text{growth} + \text{reproduction})}{\text{consumption}} \quad [4]$$

Generally, for the system to be soluble, the terms in eqn [2] for all boxes must be expressed as a set of linear equations. For this purpose the variables are usually taken to be the energy, carbon or biomass flux through each box, T_i, that is available to higher trophic levels. Linearity requiresthat the rate processes in eqn [2] are constants independent of T_i.

$$\frac{T_i}{e_i} = \sum b_{ij} \cdot T_j + c_j \qquad i = 1, n, \qquad [5]$$

where b_{ij} is the fraction of T_j consumed by T_i and c_i is the constant rate of external input to T_i; c_i would be, for example, the rate of primary production. Note that, as in **Figure 1**, all the variables are rates of throughput in the food web and the parameters are non-dimensional.These matrix inversion techniques can also be used with nutrient flows that involve recycling.

An alternative top-down approach is often used for flow calculations where emphasis is on the higher trophic levels, and fish yields are the defining input. For these situations,consumption, C_i, is expressed as:

$$\text{consumption} = \frac{(\text{consumption})}{\text{biomass}} \cdot \text{biomass} = \left(\frac{C_i}{B_i}\right) \cdot B_i. \qquad [6]$$

Then the biomass in each box, B_i, becomes the state variable. The rate process C_i/B_i is assumed constant, for each box. Then:

$$\mathrm{a}_i \cdot B_i = \sum p_{ij} \cdot B_j + \mathrm{d}_i, \qquad [7]$$

where $\mathrm{a_i}$ represents production per unit biomass, the ‘P/B’ ratio, assumed constant for any box; p_{ij} is the unit consumption rate of B_j on B_i, assumed to be independent of the magnitude of B_j or B_i; d_i is the export, assumed constant for each box.

If all the parameters are known for either eqns [2] and [7] then the set of equations canbe solved for T_i or B_i. In practice it is never as simple as that. Usually a number of parameters are unknown or ill defined; for example, as upper or lower bounds.Then iterative procedures can be used to obtain best fits. The selection of the number of boxes and the content of each box is the critical process. The assumption of a linear, steady-state ecosystem is the critical constraint for this type of analysis involving a large number of boxes. This approach is complemented by the highly nonlinear analysis used for modelling offisheries and plankton dynamics.

Examples of Carbon and Biomass Networks

The emergence of the microbial loop as asignificant feature of pelagic food webs, together with the availability of computer based inverse methods, resulted in a focus on flow analysis of the lower levels of the pelagic ecosystem; levels that were represented simply as phytoplankton-to-zooplankton in **Figure 1**. The results of analyses (**Figure 2**) illustrate how this part of the system isexpanded into seven boxes with 19 links. Two examples in **Figure 2**, from the continental shelf around Britain, show the kinds of patterns that arise from these calculations. The authors point out differences between the English Channel and the Celtic Sea. The former puts 85% of primary production through the microbial loop, whereas the latter has 40% going directly to the mesoplankton (predominantly copepods). However,for both examples, the exports from this part of the food web are very similar: 3% to higher trophic levels via the mesoplankton, and 30% to the benthos as detritus. The major change from the earlier calculations is the dominant role of the microbial loop, involving recycling of most of the primary production.Only the ‘new’ production from nitrate (NO_3) in **Figure 2A** fuels the export of detritusand mesoplankton to higher trophic levels.

The alternative approach (eqn [7]) has been used for a wide range of aquatic ecosystems, including lakes and coral reefs as well as ocean systems. One example (**Figure 3**) shows calculations for the continental shelf around the Gulf of Mexico. There is no microzooplankton box and the general flow patterns are not dissimilar to those of **Figure 1**; 30% of primary production goesdirectly to detritus, as does 40% of zooplankton food, presumably as fecal material. Detritus then feeds the demersal fish through the benthos. Note, however, that once again detritus is the dominant biomass (85% of the total).

It would appear that all the solutions of these linear systems – whether as energy or biomass, for lower or higher trophic levels – require a major role for detritus. Detritus is a difficult variable to define and measure. This box can contain phytoplankton cells, zooplankton feces, marine snow and other residues. It is assumed that detritus is broken down by bacteria but the a_i or e_i values are not well known. It is rarely sampled directly, eitheras biomass or for rate processes, and so is an empty box that can be used to balance flows.

Discussion

In terrestrial ecosystems most of our food comes directly from plants, with the remainder from herbivores – animalsthat eat plants. In the sea, the fish we eat are nearly all carnivores. Many feed on the small herbivorous crustaceans, such as krill, but some – the most highly prized, like tuna – themselves feed on carnivores, such as smaller fish. So what we hope to take from the sea, the potential fishery yield,

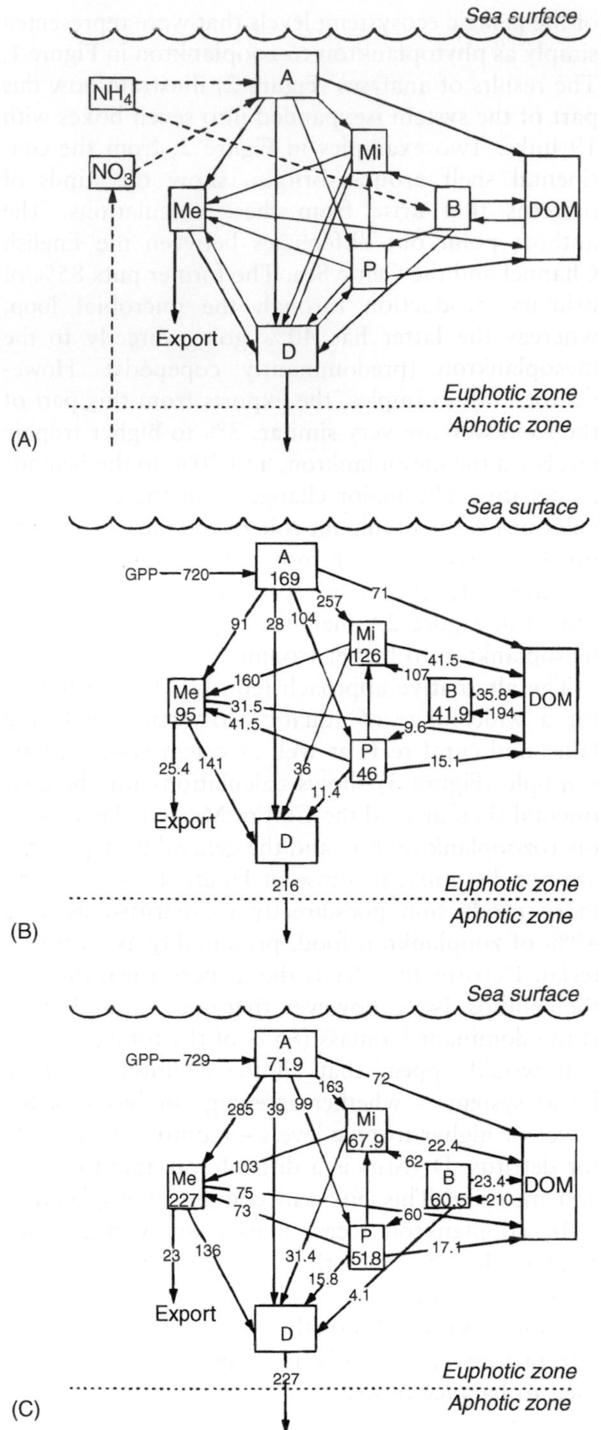

Figure 2 (A) Generic model of a plankton food web in the upper layer of a stratified water column. (B) The inverse solution for carbon fluxes in summer at a station in the English Channel. Values inside the boxes are respiration flows (mgC m^{-2} per day).(C) Flows at a station in the Celtic Sea. A, autotrophs; B, bacteria; D, detritus; DOM, dissolved organic matter; solid arrows denote intercompartmental transfer of carbon; dashed arrows indicate flows of inorganic nitrogen into the system; Me, mesozooplankton; Mi, microzooplankton; P, Protozoans. (Vezina and Platt, 1988.)

depends very much on the patterns and magnitudes of the flow of energyor carbon from primary production. Our understanding of these patterns can-therefore provide valuable estimates on the limits of what we can take from thesea, as well as increasing our insight into the ecological processes.

The selection of boxes and the arrows between them depends on our knowledge of which prey–predator interactions are quantitatively significant. It is obvious from the examples here that thereis a considerable compaction into large, often heterogeneous groups, such as mesoplankton or pelagic fish. Thus the level of organization is very different-from that for biodiversity studies, or even for descriptions of the full complexity of the food web. However, in producing a manageable number of boxes, it is important not to fold significant prey–predator interactions into the same box. Thus, in **Figure 1**, prey–predator interactions at the microbial scale were ignored. On the other hand, it has been pointed out that, for the North Sea, the inclusion of both the detrital box and the invertebrate carnivores does not provide enough energy flow for the fishery yield. The definition of boxes will remain acentral problem, but this is also the great strength of this approach: it requires attention to all aspects of the food web.

The major limitation is that the method assumes that the system is in steady state. This is usually achieved by taking yearly averages and ignoring shorter seasonal variability and longer decadal-trends. For the former, several researchers have constructed dynamic planktonmodels of the non-linear interactions between nutrients and detritus based onthe seven boxes or variables in **Figure 2**.

The longer term changes are especially important for fisheries. It is possible to transform the linearized eqn [7] for equilibrium states into a time varying system by writing:

$$\frac{dB_i}{dt} = a_i \cdot B_i - \sum p_{ij} \cdot B_j - d_i. \qquad [8]$$

This linearized approximation to an essentially non-linear system can be used to indicate the directions that changes may take when the system departs from a previous steady state, but is unlikely to be adequate for the very large switches, or regime shifts, that occur in the relative abundance of different fish stocks. An alternative is to assign very different values to the boxes for the fish stock biomass and then recalculate the network. This technique can be used to estimate the status of ecosystems before the impact of human predation on fish or marine mammals. As an example,the Gulf of Mexico flows were

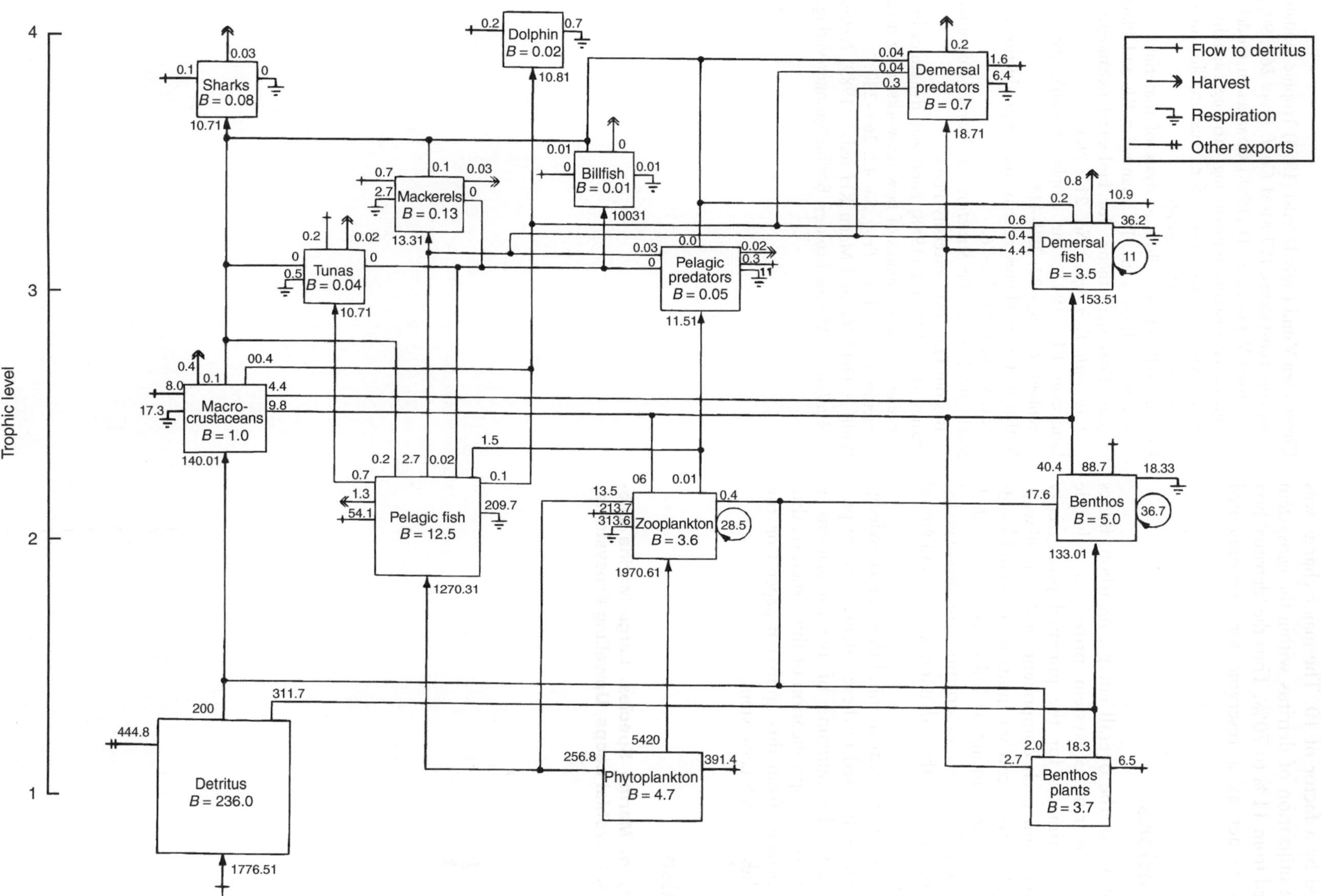

Figure 3 Gulf of Mexico ecosystem model: trophic compartments with values for biomass (g m^{-2}) and connecting flows (gm^{-2} per year^{-1}). (Christensen and Pauly, 1993.)

recalculated, with the biomass of the top predators increased by a factor of 10. The major change was that the utilization of detritus within the ecosystem increased from 11% to 70%. Thus the 'detritus' box appears to act as a reservoir for over-exploited systems.

Conclusions

Calculations of the overall energy, orcarbon, fluxes through a marine ecosystem provide a valuable check on estimates for the potential productivity associated with each component of the food web. These calculations can set limits on expected yields to humans; they can act as links between detailed, but necessarily static, descriptions of biodiversity, and models of the dynamics of individual populations.

It is necessary to bear in mind that our knowledge of the food webs used in these calculations are provisional and the outcome of the calculations is dependent on the specification of this structure; thus the full benefits from this approach depend on further studies of each ecosystem.

See also

Diversity of Marine Species. Large Marine Ecosystems. Microbial Loops. Upwelling Ecosystems.

Further Reading

Christensen V and Pauly D (eds) (1993) Trophic models of aquatic ecosystems. *ICLARM Conf. Proc.* 26: 390pp.

Christensen V and Pauly D (1998) Changes in models of aquatic ecosystems approaching carrying capacity. In: Ecosystems management for sustainable marine fisheries. *Ecol. Appl* 8 (suppl).

Fasham MJR (1985) Flow analysis of materials in the marine euphotic zone. In: Ulanowicz RE and Platt T (eds) Ecosystem theory for biological oceanography. *Can. Bull. Fish Aquat. Sci.* 213: 260pp.

Lindeman RL (1942) The trophic-dynamic aspect of ecology. *Ecology* 23: 399–418.

Ryther JH (1969) Photosynthesis and fish production in the sea. *Science* 166: 72–76.

Steele JH (1974) *The Structure of Marine Ecosystems.* Boston: Harvard University Press.

Vezina AF and Platt T (1988) Food web dynamics in the ocean. I. Best-estimates of flow networks using inverse methods. *Mar. Ecol. Prog. Ser* 42: 269–287.

Wulff F, Field JG, and Mann KH (eds.) (1989) *Network Analysis in Marine Ecology.* Berlin: Springer-Verlag.

REGIME SHIFTS, ECOLOGICAL ASPECTS

L. J. Shannon, Marine and Coastal Management, Cape Town, South Africa
A. Jarre, University of Cape Town, Cape Town, South Africa
F. B. Schwing, NOAA Fisheries Service, Pacific Grove, CA, USA

Introduction

This overview aims to introduce the reader to the ecological aspects of regime shifts by (1) introducing the current understanding of the concept, (2) summarizing the current knowledge on underlying mechanisms, (3) presenting selected case studies, and (4) examining the ecological implications of regime shifts for fisheries management.

The concept of regime shifts or 'abrupt discontinuities' was first formally referred to, in the fisheries context, in the mid-1970s and has since been the topic of much debate and discussion. In particular, the regime shift concept was debated in the larger scientific public through the works of the SCOR WG 95 on 'Worldwide large-scale fluctuations of anchovy and sardine populations', showing global synchrony in several upwelling systems (*see* Upwelling Ecosystems) characterized by alternating periods of anchovy and sardine dominance (**Figure 1**). This suggested global climatic forcing of some sort, whereby the environment may affect small pelagic fish (*see* Small Pelagic Species Fisheries) or their main prey species. There is evidence for some teleconnections (in or out of phase, and sometimes lagged by up to a decade) among shifts in periods of small pelagic fish dominance in the Pacific, North Atlantic, and the Benguela (**Figure 2**). Although the detailed mechanisms are poorly understood, these have been linked to ocean-atmospheric forcing (*see* Regime Shifts, Physical Forcing). For example, ENSO events (see below) have been associated with the Aleutian low-pressure cell, with implications for coastal ocean productivity (see North Pacific case study). Variability in ocean–atmosphere processes in the Black Sea is considered to be driven by atmospheric teleconnections with the Atlantic. Because small pelagic fish are important forage species in pelagic systems, and can also control their own prey when abundant, these environmental effects can be propagated up and down pelagic food webs, with wide-ranging implications for the structure and trophic functioning of an ecosystem.

Regime shifts in marine ecosystems have since been documented beyond upwelling areas. The current understanding of the meaning of the term 'regime shift' is "a relatively abrupt change in marine system functioning that persists on a decadal timescale, at large spatial scales, and observed in multiple aspects of the physical and biological system." This current definition stands in contrast to the concept of species alternation which describes an alternation in species dominance, but which does not upset the overall flows or functioning of an ecosystem. Further, regime shifts are to be distinguished from interannual-scale changes such as El Niño events, the effects of which typically only last 1–5 years. (El Niño events manifest as abnormally warm sea surface temperatures off Peru, usually spreading over the eastern equatorial Pacific. La Niña refers to the opposite situation of colder than normal waters in this region. ENSO events are east–west oscillations in atmospheric pressure gradients in the eastern equatorial Pacific. These events persist for several months to a year.) For example, in the North Pacific (see case study later), regime shifts overlay the ecosystem impacts of El Niño events, combining to be reflected in ecosystem variability visible on several timescales. The methods used to detect regime shifts have developed fast recently and are described in Regime Shifts: Methods of Analysis.

Mechanisms

The mechanisms underlying regime shifts are diverse and still not well understood. This makes it impossible to predict reliably how climate, fishing, or a combination of both drivers may manifest themselves as ecosystem changes. The physical, or environmental, dynamics of regime shifts are dealt with in Regime Shifts, Physical Forcing. Environmental control of regime shifts is hypothesized to operate in two ways, through continuous environmental change and/or episodic events. An example of continuous change is a prolonged period of warming, permitting expansion of a species' spawning range and thus an increased stock due to favorable habitat conditions. Alternatively, the environment can operate directly on fish recruitment (*see* Fisheries and Climate). Climatic regime shifts have been associated with changes in winds, temperature, rainfall, storm intensity, sea levels, and ice volume (affecting salinity, sea

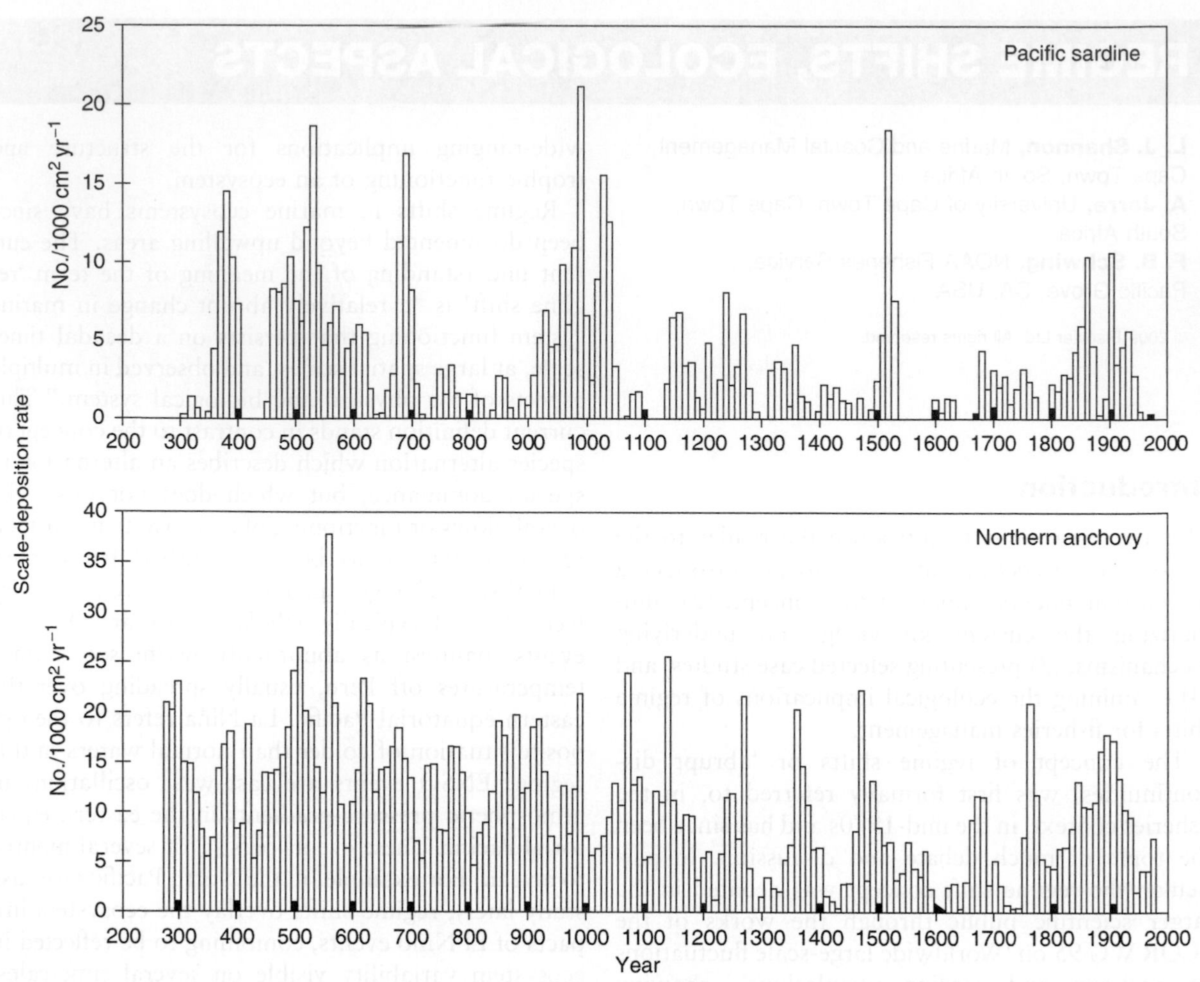

Figure 1 Time series of California sardine and northern anchovy scale deposition off southern California, a proxy for historical population biomass over the past two millennia. Series vary on *c.* 60-year cycles, similar to observed regime shift cycles. Reproduced with permission from Baumgartner TR, Soutar A, and Ferreira-Bartrina V (1992) Reconstruction of the history of Pacific sardine and northern anchovy populations over the past two millennia from sediments of the Santa Barbara Basin, California. *California Co-Operative Oceanic Fisheries Investigations Reports* 33: 24–40.

temperature, mixing, and water circulation). These climatic shifts affect biological components of the ecosystem by changing the three essential elements required for successful recruitment in fish populations, namely enrichment of the water column with nutrients, concentration of food organisms, and retention of fish larvae in favorable habitat. Environmental factors that are considered to play roles in sustaining, rather than initiating, regime shifts include changes in circulation, temperature, upwelling intensity (winds), changes in availability of plankton (related to turbulence), and changes in extent of suitable habitat for spawning or recruitment.

Environmental changes can initiate regime shifts by changing the community composition of phytoplankton and/or zooplankton. By virtue of bottom-up control mechanisms, these changes can initiate or sustain shifts in planktivores that, in turn, can propagate up the food web to predatory species. Climatic shifts cause changes in zooplankton biomass within 1–2 years, affecting fish and bird or mammal predator populations on different scales, with responses lagging several years. For example, changes in zooplankton availability have been found to affect groundfish recruitment on fairly short timescales via changing the enrichment and concentration processes (changes observed within 1–2 years), whereas responses are largely indirect and lagged for up to 9 years in spawner stock biomass. In addition, climatic shifts can have direct impacts on groundfish recruitment within 0–7 years due to change to/from favorable/unfavorable environmental conditions for larval survival and for spawners (within 0–10 years, affecting adult growth and spawning).

Complex ecological interactions come into play to sustain regimes and are evident in changes that occur

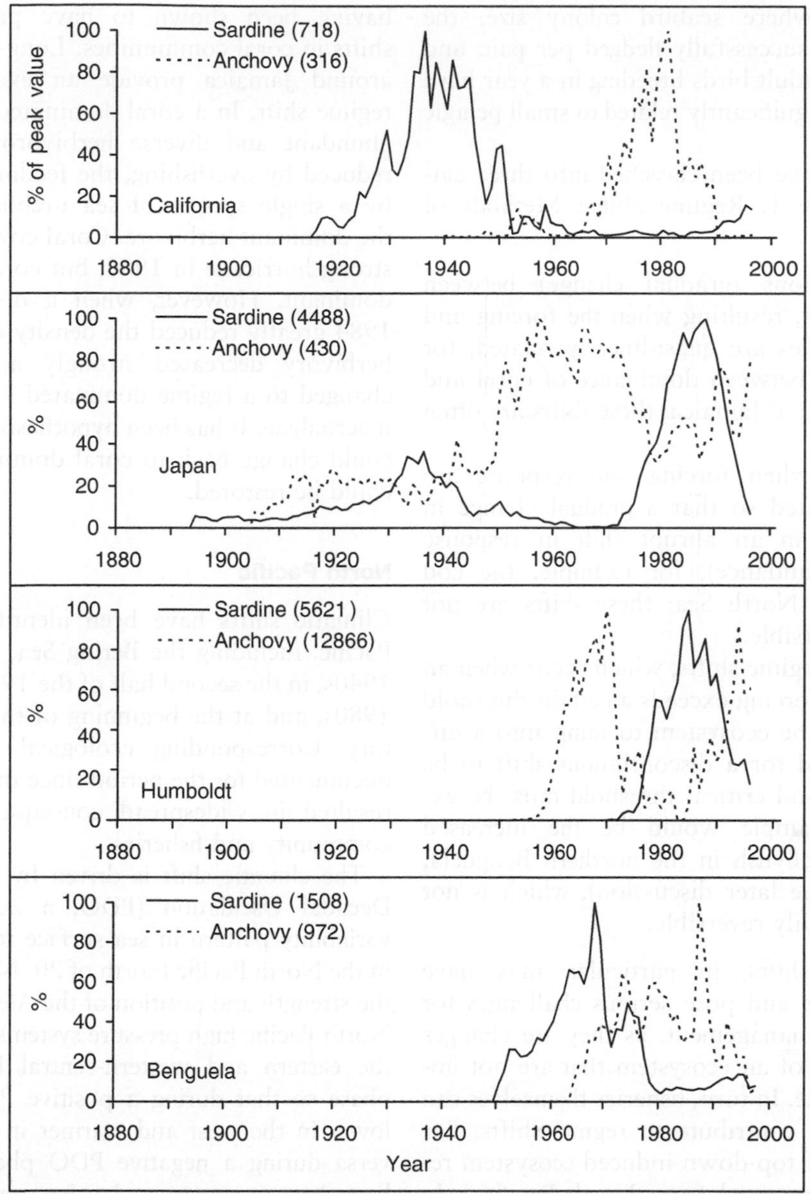

Figure 2 Catches of anchovy and sardine relative to maximum peak values (given in parentheses) of each series over the last century in the Pacific Ocean (California Current system, Kuroshia Current system off Japan, Humboldt Current system off Peru) and in the Atlantic Ocean (Benguela Current system). Figure used with permission of Lehodey *et al.*, redrawn from Schwartzlose RA, Alheit J, Bakun A, *et al.* (1999) Worldwide large-scale fluctuations of sardine and anchovy populations. *South African Journal of Marine Science* 21: 289–347.

throughput the food web. As already discussed, small pelagic fish are influenced by zooplankton prey availability, but in turn, by virtue of wasp-waist trophic control in the food web, they impact their zooplankton prey and their fish, bird, and mammal predators too. (Wasp-waist flow control refers to the trophic control exerted by small pelagic fish, which are abundant yet comprised of a few dominant species (the wasp waist), and which, by virtue of their position at the center of the food web, exert trophic control on their predators (bottom-up or donor control) and their prey (top-down or predator control).) By means of this mechanism, it has been suggested that small pelagic fish are key species influencing ecosystem dynamics. Thus, the impacts of regime shifts are not restricted to fish and their prey but extend across the whole ecosystem, including top predators such as birds, seals, and predatory fish. Seabirds, many of which are specialist feeders, are strongly affected by regime shifts. The diets of several seabird species have been found to reflect shifts in small pelagic fish prey, and there are many examples where seabird colony size, the number of chicks successfully fledged per pair, and the proportion of adult birds breeding in a year have been found to be significantly related to small pelagic fish availability.

Regime shifts have been classified into three categories (*see* **Figure 1**, Regime Shifts: Methods of Analysis):

- smooth transitions (gradual change) between ecosystem states, resulting when the forcing and response variables are quasi-linearly related, for example, shifts between dominance of coral and macroalgae around Jamaica: these shifts are often reversible.
- abrupt shifts, when forcing and response are nonlinearly related so that a gradual change in forcing results in an abrupt shift in response (population abundance), for example, the cod collapse in the North Sea: these shifts are not necessarily reversible.
- discontinuous regime shifts, which occur when an environmental forcing exceeds a certain threshold value, causing the ecosystem to jump into a different state, and for a discontinuous shift to be reversed, a second critical threshold must be exceeded: an example would be the increased abundance of jellyfish in the northern Benguela, off Namibia (see later discussion), which is not likely to be readily reversible.

Discontinuous shifts, in particular, may have major implications and pose serious challenges for fisheries resource management, as they are changes in the functioning of an ecosystem that are not immediately reversible. In turn, fisheries themselves can induce or at least contribute to regime shifts. For example, a case of top-down induced ecosystem regime shift is documented from North Pacific kelp forests, where hunting of sea otters during the nineteenth and the beginning of the twentieth century, and consequently the lack of their predation in kelp forests, led to the dominance of sea urchins in previously diverse ecosystems.

Case Studies

A selection of case studies is presented. The focus here is on regime shifts in large marine ecosystems for which there is some understanding of the causal mechanisms. See also Regime Shifts, Physical Forcing on the physical forcing of regime shifts.

Coral Reef Systems

Coral reef systems are particularly susceptible to environmental change and fishing, both factors having been shown to have precipitated regime shifts in coral communities. Long-term observations around Jamaica provide an example of a clear regime shift. In a coral-dominated ecosystem, once-abundant and diverse herbivorous fish had been reduced by overfishing, the feeding niche was filled by a single species of sea urchin, which acted as the dominant herbivore. Coral cover declined after a strong hurricane in 1980, but coral continued to be dominant. However, when a disease outbreak in 1984 greatly reduced the density of the sea urchins, herbivory decreased strongly and the ecosystem changed to a regime dominated by fleshy species of macroalgae. It has been hypothesized that the regime could change back to coral dominance if herbivory could be restored.

North Pacific

Climatic shifts have been identified in the North Pacific, including the Bering Sea, during the 1920s, 1940s, in the second half of the 1970s, the end of the 1980s, and at the beginning of the twenty-first century. Corresponding ecological regime shifts are documented for the period since the 1970s, and have resulted in widespread consequences for the fish community and fisheries.

The climatic shift is driven by the Pacific (Inter-) Decadal Oscillation (PDO, a 20–30-year climatic variability pattern in sea surface temperatures (SSTs) in the North Pacific (north of 20° N)), which relates to the strength and position of the Aleutian Low and the North Pacific high-pressure systems. SST anomalies in the eastern and western-central Pacific are out of phase so that during a positive PDO phase, SST is lower in the west and warmer in the east, and vice versa during a negative PDO phase. PDO patterns have been reconstructed using tree rings, where 11 reversals between the mid-seventeenth and the late twentieth centuries were identified. Phases of a strong Aleutian low-pressure system (positive PDO) are connected with enhanced biological productivity in Alaskan waters. The increased winds during strong Aleutian lows cause the ocean pycnocline to shallow, which reduces the depth of the upper mixed layer, generating higher productivity of plankton. This in turn favors recruitment of several groundfish species, and enhances groundfish recruitment and the production of several salmon species in Alaska and northern British Columbia. In contrast, production of salmon in southern British Columbia and the northern states of the Pacific USA is reduced (**Figure 3**). When the Aleutian low is weak (negative PDO), the North-Pacific high-pressure system strengthens, resulting in stronger upwelling-favorable winds and greater

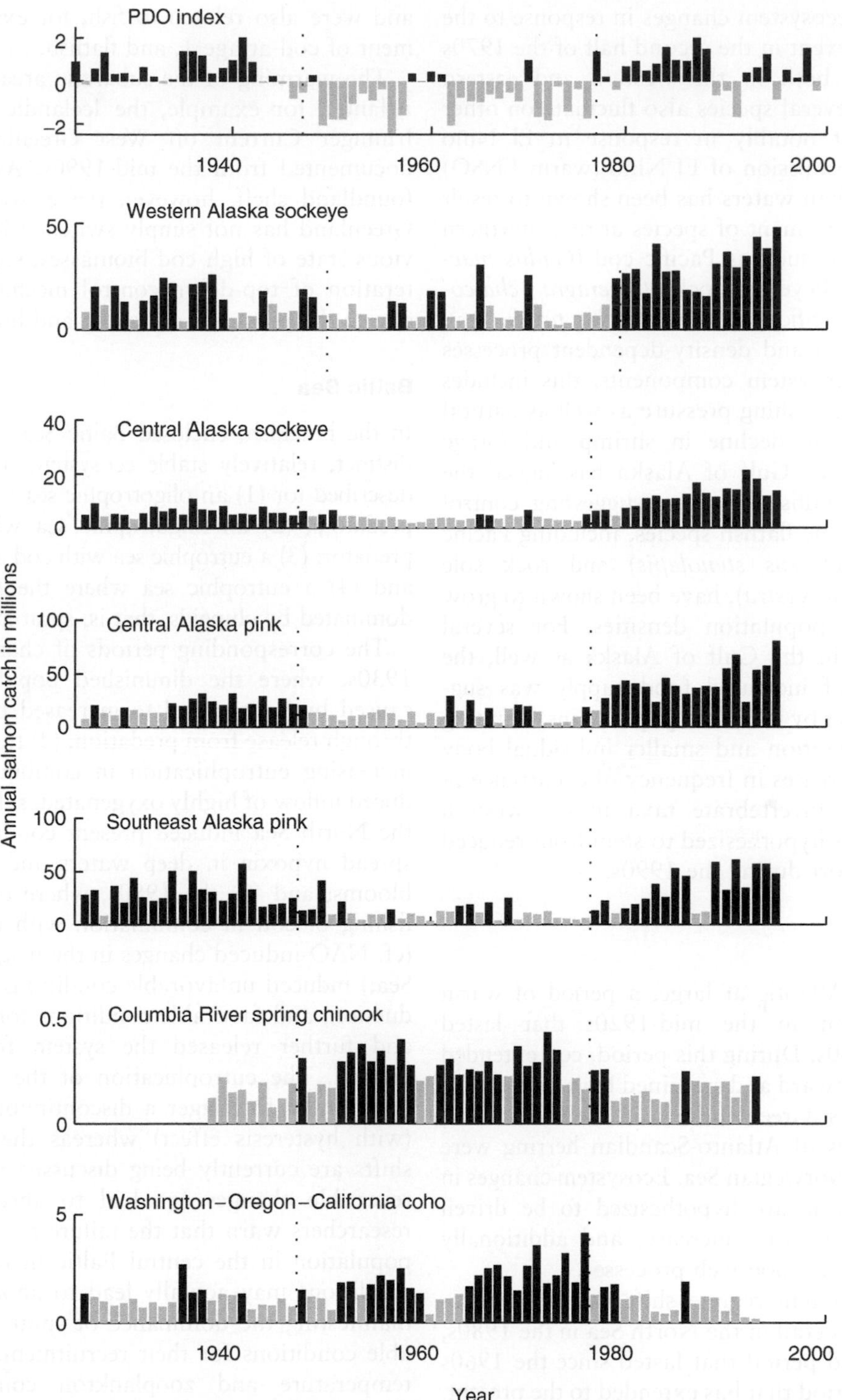

Figure 3 Fluctuations in catches of salmon in various parts of the North Pacific. Black/gray bars denote catches higher/lower than median catches for Alaskan salmon stocks. Vertical dotted lines denote regime shifts in the Pacific Decadal Oscillation (PDO) in 1925, 1947, and 1977. Courtesy of Nathan Mantua.

productivity off California (sardine-dominated regime; see Figure 2). Interestingly, this pattern appears to have broken down in the late 1990s, when salmon species were doing well in both areas – and at the same time, when a synchronous upsurge of the abundance of small pelagic fish was observed across the globe, in the Benguela system of the Southeast Atlantic Ocean.

Whereas the ecosystem changes in response to the strong climatic event in the second half of the 1970s are unmatched both in the western and eastern North Pacific, several species also fluctuate on other timescales, most notably in response to El Niño events. The transmission of El Niño (warm ENSO) events into Alaskan waters has been shown to result in improved recruitment of species at their northern distribution limit, such as Pacific cod (*Gadus macrocephalus*), walleye pollock (*Theragra chalcogramma*), and Pacific hake (*Merluccius productus*). Further, top-down and density-dependent processes affect several ecosystem components; this includes the effects of high fishing pressure as well as natural effects. The strong decline in shrimp and forage fish in the western Gulf of Alaska has lagged the increase in groundfish biomass, suggesting control by predation. Some flatfish species, including Pacific halibut (*Hippoglossus stenolepis*) and rock sole (*Lepidopsetta polyxystra*), have been shown to grow slowly at high population densities. For several salmon species in the Gulf of Alaska as well, the positive effect of increased food supply was suggested to be offset by increased populations, resulting in greater competition and smaller individual body sizes. Recent increases in frequency of occurrence of noncommercial, invertebrate taxa in the western Gulf of Alaska is hypothesized to stem from reduced trawl-fishing effort during the 1990s.

North Atlantic

For the North Atlantic at large, a period of warm conditions began in the mid-1920s that lasted through the 1960s. During this period, cod extended their range northward and sustained high catches, for example, off West Greenland and in the Barents Sea, and high catches of Atlanto-Scandian herring were obtained in the Norwegian Sea. Ecosystem changes in the North Atlantic are hypothesized to be driven directly by temperature increase, and additionally through bottom-up food web processes.

Another ecosystem regime shift has been documented in more detail in the North Sea in the 1980s, separating a cold period that lasted since the 1960s from a warm period that has extended to the present. Changes in the North Atlantic Oscillation have been related to this regime shift, inducing changes in wind intensity and direction, as well as an increase in sea temperature. These in turn act on current intensity and direction, the magnitude of the oceanic inflow into the North Sea, and the stratification of the water column. The increase in sea temperature triggered a change in the phytoplankton community, as well as calanoid copepod species composition and diversity, and were also related to fish, for example, recruitment of cod at age 1, and flatfish.

The warming of the subarctic areas of the North Atlantic, for example, the Icelandic Shelf and the Irminger Current off West Greenland, has been documented from the mid-1990s. As on the Newfoundland shelf, however, the ecosystem off West Greenland has not simply switched back to its previous state of high cod biomasses, suggesting an alteration of top-down control mechanism, possibly due to the pressures of fishing and hunting.

Baltic Sea

In the brackish, enclosed Baltic Sea, four periods of distinct, relatively stable ecosystem states have been described for (1) an oligotrophic sea with seals as top predators; (2) an oligotrophic sea with cod as top predator; (3) a eutrophic sea with cod as top predator; and (4) a eutrophic sea where the fish biomass is dominated by clupeids, that is, sprat and herring.

The corresponding periods of change are (1) the 1930s, where the diminished population of seals caused by hunting led to increased biomass of cod through release from predation; (2) the 1950s, where increasing eutrophication in combination with reduced inflow of highly oxygenated, saline water from the North Sea induced present conditions of widespread hypoxia in deep waters and frequent algal blooms; and (3) the 1980s, where continued overfishing of cod in combination with reduced inflow (cf. NAO-induced changes in the neighboring North Sea;) induced unfavorable conditions for cod reproduction, and favorable conditions for sprat feeding, and further released the system from top-down control. The eutrophication of the Baltic Sea has been found to trigger a discontinuous regime shift (with hysteresis effect) whereas the previous two shifts are currently being discussed as having been reversible changes (gradual to abrupt). However, researchers warn that the failure to rebuild the cod population in the central Baltic in relatively warm conditions may actually lead to another hysteresis, manifesting the dominance of sprat through favorable conditions for their recruitment, both through temperature and zooplankton community composition, and by predation by sprat on cod offspring.

Black Sea

In the Black Sea ecosystem, a regime shift to a more productive state is reported to have resulted from a combination of several factors: climate, fishing, and eutrophication. Evidence suggests that favorable ocean climate conditions enhanced plankton and plantivorous fish productivity in the 1970s and 1980s. SST declined after 1965, causing the upper water layer to become less stable, resulting in intensified upwelling and mixing. Thus, enrichment of the upper layer was enhanced. In addition, river runoff increased, contributing to enhanced productivity in the area. These favorable oceanic conditions lasted until the end of the 1980s, whereafter productivity again declined. Fishing has also been shown to have played an important role in the observed regime shift in the Black Sea. In the 1970s, heavy industrialized fishing began on pelagic predators such as bonito, mackerel, bluefish, and dolphins, resulting in trophic effects cascading down through the food web as follows: reduced predator populations resulted in large planktivorous fish stocks forming, which depleted zooplankton, in turn leading to an increase in phytoplankton productivity. However, fishing and natural forcing were insufficient to explain the large increase in productivity in the Black Sea during the 1980s, and anthropogenic eutrophication (e.g., phosphorus discharge into the Black Sea) was found to have contributed to enhancing productivity across the food web.

Toward the end of the 1980s, the Black Sea ecosystem again shifted; there was a collapse of small pelagic fish stocks, the jellyfish *Aurelia aurita* declined after 1985, around the time at which an alien ctenophore, *Mnemiopsis leidyi*, proliferated noticeably, and phytoplankton productivity in the early 1990s was much higher than in the late 1970s. It has been shown that the decline in small pelagic fish in the Black Sea was primarily due to overfishing and that the outburst of the alien gelatinous *M. leidyi* in the 1990s only played a small role in this decline. Rather, it appears that gelatinous zooplankton proliferate in response to vacation of an ecological niche when small pelagic fish stocks collapse. In the Black Sea, collapse of small pelagics afforded the alien ctenophore the opportunity to take advantage of increased plankton production that had occurred in response to eutrophication in the 1980s. This situation in which gelatinous species proliferate in response to collapses in small pelagic fish stocks also appears to occur in other ecosystems, for example, the northern Benguela (see below).

Benguela

Regime shifts have been recorded in a number of upwelling ecosystems. Here we use the well-documented case of the Benguela upwelling system as an example.

South African and Namibian fisheries target similar species (Cape hake *Merluccius* spp., sardine *Sardinpos sagax*, anchovy *Engraulis encrasicolus*, and horse mackerel *Trachurus trachurus capensis*). Before human intervention, the South African (southern Benguela) and Namibian (northern Benguela) ecosystems may have been fairly similar in structure and functioning; however, the two have undergone very different fishing histories and different environmental perturbations have had large effects in the two ecosystems. Thus, northern and southern ecosystems now contrast in their species compositions, community structures, and abundance trends (**Figure 4**).

Catches of hake, off South Africa, increased from the 1950s, had peaked by 1977, when a 200-mile Fishing Zone was proclaimed by South Africa, and have since remained fairly stable. Sardine dominated the South African catch from 1950 to 1965, and anchovy from then until the mid-1990s. Sardine again began to recover in the late 1990s, and in the early 2000s, both anchovy and sardine were relatively abundant. It has been hypothesized that in the southern Benguela, the water column stabilizes during warm periods as upwelling events are weak and less frequent, phytoplankton is dominated by small cells, leading to a zooplankton community dominated by small copepods that favor filter-feeding sardine. In cool periods when upwelling is stronger, phytoplankton biomass is high and dominated by large chain-forming diatoms, and zooplankton are larger and ideal for the biting feeding behavior of anchovy. A single or a series of good year-classes may result from an episodic, favorable environmental event, and may be adequate to shift the stock into a high-abundance/productivity regime. Off South Africa, two favorable episodic environmental events, resulting from wind-induced upwelling in the summer of 1999–2000, have been proposed as the reason for the sudden upsurge in abundance of small pelagic fish in the southern Benguela in the early 2000s.

Similar to events observed in the Black Sea, the northern Benguela has suffered collapses of anchovy and sardine stocks while a proliferation of jellyfish *Chrysaora hysoscella* and *Aequorea aequorea* has been observed since the 1980s. These changes have been attributed to a series of unfavorable environmental events having been exacerbated by fishing at levels that were not sustainable at such low stock biomasses. The northern Benguela has seen a sequential depletion of sardine, anchovy, and hake, attributed to ineffective management measures prior to Namibia's independence and proclamation of a 200-mile EEZ in 1990. Since the 1990s, hake has recovered slightly but sardine has remained depleted. Horse mackerel dominated catches since the 1970s, declining slightly since the late 1980s. Total catches off Namibia have

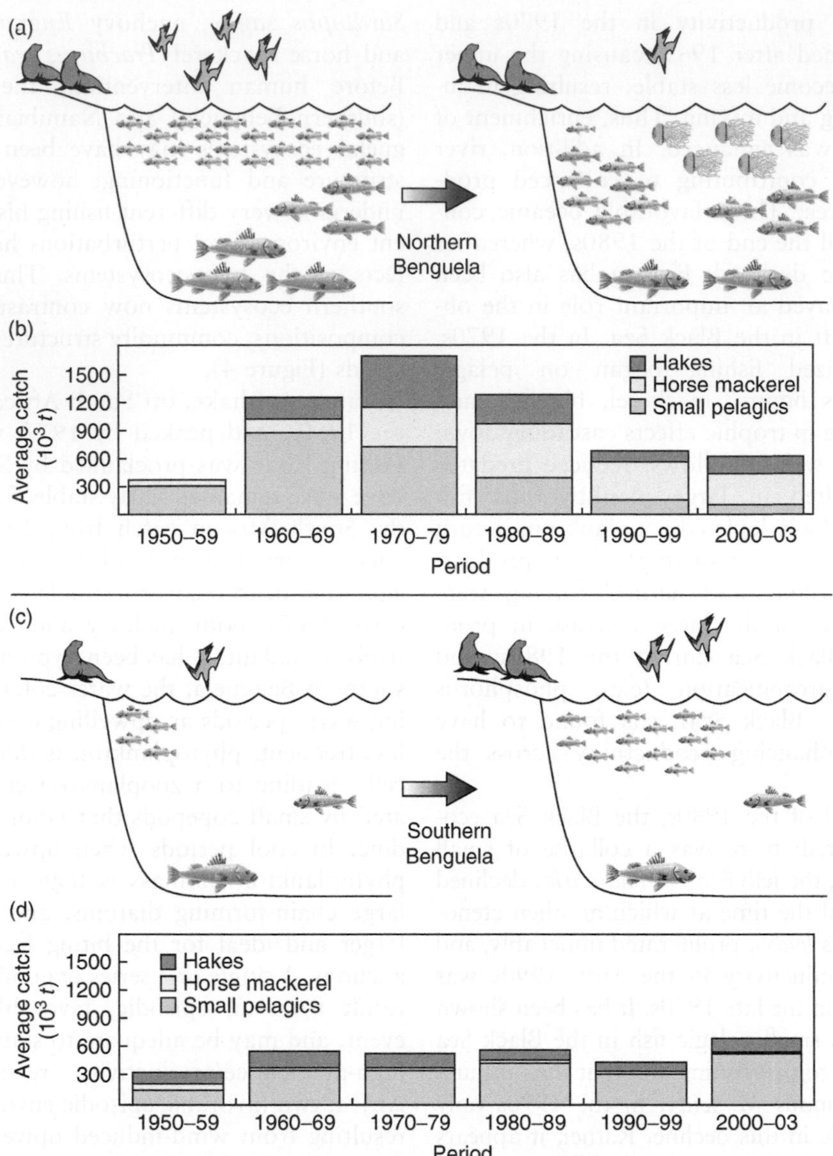

Figure 4 Diagrammatic overview of the changes in ecosystem structure and relative abundance of dominant species (small pelagic fish, hake, goby, jellyfish, Cape gannet, and Cape fur seal; number of individuals approximately represents relative abundance) in the northern and southern Benguela between the 1970s (left) and the early 2000s (right). Reprinted from *Large Marine Ecosystems Series 14: Benguela – Predicting a Large Marine Ecosystem*, Shannon LV, Hempel G, Malanotte-Rizzoli P, Moloney CL, and Woods J (eds.), ch. 8, Van der Lingen CD, Shannon LJ, Cury P, *et al.*, Resource and ecosystem variability, including regime shifts, in the Benguela Current system, pp. 147–184, copyright (2006), with permission from Elsevier and Dr. Carl van der Lingen.

decreased steadily since the late 1960s. As off South Africa, sardine was the dominant small pelagic fish in the early 1960s but year-classes were poor from 1965. However, large catches of sardine were still taken in the late 1960s, and sardine abundance decreased. In contrast to the southern Benguela, where sardine was replaced by anchovy, in the northern Benguela sardine was replaced by a guild of planktivores comprising horse mackerel, bearded goby *Sufflogobius bibarbatus*, and to a small extent also anchovy. In the late 1980s, sardine and hake increased again off Namibia, but two unfavorable environmental events occurred in the northern Benguela in the mid-1990s (a low oxygen event in 1993–94 and an intrusion of warm-water Benguela Niño in 1995). Subsequently, several fish species underwent geographical shifts in distribution, sardine recruitment was low, and most commercial species (hake, horse mackerel, etc.) declined. For example, hake migrated offshore in the mid-1990s, corresponding to reduced catch rates. Reduced availability of prey for top predators was evident in observed starvation of seals causing the seal population to be reduced by one-third in the mid-1990s, as well as seabird species of conservation concern, such as African penguin and bank cormorant.

Management Implications

Current management strategies worldwide do not account for environmental variability, or more specifically for multi-year changes in the environment linked to regime shifts, and how this variability may affect recruitment into commercially important fish stocks. On the other hand, fishing itself affects stock fluctuations by increasing mortality at various spatial and temporal scales. Overfishing can exacerbate the effects of natural regime shifts in an ecosystem, as detailed in the case studies of the northern Benguela, and the Black Sea (see above).

If generalized feeding is the norm, the collapse of a large stock may benefit a suite of other species, the outcome being influenced largely by fishing and predation during and post stock collapse. The uncertainty around predicting regime shifts and ecosystem changes is highlighted in plausible yet so-far unsuccessful attempts to manipulate ecosystems. Under the shaky assumption that removal of the dominant pelagic species should favor the subordinate one (presumed to be its competitor), if the latter is only lightly fished, management options along these lines have been considered in some areas. For example, off California, it is possible that anchovy and sardine compete for zooplankton prey, and thus it has been suggested that heavily fishing anchovy may allow the sardine stock to increase in size. In the northern Benguela, the management decision was in fact taken to fish heavily on anchovy in the 1970s when sardine, the more commercially valuable small pelagic species, began to collapse, in an attempt to reduce competition between the two species. This management strategy was proven ineffective as both Namibian anchovy and sardine stocks collapsed in the late 1970s. By comparison, anchovy off South Africa were conservatively managed when South African sardine showed signs of collapse in the late 1960s, enabling anchovy to proceed with its natural cycle of increase and to support a healthy fishery in the 1980s. For the upwelling ecosystem off northern-central Peru, alternation between pelagic fisheries for reduction and consumption has been suggested, aiming at maintaining a viable pelagic fishery in a highly variable environment. This scheme would have implied fishing anchoveta in cold years, and mackerel and horse mackerel in warm years. It was hypothesized that in this way, both fishing and predation pressures on anchoveta would be reduced in warm years, when mackerel and horse mackerel tend to be distributed close to the coast, thus avoiding anchoveta stock depletion. In cold years, mackerel and horse mackerel are distributed further offshore, outside the upwelling zone where anchoveta are found. Cold conditions are also favorable for anchoveta recruitment, and the anchoveta stock preserved under warm conditions could then theoretically bounce back quickly to high biomasses, which in turn could sustain a considerable reduction fishery. To our knowledge, however, this hypothesis has not yet been taken further in the management process.

Our ecosystems may be becoming even more sensitive to environmentally driven regime shifts as a result of fishing. The worldwide phenomenon of 'fishing down the food web' to progressively remove fish at lower and lower trophic levels is apparent in the observed shifts from ecosystems dominated by large, predatory fish to those dominated by pelagic fish, which by virtue of their short life span and high population-turnover rates generally appear highly sensitive to environmental changes. (Trophic level refers to a species' trophic position in the food web. Species may be aggregated into discrete trophic levels, which provide a measure of the number of steps from producer to top predator within a food chain. Alternatively, species in a food web can be assigned noninteger trophic levels on the basis that producers are at trophic level 1, and consumers at trophic level 1 plus the average trophic levels of their prey, weighted by the proportion of each prey species in the diet of the given predator. Readers are also referred to Fisheries and Climate dealing in more detail with fisheries and climate.)

Fisheries managers have to deal with a high degree of uncertainty when a sudden decline in an exploited stock is observed, as this could be interannual variability, a sign of overfishing, or a shift to a new, possibly less-productive regime which may not be easily reversible. If quantification of regime shifts were fully possible, their early identification would likely also be possible, which would be highly beneficial for fisheries and ecological management purposes. Early identification of regime shifts may be achievable by means of ecosystem monitoring, in addition to monitoring of climatic events. Further, monitoring of large-scale climatic teleconnections may assist in forecasting probably regional-scale environmental changes.

An ideal management system would be a precautionary, adaptive one that enabled fisheries to take advantage of regimes of high abundance and to restrict fishing pressure during years of change or regimes of low production. A shift in current management approaches is needed to consider regime-specific harvest rates so as to optimize catches while also accommodating periods of low productivity. This is a particular challenge of an ecosystem approach to fisheries management. Several researchers highlight the need for immediate management action already now, even while the scientific concept of long-term changes continues to be refined, time series analysis continues, and our knowledge of the functioning of marine systems improves.

See also

Fisheries and Climate. Regime Shifts: Methods of Analysis. Regime Shifts, Physical Forcing. Small Pelagic Species Fisheries. Upwelling Ecosystems.

Further Reading

Baumgartner TR, Soutar A, and Ferreira-Bartrina V (1992) Reconstruction of the history of Pacific sardine and northern anchovy populations over the past two millennia from sediments of the Santa Barbara Basin, California. *California Co-Operative Oceanic Fisheries Investigations Reports* 33: 24–40.

Beaugrand G (2004) The North Sea regime shift: Evidence, causes, mechanisms and consequences. *Progress in Oceanography* 60(2–4): 245–262.

Collie JS, Richardson K, and Steele J (2004) Regime shifts: Can ecological theory illuminate the mechanisms? *Progress in Oceanography* 60: 281–302.

Collie JS, Richardson K, Steele JH, and Thybo Mouritsen L (2004) *Physical Forcing and Ecological Feedbacks in Marine Regime Shifts*. ICES CM 2004/M:06, 26pp.

Cury PM and Shannon LJ (2004) Regime shifts in upwelling ecosystems: Observed changes and possible mechanisms in the northern and southern Benguela. *Progress in Oceanography* 60: 223–243.

Daskalov GM (2003) Long-term changes in fish abundance and environmental indices in the Black Sea. *Marine Ecology Progress Series* 255: 259–270.

de Young B, Harris R, Alheit J, Beaugrand G, Mantua N, and Shannon L (2004) Detecting regime shifts in the ocean: How much data are enough? *Progress in Oceanography* 60: 143–164.

Drinkwater K (2006) The regime shift of the 1920s and 1930s in the North Atlantic. *Progress in Oceanography* 68: 134–151.

Folke C, Carpenter S, Walter B, *et al.* (2004) Regime shifts, resilience and biodiversity in ecosystem management. *Annual Review of Ecology, Evolution and Systematics* 35: 557–581.

Hare SR and Mantua NJ (2000) Empirical evidence for North Pacific regime shifts in 1977 and 1989. *Progress in Oceanography* 47: 103–145.

Hughes TP (1994) Catastrophes, phase shifts, and large-scale degradation of a Carribean coral reef. *Science* 265: 1547–1551.

Lees K, Pitois S, Scot C, Frid C, and Mackinson S (2006) Characterizing regime shifts in the marine environment. *Fish and Fisheries* 7: 104–127.

Lehodey P, Alheit J, Barange M, *et al.* (2006) Climate variability, fish and fisheries. *Journal of Climate* 19: 5009–5030.

Manuta NJ, Hare SR, Zhang Y, Wallace JM, and Francis RC (1997) A Pacific interdecadal climate oscillation with impacts on salmon production. *Bulletin of the American Meteorological Society* 78(6): 1069–1079.

Österblom H, Hansson S, Larsson U, *et al.* (2007) Human-induced trophic cascades and ecological regime shift in the Baltic Sea. *Ecosystems* 10: 877–889.

Overland J, Percival DB, and Mojfjeld HO (2006) Regime shifts and red noise in the North Pacific. *Deep-Sea Research I* 53: 582–588.

Rothschild B and Shannon LJ (2004) Regime shifts and fishery management. *Progress in Oceanography* 60: 397–402.

Schwartzlose RA, Alheit J, Bakun A, *et al.* (1999) Worldwide large-scale fluctuations of sardine and anchovy populations. *South African Journal of Marine Science* 21: 289–347.

Van der Lingen CD, Shannon LJ, Cury P, *et al.* (2006) Resource and ecosystem variability, including regime shifts, in the Benguela Current system. In: Shannon LV, Hempel G, Malanotte-Rizzoli P, Moloney CL and Woods J (eds.) *Large Marine Ecosystems Series 14: Benguela – Predicting a Large Marine Ecosystem*, ch. 8, pp. 147–184. New York: Elsevier.

REGIME SHIFTS, PHYSICAL FORCING

F. B. Schwing, NOAA Fisheries Service,
Pacific Grove, CA, USA

Introduction

In the ocean observational record, there appear to be multidecadal periods of relatively stable environmental states separated by short, abrupt transitions. These climate regimes, and their associated regime shifts, are key in regulating the productivity and structure of marine ecosystems and their populations, including many economically important fish stocks, marine mammals, and protected species. They also play an important role in regulating large-scale weather patterns, thus determining decadal continental precipitation and drought patterns, which also have considerable terrestrial ecological, social, and economic impacts.

Understanding and quantifying regime shifts is also important to separate out and quantify longer-term and anthropogenic climate change, whose signals may be obscured by the oscillatory nature of regimes in the relatively short observational record. The distinction between regime shifts – alternating equilibrium states in ecosystems – versus a steady baseline state is a critical one for projecting future climate states and their impacts.

Their signals may be similar in morphology and impact to the more familiar interannual El Niño–Southern Oscillation (ENSO) cycle, but on a much longer timescale which has major social and economic impacts on fishery-dependent communities and on the intrinsic benefits of marine ecosystems. Our current understanding of regime shifts is analogous to that of ENSO about two decades ago. We have observed them, but the understanding of their causes and the ability to forecast them is still speculative.

A Functional Definition of Regime Shifts

The term regime shift is commonly used in political and economic sciences, and generally describes any rapid transition to a new, relatively stable state. Its use in environmental science is based on principles of theoretical ecology and was initially applied by biologists to observe rapid ecosystem changes. Ecologists refer to punctuated equilibrium, where ecological communities switch suddenly between distinct equilibrium states. Others refer to cycles, epochs, episodes, even El Viejo (old man in Spanish) – an allusion to their morphological similarity to El Niño events.

Because of our limited experience in observing and understanding regime shifts in the ocean, there is no established or commonly accepted definition. Most attempts have focused on infrequent, large changes or reorganizations in marine population abundance, productivity, or distribution throughout one or more ecosystems. There remains some debate about whether regime shifts are physical or biological, or must be reflected in both the living and abiotic components of marine ecosystems to qualify.

In general, regime shifts in the marine environment are characterized by a relatively rapid (over a few years) persistent (decades) change in ocean state that is associated with large-scale alterations in the climate system, with corresponding changes in community structure or at multiple trophic levels. A regime shift results in a change in the dynamics (mean, variance, extrema, seasonality) of a number of key variables that characterize the ecosystem's environment and populations. Shifts are not necessarily simple and instantaneous. Their timing may not be simultaneous in physical and biological series, and their evolution may appear smooth, cyclic, abrupt, chaotic, or discontinuous.

As a working definition, regime shifts are relatively abrupt changes in marine system functioning that occur at large spatial scales and are observed in multiple aspects of the physical and biological system.

History

The first relatively well-observed and identified ocean regime shift was in the North Pacific in the 1970s. However indications of multidecadal periods of atmospheric pressure, ocean temperature, and fish catch have been known for much longer. Shifts in North Pacific observations, most of which can be identified in North Atlantic time series as well, appeared in around 1850, 1880, 1925, 1945, 1958, 1976, 1989, and 1998 (**Figure 1**). A few of these were short-lived and regional in nature. Japanese fishery records extending back to the fourteenth century show large swings in herring, sardine, and

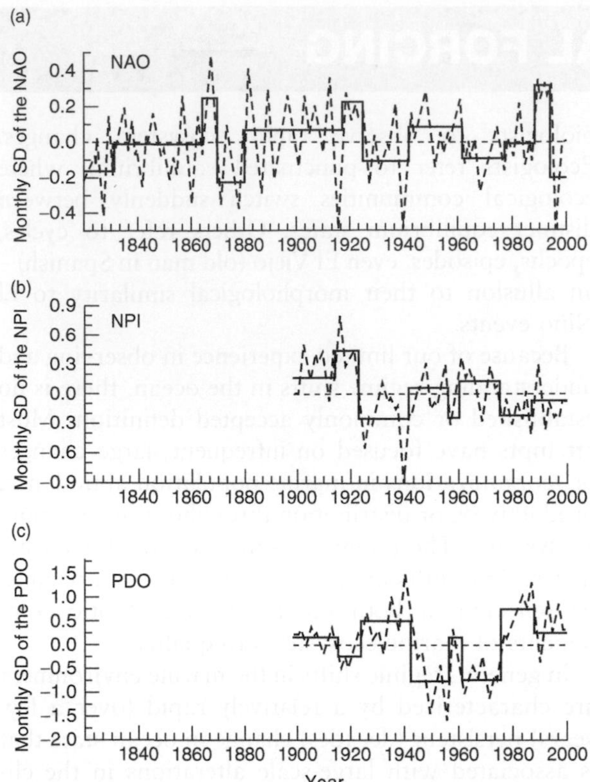

Figure 1 Climate regimes and regime shifts based on climate indices (a) North Atlantic Oscillation, (b) North Pacific Index, and (c) Pacific Decadal Oscillation. Regimes noted by thick solid lines, smoothed time series by dashed lines, and series mean by dotted lines. Regime shifts are determined objectively from abrupt change analysis, (Schwing *et al.*, 2003). Synchrony of regimes indicates global coherence of regime shifts.

tuna catch on multidecadal scales. Fluctuations of 20–50 years in migration and fishing of herring off Sweden date back to the tenth century. Proxy chronologies derived from marine sediments, ice cores, corals, and tree rings support the existence of physical and biological regime shifts for several centuries prior to the instrumental record.

Regime shifts in the ocean were noted initially in biological observations, and are generally more distinct in biological than physical indicators. However, the concept of regime shifts in ocean systems took some time to be accepted. The ocean environment and marine ecosystems had long been considered by many to be quasi-stable with little temporal variability, certainly in comparison to terrestrial and freshwater systems. While fishery catches worldwide were known to vary, sometimes declining dramatically within a few years, these fluctuations were not linked to environmental change. This was at odds with theoretical ecologists who embraced the punctuated equilibrium concept. This paradox was somewhat driven by the theoretical view of pristine communities versus the worldly realities of overfishing and other human effects that could drive natural ecosystems into a new state.

As systematic long-term observations accumulated in regions such as the Kuroshio Current, California Current, Gulf of Alaska, and North Sea, biologists began to relate the collapse (growth) of major fisheries on decadal scales to a general decline (increase) in ecosystem productivity and an associated shift in physical conditions. The realization that shifts in fisheries occur with apparent global synchrony greatly advanced the idea of climate-driven ecological variability, and the general concept of regime shifts in the ocean.

A series of scientific conferences that assembled the leading experts on climate, oceanography, and fisheries developed the notion that not only do fishery populations fluctuate on decadal and longer timescales coincident with apparent swings in physical climate forcing and ecosystem productivity and structure, but that these fluctuations seem to be synchronous across disparate ecosystems, indicative of a common global physical forcing. The term regime shift began to be used as a common descriptor for these events. The term has been adopted widely in management, policy, and public circles, even as the scientific community continues to debate their existence and cause.

Observations of Regime Shifts

In the 1970s, scientists noted a major shift in North Pacific atmospheric pressure and wind patterns. The Aleutian Low, which dominates winter winds, deepened and shifted toward the east. Eastward wind stress doubled and extended southward in a broad area across the central North Pacific (**Figure 2**). This brought cooler winds over much of the North Pacific, and a warmer, moist flow along the North American west coast. Turbulent wind mixing of the upper ocean increased as well throughout the North Pacific, with the greatest increase in a broad region between 30° and 40° N.

These atmospheric shifts produced changes in the North Pacific circulation on regional to basin scales. Upwelling in the California Current was both weaker and less productive. Transport in the North Pacific Current increased by nearly 10%. Southward surface Ekman transport of subarctic water into the North Pacific Current increased. The region of Ekman divergence (upward vertical transport, or

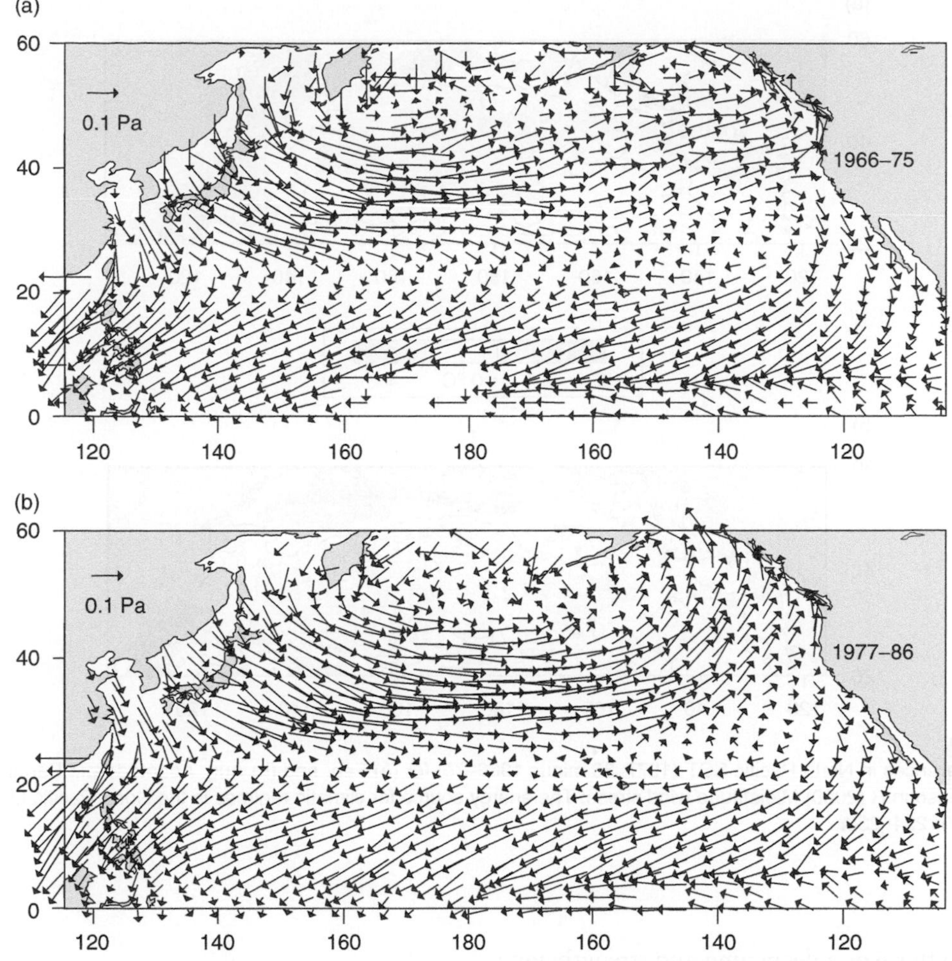

Figure 2 Winter wind stress over the North Pacific for (a) 1966–75 and (b) 1977–86. The post-1976 decade featured a mid-basin doubling of wind stress, and a more northward stress in the northeastern Pacific. Adapted from Parrish RH, Schwing FB, and Mendelssohn R (2000) Midlatitude wind stress: The energy source for climatic regimes in the North Pacific Ocean. *Fisheries Oceanography* 9: 224–238.

Ekman pumping) strengthened in the central basin and expanded southward.

Adjustments between the atmosphere and upper ocean due to heat exchange, wind-driven Ekman ocean currents, and basin-scale Sverdrup circulation patterns contributed to anomalously cool sea surface temperatures (SSTs) in the central portion of the basin, and warmer SSTs in a horseshoe-shaped region from the Bering Sea and Gulf of Alaska, along the North American coast, and extending southwest from Baja Mexico (roughly along the axis of the trade winds) (**Figure 3**). The core of the anomalously cool area was located along the northern, subarctic edge of the eastward flowing Kuroshio Extension and North Pacific Current that feeds the eastern boundary currents adjacent to North America. SSTs along the North American coast were 1–2 °C warmer than before, while a similar level of cooling in the central North Pacific extended as far as 200–400 m deep. Cooler surface water subducted from the central North Pacific in the thermocline, and propagated westward to the Kuroshio–Oyashio extension region.

Shifts in the North Pacific Ocean coincided with changes in nutrient availability and biological productivity at multiple trophic levels. While basin and global climate variations set the conditions for decadal regime shifts, the ecosystem responded to local to regional perturbations in the environment due to different processes controlling the biology in different areas. Enhanced mixing and adjustments in basin circulation contributed to a 30–80% deeper mixed layer in the subtropical North Pacific after the 1970s regime shift, which led to greater nutrient availability and a dramatic increase in phytoplankton in the subsurface chlorophyll maximum. The light-limited subarctic shoaled simultaneously by 20–30%, contributing to a doubling of zooplankton biomass.

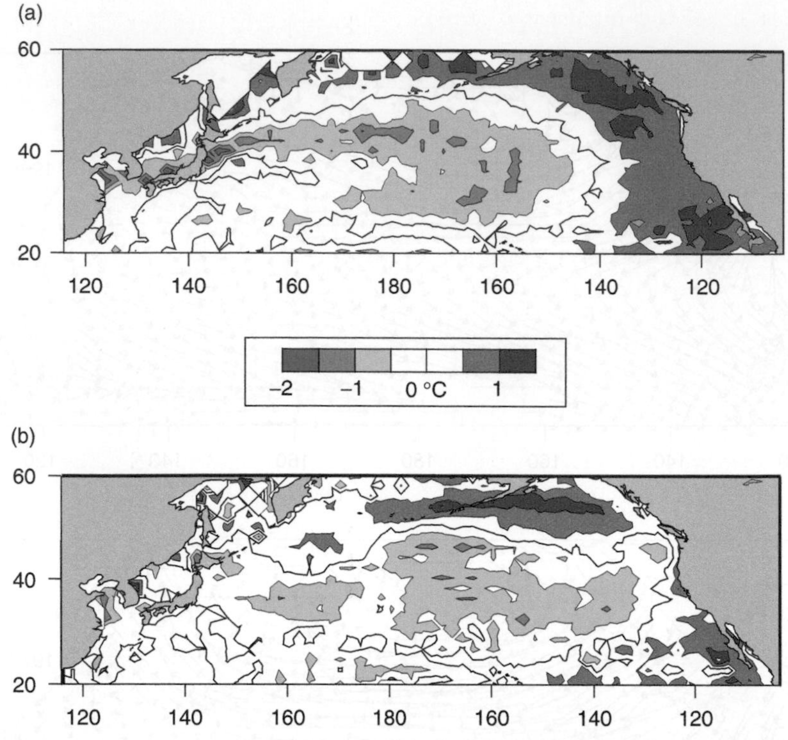

Figure 3 Differences in North Pacific SST (1977–86 minus 1966–75) for (a) Feb. and (b) Aug. Adapted from Parrish RH, Schwing FB, and Mendelssohn R (2000) Midlatitude wind stress: The energy source for climatic regimes in the North Pacific Ocean. *Fisheries Oceanography* 9: 224–238.

Coincident to this was a deepening and strengthening of the thermocline and nutricline, and a sharp decline in zooplankton biomass, in the California Current.

After 1989, the anomalously warm region in the northeast Pacific cooled by about 0.2–0.5 °C, with the greatest change in the Bering Sea and Gulf of Alaska. SSTs in the central basin, on the other hand, returned to levels seen prior to the mid-1970s. Retrospective analyses of atmospheric and oceanic fields suggested that the shifts in the 1970s and late 1980s were part of a bimodal pattern, termed the Pacific Decadal Oscillation (PDO), where SST anomalies in the eastern and central Pacific are out of phase (**Figure 4**). But a more recent shift in the late 1990s appeared as a north–south SST mode. California Current productivity has appeared to increase, while the Bering Sea was unaffected.

Meanwhile, enhanced inflow beginning in the 1920s warmed the North Atlantic dramatically. This warming continued through the 1960s, reducing boreal sea ice. A declining inflow of Atlantic water beginning in the 1960s returned cooler conditions to the region. Atlantic inflow increased again after 1980. Similar multidecadal variability has been reconstructed from a 450-year Arctic ice core record.

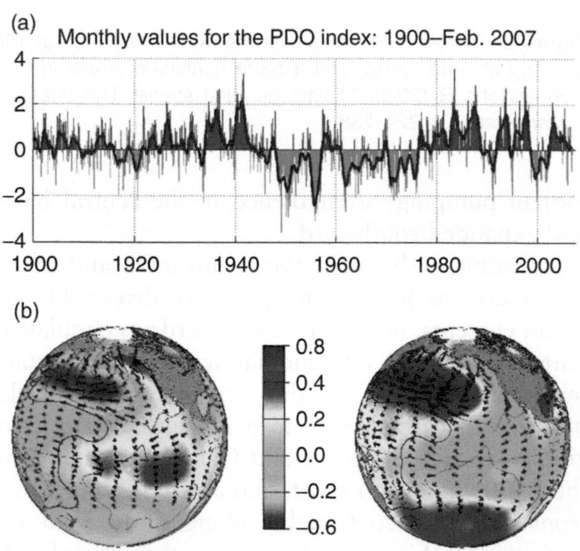

Figure 4 (a) Time series of the Pacific Decadal Oscillation (PDO) index, defined as the leading principal component of North Pacific monthly SST. (b) Typical winter SST (colors), sea level pressure (contours), and surface wind stress (arrows) anomaly patterns during warm (positive) and cool (negative) phases of the PDO. The PDO cycle is roughly 50–70 years, with reversals occurring around 1925, 1947, 1977, and 1999. Courtesy of Nathan Mantua.

Synchronous fluctuations in the intensity of the Iceland Low and Azores High, termed the North Atlantic Oscillation (NAO), control climate variability of the North Atlantic. Shifts in the NAO are thought to lead to the major shifts in ocean circulation and large-scale water mass shifts seen in the twentieth century. An index based on North Atlantic SST, termed the Atlantic Multidecadal Oscillation (AMO), is thought to characterize the upper ocean's reflection of thermohaline circulation fluctuations that are driven by large-scale atmosphere–ocean interactions.

The plankton community structure in the English Channel also switched in the 1920s and 1960s. Associated physical and biological changes, referred to as the Russell Cycle, have been linked to variations in Atlantic flow as well as fluctuations in marine populations throughout the Atlantic. This cycle traces back over a century of North Sea mollusk growth variations and 400 years of fishery records.

Physical Mechanisms

Regime shifts in marine ecosystems are thought to occur when large-scale changes in ocean conditions, including factors such as temperature, stratification, circulation, freshwater inflow, and geographic shifts in water mass types and biophysical features and boundaries, affect biological production and community composition. For example, alternating species dominance and catch of fish in many coastal marine ecosystems is thought to be a response to changing food type, thermal, and hydrodynamic conditions.

Much of the basis of the theory behind these mechanisms is the analysis of long physical and biological time series that feature fluctuations in level and variance energy on decadal scales. Determining whether regimes are cyclic, regular, or random, and whether they occur as abrupt switches or periodic oscillations, is critical in isolating the forces that drive them.

Initial descriptions of regime shifts considered this phenomenon to be a single mode of variability, alternating between two opposing states (e.g., cool vs. warm, low vs. high productivity), suggesting that their forcing occurs as a standing wave. The precedence of periodic phenomena, from seasonal and interannual (ENSO) to decadal (solar and tidal) to centennial and millenial (astronomical and Earth) gives reason to expect similar cyclicity in regime shifts. Some analyses and models suggest that regimes may be cyclic, but composed of multiple modes. Spectral analysis of many physical and biological time series have found cycles with roughly 20-, 50-, 70-, and 100-year periods. The interactions of these dominant periods reproduce fairly well the long-term ocean variability.

Interventions and other related statistical analyses have identified abrupt shifts in several physical and biological time series in the Pacific, and indicate a close coupling between the atmosphere and ocean. However, other analyses have concluded that detecting shifts by these methods do not conclusively demonstrate significant shifts between stable states, and could with equal likelihood identify shifts in red noise series.

Despite similarities in their anomaly patterns, the mechanisms behind ENSO and decadal regime shifts are different. The mechanisms driving regime shifts involve stochastic atmospheric forcing and teleconnections between distant regions, planetary ocean waves and internal adjustments, and air–ocean feedbacks. Stochastic atmospheric forcing, roughly reproducing the weather-scale variance, explains mid-latitude ocean decadal variability to first order. However, internal ocean processes and ocean–atmosphere coupling are required to reproduce the observed sequence of decadal cycling. Additional adjustments of upper ocean forcing (via Ekman and Sverdrup processes) on interannual and longer scales are associated with the transfer of momentum, heat, and moisture via atmospheric teleconnections from the Tropics, zonally (linking Asia, the Pacific, North America, and the Atlantic), and possibly Arctic to mid-latitude ocean regions.

Other evidence suggests that regime shifts in the North Pacific are a direct response to the stronger eastward wind stress due to an intensified Aleutian Low. The central basin adjusts locally to enhanced heat-flux cooling, more southward Ekman advection of cool water, wind mixing of the preconditioned temperature, and vertical shifts in the thermocline through greater Ekman divergence (**Figure 5**). Ocean basin gyre spin-up associated with changes in large-scale wind patterns alters the geostrophic flow field and the source waters. Upper ocean inertia damps these atmospheric forcings, creating an integrated signal that is more energetic at decadal and longer periods (red spectrum).

Upper ocean conditions in the Kuroshio–Oyashio Extension appear much less sensitive to direct atmospheric forcing on decadal scales. Shifts in the central North Pacific Ocean propagate west along the thermocline via Rossby waves forced by the subtropical wind stress curl. Variations in temperature and salinity travel along isopycnal surfaces, finally perturbing surface waters in the Kuroshio region with a *c.* 5-year lag. There, surface heat

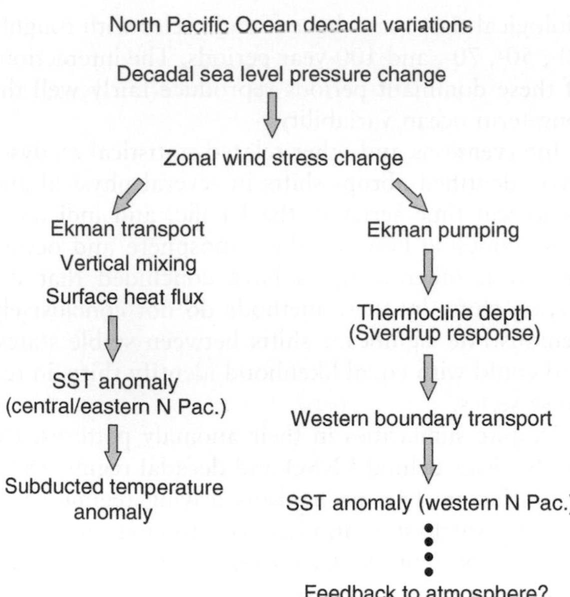

Figure 5 Schematic diagram of the mechanisms for regime shifts in the upper ocean of the North Pacific. The left-hand side describes SST anomalies in the central North Pacific, which are driven by perturbations of the surface heat budget and subducted into the main thermocline. The right-hand side depicts the generation of SST anomalies in the western North Pacific through a Sverdrup response to changes in Ekman pumping. The sources of these ocean mechanisms are decadal variations in atmospheric pressure, which change surface wind stress patterns. From Miller A and Schneider N (2000) Interdecadal climate regime dynamics in the North Pacific Ocean: Theories, observations and ecosystem impacts. *Progress in Oceanography* 47: 355–380.

exchange provides a feedback to the western Pacific atmosphere, creating a decadal equivalent of the tropical ENSO delayed oscillator. The perturbation in the western Pacific atmosphere feeds into the pressure pattern (e.g., Aleutian Low) that resets the wind stress forcing of the ocean, an order 20-year cycle. A key question remains – what is the relative importance of the atmosphere–ocean coupling vs. stochastic forcing, and does this dominance change between regime shifts?

In the eastern Pacific, upper ocean temperature shifts are largely controlled by wind-driven heat gains or losses. A stronger Aleutian Low brings warmer, moister air over the Northeast Pacific, reducing the surface heat flux from the upper ocean and increasing ocean advection of warmer water into the region. Adjustments in Ekman processes also reduce upwelling, contributing to the warm SST anomalies that contrast with the cooler SSTs in the central Pacific. Shifts in precipitation patterns affect coastal runoff, contributing to physical and biological variability through adjustments to buoyancy, temperature, circulation, and nutrients.

In the positive mode of the NAO, a deeper Icelandic Low leads to stronger storms and their more northerly track across the Atlantic. Stronger winds mean more intense vertical mixing and reduced stratification. The result is less available productivity in subpolar regions, but more in the subtropical gyre. On decadal scales, adjustments in the basin ocean circulation and a greater Gulf Stream transport are thought to drive warm nutrient-rich slope water at depth, with a 1-year lag, onto the shelf of Canada and the United States. Cool regions of the North Atlantic, such as the Barents Sea, are warmed by a positive heat exchange with the atmosphere and the advection of warmer, saline Atlantic ocean water inflow. Thus the direction of regime shifts in the Atlantic and their ecological effects depend on the region and organism considered.

Equatorial symmetry in atmosphere and ocean anomaly patterns indicates that similar mechanisms are responsible for decadal variability in the South Pacific and Atlantic, although the difference in their spatial scales results in natural basin modes and basin-specific regime shift patterns.

Ocean regime shifts appear to be global phenomena; they are seen in storm and drought cycles over land, and in forest ecosystem variability. Atmospheric teleconnections link decadal variability between the Pacific and Atlantic, and may be important in regulating regime shift synchrony across ecosystems. However, teleconnections between the Pacific and Atlantic appear at this point to be either weak or a nonstationary process that fluctuates over multiple regimes, such that the phase relationship between the basins appears to change.

Communication between the Tropics and extratropical ocean regions through atmospheric teleconnections, ocean advection, and planetary ocean waves create a coupling between the atmosphere and ocean that drives decadal regime shifts. Other studies, however, indicate that the Tropics have only a minor influence on the mid-latitudes on decadal scales, and that the local atmosphere–ocean coupling drives regime shifts. The ocean is thought to integrate atmospheric weather, which is roughly a white noise (random) process, creating increased variability at longer periods (red noise).

The interaction of equatorial ENSO forcing with the integrated atmospheric signal may be the source of decadal variability in the Pacific. This pattern is not predictable, and may not be apparent until years after a regime shift has occurred. On the other hand, if the processes responsible for these decadal signals are a sequence of lagged couplings and propagations explained by nonlinear physical concepts, then improved understanding of these processes will allow nowcasting of the existing climate state, hence a prediction of the slow evolution of the atmosphere–ocean cycle.

Strong atmosphere–ocean coupling occurs in the North Atlantic, especially on short-term (<5 year) scales. Quasi-decadal variability in the North Atlantic can be described with modulated oscillatory modes coupled between the ocean and atmosphere, although there is some doubt about the significance of decadal spectral peaks. Other analyses suggest this variability is better expressed as an autoregressive process, or as some combination of varying seasonal and nonseasonal trends. Some time series display nonstationary behavior on decadal scales, although it is not possible to differentiate clearly between nonstationarity and random fluctuations.

The reorganization of energy as it is transferred from the atmosphere to the ocean is reflected in the coupling between physics and biology. In general, physical and biological time series have fundamentally different dynamics. Key physical time series in the North Pacific are best modeled as linear but composed of random fluctuations and thus stochastic. Biological time series exhibit low-order nonlinear properties, suggesting a less random amplification of physical environmental variability. In summary, as variability propagates from large-scale atmospheric forcing to ocean structure and into the biotic marine world, it becomes more organized and thus more likely to appear as cyclic or regime-like.

Regional responses to a particular regime shift may not be of the same magnitude, or even in the same direction. The same physical change may result in opposing responses in different regions. For example, warmer SSTs and a shallower mixed layer depth following the 1970s shift led to higher productivity in the subarctic Pacific, which is light-limited. However, the shallower mixed layer lowered productivity in the California Current because of lower nutrient availability in the photic zone.

Glossary

Delayed oscillator A proposed mechanism to explain the evolution of ENSO. Because the response time of the ocean is much slower than the atmosphere, ocean anomalies developing due to anomalous wind forcing also feed back to the atmosphere in a way that further reinforces ENSO conditions in the ocean.

Ekman transport The process by which wind stress balances the effect of Earth's rotation (the Coriolis force) to create a transport of water in the surface layer roughly 90° to the right (left) of the wind in Northern (Southern) Hemisphere. Along a continental west coast, this transport away from the coast results in upwelling of cool, nutrient-rich water. A divergent Ekman transport (positive wind stress curl) in the open ocean will cause Ekman pumping, or open ocean upwelling.

Red noise Variability with energy decreasing uniformly with increasing frequency. The result is a time series pattern that appears to have a coherent cycle at increasingly long periods as the series gets longer.

Rossby wave Freely traveling long planetary waves (100–1000 km wavelength, *c.* 10 cm s^{-1} phase velocity in the internal ocean) that always propagate to the west due to the beta-effect. It is a dynamic mechanism for ocean adjustment to large-scale perturbations, such as atmospheric forcing. Rossby waves are responsible for the intensification of western boundary currents and the propagation of decadal signals across ocean basins.

Sverdrup transport Ocean basin-scale mass transport in balance with the wind stress curl. Sverdrup processes are responsible for maintaining anticyclonic gyre circulations in ocean basins.

White noise Random variability (noise) that has uniform spectral energy at all frequencies.

See also

Fisheries and Climate. Fisheries Overview. Marine Fishery Resources, Global State of. Molluskan Fisheries. Regime Shifts, Ecological Aspects. Regime Shifts: Methods of Analysis. Upwelling Ecosystems.

Further Reading

Bond NA, Overland JE, Spillane M, and Stabeno P (2003) Recent shifts in the state of the North Pacific. *Geophysical Research Letters* 30: 2183–2186.

Gedalof Z, Mantua NJ, and Peterson DL (2002) A multi-century perspective of variability in the Pacific Decadal Oscillation: New insights from tree rings and corals. *Geophysical Research Letters* 29 (doi: 10.1029/2002 GL015824).

Latif M and Barnett TP (1994) Causes of decadal climate variability over the North Pacific and North America. *Science* 266: 634–637.

Mantua NJ, Hare SR, Zhang Y, Wallace JM, and Francis RC (1997) A Pacific interdecadal climate oscillation with impacts on salmon production. *Bulletin of the American Meteorological Society* 78: 1069–1079.

Miller A and Schneider N (2000) Interdecadal climate regime dynamics in the North Pacific Ocean: Theories, observations and ecosystem impacts. *Progress in Oceanography* 47: 355–380.

Minobe S (1999) Resonance in bidecadal and pentadecadal climate oscillations over the North Pacific: Role in climatic regime shifts. *Geophysical Research Letters* 26: 855–858.

Newman M, Compo GP, and Alexander MA (2003) ENSO-forced variability of the Pacific Decadal Oscillation. *Journal of Climate* 16: 3853–3857.

Parrish RH, Schwing FB, and Mendelssohn R (2000) Midlatitude wind stress: The energy source for climatic regimes in the North Pacific Ocean. *Fisheries Oceanography* 9: 224–238.

Percival DB, Overland JE, and Mofjeld HO (2001) Interpretation of North Pacific variability as a short- and long-memory process. *Journal of Climate* 14: 4545–4559.

Peterson WT and Schwing FB (2003) A new climate regime in northeast Pacific ecosystems. *Geophysical Research Letters* 30(17): 1896 (doi:10.1029/2003GL017528).

Pierce DW (2001) Distinguishing coupled ocean–atmosphere interactions from background noise in the North Pacific. *Progress in Oceanography* 49: 331–352.

Schwing FB, Jiang J, and Mendelssohn R (2003) Coherency of regime shifts between the NAO, NPI, and PDO. *Geophysical Research Letters* 30: 1406 (doi:10.1029/2002GL016535).

Steele JH (2004) Regime shifts in the ocean: Reconciling observations and theory. *Progress in Oceanography* 60: 135–141.

Trenberth KE and Hurrell JW (1994) Decadal atmospheric–ocean variations in the Pacific. *Climate Dynamics* 9: 303–319.

Zhang RH and Liu Z (1999) Decadal thermocline variability in the North Pacific Ocean: Two pathways around the subtropical gyre. *Journal of Climate* 12: 3273–3296.

REGIME SHIFTS: METHODS OF ANALYSIS

B. deYoung, Memorial University, St. John's, NL, Canada
A. Jarre, University of Cape Town, Cape Town, South Africa

Introduction: Drivers and Response of Regime Shifts

Regime shifts as understood here are relatively abrupt changes in marine system functioning that persist on a decadal timescale, at large spatial scales, and are observed in multiple aspects of the physical and biological system. From a systems analysis perspective, one approach to analyzing regime shifts is to separate out the drivers of the shift and the response (*see* Regime Shifts, Ecological Aspects). Distinguishing the driver and the response is helpful because one expects them to have differing characteristics and so mixing them up would bias the analysis. This separation also supports the view that regime shifts in the marine ecosystem are often driven by external forcing from the physical system. The primary drivers for a regime shift are generally considered to be abiotic processes, for example, changes in ocean temperature, salinity, or circulation. However, in principle, following the nonlinear systems approach, these could include biotic processes including, for example, internal population dynamics, or changes to structural habitat, primarily considered for nearshore coastal ecosystems. Abiotic factors are typically the most easily identified since there are generally more physical than biological data and we normally have greater confidence in measurements of things such as temperature than fish landings or zooplankton abundance (*see* Regime Shifts, Physical Forcing). The response typically focuses on the biological organisms. It is also possible, however, to consider purely physical state regime shifts, perhaps the most obvious one being the ice-age cycles, but here the focus is on marine ecosystems, in which biological and physical dynamics are coupled.

Labeling variables as drivers and responses is somewhat like being identified at a wedding where you are asked whether your connection is to the family of the bride or the groom. At times, the categories are mixed, you may be connected to both the bride and the groom. Oceanic variables such as nutrient concentration may also sit in both camps, not only as a driver of biological changes but also responding to other changes in the physical environment such as temperature or the mixed layer depth. This mixing of the dynamics is one of the challenges in selecting variables for quantitative regime shift analysis since many of the techniques mix the variables together.

What are the characteristics of the drivers and response that we should consider in analyzing a regime shift? The most basic question is whether or not a shift has taken place. One can try to determine when the shift took place and the duration of the transition. Another timescale to consider is how long the system was in the state before or after the shift, or the persistence. Of course, these timescales could be different for the driver and the response and so yet a third timescale might need to be considered, the time lag between the driver and the response. The spatial scale of the drivers and response also provide a means to characterize the regime state and shift. Again, we might expect that each variable, whether under the umbrella of driver or that of response, may not exhibit uniform spatial characteristics. There may also be particular characteristics for each variable that relate to the interpretation of the regime state. While it may be as simple as a presence or absence differentiation, as in the case of algae and coral in coral reef regime shifts, in other situations the differences may be more subtle and relate to changing abundance or changing characteristics (*see* also Regime Shifts, Ecological Aspects). It may also be that a regime shift corresponds not to a change in the abundance but to a change in the timing of presence, as in a change in the phase of the seasonal cycle of abundance.

One can conduct numeric the analysis on the mean, the variance, or the frequency structure of the data. To date most of the analysis has focused on the mean or the variance of the data. It has been suggested that regime shifts can appear as a result of the statistical aggregation of the many different time series that are fed into the analysis (**Figure 1**). Our ocean time series of direct measurements now extend in length from many decades to a century for some variables and in some regions. As we have discovered these growing time series, we have seen growing evidence of interannual and decadal variability. It has been suggested that there is no real abrupt change in the ecosystem dynamics but that the nonlinear long cycles that we discover are an artifact of our analysis.

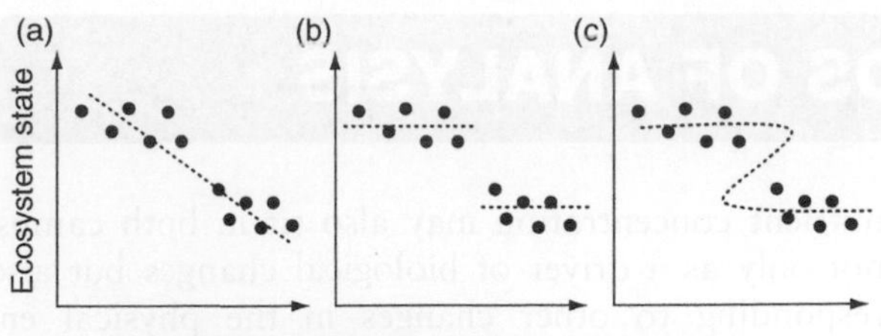

Figure 1 Schematic showing different possible interpretations of the same data. The vertical axis defines the ecosystem state, perhaps through the average of the normalized abundance of several different species. The horizontal axis could be time, in which case we are just looking at the biotic part of the marine ecosystem or perhaps is the average of several different physical environmental indices, for example, the sea surface temperature and the pressure difference across the ocean basin. The dashed lines illustrate the conceptual models that are fitted to the data: (a) a linear evolution of a declining abundance, (b) an abrupt but potentially reversible change in abundance, and (c) an abrupt change that is not easily reversible as the system exhibits a highly nonlinear and nonfunctional response. Given the complexity and multidimensionality of the data, including both biological and physical data, there are many different such plots that could be generated for a particular ecosystem.

The suggestion is that artificial patterns appear out of the 'soup' of variables. Nonetheless, tests of the statistical character of the time series suggest that marine ecosystems are highly nonlinear and quite susceptible to dramatic change. While there may be times when our analysis will produce false positives, it does appear that regime shifts do indeed take place in the ocean (*see* Regime Shifts, Ecological Aspects and Regime Shifts, Physical Forcing).

Approaches to the Analysis of Regime Shifts

The goal of regime shift analysis is to demonstrate a shift in a marine ecosystem, typically through analysis of its mean or variance. While there have been many papers on regime shifts in the ocean, there is not yet an agreed-upon test case for the testing of different analysis approaches. Indeed, the diversity of analysis techniques suggests that we have yet to discover a single unambiguous tool for regime shift detection. In view of the diversity of marine ecosystems, it may not even exist, and consequently, a carefully composed set methods that have performed well in several applications – a 'toolbox' – may be most appropriate. Another challenge is to find an appropriate test case, such as the coral-algal system. Like some of the freshwater examples, they have shown clear shifts in structure. As in freshwater systems they have relatively few independent variables to consider, making their recognition more straightforward, but this simplicity also makes their analysis unlike that of other complex open ocean ecosystems such as the North Pacific.

Regime shift analysis is a particular case of time-series analysis. The first step in the analysis is the selection of the time series. Some analysis involves the exploration of hundreds of different time series, which is of course one of the difficulties. The goal is to find a pattern that represents the ecosystem structure, and can illustrate the shift in structure associated with the regime shift. One reason for using many different time series is that one expects the ecosystem structure to include such complexity, although using hundreds of variables is arguably an extreme. But there is another reason for selecting so many time series, and that is that by bringing together so many different variables one hopes that the noise, that is the things that are not connected to the pattern, or to the shift, will average out allowing the pattern to be revealed. Unfortunately, bringing together so many time series may also produce false positives in which we see a regime shift that is not really there. It has been suggested that since these oceanographic data have so much low-frequency variability, interannual and decadal, it is quite likely that analyzing them together, that is combining the different time series, will occasionally yield coherent jumps that look like a regime shift. This is a potential concern which really just highlights the problem that even when a regime shift is identified through an analysis technique we have great difficulty in determining the dynamics of the state of the two separated regimes. If we could confirm, through understanding, the character of the defined regimes, then we would have greater confidence in the analysis. Ultimately, this numerical analysis needs to be supported by ecosystem understanding.

With the time series in place, there are several different options, to consider. One approach, now common was applied to North Pacific data (**Figure 2**). It involves looking at the variability of the time series, the variance, through an analysis in which all the different time series are first normalized by removing the mean and then dividing by the standard deviation. Normalizing each of the time series removes the differences between the units. It is necessary to guess at a date for the regime shift as the series are adjusted so that the series show the same direction of change across the regime-shift date. The time series are then averaged together and the standard errors calculated ($s/\sqrt{n}$), where s is the standard deviation and n is the number of samples. If the shift shows a significant change across the selected time, that is, the jump exceeds the standard error and the change is stable for

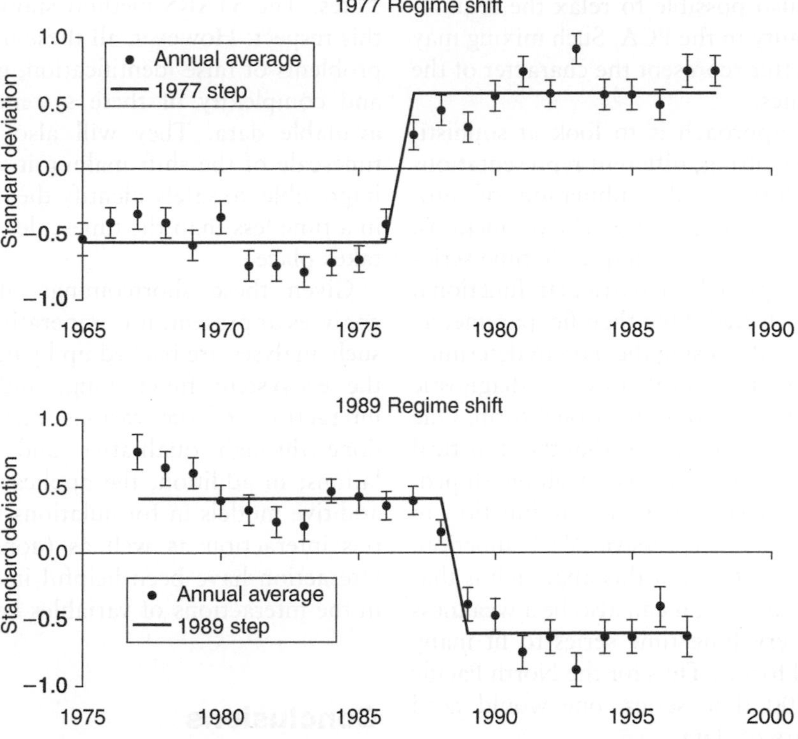

Figure 2 Hare and Mantua looked at more than 100 variables around the North Pacific to illustrate two regime shifts in the Northeast Pacific, one in 1977 and the other in 1989. These two figures have been reproduced many times as icons of ocean regime shifts. The approach to analysis taken in this figure was to sum the data that is both normalized, adjusted and shifted. Adapted from Hare S and Mantua N (2000) Empirical evidence for North Pacific regime shifts in 1977 and 1989. *Progress in Oceanography* 47: 103–145, with permission.

some period of time that exceeds the period of the change, then a regime shift has been identified.

Another approach has been developed that is tied to the statistical Student *t*-test, called sequential *t*-test algorithm for analyzing regime shifts (STARS). This approach has the advantage that it does not require selecting a time of the shift as part of the analysis. One steps through the time series to determine when there is a statistically significant change in the mean state. It is also said that unlike many of the other methods described here, it detects changes toward the end of a time series. A variant on this approach is to analyze the variance through time to determine when there is a significant change of variance. The algorithm has been applied to the Gulf of Alaska and Bering Sea ecosystems. Recent exploratory application to the southern Benguela upwelling ecosystem generally confirmed the ability of the method to detect changes toward the end of the time series. However, the performance of the algorithm for autocorrelated time series still needs to be enhanced.

Principal-component analysis (PCA) is a technique well used both by biological and physical scientists. In physical oceanography, the technique is called empirical orthogonal functional (EOF) analysis. The approach is quite straightforward, essentially one is looking for the eigenvalues and eigenvectors of the covariance matrix of the data. The covariance matrix defines the relation between the different variables. The eigenvectors should define the most effective representation of the variability of the variable dynamics. Thus, they define the pattern of the dynamics. The resulting patterns from PCA are called principal components, eigenvalues, or empirical functions, depending on who is doing the analysis. It is both a strength and a weakness that these functions are orthogonal. In physical systems, it is generally expected that dynamical modes will be independent, or orthogonal, and so this characteristic is seen as a strength. For this problem, it may not be realistic to assume that the regime states will be independent, or orthogonal, and it certainly seems doubtful that they will follow linear relationships. Thus while there are clear strengths associated with this approach, there are also some limitations. While the standard approach to PCA is linear, a limitation to the analysis, nonlinear techniques using neural networks have also been developed although they have not yet been tried for regime shifts. It will be interesting to see what difference a nonlinear analysis might make. It is also possible to relax the requirement for orthogonality in the PCA. Such mixing may yield results that better represent the character of the regime shift dynamics.

Another general approach is to look at sophisticated forms of curve fitting, different representations of basic functional forms, and combinations of autoregressive (AR) and moving average (MA) functions. This is a commonly used technique in time-series analysis. For this approach, a particular functional form must first be selected. One then fits parameters for the time-series model, using the data to determine the parameters, and then evaluates the diagnostic characteristics for the resulting functional form. One clear advantage of this approach is that the statistical analysis for such modeling has been well developed. A rather similar approach uses an AR function in vector form, a vector-autoregressive (VAR) function. While one of the advantages of this approach is that it includes a vector analysis, it can also be a weakness as it will require very long time series to fit many different functional forms. Thus for the North Pacific data, with over 100 time series, one would need more than 100 years of data.

An information theory approach has been suggested that requires an estimate of the system's trajectory in state-space. The idea is quite straightforward. The derivative of the time series is a measure of its changing state. The acceleration, a measure of the changing state of the individual time series, is calculated. The maxima of the averaged accelerations should then provide a measure of the changing state of the system. These averages are called Fisher information (FI) indices. One should see a drop in the index as we move through a period of disorder and a gain as the system moves through a period of order. One weakness of this approach is that there does not appear to be a clear statistical measure of the significance levels for the changes in the Fisher index.

Continuous wavelet transformation is a methodology that has proven successful in detecting climate oscillations in several applications, ranging from paleo-climatic analyses to temperature and wind time series of only a few decades. The applicability of this method for detecting decadal-scale changes in biological time series, the length of which typically is limited to a few decades, is currently under investigation.

The analysis techniques presented here can only be conducted retrospectively and typically require long time series. With the need for an understanding of regime shifts for application to ocean management, there is a need for a tool that offers relatively quick identification of a change in ocean ecosystem regime states. The STARS method shows some promise in this respect. However, all these methods are open to problems of false identification, given the variability and complexity of these systems, and the limited available data. They will also be limited by the timescale of the shift making it very difficult, if not impossible, to safely identify the occurrence of a shift in a time less than the timescale over which the shift takes place.

Given these shortcomings of purely statistical analyses at present, it is imperative that the results of such analyses are backed up by our understanding of the ecosystem functioning, and modeling of the interactions of the various variables. This can be done through qualitative and quantitative simulations; in addition, the applications of generalized additive models in formulations that allow continuous interaction as well as (nonadditive) threshold interaction have been helpful in explaining changes in the interactions of variables between regimes.

Conclusions

Our understanding of regime shift dynamics will grow with new data and new analyses. Continued studies of ocean ecosystems are extending the length of the time series with which we have to work. These extended data not only offer more degrees of freedom for analysis, hence improving the statistical confidence of our understanding, but also cover a growing number of behaviors of marine ecosystems. We need to learn more about the true dynamical relations among variables and we need to develop explicit rather than the statistical models so that we have a clearer model of the relation between the drivers and the response.

See also

Regime Shifts, Ecological Aspects. Regime Shifts, Physical Forcing.

Further Reading

Box GEP and Jenkins GM (1976) *Time Series Analysis: Forecasting and Control*, rev. edn. San Francisco, CA: Holden-Day.

Cianelli L, Chan KS, Bailey KM, and Stenseth NC (2004) Non-additive effects of the environment on the survival of a large marine fish population. *Ecology* 85(12): 3418–3427.

deYoung B, Harris R, Alheit J, Beaugrand G, Mantua N, and Shannon L (2004) Detecting regime shifts in the oceans: Data considerations. *Progress in Oceanography* 60: 143–164.

Fath BD, Cabezas H, and Pawlowski CW (2003) Regime changes in ecological systems: An information theory approach. *Journal of Theoretical Biology* 222: 517–530.

Hare S and Mantua N (2000) Empirical evidence for North Pacific regime shifts in 1977 and 1989. *Progress in Oceanography* 47: 103–145.

Hsieh CH, Glaser SM, Lucas AJ, and Sugihara G (2005) Distinguishing random environmental fluctuations from ecological catastrophes for the North Pacific Ocean. *Nature* 435: 336–340.

Mantua N (2004) Methods for detecting regime shifts in large marine ecosystems: A review with approaches applied to North Pacific data. *Progress in Oceanography* 60(2–4): 165–182.

Overland JE, Percival DB, and Mofjeld HO (2006) Regime shifts and red noise in, the North Pacific. *Deep-Sea Research I* 53(4): 582–588.

Rodionov SN (2004) A sequential algorithm for testing climate regime shifts. *Geophysical Research Letters* 31 (doi:10.1029/2004GL019448).

Rodionov SN (2006) Use of pre-whitening in climate regime shift detection. *Geophysical Research Letters* 33: L12707.

Rudnick DL and Davis RE (2003) Red noise and regime shifts. *Deep-Sea Research* 50: 691–699.

Sheffer M, Carpenter S, Foley JA, Folkes C, and Walker B (2001) Catastrophic shifts in ecosystems. *Nature* 413: 591–596.

Solow AR and Beet AR (2005) A test for regime shifts. *Fisheries Oceanography* 14: 236–240.

MICROBIAL LOOPS

M. Landry, University of Hawaii at Manoa, Department of Oceanography, Honolulu, HI, USA

Introduction

The oceans contain a vast reservoir of dissolved, organically complex carbon and nutrients. At any given point in time, most of this dissolved organic matter (DOM) is refractory to biological utilization and decomposition. However, a significant flow of material cycles rapidly through a smaller labile pool and supports an important component of the food web based on bacterial production. The recapturing of this otherwise lost dissolved fraction of production by bacteria and its subsequent transfer to higher trophic levels by a chain of small protistan grazers was initially called the microbial loop by Farooq Azam and co-workers in the early 1980s. The following decades have been charecterized by remarkable discoveries of previously unknown life-forms and by major advances in our understanding of trophic pathways and concepts involving the seas' smallest organisms. In modern usage, the term microbial food web incorporates the original notion of the microbial loop within this broader base of microbially mediated processes and interactions.

Perspectives on an Evolving Paradigm

Although early studies date back more than a century, our understanding of the microbial ecology of the seas initially advanced slowly relative to other aspects of biological oceanography largely because of inadequate methods. Prior to the mid-1970s, for example, the simple task of assessing bacterial abundance in sea water was done indirectly, by counting the number of colonies formed when sea water was spread thinly over a nutritionally supplemented agar plate. We know now that about one marine bacterium in a thousand is 'culturable' by such methods. At the time, however, the low counts on media plates, typically tens to hundreds of cells per ml, were consistent with the then held view that sea water would not support a large and active assemblage of free-living bacteria. The role of bacteria was therefore assumed to be that of decomposers of organically rich microhabitats such as fecal pellets or detrital aggregates.

Coincidentally, early deficiencies in phytoplankton production estimates, due to trace metal contaminates and toxic rubber springs in water collection devices, were giving systematic underestimates of primary production, particularly in the low-nutrient central regions of the oceans that we now know to be dominated by microbial communities. For such regions, the low estimates of bacterial standing stocks and primary production were mutually consistent, reinforcing the notion of the central oceans as severely nutrient-stressed 'biological deserts' with sluggish rates of community growth and activity.

Even so, there were early signs from both coastal and open-ocean studies of much greater microbial potential – one from newly developed sea water analyses of ATP (adenosine triphosphate), the short-lived energy currency of all living organisms; another from respiration (oxygen utilization) measurements of whole and size-fractioned sea water samples. Such measurements implied much larger concentrations of life-forms and metabolic rates than could be explained by the planktonic plants and animals that were being studied intensively in the 1970s. Recently developed methods for measuring bacterial production in the oceans, based on the uptake of radioisotope-labeled thymidine precursor for nucleic acid synthesis, were also giving results inconsistent with conventional wisdom. Thus, by the late 1970s, the stage was set for a revolutionary new paradigm, the microbial loop.

Epifluorescence microscopy was perhaps the tool that most facilitated scientific discovery and general acceptance of the new paradigm. Here, for the first time, one could directly visualize, with the aid of fluorescent stains, the multitude of coccoid, rod, and squiggle-shaped forms that comprised the typical bacterial assemblage of about a million cells per ml. Autoradiography, a technique by which the cellular uptake of radioisotope-labeled substrates is developed on sensitive films, confirmed that most of these cells were living and active.

Epifluorescence microscopy (EPI) also opened the door to discovering and studying other components of the microbial community. *Synechococcus*, a genus of small photosynthetic coccoid cyanobacteria, was soon recognized as a ubiquitous and important component of the marine phytoplankton from its characteristic shape and phycobilin accessory pigments, which glow orange under the EPI standard blue-light excitation. The same excitation wavelength causes the chlorophyll in photosynthetic cells to fluoresce bright red, allowing purely heterotrophic cells to be easily distinguished from chlorophyll-containing cells of similar size and shape. This distinction was critical in demonstrating a clear food web coupling for bacterial production via small phagotrophic (i.e., particle-consuming) colorless flagellates, and it provided an important technique for identifying and quantifying trophic connections using fluorescently labeled beads and cells as tracers to assess grazer uptake rates.

Such studies also had the unintended effect of illustrating the blurry distinction between pure autotrophy and heterotrophy within the microbial assemblage. For example, many of the small photosynthetic flagellates (containing chlorophyll) in the oceans have been observed to consume bacterial-sized particles. Similarly, many, if not most, of the larger pigmented dinoflagellates follow a 'mixed' mode of nutrition involving photosynthesis and phagotrophy. In a phenomena known as kleptoplastidy, common forms of ciliated protozoa have also been shown to retain (literally 'steal') the chlorophyll-containing plastids of their prey and use them as functional photosynthetic units for a day or longer. The widespread occurrence of mixotrophy, in all of its various forms, has consequently emerged as one of the important findings related to the microbial food web.

One notable discovery of the mid-1980s was that of *Prochlorococcus*, a tiny photosynthetic bacterium now known to be one of the most important primary producers in the tropical oceans, and probably on the planet. Although it seems remarkable that such an important organism could have escaped detection for more than a century of oceanographic investigation, this advance was again only made possible by new methods. In this case, the facilitating technology was a laser-based optical instrument, the flow cytometer, developed by medical research for the rapid analysis of individual cells in a narrowly focused fluid stream. The application of this new approach in the ocean sciences was quick to reveal high concentrations of the dimly red-fluorescing (chlorophyll-containing) *Prochlorococcus*, which could not be distinguished from nonpigmented bacteria by standard EPI techniques. Herein lies one of the problems of epifluorescence microscopy, the confounding of significant populations of the autotrophic *Prochlorococcus* cells with heterotrophic bacteria. Some early reports of heterotrophic biomass greatly exceeding autotrophs in surface waters of the tropical oceans were a consequence of this methodological artifact.

Even among the heterotrophic bacteria, there have been discoveries of fundamental importance. For example, kingdom-specific molecular probes have recently shown that a significant fraction of the 'bacteria' from EPI counts are not true Bacteria at all, but lesser-known prokaryotes of the Kingdom Archaea. These organisms have been known to inhabit extreme environments such as hot springs and the interstices of salt crystals, but their presence in more typical oceanic habitats raises many interesting questions about the roles of their unique metabolic systems in ocean biogeochemistry. The application of powerful molecular methods to the ocean sciences in the 1990s has signaled the beginning a new era in marine microbial ecology with the potential to reveal the full spectrum of microbial population and physiological and metabolic diversity.

As a natural extension of new methods to visualize and characterize community components of decreasing size, the 1990s have also seen a growing recognition of the importance of viruses in marine microbial communities. Based on a combination of fluorescent staining methods and electron microscopy, viruses are now clearly the most numerous component of the microbial community, exceeding bacteria in most cases by an order of magnitude. While not technically 'real' organisms, in the sense of having independent metabolic or reproductive capabilities, viruses can be significant vectors of bacterial and algal mortality and therefore have important implications for the functioning and energy flows in marine microbial communities. Recent studies have shown rates of microbe infection and viral turnover that would account for the loss of about half of bacterial production. The typical host-specificity of viruses and their spread by density-dependent encounter frequency suggest that they also have a major role in maintaining bacterial diversity, by selectively punishing the most successful competitors.

Organization of the Microbial Food Web

For the sake of representing diagrammatically the trophic connections among marine microbes, and indeed as a practical strategy for most ecological studies, it is necessary to compress the known complexities of microbial communities into a few functional categories (**Figure 1**). There is no clear cut-off between particulate and dissolved organic matter in the oceans, for example, but rather a spectrum of material ranging from low molecular weight (low-MW) amino acids and sugars to large complex

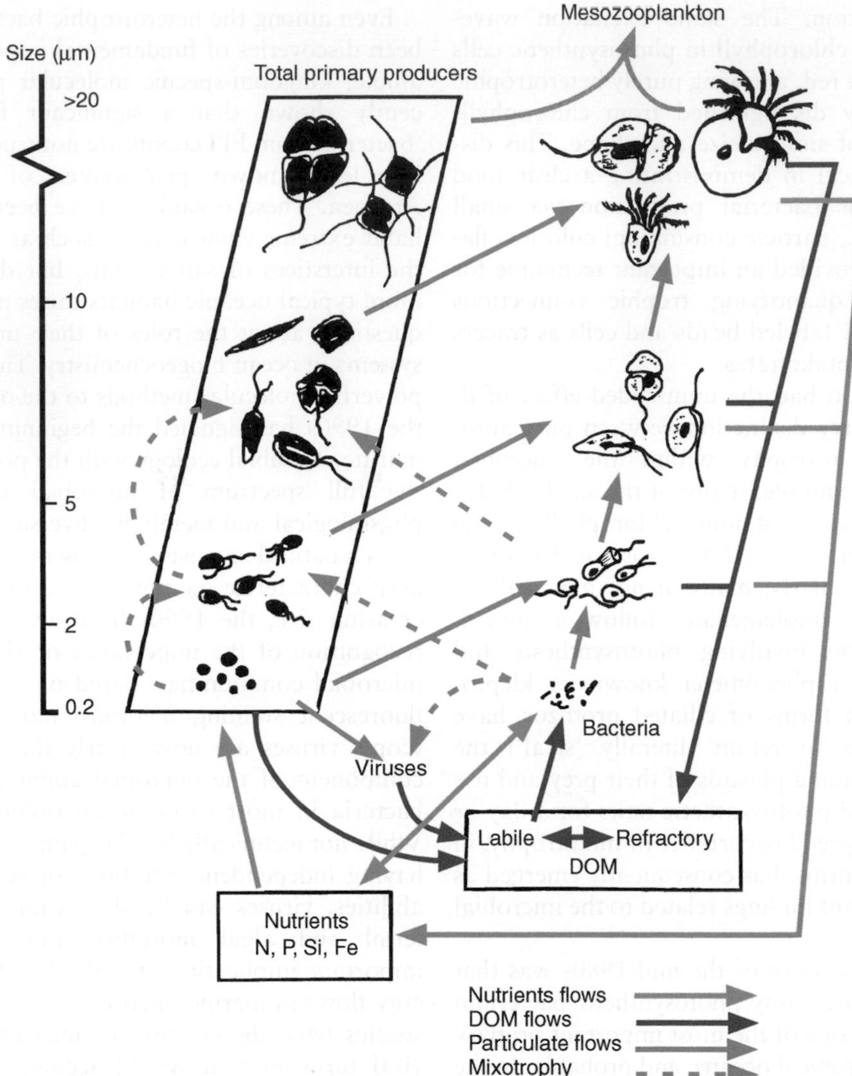

Figure 1 Conceptual representation of the microbial food web showing flows of nutrients and dissolved organics and interactions among various size classes of bacteria and protists. Primary producers (left side) are shown as chlorophyll-containing cells and heterotrophic (right side) cells are drawn without pigments. Mixotrophy is shown as feeding of some of the pigmented cells on smaller prey organisms.

molecules, to colloids, viruses and submicrometer remnants of previously consumed biota, and to wispy strands that link a fragile gelatinous matrix of living and dead material. Labeled simply as DOM in **Figure 1**, this material can cycle between refractory and labile pools by slow physical winnowing and leaching, extracellular enzymatic cleavage, or accelerated photochemical oxidation, the latter principally from enhanced ultraviolet radiation in near-surface waters. Although the relative magnitudes are debatable and likely to vary seasonally and regionally, inputs to the DOM box come from virtually all components of the marine plankton. Phytoplankton leak low-MW compounds across porous membranes, and they also produce the sugary products of photosynthesis in excess when nutrients are insufficient for cell growth. Many larger consumers feed sloppily, producing DOM and particulate fragments as they grind and rip their food with silica-tipped teeth. Both large and small consumers excrete low-MW organics, often as a significant fraction of their total metabolism, and release organics in the form of incompletely digested material. Lastly, whole cells are fragmented into numerous components in the operationally defined DOM size range during the final stages of the viral lytic cycle. DOM from all of these various sources provides the substrate that fuels the growth of marine heterotrophic bacteria.

The original microbial loop was patterned on the notion that consumers typically feed on prey organisms that are a factor of 10 less in cell size (length or equivalent spherical diameter), a view that fitted nicely with the traditional decadal size classifications of marine plankton. Accordingly, picoplankton (0.2–2 μm cells, including most prokaryotes) would be fed upon by nano-sized protists (2–20 μm cells, typically flagellates), and they, in turn, by microheterotrophs (20–200 μm cells, typically ciliates). From laboratory feeding experiments as well as from size-fraction manipulations of natural communities, however, it has become increasingly clear that many protists feed optimally on prey much closer to their own body size. The most common bacterivores in the oceans, small naked flagellates, are typically only 3–6 times the size of their average bacterial prey. Heterotrophic and mixotrophic dinoflagellates generally prey on cells of even larger relative size, some using extracellular capture and digestion to handle preferred prey larger than their own body size. Compared to the original microbial loop paradigm, the effect of compacting mean predator–prey size relationships for flagellates can add at least one more level of intermediate consumer between bacteria and the largest protozoa in the grazing chain.

As will be presented in more detail below, complex relationships and feedbacks among bacteria and protists and their different degrees of availability to higher-order consumers argue for a broad view of the microbial food web, rather than a narrow focus on a single loop element. This brings us to the matter of definition. Are all single-celled organisms to be included in this web, or are there size or functional reasons to exclude some? **Figure 1** is organized from the perspective of organisms described in the original microbial loop paradigm, the bacteria and the grazer chain. It therefore includes the autotrophic organisms that would constitute the main food items of some protistan grazers. Notably absent are the very large phytoplankton, principally large solitary cells or long diatom chains that figure so prominently in classical descriptions of the seasonal bloom cycles of temperate and boreal oceans. Such cells would not be readily available to protistan grazers because of their size, spines or other defensive strategies. They also function differently from smaller primary producers, being more intimately related to export flux from the euphotic zone by aggregate formation and direct cell sinking or by incorporation into the fast-sinking fecal pellets of large metazoan consumers. Thus, while all planktonic organisms are related in a sense by trophic linkages and feedbacks to dissolved organics and nutrients, we specifically exclude these larger primary producers from the microbial assemblage. The division is functional and roughly follows size, distinguishing the direct flow of primary production to a network of metazoan consumers as opposed to that going primarily to protists. It is important to observe that the division is only loosely related to taxonomic groupings. For example, the dominant diatoms in many open ocean regions are tiny (<10 μm) pennate cells and well within the size range that can be grazed efficiently by ciliates and large flagellates. By the same token, large clump-forming filaments of the nitrogen-fixing cyanobacteria *Trichodesmium* spp. appear to be directly utilized only by certain harpacticoid copepods.

Food Web Transfers

The original descriptions of the microbial loop by Williams and Azam and co-workers carefully observed that the transfer of bacterial production to higher trophic levels by a chain of protistan consumers was likely to be inefficient. Nonetheless, there was early speculation that this newly discovered pathway could represent a significant link or energy bonus to higher levels as opposed to a metabolic sink. This view was bolstered by evidence of high gross growth efficiencies (e.g., carbon growth = 50–70% of carbon substrate uptake) for bacteria under optimal laboratory conditions and observations that the growth efficiencies of small protists were also much higher and less sensitive to food concentration that those of metazoan consumers, like copepods.

There is little support these days for significant transfer of DOM uptake by bacteria through the protistan grazing chain. For one, experimental studies, now available from many marine ecosystems, suggest an average bacterial growth efficiency of about 20% on naturally occurring organic substrates. In other words, 5 moles of dissolved organic carbon are needed to produce 1 mole C of bacteria, the remainder being metabolized to inorganic carbon. Virus-induced cell lysis, the so-called viral shunt to DOM, represents a further loss of potential production to trophic transfer. Lastly, each step in the grazing chain, operating at about 30% efficiency, takes its toll. Assuming half of bacterial mortality goes to viral lysis and half to small bacterivorous flagellates, less than 0.3% ($0.2 \times 0.5 \times 0.3 \times 0.3 \times 0.3 = 0.0027$) of the DOM carbon uptake would be transferred past the largest protistan consumers in **Figure 1**.

While such calculations can diminish one's expectations for supporting significant fish production from bacteria *per se*, we must take a broader view of microbial contributions to plankton energy flows. In **Figure 1**, for example, bacteria do not constitute the single source of materials to large heterotrophic protists and the animals that ultimately feed on them. At each step along the grazing chain, production of flagellates and ciliates is well supplemented by consumption of appropriate sizes of photosynthetic organisms. Two-way flows between autotrophic and heterotrophic compartments due to mixotrophs further complicate these interactions, making it difficult to view the contribution of individual components and loops in isolation of the others.

To make matters even more complex, metazoan consumers do not as a rule wait patiently to siphon off only those resources that make it to the largest size categories of the microbial grazing chain. Mucus-net feeding pelagic tunicates, like appendicularians and salps, short-circuit the grazing chain by efficiently exploiting bacterial-sized particles, or at least the smallest size categories of nanoplankton. Somewhat less appreciated, the early developmental stages of planktonic crustaceans like copepods and euphausiids, as well as the larvae of benthic invertebrates in coastal areas, typically feed efficiently on the smallest size fractions of the microbial assemblage, graduating to larger prey as they grow. Regardless of the feeding habits of the adults, therefore, the developmental success of these organisms may depend on interactions with the smallest microbial size fractions.

Nutrient Cycling

The lack of efficient transfer of bacterial production through the long protistan grazing chain suggests that the microbial loop must be extremely important in remineralization and nutrient cycling. This has clearly emerged as one of the primary functions of the microbial food web and has brought particularly a new understanding to the ecology of the open oceans where microbial interactions predominate. Such systems are often limited by primary nutrients such as nitrogen or phosphorus, and in some cases trace elements like iron. Their common characteristic is the relatively high turnover rate of small primary producers supported by the efficient recycling of nutrients. Without this positive feedback of nutrient recycling, the available resources would rapidly be locked into the standing biomass of plankton, and new growth and production would stop.

Even though bacteria grow inefficiently on naturally available DOM and help to solubilize and degrade particulate organics with extracellular enzymes, they are typically not the major remineralizers in the seas. In fact, they often compete significantly with phytoplankton for the uptake of dissolved nutrients. Their superior competitive ability comes, of course, from the high surface area to volume ratios of bacterial cells. Their demand derives from their relatively high nutrient requirements for cell growth compared to phytoplankton. For example, while phytoplankton grow optimally with a carbon to nitrogen ratio (C:N) of about 7, the typical ratio for bacteria is 5. Compared to phytoplankton, bacteria also seem to require iron at about twice the concentration relative to carbon. We can appreciate the interplay among growth efficiencies, nutrient richness of available substrates, and bacteria as a source or sink for recycled nutrients from **Figure 2**. High growth efficiency allows bacteria to use more nutrients for growth, with less recycled. Reducing the nutrient content (increasing C:N) of available DOM has a similar effect. Where the dissolved substrates are nutritionally rich, growth efficiency is likely to be high, so nutrient release will be positive but depressed. On the other hand, when the C:N ratio of DOM is too high to satisfy nutrient needs for growth, bacteria will seek limiting elements in inorganic pools.

It is precisely in this latter case that we run across an interesting paradox of microbial interrelationships. Phytoplankton respond to nutrient-limited conditions by producing photosynthate sugars in excess of their growth needs and excreting them to the external environment. Since these sugars are

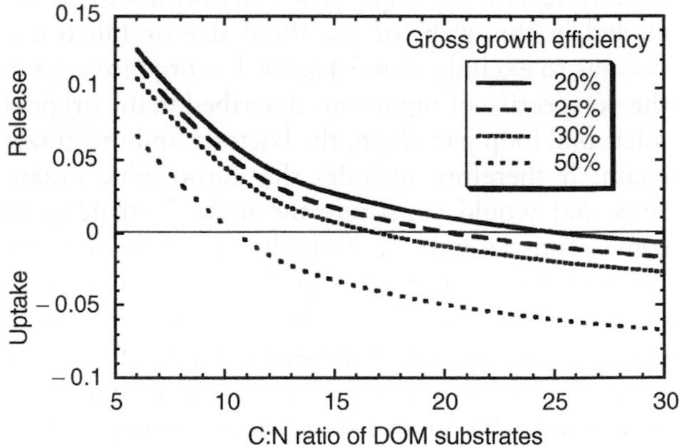

Figure 2 Uptake or release of dissolved inorganic nitrogen (DIN) by bacteria as a function of gross growth efficiency (GGE) and the carbon:nitrogen ratio (C:N) of dissolved organic substrates. Figure shows that bacteria can act as decomposers (releasing nutrients) when substrates are nutrient rich (low C:N). With increasing C:N or increasing carbon growth efficiency, bacteria will show a deficit (negative) in nutrients from DOM and will compete with phytoplankton in nutrient uptake. In comparision, bacterivorous flagellates feeding on relatively nutrient-rich bacteria should always serve as significant decomposers, recycling about 75% of ingested N as DIN or small particulates at GGE = 30%.

easily assimilated by bacteria but are devoid of associated nutrients (i.e., high C:N), the effect is to enhance bacterial demand for inorganic nutrients and hence their competition for limiting substrate with phytoplankton. According to some analyses, if one considers the many indirect consequences of this enhancement effect on the microbial network, phytoplankton could 'win' in the end by stimulating grazing on bacteria and subsequent nutrient remineralization. However, bacterivorous mixotrophs have an inherent advantage under such conditions since they benefit both from direct consumption of nutrient-rich bacteria and by the stimulation of bacterial growth and nutrient cycling from the DOM released by true autotrophs. At the same time, photosynthetic carbon production allows mixotrophs to utilize ingested nutrients for growth at very high efficiency, releasing little back to the environment compared to pure heterotrophs. As one might imagine, bacterivorous mixotrophs become more important and can even dominate over pure autotrophic or heterotrophic flagellates in systems of increasing oligotrophy.

Regional Patterns and Variations

Bacterial abundance and biomass vary in the oceans, within and between regions, but the range of variability is generally less than that observed for larger components of the food web. This is because bacteria and small algae in the microbial food web can be contained within certain limits by fast-growing protistan predators. In contrast, larger phytoplankton, such as diatoms, enjoy a substantial growth rate advantage over slow-responding mesozooplankton consumers, allowing them to increase explosively when light and nutrient conditions become optimal. Such cells give many regions of the oceans their characteristic seasonal blooms.

Plankton blooms are generally short-lived phenomena because the larger components of the food web are not good at retaining nutrients in the surface waters once the water column is seasonally stratified. Large cells aggregate and sink when nutrients are exhausted, and mesozooplankton export nutrient-rich material from surface waters as compact, fast-sinking fecal pellets. Increasing nutrient stress naturally favors smaller competitors for the limiting resource and the smaller consumers that feed on them. Therefore, a declining bloom will evolve through various successional states toward a microbially dominated community in which production, grazing, and nutrient remineralization are more tightly coupled. One such transitional state is likely to be a period in which the bacterial carbon demand overshoots the concurrent production of phytoplankton. This occurs during the declining stages of the bloom, when senescent phytoplankton produce carbohydrates in excess, when sick phytoplankton cells lyse, or when diminished antibacterial chemical defenses of phytoplankton allow bacteria to more effectively exploit the DOM accumulated during the bloom. Reduced carbon supply and enhanced mortality to bacterivorous protists and viruses will rapidly bring the bacteria back into balance.

If we consider the full range of variability in the oceans, it is possible to find very rich coastal ecosystems or events with chlorophyll concentrations of $30\,\mu g\,l^{-1}$ or more and bacterial abundances exceeding 10^7 cells ml^{-1}. However, such extremes are relatively rare. Average concentrations are about two orders of magnitude lower for phytoplankton chlorophyll and 10-fold lower for bacteria. Particularly in the open oceans, many regions share similar characteristics with regard to general low levels of bacterial and phytoplankton standing stocks.

If one looks, for example, at mean levels of bacterial abundance and biomass in tropical and subtropical seas, they vary quite little, regardless of whether the regions are relatively rich and productive (Arabian Sea), iron-limited (equatorial Pacific), or extremely oligotrophic (subtropical Pacific) (**Table 1**). As an indication of the selective pressures for small primary producers in such systems, photosynthetic bacteria (*Prochlorococcus* and *Synechococcus*) usually account for 40–50% of total bacterial biomass, with the relative abundance shifting toward *Synechococcus* in richer more-productive systems or seasons. As we move toward higher latitudes, photosynthetic bacteria decline in importance relative to eukaryotic phytoplankton, and bacteria as a group become increasingly more heterotrophic. *Prochlorococcus* are largely absent from the plankton at water temperatures below 12°C, while *Synechococcus* are rare in polar waters.

The importance of the microbial communities in the ecology of the oceans derives from the flow of nutrients and fixed carbon through them. This is in addition to the many roles that bacteria have with respect to chemical transformations and ocean geochemistry. When one takes into account the low growth efficiencies of bacteria and the deficiencies of the carbon-14 method, typical bacterial production estimates on the order of 5–20% of ^{14}C-bicarbonate uptake (**Table 1**) are consistent with a carbon demand of about 50% of primary production. In addition, protistan grazers directly consume 50–90% of phytoplankton cellular growth in the open oceans, and often half or more in coastal waters. Through

Table 1 Representative estimates of bacterial cell abundances, carbon biomass, and percentage heterotrophic cell biomass for several regions of the open oceans according to recent studies

Region		Abundance (10^3 cells ml^{-1})			Biomass ($\mu g\ C\ l^{-1}$)	% Hetero	Chla ($\mu g\ l^{-1}$)	BP/PP
	T (°C)	Hbact	PRO	SYN				
Central Equatorial Pacific, 1°S–1°N	28	670	140	7	14	59	0.26	0.17
Subtropical Pacific, 23°N	25	550	210	1	14	47	0.06	
Arabian Sea, NE Monsoon	27	850	70	90	22	47	0.40	0.22
Tropical Atlantic, 5–20°N	27	750	210	12	18	51	0.41	
Subtropical Atlantic, 30°N	22	500	70	7	9	66	0.07	0.05
Subarctic Atlantic, 50–60°N	12	1500	20	40	23	79	0.87	0.15
Southern Ocean bloom, 47°S, 6°W	4	2000	ND	ND	24	–	3.0	0.16
Southern Ocean, 60°S, 170°W	0	300	0	0	4	100	0.4	

T (°C) and Chl*a* are mean environmental temperature and total phytoplanton chlorophyll *a*. BP/PP is the ratio of bacterial (heterotrophic) production to total primary production. Total bacterial biomass is for combined populations of heterotrophic cells (Hbact), *Prochlorococcus* (PRO) and *Synechococcus* (SYN) determined from cell counts and mean carbon contents of 12, 35 and 100×10^{-15} g C per cell.
ND, not determined.

these two routes, most of the organic production of the oceans is dissipated in the microbial food web.

Conclusion

The last 20–30 years have been a period of unprecedented discovery relating to the microbial ecology of the oceans. Largely ignored only a short while ago, microbial food web interactions are now central to our understanding of energy and nutrient flows in the oceans. The ubiquitous and self-regulating microbial community provides the foundation upon which the rest of the food web operates. Though sometimes overridden by the dynamics of larger bloom-forming organisms, it emerges as the dominant trophic pathway in most open-ocean regions and the end point of community succession when nutrients become limiting. In contrast to the ecology of larger organisms in the seas, we know little about the dynamics and unique contributions of microbes at the species level. This remains an exciting area of research for the future.

Glossary

Autotrophs Primary producers, organisms that utilize only inorganic carbon for metabolic synthesis.

Bacterivory Consumption of bacteria.

DOM Dissolved organic matter; operationally, all organic matter that can pass through a filter with 0.2 μm pores.

Gross growth efficiency For a given organism, the efficiency of conversion of carbon intake into new carbon growth.

Heterotrophs Organisms that utilize organic sources of carbon (particulate or dissolved) for metabolic synthesis.

Microplankton Planktonic organisms in the size range 20–200 μm; includes single-celled as well as multicellular organisms.

Mixotrophic Organisms with a mixed mode of nutrition, typically combining the ability to derive significant nutrition from photosynthesis as well as feeding directly on other organisms (or dissolved substrates).

Nanoplankton Planktonic singled-celled organisms in the size range 2–20 μm.

Oligotrophic System characterized by low concentrations of nutrients and plankton biomass.

Picoplankton Planktonic singled-celled organisms in the size range 0.2–2 μm.

See also

Phytoplankton Blooms.

Further Reading

Azam F, Fenchel T, Gray JG, Meyer-Reil LA, and Thingstad T (1983) The ecological role of water-column microbes in the sea. *Marine Ecology Progress Series* 10: 257–263.

Fenchel T (1987) *Ecology of Protozoa. The Biology of Free-living Phagotrophic Protists*. Madison, WI: Brock/ Springer Science Tech.

Fuhrman JA (1999) Marine viruses and their biogeochemical and ecological effects. *Nature* 399: 541–548.

Hobbie JE and Williams PJ leB (eds.) (1984) *Heterotrophic Activity in the Sea*, NATO Conference Series. IV, vol. 15. New York: Plenum Press.

Partensky F, Hess WR, and Vaulot D (1999) *Prochlorococcus*, a marine photosynthetic prokaryote of global significance. *Microbiology and Molecular Biology Reviews* 63: 106–127.

Raven JA (1997) Phagotrophy in phototrophs. *Limnology and Oceanography* 42: 198–205.

Thingstad FT (1998) A theoretical approach to structuring mechanisms in the pelagic food web. *Hydrobiologia* 363: 59–72.

Williams PJ leB (1981) Incorporation of microheterotrophic processes into the classical paradigm of the planktonic food web. *Kieler Meeresforsch. Sondh* 5: 1–28.

PATCH DYNAMICS

K. L. Denman, University of Victoria, Victoria, BC, Canada
J. F. Dower, University of British Columbia, Vancouver, BC, Canada

Introduction

Once out of sight of land, the vastness and apparently unchanging nature of the open ocean might lead one to think that the plankton would be distributed evenly in space. In fact, this is rarely the case. Rather, planktonic organisms are generally distributed unevenly, in clumps of all shapes and sizes usually referred to as patches. As oceanographic platforms and various sensors for observing plankton have improved, we have discovered that plankton patches exist on scales of less than a meter (microscales) right up to scales of hundreds of kilometers (called the mesoscale) that are characteristic of the oceanic equivalent of storms.

We have also learned that the two main groups of plankton – phytoplankton (the microscopic plants in the ocean) and zooplankton (the small animals that usually feed on phytoplankton) – are not patchy in the same way. Rather, zooplankton seem to be organized into many more patches at smaller scales, such that their distributions sometimes approach being random. This characteristically smaller patch size of zooplankton is believed to reflect their greater ability for swimming, swarming, or other directed motions, perhaps in response to patchiness in their food sources or other environmental cues. Similarly, the spatial distribution of larval fish, which have an even greater capability for directed motion and are capable of swimming and orienting themselves in groups in response to cues in their environment, is usually even more patchy.

What does all this mean? What causes plankton patchiness? Are the causes physical, biological, or both? Are there ecological advantages or disadvantages to plankton patchiness? Does such patchiness promote or interfere with the transfer of energy and biomass from phytoplankton to zooplankton and larval fish and on to higher trophic levels?

The related issue of how small-scale physical processes influence the biology of planktonic organisms is addressed elsewhere (*see* Small-Scale Physical Processes and Plankton Biology). There the focus is on how individual predators interact with individual prey within their physical environment. Observations of interactions between individual planktonic organisms are possible primarily in the laboratory. Here we consider patches as being comprised of groups of individual planktonic organisms: how they are actually observed by oceanographers in the ocean, how we then analyze such observations and assess their potential significance to food web dynamics.

The two approaches (patch-based versus individual-based observations) are connected as follows. Better understanding and prediction of patch dynamics and their importance to food web dynamics requires understanding of the processes by which individual organisms interact with each other and with the fluid flow. Generally, we are not yet able to make detailed observations of interactions between individual organisms in the ocean, and so we must rely on laboratory observations of individual organisms. To put these individual-based observations into an appropriate ecological context, however, it is also necessary to formulate or generalize how such individual interactions contribute to the behavior and responses at the level of a patch, which may contain millions of individuals. The other point of contact between the two approaches is to consider the case of an individual predator interacting with a patch of prey. One might reasonably ask whether there is some threshold difference in size between predator and prey at which an individual predator begins to interact with a patch of prey organisms (e.g., a gray whale feeding on a swarm of mysids) rather than with individual prey.

Because the observational technologies and analytic techniques used to study plankton patchiness are both complex and fascinating in their own right, it is wise during the discussion to follow to remain conscious of the underlying question being addressed: What are the implications of plankton patchiness to foodweb dynamics?

History

The existence of plankton patchiness was known as far back as the 1930s from observations that net tows taken simultaneously from two sides of the ship did not capture identical or often even similar collections of zooplankton. Subsequent studies and statistical analyses into the 1960s determined that this observed variability was not random but had structure, and that patches or clumps existed at sizes of less than a meter to sizes spanning hundreds of kilometers. In the 1970s continuous-flow fluorometers allowed high-resolution measurements of chlorophyll fluorescence (an indicator of the concentration of phytoplankton) continuously and simultaneously with comparable temperature observations. Similarities in the statistical distributions of fluorescence versus temperature quickly led to the realization that plankton patchiness was in large part controlled by physical turbulence. However, the fact that fluorescence distribution patterns did not exactly mimic that of the temperature field signaled that phytoplankton are not just 'passive tracers' of the flow, but that they are capable of changing their spatial distribution by growing. Comparable high-resolution observations of zooplankton distributions were more difficult to obtain, but their statistical properties differed even more from those of the temperature field, suggesting that, in addition to growth, zooplankton are also not passive tracers of the flow field because of their ability to swim and aggregate.

In the 1980s, the development of new observational techniques – moored fluorometers, instrumented multiple-net systems, multifrequency acoustics, optical plankton counters, and satellite-borne multichannel remote color sensors (to name but a few) – allowed a three-dimensional picture of the structure of patchiness and its relationship to the physical environment to emerge. In contrast, during the 1990s there was a shift toward the development of observational techniques that focused on individual organisms, usually under laboratory conditions. An exception to this trend was the development of the video plankton recorder (VPR), a towed oceanographic instrument that allowed *in situ* detection, identification, and counting of individual planktonic organisms, albeit with great manual labor in the analysis. Together with sensors of related oceanographic properties, the VPR allows construction of three-dimensional images of plankton patchiness simultaneously with a three-dimensional description of the immediate ocean environment.

In concert with the continuing emergence of an observational description of plankton patchiness in the context of the physical environment, there has also been a parallel development in our conceptual understanding of (i) the factors that create and control patchiness and (ii) the significance of patchiness to planktonic food web dynamics. Taking phytoplankton patchiness as an example, the 1970s view was that patch formation and decay was a balance between phytoplankton growth (which would act to stabilize the structure of a patch) versus small-scale turbulent mixing and diffusion (which combine to erode patch structure). The idea was that, in patches larger than some characteristic size, phytoplankton growth would win out over mixing and diffusion and the patch would be stable, but that, for patches smaller than this characteristic size, erosion by small-scale turbulent diffusion would be more efficient and the patch would be smoothed out. Around 1980, however, there was a realization that grazing by zooplankton could also affect the patchiness of phytoplankton. In addition, it was recognized that zooplankton can respond to variability in both phytoplankton abundance and physical structure by actively swimming to different positions to graze selectively in certain areas. Although recognized as an important process, the actual means by which zooplankton respond to cues in their environment and prey (both in terms of changing their distribution through swimming and by varying the intensity of their grazing) remains largely unknown.

Since the 1980s we have come to appreciate that the action of ocean currents and turbulence is much more complex than simply smoothing out smaller-scale structure through diffusive mixing. We now think more in terms of 'geostrophic turbulence', where variable currents and eddies (with structure from hundreds of meters down to submeter scales) create, distort, and evolve patches in various complex ways. Variable currents interacting with gradients in phytoplankton concentration can stretch, distort, and sharpen the boundaries of patches, sometimes creating very clear boundaries (often referred to as fronts) that separate patches from their background. Depending on the associated physical motions, these sharp frontal boundaries can increase or decrease the transfer of planktonic organisms across the front, or they can expose those predators that are capable of swimming to prey either of different concentration or of different community structure. We have also come to realize that perhaps the most important patches in terms of food web dynamics are those that tend either to recur or to persist at certain locations owing to features of coastal geometry or bottom topography interacting with the flow field. In fact, it is the very 'predictability' of these patches that makes them so important. These can be sites of enhanced tidal mixing, of increased retention of zooplankton and fish larvae due to convergent circular currents, or of increased dispersal due to the enhancement or divergence of currents. That such sites are important for the transfer of energy to higher trophic levels is evidenced by the fact that they are often sites where larger fish, as well as sea birds and marine mammals also congregate.

(A)

Figure 1 (A) Derived phytoplankton pigment concentration from the SeaWiFS satellite on 8 March 2000. The image (about 800 km in the vertical dimension) shows phytoplankton patchiness (red indicates highest pigment concentration) associated with filaments of nutrient-rich upwelled waters flowing offshore from the western coast of South Africa. (B) Sea surface temperature for the same day. (Images provided by the SeaWiFS Project, NASA/Goddard Space Flight Center and ORBIMAGE, courtesy of Gene Feldman (NASA) and Scarla Weeks (OceanSpace).)

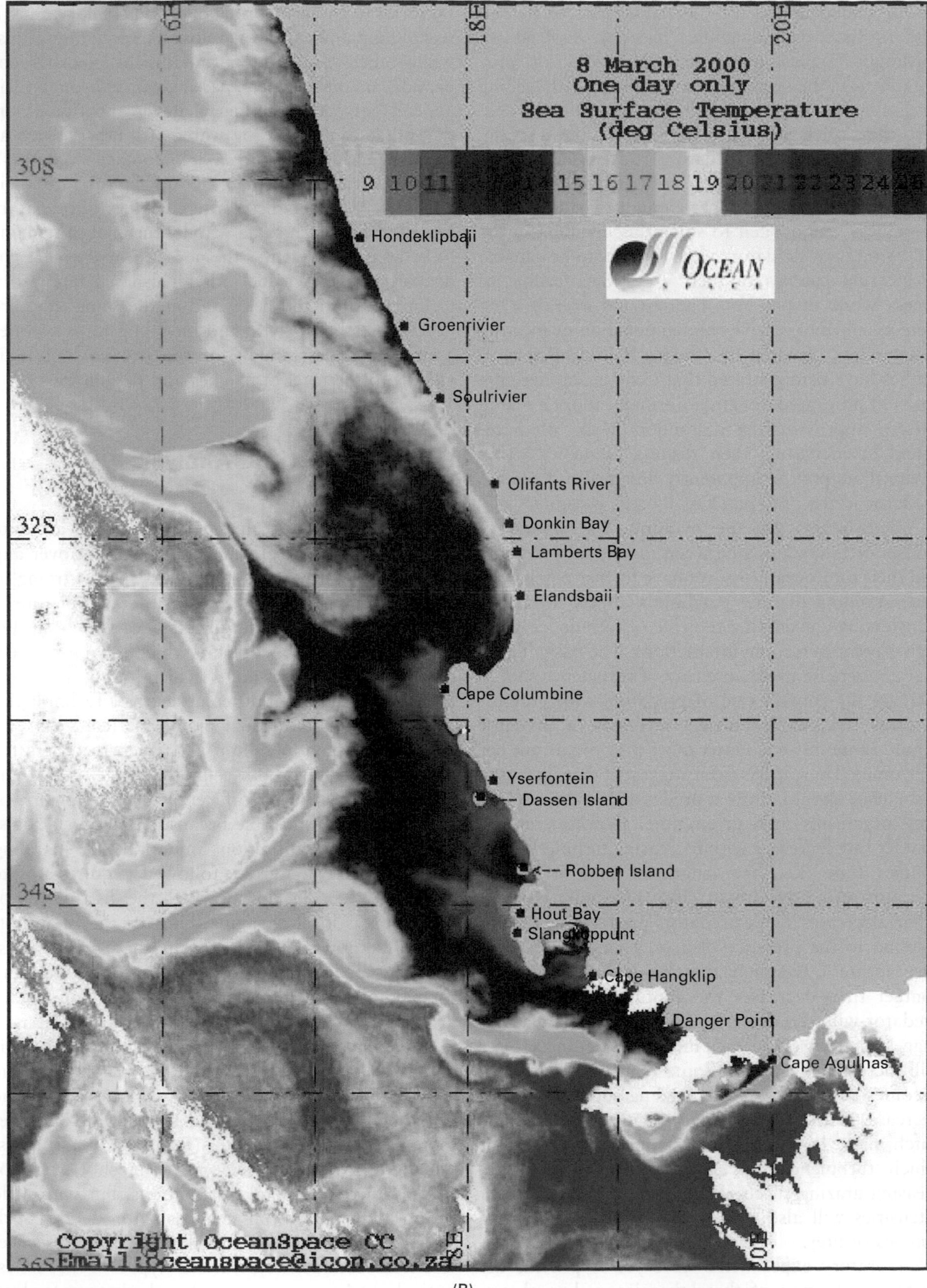

(B)

Figure 1 *Continued*

Our understanding of the importance of patchiness to food web dynamics has also evolved accordingly. Typical mean concentrations of prey organisms in the ocean, when compared with feeding studies conducted under laboratory conditions, suggest that predators generally should be starving. Other observations of predators, such as determining the contents of their guts, seem to indicate that generally they somehow get enough to eat. Initially, patchiness was evoked to solve this dilemma in the following way. Predators were thought to be capable of locating patches of their prey, then feeding to excess while in patches. Thereby, on average they were assumed to receive enough nutrition by locating patches and feeding in them. Indeed, empirical studies have demonstrated that upon encountering a patch of prey, predators that normally adopt a search pattern approximating a 'random walk' often respond by increasing their turning frequency, a behavioral adaptation apparently designed to keep the predator in the prey patch as long as possible.

Aided largely by the evolution of laboratory studies of interactions between individual predators and their prey organisms, we have further refined our understanding of how patchiness affects food web transfers by considering the role of turbulence as an active agent in feeding interactions. The basic idea is that an increase in the intensity of turbulence should increase the contact rate of predators with prey organisms (i.e., as turbulence increases, a predator should randomly encounter more prey organisms per unit time). Thus, it has been suggested that increased turbulence should allow a predator to capture more prey organisms in a given time. However, as the contact rate increases with increasing turbulence, the contact time (i.e., the length of time when the predator and prey are in close proximity to each other) decreases. Since predators need a finite time to respond to the presence of a prey item before they can capture it, we might therefore expect that as the contact time decreases the capture success of the predator will also decrease. Some model simulations suggest that, as turbulence increases, feeding success will increase, but that at some point the turbulence becomes so disruptive that feeding success starts to decrease. Whether the predator is inside or outside a patch might be expected to change the threshold at which turbulence moves from enhancing to decreasing grazing success. In addition, high turbulent intensities will also tend to spread out or disperse patches of prey.

Laboratory experiments have yet to provide definitive evaluation of these theories or hypotheses, and field observations are even more ambiguous. To add further complexity, some zooplankton are raptorial (e.g., carnivorous copepods or larval fish), seeking out and capturing individual prey organisms, while other zooplankton (notably herbivorous copepods) are filter feeders, passing water through their appendages while filtering whatever nutritional particles they can out of the water. Still others, such as jellyfish and ctenophores, are 'contact predators', which capture prey that literally bump into their tentacles. Whether these different types of predators are all affected the same way by turbulence remains to be seen. As we enter the new century, we are endeavoring to move the individual-based laboratory techniques into the field, and at the same time are developing the analytic and modeling tools to integrate our understanding of the behavior of individuals to the level of patches and populations.

Observations and Analysis Techniques

It is not yet possible in the sea to obtain accurate 'snapshots' of the distribution of plankton over any large volume. The basic problem is that of trying to construct a synoptic (yet necessarily static) picture of the distribution of plankton, many of which display species-specific behaviors and which inhabit a three-dimensional moving fluid (which often moves in different directions at different depths). Technologies that can count individual organisms (e.g., cell cytometers for phytoplankton and video recorders for zooplankton) have both a limited range and a narrow field of view. A sufficiently dense network of sensors is rarely possible because of resource, logistical, and intercalibration limitations. A mapping strategy, whereby a grid is followed, requires a finite time for execution, during which the distribution of organisms is itself changing, blurring the 'snapshot' to an unknown degree. Within these constraints, oceanographers employ two basic types of sampling to maximize the information that they can obtain about the patchy nature of plankton distributions.

The first strategy is to tow sensors along horizontal or vertical straight lines, or along oblique 'sawtooth' paths, over a grid of transects in the horizontal plane. The second strategy involves remote sensing from satellites (**Figure 1**) or airplanes, which can give nearly 'synoptic' (snapshot) views of near surface horizontal distributions, but with little or no information on changes with depth. Such sensors include stimulated chlorophyll fluorescence, electrical conductivity, various types of acoustics, optical plankton counters, video plankton recorders, and various combinations of instrumented nets. Some of these sensors are used to count individual

Table 1 General properties of the most common sensors used to observe plankton patchiness. Size range of typical phytoplankton organisms is 1 μm to 20 μm; size range of typical zooplankton organisms is 1 mm to 1 cm

Sensor	*Organism*	*Range (horizontal)*	*Determine size and identity*
Fluorometer	Phytoplankton	0.1–100 km	No
Satellite color sensor	Phytoplankton	1–1000 km	No
Towed nets	Zooplankton	100 m–10 km	Yes
Optical plankton counter	Zooplankton	1 m–100 km	Size
Acoustic sounding	Zooplankton	10 m–100 km	Approximate size
Video plankton recorder	Zooplankton and phytoplankton chains	10 cm–100 km	Yes

organisms, while others are designed to estimate plankton biomass in a given sampling volume. **Table 1** summarizes the most common sensors in terms of their spatial resolution and range, their ability to detect and count individual organisms, and the target planktonic group. Usually, oceanographers employ a combination of these sensors to obtain the best picture of the plankton distributions (**Figure 2**).

Increasingly, we want simultaneous high-resolution observations of both phytoplankton and zooplankton as well as related environmental variables (e.g., incident light, temperature, salinity, nutrients, turbulent mixing, local flow phenomena, etc.) in an attempt to identify and quantify mechanisms relating causes and consequences of patchy distributions in the planktonic ecosystem. Of course, logistic problems aside, the problem with attempting to tie all this information together is that the plankton distribution observed at some point in time may correlate better with environmental conditions at some time in the recent past (hours or days ago) rather than with the environmental conditions at the time of capture. This delay may occur because, being living organisms, plankton require time to respond to changes in their physical environment. For planktonic organisms that reproduce quickly (e.g., phytoplankton and microbes) this 'lag time' might be of the order of hours to a day. However, for organisms that take longer to grow and/or reproduce (e.g., zooplankton and larval fish) the lag time between a favorable change in the environment and a measurable response in the plankton may be of the order of days to weeks, or even longer. Thus, although we now have the technology to sample both physics and biology at very high resolutions in the ocean, we must be careful how we interpret the masses of data we collect.

Direct observations of patchiness, such as visual images from satellites and video plankton recorders, can be compelling, especially when they point to previously unknown or unexpected phenomena. But how do we proceed from 'pretty pictures' to mechanisms and then to prediction? Usually, we try to apportion the observed variability or variance into characteristic size scales: are there larger changes over kilometers or over meters? In addition, to determine interactions between different organisms or groups of organisms, or between plankton and their environment, it is necessary to perform various types of correlative analyses: are changes in phytoplankton concentrations associated mostly with changes in populations of zooplankton or with changes in temperature or nutrients? For observations obtained in a regular fashion (at regular intervals in time or space), spectrum analysis has proved to be useful, both to apportion variance according to patch size and to identify correlative structures between different variables. More recently, fractal analysis, which requires fewer assumptions regarding the 'well-behavedness' of the statistics of the observations, promises to reveal more insights into the spatial variability of plankton.

Food Web Implications

From the perspective of food web dynamics, it is interesting to consider whether plankton patchiness can be caused by predator–prey interactions. Certainly, empirical observations and some field-based work (particularly in freshwater ecosystems) have demonstrated that, under certain conditions, predation can alter plankton community structure. However, although this concept is appealing, the bulk of observational evidence in marine ecosystems strongly suggests that most plankton patchiness results from environmental factors. Examples include locally enhanced phytoplankton growth in regions of favorable light and nutrient conditions; turbulent mixing eroding patches; variable circulation patterns

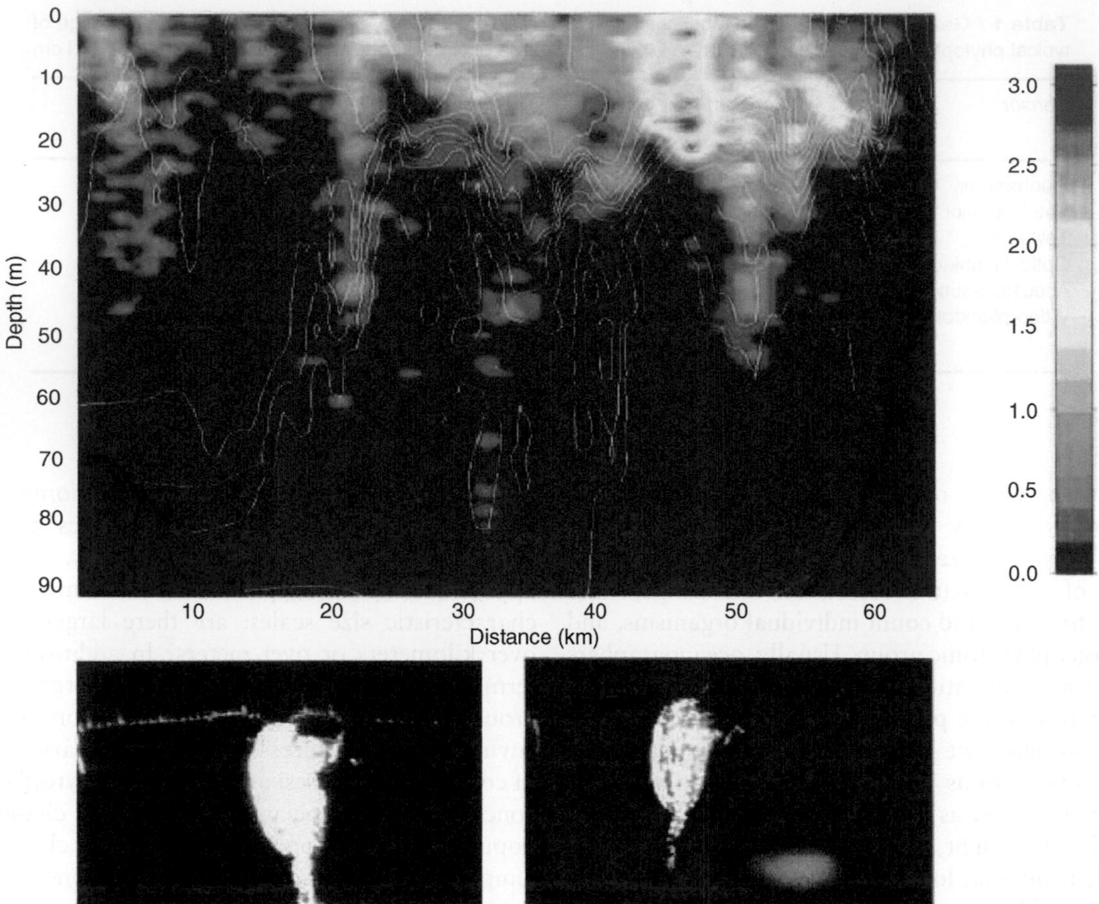

Figure 2 Distribution, abundance, and two representative images of the copepod *Calanus finmarchicus* along a transect across the Great South Channel east of Woods Hole, USA, with the video plankton recorder. Lines represent contours of constant temperature (0.5°C intervals). (From Gallager SM *et al.* (1996) High-resolution observations of plankton spatial distributions correlated with hydrography in the Great South Channel, Georges Bank. *Deep-Sea Research II* 43: 1627–1663 © 1996, with permission of Elsevier Science.

creating and distorting spatial structure in organism distributions; zooplankton swarming in response to environmental cues; or phytoplankton altering their buoyancy in response to variability in light, gravitational, or chemical cues. To be fair, however, it should also be noted that observations in the ocean have generally been inadequate to detect without ambiguity evidence of patchiness generated through grazing interactions. In fact, such 'top-down' regulation of planktonic concentration patterns is expected on ecological grounds. Thus, its existence should not yet be dismissed on the basis of contemporary observational capabilities.

Evidence is more clear with regard to flow patterns that can cause nutrient injection and/or the retention or even concentration of planktonic organisms for sufficient duration that ecologically significantly enhanced feeding by predators on their prey can occur. Examples of such flow patterns are fronts, rotating rings or eddies, and regular or recurring tidal patches associated with headlands, subsurface banks, or seamounts. Persistent or regularly recurring fronts, eddies, or convergent zones may also induce behavioral adaptation in predators, especially fish populations, such that they return to sites of recurring enhanced prey. For 'passive' plankton without the capability for directed motion, it is not always easy to determine whether the higher concentrations or numbers of organisms associated with these flow features result from locally enhanced growth or from concentration by convergent flow. Generally, concentration can only result when there is convergent flow and the organisms have a density different from that of the local fluid.

In recent years, our thinking has generally moved away from the concept that random encounters of patches of predators and prey result in enhanced food web transfers, and toward the concept that persistent or recurring patches (usually associated with certain flow or mixing regimes caused by particular geographical features) have the most potential to affect food web dynamics. The exception to this is the concept that turbulent motions alter the 'contact rate' between individual predator and prey organisms, an area treated in more detail in Small-Scale Physical Processes and Plankton Biology.

What Remains 'Unknown'?

To what extent has our understanding of the importance of plankton patchiness to food web dynamics been determined by sampling inadequacies? The question can only be answered in the context of how changes in our conceptual understanding have paralleled changes in our sampling capabilities over the last several decades. As sampling capabilities have improved both under laboratory conditions and at sea, we have started to move away from considering patch–patch food web interactions. Increasingly we consider these interactions more in terms of individual predators interacting with individual prey, and determining whether 'contact rate' between predator and prey organisms is regulated mostly by the turbulent flow field or by the actions of individual predators in reponse to their sensing of prey organisms. At sea, even guided by the results of laboratory observations, we can seldom observe individual phytoplankton organisms, so conceptual models have evolved toward consideration of individual predators interacting with patches of prey. Is this the ecologically meaningful view of plankton patchiness? To resolve our conceptual uncertainties we must improve our sampling capabilities such that we can observe individual prey and predator organisms simultaneously over the dimensions of observed patches. The challenge will then be to scale up from the observations of how individual organisms interact with their environment and with other individual organisms to an understanding of how populations of organisms collectively interact with their environment and with other populations of both their predators and their prey.

See also

Small-Scale Physical Processes and Plankton Biology.

Further Reading

Davis CS, Flierl GR, Wiebe PH, and Franks PJ (1991) Micropatchiness, turbulence and recruitment in plankton. *Journal of Marine Research* 49: 109–151.

Denman KL (1994) Physical structuring and distribution of size in oceanic foodwebs. In: Giller PS, Hildrew AG, and Raffaelli DG (eds.) *Aquatic Ecology: Scale, Pattern and Process*. Oxford: Blackwell Scientific.

Levin SA, Powell TM, and Steele JH (eds.) (1993) *Patch Dynamics, Lecture Notes in Biomathematics, 96*. Berlin: Springer-Verlag.

Mackas DL, Denman KL, and Abbott MR (1985) Plankton patchiness: biology in the physical vernacular. *Bulletin of Marine Science* 37: 652–674.

Okubo A (1980) *Diffusion and Ecological Problems: Mathematical Models*. Berlin: Springer-Verlag.

Platt T (1972) Local phytoplankton abundance and turbulence. *Deep-Sea Research* 19: 183–188.

Platt T and Denman KL (1975) Spectral analysis in ecology. *Annual Review of Ecology and Systematics* 6: 189–210.

Powell TM and Okubo A (1994) Turbulence, diffusion and patchiness in the sea. *Philosophical Transactions of the Royal Society of London B* 343: 11–18.

Seuront L, Schmitt F, Lagadeuc Y, Schertzer D, and Lovejoy S (1999) Universal multifractal analysis as a tool to characterize multiscale intermittent patterns: example of phytoplankton distribution in turbulent coastal waters. *Journal of Plankton Research* 21: 877–922.

Steele JH (ed.) (1978) *Spatial Pattern in Plankton Communities*. New York: Plenum Press.

DIVERSITY OF MARINE SPECIES

P. V. R. Snelgrove, Memorial University of Newfoundland, Newfoundland, Canada

Overview

The oceans comprise the largest habitat on Earth, both in area and in volume. Some 70.8% of the Earth's surface is covered by sea water, making marine benthos (bottom-living organisms) the most widespread collection of organisms on the planet. Pelagic organisms (organisms that live in the water column above the bottom) occupy the sea water that fills the ocean basins, which represent some 99.5% of occupied habitat on Earth. Within these benthic and pelagic environments, there is a wide variety of habitat types from tropical to polar latitudes, that ranges from the narrow band of rocky intertidal to open ocean surface waters to vast plains of muddy sediments on the deep-sea floor. Patterns of species composition and diversity vary considerably among habitats (**Figure 1**), although our understanding of these patterns is limited. Given the range and size of the habitats that occur in the oceans, this article will focus primarily on habitat and species diversity, while acknowledging the importance of genetic diversity.

Within the oceans, the major variables that delimit distributions of organisms include tidal exposure, temperature, salinity, oxygen, light availability, productivity, biotic interactions, and pressure. For benthic organisms, substrate composition (rock, gravel, sand, mud, etc.) also plays a key role. Not all of these variables play the same role in different environments but the sum total of their interactions determines biological pattern in the oceans.

Within the pelagic realm, organisms such as fish and whales are strong swimmers and can make

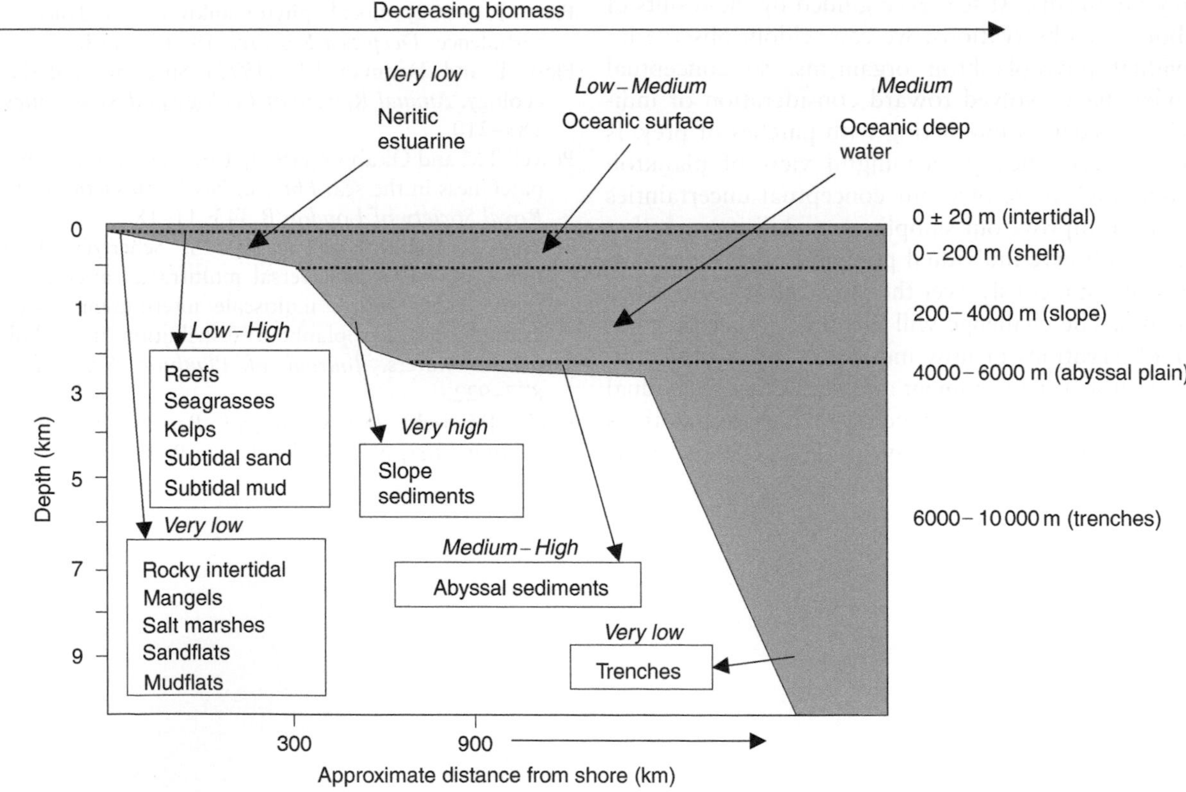

Figure 1 Diagram indicating different marine habitats. In relative terms, the horizontal axis is greatly compressed; it should also be noted that some continental margins are much narrower than the example shown. Bottom habitats are shown in the lower left boxes and water column habitats are shown in boxes near the top of the figure. Above each habitat name, a relative estimate of diversity within the different habitats is shown, but these comparisons are only generalities for which there are important exceptions. Coral reefs are almost always highly diverse.

significant headway against currents (nekton), whereas planktonic organisms, such as single celled and chain-forming algae (phytoplankton) and gelatinous forms (jellyfish, comb jellies, salps) drift passively with the predominant currents. Distributions of organisms are regulated by water mass characteristics such as salinity and temperature, water depth, and productivity; productivity varies with geography and depth in the water column. Life in the benthic environment is very different. Many benthic organisms have either limited mobility or are completely sessile. Much of their dispersal process occurs at the larval stage, which is planktonic for many species. The nature of the ocean bottom is a critical factor in determining the species that reside upon it. Species that attach to rocky substrate rarely occur in sediments, and species that occur in sands are usually relatively rare in mud. Shallow areas of the nearshore may be densely vegetated whereas deep-sea bottoms completely lack plant structure or photosynthetic primary producers of any sort.

The Organisms

The number of described plant species in the oceans is relatively modest. Vascular plants, such as those found in seagrass beds, mangrove forests (mangels) and salt marshes, are few in number (~45 described species of seagrasses, for example), and the numbers of these species that co-occur at a given site are few. Described species of phytoplankton (~3500–4500) are considerably higher than for marine vascular plants, but by comparison with terrestrial plants (~250 000) the numbers are also modest. There has been some suggestion that phytoplankton taxonomists have tended to lump species, and that the number of described species in some groups is significantly less than the actual number of species, but this possibility has not yet been resolved.

Defining species is even more problematic in marine microbes, a group for which it is believed that we have not yet even begun to describe their species numbers. In the past, species were defined based on characteristics of populations grown in culture. Recent advances in molecular biology have indicated that there are vast numbers of species that are completely missed with this approach, and that a small volume of sea water or sediment may contain tremendous microbial diversity.

Invertebrates comprise most of the described diversity in the oceans. Some field researchers studying invertebrate communities subdivide the fauna into size groupings, rather than along strict taxonomic lines. Benthic ecologists define megafauna as animals that can be identified from bottom photographs; macrofauna are those that are retained on a 300 μm sieve; meiofauna are those that are retained on a 44 μm sieve; and microbes are organisms that pass through a 44 μm sieve. Similar size groupings are used by plankton biologists. Examples of megafauna are fish and crabs, whereas macrofauna include polychaete worms, amphipod crustaceans, and small clams. Meiofauna include nematodes, small crustaceans, and foraminifera. Microbes include bacteria and protistans. In the past, larger organisms such as vertebrates have generated the greatest interest in terms of marine biodiversity and conservation, but in recent years the concern for smaller invertebrate and single-celled organisms has increased with the recognition that they form the bulk of the global species pool.

Sampling Biodiversity

Organisms are subdivided on size out of practical necessity, in that the sampling approach and sample size that are appropriate for one group are often inappropriate for another. For megafauna living in the bottom or in the water column, trawls and nets of various sorts are used. In some cases, photographic identification is also used, but if a high degree of taxonomic resolution is desired then this approach is only useful for larger organisms. Similarly, acoustic devices use sound waves to estimate abundances of megafauna over comparatively large swaths of the ocean (hundreds of square kilometers can be covered in a day), but the taxonomic resolution is again poor and the approach has only limited utility for benthic habitats. For macrofaunal and meiofaunal groups, net samplers or pumps are used in the water column, and various types of grabs or corers are used to sample marine sediments. For microbial groups, a single cubic centimeter of sediment may typically contain 10^6 bacteria and it is clear that different sampling approaches are needed for different groups of organisms; typically, small water samples are collected for pelagic microbes whereas subsamples of bottom cores are typically taken for benthic microbes. The disparity in appropriate techniques for different size groups of organisms has contributed greatly to the paucity of studies on more than one taxonomic size grouping at a given locale.

Unfortunately, where conflicting conclusions have been drawn about patterns in different groups of organisms, it is rarely possible to know whether the patterns truly vary among groups or merely reflect differences in sampling efforts.

Coastal Environments

The shallowest marine environments are those forming the transition between land and sea, and include intertidal habitats that are regularly exposed to air, full sunlight, rapid changes in salinity and temperature and, at higher latitudes, ice abrasion. The best studied of these environments are rocky intertidal habitats, but sand- and mudflats, salt marshes, and mangels (mangrove communities) are other intertidal environments that occur at the land–sea interface and are alternately flooded with sea water, and then exposed to high temperatures, desiccation, and potentially hypersaline conditions. Intertidal habitats, because they are physically harsh environments that require specialized adaptation, are relatively low in diversity, although the modest numbers of species that are present are often represented by many individuals.

Mangroves comprise globally only ~50 species, and within a given area only one or two mangrove species may be present, but mangels contribute to local diversity patterns by creating habitat for marine and terrestrial species. Mangroves are limited to tropical waters, and in temperature and boreal latitudes a similar niche is filled by salt marshes. Marshes and mangroves are also extremely productive habitats, and although some of the detritus resulting from breakdown of the plant materials is exported to, and diluted in, the adjacent shelf system, decomposition of large amounts of this detrital organic matter can occur in bottom sediments within the marsh or mangel. As a general rule, near-shore environments that are highly productive are often relatively low in diversity. This is true of primary producers, which are often dominated by only a few species as occurs in marshes and mangroves, but also in near-shore areas where phytoplankton thrive under high nutrient conditions. The benthic communities in these environments are often low in diversity because a few species are able to take advantage of the abundance of food and are tolerant of the hypoxic conditions that are often associated with high organic input.

All of these harsh environmental variables require specific adaptations for survival, and the relatively few species that are able to live under these conditions often occur in very high densities. This diversity is, nonetheless, important. The relative simplicity of intertidal habitats in terms of the numbers of species present and their accessibility has made them particularly conducive to experimental manipulation, and our understanding of biodiversity regulation in intertidal environments is probably greater than in any other marine habitat. Indeed, intertidal systems provide a useful model community that has generated many of the major paradigms for marine ecosystems. Limited data from sedimentary communities suggest that paradigms developed in the rocky intertidal are not necessarily transferable to sedimentary habitats, but they do provide a useful framework of ideas for other environments.

Intertidal habitats are important not only for the ecosystem services they provide (summarized below) but also because they represent a key ecotone (a transition zone between different communities) between land and ocean. The plants in mangels and salt marshes, for example, support a terrestrial fauna in their upper branches but a marine fauna in the sediments at their roots. These faunas may also interact with one another, and with other groups of organisms such as migratory birds that pass through the environment and feed while in transit.

Estuaries represent another group of specialized coastal habitats, where fresh water and sea water meet and mix to form a gradient in salinity. Estuaries can encompass the intertidal habitats described above but they also contain a subtidal habitat and their influence may be felt well beyond the land–sea interface. Typically, estuaries are relatively low in diversity because the salinity is too high or variable for freshwater species or too dilute for most marine species. Thus, most estuarine species have physiological adaptations to cope with the salinity problem. Depending on the hydrography of the estuary, salinity can also vary considerably on a seasonal or even a tidal basis, and variation can be even more limiting to species diversity than mean salinity.

Although some estuarine bottoms are rocky, most are at least partially covered by sediments transported from land by rivers and runoff. As is the case with most sedimentary bottoms, the majority of species and individuals live among the sediment grains, concentrated in the upper few centimeters below the sediment–water interface where oxygen is available. Further down in the sediment, oxygen is typically reduced or absent, so that organisms living deeper in the sediment (to tens of centimeters) must be able to tolerate low oxygen or use a long siphon or burrow to maintain oxygen exchange with the surface. Thus, diversity at depth is usually reduced relative to that near the sediment–water interface. Many estuaries are very productive, in part because they are often relatively nutrient rich as a result of the freshwater inflow and terrestrial runoff. As a result, these estuaries often support a very high biomass of a relatively species-poor community.

Some coastal habitats are vegetated with photosynthetic plants. Seagrasses are true grasses with roots that occur in shallow subtidal areas, and are usually most abundant in estuaries. Globally, there are only 45 species, but seagrasses provide habitat for other species; these species may include epibionts that live on the seagrass blades, or a variety of fish and sedimentary invertebrates that live below or between the grass blades. Seagrass beds are more diverse, for example, than adjacent sandy bottoms, likely because they provide a predator refuge that a sand bottom cannot.

Another common form of vegetation along the seashore is seaweeds. Seaweeds, like seagrasses, require light for photosynthesis and are typically attached to a shallow hard bottom; they are therefore confined to nearshore environments. Some seaweeds are intertidal, showing strong zonation patterns with tidal exposure. Species that are tolerant to prolonged exposure have adaptations that allow them to tolerate the harsh conditions, but again the harsh conditions mean that species numbers in a given intertidal area are relatively few.

Kelps make up another dense vegetation habitat that occurs subtidally in cold, clear, nutrient-rich water less than ~30m depth. The kelps are attached to the hard bottom, but in some cases sediments may accumulate in areas around the holdfast that attaches the kelp to the substrate. As is the case with seagrasses, a kelp bed is typically dominated by one or two kelp species, but provides critical habitat for many other species. Another feature common to kelps and seagrasses is that they are extremely productive habitats and often support a high biomass of primary and secondary producers.

The most spectacular marine habitat in terms of diversity of organisms is the coral reef. Coral reefs are formed by hard corals, which are limited to areas with average surface temperatures above ~20°C. Reefs offer a wide variety of three-dimensional habitat that is utilized by a variety of other species, resulting in the most diverse marine habitat in terms of number of species per unit area. Many of the most productive marine environments are low in diversity, but reefs are an interesting exception. They are very productive (tight nutrient cycling) as a result of photosynthesizing dinoflagellates called zooxanthellae that are symbionts with reef-building corals. Thus, coral reefs support a high biomass of organisms from corals to fishes.

Beyond the immediate near-shore environment lies the continental shelf, which is largely sediment-covered and extends to approximately 200 m depth. The continental shelf varies in width from tens of kilometers to ~200 km. Depending on the current and wave regime, sediments may be sandy or muddy, and the sediment composition will influence species composition. Most of the primary production in the shelf environment is provided by phytoplankton in near-surface waters, and light penetration is typically reduced relative to the open ocean because of increased turbidity and phytoplankton abundance. Benthic vegetation is lacking, but benthic algal diatom mats can sometimes be important where light penetration is sufficient to support them. These mats provide a potential food source for benthic organisms, but contribute little in the form of structural habitat. Epifaunal organisms (those living upon rather than within the seafloor) can provide a habitat that can be important for some species. The productivity of shelf environments is extremely variable geographically, but most of the world's most-productive fisheries occur on the shelf, particularly where nutrient-rich deep waters are upwelled to the surface. Relative to the near-shore, shelf habitats are usually more diverse, but the level of diversity on shelf environments can vary considerably with geography.

For groups as varied as shallow-water molluscs, hard substrate fauna, and deep-sea macrofauna, it has been suggested that species number generally declines from the tropics to the poles. This gradient is much more obvious in the northern hemisphere, where relatively recent glaciation-related extinctions contribute greatly to the pattern. The latitudinal gradient has also been described in the southern hemisphere, though the pattern is less striking. Not surprisingly, given their relative ages, the older Antarctic benthos is more diverse than its Arctic counterpart. But a simple latitudinal gradient is not evident in all taxa. Some groups, such as macroalgae, appear to be most diverse at temperate latitudes, and emerging evidence suggests that shallow-water and deep-sea nematodes may also peak in diversity at temperate latitudes.

The Open Ocean

Beyond the continental shelf lies the open ocean, or the pelagic environment. Pelagic communities are delineated primarily by water masses, so that assemblages are often broadly distributed over ranges that may be characterized by differences in temperature, salinity, and nutrient values. Thus, surface circulation and wind effects can interact with thermohaline circulation patterns to create distinct water masses with distinct faunas. Interesting processes occur where these water masses meet, creating complex spatial patterns at the boundaries. Clearly these environmental variables play a major role in regulating pattern, but the geological history of a habitat can also play a role by determining the regional species pool. Major geological events such as the elevation of the Panama Isthmus, the opening of the Drake Passage, and Pleistocene glaciation all had profound effects on circulation, which in turn had major effects on marine distribution patterns that are reflected in modern communities. Unlike shelf and coastal environments, the influence of benthic communities on water column processes in the open ocean is minimal and indirect.

Vertically migrating species play a significant role in oceanic pelagic communities and provide a conduit between surface and deep waters. Some species migrate many hundreds of meters on a daily basis, thus complicating efforts to evaluate biodiversity in a given water mass. These migrating species provide a means of energy transfer between waters at different depths and also provide a mechanism by which regulation of diversity pattern in surface waters could be related to diversity patterns in deeper waters or the benthos. Many benthic species also produce larval stages that may contribute to the biodiversity of surface waters, often on a seasonal basis. Surface waters may also have a major impact on benthic communities in the deep sea because the deep sea depends largely on surface primary production.

Photosynthesis is limited to the upper portion of the water column ($\sim$200 m) where there is sufficient light penetration. Beneath these waters are the continental slope (200–3000 m), the continental rise (3000–4000 m), the abyssal plains (4000–6500 m), and trenches (6500–10 000 m) of the deep ocean, which will be grouped here as 'deep-sea' habitats. Like the shelf environment, most of the deep-sea bottom is covered in sediments, some of which are geologically derived and others that have formed from sinking skeletons of pelagic organisms.

As described earlier, pelagic communities are delineated primarily by water masses, and a number of biogeographic provinces have been described for the world's oceans. Shelf and offshore communities are markedly different in composition and abundance. Local shelf communities are less species rich than offshore communities, but greater spatial heterogeneity in near-shore environments typically results in greater total species richness in neritic environments. One major variable that affects diversity is productivity; species richness tends to be depressed in areas where productivity is high and seasonally variable. This pattern may explain the general pattern of decreasing species number with increased latitude. Characteristics of the regional circulation can offset this general pattern, particularly in coastal environments where productivity does not show as clear a relationship with latitude. Variation in diversity has also been observed with depth in the water column, which is also consistent with primary productivity being restricted to the lighted surface layer of the ocean. The shallowest areas of the oligotrophic North Pacific are more productive in terms of phytoplankton than deeper waters, and phytoplankton diversity is higher in deeper waters than near the surface. Zooplankton, by contrast, show a slightly different pattern where species numbers are higher in surface waters. Studies from the North Atlantic suggest that species richness increases and then peaks at approximately 1000 m. One critical variable that may contribute to these differences in pattern is the pulsed nature of organic input in some systems; highly variable organic flux may represent a strong disturbance, and therefore depress diversity.

Because water depth is so great, much of the water column and benthic environment is devoid of light, and the food source is the material sinking from surface waters and advected from the adjacent shelf habitat. The great depths also result in ambient pressures far in excess of those in shallow water, and water temperatures are relatively low (<4°C) and are seasonally and spatially much less variant than in shallow-water systems. The seeming inhospitable nature of the deep-sea environment led some earlier investigators to speculate that it was azoic, or devoid of life. Work by Hessler and Sanders in the 1960s and more recent work by Grassle and Maciolek in the 1980s has dramatically changed this view. Although the densities of organisms that live in deep-sea sediments are very low and individuals tend to be very small, the numbers of species present is usually very high. Thus, a given sample will contain few individuals, but many of them will represent different species. This generalization is true for most deep-sea habitats, but areas such as trenches, upwelling areas, areas with intense currents, and high latitudes can be low in diversity. Low diversity in these areas results from some overwhelming environmental variable, such as low oxygen, resulting in the exclusion of many species.

Depth-related patterns have also been described in benthic communities. A peak in biodiversity has been described at continental slope depths, with lower diversity at shelf and abyssal plain depths. This pattern depends on how diversity is defined. In terms of total numbers of species per unit area, shallow water habitats sometimes have higher values because they support much higher densities of individuals. Shallow-water habitats are also patchier over spatial scales of tens to hundreds of kilometers in terms of sediment type, and other habitat variables that change species composition. Thus, pooling of samples from coastal environments can sometimes produce greater total species numbers than pooling over similar distances in deep-sea sediments. For most areas that have been sampled, there is another key difference between deep-sea and shallow-water samples. An individual deep-sea sample is typically characterized by low dominance and greater dissimilarity between proximate samples than would be found in shallow-water samples. One other key difference is total area; although densities of organisms in the deep sea are much lower than in shallow water, the tremendous area of the deep sea alone is enough to support very large numbers of species. But some recent evidence from Australian shelf samples suggests that some shallow-water communities may rival deep-sea communities even at the sample scale. Thus, poorly sampled areas such as tropical shorelines and the southern hemisphere need to be better understood before definitive 'rules' on biodiversity patterns can be established.

The one major exception to the generality of low productivity in the deep sea is hydrothermal vents, which were first discovered in 1977. Their discovery came as a great surprise to deep-sea scientists because they supported high abundances of novel megafaunal species. The size and numbers of vent organisms is in sharp contrast to most deep-sea habitats, and is possible only because of the chemosynthetic bacteria that form mats or live symbiotically with several vent species. These bacteria are dependent on hydrogen sulfide and other reduced compounds emitted at vents. Since the discovery of vents over 20 new families, 100 new genera, and 200 new species have been described from these communities, but diversity is very low as a result of the toxic hydrogen sulfide. Vent habitats have extremely high levels of endemism resulting from the evolution of forms that are able to thrive in the toxic conditions and take advantage of the high levels of bacterial chemosynthetic production that drives the food chain at vents. Moreover, the fauna at vents is very distinct from other habitats, sometimes at the family level or higher.

Among deep-sea habitats, there are other low-diversity communities. Low diversity is also observed beneath upwelling regions, where high levels of organic matter sinking from surface water to bottom sediments may create hypoxic conditions that eliminate many species. Deep-sea trenches are subject to slumping events that contribute to relatively low numbers of species. Deep-sea areas in the Arctic are also still rebounding from loss of much of the fauna during glaciation and anoxic periods that were associated with that time period.

Regulation of Pattern and Linkages Between Organisms and Habitats

One pattern common to marine communities and terrestrial communities alike is that habitats characterized by high levels of disturbance, or conditions that are extremely challenging from a physiological perspective, often support relatively few species.

Aside from habitats that are strongly influenced by overriding environmental variables that depress diversity, regulation of diversity in marine ecosystems is not fully understood. As a general rule in ecology, high habitat complexity is thought to support the highest diversity because of the many available habitats and niches. Certainly the high level of habitat complexity observed in reefs contributes to their diversity, but pelagic habitats and deep-sea sediments would appear at first glance to be among the least spatially complex habitats in nature. But in these environments, complexity may occur at small temporal and spatial scales. In the pelagic realm, microstructure of nutrients is thought to be an important aspect of species coexistence. Thus, where nutrient levels are relatively low and patchy, primary producers are most diverse. Patchiness is also thought to be important in maintaining diversity in coral reefs and deep-sea communities, the two most speciose community types in the oceans. In both cases, it is thought that intermediate levels of disturbance may be important in preventing the strongest competitors from taking over and eliminating the weaker competitors. In reef habitats, periodic small-scale disturbance in the form of storms keeps any one species from taking over. In the deep sea, it is thought that small-scale disturbances in the form of food patches, biological structures, and sediment topography, can all create microhabitats that allow species to coexist. Determining the factors that regulate biodiversity in marine systems is, nonetheless, an ongoing research question.

Estimates of Total Species Numbers

Because such a large portion of the ocean is poorly sampled, estimates of total species numbers vary considerably depending on the specific assumptions used to extrapolate from those areas that have been well sampled. The number of described species from the oceans is approximately 300 000. For some areas of the ocean, such as coastal environments of the North Atlantic, most faunal groups are reasonably well described. However, major gaps in our knowledge exist for some taxonomic groups and some marine habitats. In terms of taxonomic groups, microbes are very poorly known but new evidence based on molecular approaches suggests that the pelagic and sedimentary realms may both support numbers of bacterial species that would add greatly to the approximately 1.75 million species of non-microbial species presently described globally. Protistan diversity is also poorly known. Surprisingly, even for relatively well-sampled areas of the oceans, diversity in some groups of organisms, such as the nematodes, remains poorly known. The specific habitats that are very poorly known include tropical sediments, coral reefs, and deep-sea sediments. Reaka-Kudla has estimated that up to 9 million species may inhabit coral reefs, and Grassle and Maciolek estimated that there may be 10 million macrofaunal species in the deep sea. Others have extrapolated from the ratio of known to unknown species in specific areas to generate estimates of 500 000 to 5 million macrofaunal species in deep-sea sediments alone. Based on the typical ratio of nematode species to macrofaunal species, Lambshead hypothesized that there may be as many as 100 million nematode species in the deep sea. The problem with these estimates is that they are based on a relatively small area, so that the error in extrapolating to the whole of the deep sea, or all oceans, is quite large. For example, it has been estimated that deep-sea sediments (defined here as shelf edge and greater depths) cover approximately 65% of the Earth's surface yet globally only $\sim 2\,km^2$ of ocean floor has been sampled for macrofauna, and $\sim 5\,m^2$ has been sampled for meiofauna.

Threats

Different habitats presently face different levels of threat as a result of human activity. Open ocean habitats, both pelagic and benthic, have been least impacted by human activity. The greater distance from human populations and their influences, and the shear size of the habitats themselves, make them much less vulnerable than the shoreline and shelf habitat where most marine habitat destruction and local species loss is occurring.

Most of the world's fisheries are concentrated in shelf or near-shore environments, although the capacity to fish deeper waters is increasing all the time. It is estimated that >65% of global fisheries are either fully or overexploited. There are a variety of mechanisms by which this fishing activity may affect biodiversity of marine systems. One of the greatest concerns in terms of benthic habitats is the habitat destruction caused by fishing gear that is dragged across the sea floor, damaging organisms that live in and upon the seabed, homogenizing bottom habitats, and disrupting the sediment fabric and its geochemistry and microhabitats. Recent estimates suggest that some major fishing grounds, such as Georges Bank, have experienced trawl coverage that exceeds 200–400% annually. Habitat damage is not an issue for the fluid pelagic habitat, but the removal of nontarget species as by-catch remains a problem in bottom and pelagic fisheries. Organisms ranging from sea birds to marine mammals to fish to invertebrates are all known to suffer high by-catch mortality; by-catch levels can often rival or even exceed the biomass of the target species removed from an ecosystem. Another major concern with fisheries is that they are often very effective at removing large numbers of individuals of the target species, which is often a top predator within the ecosystem. The potential for alteration of food chains is great, both for pelagic and benthic species. Evidence is accumulating that the trophic structure of many heavily fished ecosystems is changing, with upper trophic levels sometimes being eliminated and the transfer of energy through remaining species altered substantially. One final aspect of fishing activity that is of particular concern for pelagic habitats is lost fishing gear and debris that may continue to 'ghost fish' and capture target and nontarget organisms for years after the gear has been lost.

One chronic problem with studies of fisheries impacts is that we lack good 'control' sites where fishing impacts are minimal. Advances in fishing technology have made most habitats accessible to fishing gear, and the few areas that remain inaccessible are typically poor 'control' sites because they are fundamentally different in physical topography and species composition than the areas that are fished.

Coral reef fisheries have their own unique problems. Dynamite, cyanide, and bleach fishing are all used to get around the problem of fishing a topographically complex habitat, but they are also methods that destroy many nontarget species including the corals that make up the habitat itself. One of the discouraging facts about reef systems is that in most instances, reefs were considerably altered by removal of large and potentially important species by early settlers, and we have little idea of what the pristine systems looked like in terms of species relationships.

Aquaculture represents a special case in terms of fishing activity, in that impacts are typically more localized and easily seen. The issue of ownership is also less contentious, stocks are more easily managed, and in some cases it is also easier to assign accountability for environmental damage. But aquaculture can be extremely destructive, in that some activities such as shrimp farming are often achieved by destroying other coastal habitats such as mangroves. Aquaculture also often involves moving organisms around as brood stock, potentially allowing invasion of nonnative habitat by the species being cultured, or parasites and diseases that the organism may carry into the new environment. Even for those organisms that have been deliberately moved, there are potential consequences in terms of alteration of local genetic structure of natural populations.

Many coastal environments are also threatened by pollution resulting from the dumping of various toxic wastes such as heavy metals, organic compounds (e.g., polychlorinated biphenyls or PCBs), metals, and eutrophication resulting from increased input of macronutrients from sewage and agricultural runoff. These excess nutrients result in blooms of a low diversity phytoplankton community (sometimes favoring toxic species) that sink to the bottom and undergo microbial decomposition. In some instances, microbial respiration will deplete bottom waters of oxygen, and benthic communities may subsequently be wiped out. Whether the cause is eutrophication or toxic chemicals, both forms of pollution depress diversity and favor a few weedy species. On coral reefs, increased nutrients will typically lead to macroalgal blooms and the loss of corals. In this instance, the entire habitat may be destroyed.

The effect of fisheries operations on habitat alteration has already been mentioned, but marine habitats may also be altered for other activities. The demand for coastal real estate, both for industrial and residential use, has eliminated large areas of salt marsh and mangrove. Expansion of coastal waterways by dredging and replenishing of eroded beaches by mining subtidal sands are two mechanisms by which habitats may be altered. Invariably, the loss of habitat means the loss of species, at least on a local scale. Unfortunately, the amount of habitat being lost is becoming so extensive that the number of habitat refugia remaining are becoming fewer and fewer and more widely separated.

There has been great concern over the introduction of non-native ('exotic') species into marine habitats in recent years, a phenomenon that has been ongoing for some time. Indeed, it has been estimated that between the years 1500 and 1800, more than 1000 intertidal and subtidal species worldwide may have been transported and introduced into nonnative habitats. As many as 3000 species may be in transit in the ballast water of ships on a given day. In some instances it has been documented that invasive species have greatly altered the species composition and ecological processes in the areas they have invaded. San Francisco Bay, for example, now harbors hundreds of exotic species, including Asian clams that have become so abundant that they have altered the seasonal phytoplankton production cycle within the bay. The problem of invasive species is thought to be most severe in coastal and estuarine regions, where open ocean waters have historically provided a barrier to dispersal. Ballast water is not, however, the only culprit in facilitating the movement of invasive species. Fouling organisms on the hulls of ships provide another mechanism, and aquaculture and scientific study are additional transport vectors.

One other threat to marine biodiversity is global climate change. The effects of climate change are difficult to know given the different trajectories predicted by different models, but several general categories of effect may be expected. First, as air temperatures increase, so will ocean surface temperatures, presumably shifting faunas toward the poles. Such an effect has already been documented in California intertidal and pelagic communities. What is less obvious is that some species will be unable to simply shift poleward because other habitat requirements, such as the presence of a shallow bank, may not be met at another latitude. Coral reefs provide an excellent example of this phenomenon. Recently, large areas of coral reefs have experienced increased occurrences of bleaching, a phenomenon where corals expel their symbiotic dinoflagellates and die. Bleaching has been linked to the increased frequency of El Niño events, higher water temperature, and elevated ultraviolet radiation; in this case, corals cannot simply colonize an adjacent habitat because it is typically too deep and warming trends are much faster than potential colonization rates. A second category of effect is rising sea level. In some instances it may be possible that intertidal organisms may simply shift upward in response to rising waters, but in areas where human populations have developed areas in the landward direction, mangroves, salt marshes, and intertidal habitats may be prevented from advancing by seawalls or other physical barriers. Perhaps the most difficult aspect of global change to predict is the effect that alteration of temperature and rainfall pattern will have on ocean circulation. Some global warming models have suggested that even relatively modest temperature increases may alter ocean circulation patterns. Because ocean circulation is a critical variable in terms of surface productivity and transport of reproductive propagules, any change to ocean circulation could potentially affect every marine community from shallow surface waters to the deepest areas of the ocean.

The Importance of Marine Biodiversity

Marine organisms contribute to many critical processes that have direct and indirect effects on the health of the oceans and humans. In the majority of instances there are few data to demonstrate that total numbers of species are important, but data on this question are only starting to be assembled. What is obvious is that there are specific species and functional groups that play very critical roles in important ecosystem processes, and the loss of these species may have significant repercussions for the whole ecosystem.

Primary and secondary production are critical mechanisms by which marine communities contribute to global processes. It has been estimated that approximately half the primary production on Earth is attributable to marine organisms. It has also been estimated that approximately 20% of animal protein consumed by humans is from the oceans. Perhaps more importantly, this consumption is much higher in some countries than in others, making it a critical staple of many diets. Marine organisms are also harvested for extractable products, including medicines and various industrial products.

The global cycling of nutrients and even carbon depend on marine communities. Without primary producers in surface waters, the oceans would quickly run out of food, but without planktonic and benthic organisms to facilitate nutrient cycling, the primary producers would quickly become nutrient limited.

Benthic marine organisms can contribute to sediment and shoreline stability. Mangroves, salt marsh plants, and seagrasses all bind sediments together and thereby reduce shoreline erosion. Sedimentary organisms can either stabilize or destabilize sediments, thus affecting coastal sediment budgets. One direct ramification of these activities is that sediment bound pollutants will be greatly affected by binding and destabilization of sediments, creating a situation where sedimentary bacteria, diatoms, and invertebrates can influence whether pollutants are buried or resuspended. Some bacteria and invertebrates are also able to metabolize certain pollutants and reduce or eliminate their toxicity.

In coastal environments, salt marshes, mangroves, and seagrasses can help trap sediments and absorb nutrients from sewage and agricultural sources, thereby filtering coastal runoff and helping to maintain relatively non-turbid, non-eutrophied coastal waters. In a related manner, benthic organisms may be important filter feeders, removing particles that would reduce water clarity. In addition to allowing light penetration to greater depths, this filtering activity can contribute to coastal aesthetics and clear water. Other aesthetic services are provided by coastal wetlands and coral reefs, both of which generate tremendous tourist interest.

Concluding Remarks

Although documented extinctions are relatively few, there are good reasons for improving our understanding of biodiversity pattern and regulation, and the role that biodiversity plays in key ecosystem processes. One problem is that biodiversity in the oceans is poorly described, and the handful of seabirds, marine mammals, and marine snails that have become extinct may represent the visible few of a much larger number of organisms that have been lost without our knowing it. It has been estimated, for example, that 50 000 undescribed species may have already been eliminated from coral reefs. A second concern is that some of the species we have treated as cosmopolitan may, in fact, represent a sibling species group, and when we eliminate a species from an area, it may be a different species than occurs elsewhere. A third concern is that even if the elimination of species from a given area does not represent a global extinction, it may represent a unique genotype that cannot be replaced from a surviving population elsewhere.

Although several general patterns of biodiversity have been described in the oceans, and our understanding of how biodiversity is regulated and maintained is still limited, several paradigms point to the importance of disturbance and habitat heterogeneity. Because vast areas of the oceans remain unsampled, our estimates of species numbers are crude and based on a very small portion of the marine habitat, but they do suggest that the oceans contain a significant portion of the global species pool. Unfortunately, we know little about how biodiversity contributes to the many critical ecosystem services that marine communities provide. In many instances, it is possible to generate very clear examples of a single species having a major impact on its ecosystem, and based on this observation it is safe to say that loss of biodiversity may have very negative effects on marine ecosystems if the species lost is one of these key players. But we have little idea of whether numbers of species matter, or which species are most important in maintaining a properly functioning ecosystem. An improved understanding of biodiversity and regulation will give us better tools to predict where biodiversity changes are likely to occur and why, how these changes will affect other components of marine biodiversity, and the best strategy to mitigate such changes.

Further Reading

Angel MV (1997) Pelagic biodiversity. In: Ormond RFG, Gage JD, and Angel MV (eds.) *Marine Biodiversity. Patterns and Processes*. Cambridge: Cambridge University Press.

Birkeland C (ed.) (1997) *Life and Death of Coral Reefs*. New York: Chapman and Hall.

Costanza R, d'Arge R, de Groot R, *et al.* (1997) The value of the world's ecosystem services and natural capital. *Nature* 387: 253–260.

Dayton PK, Thrush SF, Agardy MT, and Hofman RJ (1995) Environmental effects of marine fishing. *Aquatic Conservation: Marine and Freshwater Ecosystems* 5: 205–232.

Gage JD and Tyler PA (1991) *Deep-Sea Biology: A Natural History of Organisms at the Deep-Sea Floor.* Cambridge: Cambridge University Press.

Grassle JF, Lasserre P, McIntyre AD, and Ray GC (1991) Marine biodiversity and ecosystem function: a proposal for an international programme of research. *Biology International Special Issue 23*: 1–19.

Grassle JF and Maciolek NJ (1992) Deep-sea species richness: regional and local diversity estimates from quantitative bottom samples. *American Naturalist* 139: 313–341.

Gray JS (1997) Marine biodiversity: patterns, threats and conservation needs. *Biodiversity and Conservation* 6: 153–175.

Gray J, Poore G, and Ugland K (1997) Coastal and deep-sea benthic diversities compared. *Marine Ecology Progress Series* 159: 97–103.

Hall SJ (1999) *The Effects of Fishing on Ecosystems and Communities*. Oxford: Blackwell Science.

Lambshead PJD (1993) Recent developments in marine benthic biodiversity research. *Oceanis* 19: 5–24.

May R (1992) Bottoms up for the oceans. *Nature* 357: 278–279.

McGowan JA and Walker PW (1985) Dominance and diversity maintenance in an oceanic ecosystem. *Ecological Monographs* 55: 103–118.

National Research Council (1995) *Understanding Marine Biodiversity: a Research Agenda for the Nation.* Washington, DC: National Academy Press.

Norse EA (ed.) (1993) *Global Marine Biological Diversity: a Strategy for Building Conservation into Decision Making*. Washington, DC: Island Press.

Rex MA, Etter RJ, and Stuart CT (1997) Large-scale patterns of biodiversity in the deep-sea benthos. In: Ormond RFG, Gage JD, and Angel MV (eds.) *Marine Biodiversity. Patterns and Processes.* Cambridge: Cambridge University Press.

Snelgrove PVR, Blackburn TH, Hutchings PA, *et al.* (1997) The importance of marine sedimentary biodiversity in ecosystem processes. *Ambio* 26: 578–583.

SMALL-SCALE PHYSICAL PROCESSES AND PLANKTON BIOLOGY

J. F. Dower, University of British Columbia, Vancouver, BC, Canada
K. L. Denman, University of Victoria, Victoria BC, Canada

Introduction

By definition, plankton are aquatic organisms (including plants, animals, and microbes) that drift in the water and which cannot swim against any appreciable current. Most plankton are also very small, usually much less than 1 cm in size. Thus, it should come as no surprise that the behavior of, and the interactions between, individual plankton are strongly influenced by small-scale physical processes. Although this may seem intuitive, the fact is that oceanographers have only been aware of the importance of small-scale physical processes to plankton ecology for about the past 20 years, and have only been able to directly study plankton biology at these scales for the past 10 years or so. The primary reason for this is that the space and timescales relevant to individual plankton (millimeters → meters, and seconds → hours) are quite small, making it extremely difficult to sample properly in the oceans. Thus, much of what we know about the effect of small-scale physical processes on plankton biology is necessarily based on empirical and theoretical studies.

One might well ask why we need to understand the behavior of individual plankton at all, especially given that traditional plankton ecology (at least as conducted in field studies) generally involves comparisons between averages. For instance, we usually compare differences in the average zooplankton density or the average rate of primary productivity at several sites. However, in comparing averages, researchers make many implicit (though often unstated) assumptions about the behavior of the individual plankton that comprise these populations. Specifically, researchers assume that differences in average population-level responses are merely the sum of many individual responses. But is this a realistic assumption? To explore this idea, let us consider what individual planktonic organisms actually 'do' and examine how they are affected by small-scale physical processes.

Life in the Plankton

First, let us consider a typical phytoplankton cell. We will assume it is a large (e.g., 100 μm) diatom. In the simplest sense, the biological processes of prime importance to this organism are that of finding sufficient light and nutrients to photosynthesize, grow, and reproduce, while at the same time trying to avoid sinking and predation. Next, let us consider a typical zooplanktonic organism. We will assume it is a copepod. What does it actually 'do'? On a daily basis it spends much of its time searching for food. Once food is encountered it must then be captured and ingested. In the meantime, this copepod must also try to avoid its own predators. Toward these goals of predator avoidance and feeding, certain zooplankton (including many copepods) also undertake diel vertical migrations of hundreds of meters (for more details on diel vertical migrations). Assuming our copepod reaches maturity (having successfully avoided being eaten) it must then find a mate in order to reproduce. Finally, like the diatom cell, our copepod must also try to avoid sinking. How might these various biological processes be affected by small-scale physics? There are two main factors that must be considered: viscosity and turbulence. Interestingly, both are related to the interaction between the very small size of most plankton, and the nature of the fluid environment that they inhabit.

Effects of Viscosity

The first thing to consider is that, although they live in open water and are transported by the background flow, the world probably feels quite 'sticky' to most plankton. This is due to the fact that at very small spatial scales and at the relatively slow swimming speeds or sinking rates of most plankton (i.e., of order mm s^{-1}), viscous forces dominate over inertial forces. The relative importance of viscous to inertial forces can be determined by calculating a dimensionless quantity known as the Reynolds number:

$$Re = \ell U/v \quad [1]$$

Table 1 Approximate value of Reynolds numbers for the swimming speeds of various animals. Note that the kinematic viscosity was taken as 1.05×10^{-6} m^2 s^{-1}, which approximates that of sea water at 20°C

Animal	*Length (m)*	*Speed (m s^{-1})*	*Re*
Whale	20	10	200 000 000
Tuna	2	10	20 000 000
Human	2	2	4 000 000
Small adult fish	0.5	0.5	240 000
Post-metamorphosis fish	0.05	0.05	2400
Larval fish	0.005	0.005	24
Adult copepod	0.002	0.02	4

where is the characteristic length (m) of the object in question. U is the speed (m s^{-1}) at which the object is moving, and υ is the kinematic viscosity (m^2 s^{-1}) of the medium in which the object is moving. For our purposes we will take $\upsilon = 1.05 \times 10^{-6}$, a value typical of sea water at about 20°C.

When $Re > 2000$, inertial forces dominate and the flow is turbulent. For values of $Re < 2000$ viscous forces become progressively more important. Note that *Re* is not meant to be a precise measure, serving merely as a 'ballpark measure' for comparing the flow regimes that apply to different objects. **Table 1** lists some estimates of *Re* that apply to the swimming motions of various animals. Note that for most plankton, *Re* is generally less than 100, meaning that flow conditions are strongly viscous. Thus, once active swimming ceases, instead of continuing to glide (as does an adult fish once it ceases swimming), the low Reynolds numbers that apply to most plankton ensure that they come to an almost immediate stop.

There is at least one notable exception to this, however; the escape responses initiated by certain copepods. Copepods are equipped with antennae that serve as highly sensitive mechanoreceptors. In addition to detecting food and potential mates, these mechanoreceptors can also warn of approaching predators. Recent work has shown that some copepods can initiate escape responses (a series of rapid hops in a direction away from the perceived threat) in <10 ms. Moreover, they can achieve burst speeds of several hundred body lengths per second. In some cases, the acceleration achieved is sufficient to enable the copepod to break through the 'viscous barrier' and enter the inertial world, albeit temporarily. Some copepod species have even evolved myelinated axons in their antennae to boost signal conduction along the nerves responsible for initiating the escape response.

Effects of Turbulence

Although occasionally occurring in very dense patches most zooplankton are actually rather dilute. Concentrations of 10–100 l^{-1} are typical for many neritic copepods, and concentrations of planktonic organisms such as jellyfish and larval fish are usually orders of magnitude lower still (e.g. 1 per 10–100 m^3). Thus, given these low concentrations, it has long been (and continues to be) widely held that food concentrations in the ocean are limiting to growth and biological production. Throughout the 1960s and 1970s this belief was strengthened by the observation that successfully rearing copepods and larval fish in the lab often required food concentrations several times higher than those encountered in the oceans.

The point to consider, however, is that at the submeter scales relevant to most plankton, water motions tend to be dominated by random turbulence rather than directional flow. Turbulence is a ubiquitous feature of the ocean, and exists at all scales. In the surface layer of the ocean turbulent energy usually comes from the mixing effects of winds, or from the current shear between layers of water moving in different directions. In shallow coastal waters a second source of turbulence is tidal friction with the seafloor, which stirs the water column from the bottom up. From the perspective of our individual plankton, the key point is that in the surface layer of the ocean these small-scale turbulent velocities tend to be of the same order of magnitude as (or larger than) the typical swimming and/or sinking velocities of most plankton. How will this affect interactions between individual plankton?

Turbulence and Predator–Prey Interactions

Until the late 1980s, the contact rate between planktonic predators and prey was usually expressed as a simple function of (1) the relative swimming velocities of the predator and its prey, and (2) the prey concentration. Put simply, the faster predator and prey swim, and the higher the prey concentration, the more often should the predator and prey randomly encounter each other. Numerically, this can be represented as:

$$Z = D \times A \qquad [2]$$

where Z is the encounter rate, D is the prey density, and A is the relative velocity term. However, in 1988 researchers first theorized that small-scale turbulence might play an important role in predator–prey interactions in the plankton. Specifically, it was hypothesized that small-scale turbulent motions can randomly bring predator and prey together in the water column and, thus, that D and A from eqn. [2] should be rewritten as follows:

$$D = \pi R^2 N \qquad [3]$$

$$A = \left(u^2 + 3v^2 + 4w^4\right)/\sqrt{3}\left(v^2 + w^2\right) \qquad [4]$$

where R is the distance at which the predator can detect the prey (m), N is the prey concentration (m^{-3}), u and v are the prey and predator swimming speeds ($m\ s^{-1}$), and w is the turbulent velocity ($m\ s^{-1}$).Eqn. [3] assumes that the predator searches a circular area in front of itself, the result being that as it swims forward its 'search volume' assumes the shape of a cylindrical tube of radius R. In fact, this is only one of many search geometries displayed by visual planktonic predators. For instance, although some larval fish (e.g., herring larvae) are 'cruise predators' and scan a cylinder of water as described above, laboratory experiments have also shown that other species (e.g., cod larvae) are better described as 'pause–travel' predators. In this case, the predator searches for prey only during short pauses. These are followed by short bursts of swimming, during which there is no searching. Furthermore, the geometry of the volume searched also appears to be species specific, although in general it assumes the shape of a 'pie-shaped wedge' centered along the predator's line of vision.

There are also many predatory zooplankton that do not rely on vision at all. These include mechanoreceptor predators (e.g., raptorial copepods, chaetognaths) which detect the vibrations and hydrodynamic disturbances created by approaching prey, and contact predators (e.g., jellyfish, ctenophores) armed with stinging or entangling tentacles and which essentially rely on prey bumping into them. There are also a wide variety of filter-feeding (e.g., copepods, larvaceans) and suspension feeding zooplankton (e.g., heteropods) which generally feed on phytoplankton and protozoans that are either strained out of a feeding current or which are ingested after becoming trapped on mucus coated surfaces.

Regardless of the search geometry and mode of feeding, however, the general hypothesis is that as turbulence increases so, too, should encounter rates between predator and prey. Among the chief reasons plankton ecologists are interested in this idea is the longstanding belief that food availability is an important regulator of the growth and survival of zooplankton and larval fish. Throughout the 1990s, researchers sought empirical evidence of this phenomenon. A number of laboratory studies did show that copepods encounter more prey under increased turbulence. These experiments also showed that many copepods initially respond to increased turbulence by initiating escape responses. The explanation of this result was that, being mechanoreceptor (rather than visual) predators, copepods initially interpret an abrupt increase in turbulence as signaling the approach of a potential predator. Other laboratory studies have since demonstrated that larval fish also initiate more attacks and have higher levels of gut fullness under increased turbulence.

Of course, there are limitations to what can be modeled realistically in any laboratory study, especially those studying small-scale physical processes. For instance, the experiments described above typically involved videotaping the behavior of individual copepods that had been tethered in a flow field in which the turbulence could be varied. It remains to be seen whether the behavior of such tethered animals is the same as that of free-swimming copepods. Likewise, experiments on the effect of turbulence on larval fish feeding ecology usually involve offering only a single type and size of prey at a time, and usually at unrealistically high prey concentrations (e.g. 1000s of prey per liter). In the ocean, of course, larval fish encounter a wide range of prey types and prey sizes, many of which typically occur at relatively low concentrations.

Logistically, the biggest challenge is that of trying to create a realistic turbulent field under laboratory conditions. The small size of most plankton necessitates the use of rather small volume aquaria for experimentation, especially if (as is often the case) the goal is to use videographic techniques to follow individual plankton. Such experiments generally rely on variable speed oscillating grids or paddles to generate different levels of turbulence. Thus, although empirical studies have taught us a lot about the potential importance of turbulence in plankton ecology, the question remains as to whether this process is actually important in the ocean.

Further refinements of the theory have since suggested that the relationship between turbulence and feeding success should be dome-shaped (rather than linear) since, at some point, the predator will be

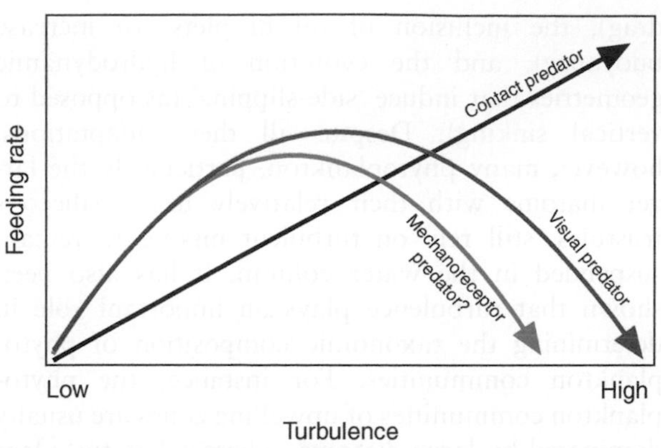

Figure 1 Possible responses of a visual predator, a mechanoreceptor predator, and a contact predator to turbulence. The domed response of visual predators such as larval fish has been supported by laboratory studies. The other two lines are more speculative. The linear increase proposed for contact predators (e.g. jellyfish and ctenophores) is based on the premise that these predators rely on prey bumping into them. However, given that mechanoreceptor predators such as raptorial copepods actually 'hear' the turbulence, it may well be that their ability to separate the signal of an approaching prey item from that of the background turbulent noise declines at higher turbulence levels, leading to a lower optimum level of turbulence than for the other types of predators.

unable to react to prey before they are carried away by high levels of turbulence (**Figure 1**). However, this theory is based largely on the assumption of a visual predator. It remains to be seen how other types of planktonic predators should respond to turbulence. For instance, will the functional response of a mechanoreceptor predator (e.g., a raptorial copepod) be dome-shaped, too? Perhaps such a predator should have a lower 'optimum' level of turbulence than a visual predator since, at high turbulence levels, the background turbulence might make it more difficult to sense prey? Similarly, perhaps a linear response should be expected for 'contact predators' such as jellyfish or ctenophores. Further laboratory work will be needed to clarify this matter.

There have been some attempts to quantify the effect of turbulence on zooplankton and larval fish in the field. To date, although the evidence broadly suggests that turbulence increases encounter rates and even gut fullness (**Figure 2**), its effect on the growth and survival rates of larval fish and zooplankton remains uncertain. What is known is that not all species respond to turbulence in the same way. For instance, although gut fullness in larval Atlantic cod (*Gadus morhua*) and radiated shanny (*Ulvaria subbifurcata*) increases in response to increased turbulence, other species such as herring (*Clupea harengus*) and walleye pollock (*Theragra chalcogramma*) actively move deeper in the water column to avoid turbulence during windy conditions. Similarly, field observations demonstrate that the vertical segregation of at least two congeneric copepod species (*Neocalanus cristatus* and *Neocalanus plumchrus*) is determined by species-specific preferences for different turbulent regimes. Such responses make it extremely difficult to generalize whether turbulence is of net positive benefit to zooplanktonic predators.

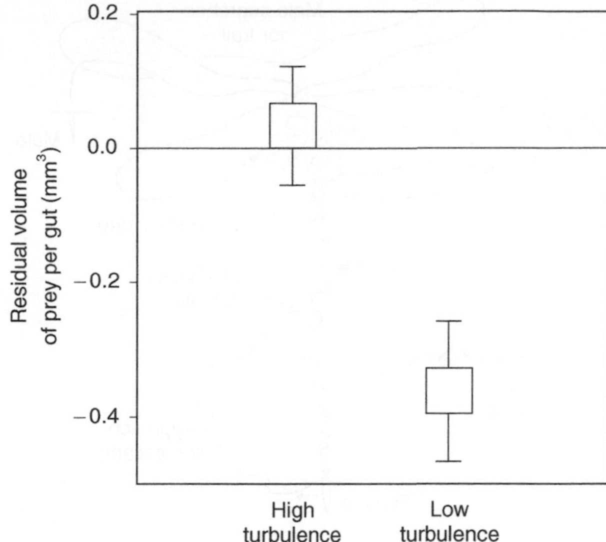

Figure 2 The effect of turbulence on gut fullness in larval radiated shanny *(Ulvaria subbifurcata)* during a 23-day time-series in July/August 1995 in coastal Newfoundland. Despite the fact that prey concentrations remained relatively constant throughout, the figure shows that the average volume of food in the guts of larval fish was significantly higher on high turbulence days than on low turbulence days. Note that the data have been detrended, since larger larvae will generally always have more food in their guts than smaller larvae. The figure thus shows the residual volume of food per gut (i.e., after the size effect has been removed). (Redrawn from Dower *et al.*, 1998.).

Turbulence and Reproductive Ecology

In addition to its effect on feeding ecology, turbulence also appears to play a role in zooplankton reproduction. Recent work has shown that the males of some copepod species find mates by following chemical 'odor trails' left by the females. High-resolution videographic observations show the male swimming back and forth until he crosses the odor trail, at which point he immediately reverses direction to pick up the trail again. Having locked onto the trail the male then follows it back to the female (**Figure 3**). Should the male initially head the wrong way down the trail, he quickly reverses direction and goes the right way until reaching the female.

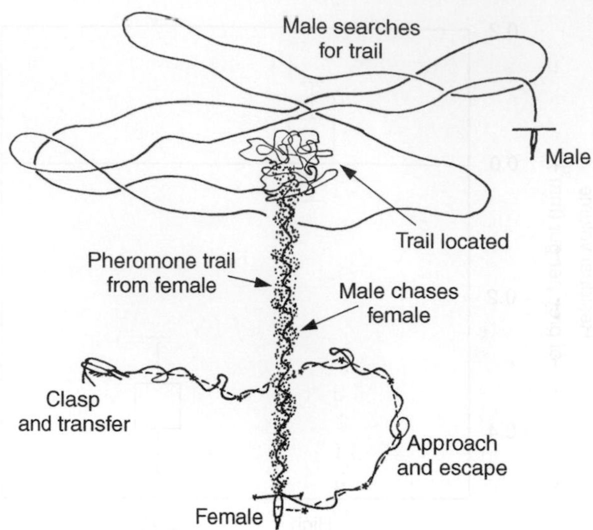

Figure 3 Conceptual interpretation of mate-attraction and mate-searching behavior in the copepod *Calanus marshallae*. The sequence of events first involves the female producing a pheromone trail, which alerts the male that a female is in the area. The male initially swims in smooth horizontal loops until he crosses the trail. At this point the male follows the trail back to the female. (Redrawn from Tsuda and Miller (1998) and used with the kind permission of the authors.).

Turbulence comes into play because the odor trails persist for only a few seconds before being dissipated by small-scale turbulent motions, making it harder for the males to find females. Although it has yet to be confirmed in the lab, this may partly explain why many zooplankton form dense swarms during mating. If the males are to successfully find females via odor trails it might be predicted that mating swarms should be dense enough that the time taken to randomly encounter an odor trail is shorter than the time taken for the trails to dissipate. It is also known that copepod species that usually inhabit the near-surface layers of the ocean often descend to depths of hundreds of meters to reproduce. Is it possible that this behavior is partly a response to the need for relatively quiescent waters in order to allow males to track females before odor trails are dissipated by small-scale turbulent motions?

Turbulence and Phytoplankton Ecology

Recall our typical diatom. Like other phytoplankton, it needs to stay relatively close to the surface in order to capture sufficient light for photosynthesis. Phytoplankton have evolved a variety of adaptations to reduce sinking, including the production of spines and other external ornamentations (to increase drag), the inclusion of oil droplets (to increase buoyancy), and the evolution of hydrodynamic geometries that induce 'side-slipping' (as opposed to vertical sinking). Despite all these adaptations, however, many phytoplankton, particularly the larger diatoms with their relatively heavy siliceous frustules, still rely on turbulent mixing to remain suspended in the water column. It has also been shown that turbulence plays an important role in determining the taxonomic composition of phytoplankton communities. For instance, the phytoplankton communities of upwelling zones are usually dominated by large diatoms, whereas less turbulent regions (e.g., open ocean) are more often dominated by much smaller phytoplankton taxa. Similarly, and particularly at temperate latitudes where there is a strong seasonality in wind-mixing, there is often a seasonal succession of phytoplankton species in coastal waters whereby large diatoms dominate during the spring bloom when winds, and thus near-surface turbulent mixing, are strongest. As the summer progresses, and stratification of the upper water column proceeds, large diatoms are often succeeded by more motile forms (e.g., dinoflagellates and coccolithophores) and smaller phytoplankton (e.g., cyanobacteria) that rely less on turbulence to remain near the surface.

A growing body of literature suggests that turbulence can also have novel effects on dinoflagellates. Many dinoflagellate species are bioluminescent, producing a flash of light when disturbed. Recent work has revealed that such bioluminescence may provide an indirect means of reducing predation. Experiments show that some dinoflagellates only bioluminesce in response to strong current shear, such as that induced by breaking surface waves or by predators attempting to capture them. The so-called 'burglar alarm' theory suggests that, by flashing in response to strong shear and lighting up the water around themselves, the dinoflagellates make their potential predators more visible, and thereby more prone to predation from *their* predators. This effect has been demonstrated in the lab, where species of squid and fish (feeding on mysids, shrimp, and other fish species) have been observed to increase their rates of attack and capture success when the water in which they are foraging contains bioluminescent dinoflagellates. Apparently, prey movements induce the dinoflagellates to flash, thereby illuminating the prey and making it easier for the predator to capture them.

Turbulence has also been shown to have negative effects on dinoflagellates. It has long been known that dinoflagellate blooms are most common when conditions are calm. However, recent work has shown that even moderate amounts of turbulent mixing can actually inhibit the growth of certain dinoflagellates. The effect seems to be that of preventing cell division, since other cellular processes (e.g., photosynthesis, pigment synthesis, nucleic acid synthesis) appear to continue as usual. If the period of turbulent mixing is short-lived, the cells recover and quickly begin dividing, possibly going on to form a bloom. If, however, the turbulence continues for longer periods, the cells will be prevented from dividing, or they may even die. Of particular interest is that the growth of at least two dinoflagellate species that form harmful algal blooms *(Lingulodinium polyedrum* and *Heterosigma carterae*) seems to be inhibited by turbulence.

Conclusions

Although our basic view of plankton as being organisms that 'go with the flow' still holds true, over the past 20 years we have also learned that interactions between plankton and their physical environment can be quite complex. Perhaps the most important result of this research has been the realization that small-scale physical processes affect so many different aspects of plankton ecology including feeding, predator–prey interactions, swimming and buoyancy, nutrient diffusion, mate selection, and even patterns of community composition. That many of these discoveries are rather new is largely due to the fact that, until relatively recently, oceanographers were simply unable to conduct experiments and observe plankton at appropriately small scales. Now that we have that capability, however, the challenge in the coming years will be to find ways to integrate what has been learned about the behavior of individual plankton to further develop our understanding of population-level processes. Are population-level processes merely the sum of innumerable individual interactions, or are there other physical processes that affect populations at larger space and timescales? Alas, we do not yet know the answer to this question. However, finding new ways to extend what has been learned in the laboratory into more realistic field settings may prove one step in the right direction.

Further Reading

Berdalet E (1992) Effects of turbulence on the marine dinoflagellate Gymnodinium nelsonii. *Journal of Phycology* 28: 267–272.

Dower JF, Miller TJ, and Leggett WC (1997) The role of microscale turbulence in the feeding ecology of larval fish. *Advances in Marine Biology* 31: 170–220.

Dower JF, Pepin P, and Leggett WC (1998) Enhanced gut fullness and an apparent shift in size selectivity by radiated shanny (*Ulvaria subbifurcata*) larvae in response to increased turbulence. *Canadian Journal of Fisheries and Aquatic Sciences* 55: 128–142.

Fuiman LA and Batty RS (1997) What a drag it is getting cold: partitioning the physical and physiological effects of temperature on fish swimming. *Journal of Experimental Biology* 200: 1745–1755.

Kiorboe T (1993) Turbulence, phytoplankton cell size, and the structure of pelagic food webs. *Advances in Marine Biology* 29: 1–72.

Mensinger AF and Case JF (1992) Dinoflagellate luminescence increases susceptibility of zooplankton to teleost predation. *Marine Biology* 112: 207–210.

Rothschild BJ and Osborn TR (1988) Small-scale turbulence and plankton contact rates. *Journal of Plankton Research* 10: 465–474.

Tsuda A and Miller CB (1998) Mate-finding behaviour in *Calanus marshallae* Frost. *Philosophical Transactions of the Royal Society of London* B 353: 713–720.

Vogel S (1989) *Life in Moving Fluids: The Physical Biology of Flow* 3rd edn. Princeton, NJ: Princeton University Press.

PHYTOPLANKTON BLOOMS

D. M. Anderson, Woods Hole Oceanographic Institution, Woods Hole, MA, USA

Introduction

Among the thousands of species of microscopic algae at the base of the marine food chain are a few dozen that produce toxins. These species make their presence known in many ways, ranging from massive 'red tides' that discolor the water, to dilute, inconspicuous concentrations of cells noticed only because of the harm caused by their highly potent toxins. Impacts include mass mortalities of wild and farmed fish and shellfish, human illness and death, alterations of marine trophic structure, and death of marine mammals, sea birds, and other animals.

'Blooms' of these algae are commonly called red tides, since, in some cases, the tiny plants increase in abundance until they dominate the planktonic community and change the color of the water with their pigments (**Figure 1**). The term is misleading, however, since nontoxic species can bloom and harmlessly discolor the water, or can cause ecosystem damages as severe as those linked to toxic organisms. Adverse effects can also occur when toxic algal cell concentrations are low and the water is not discolored. Given the confusion surrounding the meaning of 'red tide', the scientific community now prefers the term 'harmful algal bloom' or HAB. This new descriptor includes algae that cause problems because of their toxicity, as well as nontoxic algae that cause problems in other ways. It also applies to macroalgae (seaweeds) which can cause major ecological impacts as well (**Figure 2**).

Impacts

Toxic Algae

HAB phenomena take a variety of forms, with multiple impacts (**Table 1**). One major category of impact occurs when toxic phytoplankton are filtered from the water as food by shellfish which then accumulate the algal toxins to levels which can be lethal to humans or other consumers. The poisoning syndromes have been given the names paralytic, diarrhetic, neurotoxic, amnesic, and azaspiracid shellfish poisoning (PSP, DSP, NSP, ASP, and AZP, respectively). The symptomology and exposure route for each of these are presented in **Table 2**. Except for ASP, all are caused by biotoxins synthesized by a class of marine algae called dinoflagellates. A fifth human illness, ciguatera fish poisoning (CFP) is caused by toxins produced by dinoflagellates that attach to surfaces in many coral reef communities. Ciguatoxins are transferred through the food chain from herbivorous reef fishes to larger carnivorous, commercially valuable finfish. The final human illness linked to toxic algae is called possible estuary-associated syndrome (PEAS). This vague term reflects the poor state of knowledge of the human health effects of the dinoflagellate *Pfiesteria piscicida* and related organisms that have been linked to symptoms such as deficiencies in learning and memory, skin lesions, and acute respiratory and eye irritation – all after exposure to estuarine waters where *Pfiesteria*-like organisms have been present.

Another type of HAB impact occurs when marine fauna are killed by algal species that release toxins and other compounds into the water, or that kill without toxins by physically damaging gills or by creating low oxygen conditions as bloom biomass decays. Fish and shrimp mortalities from these types of HABs at aquaculture sites have increased considerably in recent years. HABs also cause mortalities of wild fish, sea birds, whales, dolphins, and other marine animals. To understand the breadth of these ecosystem impacts, think of the transfer of toxins through the food web as analogous to the flow of carbon. Just as phytoplankton are the source of fixed carbon that moves through the food web, they can also be the source of toxins which cause adverse effects either through toxin transmitted directly from the algae to the affected organism or indirectly through food web transfer.

A prominent example of direct toxin transfer was the death of 19 whales in Massachusetts in 1987 due to saxitoxin that had accumulated in mackerel that the whales consumed. Similar events occurred in Monterey Bay, California in 1998 and again in 2000 when hundreds of sea lions died from domoic acid (the ASP toxin) vectored to them via anchovies. In both cases, the food fish accumulated algal toxins through the food web and passed those toxins to the marine mammals.

Adult fish can be killed by the millions in a single outbreak, with obvious long- and short-term ecosystem impacts (**Figure 3**). Likewise, larval or juvenile stages of fish or other commercially important fisheries species can experience mortalities from algal

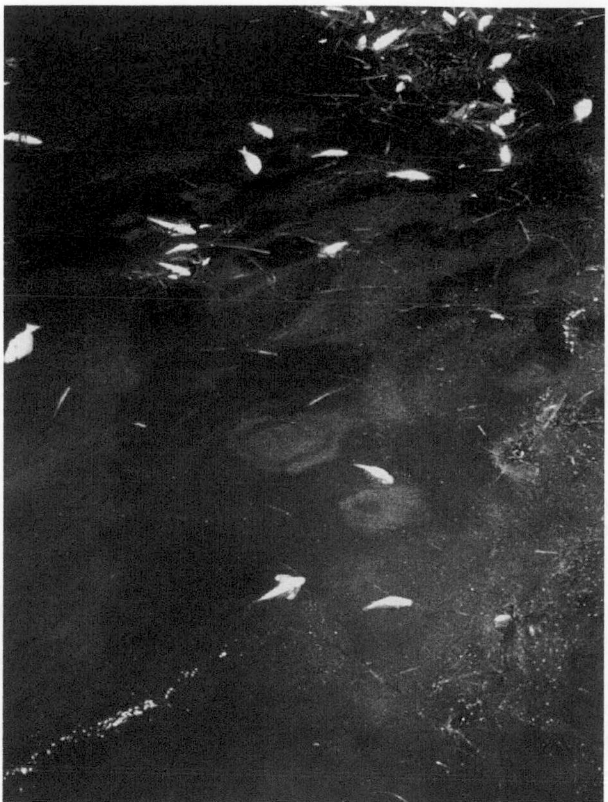

Figure 1 Dead fish and discolored water during a Florida red tide. (Photo credit: Florida Department of Environmental Protection.)

Figure 2 Seaweed washed up on a Florida beach. Macroalgal blooms, although not toxic, can be quite harmful to coastal ecosystems and cause major economic problems to the tourist industry. (Photo credit: B. Lapointe).

toxins. Impacts of this type are far more difficult to detect than the acute poisonings of humans or higher predators, since exposures and mortalities are subtle and often unnoticed. Impacts might not be apparent until years after a toxic outbreak, such as when a year class of commercial fish reaches harvesting age but is in low abundance. Chronic toxin exposure may therefore have long-term consequences that are critical with respect to the sustainability or recovery of natural populations at higher trophic levels. Many believe that ecosystem-level effects from toxic algae are more pervasive than are realized, affecting multiple trophic levels, depending on the ecosystem and the toxin involved.

Nontoxic Blooms

Nontoxic blooms of algae can cause harm in a variety of ways. One prominent mechanism relates to the high biomass that some blooms achieve. When

Table 1 Some harmful effects caused by algae in coastal and brackish waters

Effect	*Causative organisms*
Human health syndrome	
Paralytic shellfish poisoning (PSP)	*Alexandrium* spp., *Pyrodinium bahamense* var. *compressum, Gymnodinium catenatum, Anabena circinalis, Aphanizomenon* spp.
Diarrhetic shellfish poisoning (DSP)	*Dinophysis* spp., *Prorocentrum* spp.
Neurotoxic shellfish poisoning (NSP)	*Gymnodinium breve*
Azaspiracid shellfish poisoning (AZP)	*Protoperidinium crassipes*
Amnesic shellfish poisoning (ASP)	*Pseudo-nitzschia* spp.
Ciguatera fish poisoning (CFP)	*Gambierdiscus toxicus*
Respiratory problems and skin irritation	*Gymnodinium breve, Pfiesteria piscicida, Nodularia spumigena*
Possible estuary-associated syndrome (PEAS)	*Pfiesteria piscicida, P. shumwayae*, and possibly other *Pfiesteria*-like organisms
Hepatotoxicity	*Microcystis aeruginosa, Nodularia spumigena*
Mortality of wild and cultured marine resources	
Hemolytic, hepatotoxic, osmoregulatory effects and other unspecified toxicity	*Gymnodinium* spp., *Cochlodinium polykrikoides, Pfiesteria piscicida, Heterosigma carterae, Chattonella* spp., *Fibrocapsa japonica, Chrysochromulina* spp., *Prymnesium* spp., *Aureococcus anophagefferens, Microcystis* spp.
Mechanical damage to gills	*Chaetoceros* spp., *Leptocylindrus* spp.
Gill clogging and necrosis	*Phaeocystis* spp., *Thalassiosira* spp.
Inhibition of tourism and recreational activities	
Production of foams, mucilage, discoloration, repellent odors	*Noctiluca scintillans, Phaeocystis* spp., *Cylindrotheca closterium, Aureococcus anophagefferens, Nodularia spumigena*
Seaweed accumulation on beaches; coral reef overgrowth	*Cladophera, Codium, Ulva*, and many other species
Adverse effects on marine ecosystems	
Hypoxia, anoxia	*Noctiluca scintillans, Ceratium* spp., many others
Negative effects on feeding behavior, shading, reduction of water clarity	*Aureococcus anophagefferens, Aureoumbra lagunensis, Phaeocystis* spp.
Seaweed accumulation on beaches; coral reef overgrowth	*Cladophera, Codium, Ulva*, and many other species
Toxicity to marine organisms, including invertebrates, fish, mammals, and birds	*Gymnodinium breve, Alexandrium* spp., *Pseudo-nitzschia australis, Cocchlodinium polykrikoides*

(Modified from **Zingone and Enevoldsen, 2000.**)

this biomass begins to decay as the bloom terminates, oxygen is consumed, leading to widespread mortalities of all plants and animals in the affected area. These 'high biomass' blooms are sometimes linked to excessive pollution inputs (see below), but can also occur in relatively pristine waters.

Large, prolonged blooms of nontoxic algal species can reduce light penetration to the bottom, decreasing densities of submerged aquatic vegetation (SAV). Loss of SAV can have dramatic impacts on coastal ecosystems, as these grass beds serve as nurseries for the food and the young of commercially important fish and shellfish. Such indirect ecosystem effects from a nontoxic HAB were seen with a dense brown tide that lasted for over 7 years in the Laguna Madre section of southern Texas. The dense, coffee-colored bloom reduced light penetration and dramatically altered the abundance of seagrasses over a wide area.

Macroalgae (seaweeds) also cause problems. Over the past several decades, blooms of macroalgae have been increasing along many of the world's developed coastlines. Macroalgal blooms occur in nutrient-enriched estuaries and nearshore areas that are shallow enough for light to penetrate to the seafloor. These blooms have a broad range of ecological effects, and often last longer than 'typical' phytoplankton HABs. Once established, macroalgal blooms can remain in an environment for years unless the nutrient supply decreases. They can be particularly harmful to coral reefs. Under high nutrient conditions, opportunistic macroalgal species outcompete, overgrow, and replace the coral.

Nontoxic phytoplankton can also kill fish. The diatom genus *Chaetoceros* includes several species which have been associated with mortalities of farmed salmon, yet no toxin has ever been identified in that group. These species have long, barbed spines

Table 2 Human illnesses associated with harmful algal blooms

Syndrome	*Causative organisms*	*Toxin produced*	*Route of acquisition*	*Clinical manifestations*
Ciguatera fish poisoning (CFP)	*Gambierdiscus toxicus* and others	Ciguatoxin, maitotoxin	Toxin passed up marine food chain; illness results from eating large, carnivorous reef fish	Acute gastroenteritis, paresthesias and other neurological symptoms
Paralytic shellfish poisoning (PSP)	*Alexandrium* species, *Gymnodinium catenatum, Pyrodinium bahamense* var. *compressum*, and others	Saxitoxins	Eating shellfish harvested from affected areas	Acute paresthesias and other neurological manifestations; may progress rapidly to respiratory paralysis and death
Neurotoxic shellfish poisoning (NSP)	*Gymnodinium breve* and others	Brevetoxins	Eating shellfish harvested from affected areas; toxins may be aerosolized by wave action	Gastrointestinal and neurological symptoms; respiratory and eye irritation with aerosols
Diarrhetic shellfish poisoning (DSP)	*Dinophysis* species	Okadaic acid and others	Eating shellfish harvested from affected areas	Acute gastroenteritis
Azaspiracid shellfish poisoning (AZP)	*Protoperidinium crassipes*	Azaspiracids	Eating shellfish harvested from affected areas	Neurotoxic effects with severe damage to the intestine, spleen, and liver tissues in test animal
Amnesic shellfish poisoning (ASP)	*Pseudo-nitzchia* species	Domoic acid	Eating shellfish (or, possibly, fish) harvested from affected areas	Gastroenteritis, neurological manifestations, leading in severe cases to amnesia, coma, and death
Possible estuary-associated syndrome (PEAS)	*Pfiesteria piscicida* and others	Unidentified	Exposure to water or aerosols containing toxins	Deficiencies in learning and memory; acute respiratory and eye irritation, acute confusional syndrome

(Modified from Morris, 1999.)

Figure 3 Toxic blooms can kill wild and aquacultured fish by the millions. (Photo credits: G. Pitcher and M. Aramaki).

that lodge in gill tissues, leading to massive discharge of mucus that eventually results in lamellar degeneration and death due to the reduction in oxygen exchange. Why these diatoms have barbed spines is unknown. It is improbable that they have evolved specifically to kill fish, since the only mortalities from these species are caged fish that cannot escape the blooms. The problems now faced by the fish-farming industry are most likely the unfortunate result of an evolutionary strategy by certain *Chaetoceros* species to avoid predation or to stay afloat.

The Toxins

The toxins produced by some HAB species are among the most potent natural poisons known. Saxitoxin, for example, is 1000 times more potent than cyanide, and 50 times stronger than curare. Many of the toxin classes are not single chemical entities, but instead represent families of compounds of similar chemical structure (**Table 2**). These toxins bind with high affinity to specific receptor sites such as ion channels of excitable cells (**Table 3**). Most binding is reversible, but dissociation times can be quite prolonged. Each derivative of the parent toxin is slightly different in structure, and thus binds to the target receptor with different affinity. Weaker toxins tend to dissociate more quickly, accounting for their reduced potency. The saxitoxin family binds to sodium channels, which are the protein 'tunnels' responsible for the flux of sodium in and out of nerve and muscle cells. The brevetoxins also bind to the sodium channel, but to a different site. Their effect is to cause continuous activation of the channel, rather than a blockage, as with saxitoxin. Domoic acid disrupts normal neurochemical transmission in the brain by binding to kainate receptors in the central nervous system. This results in sustained depolarization of the neurons, and eventually in cell degeneration and death. One unfortunate symptom in ASP victims is permanent, short-term memory loss due to lesions that form in portions of the brain such as the hippocampus where there is a high density of kainate receptors.

Man is exposed to algal toxins principally by consumption of contaminated seafood products, although one type of toxin (brevetoxin) also causes respiratory asthma-like symptoms because of aerosol formation due to wave action. Acute single-dose lethality of seafood toxins has been extensively studied, but chronic and/or repeated exposure to low toxin concentrations, which undoubtedly occurs, has not been adequately examined. There is also a serious lack of knowledge as to how the toxins are distributed throughout the body and eliminated. Other important questions include how long the toxins circulate before elimination, and how they are metabolized by living organisms. These knowledge gaps prevent researchers from devising antidotes or effective treatments which may alleviate or lessen the symptoms. Therapeutic intervention is primarily limited to symptomatic treatment and life support.

The mechanisms by which fish are killed by toxic HABs are not well understood. Neurotoxins can rapidly act on specific fish tissues, significantly reducing heart rate and resulting in reduced blood flow and death from lack of oxygen. Other classes of compounds released by HABs such as reactive oxygen species, polyunsaturated fatty acids, and galactolipids are not toxins *per se*, but still can kill fish and other animals rapidly. Some have hemolytic activity,

Table 3 Properties of major algal toxins (values in parentheses (n) indicate the number of toxin derivatives)

Toxin family (n)	*Syndrome*	*Solubility*	*Action on*	*Pharmacology*
Saxitoxin (18)	PSP	Water soluble	Nerve, brain	Blockage of sodium channel
Brevetoxin (9)	NSP	Lipid soluble	Nerve, muscle, lung, brain	Selective sodium channel activators
Azaspiracid (3)	AZP	Lipid soluble	Nerve, liver, intestine, spleen	Not yet characterized
Okadaic acid (12)	DSP	Lipid soluble	Enzymes	Diarrhea by stimulating phosphorylation of a protein that controls sodium excretion of intestinal cells
Domoic acid (11)	ASP	Hot acid extraction	Brain	Binds to glutamate receptors in the brain, causing continuous stimulation of brain cells; lesions formed
Maitotoxin/ciguatoxin	CFP	Water/lipid soluble	Nerve, muscle, heart, brain	Activation of calcium/sodium channels

and others cause gill rupture, excessive production of mucus, hypoxia, and edema.

Scientists are searching for biochemical pathways within the algae to explain the metabolic role of the HAB toxins, but the search thus far has been fruitless. The toxins are not proteins, and all are synthesized in a series of chemical steps requiring multiple genes. Biosynthetic pathways have been proposed, but no chemical intermediates have been identified, nor have unique enzymes used only in toxin production yet been isolated. All we have are tantalizing clues that relate to toxin variability in individual cells, such as a 10-fold increase in saxitoxin production in cultured *Alexandrium* cells that are limited by phosphorus rather than nitrogen, or changes in the relative abundance of the different toxin derivatives produced by a particular dinoflagellate strain when growth conditions are varied. These observations indicate that the metabolism of the toxins is a dynamic process within the algae, but it is still not clear whether they have a specific biochemical role or are simply secondary metabolites. As with the nontoxic, spiny *Chaetoceros* species that kill fish, the illnesses and mortalities caused by algal 'toxins' may simply be the result of the 'accidental' affinity of those compounds for receptor sites on ion channels or tissues humans and in higher animals.

Ecology and Population Dynamics

Growth Features, Bloom Mechanisms

Impacts from HABs are necessarily linked to the population size and distribution of the causative algae. The growth and accumulation of individual algal species in a mixed assemblage of marine organisms are exceedingly complex processes involving an array of chemical, physical, and biological interactions. Given the diverse array of algae from several different classes that produce toxins or cause problems in a variety of oceanographic systems, attempts to generalize the bloom dynamics of HAB species are doomed to failure. Some common mechanisms can nevertheless be highlighted.

HABs are seldom caused by the explosive growth of a single species that rapidly dominates the water column. As stated by Ryther and co-workers many years ago, '...there is no necessity to postulate obscure factors which would account for a prodigious growth of dinoflagellates to explain red water. It is necessary only to have conditions favoring the growth and dominance of a moderately large population of a given species, and the proper hydrographic and meteorological conditions to permit the accumulation of organisms at the surface and to effect their future concentrations in localized areas.' In other words, winds, tides, currents, fronts, and other features can create discrete patches of cells of streaks of red water at all scales.

HABs in temperate and high latitudes are predominantly summer, coastal phenomena. They commonly occur during periods when heating or freshwater runoff create a stratified surface layer overlying colder, nutrient rich waters. Since the upper layer is quickly stripped of nutrients by other fast-growing algae, the onset of stratification often means that the only significant source of major plant nutrients lies below the interface between the layers – the pycnocline. This situation favors dinoflagellates and other motile algae, since nonmotile phytoplankton are unable to remain in suspension in the upper layer, and thus sink out of the zone where light permits photosynthesis. In contrast, motile algae are able to regulate their position and access the nutrients below the pycnocline. Many HAB species can swim at speeds in excess of $10\,\mathrm{m\,d^{-1}}$, and some undergo marked vertical migration, in which they reside in surface waters during the day to harvest the sunlight and then swim to the pycnocline and below to take up nutrients at night. This strategy can explain how dense accumulations of cells can appear in surface waters that are devoid of nutrients and which would seem to be incapable of supporting such prolific growth.

Swimming can also allow a species to persist at some optimum depth in the presence of vertical currents. One striking observation about some HAB blooms is that highly concentrated subsurface layers of cells can sometimes be only tens of centimeters thick – so called 'thin layers' of cells. This is another example of the importance of physical processes in determining organism distributions in stratified coastal systems. **Figure 4** shows a vertical profile of temperature and the corresponding thin layer of dinoflagellates that formed under those conditions along the coast of France. Similar thin layers are found throughout the world with scales as small as 10 cm in the vertical and 10 km in the horizontal.

Horizontal transport of blooms is also an important feature of some HABs, often over large distances. Major toxic outbreaks can suddenly appear at a site due to the transport of established blooms from other areas by ocean currents. This is the case in the western Gulf of Maine in the USA, where blooms of toxic dinoflagellates are transported hundreds of kilometers within a buoyant coastal current formed by the outflow of a major river system. An NSP outbreak in North Carolina in 1987 is another example of such long distance transport. The toxic *Gymnodinium breve* population that contaminated

North Carolina shellfish that year and drove tourists and residents from beaches originated in a bloom off the south-west coast of Florida, nearly 1500 km away. That bloom was carried out of the Gulf of Mexico and up the south-east coast of the USA by a combination of current systems, culminating in the

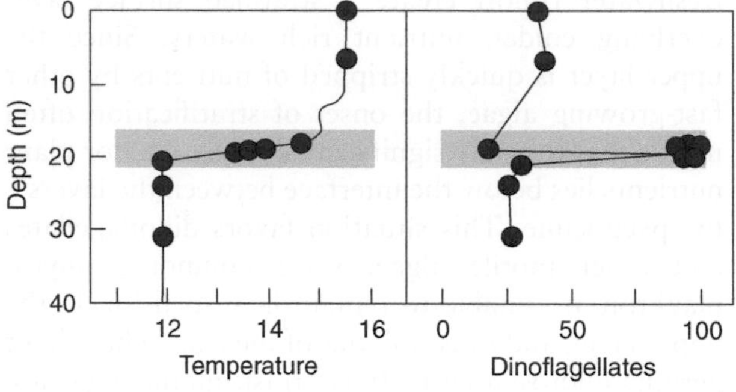

Figure 4 Subsurface accumulation of 'thin layers' of HAB species are common, as shown in this dataset from La Rochelle, France. (Adapted with permission from Gentien *et al.* (1995)).

Gulf Stream. After approximately 30 days of transport, a filament of water separated from the Gulf Stream and moved onto the narrow continental shelf of North Carolina, carrying toxic *G. breve* cells with it. The warm water mass remained in nearshore waters and was identifiable in satellite images for nearly 3 weeks (**Figure 5**).

Another important aspect of physical/biological coupling relates to the enhanced phytoplankton biomass that can occur at hydrographic features such as fronts. This enhanced biomass is the result of the interaction between physical processes such as upwelling, shear, and turbulence, and physiological processes such as swimming and enhanced nutrient uptake. One example is the linkage between tidally generated fronts and the sites of massive blooms of the toxic dinoflagellate *Gyrodinium aureolum* (now called *Gymnodinium mikimotoi*) in the North Sea. The pattern generally seen is a high surface concentration of cells at the frontal convergence, contiguous with a subsurface chlorophyll maximum which

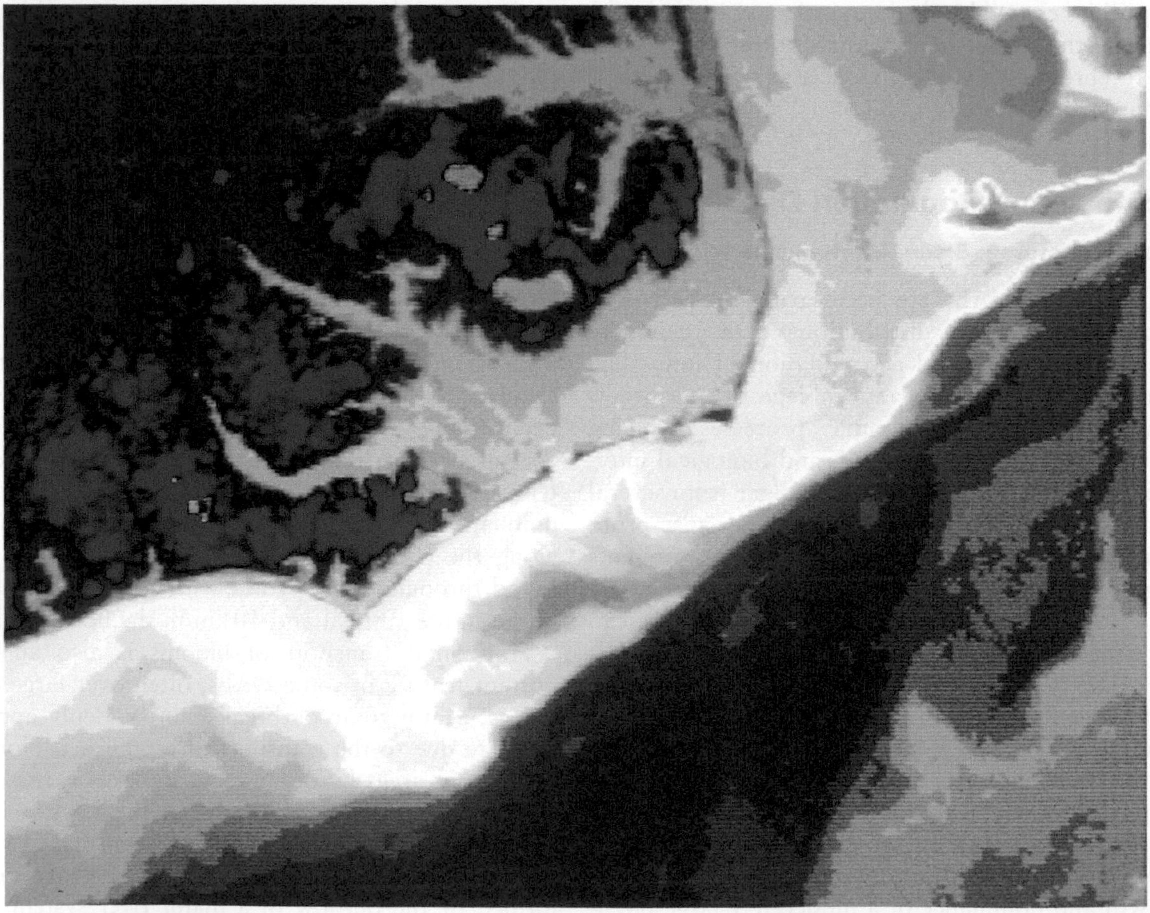

Figure 5 Satellite image of sea surface temperature, showing the warm Gulf Stream (blue) off the coast of North Carolina, A filament of the Gulf Stream can be seen extending towards shore. This carried a toxic population of *Gymnodinium breve*; that population originated in the Gulf of Mexico, >1500 km away. (Photo credit: P. Tester.).

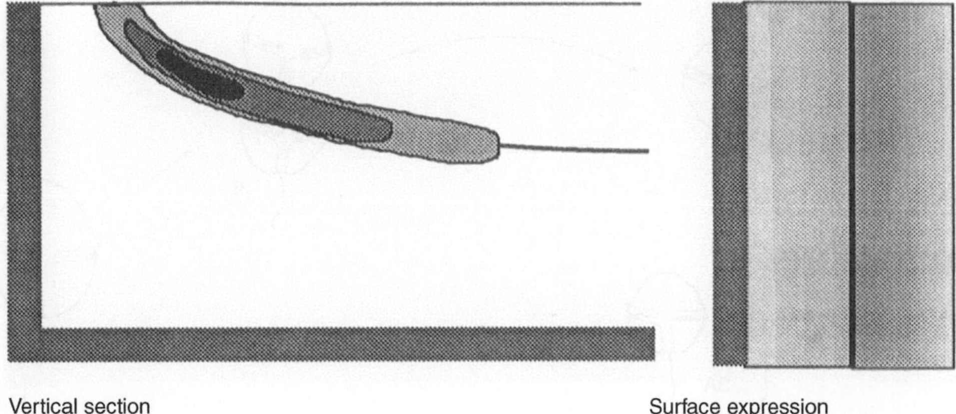

Figure 6 Accumulation of HAB cells near a front. This schematic demonstrates how cells can accumulate at a frontal convergence, with a surface manifestation of the bloom at the frontal convergence, and a subsurface extension of the bloom that extends along the sloping pycnocline. (Adapted with permission from Franks (1992).)

follows the sloping interface between the two water masses beneath the stratified side of the front (**Figure 6**). The surface signature of the chlorophyll maximum (sometimes a visible red tide) may be 1–30 km wide. Chlorophyll concentrations are generally lower and uniform on the well-mixed side of the front. Lateral transport of the front and its associated cells brings toxic *G. aureolum* populations into contact with nearshore fish and other susceptible resources, resulting in massive mortalities. This is a case where small-scale physical/biological coupling results in biomass accumulation, and larger-scale advective mechanisms cause the biomass to become harmful.

Life Histories

An important aspect of many HABs is their reliance on life history transformations for bloom initiation and decline. A number of key HAB species have dormant, cyst stages in their life histories, including *Alexandrium* spp., *Pyrodinium bahamense*, *Gymnodinium catenatum*, *Cocchlodinium polykrikoides, Gonyaulax monilata, Pfiesteria piscicida*, *Chattonella* spp., and *Heterosigma carterae*. Resting cysts remain in bottom sediments (sometimes termed 'seedbeds') when conditions in the overlying waters are unsuitable for growth. When conditions improve, such as with seasonal warming, the cysts germinate, inoculating the water column with a population of cells that begins to divide asexually via binary fission to produce a bloom. At the end of the bloom, often in response to nutrient limitation, vegetative growth ceases and the cells undergo sexual reproduction, whereby gametes are formed that fuse to form the swimming zygotes that ultimately become dormant cysts. **Figure 7** shows an example of the *Alexandrium tamarense* life history. Clearly, the location of cyst seedbeds can be an important determinant of the location of resulting blooms, and the size of the cyst accumulations can affect the magnitude of the blooms as well. It is generally believed, however, that environmental regulation of cell division is more important to eventual bloom magnitude than the size of the germination inoculum from cysts.

Cysts are also important in species dispersal. Natural or human-assisted transport of cysts from one location to another (e.g. via ballast water discharge or shellfish seeding) can allow a species to colonize a region and extend its geographic range. In 1972, a hurricane introduced *Alexandrium fundyense* into southern New England waters from established populations in the Bay of Fundy. Since that time, PSP has become an annually recurrent phenomenon in the region. Another example of human-assisted species introductions is the appearance of *Gymnodinium catenatum* in Tasmania in the 1970s, coincident with the development of a wood chip industry involving commercial vessels and frequent ballast water discharge.

Grazing Interactions

HAB cells can be food for herbivorous marine organisms (grazers). The extent to which grazing can control HABs depends upon the abundance of zooplankton, their ability to ingest the harmful algal species, and the effects of the HAB species on the grazers. In this context, one of the least understood aspects of toxic algal physiology concerns the metabolic or ecological roles of the toxins. Some argue that toxins evolved as a defense mechanism against grazing. Indeed, experimental studies demonstrate that zooplankton are affected to some degree by the

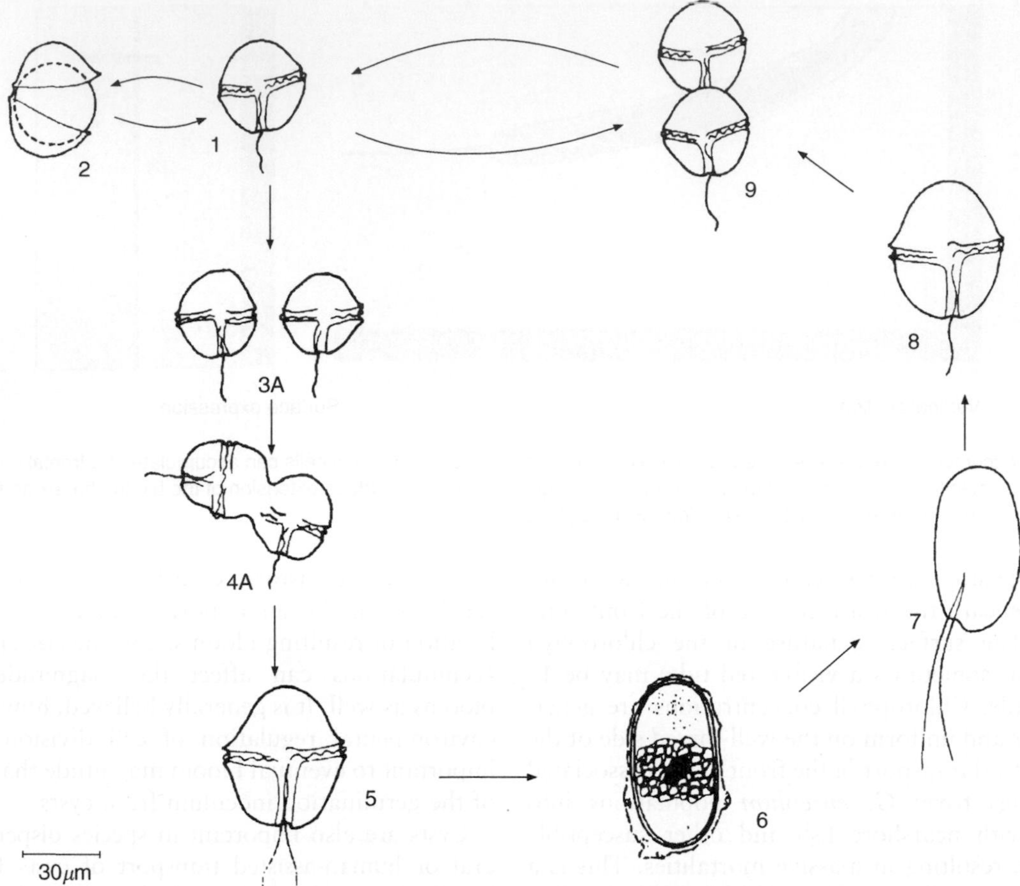

Figure 7 Life cycle diagram of *Alexandrium tamarense*. Stages are identified as follows: (1) vegetative, motile cell; (2) temporary or pellicle cyst; (3) 'female' and 'male' gametes; (4) fusing gametes; (5) swimming zygote or planozygote; (6) resting cyst or hypnozygote; (7,8) motile, germinated cell or planomeiocyte; and (9) pair of vegetative cells following division. Redrawn from Anderson *et al.* 1998.).

algal toxins they ingest. Two different effects have been noted. The first is a gradual incapacitation during feeding, as if the zooplankton are gradually paralyzed or otherwise impaired. (One study even showed that a tintinnid ciliate could only swim backwards, away from its intended algal prey, after exposure to a toxic dinoflagellate culture.) The second response is an active rejection or regurgitation of the toxic algae by a grazing animal, as if it had an unpleasant taste. Either of these mechanisms would result in reduced grazing pressure on toxic forms, which would facilitate the formation of a bloom. However, this cannot be the sole explanation for the presence of the toxins, as nontoxic phytoplankton are abundant, and form massive blooms.

An example of these mechanisms is seen in **Figure 8**, which depicts the feeding behavior of the copepod *Acartia tonsa* in mixtures containing equal concentrations of toxic and nontoxic *Alexandrium* spp. dinoflagellates. The copepods consumed nontoxic *Alexandrium* sp. cells at significantly higher rates than toxic cells, indicating chemosensory discrimination of toxic cells by the grazers.

Although there is clear evidence of the avoidance of certain HABs by grazers, or of incapacitation of the grazers themselves during feeding, it is also apparent that some toxins can be vectored through the food web via grazers. Fish kills of herring in the Bay of Fundy have been linked to PSP toxins from *Alexandrium* species that had been grazed by pteropods (small planktonic snails that are a favorite food of herring). Fortunately, in terms of human health, subsequent laboratory studies found that herring and other fish are very sensitive to these toxins and, unlike shellfish, die before they accumulate the toxins at dangerous levels in their flesh. Through similar food web transfer events, *Alexandrium* toxins have been implicated in kills of menhaden, sand lance, bluefish, dogfish, skates, monkfish, and even whales. Similar arguments can be made for other HAB toxins, especially the brevetoxins which have been linked to deaths of dolphins, and domoic acid (the ASP toxin),

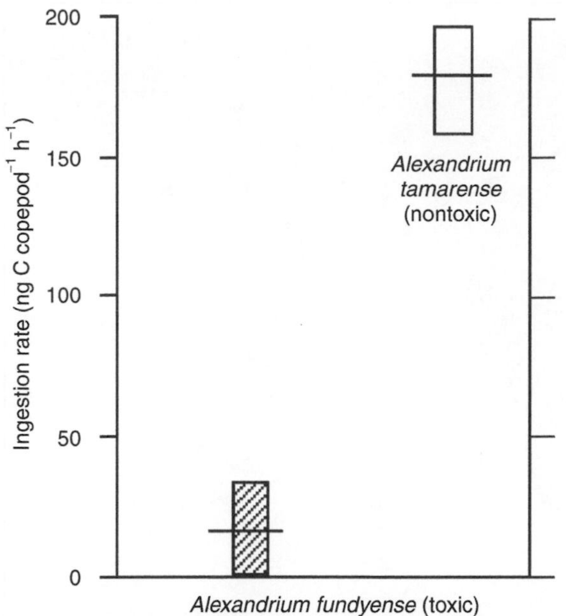

Figure 8 Feeding behavior of the copepod *Acartia tonsa* in mixtures containing equal concentrations of toxic and non-toxic *Alexandrium* spp. dinoflagellates. Central bars denote mean rates, boxes indicate standard deviation. Redrawn with permission from Teegarden 1999.

which has caused extensive bird and sea lion mortalities along the eastern Pacific coast of the USA and Mexico.

Trends

The nature of the global HAB problem has changed considerably over the last several decades (**Figure 9**). Virtually every coastal country is now threatened by harmful or toxic algal species, often in many locations and over broad areas. Thirty years ago, the problem was much more scattered and sporadic. The number of toxic blooms, the economic losses from them, the types of resources affected, and the number of toxins and toxic species have all increased dramatically in recent years. Disagreement only arises with respect to the reasons for this expansion, of which there are several possibilities.

New bloom events may simply reflect indigenous populations that are discovered because of improved chemical detection methods and more observers. The discovery of ASP along the west coast of the USA in 1991 is a good example of this, as toxic diatom species were identified and their toxin detected as a direct result of communication with Canadian scientists who had discovered the same toxin 4 years earlier and developed new chemical detection methods for domoic acid.

Several other 'spreading events' are most easily attributed to natural dispersal via currents, rather than human activities. As described above, the first NSP event ever to occur in North Carolina was shown to be a Florida bloom transported over 1500 km by the Gulf Stream – a totally natural phenomenon with no linkage to human activities. Some believe that humans may have contributed to the global HAB expansion by transporting toxic species in ship ballast water (e.g. *Gymnodinium catenatum* in Tasmania) or by shellfish relays and seeding. Another factor underlying the global increase in HABs is that we have dramatically increased aquaculture activities worldwide (**Figure 10**), and this inevitably leads to increased monitoring of product quality and safety, revealing indigenous toxic algae that were probably always there.

Nutrient enrichment is another cause for the increasing frequency of HAB events worldwide. Manipulation of coastal watersheds for agriculture, industry, housing, and recreation has drastically increased nutrient loadings to coastal waters. Just as the application of fertilizer to lawns can enhance grass growth, marine algae grow in response to nutrient inputs. Shallow and restricted coastal waters that are poorly flushed appear to be most susceptible to nutrient-related algal problems. Nutrient enrichment of such systems often leads to excessive production of organic matter, a process known as eutrophication, and increased frequencies and magnitudes of phytoplankton blooms, including HABs. There is no doubt that this is true in certain areas of the world where pollution has increased dramatically. It is perhaps real, but less evident in areas where coastal pollution is more gradual and unobtrusive. A frequently cited dataset from an area where pollution has been a significant factor in HAB incidence is from the Inland Sea of Japan, where visible red tides increased steadily from 44 per year in 1965 to over 300 a decade later, matching the pattern of increased nutrient loading from pollution (**Figure 11**). Effluent controls were instituted in the mid-1970s, resulting in a 70% reduction in the number of red tides that has persisted to this day. A related data set for the Black Sea documents a dramatic increase in red tides up to the mid-1990s, when the blooms began to decline. That reduction, which has also continued to this day, has been linked to reductions in fertilizer usage in upstream watersheds by former Soviet Union countries no longer able to afford large, state-subsidized fertilizer applications to agricultural land.

Anthropogenic nutrients can stimulate or enhance the impact of toxic or harmful species in several ways. At the simplest level, toxic phytoplankton may increase in abundance due to nutrient enrichment, but remain as the same relative fraction of the total phytoplankton biomass (i.e. all phytoplankton

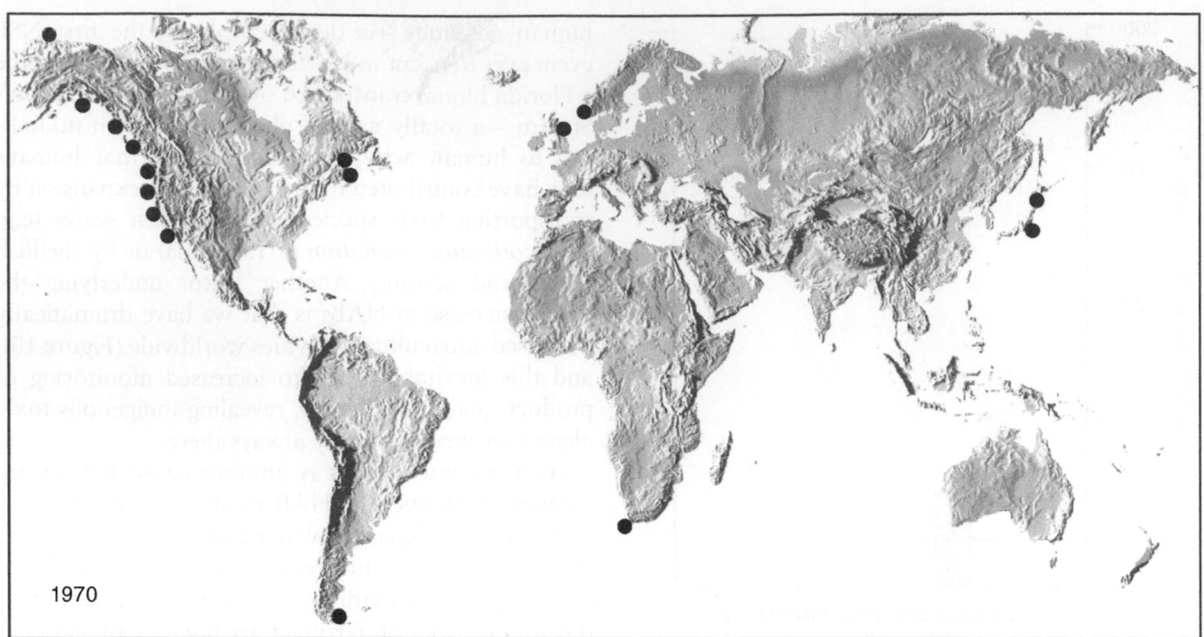

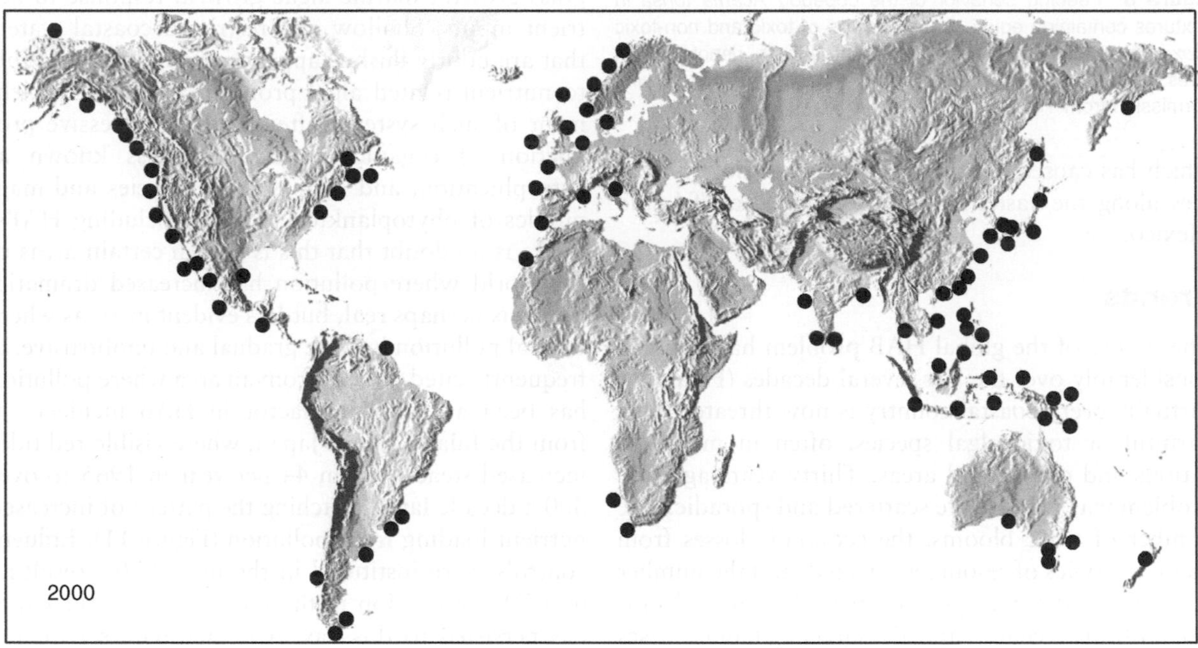

Figure 9 Expansion of the PSP problem over the past 30 years. Sites with proven records of PSP-causing organisms are noted in 1970, and again in 2000. Modified from Hallegraeff 1993. •, PSP.

species are affected equally by the enrichment). Alternatively, some contend that there has been a selective stimulation of HAB species by pollution. This view is based on the 'nutrient ratio hypothesis' which argues that environmental selection of phytoplankton species is associated with the relative availability of specific nutrients in coastal waters, and that human activities have altered these nutrient supply ratios in ways that favor harmful forms. For example, diatoms, the vast majority of which are harmless, require silicon in their cell walls, whereas most other phytoplankton do not. Since silicon is not abundant in sewage effluent but nitrogen and phosphorus are, the N:Si or P:Si ratios in coastal waters have increased through time over the last several decades. Diatom growth in these waters will cease when silicon supplies are depleted, but other phytoplankton classes (including toxic or harmful species) can continue to proliferate using the 'excess' nitrogen and phosphorus.

Figure 10 One reason for the global expansion in HABs relates to the increase in aquaculture activities, as shown with the dense concentration of mussel rafts in Spain. Not only do these intense farming operations alter the nutrient loads and plankton composition of local waters, but the intense scrutiny of the product for quality control purposes can reveal cryptic toxins that may have always been present in the region. (Photograph courtesy of D.M. Anderson.).

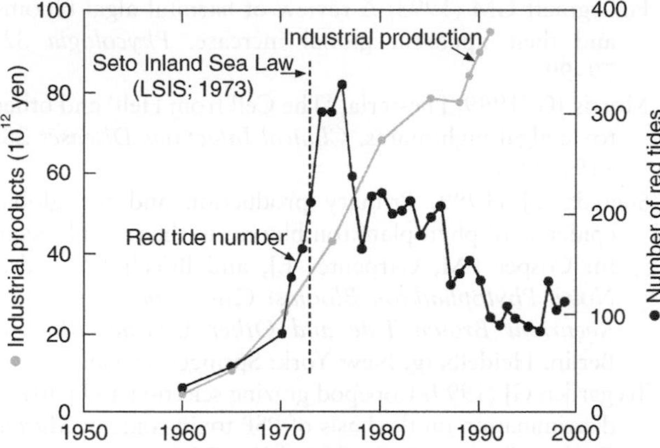

Figure 11 Human pollution can directly affect the number of red tides and HABs. This figure shows how blooms increased dramatically during the 1960s and 1970s in the Seto Inland Sea of Japan, but declined after pollution control legislation was enacted. Redrawn from Okaichi 1998.

This concept is controversial, but is supported by a 23-year time series off the German coast which documents the general enrichment of coastal waters with nitrogen and phosphorus, as well as a fourfold increase in N:Si and P:Si ratios. This was accompanied by a striking change in the composition of the phytoplankton community, as diatoms decreased and flagellates increased more than 10-fold.

Similar arguments hold for changes in N:P ratios, which have also changed due to the differences in removal of these nutrients with standard wastewater treatment practices. Some HAB species, such as *Phaeocystis pouchetii*, the species responsible for massive blooms that foul beaches with thick layers of smelly foam and slime, are not good competitors for phosphorus. As the N:P ratio of Dutch coastal waters decreased over time due to pollution effects, the size

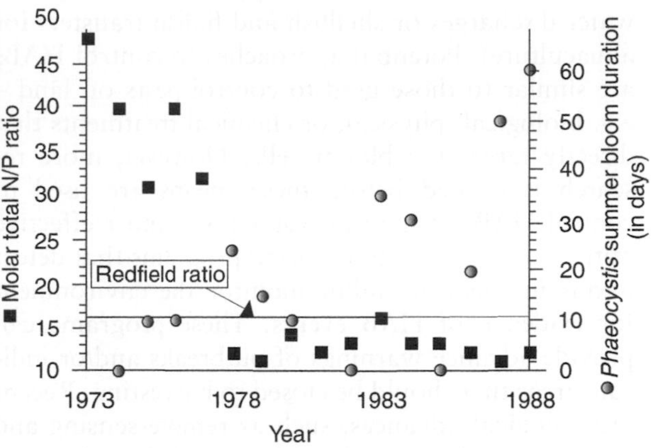

Figure 12 Changes in the duration of *Phaeocystis* blooms as the N:P ratio of Dutch coastal waters declined. Redrawn from Riegman *et al.* 1992.

and duration of *Phaeocystis* blooms increased (**Figure 12**).

The global increase in HAB phenomena is thus a reflection of two factors – an actual expansion of the problem, and an increased awareness of its size or scale. It is expanding due to pollution or other global change issues, but improved methods and enhanced scientific inquiry have also led to a better appreciation of its size or scale. The fact that part of the expansion is a result of increased awareness should not negate our concern, nor should it alter the manner in which we mobilize resources to attack it. The fact that it is also growing due to human activities makes our concerns even more urgent.

Management Issues

Management options for dealing with the impacts of HABs include reducing their incidence and extent (prevention), stopping or containing blooms (control), and minimizing impacts (mitigation). Where possible, it is preferable to prevent HABs rather than to treat their symptoms. Since increased pollution and nutrient loading may cause an increased incidence of outbreaks of some HAB species, these events may be prevented by reducing pollution inputs to coastal waters, particularly industrial, agricultural, and domestic effluents high in plant nutrients. This is especially important in shallow, poorly flushed coastal waters that are most susceptible to nutrient-related algal problems. Other strategies that may prevent HAB events include: regulating the siting of aquaculture facilities to avoid areas where HAB species are present, modifying water circulation for those HABs where restricted water exchange is a factor in bloom development, and restricting species introductions (e.g. through regulations on ballast water discharges or shellfish and finfish transfers for aquaculture). Potential approaches to control HABs are similar to those used to control pests on land – e.g. biological, physical, or chemical treatments that directly target the bloom cells. However, more research is needed before these means are used to control HABs in natural waters. The most effective mitigation tools are monitoring programs that detect toxins in shellfish and/or monitor the environment for evidence of HAB events. These programs can provide advance warnings of outbreaks and/or indicate areas that should be closed to harvesting. Recent technological advances, such as remote-sensing and molecular techniques, have increased detection and characterization of HAB species and blooms, and are playing an increasing role in monitoring programs worldwide. A long-term goal of these HAB monitoring programs and tools is to develop the ability to forecast bloom development and movement.

Summary

The HAB problem is significant and growing worldwide and poses a major threat to public health and ecosystem health, as well as to fisheries and economic development. The problems and impacts are diverse, as are the causes and underlying mechanisms controlling the blooms. The signs are clear that pollution and other alterations in the coastal zone have increased the abundance of algae, including harmful and toxic forms. All new outbreaks and new problems cannot be blamed on pollution, however, as there are numerous other explanations, some of which involve human activities, and some of which do not. As a growing world population increases its use of the coastal zone and demands more fisheries and recreational resources, there is a clear need to understand HAB phenomena and to develop scientifically sound management and mitigation policies.

See also

Molluskan Fisheries. Network Analysis of Food Webs.

Further Reading

Anderson DM (1994) Red tides. *Scientific American* 271: 52–58.

Anderson DM, Cembella AD, and Hallegraeff GM (eds.) (1998) *The Physiological Ecology of Harmful Algal Blooms*. Heidelberg: Springer-Verlag.

Burkholder JM and Glasgow HB Jr (1997) Pfiesteria piscicida and other Pfiesteria-like dinoflagellates: Behavior, impacts, and environmental controls. *Limnology and Oceanography* 42: 1052–1075.

Carmichael WW (1997) The cyanotoxins. *Advances in Botanical Research* 27: 211–256.

Fogg GE (1991) The phytoplanktonic ways of life. *New Phytologist* 118: 191–232.

Franks PJS (1992) Phytoplankton blooms at fronts: patterns, scales, and physical forcing mechanisms. *Reviews in Aquatic Sciences* 6(2): 121–137.

Gentien P, Lunven M, Lehaitre M, and Duvent JL (1995) In situ depth profiling of particle sizes. *Deep-Sea Research* 42: 1297–1312.

Hallegraeff GM (1993) A review of harmful algal blooms and their apparent global increase. *Phycologia* 32: 79–99.

Morris JG (1999) Pfiesteria, 'The Cell from Hell' and other toxic algal nightmares. *Clinical Infectious Diseases* 28: 1191–1198.

Smayda TJ (1989) Primary production and the global epidemic of phytoplankton blooms in the sea: a linkage In: Cosper EM, Carpenter EJ, and Bricelj VM (eds.) *Novel Phytoplankton Blooms: Causes and Impacts of Recurrent Brown Tide and Other Unusual Blooms*. Berlin, Heidelberg, New York: Springer-Verlag.

Teegarden GJ (1999) Copepod grazing selection and particle discrimination on the basis of PSP toxin content. *Marine Ecology Progress Series* 181: 163–176.

Yasumoto T and Murata M (1993) Marine toxins. *Chemical Reviews* 93: 1897–1909.

Zingone A and Enevoldsen HO (2000) The diversity of harmful algal blooms: A challenge for science and management. *Ocean and Coastal Management* 43: 725–748.

HYPOXIA

N. N. Rabalais, Louisiana Universities Marine Consortium, Chauvin, LA, USA

Introduction: Definitions

Hypoxic (low-oxygen) and anoxic (no-oxygen) waters have existed throughout geologic time. Presently, hypoxia occurs in many of the ocean's deeper environs, open-ocean oxygen-minimum zones (OMZs), enclosed seas and basins, below western boundary current upwelling zones, and in fjord. Hypoxia also occurs in shallow coastal seas and estuaries, where their occurrence and severity appear to be increasing, most likely accelerated by human activities (**Figure 1**). A familiar term used in the popular press and literature, 'dead zone', used for coastal and estuarine hypoxia, refers to the fish and shellfish killed by the suffocating conditions or the failure to catch these animals in bottom waters when the oxygen concentration in the water covering the seabed is below a critical level.

Based on laboratory or field observations or both, the level of oxygen stress and related responses of invertebrate and fish faunas vary. The units are often determined by oxygen conditions that are physiologically stressful, but these levels also differ depending on the organisms considered, and the pressure, temperature, and salinity of the ambient waters. The numerical definition of hypoxia varies as do the units used, but hypoxia has mostly been defined as dissolved oxygen levels lower than a range of 3–2 $ml\,l^{-1}$, with the consensus being in favor of 1.4 $ml\,l^{-1}$ (= 2 $mg\,l^{-1}$ or ppm). This value is approximately equivalent to 30% oxygen saturation at 25 °C and 35 salinity (psu). Below this concentration, bottom-dragging trawl nets fail to capture fish, shrimp, and swimming crabs. Other fishes, such as rays and sharks, are affected by oxygen levels below 3 $mg\,l^{-1}$, which prompts a behavioral response to evacuate the area, up into the water column and shoreward. Water-quality standards in the coastal waters of Long Island Sound, New York, and Connecticut, USA, consider that dissolved oxygen conditions below 5 $mg\,l^{-1}$ result in behavioral effects in marine organisms and fail to support living resources at sustainable levels.

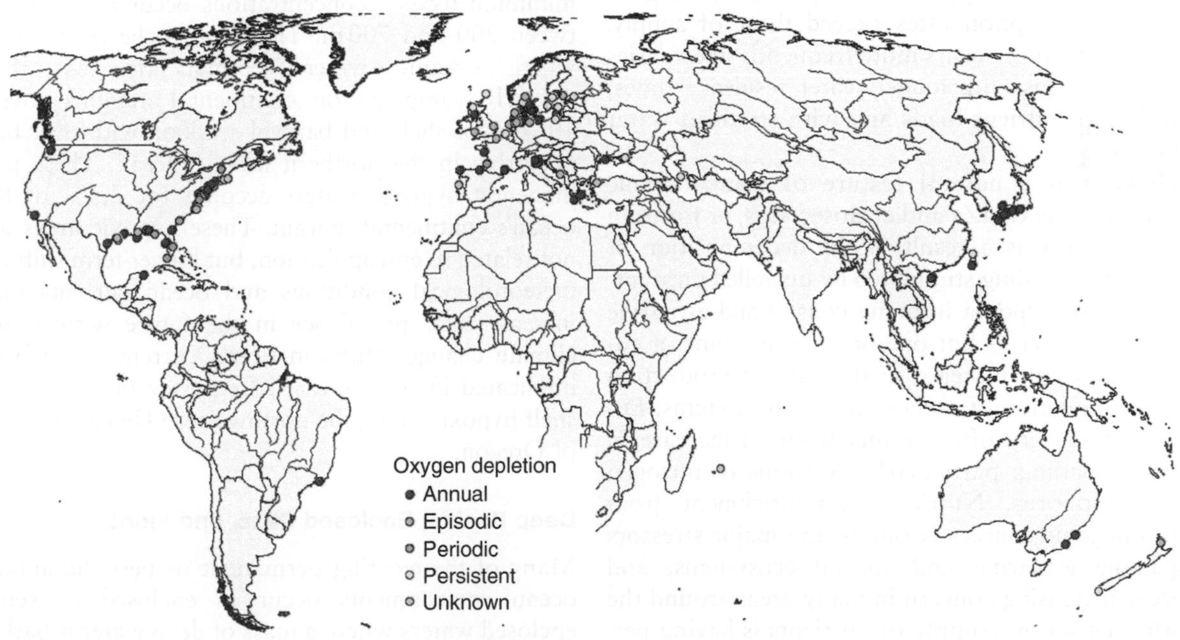

Figure 1 Distribution of coastal ocean hypoxic areas; excludes naturally occurring hypoxia, such as upwelling zones and OMZs. Reproduced from Díaz RJ, Nestlerode J, and Díaz ML (2004) A global perspective on the effects of eutrophication and hypoxia on aquatic biota. In: Rupp GL and White MD (eds.) *Proceedings of the 7th International Symposium on Fish Physiology, Toxicology and Water Quality, Tallinn, Estonia, May 12-15, 2003. EPA 600/R-04/049*, pp. 1–33. Athens, GA: Ecosystems Research Division, US Environmental Protection Agency, with permission from Robert J. Díaz.

The most commonly used definition for oceanic waters is dissolved oxygen content less than $1\,\mathrm{ml\,l^{-1}}$ (or $0.7\,\mathrm{mg\,l^{-1}}$). Disoxyic or disaerobic refers to oxygen levels between 0.1 and $1.0\,\mathrm{ml\,l^{-1}}$. OMZs are usually defined as waters less than $0.5\,\mathrm{ml\,l^{-1}}$ dissolved oxygen.

Causes

Hypoxia occurs where the consumption of oxygen through respiratory or chemical processes exceeds the rate of supply from oxygen production via photosynthesis, diffusion through the water column, advection, or mixing. The biological and physical water-column characteristics that support the development and maintenance of hypoxia include (1) the production, flux, and accumulation of organic-rich matter from the upper water column; and (2) water-column stability resulting from stratification or long residence time. Dead and senescent algae, zooplankton fecal pellets, and marine aggregates contribute significant amounts of organic detritus to the lower water column and seabed. Aerobic bacteria consume oxygen during the decay of the carbon and deplete the oxygen, particularly when stratification prevents diffusion of oxygen. Stratification is the division of the water column into layers with different densities caused by differences in temperature or salinity or both. Hypoxia will persist as long as oxygen consumption rates exceed those of supply. Oxygen depletion occurs more frequently in estuaries or coastal areas with longer water residence times, with higher nutrient loads and with stratified water columns.

Hypoxia is a natural feature of many oceanic waters, such as OMZs and enclosed seas, or forms in coastal waters as a result of the decomposition of high carbon loading stimulated by upwelled nutrient-rich waters. Hypoxia in many coastal and estuarine waters, however, is but one of the symptoms of eutrophication, an increase in the rate of production and accumulation of carbon in aquatic systems. Eutrophication very often results from an increase in nutrient loading, particularly by forms of nitrogen and phosphorus. Nutrient over-enrichment from anthropogenic sources is one of the major stressors impacting estuarine and coastal ecosystems, and there is increasing concern in many areas around the world that an oversupply of nutrients is having pervasive ecological effects on shallow coastal waters. These effects include reduced light penetration, increased abundance of nuisance macroalgae, loss of aquatic habitat such as seagrass or macroalgal beds, noxious and toxic algal blooms, hypoxia and anoxia, shifts in trophic interactions and food webs, and impacts on living resources.

Hypoxic Systems

Oxygen-Minimum Zones

Persistent hypoxia is evident in mid-water OMZs, which are widespread in the world oceans where the oxygen concentrations are less than $0.5\,\mathrm{ml\,l^{-1}}$ (or about 7.5% oxygen saturation, $<22\,\mu\mathrm{M}$). They occur at different depths from the continental shelf to upper bathyal zones (down to 1300 m). Many of the OMZs form as a result of high primary production associated with coastal upwelled nutrient-rich waters. Their formation also requires stagnant circulation, long residence times, and the presence of oxygen-depleted source waters. The extensive OMZ development in the eastern Pacific Ocean is attributed to the fact that intermediate depth waters of the region are older and have overall oxygen concentrations lower than other water masses. The largest OMZs are at bathyal depths in the eastern Pacific Ocean, the Arabian Sea, the Bay of Bengal, and off southwest Africa. The upper boundary of an OMZ may come to within 10 or 50 m of the sea surface off Central America, Peru, and Chile. The OMZ is more than 1000-m thick off Mexico and in the Arabian Sea, but off Chile, the OMZ is <400-m thick. Along continental margins, minimum oxygen concentrations occur typically between 200 and 700 m. The area of the ocean floor where oceanic waters permanently less than $0.5\,\mathrm{ml\,l^{-1}}$ impinge on continental margins covers $10^6\,\mathrm{km^2}$ of shelf and bathyal seafloor, with over half occurring in the northern Indian Ocean. These permanently hypoxic waters account for 2.3% of the ocean's continental margin. These hypoxic areas are not related to eutrophication, but longer-term shifts in meteorological conditions and ocean currents may increase their prevalence in the future with global climate change. Shifts in ocean currents have been implicated in the increased frequency of continental shelf hypoxia along the northwestern US Pacific coast of Oregon.

Deep Basins, Enclosed Seas, and Fjord

Many of the existing permanent or periodic anoxic ocean environments occur in enclosed or semi-enclosed waters where a mass of deep water is bathymetrically isolated from main shelf or oceanic water masses by surrounding landmasses or one or more shallow sills. In conjunction with a pycnocline, the bottom water volume is restricted from exchange with deep open water. Examples of hypoxic and anoxic basins include anoxic deep water fjord, such as Saanich Inlet, the deeper basins of the Baltic, the basin of the Black Sea, the Japanese Seto Inland Sea, deep waters of the Sea of Cortez, Baja California, and Santa Barbara Basin in the southern California borderland.

Coastal Seas and Estuaries

Periodic hypoxia or anoxia also occurs on open continental shelves, for example, the northern Gulf of Mexico and the Namibian and Peruvian shelves where upwelling occurs. More enclosed shelves such as the northern Adriatic and the northwestern shelf of the Black Sea also have periodic hypoxia or anoxia. In these instances, there is minimal exchange of shelf-slope water and/or high oxygen demand on the shallow shelf. Estuaries, embayments, and lagoons are susceptible to the formation of hypoxia and anoxia if the water residence time is sufficiently long, especially where the water column is stratified. Light conditions are also important in these coastal habitats as a limiting factor on phytoplankton growth, which, if excessive, contributes to high organic loading within the confined waters.

Coastal ecosystems that have been substantially changed as a result of eutrophication exhibit a series of identifiable symptoms, such as reduced water clarity, excessive, noxious, and, sometimes, harmful algal blooms, loss of critical macroalgal or seagrass habitat, development or intensification of oxygen depletion in the form of hypoxia or anoxia, and, in some cases, loss of fishery resources. More subtle responses of coastal ecosystems to eutrophication include shifts in phytoplankton and zooplankton communities, shifts in the food webs that they support, loss of biodiversity, changes in trophic interactions, and changes in ecosystem functions and biogeochemical processes.

In a review of anthropogenic hypoxic zones in 1995, Díaz and Rosenberg noted that no other environmental variable of such ecological importance to estuarine and coastal marine ecosystems around the world has changed so drastically, in such a short period of time, as dissolved oxygen. For those reviewed, there was a consistent trend of increasing severity (either in duration, intensity, or size) where hypoxia occurred historically, or hypoxia existed presently when it did not occur before. While hypoxic environments have existed through geologic time and are common features of the deep ocean or adjacent to areas of upwelling, their occurrence in estuarine and coastal areas is increasing, and the trend is consistent with the increase in human activities that result in nutrient over-enrichment.

The largest human-caused hypoxic zone is in the aggregated coastal areas of the Baltic Sea, reaching 84 000 km^2. Hypoxia existed on the northwestern Black Sea shelf historically, but anoxic events became more frequent and widespread in the 1970s and 1980s, reaching over areas of the seafloor up to 40 000 km^2 in depths of 8–40 m. There is also evidence that the suboxic zone of the open Black Sea enlarged toward the surface by about 10 m since 1970. The condition of the northwestern shelf of the Black Sea, in which hypoxia covered up to 40 000 km^2, improved over the period 1990–2000 when nutrient loads from the Danube River decreased, but may be experiencing a worsening of hypoxic conditions more recently.

Similar declines in bottom water dissolved oxygen have occurred elsewhere as a result of increasing nutrient loads and cultural eutrophication, for example, the northern Adriatic Sea, the Kattegat and Skaggerak, Chesapeake Bay, Albemarle-Pamlico Sound, Tampa Bay, Long Island Sound, New York Bight, the German Bight, and the North Sea. In the United States, over half of the estuaries experience hypoxia at some time over an annual period and many experience hypoxia over extensive areas for extended periods on a perennial basis. The number of estuaries with hypoxia or anoxia continues to rise.

Historic data on Secchi disk depth in the northern Adriatic Sea in 1911 through the present, with few interruptions of data collection, provide a measure of water transparency that could be interpreted to depict surface water productivity. These data coupled with surface and bottom water dissolved oxygen content determined by Winkler titrations and nutrient loads outline the sequence of eutrophication in the northern Adriatic Sea. Similar historical data from other coastal areas around the world demonstrate a decrease in water clarity due to phytoplankton production in response to increased nutrient loads that are paralleled by declines in water column oxygen levels.

There are strong relationships between river flow and nutrient flux into the Chesapeake Bay and northern Gulf of Mexico and phytoplankton production and biomass and the subsequent fate of that production in spring deposition of chlorophyll *a*. Further there is a strong relationship between the deposited chlorophyll *a* and the seasonal decline of deep-water dissolved oxygen. Excess nutrients in many watersheds are driven by agricultural activities and atmospheric deposition from burning of fossil fuels. The link with excess nutrients in more urban areas, such as Long Island Sound, is with the flux of nutrients associated from numerous wastewater outfalls.

Swift currents that move materials away from a river delta and that do not permit the development of stratification are not conducive to the accumulation of biomass or depletion of oxygen, for example in the Amazon and Orinoco plumes. Similar processes off the Changjiang (Yantze River) and high turbidity in the plume of the Huanghe (Yellow River) were once thought to be reasons why hypoxia did not develop in those coastal systems. Incipient indications of the beginning of symptoms of cultural eutrophication were becoming evident at the terminus of both these systems as nutrient loads increased. The severely reduced, almost minimal, flow of the Huanghe has prevented the formation of hypoxia, but other coastal ecosystem problems remain. There is, however, now a hypoxic area off the Changjiang Estuary and harmful algal blooms are more frequent in the East China Sea. The likelihood that more and more coastal systems, especially in developing countries, where the physical conditions are appropriate will become eutrophic with accompanying hypoxia is worrisome.

Northern Gulf of Mexico

The hypoxic zone on the continental shelf of the northern Gulf of Mexico is one of the largest hypoxic zones in the world's coastal oceans, and is representative of hypoxia resulting from anthropogenic activities over the last half of the twentieth century (**Figure 2**). Every spring, the dissolved oxygen levels in the coastal waters of the northern Gulf of Mexico decline and result in a vast region of oxygen-starved water that stretches from the Mississippi River westward along the Louisiana shore and onto the Texas coast. The area of bottom covered by hypoxic water can reach 22 000 km^2, and the volume of hypoxic waters may be as much as 10^{11} m^3. Hypoxia in the Gulf of Mexico results from a combination of natural and human-influenced factors. The Mississippi River, one of the 10 largest in the world, drains 41% of the land area of the lower 48 states of the US and delivers fresh water, sediments, and nutrients to the Gulf of Mexico. The fresh water, when it enters the Gulf, floats over the denser saltier water, resulting in stratification, or a two-layered system. The stratification, driven primarily by salinity, begins in the spring, intensifies in the summer as surface waters warm and winds that normally mix the water subside, and dissipates in the fall with tropical storms or cold fronts.

Hypoxic waters are found at shallow depths near the shore (4–5 m) to as deep as 60 m. The more typical depth distribution is between 5 and 35 m. The hypoxic water is not just located near the seabed, but may rise well up into the water column, often occupying the lower half of a 20-m water column (**Figure 3**). The inshore/offshore distribution of hypoxia on the Louisiana shelf is dictated by winds and currents. During typical winds from the southeast, downwelling favorable conditions force the hypoxic bottom waters farther offshore. When the wind comes from the north, an upwelling favorable current regime promotes the movement of the hypoxic bottom waters close to shore. When the hypoxic waters move onto the shore, fish, shrimp, and crabs are trapped along the beach, resulting sometimes in a 'jubilee' when the stunned animals are easily harvested by beachgoers. A more negative result is a massive fish kill of all the sea life trapped without sufficient oxygen.

Hypoxia occurs on the Louisiana coast west of the Mississippi River delta from February through November, and nearly continuously from mid-May through mid-September. In March and April, hypoxic water masses are patchy and ephemeral. The hypoxic

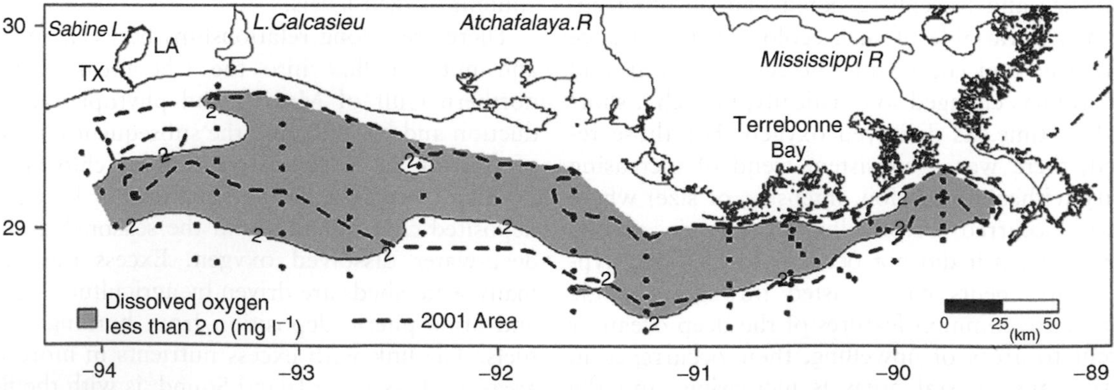

Figure 2 Similar size and expanse of bottom water hypoxia in mid-July 2002 (shaded area) and in mid-July 2001 (outlined with dashed line). Data source: N. N. Rabalais, Louisiana Universities Marine Consortium.

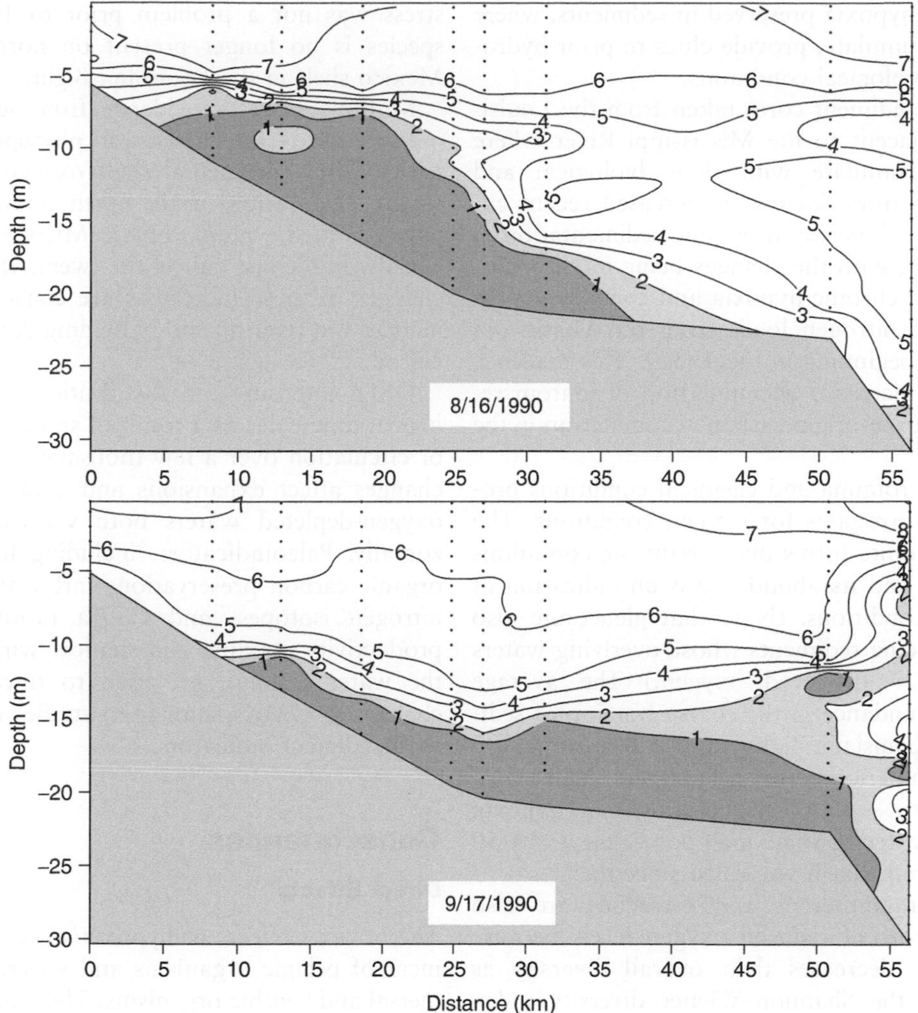

Figure 3 Contours of dissolved oxygen (mg l^{-1}) across the continental shelf of Louisiana approximately 200 km west of the Mississippi River delta in summer. The distribution across the shelf in August is a response to an upwelling favorable oceanographic regime and that of September to a downwelling favorable oceanographic regime. These contours also illustrate the height above the seabed that hypoxia can reach, i.e., over half the water column. Data source: N. N. Rabalais, Louisiana Universities Marine Consortium.

zone is most widespread, persistent, and severe in June, July, and August, and often well into September, depending on whether tropical storm activity begins to disrupt the stratification and hypoxia. Anoxic waters occur periodically in midsummer.

The midsummer size of the hypoxic zone varies annually, and is most closely related to the nitrate load of the Mississippi River in the 2 months prior to the typically late-July mapping exercise. The load of nitrate is determined by the discharge of the Mississippi River multiplied by the concentration of the nitrate, so that the amount of water coming into the Gulf of Mexico is also a factor. The relationship of the size of hypoxia, however, is stronger with the load of nitrate than with the total river water discharge or any other nutrient or combination of nutrients. Changes in the severity of hypoxia over time are related mostly to the change in nitrate concentration in the Mississippi River (80%), the remainder to changes in increased discharge (20%).

Historical Change in Oxygen Conditions

Historical dissolved oxygen data such as those for the northern Adriatic Sea beginning in 1911 are not commonly available. A solution is to turn to the sediment record for paleoindicators of long-term transitions related to eutrophication and oxygen deficiency. Biological, mineral, or chemical indicators of plant communities, level of productivity, or conditions of hypoxia preserved in sediments, where sediments accumulate, provide clues to prior hydrographic and biological conditions.

Data from sediment cores taken from the Louisiana Bight adjacent to the Mississippi River where sediments accumulate with their biological and chemical indicators document increased recent eutrophication and increased organic sedimentation in bottom waters, with the changes being more apparent in areas of chronic hypoxia and coincident with the increasing nitrogen loads from the Mississippi River system beginning in the 1950s. This evidence comes as an increased accumulation of diatom remains and marine-origin carbon accumulation in the sediments.

Benthic microfauna and chemical conditions provide several surrogates for oxygen conditions. The mineral glauconite forms under reducing conditions in sediments, and its abundance is an indication of low-oxygen conditions. (Note that glauconite also forms in reducing sediments whose overlying waters are $>2\,\mathrm{mg\,l^{-1}}$ dissolved oxygen.) The average glauconite abundance in the coarse fraction of sediments in the Louisiana Bight was $\sim 5.8\%$ from 1900 to a transition period between 1940 and 1950, when it increased to $\sim 13.4\%$, suggesting that hypoxia 'may' have existed at some level before the 1940–50 time period, but that it worsened since then.

Benthic foraminiferans and ostracods are also useful indicators of reduced oxygen levels because oxygen stress decreases their overall diversity as measured by the Shannon–Wiener diversity index (SWDI) and shifts community composition. Foraminiferan and ostracod diversity decreased since the 1940s and early 1950s, respectively. While present-day foraminiferan diversity is generally low in the Louisiana Bight, comparisons among assemblages from areas of different oxygen depletion indicate that the dominance of *Ammonia parkinsoniana* over *Elphidium* spp. (A–E index) was much more pronounced in oxygen-depleted compared to well-oxygenated waters. The A–E index has also proven to be a strong, consistent oxygen-stress signal in other coastal areas, for example, Chesapeake Bay and Long Island Sound. The A–E index from sediment cores increased significantly after the 1950s, suggesting increased oxygen stress (in intensity or duration) in the last half century. *Buliminella morgani*, a hypoxia-tolerant species, known only from the Gulf of Mexico, dominates the present-day population ($>50\%$) within areas of chronic seasonal hypoxia, and has also increased markedly in recent decades. *Quinqueloculina* sp., a hypoxia-intolerant foraminiferan, was a conspicuous member of the fauna from 1700 to 1900, indicating that oxygen stress was not a problem prior to 1900, but this species is no longer present on northern Gulf of Mexico shelf in the Louisiana Bight.

Multiple lines of evidence from sediment cores indicate an overall increase in phytoplankton productivity and continental shelf oxygen stress (in intensity or duration) in the northern Gulf of Mexico adjacent to the plume of the Mississippi River, especially in the last half of the twentieth century. The changes in these indicators are consistent with the increases in river nitrate-N loading during that same period.

OMZ intensity and distribution vary over geological timescales as a result of shifts in productivity or circulation over a few thousands to 10 ky. These changes affect expansions and contractions of the oxygen-depleted waters both vertically and horizontally. Paleoindicators, including foraminiferans, organic carbon preservation, carbonate dissolution, nitrogen isotopes, and Cd:Ca ratios that reflect productivity maxima and shallow winter mixing of the water column, are used to trace longer-term changes in OMZs, similar to studies of continental shelf sediment indicators.

Consequences

Direct Effects

The obvious effects of hypoxia/anoxia are displacement of pelagic organisms and selective loss of demersal and benthic organisms. These impacts may be aperiodic so that recovery occurs; may occur on a seasonal basis with differing rates of recovery; or may be permanent so that a shift occurs in long-term ecosystem structure and function. As the oxygen concentration falls from saturated or optimal levels toward depletion, a variety of behavioral and physiological impairments affect the animals that reside in the water column or in the sediments or that are attached to hard substrates (**Figure 4**). Hypoxia acts as an endocrine disruptor with adverse effects on reproductive performance of fishes, and loss of secondary production may therefore be a widespread environmental consequence of hypoxia. Mobile animals, such as shrimp, fish, and some crabs, flee waters where the oxygen concentration falls below $3\text{–}2\,\mathrm{mg\,l^{-1}}$.

As dissolved oxygen concentrations continue to fall, less mobile organisms become stressed and move up out of the sediments, attempt to leave the seabed, and often die (**Figure 5**). As oxygen levels fall from 0.5 toward $0\,\mathrm{mg\,l^{-1}}$, there is a fairly linear decrease in benthic infaunal diversity, abundance, and biomass. Losses of entire higher taxa are features of the

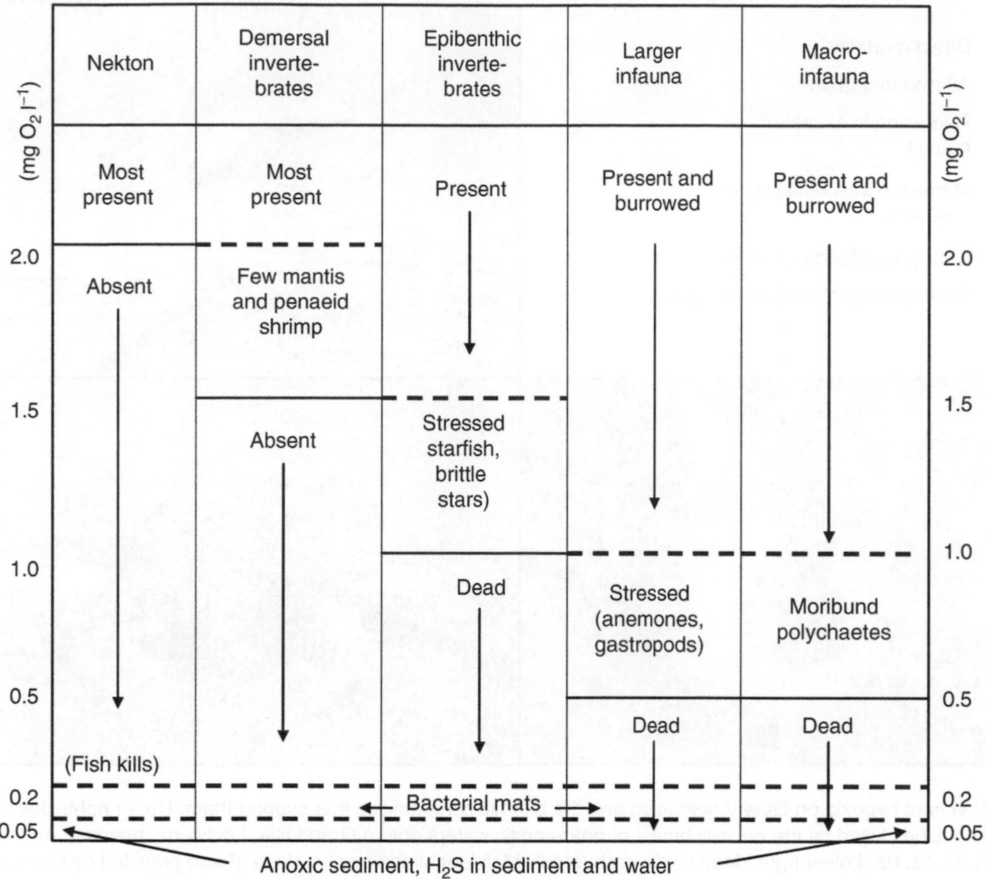

Figure 4 Progressive changes in fish and invertebrate fauna as oxygen concentration decreases from 2 mg l^{-1} to anoxia. From Rabalais NN, Harper DE, Jr., and Tuner RE (2001) Responses of nekton and demersal and benthic fauna to decreasing oxygen concentrations. In: Rabalais NN and Turner RE (eds.) *Coastal and Estuarine Studies 58: Coastal Hypoxia: Consequences for Living Resources and Ecosystems*, pp. 115–128. Washington, DC: American Geophysical Union.

depauperate benthic fauna in the severely stressed seasonal hypoxic/anoxic zone of the Louisiana inner shelf in the northern Gulf of Mexico. Larger, longer-lived burrowing infauna are replaced by short-lived, smaller surface deposit-feeding polychaetes, and certain typical marine invertebrates are absent from the fauna, for example, pericaridean crustaceans, bivalves, gastropods, and ophiuroids. Long-term trends for the Skagerrak coast of western Sweden in semi-enclosed fiordic areas experiencing increased oxygen stress showed declines in the total abundance and biomass of macroinfauna, abundance and biomass of mollusks, and abundance of suspension feeders and carnivores. These changes in benthic communities result in an impoverished diet for bottom-feeding fish and crustaceans and contribute, along with low dissolved oxygen, to altered sediment biogeochemical cycles. In waters of Scandinavia and the Baltic, there was a reduction of 3 million t in benthic macrofaunal biomass during the worst years of hypoxia occurrence. This loss, however, may be partly compensated by the biomass increase that occurred in well-flushed organically enriched coastal areas not subject to hypoxia.

Where oxygen minimum zones impinge on continental margins or sea mounts, they have considerable effects on benthic assemblages. The benthic fauna of OMZs consist mainly of smaller-sized protozoan and meiofaunal organisms, with few or no macrofauna or megafauna. The few eukaryotic organisms are nematodes and foraminiferans. Meiofauna appear to be more broadly tolerant of oxygen depletion than are macrofauna. The numbers of metazoan meiofaunal organisms, primarily nematodes, are not reduced in OMZs, presumably due to abundant particulate food and reduced predation pressure. In hypoxic waters of the northern Gulf of Mexico, harpacticoid copepod meiofauna are reduced at low oxygen levels, but the nematodes maintain their densities. Benthic macrofauna are found in all hypoxic sediments of the northern Gulf of Mexico, although the density is severely reduced

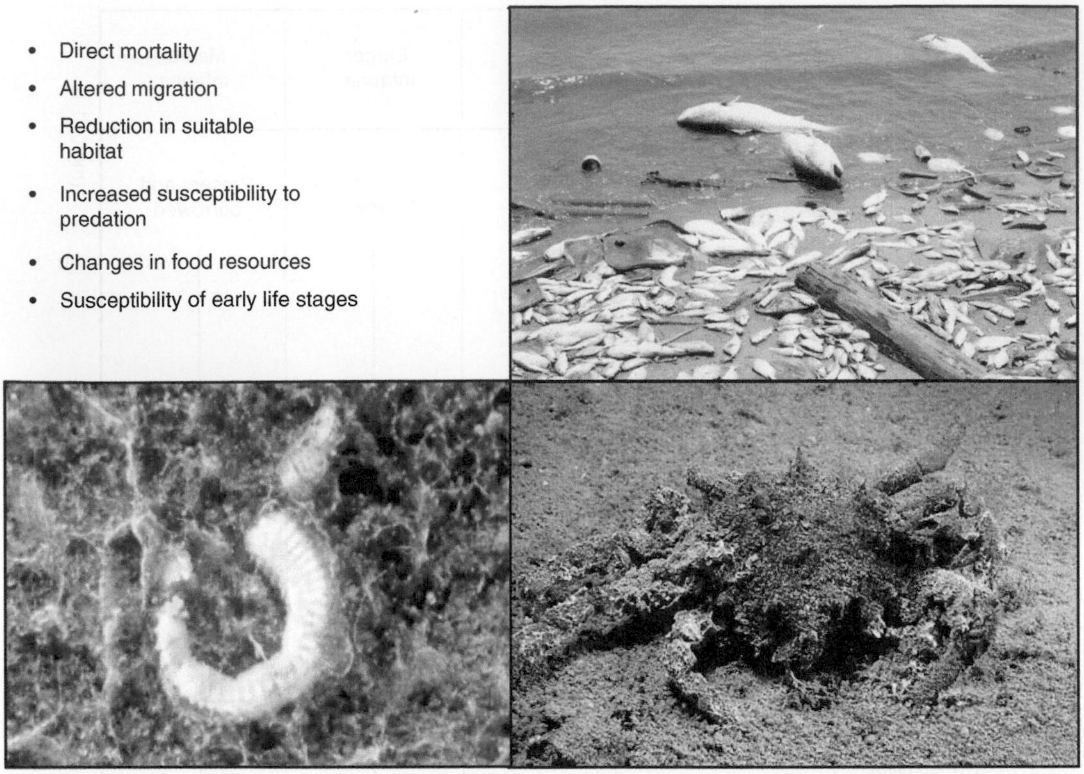

- Direct mortality
- Altered migration
- Reduction in suitable habitat
- Increased susceptibility to predation
- Changes in food resources
- Susceptibility of early life stages

Figure 5 Effects of hypoxia on fishery resources and the benthic communities that support them. Upper right: Dead demersal and bottom-dwelling fishes killed by the encroachment of near-anoxic waters onto a Grand Isle, Louisiana, beach in August 1990. Photo provided by K. M. St. Pé. Lower right: dead spider crab (family Majidae) at sediment surface. Photo provided by Franklin Viola. Lower left: dead polychaete (family Spionidae) and filamentous sulfur bacteria. Photo provided by Franklin Viola.

below $0.5\,\text{mg}\,\text{l}^{-1}$, and the few remaining organisms are polychaetes of the families Ampharetidae and Magelonidae and some sipunculans.

While permanent deep-water hypoxia that impinges on 2.3% of the ocean's continental margin may be inhospitable to most commercially valuable marine resources, they support the largest, most continuous reducing ecosystems in the world oceans. Large filamentous sulfur bacteria, *Thioploca* and *Beggiatoa*, thrive in hypoxic conditions of $0.1\,\text{ml}\,\text{l}^{-1}$. OMZ sediments characteristically support large bacteria, both filamentous sulfur bacteria and giant spherical sulfur bacteria with diameters of 100–300 μm. The filamentous sulfur bacteria are also characteristic of severely oxygen depleted waters in the northern Gulf of Mexico.

Secondary Production

An increase in nutrient availability results in an increase of fisheries yield to a maximal point; then there are declines in various compartments of the fishery as further increases in nutrients lead to seasonal hypoxia and permanent anoxia in semi-enclosed seas. Documenting loss of fisheries related to the secondary effects of eutrophication, such as the loss of seabed vegetation and extensive bottom water oxygen depletion, is complicated by poor fisheries data, inadequate economic indicators, increase in overharvesting that occurred at the time that habitat degradation progressed, natural variability of fish populations, shifts in harvestable populations, and climatic variability.

Eutrophication often leads to the loss of habitat (rooted vegetation or macroalgae) or low dissolved oxygen, both of which may lead to loss of fisheries production. In the deepest bottoms of the Baltic proper, animals have long been scarce or absent because of low oxygen availability. This area was $20\,000\,\text{km}^2$ until the 1940s. Since then, about a third of the Baltic bottom area has intermittent oxygen depletion. Lowered oxygen concentrations and increased sedimentation have changed the benthic fauna in the deeper parts of the Baltic, resulting in an impoverished diet for bottom fish. Above the halocline in areas not influenced by local pollution, benthic biomass has increased due mostly to an increase in mollusks. On the other hand, many reports document instances where local pollution resulting in severely depressed oxygen levels has greatly impoverished or even annihilated the soft-bottom macrofauna.

Eutrophication of surface waters accompanied by oxygen-deficient bottom waters can lead to a shift in dominance from demersal fishes to pelagic fishes. In the Baltic Sea and Kattegatt where eutrophication-related ecological changes occurred mainly after World War II, changes in fish stocks have been both positive (due to increased food supply; e.g., pike perch in Baltic archipelagos) and negative (e.g., oxygen deficiency reducing Baltic cod recruitment and eventual harvest). Similar shifts are inferred with limited data on the Mississippi River-influenced shelf with the increase in two pelagic species in bycatch from shrimp trawls and a decrease in some demersal species. Commercial fisheries in the Black Sea declined as eutrophication led to the loss of macroalgal habitat and oxygen deficiency, amid the possibility of overfishing. After the mid-1970s, benthic fish populations (e.g., turbot) collapsed, and pelagic fish populations (small pelagic fish, such as anchovy and sprat) started to increase. The commercial fisheries diversity declined from about 25 fished species to about five in 20 years (1960s to 1980s), while anchovy stocks and fisheries increased rapidly. The point on the continuum of increasing nutrients versus fishery yields remains vague as to where benefits are subsumed by environmental problems that lead to decreased landings or reduced quality of production and biomass.

Future Expectations

The continued and accelerated export of nitrogen and phosphorus to the world's coastal ocean is the trajectory to be expected unless societal intervention takes place (in the form of controls or changes in culture). The largest increases are predicted for southern and eastern Asia, associated with predicted large increases in population, increased fertilizer use to grow food to meet the dietary demands of that population, and increased industrialization. The implications for coastal eutrophication and subsequent ecosystem changes such as worsening conditions of oxygen depletion are significant.

See also

Ecosystem Effects of Fishing. Fiordic Ecosystems. Upwelling Ecosystems.

Further Reading

Díaz RJ, Nestlerode J, and Díaz ML (2004) A global perspective on the effects of eutrophication and hypoxia on aquatic biota. In Rupp GL and White MD (eds.) *Proceedings of* the 7th International *Symposium on Fish Physiology, Toxicology and Water Quality*, EPA 600/R-04/049, pp. 1–33. Tallinn, Estonia, 12–15 May 2003. Athens, GA: Ecosystems Research Division, US EPA.

Díaz RJ and Rosenberg R (1995) Marine benthic hypoxia: A review of its ecological effects and the behavioural responses of benthic macrofauna. *Oceanography and Marine Biology Annual Review* 33: 245–303.

Gray JS, Wu RS, and Or YY (2002) Review. Effects of hypoxia and organic enrichment on the coastal marine environment. *Marine Ecology Progress Series* 238: 249–279.

Hagy JD, Boynton WR, and Keefe CW (2004) Hypoxia in Chesapeake Bay, 1950–2001: Long-term change in relation to nutrient loading and river flow. *Estuaries* 27: 634–658.

Helly J and Levin LA (2004) Global distributions of naturally occurring marine hypoxia on continental margins. *Deep-Sea Research* 51: 1159–1168.

Mee LD, Friedrich JJ, and Gomoiu MT (2005) Restoring the Black Sea in times of uncertainty. *Oceanography* 18: 100–111.

Rabalais NN and Turner RE (eds.) (2001) *Coastal and Estuarine Studies 58: Coastal Hypoxia – Consequences for Living Resources and Ecosystems*. Washington, DC: American Geophysical Union.

Rabalais NN, Turner RE, and Scavia D (2002) Beyond science into policy: Gulf of Mexico hypoxia and the Mississippi River. *BioScience* 52: 129–142.

Rabalais NN, Turner RE, Sen Gupta BK, Boesch DF, Chapman P, and Murrell MC (2007) Characterization and long-term trends of hypoxia in the northern Gulf of Mexico: Does the science support the Action Plan? *Estuaries and Coasts* 30(supplement 5): 753–772.

Turner RE, Rabalais NN, and Justić D (2006) Predicting summer hypoxia in the northern Gulf of Mexico: Riverine N, P and Si loading. *Marine Pollution Bulletin* 52: 139–148.

Tyson RV and Pearson TH (eds.) *Geological Society Special Publication No. 58: Modern and Ancient Continental Shelf Anoxia*, 470pp. London: The Geological Society.

Relevant Website

http://www.gulfhypoxia.net
– Hypoxia in the Northern Gulf of Mexico.

POPULATION GENETICS OF MARINE ORGANISMS

D. Hedgecock, University of Southern California, Los Angeles, CA, USA

Introduction

This article provides a brief overview of the principles of population genetics and its applications in ocean science. The specialized vocabulary of genetics is defined, and central concepts and approaches are summarized in an abbreviated historical context. Finally, specific topics that have been addressed in the marine biological literature illustrate major areas of application of population genetics in ocean science.

Definitions and Historical Approaches

Population genetics is the branch of genetics that explores the consequences of Mendelian inheritance at the level of populations, rather than families. Population genetics seeks to explain the most elementary step of Darwinian evolution, change in population genetic composition over time. Genetic composition is described by the frequencies, in population samples, of 'alleles' (alternative states or forms of a particular gene or genetic marker that arise by mutation or recombination), 'phenotypes' (the perceivable properties of organisms, such as their color or the banding patterns of their macromolecules following electrophoretic separation), or 'genotypes' (the hereditary constitutions of individuals, usually inferred from their phenotypes), at particular genes or 'loci' (singular 'locus', the place on a chromosome where a genetic element is located). 'Genes' are functional units of inheritance, whether they code for the synthesis of proteins or RNA or serve some regulatory or structural purpose, but nonfunctional portions of the genome are often used as genetic markers. The zygotes of most animals and plants are 'diploid', inheriting one copy of each gene from their parents. A diploid genotype, comprising two alleles that are identical either phenotypically or by descent, is said to be 'homozygous'. A genotype comprising two different alleles is 'heterozygous'. The proportion of heterozygous genotypes in a population, the 'heterozygosity', is an important measure of genetic diversity. Genetically determined traits or markers are said to be 'polymorphic' when two or more forms or alleles occur in a population at frequencies greater than expected from recurrent mutation. Conversely, a 'monomorphic' or 'fixed' genetic trait or marker is represented by a single phenotype within a population comprising only a single allele (homozygotes).

The primary forces that change genetic composition over time are recurrent mutations, natural or artificial selection, gene flow among subpopulations, and random changes resulting from sampling errors in finite populations. In the absence of these forces, the genetic composition of a 'randomly mating' population (i.e., one in which all possible matings or unions of gametes are equally likely) is stable through time and is described by the Hardy–Weinberg (H–W) principle. Under this basic principle, the frequencies of genotypes at a single, polymorphic locus are given by the expansion of the binomial (with two alleles) or multinomial (with three or more alleles) function of allelic frequencies. For example, at a locus with two alleles of equal selective value, A and a, having relative frequencies p and q ($p+q=1.0$) in a large population, the frequencies of the genotypes, AA, Aa, and aa, are given by the binomial expansion, $(p+q)^2 = p^2+2pq+q^2$, respectively. Populations conforming to this principle are said to be in H–W equilibrium. The H–W principle, which is easily extended to multiple alleles, simplifies the description of populations, by reducing the number of parameters to n alleles at a locus, rather than the $n(n+1)$ genotypes formed by n alleles in diploid organisms.

There are three distinct and complementary approaches in population genetics. Empirical studies seek to document patterns of genetic variation in natural populations and to explain these patterns in terms of the primary forces acting on population genetic variation and the population biology of the species in question. Experimental studies seek to test the effects of inbreeding and crossbreeding, natural and artificial selection, random genetic drift owing to population bottlenecks or finite population size, or to estimate the frequency and fitness effects of spontaneous mutations. Finally, theoretical studies continue to develop and elaborate the mathematical backbone of population genetics set down in the seminal works of Sewall Wright, Sir Ronald Fisher, and J.B.S. Haldane published in the early 1930s.

In the marine sciences, the bulk of population genetic studies fall into the empirical category. The earliest population genetic studies, carried out by Charles Bocquet in France and Bruno Battaglia in Italy in the 1950s and 1960s, described color polymorphisms in natural populations of copepods and isopods. The introduction, in the late 1960s, of methods for electrophoretic separation of allelic protein variants (or 'allozymes') allowed investigation of genetic variation in organisms not amenable to laboratory culture or lacking visible polymorphisms. Among the first allozyme studies of marine organisms were those by Fred Utter and colleagues on Pacific salmon, which provided a scientific basis for management of salmon ocean fisheries. In the 1980s, methods for the analysis of polymorphism in DNA sequences were developed, first for mitochondrial DNA but eventually for single copy nuclear genes. Today, the highly abundant and polymorphic, di-, tri-, and tetra-nucleotide tandem repeat elements known as microsatellites are the most widely used DNA markers, though less polymorphic but even more abundant single nucleotide polymorphisms (SNPs), which are amenable to high-throughput genotyping technologies, promise genome-wide coverage in the near future.

Experimental genetic studies were initiated by Bocquet, Battaglia, and others working on small and easily cultured crustaceans but have exploded over the past two decades in conjunction with the growth of aquaculture. In this arena, molecular markers are being used not only to document variation in populations but also to construct linkage maps, which when combined with experimental analysis of complex phenotypes, such as growth and survival, permit the mapping of genes controlling variance in these quantitative traits (known as quantitative trait loci or QTL mapping). Application of technologies for gene-expression profiling and even whole genome sequencing in species with well-developed experimental methods are likely to increase greatly our understanding of phenotypic evolution in marine populations in the near future.

Over the past three decades, biological oceanographers and fisheries scientists have had a growing interest in the theory and practice of population genetics, particularly as enhanced by molecular methods. Major scientific questions being addressed by population geneticists working with marine animals and plants can be grouped under five headings: (1) identification of morphologically cryptic, sibling species; (2) amount and spatial structure of genetic diversity within species; (3) temporal genetic change, particularly in relation to recruitment; (4) retrospective analyses of historical collections; and (5) phylogenetic and phylogeographic analyses, which aim, respectively, to determine the evolutionary relationships among species or higher taxa and to reveal the geographical distribution of genes and genealogies within species. Some of the prominent applications of population genetics in the marine environment concern the connectivity among populations, design of marine reserves, aquaculture and stock enhancement and their ecological affects, and mixed-stock analyses in fisheries.

Morphologically Cryptic, Sibling Species

A major contribution of population genetic studies to marine biology has been the identification of sibling or cryptic biological species within morphologically defined taxa. 'Sympatric' (occupying the same geographical area), cryptic species are most often recognized by obvious failures of the H–W principle at one or more markers. If the cryptic species are fixed for alternative alleles at a marker, one observes AA and BB homozygotes only and no or few AB heterozygotes, contrary to expectations for random mating. When cryptic species are 'allopatric' (occupying nonoverlapping areas), inference of biological species status is less clear. Fixed differences at multiple genetic markers suggest an absence of genetic exchange, but additional evidence is needed to ascertain whether the populations are reproductively isolated. Morphologically, cryptic speciation might also be suggested by large divergence in DNA sequences, but, again, ancillary evidence is required to ascertain whether this variation is concordant with reproductive isolation. The percent nucleotide difference maintained within conspecific populations for mitochondrial DNA ranges from a few percent in many species to as much as 18% in the copepod *Tigriopus californicus*.

Cryptic species have been discovered in several well-studied taxa, including the blue mussel *Mytilus*, the worm *Capitella*, and the copepod *Calanus*. These discoveries, mostly serendipitous byproducts of research directed at other questions, suggest that a systematic investigation of the frequency of sibling species in various taxa could eventually increase estimates of marine species diversity by an order of magnitude. Prudence dictates that the taxonomy of all target organisms of oceanographical study should be confirmed by both traditional and molecular methods and that voucher specimens be kept. A major advantage of using DNA markers is that voucher specimens can often be preserved simply in ethanol.

Molecular diagnosis of cryptic species or of the cryptic stages of species should enable the extension of species diagnosis to early life stages. Correct species identification of larval stages should enable, in turn, the acquisition of data on species-specific patterns in dispersal and recruitment.

Amount and Spatial Structure of Genetic Diversity Within Species

Early studies of protein polymorphism revealed widely varying levels of genetic diversity among marine taxa. Comparative studies of taxa differing in life history or ecological traits have provided few compelling general explanations for what maintains different levels of genetic diversity in different taxonomic groups. Certainly, the amount of diversity measured for a particular taxon depends largely on the particular type and set of genetic markers studied. However, the amount of genetic diversity in a species depends more fundamentally on the mating system and the balance among the forces of mutation, migration, selection, and random genetic drift owing to finite population size. We focus on the last factor first.

The H–W principle applies to infinitely large or very large populations. However, this important assumption of the H–W principle is often violated in real populations, which are finite in size. Therefore, to understand the maintenance and structure of genetic diversity in finite natural populations, we require an understanding of the concepts of random genetic drift and 'effective population size' (N_e). Random genetic drift is defined as change in allelic frequency resulting from the sampling of finite numbers of gametes from generation to generation. One way to illustrate genetic drift is to consider the transition matrix that describes the probability of having i A alleles in generation $t+1$, given j A alleles in generation t. The elements of this matrix are calculated from the binomial probability:

$$x_{ij} = \frac{(2N)!}{(2N-i)!\,i!}\left(1-\frac{j}{2N}\right)^{2N-i}\left(\frac{j}{2N}\right)^{i}$$

where $2N$ is the number of alleles and the frequency of A at generation t is $j/2N$.

With this transition matrix, one can calculate how the distribution of allelic frequencies among a collection of populations will evolve. The allele frequency distribution is a vector, giving the proportion, $y_{j,t}$, of populations with j A alleles at generation t, the sum of which is 1.0. We can calculate the change in the allele frequency distribution from one generation to the next by multiplying the matrix, X, of probability transition values (i.e., the x_{ij} values) by the vector, Y, of population states (i.e., the $y_{j,t}$ values), such that $Y_{t+1}=XY_t$. This recursive relationship can be generalized as $Y_t=X_tY_0$, assuming an initial allele frequency distribution Y_0. How allele frequency and heterozygosity evolve by random genetic drift in a collection of small populations, each of size three ($2N=6$), is illustrated in **Table 1**. The initial frequency of the A allele in all populations is 0.5 (i.e., $y_{3,0}=1.0$).

Several important features of genetic drift are illustrated by **Table 1**. (1) The proportion of polymorphic populations (those with one to five A alleles) rapidly declines, until all populations become fixed with either zero or six A alleles. (2) The frequency of A across the collection of all subpopulations remains at the mean of 0.5, so that allelic diversity is conserved across groups. (3) Finally, heterozygosity for the total population declines from the initial H–W equilibrium value, $2pq=2\cdot(0.5)\cdot(0.5)=0.5$, to zero, as subpopulations become fixed. The heterozygosity at generation t can be calculated as

$$H_t = 2\sum_{j=0}^{2N}\left(1-\frac{j}{2N}\right)\left(\frac{j}{2N}\right)y_{j,t}$$

The recursive relationship between heterozygosity at generation t and heterozygosity at generation $t+1$ is readily solved, yielding the important fact that heterozygosity declines from one generation to the next by

Table 1 The evolution by random genetic drift of allele frequencies and heterozygosity (H) in a collection of small populations, each of size three ($2N=6$) with initial frequency of allele A equal to 0.5 (see text for complete discussion)

Number of A alleles	*Generation*						
	0	*1*	*2*	*4*	*8*	...	∞
0	0.0	0.0156	0.0728	0.196	0.353	...	0.5
1	0.0	0.0938	0.1326	0.1094	0.0538	...	0.0
2	0.0	0.2344	0.1893	0.1284	0.0617	...	0.0
3	1.0	0.3125	0.2106	0.1325	0.0631	...	0.0
4	0.0	0.2344	0.1893	0.1284	0.0617	...	0.0
5	0.0	0.0938	0.1326	0.1094	0.0538		0.0
6	0.0	0.0156	0.0728	0.196	0.353		0.5
Mean frequency	0.5	0.5	0.5	0.5	0.5	...	0.5
H	0.5	0.4167	0.3472	0.2411	0.1163	...	0.0

the factor $1/(2N)$. This is what makes population size such an important parameter in population and conservation genetics.

The finite populations discussed above are mathematically ideal populations that differ from real populations in a number of important ways. In the mathematically ideal population, there are equal numbers of both sexes, adults mate at random, and variance in number of offspring per adult is binomial or Poisson. In actual populations, the sexes may not be equal in number, mating may not be at random, or the variance in offspring number may be larger than binomial or Poisson. The effective size, N_e, is the size of a mathematically ideal population that has either the same rate of random genetic drift or the same rate of inbreeding as the actual population under study. Variance and inbreeding effective sizes are similar unless the population experiences a rapid change in size. Here, we are concerned primarily with the variance effective size, which determines the rate at which genetic variants are lost ($p = 0.0$) or fixed ($p = 1.0$) by random genetic drift. The number of adults in the ideal population (N) is, by definition, equal to the effective size, and the ratio, $N_e/N = 1.0$ in the ideal case. Owing to the nonideal properties of real populations, the N_e/N ratio for most terrestrial vertebrate populations is thought to lie between 0.25 and 0.75. However, as we shall see in the next section, this ratio may be much smaller in highly fecund marine species. In actual populations, N_e is substituted for N in theoretical calculations; for example, heterozygosity is lost at a rate of $1/(2N_e)$ per generation.

The spatial structure of genetic variation within a species is, perhaps, the most thoroughly studied aspect of marine population genetics. The question of how variation in mode of larval development, therefore larval dispersal potential, affects the genetics and evolution of marine species was raised over a quarter of a century ago. To investigate this question, it is useful to review the theory of population divergence developed largely by Sewall Wright. We focus primarily on the balance between random genetic drift and migration. The role of diversifying selection in causing populations to diverge is not easily generalized; moreover, the specific consequences of any selection regime can depend on the underlying balance of drift and migration.

The genetic diversity of a species can be partitioned into components within and among population units, ranging from local, randomly mating populations (or demes) to subpopulations to the total species. Wright partitioned genetic variation within a species, using F-statistics, which measure the average genetic correlation between pairs of gametes derived from different levels in a population hierarchy. At the basal level of this hierarchy, the correlation between the two gametes drawn from the same individual is symbolized as F_{IS}. F_{IS} is zero in a randomly mating subpopulation but is positive when there are excesses of homozygotes relative to H–W expectations. Mating among related individuals can cause an excess of homozygotes, in which case F_{IS} is equivalent to the coefficient of inbreeding, f. Let us return to the example of a single locus with two alleles, A and a, having relative frequencies p and q in a particular local population. With partial inbreeding, the frequencies of genotypes, AA, Aa, and aa, are $p^2 + fpq$, $2pq(1\text{-}f)$, and $q^2 + fpq$, respectively. This is a generalization of the H–W principle, in which f (or F_{IS}) governs how alleles associate into genotypes. When f (or F_{IS}) is zero, the proportions of each of the three genotypes are the same as the H–W equilibrium; when f is 1.0, the population contains no heterozygotes. Most sexually reproducing marine populations conform to H–W equilibrium, so that F_{IS} is close to zero. Clonal marine populations, such as some sea anemones and corals, often have nonzero F_{IS}, owing to a mixture of sexual and vegetative propagation of genotypes.

If a species is subdivided into partially isolated, finite subpopulations, mating among individuals in the total population cannot take place at random. At the same time, there is genetic drift within each subpopulation. The effect on the proportion of genotypes in the species is analogous to the effect of inbreeding. Local populations tend toward fixation, with a decline in heterozygosity, but genetic diversity is preserved among rather than within subpopulations (**Table 1**). The genetic correlation between gametes drawn from different individuals within a subpopulation, with respect to allelic frequencies in the total population, is given by $F_{ST} = \sigma_p^2/\bar{p}(1 - \bar{p})$. In this equation, $\bar{p}$ is the average frequency of allele A in the total population and σ_p^2 is the variance of p among subpopulations. Thus, F_{ST} is the ratio of the variance of allelic frequencies among subpopulations to the maximum variance that would be obtained if each subpopulation were fixed for one of the alternative alleles without change in mean allelic frequency. When local populations diverge from one another, there is an excess of homozygotes and a deficiency of heterozygotes, with respect to random mating expectations in the total population. The frequencies of AA, Aa, and aa in the total population are, thus, $\bar{p}^2 + \sigma_p^2$, $2\bar{p}\bar{q} - 2\sigma_p^2$, and $\bar{q}^2 + \sigma_p^2$, respectively. Note the resemblance of these genotypic frequencies to those in an inbreeding population. The principle is readily understood at the extreme, in which each subpopulation is fixed for one allele or another; in this case, there are no heterozygotes in the total population. This partitioning of genetic diversity can be extended to any number of hierarchical levels.

In the absence of gene exchange among local populations of finite size, F_{ST} increases through time as local populations become fixed through the process of random genetic drift (**Table 1**). Local fixation can be prevented, however, by 'gene flow', the exchange of gametes or individuals among local populations. Suppose that a large population is subdivided into many subpopulations, with an average allele frequency, $\bar{p}$, and that each subpopulation exchanges a proportion, m, of its population with a random sample of the whole population every generation. This is Wright's well-known "island" model of population structure. If we consider a single subpopulation, the change in allelic frequency per generation will be $\Delta p = -m(p - \bar{p})$. Although this model is not realistic (e.g., exchange among all subpopulations is not likely to be equal), it illustrates the principle that the frequency of an allele in a subpopulation will be pulled toward the mean frequency of immigrants. When m is small, the equilibrium between genetic drift and gene flow in the island model is approximated by $F_{ST} = 1/(4Nm + 1)$, which leads to a simple estimate of Nm, the absolute number of migrants exchanged among demes. Such estimates from data on real populations must be regarded as provisional, however, as the assumption of equilibrium underlying the estimate are likely rarely, if ever, met.

Gene flow is a powerful cohesive force holding populations together. Indeed, in the island model, the exchange of one migrant every other generation ($Nm = 0.5$) is theoretically sufficient to prevent the chance loss or fixation of an allele with an initial mean frequency of 0.5 in most subpopulations. However, owing to departures from the ideal populations in the island model, the exchange of thousands of individuals per generation between neighboring subpopulations of a widely distributed species could be insufficient to prevent considerable random drift. Moreover, the balance between genetic drift and gene flow is established over many generations. Divergence (or convergence) following an interruption (or resumption) of gene flow can take hundreds of generations, so that genetic similarity is not sufficient evidence for current gene flow. On the other hand, genetic similarity owing to high gene flow does not mean that subpopulations are demographically linked. Ecological and genetic processes operate over very different time scales.

Spatial structure of genetic diversity or population subdivision is the topic of most interest to oceanographers, because it is tied to the hope that the geographic sources of recruitment to marine animal populations might be identified by their genetic makeup. However, for most species of interest, those that comprise the zooplankton broadly speaking, this hope is not well founded in logic or fact. Dispersal among geographic populations, perhaps even low levels of dispersal, can eliminate the very genetic differences among populations that would permit identification of provenance. The marine population genetics literature supports the generalization that species with dispersing, planktotrophic (feeding) larvae have, as expected, much less geographic variation or population subdivision than species with poorly dispersing, lecithotrophic (nonfeeding) larvae. F_{ST} is generally much less than 0.05 for marine species with planktonic larvae.

The oceanographer's problem is to detect if a sample of zooplankton is a genetic mixture and if so, to determine the contributions of different geographic populations to the mix. This problem has been solved in mixed-stock fisheries, particularly for anadromous species. Sophisticated statistical analysis of genetic data can identify, with accuracy and precision, the relative contributions of discrete salmon stocks to ocean catches. With the advent of highly polymorphic microsatellite DNA markers, it is even possible to assign individuals to their population of origin. However, these methods work well for species like salmon because the source populations are identifiable in space and are genetically distinct, with F_{ST} values generally above 0.05 and as high as 0.5. These same methods are not likely to work for the many marine species that lack obvious spatial genetic structure. Although identification of sources and sinks of zooplanktonic populations is critical to understanding their distribution and abundance, it is unlikely to be achieved with indirect genetic methods alone. The fundamental limitation may be the dispersal biology of such species, which can easily homogenize the frequencies of alleles at all loci, whether allozymes or microsatellite DNA markers.

Studies of spatial genetic variation should be made as a part of any baseline population genetic description of a marine species. As a rule, one should not expect such studies to yield conclusive information about the sources of recruits or the water masses bearing them. Nevertheless, any given study might reveal an exception to this rule or a previously unrecognized barrier to dispersal that has resulted in a major genetic subdivision.

Temporal Genetic Change

Despite the generalization that marine species with planktonic dispersal tend to be genetically homogeneous over large geographic regions, allelic frequencies among samples, taken sometimes on scales of meters, often show unpredictable patchiness. Planktonic larvae and patterns of recruitment are thought to play a role in such paradoxical observations. Temporal genetic studies, which are few, further suggest that as much genetic variation can be observed among recruits to a single place as can be observed among adult populations on spatial scales of 100s to 1000s of kilometer. Thus, the population genetics of marine planktonic larvae or new recruits are spatially and temporally much more dynamic than expected, given that their adult populations are large and well connected by larval dispersal. This anomalous feature of marine population genetics, like patterns of larval settlement themselves, may be attributed to the dynamics of the ocean environment.

Genetic heterogeneity on fine spatial scales and temporal change may both be explained by the hypothesis that highly fecund marine organisms may have a large variance in individual reproductive success. This variance in reproductive success may result from a sweepstakes-chance matching of reproductive effort with oceanographic conditions conducive to spawning, fertilization, larval survival, and recruitment. According to this hypothesis, only small fractions (from 1/100 to 1/100 000) of spawning adults effectively reproduce and replace standing adult populations each generation. In this case, ratios of N_e/N may be far less than those observed in terrestrial animals (0.25–0.75), and random genetic drift of allelic frequencies should be measurable in some populations. This prediction has been borne out by temporal studies, in which observed genetic drift implies effective population sizes that are many orders of magnitude less than the simple abundance of adults. Studies of larval populations and comparisons of recruits and adults confirm a second prediction, that specific cohorts of larvae or recruits should show genetic evidence of having been produced by only a segment of the potential parental pool and may even show significant levels of full- or half-sib relatedness.

This hypothesis establishes a connection between oceanography and population genetics in the study of recruitment and may explain how local adaptations and speciation can occur in seemingly large and well-mixed marine populations. Temporal genetic studies are therefore required to make sense of population genetic structure.

Retrospective Analyses of Historical Collections

Enzymatic amplification of specific DNA sequences by the polymerase chain reaction (PCR) can potentially be applied to the study of preserved zooplankton or fish scale collections. Molecular methods could facilitate rapid systematic treatment of these collections and establish genetic histories for particular species of interest. The power of such an approach is illustrated by a genetic study of Georges Bank on haddock populations, which revealed significant heterogeneity in the frequencies of mitochondrial DNA genotypes between the 1975 and 1985 year-classes. Temporal change in this case was attributed to immigration rather than variance in reproductive success. On the other hand, in a study of New Zealand red snapper, for which dried scales were available from 1950 to 1986 and fresh material from 1998, microsatellite DNA analysis showed that loss of genetic diversity and changes in allele frequencies in the Tasman Bay stock of nearly 7 million fish were consistent with an effective population size of 176 individuals. The authors attribute this very low N_e/N ratio to successful breeding by relatively few adults.

Phylogenetics, Phylogeography, and Paleoceanography

Molecular genetic analyses are being used to reconstruct phylogenies or evolutionary relationships at all taxonomic levels and for many phyla. The application of phylogenetic methods to molecular as well as organismal traits, such as morphology, life history, and behavior, will undoubtedly shed new light on the evolution and systematics of marine organisms. Comparisons of genetic divergence within and between closely related species separated by barriers of known age (e.g., the Bering Strait or the Isthmus of Panama) are useful for calibrating rates of molecular evolution and reconstructing the history of faunal exchanges. Ultimately such studies may provide a basis for the development of biotechnological tools for rapid and automated classification of oceanographic samples and collections.

Within the past decade, the application of phylogenetic approaches to molecular variation within species has yielded new insights into population histories and a synthesis of the traditionally separate disciplines of systematics and population genetics. Studies of mitochondrial DNA in several species living along the Gulf of Mexico and Atlantic coasts of the US have revealed concordant patterns of major genetic subdivisions. In all of these species, Gulf genotypes give way to Atlantic genotypes across a previously recognized biogeographical boundary in southeastern Florida. Gulf and Atlantic genotypes represent clades that separated over a million years ago. Thus, 'phylogeographic' patterns reflect the persistence of historical events in the gene pools of organisms. Such information is relevant to management and conservation efforts and begs for interdisciplinary, paleoceanographic explanation. Other biogeographic boundaries should be similarly studied to assess whether they generally coincide with genetic discontinuities. A review of genetic differences between conspecific populations on either side of Point Conception, California, failed to find a consistent pattern of discontinuity like that associated with the Florida peninsula. Nevertheless, past dispersal, rather than present oceanographic connections, appears to be the message written into the genetics of contemporary populations.

See also

Population Dynamics Models. Population Genetics of Marine Organisms.

Further Reading

Avise JC (2004) *Molecular Markers, Natural History, and Evolution*, 2nd edn., 684pp. Sunderland, MA: Sinauer.

Hauser L, Adcock GJ, Smith PJ, Bernal Ramirez JH, and Carvalho GR (2002) Loss of microsatellite diversity and low effective population size in an overexploited population of New Zealand snapper (*Pagrus auratus*). *Proceedings of the National Academy of Sciences of the United States of America* 99: 11742–11747.

Hedgecock D, Launey S, Pudovkin AI, Naciri Y, Lapegue S, and Bonhomme F (2007) Small effective number of parents (N_b) inferred for a naturally spawned cohort of juvenile European flat oysters *Ostrea edulis*. *Marine Biology* 150: 1173–1182.

Hedrick PW (2005) *Genetics of Populations*, 3rd edn., 737pp. Boston, MA: Jones and Bartlett.

Hellberg ME, Burton RS, Neigel JE, and Palumbi SR (2002) Genetic assessment of connectivity among marine populations. *Bulletin of Marine Science* 70(supplement S): 273–290.

Luikart G and England PR (1999) Statistical analysis of microsatellite DNA data. *Trends in Ecology and Evolution* 14: 253–256.

Luikart G, England PR, Tallmon D, Jordan S, and Taberlet P (2003) The power and promise of population genomics: From genotyping to genome typing. *Nature Reviews Genetics* 4: 981–994.

Whitlock MC and McCauley DE (1999) Indirect measures of gene flow and migration: $F_{ST} \neq 1/(4Nm+1)$. *Heredity* 82: 117–125.

MARINE ALGAL GENOMICS AND EVOLUTION

A. Reyes-Prieto, H. S. Yoon, and D. Bhattacharya,
University of Iowa, Iowa City, IA, USA

Introduction: The Genomic Perspective

The understanding of oceanic biodiversity and ecology has been revolutionized by the study of complete nuclear genome sequences. Genomic approaches have helped to elucidate ecological-geochemical processes like the carbon cycle and the metabolism of microorganisms that are the key players in oceanic ecosystems. Genomics also has been used to understand gene expression patterns during massive algal 'blooms', the evolution, diversification, and ecology of photosynthetic eukaryotes, and to provide insights into niche adaptation at the genomic level. However, the vast majority of sequenced genomes of marine photosynthetic microorganisms comprises cyanobacteria (prokaryotes) and the genomes of only a handful of marine algae such as the diatoms *Thalassiosira pseudonana* and *Phaeodactylum tricornutum*, the pelagophyte *Aureococcus anophagefferens*, the haptophyte *Emiliania huxleyi*, and the green algae *Ostreococcus tauri* and *Ostreococcus lucimarinus* are available in the public databases. Genomes of nonmarine algae that have been sequenced include the thermophilic red alga *Cyanidioschyzon merolae* and the green algal genetic model *Chlamydomonas reinhardtii*. Numerous algal genome projects are however currently underway, such as for the red algae *Galdieria sulphuraria*, *Porphyra purpurea*, *Chondrus crispus*, the glaucophyte *Cyanophora paradoxa*, the green algae *Volvox carteri*, *Dunaliella salina*, *Micromonas pusilla* (two ecotypes), and *Bathycoccus* sp., the heterokont alga *Ectocarpus siliculosus*, the cryptophyte *Guillardia theta*, the chlorarachniophyte amoeba *Bigelowiella natans*, a number of marine diatoms (*Pseudonitzchia multiseries* and *Fragilariopsis cylindrus*), and the haptophytes *Emiliania huxleyi*, and *Phaeocystis* (two species). All of these taxa were chosen because of their fundamental roles in the marine phytoplankton, their toxicity, or central position in understanding algal evolution.

Photosynthesis in Eukaryotes: The Origin of Plastids

It has now been clearly documented that the eukaryotic photosynthetic organelle (plastid) originated through endosymbiosis, whereby a single-celled protist (host) engulfed and retained a photosynthetic cyanobacterium (endosymbiont). This primary endosymbiosis is believed to have given rise to the plastid in the common ancestor of the Plantae, which comprises red, green (including plants), and glaucophyte algae (**Figure 1**). Molecular divergence time estimations indicate that the primary endosymbiosis is an ancient event in eukaryote evolution, likely having occurred in the late Paleoproterozoic (*c.* 1.5×10^9 years ago). The establishment of the primary plastid involved critical evolutionary steps, such as the emergence of metabolic connections between the host cytoplasm and the endosymbiont. Molecular phylogenetic analyses suggest that some host proteins involved in the transport of nutrients across biological membranes (translocators) were relocated to the endosymbiont inner membrane and allowed the movement of photosynthesis products (sugars, nutrients) to the host. This was an essential step toward establishment of the symbiotic relationship (**Figure 2**). Another key step for plastid establishment was the

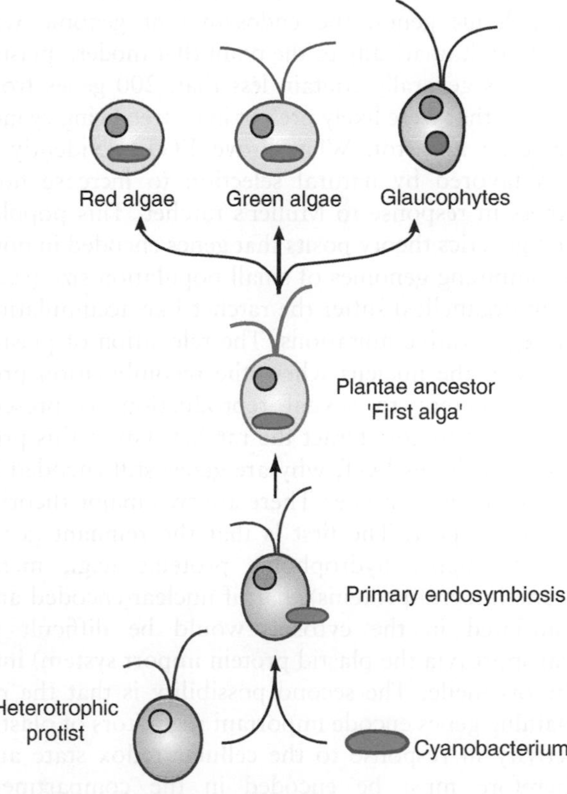

Figure 1 Origin of the plastid in the common ancestor of the Plantae through primary endosymbiosis.

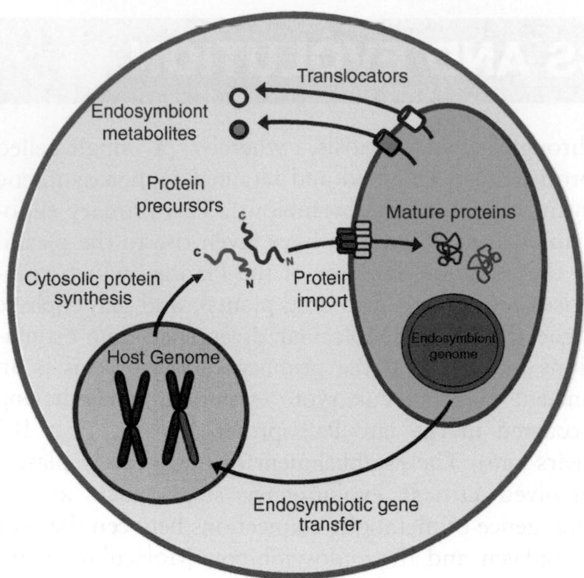

Figure 2 Key processes that occur after primary plastid endosymbiosis include endosymbiotic gene transfer (EGT), the evolution of a plastid protein import machinery (TIC-TOC translocons), and a system (translocators) for the regulated movement of metabolites across the plastid membranes.

transfer and integration of genes from the cyanobacterium to the host nucleus (endosymbiotic gene transfer, EGT). As a result of EGT and the loss of superfluous genes, the endosymbiont genome was reduced dramatically to the point that modern plastid genomes generally contain less than 200 genes from *c.* 4000 that were likely present in its free-living cyanobacterial ancestor. What drove EGT? Evidently it was favored by natural selection to increase host fitness in response to Muller's ratchet. This population genetics theory posits that genes encoded in nonrecombining genomes of small population size (such as in organelles) suffer the ratchet-like accumulation of degenerative mutations. The relocation of plastid genes to the nucleus where the recombination process, inherent to sexual reproduction, is present would act to counteract the ratchet. Given this process that drives EGT, why are genes still encoded in the organelle genome? There are two major theories in this respect. The first is that the remnant genes encode highly hydrophobic proteins (e.g., membrane-integral proteins) that, if nuclear-encoded and translated in the cytosol, would be difficult to transport (via the plastid protein import system) into the organelle. The second possibility is that the remaining genes encode important regulators of plastid activity in response to the cellular redox state and therefore must be encoded in the compartment where their gene products carry out their function. This need for co-localization of genes and their function (the *co*-localization for *r*edox *r*egulation (CORR) hypothesis) may place limits on EGT for core redox regulatory proteins such as the reaction center subunits of photosystems I and II (e.g., plastid-encoded *psa*A, *psb*A).

Another fundamental issue that needs to be clarified to understand early algal genome evolution is quantifying cyanobacterial EGT to the nucleus of the first photosynthetic eukaryotes. The first systematic analysis of EGT was done using complete genome data from the flowering plant *Arabidopsis thaliana*. Molecular evolutionary analyses of the *c.* 25 000 predicted *Arabidopsis* nuclear proteins suggested that EGT left a sizeable mark on the Plantae that far exceeds the lateral transfer of photosynthetic capacity. Apparently, *c.* 18% (4500/25 000) of *Arabidopsis* nuclear genes originated from the ancestral cyanobacterium through EGT and, remarkably, many of these genes have evolved nonphotosynthetic or nonplastid functions. A recent comparative genomic analysis of partial genome data from the early diverging glaucophyte algae *Cyanophora paradoxa*, considered a 'living fossil' among Plantae, provided results regarding EGT that are markedly different than in *Arabidopsis*. In *Cyanophora*, only about 11% (*c.* 1500 genes) of the 12 000–15 000 estimated genes have a cyanobacterial (endosymbiotic) origin. Most of the identified cyanobacterial derived genes in *Cyanophora* have a plastid-related function (90%). These results indicate that the early cyanobacterial contribution to the nuclear genome was shaped by selective forces to retain the photosynthetic endosymbiont. A concomitant key event in endosymbiois was the evolution of the plastid import machinery to transport into the organelle the protein products of those endosymbiont genes relocated to the nucleus by EGT (**Figure 2**). It is likely that the first protein import mechanism evolved from the host endomembrane system (derived from the secretory pathway), and later in evolution the sophisticated and well-known translocon complexes of the modern plastids appeared (TIC-TOC). It seems that once these (and other) intricate events had occurred and the first populations of photosynthetic eukaryotes were established, the path was set for the rise and diversification of the algae.

Secondary Endosymbiosis and the Chromalveolate Algae

The evolution of oxygenic photosynthesis in eukaryotes resulted in an enormous collection of descendant lineages that are critical for primary production in terrestrial and oceanic ecosystems.

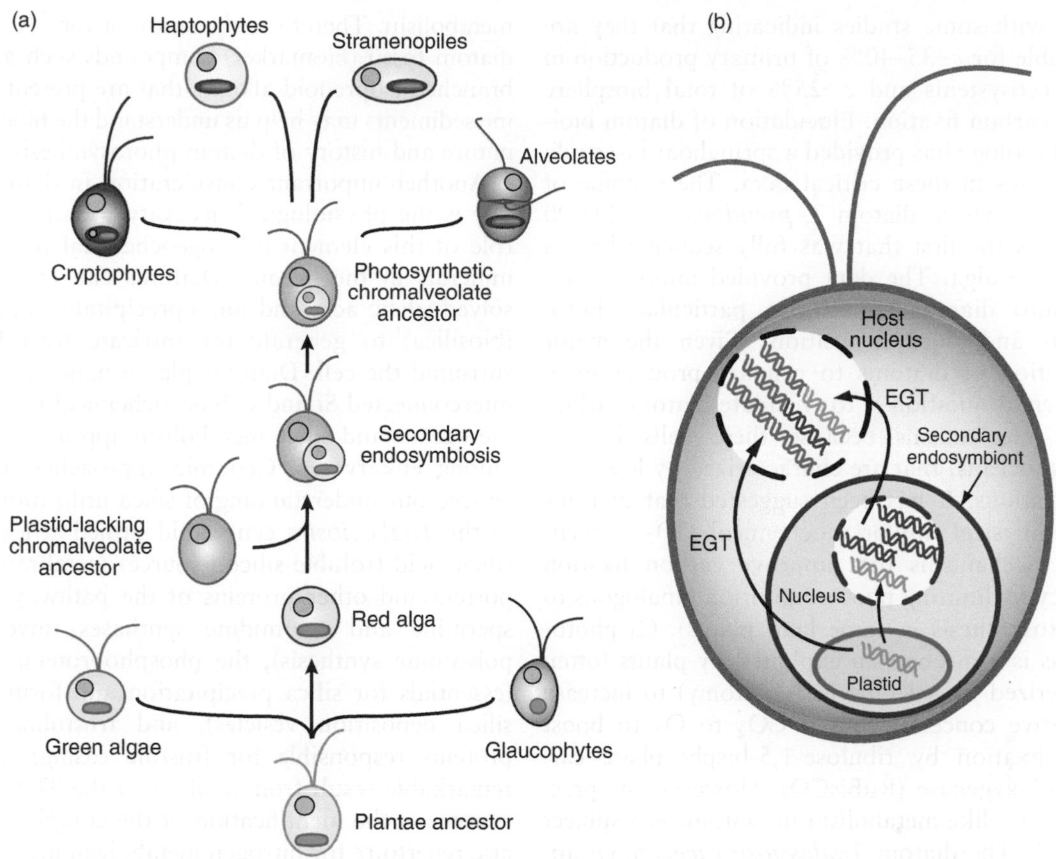

Figure 3 Evolution of chromalveolate plastids. (a) Origin of the plastid in the putative common ancestor of chromalveolates through secondary endosymbiosis. (b) The different types of gene transfers that occurred after the origin of the secondary plastid in chromalveolates.

Primary plastid origin set the wheels in motion for another major event in eukaryotic evolution – red algal secondary endosymbiosis. This is a different type of plastid acquisition in which a non-photosynthetic protist engulfed a eukaryotic red alga and retained the plastid (**Figure 3(a)**). This event is believed to have occurred *c.* 1.2×10^9 years ago in the photosynthetic ancestor of the Chromalveolata (see below). The chromalveolates are a putative monophyletic group of eukaryotes that originally comprised the following chlorophyll-*c*-containing algae and their nonphotosynthetic or plastid-lacking descendants: cryptophytes, haptophytes, stramenopiles, dinoflagellates, ciliates, and apicomplexans. Recent studies have however identified other protist lineages (e.g., katablepharids, telonemids, and Rhizaria) that also appear to be members of the Chromalveolata.

Secondary endosymbiosis also involved EGT, but now with a new level of complexity: genes were transferred from the nucleus of the red alga (secondary endosymbiont) to the host genome (**Figure 3(b)**). The massive loss of unnecessary or redundant genes and EGT to the host nucleus resulted in the diminution or complete disappearance of the red-algal endosymbiont nucleus. Only the cryptophyte algae retain a remnant of this diminished nucleus that is located between the plastid membranes and is referred to as the nucleomorph. The gain of photosynthesis by chromalveolates was a key point in the Earth's history because many of these algae now comprise the ecologically dominant photosynthetic eukaryotes in the oceans (i.e., diatoms, haptophytes, and dinoflagellates).

The Diatoms

Diatoms (Bacilliarophyta) with *c.* 20 000 described and as many as 100 000 estimated species are photosynthetic unicellular organisms characterized by the silica cell wall called a frustule. They are the most diverse group within stramenopiles and are phylogenetically related to brown algae, kelps, and nonphotosynthetic oomycetes. The diatoms play a fundamental role in energy webs in the worlds' oceans, with some studies indicating that they are responsible for *c.* 35–40% of primary production in marine ecosystems and *c.* 25% of total biosphere organic carbon fixation. Elucidation of diatom biology and ecology has provided a springboard to study the genomes in these critical taxa. The genome of the centric marine diatom *T. pseudonana* (>11 000 genes) was the first that was fully sequenced from any marine alga. The data provided important insights into diatom biology, in particular photosynthesis and silica deposition. Given the major contribution of diatoms to primary production, a fundamental question is to elucidate diatom carbon fixation mechanisms because these cells live in habitats (oceans) that are characterized by low CO_2 concentrations. It has been suggested that diatoms possess physical and/or biochemical CO_2 concentration mechanisms that improve carbon fixation efficiency by limiting photorespiration (analogous to C_4 photosynthesis in some land plants). C_4 photosynthesis is a mechanism exploited by plants (often characterized by a Kranz leaf anatomy) to increase the relative concentration of CO_2 to O_2 to boost carbon fixation by ribulose-1,5-bisphosphate carboxylase/oxygenase (RuBisCO). However, the presence of a C_4-like metabolism in diatoms is a subject of debate. The diatom *Thalassiosira weissflogii* apparently possesses a special form of C_4-like photosynthesis that has a central role under low CO_2 concentrations. The genomic data are however ambiguous in this respect because *T. pseudonana* contains the complete set of genes required for C_4-like photosynthesis, but the localization of essential enzymes for concentrating CO_2 in C_4-type photosynthesis, the carbonic anhydrases (α and γ types), in the cytoplasm is not consistent with typical C_4 photosynthesis. Recent experimental studies suggest that photosynthetic metabolism in diatoms is diverse and some diatoms perform exclusively C_3-like metabolism (more efficient in nonlimiting CO_2 conditions) and others are capable of 'intermediate' C_3–C_4 metabolism. In this context, comparative genomics has provided some important clues that can be used to direct experiments aimed at clarifying the details of photosynthesis in this group of marine algae.

Marine geosciences are also critical to understanding the history of C_3–C_4 metabolism. A typical indicator of photosynthetic activity is the carbon isotopic discrimination (fractionation) of the heavier but rare carbon isotopes ^{13}C and ^{14}C in contrast to the lightest and most abundant ^{12}C. On average, the $^{13}C/^{12}C$ ratio decreases in biologically fixed carbon compounds. Moreover, in C_3 photosynthesis, the ^{13}C depletion is significantly increased with respect to C_4 metabolism. Therefore, the study of the ^{13}C ratio in diatom fossil (biomarker) compounds such as highly branched isoprenoid alkenes that are present in marine sediments may help us understand the biochemical nature and history of diatom photosynthesis.

Another important consideration in diatom biology is the physiological necessity for silica and the role of this element in biogeochemical recycling of minerals in the oceans. Diatoms can take up dissolved silicic acid and microprecipitate silica (SiO_2) (biosilica) to generate the intricate frustules that surround the cell. Diatoms play a major role in the interconnected Si and C biogeochemical recycling in the oceans and silica metabolism appears to be rare among eukaryotes. Genomic approaches have advanced our understanding of silica utilization. Study of the *Thalassiosira* genome identified at least three silicic acid (soluble silicon source) membrane transporters and other proteins of the pathway such as spermine and spermidine synthases (involved in polyamine synthesis), the phosphoproteins silaffins (essentials for silica precipitation and formation of silica deposition vesicles), and frustulins (glycoproteins responsible for frustule casing). Another remarkable result from analysis of the *Thalassiosira* genome is the identification of the complete enzymatic repertoire for nitrogen metabolism involving the urea–ornithine cycle, suggesting that pyrimidine metabolism occurs in the cytoplasm as in heterotrophic eukaryotes. Recent advances and upcoming data in diatom genomics should significantly increase our understanding of the biological processes that govern primary production in the oceans.

The Haptophytes

The Haptophyta comprises *c.* 500 mostly marine algal species, many of which are characterized by calcite (calcium carbonate) scales that cover the cell (coccoliths). In particular, *E. huxleyi*, which is a highly abundant haptophyte in the ocean, has been considered a critical component of marine environments because of its dual capacity to fix environmental carbon via biomineralization (calcium carbonate, calcite) and through photosynthesis. The haptophytes having coccoliths are responsible for *c.* 50% of the calcium carbonate precipitation in oceans. The impact from *Emiliania* blooms is great enough to have a possible influence on global climate. Occasionally these blooms reach a size of *c.* 100 000 km^2 with cell accumulations that can affect local climate. This is due to the optical properties of coccoliths that can reflect a significant amount of sunlight and heat from the water. The study of coccolith mineralization (coccolithogenesis) has also been analyzed using gene expression patterns in *E. huxleyi* cultures raised under calcifying and noncalcifying conditions. This work has identified proteins that are putatively related to coccolithogenesis. Much work still has to be done with candidate proteins to verify their roles in coccolith formation because many lack identifiable homologs in the public sequence databases. Like other chromalveolates (e.g., diatoms), haptophytes contain a secondary plastid derived from a red algal endosymbiont with chlorophyll *c* as the principal photopigment. Despite their biological importance, genome-scale data have only now started to emerge with the *E. huxleyi* genome that is currently being sequenced to completion. This information will be critical for bioinformatically annotating *Emiliania* genes with their putative functions and testing these results using functional genomics approaches in this nascent algal model.

Dinoflagellates

Dinoflagellates, along with ciliates and apicomplexans (intracellular parasites; e.g., *Plasmodium*) form the protist group Alveolata that is subsumed into the Chromalveolata. Ciliates have apparently lost outright the ancestral plastid they shared with other chromalveolates, whereas the apicomplexans lost the photosynthetic capacity but retain a vestigial plastid (apicoplast) as a compartment for carrying out other plastid functions such as fatty acid biosynthesis. The dinoflagellates comprise *c.* 2000 species and the vast majority are free-living cells, with autotrophic and heterotrophic lifestyles, although many parasitic taxa also exist.

The photosynthetic dinoflagellates play a fundamental role as primary producers in coastal waters. One well-known example is the mutualistic association between some dinoflagellates of the genus *Symbiodinium* and the reef-forming corals. Under optimal environmental conditions, symbiotic dinoflagellates provide >90% of their total photosynthetic production to the coral host. It is believed that this mutualistic association influenced the diversification of cnidarian species since the late Triassic. The collapse of the symbiosis produces the so-called coral bleaching that is due to the loss of the dinoflagellate photopigments in the coral cells. Bleaching is a major cause of coral mortality, reduced fecundity, and increased disease vulnerability, with serious effects on reef ecosystems. The success of the symbiotic relationship is influenced by several environmental factors, and it has been suggested that elevated temperature, irradiance, and pathogen infections disrupt the symbiosis. Understanding the physiological process that governs the dinoflagellate–coral symbiosis is of high priority in marine ecology and conservation biology. An obvious goal in the genomics era is to identify the key molecular processes that lead to the establishment, maintenance, and loss of the dinoflagellate–coral symbiosis, with the ultimate goal being to identify the genes that underlie this interaction. Preliminary results of gene expression profiles with the model anemone *Anthopleura elegantissima* suggest that the mutualism is not governed by a few symbiotic genes in the coral cell, but by particular expression patterns of common major cellular processes. Proteins mediating cell–cell interactions (e.g., sym32) have been proposed to play a role in the symbiosis by facilitating the engulfment of the dinoflagellate by cnidarian gastrodermal cells. What about the contribution of the symbiont? There are numerous questions to be addressed in this regard but these require extensive genomic data from dinoflagellates, which as we discuss below is not a trivial DNA sequencing task.

Dinoflagellate marine species like *Alexandrium* and *Karenia* are infamous for producing potent neurotoxins in so-called 'red tides'. Red tides are one form of harmful algal bloom (HAB) and pose serious threats to humans, marine mammals, seabirds, fish, and many other organisms that ingest the toxin. The presence of toxins is not however always related to photopigment concentration and high cell densities that can cause the red coloring of seawater. One of the most common dinoflagellate toxic compounds is the saxitoxin (1000 times more potent than cyanide) produced typically by members of the genus *Alexandrium*. The genetic and biochemical nature of the saxitoxin biosynthetic pathway is controversial because there are closely related toxic and nontoxic strains of *Alexandrium*. It has been proposed that the ability to produce saxitoxins stems from symbiotic bacteria and is not intrinsic to the dinoflagellate genic repertoire. However, saxitoxin production apparently remains when symbiotic bacteria are removed from culture. This issue that has remarkable ecological and economic consequences remains a fundamental unresolved aspect of dinoflagellate biology. Ongoing functional genomic analysis of *Alexandrium*, *Karenia*, and other HAB-forming dinoflagellates will ultimately clarify the environmental and genetic factors that underlie bloom formation and toxin production.

However, any attempt to study dinoflagellates at the genomic level has to deal with the large amount of DNA present in their nucleus. Dinoflagellates contain 3–250 pg DNA/cell corresponding to approximately

3000–215 000 megabases (millions of base pairs, Mb), which is 1 or 2 orders of magnitude larger than the typical model organism; for example, the haploid human genome is of 3180 Mb. It has been suggested that polyploidy or polyteny may account for the massive cellular DNA content in dinoflagellates, but molecular studies do not support this hypothesis. It is therefore unclear why these algae possess such large genomes, in particular because their genetic complexity does not appear to be commensurate with their DNA content. Flow cytometry investigations suggest some members of the genus *Symbiodinium* might be targets to genome sequencing in the future. Other promising windows for algal researchers are the picoeukaryote-size dinoflagellates (<3 μm), which probably have genomes more adequate for exhaustive sequencing. Finally, another reasonable approach to take with dinoflagellate genomics is to sequence single chromosomes. It is evident therefore that there are serious practical limitations to genomic studies in dinoflagellates. However, numerous smaller-scale (e.g., expressed sequence tags, ESTs) genomic approaches have been applied to these taxa and have provided important insights into dinoflagellate evolution.

Dinoflagellate Plastid Evolution

Here we discuss some recent insights into dinoflagellate plastid evolution that have resulted from EST investigations. If we assume that there was a single ancestral red algal plastid that unites the chromalveolates, then this organelle has been independently lost in some lineages (e.g., ciliates, oomycetes, telonemids). Plastid loss is not rare in nature, but dinoflagellates show multiple examples of this phenomenon as well as the unique ability to recruit new algal plastids through tertiary endosymbiosis (see below).

The most common type of plastid in dinoflagellates contains peridinin as the major carotenoid photopigment. Remarkably, the plastid genome of peridinin-containing dinoflagellates is highly reduced and broken up into minicircles typically containing a single gene. The minicircles encode core subunits of the photosynthetic machinery (*atp*A, *atp*B, *pet*B, *pet*D, *psa*A, *psa*B, *psb*A–E (consistent with the CORR hypothesis described above)) and other proteins of unknown function (ycf16, ycf24, rpl28, and rpl23) as well as rRNA and tRNA. This unique arrangement is in stark contrast to typical algal and plant genomes that encode 100–200 genes. Analysis of an EST data set from the peridinin-containing dinoflagellate *Alexandrium tamarense* identified 15 genes in the nucleus of this species that are always plastid-encoded in other photosynthetic eukaryotes. Clearly these nuclear genes are the result of EGT from the plastid to the nuclear genome. The forces behind this unprecedented migration of universally conserved components of the plastid genome to the nucleus are not yet fully understood. Unlike other eukaryotes that have reduced plastid genomes because of a loss of photosynthesis due to a parasitic lifestyle, the peridinin dinoflagellates have drastically reduced their plastid genome but are generally free-living and retain photosynthetic capacity.

Equally surprising is the observation that a handful of *Alexandrium* genes encoding plastid-targeted proteins (*hem*B, *tuf*A) trace their origin to green algae, and not to red algae as expected under the chromalveolate hypothesis. These data may indicate a 'hidden' green algal endosymbiosis in the dinoflagellates (and potentially other chromalveolates) or, alternatively, multiple independent cases of horizontal gene transfer (HGT) from green algae. The phagotrophic capacity of dinoflagellates would provide an explanation for the latter hypothesis. Under this scenario the dinoflagellates recruit genes via HGT from algal (or other) food sources that are brought into their cells. Ongoing comparative genomics and phylogenomic studies demonstrate that endosymbiosis and HGT have played a critical role in shaping the genomes of dinoflagellates. Although fascinating in terms of genome evolution, these processes complicate interpretation of gene data and necessitate great care when assigning dinoflagellate DNA sequences to their putative sources of origin.

Plastid Replacement through Tertiary Endosymbiosis

The dinoflagellates are also unique in their ability to take up plastids through repeated endosymbioses. There are several dinoflagellates that have replaced the ancestral chromalveolate plastid of red algal origin with one from other chromalveolates (i.e., cryptophytes, diatoms, or haptophytes) or green algae through a process termed tertiary endosymbiosis (**Figure 4**). A prominent example of tertiary endosymbiosis is the red tide dinoflagellate *Karenia brevis* that has acquired a plastid from a haptophyte endosymbiont. A direct consequence of this plastid acquisition is that *Karenia* now has fucoxanthin (uniquely derived from the haptophyte) as a plastid photopigment, instead of the ancestral peridinin. Comparative genomic analysis of the fucoxanthin-containing dinoflagellates *Karenia* and *Karlodinium micrum* provide key insights into tertiary endosymbiosis. One important observation is that most (and perhaps all) of the nuclear genes that encode plastid-targeted proteins that were inherited from the

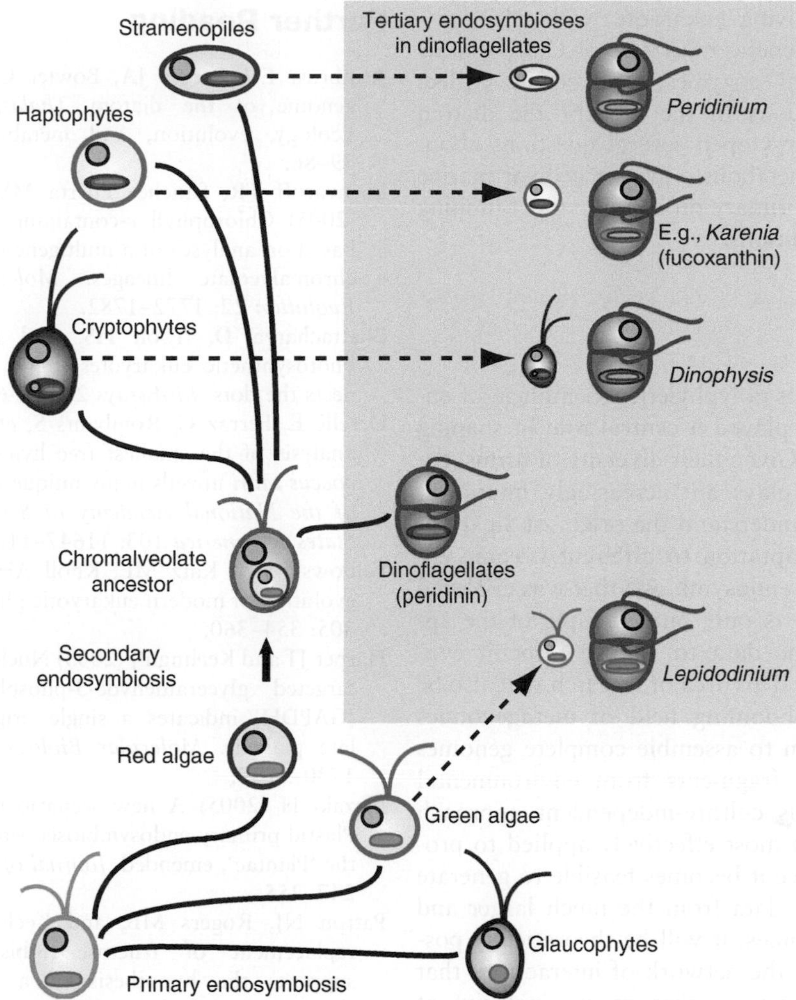

Figure 4 Tertiary endosymbiosis in dinoflagellates occurred independently in different lineages and involved evolutionarily distantly related algal endosymbionts.

red algal secondary endosymbiont via EGT have been replaced with the haptophyte homologs. This underlines the dynamic evolutionary history of dinoflagellate genomes that may be an adaptation to generate genic diversity through their mixotrophic lifestyle.

Picoeukaryotes: 'Hidden' Biodiversity in the World's Oceans

In recent years it has become apparent that a vast unexplored diversity of algal forms exist in oceans as prokaryote-sized cells (0.2–3 μm of cell mean diameter) as part of the eukaryotic picoplankton (picoeukaryotes). Marine picoeukaryotes are found in all of the major algal groups (e.g., green algae, haptophytes, stramenopiles, and dinoflagellates) and have often been uncovered using environmental (meta)genomics or environmental polymerase chain reaction (PCR) approaches in which DNA sequences are determined from environmental samples without culturing particular isolates. These data demonstrate the existence of an extraordinary biodiversity of oceanic picoeukaryotes although their potential contribution to major biogeochemical processes is largely unknown. Studies of the global distribution of the ubiquitous marine green algal picoeukaryotes *Ostreococcus* spp. and *Micromonas pusilla* (both are early-diverging prasinophyte green algae) suggest the existence of ecotypes (populations adapted to a particular ecological niche). These isolates are excellent models to study genome-wide adaptation to particular habitats with the added benefit that they have highly reduced genomes (i.e., resulting in reduced sequencing costs). Taking advantage of these aspects, the nuclear genome has been recently sequenced from three different *Ostreococcus* isolates. *Ostreococcus tauri* possesses a genome of *c.* 12.6 Mb, which is one of the smallest known for a free-living eukaryote, with 8166 predicted genes. The genetic repertoire of *Ostreococcus* includes the typical enzymes involved in C_4-like photosynthesis, and as in the case of the diatom *T. pseudonana*, leaves open several questions about the details of this metabolism in a unicellular marine alga and its role in primary production under limiting light and CO_2 conditions.

Summary

Algae are an ancient polyphyletic assemblage of eukaryotes that have played a central role in shaping the Earth's history. Given their diversity of forms, the study of genomes plays an increasingly important role in helping us understand the origin of algal lineages and their adaptation to different oceanic environments. Plastid endosymbiosis that was explored in most detail here is only one example of the application of genome data to resolve difficult evolutionary problems. This area of research will also be facilitated by the booming field of metagenomics that has as one aim to assemble complete genomes and large genome fragments from environmental DNA samples. This culture-independent approach has until now been most effectively applied to prokaryotic DNA. Once it becomes feasible to generate useful metagenome data from the much larger and complex algal genomes, it will be theoretically possible to reconstruct the network of interactions that define particular marine environments, inclusive of prokaryotes, eukaryotes, and viruses. The current acceleration in DNA-sequencing technologies that will soon provide individual researchers with hundreds of billions of base pairs of sequence information promises to make this dream a reality. Given therefore the breathtaking pace of advances in the field of genomics it is hard to predict where we will be 10 years from now. It is clear however that marine biologists and oceanographers will have unparalleled opportunities to study the ecology, diversification, and evolution of photosynthetic eukaryotes.

See also

Phytoplankton Blooms.

Further Reading

Armbrust EV, Berges JA, Bowler C, *et al.* (2004) The genome of the diatom *Thalassiosira pseudonana*: Ecology, evolution, and metabolism. *Science* 306: 79–86.

Bachvaroff TR, Sanchez Puerta MV, and Delwiche CF (2005) Chlorophyll *c*-containing plastid relationships based on analyses of a multigene data set with all four chromalveolate lineages. *Molecular Biology and Evolution* 22: 1772–1782.

Bhattacharya D, Yoon HS, and Hackett JD (2004) Photosynthetic eukaryotes unite: Endosymbiosis connects the dots. *BioEssays* 26: 50–60.

Derelle E, Ferraz C, Rombauts S, *et al.* (2006) Genome analysis of the smallest free living eukaryote *Ostreococcus tauri* unveils many unique features. *Proceedings of the National Academy of Sciences of the United States of America* 103: 11647–11652.

Falkowski PG, Katz ME, Knoll AH, *et al.* (2004) The evolution of modern eukaryotic phytoplankton. *Science* 305: 354–360.

Harper JT and Keeling PJ (2003) Nucleus-encoded, plastid-targeted glyceraldehyde-3-phosphate dehydrogenase (GAPDH) indicates a single origin for chromalveolate plastids. *Molecular Biology and Evolution* 20: 1730–1735.

Nozaki H (2005) A new scenario of plastid evolution: Plastid primary endosymbiosis before the divergence of the 'Plantae', emended. *Journal of Plant Research* 118: 247–255.

Patron NJ, Rogers MB, and Keeling PJ (2004) Gene replacement of fructose-1,6-bisphosphate aldolase supports the hypothesis of a single photosynthetic ancestor of chromalveolates. *Eukaryotic Cell* 3: 1169–1175.

Reyes-Prieto A, Weber AP, and Bhattacharya D (2007) The origin and establishment of the plastid in algae and plants. *Annual Review of Genetics* 41: 147–168.

Rodriguez-Ezpeleta N, Brinkmann H, Burey SC, *et al.* (2005) Monophyly of primary photosynthetic eukaryotes: Green plants, red algae, and glaucophytes. *Current Biology* 15: 1325–1330.

Takishita K, Ishida K, and Maruyama T (2004) Phylogeny of nuclear-encoded plastid-targeted GAPDH gene supports separate origins for the peridinin- and the fucoxanthin derivative-containing plastids of dinoflagellates. *Protist* 155: 447–458.

Yoon HS, Hackett JD, Pinto G, and Bhattacharya D (2002) The single, ancient origin of chromist plastids. *Proceedings of the National Academy of Sciences of the United States of America* 99: 15507–15512.

CLIMATE IMPACTS

PLANKTON AND CLIMATE

A. J. Richardson, University of Queensland, St. Lucia, QLD, Australia

Introduction: The Global Importance of Plankton

Unlike habitats on land that are dominated by massive immobile vegetation, the bulk of the ocean environment is far from the seafloor and replete with microscopic drifting primary producers. These are the phytoplankton, and they are grazed by microscopic animals known as zooplankton. The word 'plankton' derives from the Greek *planktos* meaning 'to drift' and although many of the phytoplankton (with the aid of flagella or cilia) and zooplankton swim, none can progress against currents. Most plankton are microscopic in size, but some such as jellyfish are up to 2 m in bell diameter and can weigh up to 200 kg. Plankton communities are highly diverse, containing organisms from almost all kingdoms and phyla.

Similar to terrestrial plants, phytoplankton photosynthesize in the presence of sunlight, fixing CO_2 and producing O_2. This means that phytoplankton must live in the upper sunlit layer of the ocean and obtain sufficient nutrients in the form of nitrogen and phosphorus for growth. Each and every day, phytoplankton perform nearly half of the photosynthesis on Earth, fixing more than 100 million tons of carbon in the form of CO_2 and producing half of the oxygen we breathe as a byproduct.

Photosynthesis by phytoplankton directly and indirectly supports almost all marine life. Phytoplankton are a major food source for fish larvae, some small surface-dwelling fish such as sardine, and shoreline filter-feeders such as mussels and oysters. However, the major energy pathway to higher trophic levels is through zooplankton, the major grazers in the oceans. One zooplankton group, the copepods, is so numerous that they are the most abundant multicellular animals on Earth, outnumbering even insects by possibly 3 orders of magnitude. Zooplankton support the teeming multitudes higher up the food web: fish, seabirds, penguins, marine mammals, and turtles. Carcasses and fecal pellets of zooplankton and uneaten phytoplankton slowly yet consistently rain down on the cold dark seafloor, keeping alive the benthic (bottom-dwelling) communities of sponges, anemones, crabs, and fish.

Phytoplankton impact human health. Some species may become a problem for natural ecosystems and humans when they bloom in large numbers and produce toxins. Such blooms are known as harmful algal blooms (HABs) or red tides. Many species of zooplankton and shellfish that feed by filtering seawater to ingest phytoplankton may incorporate these toxins into their tissues during red-tide events. Fish, seabirds, and whales that consume affected zooplankton and shellfish can exhibit a variety of responses detrimental to survival. These toxins can also cause amnesic, diarrhetic, or paralytic shellfish poisoning in humans and may require the closure of aquaculture operations or even wild fisheries.

Despite their generally small size, plankton even play a major role in the pace and extent of climate change itself through their contribution to the carbon cycle. The ability of the oceans to act as a sink for CO_2 relies largely on plankton functioning as a 'biological pump'. By reducing the concentration of CO_2 at the ocean surface through photosynthetic uptake, phytoplankton allow more CO_2 to diffuse into surface waters from the atmosphere. This process continually draws CO_2 into the oceans and has helped to remove half of the CO_2 produced by humans from the atmosphere and distributed it into the oceans. Plankton play a further role in the biological pump because much of the CO_2 that is fixed by phytoplankton and then eaten by zooplankton sinks to the ocean floor in the bodies of uneaten and dead phytoplankton, and zooplankton fecal pellets. This carbon may then be locked up within sediments.

Phytoplankton also help to shape climate by changing the amount of solar radiation reflected back to space (the Earth's albedo). Some phytoplankton produce dimethylsulfonium propionate, a precursor of dimethyl sulfide (DMS). DMS evaporates from the ocean, is oxidized into sulfate in the atmosphere, and then forms cloud condensation nuclei. This leads to more clouds, increasing the Earth's albedo and cooling the climate.

Without these diverse roles performed by plankton, our oceans would be desolate, polluted, virtually lifeless, and the Earth would be far less resilient to the large quantities of CO_2 produced by humans.

Beacons of Climate Change

Plankton are ideal beacons of climate change for a host of reasons. First, plankton are ecthothermic (their body temperature varies with the surroundings), so their physiological processes such as nutrient uptake, photosynthesis, respiration, and reproductive development are highly sensitive to temperature, with their speed doubling or tripling with a 10 °C temperature rise. Global warming is thus likely to directly impact the pace of life in the plankton. Second, warming of surface waters lowers its density, making the water column more stable. This increases the stratification, so that more energy is required to mix deep nutrient-rich water into surface layers. It is these nutrients that drive surface biological production in the sunlit upper layers of the ocean. Thus global warming is likely to increase the stability of the ocean and diminish nutrient enrichment and reduce primary productivity in large areas of the tropical ocean. There is no such direct link between temperature and nutrient enrichment in terrestrial systems. Third, most plankton species are short-lived, so there is tight coupling between environmental change and plankton dynamics. Phytoplankton have lifespans of days to weeks, whereas land plants have lifespans of years. Plankton systems will therefore respond rapidly, whereas it takes longer before terrestrial plants exhibit changes in abundance attributable to climate change. Fourth, plankton integrate ocean climate, the physical oceanic and atmospheric conditions that drive plankton productivity. There is a direct link between climate and plankton abundance and timing. Fifth, plankton can show dramatic changes in distribution because they are free floating and most remain so their entire life. They thus respond rapidly to changes in temperature and oceanic currents by expanding and contracting their ranges. Further, as plankton are distributed by currents and not by vectors or pollinators, their dispersal is less dependent on other species and more dependent on physical processes. By contrast, terrestrial plants are rooted to their substrate and are often dependent upon vectors or pollinators for dispersal. Sixth, unlike other marine groups such as fish and many intertidal organisms, few plankton species are commercially exploited so any long-term changes can more easily be attributed to climate change. Last, almost all marine life has a planktonic stage in their life cycle because ocean currents provide an ideal mechanism for dispersal over large distances. Evidence suggests that these mobile life stages known as meroplankton are even more sensitive to climate change than the holoplankton, their neighbors that live permanently in the plankton.

All of these attributes make plankton ideal beacons of climate change. Impacts of climate change on plankton are manifest as predictable changes in the distribution of individual species and communities, in the timing of important life cycle events or phenology, in abundance and community structure, through the impacts of ocean acidification, and through their regulation by climate indices. Because of this sensitivity and their global importance, climate impacts on plankton are felt throughout the ecosystems they support.

Changes in Distribution

Plankton have exhibited some of the fastest and largest range shifts in response to global warming of any marine or terrestrial group. The general trend, as on land, is for plants and animals to expand their ranges poleward as temperatures warm. Probably the clearest examples are from the Northeast Atlantic. Members of a warm temperate assemblage have moved more than 1000-km poleward over the last 50 years (**Figure 1**). Concurrently, species of a subarctic (cold-water) assemblage have retracted to higher latitudes. Although these translocations have been associated with warming in the region by up to 1 °C, they may also be a consequence of the stronger northward flowing currents on the European shelf edge. These shifts in distribution have had dramatic impacts on the food web of the North Sea. The cool water assemblage has high biomass and is dominated by large species such as *Calanus finmarchicus*. Because this cool water assemblage retracts north as waters warm, *C. finmarchicus* is replaced by *Calanus helgolandicus*, a dominant member of the warm-water assemblage. This assemblage typically has lower biomass and contains relatively small species. Despite these *Calanus* species being indistinguishable to all but the most trained eye, the two species contrast starkly in their seasonal cycles: *C. finmarchicus* peaks in spring whereas *C. helgolandicus* peaks in autumn. This is critical as cod, which are traditionally the most important fishery of the North Sea, spawn in spring. As cod eggs hatch into larvae and continue to grow, they require good food conditons, consisting of large copepods such as *C. finmarchicus*, otherwise mortality is high and recruitment is poor. In recent warm years, however, *C. finmarchicus* is rare, there is very low copepod biomass during spring, and cod recruitment has crashed.

Changes in Phenology

Phenology, or the timing of repeated seasonal activities such as migrations or flowering, is very sensitive to global warming. On land, many events in spring are happening earlier in the year, such as the arrival of swallows in the UK, emergence of butterflies in the US, or blossoming of cherry trees in Japan. Recent

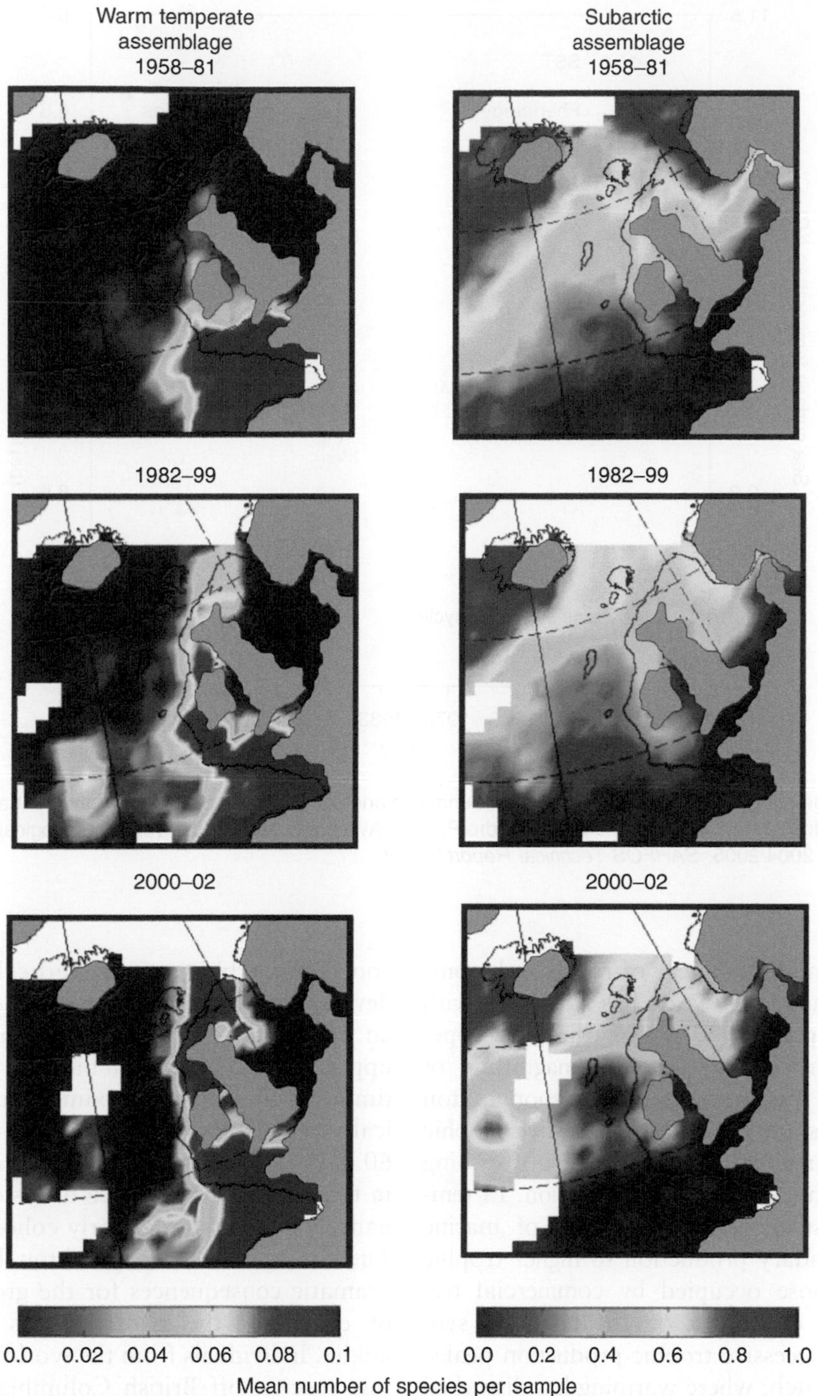

Figure 1 The northerly shift of the warm temperate assemblage (including *Calanus helgolandicus*) into the North Sea and retraction of the subarctic assemblage (including *Calanus finmarchicus*) to higher latitudes. Reproduced by permission from Gregory Beaugrand.

evidence suggests that phenological changes in plankton are greater than those observed on land. Larvae of benthic echinoderms in the North Sea are now appearing in the plankton 6 weeks earlier than they did 50 years ago, and this is in response to warmer temperatures of less than 1 °C. In echinoderms, temperature stimulates physiological developments and larval release. Other meroplankton such as larvae of fish, cirrepedes, and decapods have also responded similarly to warming (**Figure 2**).

Timing of peak abundance of plankton can have effects that resonate to higher trophic levels. In the

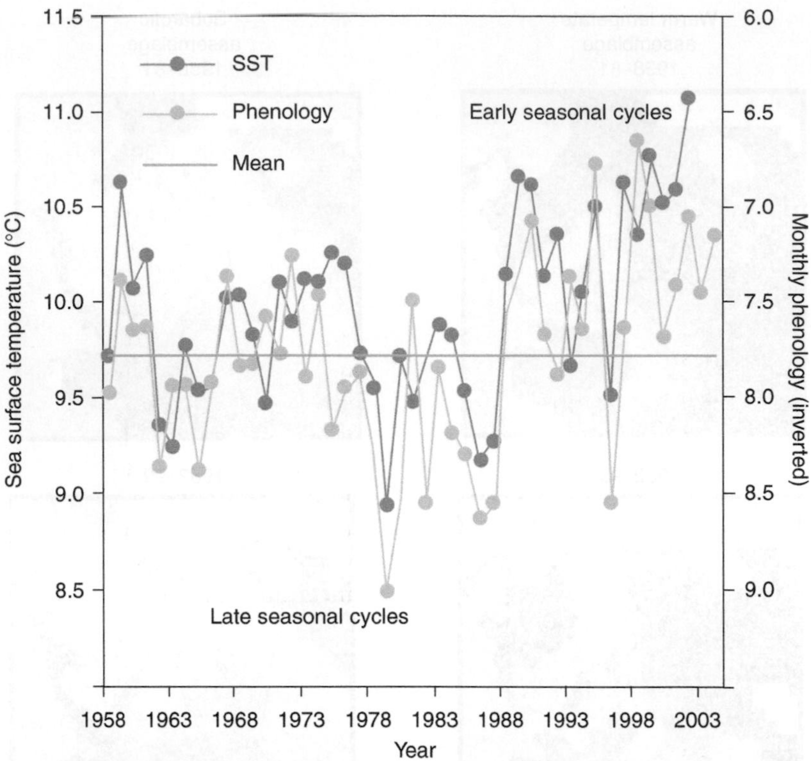

Figure 2 Monthly phenology (timing) of decapod larval abundance and sea surface temperature in the central North Sea from 1958 to 2004. Reproduced from Edwards M, Johns DG, Licandro P, John AWG, and Stevens DP (2006) Ecological status report: Results from the CPR Survey 2004/2005. *SAHFOS Technical Report* 3: 1–8.

North Sea, the timing each year of plankton blooms in summer over the last 50 years has advanced, with phytoplankton appearing 23 days earlier and copepods 10 days earlier. The different magnitude of response between phytoplankton and zooplankton may lead to a mismatch between successive trophic levels and a change in the synchrony of timing between primary and secondary production. In temperate marine systems, efficient transfer of marine primary and secondary production to higher trophic levels, such as those occupied by commercial fish species, is largely dependent on the temporal synchrony between successive trophic production peaks. This type of mismatch, where warming has disturbed the temporal synchrony between herbivores and their plant food, has been noted in other biological systems, most notably between freshwater zooplankton and diatoms, great tits and caterpillar biomass, flycatchers and caterpillar biomass, winter moth and oak bud burst, and the red admiral butterfly and stinging nettle. Such mismatches compromise herbivore survival.

Dramatic ecosystem repercussions of climate-driven changes in phenology are also evident in the subarctic North Pacific Ocean. Here a single copepod species, *Neocalanus plumchrus*, dominates the zooplankton biomass. Its vertical distribution and development are both strongly seasonal and result in an ephemeral (2-month duration) annual peak in upper ocean zooplankton biomass in late spring. The timing of this annual maximum has shifted dramatically over the last 50 years, with peak biomass about 60 days earlier in warm than cold years. The change in timing is a consequence of faster growth and enhanced survivorship of early cohorts in warm years. The timing of the zooplankton biomass peak has dramatic consequences for the growth performance of chicks of the planktivorous seabird, Cassin's auklet. Individuals from the world's largest colony of this species, off British Columbia, prey heavily on *Neocalanus*. During cold years, there is synchrony between food availability and the timing of breeding. During warm years, however, spring is early and the duration of overlap of seabird breeding and *Neocalanus* availability in surface waters is small, causing a mismatch between prey and predator populations. This compromises the reproductive performance of Cassin's auklet in warm years compared to cold years. If Cassin's auklet does not adapt to the changing food conditions, then global warming will place severe strain on its long-term survival.

Changes in Abundance

The most striking example of changes in abundance in response to long-term warming is from foraminifera in the California Current. This plankton group is valuable for long-term climate studies because it is more sensitive to hydrographic conditions than to predation from higher trophic levels. As a result, its temporal dynamics can be relatively easily linked to changes in climate. Foraminifera are also well preserved in sediments, so a consistent time series of observations can be extended back hundreds of years. Records in the California Current show increasing numbers of tropical/subtropical species throughout the twentieth century reflecting a warming trend, which is most dramatic after the 1960s (**Figure 3**). Changes in the foraminifera record echo not only increase in many other tropical and subtropical taxa in the California Current over the last few decades, but also decrease in temperate species of algae, zooplankton, fish, and seabirds.

Changes in abundance through alteration of enrichment patterns in response to enhanced stratification is often more difficult to attribute to climate change than are shifts in distribution or phenology, but may have greater ecosystem consequences. An illustration from the Northeast Atlantic highlights the role that global warming can have on stratification and thus plankton abundances. In this region, phytoplankton become more abundant when cooler regions warm, probably because warmer temperatures boost metabolic rates and enhance stratification in these often windy, cold, and well-mixed regions. But phytoplankton become less common when already warm regions get even warmer, probably because warm water blocks nutrient-rich deep water from rising to upper layers where phytoplankton live. This regional response of phytoplankton in the North Atlantic is transmitted up the plankton food web. When phytoplankton bloom, both herbivorous and carnivorous zooplankton become more abundant, indicating that the plankton food web is controlled from the 'bottom up' by primary producers, rather than from the 'top down' by predators. This regional response to climate change suggests that the distribution of fish biomass will change in the future, as the amount of plankton in a region is likely to influence its carrying capacity of fish. Climate change will thus have regional impacts on fisheries.

There is some evidence that the frequency of HABs is increasing globally, although the causes are uncertain. The key suspect is eutrophication, particularly elevated concentrations of the nutrients nitrogen and phosphorus, which are of human origin and discharged into our oceans. However, recent evidence from the North Sea over the second half of the twentieth century suggests that global warming may also have a key role to play. Most areas of the North Sea have shown no increase in HABs, except off southern Norway where there have been more blooms. This is primarily a consequence of the enhanced stratification in the area caused by warmer temperatures and lower salinity from meltwater. In the southern North Sea, the abundance of two key HAB species over the last 45 years is positively related to warmer ocean temperatures. This work

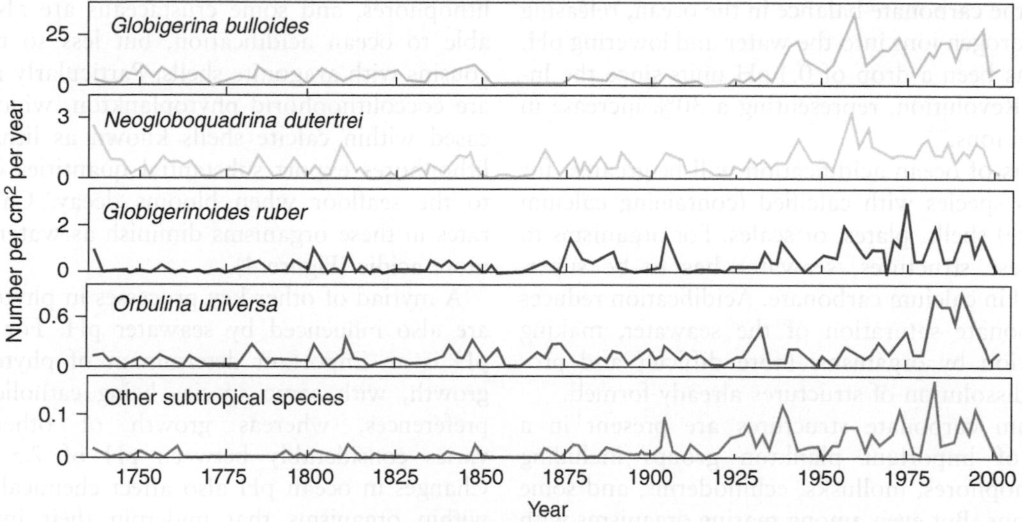

Figure 3 Fluxes of tropical/planktonic foraminifera in Santa Barbara Basin (California Current). Tropical/subtropical foraminifera showing increased abundance in the twentieth century. Reproduced from Field DB, Baumgartner TR, Charles CD, Ferreira-Bartrina V, and Ohman M (2006) Planktonic foraminifera of the California Current reflect 20th century warming. *Science* 311: 63–66.

supports the notion that the warmer temperatures and increased meltwater runoff anticipated under projected climate change scenarios are likely to increase the frequency of HABs.

Although most evidence for changes in abundance in response to climate change are from the Northern Hemisphere because this is where most (plankton) science has concentrated, there is a striking example from waters around Antarctica. Over the last 30 years, there has been a decline in the biomass of krill *Euphausia superba* in the Southern Ocean that is a consequence of warmer sea and air temperatures. In many areas, krill has been replaced by small gelatinous filter-feeding sacs known as salps, which occupy the less-productive, warmer regions of the Southern Ocean. The decline in krill is likely to be a consequence of warmer ocean temperatures impacting sea ice. It is not only that sea ice protects krill from predation, but also the algae living beneath the sea ice and photosynthesizing from the dim light seeping through are a critical food source for krill. As waters have warmed, the extent of winter sea ice and its duration have declined, and this has led to a deterioration in krill density since the 1970s. As krill are major food items for baleen whales, penguins, seabirds, fish, and seals, their declining population may have severe ramifications for the Southern Ocean food web.

Impact of Acidification

A direct consequence of enhanced CO_2 levels in the ocean is a lowering of ocean pH. This is a consequence of elevated dissolved CO_2 in seawater altering the carbonate balance in the ocean, releasing more hydrogen ions into the water and lowering pH. There has been a drop of 0.1 pH units since the Industrial Revolution, representing a 30% increase in hydrogen ions.

Impacts of ocean acidification will be greatest for plankton species with calcified (containing calcium carbonate) shells, plates, or scales. For organisms to build these structures, seawater has to be supersaturated in calcium carbonate. Acidification reduces the carbonate saturation of the seawater, making calcification by organisms more difficult and promoting dissolution of structures already formed.

Calcium carbonate structures are present in a variety of important plankton groups including coccolithophores, mollusks, echinoderms, and some crustaceans. But even among marine organisms with calcium carbonate shells, susceptibility to acidification varies depending on whether the crystalline form of their calcium carbonate is aragonite or calcite. Aragonite is more soluble under acidic conditions than calcite, making it more susceptible to dissolution. As oceans absorb more CO_2, undersaturation of aragonite and calcite in seawater will be initially most acute in the Southern Ocean and then move northward.

Winged snails known as pteropods are probably the plankton group most vulnerable to ocean acidification because of their aragonite shell. In the Southern Ocean and subarctic Pacific Ocean, pteropods are prominent components of the food web, contributing to the diet of carnivorous zooplankton, myctophids, and other fish and baleen whales, besides forming the entire diet of gymnosome mollusks. Pteropods in the Southern Ocean also account for the majority of the annual flux of both carbonate and organic carbon exported to ocean depths. Because these animals are extremely delicate and difficult to keep alive experimentally, precise pH thresholds where deleterious effects commence are not known. However, even experiments over as little as 48 h show shell deterioration in the pteropod *Clio pyrimidata* at CO_2 levels approximating those likely around 2100 under a business-as-usual emissions scenario. If pteropods cannot grow and maintain their protective shell, their populations are likely to decline and their range will contract toward lower-latitude surface waters that remain supersaturated in aragonite, if they can adapt to the warmer temperature of the waters. This would have obvious repercussions throughout the food web of the Southern Ocean.

Other plankton that produce calcite such as foraminifera (protist plankton), mollusks other than pteropods (e.g., squid and mussel larvae), coccolithophores, and some crustaceans are also vulnerable to ocean acidification, but less so than their cousins with aragonite shells. Particularly important are coccolithophorid phytoplankton, which are encased within calcite shells known as liths. Coccolithophores export substantial quantities of carbon to the seafloor when blooms decay. Calcification rates in these organisms diminish as water becomes more acidic (**Figure 4**).

A myriad of other key processes in phytoplankton are also influenced by seawater pH. For example, pH is an important determinant of phytoplankton growth, with some species being catholic in their preferences, whereas growth of other species varies considerably between pH of 7.5 and 8.5. Changes in ocean pH also affect chemical reactions within organisms that underpin their intracellular physiological processes. pH will influence nutrient uptake kinetics of phytoplankton. These effects will have repercussions for phytoplankton community

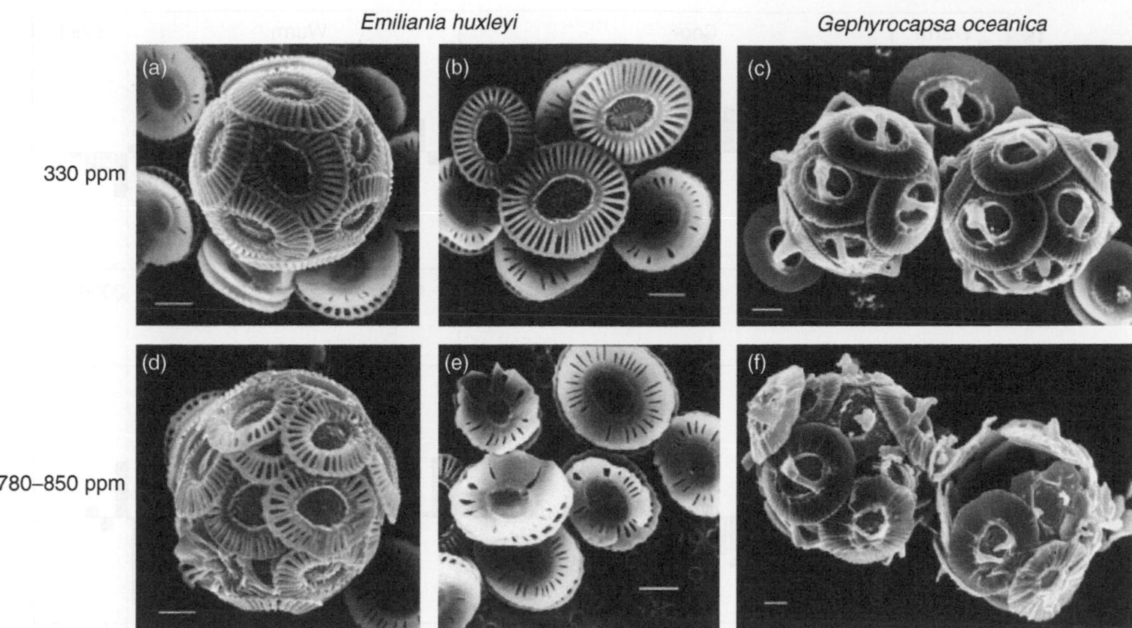

Figure 4 Scanning electron microscopy photographs of the coccolithophores *Emiliania huxleyi* and *Gephyrocapsa oceanica* collected from cultures incubated at CO_2 levels of about 300 and 780–850 ppm. Note the difference in the coccolith structure (including distinct malformations) and in the degree of calcification of cells grown at normal and elevated CO_2 levels. Scale bar = 1 mm. Repinted by permission from Macmillan Publishers Ltd., *Nature*, Riebesell U, Zondervan I, Rost B, Tortell PD, Zeebe RE, and Morel FMM, Reduced calcification of marine plankton in response to increased atmospheric CO_2, 407: 364–376, Copyright (2000).

composition and productivity, with flow-on effects to higher trophic levels.

Climate Variability

Many impacts of climate change are likely to act through existing modes of variability in the Earth's climate system, including the well-known El Niño/Southern Oscillation (ENSO) and the North Atlantic Oscillation (NAO). Such large synoptic pressure fields alter regional winds, currents, nutrient dynamics, and water temperatures. Relationships between integrative climate indices and plankton composition, abundance, or productivity provide an insight into how climate change may affect ocean biology in the future.

ENSO is the strongest climate signal globally, and has its clearest impact on the biology of the tropical Pacific Ocean. Observations from satellite over the past decade have shown a dramatic global decline in primary productivity. This trend is caused by enhanced stratification in the low-latitude oceans in response to more frequent El Niño events. During an El Niño, upper ocean temperatures warm, thereby enhancing stratification and reducing the availability of nutrients for phytoplankton growth. Severe El Niño events lead to alarming declines in phytoplankton, fisheries, marine birds and mammals in the tropical Pacific Ocean. Of concern is the potential transition to more frequent El Niño-like conditions predicted by some climate models. In such circumstances, enhanced stratification across vast areas of the tropical ocean may reduce primary productivity, decimating fish, mammal, and bird populations. Although it is unknown whether the recent decline in primary productivity is already a consequence of climate change, the findings and underlying understanding of climate variability are likely to provide a window to the future.

Further north in the Pacific, the Pacific Decadal Oscillation (PDO) has a strong multi-decadal signal, longer than the ENSO period of a few years. When the PDO is negative, upwelling winds strengthen over the California Current, cool ocean conditions prevail in the Northeast Pacific, copepod biomass in the region is high and is dominated by large cool-water species, and fish stocks such as coho salmon are abundant (**Figure 5**). By contrast, when the PDO is positive, upwelling diminishes and warm conditions exist, the copepod biomass declines and is dominated by small less-nutritious species, and the abundance of coho salmon plunges.

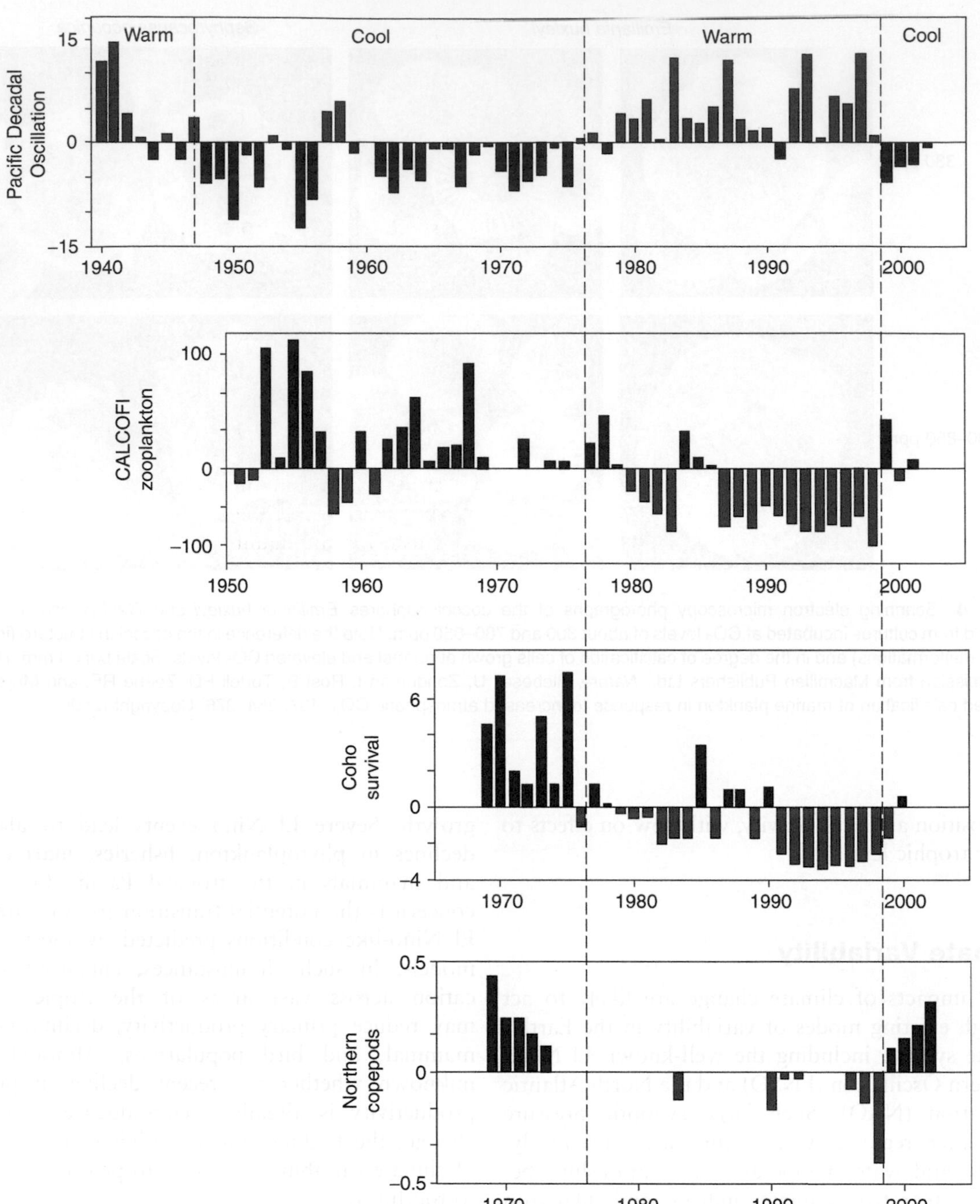

Figure 5 Annual time series in the Northeast Pacific: the PDO index from May to September; anomalies of zooplankton biomass (displacement volumes) from the California Current region (CALCOFI zooplankton); anomalies of coho salmon survival; and biomass anomalies of cold-water copepod species (northern copepods). Positive (negative) PDO index indicates warmer (cooler) than normal temperatures in coastal waters off North America. Reproduced from Peterson WT and Schwing FB (2003) A new climate regime in Northeast Pacific ecosystems. *Geophysical Research Letters* 30(17): 1896 (doi:10.1029/2003GL017528).

These transitions between alternate states have been termed regime shifts. It is possible that if climate change exceeds some critical threshold, some marine systems will switch permanently to a new state that is less favorable than present.

In Hot Water: Consequences for the Future

With plankton having relatively simple behavior, occurring in vast numbers, and amenable to experimental manipulation and automated measurements, their dynamics are far more easily studied and modeled than higher trophic levels. These attributes make it easier to model potential impacts of climate change on plankton communities. Many of our insights gained from such models confirm those already observed from field studies.

The basic dynamics of plankton communities have been captured by nutrient–phytoplankton–zooplankton (NPZ) models. Such models are based on a functional group representation of plankton communities, where species with similar ecological function are grouped into guilds to form the basic biological units in the model. Typical functional groups represented include diatoms, dinoflagellates, coccolithophores, microzooplankton, and mesozooplankton. There are many global NPZ models constructed by different research teams around the world. These are coupled to global climate models (GCMs) to provide future projections of the Earth's climate system. In this way, alternative carbon dioxide emission scenarios can be used to investigate possible future states of the ocean and the impact on plankton communities.

One of the most striking and worrisome results from these models is that they agree with fieldwork that has shown general declines in lower trophic levels globally as a result of large areas of the surface tropical ocean becoming more stratified and nutrient-poor as the oceans heat up (see sections titled 'Changes in abundance' and 'Climate variability'). One such NPZ model projects that under a middle-of-the-road emissions scenario, global primary productivity will decline by 5–10% (**Figure 6**). This trend will not be uniform, with increases in productivity by 20–30% toward the Poles, and marked declines in the warm stratified tropical ocean basins. This and other models show that warmer, more-stratified conditions in the Tropics will reduce nutrients in surface waters and lead to smaller phytoplankton cells dominating over larger diatoms. This will lengthen food webs and ultimately support fewer fish, marine mammals, and seabirds, as more trophic linkages are needed to transfer energy from small phytoplankton to higher trophic levels and 90% of the energy is lost within each trophic level through respiration. It also reduces the oceanic uptake of CO_2 by lowering the efficiency of the

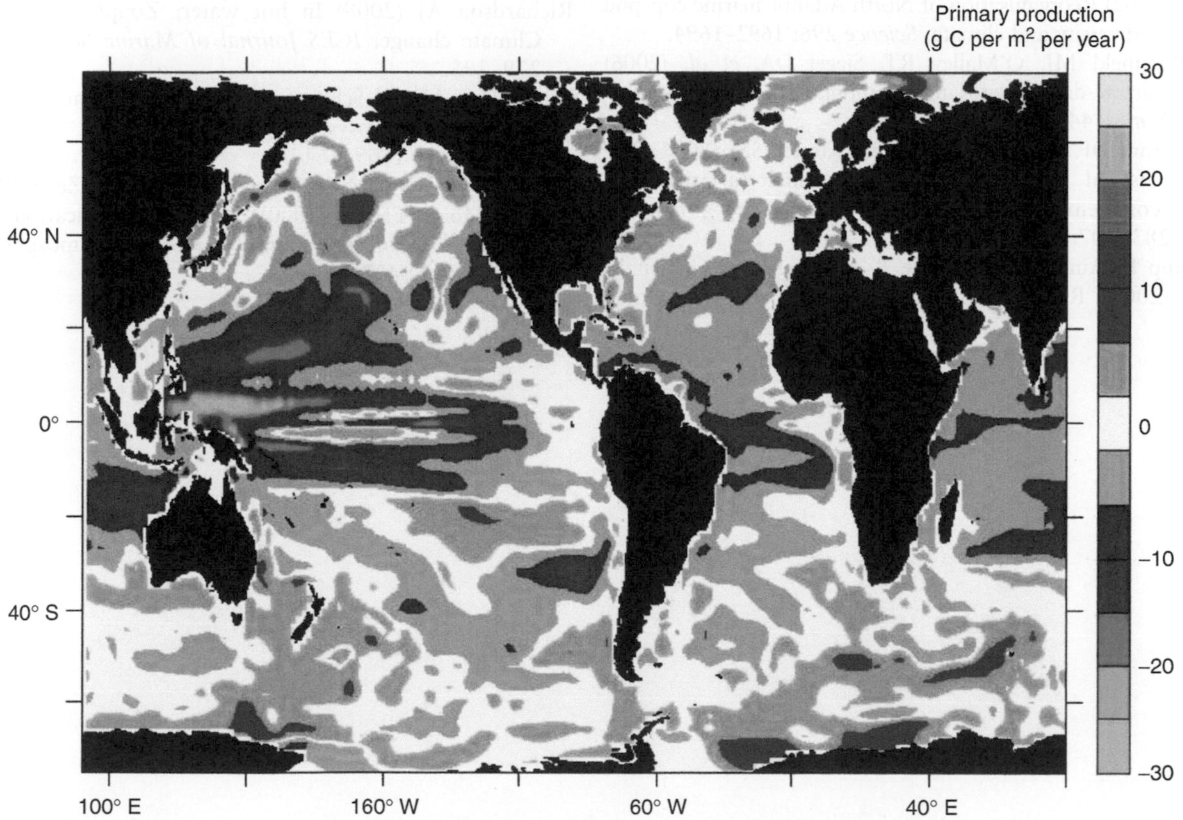

Figure 6 Change in primary productivity of phytoplankton between 2100 and 1990 estimated from an NPZ model. There is a global decline in primary productivity by 5–10%, with an increase at the Poles of 20–30%. Reproduced from Bopp L, Monfray P, Aumont O, *et al.* (2001) Potential impact of climate change on marine export production. *Global Biogeochemical Cycles* 15: 81–99.

biological pump. This could cause a positive feedback between climate change and the ocean carbon cycle: more CO_2 in the atmosphere leads to a warmer and more stratified ocean, which supports less and smaller plankton, and results in less carbon being drawn from surface ocean layers to deep waters. With less carbon removed from the surface ocean, less CO_2 would diffuse into the ocean and more CO_2 would accumulate in the atmosphere.

It is clear that plankton are beacons of climate change, being extremely sensitive barometers of physical conditions. We also know that climate impacts on plankton reverberate throughout marine ecosystems. More than any other group, they also influence the pace and extent of climate change. The impact of climate change on plankton communities will not only determine the future trajectory of marine ecosystems, but the planet.

Further Reading

Atkinson A, Siegel V, Pakhomov E, and Rothery P (2004) Long-term decline in krill stock and increase in salps within the Southern Ocean. *Nature* 432: 100–103.

Beaugrand G, Reid PC, Ibanez F, Lindley JA, and Edwards M (2002) Reorganisation of North Atlantic marine copepod biodiversity and climate. *Science* 296: 1692–1694.

Behrenfield MJ, O'Malley RT, Siegel DA, *et al.* (2006) Climate-driven trends in contemporary ocean productivity. *Nature* 444: 752–755.

Bertram DF, Mackas DL, and McKinnell SM (2001) The seasonal cycle revisited: Interannual variation and ecosystem consequences. *Progress in Oceanography* 49: 283–307.

Bopp L, Aumont O, Cadule P, Alvain S, and Gehlen M (2005) Response of diatoms distribution to global warming and potential implications: A global model study. *Geophysical Research Letters* 32: L19606 (doi:10.1029/2005GL023653).

Bopp L, Monfray P, Aumont O, *et al.* (2001) Potential impact of climate change on marine export production. *Global Biogeochemical Cycles* 15: 81–99.

Edwards M, Johns DG, Licandro P, John AWG, and Stevens DP (2006) Ecological status report: Results from the CPR Survey 2004/2005. *SAHFOS Technical Report* 3: 1–8.

Edwards M and Richardson AJ (2004) The impact of climate change on the phenology of the plankton community and trophic mismatch. *Nature* 430: 881–884.

Field DB, Baumgartner TR, Charles CD, Ferreira-Bartrina V, and Ohman M (2006) Planktonic forminifera of the California Current reflect 20th century warming. *Science* 311: 63–66.

Hays GC, Richardson AJ, and Robinson C (2005) Climate change and plankton. *Trends in Ecology and Evolution* 20: 337–344.

Peterson WT and Schwing FB (2003) A new climate regime in Northeast Pacific ecosystems. *Geophysical Research Letters* 30(17): 1896 (doi:10.1029/2003GL017528).

Raven J, Caldeira K, Elderfield H, *et al.* (2005) *Royal Society Special Report: Ocean Acidification Due to Increasing Atmospheric Carbon Dioxide*. London: The Royal Society.

Richardson AJ (2008) In hot water: Zooplankton and Climate change. *ICES Journal of Marine Science* 65: 279–295.

Richardson AJ and Schoeman DS (2004) Climate impact on plankton ecosystems in the Northeast Atlantic. *Science* 305: 1609–1612.

Riebesell U, Zondervan I, Rost B, Tortell PD, Zeebe RE, and Morel FMM (2000) Reduced calcification of marine plankton in response to increased atmospheric CO_2. *Nature* 407: 364–367.

EFFECTS OF CLIMATE CHANGE ON MARINE MAMMALS

I. Boyd and N. Hanson, University of St. Andrews, St. Andrews, UK

Introduction

The eff'ects of climate change on marine mammals will be caused by changes in the interactions between the physiological state of these animals and the physical changes in their environment caused by climate change. In this article, climate change is defined as a long-term (millennial) trend in the physical climate. This distinguishes it from short-term, regional fluctuations in the physical climate. Marine mammals are warm-blooded vertebrates living in a highly conductive medium often with a steep temperature gradient across the body surface. They also have complex behavioral repertoires that adapt rapidly to changes in the conditions of the external environment. In general, we would expect the changes in the physical environment at the scales envisaged under climate change scenarios to be well within the homeostatic capacity of these species. Effects of climate upon the prey species normally eaten by marine mammals, most of which do not have the same level of homeostatic control to stresses in their physical environment, may be the most likely mechanism of interaction between marine mammals and climate change. However, we should not assume that effects of climate change on marine mammals should necessarily be negative.

Responses to Normal Environmental Variation

Marine mammals normally experience variation in their environment that is very large compared with most variance predicted due to climate change. Examples include the temperature gradients that many marine mammals experience while diving through the water column and the extreme patchiness of the prey resources for marine mammals. Marine mammals in the Pacific have life-histories adapted to transient climatic phenomena such as El Niño, which oscillate every 4 years or so. Consequently, the morphologies, physiologies, behaviors, and life histories of marine mammals will have evolved to cope with this high level of variance. However, it is generally accepted that climate change is occurring too rapidly for the life histories of marine mammals to adapt to longer periods of adverse conditions than are experienced in examples like El Niño.

Marine mammals appear to cope with other longer wavelength oscillations including the North Pacific and North Atlantic Oscillations and the Antarctic Circumpolar Wave and it is possible that their life histories have evolved to cope with this type of long wavelength variation. Nevertheless, nonoscillatory climate change could result in nonlinear processes of change in some of the physical and biological features of the environment that are important to some marine mammal species. Although speculative, obvious changes such as the extent of Arctic and Antarctic seasonal ice cover could affect the presence of essential physical habitat for marine mammals as well as food resources, and there may be other changes in the structure of marine mammal habitats that are less obvious and difficult to both identify and quantify. Changes in the trophic structure of the oceans in ice-bound regions, where the ecology is very reliant on sea ice, may lead the trophic pyramid that supports these top predators to alter substantially. The polar bear is particularly an obvious example of this type of effect where both loss of hunting habitat in the form of sea ice and the potential effects on prey abundance are already having measurable effects upon populations. Similarly, changes in coastal habitats resulting from changes in sea level, changes in run-off and salinity, and changes in nutrient and sediment loads are likely to have important effects on some species of small cetaceans with localized distributions. Many sirenians rely upon seagrass communities and anything that affects the sustainability of this food source is likely to have a negative effect on these species.

Many marine mammal species have already experienced range retraction and population depletion because of direct interaction with man. Monk seals appear to be particularly vulnerable because they rely upon small pockets of beach habitat, many of which are threatened directly by man and also by rising sea level.

Marine mammals are known to be vulnerable to the effects of toxic algal blooms. Toxins may lead to sublethal effects, such as reduced rates of reproduction as well as direct mortality. Several mortality events including coastal whale and dolphin species as well as seals have been attributed to these effects.

Climate change could result in increased frequency of the conditions that lead to such effects, perhaps as a result of interactions between temperature and eutrophication of coastal habitats.

Classifying Effects

A common approach to the assessment of the effects of climate change is to divide these into 'direct' and 'indirect' effects. In this case, 'direct' effects are those associated with changes in the physical environment, such as those that affect the availability of suitable habitat. 'Indirect' effects are those that operate through the agency of food availability because of changes in ecology, susceptibility to disease, changed exposure to pollution, or changes in competitive interactions. Würsig *et al.* in 2002 added a third level of effect which was the result of human activities occurring in response to climate change that tend to increase conflicts between man and marine mammals. This division has little utility in terms of rationalizing the effects of climate change because, in simple terms, the effects will operate ultimately through the availability of suitable habitat. Assessing the effects of climate change rests upon an assessment of whether there is a functional relationship between the availability of suitable habitat and climate and the form of these functional relationships, which will differ between species, has not been determined.

The expansion and contraction of suitable habitat can be affected by a broad range of factors and some of these can operate on their own but others are often closely related and synergistic, such as the combined effect of retraction of sea ice upon the availability of breeding habitat for seals and also for the food chains that support these predators.

Evidence for the Effects of Climate Change on Marine Mammals

There is no strong evidence that current climate change scenarios are affecting marine mammals although there are studies that suggest some typical effects of climate change could affect marine mammal distribution and abundance. There is an increasing body of literature that links apparent variability in marine mammal abundance, productivity, or behavior with climate change processes. However, with the probable exception of those documenting the changes occurring to the extent of breeding habitat for ringed seals within some sections of the Arctic, and the consequences of this also for polar bears, most of these studies simply reflect a trend toward the interpretation of responses of marine mammals to large-scale regional variability in the physical environment, as has already been well documented in the Pacific for El Niño, in terms of climate trends. Long-term trends in the underlying regional ecosystem structure are sometimes extrapolated as evidence of climate change. In few, if any, of these cases is there strong evidence that the physical environmental variability being observed is derived from irreversible trends in climate. Some of the current literature confounds understanding of the responses of marine mammals to regional variability with that of climate change, albeit that an understanding of one may be useful in the interpretation and prediction of the effects of the other.

Based upon records of species from strandings, MacLeod *et al.* in 2005 have suggested that the species diversity of cetaceans around the UK has increased recently and that this may be evidence of range expansion in some species. However, the sample sizes involved are small and there are difficulties in these types of studies accounting for observer effort. This is a common story for marine mammals, and many other marine predators including seabirds, in that, there is a great deal of theory about what the effects of climate change might be but little convincing evidence that backs up these suggestions. Even process studies, involving research on the mechanisms underlying how climate change could affect marine mammals, when considered in detail make a tenuous linkage between the physical variables and the biological response of the marine mammals.

Is Climate Change Research on Marine Mammals Scientific?

Although it is beyond dispute that marine mammals respond to physical changes in habitat suitability, the relationship between a particular effect and the response from the marine mammal is seldom clear. Where data from time series are analyzed, as in the case of Forcada *et al.*, they are used to test *post hoc* for relationships between climate and biological variables. There is a tendency in these circumstances to test for all possible relationships using a range of physical and biological variables. Such *post hoc* testing is fraught with pitfalls because invariably the final apparently statistically significant relationships are not downweighted in their significance by all the other nonsignificant relationships that were investigated alongside those that proved to be statistically significant. Of course, there may be *a priori* reasons for accepting that a particular relationship is true, but the approach to examining time series rarely provides an analysis of the relationships that were not statistically significant or the *a priori* reasons there might be for rejection of these. Consequently, current suggestions from the literature about the potential effects of climate change may be exaggerated because of the strong possibility of the presence of type I and type II statistical error in the assessment process. Moreover, in the great majority of examples, it will be almost impossible to clearly demonstrate effects of climate change, as has been the case with partitioning the variance between a range of causes of the decline of the Steller sea lion (*Eumetopias jubatus*) in the North Pacific and Bering Sea.

Identifying Situations in Which Climate Change is Likely to Have a Negative Effect on Marine Mammals: Future Work

To date, little has been done to build predictive frameworks for assessing the effects of climate change of marine mammals. There have been broad assessments and focused ecological studies but these are a fragile foundation for guiding policy and management, and for identifying populations that are at greatest risk. The resilience of marine mammal populations to climate change will reflect resilience to any other change in habitat quality, that is, it will depend upon the extent of suitable habitat, the degree to which populations currently fill that habitat, the dispersal capacity of the species, and the structure of the current population, including its capacity for increase and demographics. Clearly, populations that are already in a depleted state, or that are dependent upon habitat that is diminishing for reasons other than climate change, will be more vulnerable to the effects of climate change. There are also some, as yet unconvincing, suggestions that habitat degradation may occur through effects of climate upon pollutant burdens.

The general demographic characteristics of marine mammal populations are relatively well known so there are simple ways of assessing the risk to populations under different scenarios of demographic stochasticity, population size, and isolation. An analysis of this type could only provide a very broad guide to the types of effects that could be expected but, whereas no such analysis has been carried out to date, this should be seen as a first step in the risk-assessment process.

The metapopulation structure of many marine mammal populations will affect resilience to climate change and will be reflected in the dispersal capacity of the population. Again, this type of effect could be included within an analysis of the sensitivity of marine mammal populations to climate change under different metapopulation structures. A feature of climate change is that it is likely to have global as well as local effects and the sensitivity to the relative contribution from these would be an important feature of such an analysis.

Further Reading

Atkinson A, Siegel V, Pakhamov E, and Rothery P (2004) Long-term decline in krill stocks and increase in salps within the Southern Ocean. *Nature* 432: 100–103.

Cavalieri DJ, Parkinson CL, and Vinnikov KY (2003) 30-year satellite record reveals contrasting Arctic and Antarctic decadal sea ice variability. *Geophysical Research Letters* 30: 1970 (doi:10.1029/2003GL018031).

Derocher E, Lunn N, and Stirling I (2004) Polar bears in a warming climate. *Integrative and Comparative Biology* 44: 163–176.

Ferguson S, Stirling I, and McLoughlin P (2005) Climate change and ringed seal (*Phoca hispida*) recruitment in western Hudson Bay. *Marine Mammal Science* 21: 121–135.

Forcada J, Trathan P, Reid K, and Murray E (2005) The effects of global climate variability in pup production of Antarctic fur seals. *Ecology* 86: 2408–2417.

Grebmeier J, Overland J, Moore S, *et al.* (2006) A major ecosystem shift in the northern Bering Sea. *Science* 311: 1461–1464.

Green C and Pershing A (2004) Climate and the conservation biology of North Atlantic right whales: The right whale at the wrong time? *Frontiers in Ecology and the Environment* 2: 29–34.

Heide-Jorgensen MP and Lairde KL (2004) Declining extent of open water refugia for top predators in Baffin Bay and adjacent waters. *Ambio* 33: 487–494.

Hunt G, Stabeno P, Walters G, *et al.* (2002) Climate change and control of southeastern Bering Sea pelagic ecosystem. *Deep Sea Research II* 49: 5821–5853.

Laidre K and Heide-Jorgensen M (2005) Artic sea ice trends and narwhal vulnerability. *Biological Conservation* 121: 509–517.

Leaper R, Cooke J, Trathan P, Reid K, Rowntree V, and Payne R (2005) Global climate drives southern right whale (*Eubaena australis*) population dynamics. *Biology Letters* 2 (doi:10.1098/rsbl.2005.0431).

Lusseau RW, Wilson B, Grellier K, Barton TR, Hammond PS, and Thompson PM (2004) Parallel influence of climate on the behaviour of Pacific killer whales and Atlantic bottlenose dolphins. *Ecology Letters* 7: 1068–1076.

MacDonald R, Harner T, and Fyfe J (2005) Recent climate change in the Artic and its impact on contaminant pathways and interpretation of temporal trend data. *Science of the Total Environment* 342: 5–86.

MacLeod C, Bannon S, Pierce G, *et al.* (2005) Climate change and the cetacean community of north-west Scotland. *Biological Conservation* 124: 477–483.

McMahon C and Burton C (2005) Climate change and seal survival: Evidence for environmentally mediated changes in elephant seal *Mirounga leonina* pup survival. *Proceedings of the Royal Society B* 272: 923–928.

Robinson R, Learmouth J, Hutson A, *et al.* (2005) Climate change and migratory species. *BTO Research Report 414*. London: Defra. http://www.bto.org/research/reports/researchrpt_abstracts/2005/RR414%20_summary_report.pdf (accessed Mar. 2008).

Sun L, Liu X, Yin X, Zhu R, Xie Z, and Wang Y (2004) A 1,500-year record of Antarctic seal populations in response to climate change. *Polar Biology* 27: 495–501.

Trillmich F, Ono KA, Costa DP, *et al.* (1991) The effects of El Niño on pinniped populations in the eastern Pacific. In: Trillmich F and Ono KA (eds.) *Pinnipeds and El Niño: Responses to Environmental Stress*, pp. 247–270. Berlin: Springer.

Trites A, Miller A, Maschner H, *et al.* (2006) Bottom up forcing and decline of Stellar Sea Lions in Alaska: Assessing the ocean climate hypothesis. *Fisheries Oceanography* 16: 46–67.

Walther G, Post E, Convey P, *et al.* (2002) Ecological responses to recent climate change. *Nature* 416: 389–395.

Würsig B, Reeves RR, and Ortega-Ortiz JG (2002) Global climate change and marine mammals. In: Evans PGH and Raga JA (eds.) *Marine Mammals – Biology and Conservation*, pp. 589–608. New York: Kluwer Academic/Plenum Publishers.

SEABIRD RESPONSES TO CLIMATE CHANGE

David G. Ainley, H.T. Harvey and Associates, San Jose, CA, USA
G. J. Divoky, University of Alaska, Institute of Arctic Biology, Fairbanks, AK, USA

Introduction

This article reviews examples showing how seabirds have responded to changes in atmospheric and marine climate. Direct and indirect responses take the form of expansions or contractions of range; increases or decreases in populations or densities within existing ranges; and changes in annual cycle, i.e., timing of reproduction. Direct responses are those related to environmental factors that affect the physical suitability of a habitat, e.g., warmer or colder temperatures exceeding the physiological tolerances of a given species. Other factors that can affect seabirds directly include: presence/absence of sea ice, temperature, rain and snowfall rates, wind, and sea level. Indirect responses are those mediated through the availability or abundance of resources such as food or nest sites, both of which are also affected by climate change.

Seabird response to climate change may be most apparent in polar regions and eastern boundary currents, where cooler waters exist in the place of the warm waters that otherwise would be present. In analyses of terrestrial systems, where data are in much greater supply than marine systems, it has been found that range expansion to higher (cooler but warming) latitudes has been far more common than retraction from lower latitudes, leading to speculation that cool margins might be more immediately responsive to thermal variation than warm margins. This pattern is evident among sea birds, too. During periods of changing climate, alteration of air temperatures is most immediate and rapid at high latitudes due to patterns of atmospheric circulation. Additionally, the seasonal ice and snow cover characteristic of polar regions responds with great sensitivity to changes in air temperatures. Changes in atmospheric circulation also affect eastern boundary currents because such currents exist only because of wind-induced upwelling.

Seabird response to climate change, especially in eastern boundary currents but true elsewhere, appears to be mediated often by El Niño or La Niña. In other words, change is expressed stepwise, each step coinciding with one of these major, short-term climatic perturbations. Intensive studies of seabird populations have been conducted, with a few exceptions, only since the 1970s; and studies of seabird responses to El Niño and La Niña, although having a longer history in the Peruvian Current upwelling system, have become commonplace elsewhere only since the 1980s. Therefore, our knowledge of seabird responses to climate change, a long-term process, is in its infancy. The problem is exacerbated by the long generation time of seabirds, which is 15–70 years depending on species.

Evidence of Sea-bird Response to Prehistoric Climate Change

Reviewed here are well-documented cases in which currently extant seabird species have responded to climate change during the Pleistocene and Holocene (last 3 million years, i.e., the period during which humans have existed).

Southern Ocean

Presently, 98% of Antarctica is ice covered, and only 5% of the coastline is ice free. During the Last Glacial Maximum (LGM: 19 000 BP), marking the end of the Pleistocene and beginning of the Holocene, even less ice-free terrain existed as the ice sheets grew outward to the continental shelf break and their mass pushed the continent downward. Most likely, land-nesting penguins (Antarctic genus *Pygoscelis*) could not have nested on the Antarctic continent, or at best at just a few localities (e.g., Cape Adare, northernmost Victoria Land). With warming, loss of mass and subsequent retreat of the ice, the continent emerged.

The marine-based West Antarctic Ice Sheet (WAIS) may have begun to retreat first, followed by the land-based East Antarctic Ice Sheet (EAIS). Many Adélie penguin colonies now exist on the raised beaches remaining from successive periods of rapid ice retreat. Carbon-dated bones from the oldest beaches indicate that Adélie penguins colonized sites soon after they were exposed. In the Ross Sea, the WAIS receded south-eastward from the shelf break to its present position near Ross Island approximately 6200 BP. Penguin remains from Cape Bird, Ross Island (southwestern Ross Sea), date back to

7070±180 BP; those from the adjacent southern Victoria Land coast (Terra Nova Bay) date to 7505±230 BP. Adélie penguin remains at capes Royds and Barne (Ross Island), which are closest to the ice-sheet front, date back to 500 BP and 375 BP, respectively. The near-coast Windmill Islands, Indian Ocean sector of Antarctica, were covered by the EAIS during the LGM. The first islands were deglaciated about 8000 BP, and the last about 5500 BP. Penguin material from the latter was dated back to 4280–4530 BP, with evidence for occupation 500–1000 years earlier. Therefore, as in Victoria Land, soon after the sea and land were free from glaciers, Adélie penguins established colonies.

The study of raised beaches at Terra Nova Bay also investigated colony extinction. In that area several colonies were occupied 3905–4930 BP, but not since. The period of occupancy, called 'the penguin optimum' by geologists, corresponds to one of a warmer climate than at present. Currently, this section of Victoria Land is mostly sea-ice bound and penguins nest only at Terra Nova Bay owing to a small, persistent polynya (open-water area in the sea ice).

A study that investigated four extinct colonies of chinstrap penguin in the northern part of the Antarctic Peninsula confirmed the rapidity with which colonies can be founded or deserted due to fluctuations in environmental conditions. The colonies were dated at about 240–440 BP. The chinstrap penguin, an open-water loving species, occupied these former colonies during infrequent warmer periods indicated in glacial ice cores from the region. Sea ice is now too extensive for this species offshore of these colonies. Likewise, abandoned Adélie penguin nesting areas in this region were occupied during the Little Ice Age (AD 1500–1850), but since have been abandoned as sea ice has dissipated in recent years (see below).

South-east Atlantic

A well-documented avifaunal change from the Pleistocene to Recent times is based on bone deposits at Ascension and St Helena Islands. During glacial maxima, winds were stronger and upwelling of cold water was more pronounced in the region. This pushed the 23°C surface isotherm north of St Helena, thus accounting for the cool-water seabird avifauna that was present then. Included were some extinct species, as well as the still extant sooty shearwater and white-throated storm petrel. As the glacial period passed, the waters around St Helena became warmer, thereby encouraging a warm-water avifauna similar to that which exists today at Ascension Island; the cool-water group died out or decreased substantially in representation. Now, a tropical avifauna resides at St Helena including boobies, a frigatebird not present earlier, and Audubon's shearwater. Most recently these have been extirpated by introduced mammals.

North-west Atlantic/Gulf of Mexico

Another well-documented change in the marine bird fauna from Plio-Pleistocene to Recent times is available for Florida. The region experienced several major fluctuations in sea level during glacial and interglacial periods. When sea level decreased markedly to expose the Isthmus of Panama, thus, changing circulation, there was a cessation of upwelling and cool, productive conditions. As a result, a resident cool-water avifauna became extinct. Subsequently, during periods of glacial advance and cooler conditions, more northerly species visited the area; and, conversely, during warmer, interglacial periods, these species disappeared.

Direct Responses to Recent Climate Change

A general warming, especially obvious after the mid-1970 s, has occurred in ocean temperatures especially west of the American continents. Reviewed here are sea-bird responses to this change.

Chukchi Sea

The Arctic lacks the extensive water–ice boundaries of the Antarctic and as a result fewer seabird species will be directly affected by the climate-related changes in ice edges. Reconstructions of northern Alaska climatology based on tree rings show that temperatures in northern Alaska are now the warmest within the last 400 years with the last century seeing the most rapid rise in temperatures following the end of the Little Ice Age (AD 1500–1850). Decreases in ice cover in the western Arctic in the last 40 years have been documented, but the recent beginnings of regional ornithological research precludes examining the response of birds to these changes.

Changes in distribution and abundance related to snow cover have been found for certain cavity-nesting members of the auk family (Alcidae). Black guillemots and horned puffins require snow-free cavities for a minimum of 80 and 90 days, respectively, to successfully pair, lay and incubate eggs, and raise their chicks(s). Until the mid-1960s the snow-free period in the northern Chukchi Sea was usually

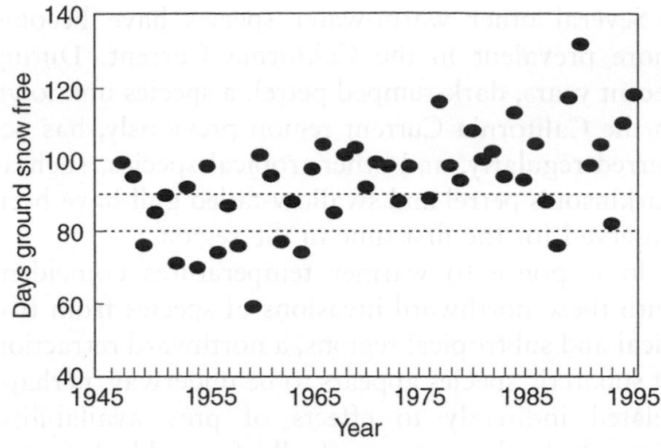

Figure 1 Changes in the length of the annual snow-free period at Barrow, Alaska, 1947–1995. Dashed lines show the number of days that black guillemots and horned puffins require a snow-free cavity (80 and 90 days, respectively). Black guillemots first bred near Barrow in 1966 and horned puffins in 1986. (Redrawn from Divoky, 1998.)

shorter than 80 days but with increasing spring air temperatures, the annual date of spring snow-melt has advanced more than 5 days per decade over the last five decades (**Figure 1**). The annual snow-free period now regularly exceeds 90 days, which reduces the likelihood of chicks being trapped in their nest sites before fledging. This has allowed black guillemots and horned puffins to establish colonies (range expansion) and increase population size in the northern Chukchi.

California Current

Avifaunal changes in this region are particularly telling because the central portion of the California Current marks a transitional area where subtropical and subarctic marine faunas meet, and where north–south faunal shifts have been documented at a variety of temporal scales. Of interest here is the invasion of brown pelicans northward from their 'usual' wintering grounds in central California to the Columbia River mouth and Puget Sound, and the invasion of various terns and black skimmers northward from Mexico into California. The pelican and terns are tropical and subtropical species.

During the last 30 years, air and sea temperatures have increased noticeably in the California Current region. The response of seabirds may be mediated by thermoregulation as evidenced by differences in the amount of subcutaneous fat held by polar and tropical seabird species, and by the behavioral responses of seabirds to inclement air temperatures.

The brown pelican story is particularly well documented and may offer clues to the mechanism by which similar invasions come about. Only during the very intense El Niño of 1982–83 did an unusual number of pelicans move northward to the Columbia River mouth. They had done this prior to 1900, but then came a several-decade long period of cooler temperatures. Initially, the recent invasion involved juveniles. In subsequent years, these same birds returned to the area, and young-of-the-year birds followed. Most recently, large numbers of adult pelicans have become a usual feature feeding on anchovies that have been present all along. This is an example of how tradition, or the lack thereof, may facilitate the establishment (or demise) of expanded range, in this case, compatible with climate change.

The ranges of skimmers and terns have also expanded in pulses coinciding with El Niño. This pattern is especially clear in the case of the black skimmer, a species whose summer range on the east coast of North America retracts southward in winter. On the west coast, almost every step in a northward expansion of range from Mexico has coincided with ocean warming and, in most cases, El Niño: first California record, 1962 (not connected to ocean warming); first invasion *en masse*, 1968 (El Niño); first nesting at Salton Sea, 1972 (El Niño); first nesting on coast (San Diego), 1976 (El Niño); first nesting farther north at Newport Bay and inland, 1986 (El Niño). Thereafter, for Southern California as a whole, any tie to El Niño became obscure, as (1) average sea-surface temperatures off California became equivalent to those reached during the intense 1982–83 El Niño, and (2) population increase became propelled not just by birds dispersing north from Mexico, but also by recruits raised locally. By 1995, breeding had expanded north to Central California. In California, with warm temperatures year round, skimmers winter near where they breed.

The invasion northward by tropical/subtropical terns also relates to El Niño or other warm-water incursions. The first US colony (and second colony in the world) of elegant tern, a species usually present off California as post-breeders (July through October), was established at San Diego in 1959, during the strongest El Niño event in modern times. A third colony, farther north, was established in 1987 (warm-water year). The colony grew rapidly, and in 1992–93 (El Niño) numbers increased 300% (to 3000 pairs). The tie to El Niño for elegant terns is confused by the strong correlation, initially, between numbers breeding in San Diego and the biomass of certain prey (anchovies), which had also increased. Recently, however, anchovies have decreased. During the intense 1997–98 El Niño, hundreds of elegant terns were observed in courtship during spring even farther north (central California). No colony formed.

Climate change, and El Niño as well, may be involved in the invasion of Laysan albatross to breed on Isla de Guadalupe, in the California Current off northern Mexico. No historical precedent exists for the breeding by this species anywhere near this region. First nesting occurred in 1983 (El Niño) and by 1988, 35–40 adults were present, including 12 pairs. Ocean temperatures off northern Mexico are now similar to those in the Hawaiian Islands, where nesting of this species was confined until recently. In the California Current, sea temperatures are the warmest during the autumn and winter, which, formerly, were the only seasons when these albatross occurred there. With rising temperatures in the California Current, more and more Laysan albatross have been remaining longer into the summer each year. Related, too, may be the strengthening of winds off the North American west coast to rival more closely the trade winds that buffet the Hawaiian Islands. Albatross depend on persistent winds for efficient flight, and such winds may limit where albatrosses occur, at least at sea.

Several other warm-water species have become more prevalent in the California Current. During recent years, dark-rumped petrel, a species unknown in the California Current region previously, has occurred regularly, and other tropical species, such as Parkinson's petrel and swallow-tailed gull have been observed for the first time in the region.

In response to warmer temperatures coincident with these northward invasions of species from tropical and subtropical regions, a northward retraction of subarctic species appears to be underway, perhaps related indirectly to effects of prey availability. Nowadays, there are markedly fewer black-footed albatross and sooty and pink-footed shearwaters present in the California Current system than occurred just 20 years ago (**Figure 2**). Cassin's auklet is becoming much less common at sea in central California, and its breeding populations have also been declining. Similarly, the southern edge of the breeding range of common murres has retreated north. The species no longer breeds in Southern California (Channel Islands) and numbers have become greatly reduced in Central California (**Figure 3**). Moreover, California Current breeding populations have lost

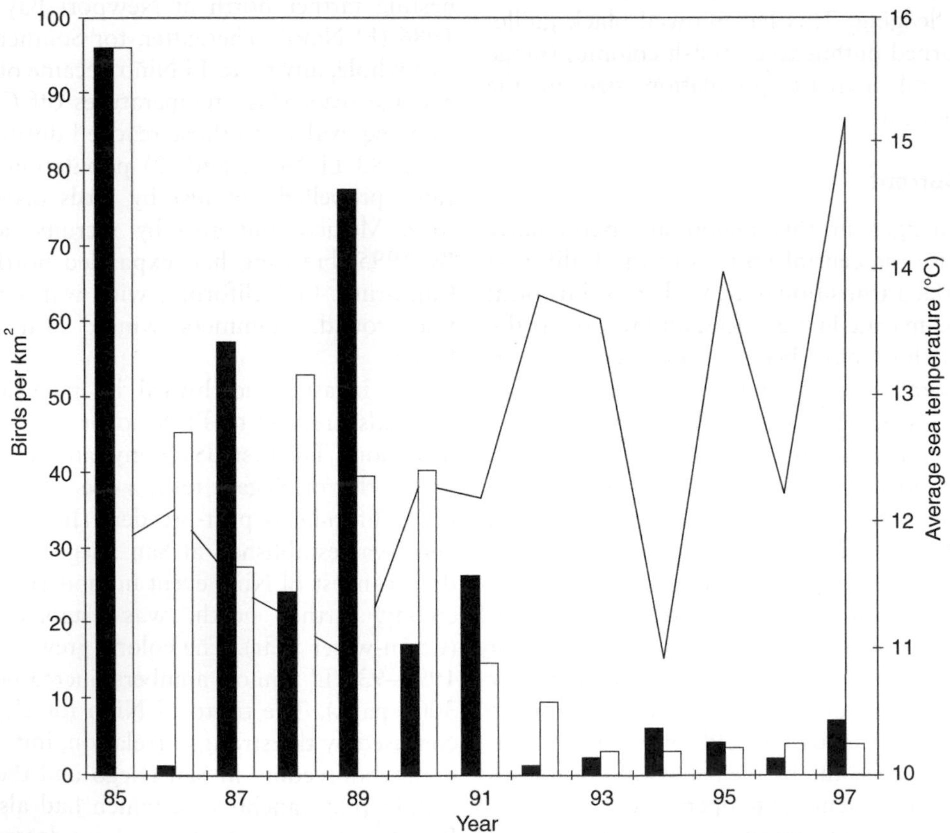

Figure 2 The density (■) (plus 3-point moving average, □) of a cool-water species, the sooty shearwater, in the central portion of the California Current, in conjunction with changes in marine climate (sea surface temperature, (—)), 1985–1997.

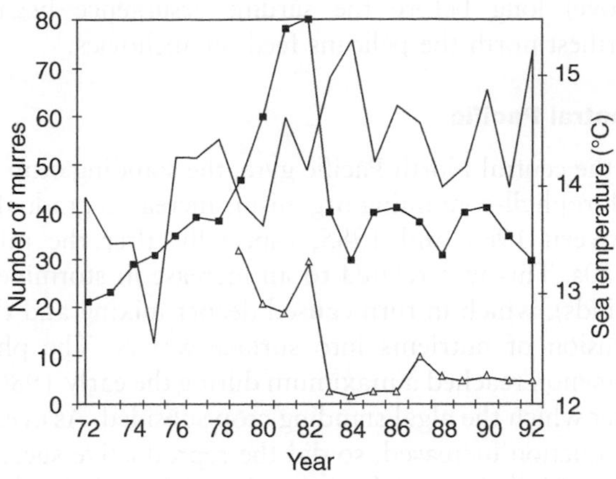

Figure 3 Changes in the number of breeding common murres in central (California, ■) and northern (Washington, △) portions of the California Current during recent decades. Sea surface temperature (—) from central California shown for comparison. During the 1970s, populations had the capacity to recover from reductions due to anthropogenic factors (e.g., oil spills). Since the 1982–83 El Niño event and continued higher temperatures, however, the species' capacity for population recovery has been lost (cf. **Figure 5**)

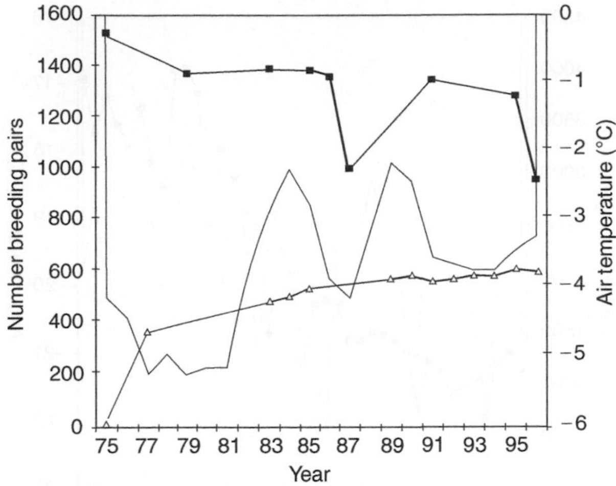

Figure 4 Changes in the number of breeding pairs of two species of penguins at Arthur Harbor, Anvers Island, Antarctica (64°S, 64°W), 1975–1996. The zoogeographic range of Adélie penguins (■) is centered well to the south of this site; the range of chinstrap penguins (△) is centered well to the north. Arthur Harbor is located within a narrow zone of overlap (200 km) between the two species and is at the northern periphery of the Adélie penguins' range.

much of their capacity, demonstrated amply as late as the 1970s, to recover from catastrophic losses. The latter changes may or may not be more involved with alterations of the food web (see below).

Northern Bellingshausen Sea

Ocean and air temperatures have been increasing and the extent of pack ice has been decreasing for the past few decades in waters west of the Antarctic peninsula. In response, populations of the Adélie penguin, a pack-ice species, have been declining, while those of its congener, the open-water dwelling chinstrap penguin have been increasing (**Figure** 4). The pattern contrasts markedly with the stability of populations on the east side of the Antarctic Peninsula, which is much colder and sea-ice extent has changed little.

The reduction in Adélie penguin populations has been exacerbated by an increase in snowfall coincident with the increased temperatures. Deeper snow drifts do not dissipate early enough for eggs to be laid on dry land (causing a loss of nesting habitat and eggs), thus also delaying the breeding season so that fledging occurs too late in the summer to accommodate this species' normal breeding cycle. This pattern is the reverse of that described above for black guillemots in the Arctic. The penguin reduction has affected mostly the smaller, outlying subcolonies, with major decreases being stepwise and occurring during El Niño.

Similar to the chinstrap penguin, some other species, more typical of the Subantarctic, have been expanding southward along the west side of the Antarctic Peninsula. These include the brown skua, blue-eyed shag, southern elephant seal and Antarctic fur seal.

Ross Sea

A large portion (32%) of the world's Adélie penguin population occurs in the Ross Sea (South Pacific sector), the southernmost incursion of ocean on the planet (to 78°S). This species is an obligate inhabitant of pack ice, but heavy pack ice negatively affects reproductive success and population growth. Pack-ice extent decreased noticeably in the late 1970s and early 1980s and air temperatures have also been rising. The increasing trends in population size of Adélie penguins in the Ross Sea are opposite to those in the Bellingshausen Sea (see above; **Figure** 5). The patterns, however, are consistent with the direction of climate change: warmer temperatures, less extensive pack ice. As pack ice has become more dispersed in the far south, the penguin has benefited.

As with the Antarctic Peninsula region, subantarctic species are invading southward. The first brown skua was reported in the southern Ross Sea in 1966; the first known breeding attempt occurred in 1982; and the first known successful nesting occurred in 1996. The first elephant seal in the Ross Sea was reported in 1974; at present, several individuals occur there every year.

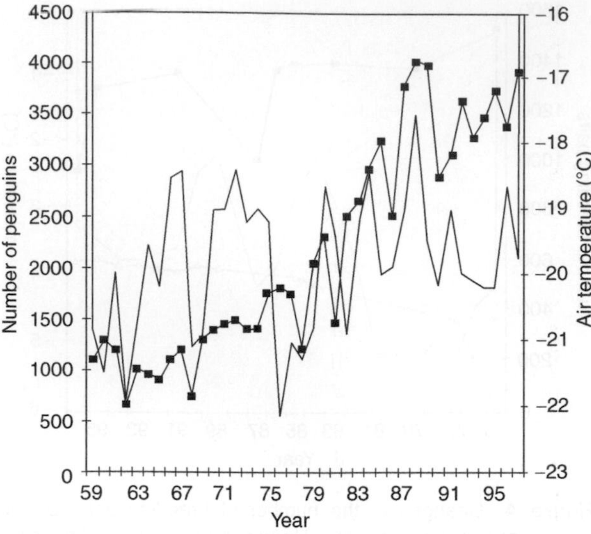

Figure 5 Changes in numbers of breeding pairs of Adélie penguins (■) at Cape Royds, Ross Island, Antarctica (77°S, 166°E), during the past four decades. This is the southernmost breeding site for any penguin species. Although changes in sea ice (less extensive now) is the major direct factor involved, average air temperatures (—) (indirect factor) of the region are shown.

Indirect Responses to Recent Climate Change

California Current

The volume of zooplankton has declined over the past few decades, coincident with reduced upwelling and increased sea-surface temperatures. In response, numbers of sooty shearwaters, formerly, the most abundant avian species in the California Current avifauna, have declined 90% since the 1980s (**Figure 2**). The shearwater feeds heavily on euphausiids in the California Current during spring. The decline, however, has occurred in a stepwise fashion with each El Niño or warm-water event. Sooty shearwaters are now ignoring the Peru and California currents as wintering areas, and favoring instead those waters of the central North Pacific transition zone, which have been cooling and increasing in productivity (see below).

The appearance of the elegant and royal terns as nesting species in California (see above) may in part be linked to the surge in abundance in northern anchovy, which replaced the sardine in the 1960s–1980s. More recently, the sardine has rebounded and the anchovy has declined, but the tern populations continue to grow (see above). Similarly, the former breeding by the brown pelican as far north as Central California was linked to the former presence of sardines. However, the pelicans recently invaded northward (see above) long before the sardine resurgence began. Farthest north the pelicans feed on anchovies.

Central Pacific

In the central North Pacific gyre, the standing crop of chlorophyll-containing organisms increased gradually between 1965 and 1985, especially after the mid-1970s. This was related to an increase in storminess (winds), which in turn caused deeper mixing and the infusion of nutrients into surface waters. The phenomenon reached a maximum during the early 1980s, after which the algal standing crop subsided. As ocean production increased, so did the reproductive success of red-billed tropicbirds and red-footed boobies nesting in the Leeward Hawaiian Islands (southern part of the gyre). When production subsided, so did the breeding success of these and other species (lobsters, seals) in the region. Allowing for lags of several years as dictated by demographic characteristics, the increased breeding success presumably led to larger populations of these seabird species.

Significant changes in the species composition of seabirds in the central Pacific (south of Hawaii) occurred between the mid-1960s and late 1980s. Densities of Juan Fernandez and black-winged petrels and short-tailed shearwaters were lower in the 1980s, but densities of Stejneger's and mottled petrels and sooty shearwaters were higher. In the case of the latter, the apparent shift in migration route (and possibly destination) is consistent with the decrease in sooty shearwaters in the California Current (see above).

Peru Current

The Peruvian guano birds – Peruvian pelican, piquero, and guanay – provide the best-documented example of changes in seabird populations due to changes in prey availability. Since the time of the Incas, the numbers of guano birds have been strongly correlated with biomass of the seabirds' primary prey, the anchoveta. El Niño 1957 (and earlier episodes) caused crashes in anchoveta and guano bird populations, but these were followed by full recovery. Then, with the disappearance of the anchoveta beginning in the 1960s (due to over-fishing and other factors), each subsequent El Niño (1965, 1972, etc.) brought weaker recovery of the seabird populations.

Apparently, the carrying capacity of the guano birds' marine habitat had changed, but population decreases occurred stepwise, coinciding with mortality caused by each El Niño. However, more than just fishing caused anchoveta to decrease; without fishing pressure, the anchoveta recovered quickly (to its lower level) following El Niño 1982–83, and trends in the sardine were contrary to those of the anchoveta beginning in the late 1960s. The seabirds that remain have shifted their breeding sites southward to southern Peru and northern Chile in response to the southward, compensatory increase in sardines. A coincident shift has occurred in the zooplankton and mesopelagic fish fauna. All may be related to an atmospherically driven change in ocean circulation, bringing more subtropical or oceanic water onshore. It is not just breeding seabird species that have been affected, nonbreeding species of the region, such as sooty shearwater, have been wintering elsewhere than the Peru Current (see above).

Trends in penguin populations on the Galapagos confirm that a system-wide change has occurred off western South America (**Figure 6**). Galapagos penguins respond positively to cool water and negatively to warm-water conditions; until recently they recovered after each El Niño, just like the Peruvian guano birds. Then came El Niño 1982–83. The population declined threefold, followed by just a slight recovery. Apparently, the carrying capacity of the habitat of this seabird, too, is much different now than a few decades ago. Like the diving, cool-water species of the California Current (see above), due to climate change, the penguin has lost its capacity for population growth and recovery.

Gulf of Alaska and Bering Sea

A major 'regime' shift involving the physical and biological make-up of the Gulf of Alaska and Bering Sea is illustrated amply by oceanographic and fisheries data. Widespread changes and switches in populations of ecologically competing fish and invertebrate populations have been underway since the mid-1970s. Ironically, seabird populations in the region show few geographically consistent patterns that can be linked to the biological oceanographic trends. There have been no range expansions or retractions, no doubt because this region, in spite of its great size, does not constitute a faunal transition; and from within the region, species to species, some colonies have shown increases, others decreases, and others stability. Unfortunately, the picture has been muddled by the introduction of exotic terrestrial mammals to many seabird nesting islands. Such introductions have caused disappearance and serious declines in sea-bird numbers. In turn, the subsequent eradication of mammals has allowed recolonization and increases in seabird numbers.

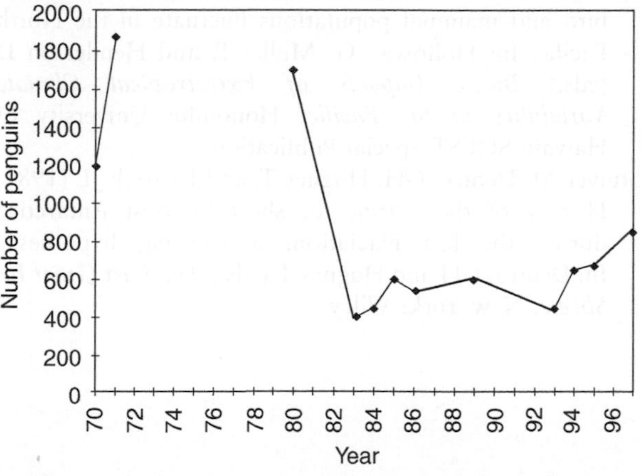

Figure 6 Changes in numbers of breeding Galapágos penguins, 1970–1997. With the 1982–83 El Niño event, the species lost the capacity to recover from periodic events leading to increased mortality. Compare to **Figure 3**, which represents another cool-water species having a similar life-history strategy, but which resides in the other eastern Pacific boundary current (i.e., the two species are ecological complements in the Northern and Southern Hemispheres).

In the Bering Sea, changes in the population biology of seabirds have been linked to decadal shifts in the intensity and location of the Aleutian Low Pressure System (the Pacific Decadal Oscillation), which affects sea surface temperatures among other things. For periods of 15–30 years the pressure (North Pacific Index) shifts from values that are above (high pressure state) or below (low pressure state) the long-term average. Kittiwakes in the central Bering Sea (but not necessarily the Gulf of Alaska) do better with warmer sea temperatures; in addition, the relationship of kittiwake productivity to sea surface temperature changes sign with switches from the high to low pressure state. Similarly, the dominance among congeneric species of murres at various sympatric breeding localities may flip-flop depending on pressure state. Although these links to climate have been identified, cause–effect relationships remain obscure. At the Pribilof Islands in the Bering Sea, declines in seabird numbers, particularly of kittiwakes, coincided with the regime shift that began in 1976. Accompanying these declines has been a shift in diets, in which lipid-poor juvenile walleye pollock have been substituted for lipid-rich forage fishes such as sand lance and capelin. Thus, the regime shifts may have altered trophic pathways to forage fishes and in turn the availability of these fish to seabirds. Analogous patterns are seen among the seabirds of Prince William Sound.

Chukchi Sea

Decrease of pack ice extent in response to recent global temperature increases has been more pronounced in the Arctic than the Antarctic. This decrease has resulted in changes in the availability of under-ice fauna, fish, and zooplankton that are important to certain arctic seabirds. The decline of black guillemot populations in northern Alaska in the last decade may be associated with this pack ice decrease.

North Atlantic

The North Atlantic is geographically confined and has been subject to intense human fishery pressure for centuries. Nevertheless, patterns linked to climate change have emerged. In the North Sea, between 1955 and 1987, direct, positive links exist between the frequency of westerly winds and such biological variables as zooplankton volumes, stock size of young herring (a seabird prey), and the following parameters of black-legged kittiwake biology: date of laying, number of eggs laid per nest, and breeding success. As westerly winds subsided through to about 1980, zooplankton and herring decreased, kittiwake laying date retarded and breeding declined. Then, with a switch to increased westerly winds, the biological parameters reversed.

In the western Atlantic, changes in the fish and seabird fauna correlate to warming sea surface temperatures near Newfoundland. Since the late 1800s, mackerel have moved in, as has one of their predators, the Atlantic gannet. Consequently, the latter has been establishing breeding colonies farther and farther north along the coast.

South-east Atlantic, Benguela Current

A record of changes in sea-bird populations relative to the abundance and distribution of prey in the Benguela Current is equivalent to that of the Peruvian upwelling system. As in other eastern boundary currents, Benguela stocks of anchovy and sardine have flip-flopped on 20–30 year cycles for as long as records exist (to the early 1900s). Like other eastern boundary currents, the region of concentration of prey fish, anchovy versus sardine, changes with sardines being relatively more abundant poleward in the region compared to anchovies. Thus, similar to the Peruvian situation, seabird populations have shifted, but patterns also are apparent at smaller timescales depending on interannual changes in spawning areas of the fish. As with all eastern boundary currents, the role of climate in changing pelagic fish populations is being intensively debated.

See also

Upwelling Ecosystems.

Further Reading

Aebischer NJ, Coulson JC, and Colebrook JM (1990) Parallel long term trends across four marine trophic levels and weather. *Nature* 347: 753–755.

Crawford JM and Shelton PA (1978) Pelagic fish and seabird interrelationships off the coasts of South West and South Africa. *Biological Conservation* 14: 85–109.

Decker MB, Hunt GL, and Byrd GV (1996) The relationship between sea surface temperature, the abundance of juvenile walleye pollock (*Theragra chalcogramma*), and the reproductive performance and diets of seabirds at the Pribilof islands, southeastern Bering Sea. *Canadian Journal of Fish and Aquatic Science* 121: 425–437.

Divoky GJ (1998) *Factors Affecting Growth of a Black Guillemot Colony in Northern Alaska.* PhD Dissertation, University of Alaska, Fairbanks.

Emslie SD (1998) *Avian Community, Climate, and Sea-level Changes in the Plio-Pleistocene of the Florida Peninsula.* Ornithological Monograph No. 50. Washington, DC: American Ornithological Union.

Furness RW and Greenwood JJD (eds.) (1992) *Birds as Monitors of Environmental Change.* London, New York: Chapman and Hall.

Olson SL (1975) *Paleornithology of St Helena Island, South Atlantic Ocean.* Smithsonian Contributions to Paleobiology, No. 23. Washington, Dc.

Smith RC, Ainley D, Baker K, *et al.* (1999) Marine ecosystem sensitivity to climate change. *BioScience* 49(5): 393–404.

Springer AM (1998) Is it all climate change? Why marine bird and mammal populations fluctuate in the North Pacific. In: Holloway G, Muller P, and Henderson D (eds.) *Biotic Impacts of Extratropical Climate Variability in the Pacific.* Honolulu: University of Hawaii: SOEST Special Publication.

Stuiver M, Denton GH, Hughes T, and Fastook JL (1981) History of the marine ice sheet in West Antarctica during the last glaciation: a working hypothesis. In: Denton GH and Hughes T (eds.) *The Last Great Ice Sheets.* New York: Wiley.

FISHERIES AND CLIMATE

K. M. Brander, DTU Aqua, Charlottenlund, Denmark

Introduction

Concern over the effects of climate has increased in recent years as concentrations of greenhouse gases have risen in the atmosphere. We all observe changes which are attributed, with more or less justification, to the influence of climate. Anglers and fishermen in many parts of the world have noticed a steady increase in warm-water species. The varieties of locally caught fish which are sold in markets and shops are changing. For example, fish shops in Scotland now sell locally caught bass (*Dicentrarchus labrax*) and red mullet (*Mullus surmuletus*) – species which only occurred in commercial quantities south of the British Isles (600 km to the south) until the turn of the millennium. The northward spread of two noncommercial, subtropical species is shown in **Figure 1**.

They were first recorded off Portugal in the 1960s and were then progressively found further north, until by the mid-1990s they occurred over 1000 km north of the Iberian Peninsula. *Zenopsis conchifer* was recorded at Iceland for the first time in 2002.

In addition to distribution changes, the growth, reproduction, migration, and seasonality of fish are affected by climate. The productivity and composition of the ecosystems on which fish depend are altered, as is the incidence of pathogens. The changes are not only due to temperature; but winds, ocean currents, vertical mixing, sea ice, freshwater runoff, cloud cover, oxygen, salinity, and pH are also part of climate change, with effects on fish. The processes by which these act will be explored in the examples cited later.

The effect of climate on fisheries is not a new phenomenon or a new area of scientific investigation. Like the effects of climate on agriculture or on hunted and harvested animal populations, it has probably been systematically observed and studied since humans began fishing. However, the anthropogenic component of climate change is a new phenomenon, which is gradually pushing the ranges of atmospheric and oceanic conditions outside the envelope experienced during human history. This article begins by describing some of the past effects of climate on fisheries and then goes on to review expected impacts of anthropogenic climate change.

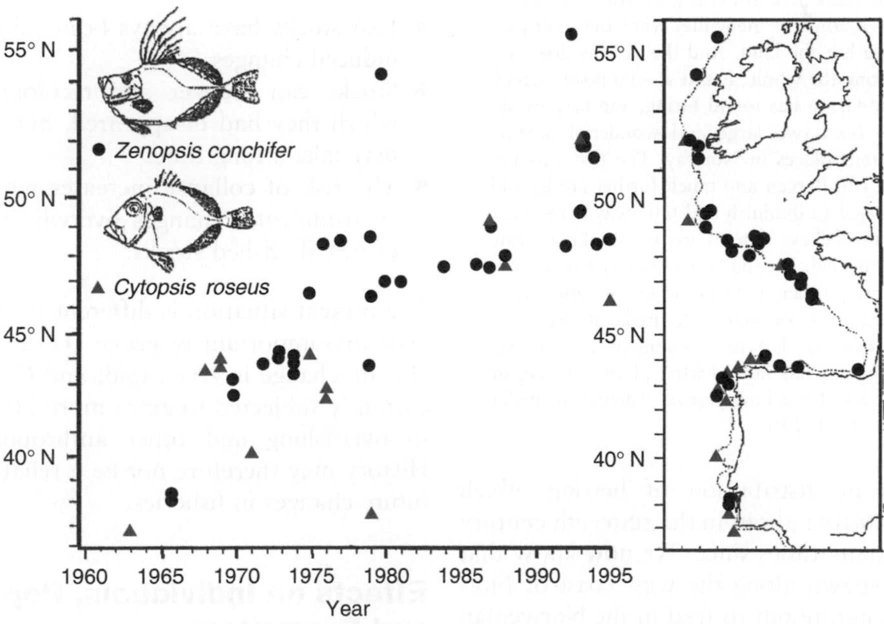

Figure 1 First records of two subtropical fish species (silvery John Dory and rosy Dory) along the NW European continental shelf. From Quéro JC, Buit HD, and Vayne JJ (1998) Les observations de poissons tropicaux et le réchauffement des eaux dans l'Atlantique européen. *Oceanologica Acta* 21: 345–351.

Climate Timescales and Terminology

A wide range of timescales of change in the physical and chemical environment may be included in the term 'climate'. In this article, 'climate variability' refers to changes in temperature, wind fields, hydrological cycles, etc., at annual to decadal timescales and 'climate change' denotes longer-term shifts in the mean values. There are both natural and anthropogenic causes of climate variability and it is not always easy to distinguish the underlying causes of a particular observed effect. Changes in the physical and chemical environment occur naturally on daily, seasonal, and longer-term (e.g., 18.6-year nodal tide) cycles, which can be related to planetary motion. Natural variability in the environment overlies these cycles, so that one can, for example, speak of a windy month or a wet year. Underlying such statements is the idea of a 'normal' month or year, which is generally defined in relation to a climatology, that is, by using a long-term mean and distribution of the variable in question. Volcanic activity and solar fluctuations are other nonanthropogenic factors which affect climate.

History

The Norwegian spring spawning herring (*Clupea harengus* L.) has been a major part of the livelihood of coastal communities in western and northern Norway for over 1000 years.

> It comes up to the shore here from the great fish pond which is the Icelandic Sea, towards the winter when the great part of other fish have left the land. And the herring does not seek the shore along the whole, but at special points which God in his Good Grace has found fitting, and here in my days there have been two large and wonderful herring fisheries at different places in Norway. The first was between Stavanger and Bergen and much further north, and this fishery did begin to diminish and fall away in the year 1560. And I do not believe there is any man to know how far the herring travelled. For the Norwegian Books of Law show that the herring fishery in most of the northern part of Norway has continued for many hundreds of years, although it may well be that in punishment for the unthankfulness of men it has moved from place to place, or has been taken away for a long period. (Clergyman Peder Claussøn Friis (1545–1614))

The changes in distribution of herring which Clergyman Friis wrote about in the sixteenth century have occurred many times since. We now know that herring, which spawn along the west coast of Norway in spring, migrate out to feed in the Norwegian Sea in summer, mainly on the copepod *Calanus finmarchicus*. The distribution shifts in response to decadal changes in oceanic conditions.

Beginning in the 1920s much of the North Atlantic became warmer, the polar front moved north, and the summer feeding migration of this herring stock expanded along the north coast of Iceland. For the next four decades, the resulting fishery provided economic prosperity for northern Iceland and for the country as a whole. When the polar front, which separates cold, nutrient-poor polar water from warmer, nutrient-rich Atlantic water, shifted south and east again, in 1964–65, the herring stopped migrating along the north coast of Iceland. At the same time, the stock collapsed due to overfishing and the herring processing business in northern Iceland died out (**Figure 2**).

There are many similar examples of fisheries on pelagic and demersal fish stocks, which have changed their distribution or declined and recovered as oceanic conditions switched between favorable and unfavorable periods. The Japanese Far Eastern sardine has undergone a series of boom and bust cycles lasting several decades as has the Californian sardine and anchovy. Along the west coast of South America the great Peruvian anchoveta fishery is subject to great fluctuations in abundance, which are not only driven by El Niño/Southern Oscillation (ENSO), but also by decadal-scale variability in the Pacific circulation. In the North Atlantic, the economy of fishing communities in Newfoundland, Greenland, and Faroe has been severely affected by climate-induced fluctuations in the cod stocks (*Gadus morhua* L.) on which they depend.

A number of lessons can be learned from history:

- Fish stocks have always been subject to climate-induced changes.
- Stocks can recover and recolonize areas from which they had disappeared, but such recoveries may take a long time.
- The risk of collapse increases when unfavorable environmental changes overwhelm the resilience of heavily fished stocks.

The present situation is different from the past in at least two important respects: (1) the current rate of climate change is very rapid; and (2) fish stocks are currently subjected to extra mortality and stress due to overfishing and other anthropogenic impacts. History may therefore not be a reliable guide to the future changes in fisheries.

Effects on Individuals, Populations, and Ecosystems

Climatic factors act directly on the growth, survival, reproduction, and movement of individuals and

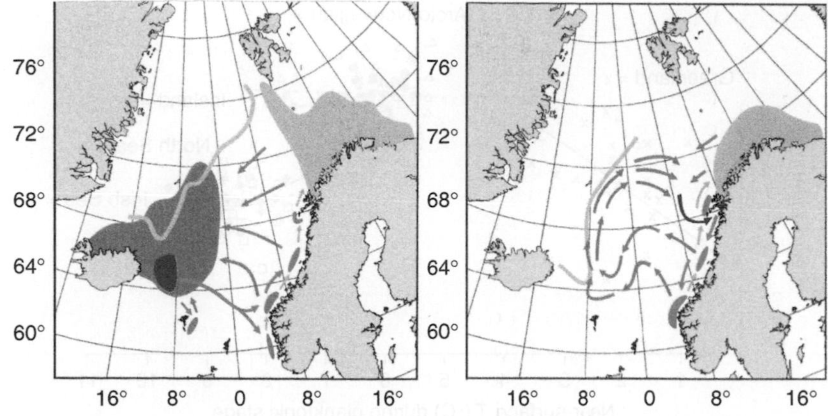

Figure 2 Left panel shows the distribution of Norwegian spring spawning herring between the 1920s and 1965, with areas of spawning (dark blue), nursery (light blue), feeding (green), and overwintering (red). Right panel is the distribution after 1990, when the stock had recovered, but had not reoccupied the area north of Iceland. The polar front separating warm, nutrient-rich Atlantic water from cooler, nutrient-depleted polar water is shown as a light blue line. From Vilhjalmsson H (1997) Climatic variations and some examples of their effects on the marine ecology of Icelandic and Greenland waters, in particular during the present century. *Rit Fiskideildar* XV(1): 9–27.

some of these processes can be studied experimentally, as well as by sampling in the ocean. The integrated effects of the individual processes, are observed at the level of populations, communities, or ecosystems.

In order to convincingly attribute a specific change to climate, one needs to be able to identify and describe the processes involved. Even if the processes are known, and have been studied experimentally, such as the direct effects of temperature on growth rate, the outcome at the population level, over periods of years, can be complex and uncertain. Temperature affects frequency of feeding, meal size, rate of digestion, and rate of absorption of food. Large-scale experiments in which a range of sizes of fish were fed to satiation (Atlantic cod, *G. morhua* L. in this case; see **Figure 3**) show that there is an optimum temperature for growth, but that this depends on the size of the fish, with small fish showing a higher optimum. The optimum temperature for growth also depends on how much food is available, since the energy required for basic maintenance metabolism increases at higher temperature. This means that if food is in short supply then fish will grow faster at cooler temperature, but if food is plentiful then they will grow faster at higher temperatures. A further complication is that temperature typically has a seasonal cycle so the growth rate may decline due to higher summer temperature, but increase due to higher winter temperature.

Recruitment of young fish to an exploited population is variable from year to year for a number of reasons, including interannual and long-term climatic variability. The effects of environmental

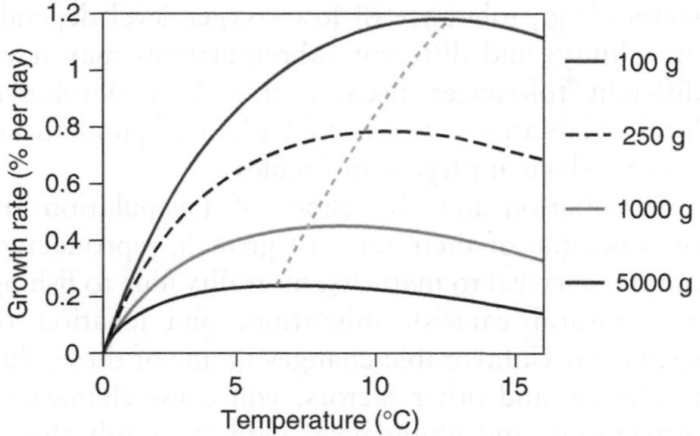

Figure 3 Growth rate of four sizes of Atlantic cod (*Gadus morhua* L.) in rearing experiments at different temperatures in which they were provided unlimited food. The steep dashed line intersects the growth curves at their maximum values, to show how the temperature for maximum growth rate declines as fish get bigger.

variability on survival during early life, when mortality rates in the plankton are very high, is thought to be critical in determining the number of recruits. When recruitment is compared across a number of cod populations, a consistent domed pattern emerges (**Figure 4**).

The relationship between temperature and recruitment for cod has an ascending limb from *c.* 0 to 4 °C and a descending limb above *c.* 7 °C.

In addition to temperature, the distribution of fish depends on salinity, oxygen, and depth, which is only affected by climate at very long timescales. A 'bioclimate envelope', defining the limits of the range, can be constructed by taking all climate-related variables together. Such 'envelopes' are in reality

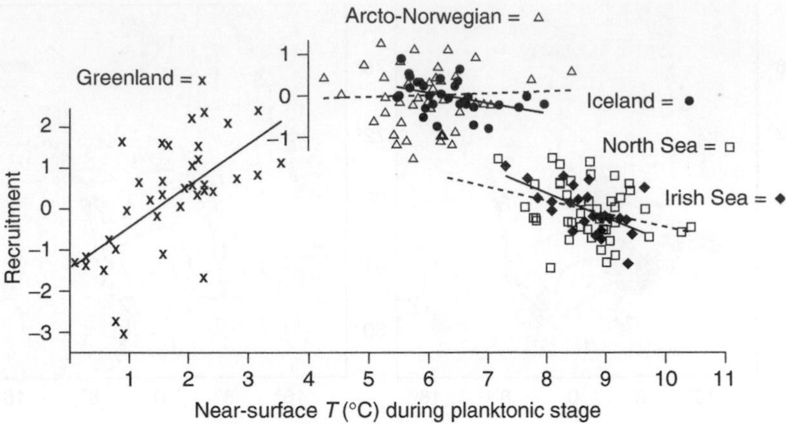

Figure 4 Composite pattern of recruitment for five of the North Atlantic cod stocks to illustrate the effect of temperature during the planktonic stage of early life on the number of recruiting fish. The scales are $\log_e$ (number of 1-year-old fish) with the means adjusted to zero. The axes for the Arcto-Norwegian and Iceland stocks have been displaced vertically.

quite complex because the tolerance and response of fish may vary with size (as shown for growth in **Figure 3**); there may be interactions between the factors (e.g., tolerance of low oxygen level depends on salinity) and different subpopulations may have different tolerances because they have developed local adaptations (e.g., cold-adapted populations may produce antifreeze molecules).

Distribution and abundance of a population are the outcome of their rates of growth, reproductive output, survival to maturity, mortality (due to fishing and natural causes), migration, and location of spawning. Unfavorable changes in any of these, due to climate and other factors, will cause changes in distribution and abundance, with the result that a species may no longer be able to maintain a population in areas which are affected. Favorable changes allow a species to increase its population or to colonize previously unsuitable areas.

Climate and fishing can both be considered as additional stresses on fish populations. In order to manage fisheries in a sustainable way the effects of both (and other factors, such as pollution or loss of essential habitat) need to be recognized, as well as the interactions between them. When the age structure and geographic extent of fish populations are reduced by fishing pressure, then they become less resilient to climate change. Conversely, if climate change reduces the surplus production of a population (by altering growth, survival, or reproductive success), then that population will decline at a level of fishing which had previously been sustainable. Because of these interactions it is not possible to deal with adaption to climate change and fisheries management as separate issues. The most effective strategy to assist fish stocks in adapting to climate change is to reduce the mortality due to fishing. Sustainable fisheries require continuous monitoring of the consequences of climate change.

Regional Effects of Climate

Tropical Pacific

The tuna of the Pacific provide one of the very few examples in which the consequences of climate change have been modeled to include geographic and trophic detail all the way through from primary and secondary production to top fish predators. The tuna species skipjack (*Katsuwonus pelamis*) and yellowfin (*Thunnus albacares*) are among the top predators of the tropical pelagic ecosystem and produced a catch of 3.6 million t in 2003, which represents *c.* 5.5% of total world capture fisheries in weight and a great deal more in value. Their forage species include the Japanese sardine *Engraulis japonicus*, which itself provided a catch of over 2 million t in 2003. The catches and distribution of these species and other tuna species (e.g., albacore *Thunnus alalunga*) are governed by variability in primary and secondary production and location of suitable habitat for spawning and for adults, which in turn are linked to varying regimes of the principal climate indices, such as the El Niño–La Niña Southern Oscillation index (SOI) and the related Pacific Decadal Oscillation (PDOI). Statistical and coupled biogeochemical models have been developed to explore the causes of regional variability in catches and their connection with climate. The model area includes the Pacific from 40° S to 60° N and timescales range from short-term to decadal regime shifts. The model captures the slowdown of Pacific meridional overturning circulation and decrease of equatorial upwelling, which has caused primary production and biomass to decrease by about 10% since 1976–77 in the equatorial Pacific. Further climate change will affect the distribution and production of tuna fisheries in rather complex ways. Warmer surface waters and lower primary production in the central and eastern Pacific may result in a redistribution of tuna to higher latitudes (such as Japan) and toward the western equatorial Pacific.

North Pacific

Investigations into the effects of climate change in the North Pacific have focused strongly on regime shifts. The physical characteristics of these regime shifts and the biological consequences differ between the major regions within the North Pacific.

The PDO tracks the dominant spatial pattern of sea surface temperature (SST). The alternate phases of the PDO represent cooling/warming in the central subarctic Pacific and warming/cooling along the North American continental shelf. This 'classic' pattern represents change along an east–west axis, but since 1989 a north–south pattern has also emerged. Other commonly used indices track the intensity of the winter Aleutian low-pressure system and the sea level pressure over the Arctic. North Pacific regime shifts are reported to have occurred in 1925, 1947, 1977, 1989, and 1998, and paleo-ecological records show many earlier ones. The duration of these regimes appears to have shortened from 50–100 years to *c.* 10 years for the two most recent regimes, although whether this apparent shortening of regimes is real and whether it is related to other aspects of climate change is a matter of current debate and concern. The SOI also has a large impact on the North Pacific, adding an episodic overlay with a duration of 1 or 2 years to the decadal-scale regime behavior.

Regime shifts, such as the one in 1998, have well-documented effects on ocean climate and biological systems. Sea surface height (SSH) in the central North Pacific increased, indicating a gain in thickness of the upper mixed layer, while at the same time SSH on the eastern and northern boundaries of the North Pacific dropped. The position of the transition-zone chlorophyll front, which separates subarctic from subtropical waters and is a major migration and forage habitat for large pelagic species, such as albacore tuna, shifted northward. In addition to its effects on pelagic fish species, shifts in the winter position of the chlorophyll front affect other species, such as Hawaiian monk seals, whose pup survival rate is lower when the front, with its associated production, is far north of the islands. Spiny lobsters (*Panulirus marginatus*) recruitment in the Northwestern Hawaiian Islands is also affected.

In the California Current System (CCS), zooplankton species characteristic of shelf waters have since 1999 replaced the southerly, oceanic species which had been abundant since 1989 and northern fish species (Pacific salmon, cod, and rockfish species) have increased, while the southern migratory pelagics such as Pacific sardines, have declined. The distribution of Pacific hake (*Merluccius productus*), which range from Baja California to the Gulf of Alaska, is closely linked to hydrographic conditions. During the 1990s, the species occurred as far north as the Gulf of Alaska, but following a contraction of range by several hundred kilometers in 2000 and 2001, its northern limit has reverted to northern Vancouver Island, a return to the distribution observed in the 1980s.

The biological response to the 1998 regime shift was weaker in the Gulf of Alaska and the Bering Sea than in the central North Pacific and CCS. The northern regions of the western North Pacific resembled the southern regions of the eastern North Pacific in showing an increase in biological production. Zooplankton biomass increased in the Sea of Okhotsk and the previously dominant Japanese sardine was replaced by herring, capelin, and Japanese anchovy.

North Atlantic

Some of the consequences of the warming of the North Atlantic from the 1920s to the 1960s have already been described. There are numerous excellent publications on the subject, dating back to the 1930s and 1940s, which show how much scientific interest there was in the effects of climate change 65 years ago.

The history of cod stocks at Greenland since the early 1900s is particularly well documented, showing how rapidly a species can extend its range (at a rate of $50\,\mathrm{km\,yr^{-1}}$) and then decline again. This rate of range extension is matched by the examples shown in **Figure 1** and also by the plankton species which have been collected systematically since the 1930s by ships of opportunity towing continuous plankton recorder on many routes in the North Atlantic (*see* Continuous Plankton Recorders). This and other evidence indicate that the rates of change in distribution in response to climate are much more rapid in the sea than they are on land. There are fewer physical barriers in the sea; many marine species have dispersive planktonic life stages; the diverse, immobile habitats, which are created on land by large, long-lived plants, do not occur in the

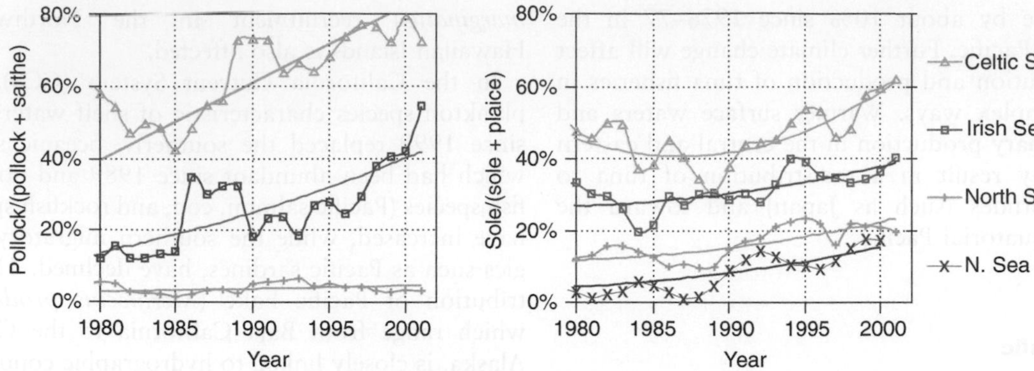

Figure 5 Increasing abundance of warm-adapted species (pollock: Pollachius pollachius and sole: *Solea solea*) relative to similar cold-adapted species (saithe: *Pollachius virens* and plaice: *Pleuronectes platessa*) as shown by catch ratios in the Celtic Sea, Irish Sea, and North Sea. The SST for the North Sea is also shown.

sea, with the exception of kelp forests in coastal areas. Coral reefs are another exception.

One of the difficulties in ascribing observed changes in fish stocks to climate is that fishing now has such a pervasive effect and exerts such high mortalities on most species. Where similar warm and cold-adapted species co-occur in a particular fishing area, the ratio of their catches gives an indication of which of them is being favored by climate trends (see **Figure 5**).

Baltic Sea

The Baltic Sea is the largest brackish water sea in the world, with much lower species diversity than the adjacent North Sea. The fisheries depend on just three marine species (cod, herring, and sprat), which are at the extreme limits of their tolerance of low salinity and oxygen. Over time, the Baltic has become fresher and less oxygenated, but there are periodic inflows of high-salinity, oxygen-rich water from the Skagerrak (North Sea) which flush through the deep basins and restore more favorable, saline, oxygenated conditions for the fish to reproduce. A specific, short-term weather pattern is required to generate an inflow: a period of easterly wind lowers the sea level in the Baltic, then several weeks of westerly wind force water from the Skagerrak through the shallow Danish Straits and into the deeper areas to the east.

On average, there has been approximately one major inflow per year since 1897 and the benefit to the fish stocks lasts for more than 1 year, but the frequency with which the required short-term weather pattern occurs depends on longer-term climatic factors. Inflows have been much less frequent over the past two decades, the last two major inflows happening in 1993 and 2003. The changes in windfields associated with climate change may continue to reduce the frequency of inflows to the Baltic.

The reproductive potential of cod in particular has been badly affected by the reduced volume of suitable water for development of its eggs. Cod spawn in the deep basins and if the salinity and density are too low, then their eggs sink into the anoxic layers near the bottom, where they die. This is one of the very few examples in which laboratory studies on individual fish can be applied almost directly to make inferences about the effects of climate. The buoyancy of cod eggs can be measured in a density gradient and their tolerance of low oxygen and salinity studied in incubation chambers. With this information, it is possible to determine what depth they will sink to under the conditions of temperature and salinity found in the deep basins and hence whether they remain in sufficiently oxygenated water for survival.

Coral Reef Fisheries

Coral reefs have suffered an increasing frequency of bleaching events due to loss of symbiotic algae. This occurs when SST remains $\sim 1\,^{\circ}\mathrm{C}$ above the current seasonal maxima. Mortality of corals occurs when the increment in SST is $>2\,^{\circ}\mathrm{C}$. Extensive and extreme bleaching occurred in 1998, associated with the strong El Niño and the hottest year on record.

The mass coral bleaching in the Indian Ocean in 1998 apparently did not result in major short-term impacts on coastal reef fisheries. However, in the longer term, the loss of coral communities and reduced structural complexity of the reefs are expected to have serious consequences for fisheries production with reduced fish species richness, local extinctions, and loss of species within key functional groups of reef fish.

Indirect Effects and Interactions

Spread of Pathogens

Pathogens have been implicated in mass mortalities of many aquatic species, including plants, fish, corals, and mammals, but lack of standard epidemiological data and information on pathogens generally make it difficult to attribute causes. An exception is the northward spread of two protozoan parasites (*Perkinsus marinus* and *Haplosporidium nelsoni*) from the Gulf of Mexico to Delaware Bay and further north, where they have caused mass mortalities of Eastern oysters (*Crassostrea virginica*). Winter temperatures consistently lower than 3 °C limit the development of the MSX disease caused by *Perkinsus* and the poleward spread of this and other pathogens can be expected to continue as such winter temperatures become rarer. This example also illustrates the relevance of seasonal information when considering the effects of climate change, since in this case it is winter temperature which controls the spread of the pathogen.

Effects of Changing Primary Production

Changes in primary production and in food-chain processes due to climate will probably be the major cause of future changes in fisheries production. Reduced nutrient supply to the upper ocean due to slower vertical mixing and changes in the balance between primary production and respiration at higher temperature will result in less energy passing to higher trophic levels. Many other processes will also have an effect, which makes prediction uncertain. For example, nutrient-depleted conditions favor small phytoplankton and longer food chains at the expense of diatoms and short food chains. Altered seasonality of primary production will cause mismatches with zooplankton whose phenology is adapted to the timing of the spring bloom. As a result, a greater proportion of the pelagic production may settle out of the water column to the benefit of benthic production.

There is evidence from satellite observations and from *in situ* studies that primary production in some parts of the ocean has begun to decline, but the results of modeling studies suggest that primary production will change very little up to 2050. Within this global result, there are large regional differences. For example, the highly productive sea–ice margin in the Arctic is retreating, but as a result a greater area of ocean is exposed to direct light and therefore becomes more productive.

Impacts of Changes in Fisheries on Human Societies

Fluctuations in fish stocks have had major economic consequences for human societies throughout history. Fishing communities which were dependent on local resources of just a few species have always been vulnerable to fluctuations in stocks, whether due to overfishing, climate, or other causes. The increase in distant water fleets during the last century reduced the dependence of that sector of the fishing industry on a particular area or species, but the resulting increase in rates of exploitation also reduced stock levels and increased their variability.

Many examples can be cited to show the effects of fish-stock fluctuations. The history of herring in European waters over the past 1000 years influenced the economic fortunes of the Hanseatic League and had a major impact on the economy of northern Europe. Climate-dependent fluctuations in the Far Eastern sardine population influenced their fisheries and human societies dependent on them. Variability in cod stocks at Newfoundland, Greenland, and the Faroe Islands due to a combination of climate-related effects and overfishing had major impacts on the economies and societies, resulting in changes in migration and human demography. The investigation of economic effects of climate change on fisheries is a rapidly developing field, which can be expected to help considerably when planning strategies for adaptation or, in some cases, mitigation of future impacts.

Given the uncertainties over future marine production and consequences for fish stocks, it is not surprising that projections of impacts on human societies and economies are also uncertain. Global aquaculture production increased by *c.* 50% between 1997 and 2003, while capture production decreased by *c.* 5% and the likelihood that these trends will continue also affects the way in which climate change will affect fisheries production.

Some areas, such as Greenland, which are strongly affected by climate variability and which have been undergoing a relatively cold period since the 1960s, can be expected to benefit from warmer oceanic conditions and changes in the marine ecosystem are occuring there quite rapidly. In other areas, such as Iceland, the positive and negative impacts are more finely balanced.

It is very difficult to judge at a global level who the main losers and winners will be from changes in fisheries as a result of climate change, aside from the obvious advantages of being well informed, well capitalized, and able to shift to alternative areas or kinds of fishing activity (or other nonfishery activities) as circumstances change. Some of the most vulnerable systems may be in the mega-deltas of rivers in Asia, such as the Mekong, where 60 million people are active in fisheries in some way or other. These are mainly seasonal floodplain fisheries, which, in addition to overfishing, are increasingly threatened by changes in the hydrological cycle and in land use, damming, irrigation, and channel alteration. Thus the impact of climate change is just one of a number of pressures which require integrated international solutions if the fisheries are to be maintained.

See also

Continuous Plankton Recorders. Ecosystem Effects of Fishing. Fisheries Overview. Fishery Manipulation through Stock Enhancement or Restoration. Marine Fishery Resources, Global State of. Open Ocean Fisheries for Deep-water Species. Open Ocean Fisheries for Large Pelagic Species. Small Pelagic Species Fisheries.

Further Reading

ACIA (2005) *Arctic Climate Impact Assessment Scientific Report*. Cambridge, UK: Cambridge University Press.

Cushing DH (1982) *Climate and Fisheries*. London: Academic Press.

Drinkwater KF, Loeng H, Megrey BA, Bailey N, and Cook RM (eds.) (2005) The influence of climate change on North Atlantic fish stocks. *ICES Journal of Marine Science* 62(7): 1203–1542.

German Advisory Council on Global Change (2006) *The Future Oceans – Warming up, Rising High, Turning Sour*. Special Report. Berlin: German Advisory Council on Global Change (WBGU). http://www.wbgu.de/wbgu_sn2006_en.pdf (accessed Mar. 2008).

King JR (ed.) (2005) *Report of the Study Group on Fisheries and Ecosystem Responses to Recent Regime Shifts*. PICES Scientific Report 28, 162pp.

Lehodey P, Chai F, and Hampton J (2003) Modelling climate-related variability of tuna populations from a coupled ocean biogeochemical-populations dynamics model. *Fisheries Oceanography* 12: 483–494.

Quéro JC, Buit HD, and Vayne JJ (1998) Les observations de poissons tropicaux et le réchauffement des eaux dans l'Atlantique européen. *Oceanologica Acta* 21: 345–351.

Stenseth Nils C, Ottersen G, Hurrell JW, and Belgrano A (eds.) (2004) *Marine Ecosystems and Climate Variation: The North Atlantic. A Comparative Perspective*. Oxford, UK: Oxford University Press.

Vilhjalmsson H (1997) Climatic variations and some examples of their effects on the marine ecology of Icelandic and Greenland waters, in particular during the present century. *Rit Fiskideildar* XV(1): 9–27.

Wood CM and McDonald DG (eds.) (1997) *Global Warming: Implications for Freshwater and Marine Fish*. Cambridge, UK: Cambridge University Press.

Relevant Website

http://www.ipcc.ch
– IPCC Fourth Assessment Report.

FISHERIES

FISHERIES OVERVIEW

M. J. Fogarty, Northeast Fisheries Science Center, National Marine Fisheries Service, Woods Hole, MA, USA
J. S. Collie, University of Rhode Island, Narragansett, RI, USA

Introduction

The long-standing importance of fishing as a human enterprise can be traced through the diversity of harvesting implements found in ancient archeological sites, artistic depictions throughout prehistory and antiquity, and the recorded history of many civilizations. Sustenance from the sea has been essential throughout human history and has formed the basis for trade and commerce in coastal cultures for millennia. The remains of fish and shellfish in extensive middens throughout the world attest to the prominence of these resources in the diets of early coastal peoples. In Northern Europe, the fortunes of the Hanseatic League during the medieval period were linked to trade in fishery resources, demonstrating a dominant role of fisheries in the trade of nations that extends over many centuries. Today, the critical importance of food resources from the sea has been further highlighted by increased demand related to the burgeoning human population and the recognized benefits of seafood as a high-quality source of protein. In turn, this importance is reflected in the diversity of fishery-related topics included in this encyclopedia. Here, we introduce the topics to be covered in the individual sections on fishing and fisheries resources and provide further background information to set the stage for these contributions.

The fishery resources of the oceans were long thought to be boundless and the high fertility of fishes was thought to render them impervious to human depredation. Thomas Henry Huxley, the preeminent Victorian naturalist, wrote in 1884 "... the cod fishery, the herring fishery, the pilchard fishery, the mackerel fishery, and probably all the great sea-fisheries, are inexhaustible; that is to say that nothing we do seriously affects the number of fish ... given our present mode of fishing. And any attempt to regulate these fisheries consequently ... seems to be useless." (quoted in Smith, 1994). Although Huxley qualified his remarks and limited them to the harvesting methods of his day and to ocean fisheries, the paradigm of inexhaustibility was broadly accepted, shaping the attitudes of fishers, scientists, managers, and politicians and complicating efforts to establish effective restrictions on fishing activities. Experimental studies of the effects of fishing on marine populations were conducted in Scotland as early as 1886. In a decade-long study, bays open and closed to fishing were compared, demonstrating that harvesting did result in declines in the abundance and average size of exploited fishes. However, the results were deemed controversial and the debate concerning the impact of fishing on marine populations continued through the middle of the twentieth century. By the second half of the last century with the development of large-scale industrial fisheries and distant-water fleets, it was abundantly clear that humans have the capacity to outstrip the production capacity of exploited marine populations, resulting in resource depletion and loss in yield.

In contrast to the long history of human harvest of the oceans, scientific endeavors in support of resource management (and the understanding of underlying basic ecological principles) are comparatively recent. Indeed, the term ecology (*oecologie*) itself was not coined until 1866, while written records of large-scale marine fisheries predate this landmark by several centuries. In many instances, fish and shellfish populations had already been substantially altered by fishing prior to the development of a scientific framework within which to evaluate these changes and true baseline conditions can only be inferred.

Because of the broad spatial and temporal scales over which fisheries operate, institutions dedicated to monitoring fisheries and resource species and providing scientific advice in support of management have been established. In Western countries, many of these institutions were formed in the latter nineteenth and early twentieth centuries. For example, the US Fish Commission was established in 1871 in response to concerns over declines in coastal fishery resources at that time. The Fishery Board of Scotland was established in 1883 and the International Council for Exploration of the Sea followed in 1902. These institutions and others such as the Marine Biological Association of the United Kingdom, established in 1884, approached the problem of understanding fluctuations of exploited fish and shellfish species from a broad scientific perspective, including consideration of the physical and biological environments of the organisms. Spencer Fullerton Baird, first US Commissioner of Fisheries, wrote that studies of fish "... would not be complete without a thorough knowledge of their associates in the sea, especially of such as prey upon them or constitute their food" (Baird, 1872). Baird further noted with respect to the importance of understanding the physical setting in the ocean that "... the temperature taken at different depths, its varying transparency, density, chemical composition, percentage of saline matter, its surface- and undercurrents, and other features of its physical condition ... throw more or less light on the agencies which exercise and influence upon the presence or absence of particular fishes." This broad multidisciplinary perspective remains an important component of fisheries investigations. Recently, the importance of an ecosystem perspective has been reemphasized and efforts to understand the potential impact of climate change on fishery resources have assumed high priority.

Attempts to estimate the production potential of the seas, based on energy flow in marine ecosystems, initially indicated that the coastal ocean could sustain yields of approximately 100 million tons of fish and shellfish on a global basis. Worldwide marine landings in 2004 were 86 million tons (*see* Marine Fishery Resources, Global State of). Globally, 3% of the stocks for which information is available are classified as underexploited, 20% are moderately exploited, 52% are considered fully exploited, 17% are overexploited, 7% are depleted, and 1% are listed as recovering. It is clear that we are at or near the limits to production for many exploited populations and have exceeded sustainable levels for many others. Improved management does hold the promise of increasing potential production in overexploited and depleted populations; carefully controlled increases in exploitation of currently underutilized species also may result in some increase in yield. As noted above, the pressures on fishery resources on a global basis are directly related to increases in human population size and the resulting increased demand for protein from the sea. Further, the recently emphasized health benefits of seafood consumption have resulted in increases in per capita consumption of fish in many Western countries that traditionally had comparatively low consumption levels.

Fishing and Fishery Resources

The articles concerning fishery resources in this encyclopedia document the broad spectrum of species taken and modes of capture in fisheries worldwide. The descriptions of fisheries for those species groups summarized herein are linked to overviews of their biological and ecological features elsewhere in the encyclopedia. Reviews are provided for major species groups supporting important fisheries. These reviews cover small-bodied fishes inhabiting the open water column (small pelagic fishes), larger-bodied pelagic fishes, bottom-dwelling organisms (demersal species), salmon, and shellfish (including crustaceans and mollusks). Regions (and associated species) requiring special consideration such as the vulnerable Antarctic marine ecosystem(s), coral reefs, and deepwater habitats are accorded separate treatment. In addition, articles dealing with a number of overarching issues are included to provide a broader context for understanding the importance of fisheries to society and their impacts on natural systems. An overview of fishing methods and techniques employed throughout the world is provided as are review articles on the global status of marine fishery resources, factors controlling the dynamics of exploited marine species, harvesting multiple species assemblages, and the ecosystem effects of fishing. Key considerations in the management of marine resources are documented and the intersection between human interests and motivations and resource management are explored.

The biological and ecological characteristics of the species sought in different fisheries and their behavioral patterns play a dominant role in harvesting methods, vulnerability to exploitation, and overall yields. The general strategies involved in fish capture include the use of entangling gears, trapping, filtering with nets, and hooking or spearing. Based on their experience and that of others gained over time, fishermen use detailed information on distribution, seasonal movement patterns, and other aspects of behavior of different species in the capture process. Recently, advances in electronic equipment, ranging from sophisticated hydroacoustic fishfinders to satellite navigation systems, have allowed fishermen to refine their understanding of these characteristics, greatly increasing the efficiency of fishing operations and the consequent impact on fishery resources.

Small-bodied fish such as herring, mackerel, and sardines that characteristically form large schools are often taken by surrounding nets or purse seines in large quantities (*see* Small Pelagic Species Fisheries). These pelagic species typically inhabit mid- to near-surface water depths. Schooling behavior can increase the detectability of these species and therefore their vulnerability to capture, a problem which has resulted in overharvesting of small pelagic species in many areas. Similar considerations hold for larger-bodied pelagic fishes such as the tunas (*see* Open Ocean Fisheries for Large Pelagic Species). The harvesting pressure on these larger species is fueled by their high unit value; management is complicated by their extensive movements and migrations, necessitating international management protocols and agreements.

A diverse assemblage of fish typically inhabiting near-bottom or bottom waters support fisheries of long-standing importance such as the cod fisheries throughout the North Atlantic, halibut fisheries in both the Atlantic and Pacific, and many others (*see* Demersal Species Fisheries).

The exploited demersal species exhibit a wide range of body sizes and life history characteristics that influence their response to exploitation. These species are often captured with nets dragged over the seabed, although traps, entangling nets, and other devices are also used. Such fishing gear often captures many species not specifically targeted by the fishery and may also disrupt the bottom habitat and associated species. As a result, substantial concern has been expressed over both the direct and indirect effects of bottom fishing practices with respect to alterations of food web structure and disturbance to critical habitat (*see* Ecosystem Effects of Fishing).

Species such as Atlantic and Pacific salmon spawn in fresh water but spend a substantial part of their life cycle in the marine environment (*see* Salmon Fisheries, Atlantic and Salmon Fisheries, Pacific). These anadromous species are impacted not only by harvesting but also by land- and freshwater-use practices affecting the part of the life cycle occurring in fresh water. Damming of rivers, deforestation, pollution, and overharvesting have all resulted in declines in these species in some areas. Salmon exhibit strong homing instincts and typically return to their natal river system to breed. The concentration of fish as they return to river systems and in the rivers and lakes themselves makes these species particularly vulnerable to capture. Pacific salmon differ considerably from most other fish species that support important fisheries in that they die after spawning only once; it is essential that a sufficient number of adults escape the capture process to replenish the population. Artificial enhancement through hatchery programs has been very widely employed for salmon stocks in an attempt to maintain viable populations (*see* Fishery Manipulation through Stock Enhancement or Restoration) although concerns have been raised about effects on the genetic structure of natural stocks and the possibility of transmission of diseases from hatchery stocks.

Fisheries for shellfishes, including those for lobsters, crabs, and shrimp (*see* Crustacean Fisheries) and for clams, snails, and squids (as well as other cephalopods; *see* Molluskan Fisheries) are among the most lucrative in the world. Mollusks also are taken for uses other than food, such as the ornamental trade in shells. Most shellfish live immediately on or near the seabed and are harvested by traps (lobsters, crabs, some cephalopods such as octopus, and some snails such as whelks), towed nets (e.g., shrimps and squid), and dredges (oysters, clams, etc.) among other devices. The ready availability of some types of shellfish in intertidal habitats has resulted in a very long history of exploitation by coastal peoples. Important commercial and recreational fisheries continue for these shellfish, which are often harvested with very simple implements. The high unit value of shellfish makes aquaculture economically feasible to supplement harvesting of natural populations.

Fisheries prosecuted in some habitats and environments require special considerations. For example, in the waters off Antarctica, the need to protect the potentially vulnerable food web, encompassing endangered and threatened marine mammal populations, has led to the development of a unique ecosystem-based approach to management (*see* Southern Ocean Fisheries). Harvesting of krill populations, a preferred prey of a number of whale, seal, and seabird populations, is regulated with specific recognition of the need to avoid disruption of the food web and impacts on these predators. In coral reef systems, the very high diversity of species found and the sensitivity of the habitat to disruption has highlighted the need for an ecosystem approach to management in these systems. Growing concern over destructive fishing practices using toxins and explosives has emphasized the need for effective management in these areas. In deep-water habitats, many resident species exhibit life history characteristics such as slow growth and delayed maturation that make them particularly vulnerable to exploitation (*see* Open Ocean Fisheries for Deep-water Species). The vulnerability of deep-sea habitats to disturbance by fishing gear is also a dominant concern in these environments.

Issues in Fishery Management

Fishery management necessarily entails consideration of resource conservation, the economic implications of alternative management strategies, and the social context within which management decisions are effected. The relative weights assigned to these diverse considerations can vary substantially in different settings, resulting in very different management decisions and outcomes. The setting of conservation standards is tied directly to understanding of basic life history characteristics such as the rate of reproduction at low population levels, growth characteristics, and factors affecting the survivorship from the early life stages to the age or size of vulnerability to the fishery (*see* Fisheries and Climate). The choice of particular harvesting strategies and levels holds both economic and social implications. Fishery management is ultimately a political process and decisions concerning allocation of fishery resources often engender intense debates. These debates are often set within the context of differing perspectives on fishing rights and privileges. In many societies, fishing is viewed as a basic right open to all citizens and fishery resources are often viewed as a form of common property.

Formal designation of fishery resources as *res nullia* (things owned by no one) can be traced to Roman law where ownership was conferred by the process of capture. Traditions of open access to fishery resources in many Western countries persist and remain a principal factor in the global escalation of fishing pressure. This legacy has led to excess capacity and overcapitalization of world fishing fleets, resulting in conflicts between conservation requirements and the social and short-term economic impacts of implementing rational and effective management. Garrett Hardin's (1968) influential statement of the "Tragedy of the commons" – a resource owned by no one is cared for by no one – has been applied to fishery resources and further honed to reflect considerations of the importance of well-defined property rights and attendant responsibilities in natural resources management. Various forms of dedicated access privileges have been implemented in fisheries around the world to reduce overcapitalization and to vest fishermen in the long-term sustainability of fisheries.

Biological reference points provide the basis for specifying objectives for fishery management in many of the major fisheries throughout the world. Limit reference points define the boundaries of a situation that could cause serious harm to a stock, while target reference points are used to determine harvest control rules that are risk-averse and have a low probability of causing serious harm. Limits are conceived as reference levels that should have a low probability of being exceeded and are designed to prevent stock declines through recruitment overfishing. Targets are reference levels providing management goals but which may not necessarily be met under all conditions. Although originally conceived as target reference points, the fishing mortality rate resulting in maximum sustainable yield and the corresponding level of equilibrium biomass are now commonly employed as limit reference points. Yield and spawning stock biomass per recruit analyses have been used to provide both limit and target reference points. It is possible to construct a two-dimensional representation of the exploitation status of a stock in relation to the estimated levels of fishing mortality and population size (**Figure 1**). When coupled with information on threshold levels of fishing mortality and population size used to define limit reference points, decision rules can be defined to assess appropriate courses of action. In instances where the limit fishing mortality reference point is exceeded, 'overfishing' is said to occur; when the stock declines below the limit biomass reference point, the stock is 'overfished' and management action is required. It is further possible to specify target exploitation levels in this context.

Although biological reference points have been widely applied on a global basis and are often required under fisheries legislation, corresponding economic reference points exist and deserve special consideration. For example, the concept of maximum economic yield has served as a cornerstone of resource economic theory. Resource economists have long recognized that in an unregulated open-access fishery, fishing effort increases to a bioeconomic equilibrium at which profits are completely dissipated. Developing ways to understand and create appropriate incentive structures to ensure appropriate and efficient economic utilization of fishery resources is critically important.

Tools available to fishery managers to control the activities of harvesters include constraints on the overall amount of fishing activity (measured as the number of vessels allowed permits, the number of days vessels can spend at sea, etc.), the total amount of the catch allowed in a specified time period and regulated by various forms of quota systems, the types of fishing gear that can be used and their

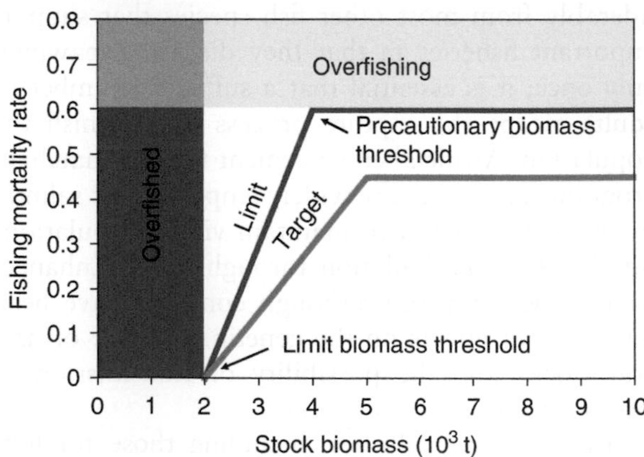

Figure 1 Control diagram for evaluating the status of a marine fish stocks in relation to biological reference points. The green line defines target levels of stock biomass and fishing mortality; the red line defines limits to fishing mortality and biomass thresholds.

characteristics (e.g., net mesh size), and closures of fishing grounds (including seasonal and year-round closures). Often, several of these tools will be used in combination to meet specified management objectives. As we move toward a paradigm of Ecosystem Approaches to Fishery Management (EAFM), these basic tools will remain the essential elements in tactical fishery management. However, because a broader suite of objectives will be embodied in EAFM, the mix of management tools applied in a given setting will undoubtedly differ relative to single-species management strategies.

Failures in fishery management can often be traced to conflicting goals and objectives in the conservation, economic, and social dimensions. For example, the needs for conservation can be compromised by desires to maintain full employment opportunities in the fishing industry if this leads to political pressure to permit high harvest levels. In the longer term, actions taken to ensure the sustainability of fishery systems also ensure the viability of the industries dependent on these resources. However, the short-term impacts (or perceived impacts) of fisheries regulations on fishers and fishing communities are often dominant considerations in whether particular regulations will be put in place.

Emerging Issues in Fisheries and Fisheries Science

The need for a more holistic view of human impacts in the marine environment is increasingly recognized. Harvesting has both direct and indirect effects on marine ecosystems. The former include removal of biomass and potential impacts on habitat and non-target species (*see* Ecosystem Effects of Fishing). The latter includes alteration in trophic structure through species-selective harvesting patterns changing the relative balance of predators and their prey . Multispecies considerations in fishery management account for interactions among harvested species and the need to consider factors such as the food and energetic requirements of protected resource species. These interactions must be further viewed in the context of external forcing such as climate change and variability as noted in Fisheries and Climate (see also below). Collectively, these factors can result in shifts in productivity states that must be accounted for in management. Further, species interactions require that we explicitly deal with trade-offs in management (e.g., between predators and their prey).

The global escalation in fishing pressure and recognition of the potential environmental impacts of fishing activities have led to an increased interest in and emphasis on an ecosystem approach to fisheries management. An ecosystem approach to fisheries management seeks to ensure the preservation of ecosystem composition, structure, and function based on an understanding of ecological interactions and processes required for ecosystem integrity. As noted above, ecosystem principles guide the management of marine resources off Antarctica (*see* Southern Ocean Fisheries), and broader application of an ecosystem approach to fisheries management approaches is under development in many other regions.

The interest in an ecosystem approach to fisheries management has led to a reevaluation of management tools and an increased emphasis on the use of strategies such as the development of marine protected areas in which harvesting and other extractive activities are strongly controlled. The use of closed areas as a fishery management tool has an extensive history and the current focus on marine reserves can be viewed as an extension of these long-standing approaches to meet broader conservation objectives. Areas in which all extractive practices are banned may reduce overall exploitation rates, protect sensitive habitats and associated biological communities, and preserve ecosystem structure and function. The use of no-take marine reserves has been also advocated as a hedge against uncertainty in our understanding of ecosystem structure and function and in our ability to control harvest rates. While there is substantial evidence of increases in biomass, mean size, and biodiversity within reserves, the effects on adjacent areas through spillover effects are less well documented, although information is accruing. The importance of reserves as source areas for adjacent sites open to harvesting and other activities can be critical for the resilience of spatially linked populations. Used in concert with other measures that restrain fishing activities in areas open to fishing, no-take marine protected areas can be a highly effective component of an ecosystem approach to fisheries management strategies.

The prospect of global climate change and its implications for terrestrial and marine systems is one of the most pressing issues facing us today. The potential impacts of global climate change in the marine environment are receiving increased attention as higher-resolution forecasting models are developed and oceanographic features are more fully represented in general circulation models. The projected climate impacts with respect to changes in temperature, precipitation, and wind fields hold important implications for oceanic current systems, mesoscale features such as frontal zones and eddies, stratification, and thermal structure. In turn, these impacts will affect marine organisms dependent on the physical geography of the sea for dispersal in currents, the size and location of feeding and spawning grounds, and basic biological considerations such as temperature effects on metabolism. Shifts in distribution patterns of fishery resource species, changes in vital rates such as survivorship and growth, and alterations in the structure of marine communities can be anticipated. Persistent changes in environmental conditions can affect the production characteristics of different systems and the production potential of individual species (*see* Fisheries and Climate). In particular, a shift in environmental states can interact synergistically with fishing pressure to destabilize an exploited population. Exploitation rates that are sustainable under a favorable environmental regime may not remain so if a shift to less favorable conditions occurs. Large-scale research programs such as the Global Ocean Ecosystem Dynamics (GLOBEC) Program have now been implemented throughout the world to assess the potential effects of global climate change on marine ecosystems, including impacts on resource species.

Collectively, the problems of overexploitation, habitat loss and degradation, alteration of ecosystem structure, and environmental change caused by human activities point to the need to consider humans fully as part of the ecosystem and not somehow apart, and to manage accordingly. Notwithstanding the depleted status of many world fisheries, several important fisheries have recovered from overexploitation in response to management regulations. Wisely managed, fisheries can continue to meet important human needs for food resources from the sea while meeting our obligations to future generations.

See also

Crustacean Fisheries. Demersal Species Fisheries. Ecosystem Effects of Fishing. Fisheries and Climate. Fishery Manipulation through Stock Enhancement or Restoration. Marine Fishery Resources, Global State of. Molluskan Fisheries. Open Ocean Fisheries for Deep-water Species. Open Ocean Fisheries for Large Pelagic Species. Salmon Fisheries, Atlantic. Salmon Fisheries, Pacific. Small Pelagic Species Fisheries. Southern Ocean Fisheries.

Further Reading

Baird SF (1872) Report on the conditions of the sea fisheries, 1871. *Report of the US Fish Commission 1*. Washington, DC: US Fish Commission.

Cushing DH (1988) *The Provident Sea*. Cambridge, UK: Cambridge University Press.

FAO (2006) *The State of World Fisheries and Aquaculture*. Rome: Food and Agriculture Organization of the United Nations. http://www.fao.org/sof/sofia/index_en.htm (accessed Mar. 2008).

Fogarty MJ, Bohnsack J, and Dayton P (2000) Marine reserves and resource management. In: Sheppard C (ed.) *Seas at the Millennium: An Environmental Evaluation*, ch. 134, pp. 283–300. Amsterdam: Elsevier.

Hardin G (1968) The tragedy of the commons. *Science* 162: 1243–1247.

Jennings S, Kaiser MJ, and Reynolds JD (2001) *Marine Fisheries Ecology*. Oxford, UK: Blackwell Science.

Kingsland SE (1994) *Modeling Nature*. Chicago, IL: University of Chicago Press.

Pauly D (1996) One hundred million tons of fish, and fisheries research. *Fisheries Research* 25: 25–38.

Quinn TJ, II and Deriso RB (1999) *Quantitative Fish Dynamics*. Oxford, UK: Oxford University Press.

Ryther J (1969) Photosynthesis and fish production in the sea. *Science* 166: 72–76.

Sahrhage D and Lundbeck J (1992) *A History of Fishing*. Berlin: Springer.

Smith TD (1994) *Scaling Fisheries: The Science of Measuring the Effects of Fishing, 1855–1955*. Cambridge, UK: Cambridge University Press.

DEMERSAL SPECIES FISHERIES

K. Brander, International Council for the Exploration of the Sea (ICES), Copenhagen, Denmark

Introduction

Demersal fisheries use a wide variety of fishing methods to catch fish and shellfish on or close to the sea bed. Demersal fisheries are defined by the type of fishing activity, the gear used and the varieties of fish and shellfish which are caught. Catches from demersal fisheries make up a large proportion of the marine harvest used for human consumption and are the most valuable component of fisheries on continental shelves throughout the world.

Demersal fisheries have been a major source of human nutrition and commerce for thousands of years. Models of papyrus pair trawlers were found in Egyptian graves dating back 3000 years. The intensity of fishing activity throughout the world, including demersal fisheries, has increased rapidly over the past century, with more fishing vessels, greater engine power, better fishing gear and improved navigational and fish finding aids. Many demersal fisheries are now overexploited and all are in need of careful assessment and management if they are to provide a sustainable harvest.

Demersal fisheries are often contrasted with pelagic fisheries, which use different methods to catch fish in midwater and close to the water surface. Demersal species are also contrasted with pelagic species (see relevant sections), but the distinction between them is not always clear. Demersal species frequently occur in mid-water and pelagic species occur close to the seabed, so that 'demersal' species are frequently caught in 'pelagic' fisheries and 'pelagic' species in demersal fisheries. For example Atlantic cod (*Gadus morhua*), a typical 'demersal' species, occurs close to the seabed, but also throughout the water column and in some areas is caught equally in 'demersal' and 'pelagic' fishing gear. Atlantic herring (*Clupea harengus*), a typical pelagic species, is frequently caught on the seabed, when it forms large spawning concentrations as it lays its eggs on gravel banks.

Total marine production rose steadily from less than 20 million tonnes (Mt) in 1950 to around 100 Mt during the late 1990s (**Figure 1**). The demersal fish catch rose from just over 5 Mt in 1950 to around 20 Mt by the early 1970s and has since fluctuated around that level (**Figure 2**). The proportion of demersal fish in this total has therefore declined over the period 1970–1998.

The products of demersal fisheries are mainly used for human consumption. The species caught tend to be relatively large and of high value compared with typical pelagic species, but there are exceptions to such generalizations. For example the industrial (fishmeal) fisheries of the North Sea, which take over

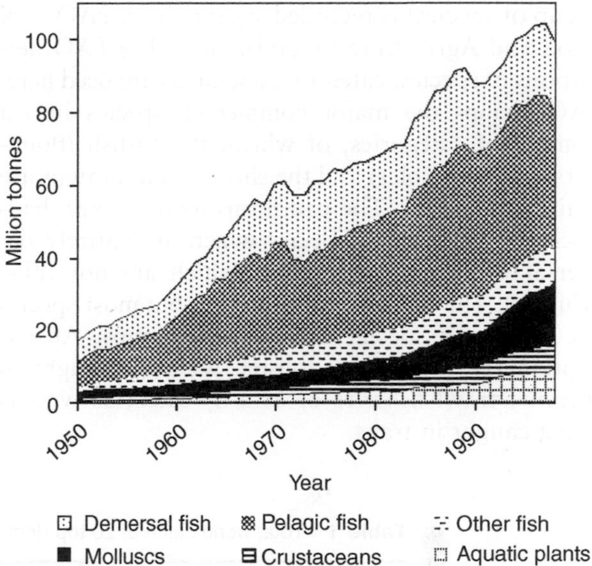

Figure 1 Total marine landings.

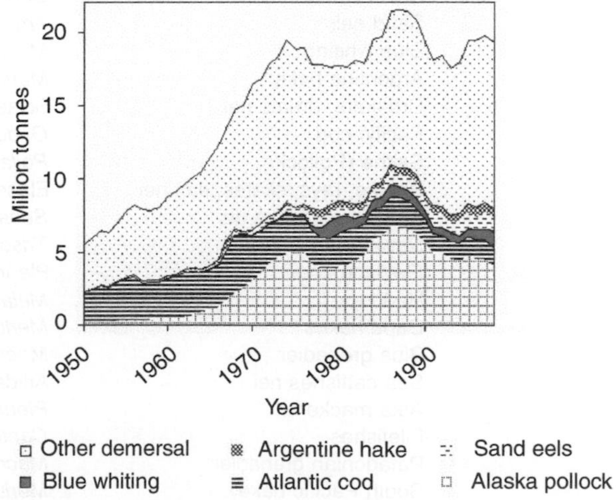

Figure 2 Total demersal fish landings.

half of the total fish catch, are principally based on low-value demersal species which live in or close to the seabed (sand eels, small gadoids).

Demersal fisheries are also often known as groundfish fisheries, but the terms are not exact equivalents because 'groundfish' excludes shellfish, which can properly be considered a part of the demersal catch. Shellfish such as shrimps and lobsters constitute the most valuable component of the demersal trawl catch in some areas.

Principal Species Caught in Demersal Fisheries

For statistical, population dynamics and fisheries management purposes the catch of each species (or group of species) is recorded separately by FAO (UN Food and Agriculture Organization). The FAO definitions of species, categories and areas are used here. FAO groups the major commercial species into a number of categories, of which, the flatfish (flounders, halibuts, soles) and the shrimps and prawns are entirely demersal. The gadiforms (cods, hakes, haddocks) include some species which are entirely demersal (haddock) and others which are not (blue whiting). The lobsters are demersal, but most species are caught in special fisheries using traps. An exception is the Norway lobster, which is caught in directed trawl fisheries or as a by-catch, as well as being caught in traps.

The five demersal marine fish with the highest average catches over the decade 1989–98 are Alaska (walleye) pollock, Atlantic cod, sand eels, blue whiting and Argentine hake. These species all spend a considerable proportion of their time in mid-water. Sand eels spend most of their lives on or in the seabed, but might not be regarded as a typical demersal species, being small, relatively short-lived and of low value. Sand eels and eight of the other 20 top 'species' in fact consist of more than one biological species, which are not identified separately in the FAO classification (**Table 1**).

The vast majority of demersal fisheries take place on the continental shelves, at depths of less than 200 m. Fisheries for deep-sea species, down to several thousand meters, have only been undertaken for the past few decades, as the technology to do so developed and it became profitable to exploit other species.

The species caught in demersal fisheries are often contrasted with pelagic species in textbooks and described as large, long-lived, high-value fish species, with relatively slow growth rates, low variability in recruitment and low mortality. There are so many exceptions to such generalizations that they are likely to be misleading. For example tuna and salmon are large, high-value pelagic species. Three of the main pelagic species in the North Atlantic (herring, mackerel and horse mackerel) are longer lived, have lower mortality rates and lower variability of recruitment than most demersal stocks in that area (**Table 2**).

Table 1 Total world catch of 20 top demersal fish species (averaged from 1989–1998)

Common name	*Scientific name*	*Tonnes*
Alaska pollock	*Theragra chalcogramma*	3 182 645
Atlantic cod	*Gadus morhua*	2 317 261
Sand eels	*Ammodytes* spp.	1 003 343
Blue whiting	*Micromesistius poutassou*	628 918
Argentine hake	*Merluccius hubbsi*	526 573
Croakers, drums nei	Sciaenidae	492 528
Pacific cod	*Gadus macrocephalus*	425 467
Saithe (Pollock)	*Pollachius virens*	385 227
Sharks, rays, skates, etc. nei	Elasmobranchii	337 819
Atlantic redfishes nei	*Sebastes* spp.	318 383
Norway pout	*Trisopterus esmarkii*	299 145
Flatfishes nei	Pleuronectiformes	289 551
Haddock	*Melanogrammus aeglefinus*	273 459
Cape hakes	*Merluccius capensis, M. paradox*	266 854
Blue grenadier	*Macruronus novaezelandiae*	255 421
Sea catfishes nei	Ariidae	242 815
Atka mackerel	*Pleurogrammus azonus*	237 843
Filefishes	*Cantherhines* (= *Navodon*) spp.	234 446
Patagonian grenadier	*Macruronus magellanicus*	230 221
South Pacific hake	*Merluccius gayi*	197 911
Threadfin breams nei	*Nemipterus* spp.	186 201

Table 2 The intensity of fishing is expressed as the average (1988–1997) probability of being caught during the next year. The interannual variability in number of young fish is expressed as the coefficient of variation of recruitment. Species shown are some of the principal demersal and pelagic fish caught in the north-east Atlantic

	Probability of being caught during next year	*Coefficient of variation of recruitment*
Demersal species		
Cod	33–64%	38–65%
Haddock	19–52%	70–151%
Hake	27%	33%
Plaice	32–48%	35–56%
Saithe	29–42%	45–56%
Sole	28–37%	15–94%
Whiting	47–56%	41–61%
Pelagic species		
Herring	12–40%	56–63%
Horse mackerel	16%	40%
Mackerel	21%	41%

Fishing Gears and Fishing Operations

A very wide range of fishing gear is used in demersal fisheries, the main ones being bottom trawls of different kinds, which are dragged along the seabed behind a trawler. Other methods include seine nets, trammel nets, gill nets, set nets, baited lines and longlines, temporary or permanent traps and barriers.

Some fisheries and fishermen concentrate exclusively on demersal fishing operations, but many alternate seasonally, or even within a single day's fishing activity, between different methods. Fishing vessels may be designed specifically for demersal or pelagic fishing or may be multipurpose.

Effects of Demersal Fisheries on the Species They Exploit

Most types of demersal fishing operation are nonselective in the sense that they catch a variety of different sizes and species, many of which are of no commercial value and are discarded. Stones, sponges, corals and other epibenthic organisms are frequently caught by bottom trawls and the action of the fishing gear also disturbs the seabed and the benthic community on and within it. Thus in addition to the intended catch, there is unintended disruption or destruction of marine life (*see also* Ecosystem Effects of Fishing)

The fact that demersal fishing methods are nonselective has important consequences when trying to limit their impact on marine life. There are direct impacts, when organisms are killed or disturbed by fishing, and indirect impacts, when the prey or predators of an organism are removed or its habitat is changed.

The resilience or vulnerability of marine organisms to demersal fishing depends on their life history. In areas where intensive demersal fisheries have been operating for decades to centuries the more vulnerable species will have declined a long time ago, often before there were adequate records of their occurrence. For example, demersal fisheries caused a decline in the population of common skate (*Raia batis*) in the north-east Atlantic and barndoor skate (*Raia laevis*) in the north-west Atlantic, to the point where they are locally extinct in areas where they were previously common. These large species of elasmobranch have life histories which, in some respects, resemble marine mammals more than they do teleost fish. They do not mature until 11 years old, and lay only a small number of eggs each year. They are vulnerable to most kinds of demersal fishery, including trawls, seines, lines, and shrimp fisheries in shallow water.

The selective (evolutionary) pressure exerted by fisheries favors the survival of species which are resilient and abundant. It is difficult to protect species with vulnerable life histories from demersal fisheries and they may be an inevitable casualty of fishing. Some gear modifications, such as separator panels may help and it may be possible to create refuges for vulnerable species through the use of large-scale marine protected areas. Until recently fisheries management ignored such vulnerable species and concentrated on the assessment and management of a few major commercial species.

In areas with intensive demersal fisheries the probability that commercial-sized fish will be caught within one year is often greater than 50% and the fisheries therefore have a very great effect on the level and variability in abundance (**Table 2**). The effect of fishing explains much of the change in abundance of commercial species which has been observed during the few decades for which information is available and the effects of the environment, which are more difficult to estimate, are regarded as introducing 'noise', particularly in the survival of young fish. As the length of the observational time series increases and information about the effects of the environment on fish accumulates, it is becoming possible to turn more of the 'noise' into signal. It is no longer credible or sensible to ignore environmental effects when evaluating fluctuations in demersal fisheries, but a considerable scientific effort is still needed in order to include such information effectively.

Effects of the Environment on Demersal Fisheries

The term 'environment' is used to include all the physical, chemical and biological factors external to the fish, which influence it. Temperature is one of the main environmental factors affecting marine species. Because fish and shellfish are ectotherms, the temperature of the water surrounding them (ambient temperature) governs the rates of their molecular, physiological and behavioral processes. The relationship between temperature and many of these rates processes (growth, reproductive output, mortality) is domed, with an optimum temperature, which is species and size specific (**Figure 3**). The effects of variability in temperature are therefore most easily detected at the extremes and apply to populations and fisheries as well as to processes within a single organism. Temperature change may cause particular species to become more or less abundant in the demersal fisheries of an area, without necessarily affecting the aggregate total yield.

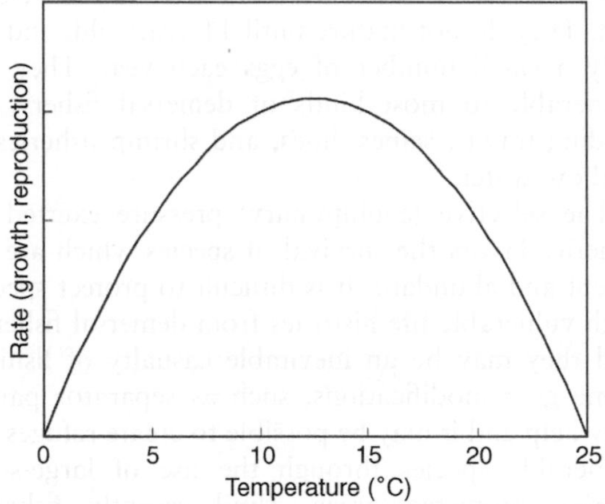

Figure 3 The relationship between temperature and many rate processes (growth, reproductive output, mortality) is domed. The optimum temperature is species and size specific.

The cod (*Gadus morhua*) at Greenland is at the cold limit of its thermal range and provides a good example of the effects of the environment on a demersal fishery; the changes in the fishery for it during the twentieth century are mainly a consequence of changes in temperature (**Figure 4**). Cod were present only around the southern tip of Greenland until 1917, when a prolonged period of warming resulted in the poleward expansion of the range by about 1000 km during the 1920s and 1930s. Many other boreal marine species also extended their range at the same time and subsequently retreated during the late 1960s, when colder conditions returned.

Changes in wind also affect demersal fish in many different ways. Increased wind speed causes mixing of the water column which alters plankton production. The probability of encounter between fish larvae and their prey is altered as turbulence increases. Changes in wind speed and direction affect the transport of water masses and hence of the planktonic stages of fish (eggs and larvae). For example, in some years a large proportion of the fish larvae on Georges Bank are transported into the Mid-Atlantic Bight instead of remaining on the Bank. In some areas, such as the Baltic, the salinity and oxygen levels are very dependent on inflow of oceanic water, which is largely wind driven. Salinity and oxygen in turn affect the survival of cod eggs and larvae, with major consequences for the biomass of cod in the area. These environmental effects on the early life stages of demersal fish affect their survival

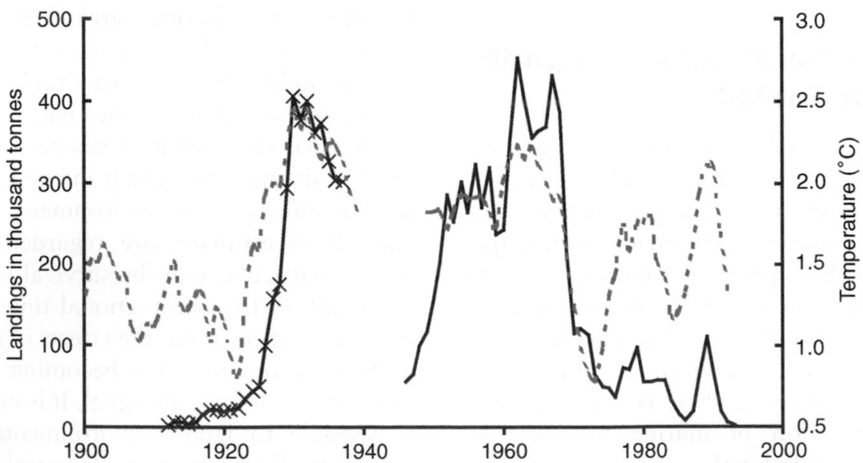

Figure 4 Cod catch and water temperature at West Greenland. Temperature is the running five-year mean of upper layer (0–40 m) values. ×, local catch × 20; ——, international catch; – – –, temperature.

and hence the numbers which recruit to the adult population.

World Catches from Demersal Fisheries and the Limits

Demersal fisheries occur mainly on the continental shelves (i.e., at depths less than 200 m). This is because shelf areas are much more productive than the open oceans, but also because it is easier to fish at shallower depths, nearer to the coast. The average catch of demersal fish per unit area on northern hemisphere temperate shelves is twice as high as on southern hemisphere temperate shelves and more than five times higher than on tropical shelves (**Figure 5**). The difference is probably due to nutrient supply. The effects of differences in productive capacity of the biological system on potential yield from demersal fisheries are dealt with elsewhere (*see also* Ecosystem Effects of Fishing).

Demersal fisheries provide the bulk of fish and shellfish for direct human consumption. The steady increase in the world catch of demersal fish species ended in the early 1970s and has fluctuated around 20 Mt since then (**Figure 2**). Many of the fisheries are overexploited and yields from them are declining. In a few cases it would seem that the decline has been arrested and the goal of managing for a sustainable harvest may be closer.

A recent analysis classified the top 200 marine fish species, accounting for 77% of world marine fish production, into four groups – undeveloped, developing, mature and declining (senescent). The proportional change in these groups over the second half of the twentieth century (**Figure 6**) shows how fishing has intensified, so that by 1994 35% of the fish stocks were in the declining phase, compared with 25% mature and 40% developing. Other analyses reach similar conclusions – that roughly two-thirds of marine fish stocks are fully exploited or overexploited and that effective management is needed to stabilize current catch levels.

Fish farming (aquaculture) is regarded as one of the principal means of increasing world fish production, but one should recall that a considerable proportion of the diet of farmed fish is supplied by demersal fisheries on species such as sand eel and Norway pout. Fishmeal is also used to feed terrestrial farmed animals.

Management of Demersal Fisheries

The purposes of managing demersal fisheries can be categorized as biological, economic and social. Biological goals used to be set in terms of maximum sustainable yield of a few main species, but a broader and more cautious approach is now being introduced, which includes consideration of the ecosystem within which these species are produced and which takes account of the uncertainty in our assessment of the consequences of our activities. The formulation of biological goals is evolving, but even the most basic, such as avoiding extinction of species, are not being achieved in many cases. At a global level it is evident that economic goals are not being achieved, because the capital and operating costs of marine fisheries are about 1.8 times higher than the gross revenue. There are innumerable examples of adverse social impacts of

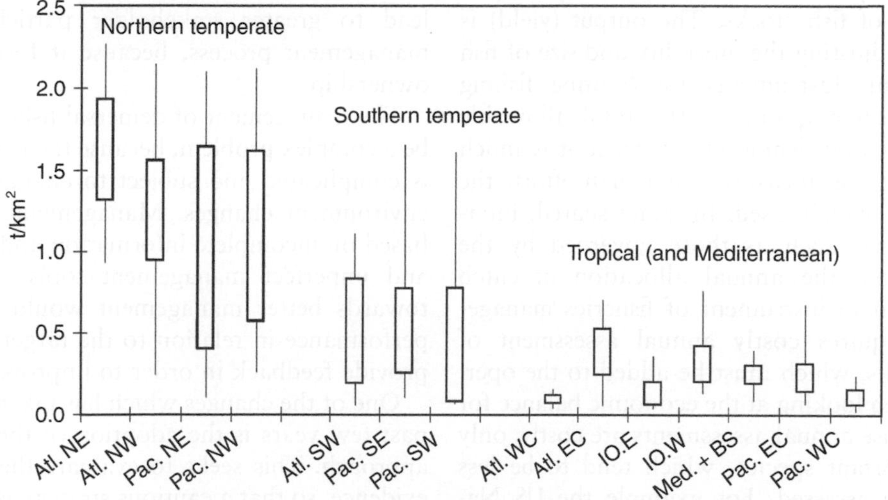

Figure 5 Demersal fish landings per unit area of continental shelf <200 m deep for the main temperate and tropical areas. The boxes show the spread between the upper and lower quartiles of annual landings and the whiskers show the highest and lowest annual landings 1950–1998. Atl, Atlantic; Pac, Pacific; IO, Indian Ocean; Med, Mediterranean; BS, Black Sea.

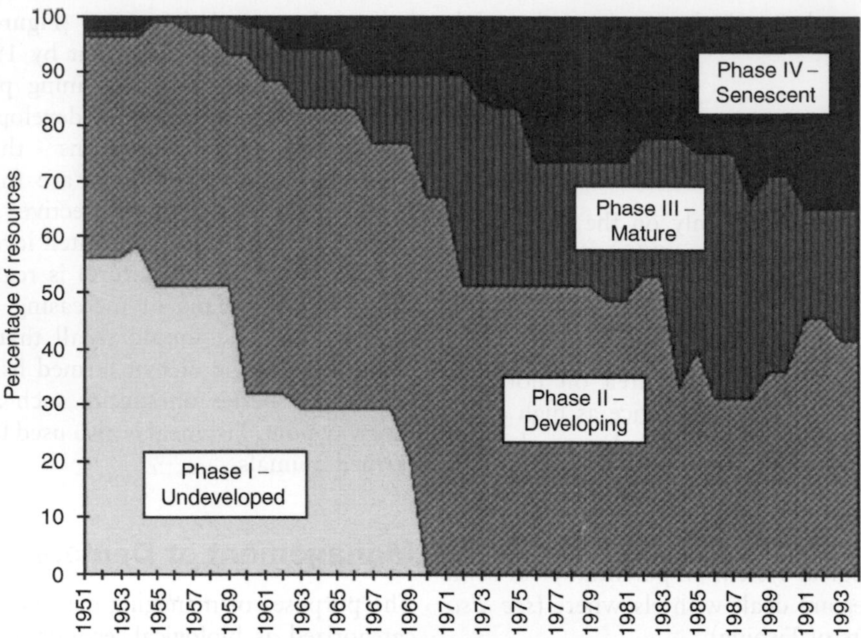

Figure 6 Temporal change in the level of exploitation of the top 200 marine fish species, showing the progression from mainly undeveloped or developing in 1951 to mainly fully exploited or overexploited in 1994.

changes in fisheries, often caused by the effects of larger, industrial fishing operations on the quality of life and standard of living of small-scale fishing communities. Clearly there is scope for improvement in fisheries management.

From this rather pessimistic analysis of where fisheries management has got us to date, it follows that a description of existing management regimes is a record of current practice rather than a record of successful practice.

Biological management of demersal fisheries has developed mainly from a single species 'yield-per-recruit' model of fish stocks. The output (yield) is controlled by adjusting the mortality and size of fish that are caught. Instruments for limiting fishing mortality are catch quotas (TAC, total allowable catch) and limits on fishing effort. Since it is much easier to define and measure catch than effort, the former is more widely used. In many shared, international fisheries, such as those governed by the European Union, the annual allocation of catch quotas is the main instrument of fisheries management. This requires costly annual assessment of many fish stocks, which must be added to the operating costs when looking at the economic balance for a fishery. Because annual assessments are costly, only the most important species, which tend to be less vulnerable, are assessed. For example the US National Marine Fisheries Service estimate that the status of 64% of the stocks in their area of responsibility is unknown.

The instruments for limiting the size of fish caught are mesh sizes, minimum landing sizes and various kinds of escape panels in the fishing gear. These instruments can be quite effective, particularly where catches are dominated by a single species. In multispecies fisheries, which catch species with different growth patterns, they are less effective because the optimal mesh size for one species is not optimal for all.

There are two classes of economic instruments for fisheries management: (1) property rights and (2) corrective taxes and subsidies. The former demands less detailed information than the latter and may also lead to greater stakeholder participation in the management process, because it fosters a sense of ownership.

The management of demersal fisheries will always be a complex problem, because the marine ecosystem is complicated and subject to change as the global environment changes. Management will always be based on incomplete information and understanding and imperfect management tools. A critical step towards better management would be to monitor performance in relation to the target objectives and provide feedback in order to improve the system.

One of the changes which has taken place over the past few years is the adoption of the precautionary approach. This seeks to evaluate the quality of the evidence, so that a cautious strategy is adopted when the evidence is weak. Whereas in the past such balance of evidence arguments were sometimes applied in order to avoid taking management action unless the evidence was strong (in order to avoid possible unnecessary disruption to the fishing industry), the presumption now is that in case of doubt it is the fish stocks rather than the short-term interests of the fishing industry which should be protected. This is a very significant change in attitude, which gives some grounds for optimism in the continuing struggle to achieve sustainable fisheries and healthy ecosystems.

Demersal Fisheries by Region

The demersal fisheries of the world vary greatly in their history, fishing methods, principal species and management regimes and it is not possible to review all of these here. Instead one heavily exploited area with a long history (New England) and one less heavily exploited area with a short history (the south-west Atlantic) will be described.

New England Demersal Fisheries

The groundfish resources of New England have been exploited for over 400 years and made an enormous contribution to the economic and cultural development of the USA since the time of the first European settlements. Until the early twentieth century, large fleets of schooners sailed from New England ports to fish, mainly for cod, from Cape Cod to the Grand Banks. The first steam-powered otter trawlers started to operate in 1906 and the introduction of better handling, preservation and distribution changed the market for fish and the species composition of the catch. Haddock became the principal target and their landings increased to over 100 000 t by the late 1920s. The advent of steam trawling raised concern about discarding and about the damage to the seabed and to benthic organisms.

From the early 1960s fishing fleets from European and Asian countries began to take an increasing share of the groundfish resources off New England. The total groundfish landings rose from 200 000 t to 760 000 t between 1960 and 1965. This resulted in a steep decline in groundfish abundance and in 1970 a quota management scheme was introduced under the International Commission for Northwest Atlantic Fisheries (ICNAF). Extended jurisdiction ended the activities of distant water fleets, but was quickly followed by an expansion of the US fleet, so that although the period 1974–78 saw an increase in groundfish abundance, the decline subsequently continued. Most groundfish species remain at low levels and, even with management measures intended to rebuild the stocks, are likely to take more than a decade to recover.

The changes in abundance of ‘traditional’ groundfish stocks (cod, haddock, redfish, winter flounder, yellowtail flounder) have to some extent been offset by increases in other species, including some elasmobranchs (sharks and rays). However, prolonged high levels of fishing have resulted in severe declines in the less resilient species of both elasmobranchs (e.g. barndoor skate) and teleosts (halibut, redfish). This is covered more fully elsewhere (*see* Ecosystem Effects of Fishing).

The second half of the twentieth century saw major changes in the species composition of the US demersal fisheries. By the last decade of the century catches of the two top fish species during the decade 1950–59, redfish and haddock, had declined to 0.6% and 2.3% of their previous levels, respectively. Shellfish had become the main element of the demersal catch (**Table 3**).

Demersal Fisheries of the South-West Atlantic (FAO Statistical Area 41)

The catch from demersal fisheries in the SW Atlantic increased steadily from under 90 000 tonnes in 1950 to over 1.2 Mt in 1998 (**Figure 7**). The demersal catch consists mainly of fish species, of which Argentine hake has been predominant throughout the 50-year record. The catch of shrimps, prawns, lobsters and crabs is almost 100 000 t per year and they have a relatively high market value.

Thirty countries have taken part in demersal fisheries in this area, the principal ones being Argentina, Brazil, and Uruguay, with a substantial and continuing component of East European effort. A three year ‘pulse’ of trawling by the USSR fleet resulted in a catch over 500 000 t of Argentine hake and over 100 000 t of demersal percomorphs in 1967. The fisheries in this area are mostly industrialized and long range. As the stocks on the continental shelf have become fully exploited, the fisheries have extended into deeper water, where they take pink cusk eel and Patagonian toothfish. The main coastal demersal species are whitemouth croaker, Argentine croaker and weakfishes.

The Argentine hake fishery extends over most of the Patagonian shelf. It is now fully exploited and possibly even overexploited. Southern blue whiting and Patagonian grenadier are also close to full exploitation. Hake and other demersal species are regulated by annual TAC and minimum mesh size regulations.

Conclusions

Demersal fisheries have been a major source of protein for people all over the world for thousands of years. World catches increased rapidly during the

Table 3 Average US catch from the north-west Atlantic region

Species	*Scientific name*	*Average catch (t)*	
		1950–59	*1989–98*
Atlantic redfishes nei	*Sebastes* spp	77 923	518
Haddock	*Melanogrammus aeglefinus*	64 489	1487
Silver hake	*Merluccius bilinearis*	47 770	16 469
Atlantic cod	*Gadus morhua*	18 748	24 112
Scup	*Stenotomus chrysops*	17 904	3846
Saithe (Pollock)	*Pollachius virens*	11 127	6059
Yellowtail flounder	*Limanda ferruginea*	9103	5085
Winter flounder	*Pseudopleuronectes americanus*	7849	5700
Dogfish sharks nei	Squalidae	502	17 655
Raja rays nei	*Raja* spp.	73	10 304
American angler	*Lophius americanus*	41	19 136
American sea scallop	*Placopecten magellanicus*	77 650	81 320
Northern quahog (Hard clam)	*Mercenaria mercenaria*	52 038	25 607
Atlantic surf clam	*Spisula solidissima*	43 923	160 795
Blue crab	*Callinectes sapidus*	28 285	57 955
American lobster	*Homarus americanus*	12 325	29 829
Sand gaper	*Mya arenaria*	12 324	7739
Ocean quahog	*Arctica islandica*	1139	179 312

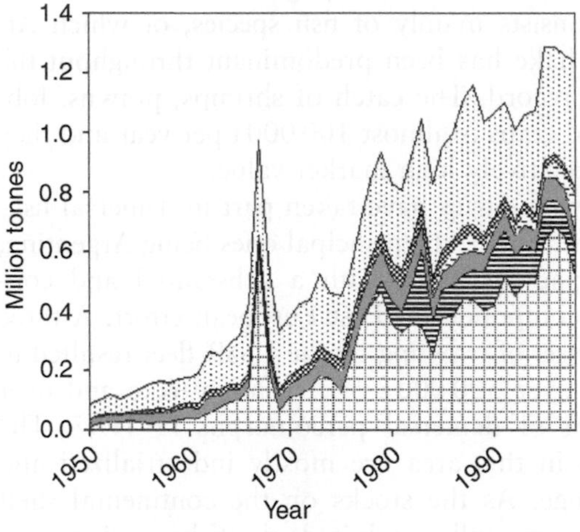

Figure 7 Total demersal fish landings from the south-west Atlantic (FAO Statistical area 41).

first three-quarters of the twentieth century. Since then the catches from some stocks have declined, due to overfishing, while other previously underexploited stocks have increased their yields. The limits of biological production have probably been reached in many areas and careful management is needed in order to maintain the fisheries and to protect the ecosystems which support them.

See also

Ecosystem Effects of Fishing.

Further Reading

Cochrane KL (2000) Reconciling sustainability, economic efficiency and equity in fisheries: the one that got away? *Fish and Fisheries* 1: 3–21.

Cushing DH (1996) *Towards a Science of Recruitment in Fish Populations*. Ecology Institute, D-21385 Oldendorf/Luhe, Germany.

Gulland JA (1988) *Fish Population Dynamics: The Implications for Management*, 2nd edn Chichester: John Wiley.

Kurlansky M (1997) *Cod. A Biography of the Fish That Changed the World*. London: Jonathan Cape.

CRUSTACEAN FISHERIES

J. W. Penn, N. Caputi, and R. Melville-Smith,
Fisheries WA Research Division, North Beach, WA, Australia

Introduction

The Crustacea are one of the most diverse groups of aquatic animals, occupying a wide variety of habitats from the shore to the deep ocean and the tropics to Arctic waters, and extending into fresh water and in some cases on to land for part of their life history.

Crustacean species contribute in the order of 7 million tonnes annually, or about 6–8% of the total world supply of fish, according to FAO statistics. Approximately 75% of this volume is from harvesting wild stocks, with the remainder from aquaculture, dominated by the tropical marine and freshwater shrimps, crayfish, and crab species. Owing to their high market value as a sought-after high-protein food, crustaceans make up a disproportionate share of the value of the world's seafoods. As a result of their high value, crustacean fisheries are generally heavily exploited and require active management to be sustained. Research to underpin management of these resources has been undertaken in many parts of the world, and particularly Australia where, unusually, lobsters and shrimps are the dominant fisheries.

Crustacean fisheries are focused on the more abundant species, particularly those in relatively shallow, accessible areas. Shrimps are the most important wild fishery products, followed by the crabs, lobsters, and krill.

Biology and Life History

Fisheries research on crustacean stocks is significantly influenced by their unusual life history and biology. A unique feature of crustaceans is that they all must undergo a regular process of molting (casting off their outer shell or exoskeleton) to grow. Once the old shell is cast off, the animal absorbs water to swell or 'grow' to a larger size before the shell hardens. Volume increase at a molt varies between species, but typically results in a gain in the range of 10–60%. The molt also serves as an opportunity to regenerate damaged limbs, such as legs or antennae, although regeneration results in lower or even negative growth increments. This molting process occurs throughout all stages of the life history and is often correlated with environmental factors such as temperature, moon phase, or tidal cycles. Because of this mechanism, growth occurs as a series of discrete 'steps' rather than a 'smooth' increase over time and is complicated to measure. Growth rates are highly dependent on water temperature, with tropical and shallow-water crustaceans tending generally to grow much faster than those in cooler and deeper waters. As a result of the molting process, it is not possible to 'age' crustaceans using any of the usual methods (growth rings on bones or shells) applied to other fished species.

The molting process also significantly influences feeding activity and hence catch rates for all crustaceans. Prior to a molt, feeding activity is generally reduced, then ceases in the lead-up to the actual molt process. Following the molt the animals are particularly hungry, begin active feeding, and are more easily caught in baited traps or trawls, but contain relatively little meat for their shell size and are of lower market value.

These cyclic catches in crustacean fisheries are well known to fishers and are also crucial knowledge for fisheries stock assessment and industry management. Significant short-term gains in value of the catch can be achieved where the fishery management arrangements take them into account.

The second very important feature of crustaceans is that exploited marine species are generally highly fecund, producing large numbers of eggs (millions in some species) which hatch into pelagic larvae which in turn can be widely distributed by ocean currents. Typically, the early larval stages are of a different form to the adults, but after a series of larval stages the individual molts into a form resembling the adults. Larval stages generally have limited swimming ability but are able to migrate up and down within the water column, and often have behaviors which, in combination with tides and currents, result in active dispersal into 'nursery' areas suitable for the later juvenile and adult stages. The large numbers of eggs and larvae produced, together with widespread dispersal mechanisms common in the major marine crustacean species, make them relatively resilient to fishing pressure compared with the freshwater species.

The typical marine larval life history does not generally apply to the freshwater species, where some or all larval development stages occur within a much larger egg and live young are often produced to minimize downstream losses due to river flow. These alternative larval strategies adopted by freshwater crustacea are efficient but, owing to the relatively low numbers produced, make such species more susceptible to overfishing than their marine counterparts.

Marine Shrimps and Prawns

This group contains by far the most important crustacean fisheries. In marine waters two major families, the Penaeidae in tropical waters and the Caridea in cold waters, support most of the significant export fisheries. In tropical fresh waters, paleomonid shrimps of the genus *Macrobrachium* support the major commercial production.

The terms 'shrimp' and 'prawn' have no scientific basis and are used interchangeably in different parts of the world. For the purposes of this chapter, the more commonly applied term 'shrimp' will be used for simplicity.

A wide array of penaeid species are harvested from tropical to subtropical waters. These species have a complex life cycle where mated females spawn generally in coastal marine waters where eggs (hundreds of thousands per spawning) are broadcast freely and hatch as free-swimming planktonic nauplius larvae. After a series of larval molts, postlarvae actively move into estuaries and coastal embayments where they develop into the juvenile stage. Juveniles and subadults then actively migrate offshore using tidal flows to further develop, mate, and spawn at 6–12 months of age.

Coupled with these relatively short life cycles is rapid growth, but also high levels of natural mortality such that few individuals survive to more than 12 months of age, although some species may live to 2 or 3 years without fishing. High natural mortality does, however, allow for high but sustainable exploitation rates for most of this group of commercially important species, although there are some exceptions noted later.

Fishing for these species is generally by means of fixed nets in estuary mouths during the offshore migration of subadults, or by vessels otter trawling in waters offshore from the estuarine nursery areas. Penaeid shrimps are generally not catchable in traps.

Powered otter trawling for shrimp, which evolved in the Gulf of Mexico, has now been adopted worldwide as the main method for industrial-scale catching of the more valuable export market-sized adult shrimps. Otter trawling can only occur on smooth bottoms, usually sand or mud adjacent to nursery areas, and typically harvests approximately 20–50% of the shrimps in the path of the net. The remainder are generally buried in the sediments, particularly during the day. The exception to this is where some shrimp species (e.g. *Penaeus merguiensis*) form dense schools and generate turbid mud 'boils' as a defense against predators. This behavior, including mid-water swimming, allows very high exploitation rates and catches on some occasions, but has been noted to break down at high levels of exploitation and in areas where river/estuarine habitats and adjacent waters have become increasingly turbid.

Major fisheries for penaeid shrimps occur through the Gulf of Mexico and Central/South American coasts (*P. aztecus, P. setiferus, P. duorarum, P. braziliensis, P. californiensis and P. vannamei*), off the Chinese river deltas (*P. orientalis*), through southeast Asia (various *Metapenaeus* and *Penaeus* species), Indonesia–Papua New Guinea (*P. merguiensis*), Australia (*P. latisulcatus, P. esculentus/semi-sulcatus, P. merguiensis*), and the African coasts (*P. indicus, P. notialis*).

In addition to the large or more valuable penaeids, very large quantities of very small *Acetes* and sergestid shrimps are harvested, particularly in Asian coastal waters, by small-scale coastal fisheries.

The second commercially important group of shrimps comprises the caridean species, which occur predominantly in the Northern Hemisphere, in temperate to Arctic waters. Where they extend into more tropical waters, they do so only at greater depths with cold temperatures corresponding to Arctic waters. This group of shrimps is relatively long-lived (up to 4–6 years), spawning at several years of age. These species are also typically protandric hermaphrodites, growing into functional males before undergoing a series of molts to become female for the remainder of their life. Females produce larger but fewer eggs than the penaeid species, and carry them after spawning attached under their tail. The eggs remain attached for an extended period, undergoing some developmental stages within the egg before hatching into pelagic larvae which grow for several months before settling onto a wide range of habitat types. *Pandalus borealis* is a typical caridean shrimp for which the life cycle has been well studied and represents the general life history pattern for this important group.

Fishing occurs by both otter trawling and trapping with baited traps which are particularly effective for these species, unlike penaeids which do not trap easily. Major fisheries for these pandalid species occur in the northern Atlantic and north Pacific.

In the Antarctic zone, the major equivalent crustacean fishery is for euphausiids or krill. These krill species are particularly abundant in the nutrient-rich Southern Ocean, where it is estimated that a biomass of 20 or more million tonnes occurs. Krill are small pelagic species which swim by way of modified walking legs (swimmerets), and generally undertake a diurnal migration between the surface and significant depths. They form dense schools on the surface, particularly at night, where they are a major source of food for Antarctic whales, seals, and fish stocks. Estimates of potential sustainable yield range into millions of tonnes per year, but the catch has been limited by processing difficulties to about 100 000 tonnes.

Crabs

Most commercially significant crabs belong to the Brachyara (true crabs) or Anomura (hermit crabs and king crabs) within the order Decapoda. They are generally characterized by a pair of claws, three pairs of walking legs, and a wide, flattened body. Crabs are probably the most highly developed, successful, and diverse of the crustaceans. They occupy a wide range of environments, from shallow tropical seas to deep ocean trenches, estuarine and fresh waters, and some species spend the majority of their life on land, only returning to the water to reproduce.

Reproductive patterns in crabs are diverse and often involve intricate courtship behaviors where the male protects the female before mating. Following copulation, crabs retain spermatozoa until egg laying, at which time fertilization takes place as the eggs are extruded. Spermatozoa can be retained by the female in a viable condition for considerable periods of time – more than a year in some species.

The important swimming crab species are generally resilient to heavy fishing pressure due to their often complex but efficient reproductive behavior and high levels of fecundity. Some species carry multiple broods of eggs, which are extruded, fertilized, and attached to the underside of the female during the early development stages. Numbers of eggs produced per year are frequently in the order of 50 000–500 000 per female, and over a million eggs are achieved by some species. Crab larval stages are known as zoea and most marine species have four or five zoeal stages before molting into a megalopa, which generally settles out of its planktonic existence.

In a number of cold-water crab fisheries (e.g. the important snow, tanner, king, and Dungeness fisheries) where breeding is more restricted, managers have elected to allow harvesting of males only, thereby giving complete protection to the brood stock. This precautionary approach, whilst useful for these species, imposes unusual constraints on research due to the inability to monitor female crabs in the commercial catch.

Most crab fishing worldwide is by use of traps. This method is preferred to most others because traps are simple to use (particularly in deep water) and labor-efficient, and the crabs are less likely to be injured. This latter fact is particularly important because it allows the product to be sold live, guaranteeing a better market price than frozen forms. Many other methods are used to catch crabs, including trawling, tangle netting, dredges, trotlines, and drop nets.

Interestingly, the majority of the large crab fisheries operate in tropical and Northern Hemisphere temperate and Arctic waters.

Table 1 shows that the three most important commercial crab species are all fast-growing 'swimming crabs', found in shallow tropical or temperate waters and embayments. These are a family of crabs which have a flattened, paddle-like hindmost leg used to burrow in sand and mud, or to propel them

Table 1 World landings in order of quantity reported by FAO catch statistics for 1996

Common name	*Species name*	*1996 catch (tonnes)*
Gazami crab	*Portunus trituberculatus*	303 000
Blue crab	*Callinectes sapidus*	116 000
Blue swimmer crab	*Portunus pelagicus*	112 000
Snow and tanner crab	*Chionoecetes* spp.	100 000
King crab	*Paralithodes* spp.	81 000
Dungeness crab	*Cancer magister*	34 000
Edible crab	*Cancer pagurus*	29 000
Red crab	*Geryon/Chaceon* spp.	7000

The table excludes landings of mud crab (*Scylla* spp.), as the majority of that is produced by aquaculture.

through the water during infrequent occasions when they 'swim' over short distances. They reach maturity and are harvested between 1 and 3 years of age.

The largest crab catches landed worldwide are those of gazami crab; however, a very substantial, but unspecified portion of these reported landings are from aquaculture operations. China alone produced 80 000 tonnes of gazami crab by aquaculture in 1997. The species has a wide distribution through the western Pacific and lives in shallow inshore waters in sheltered embayments. Stocking of waters with juvenile gazami crab has become widespread, particularly off the Japanese coast, and is considered to be economically effective.

Blue crabs occur in the western and central western Atlantic. The vast majority of the landings are made off the US coastline from states in the Gulf of Mexico and mid-Atlantic. The commercial fishery targets both hard crabs and peeler/soft crabs, soft-shelled crabs being considered a delicacy in the USA. Soft crabs have very recently molted and have a shell that has yet to become hard. While some of the peeler/soft crab product is taken with crab scrapes and other specialized methods capable of taking nonfeeding animals, the majority of the product is produced in operations which hold peelers in shedding tanks until molting occurs.

The snow, tanner, and king crabs, which are high on the list of important species in **Table 1**, are examples of moderately deep-water species occurring in cold water conditions. The distributional range of these species encompasses water <400 m deep (and, particularly for king and snow crabs, usually <200 m and colder than 10°C). These species are very slow growing when compared with the inshore warmer water species mentioned earlier. Their age at maturity is generally upward of 5 years and in most cases they enter into the commercial fishery over 8 years after settlement.

Over the long history of crab production in the north Pacific, large catches of king, tanner, and snow crabs have been made. Despite stock collapses of some species, this area is still important for its crab production and for the research efforts that have been made to understand the biology and management of these important stocks.

Crabs belonging to the *Geryon* and *Chaceon* genus (**Table 1**) are commercially important deep-water crabs. They have a wide depth range, but most of the commercially exploited populations tend to be in the 500–1000 m depth range. Water temperatures at these depths are typically less than 10°C and these animals are therefore slow growing. In Namibia, where the largest and one of the longest-standing *Chaceon* fisheries exists, the crabs take approximately 8 years to reach maturity.

Lobsters and Crayfish

Lobster and crayfish species support significant and high-value fisheries. The major commercial lobster species are marine and taken from tropical to cold temperate waters, while freshwater crayfish are mostly taken from tropical and subtropical regions. Most of the marine species have similar life history patterns, where females carry fertilized eggs externally under their abdomens. Following hatching, the larvae undergo a series of molts before taking up a benthic habitat and growing to adulthood, a process which can take many years. Freshwater crayfish species generally have a reduced larval life and hatch as small juveniles.

Fisheries are dominated by three groups, the *Homarus* species (large-clawed lobsters), the *Nephrops* species (small-clawed lobsters), and the palinurid group (spiny or rock lobsters, without claws). Catches of each of these groups are in the order of 60 000–80 000 tonnes annually.

Fishing is generally by baited traps (*Homarus* and most palinurid species), although some are taken by trawl (*Nephrops*), and diving (tropical *Panulirus* species). Freshwater crayfish are also taken by baited traps.

The major clawed lobster fishery is for *Homarus americanus* off eastern Canada and the USA. A similar-sized fishery for *Nephrops norvegicus* occurs off the European Atlantic coasts and through the Mediterranean. Spiny lobster fisheries for the *Panulirus* species occur through the tropics, with major fisheries in the Caribbean (*P. argus*) and Western Australia (*P. cygnus*). Smaller but significant fisheries for *Jasus* species occur in the temperate waters off southern Australia, New Zealand, and South Africa. The major freshwater crayfish fishery occurs in the southern states of the USA.

Stocks of these species have generally been resilient to fishing, with the exception of some *Jasus* species off Africa which have been significantly reduced over time.

Fishery Assessment Research

There are two fundamental biological issues to be addressed in the management of fish stocks (including crustaceans). The most important problem is to control the level of fishing such that there is sufficient breeding stock to continue to provide adequate supply of new recruits to the fishery. This is generally tackled using the relationship between breeding stock and recruitment. The second issue is to maximize the overall catch (and value). This problem is traditionally examined using a yield-per-recruit model which examines the trade-off between the increase in biomass through growth over time and the decrease in survival through natural and fishing mortality. The other biological studies undertaken, such as growth, migration, reproduction, and mortality, are generally the building blocks to enable the assessment of these two key issues.

There are three main differences in the biological assessment of crustacean fisheries compared with many finfish fisheries; they are growth, migration, and catchability. Growth by molting is probably the key difference between crustaceans and other marine species. Thus stock assessment needs to take into account the timing of the growth and the size increment at the molt. The frequency and size increment of molting usually decreases with age, especially after reaching maturity.

Crustacean growth contrasts with that of finfish populations, which can generally be modeled using a continuous growth model. Because crustaceans totally replace their outer shell with each molt they cannot be aged in this way, creating a major problem for stock assessment. This process also makes tag recapture less reliable for these species. As a consequence, age is usually estimated by following length frequencies of particular year-classes, although this is often possible for only the younger year-classes. Some recent work on measuring the age pigment, lipofuscin, which generally increases linearly with age and is not lost at the molt, may provide an opportunity in the future to regularly utilize age information in the stock assessment of crustaceans.

The second feature of crustaceans which sets them apart from finfish and affects their stock assessment is migration. While generally poor in swimming ability, crustacean species often undergo significant migrations linked to specific stages in their life cycle. For example, tropical shrimps and swimming crabs have specific behaviors which enable them to actively migrate offshore, utilizing ebb tidal flows, as they approach sexual maturity. Many spiny lobster species also undergo extensive directional migrations at a particular age, usually following a coordinated molt, marching in columns from shallow nursery areas to offshore spawning areas before reaching sexual maturity.

These migration 'events' are often of short duration and usually unidirectional. Such rapid, short-term interruptions to the normal, relatively sedentary behavior of crustaceans pose special constraints on stock assessment. That is, crustacean migration typically causes erratic changes in stock distribution and catches, contrasting with most finfish fisheries where the regular, more consistent swimming movements of the fish result in continuous redistribution of the stock.

Because molting is often synchronized and related to growth and migration, the catchability of many crustaceans is also typically inconsistent and often cyclic. For this reason, crustacean catch rates do not directly reflect the abundance of the stock and must be corrected for in the data sets utilized in stock assessments. To assess the status of exploited crustacean stocks which present these particular problems, long data series are extremely valuable, especially where they can be used to refine the catch rate–abundance relationship and in establishing the linkage between different life history stages. Such data can be used to assess the relationship between spawning stock, environmental factors, and recruitment to the fishery for managing the critical effects of fishing on the spawning stocks.

For example, the long time-series of data (**Figure 1**) was fundamental in assessing the cause of the collapse of tiger prawn stocks in Shark Bay, Western Australia. These data were used to evaluate the impact of fishing effort on the spawning stock and assess the reduction in fishing effort required for the fishery to recover to its optimal level.

Figure 2, derived from the historical data, shows the relationship between spawning stock levels and subsequent recruitment to the Shark Bay tiger prawn stock. This relationship, together with the reverse relationship between recruitment and surviving spawner abundance (in the same year) relative to variations in fishing effort targeting the stock, has been used to construct a simple model (**Figure 3**) to determine optimal levels of tiger prawn fishing effort. Management changes to redirect effort away from the species based on this modeling have resulted in a recovery of the tiger prawn stock (**Figure 1**).

Similarly, the use of catch predictions in the western rock lobster (*Panulirus cygnus*) fishery up to 4 years ahead using an index of abundance of settling puerulus (first post-larval stage) and juveniles entering the fishery has enabled fisheries management to be proactive rather than reactive to changes in stock abundance. This relationship, presented in **Figure 4**, shows that catch is determined by variations in puerulus settlement 3 and 4 years previously, and fishing effort during the year of recruitment to the fishery. Such predictive relationships have been used in Western Australia to adjust fishing levels in

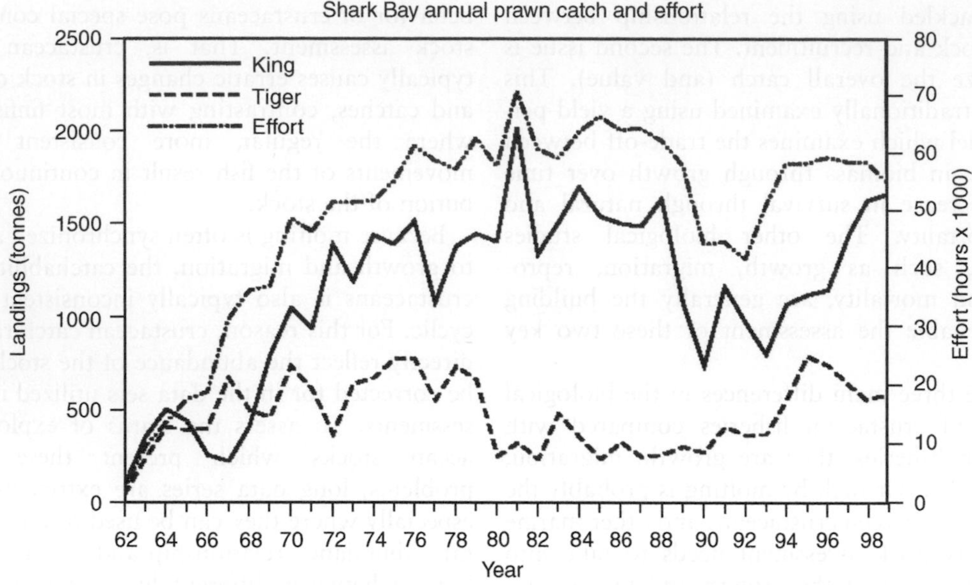

Figure 1 The time-series data on catch and fishing effort (hours trawled) for tiger prawns (*Penaeus esculentus*) and western king prawns (*P. latisulcatus*) since the inception of the Shark Bay (Western Australia) prawn fishery in 1962.

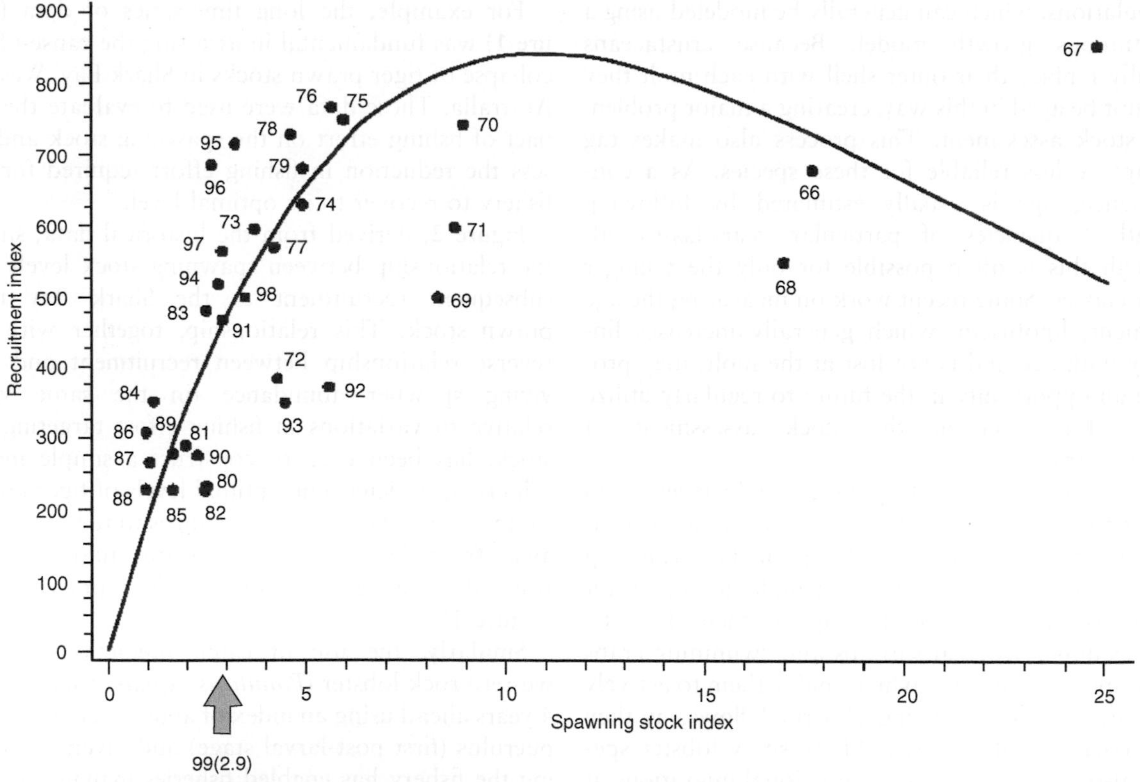

Figure 2 The relationship between spawner abundance and recruitment (1 year later) for the tiger prawn (*P. esculentus*) stock in Shark Bay (Western Australia). Year of recruitment is shown against each data point.

advance to ensure that breeding stock levels are maintained.

This development of predictive relationships using long-run data sets also enables environmental factors which may influence survival of larval stages, and catchability in crustacean stocks, to be examined. Figure 5, showing the relationship between rock lobster puerulus settlement and Fremantle sea level

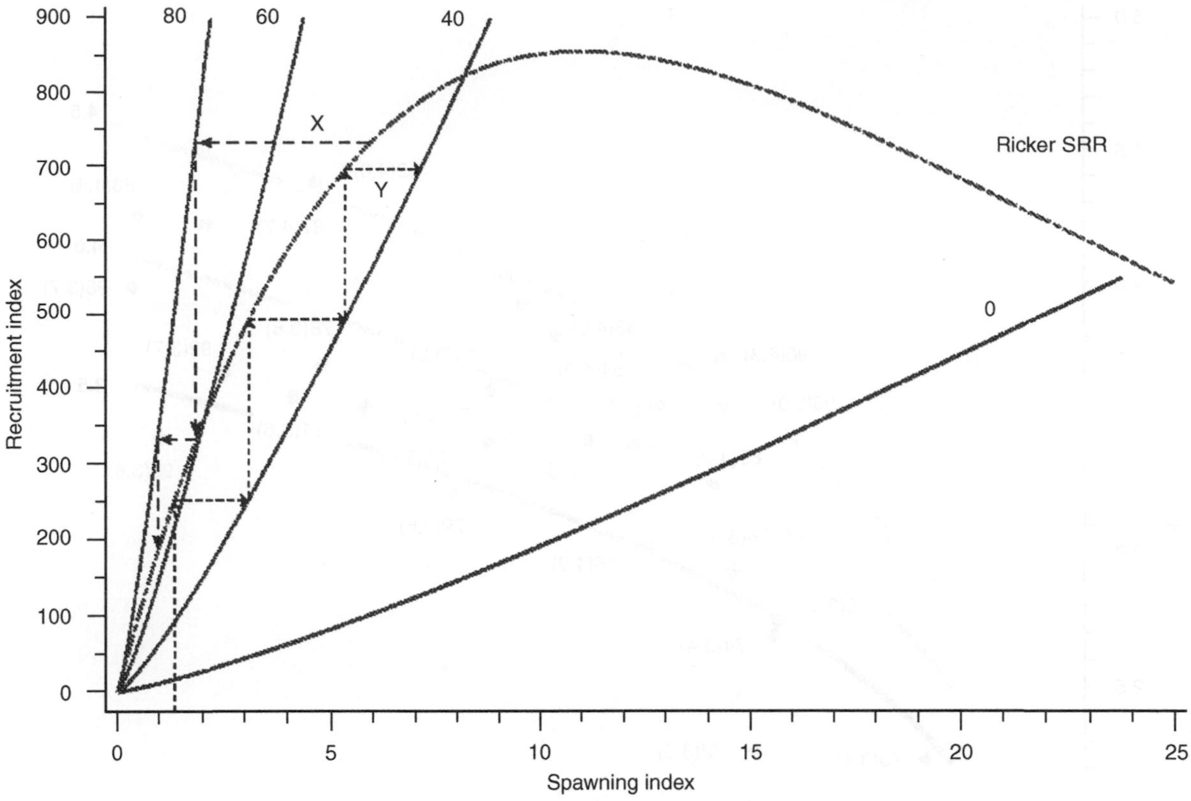

Figure 3 A model combining the spawner–recruit relationship SRR and recruitment to spawner (as affected by fishing effort) for the Shark Bay tiger prawn stock, which has been utilized to estimate optimal fishing effort levels. Trajectory 'X' shows the expected annual decline in recruitment and spawning stock at an unsustainable level of fishing effort (80 000 trawling hours). Trajectory 'Y' shows the converse stock recovery from low levels when fishing occurs at optimal levels of about 40 000 hours of trawling effort.

as an index of flow of the Leeuwin Current along the WA coastline, is an example of this type of analysis.

The availability of this type of relationship is particularly valuable to researchers attempting to distinguish between the effects of fishing and short-term 'natural' variations in recruitment to the fishery caused by environmental influences. This is an important distinction, as a poor year-class due to environmental factors, coupled with high fishing effort, can combine to produce a very low breeding stock and trigger a long-term stock decline.

The second most important fisheries problem, optimizing yield per recruit to the fishery, is also particularly difficult to assess in crustaceans owing to the molting process. The resulting inability to age or reliably tag these species makes estimation of natural mortality and growth of pre-recruit year-classes relatively unreliable.

This has led to an 'adaptive' management approach using adjustments to sizes at first capture over a number of years to directly assess the resulting impact on catch. The alternative approach has been to develop complex simulation models based on length rather than age. These model-based assessments have improved significantly the ability to manage crustacean fisheries, but again where successful have relied heavily on long-run, detailed fishery databases for their testing and validation.

Crustacean Management Techniques

Management techniques applied to the significant tropical shrimp fisheries focus mainly on minimum trawl mesh sizes, accompanied by area and seasonal closures to optimize the quantity and size of shrimps caught. Many of these trawl fisheries also involve specific gear regulations to minimize unwanted bycatch. Owing to the highly variable annual recruitment to these fisheries, the most common and successful management approaches have involved fishing effort controls and, more recently, transferable effort quotas. For the longer-lived, more consistent cold-water pandalid shrimp and krill fisheries, catch quotas are more applicable and often utilized.

Because of their larger individual size (and value) and the dominant method of capture (trapping) which facilitates live discarding of unwanted catch, the

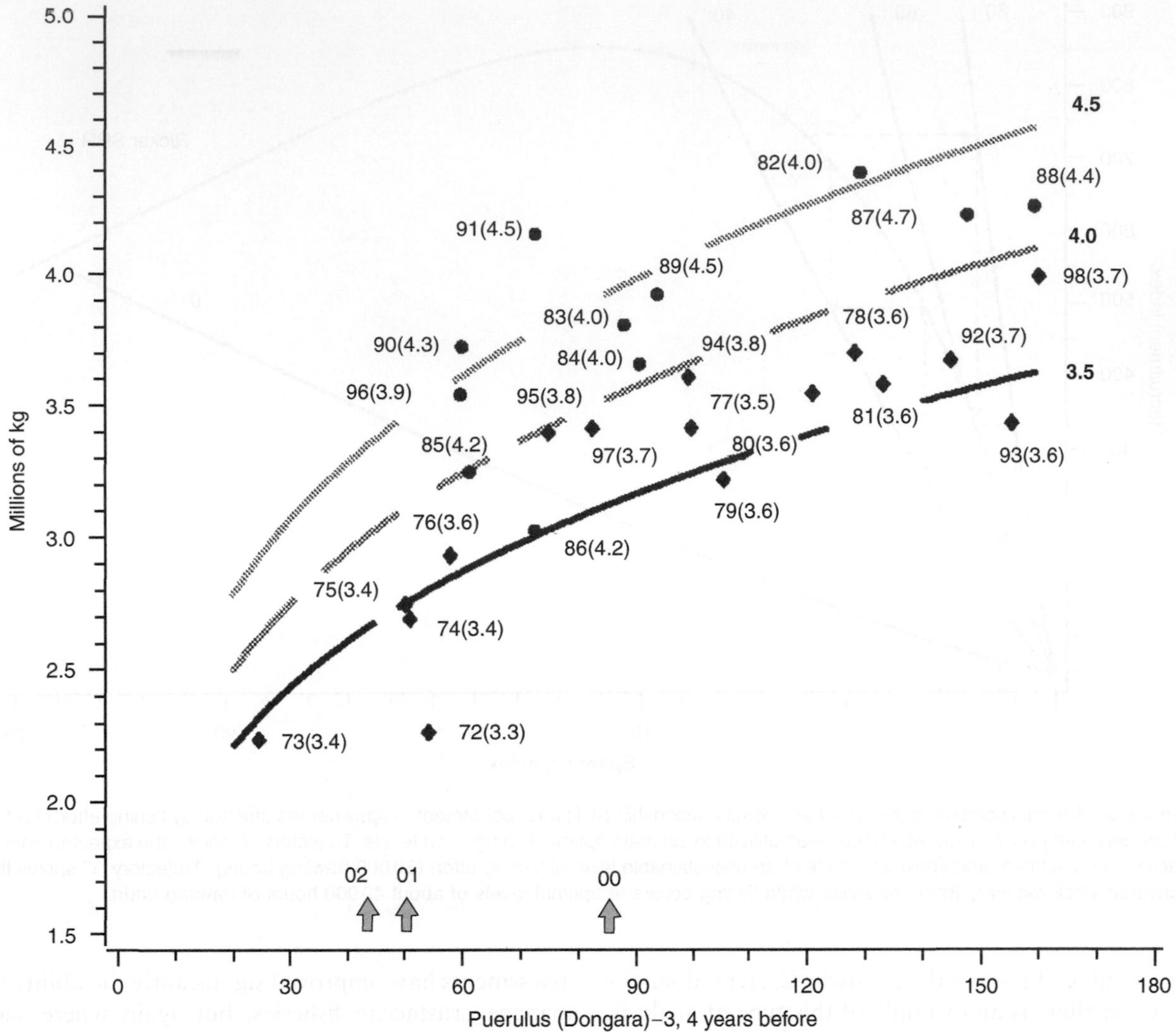

Figure 4 Catch forecast for western rock lobster in the northern part of its range, based on the level of puerulus settlement at Dongara 3–4 years earlier and the number of pot lifts. The catch year is shown with millions of pot lifts in brackets.

common management focus in crab fisheries is on legal minimum size regulations and protection for spawning females. Gear design rules specifying 'escape gaps' to reduce the capture of undersize crabs have become a common management tool for these fisheries. Overall management of the longer-lived temperate or deep-water crabs often involves catch quotas to ensure maintenance of breeding stocks and economic performance of fisheries. This methodology is less relevant to the faster-growing, more variable tropical crab fisheries, which lack the predictable recruitment and longevity necessary for effective catch quota management. In these fisheries, effort controls through limited entry are more useful and common.

The management techniques for high-value lobster stocks are generally similar to those for crabs, focusing on legal minimum sizes, associated gear controls, and female protection in the trap fisheries which dominate this crustacean sector. Historically, limited entry arrangements have been the most common overall management strategy for sustaining lobster fisheries, with 'individually transferable trap quotas' first applied in the 1960s to the Australian spiny lobster fisheries. Total catch quotas, applied more recently through 'individually transferable quotas', have also been utilized to control fishing, particularly for the more consistent, longer-lived cold-water lobster fisheries.

Conclusions

The high value of crustaceans has led to increasing exploitation pressures and the need for improved research assessments to underpin management. Stock assessment methods for crustacean fisheries, however,

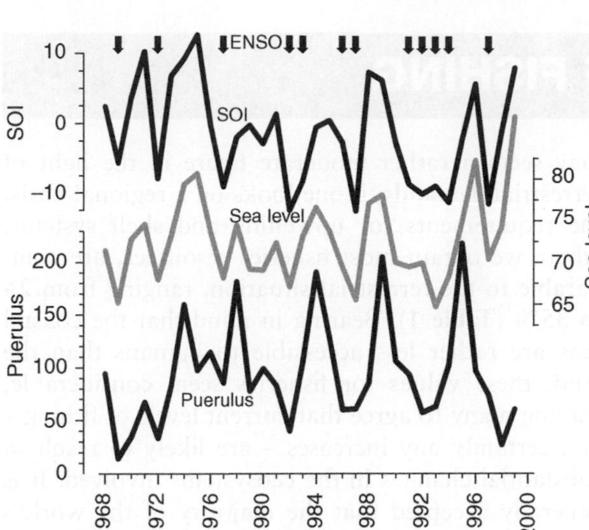

Figure 5 Annual mean values of Southern Oscillation Index (SOI), Fremantle sea level, and puerulus settlement at Dongara. ENSO (El Niño/Southern Oscillation) periods are indicated with arrows.

provide significant scientific challenges owing to the unique crustacean method of growth through molting. This mode of growth prevents the use of the long-established age-based methods applied to the more generic finfish and some molluscan fisheries.

The stock assessment approach adopted has therefore focused on direct measurement of recruitment and spawning stocks and the use of long-run fishery databases. The importance of using long-run data sets to validate and test length-based models has been critical to the more recent improvements in crustacean fisheries assessment for management.

See also

Demersal Species Fisheries. Population Dynamics Models.

Further Reading

Caddy JF (1989) *Marine Invertebrate Fisheries: Their Assessment and Management*. New York: John Wiley & Sons.

Caputi N, Penn JW, Joll LM, and Chubb CF (1998) Stock-recruitment-environment relationships for invertebrate species of Western Australia. In: Jamieson GS and Campbell A (eds.) Proceedings of the North Pacific Symposium on Invertebrate Stock Assessment and Management. *Canadian Special Publication on Fisheries and Aquatic Science* 125: 247–255.

Cobb JS and Phillips BF (1980) *The Biology and Management of Lobsters*, vol. II: *Ecology and Management*. New York: Academic Press.

Debelius H (1999) *Crustacea Guide of the World*. Frankfurt: IKAN-Unterwasserarchiv.

Gulland JA and Rothschild BJ (eds.) (1984) *Penaeid Shrimps: Their Biology and Management*. Farnham, Surrey: Fishing News Books.

Phillips BF and Kittaka J (eds.) (2000) *Spiny Lobsters: Fisheries and Culture*. Oxford: Fishing News Books.

Provenzano AJ (1985) *The Biology of Crustacea*, vol. 10: *Economic Aspects: Fisheries and Culture*. Orlando, FL: Academic Press.

ECOSYSTEM EFFECTS OF FISHING

S. J. Hall, Flinders University, Adelaide, SA, Australia

Introduction

In comparison with conventional fisheries biology, which examines the population dynamics of target stocks, there have been relatively few research programs that consider the wider implications of fishing activity and its effects on ecosystems. With growing recognition of the need to conduct and manage our activities within a wider, more environmentally sensitive framework, however, the effects of fishing on ecosystems is increasingly being debated by scientists and policy makers around the world. As with many other activities such as waste disposal, chemical usage or energy policies, scientists and politicians are being asked whether they fully understand the ecological consequences of fishing activity.

The scale of biomass removals and its spatial extent make fishing activity a strong candidate for effecting large-scale change to marine systems. Coarse global scale analyses provide a picture of our fish harvesting activities as being comparable to terrestrial agriculture, when expressed as a proportion of the earth's productive capacity. It has been estimated that 8% of global aquatic primary production was necessary to support the world's fish catches in the early 1980s, including a 27 million tonne estimate of discards (see below). Perhaps the most appropriate comparison is with terrestrial systems, where almost 40% of primary productivity is used directly or indirectly by humans. Although 8% for marine systems may seem a rather moderate figure in the light of terrestrial demands, if one looks on a regional basis, the requirements for upwelling and shelf systems, where we obtain most fisheries resources, are comparable to the terrestrial situation, ranging from 24 to 35% (**Table 1**). Bearing in mind that the coastal seas are rather less accessible to humans than the land, these values for fisheries seem considerable, leading many to agree that current levels of fishing – and certainly any increases – are likely to result in substantial changes in the ecosystems involved. It is generally accepted that the majority of the world's fish stocks are fully or overexploited.

When considering ecosystem effects it is useful to distinguish between the direct and indirect effects of fishing. Direct effects can be summarized as follows:

1. fishing mortality on species populations, either by catching them (and landing or throwing them back), by killing them during the fishing process without actually retaining them in the gear or by exposing or damaging them and making them vulnerable to scavengers and other predators;
2. increasing the food available to other species in the system by discarding unwanted fish, fish offal and benthos;
3. disturbing and/or destroying habitats by the action of some fishing gears.

In contrast, indirect effects concern the knock-on consequences that follow from these direct effects, for example, the changes in the abundances of predators, prey and competitors of fished species that might occur due to the reductions in the abundance of target species caused by fishing, or by the provision of food through discarding of unwanted catch.

Table 1 Global estimates of primary production and the proportion of primary production required to sustain global fish catches in various classes of marine system

Ecosystem type	*Area (10^6 km^2)*	*Primary production (g C m^{-2}y^{-1})*	*Catch (g m^{-2}y^{-1})*	*Discards (g m^{-2}y^{-1})*	*Mean % of primary production*	*95% CI*
Open ocean	332.0	103	0.01	0.002	1.8	1.3–2.7
Upwellings	0.8	973	22.2	3.36	25.1	17.8–47.9
Tropical shelves	8.6	310	2.2	0.671	24.2	16.1–48.8
Nontropical shelves	18.4	310	1.6	0.706	35.3	19.2–85.5
Coastal reef systems	2.0	890	8.0	2.51	8.3	5.4–19.8

Reproduced from Hall (1999).

By-catch and Discards

In many areas of the world a wide variety of fishing gears are used, each focusing on one or a few species. Unfortunately, this focus does not mean that non-target species, sexes or size-classes are excluded from catches. Target catch is usually defined as 'the catch of a species or species assemblage that is primarily sought in a fishery' – nontarget catch, or by-catch as it is usually called, is the converse. By-catch can then be further classified as incidental catch, which is not targeted but has commercial value and is likely to be retained if fishing regulations allow it and discard catch, which has no commercial value and is returned to the sea.

The problem of by-catch and discarding is probably one of the most important facing the global fishing industry today. The threat to species populations, the wastefulness of the activity and the difficulties undocumented discarding poses for fish stock assessment are all major issues. A recent published estimate of the annual total discards was approximately 27 million tonnes, based on a target catch of 77 million tonnes. This figure, however, did not include by-catch from recreational fisheries, which could add substantially to the total removals, and the estimate is subject to considerable uncertainty. **Figure 1** shows how these discard figures break down on a regional basis. Just over one-third of the total discards occur in the Northwest Pacific, arising from fisheries for crabs, mackerels, Alaskan pollock, cod and shrimp, the latter accounting for about 45% of the total. The second ranked region is the Northeast Atlantic where large whitefish fisheries for haddock, whiting, cod, pout, plaice and other flatfish are the primary sources. Somewhat surprisingly, capelin is also a rather important contributor to the total, primarily because capelin are discarded due to size, condition and other market-related factors. The third place in world rankings is the West Central Pacific, arising largely through the action of shrimp fisheries. These fisheries, prosecuted mainly off the Thai, Indonesian and Philippine coasts, accounted for 50% of the total by-catch for the region, although fisheries for scad, crab and tuna are also substantial contributors. Interestingly, the South East Pacific ranks fourth, not because the fisheries in the area have high discard ratios (on the contrary, the ratios for the major anchoveta and pilchard fisheries are only 1–3%), but simply due to the enormous size of the total catch. For the remaining tropical regions, by-catch is again dominated by the actions of shrimp fisheries, although some crab fisheries are also significant.

One characteristic difference between temperate and tropical fishery discards is worthy of note. In the tropics, where shrimp fisheries dominate the statistics, discards mainly comprise small-bodied species which mature at under 20 cm and weigh less than 100 g. In contrast, for the temperate and subarctic

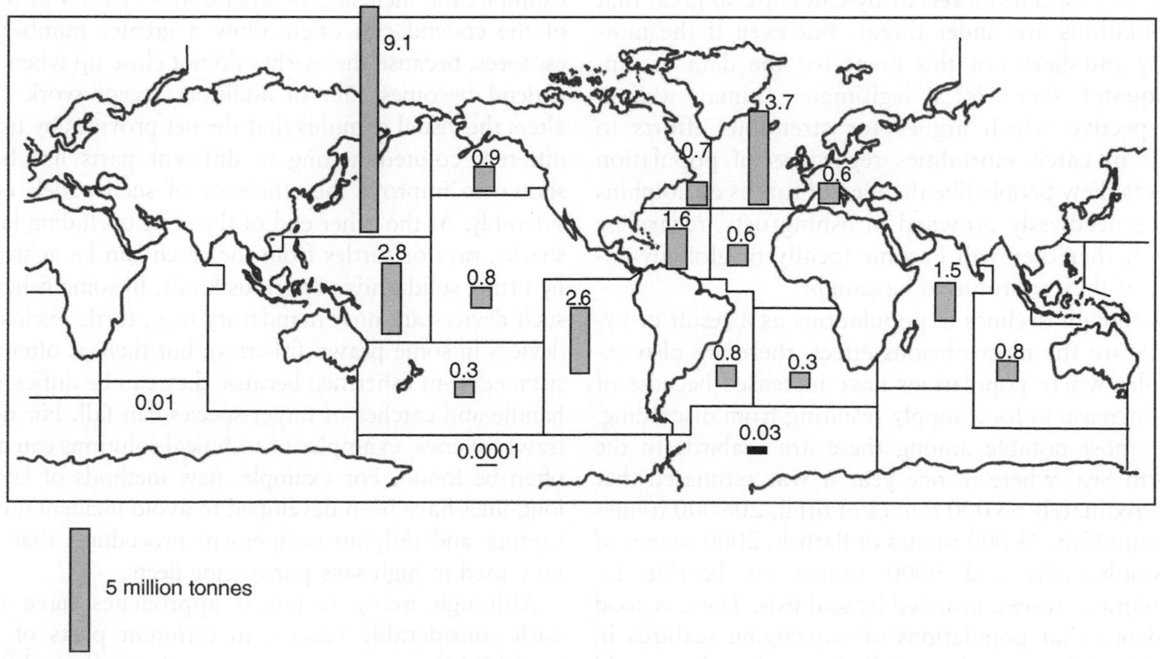

Figure 1 The regional distribution of discards. (Reproduced from Hall, 1999.)

regions discards are generally dominated by sublegal and legal sizes of commercially important, larger-bodied species. Thus, in the temperate zone discarding is not only an ecological issue, it is also a fisheries management issue in the strictest sense. Fish are being discarded which, if left alone, would form part of the future commercial catch.

A cause of particular concern is the incidental catch of larger vertebrate fauna such as turtles, elasmobranchs and marine mammals. Catch rates for these taxa are generally highest in gillnet fisheries, which increased dramatically in the 1970s and 1980s, particularly for salmonids, squid and tuna. In some fisheries, the numbers caught can be very substantial. In the high seas longline, purse seine and driftnet fisheries for tuna and billfish, for example, migratory sharks form a large component of the catch, with some 84 000 tonnes estimated to have been caught from the central and south Pacific in 1989.

As with the other nonteleost taxa for which by-catch effects are a concern, the life-history characteristics of sharks make them particularly vulnerable to fishing pressure. Slow growth, late age at maturity, low fecundity and natural mortality, and a close stock recruitment relationship all conspire against these taxa. Such life-history attributes have also led to marked alterations in the absolute and relative abundance of ray species in the North and Irish Sea, which are subject to by-catch mortality from trawl fisheries. In the Irish Sea for example, the 'common skate' (*Raja batis*) is now rarely caught.

For some species (e.g., some species of albatross and turtle species) levels of by-catch are so great that populations are under threat. But even if the mortality rates are not this great (or the data are inadequate) there is a legitimate animal welfare perspective which argues for strenuous efforts to limit by-catch mortalities regardless of population effects. Few people like the idea of turtles or dolphins being needlessly drowned in fishing nets, regardless of whether they will become locally or globally extinct if they continue to be caught.

Although declines in populations as a result of by-catch are the most obvious effect, there are also examples where populations have increased because of the increase in food supply resulting from discarding. The most notable among these are seabirds in the North Sea, where in one year it was estimated that approximately 55 000 tonnes of offal, 206 000 tonnes of roundfish, 38 000 tonnes of flatfish, 2000 tonnes of elasmobranchs and 9000 tonnes of benthic invertebrates were consumed by seabirds. There is good evidence that populations of scavenging seabirds in the North Sea are substantially larger than they would be without the extra food provided by discards.

Solving the By-catch and Discard Problem

There is no universally applicable solution for mitigating by-catch and discard problems. Each fishery has to be examined separately (often with independent observers on fishing vessels) and the relative merits of alternative approaches assessed. One obvious route to reducing unwanted catch, however, is to increase the selectivity of the fishing method in some way. In trawl fisheries, in particular, technical advances, combined with a greater understanding of the behavior of fish in nets has led to the development of new methods to increase selectivity. These methods adopt one of two strategies. The first is to exploit behavioral differences between the various fished species, using devices such as separator trawls, modified ground gear (i.e., the parts of the net that touch the seabed) or modifications to the sweep ropes and bridles that attach to the trawl doors. For example, separator trawls in the Barents Sea have been shown successfully to segregate cod and plaice into a lower net compartment from haddock, which are caught in an upper compartment. In Alaska this approach has been used to allow 40% of bottom-associated halibut to escape while retaining 94% of cod, the target species.

The second approach is to exploit the different sizes of species. In many fisheries it is the capture of undersized fish that is the main problem and regulation of minimum permissible mesh size is of course a cornerstone of most fisheries management regimes. Such a measure can often, however, be improved upon. For example, the inclusion of square mesh panels in front of the codend can often allow a greater number of escapees, because the meshes do not close up when the codend becomes full. In addition, recent work that alters the visual stimulus that the net provides by using different colored netting in different parts has been shown to improve the efficiency of such panels considerably. At the other end of the scale excluding large sharks, rays or turtles from the catch can be achieved by fitting solid grids of various kinds. In some fisheries such devices are now mandatory (e.g., turtle exclusion devices in some prawn fisheries), but there is often resistance from fishermen because they can be difficult to handle and catches of target species can fall. For non-trawl fisheries, examples of technical solutions can also often be found. For example, new methods of laying long-lines have been developed to avoid incidental bird capture and dolphin escapement procedures that are now used in high seas purse seine fleets.

Although many technical approaches have met with considerable success in different parts of the world it is important to recognize that technical fixes are only part of the solution – the system in which they have to operate must also be considered. The regulations that govern fisheries and the vagaries of the marketplace often create a complex web of incentives and disincentives that drive the discarding practices of fishermen. The situation can be especially complicated in multispecies fisheries.

The Effects of Trawling and Dredging on the Seabed

Disturbance of benthic communities by mobile fishing gears is the second major cause for concern over possible ecosystem effects of fishing, threatening nontarget benthic species and perhaps also the longer-term viability of some fisheries themselves if essential fish habitat is being destroyed. With continuing efforts to find unexploited fish resources, hitherto untouched areas are now becoming accessible as new technologies such as chain mats, which protect the belly of the net, are developed. In Australia, for example, new fisheries are developing in deeper water down to depths of 1200 m.

A prerequisite for a rational assessment of fishing effects on benthos is an understanding of the distribution, frequency and temporal consistency of bottom trawling. On a global basis, recent estimates obtained using Food and Agriculture Organization catch data from fishing nations suggest that the continental shelves of 75% of the countries of the world which border the sea were exposed to trawling in 1996. It would appear, therefore, that few parts of the world's continental shelf escape trawling, although it should be borne in mind that in many fisheries trawl effort is highly aggregated. Although we have an appreciation of average conditions, these are derived from a mosaic of patches, some heavily trawled along preferred tows, others avoided by fishermen because they are unprofitable or might damage the gear. Unfortunately, lack of data on the spatial distribution of fishing effort prevents estimates of disturbance at the fine spatial resolution required to obtain a true appreciation of the scale of trawl impacts. Nevertheless, there is little doubt that substantial areas of the world's continental shelf have been altered by trawling activity.

For the most part the responses of benthic communities to trawling and dredging is consistent with the generalized model of how ecologists expect communities to respond, with losses of erect and sessile epifauna, increased dominance by smaller faster-growing species and general reductions in species diversity and evenness. This agreement with the general model is comforting, but we have also learnt that not all communities are equally affected. For example, it is much more difficult to detect effects in areas where sediments are highly mobile and experience high rates of natural disturbance, whereas boulder or pebble habitats, those supporting rich epifaunal communities that stabilize sediments, reef forming taxa or fauna in habitats experiencing low rates of natural disturbance, seem particularly vulnerable. However, despite the body of experimental data that has examined the impacts of trawling on benthic communities, it is often not possible to deduce the original composition of the fauna in places where experiments have been conducted because data gathered prior to the era of intensive bottom-fishing are sparse. This is an important caveat because recent analyses of the few existing historical datasets suggest that larger bodied organisms (both fish and benthos) were more prevalent prior to intensive bottom trawling. Moreover, in general, epifaunal organisms are less prevalent in areas subjected to intensive bottom fishing. Communities dominated by sponges, for example, may take more than a decade to recover, although growth data are notably lacking. Such slow recovery contrasts sharply with habitats such as sand that are restored by physical forces such as tidal currents and wave action.

Habitat Modification

An important consequence of trawling and dredging is the reduction in habitat complexity (architecture) that accompanies the removal of sessile epifauna. There is compelling evidence from one tropical system, for example, that loss of structural epibenthos can have important effects on the resident fish community, leading to a shift from a high value community dominated by Lethrinids and Lutjanids to a lower value one dominated by Saurids and Nemipterids. Similar arguments have also been made for temperate systems where structurally rich habitats may support a greater diversity of fish species. Importantly, such effects may not be restricted to the large biotic or abiotic structure provided by large sponges or coral reefs. One could quite imagine, for example, that juveniles of demersal fish on continental shelves might benefit from a high abundance of relatively small physical features (sponges, empty shells, small rocks, etc.) but that over time trawling will gradually lower the physical relief of the habitat with deleterious consequences for some fish species. Such effects may account for notable increases in the dominance of flatfish in both tropical and temperate systems. Our current understanding of the functional role of many of the larger-bodied long-lived species (e.g., as habitat features, bioturbators, etc.) is limited and needs to be addressed to predict the outcome of permitting chronic fishing disturbance in areas where these animals occur.

Although fishing-induced habitat modification is probably most widely caused by mobile gears, it is important to recognize that other fishing methods can also be highly destructive. For coral reef fisheries, dynamite fishing and the use of poisons represent major threats in some parts of the world.

Perhaps the only effective approach for mitigating the effects of trawling in vulnerable benthic habitats is to establish marine protected areas in which the activity is prohibited. Given the widespread distribution of trawling, it is not surprising that the establishment of marine protected areas is a key goal for many sectors of the marine conservation movement, although it should be borne in mind that it is not only trawling effects that can be mitigated by the approach. A key driver for the establishment of marine protected areas has come from The World Conservation Union (IUCN) and others who have called for a global representative system of marine protected areas and for national governments to also set up their own systems. A number of nations have already taken such steps, including Australia, Canada and the USA, with other nations likely to follow suit in the future.

Species Interactions

Even species that are not directly exploited by a fishery are likely to be affected by the removal of a substantial proportion of their prey, predator or competitor biomass and there are certainly strong indications that interactions with exploited species should be strong enough to lead to population effects elsewhere. For example, an analysis of the energy budgets for six major marine ecosystems found that the major source of mortality for fish is predation by other fish. Predatory interactions may, therefore, be important regulators for marine populations and removing large numbers of target species may lead to knock-on effects. Unfortunately, however, gathering the data necessary to demonstrate such controls is a major task that has rarely been achieved. Without studies directed specifically at the processes underlying the population dynamics of specific groups of species, it is difficult to evaluate the true importance of the effects of fisheries acting through species interactions in marine systems. Despite this caveat, some general effects appear to be emerging.

Removing Predators

For communities occupying hard substrata, there is good evidence that some fisheries have reduced predator abundances and that this has led to marked changes lower in the food web. Both temperate hard substrates and coral reefs provide good examples where reductions in predator numbers have led to change in the abundance of prey species that compete for space (e.g., mussels or algae), or in prey that themselves graze on sessile species. Such changes have led in turn to further cascading changes in community composition. For example, in some coral reef systems, removal of predatory fish has led to increases in sea urchin abundance and consequent reductions in coral cover.

Examples of strong predator control are much less easy to find in pelagic systems than they are in hard substratum communities. This perhaps suggests that predator control is less important in the pelagos. Alternatively, the lack of evidence may simply reflect our weak powers of observation; it is much harder to get data that would support the predator control theory in the pelagic than it is on a rocky shore.

Removing Prey

Fluctuations in the abundance of prey resources can affect a predator's growth and breeding success. Thus, if prey population collapses are sustained over the longer term due to fishing this will translate into a population decline for the predator. Examples of such effects can be found, particularly for bird species, but also for other taxa such as seals. Since many people have strong emotional attachments to such taxa, there is often intense interest when breeding failures or population declines occur. In the search for a culprit fishing activity is often readily offered as an explanation for the prey decline, or at least as an important contributory factor. In assessing the effect of prey removal, however, one must consider whether the fishery and predator compete for the same portion of the population, either in terms of spatial location or stage in the life cycle. For example, if the predator eats juveniles whose abundance is uncorrelated with the abundance of the fishable stock, the potential for interactions is greatly reduced. Such a feature seems rather common and probably needs to be examined closely in cases where a fishery effect is implicated. Nevertheless, there can be little doubt that unrestrained exploitation increases the likelihood of fisheries collapses and this is turn will take its toll on predator populations.

Removing Competitors

Unequivocal demonstrations of competition in most marine systems are rare. Perhaps the only exception to this is for communities occupying hard substrates where competition for space has been demonstrated and can be important in determining community responses to predators (see above). For other systems (e.g., the pelagic or soft-sediment benthos), we can only offer opinions, based on our assessment of the importance of other factors (e.g., predation, low quality food, environmental conditions). One system where fishing activity has been generally accepted to have an impact through competitive effects is the Southern Ocean, where massive reductions in whale populations by past fishing activity has led to apparent increases in the population size, reproduction or growth of taxa such as seals and penguins. A recent assessment, however, has even cast doubt on this interpretation, concluding that there is little evidence that populations have responded to an increase in available resources resulting from a decline in competitor densities.

Species Replacements

Despite the difficulties of clearly identifying the ecological mechanism responsible for the changes, there are some examples where fishing is heavily implicated in large-scale shifts in the species composition of the system and apparent replacement of one group by another. The response of the fish assemblage in the Georges Bank/Gulf of Maine area is, perhaps, the clearest example (**Figure 2**). During the 1980s the principal groundfish species, flounders and other finfish, declined markedly in abundance after modest increases in the late 1970s. It seems almost certain that the subsequent decline was a direct result of overexploitation by the fishery. In contrast, the elasmobranchs (skates and spiny dogfish) continued to increase during the 1980s. It would appear, therefore, that the elasmobranchs have responded opportunistically to the decline in the other species in the system, perhaps by being able to exploit food resources that were no longer removed by target species. Other possible examples of species replacements are the apparent increase in cephalopod species in the Gulf of Thailand, which coincided with the increase in trawl fishing activity and reduction in the abundance of demersal fish, and the increase in flatfish species that seems to have occurred in the North Sea and elsewhere.

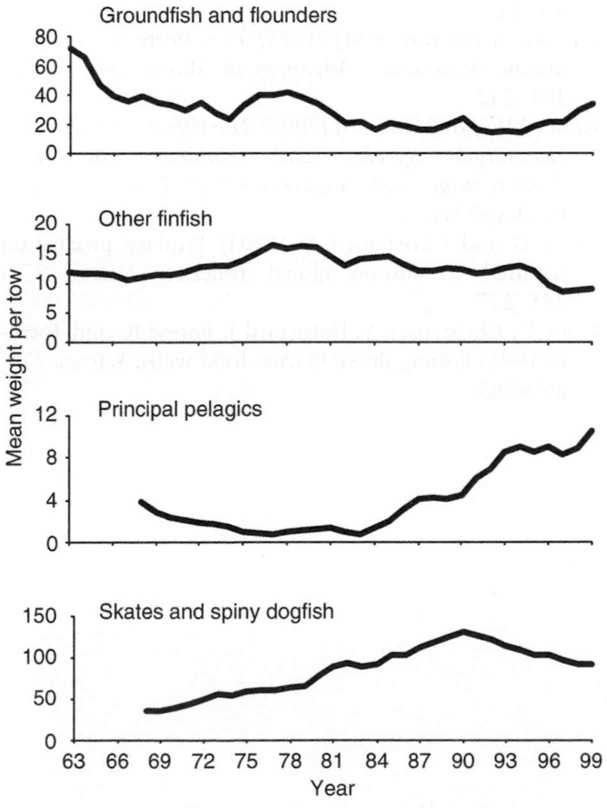

Figure 2 Trends in the relative contribution to total biomass (numbers) made by major taxonomic fish groups. Data from National Oceanographic and Atmospheric Administration, National Marine Fisheries Service, Woods Hole, MA, USA (personal communication).

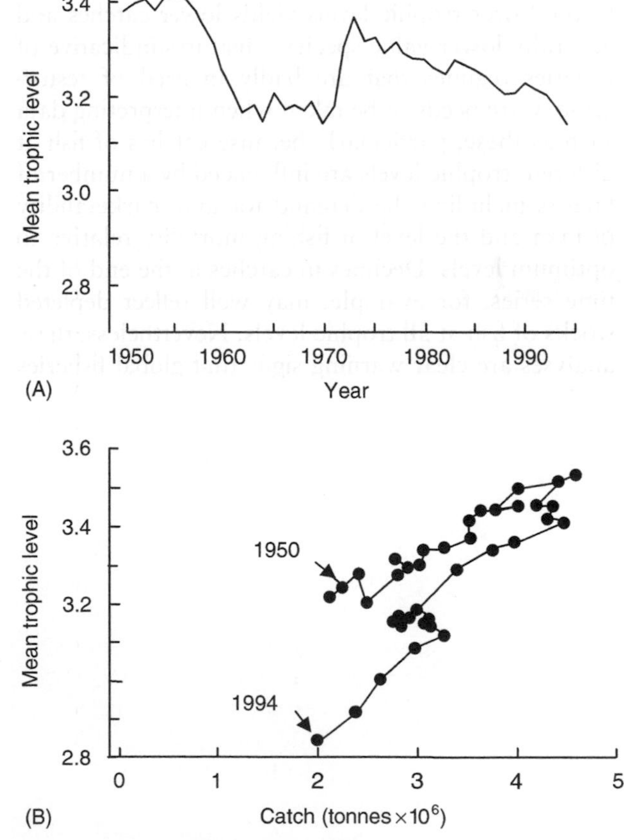

Figure 3 (A) Global trends in mean trophic level of fisheries landings from 1950 to 1994. (B) Plot of mean trophic level versus catch for the north-west Atlantic. (Reproduced from Hall, 1999.)

Conclusion

A final perspective on the system-level effects of fisheries come from an examination of changes over the last 45 years in the average trophic level at which landed fish were feeding (**Figure 3A**). This analysis indicates that there has been a decline in mean trophic level from about 3.3 in the early 1950s to 3.1 in 1994. Very large landings of Peruvian anchoveta, which feeds at a low trophic level, account for the marked dip in the time series in the 1960s and early 1970s. When this fishery crashed in 1972–73 the mean trophic level of global landings rose again. For particular regions, where fisheries have been most developed there have been generally consistent declines in trophic level over the last two decades.

Plots of mean trophic level against catches give a more revealing insight into the system-level dynamics of fisheries (**Figure 3B**). Contrary to expectations from simple trophic pyramid arguments, highest catches are not associated with the lowest trophic levels. This is important because it has been suggested in the past that fishing at lower trophic levels will give greater yields because energy losses from transfers up the food chain will be less. It appears, however, that the global trend towards fishing down to the lower trophic levels yields lower catches and generally lower value species – features indicative of fisheries regimes that are badly in need of restoration. Care needs to be taken when interpreting data such as these, particularly because catches of fish at different trophic levels are influenced by a number of factors including the demand for and marketability of taxa and the level of fishing mortality relative to optimum levels. Declines in catches at the end of the time series, for example, may well reflect depleted stocks of fish at all trophic levels. Nevertheless, these analyses are clear warning signs that global fisheries are operating at levels that are certainly inefficient and probably beyond those that are prudent if we wish to prevent continuing change in the trophic structure of marine ecosystems.

See also

Fisheries Overview. Large Marine Ecosystems. Network Analysis of Food Webs.

Further Reading

Alverson DL, Freeberg MH, Murawski SA, and Pope JG (1994) A global assessment of bycatch and discards. *FAO Fisheries Technical Paper 339*, 233pp, Rome.

FAO (1996) Precautionary approach to fisheries. Part 1: Guidelines on the precautionary approach to capture fisheries and species introductions. *FAO Fisheries Technical Paper 350/1*, Rome.

FAO (1997) Review of the state of the world fishery resources: marine fisheries. *FAO Fisheries Circular no. 920*. Rome.

Hall SJ (1999) *The Effects of Fishing on Marine Ecosystems and Communities*. Oxford: Blackwell Science.

Jennings S and Kaiser MJ (1998) The effects of fishing on marine ecosystems. *Advances in Marine Biology* 34: 201–352.

Kaiser MJ and deGroot SJ (2000) *The Effects of Fishing on Non-target Species and Habitats: Biological, Conservation and Socio-economic Issues*. Oxford: Blackwell Science.

Pauly D and Christensen V (1995) Primary production required to sustain global fisheries. *Nature* 374: 255–257.

Pauly D, Christensen V, Dalsgaard J, Forese R, and Torres F (1998) Fishing down marine food webs. *Science* 279: 860–863.

FISHERY MANIPULATION THROUGH STOCK ENHANCEMENT OR RESTORATION

M. D. J. Sayer, Dunstaffnage Marine Laboratory, Oban, Argyll, UK

Introduction

The continuing advance in the technology associated with the rearing and ongrowing of the early life stages of a number of marine species with commercial importance has resulted in mounting interest in the artificial manipulation of some marine fishery stocks. This manipulation can take two main forms: (1) enhancement of pre-existing or declining stocks and (2) restoration of damaged or extinct stocks.

Enhancement itself can define three forms of fishery manipulation: as being

1. A traditional method of marine ranching for many shellfish species (e.g., oysters or scallops);
2. A method for augmenting existing low volume/high value stocks where the habitat is perceived as being below the potential carrying capacity (e.g., lobster stock enhancement programs); and
3. A reversal of a trend of declining harvests, possibly at reduced catch per unit effort rates, from a fishery which may be recruit-limited (e.g., cod and salmonid stock enhancement programs).

Restoration only occurs where the decline in fishery status is so great that an active fishery no longer exists. Complete fishery collapse has been attributed to natural variation in population shifts but invariably is more often caused, either directly or indirectly, by anthropogenic influence such as nursery habitat destruction or alteration, migratory barriers, overfishing and acute or chronic pollution episodes. In these cases mitigation activities can attempt to reconstruct the previous fishery or replace it with other species of commercial importance that may be better suited to the altered conditions that subsequently exist.

Within both enhancement and restoration programs fishery managers may identify potential benefits associated with basing the manipulation on alternative species to those lost, in decline or to be augmented. These alternative species can be native to the waters of introduction but there are plenty of examples of exotic or closely related marine species being transplanted from different areas of the globe because of the potential for higher yields or better survival potential (e.g., the Pacific oyster, *Crassostrea gigas* (Thunberg)). In addition, artificially increasing the relative numbers of any one species, either naturally occurring or introduced, may alter the ecological balance of an existing ecosystem. Both of these scenarios have been defined previously in the literature as community change marine ranching but, although community change will result, the driving forces for introductions were either enhancement or restoration and so it is not a valid form of manipulation in its own right. **Table 1** summarizes the types and subtypes of fishery manipulation, the reasons for manipulation, the type of stock species used and the key assumptions for artificial fishery intervention. **Table 1** also introduces the concepts of ownership, value, habitat carrying capacity, and recruit limitation which are all contributory factors influencing the scale of the type and value of the intervention.

A Global Perspective of Fishery Manipulation

Over the past few decades, the harvest from the global marine fishery has been maintained with the trend towards a steady but slight increase. However, there have been marked and dramatic declines in some fisheries that have been traditionally exploited possibly caused by overfishing, environmental or ecological change, inadequate fishery management, or combinations of all three. Sometimes the specific fishery or an individual stock have provided a historical basis for the development and maintenance of dependent human communities and so the often sudden reduction in yields can result in significant deleterious socioeconomic degradation.

The history of stock enhancement using hatchery-reared juveniles began in the late 1870s with releases of Atlantic cod (*Gadus morhua* L.) and salmonids principally in the United States and Norway and salmonids only in Japan. Enhancements continued for almost 90 years on varying scales with inconclusive effectiveness. In total it is estimated that 27 countries have been involved with the stocking or ranching of marine species. However, it is Japan that has pursued stock enhancement with long-term vigour. Since 1963, massive stock enhancement

Table 1 Summary of the major types of fishery stock manipulation

Type of manipulation	*Subtype*	*Reasons for manipulation*	*Species stocked*	*Key assumptions*
Enhancement	Traditional marine ranching	To improve harvests on a semi-intensive scale through the captive on-growth of juveniles utilizing natural food availability To increase the stock above historical levels	Natives Close relatives Exotics	That the lease of the marine habitat and ownership of the seeded animal is conveyed and thus private commercial status is possible That the natural fishery is low volume/low value[a] That the carrying capacity of the habitat can be significantly increased That the natural fishery is not recruit-limited
	Augmentation of existing stocks	To improve harvests on an extensive scale through noncaptive on-growth of juveniles utilizing natural food availability. To increase the stock above historical levels.	Natives Close relatives	That there is limited access to ownership and cooperative status or government funding is required That the natural fishery is low volume/high value[a] That the carrying capacity of the habitat can be increased. That the natural fishery is not recruit-limited
	Enhancement of declining stocks	To reverse declines in the harvests on an extensive scale through noncaptive on-growth of juveniles utilizing natural food availability. To maintain the stock at historical levels.	Native	That there is no ownership over and above existing regional or national restrictions to the fishery and centralized funding is required That the natural fishery is high volume/low value[a] That the carrying capacity of the habitat can be increased That the natural fishery is recruit-limited
Restoration	Reconstruction of damaged fishery	To re-establish a historical fishery that had declined below levels of sustainability through noncaptive on-growth of juveniles utilizing natural food availability. To return the stock to historical levels	Natives	That there is no ownership over and above existing regional or national restrictions to the fishery and centralized funding is required That the natural fishery is high volume/low value[a] That the carrying capacity of the habitat can be increased That the natural fishery is recruit-limited
	Replacement of damaged fishery	To replace a historical fishery with a new species that can adapt better to the altered conditions through noncaptive on-growth of juveniles utilizing natural food availability. To establish a completely new fishery	Natives Close relatives Exotics	Ownership and volume/value parameters are likely to mimic extinct fishery That the carrying capacity of the habitat can be increased There is no natural fishery

[a]Volume refers to the proportionate contribution of the fishery to the total national harvest; value refers to the individual animal. Therefore a high volume/low value fishery will have a significantly greater socioeconomic importance compared with a low volume/high value fishery.

Adapted from Barthey (1999).

schemes in Japan have been supported by the national government. In the 1960s there was widespread destruction of Japanese coastal areas, accelerated by land-reclamation projects and industrial pollution. In addition, overfishing had contributed to seriously low stock levels of major traditional fisheries for red sea bream (*Pagurus major* (Temminck & Schlegel)), kuruma prawn (*Penaeus japonicus* (Bate)) and swimming crab (*Portunus trituberculatus* (Miers)). A program of stock enhancement was therefore initiated by the Japanese Government in order to improve fishery resources in coastal areas. This program has steadily increased over the years and, at present, there are over 70 national and local government hatchery facilities contributing or developing almost 80 species for stock release (1995 figures: 33 fish species, 13 crustacean, 24 molluscan, six sea urchin, one sea cucumber, and one octopus). The scales of release have been massive in some cases with, for example, 3 billion scallop, 300 million prawns and 70 million crabs released in total since the program was initiated. Over the years there have been marked variations in the effectiveness of this large-scale enhancement program, but long-term results for chum salmon (*Onchorhynchus keta* (Walbaum)) and the scallop (*Patinopecten yessoensis* (Jay)) indicate a proven augmentation of net fishery production at acceptable economic rates. However, distinct positive economic benefits for these programs are not observed routinely. It is notable that the Japanese stock enhancement program co-evolved alongside large artificial reef and seaweed bed restoration programs. Similar enhancement programs for marine vascular plants and seaweeds, some in addition to the construction of artificial reefs, are globally widespread. Habitat restoration or manipulation by itself can have a positive effect on fishery status without the additional requirement of hatchery-reared animal introductions.

The release of Atlantic cod in north Atlantic waters has a long history. Large numbers continue to be released in programs in Norway, but also to a lesser extent in the Faroe Islands and Iceland. Although evidence has been collected to suggest that the condition and growth rates of released cod are better than those of wild stocks on recapture, the overall effect, though variable, was not to produce a significant increase in the fishery and certainly not within any limits of economic viability. Just as enhancement schemes for Atlantic salmon were eventually occluded by intensive aquaculture of that species, there is growing interest in the intensive farming of cod in north temperate waters.

In the 1990s there were widespread releases of microtagged juvenile European lobsters (*Homarus gammarus* (L.)) in Norway. Following the pioneering development work carried out in the UK during the 1970s and 1980s the large-scale production of juvenile European lobsters has become technically straightforward and low-cost. Seven years after initial releases, over 40% of the commercial catches and over 70% of the sublegal-sized catches in south-western Norway were from the released stock. This scale of enhancement was achieved with a total recovery rate of approximately 8% of the released stock. However, even at this rate of return (which approximates to other recapture rates for the European lobster) it is concluded that the continued enhancement of a depleted lobster population in Norway is economically feasible.

There are many other examples of manipulation programs occurring over the globe at varying levels of size and success. Some concentrate on enhancement, others on restoration. In all cases, the measurement of success is complex. For example, a large fishery of significant social importance can be maintained at an economic loss through stock enhancement schemes, but can still be measured as a success by the funding agency if the losses sustained through enhancement outweigh the socio-economic costs of societal collapse. In another case, enhancement or restoration may not increase the numbers of animals harvested but could increase the unit price through improved quality. In a similar way, there have been examples of artificially moving the fishery closer to the market, thereby increasing the unit price and restoring an economically viable fishery even though catch volumes were not improved.

Quality and Survival of the Release Stock

The anticipated proportion of released individuals recruiting to the fishery and eventually to harvest will differ between species and the intended type of manipulation. However, an essential objective in the production of hatchery-reared animals is that they should possess similar physical and behavioral capabilities to their wild counterparts in order to minimize differences that would compromise their survival in a natural environment. Many stock enhancement programs have reported high mortality rates in released individuals over timescales of days postrelease. Through subsequent laboratory experimentation significant progress has been made into identifying approaches to the ways the enhancement stocks are cultured, prepared for release, and eventually liberated.

There are a significant number of marine species that, when reared artificially, present differences in structure and coloration compared with wild individuals. Examples of this are hatchery-reared halibut (*Hippoglossus hippoglossus* (L.)) where a significant proportion of the cultured individuals are non-pigmented, and physical jaw deformities characteristically caused during the rearing of Atlantic cod and herring (*Clupea harengus* L.). Many of these abnormalities can be corrected for through dietary improvement. In addition abnormal colorations are sometimes a result of being reared in bare unnaturally colored culture tanks and only a short prerelease exposure period to simulated natural conditions is sufficient to improve coloration. The culture environment itself is likely to lack many of the physicochemical attributes of the wild. Numerous fish species depend on tidal and diel variations in parameters such as light and pressure to drive short-term migratory patterns that may be essential in optimizing foraging and antipredation behaviors. The ability to learn potentially inherent rhythmic behavior cycles over relatively short conditioning periods has been shown to reduce initial levels of postrelease vulnerability.

In intensive fish farming the removal of many natural behavioral traits that can potentially result in intraspecific damage can be advantageous. However, the retention of aggressive, predatory and antipredatory behavior is essential in many species that are intended for wild-release in manipulation programs. A number of studies have shown that cultured juveniles rarely possess the same abilities as wild fish of similar age to detect and/or react to potential predators or prey, or react in the same way to different environmental cues and clues. Systematic approaches have been taken to dissipate the effects of hatchery culture including prerelease exposures to predators, prey organisms (weaning hatchery-reared individuals from artificial to live diets), and simulated natural environments. Some of the juvenile cod intended for release in the Norwegian cod enhancement program were cultured in extensive systems that potentially preconditioned the juveniles to a suite of behavioral and physical conditions that were similar to those expected in the wild. The same enhancement program also produced 0-group juveniles that at the time of liberation had achieved a higher growth rate and were in a higher condition than their wild counterparts. It was considered that this advanced growth and condition allowed the culture animals to withstand a period of poor feeding after release.

The actual methodological approach to the practical task of liberation of the reared stock will vary considerably between species. However, it is essential that release is based on a sound knowledge of the biology, environmental requirements, and stocking densities of each species. Some release strategies are plainly obvious; significant mortality levels would be expected if rock-dependent fish were released onto a sand-dominated habitat and vice versa. Detailed knowledge of the animal's life history may indicate that a range of habitats may be required for the successful on-growth of the released juveniles. One method of optimizing the habitat requirements of ranched species may be to deploy artificial habitats at, or in the vicinity of, the site of release. As well as potentially improving survival, artificial habitats can increase the carrying capacity of an environment and help in designating ownership within an open system environment (see below).

The method of release also has to consider the stocking density, the early life-history tendencies and any ontogenic shifts in habitat utilization. Research has indicated the benefits of introducing shoaling species initially into cages so that the individuals can recover from the stresses associated with release and develop strong shoaling tendencies prior to eventual liberation. Conversely, dispersal methods are essential in species that are strongly territorial and potentially cannibalistic. Ontogenic changes in habitat requirements may also have to be considered if maintenance of the species within set geographical limitations is an objective of the enhancement program.

Genetic Considerations and the Introduction of Exotics

The maintenance of native gene pools and the preservation of genetic variation and adaptive gene combinations in natural populations have the potential to be compromised through the deliberate release of hatchery-reared fishery stocks of different or restricted genetic profiles. Conserving natural genetic variation is important for continued evolution in wild stocks, but also has direct economic implications for use in aquaculture. However, genetic variability also represents an opportunity to improve stocks for enhancement or restoration by selective breeding. Selective breeding is at one end of a scale of genetic modification that could potentially result in the released animals out-competing the natural stocks and eventually completely replacing, rather than augmenting, the target fishery. A cautionary approach has always been urged with regard to improving strains for eventual release by selection.

The genetical ethics involved with the deliberate release of fish into the wild have received a lot of attention and most large-scale, public-funded release schemes have in place guidelines for the selection of broodstock, release sites and the health status of the released fish. Unfortunately, what is largely lacking in many programs is genetic information tracking the interactions of released and wild stocks, and so the analysis of the potential risks posed by large-scale releases is largely incomplete. What is known is that selective breeding can result in improved return rates and is, therefore, of significant economic importance in manipulation programs. In general, most proposals for minimizing genetic pollution within selective breeding programs suggest using local broodstock where possible to produce the juveniles at each release point and preserving genetic diversity through the maintenance of large broodstock numbers.

The process of fishery replacement often occurs where, for a variety of potential reasons, the historical fishery has become extinct. Often in these cases it is identified that the re-establishment of the fishery using the same species would not be successful. In this situation, and where an identified basis for a new fishery to replace the failed one exists, introductions of different native species or in some cases nonnative species, or exotics, has taken place. Often the greatest care is taken with such introductions and many countries have strict laws governing the movement and introduction of nonnative species in order to minimize risks of disease transfer, to which native wild populations may have no resistance, and to prevent adverse competition or interbreeding. Clearly such introductions, and on scales likely to be economically viable, create special concerns. As a consequence large-scale introductions of nonnatives are unlikely to form a significant proportion of future manipulation programs. Current practices do involve important safeguards to protect the ecological integrity of systems in which enhancement or restoration efforts are carried out. However, many such programs were initiated at a time when modern environmental ethics were unrecognized. As a consequence, irreversible changes have occurred in some systems.

Ecological Balance and Carrying Capacity

Crucial to any form of fishery manipulation is the question of whether or not the ecosystem that hatchery-reared fish are being introduced to can sustain and support the new introductions. Quite often it is assumed that because stocks of the main commercially relevant species are in decline, the ecosystem can support reintroduction with the aim of at least attaining historical levels. However, this ignores how the total biomass has changed and whether there is a concomitant decline. A decline in the density and abundance of one species can result in one or more other species increasing in volume because there is now spare capacity in the system. So, if the commercial species has declined to a level at which manipulation is being considered, but the spare capacity in the system once occupied by that species has now been filled by expansion from other species, the carrying capacity may be limited and, irrespective of the numbers of introductions, the target species entering the fishery will not increase.

Measurement of total biomass may, therefore, give indications as to the potential success of a fishery manipulation program. However, estimating total biomass is very difficult in what are usually dynamic trophic situations and rarely in manipulation programs do total biomass estimates exist prior to the measured decline in the target species. A potentially more practical methodology is to examine the ecosystem as a whole and identify what species are inhabiting the trophic niche that the target species would be expected to occupy on introduction. If there is a possibility of interspecific competition occurring with the target species being successful against trophic niche competitors then an introduction may be successful. This is, of course, an extremely simplistic approach and totally ignores the questions of food supply and higher order predation. If a manipulation is taken to the extreme then enhancements of food supply, reductions in predators, reductions in competitors, and increased habitat availability should also be considered. Artificial reef deployments that are designed to optimize habitat requirements, enhance lower order productivity, and minimize higher order predation and interspecific competition have proved successful in small-scale localized manipulations.

Recapture and Monitoring Performance

All fishery manipulations will require some degree of performance auditing. As well as yielding feedback as to how well the manipulation has worked there will be real economic data to be obtained in order to assess the socioeconomic and commercial validity of the manipulation. In addition, fishery managers will also obtain an indication as to any additional manipulations that may be required. Performance monitoring usually takes the form of recapture. Simple ratios of wild compared with introduced individuals give an initial indication of how the manipulation has worked, though these can be further modified through condition indices and age/growth functions. The criteria against which a successful manipulation can be assessed will vary markedly between target species and the economic reasons for the initial intervention. In most cases it is the target stock mass that will be of central importance with the objective of maximizing conversion rates of food to flesh in order to produce maximum sustainable yield. However, increasing numbers may have equal importance particularly in cases where the target species are for sport fishing.

The methods for recording the incidence of released individuals will also vary between fisheries. Large-scale introductions will invariably have to rely on return information gathered from the commercial fishery. Often these data can be obtained through a reward system to the fishermen if the method of tagging is visible enough to be identified easily. Manipulations that have employed the use of small microtag injections have required a much more active role for the assessors who have to monitor all or parts of the fishery landings in order to estimate the efficacy of the manipulation. Where intervention has occurred on a smaller scale, the active fishery may be small enough to allow for all catches to be assessed.

Ownership, Exploitation Rights, and Operational Controls

Fundamental to the form and viability of any manipulation program is the legal framework on which the ownership rights to the resource and its exploitation are based. Unless traditional rights to the existing or damaged fishery exist or legal provisions are successfully made to the contrary, the default position is usually open-access exploitation. Unlimited access to the resource, within possible national or international quota controls, means that ownership of the released stock cannot be retained and, therefore, the instigation of such an enhancement or restoration program can only come from supraregional, national, or international initiatives. Only in programs where ownership of the released stock is conveyed and can be managed will private or cooperative programs be assured. In a similar way, the pattern of exploitation rights will influence heavily the investment procedures in the manipulation program. There is an established history of private, cooperative, and centralized public funding in fishery enhancement and restoration programs where, in general, private and cooperative investment only occurs where some legally protected proprietary rights exist. A fishery that remains open-access will only ever attract centralized public investment.

The source of investment will also dictate the type of fishery that is to be enhanced or restored. Centralized public funding has traditionally targeted fisheries that have large socioeconomic impacts. These fisheries are invariably ones of high volume but where the unit value of the fishery is low. Invariably these high volume/low value fisheries are ones with significant historical resonance but are declining through recruit limitations. Stock enhancement in these cases will be as much to maintain social structure and tradition than to rescue the fishery *per se*. Private and cooperative investment will be on a smaller scale and will, therefore, be attracted to low-volume fisheries, where the unit value of the fishery is high. This latter form of fishery also tends to be the type where ownership and exploitation rights are more easily established because of the smaller areas involved. In cases of private restricted ownership the ability to identify and retain the released stock is important. Also through artificial fishery manipulation the location of the fishery can be altered to the advantage of the fishermen by, for example, bringing the fishery closer to the markets or moving it to more protected waters. Location manipulation, retention of the released stock, habitat optimization, and fishery identification have all been achieved through the construction of artificial structures, that either mimic the habitat provided by a natural reef or act as fish attraction devices. In many countries the deployment of artificial structures in association with fishery enhancement, although potentially advantageous for the above reasons, can, in itself, carry a high level of legal burden even before the legal provisions for ownership regulation of the released stock are attained.

Conclusions

The manipulation of some marine fisheries has been considered to be an important tool available to fishery managers to prevent or reverse declining fisheries, or to restore or replace lost ones. However, many manipulation schemes have attracted both controversy and critisism and it has not always been possible to quantify the efficacy of all programs. It has been proposed that future manipulation programs follow a two-staged approach. This approach suggests that managers should firstly quantify the existing status of the fishery and the environment (to include other species present and the ecological carrying capacity) prior to considering manipulation and then undertake a detailed premanipulation study that estimates the expected returns (numbers and value), identifies the expected beneficiaries, assigns ownership, and introduces a legal framework of regulation. Once manipulation is embarked on then a precautionary approach should be adopted. This entails adherence to agreed and planned manipulation protocols and the continuing evaluation of potential impacts with contingency plans to either adapt or end the manipulation if adverse impacts are detected. In addition, there are now case studies from around the world that highlight decades of past manipulation research, dozens of species released and many fisheries targeted. In general, there does appear to be a trend emerging, which is that successful manipulation tends only to occur where the species is not migratory on a large scale, is part-contained by habitat availability, and is dependent on relatively low levels of recapture in order to be economically or socially viable. These type of criteria are best represented by low volume, high value fisheries where environmental carrying capacity can be increased. In addition, manipulations that have been undertaken in parallel with habitat enhancement schemes (for example, artificial reefs, nursery ground restoration or protection) have been among the most successful.

See also

Crustacean Fisheries. Demersal Species Fisheries. Marine Fishery Resources, Global State of. Molluskan Fisheries. Open Ocean Fisheries for Deep-water Species. Open Ocean Fisheries for Large Pelagic Species. Salmon Fisheries, Atlantic. Salmon Fisheries, Pacific. Small Pelagic Species Fisheries. Southern Ocean Fisheries.

Further Reading

Addison JT and Bannister RCA (1994) Re-stocking and enhancement of clawed lobster stocks: a review. *Crustaceana* 67: 131–155.

Bartley DM (1999) Marine ranching: a global perspective. In: Howell BR, Moksness E, and Svåsand T (eds.) *Stock Enhancement and Sea Ranching*, pp. 79–90. Oxford: Fishing News Books.

Blaxter JHS (2000) The enhancement of marine fish stocks. *Advances in Marine Biology* 38: 1–54.

Carvalho GR and Pitcher TJ (eds.) (1995) *Molecular Genetics in Fisheries*. London: Chapman and Hall.

Cowx IG (ed.) (1998) *Stocking and Introductions of Fish*. Oxford: Fishing News Books.

Cross TF (1999) Genetic considerations in enhancement and ranching of marine and anadromous species. In: Howell BR, Moksness E, and Svåsand T (eds.) *Stock Enhancement and Sea Ranching*, pp. 37–48. Oxford: Fishing News Books.

Danielssen DS and Moksness E (eds.) (1994) Sea ranching of cod and other marine species. *Aquaculture and Fisheries Management*, (Supplement 1) 25.

Hilborn R (1998) The economic performance of marine stock enhancement projects. *Bulletin of Marine Science* 62: 661–674.

Howarth W and Lería C (1999) Legal issues relating to stock enhancement and marine ranching. In: Howell BR, Moksness E, and Svåsand T (eds.) *Stock Enhancement and Sea Ranching*, pp. 509–525. Oxford: Fishing News Books.

Howell BR (1994) Fitness of hatchery-reared fish for survival at sea. *Aquaculture and Fisheries Management* 25 (Supplement 1): 3–17.

Howell BR, Moksness E, and Svåsand T (eds.) (1999) *Stock Enhancement and Sea Ranching*. Oxford: Fishing News Books.

Isaksson A (1988) Salmon ranching: a world review. *Aquaculture* 75: 1–33.

Jensen AC, Collins KJ, and Lockwood AP (eds.) (1999) *Artificial Reefs in European Seas*. Dordrecht: Kluwer Academic Publishers.

Kitada S (1999) Effectiveness of Japan's stock enhancement programmes: current perspectives. In: Howell BR, Moksness E, and Svåsand T (eds.) *Stock Enhancement and Sea Ranching*, pp. 103–131. Oxford: Fishing News Books.

Pickering H (1999) Marine ranching: a legal perspective. *Ocean Development and International Law* 30: 161–190.

Svåsand T, Kristiansen TS, Pedersen T, *et al.* (2000) The enhancement of cod stocks. *Fish and Fisherie* 1: 173–205.

MARINE FISHERY RESOURCES, GLOBAL STATE OF

J. Csirke and S. M. Garcia, Food and Agriculture Organization of the United Nations, Rome, Italy

Introduction

The Fisheries Department of the Food and Agriculture Organization of the United Nations (FAO) monitors the state of world marine fishery resources and produces a major review every 6–8 years, with shorter updates presented every 2 years to the FAO Committee on Fisheries (COFI) as part of a more general report *The State of Fisheries and Aquaculture* (*SOFIA*). The latest major FAO review of the state of world marine fishery resources was issued in 2005 with a shorter update in *SOFIA 2006* and a further updating focusing on fishery resources that can be found partly or entirely on the high seas also published in 2006. This article draws significantly from sections of the above FAO publications and uses catch information available from 1950 to 2004 (the last year for which global catch statistics are available).

With a view to offering a comprehensive description of the global state of world fish stocks, the short analysis provided below considers successively: (1) the relation between 2004 and historical production levels; (2) the state of stocks, globally and by regions according to information from the FAO reports above; and (3) the trends in state of stocks since 1974, globally and by region.

Relative Production Levels

The catch data available for the 19 FAO statistical areas (**Table 1**) or regions of the world's oceans indicate that four of them are at or very close to their maximum historical level of production: the Eastern Indian Ocean and the Western Central Pacific Oceans reached maximum production in 2004 while the Western Indian Ocean and the NE Pacific produces 95% and 90% of the maximum, produced in 2003 and 1987, respectively. All other regions are presently producing less than 90% of their historical maximum, for various reasons. This may result, at least in part, from natural oscillations in productivity caused by ocean climate variability or by the fact that short-term exceptionally high initial catches (a phenomenon known as 'overshooting') can be obtained when a new fishery on a previously unexploited stock develops too fast. However, the lower catch values in recent years may also be indicative of an increase in the number of resources being overfished or depleted.

Global Levels of Exploitation

The recent FAO reviews report on 584 stock or species groups (stock items) being monitored. For 441 (or 76%) of them, there is some more-or-less recent information allowing some estimates of the state of exploitation. These 'stock' items are classified as underexploited (U), moderately exploited (M), fully exploited (F), overexploited (O), depleted (D), or recovering (R), depending on how far they are from 'full exploitation' in terms of biomass and fishing pressure. 'Full exploitation' is used by FAO as loosely equivalent to the level corresponding to maximum sustainable yield (MSY) or maximum long-term average yield (MLTAY).

1. U and M stocks could yield higher catches under increased fishing pressure, but this does not imply any recommendation to increase fishing pressure.
2. F stocks are considered as being exploited close to their MSY or MLTAY and could be slightly under or above this level because of natural variability or uncertainties in the data and in stock assessments. These stocks are usually in need of (and in some cases already have) effective control on fishing capacity in order to remain as fully exploited, and avoid falling in the following category.
3. O or D stocks are exploited beyond MSY or MLTAY levels and have their production levels reduced; they are in need of effective strategies for capacity reduction and stock rebuilding.
4. R stocks are usually at very low abundance levels compared to historical levels. Directed fishing pressure may have been reduced by management or because of profitability being lost, but may nevertheless still be under excessive fishing pressure. In some cases, their indirect exploitation as bycatch in another fishery might be enough to keep them in a depressed state despite reduced direct fishing pressure.

According to information available in 2004 (**Figure 1**), 3% of the world fish stocks (all included)

Table 1 Ratio between recent (2004) total catch and maximum reached catch, by FAO statistical areas

FAO area (code and name)	*Year when maximum catch was reached*	*Total catch in 2004 (10^3)*	*Ratio of 2004 over maximum catch (%)*
18 – Arctic Sea	1968	0	0
21 – Atlantic Northwest	1968	2353	52
27 – Atlantic Northwest	1976	9952	77
31 – Atlantic, Western Central	1984	1652	66
34 – Atlantic, Eastern Central	1990	3392	82
37 – Mediterranean and Black Sea	1988	1528	77
41 – Atlantic, Southwest	1997	1745	66
47 – Atlantic, Southwest	1978	1726	53
51 – Indian Ocean, Western	2003	4147	95
57 – Indian Ocean, Western	2004	5625	100
61 – Pacific, Northwest	1998	21 558	87
67 – Pacific, Northwest	1987	3050	90
71 – Pacific, Western Central	2004	11 011	100
77 – Pacific, Eastern Central	2002	1701	87
81 – Pacific, Southwest	1992	736	80
87 – Pacific, Southwest	1994	15 451	76
48 – Atlantic, Antarctic	1987	125	25
58 – Indian Ocean, Antarctic	1972	8	4
88 – Pacific, Antarctic	1983	3	28

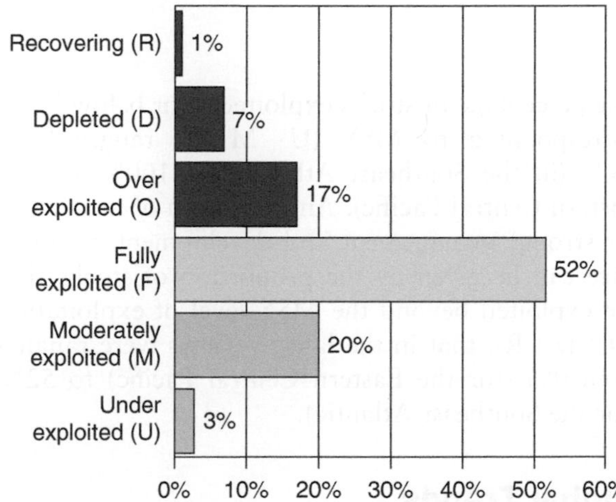

Figure 1 State of world fish stocks in 2004. Reproduced with permission from FAO (2005) Review of the State of World Marine Fishery Resources. *FAO Fisheries Technical Paper, 457*, 235pp, figure A2.1. Rome: FAO.

appeared to be underexploited, 20% moderately exploited, 52% fully exploited, 17% overfished, 7% depleted, and 1% recovering.

On the one hand, this indicates that 25% of the world stocks (O + D + R) for which some data are available are below the level of abundance corresponding to MSY or are exploited with a fishing capacity well above this level. They require management to rebuild them at least to the level corresponding to MSY as provided by the 1992 UN Convention on the Law of the Sea (UNCLOS). As 52% of the stocks appear to be exploited around MSY and most, if not all, also require that capacity control measures be applied to avoid the negative effects of overcapacity, it appears that 77% (F + O + D + R) of the world stocks for which data are available require that strict capacity and effort control be applied in order to be stabilized or be rebuilt around the MSY biomass levels, and possibly beyond. Some of the fisheries concerned may already be under such management schemes.

On the other hand, **Figure 1** also indicates that 23% (U + M) of the world stocks for which some data are available are above the level of abundance corresponding to MSY, or are exploited with a fishing capacity below this level. Considering again that 52% of the stocks are exploited around MSY, this means that 75% of the stocks (U + M + F) are at or above MSY level of abundance, or are exploited with a fishing capacity at or below this level, and should be therefore considered as compliant with UNCLOS basic requirements.

These two visions of the global situation of fishery stocks indicate that the 'glass is half full or half empty' and both are equally correct depending on which angle one takes. From the 'state of stocks' point of view, it is comforting to see that 75% of the world resources are still in a state which could produce the MSY, as provided by UNCLOS. From the management point of view, it should certainly be noted that 77% of the resources require stringent management and control of fishing capacity. As mentioned above, some of these (mainly in a few

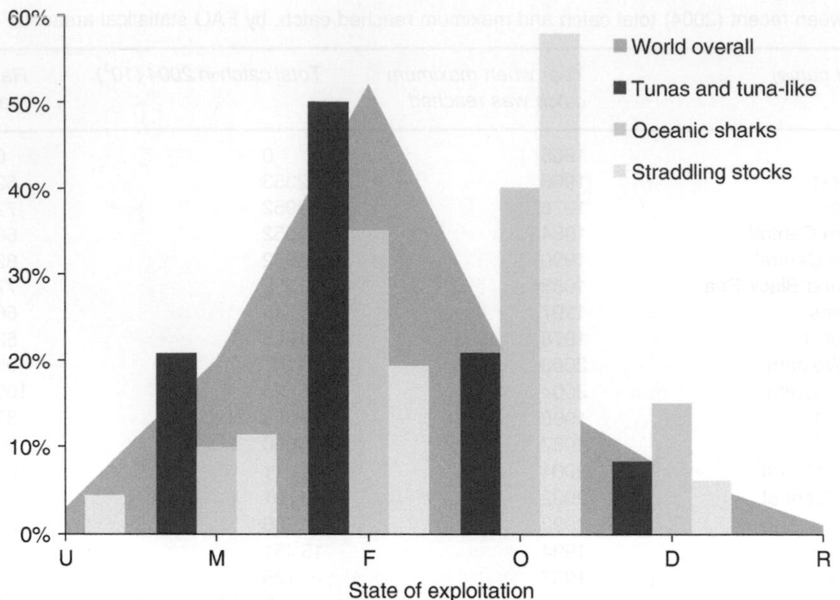

Figure 2 State of exploitation of world highly migratory tuna and tuna-like species, highly migratory oceanic sharks, and straddling stocks (including high seas fish stocks). Reproduced with permission from Maguire J-J, Sissenwine M, Csirke J, Grainger R, and Garcia S (2006) The State of World Highly Migratory, Straddling and Other High Seas Fishery Resources and Associated Species. *FAO Fisheries Technical Paper*, *495*, 84pp, figure 59. Rome: FAO.

developed countries) are already under some form of capacity management. Many, however, would require urgent action to stabilize or improve the situation. For 25% of them, energetic action is required for rebuilding.

The situation appears more critical in the case of fish stocks that occur partly or entirely, and can be fished in the high seas (**Figure 2**).

While the state of exploitation of highly migratory tuna and tuna-like species is very similar to that of the world overall, the state of exploitation of highly migratory oceanic sharks appears to be more problematic, with more than half of the stocks listed as overexploited or depleted. The state of straddling stock (including high seas stocks) is even more problematic with nearly two-thirds being classified as overexploited or depleted.

State of Stocks by Region

When the available information is examined by regions (**Figure 3**), the percentage of stocks exploited at or beyond levels of exploitation corresponding to MSY (F + O + D + R) and needing fishing capacity control ranges from 43% (for the Eastern Central Pacific) to 100% (in the Western Central Atlantic). Overall, in most regions, 70% of the stocks at least are already fully exploited or overexploited. The percentage of stocks exploited at or below levels corresponding to MSY (U + M + F) ranges from 48% (in the Southeast Atlantic) to 100% (in the Eastern Central Pacific). An indication of how weak (or strong) management and development performance can be given by the proportion of stocks that are exploited beyond the MSY level of exploitation (O + D + R), that in the latest reviews were ranging from 0% (for the Eastern Central Pacific) to 52% (for the Southeast Atlantic).

Global Trends

The trends in the proportion of stocks in the various states of exploitation as taken from the various FAO reviews since 1974 (**Figure 4**) shows that the percentage of stocks maintained at MSY level, or fully exploited (F) has slightly increased since 1995, reversing the previous decreasing trend since 1974. The underexploited and moderately exploited stocks (U + M), offering some potential for expansion, continue to decrease steadily, while the proportion of stocks exploited beyond MSY levels (O + D + R) increased steadily until 1995, but has apparently leveled off and remained more or less stable at around 25% since. The number of 'stocks' for which information is available has also increased during the same period, from 120 to 454.

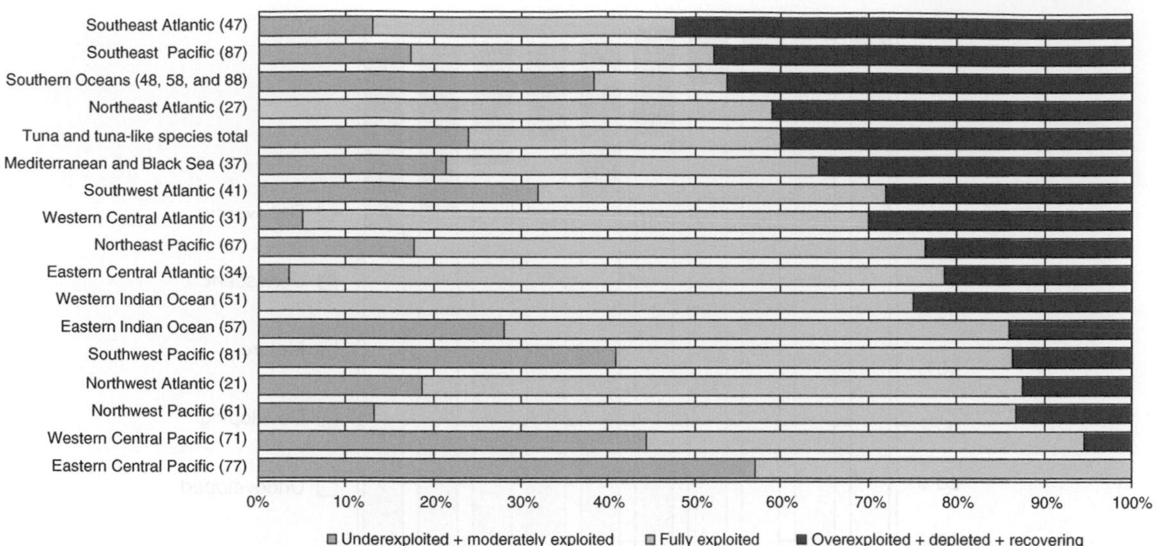

Figure 3 Percentage by FAO fishing areas of stocks exploited at or beyond MSY levels (F + O + D + R) and below MSY levels (U + M). Reproduced with permission from FAO (2005) Review of the State of World Marine Fishery Resources. *FAO Fisheries Technical Paper, 457*, 235pp, figure A2.2. Rome: FAO.

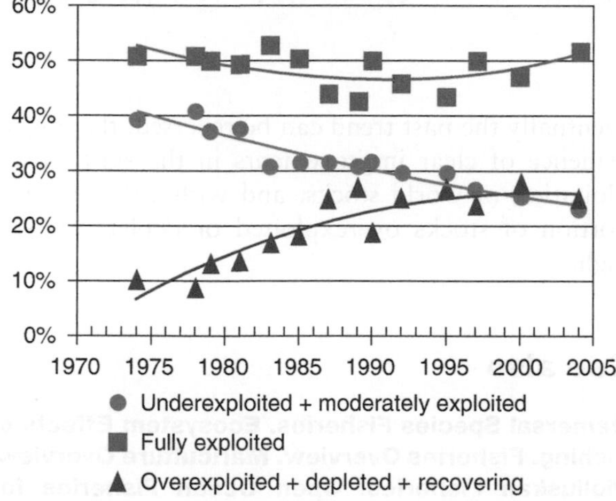

Figure 4 Global trends in the state of exploitation of world stocks since 1974. Reproduced with permission from FAO (2005) Review of the State of World Marine Fishery Resources. *FAO Fisheries Technical Paper, 457*, 235pp, figure A2.3. Rome: FAO.

Discussion

The perspective view of the state of world stocks obtained from the series of FAO fishery resources reviews indicates clearly a number of trends. Globally, between 1974 and 1995, there was a steady increase in the proportion of stocks classified as 'exploited beyond the MSY limit', that is, overfished, depleted, or recovering (after overexploitation and depletion). These conclusions are in line with earlier findings summarized by Garcia, de Leiva, and Grainger in **Figure 5**. Since the findings by Garcia and Grainger in 1996 were based on a sample of the world stocks, severely constrained by availability of information to FAO, the conclusions are considered with some caution. A key question is: To what extent does the information available to FAO reflect reality? There are many more stocks in the world than those referred to by FAO. In addition, some of the elements of the world resources referred to by FAO as 'stocks' are indeed conglomerates of stocks (and often of species). One should therefore ask what validity a statement made for the conglomerate has for individual stocks (*stricto sensu*). There is no simple reply to this question and no research has been undertaken in this respect.

However, while recognizing that the global trends observed reflect trends in the monitored stocks, it is also noted that the observations generally coincide with reports from studies conducted at a 'lower' level, usually based on more insight and detailed data. For instance, an analysis on Cuban fisheries using the same approach as used by Garcia and Grainger for the whole world leads to surprisingly similar conclusions, using less coarse aggregations with an even longer time series.

There is of course the possibility that stocks become 'noticed' and appear in the FAO information base as 'new' stocks only when they start getting into trouble and scientists having accumulated enough data start dealing with them, generating reports that FAO can access. This could explain the increase in the percentage of stocks exploited beyond MSY since

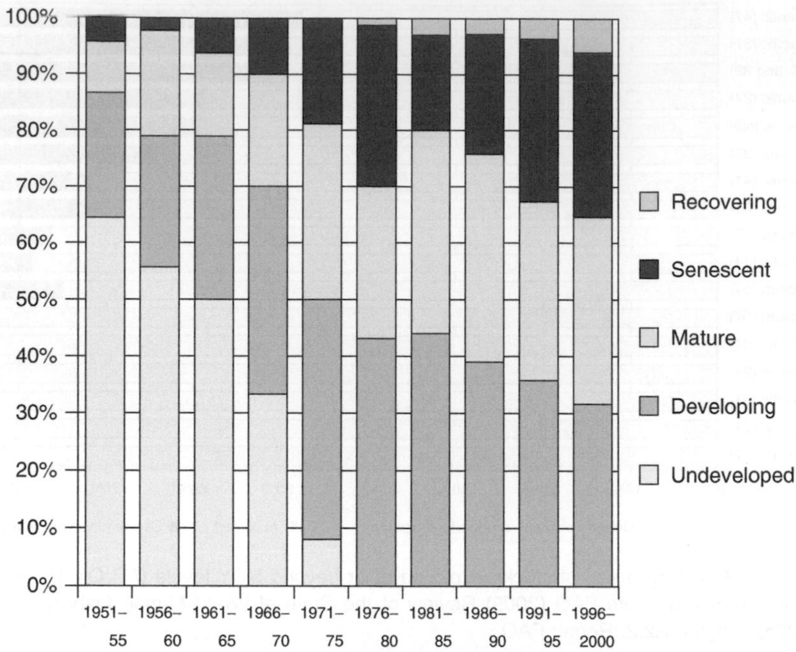

Figure 5 Stage of development of the 200 major marine fishery resources: 1950–2000. Reproduced with permission from FAO (2005) Review of the state of world marine fishery resources. *FAO Fisheries Technical Paper, 457*, 235pp, figure A2.4. Rome: FAO.

1974. This assumption, however, does not hold for at least two reasons.

1. The number of 'stock items' identified by FAO but for which there is not enough information has also increased significantly with time, from seven in 1974 to 149 in 1999, clearly showing that new entries in the system are not limited to well-developed fisheries or fish stocks in trouble.
2. From the 1980s, based on the recognition of the uncertainties behind identification of the MSY level, and recognizing also the declines due to decadal natural fluctuations, scientists have become more and more reluctant to definitely classify stocks as 'overfished'. The apparent 'plateauing' of the proportion of stocks with excessive exploitation in the world oceans may in part be due to this new trend.

While the trend analyses of world fisheries landings (**Figure 5**) tend to suggest that the proportion of senescent and recovery fisheries (taken as grossly corresponding to those being exploited beyond MSY) is still on the increase, the resources analyses summarized in **Figure 4** indicate that there is a 'flattening' in the proportion of fish stocks that are overexploited (or exploited beyond MSY). Even if the latter may be indicative that the 'deterioriation' process leading to the overexploitation of marine fishery resources has slowed down and that eventually the past trend can be reversed, there is no evidence of clear improvements in the state of exploitation of world stocks, and with 25% the proportion of stocks overexploited or depleted is still high.

See also

Demersal Species Fisheries. Ecosystem Effects of Fishing. Fisheries Overview. Mariculture Overview. Molluskan Fisheries. Open Ocean Fisheries for Deep-water Species. Open Ocean Fisheries for Large Pelagic Species. Salmon Fisheries, Atlantic. Salmon Fisheries, Pacific. Small Pelagic Species Fisheries. Southern Ocean Fisheries.

Further Reading

Baisre JA (2000) Chronicles of Cuban Marine Fisheries (1935–1995): Trend Analysis and Fisheries Potential. *FAO Fisheries Technical Paper, 394*. Rome: FAO.

Csirke J Vasconcellos M (2005) Fisheries and long-term climate variability. In: Review of the State of World Marine Fishery Resources. *FAO Fisheries Technical Paper, 457*, pp. 201–211. Rome: FAO.

FAO (2005) Review of the State of World Marine Fishery Resources. *FAO Fisheries Technical Paper, 457*, 235pp. Rome: FAO.

FAO (2007) *The State of World Fisheries and Aquaculture 2006*, 162pp. Rome: FAO.

Garcia SM and De Leiva Moreno I (2000) Trends in world fisheries and their resources: 1974–1999. *The State of World Fisheries and Aquaculture 2000*. Rome: FAO.

Garcia SM and Grainger R (1996) Chronicles of Marine Fishery Landings (1950–1994): Trend Analysis and Fisheries Potential. *FAO Fisheries Technical Paper, 359*. Rome: FAO.

Maguire J-J, Sissenwine M, Csirke J, Grainger R, and Garcia S (2006) The State of World Highly Migratory, Straddling and Other High Seas Fishery Resources and Associated Species. *FAO Fisheries Technical Paper, 495*, 84pp. Rome: FAO.

MOLLUSKAN FISHERIES

V. S. Kennedy, University of Maryland, Cambridge, MD, USA

Introduction

The Phylum Molluska, the second largest phylum in the animal kingdom with about 100 000 named species, includes commercially important gastropods, bivalves, and cephalopods. All are soft-bodied invertebrates and most have a calcareous shell secreted by a phylum-specific sheet of tissue called the mantle. The shell is usually external (most gastropods, all bivalves), but it can be internal (most cephalopods). Gastropods live in salt water, fresh water, and on land; bivalves live in salt water and fresh water; cephalopods are marine. Gastropods include land and sea slugs (these have a reduced or no internal shell and are rarely exploited by humans) and snails (including edible periwinkles, limpets, whelks, conchs, and abalone). Bivalves include those that cement to hard substrates (oysters); that attach to hard substrates by strong, beard-like byssus threads (marine mussels, some pearl oysters); that burrow into hard or soft substrates (clams, cockles); or that live on the sea bottom but can move into the water column if disturbed (scallops). Cephalopods are mobile and include squid, cuttlefish, octopus, and the chambered nautilus (uniquely among cephalopods, the nautilus has a commercially valuable external shell). Most gastropods are either carnivores or grazers on algae, most commercial bivalves 'filter' suspended food particles from the surrounding water, and cephalopods are carnivores.

Humans have exploited aquatic mollusks for thousands of years, as shown worldwide by shell mounds or middens produced by hunter-gatherers on sea coasts, lake margins, and riverbanks. Some mounds are enormous. In the USA, Turtle Mound in Florida's Canaveral National Seashore occupies about a hectare of land along about 180 m of shoreline, contains nearly 27 000 m^3 of oyster shell, and was once an estimated 23 m high (humans have mined such mounds worldwide for the shells, which serve as a source of agricultural and building lime, as a base for roads, and for crumbling into chicken grit). A group of oystershell middens near Damariscotta, Maine, USA stretches along 2 km of shoreline. One mound was about 150 m long, 70 m wide, and 9 m high before it was mined; a smaller mound was estimated to contain about 340 million individual shells. Studies by anthropologists have shown that many middens were started thousands of years ago (e.g. 4000 years ago in Japan, 5000 years ago in Maine, 12 000 years ago in Chile, and perhaps up to 30 000 years ago in eastern Australia; sea level rise may have inundated even older sites).

Mollusks have been exploited for uses other than food. More than 3500 years ago, Phoenicians extracted 'royal Tyrian purple' dye from marine whelks in the genus *Murex* to color fabrics reserved for royalty, and other whelks in the Family Muricidae have long been used in Central and South America to produce textile dyes. From Roman times until the early 1900s, the golden byssus threads of the Mediterranean pen shell *Pinna nobilis* were woven into fine, exceedingly lightweight veils, gloves, stockings, and shawls. Wealth has been displayed worldwide on clothing or objects festooned with cowry (snail) shells. Cowries, especially the money cowry *Cypraea moneta*, have been widely used as currency, perhaps as early as 700 BC in northern China and the first century AD in India. Their use spread to Africa and North America. Also in North America, East Coast Indians made disk-shaped beads, or wampum, from the shells of hard clams *Mercenaria mercenaria* for use as currency.

Mollusk shells have been used as fishhooks and octopus lures (the latter often made of cowry shells), household utensils (bowls, cups, spoons, scrapers, knives, boring devices, adzes, chisels, oil lamps), weapons (knives, axes), signaling devices (trumpets made from large conch shells), and decorative objects (beads, lampshades made of windowpane oyster shell, items made from iridescent 'mother-of-pearl' shell of abalone). Many species of mollusks produce 'pearls' when they cover debris that is irritating their mantle with layers of nacre (mother-of-pearl; an aragonite form of calcium carbonate). The pearl oyster (not a true oyster) and some freshwater mussels (not true mussels) use an especially iridescent nacre that results in commercially valuable pearls. Some religious practices involve shells, including scallop shells used as symbols by pilgrims trekking to Santiago de Compostela in Spain to honor St James (the French name for the scallop is Coquille St-Jacques).

As human populations and the demand for animal protein have grown, harvests of wild mollusks have

Figure 1 Post-harvest activities on shallow-draft tonging boats in Chesapeake Bay, Maryland, USA. Note the array of tongs, and the sorting or culling platform in the boat in the left foreground. (Photograph by Skip Brown, courtesy of Maryland Sea Grant.)

expanded. However, overharvesting and pollution of mollusk habitat have depleted many wild populations. To meet the demand for protein and to combat these losses, aquaculture has increased to supplement wild harvests. Unfortunately, some harvesting and aquaculture practices can have detrimental effects on the environment (see below).

Harvesting Natural Populations

Historical exploitation of mollusks occurred worldwide in shallow coastal and freshwater systems, and artisanal fishing still takes place there. The simplest fisheries involve harvesting by hand or with simple tools. Thus marine mussels and various snails exposed on rocky shores at low tides are harvested by hand, with oysters, limpets, and abalone pried from the rocks. Low tide on soft-substrate shores allows digging by hand to capture many shallow-dwelling species of burrowing clams. Some burrowing clams live more deeply or can burrow quickly when disturbed. Harvesting these requires a shovel, or a modified rake with long tines that can penetrate sediment quickly and be rotated so the tines retain the clam as the rake is pulled to the surface.

Mollusks living below low tide are usually captured by some sort of tool, the simplest being rakes and tongs. For example, oyster tongs in Chesapeake Bay (**Figure 1**) have two wooden shafts, each with a half-cylinder, toothed, metal-rod basket bolted at one end and with a metal pin or rivet holding the shafts together like scissors. A harvester standing on the side of a shallow-draft boat lets the shafts slip through his hands (almost all harvesters are male) until the baskets reach the bottom and then moves the upper ends of the shafts back and forth to scrape oysters and shells into a pile. He closes the baskets on the pile and hoists the contents to the surface manually or by a winch, opening the baskets to dump the scraped material onto a sorting platform on the boat (see **Figure 1**). Harvesters use hand tongs at depths up to about 10 m. Also in Chesapeake Bay, harvesters exploiting oysters living deeper than 10 m deploy much larger, heavier, and more efficient tongs from their boat's boom, using a hydraulic system to raise and lower the tongs and to close them on the bottom. In addition to capturing bivalves, rakes and hand tongs are used worldwide to harvest gastropods like whelks, conchs, and abalone (carnivorous gastropods like whelks and conchs can also be captured in pots baited with dead fish and other animals).

Mollusks can be captured by dredges towed over the bottom by boats, some powered by sail (**Figure 2**) and others by engines (**Figure 3**). Dredges harvest attached mollusks like oysters and marine mussels, as

Figure 2 Oyster dredge coming on board a sailboat ('skipjack') in Chesapeake Bay, Maryland, USA. Note the small 'push-boat' or yawl hoisted on the stern on this sailing day. (Photograph by Michael Fincham, courtesy of Maryland Sea Grant.)

Figure 3 Harvesting vessel *Mytilus* with blue mussel dredges in Conwy Bay, Wales. (Photograph courtesy of Dr Eric Edwards.)

well as buried clams, scallops lying on or swimming just above the sea bottom, and some gastropods (whelks, conchs). Dredges are built of metal and usually have teeth on the leading edge that moves over the bottom. Captured material is retained in a sturdy mesh bag made of wear-resistant heavy metal rings linked together and attached to the dredge frame. Mesh size is usually regulated so that small mollusks can fall out of the bag and back onto the sea bottom.

Some dredges use powerful water jets to blow buried mollusks out of soft sediment and into a metal-mesh bag. Where the water is shallow enough, subtidal clams (and oysters in some regions) are harvested by such water jets, but instead of being captured in a mesh bag the clams are blown onto a wire-mesh conveyor belt that carries them to the surface alongside the boat. Harvesters pick legal-sized clams off the mesh as they move past on the belt. Mesh size is such that under-sized clams and small debris fall through the belt and return to the bottom while everything else continues up the belt and falls back into the water if not removed by the harvester.

Commercial harvesting of freshwater mussels in the USA since the late 1800s has taken advantage of the propensity of bivalves to close their shells tightly when disturbed. Harvest vessels tow 'brails' over the bottom. Brails are long metal rods or galvanized pipe with eyebolts at regular intervals. Wire lines are attached to the eyebolts by snap-swivels. Each line holds a number of 'crowfoot' hooks of various sizes and numbers of prongs, depending on the species being harvested. Small balls are formed on the end of each prong so that, when the prong tip enters between the partially opened valves of the mussel, the valves close on the prong and the ball keeps the prong from pulling free of the shell. When the brails are brought on deck the clinging mussels are removed. This method works best in river systems with few snags (tree stumps, rocks, trash) that would catch and hold the hooks.

Cephalopods are captured by trawls, drift nets, seines, scoop and cast nets, pots and traps, and hook and line. A traditional gear is the squid jig, which takes various shapes but which has an array of barbless hooks attached. Jigs are moved up and down in the water to attract squid, which grab the jig and ensnare their tentacles, allowing them to be hauled into the boat. In oceanic waters, large vessels using automated systems to oscillate the jig in the water may deploy over 100 jig lines, each bearing 25 jigs. Such vessels fish at night, with lights used to attract squid to the fishing boat. A typical vessel may carry 150 metal-halide, incandescent lamps that together produce 300 kW of light. Lights from concentrations of vessels in the global light-fishing fleet off China and south-east Asia, New Zealand, the Peruvian coast, and southern Argentina can be detected by satellites. With a crew of 20, a vessel as described above may catch 25–30 metric tons of squid per night.

Some harvesters dive for mollusks, especially solitary organisms of high market value such as pearl oysters and abalone. Breath-hold diving has been used for centuries, but most divers now use SCUBA or air-delivery (hooka) systems. Diving is efficient because divers can see their prey, whereas most other capture methods fish 'blind'. Unfortunately, although diving allows for harvesting with minimal damage to the habitat, it has led to the depletion or extinction of some mollusk populations such as those of abalone.

A variety of measures are in place around the world to regulate mollusk fisheries. A common regulation involves setting a minimum size for captured animals that is larger than the size at which individuals of the species become capable of reproducing. This regulation ensures that most individuals can spawn at least once before being captured. Size selection is often accomplished by use of a regulated mesh size in dredge bags or conveyor-belts as described earlier. If the animals are harvested by a method that is not size selective, such as tonging for oysters or brailing for freshwater mussels, then the harvester is usually required to cull undersized individuals from the accumulated catch (see **Figure 1** for the culling platform used by oyster tongers) and return them to the water, usually onto the bed from which they were taken. Oysters are measured with a metal ruler; freshwater mussels are culled by attempting to pass them through metal rings of legal diameter and keeping those that cannot pass through.

Other regulatory mechanisms include limitations as to the number of harvesters allowed to participate in the fishery, the season when harvesting can occur, the type of harvest gear that can be used, or the total catch that the fishery is allowed to harvest. There may be areas of a species' range that are closed to harvest, perhaps when the region has many undersized juveniles or when beds of large adults are thought to be in need of protection so that they can serve as a source of spawn for the surrounding region. Restrictions may spread the capture effort over a harvest season to prevent most of the harvest from occurring at the start of the season, with a corresponding market glut that depresses prices. Finally, managers may protect a fishery by regulations mandating inefficiencies in harvest methods. Thus, in Maryland's Chesapeake Bay, diving for oysters has been strictly regulated because of its efficiency. Similarly, the use of a small boat (**Figure 2**) to push sailboat dredgers ('power dredging') is allowed only on 2 days per week (the days chosen – usually days without wind – are at the captain's discretion). Dredging must be done under sail on the remaining days of the week (harvesting is not allowed on Sunday). Around the world, inspectors ('marine police') are empowered to ensure that regulations are followed, either by boarding vessels at sea or when they dock with their catch.

A great hindrance to informed management of molluskan (and other) fisheries is the lack of data on the quantity of organisms taken by noncommercial (recreational) harvesters. For example, in Maryland's Chesapeake Bay one can gather a bushel (around 45 l) of oysters per day in season without needing a license if the oysters are for personal use. Clearly it is impossible to determine how many of these bushels are harvested during a season in a sizeable body of water. If such harvests are large, the total fishing mortality for the species can be greatly underestimated, complicating efforts to use fishery models to manage the fishery.

Processing mollusks involves mainly shore-based facilities, except for deep-sea cephalopod (mostly squid) fisheries where processing, including freezing, is done on board, and for some scallop species (the large muscle that holds scallop shells shut is usually cut from the shell at sea, with the shell and remaining soft body parts generally discarded overboard). Thus the catch may be landed on the same day it is taken (oysters, freshwater and marine mussels, many clams and gastropods) or within a few days (some scallop and clam fisheries). If the catch is not frozen, ice is used to prevent spoilage at sea.

Suspension-feeding mollusks (mostly bivalves) can concentrate toxins from pollutants or poisonous algae and thereby become a threat to human health. If such mollusks are harvested, they have to be held in clean water for a period of time to purge themselves of the toxin (if that is possible). Thus they may be relaid on clean bottom or held in shore-based systems ('depuration facilities') that use ozone or UV light to sterilize the water that circulates over the mollusks. Relaying and reharvesting the mollusks or maintaining the land-based systems adds to labor, energy, and capital costs.

Aquaculture

Aquaculture involves using either natural 'seed' (small specimens or juveniles that will be moved to suitable habitat for further growth) that is harvested from the wild and reared in specialized facilities, or producing such seed in hatcheries. Most commercial bivalves and abalone can be spawned artificially in hatcheries. The techniques involve either taking adults ripe with eggs or sperm (gametes) from nature or assisting adults to ripen by providing algal food in abundance at temperatures warm enough to support gamete production. Most ripe adults will spawn when provided with a stimulus such as an increase in temperature or food or both, or by the addition to the ambient water of gametes dissected from sacrificed adults. The spawned material is washed through a series of screens to separate the gametes from debris. Eggs are washed into a container and their numbers are estimated by counting samples, then the appropriate density of sperm is added to fertilize the eggs. Depending on the species of mollusk, an adult female may produce millions of eggs in one spawning event, so hundreds of millions of larvae produced by a relatively small number of females may be available for subsequent rearing.

Larvae are reared in specialized containers that allow culture water to be changed every few days and the growing larvae to be captured on screens, counted, and replaced into clean water until they become ready to settle. As they grow, larvae are fed cultured algae at appropriate concentrations (this requires extremely large quantities of algae to support the heavily feeding larvae). Depending on the species, after 2 or more weeks many larvae are ready to settle ('set') onto a solid surface (e.g., oysters, marine mussels, scallops, pearl oysters, abalone) or onto sediment into which they will burrow (e.g. clams). The solid material on which setting occurs is called 'cultch'. Settled larvae are called 'spat'.

Spat can be reared in the hatchery if sufficient algal food is available (usually an expensive proposition given the large quantity of food required). Thus most production facilities move spat (or seed) into nature soon after they have settled. However, this exposes the seed to diverse natural predators – including flatworms, boring sponges, snails, crabs, fish, and birds. Consequently, the seed may need to be protected (e.g. by removing predators by hand, poisoning them, and providing barriers to keep them from the seed), which is labor-intensive and expensive. Mortalities of larvae, spat, and seed are high at all stages of aquaculture operations.

The energy and monetary costs of maintaining hatchery water at suitable temperatures and of rearing enormous quantities of algae means that molluskan hatcheries are expensive to operate profitably. Thus only 'high-value' mollusks are cultured in hatcheries. One option for those who cannot afford to maintain a hatchery is to purchase larvae that are ready to settle. In the supply hatchery, larvae are screened from the culture water onto fine mesh fabric in enormous densities. The densely packed larvae withdraw their soft body parts into their shells, closing them. The larvae can then be shipped by an express delivery service in an insulated container that keeps them cool and moist until they reach the purchaser, who gently rinses the larvae off the fabric into clean water of the appropriate temperature and salinity in a setting tank. Also in the tank is the cultch that the larvae will attach to or the sediment into which they will burrow. This procedure of setting purchased larvae is called remote setting.

Among bivalves, oyster larvae settle on hard surfaces, preferably the shell of adults of the species, but also on cement materials, wood, and discarded trash. The larva cements itself in place and is immobile for the rest of its life. Thus oysters have traditionally been cultured by allowing larvae to cement to cultch like wooden, bamboo, or concrete stakes; stones and cement blocks; or shells (such as those of oysters and scallops) strung on longlines hanging from moored rafts (**Figure 4C**). As noted above, the settling larvae may be either those produced in a hatchery or those living in nature. The spat may be allowed to remain where they settle, or the cultch may be moved to regions where algal production is high and spat growth can accelerate. Scallop, marine mussel, and pearl oyster larvae do not cement to a substrate, but secrete byssus threads for attachment. For these species, fibrous material like hemp rope is provided in hatcheries or is deployed in nature where larvae are abundant. As these settled spat grow, they may be transferred to containers such as single- or multi-compartment nets (**Figure 4A, B**) or to flexible mesh tubes (used for marine mussels). Most clams are burrowers and will settle onto sediment. This may be provided in ponds, sometimes including those used to grow shrimp and other crustaceans (simultaneous culture of various species is called polyculture). Abalone settle on hard surfaces to which their algal food is attached, so they are usually cultured initially on corrugated plastic plates coated with diatoms. As they grow, they may be moved to well-flushed raceways or held suspended in nature (**Figure 4C**) in net cages.

There are risks associated with aquaculture, just as with harvesting natural populations. A major problem is the one that affects many agricultural monocultures and that is the increased susceptibility to disease outbreaks among densely farmed organisms. Such diseases now affect cultured abalone and scallops in China. Another problem involves the reliance on a relatively few animals for spawning purposes, which can lead to genetic deficiencies and inbreeding, further endangering a culture program. The coastal location of aquaculture facilities makes them susceptible to damage by storms, and if such storms increase or intensify with global warming, aquaculturists risk losing their animals, facilities, and investment. Ice can cause damage in cold-winter regions, although this is more predictable than are intense storms and preventative steps can be taken (not always successfully).

Extensive use of rafts and other systems for suspension culture (**Figure 4C**) may interfere with shipping and recreational uses of the water, and in some regions, laws against navigational hazards prevent water-based culture systems. If land

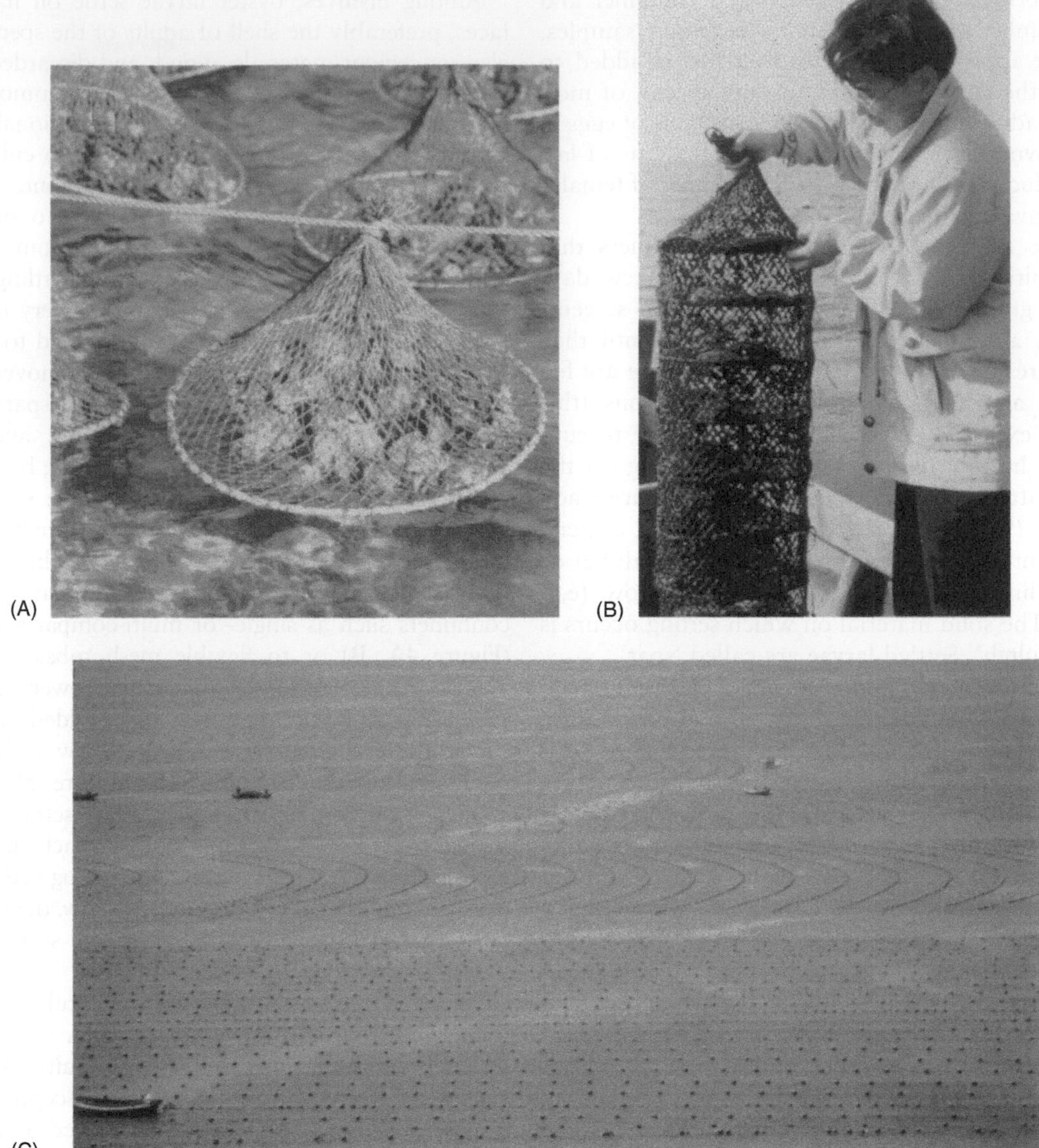

Figure 4 (A) Single-compartment pearl oyster cages attached to intertidal longlines. (B) Multi-compartment lantern net used to culture scallops. (C) Suspended longlines for growing Pacific oyster, scallops, and abalone off Rongchen, China. (Photographs courtesy of Dr Ximing Guo; reproduced with permission from Guo *et al.* (1999) *Journal of Shellfish Research* 18: 19–31.)

contiguous to the water is too expensive because of development or other land-use practices, land-based culture systems may not be economically feasible. Finally, coastal residents may consider rafts to be an eyesore that lowers the value of their property and they may seek to have them banned from the region.

Comparisons of Wild and Cultured Production

Of the 86–93 million metric tons of aquatic species harvested from wild stocks in 1996–1998 (capture landings), fish comprised 86%, mollusks 8%, and crustaceans (shrimp, lobsters, crabs) 6% (**Table 1**). Among the mollusks, the average relative proportions over the 3 years were cephalopods, 44%; bivalves, 32%; gastropods, 2%; and miscellaneous, 21%. Within the bivalves, the relative proportions over the 3 years were clams, 38% of all bivalves; freshwater mussels, 25%; scallops, 22%; marine mussels, 9%; and oysters, 6%.

Aquaculture had a greater effect on total landings (wild harvest plus aquaculture production) of mollusks than of fish and crustaceans. Of FAO's estimated 27–31 million metric tons of aquatic animals cultured worldwide from 1996 to 1998, fish comprised 64% and crustaceans comprised 5% by weight, declines from their proportions of the wild harvest (**Table 1**). In contrast, mollusks comprised 31% of culture production, about four times their proportion of the wild harvest. In addition, 20–40% more mollusks by weight were produced by aquaculture than were harvested from nature; by contrast, the quantities of cultured fish and crustaceans were a small fraction of quantities harvested in nature (**Table 1**). Of the cultured mollusks produced from 1996 to 1998, bivalves represented 87% by weight, followed by miscellaneous mollusks at 13%; cultured gastropods and cephalopods represented fractions of a percent of produced weight. When the FAO's harvest values of the wild fisheries and aquaculture production from 1984 to 1998 are combined for bivalves, cephalopods, and gastropods (**Figure 5**), the production of bivalves is seen to have risen greatly in the 1990s, with cephalopod production having increased modestly and gastropod production not at all. The differences can be attributed to the relatively greater yield of cultured bivalves compared with the other two molluskan groups.

Table 1 Worldwide capture landings (wild harvest) and aquaculture production and value of fish, mollusks, and crustaceans from 1996 to 1998

Category	*1996 (%)*	*1997 (%)*	*1998 (%)*	*Average percentage*
Capture landings (million metric tons)				
Fish	80.9 (87)	79.7 (86)	72.7 (85)	86
Mollusks	6.6 (7)	7.3 (8)	6.6 (8)	8
Crustaceans	5.5 (6)	5.9 (6)	6.4 (7)	6
Total	93.0	92.9	85.7	
Aquaculture production (million metric tons)				
Fish	17.0 (63)	18.8 (65)	20.0 (65)	64
Mollusks	8.6 (32)	8.7 (30)	9.2 (30)	31
Crustaceans	1.2 (4)	1.4 (5)	1.6 (5)	5
Total	26.8	28.9	30.8	
Aquaculture value (billion $US)Fish				
Fish	26.5 (62)	28.3 (62)	27.8 (61)	62
Mollusks	8.6 (20)	8.7 (19)	8.5 (19)	19
Crustaceans	7.8 (18)	8.5 (19)	9.2 (20)	19

From UN Food and Agriculture Organization statistics as at 21 March 2000.

The economic value of cultured animals ranged from 43 to 46 billion US dollars over the period 1996–1998, with fish comprising an average of 62%, mollusks 19%, and crustaceans 19% of this amount (**Table 1**). Thus, although production (weight) of cultured mollusks was six to seven times that of cultured crustaceans, their value just equaled that of crustaceans (this was a result of the premium value of cultured shrimps and prawns). Of the cultured mollusks, bivalves represented 93% by economic value followed by miscellaneous mollusks at 6%; cultured gastropods and cephalopods represented fractions of a percent of overall economic value.

Among cultured bivalves, the relative proportions by weight and by economic value were: Oysters, 42% and 42% respectively; clams, 26% and 32%; scallops, 15% and 20%; marine mussels, 17% and 6%; and freshwater mussels, <1% in both categories. Thus, although oysters were the least important bivalve in terms of wild harvests (6% of bivalve landings, see above), they were the most important cultured bivalve. On the other hand, freshwater mussels represented 25% of wild harvests but were relatively insignificant as a cultured item.

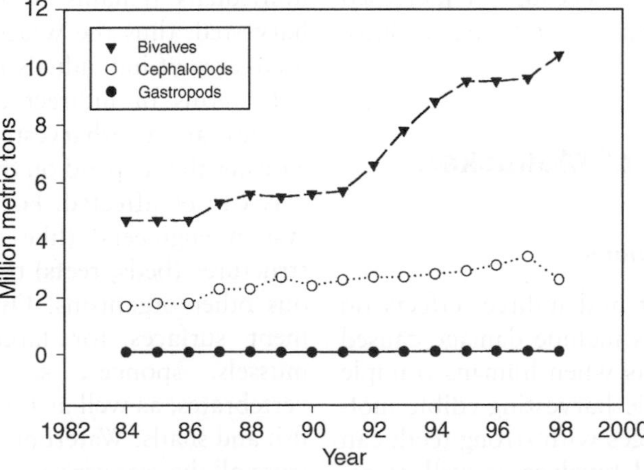

Figure 5 Worldwide production of bivalve, cephalopod, and gastropod mollusks from wild fisheries plus aquaculture activities in the period from 1984 to 1998. (From UN Food and Agriculture Organization statistics as at 21 March 2000.)

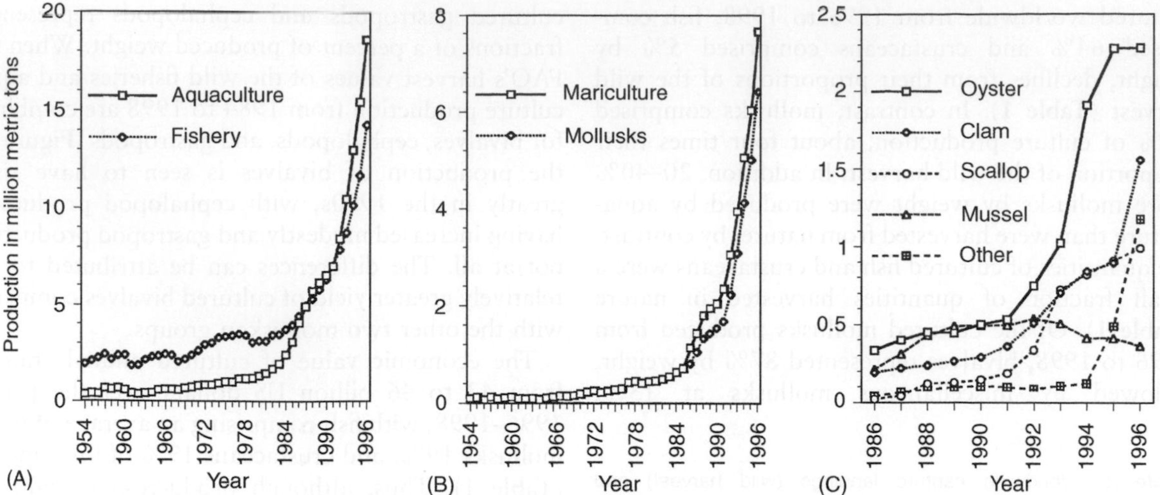

Figure 6 Production (wet, whole-body weight) in the natural fishery and in aquaculture in China. (A) Natural fishery and aquaculture. (B) Total mariculture (marine aquaculture) production and the molluskan component. (C) Comparative production among molluskan groups. (Adapted with permission from Guo *et al.* (1999) *Journal of Shellfish Research* 18: 19–31.)

China provides an excellent example of increased efforts in aquaculture, with its industry producing about 50% of the world's aquacultural output in the early 1990s, rising to about 67% by 1997 and 1998. Oyster and clam culture has been practiced in China for about two millennia, but growth of aquaculture accelerated in the early 1950s, with production outstripping wild fishery harvests by 1988 (**Figure 6A**). Mollusk culture was a substantial factor in this growth (**Figure 6B**). The intensified culture involved new species beyond the traditional oysters and clams, beginning with marine mussels, then scallops, then abalone. Over 30 species of marine mollusks are farmed along China's coasts, including 3 species of oysters, 14 of clams, 4 of marine mussels, 4 of scallops, 2 of abalone, 2 of snails, and 1 of pearl oyster. Marine mussel production has declined since 1992 (**Figure 6C**), apparently because of increased production of 'high-value' species (oysters, scallops, abalone).

Detrimental Effects of Molluskan Fisheries

Harvesting Natural Populations

Harvesting can have direct and indirect effects on local habitats. Direct effects include damage caused by harvest activities, such as when humans trample rocky shore organisms while harvesting edible mollusks. Similarly, heavy dredges with strong teeth can damage and kill undersized bivalves as well as associated inedible species. Dredges that use high-pressure water jets churn up bottom sediments and displace undersized clams and nontarget organisms that may not be able to rebury before predators find them exposed on the surface. When commercial shellfish are concentrated within beds of aquatic grasses, dredging can uproot the plants.

On the other hand, some harvesting techniques are relatively benign. For example, lures like squid jigs attract the target species with little or no 'by-catch' of nontarget species. Size selectivity can be attained by using different sizes of jigs. However, as with most fishing methods, overharvesting of the target species can still occur.

From the perspective of population dynamics, intense harvesting pressure commonly results in smaller average sizes of individual organisms and eventually most survivors may be juveniles. Care must be taken to ensure that a suitable proportion of individuals remains to reproduce before they are harvested, thus the widespread regulations on minimal sizes of individuals that can be harvested.

In terms of indirect effects, if some molluskan species are overharvested, inedible associated organisms that depend on the harvested species in some way can be affected. For example, oysters are 'ecosystem engineers' (like corals) and produce large structures (beds, reefs) that are exploited by numerous other organisms. Oyster shells provide attachment surfaces for large numbers of barnacles, mussels, sponges, sea squirts, and other invertebrates, as well as for the eggs of some species of fish and snails. Waterborne sediments as well as feces from all the reef organisms settle in interstices among the shells and become habitat for worms and other burrowing invertebrates. The presence of these organisms attracts predators, including crustaceans and fish. Consequently, the diversity of species on and around oyster reefs is higher than it is for adjacent soft-bottom systems. The same is true for marine mussel beds. Harvesting these bivalves can result in lowered biomass and perhaps lowered diversity of associated organisms.

Depleted populations of commercial bivalves may have other ecological consequences. These bivalves are 'suspension feeders' and pump water inside their shell and over their gills to extract oxygen and remove suspended food particles. When Europeans sailed into Chesapeake Bay in the 1600s, oyster reefs broke the Bay's surface and were a navigational hazard. Today, many oyster beds have been scraped almost level with the Bay bottom and annual harvests in Maryland have dropped from an estimated 14 million bushels of oysters in the 1880s to a few hundred thousand bushels today. Knowing the pumping capacity of individual oysters and the estimated abundances of pre-exploitation and present-day populations, scientists can calculate the filtering ability of those populations. Estimates are that a quantity of water equivalent to the total amount contained in the Bay could have been filtered in about a week in the early eighteenth century; today's depleted populations are thought to require nearly a year to filter the same amount of water.

This diminished filtering capacity has implications for food web dynamics. A feeding oyster ingests and digests food particles and expels feces as packets of material larger than the original food particles. Nonfood particles are trapped in mucous strings and expelled into the surrounding water as pseudofeces. Thus, oysters take slowly sinking microscopic particles from the water and expel larger packets of feces and pseudofeces that sink more rapidly to the Bay bottom. When oyster numbers are greatly diminished, this filtering and packaging activity is also diminished and more particles remain suspended in the water column. Thus ecological changes in Chesapeake Bay over the last century may have been enormous. Many scientists believe the Bay has shifted from a 'bottom-dominated' system in which oyster reefs were ubiquitous and the water was clearer than now to one dominated by water-column organisms (plankton) in which light levels are diminished. Smaller plankton serve as food for larger plankton, including jellyfish, and some scientists believe that the large populations of jellyfish present in the Bay in warmer months may be a result of overharvesting of oysters over the past century.

The reverse of this phenomenon is seen in the North American Great Lakes, where huge populations of zebra mussels *Dreissena polymorpha* and quagga mussels *Dreissena bugensis* have developed from invaders inadvertently introduced from Europe in ships' ballast water. The mussels filter the lake water so efficiently that the affected lakes are clearer than they have been for decades – they have become 'bottom-dominated' and plankton populations have decreased in abundance. Unintended consequences resulting from overharvesting other mollusk populations remain to be elucidated.

Aquaculture Practices

A number of environmental problems affect molluskan aquaculture. For one, heavy suspension feeding by densely farmed mollusks in coastal bays may outstrip the ability of the environment to supply algae, so the mollusks may starve or cease to grow. Another problem is that large concentrations of cultured organisms produce fecal and other wastes in quantities that can overwhelm the environment's ability to recycle these wastes. When this happens, eutrophication occurs, inedible species of algae may appear, and abundances of algal species that support molluskan growth may be reduced. To counter these problems, countries like Australia are using Geographic Information System technology to pinpoint suitable and unsuitable locations for aquaculture.

Introductions of exotic species of shellfish for aquaculture purposes have sometimes been counterproductive. A variety of diseases that have depleted native mollusk stocks have been associated with some introductions (e.g. MSX disease in the eastern oyster *Crassostrea virginica* in Chesapeake Bay is thought to be linked to attempts to import the Pacific oyster *Crassostrea gigas* to the bay). 'Hitch-hiking' associates that are carried to the new environment by imported mollusks have sometimes caused problems. For example, the gastropod *Crepidula fornicata* that accompanied oysters of the genus *Crassostrea* that were brought to Europe from Asia has become so abundant that it competes with the European oyster *Ostrea edulis* and blue mussel *Mytilus edulis* for food and may foul oyster and mussel beds with its wastes. Recently, the veined rapa whelk *Rapana venosa* has appeared in lower Chesapeake Bay, apparently arriving from the Black Sea or from Japan in some unknown fashion. It is a carnivore and may pose a threat to the indigenous oyster and hard clam fisheries (although it might prove to be a commercial species itself if a market can be found). As a result of these and other problems, many countries have developed stringent rules to govern movement of mollusks locally and worldwide.

See also

Corals and Human Disturbance. Ecosystem Effects of Fishing. Fisheries and Climate. Fishery Manipulation through Stock Enhancement or Restoration. Marine Fishery Resources, Global State of. Rocky Shores.

Further Reading

Andrews JD (1980) A review of introductions of exotic oysters and biological planning for new importations. *Marine Fisheries Review* 42(12): 1–11.

Attenbrow V (1999) Archaeological research in coastal southeastern Australia: A review. In: Hall J and McNiven IJ (eds.) *Australian Coastal Archaeology*, pp. 195–210. ANH Publications, RSPAS. Canberra, Australia: Australian National University.

Caddy JF (ed.) (1989) *Marine Invertebrate Fisheries: Their Assessment and Management*. New York: John Wiley Sons.

Caddy JF and Rodhouse PG (1998) Cephalopod and groundfish landings: Evidence for ecological change in global fisheries. *Reviews in Fish Biology and Fisheries* 8: 431–444.

Food and Agriculture Organization of the United Nations' website for fishery statistics (www.fao.org).

Guo X, Ford SE, and Zhang F (1999) Molluscan aquaculture in China. *Journal of Shellfish Research* 18: 19–31.

Kaiser MJ, Laing I, Utting SD, and Burnell GM (1998) Environmental impacts of bivalve mariculture. *Journal of Shellfish Research* 17: 59–66.

Kennedy VS (1996) The ecological role of the eastern oyster, *Crassostrea virginica*, with remarks on disease. *Journal of Shellfish Research* 15: 177–183.

MacKenzie CL Jr, Burrell VG Jr, Rosenfield A, and Hobart WL (eds.) (1997) *The History, Present Condition, and Future of the Molluscan Fisheries of North and Central America and Europe. Volume 1, Atlantic and Gulf Coasts. (NOAA Technical Report NMFS 127); Volume 2, Pacific Coast and Supplemental Topics (NOAA Technical Report NMFS 128); Volume 3, Europe (NOAA Technical Report NMFS 129)*. Seattle, Washington: US Department of Commerce.

Menzel W (1991) *Estuarine and Marine Bivalve Mollusk Culture*. Boca Raton, FL: CRC Press.

Newell RIE (1988) Ecological changes in Chesapeake Bay: Are they the result of overharvesting of the American oyster, *Crassostrea virginica*. In: Lynch MP and Krome EC (eds.) *Understanding the Estuary: Advances in Chesapeake Bay Research*, pp. 536–546. Chesapeake Research Consortium Publication 129. Solomons, MD: Chesapeake Research Consortium.

Rodhouse PG, Elvidge CD, and Trathan PN (2001) Remote sensing of the global light-fishing fleet: An analysis of interactions with oceanography, other fisheries and predators. *Advances in Marine Biology* 39: 261–303.

Safer JF Gill FM (1982) *Spirals from the Sea. An Anthropological Look at Shells*. New York: Clarkson N Potter.

Sanger D and Sanger M (1986) Boom and bust in the river. The story of the Damariscotta oyster shell heaps. *Archaeology of Eastern North America* 14: 65–78.

OPEN OCEAN FISHERIES FOR DEEP-WATER SPECIES

J. D. M. Gordon, Scottish Association for Marine Science, Oban, Argyll, UK

Introduction

Deep-water fisheries are considered to be those that exploit fish or shellfish that habitually live at depths greater than 400 m. With the exception of some localized line fisheries around oceanic islands, such as for *Aphanopus carbo* (black scabbardfish) at Madeira in the Atlantic and for *Ruvettus* in Polynesia, the fisheries are mostly of recent origin. The deep-water fisheries of the continental slopes only developed in the 1960s when Soviet trawlers discovered and exploited concentrations of roundnose grenadier (*Coryphaenoides rupestris*) off Canada and Greenland. At the same time, the Soviet fleet was exploiting similar resources in the North Pacific. Since then, despite a decline in Russian landings, deep-water fisheries have continued to expand on a global scale. Some of these fisheries target species that had never previously been exploited, such as orange roughy (*Hoplostethus atlanticus*). Some species have a depth range that extends from the shallow continental shelf depths into deeper water, and there is an increasing trend for the fisheries on these species to extend into deeper water and exploit all stages of the life history. Examples of such species in the North Atlantic are Greenland halibut (*Reinhardtius hippoglossoides*), anglerfish (*Lophius* spp.) and deep-water redfish (*Sebastes mentella*).

The United Nations Food and Agriculture Organization (FAO) has divided the world's oceans into statistical areas (**Figure 1**) and for some species there are data on landings. However, in many countries deep-water species are frequently landed unsorted in grouped categories that sometimes makes it difficult, if not impossible, to specify a proportion of the catch as deep-water. For example, many sharks are grouped and the term 'sharks various' can include inshore, deep-water, and oceanic pelagic species.

Deep-water fishes are generally perceived as being slow-growing and having a high age at first maturity and a low fecundity, all of which contribute to making them susceptible to overexploitation. In a new fishery the initial catch rates will be high, but, as the accumulated biomass of older fish is removed, the catch rates will decrease. For some species the sustainable yield may be only 1–2% per year of the pre-exploitation biomass. Few deep-water fisheries are managed, but where they are managed an objective has often been to maintain the stock at above 30% of the virgin biomass. Some deep-water species, such as sharks and orange roughy, have a global distribution, while others, such as roundnose grenadier, *Sebastes* spp. and Patagonian toothfish (*Dissostichus eleginoides*), are widely distributed within different oceans. Virtually nothing is known about the stock structure of most deep-water species and this can cause problems when developing management strategies. The following examples have been chosen to illustrate some of the problems associated with the sustainable exploitation and management of deep-water fisheries.

Black Scabbardfish (*Aphanopus carbo*) (Figure 2)

One of the longest-established deep-water fisheries is the fishery for black scabbardfish in the eastern North Atlantic. It began as a longline fishery around the oceanic island of Madeira and probably originates from the seventeenth century. Records of landings exist from about the 1930s, but because of the artisanal nature of the fishery these may be unreliable. At the beginning the fishery was prosecuted from open boats using vertical longlines set at about 1000 m depth and which were hauled by hand. However, in recent years the fishery has evolved and now uses larger vessels, mechanized line haulers and horizontal floating longlines. As a result the landings have been increasing, partly as a result of increasing efficiency and also because of the discovery of new fishing grounds farther from the islands. In 1983 a longline fishery for black scabbardfish developed off mainland Portugal and catches increased rapidly. When a mixed bottom-trawl fishery developed on the continental slope to the west of the British Isles in the 1990s black scabbardfish was an important bycatch and landings increased considerably. There is little doubt that the catch per unit of effort (CPUE) in the trawl fishery is decreasing and in 2000 the International Council for the Exploration of the Sea (ICES) recommended a 50% decrease in fishing effort. However, there is a lack

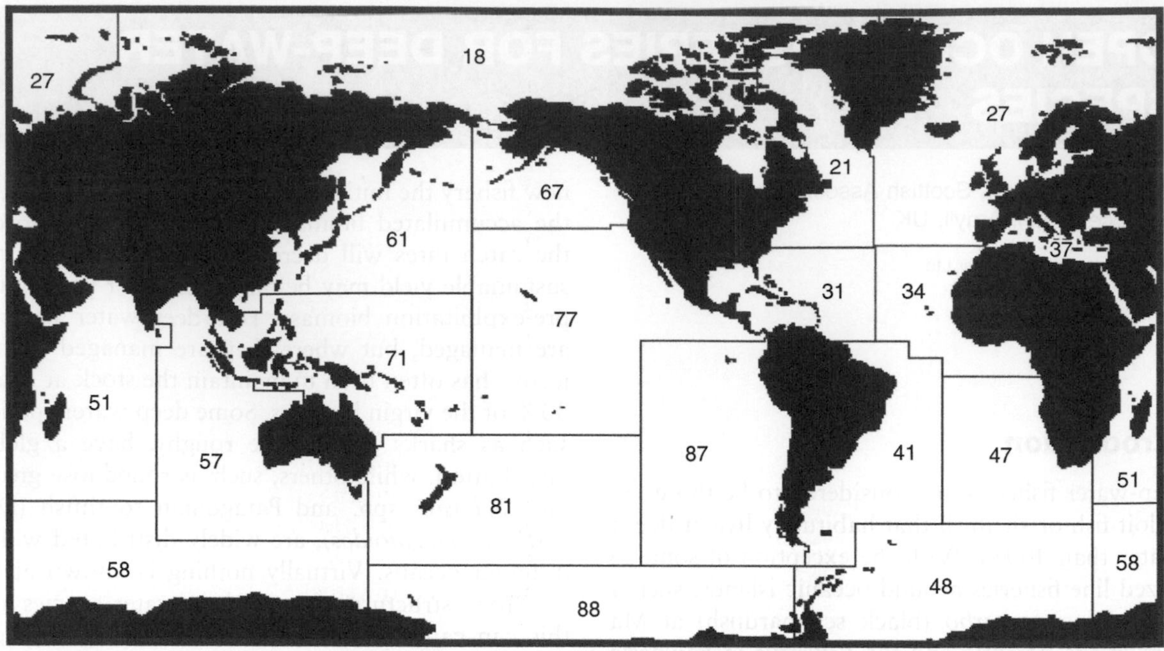

Figure 1 Map showing major fishing areas used by FAO for statistical purposes. The numbered areas are referred to by name in the text and **Figures 2–7** and can be identified by reference to **Table 1**. (From Gordon, 1999.)

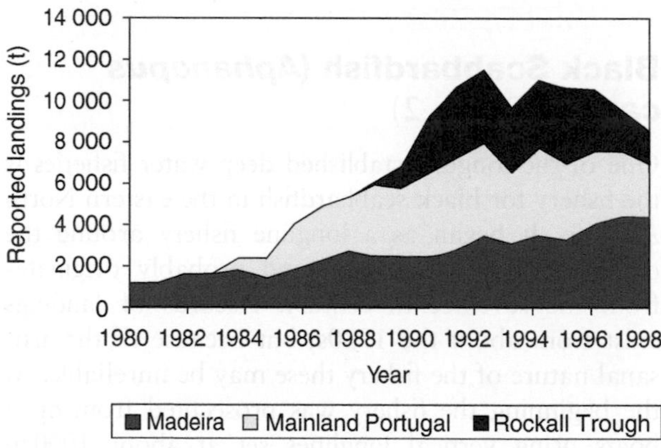

Figure 2 The reported landings (tonnes) of black scabbardfish (*Aphanopus carbo*) caught by longline around Madeira and off mainland Portugal and by bottom trawl in the Rockall Trough (all north-east Atlantic).

of basic biological knowledge on which to base any effective management. The eggs, larvae, and smallest juveniles of black scabbardfish are unknown. Estimated ages range from about 8 to over 20 years, but without information on the juvenile stages these cannot be validated. Mature female fish are found around Madeira, off mainland Portugal, and probably also from the Azores and northward along the Mid-Atlantic and Reykjanes Ridges. To the west of the British Isles the fish are all immature. One hypothesis to explain this distribution is that there is a single north-eastern Atlantic stock and that the adult fish spawn in the southern areas and the subadults migrate northward on a feeding migration. Studies of the diet of the subadult fish in the northern areas have shown that they feed on other fish such as blue whiting (*Micromesistius poutassou*) and argentine (*Argentina silus*) that shoal along the upper continental slope. If this hypothesis proves to be correct, effective management will have to take into account the three established fisheries and another that is developing in the Mid-Atlantic. An important by-catch of the longline fisheries for black scabbardfish are deep-water squalid sharks, such as *Centroscymnus coelolepis*, *Centrophorus squamosus*, and *Centrophorus granulosus*. Deep-water shark fisheries have their own problems (see below).

Roundnose Grenadier (*Coryphaenoides rupestris*) (Figure 3)

Concentrations of roundnose grenadier were discovered along the continental slope of the western North Atlantic north of 50°N in the late 1950s. This led to fishery investigations and exploratory voyages by Russia, the then German Democratic Republic, and Poland, and also coincided with the introduction of the large factory-freezer stern trawlers. The first directed fishery for roundnose grenadier was by Russia in the north-west Atlantic in the mid 1960s. This fishery began off the north-west of Newfoundland, and later extended northwards to Labrador, Baffin Island and

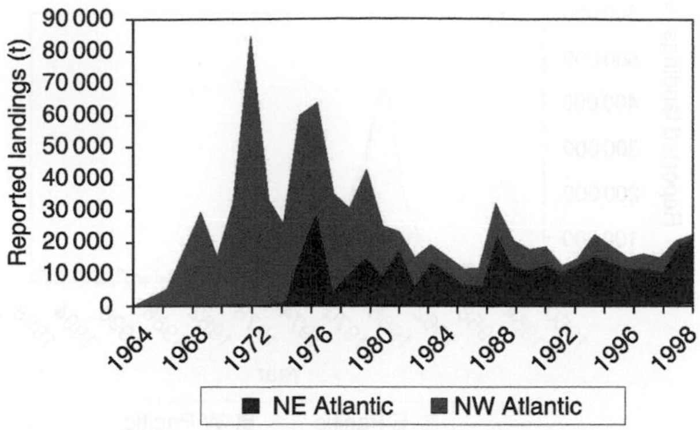

Figure 3 The reported landings (tonnes) of roundnose grenadier (*Coryphaenoides rupestris*) caught by bottom trawl in the north-east Atlantic and the north-west Atlantic (FAO statistical areas).

western Greenland. The fishery developed rapidly, reaching a peak of over 83 000 t in 1971. The subsequent decline in the fishery has often been ascribed to overexploitation and, while this might be an important factor, other contributory factors are undoubtedly the establishment of 200-mile national fishery zones, changes in the allowable by-catches in the Greenland halibut fishery, and probably some misreporting of initial catches. Scientific research on roundnose grenadier lagged far behind the fishery and it was not until 1975 that a precautionary total allowable catch (TAC) was imposed. However, the landings continued to decline and were soon well below the level of the TAC, which was also being steadily reduced. It was only in the 1990s that landings reached the level of the TAC as a result of the by-catch of the European trawlers fishing for Greenland halibut in international waters around the Flemish Cap.

A similar fishery developed in the international waters of the eastern North Atlantic (including the Mid-Atlantic and Reykjanes Ridges) in about 1973. Historically, the largest catches were from Russian vessels fishing the southern part of the Reykjanes Ridge. The landings peaked in 1975 and have fluctuated ever since. ICES has expressed concern that some of the catches in international waters are not being reported. By-catches of other species in these mixed, bottom-trawl catches have also been inadequately reported. In the early 1970s German trawlers and later French trawlers began to exploit spawning aggregations of blue ling (*Molva dypterygia*) in the northern parts of the Rockall Trough. Some French trawlers, which traditionally fish along the shelf edge for species such as saithe (*Pollachius virens*), began to move onto the upper and mid-continental slope of the Rockall Trough to exploit blue ling and by 1989 were beginning to land a by-catch of species such as roundnose grenadier, black scabbardfish, deep-water sharks, and several other less abundant species. As the fishery evolved it moved into deeper waters and soon roundnose grenadier was a target species in its own right. The landings have remained fairly constant in recent years, but there is good evidence to suggest that the CPUE has decreased. The landings have been maintained by increasing fishing effort and efficiency, moving into new areas and deeper water, and discarding fewer of the smaller fish. In 2000 ICES recommended a 50% cut in fishing effort and the European Commission proposed a precautionary TAC on this and other deep-water species in 2001. However, the fisheries ministers deferred the Commission's proposals and the fisheries remain unregulated. Almost nothing is known about the stock structure of roundnose grenadier and, although knowledge of the biology is improving, the data for analytical, age-based stock assessments is lacking. It is probable that the stocks in the area of the Rockall Trough that is under national jurisdiction are part of a larger stock around the Rockall and Hatton Banks that is in international waters. The lack of any regulation in international waters will diminish the value of any management measures introduced in waters under coastal state jurisdiction. There is also the issue whether TACs are an appropriate management tool for mixed fisheries. The problems of discarding over-quota marketable species, so-called high grading, etc., are all well documented in shelf fisheries, but the additional problem of a changing catch composition with depth in deep-water fisheries is something that will be difficult to incorporate into a management scheme.

Scorpaenid Fisheries (*Sebastes* spp. and *Sebastolobus* spp.) (Figures 4 and 5)

The trawl fisheries for redfishes are among the most important deep-water fisheries of the North Atlantic. In the north-western Atlantic the landings comprise a mixture of three species (*Sebastes marinus*, *S. fasciatus*, and *S. mentella*), but only *S. mentella* is a truly deep-water species. Because the landings are not separated into species, it is impossible to assess the importance of the deep-water component of the catch. The fishery peaked at almost 400 000 t in the mid-1950s and, except for an increase in the 1970s, has been steadily declining. The peaks in landings represent times when different nations entered the fishery but, at the present time, the fishery is dominated by Canada. It appears that this fishery is now in serious decline, which is perhaps not surprising given the high level of exploitation, the slow growth,

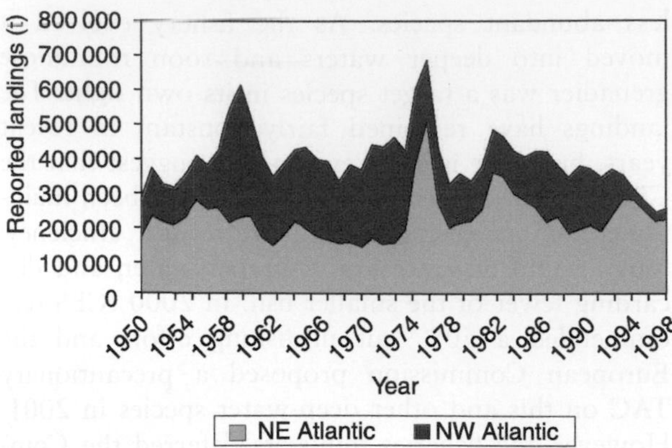

Figure 4 The reported landings (tonnes) of redfish (*Sebastes* spp.) caught by trawl in the north-east Atlantic and the north-west Atlantic (FAO statistical areas).

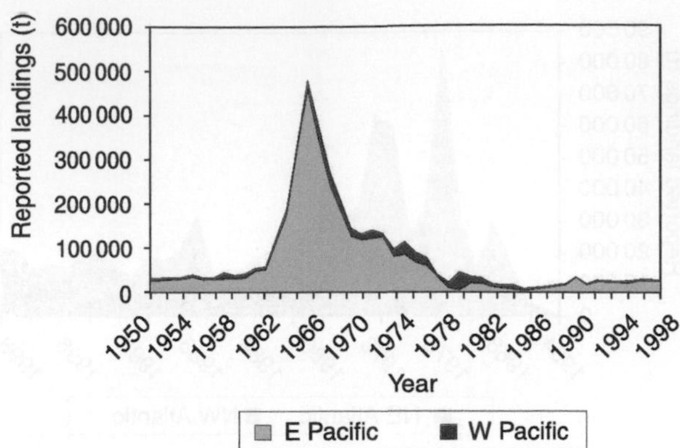

Figure 5 The reported landings (tonnes) of Pacific Ocean perch (*Sebastes alutus*) caught by trawl in the east and west Pacific (FAO statistical areas).

the high age at first maturity and the viviparous mode of reproduction. The mean length of fish caught by research vessels has been declining steadily. In the eastern North Atlantic the landings are comprised of two species (*Sebastes marinus* and *S. mentella*). Recently the landings of redfish by some countries have been partly apportioned into the two species on the basis of research surveys. *S. mentella* is the deep-water species, but these data do not give a true indication of the likely proportion of this species to the total catch of the whole region. However, there can be little doubt that the trend in recent years has been to increasingly exploit *S. mentella* in deeper water and for an overexploitation of all stocks. A longline fishery for so called 'giant redfish' developed in 1996 on the Reykjanes Ridge but only lasted for that year as a profitable fishery.

In the North East Pacific the fishery for Pacific Ocean perch (*Sebastes alutus*) began in 1946 and, after a slow period of development, the landings peaked at over 460 000 t in 1965 and then steadily declined over the next decade so that they now they range between about 10 000 to 30 000 t (**Figure 5**). In the North West Pacific the fishery for Pacific ocean perch is on a smaller scale but peaked in the 1970s and has since declined. This species has a range from the outer shelf to the upper slope and most of the fishery is on the outer shelf at about 200–300 m.

There are several scorpaenid fishes that extend from the outer shelf to the continental slope and, with the exception of the Pacific Ocean perch, the FAO statistics do not separate them at the species level. However, many of the stocks in shallower waters have become depleted and the fishery has progressively moved into deeper waters to exploit several species. These include in the east the long spine thornyheads (*Sebastolus altivelis*) and splitnose rockfish (*Sebastes diplopoa*) and in the west the longfin thornyhead (*Sebastolobus macrochir*), the shortspine thornyhead (*S. alascanus*), the rougheyed rockfish (*Sebastes aliutianus*) and the osaga (*S. iracundus*).

The Orange Roughy (*Hoplostethus atlanticus*) (Figure 6)

The orange roughy is probably the most frequently cited example of an exploited deep-water fish partly because it is an extreme example of all the features than can lead to over-exploitation; a life span of >100 y, high age at first maturity (~25 y), and a relatively low fecundity. It also has many qualities that make it a highly marketable commodity. It tends to aggregate in large numbers around seamounts, pinnacles, and steep slopes and with sophisticated fishing techniques these stocks can be fished down

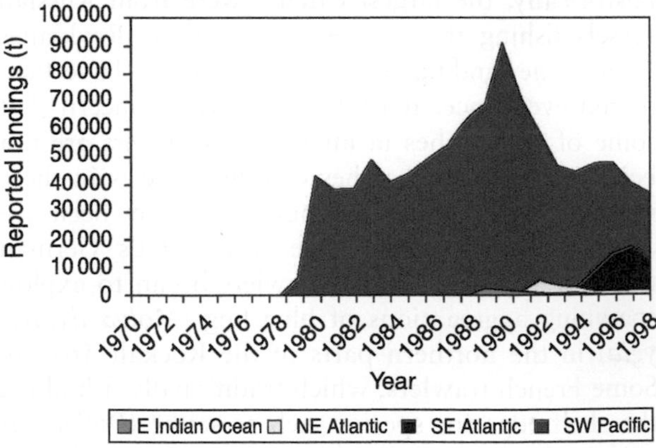

Figure 6 The reported landings (tonnes) of orange roughy (*Hoplostethus atlanticus*) caught by trawl in the south-west Pacific (New Zealand and eastern Australia), the eastern Indian ocean (Australia), the south-east Atlantic and the north-east Atlantic.

very rapidly: a type of fishing sometimes referred to as 'mining'.

The most important fishery for orange roughy is around New Zealand. This trawl fishery, at depths from about 700 to 1200 m, began in the late 1970s and expanded rapidly, with landings exceeding 60 000 t per annum in the late 1980s and early 1990s. Since then landings have declined to about 25 000 t. The dense spawning and feeding aggregations were systematically fished down and the fishery was maintained by the continual discovery of new fishing grounds, mainly on the extensive deep-water plateaus and rises that surround New Zealand. The present level of landings is composed of a reduced catch from established fisheries and higher catches from new areas.

The Australian fishery for orange roughy extends along the slope of south-eastern Australia, around Tasmania, and extends up to New South Wales. Fishing began in 1982 as part of a general slope fishery and it was not until 1986 that dense aggregations were found off Tasmania and a directed fishery was established. Between 1986 and 1988 the landings of orange roughy ranged between about 4600 and 7200 t per annum. New non-spawning aggregations and a new spawning aggregation resulted in a dramatic increase in catches from 26 000 t in 1989 to 41 000 t in 1990. Since then the landings have declined. FAO reports the landings of this area together with the New Zealand Fishery (south-west Pacific). A smaller fish fishery is reported by FAO in the eastern Indian Ocean and landings have steadily decreased from initial levels of about 2000 t in 1989.

In the north-eastern Atlantic, French trawlers began landing orange roughy from deep water to the west of the British Isles in 1991. Although there is still a degree of secrecy associated with this fishery, there is little doubt that it takes place at greater depths (down to about 1700 m) than the multispecies trawl fishery for roundnose grenadier (see above) and most probably in areas of steep slopes and seamounts. The landings peaked in 1992 and the fishery has subsequently declined in the Rockall Trough (ICES Sub-area VI). Landings from west of Ireland (ICES Sub-area VII) have remained fairly constant, probably as a result of the sequential discovery and fishing down of new aggregations. In the international waters of the north-eastern Atlantic, a Faroese vessel has been fishing on the Mid-Atlantic Ridge and other offshore banks and seamounts. There have also been spasmodic catches by Iceland on the Reykjanes Ridge.

The New Zealand orange roughy fishery, because of its economic importance, has been the subject of considerable research in recent years. Nevertheless, the rapid development of the orange roughy fishery far outpaced the scientific knowledge of the fish and the stocks. The initial quotas were essentially precautionary, to allow time for an assessment to be carried out. These assessments were based on random stratified trawl surveys but, because there was a lack of knowledge about the great age and low productivity and the stock discrimination of this fish, some inappropriate assumptions were made. As a result the quotas set in the early years of the fishery, mainly for the Chatham Rise and Challenger Plateau, were too high and the stocks were overexploited. In recent years there has been a marked improvement in the knowledge of the biology of the species, although the estimates of very high ages continue to be controversial. Improved stock assessment, using a variety of methods, and the application of genetics to stock discrimination have allowed more realistic TACs to be implemented. For example, on the Chatham Rise the TAC was decreased from 33 000 t in 1990 to 7000 t in 1996. By 1998 the stock was considered to be at 15–20% of virgin biomass but rebuilding. However, in other areas quotas continue to be reduced and on the Challenger Plateau, once a major fishery, the quota for the 2000/2001 fishing year has been reduced to 1 tonne, thereby effectively closing the fishery. The Australian fishery is also managed by quotas and these have been reduced drastically since the start of the fishery. The fishery in the north-east Atlantic remains totally unregulated.

Deep-water Sharks

Deep-water sharks are landed as a by-catch of many deep-water fisheries and there are some targeted

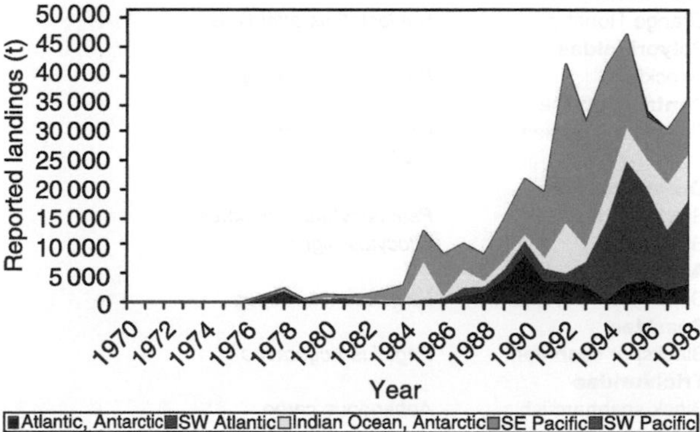

Figure 7 The reported landings (tonnes) of Patagonian toothfish (*Dissostichus eleginoides*) caught in the south-west Pacific (New Zealand and eastern Australia), the south-east Pacific, the Indian and Atlantic Ocean sectors of Antarctica (Australia) and the south-west Atlantic.

Table 1 Deep-water fish species that are listed by area in FAO fishery statistics

		Northwest Atlantic	Northeast Atlantic	Central Atlantic	Mediterranean	Southwest Atlantic	Southeast Atlantic	Atlantic Antarctic	Eastern Indian Ocean	Indian Ocean Antarctic	Northwest Pacific	Northeast Pacific	Central Pacific	Southwest Pacific	Southeast Pacific	Pacific Antarctic
	FAO Statistical Area	*21*	*27*	*31 & 34*	*37*	*41*	*47*	*48*	*57*	*58*	*61*	*67*	*71 & 77*	*81*	*87*	*88*
Common name	Scientific name															
PISCES																
Scyliorhinidae																
Blackmouth dogfish	*Galeus melastomus*		+		+											
Squalidae																
Portuguese dogfish	*Centroscymnus coelolepis*		+													
Leafscale gulper shark	*Centrophorus squamosus*		+													
Kitefin shark	*Dalatias licha*		+													
Argentinidae																
Argentine	*Argentina silus*		+													
Deep-sea smelt	*Glossanodon semifasciatus*										+					
Congridae																
Conger eel	*Conger conger*		+													
Ophidiidae																
Pink cusk eel	*Genypterus blacodes*					+									+	
Kingclip	*Genypterus capenensis*						+									
Macrouridae																
Roundnose grenadier	*Coryphaenoides rupestris*	+	+													
Roughhead grenadier	*Macrourus berglax*	+	+													
Pacific grenadier	*Coryphaenoides acrolepis*										+		+			
Longfin grenadier	*Coryphaenoides longifilis*										+					
Merlucciidae																
Hake	*Merluccius merluccius*		+		+											
Cape hakes	*Merluccius capenensis & M. paradoxus*						+									
Patagonian grenadier	*Macruronus magellanicus*					+								+		
Hoki	*Macruronus novaezelandiae*															
Gadidae																
Greater forkbeard	*Phycis blennoides*		+		+											
Blue ling	*Molva dypterygia*		+													
Blue whiting	*Micromesstius poutassou*		+		+											
Southern blue whiting	*Micromesistius australis*					+								+	+	
Moridae																
Morid	*Laemonema longips*										+					
Berycidae																
Alfonsinos	*Beryx* spp.		+				+						+			
Trachichthyidae																
Orange Roughy	*Hoplostethus atlanticus*		+				+		+					+		
Polyprionidae																
Wreckfish	*Polyprion americanus*				+											
Pentacerotidae																
Pelagic armourhead	*Pseudopentaceros wheeleri*												+			
Oreosomatidae																
Smooth oreo	*Pseudocyttus maculatus*													+		
Black oreo	*Allocytus niger*													+		
Apogonidae																
Deep-water cardinal fish	*Epigonus telescopus*		+											+		
Sparidae																
Blackspot seabream	*Pagellus bogaraveo*		+		+											
Trichiuridae																
Black scabbardfish	*Aphanopus carbo*		+	+												
Scorpaenidae																
Bluemouth	*Helicolenus dactylopterus*		+		+											
Pacific Ocean perch	*Sebastes alutus*										+	+				
Redfish	*Sebastes spp*	+	+													
Rockfishes	*Sebastolobus* spp.										+	+				

Table 1 *Continued*

		Northwest Atlantic	*Northeast Atlantic*	*Central Atlantic*	*Mediterranean*	*Southwest Atlantic*	*Southeast Atlantic*	*Atlantic Antarctic*	*Eastern Indian Ocean*	*Indian Ocean Antarctic*	*Northwest Pacific*	*Northeast Pacific*	*Central Pacific*	*Southwest Pacific*	*Southeast Pacific*	*Pacific Antarctic*
	FAO Statistical Area	*21*	*27*	*31 & 34*	*37*	*41*	*47*	*48*	*57*	*58*	*61*	*67*	*71 & 77*	*81*	*87*	*88*
Anoplopomatidae																
Sablefish	*Anoplopoma fimbria*										+	+	+			
Nototheniidae																
Patagonian toothfish	*Dissostichus eleginoides*					+		+		+				+	+	+
Pleuronectidae																
Dover sole	*Microstomus pacificus*											+				
Greenland halibut	*Reinhardtius hippoglossoides*	+	+													
CRUSTACEA																
Penaeidae																
Rose shrimp	*Parapenaeus longirostris*		+	+												
Aristeidae																
Blue and red shrimp	*Aristeus antennatus*			+												
Nephropidae																
Norway lobster	*Nephrops norvegicus*		+	+												
Geryonidae																
Deep-water red crab	*Chaceon maritae*			+												

fisheries. In some fisheries, such as in the bottom-trawl fishery for roundnose grenadier in the north-eastern Atlantic, only some of the shark species that are caught are landed. The rest are discarded, sometimes after removal of the liver and/or the fins. The reported landings to FAO are, with a few exceptions, by grouped category and it is therefore difficult to estimate the true extent of the fishery. Deep-water sharks of small adult size or the juveniles of marketable species that are discarded are unlikely to survive. No records are kept of discards and there have been few scientific studies of discarding.

Sharks are vulnerable to over-exploitation because of their longevity and slow growth. They have a high age at first maturity and a low reproductive rate per female as a result of the long embryonic development for oviparous and viviparous species, from several months up to about two years. The majority of deep-water species might have biennial or even triennial reproductive cycles.

Several fisheries that target sharks, such as the longline fishery for gulper shark (*Centrophorus granulosus*) off mainland Portugal and the kitefin shark (*Dalatias licha*) at the Azores, have declined rapidly after a few years and are good examples of the vulnerability of these stocks. The CPUE of the combined catches of leafscale gulper shark (*Centrophorus squamosus*) and the Portuguese dogfish (*Centroscymnus coelolepis*) from the trawl fishery to the west of the British Isles has declined by about 50% in less than ten years. Longlining is often presented as a more selective and environmentally friendly method of exploiting deep-water fish resources. However, many exploratory surveys have shown that there can be a high, perhaps unacceptable, by-catch of unwanted sharks.

Patagonian Toothfish (*Dissostichus eleginoides*) (Figure 7)

The main deep-water fishery in the southern oceans is for the Patagonian toothfish. It is widely distributed in the area and it is caught by longline at depths down to 2000 m or more. The reported landings are probably unreliable as there is thought to be illegal fishing in the area. The significant landings are from the Atlantic (FAO Area 48) and Indian Ocean (FAO Area 58) sectors.

In the Commission for the Conservation of Antarctic Marine Living Resources (CCAMLR) area the toothfish is managed by TAC with additional technical measures, some of which are designed to limit the accidental capture of seabirds attracted to the baited hooks as the lines are being set.

Other Deep-water Fisheries

The above examples were chosen to illustrate some of the different aspects of deep-water fisheries and the associated problem of managing what are generally considered to be fragile resources. There are many other deep-water species that are either targeted or are a by-catch of other deep-water fisheries. **Table 1** gives some of the important deep-water species that are clearly identifiable in published FAO statistics by statistical area. This list excludes many species that are recorded under grouped categories or where most of the fishery takes place on the continental shelf.

Ecosystem Effects

There is a growing concern, in part resulting from new technologies that allow direct observation, about the effects of fishing on the ecosystem and the probability that the deep-water ecosystems will be particularly susceptible to damage and will take a long time to recover. There is an expanding literature on the physical effects of fishing gear, especially trawls, on the sea bed in shallow water. In deeper water there are reports from photographic surveys of trawl marks in soft sediments, but little information on the effects at the biological level. Trawl damage to hard bottoms such as deep-water reefs and seamounts has been documented and reefs of the deep-water coral *Lophelia* off Norway and some seamounts off Australia and New Zealand are now protected. Discarding of species of no commercial value can represent a very high proportion of the catch. None of these discards will survive the trauma of being brought to the surface from great depths. Deep-water fishes generally have fragile skins and it is very probable that most fish entering the trawl and escaping through the meshes will not survive. The impacts of these discards and unseen mortalities on the ecosystem are unknown. The effect of the selective removal of top predators is also largely unknown.

See also

Demersal Species Fisheries. Open Ocean Fisheries for Large Pelagic Species. Salmon Fisheries, Pacific.

Further Reading

Clark M (2001) Are deepwater fisheries sustainable? – The example of orange roughy (*Hoplostethus atlanticus*) in New Zealand. *Fisheries Research* 52: 123–136.

Clark MR, Anderson OW, Francis RIC, and Tracey DM (2000) The effects of commercial exploitaton on orange roughy (*Hoplostethus atlanticus*) from the continental slope of the Chatham Rise, New Zealand, from 1979 to 1997. *Fisheries Research* 45: 217–238.

Food and Agriculture Organization (1994) Review of the state of world fishery resources. *FAO Fisheries Technical Paper, No. 335*, p. 136. Rome: FAO.

Food and Agriculture Organization (1997) Review of the state of world fishery resources: marine fisheries. FAO Fisheries Circular No. 920. 173 pp. Rome: FAO.

Gordon JDM (1999) Management considerations of deep-water shark fisheries. In: Shotton R (ed.) Case studies of the management of elasmobranch fisheries. FAO Fisheries Technical Paper. No. 378, pp. 774–818. Rome: FAO.

Gordon JDM (2001) Deep-water fisheries at the Atlantic frontier. *Continental Shelf Research* (In press).

Hopper AG (ed.) (1995) *Deep-water Fisheries of the North Atlantic Oceanic Slope*. Dordrecht: Kluwer Academic Publishers.

Koslow JA, Boehlert GW, Gordon JDM, *et al.* (2000) Continental slope and deep-sea fisheries implications for a fragile ecosystem. *ICES Journal of Marine Science* 57: 548–557.

Lorance P and Dupouy H (2001) CPUE abundance indices of the main target species of the French deep-water fishery in ICES sub-areas V, VI and VII. *Fisheries Research* 52: 137–150.

Merrett NR and Haedrich RL (1997) *Deep-sea Demersal Fish and Fisheries*. London: Chapman and Hall.

Moore G and Jennings S (eds.) (2000) *Commercial Fishing: The Wider Ecological Impacts*.

Randall DJ and Farrell AP (eds.) (1997) *Deep-sea Fishes*. London: Academic Press.

Uchida RN, Hayasi S and Boehlert GW (eds.) (1986) *Environment and Resources of Seamounts in the North Pacific*. NOAA Technical Report NMFS 43. 105 pp. Seattle: United States Department of Commerce.

OPEN OCEAN FISHERIES FOR LARGE PELAGIC SPECIES

J. Joseph, La Jolla, California, USA

Introduction

Open-ocean fisheries for large pelagic species target relatively large organisms that spend most of their lives in offshore waters, usually within about 100–150 m of the surface. These large pelagic species include finfish, such as scombrids (tunas) (**Figure 1**), istiophorids (billfishes: spearfish, sailfish, and marlins), xiphiids (swordfish) (**Figure 2**), coryphaenids (mahimahi, or dolphinfish), and elasmobranchs (sharks), and also cetaceans (whales, dolphins, and porpoises). Many of these species do not spend their entire lives in the open ocean but, as their biological needs dictate, make sporadic forays into the coastal zone. Therefore, although they are captured mostly in the open ocean, they are sometimes taken near shore.

This article concentrates on fisheries for tunas and billfishes, because they account for by far the greatest proportion of the total catch of all large pelagic species. (In this article catch includes fish discarded at sea, but landings do not.) The various types of fishing gear and vessels used to catch large pelagics, and the magnitude of those catches, are reviewed. Because of their great size, speed, and stamina, these large pelagics are much sought after by recreational fishers, so sportfishing is also included in this article. Fisheries for sharks and other large pelagics are discussed only briefly, and whales and whaling are discussed elsewhere.

The Animals

Tunas, and many of the billfishes, comprise most of the world's catch of large pelagic species. They have many characteristics that set them apart from most other types of fishes and which contribute to their nomadic lifestyle. They have very high metabolic rates, resulting in high energy and oxygen demands. Tunas must swim constantly in order to pass enough water over their gills to meet this high demand for oxygen. Unlike most other fishes they cannot pump water over their gills, so if they stopped swimming they would suffocate, and they would also sink because they are denser than the water that surrounds them. They possess highly developed circulatory systems that allow them to retain or dissipate heat as needed for efficient operation of their nervous, digestive, and locomotor systems, and are capable of maintaining their body temperature up to 15°C above ambient. Their fusiform shape, highly developed swimming muscles, and crescent-shaped tail, which provides maximum forward thrust, enable them to swim at extraordinary speeds. Many of the other large pelagic species have a similar propensity for travel, but their adaptations to a nomadic life style are different.

Tuna tend to form large schools, which makes it relatively easy to locate and catch them.

Figure 1 Yellowfin tuna (*Thunnus albacares*) (Source: Inter-American Tropical Tuna Commission).

Figure 2 Swordfish (*Xiphias gladius*) (Source: Inter-American Tropical Tuna Commission).

Methods of Catching the Fish

Humans have most likely been fishing since the first days they walked the earth. The first marine organisms caught were probably taken by hand from tidepools and beaches. As their fishing skills developed, early humans fashioned hooks, harpoons, traps, and nets to capture their prey. Archaeological evidence indicates that ancient man harvested large pelagic species: skeletal remains found in caves indicate that giant Atlantic bluefin (*Thunnus thynnus*) were caught near modern-day Sweden more than 6000 years ago, and similar evidence shows that giant Pacific bluefin (*Thunnus orientalis*), weighing more than 250 kg, were taken by native Americans in the region between the southern Queen Charlotte Islands, British Columbia, and Cape Flattery, Washington, more than 5000 years ago. Just how early humans caught these giant fish is uncertain, but probably harpoons or handlines with baited hooks were used.

As civilizations developed, so did trade in agricultural and natural resource products, including fish. Historical evidence indicates that nearly 3000 years ago the Phoenicians salted and dried tuna that they had caught, and traded it throughout the Mediterranean region.

Harpoons

The harpoon is simply a spear modified for fishing by attaching a line and buoy for retrieving both spear and prey. Harpoons have been used since early times for capturing large pelagics, but were perfected for whaling. For many years harpoons were the primary means of capturing swordfish, which have a tendency to bask at the surface of the ocean, and they are still used today in many parts of the world for this purpose. In recent years they have been replaced by more efficient forms of fishing, but harpoon-caught swordfish still command premium prices, apparently because of their better quality. Some marlin (*Makaira* spp. and *Tetrapturus* spp.) and giant bluefin tuna are also taken with harpoons. Vessels used for harpoon fishing have a long, narrow platform projecting from the bow, on which the harpooner stands, and from which vantage point there is a better chance of harpooning the fish before it is aware of the approach of the vessel.

Traps

Traps have been used extensively to catch tunas throughout the Mediterranean right up to the present time. The *almadraba*, a type of trap net used since the time of the Phoenicians, consists of corridors of netting, called leads, up to several kilometers long, up which migrating fish swim until they reach holding chambers, where they are harvested. Because the fish need not be pursued, vessels are not required to catch the fish, but only to transport the catch from the traps to shore.

Hook and Line

Hand lines The simplest form of hook-and-line fishing is a single hook attached to a hand-held line. The hooks, which can be baited with live, dead, or artificial bait, are generally set from a vessel or floating platform. All types and sizes of vessels, propelled by engine, sail or paddles, are used today to capture large pelagic fishes in various areas of the world. Hand-line fisheries, although primitive and, in a commercial sense, inefficient, are nevertheless widespread, and often spectacular to observe. In one such fishery, in Ecuador, the fishermen, frequently father and son, set out to sea shortly after midnight in sail-powered dugout canoes about 6 m long. Once on the fishing grounds, a single hook, baited with a small fish, is trailed several meters behind the canoe. When a fish, frequently large yellowfin tuna (*Thunnus albacares*) or black marlin (*Makaira indica*), which can weigh up to several hundred kilos, is hooked, the sails are reefed and the fishermen attempt to bring it to the boat. The struggle can last several hours, and once the fish is brought alongside it is clubbed in order to immobilize it. If it is too large to lift into the canoe, the fishermen enter the water, swamp the canoe, and roll the fish into it. They then bail out the vessel, hoist sail, and head for shore and the fish market. Scenes such as these are repeated in many artisanal fisheries throughout the world, but at diminishing rates as motorized fiberglass skiffs replace traditional canoes.

Pole and line Pole-and-line fishing is used in many parts of the world to catch large pelagics, principally tunas. The technique, which was developed independently in several separate regions of the world, is similar to that used for handlines, with the difference that the hook and line are attached to the end of a pole, giving greater reach and better leverage.

In the South Pacific, such fishing is done from canoes (**Figure 3**) and small motorized vessels using lures traditionally made from seashells (**Figure 4**). In Japan, commercial pole-and-line fishing for skipjack (*Katsuwonus pelamis*), yellowfin, and bluefin tunas was common in coastal waters for many centuries. During the twentieth century the Japanese developed larger pole-and-line vessels that were able to travel to all oceans of the world to fish for tunas. These vessels carry live bait in tanks of circulating sea water and

Figure 3 Pole-and-line fishing from a canoe in Tokelau (Courtesy of Robert Gillette, Suva, Fiji).

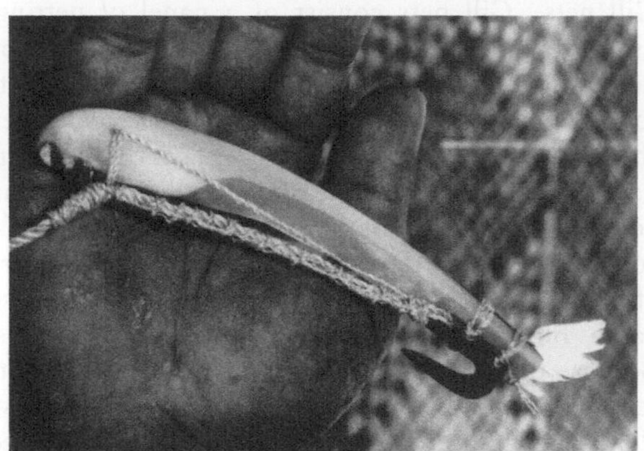

Figure 4 Pearl shell lure for catching large pelagic fish, from Tokelau (Courtesy of Robert Gillette, Suva, Fiji).

crews of more than 25; they can freeze their catches and stay at sea for several months. Some now carry automated machine-operated poles; a single crewman can tend several poles at once, thereby increasing vessel efficiency.

In the Indian Ocean, the island nation of Maldives was 'built on the backs of fish,' as the majority of its people made their living or derived their sustenance from fishing. The major form of fishing is with poles and lines, targeting skipjack, yellowfin, and wahoo (*Acanthocybium solandri*), using small vessels originally powered by sail with compartments for live bait built into the hull through which water is circulated. This method of fishing, which uses both artificial and live bait and a hook of unique design, is distinctively different in detail from that of Japan.

In southern California, after the introduction of tuna canning in the early twentieth century, pole-and-line fishing was used to supply the growing demand for tuna. The first vessels were small, used ice to preserve their catch, and fished within a few days of port. As the demand for tuna grew, larger, longer-range boats were built. This was the origin of the 'tuna clipper,' pole-and-line vessels that could pack several hundred tons of tuna in refrigerated fish holds, carry large amounts of live bait, and stay at sea for many months; they plied the eastern Pacific between California and Chile. Typically a vessel would catch bait before putting to sea to search for tuna. On sighting a school, the 'chummer' would throw the chum, or bait, to bring the fish alongside the vessel, and fishermen standing in racks at water level on the port side and stern of the vessel would catch the fish, usually using artificial lures (**Figure 5**). Once a fish was hooked, the fisherman lifted the pole, jerking the fish from the water and over his head. The pole would be stopped in the vertical position, and the fish would fall off the barbless hook onto the deck. If the school was feeding well, a fisherman could catch fish as quickly as he could get the hook into the water. If the chummer was successful in keeping a large school feeding alongside the boat, several tons could be taken in a short time. Because many fish are too large to be lifted by one fisherman on a single pole, a single hook might be attached to two, three and even four poles (**Figure 6**). For economic reasons, pole-and-line fishing is no longer predominant in the eastern Pacific, and only a few vessels operate on a regular basis.

During the 1950s pole-and-line fishing was introduced off tropical West Africa by the French and Spanish; it is still practiced there today, but to a limited extent.

Longlines Longlines, as the name implies, consist of a long mainline, kept afloat by buoys, from which

Figure 5 Pole-and-line fishing from a baitboat in Ecuador (Courtesy of Robert Olson, IATTC).

Figure 6 Catching a three-pole yellowfin tuna from a US baitboat (Courtesy of Pete Foulger, Harbor Marine Supplies, San Diego).

a number of branch lines, each terminating with a hook, are suspended (**Figure** 7). Longlines for large pelagics can have a mainline up to 125 km long, with as many as 500 buoys and 2500 hooks, and can take 8–12 h to set or retrieve. Longlines catch whatever fish happen to take the hook; a single set can capture several species of tunas, billfishes, and sharks. The distance between the buoys determines the depth of the hooks; they are normally suspended at depths between 100 and 150 m, where the water is cooler and large tunas are most likely to be. These large tunas, especially bigeye (*Thunnus obesus*), command high prices in the *sashimi* markets of Japan, and most of the larger longline fishing vessels of the world target this market.

Longline vessels vary in size. Small vessels use much shorter lines, and normally operate in coastal waters, whereas the larger vessels roam the oceans of the world in search of their prey and, supplied by tender vessels, can stay at sea for extended periods. Japanese vessels account for most of the longline catches, followed by Taiwanese and South Korean vessels.

Most billfishes are taken by longlines. During the last two decades longlining for swordfish has become very popular, and accounts for most of the landings of that species. Longline vessels targeting swordfish operate in the Atlantic, Pacific, and Indian Oceans.

Trolling In this method of fishing a number of lines to which lures are attached are towed from outrigger poles and from the stern of a vessel. When fishing for large pelagics, the lures are towed at a speed of several knots. Most troll fisheries target albacore tuna (*Thunnus alalunga*), but many other species, including bluefin, yellowfin, and wahoo, are also caught. This form of fishing is used throughout the world, usually from vessels of less than 20 m in length.

Nets

Gill nets Gill nets consist of a panel of netting, usually synthetic, held vertically in the water by floats along its upper edge and weights along its lower edge, in which fish become enmeshed when they attempt to swim through. The drift gill nets used to capture large pelagics in the open ocean consist of continuous series of such panels, sometimes more than 100 k long, and are very effective in catching the target species, but they also catch birds, sea turtles, and marine mammals. Because of these by-catches, and the fact that nets lost or abandoned at sea continue to catch fish ('ghost fishing'), in the late 1980s the United Nations recommended banning the use of drift gill nets over 2.5 km long on the high seas. However, such nets are still used within the Exclusive Economic Zones (EEZs) of some nations, particularly for catching swordfish and sharks.

Purse seines Purse seiners catch more tuna than all other types of vessels combined. Like a gill net, a purse seine is set vertically in the water, with floats attached to the upper edge; along the lower edge is a chain, for weight, and a series of rings, through which the pursing cable passes. Purse seines are constructed of heavy webbing, can be up to 1.5 km long and 150 m deep. When a school of tuna is sighted a large skiff, to which one end of the net is attached, is released from the stern of the fishing vessel (**Figure** 8). The vessel circles the school, paying out net, until it reaches the skiff, closing the circle. The pursing cable is winched aboard the vessel, closing the bottom of the net and trapping the fish; the net is then hauled back on board, concentrating the catch for loading. Purse-seine vessels fishing for large pelagics range from about 25 to 115 m in length, and can carry up to

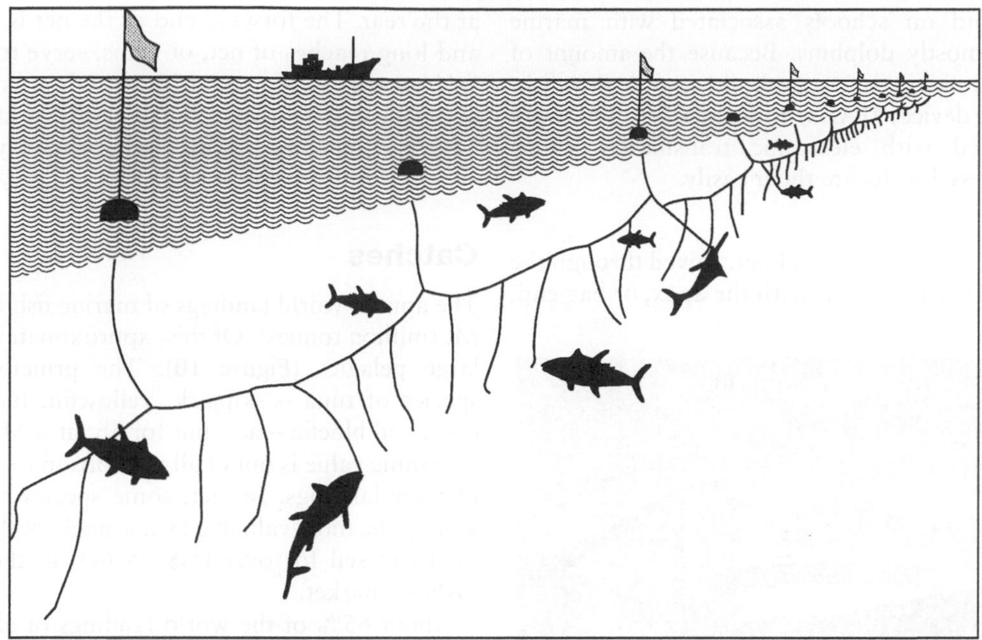

Figure 7 Diagrammatic sketch of longline fishing.

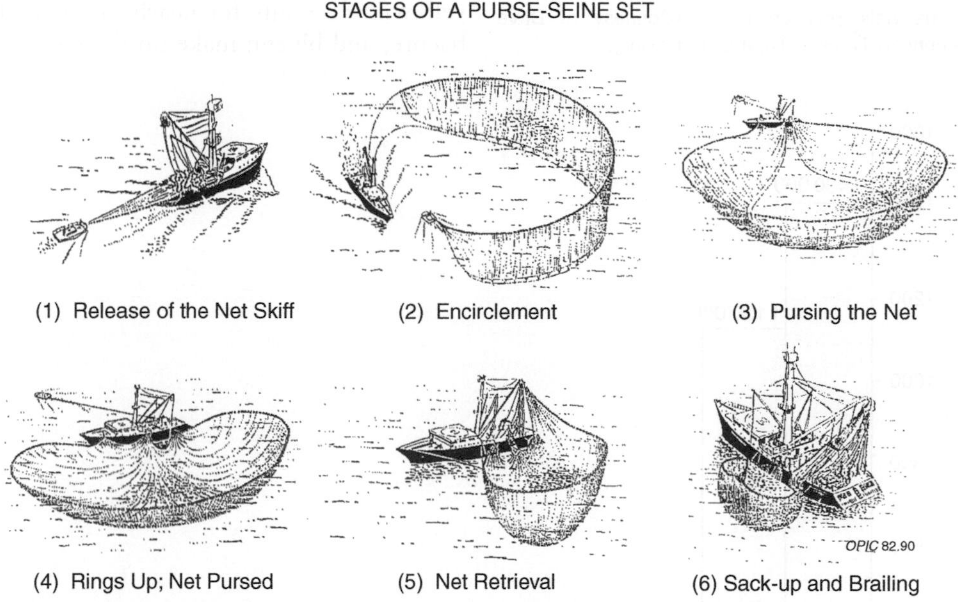

Figure 8 Setting a purse-seine tuna net (Source: Reproduced with permission from stequent and Marsac, 1991).

3000 tonnes of frozen fish, but most such vessels targeting the two principal species caught by this fishery, skipjack and yellowfin tuna, average about 75 m and 1200 tonnes (**Figure 9**). These vessels are capable of fishing all the oceans of the world and staying at sea for several months. Many carry helicopters for locating fish and directing the vessel while the net is being set, and are also equipped with sophisticated electronic devices such as specialized radar for detecting flocks of birds at great distances, current meters, satellite communication gear for receiving data on weather and ocean conditions, global positioning systems, and a variety of sonars for detecting schools of fish underwater. Purse seiners make three types of sets: on free-swimming schools of tuna, on schools associated with flotsam such as parts of trees, and on schools associated with marine mammals, mostly dolphins. Because the amount of flotsam is limited, fishermen build and deploy fish-aggregating devices (FADs) to attract fish. FADs are usually fitted with electronic transmitters which allow the vessel to locate them easily.

Trawls A trawl is a conical net, towed through the water by a trawler vessel, with the apex, or bag end, at the rear. The forward end of the net is held open, and long reaches of net, or arms, serve to guide the fish toward the bag end. Most trawls are used to fish on or near the seabed, but recently pelagic trawls have been used to fish for tuna, primarily albacore.

Figure 9 A modern purse-seine vessel of 1200 tonnes capacity; note the helicopter on board (Courtesy of Dave Bratten, Inter-American Tropical Tuna Commission).

Catches

The annual world landings of marine fish is about 85 Mt (million tonnes). Of this, approximately 4 Mt are large pelagics (**Figure 10**). The principal market species of tuna – skipjack, yellowfin, bigeye, albacore, and bluefin – account for about 3 Mt, but their economic value is out of all proportion to the volume of their landings. In fact, some species of tuna are among the most valuable fish: a single 300-kg bluefin tuna can sell for over US$ 75 000 in the Japanese *sashimi* market.

About 65% of the world landings of all large pelagic species is taken from the Pacific Ocean, 20% from the Indian Ocean, and 15% from the Atlantic Ocean. Skipjack represents the greatest proportion in all three oceans, constituting about 50% of the total landings of the principal market species of tuna; yellowfin accounts for nearly 35%, and bigeye, albacore, and bluefin make up the rest.

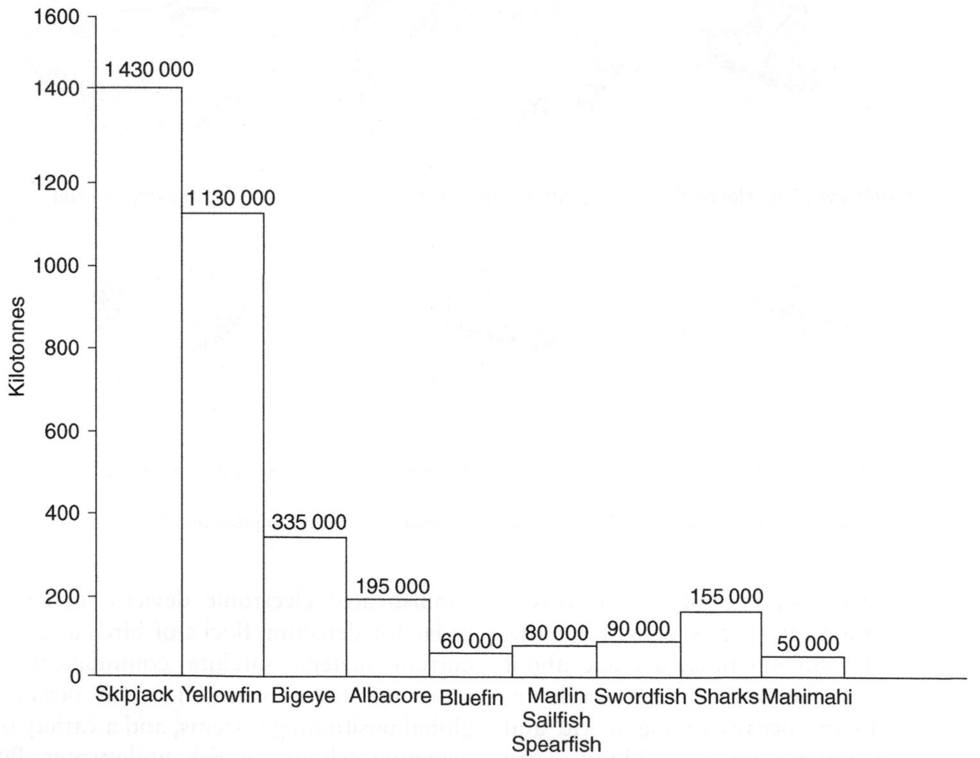

Figure 10 Recent world landings of certain large pelagic species. (Data modified from the 1996 and 1997 FAO Yearbooks of Fishery Statistics.)

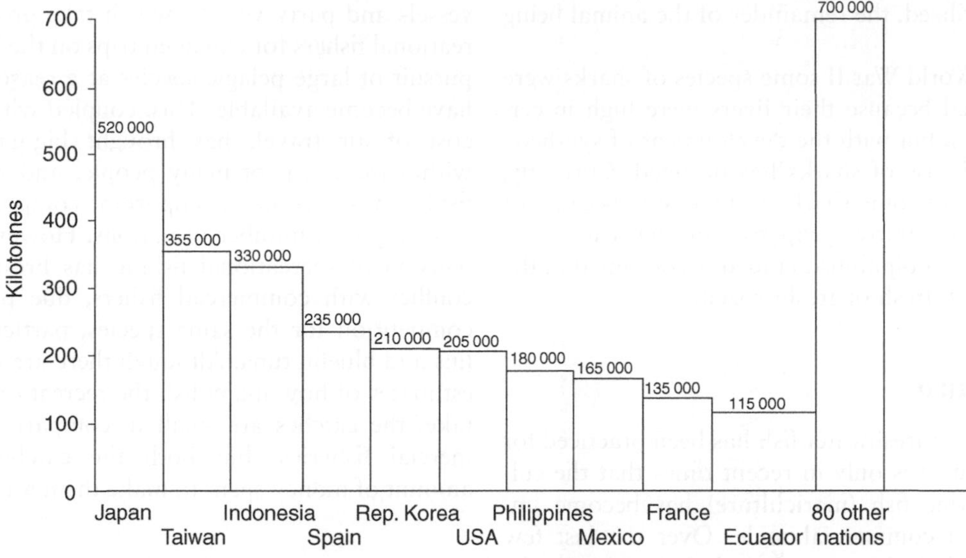

Figure 11 Recent world landings of principal market species of tunas, ranked by country of capture. (Data modified from the 1996 and 1997 FAO Yearbooks of Fishery Statistics.)

The largest fishery for tuna in the world is in the western Pacific, which produces about 35% of the world landings, mostly skipjack and yellowfin. Other large fisheries are in the eastern Pacific, the western Indian Ocean, and the eastern Atlantic. In all of these fisheries vessels of many nations fish side by side, using all types of fishing gear. Prior to about 1975 most tuna-fishing vessels were longliners or baitboats, but now purse seiners predominate.

In terms of landings, Japan is the leader, with 20% of the world total, followed by Taiwan, Indonesia, Spain, South Korea, the United States, the Philippines, Mexico, France, and Ecuador (**Figure 11**). These ten nations account for 80% of the total landings, the remainder being shared by about 80 other nations. Japan is also the principal consumer of large pelagics, accounting for about 30% of the world total, followed by Western Europe with about 25% and the United States with about 20%.

In recent years the annual landings of billfish have been about 170 000 tonnes, about half of it swordfish. About 160 000 tonnes of sharks are landed annually, but of this total probably less than 20% are pelagic species. The actual catches are almost certainly higher, but many sharks (mostly blue sharks, *Prionace glauca*) are caught only for their fins and are not reflected in world catch statistics. Japan, once again, accounts for the highest landings of billfish and, with the exception of swordfish, also consumes most of the billfish caught. Both the catch and consumption of sharks are widespread among many nations. The annual commercial landings of the common dolphinfish (*Coryphaena hippurus*) are less than 50 000 tons.

Utilization

Although tuna was first canned in Europe in the mid-1800s, prior to the twentieth century nearly all tuna and other large pelagics were eaten either fresh, salted or dried, or used to make sauce. In 1903 tuna canning was introduced to the United States in San Pedro, California. The canned product was well received by American consumers, increasing the demand for fish and ushering in the modern tuna industry. Since then the consumption of tuna has steadily increased, and today about 60% of the world landings, nearly all caught by purse seiners, is canned, yielding some 150 million cases, or about 1.5 Mt, annually. The United States consumes almost 50 million cases, Western Europe just over 50 million cases, and the remainder is consumed mostly in Asia and Latin America. Second in importance is fresh tuna, for consumption either raw, as *sashimi*, mostly in Japan, or grilled. *Katsuobushi*, lightly fermented, smoked and dried skipjack tuna, used as a condiment in Japanese cuisine, is another important use of tuna.

Of the billfishes, swordfish is consumed mostly as steaks, whereas blue marlin (*Makaira* spp.) and striped marlin (*Tetrapturus audax*) are frequently consumed fresh or smoked. Much of the catch of spearfish (*Tetrapturus* spp.) and sailfish (*Istiophorus platypterus*) is processed into imitation shrimp and crab, and other products.

With the exception of a few species, sharks are generally held in lower esteem as a food fish, and command much lower prices than tunas and some of the billfish. In many cases only the fins, which fetch high prices in Asian markets as an ingredient for soup, are utilized, the remainder of the animal being discarded.

Prior to World War II some species of sharks were heavily fished because their livers were high in certain vitamins, but with the development of synthetic vitamins this use of sharks has declined. Currently, the demand for some sharks is increasing because of the purported curative properties of their cartilage.

Most of the dolphinfish landed is consumed in the United States, fresh or fresh/frozen.

Mariculture

Propagation of freshwater fish has been practiced for centuries, but it is only in recent times that the culture of marine fish (mariculture) has become important on a commercial scale. Over the last few years mariculture has contributed about 10% to the total annual landings of all marine fishes. The pelagic habitat, migratory nature, and large size of pelagic species make rearing them in captivity difficult, and only recently have scientists been able to spawn and rear tunas artificially. In Japan there has been limited success in spawning bluefin tuna in captivity, hatching the eggs, and rearing the young. At the Inter-American Tropical Tuna Commission's Achotines Laboratory in Panama, scientists have been able to spawn yellowfin tuna held in captivity in onshore tanks on a regular basis, and have also successfully reared the hatched eggs to the juvenile stage. It is anticipated that this research will eventually make it possible to complete the life cycle of yellowfin tuna in captivity, an essential precondition to rearing tuna commercially. In the meantime, attempts at what is known as bluefin ranching are being increasingly successful. This involves catching adult bluefin tuna and transporting them to pens in inshore areas, where they are held and fed until reaching a marketable size and condition. Because of the high value of bluefin in the *sashimi* market, ranching is a growing enterprise in many countries, including Australia, Croatia, Japan, Mexico and Spain.

Recreational Fishing

Recreational fishing for large pelagics has been important since the early 1900s, and became more so after it was popularized by writers such as Zane Gray and Ernest Hemingway. The most sought-after species are the larger ones, like marlins, sailfish, and bluefin and yellowfin tuna. It was originally mostly a sport for the wealthy, because of the high cost of fishing boats and access to areas where these species abound. More recently, however, fleets of charter vessels and party vessels, which take groups of recreational fishers for extended trips on the high seas in pursuit of large pelagic species at a reasonable cost, have become available. This, coupled with the lower cost of air travel, has brought big-game fishing within the grasp of many people, and recreational fishing has become an important component of the economy of a number of nations. However, this expansion of recreational fishing has brought it into conflict with commercial fishers, due primarily to competition for the same species, particularly marlins and bluefin tuna. Although there are no accurate estimates of how many fish the recreational fisheries take, the catches are small in comparison to commercial fisheries, but both the catches and the amount of money spent to make them are increasing.

Conservation of the Resource

Large pelagics are a renewable resource whose abundance can be profoundly influenced by human activities, particularly fishing. These activities have been steadily increasing as the demand for food and recreation from the sea increases, and landings of tunas and related species have increased tenfold since 1945. These increases have resulted in overfishing of some species, particularly bluefin tuna, swordfish, and some sharks. Many of the other species of large pelagics, with the notable exception of skipjack tuna, are currently fully, or nearly fully exploited, and sustained increased landings cannot be expected. Many of these other species are in urgent need of conservation if they are to be protected from overexploitation. Widespread concern over the status of some of these stocks has led to action by coalitions of chefs throughout the United States to boycott the use of swordfish in their restaurants, and attempts by public interest groups to include sharks and bluefin tuna as species covered by the Convention on International Trade in Endangered Species (CITES). This increased fishing has also caused problems for other marine species taken incidentally as by-catch with the species targeted by the fishery. These by-catch species are often of no value to the fishers, and are thrown back to the sea, dead. By-catches occur in most fisheries for large pelagic species, and it is important to determine their impact on the populations from which they are taken and on the ecosystem to which they belong. A similar problem in some purse-seine fisheries for tuna, especially in the eastern Pacific, is the capture of dolphins. Many dolphins were killed in purse-seine nets between the inception of this type of fishing in the late 1950s and the early 1990s, but in recent years, thanks to the joint efforts of environmental organizations, the fishing industry, and governments, the mortality of dolphins caused by tuna fishing has been reduced to very low levels.

Populations of pelagic sharks, in addition to being heavily exploited for their flesh and/or fins, are also taken in large quantities as a by-catch in some fisheries. Because many species of sharks have low fecundity, are slow growing and later maturing, they are vulnerable to severe overexploitation which could lead to the collapse of some populations.

Because tunas and other large pelagic species are highly migratory, the vessels that fish for them roam the oceans of the world, and many nations are involved in their harvest, international cooperation is essential for effective conservation of these species. Article 64 of the United Nations Convention on the Law of the Sea calls on nations to work jointly through appropriate international bodies to manage and conserve such highly migratory species. There are currently four such bodies, the Inter-American Tropical Tuna Commission, the International Commission for the Conservation of Atlantic Tunas, the Commission for the Conservation of Southern Bluefin Tuna, and the Indian Ocean Tuna Commission, and negotiations for creating a similar body for the western and central Pacific are near completion. These bodies are responsible for coordinating and conducting scientific research and, on the basis of that research, making recommendations to governments for the conservation of the tunas and tuna-like species. As a result of their efforts conservation measures are in effect for many of the large pelagic species. Such measures can take a variety of forms, including catch quotas that limit the amount of fish that can be caught or landed, closed areas and seasons, limits on the size of fish that can be landed, and restrictions on the types and amounts of gear that can be used to catch fish. As fishing pressure continues to increase, there will be a need to expand such measures to other areas and species.

See also

Fishery Manipulation through Stock Enhancement or Restoration. Marine Fishery Resources, Global State of.

Further Reading

Anganuzzi AA, Stobberup KA, and Webb NJ (eds.) (1996) *Proceedings of the 6th Expert Consultation on Indian Ocean Tunas*. Colombo, Sri Lanka: Indo-Pacific Tuna Development and Management Programme.

Beckett JJ (ed.) (1998) *Proceedings of the ICCAT Tuna Symposium*. Collective Volume of Scientific Papers, Vol. 50. Madrid: International Commission for the Conservation of Atlantic Tunas.

Cort JL (1990) Biologia y pesca del atún rojo. Institut. Español Oceanographia, Pub. Esp. No. 4. Madrid.

Doumenge F (1999) ha Storia Delle Pesche Tonniere (The history of tuna fisheries). Biol Mar Medit 6(2): 1–106.

FAO Fisheries Department (1993) *World Review of High Seas and Highly Migratory Fish Species and Straddling Stocks. FAO Fisheries Circular. No. 858*. Rome: FAO.

Fonteneau A (1997) *Atlas of Tropical Tuna Fisheries – World Catches and Environment*. Paris: Orstom editions.

Goadby P (1987) *Big Fish and Blue Water – Gamefishing in the Pacific*. North Ryde, London: Angus and Robertson.

Jones S and Kumaran M (1959) The fishing industry of Minicoy Island with special reference to the tuna fishery. *Indian Journal of Fisheries* VI(1): 30–57.

Joseph J (1998) A review of the status of world tuna resources. In: Nambiar KPP and Pawiro S (eds.) *Papers of the 5th World Tuna Trade Conference*, pp. 8–21. Bangkok: INFOFISH.

Joseph J and Greenough JW (1979) *International Management of Tuna, Porpoise, and Billfish – Biological, Legal, and Political Aspects*. Seattle: University of Washington Press.

Joseph J, Klawe W, and Murphy P (1988) *Tuna and Billfish – Fish without a Country*. La Jolla: Inter-American Tropical Tuna Commission.

Orbach MK (1977) *Hunters, Seamen, and Entrepreneurs – The Tuna Seinermen of San Diego*. Berkely: University of California Press.

Shomura RS, Majkowski J, and Langi S (eds.) (1994) *Interactions of Pacific Tuna Fisheries*. FAO Fisheries Technical Paper 336, Vols 1 and 2. Rome: FAO.

Stroud RH (ed.) (1989) *Planning the Future of Billfishes: Research and Management in the 90s and Beyond vols 1 and 2*. Savannah: National Coalition for Marine Conservation, Inc.

Ward P and Elscot S (2002) *Broadbile Swordfish: Status of World Fisheries*. Australia: Bureau of Rural Sciences, Aquaculture, Fisheries and Forestry.

SALMON FISHERIES, ATLANTIC

P. Hutchinson and M. Windsor,
North Atlantic Salmon Conservation Organization,
Edinburgh, UK

Introduction

Migrating animals, concentrated in space and time, represent readily harvestable resources that have a long history of exploitation by humans. The anadromous Atlantic salmon (*Salmo salar*) is no exception. Cave paintings and stone carvings dating back 25 000 years from the Dordogne region of France confirm its long association with, and importance to, humans. Throughout its range in the North Atlantic, the Atlantic salmon has been and continues to be exploited by a variety of gear in rivers, lakes, estuaries, and the sea, providing employment and recreation, and generating considerable economic benefits, often in remote rural areas. The Atlantic salmon also has cultural, ceremonial, and symbolic significance, but it is difficult to ascribe a value to these important facets of the resource. Throughout the history of exploitation of Atlantic salmon by humans, there have been many changes in the nature and scale of the fisheries.

Description of the Salmon Fisheries

Although it is a matter of conjecture, the most ancient method of harvesting Atlantic salmon was probably by hand in rivers where adults returning to spawn may well have been an important component of the diet prior to the development of agriculture and techniques for animal husbandry. Apart from the use of clubs or stones, the spear or harpoon was probably the first fishing gear used for salmon. The snare, hook, and dip net probably followed. The earliest method of harvesting salmon in quantity was probably the fishing weir. Spears, hooks, nets, and weirs thought to have been used for catching salmon at least 8000 years ago have been discovered in Sweden. In eastern Canada, the first harvesting of Atlantic salmon is thought to have started about 8800 BC when Amerindians arrived in the area. The spear was the preferred implement.

Documentary evidence of the use of salmon weirs is available from the eleventh century. The Battle of Clontarf in Ireland in 1014 was known as the Battle of the Salmon Weir. Use of weirs and nets (probably hand nets, seine, and gill nets) by North American Indians was documented in the sixteenth century. The seine net is known to have been used for catching salmon in Scotland and Ireland in the seventeenth century and probably much earlier than that. Amerindians practiced a primitive form of angling. Angling for salmon as a hobby is known from at least the seventeenth century in some countries, although it was introduced to Norway only in the nineteenth century.

While a considerable variety of types of salmon fishing gear has been developed, on comparison they appear to be based on a few basic methods of capture, which have been categorized under four general headings: fixed gears or traps (e.g., bag nets, stake nets, set gill nets); floating gears (e.g., longlines and drift nets); seine or draft nets; and rod and line (using a variety of artificial flies, baits, and lures).

There have been many improvements to these fishing methods over the period of their deployment. One of the most significant has been the development of synthetic twines that made the gear (including recreational fishing gear) easier to handle and less visible.

Salmon fisheries are often categorized as 'recreational' and 'commercial' to distinguish between sport fishing with rod and line and fishing with other gears with the intention of selling the harvest. However, the distinction is sometimes blurred. For example: 'recreational' licenses may be issued to fish for salmon with gill nets for local consumption purposes in Greenland; in some countries the sale of rod-caught salmon is permitted; and rod-and-line fisheries may be let or sold for considerable sums of money. For the purposes of this article, recreational fisheries are considered to be sport fisheries using rod and line and a variety of artificial flies, baits, and lures; commercial fisheries are those fisheries conducted with a variety of other gears where the intention is to sell the harvest. A third category, 'subsistence fisheries', is conducted with the intention of using the harvest of salmon for consumption by the local community; for example, the fisheries by native people in Canada, Finland and Greenland. Salmon fishing may also be conducted for research purposes, in some cases using methods that would ordinarily be prohibited.

Some countries have only recreational fisheries. For example, all netting of salmon was prohibited in Spain and the salmon fishery was dedicated entirely to recreational fishing in 1942 following the Civil

Table 1 Origin of salmon caught in home water fisheries in the North-east Atlantic in 1992

	Catch by country									
Origin of stock	*Russia*	*Finland*	*Norway*	*Sweden*	*England and Wales*	*Scotland*	*Northern Ireland*	*Ireland*	*France*	*Iceland*
Wild										
Russia	100%	–	+	–	–	–	–	–	–	–
Finland	–	99%	+	–	–	–	–	–	–	–
Norway	–	+	75%	6%	–	–	–	+	–	–
Sweden	–	–	1%	46%	–	–	–	–	–	–
England and Wales	–	–	–	–	62%	+	+	10%	–	–
Scotland	–	–	–	–	38%	95%	3%	5%	–	–
Northern Ireland	–	–	–	–	+	+	92%	5%	–	–
Ireland	–	–	–	–	+	+	+	80%	–	–
France	–	–	–	–	+	+	+	+	100%	–
Iceland	–	–	–	–	–	–	–	–	–	28%
Reared										
Escapees	–	<1%	23%	2%	–	5%	1%	–	–	–
Ranched	–	–	1%	46%[a]	–	–	3%	<1%	–	72%

[a]Fish released for mitigation purposes and not expected to contribute to spawning.
Source: Report of the ICES Advisory Committee on Fishery Management 1994. NASCO Council document CNL(94)13.
+, Catches thought to occur but contribution not estimated.
–, Catches occur rarely or not at all.

War, during which the salmon populations had been heavily exploited for food. Similarly, in the United States all commercial exploitation of Atlantic salmon ceased in 1948. Other countries have a mixture of commercial and recreational fisheries (e.g., Norway, United Kingdom, Ireland, France, Iceland, and Russia). In Iceland there is no coastal netting and commercial fisheries are conducted in only two rivers in the south of the island (A. Isaksson, personal communication). Canada had a major commercial fishery, but management measures introduced since the mid-1960s have progressively reduced the fishery and in 2000 no commercial licenses were issued, with the effect that the Canadian salmon fishery is now recreational and subsistence in nature. In Russia, the fisheries were mainly commercial and angling for salmon was prohibited in all but three rivers, where it was strictly controlled by restrictions on the number of licenses issued. However, since the mid-1980s, recreational fisheries have developed in the rivers of the Kola peninsula and are popular with foreign anglers. Greenland and the Faroe Islands have only one and five salmon rivers, respectively, so the opportunities for recreational fishing are limited and fishing has mainly been either commercial or subsistence in Greenland, and commercial or research in the Faroe Islands.

Salmon fisheries have been described as single or mixed stock on the basis of whether they exploit a significant number of salmon from one or from more than one river stock, respectively. Some mixed stock fisheries may exploit salmon originating in different countries. Mixed stock fisheries have also been referred to as interception fisheries and the term is often applied specifically to the Greenland and Faroes fisheries. However, prior to the closure of the commercial fisheries in Newfoundland and Labrador in Canada, there was concern about the harvest of US fish by this fishery. Similarly, in the North-East Atlantic area there are harvests in the fisheries of one country of salmon originating in the rivers of another country (**Table 1**). Thus, many salmon fisheries are interceptory in nature, but these interceptions have declined in recent years as a result of international agreements in the North Atlantic Salmon Conservation Organization (NASCO), national or regional regulations, economic factors and other reasons.

The terms 'home water' and 'distant water' are also used in relation to salmon fisheries. Since 1983, with the implementation of the Convention for the Conservation of Salmon in the North Atlantic Ocean, which prohibits fishing beyond areas of fisheries' jurisdiction, all salmon fisheries are in effect home water fisheries. The term home water fishery is therefore more correctly used to indicate fisheries within the jurisdiction of the state of origin of the salmon. (i.e., in the country in whose rivers the salmon originated), as opposed to distant water fisheries, which harvest salmon outside the jurisdiction of the state of origin.

The Resource

Limits on production during the freshwater phase of the life cycle constrain the abundance of Atlantic salmon and result in catches that are low compared to those of Pacific salmon and pelagic marine fish species such as herring and mackerel.

A wide range of factors has already affected this freshwater production capacity, including urbanization, land drainage, overgrazing, forestry practices, infrastructure developments, water abstraction, sewage and industrial effluents, hydroelectricity generation, and the introduction of nonindigenous species. Many salmon rivers were damaged as a result of the Industrial Revolution. For example, in Canada, there has been a net loss of productive capacity of salmon of 16% since 1870, and in the state of Maine, USA, about two-thirds of the historic salmon habitat had been lost by the mid-1980s. Early attempts at enhancement through stocking programs date to the middle of the nineteenth century. These stocking programs continue and in 1999 more than 30 million Atlantic salmon eggs and juveniles were stocked in rivers around the North Atlantic. With the decline of many heavy industries there have been improvements in salmon habitat and in England and Wales, for example, there are now more salmon-producing rivers than there were 150 years ago. Much progress has also been made in recent years in improving fish passage facilities at dams. The effects of the Industrial Revolution are, however, still being felt today, through the continuing problem of acidification of rivers and lakes, for example. As the human population continues to increase, pressures on salmon habitat from domestic, industrial, and agricultural demands will increase.

Catch statistics compiled for the North Atlantic region by the International Council for the Exploration of the Sea (ICES) are available for the period from 1960, during which the total reported catch has ranged from approximately 2200 tonnes in 1999 to approximately 12 500 tonnes in 1973 (**Figure 1**). The mean reported catch in tonnes by country for each of the four decades 1960–69 to 1990–99 is shown in **Table 2**. There has been a steady decline in the total reported North Atlantic catch of salmon since the early 1970s. **Figure 2** shows for four major states of origin that there is some degree of synchronicity in the trend in catches over the 40-year period from 1960 when expressed as the percentage difference from the long-term mean reported catch. While catches in all four countries were above or close to the 40-year mean in the period from the 1960s to the late 1980s, the last decade of the twentieth century was characterized by below-average catches. Although some of the reduction in catches was the result of the introduction of management measures, which have reduced fishing effort, the abundance of both European and North American Atlantic salmon stocks has declined since the 1970s, particularly the multi-sea-winter components. This decline in abundance appears to be related to reduced survival at sea.

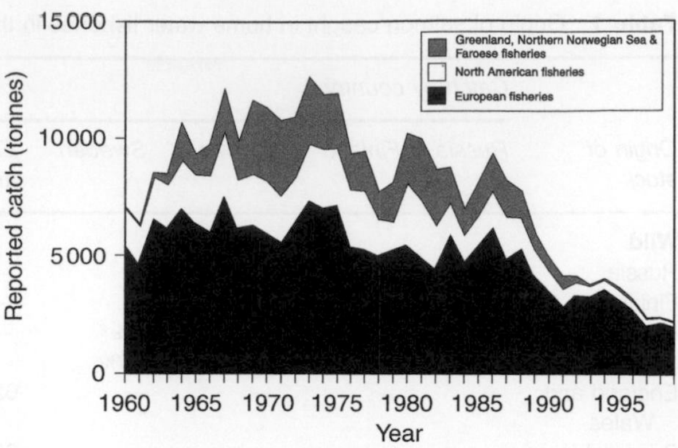

Figure 1 Reported catch of Atlantic salmon (tonnes) from the North Atlantic area, 1960–1999.

In addition to the reported catches illustrated in **Figure 1**, catches may go unreported for a variety of reasons. These include the absence of a requirement for statistics to be collected; suppression of information thought to be unfavorable; and illegal fishing. Estimates of unreported catch for the North Atlantic region for the period from 1987 have ranged between approximately 800 and 3200 tonnes, or 29–51% of the reported catch. Illegal fishing appears to be a particular problem in some countries. Associated with all forms of fishing gear is mortality generated directly or indirectly by the gear but which is not included in reported catches. This mortality may be associated with predation, discards, and escape from the gear. For salmon fishing gear the contribution of most sources of this mortality is estimated to be low (0–10%) but highly variable.

By-catch of nontarget species in salmon fishing gear is thought to be generally low. Drift nets may have a by-catch associated with their use, but this has not been fully quantified. However, as this gear is often tended by the fishermen, there may be an opportunity to release sea birds and marine mammals from the nets. 'Ghost fishing' by lost or abandoned nets is not thought to be a problem associated with salmon fishing gear. By-catch of salmon in gear set for species such as bass, lumpsucker, mackerel, herring, and cod is known to occur but it is not generally a problem. In some countries regulations have been introduced to protect salmon from capture in coastal fisheries for other species. There is, however, concern about the

Table 2 Mean reported catch (in tonnes) by country during the four decades 1960–1969 to 1990–1999

Country	*1960–1969*	*1970–1979*	*1980–1989*	*1990–1999*
Canada	2053	2142	1638	395
Denmark	138	491	152	1
England and Wales	325	384	370	224
Faroe Islands	64	152	606	47
Finland	—	42	54	57
France	—	14	23	13
Germany	2	3	—	—
Greenland	773	1300	816	119
Iceland	131	197	176	138
Ireland	1329	1676	1263	616
Northern Ireland	291	174	114	82
Norway	1822	1745	1453	840
Russia	690	559	520	158
Scotland	1684	1437	1058	468
Spain	36	27	21	8
St. Pierre and Miquelon	—	—	3	2
Sweden	50	39	35	34
USA	1	2	3	1

Notes: (1) The catch for Iceland excludes returns to commercial ranching stations. (2) The catches for Norway, Sweden, and Faroe Islands include harvests at West Greenland and in the Northern Norwegian Sea fishery. (3) The catch for Finland includes harvests in the Northern Norwegian Sea fishery. (4) The catch for Denmark includes catches in the Faroese zone, in the Northern Norwegian Sea fishery, and at West Greenland. (5) The catch for Germany is from the Northern Norwegian Sea fishery.

possible by-catch of salmon post-smolts in pelagic fisheries for herring and mackerel in the Norwegian Sea, which overlap spatially and temporally with European-origin post-smolt migration routes.

In addition to exploitation of Atlantic salmon in the North Atlantic region, there are fisheries in the Baltic Sea. Catches since 1972 have ranged from approximately 2000 to 5600 tonnes. These fisheries, which are based to a large extent on hatchery smolts released to compensate for loss of habitat following hydroelectric development, are described in detail by Christensen *et al.* (see Further Reading).

Economic Value

A wide variety of techniques have been used to assess the economic value of Atlantic salmon and, in the absence of a standardized approach, assessment of the economic value of salmon fisheries on a North Atlantic basis is not possible. Many assessments concern the expenditure associated with salmon fishing. Economic value, however, reflects willingness to pay for use of the resource, and as willingness to pay must at least be equal to actual expenditure, many assessments underestimate the full economic value. However, it is clear that throughout its range the Atlantic salmon generates considerable economic benefits that may have impacts on a regional basis or, where visiting anglers from other countries are involved or where the harvest is exported, impacts on national economies. The following examples serve to highlight the considerable economic value of salmon fisheries.

The total net economic value of salmon fisheries, both recreational and commercial, in Great Britain was estimated in 1988 to be £340 million, of which the recreational fisheries accounted for approximately £327 million.

In Canada, recreational anglers spent Can$39 million on salmon fishing in 1985 with a further Can$45 million invested on major durables and property.

In Greenland, the salmon fishery in 1980 was a substantial source of income (30–35% of total annual income to the fishermen), and 50% of fishermen could not have met their current vocational and domestic expenses at that time without the salmon fishery. Many people other than the fishermen depended on the salmon fishery for gear and equipment sales and repair and shore processing.

The expenditure by recreational salmon fishermen visiting one major Scottish salmon river, the Tweed, was estimated to be £9 million in 1996, with a total economic impact of more than £12 million. Approximately 500 full-time job equivalents depended on this activity. This is for one river and there are more than 2000 salmon rivers in the North Atlantic region, with fisheries that bring economic benefits,

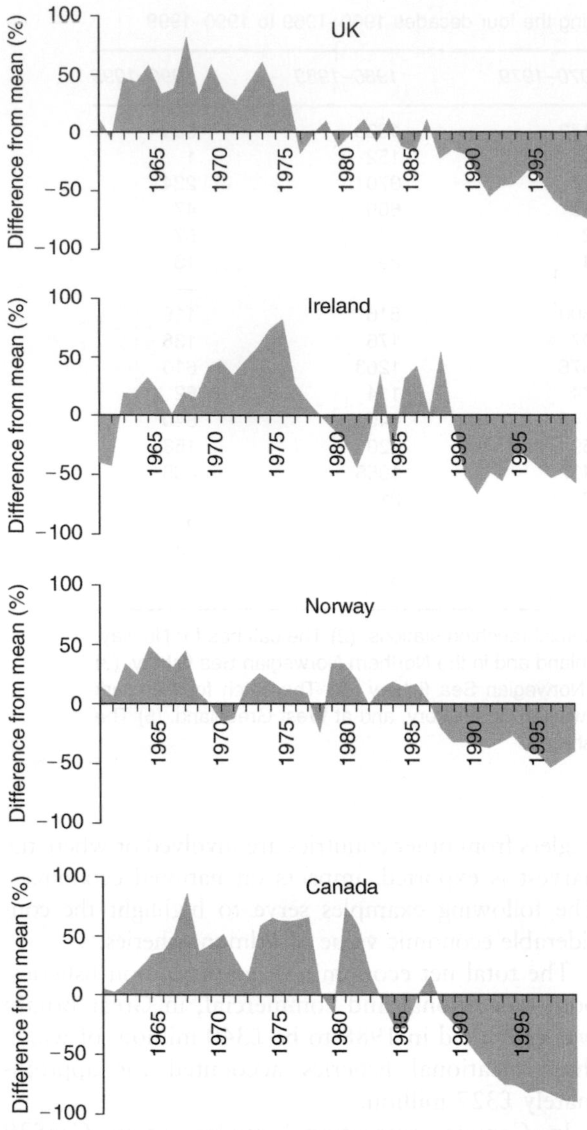

Figure 2 Catches of Atlantic salmon (tonnes), expressed as the percentage difference from the 40-year mean, for four major states of origin.

often to remote areas where job creation is otherwise very difficult.

In addition to the economic value associated with the fisheries, individuals are willing to contribute to salmon conservation even though they have no interest in fishing. Sixty percent of the New England population was found to 'care' about the Atlantic salmon restoration program and in 1987 their willingness to pay was estimated to exceed the cost of the restoration program. Economic assessments that fail to take these non-user aspects into account will considerably underestimate the economic value of the resource. The salmon has a special place in human perception and there are many nongovernment organizations dedicated to its conservation.

Management of the Fisheries

Legislation regulating the operation of salmon fisheries is known to have been introduced in Europe as early as the twelfth century. In Scotland, for example, legislation was introduced to establish a weekly close time and to prevent total obstruction of rivers by fishing weirs. Similarly, in the middle of the thirteenth century, legislation establishing close seasons was introduced in Spain. Since these early conservation measures were enacted, a wide variety of laws and regulations concerning the salmon fisheries have been developed by each North Atlantic country. These laws and regulations include those that permit or prohibit certain methods of fishing; specify permitted times and places of fishing; restrict catch by quota; prohibit the taking of young salmon and kelts; restrict or place conditions on the trade in salmon; and ensure the free passage of salmon.

The last quarter of the twentieth century witnessed dramatic changes in the exploitation of Atlantic salmon. Commercial fisheries have been greatly reduced, partly as a result of management measures taken in response to concern about abundance and partly as a result of the growth of salmon farming. Production of farmed Atlantic salmon has increased from less than 5000 tonnes in 1980 to more than 620 000 tonnes in 1999 (**Figure 3**). This rapidly growing industry has had a marked impact on the profitability of commercial fisheries for salmon. While it has been argued that the growth of salmon farming, which in 1999 produced about 300 times the harvest of the fisheries, has reduced exploitation pressure on the wild stocks, there are concerns about the genetic, disease, parasite, and other impacts the industry may be having on the wild Atlantic salmon. In some countries, escaped farm salmon frequently occur in fisheries for wild salmon and in spawning stocks.

Distant Water Fisheries

Prior to the 1960s, management of salmon fisheries in the North Atlantic region was at a local, regional, or national level. During the 1960s and early 1970s, however, distant water fisheries developed at West Greenland (harvesting both European and North American origin salmon) and in the Northern Norwegian Sea and, later, in the Faroese zone (harvesting predominantly European-origin salmon). The rational management of these fisheries required international cooperation, the forum for which was created in 1984 with the establishment of the intergovernment North Atlantic Salmon Conservation Organization (NASCO). The development and subsequent regulation of these fisheries in terms of reported catch are illustrated in **Figure 4**. The

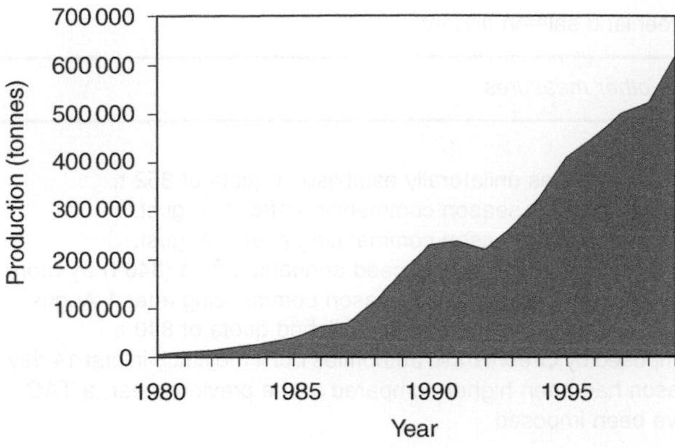

Figure 3 Production of farmed Atlantic salmon (tonnes) in the North Atlantic area.

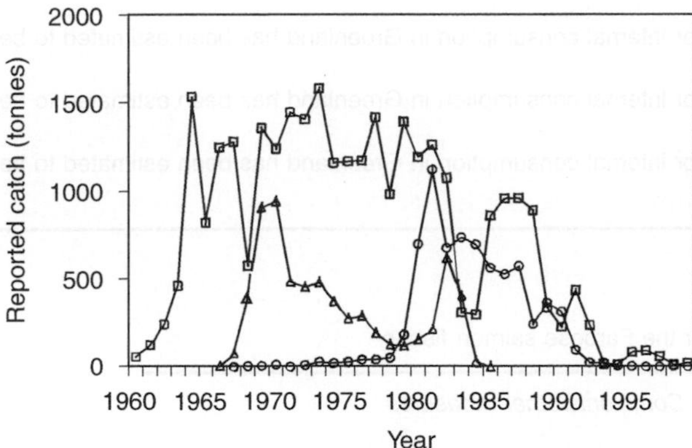

Figure 4 Reported catches (tonnes) of Atlantic salmon in the fisheries at West Greenland (□), in the Northern Norwegian Sea (△) and in the Faroese zone (○).

Newfoundland and Labrador commercial fishery in Canada, which before its closure harvested US-origin salmon in addition to salmon returning to Canadian rivers, was also subject to a regulatory measure agreed in NASCO.

West Greenland Salmon Fishery The presence of salmon off West Greenland was first reported in the late eighteenth century and a fishery for local consumption purposes has probably been conducted since the beginning of the twentieth century. From 1960 to 1964 the landings by Greenlandic vessels using fixed gill nets increased from 60 tonnes to more than 1500 tonnes and increased further from 1965 when vessels from Denmark, Sweden, the Faroe Islands, and Norway joined the fishery and monofilament gill nets were introduced. From 1975 the fishery was restricted to Greenlandic vessels. The salmon harvested at West Greenland are almost exclusively one-sea-winter salmon that would return to rivers in North America (principally Canada, but harvests of US salmon were significant in comparison to the number of fish returning to spawn) and Europe (particularly the United Kingdom and Ireland) as multi-sea-winter salmon. During the 1990s, the proportion of salmon of North American origin in the catch has increased, comprising 90% of samples in 1999.

International agreement on regulation of the harvests at West Greenland first occurred in 1972 when the International Commission for the Northwest Atlantic Fisheries (ICNAF) endorsed a US–Danish bilateral agreement to limit the catch to 1100 tonnes (adjusted to 1191 tonnes in 1974). This quota, with small adjustments to take account of delays in the start of the seasons in 1981 and 1982, applied until 1984, since when regulatory measures have been developed within NASCO. Details of these measures are given in **Table 3**.

Northern Norwegian Sea Fishery Seven years after the start of the West Greenland fishery, a salmon fishery involving, at different times, vessels from Denmark, Norway, Sweden, Finland, Germany, and the Faroe Islands commenced in the Northern Norwegian Sea. Initially drift nets were used, but the vessels soon changed to longlines. Prior to 1975, the fishery was conducted over a large geographical area between 68°N and 75°N and between the Greenwich meridian and 20°E. However, following the extension of fishery limits to 200 nautical miles, the fishery shifted westward to the area between the Norwegian fishery limit and Jan Mayen Island. The catch peaked at almost 950 tonnes in 1970. In response to the rapid escalation of this fishery, the North-East Atlantic Fisheries Commission (NEAFC) adopted a variety of measures intended to stabilize harvests, although a proposal to prohibit high-seas salmon fishing failed to obtain unanimous approval. However, this fishery ceased to exist in 1984 as a result of the prohibition on fishing for salmon beyond areas of fisheries jurisdiction in the Convention for the Conservation of Salmon in the North Atlantic Ocean (the NASCO Convention).

In the period 1989–94 vessels were identified fishing for salmon in international waters. These vessels were based mainly in Denmark and Poland and some had re-registered to Panama in order to avoid the provisions of the NASCO Convention. On the basis of information on the number of vessels, the number of trips per year, and known catches, ICES has provided estimates of the harvest (tonnes) as follows:

1990	1991	1992	1993	1994
180–350	25–100	25–100	25–100	25–100

Following diplomatic initiatives by NASCO and its Contracting Parties there have been no sightings

Table 3 Regulatory measures agreed by NASCO for the West Greenland salmon fishery

Year	*Allowable catch* (tonnes)	*Comments/other measures*
1984	870	
1985	—	Greenlandic authorities unilaterally established quota of 852 t.
1986	850	Catch limit adjusted for season commencing after 1 August.
1987	850	Catch limit adjusted for season commencing after 1 August.
1988–1990	2520	Annual catch in any year not to exceed annual average (840 t) by more than 10%. Catch limit adjusted for season commencing after 1 August.
1991	—	Greenlandic authorities unilaterally established quota of 840 t.
1992	—	No TAC imposed by Greenlandic authorities but if the catch in first 14 days of the season had been higher compared to the previous year, a TAC would have been imposed.
1993	213	
1994	159	
1995	77	
1996	—	Greenlandic authorities unilaterally established a quota of 174 t.
1997	57	
1998	Internal consumption fishery only	Amount for internal consumption in Greenland has been estimated to be 20 t.
1999	Internal consumption fishery only	Amount for internal consumption in Greenland has been estimated to be 20 t.
2000	Internal consumption fishery only	Amount for internal consumption in Greenland has been estimated to be 20 t.

TAC, total allowable catch.

Table 4 Regulatory measures agreed by NASCO for the Faroese salmon fishery

Year	*Allowable catch* (tonnes)	*Comments/other measures*
1984/85	625	
1986	—	
1987–1989	1790	Catch in any year not to exceed annual average (597 t) by more than 5%.
1990–1991	1100	Catch in any year not to exceed annual average (550 t) by more than 15%.
1992	550	
1993	550	
1994	550	
1995	550	
1996	470	No more than 390 t of the quota to be allocated if fishing licenses issued.
1997	425	No more than 360 t of the quota to be allocated if fishing licenses issued.
1998	380	No more than 330 t of the quota to be allocated if fishing licenses issued.
1999	330	No more than 290 t of the quota to be allocated if fishing licenses issued.
2000	300	No more than 260 t of the quota to be allocated if fishing licenses issued.

Note: The quotas for the Faroe Islands detailed above were agreed as part of effort limitation programs (limiting the number of licenses, season length, and maximum number of boat fishing days) together with measures to minimize the capture of fish less than 60 cm in length. The measure for 1984/85 did not set limits on the number of licenses or the number of boat fishing days.

of vessels fishing for salmon in international waters in the North-East Atlantic since 1994. NASCO is cooperating with coastguard authorities in order to coordinate and improve surveillance activities.

Faroes Salmon Fishery During the period 1967–78 exploratory fishing for salmon was conducted off the Faroe Islands using floating longlines. During this period no more than nine Faroese vessels were involved in the fishery and the catches, which were mainly of one-sea-winter salmon, did not exceed 40 tonnes. During the period 1978–85 Danish vessels also participated in the fishery and in 1980 and 1981 there was a marked increase in fishing effort and catches. As the fishery developed, it moved farther north and catches were dominated by two-sea-winter salmon. The salmon caught in the fishery are mainly of Norwegian and Russian origin. Initially negotiations on regulatory measures for the Faroese fishery were conducted on a bilateral basis between the Faroese authorities and the European Commission. Since 1984, the fishery has been regulated through NASCO. Details of these measures are given in **Table 4**.

Compensation Arrangements In the period 1991–98 the North Atlantic Salmon Fund (NASF) entered into compensation arrangements with the Faroese salmon fishermen. Similar arrangements were in place at West Greenland in 1993 and 1994. Under these arrangements the fishermen in these countries were paid not to fish the quotas agreed within NASCO. As a result of the permanent closure of the Northern Norwegian sea fishery, regulatory measures agreed by NASCO, and compensation arrangements, the proportion of the total North Atlantic catch taken in the distant water fisheries declined from an average of 21% in the 1970s to an average of only 4% in the 1990s.

Home Water Fisheries

Management measures introduced in home water fisheries partly for domestic reasons and partly under

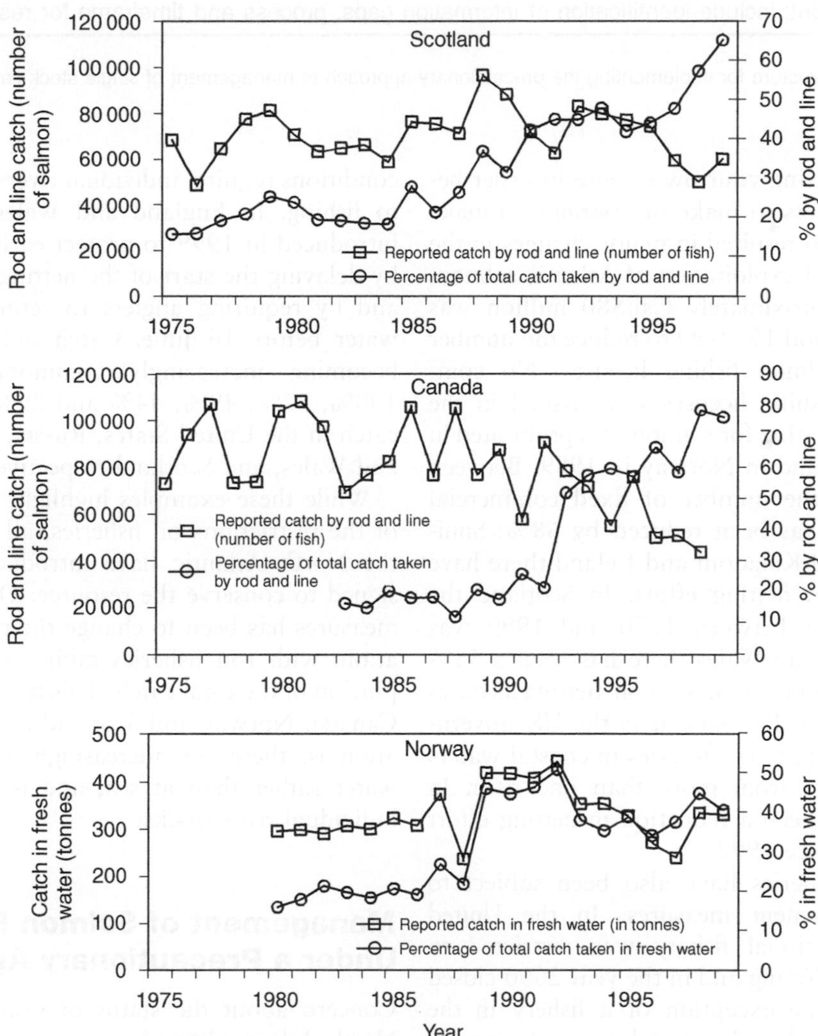

Figure 5 Reported catches (tonnes) of Atlantic salmon by rod and line (Scotland and Canada) or in fresh water (Norway) expressed as number or weight of fish and as a percentage of the total catch. Source of data: Scottish Rural Affairs Department, Edinburgh; Canadian Department of Fisheries and Oceans, Ottawa; Norwegian Directorate for Nature Management, Trondheim.

1. Is the stock threatened by external factors (e.g., acidification, disease)?
If yes, take special management action as appropriate (e.g., establish gene bank).
If no, go to (2).

2. Assess status of the stock (abundance and diversity)
(a) Have age-specific conservation limits been set?
(i) If yes, is the conservation limit being exceeded according to agreed compliance criteria (e.g., 3 out of 4 years)?
(ii) If no, assess other measures of abundance.
(b) Is the stock meeting other diversity criteria?

3. If either abundance or diversity are unsatisfactory, then seek to identify the reasons
(a) Immediately implement pre-agreed procedures to introduce appropriate measures to address reasons for failure (including stock rebuilding programs).
(b) Monitor the effect of the measures and take the results into account in future management and assessment; include identification of information gaps, process, and timeframe for resolution.

4. If both abundance and diversity are satisfactory
(a) Implement pre-agreed management actions to permit harvest of the surplus taking into account uncertainty (where appropriate use management targets to establish the exploitable surplus).
(b) Monitor the effect of the measures and take the results into account in future management and assessment; include identification of information gaps, process and timeframe for resolution.

Figure 6 Decision structure for implementing the precautionary approach to management of single stock salmon fisheries.

the process of 'putting your own house in order before expecting others to make or continue to make sacrifices' have also resulted in major changes in the level and pattern of exploitation of Atlantic salmon.

In Canada, approximately Can$80 million was invested in the period 1972–99 to reduce the number of commercial salmon fishing licenses. No commercial salmon fishing licenses were issued in the year 2000. Drift netting for salmon was prohibited in Scotland in 1962 and in Norway in 1989. Between 1970 and 1999, the number of fixed commercial gears in Norway has been reduced by 68%. Similarly, in the United Kingdom and Ireland there have been reductions in netting effort. In Scotland the reduction in effort between 1970 and 1999 was 83%. In England and Wales there has been a 53% reduction in the number of salmon netting licenses issued over the last 25 years. It is the UK government's policy to phase out fisheries in coastal waters that exploit stocks from more than one river. In Ireland, there has been a reduction in netting effort of at least 20% since 1997.

Recreational fisheries have also been subject to restrictive management measures. In the United States, the recreational fishery was restricted to catch-and-release fishing and in the year 2000 closed completely with the exception of a fishery in the Merrimack River based on releases of surplus hatchery broodstock. In Canada, daily and seasonal catch limits have been reduced, mandatory catch-and-release has been introduced, and where conditions require, individual rivers have been closed to fishing. In England and Wales, measures were introduced in 1999 to protect early running salmon by delaying the start of the netting season to 1 June and by requiring anglers to return salmon to the water before 16 June. Catch-and-release fishing is becoming increasingly commonplace. In 1999, 100%, 77%, 49%, 44% and 29% of the total rod catch in the United States, Russia, Canada, England and Wales, and Scotland respectively, was released.

While these examples highlight the severe nature of the restrictions on fisheries, all countries around the North Atlantic have introduced measures designed to conserve the resource. One result of these measures has been to change the pattern of exploitation, with rod fisheries taking an increasing proportion of the total catch. This trend is illustrated for Canada, Norway and Scotland in **Figure 5**. Exploitation is, therefore, increasingly occurring in fresh water rather than at sea, and is focused more on individual river stocks.

Management of Salmon Fisheries Under a Precautionary Approach

Concern about the status of salmon stocks in the North Atlantic has given rise to the adoption of a precautionary approach to salmon management by NASCO and its Contracting Parties. This approach, which will guide management of North Atlantic

1. **Identify river stocks that are available to the fishery**
2. **Identify stock components that are exploited by the fishery**
3. **Assess abundance and diversity of individual stocks contributing to the fishery**
4. **Are abundance and diversity satisfactory (consider the percentage of stocks that are unsatisfactory and the extent of failure for each stock)?**
   (a) If yes, go to (5).
   (b) If no, consider closing the fishery (taking into account socioeconomic factors). If the decision is made not to close the fishery, then continue to (5).
5. **Are the combined conservation limit(s) for all stocks subject to the fishery being exceeded?**
   (a) If yes, implement pre-agreed procedures for the management of the fishery based on effort or quota control:
   - *Quota control*
     – define management target based on an assessment of risk of failing conservation limits
     – predict prefishery abundance
     – determine exploitable surplus
     – apply pre-agreed rules on setting quotas
   - *Effort control* (and quota control in the absence of management targets and/or prediction of prefishery abundance)
     – evaluate effectiveness of previous effort control measures and apply appropriate changes.

   (b) If no, consider closing the fishery, taking into account socioeconomic factors. If the decision is made not to close the fishery, apply pre-agreed reserve measures to minimize exploitation.
6. **Monitor the effect of the measures and take the results into account in future management and assessment; include identification of information gaps, process, and timeframe for resolution**

Figure 7 Decision structure for implementing the precautionary approach to mixed stock salmon fisheries.

salmon fisheries in the twenty-first century, means that there is a need for caution when information is uncertain, unreliable, or inadequate and that the absence of adequate scientific information should not be used as a reason for postponing or failing to take conservation and management measures. The precautionary approach requires, *inter alia*, consideration of the needs of future generations and avoidance of changes that are not potentially reversible; prior identification of undesirable outcomes and of measures that will avoid them or correct them; and that priority be given to conserving the productive capacity of the resource where the likely impact of resource use is uncertain. A decision structure for the management of North Atlantic salmon fisheries has been adopted by NASCO on a preliminary basis. This decision structure is shown in **Figure 6** and 7 for single stock (i.e., exploiting salmon from one river) and mixed stock (i.e., exploiting salmon from more than one river) fisheries, respectively.

In short, salmon fisheries changed greatly in the last four decades of the twentieth century and the development of salmon farming had a marked effect on these fisheries. There is great concern about the future of the wild stocks and the fisheries continue to undergo critical re-examination.

See also

Salmonid Farming.

Further Reading

Anon (1991) *Salmon Net Fisheries: Report of a Review of Salmon Net Fishing in the Areas of the Yorkshire and Northumbria Regions of the National Rivers Authority and the Salmon Fishery Districts from the River Tweed to the River Ugie*. London: HMSO.

Ayton W (1998) *Salmon Fisheries in England and Wales*. Moulin, Pitlochry: Atlantic Salmon Trust.

Barbour A (1992) *Atlantic Salmon, An Illustrated History*. Edinburgh: Canongate Press.

Baum ET (1997) *Maine Atlantic Salmon. A National Treasure*. Herman, ME: Atlantic Salmon Unlimited.

Dunfield RW (1985) *The Atlantic Salmon in the History of North America*. Canadian Special Publication of Fisheries and Aquatic Sciences 80. Ottawa: Department of Fisheries and Oceans.

Hansen LP and Bielby GH (1988) *Salmon in Norway*. Moulin, Pitlochry: Atlantic Salmon Trust.

Jakupsstovu SHi (1988) Exploitation and migration of salmon in Faroese waters. In: Mills D and Piggins D (eds.) *Atlantic Salmon: Planning for the Future*, pp. 458–482. Beckenham: Croom Helm.

Mills DH (1983) *Problems and Solutions in the Management of Open Seas Fisheries for Atlantic Salmon*. Moulin, Pitlochry: Atlantic Salmon Trust.

Mills DH (1989) *Ecology and Management of Atlantic Salmon*. London: Chapman and Hall.

Møller Jensen J (1988) Exploitation and migration of salmon on the high seas in relation to Greenland. In: Mills D and Piggins D (eds.) *Atlantic Salmon: Planning for the Future*, pp. 438–457. Beckenham: Croom Helm.

NASCO (1991) *Economic Value of Atlantic Salmon*. Council document CNL(91)29. Edinburgh: North Atlantic Salmon Conservation Organization.

Taylor VR (1985) *The Early Atlantic Salmon Fishery in Newfoundland and Labrador*. Canadian Special Publication of Fisheries and Aquatic Sciences 76.

van Brandt A (1964) *Fish Catching Methods of the World*. London: Fishing News (Books) Ltd.

Vickers K (1988) *A Review of Irish Salmon and Salmon Fisheries*. Moulin, Pitlochry: Atlantic Salmon Trust.

Went AEJ (1955) *Irish Salmon and Salmon Fisheries*. London: Edward Arnold: The Buckland Lectures.

Williamson R (1991) *Salmon Fisheries in Scotland*. Moulin, Pitlochry: Atlantic Salmon Trust.

SALMON FISHERIES, PACIFIC

R. G. Kope, Northwest Fisheries Science Center, Seattle, WA, USA

Introduction

Pacific salmon comprise six species of anadromous salmonids that spawn in fresh water from central California in North America across the North Pacific Ocean to Korea in Asia: chinook salmon (*Oncorhynchus tshawytscha*), coho salmon (*O. kisutch*), sockeye salmon (*O. nerka*), chum salmon (*O. keta*), pink salmon (*O. gorbuscha*), and masu or cherry salmon (*O. masou*). Pacific salmon spawn in rivers, streams, and lakes where they die soon after spawning. Most juveniles migrate to the ocean as smolts, where they spend a significant portion of their life cycle. The length of freshwater and marine residence varies by species and the life span ranges from 2 years for pink salmon to as much as 7 or 8 years for some chinook salmon populations. Spawning runs of adult salmon have contributed an important source of protein for human cultures as well as a large influx of marine nutrients into terrestrial ecosystems. Large runs of mature fish returning from the sea every year have been highly visible to people living near rivers and salmon have historically assumed a role in the lives of people that extends beyond subsistence and commerce. Salmon became part of the social fabric of the cultures with which they interacted, and this significance continues today.

History

Salmon played an important role in the lives of people long before the arrival of Europeans on the Pacific rim. The predictable appearance of large runs of fish in the rivers emptying into the North Pacific Ocean provided a readily available source of high quality protein that could be harvested in large quantities and preserved for consumption in the winter when other sources of food were scarce. In the coastal areas of Washington, British Columbia, and south-east Alaska, Pacific salmon were a staple in the diets of tribal people and supported a level of human population density, commerce, and art unrivaled elsewhere on the continent. In Asia, salmon were harvested for subsistence by native people in Siberia, and the Japanese have harvested and dry-salted salmon in Kamchatka and Sakhalin at least since the seventeenth century.

When Europeans arrived on the Pacific rim, they were quick to take advantage of the abundant salmon. In the late eighteenth century Russian fur traders in Alaska caught and preserved salmon to provision native trappers. Distant markets developed for fresh, dried, salted, and pickled salmon, but the industry was hampered because the methods of preserving fish were inadequate, and spoilage was a recurrent problem. Commercial fisheries for Pacific salmon did not expand to an industrial scale in North America until the introduction of canning in the 1860s.

Pacific salmon were first canned on the Sacramento River in California in 1864. The canning industry then rapidly spread to the Columbia River, Puget Sound, British Columbia, and Alaska. Within 20 years canneries were established along the entire west coast from Monterey in California to Bristol Bay in Alaska, and commercial fisheries for Pacific salmon were conducted on an industrial scale. By the beginning of the twentieth century a few canneries had also been established in Asia, but the principle product produced in Asian salmon fisheries remained dry-salted salmon preferred by Japanese consumers.

With the advent of powered fishing boats in the early twentieth century, new fishing gears became prevalent and a trend developed to intercept salmon further and further from their natal streams. Two factors contributing to this trend were the possibility of harvesting salmon at times when local stocks were unavailable because of run timing or depletion, and the harvesting of fish before they became available to other fisheries. Off the coast of the United States troll fisheries targeting chinook and coho salmon developed (largely to avoid the closures imposed on river fisheries to protect stocks) and purse seine gear came into use in pink, chum, and sockeye fisheries. The Japanese began using drift gill nets which enabled the development of a high-seas mothership fleet off the coast of Kamchatka, and shore-based fisheries in the Kuril Islands.

Increasing catch of salmon stocks offshore led to a period, in the latter half of the twentieth century, of increasing international collaboration on research and management of Pacific salmon fisheries. One of the underlying principles of international management in the twentieth century has been that salmon belong to the countries in which they originate. Because of this principle, much of the research has focused on the migration and distribution of different stocks, and on methods to determine the origin of salmon encountered on the high seas. This principle of ownership has also encouraged the escalation of hatchery production.

Gear Types

Aboriginal fisheries traditionally used a variety of gear to harvest salmon. Weirs and traps were employed throughout North America and Asia and were probably the most common and efficient means of harvest. Spears and dipnets were also widely used. North American tribal fishermen also used hook and line, bow and arrow, gill nets, seines, and an elaborate gear called a reef net. The reef net was employed in northern Puget Sound to target primarily Fraser River sockeye salmon. It consisted of a rectangular net, suspended between two canoes in shallow water in the path of migrating salmon and was fished on a flood tide. Leads helped to direct fish into the net, which was raised when fish were seen swimming over it.

Commercial fisheries in Asia have primarily utilized traps in coastal waters, and seines and weirs in the rivers. In the early twentieth century the Japanese began using drift gill nets largely to avoid the uncertainty of retaining access to trap sites in Kamchatka. This allowed the Japanese to harvest salmon stocks originating from Kamchatka and the Sea of Okhotsk outside of Russian territorial waters and ultimately led to the development of the Japanese high-seas mothership fishery. Today traps remain the most prevalent and efficient gear used in Asian coastal fisheries.

In North America, early commercial fisheries primarily used haul seines and gill nets. Traps were employed very effectively in the Columbia River, Puget Sound, and Alaska, and fishwheels were also used on the Columbia River and in Alaska. Traps are no longer permitted in North America, and fishwheels are only used in Alaskan subsistence fisheries and as experimental gear in tribal fisheries in British Columbia. As internal combustion engines came into use by the fishing fleet, troll and purse seine gears were developed to target salmon in coastal waters.

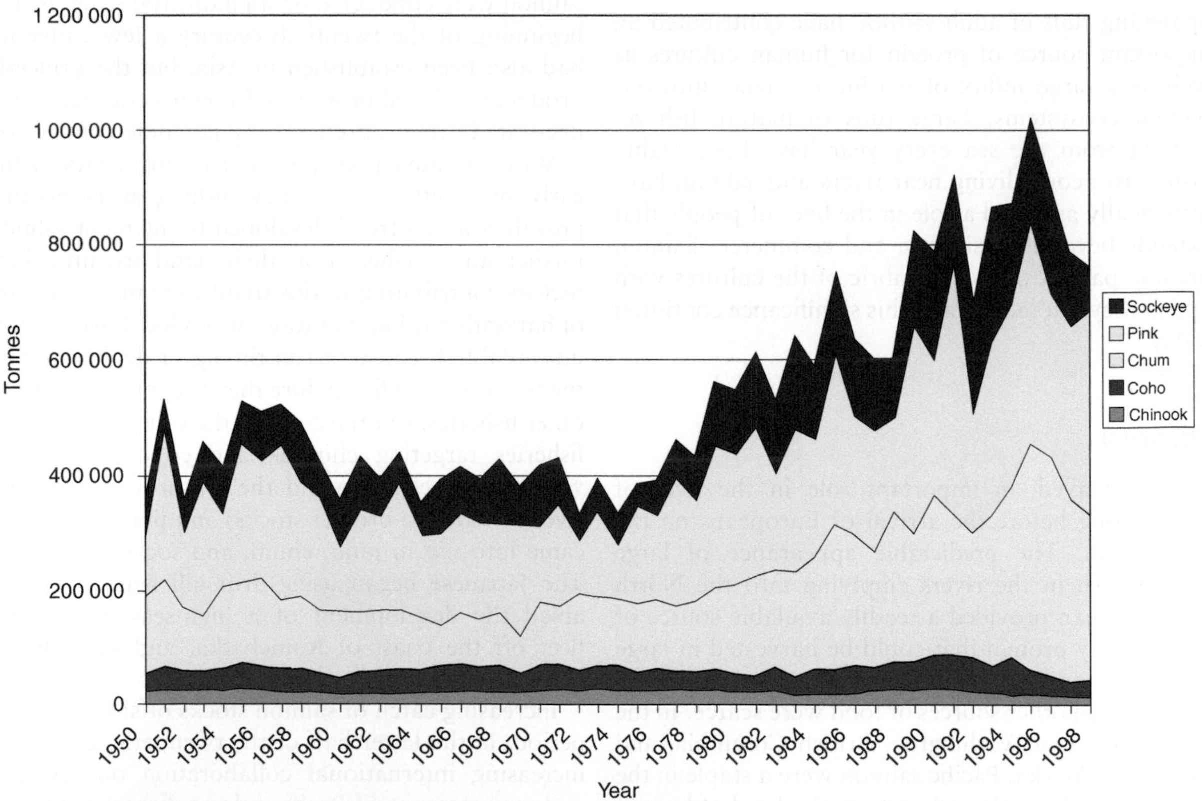

Figure 1 Commercial harvest of Pacific salmon in the North Pacific Ocean by species. Harvest includes both commercial capture fisheries and aquaculture. (Data from FAO Fishstat database.)

Catch

Since 1950, the catch of Pacific salmon has been dominated by pink and chum salmon, followed by sockeye, coho, and chinook salmon (**Figure 1**). Masu salmon are relatively uncommon, and spawn only in Japan and rivers entering the Seas of Japan and Okhotsk. They account for 1000–3000 tonnes per year, nearly all of that in Japan. Total harvest of Pacific salmon was relatively stable at around 400 000 tonnes during the 1950s, 1960s, and 1970s, but has increased dramatically since the 1970s, reaching a peak of more than 1 million tonnes in 1995. This has been due to increases in harvest of chum salmon in Japan, and in pink and sockeye salmon, primarily in Alaska (**Figure 2**). It is noteworthy that these three species have fundamentally different marine distributions to chinook and coho salmon. Chinook and coho salmon tend to have more coastal distributions while sockeye, pink, and chum undergo extensive marine migrations and have more offshore distributions. These differences in marine distributions have apparently contributed to differences in abundance trends in response to environmental changes like El Niño-Southern Oscillation (ENSO) events and regime shifts occurring on decadal scales.

Chinook Salmon

Harvest of chinook salmon in the North Pacific has been variable but has been relatively stable since 1950, at least until the 1990s (**Figure 3**). Until the last couple of years, total production has varied between 15 000 and 30 000 tonnes with the bulk of production coming from North America. Average annual total harvest from the North Pacific for the 5 years period from 1994 through 1998 has been 18 000 tonnes including production from aquaculture, with approximately 10 000 tonnes coming from capture fisheries.

Chinook salmon originate primarily from large river systems, with the Columbia, Fraser, Sacramento, and Yukon Rivers being the largest producers. Historically the majority of the chinook harvest came from the USA, but recent environmental conditions have impacted southern stocks and the harvest in Canada has surpassed that of the USA. Two pronounced dips in production, in early 1960s and 1980s, immediately followed strong ENSO events in 1959 and in the winter of 1982–83. The decline in production in the 1990s has also been associated with a series of ENSO events.

A trend which is not apparent in **Figure 3** is the shift from natural to hatchery production and

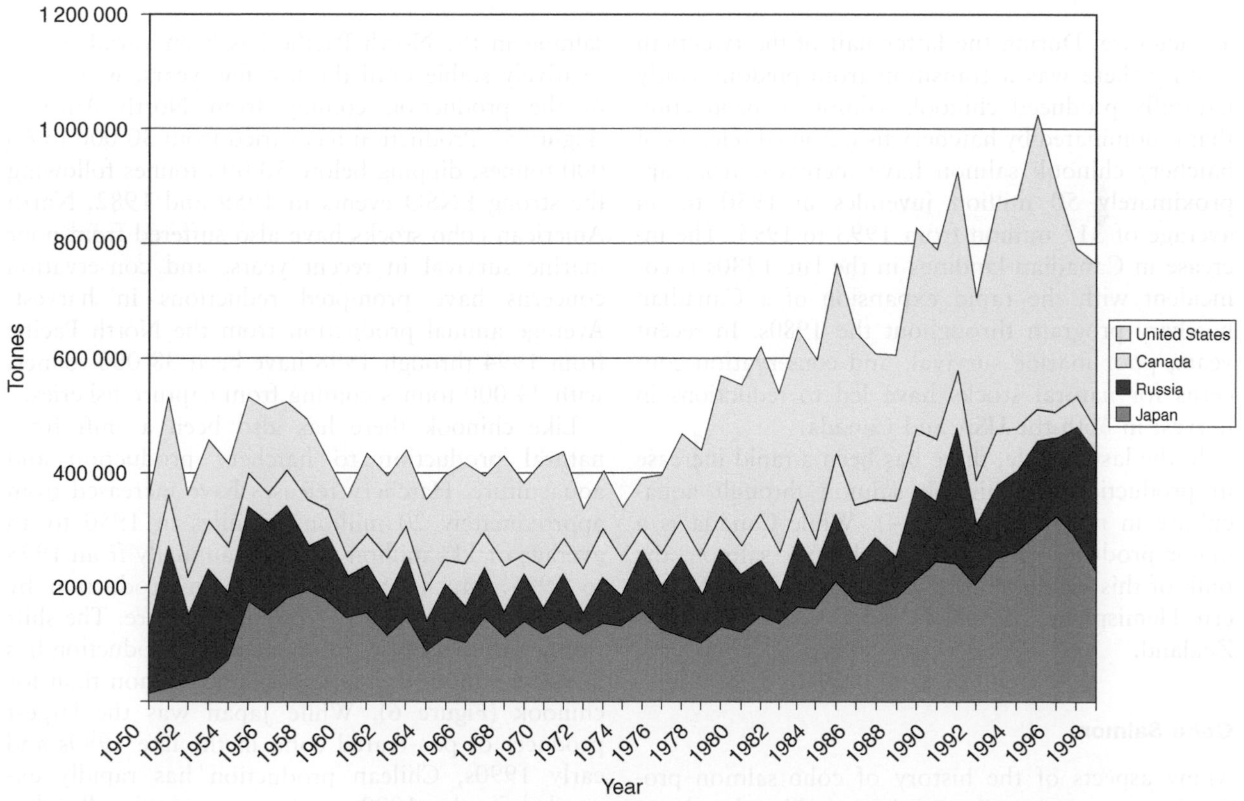

Figure 2 Commercial harvest of Pacific salmon by major fishing nations. (Data from FAO Fishstat database.)

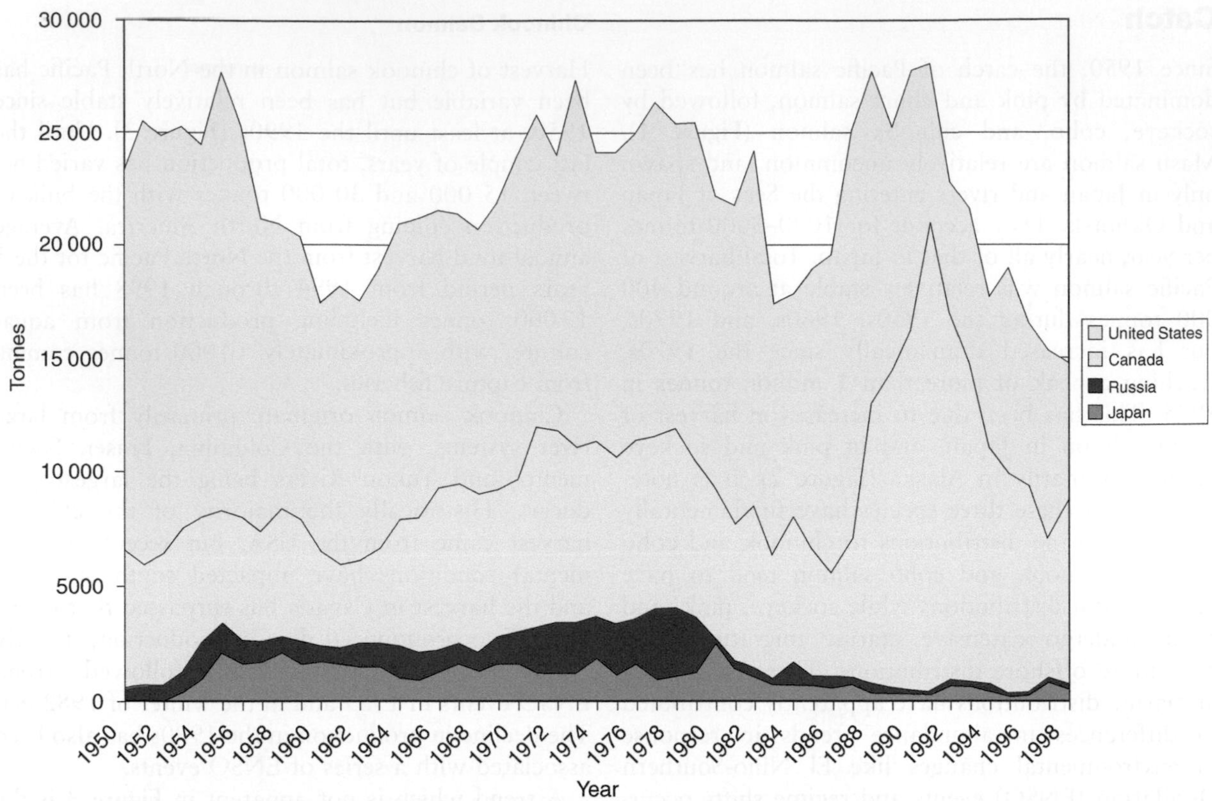

Figure 3 Commercial harvest of chinook salmon by major fishing nations in the North Pacific Ocean. Harvest includes both capture fisheries and aquaculture. (Data from FAO Fishstat database.)

aquaculture. During the latter half of the twentieth century there was a transition from predominantly naturally produced chinook salmon to production that is dominated by hatchery fish. Annual releases of hatchery chinook salmon have increased from approximately 50 million juveniles in 1950 to an average of 317 million from 1993 to 1995. The increase in Canadian landings in the late 1980s is coincident with the rapid expansion of a Canadian hatchery program throughout the 1980s. In recent years, poor marine survival, and conservation concerns for natural stocks have led to reductions in harvest in both the USA and Canada.

In the last decade, there has been a rapid increase in production of chinook salmon through aquaculture in net pens (**Figure 4**). While Canada is a major producer of pen-reared chinook salmon, the bulk of this production now comes from the Southern Hemisphere, primarily from Chile and New Zealand.

Coho Salmon

Many aspects of the history of coho salmon production are very similar to that of chinook salmon production. Like chinook salmon, harvest of coho salmon in the North Pacific has been variable, but relatively stable until the last few years, with most of the production coming from North America (**Figure 5**). Production has varied from 30 000 to 60 000 tonnes, dipping below 30 000 tonnes following the strong ENSO events in 1959 and 1982. North American coho stocks have also suffered from poor marine survival in recent years, and conservation concerns have prompted reductions in harvest. Average annual production from the North Pacific from 1994 through 1998 have been 38 000 tonnes, with 24 000 tonnes coming from capture fisheries.

Like chinook there has also been a shift from natural production to hatchery production and aquaculture. Hatchery releases have increased from approximately 20 million juveniles in 1950 to an average of 111 million juveniles annually from 1993 to 1995. Much of the increase in production by Japan in the 1980s was from aquaculture. The shift from capture fisheries to aquaculture production has been even more dramatic for coho salmon than for chinook (**Figure 6**). While Japan was the largest producer of pen-reared coho in the late 1980s and early 1990s, Chilean production has rapidly expanded in the 1990s and now exceeds all other production combined.

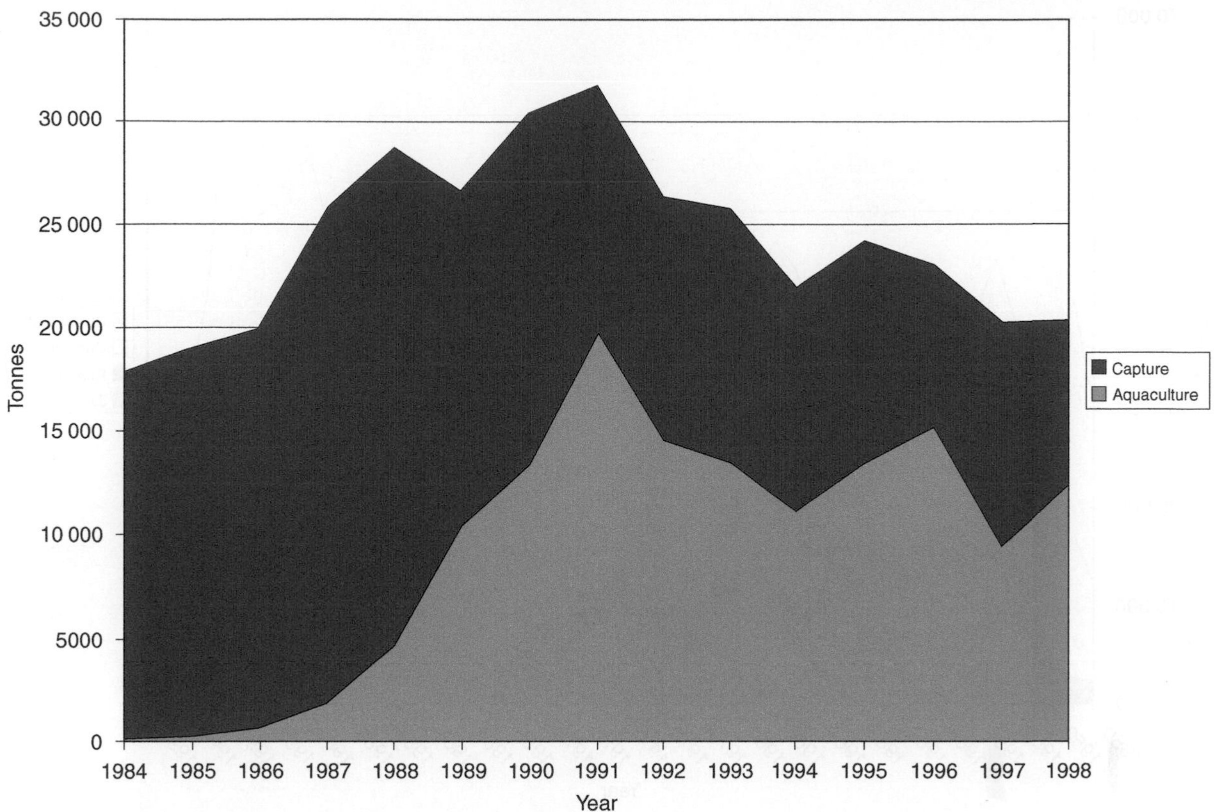

Figure 4 Total commercial production of chinook salmon in the Pacific basin. Most aquaculture production is in the Southern Hemisphere. (Data from FAO Fishstat database.)

Sockeye Salmon

Sockeye harvest was relatively stable in the 1950s, varying between 50 000 and 100 000 tonnes, but reached a low point in the 1970s of <40 000 tonnes (**Figure** 7). In the 1980s and 1990s, a combination of favorable marine conditions and harvest management that allowed sufficient spawning escapement led to dramatic increases in sockeye production in Alaska and the Fraser River in British Columbia. This has led to all-time record harvests in Alaska and the highest harvests in British Columbia since a blockage at Hell's Gate on the Fraser River devastated sockeye salmon runs in 1913 and 1914. The peak harvest exceeded 230 000 tonnes in 1993, and the average harvest from 1994 to 1998 was 148 000 tonnes annually.

Sockeye salmon have the most complex life history of any Pacific salmon. Most sockeye salmon rear in lakes for 1–3 years. They then migrate to the ocean where they spend from 1 to 4 years. Despite this variability, most spawning runs are dominated by fish of a single total age, usually either 4 or 5 years. The extended freshwater rearing period allows for interactions between successive year-classes within populations which contribute to cycles with periods of 4–5 years. While there is some evidence of cyclic dominance in the catch record, it is largely masked by aggregation of stocks in the fisheries. Differences in spawner abundance between peak and off-peak years in individual populations can be greater than an order of magnitude, and cycles can persist for decades.

Unlike chinook and coho salmon, most sockeye production is the result of natural spawning. While artificial production from 1993 to 1995 has averaged 326 million juveniles annually, 75% of this artificial production has been from natural spawning in artificial spawning channels.

Pink Salmon

Pink salmon are the most abundant species of Pacific salmon, but they also have the smallest average body size. Unlike other species of salmon, pink salmon all have a fixed 2 year life span. Because of this, the even-year and odd-year brood lines are genetically and demographically isolated, and tend to fluctuate independently with either the even-year or odd-year being dominant for long periods of time. In some streams only one of the brood lines is present. This is readily apparent in their harvest history (**Figure 8**),

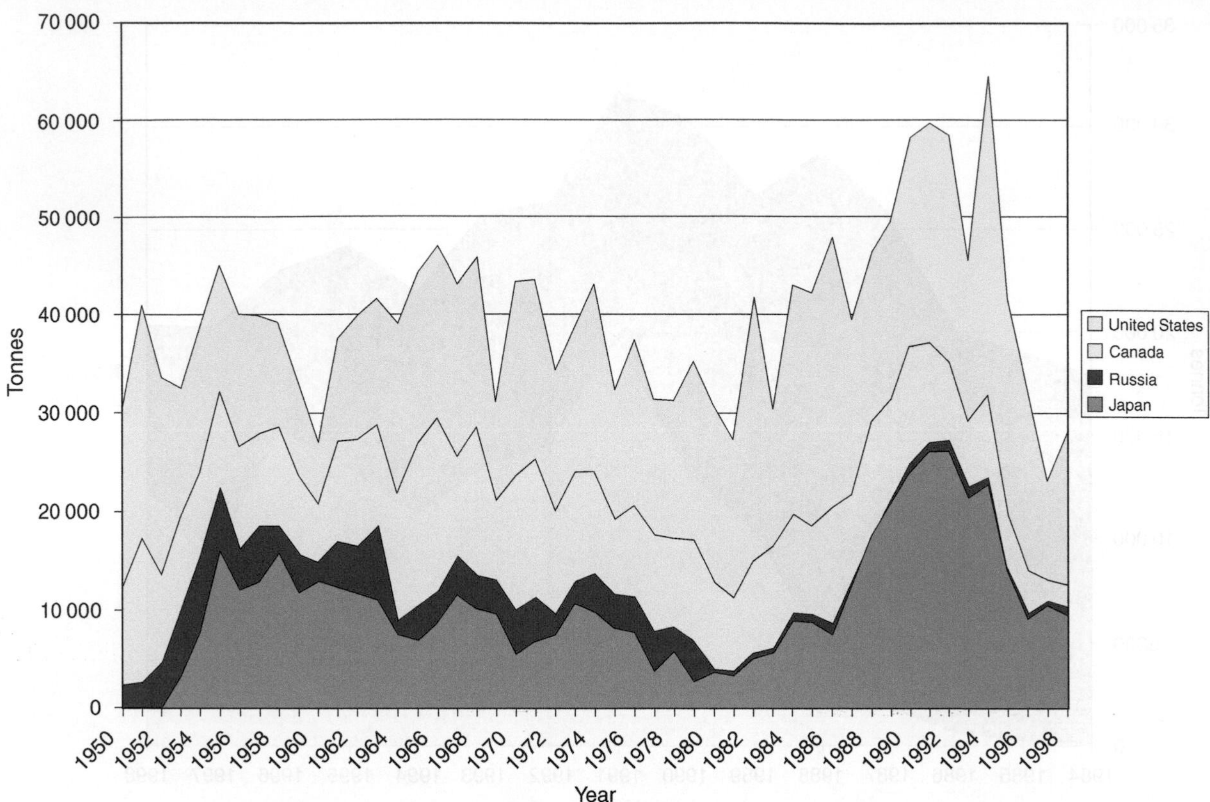

Figure 5 Commercial harvest of coho salmon by major fishing nations in the North Pacific Ocean. Harvest includes both capture fisheries and aquaculture. (Data from FAO Fishstat database.)

especially in Asian stocks where the odd-year brood line has been dominant.

The historical pattern of pink salmon harvest is similar to that of sockeye; landings were relatively stable in the 1950s and 1960s but declined slightly to reach a low in the early 1970s (**Figure 8**). Since the mid-1970s production has increased to recent record levels, reaching a peak harvest of >430 000 tonnes in 1991. This increase has occurred in Russian and Alaskan harvested while Canadian harvest has been stable and Japanese harvest has declined. The average annual harvest of pink salmon from the North Pacific from 1994 through 1998 was 327 000 tonnes.

Russia and Alaska are also where most pink salmon originate. While natural production is the major source of pink salmon, Alaska, Russia, and Japan have significant hatchery programs. Japan has had a long-standing hatchery program that has increased gradually, while Alaska did not begin hatchery production of pink salmon until the 1970s, but has expanded rapidly to surpass all others combined.

Chum Salmon

Like pink and sockeye salmon, the harvest of chum salmon declined in the 1950s and 1960s, and has increased to record levels in the 1990s (**Figure 9**). However, there is a fundamental difference between the production history of chum salmon and those of pink and sockeye salmon. North American production has been relatively stable and natural production in Asia has declined while there has been a large increase in Japanese hatchery production. The decline in Asian natural production is greater than is apparent from the catch history of Russian fisheries because the Japanese catch in the 1950s and 1960s was primarily from high-seas fisheries, while the recent Japanese catch has been predominantly from coastal fisheries targeting returning hatchery fish from Hokkaido and northern Honshu.

Issues

Hatcheries

Pacific salmon fisheries are unique among marine commercial and recreational fisheries in the scale of their dependence on artificial propagation. Because freshwater habitat is often the factor most limiting to salmon production, massive artificial propagation programs have been implemented for all species of

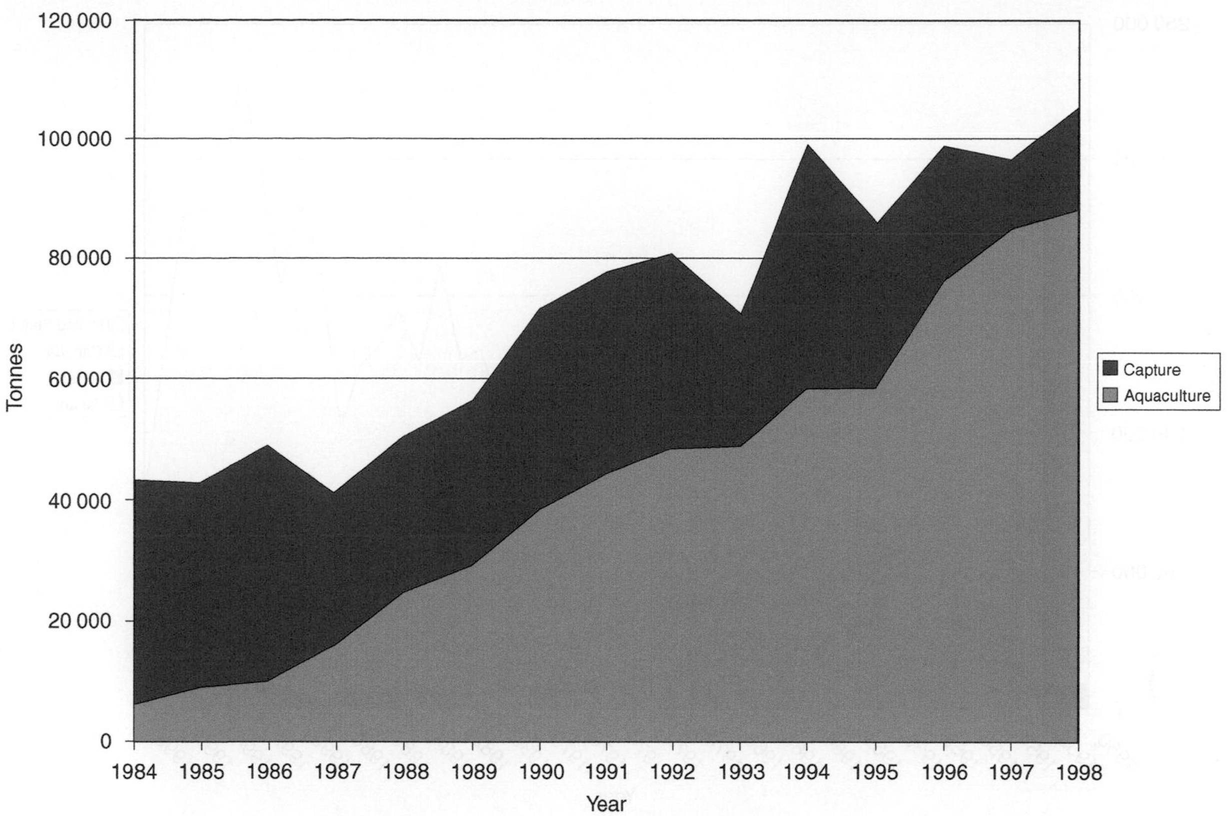

Figure 6 Total commercial production of coho salmon in the Pacific basin. Most aquaculture production is in the Southern Hemisphere. (Data from FAO Fishstat database.)

Pacific salmon to enhance fisheries, and as mitigation for freshwater habitat losses.

While hatcheries have augmented the abundance of salmon, there is increasing concern over potentially deleterious effects of hatcheries on natural salmon populations. Mixed stock fisheries targeting abundant hatchery stocks can overharvest less productive natural stocks that are intermingled with the hatchery fish. Hatchery stocks often differ genetically from the local natural stocks because the original broodstock for the hatchery may have been obtained from a nonlocal population or hatchery, hatchery breeding practices which unintentionally exert selection pressure for particular traits, and because of domestication through adaptation to hatchery rearing conditions. Stray spawners from the hatcheries can make natural stocks appear more productive than they really are, and interbreeding between natural and hatchery stocks can further reduce the productivity of natural stocks through outbreeding depression.

While artificial propagation is widespread, the focus has been on different species in different areas.

In the contiguous United States, the focus of artificial propagation has been on chinook and coho salmon. Hatchery programs exist in most rivers that support chinook and coho populations, and hatchery releases from 1993 through 1995 have averaged approximately 71 million coho and 252 million chinook annually. Alaska and Canada also have hatchery programs, but their combined releases in the same period have averaged about 38 million coho salmon and 64 million chinook salmon annually. Hatchery fish account for the majority of the harvest of these two species in the contiguous United States.

Canada has the largest program for artificial production of sockeye salmon, but the majority of this is in the form of artificial spawning channels in the Fraser River system. In these artificial channels adults spawn naturally and juveniles emigrate to lakes for rearing without being artificially fed. This program has developed rapidly since 1960 and the 1993–95 average annual production was approximately 244 million fry. Alaska has the largest hatchery program for sockeye smolts with annual production that has averaged approximately 69 million.

Alaska also has the largest hatchery program for pink salmon. This program has expanded rapidly since the 1970s. Annual releases of hatchery fry

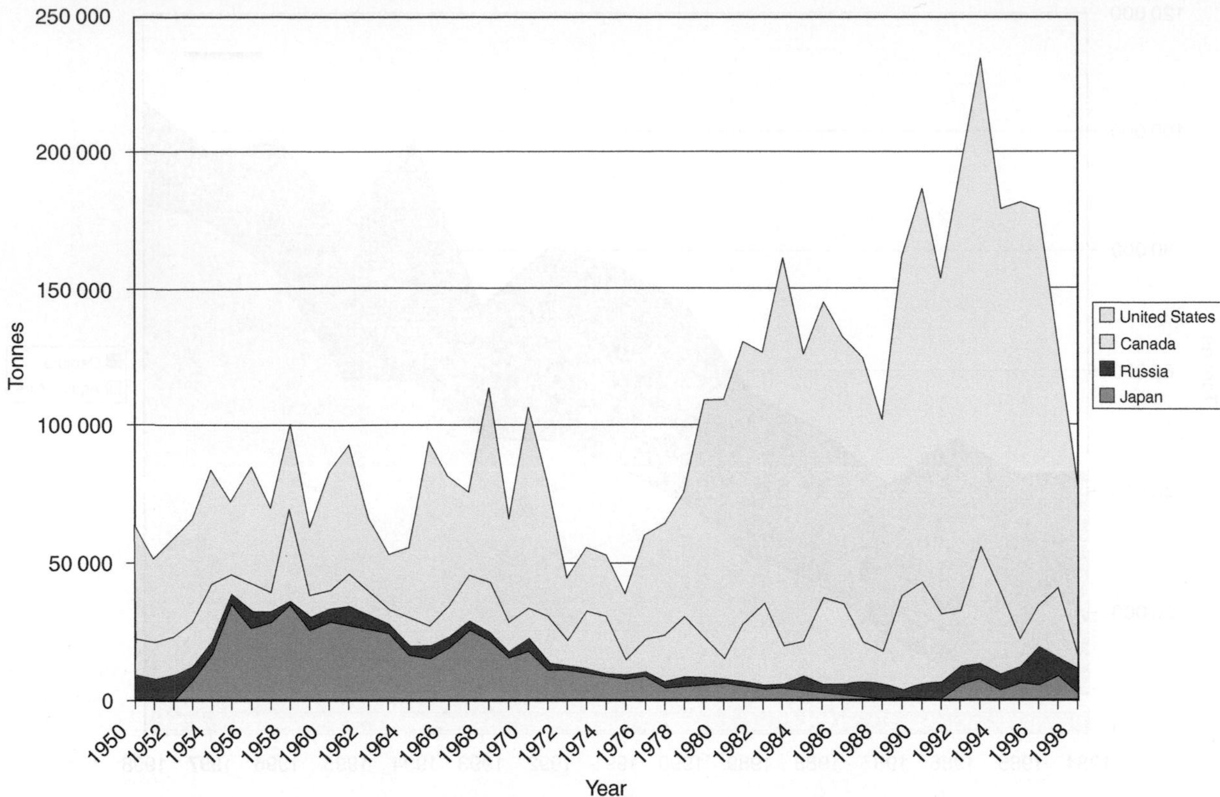

Figure 7 Commercial harvest of sockeye salmon by major fishing nations. (Data from FAO Fishstat database.)

averaged 843 million from 1993 to 1995, and returns from this program have contributed substantially to the increases in landings of pink salmon in recent years. Over the same period, Japanese hatcheries have released an annual average of 132 million, and Russian hatcheries, 264 million pink salmon fry.

The largest Pacific salmon hatchery program is operated by the Japanese for chum salmon. Japan's chum salmon hatchery program dates back to 1888 when the national salmon hatchery was built in Chitose on Hokkaido. This program released 200 000–500 000 juveniles annually in the 1950s, but expanded rapidly in the 1970s. From 1993 to 1995 Japanese hatcheries released an average of more than 2 billion chum salmon fry annually. During the same period, Canada and the Russian Federation each released an average of 220 million hatchery chum salmon annually, and annual releases from US hatcheries have averaged more than 500 million hatchery chum salmon fry. Combined, these programs have been releasing nearly 3 billion chum salmon juveniles per year, nearly 60% of the hatchery production of all Pacific salmon combined.

The recent and rapid increase in hatchery production of salmon, coupled with the increase in total landings and a concurrent decline in the average size of individuals of all salmon species has raised concerns, and prompted research into the possibility that ocean carrying capacity is currently limiting production of salmon in the North Pacific Ocean.

International Management

One of the challenges in management of Pacific salmon fisheries is the direct result of their anadromous life history. Although stocks are genetically distinct, and segregate at spawning, they undergo extensive marine migrations and co-mingle in the ocean. This characteristic makes stocks vulnerable to interception in foreign fisheries. Countries harvesting Pacific salmon have long recognized the need for international coordination of harvest management, but cooperation has been an ongoing challenge complicated by international relationships that have been less than cordial at times.

Japanese have fished for salmon along the coasts of Sakhalin, Kamchatka, and the Kuril Islands at least since the early seventeenth century. In the early nineteenth century, Russia levied a tariff on salmon exported to Japan, and in the 1890s Russia began to restrict Japanese access to trap sites. Through the

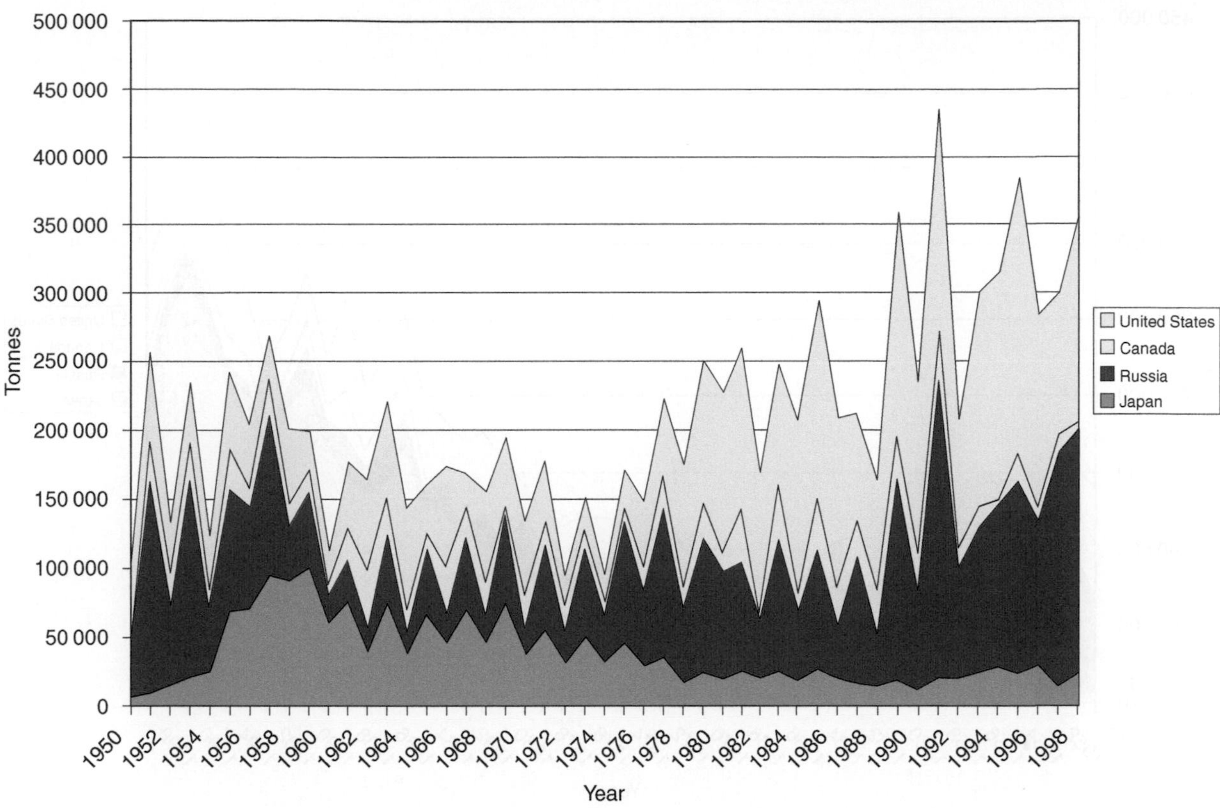

Figure 8 Commercial harvest of pink salmon by major fishing nations. (Data from FAO Fishstat database.)

early part of the twentieth century relations between Japan and Russia became increasingly strained and the Japanese fishing industry generally had increasing difficulty accessing traditional shore-based trap sites in Russia. This encouraged the development and expansion of Japanese high-seas drift gill net fisheries in the western Bering Sea and southern Sea of Okhotsk and of shore-based trap and gill net fisheries in the Kuril Islands targeting salmon stocks originating primarily from the Sea of Okhotsk. During World War II Japan lost all access to trap sites in the Soviet Union, and the Japanese high-seas fishery ceased. At the end of the war, the Soviet Union took possession of the Kuril Islands and Japanese fisheries based in the Kuril Islands ceased as well.

In 1952, the governments of Canada, Japan, and the USA signed the International Convention for the High-Seas Fisheries of the North Pacific Ocean which established the International North Pacific Commission (INPFC). Under the terms of this Convention, Japan resumed and rapidly expanded their high-seas salmon fishery in the northern Pacific Ocean, Bering Sea, and the southern portion of the Sea of Okhotsk. The primary management action of the INPFC was to set an eastern limit on the extent of high-seas fishing, but it also embarked on a large-scale research program directed at addressing issues relevant to the Commission with much of the focus on stock identification, distribution, and migration patterns. Because the Japanese high-seas fishery targeted Asian stocks, mostly from the Soviet Union, the Soviet Union responded by excluding the Japanese high-seas fishery from the Sea of Okhotsk and a large portion of the western Bering Sea. This action led to the negotiation of a Convention between Japan and the USSR in 1957 which established the Soviet–Japan Fisheries Commission to regulate the harvest of Asian salmon stocks on the high-seas.

In the late 1970s, leading up to adoption of the United Nations Convention of Law of the Sea in 1982, the USA, Canada, and the Soviet Union established 200 mile fishery zones off their coasts. This further reduced the areas accessible to high-seas fisheries. In 1993 the North Pacific Anadromous Fisheries Commission (NPAFC) replaced the INPFC with Japan, the Russian Federation, Canada, and the USA as members. Between the NPAFC Convention, which prohibits high-seas salmon fishing and trafficking in illegally caught salmon, and the United Nations General Assembly Resolution 46/215, prohibiting large-scale pelagic drift gill netting, high-seas

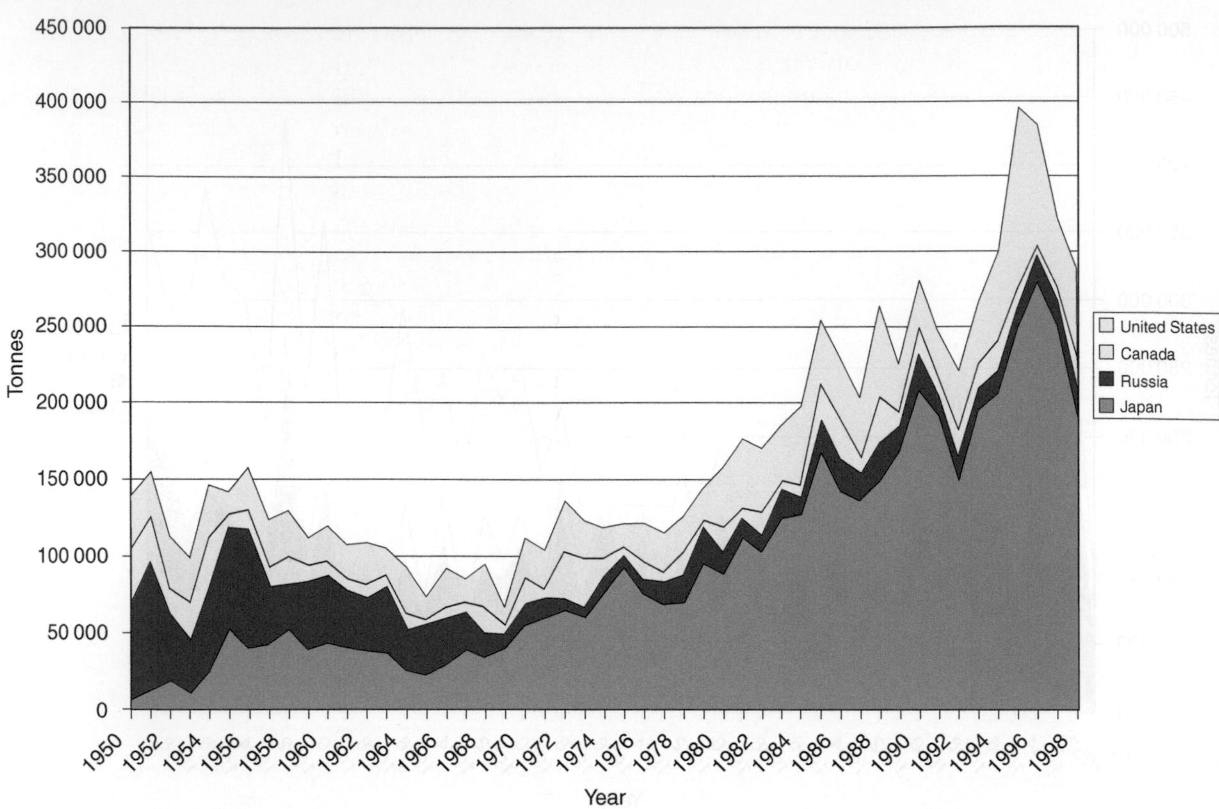

Figure 9 Commercial harvest of chum salmon by major fishing nations. (Data from FAO Fishstat database.)

fisheries for Pacific salmon have been eliminated except for illegal fishing.

In North America, much of the focus of international management has been on the Fraser River run of sockeye salmon. This was the largest run of sockeye in North America, and one of the most profitable for the canning industry in the late nineteenth century and early twentieth century. Returning adult Fraser River sockeye have two migratory approaches to the Fraser River: the southern approach through the Strait of Juna de Fuca of the south of Vancouver Island where they are vulnerable to American fisheries, and the northern approach through the Johnstone and Georgia Straits which lies entirely within Canadian waters. In most years the majority of the run has taken the southern approach where it was intercepted by US fisheries, causing persistent friction between the two countries.

The need for cooperative management of the fisheries in these southern boundary waters was recognized by both countries. Beginning in the 1890s, a series of commissions were formed to study the problem and make recommendations. In 1913 the Fraser River was obstructed by debris from railroad construction at Hell's Gate in the Fraser River Canyon. This passage problem was exacerbated by a rock slide caused by the construction in 1914 which effectively destroyed the largest, most productive sockeye run in North America. This had devastating impacts on inland native tribes who depended on salmon as a dietary staple, and had lasting impacts on the commercial fisheries as well. The desire to restore Fraser River sockeye production increased the incentive for cooperation, but it has been a long and difficult process.

Bilateral treaties were signed in 1908, 1919, and 1929, but enabling legislation was never passed by the US Congress. The 1929 treaty was amended and again signed in 1930 but was not ratified until 1937, 45 years after negotiations had started. The treaty established the International Pacific Salmon Fisheries Commission (IPSFC) to study and restore Fraser River sockeye salmon, and manage sockeye fisheries in Convention waters of the southern approach to the Fraser River, but postponed any regulatory authority for another 8 years. The IPSFC constructed many fish passage structures in the Fraser River basin, and its authority was extended to include pink salmon with the addition of a pink salmon protocol in 1957.

In 1985 a new treaty was signed and ratified that replaced the IPSFC with the Pacific Salmon

Commission (PSC). The PSC has expanded authority, which includes all five species of Pacific salmon harvested in North America, and encompasses marine boundary waters of the northern and southern Canada–US borders, and transboundary rivers passing through Canada and south-east Alaska. However, the effectiveness of the PSC has been hampered by the inability to come to unanimous agreement on allocation issues in some years.

Endangered Species

Because of their dependence on freshwater habitat, anadromous fish are vulnerable to habitat degradation resulting from human activities in inland areas. The requirements of Pacific salmon for spawning gravel free of fine sediments and cold oxygenated water for spawning and juvenile rearing make them particularly vulnerable to habitat loss and degradation from nearly all human activities where salmon and people co-occur. Construction of impassable dams for flood control, hydropower, and irrigation has eliminated access to much historic habitat. Stream channelization and levee construction for flood control has reduced habitat availability and complexity. Logging has resulted in scoured stream channels, increased runoff and sediment loads, and has diminished the availability of shade and large woody debris which provides rearing habitat. Mining has directly destroyed stream channels and choked streams with sediment, as well as contaminating the water with heavy metals. Water diversions for agricultural, industrial, and domestic use have reduced the water available and directly removed juvenile salmon. Grazing has increased sedimentation and destroyed streamside vegetation. Home construction and urbanization have contributed to sedimentation and chemical pollution, and have increased the amount of impervious surface, increasing the variability in stream flow.

Collectively these impacts have eliminated many populations of Pacific salmon and have compromised the productivity of many remaining ones. Reduced productivity of natural stocks has increased their susceptibility to overharvest, and the construction of hatcheries to mitigate the impacts of water development projects and enhance fisheries has often exacerbated the problem by increasing competition between natural and hatchery fish and increasing the harvest pressure on all fish in the attempt to harvest hatchery fish. As a result, many natural populations are at critically low abundance where they are at higher risk of extirpation from random environmental and demographic variability or from catastrophic events. As a result, a number of distinct population segments of Pacific salmon and steelhead trout in the contiguous United States have been listed as threatened or endangered under the US Endangered Species Act. At the time of writing (2000) the listings include 17 distinct population segments of chinook, coho, chum, and sockeye salmon, and another 10 distinct population segments of steelhead trout. The additional regulatory complexities of dealing with listed species has greatly complicated the management of Pacific salmon fisheries in the USA.

In response to these listings, and critically low abundance of some Canadian stocks, harvest impacts on depressed stocks have been substantially reduced in the contiguous United States and British Columbia. Efforts are being made to reduce the combined negative impacts of habitat loss and degradation, overharvest, and the negative impacts of hatchery production. Fishery scientists and managers are exploring changes in harvest practices to allow more selective harvest of hatchery stocks and healthy natural stocks while reducing impacts on listed stocks.

See also

Ocean Ranching. Salmonid Farming.

Further Reading

Cobb JN (1917) *Pacific Salmon Fisheries*. Bureau of Fisheries Document No. 839. Washington, DC: Government Printing Office.

Groot C and Margolis L (eds.) (1991) *Pacific Salmon Life Histories*. Vancouver, BC: UBC Press.

McNeil WJ (ed.) (1988) *Salmon Production, Management and Allocation*. Corvallis, OR: Oregon State University Press.

NMFS (1999) *Our Living Oceans. Report on the Status of US Living Marine Resources, 1999*. US Department of Commerce, NOAA Technical Memo. NMFS-F/SPO-41.

Shepard MP, Shepard CD, and Argue AW (eds.) (1985) *Historic Statistics of Salmon Production Around the Pacific Rim*. Canadian Manuscript Report of Fisheries and Aquatic Sciences No. 1819. 297 pages.

SMALL PELAGIC SPECIES FISHERIES

R. L. Stephenson, St. Andrews Biological Station, St. Andrews, New Brunswick, Canada
R. K. Smedbol, Dalhousie University, Department of Oceanography, Halifax, Nova Scotia, Canada

Introduction

The so-called 'pelagic' fish are those that typically occupy the midwater and upper layers of the oceans, relatively independent of the seabed. Small pelagic species include herrings and sprats, pilchards and anchovies, sardines, capelin, sauries, horse mackerel, mackerels, and whiting. Most of these fish have a high oil content and are characterized by a strong tendency to school and to form large shoals. These features have contributed to the development of large fisheries, using specialized techniques, and to a variety of markets for small pelagic species.

Two of the major pelagic species (herring and capelin) are found in polar or boreal waters, but most are found in temperate or subtropical waters. The sardine, anchovy, mackerel, and horse mackerel are each represented by several species around the world, with the largest concentrations being found in the highly productive coastal upwelling areas typically along western continental coasts. Intense fisheries have developed on a few small pelagic species, which occur in very dense aggregations in easily accessible areas near the coast. In recent years the largest fisheries have been on Atlantic herring, Japanese anchovy and Chilean Jack mackerel, although the dominance of individual fisheries has changed as the abundance of these species has fluctuated due to both natural factors and intense fishing. Small pelagic fish species with recent annual landings exceeding 100 000 t are listed in **Table 1**. Over the past three decades, fisheries for small pelagic species have made up almost half the total annual landings of all marine fin-fish species (**Figure 1**).

Catch Techniques in Pelagic Fisheries

Pelagic fish, by definition, are found near the ocean surface or in middle depths. As a result, pelagic fisheries must search larger volumes than demersal fisheries. However, most pelagic fish species exhibit behaviors that increase their catchability. The most import characteristic is shoaling, in which individuals of the same species form and travel in aggregations. Several pelagic species also exhibit clear patterns of vertical migration, often staying deep in the water column, or near bottom, by day but migrating to surface waters at dusk. In some cases fishers have used techniques such as artificial light sources to enhance shoaling behavior and improve fishing.

Pelagic fish shoals can be very large, which greatly increases their detectability. Surface shoals can be located visually, using spotters from shore or aboard vessels, and aerial searches are used in the fishery for some species. The location and depth of shoals can be determined using hydroacoustics, which has developed greatly in association with pelagic fisheries in recent decades. Echo sounders and sonars transmit an acoustic signal from a transducer associated with the vessel and receive echoes from objects within the path of the beam such as fish and the seafloor. When sector scanning and multibeam sonars are mounted on vessels within a moveable housing, they can be used to search the water column ahead and to the sides of the vessel. This permits improved detection of schools and targeting of shoals throughout the water column.

Methods for catching small pelagic fishes range from very simple to highly sophisticated. This section deals only with those catch methods and specialized gear that have been developed specifically for pelagic fisheries, and are widely used around the globe. In general, these methods can be placed in several broad categories:

1. static gear (trap, lift net and gill net)
2. towed gear (trawl net)
3. surrounding gear (purse seine, ring net and lampara net).

Static Gear (Trap, Lift Net and Gill Net)

The earliest fisheries for small pelagic species relied on shoals of fish encountering and becoming entrapped in nets that were fixed to the bottom or floating passively. A variety of forms of trap nets have been used historically where small pelagic species such as herring and mackerel occur near shore. Typically these nets have a lead and wings which intercept and direct a school of fish into a pot or pound where they remain alive until dipped or seined from the net. In the Bay of Fundy and Gulf of Maine, for example, juvenile herring have been caught in

Table 1 Main species of pelagic fish with world catches greater than 100 000 tonnes in 1998. Catch information from FAO 1999

Order	*Family*	*Species*		*Catch (10^3tonnes)*
Beloniformes	Scomberesocidae	Pacific saury	*Coloabis saira*	181
Clupeoidei	Clupeidae	Gulf menhaden	*Brevoortia patronus*	497
		Atlantic menhaden	*Brevoortia tyrannus*	276
		Atlantic herring	*Clupea harengus*	2419
		Pacific herring	*Clupea pallasi*	498
		Bonga shad	*Ethmalosa fimbriata*	157
		Round sardinella	*Sardinella aurita*	664
		Goldstripe sardinella	*Sardinella gibbosa*	161
		Indian oil sardine	*Sardinella longiceps*	282
		California pilchard	*Sardinops caeruleus*	366
		Japanese pilchard	*Sardinops melanostictus*	296
		Southern African pilchard	*Sardinops ocellatus*	197
		South American pilchard	*Sardinops sagax*	937
		Araucanian herring	*Strangomera bentincki*	318
	Engraulidae	Pacific anchoveta	*Cetengraulis mysticetus*	181
		European anchovy	*Engraulis encrasicolus*	492
		Japanese anchovy	*Engraulis japonicus*	2094
		Peruvian anchoveta	*Engraulis rigens*	1792
		European pilchard	*Sardina pilchardus*	941
		European sprat	*Sprattus sprattus*	696
Gadiformes	Gadidae	Southern blue whiting	*Micromesistius australis*	184
		Blue whiting	*Micromesistius poutassou*	1191
Percoidei	Carangidae	Cape horse mackerel	*Trachurus capensis*	184
		Japanese jack mackerel	*Trachurus japonicus*	341
		Chilean jack mackerel	*Trachurus murphyi*	2056
		Atlantic horse mackerel	*Trachurus trachurus*	388
Salmonoidei	Osmeridae	Capelin	*Mallotus villosus*	988
Scombroidei	Scombridae	Indian mackerel	*Rastrelliger kanagurta*	284
		Chub mackerel	*Scomber japonicus*	1910
		Atlantic mackerel	*Scomber scombrus*	657
		Japanese Spanish mackerel	*Scomberomorus niphonius*	552
Scorpaeniformes	Hexagrammidae	Atka mackerel	*Pleurogrammus azonus*	344

heart-shaped traps known locally as weirs (made of fine mesh covering poles driven into the seabed), or even trapped in coves by closing off the entrance with a seine. These catches have formed the basis of the canned sardine industry for over a century.

Unlike trap nets, a lift net functions as a moveable container that traps the target fish. The shoal swims over the net, and the fish are caught by lifting the net and thereby trapping the shoal within. Fish are then usually removed by dipnet. This type of gear is used commonly in South-east Asia. The nets may be operated from shore or from specialized vessels. These vessels often use bright lights to lure the target shoals over their lift nets.

The gill net catches fish by serving as a barrier or screen through which fish larger than the mesh size cannot pass, but become entangled, usually by their gills, and are removed when the net is retrieved. The gill net is constructed of a section of fine netting with a border of stronger netting. The bottom of the net is weighted with a leadline, and the top or headline has floats so that the net hangs vertically in the water. Fishing typically involves a string of nets that can be set to a desired depth. Gillnets may be moored in a specific location (set nets) or may be allowed to float with the prevailing winds and currents (drift nets).

Towed Gear (Trawl Net)

The midwater trawl was developed specifically to capture pelagic species. Like the bottom trawl commonly used in demersal fisheries, it is towed behind the vessel. Trawling may be undertaken from either single or paired vessels, although trawling by one vessel is much more common. The net is set to the depth for a particular, targeted school of fish, and is towed only long enough to pass through (or to encompass) the school.

The trawl net is usually conical in form and may be very large. Typically, the net mouth, which may be circular or square, is held open during a tow by doors which spread the wings of the net, together

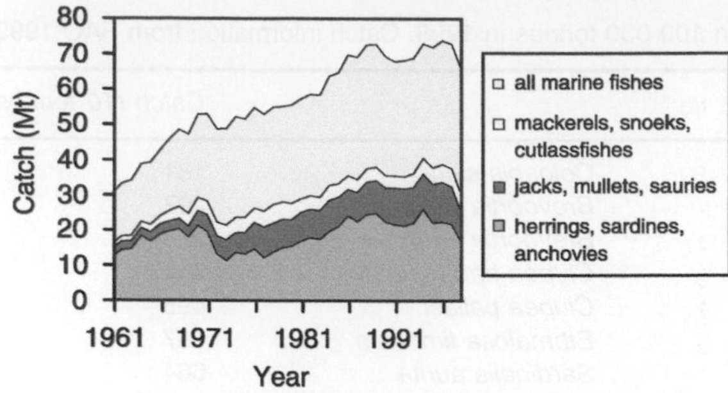

Figure 1 World catch (million tonnes, Mt) of groups of small pelagics in relation to the total world catch of all marine fish species. Catch information from FAO (1999).

with the buoyancy of floats placed on the headline, and a weighted footrope attached to the bottom of the net mouth. A transducer is placed on the headline so that the depth of the net can be monitored during a tow, and the school of fish can be seen passing the mouth into the net. Additional sensors on the trawl provide information about the state of the gear and potential catch.

Surrounding Gears (Purse Seine, Ring Net and Lampara Net)

The characteristic of pelagic fishes to occur in large shoals, right at the surface, makes them vulnerable to capture with surrounding gears. The purse seine is used worldwide and accounts for an appreciable fraction of the total annual catch by global fisheries. In general, this fishing method involves setting out a long net hanging vertically in the water to surround a school of fish. The top of the net usually floats at the surface. Following envelopment of the school, the bottom of the net is pulled together to trap the fish in a cup or 'purse' of netting. The net is hauled gradually, such that the purse shrinks in size until the school of fish is alongside the vessel. The fish are then brought aboard using pumps or a lifting net. The seine may be very large, up to 730 m long and 180 m deep. The main area of the net is usually of constant mesh size and material. Only the section of the gear wherein the catch will be concentrated during retrieval is strengthened. Purse seining usually targets fish found at the surface to a depth of about 130 m.

Purse seining may use one or two vessels. The gear is set beginning with one end and the vessel deploys the gear as it slowly steams around the fish school. When two vessels are used, the tasks of setting and retrieving the seine are shared between the vessels, and each vessel carries part of the gear. During deployment, each vessel sets the gear, beginning with the middle of the seine net, and then they proceed to envelope the school. Most often retrieval is with the aid of a power block.

A variation of the purse seine is the ring net. This encircling net is generally smaller than a purse seine and is used in relatively shallow waters. Ring net gear is usually used by pairs of vessels in the 12–19 m range.

A third type of surrounding gear is the lampara net. Although similar in shape and use to a purse seine, the bottom of this net is not pursed during retrieval. The float line on a lampara net is much longer than the weighted bottom line, which causes the middle portion of the net to form a bag. When the net is hauled, both sides (wings) of the gear are retrieved evenly. The leadlines meet, effectively closing off the bag of the net. The bag is brought alongside the vessel and the catch transferred aboard.

Products Derived from Pelagic Fisheries

Some small pelagic species (notably Atlantic herring) have been important as food, and have sustained coastal communities for centuries. Principal products (those used for human consumption) are listed in Table 2. The main reason for the underutilization of small pelagics as food is a lack of consumer demand. The development of new products and uses may increase this demand. One such innovation is the use of small pelagics as raw material in processed seafoods such as surimi.

More than half (approximately 65% during the period 1994–1998) of the harvest of small pelagic species in recent years has not been used directly for food, but rather has been processed into a variety of by-products. Some of these by-products, such as fish protein concentrate, may then be used in the production of foodstuffs for human consumption. The main by-products of the pelagic fisheries are:

1. fish meal and oil
2. fertilizer
3. silage and hydrolysate
4. compounds for industrial, chemical and pharmaceutical products.

Fish Meal and Fish Oil

Fish meal and fish oil are the most important by-products produced from pelagic fish. Meal is used mainly as a protein additive in animal feeds in both agriculture and aquaculture. During the 1980s the annual world production of fish meal was in the order of 6–7 million tonnes (Mt), requiring landings of 35–40 Mt of whole fish. Production has declined

Table 2 World production of the main primary products derived from small pelagics

	Live-weight equivalent of product (10³ tonnes)			
Product	*1995*	*1996*	*1997*	*1998*
Fresh or chilled	836	948	950	1084
Frozen	6138	6935	7481	6911
Prepared or preserved	1759	1743	2072	2013
Smoked	61	66	65	63
Dried or salted	464	433	391	360
Fats and oils of fish	3399	3090	2732	1709
Fish meal fit for human consumption	162	118	182	140

Source: FAO.

somewhat during the mid 1990s due to fluctuations in abundance of key target species.

Fish meal and oil are usually produced in tandem. The fish is cooked and then pressed to separate the solids (presscake) and liquids (pressliquor). After separation of the oil from the pressliquor, the rest of the soluble components are remixed with the presscake and this mixture is dried to form the finished solid product.

Fish oil is an important by-product of the production process for fish meal. Annual production of fish oil is approximately 1.5 Mt, making up 1–2% of the total production of fats and oils worldwide. The oil is used mainly in margarines and shortenings, but there are also technical and industrial uses, such as in detergents, lubricants, water repellents, and as fuel. Fish oil consists mainly of triglycerides. The oils derived from pelagics comprise two main groups. The first group of oils contains a relatively large amount of monoenic fatty acids and a moderate proportion of polyunsaturated fatty acids. This group of oils is derived from fish caught in the North Atlantic. The second group has a high iodine number and exhibits a relatively high content of polyunsaturated fatty acids. These oils are from fish originating in the Pacific Ocean, and tropical and southern region of the Atlantic Ocean.

Fertilizer

Fish offal and soluble compounds are used as fertilizer on farms in coastal areas. Fertilizers are also produced through industrial processes, but the industry is relatively small compared to the economic value of other fish by-products.

Fish Silage and Hydrolysates

Fish silage is used as an ingredient in animal feed, and is treated separately from fish meal due mainly to differences in the production process. Fish silage is used mainly in fish feeds and moist feed pellets. During production, the fish is usually minced and mixed with materials that inhibit bacterial growth. The product is stored for a period during which the fish is liquefied by digestive enzymes. Some of the water may be removed via evaporation.

The production process for hydrolysates is similar to that of fish silage. Water is added to the minced fish, and suspended solids are liquefied through the action of proteolytic enzymes. Oils and the remaining solids are removed, leaving a hydrolysate. The product may be treated further, depending on how the product is to be used. Fish hydrolysate may be used as an ingredient in compound feeds much like fish silage. It is also used as a flavor additive. A third use for fish hydrolysate is as a component of growth media for industrial fermentation.

Pelagic Fish as a Resource for Biotechnological Products

A number of organic compounds found within small pelagic fish are of interest to the chemical and pharmaceutical industries. These compounds include, among others, fatty acids, lipids, hormones, nucleic acids, and organometallics. Scales have been used as a source of pearl essence.

Management Issues for Small Pelagic Fisheries

The exploitation of many stocks of pelagic fishes has exhibited a pattern of sharply increasing catches followed by an even more rapid decline, leading in several cases to closure of the fishery (**Figure 2**). The rapid increases in catches have been largely due to increasing technical developments, increasing markets and movement into new fishing areas of high

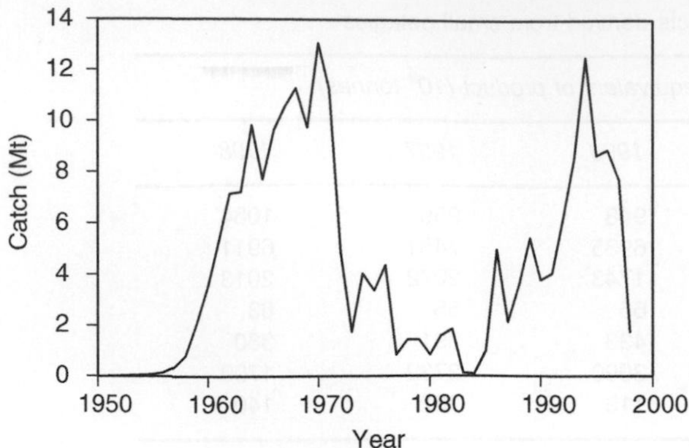

Figure 2 Catch (million tonnes; Mt) of Peruvian anchoveta (*Engraulis ringens*) showing fluctuations in abundance attributed largely to fishing. (Data from FAO, 1999.)

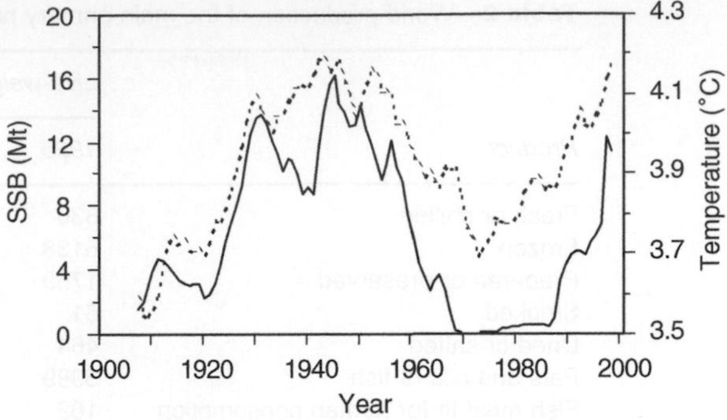

Figure 3 An example of possible environmentally induced biomass changes. Estimated spawning stock biomass (million tonnes; Mt) of Norwegian spring spawning herring (*Clupea harengus*), 1907–1998 and associated long-term temperature fluctuations. —, spawning stock, - - -, temperature. Adapted from Toresen and Østvedt (2000).

abundance. The rapid decline has been considered, in some cases, to have a link with environmental change, but has been attributed generally to heavy fishing pressure, recruitment failure, and ineffective management. Most stocks have recovered after reduction/termination of fishing, and some fisheries have resumed following recovery.

Assessment and management of pelagic fisheries has been complicated by a number of issues. The long history of information of some pelagic stocks prior to mechanized fisheries demonstrates that these stocks are prone to pronounced fluctuations in abundance in the absence of fishing (**Figure 3**). Although such natural fluctuations are undoubtedly linked to the environment, there are likely several mechanisms and these are not well understood. In some cases (e.g., sardines and anchovies, herring, and pilchard) there have been apparent shifts in dominance over time, where periods of high abundance have alternated for the different species groups (**Figure 4**). It has been hypothesized that these switches in dominance relate to shifts in environmental conditions favoring one species over another, possible interactions with other species, and the effects of harvesting. Some species or species groups have exhibited coherence in high abundance periods over broad geographical scales (e.g., high abundance of sardines throughout the North Pacific basin during the same time periods).

The shoaling behavior of pelagic species can result in high fishery catch rates when there are relatively few fish remaining in the stock. As a result catch per unit effort, which is commonly used as an indicator of stock decline in demersal fisheries, is less useful in evaluating the state of pelagic fisheries. High catch rates can be maintained, in spite of declining stock size, and this has contributed to sudden stock collapse.

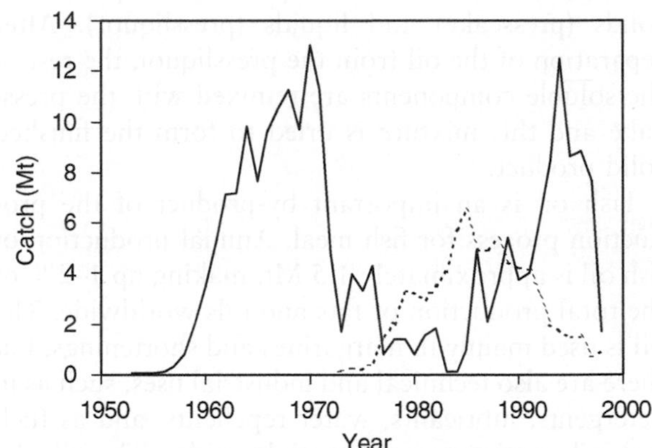

Figure 4 Catch (million tonnes; Mt) of Peruvian anchoveta (*Engraulis ringens*; —), and South American sardine (or pilchard) (*Sardinops sagax*; - - -), showing a shift in species dominance. Data from FAO (1999).

The biological basis for assessment and estimation of stock size for many small pelagic species is complicated by migration, mixing of stocks, lack of fishery-independent abundance indices, and poor analytical performance of common assessment methods. Management attempts have been further complicated by the rapid rate of decline of some small pelagic stocks.

The economic importance of herring to coastal European nations contributed to the prominent consideration of small pelagic fisheries in the evolution of fisheries science and management. Several developments, including the stock concept, recognition of year-to-year differences in year-class strength, development of hydroacoustic survey methods, and early fishery management systems have been based on fisheries for small pelagic species.

Most fisheries for small pelagic species are regulated by some form of quota, although in some cases management is complicated by the occurrence of these fisheries outside or across areas of national jurisdiction.

Continued increase in mechanization of fisheries and expansion of markets increases the fishing pressure on several pelagic stocks. Technical developments such as improved hydroacoustic (sonars and sounders) detection of fish, global positioning systems, improved net design and materials, and improved vessel characteristics (e.g., larger size and refrigeration) have increased the efficiency of the fishery. These developments have contributed to increasing fishing effort, in spite of management attempts in some cases to regulate fishing capacity. The historical exploitation of several pelagic fish stocks can be shown to be directly linked to changes in markets. The increasing demand for fishmeal, for example, resulted in large increases in the landings from some pelagic fish stocks in the 1970s and 1980s.

Current research areas of particular interest for small pelagic species include the link between stock abundance, environmental fluctuation (particularly in areas of fronts and upwelling) and climate change. Most small pelagic species feed on zooplankton, and are in turn important food items for other fish, marine mammals, and seabirds. Assessment and management of fisheries for small pelagic fisheries are increasingly becoming concerned with multispecies interactions, including both the estimation of mortality of the pelagic species caused by predation and maintaining sufficient stock size of small pelagic fish as forage for other species.

See also

Demersal Species Fisheries. Fisheries and Climate.

Further Reading

Burt JR, Hardy R, and Whittle KJ (eds.) (1992) *Pelagic Fish: The Resource and its Exploitation.* Oxford: Fishing News Books.

FAO (1999) *The State of World Fisheries and Aquaculture 1998*. Rome: Food and Agriculture Organization of the United Nations.

FAOSTAT Website. http://apps.fao.org/default.htmThe FAO website contains information from the FAO Yearbook of Fishery Statistics.

Patterson K (1992) Fisheries for small pelagic species: an empirical approach to management targets. *Review of Fish Biology and Fisheries* 2: 321–338.

Saville A (ed.) (1980) *The assessment and management of pelagic fish stocks. Rapp. P-v Reun., Cons. Int. Explor. Mer*, 177–517.

Toresen R and Østvedt OJ (2000) Variation in abundance of Norwegian spring-spawning herring (*Clupea harengus*, Clupeidae) throughout the 20th century and the influence of climatic fluctuations. *Fish and Fisheries* 1: 231–256.

SOUTHERN OCEAN FISHERIES

I. Everson, British Antarctic Survey Cambridge, UK

The history of harvesting living resources in the Southern Ocean goes back two centuries. Soon after South Georgia was discovered by Captain Cook, hunters arrived in search of fur seals (*Arctocephalus gazella*). Between 1801 and 1822, over 1 200 000 skins were taken there before the sealers moved on to the South Shetlands. The extent of the carnage was such as to bring the species close to extinction by the 1870s. In the early twentieth century elephant seals (*Mirounga leonina*) were harvested at South Georgia because their blubber provided higher-grade oil than that from whales. A strong management regime was instituted in 1952, which allowed the population to recover from earlier overexploitation. Elephant seals were also harvested at Macquarie Island, although there was less of a link to the whaling industry.

Whaling has been the largest fishing operation in the Southern Ocean. Initially this was shore-based but, with the introduction of floating factories in 1925, catching vessels could search much of the Southern Ocean. Overcapitalization of the industry and, in the early years, a lack of robust population models to provide management advice, meant that the industry declined. Research at that time, particularly by Discovery Investigations, provided much valuable information not only on whales but also on their food, krill.

With the decline in shore-based whaling, fishing fleets from Japan and the former Soviet Union turned their attention to harvesting fish and krill. Scientists expressed several concerns over this development: First that there was no international fishery regime in place for the region; second that an unregulated fishery on krill could lead to overexploitation as had happened with the whales; and third that significant fishing of krill might adversely affect the recovery of baleen whales.

Member states of the Antarctic Treaty, an agreement that arose directly from scientific collaboration during the International Geophysical Year, agreed that a management regime needed to be established for Southern Ocean resources. The first step was the Convention for the Conservation of Antarctic Seals, which came into force in 1972 even though there was at that time no harvesting of seals or any intention voiced to that end. Scientific advice underpinning this agreement come from the Scientific Committee for Antarctic Research (SCAR). The establishment of the BIOMASS (Biological Investigations of Marine Antarctic Systems and Stocks) program in 1976, again through support from SCAR, in collaboration with the Scientific Committee for Oceanic Research (SCOR), International Association of Biological Oceanography (IABO) and Advisory Committee on Marine Resources Research (ACMRR), led naturally to the Convention for the Conservation of Antarctic Marine Living Resources, negotiated in 1980 coming into force in 1982.

Commission for the Conservation of Antarctic Marine Living Resources (CCAMLR)

The overriding concern leading to the establishment of CCAMLR was that overexploitation of krill might adversely affect dependent species. This is encapsulated in Article II as a series of general principles of conservation that in summary form set out in paragraph 3 state that any harvesting shall be conducted in such a way as to

(a) prevent the decrease in the size of the harvested population to levels below those which ensure its stable recruitment;
(b) maintain the ecological relationships between harvested, dependent, and related populations and;
(c) prevent changes or minimize the risk of changes in the marine ecosystem that are not reversible over two or three decades.

Subparagraph (a) is essentially a statement of the traditional single-species approach to fisheries management and setting an upper limit to the total catch that can be taken for any individual species. This maximal level – which, it should be noted, is a limit and not a target – must then be reduced to take account of the dependent species as required by subparagraph (b). Finally, there is the requirement in subparagraph (c) that any ecological changes should be reversible within a finite and limited period.

These three components of Article II encapsulate what has come to be known as the ecosystem approach to management of marine living resources. CCAMLR was the first international fisheries commission to make this a central plank of its management regime, and even though the approach has received wide approval it is only now, twenty years later, that it is beginning to be considered for implementation in other forums.

The Antarctic Treaty covers the land and ice-shelves south of 60°S with an agreement that all territorial claims within that region are 'frozen.' Essentially this means that anything other than sovereignty can be discussed. At the time of the negotiation of CCAMLR it was known that the distribution of Antarctic krill and some finfish stocks extended some way north of the Treaty zone. Accordingly, the CCAMLR region was designated to cover the area south of the Antarctic Convergence (now called the Antarctic Polar Frontal Zone, APFZ). This area is designated by lines joining parallels of latitude and meridians of longitude, that are set out in Article I paragraph 5, which approximate the Antarctic Convergence. For the most part the fit is good, but for political reasons some small parts were excluded. Within the region between the Convergence and latitude 60° S lie several Sub-Antarctic islands for which sovereignty is claimed by member states of CCAMLR. Prior to the conclusion of negotiation of the CCAMLR Convention, France had declared a 200 mile exclusive economic zone (EEZ) around Kerguelen and Crozet and obtained agreement that, through what has become to be known as 'The Chairman's Statement' of 19 May 1980, within the zone France would take note of CCAMLR advice but would be free to implement its own policy. Subsequently, other claimant states of Sub-Antarctic islands have implemented a similar policy to France.

Krill

Although originally applied to 'fish fry,' the term krill is now taken to refer to euphausiids, of which six species occur in the Southern Ocean. Of these the only one that is targeted commercially is the Antarctic krill, *Euphausia superba*. Antarctic krill are small shrimplike crustacea that grow to a maximum length of 65 mm. They are widely distributed in the Southern Ocean south of the APFZ. On the continental shelf they are replaced by their smaller congener *E. crystallorophias*. Concentrations are normally found close to islands and the continental shelf break and have in the past been considered as constituting separate populations or management units. Although they are carried around the continent on the Antarctic Circumpolar Current (ACC), the extent of mixing between groups of krill from different regions is not clear.

The sheer size of the Southern Ocean, allied to the fact that much of it is covered in sea ice, means that it

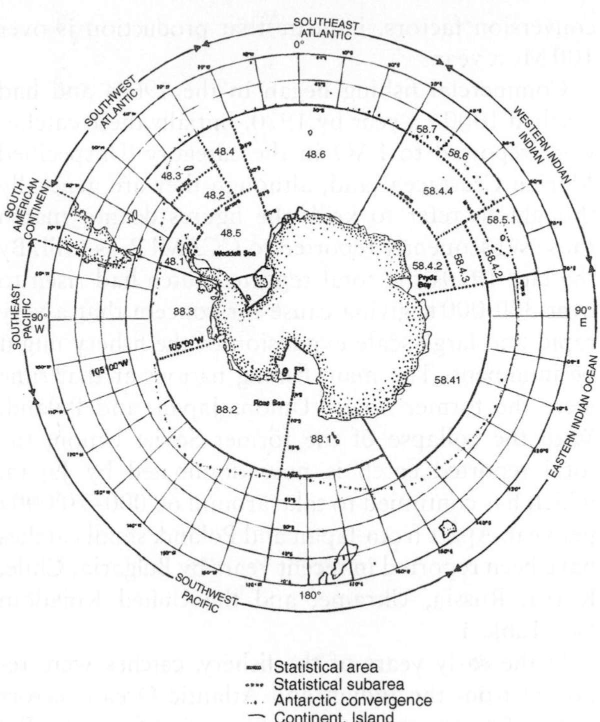

Figure 1 CCAMLR Statistical Areas, subareas, and divisions.

has never been possible to estimate the standing stock over the whole region. Synoptic surveys have been undertaken over part of the region in the Atlantic and Indian Ocean sectors. The most recent in the Atlantic sector, CCAMLR Area 48 (see **Figure 1**), gave an estimated standing stock of 44 Mt, while a survey in the Indian Ocean sector in 1996 gave estimated values of 3.04 Mt between 80° and 115°E and 1.79 Mt between 115° and 150°E.

In common with many zooplankters, krill undergo diurnal vertical migration within the top 200 m, although the pattern is not consistent with time or locality. They are often found at the surface at night, although daytime surface concentrations are not infrequent. They frequently occur in swarms with a density of several thousand individuals per cubic meter and it is these swarms that are targeted by commercial fishers.

Antarctic krill are filter feeders dependent primarily on phytoplankton. Arising from this, in the early season 'green' krill, with large amounts of chlorophyll in the hepatopancreas, predominate. These green krill are associated with a 'grassy' flavour to products manufactured for human consumption. Later in the season, colorless 'white' krill predominate.

Estimates of production have been based on consumption of krill by dependent species and, even allowing for considerable uncertainty over the conversion factors, indicate that production is over 100 Mt a year.

Commercial fishing began in the 1960s and had reached 1000 t a year by 1970. Initially these catches were reported to FAO in the category 'Unspecified Marine Crustacea' and, although they are generally thought to refer to krill, the figures do not match those subsequently reported to CCAMLR as krill. By the late 1970s the total reported catch had risen to over 300 000 t, giving cause for concern that a very rapid and large-scale expansion of the fishery might be imminent. The main fishing nations at that time were the former Soviet Union, Japan, and Poland. With the collapse of the former Soviet Union, the total reported catch is now dominated by Japan, which has continued to take around 60 000–70 000 t per year. Apart from Japan and Poland, small catches have been reported in recent years by Bulgaria, Chile, Korea, Russia, Ukraine, and the United Kingdom (see **Table 1**

In the early years of the fishery, catches were reported from the Indian and Atlantic Ocean sectors with relatively small amounts coming from the Pacific sector. During the 1990s the pattern has changed such that virtually all catches are taken from the Atlantic sector. In the summer months this is predominantly the Antarctic Peninsula (Subarea 48.1) and South Orkneys (Subarea 48.2); and in the winter, because the more southerly grounds are closed owing to sea ice, around South Georgia (Subarea 48.3).

In the early years a variety of fishing methods were attempted. Early attempts at developing surface trawls, with a codend pump, towed alongside the fishing vessel failed because of the cumbersome nature of the gear and the relative infrequency of daytime surface swarms. Likewise, purse seines have proved impractical. Currently, midwater trawls with a mouth area of 500–700 m^2 and a fine mesh throughout are used. Krill swarms are detected using directional sonar and the fishing depth of the net is adjusted by reference to a netsounder. When regions of high concentration have been found, the vessels continue to fish on them without spending a great deal of time in searching farther afield. Haul duration is generally controlled to ensure a maximum catch of 7–10 t. This is because with large catches the krill tend to became crushed in the cod end and product quality suffers.

Over several seasons Japanese fleets have changed their fishing patterns, delaying the commencement of fishing until late in the austral summer and continuing through to the winter so as to avoid catching 'green' krill; products made from green krill have a lower market value than white krill. There is also a preference for larger krill, which, owing to diurnal vertical migration, tend to be deeper by day than small krill, a factor reflected in the fishing depth.

Much effort has been expended in developing krill products for direct human consumption, but much of the current catch has been used for domestic animal feed and, particularly in recent years, for aquaculture feed. The Japanese fishery produces four types of product: fresh frozen (46% of catch), boiled frozen (10% of catch), peeled krill meat (10% of catch), and meal (34% of catch). These are used for aquaculture and aquarium feed (43% of catch), for sport fishing bait (45% of catch), and for human consumption (12% of catch). There are plans for further products such as freeze-dried krill and krill hydrolysates, although currently these are only at an early developmental stage.

In the 1970s high contents of fluoride were reported from krill. This was found to be localized to the exoskeleton, where it can reach 3500 μg F per g dry mass, although concentrations in the muscle appear to be less than 100 μg F per g dry mass. Providing the krill are peeled soon after capture or the whole krill are frozen at a temperature lower than -30°C, the fluoride concentration in the muscle remains low.

Finfish

Fisheries for finfish developed very quickly. In 1969 and 1970 around 500 000 t of fish were taken from the South Georgia, and in 1971 and 1972 a further 300 000 t were taken from Kerguelen. Although the catches were initially reported to FAO as 'Unspecified Demersal Percomorphs' it is widely assumed that the dominant species in the catches was *Notothenia rossii*. Smaller catches in following years were reported from South Shetland and South Orkney Islands. These catch rates were unsustainable and the species remains at a low stock level. Other species that were taken at this time included grey rockcod, *Lepidonotothen (Notothenia) squamifrons*, and mackerel icefish, *Champsocephalus gunnari*. During the 1980s and coincident with the start of CCAMLR, catches of other species were reported including *Chaenocephalus aceratus, Chaenodraco wilsoni, Channichthys rhinoceratus, Chionodraco rastrospinosus Pseudochaenichthys georgianus, Dissostichus eleginoides, Gobionotothen (Notothenia) gibberifrons* and *Patagonotothen guntheri*. There was also a fishery for Myctophidae, principally *Electrona carlsbergi*, which reached a peak of 78 000 t in 1991 but ceased two years later coincident with the collapse of the former Soviet Union. In recent years the only target species have been mackerel icefish and Patagonian toothfish.

Table 1 Annual reported catches (tonnes) of Antarctic krill by nation

Year	Japan	Poland	USSR	Russia	Ukraine	Argentina	Bulgaria[a]	Chile	E. Germany	Korea	Spain	Latvia[a]	Panama[a]	UK
1970														
1971														
1972														
1973	59													
1974	646		19 139											
1975	2 677		41 352											
1976	4 750		609					276						
1977	12 802	6 966	71 656					92						
1978	25 219		106 991				94		8					
1979	36 961		295 508				46		102	511				
1980	36 275	226	440 516					0	0	0				
1981	27 698	0	420 434					0	0	0				
1982	35 116	0	491 656					0	0	1 429				
1983	42 282	360	180 290					3 752	0	1 959				
1984	49 531	0	74 381					1 649	0	5 314				
1985	38 274	0	150 538					2 598	0	0				
1986	61 074	2 065	379 270					3 264	0	0				
1987	78 360	1 726	290 401					4 063	0	1 527	379			
1988	73 112	5 215	284 873					5 938	0	1 525				
1989	78 928	6 997	301 498					5 329	0	1 779				
1990	62 187	1 275	302 376					4 501	396	4 040				
1991	67 582	9 571	275 495					3 679		1 211				
1992	74 325	8 607		151 725	61 719			6 066		519				
1993	59 272	15 911		4 249	6 083			3 261		0				
1994	62 322	7 915		965	8 852			3 834		0		71		
1995	60 303	9 384			48 886			0		0			141	
1996	60 546	20 610			20 056			0		0			495	
1997	58 798	19 156			4 246			0		0			0	308
1998	63 233	15 312			0			0		1 618			0	634
1999	71 318	18 554			5 694	6 524		0		1 228			0	

[a]Not a member state of CCAMLR when the catches were made.

Mackerel Icefish

The mackerel icefish is a Channichthyid; these 'white blooded fish' are the only group of vertebrates that do not have erythrocytes or any effective blood pigment. They occur on the shelf in waters rarely more than 500 m deep. Spawning is thought to occur in fiords and bays during April and May and the eggs, >3 mm diameter, hatch in the spring coincident with the spring primary production bloom and new generations of copepods in the plankton. The fish become sexually mature at around 3 years of age when they are 25 cm in total length. Their diet is predominantly krill, although they will take a variety of other species, particularly when krill are scarce.

At Kerguelen it has been possible to follow growth through several years by analysis of length distributions. Of particular interest is a three year cycle in the population. It is not clear whether this is a natural phenomenon or one induced by fishing pressure. At South Georgia there have been similar fluctuations in stock size, although in that case they have been attributed to enhanced predation pressure by fur seals at times of krill scarcity.

Initially bottom trawls were used in the fishery but these were banned by CCAMLR in the early 1990s to reduce damage to the benthic environment. The use of midwater trawls, as well as having a much reduced impact on benthic biota, is also accompanied by reduced by-catches. Currently there are limited trawl fisheries at Kerguelen and South Georgia. The fish are generally headed and gutted and then frozen; they fetch a price similar to that of medium-small to medium sized cod. The fillets contain slightly more fat than average whitefish.

At Kerguelen the management regime has been operated exclusively through the French authorities. Around South Georgia a Total Allowable Catch (TAC) has been set based on production estimates using trawl survey results. The most recent TACs have been based on the lower 95% confidence limit from the surveys.

Toothfish

There are two species of the genus *Dissostichus*, which are broadly separated geographically. *D. mawsoni* is present in the high Antarctic close to the continent, whereas *D. eleginoides* is present around sub-Antarctic islands and extends outside of the Antarctic zone to Patagonia and the west coast of South America. Toothfish are the largest Nototheniid fish in the Southern Ocean, growing to a maximum length of over 2 m and mass of around 100 kg. They are relatively slow-growing reaching sexual maturity at around 90–100 cm. Spawning is thought to occur in the latter part of the winter.

At Kerguelen a trawl fishery has become established in various parts of the shelf, with trawlers mainly from France and Russia. The management regime is implemented by French authorities.

In the early years at South Georgia, toothfish appeared as a by-catch in the trawl fishery, sometimes to the extent of a tonne of fish in a haul. Toward the end of the 1980s, long-liners from the former Soviet Union began fishing at the shelf slope around South Georgia and obtained good catch rates. With the increasing fishing effort being applied, concern was expressed at the sustainability of the resource. Owing to the high value of the catch, other nations have joined the fishery; these include Argentina, Bulgaria, Chile, Korea, Ukraine, and the United Kingdom. Two types of long-line system are used: Mustad Autoliner and the Spanish system with a ground line and parallel fishing line. The lines are baited with squid, horse mackerel or sprats and fished on the bottom for around 12 h in water around 1000 m deep.

It is possible that the by-catch of rays and grenadiers may be significant, although at present data are scarce. Of much greater concern is the incidental mortality caused to seabirds that attack the baits as the lines are shot. Very low recruitment of albatrosses to the breeding populations at South Georgia has been directly attributed to mortality caused by long-line fishing, not only for toothfish within the Southern Ocean but also due to tuna long-lining over much of the Southern Hemisphere. It has been demonstrated that restricting the setting of lines to the hours of darkness together with weighting the lines is adequate to eliminate this mortality, if applied diligently.

Toothfish fetch a high price on international markets and this has been a major cause of illegal and unregulated fishing within the Southern Ocean. This reached a peak in 1997 when over 90% of catches in some areas were thought to have been taken outside of CCAMLR regulations. To combat this activity, CCAMLR has introduced a catch documentation scheme whereby only fish that have been taken in compliance with CCAMLR Conservation Measures can be issued with a valid certificate. This measure is having a positive effect as currently certificated catches attract a price of around $10 000 per tonne whereas uncertificated catches fetch only around $3000 per tonne (see **Table 2**).

Crabs

There are two species of crab in the waters around South Georgia, *Paralomis spinosissima* and *P. formosa*.

Table 2 Toothfish catches (tonnes) by area[a]

Subarea/Division	*Estimated total catch*	*Reported catch 1996/97*	*Estimated unreported catch from catch/effort data*	*Unreported catch as % of the estimated total catch*
South Georgia (48.3)	2 389	2 389	Probably low	Probably low
Prince Edward Is. (58.7)	14 286	2 386	11 900	83.3
Crozet Is. (58.6)	19 233	333	18 900	98.2
Kerguelen (58.5.1)	6 681	4 681	2 000	29.9
Heard Is. (58.5.2)	8 037–12 837	837	7 200–12 000	89.6–93.4
All subareas	48 856–53 656	10 856	38 000–42 800	77.8–79.8

[a]Reported in relation to unreported catch of toothfish during the 1996/97 season. The unreported catch is estimated from market landings, numbers of vessels seen, and observed catch rates. (*Anon, 1997.*)

In the early 1990s there was an exploratory study to fish them. In view of the lack of knowledge of either species, CCAMLR imposed a rigorous fishing plan for the study. Following fishing in the area during 1993, when 299 t were caught, and 1996, when 497 t were caught, the fishing company decided that these catches were uneconomic and did not pursue the matter. Subsequently there have been expressions of interest by other companies, but no further fishing activity.

Squid

For many years it has been known that squid figure in the diets of many predators in the Southern Ocean. Regurgitated stomach content samples from albatrosses and petrels indicated that in many cases these squid were quite fresh, leading to the conclusion that they had been taken from adjacent waters. Exploratory ventures in 1989 and 1996 by Korean squid jiggers found small concentrations of *Martialia hyadesi* close to the APFZ in summer and on the South Georgia shelf in winter. World market prices of squid have meant that further ventures are likely to be only marginally profitable. However, advances in processing whereby the second tunic can be easily removed may change this situation in the near future.

Southern Ocean Management Regime

The sections of Article II of the CCAMLR Convention provide a clear framework for the management regime. In the first instance, harvested species need to be managed such that their long-term sustainability is not threatened by commercial activities. This requirement can be met through the application of traditional fisheries models and others developed within CCAMLR such as the Krill Yield Model (KYM) and Generalized Yield Model, the underlying principles of which are set out below.

1. To aim to keep the krill biomass at a level higher than would be the case for single-species harvesting considerations.
2. Given that krill dynamics have a stochastic component, to focus on the lowest biomass that might occur over a future period, rather than on the average biomass at the end of that period, as might be the case in a single species context.
3. To ensure that any reduction of food to predators that might arise out of krill harvesting is not such that land-breeding predators with restricted foraging ranges are disproportionately affected compared to predators in pelagic habitats.
4. To examine what levels of krill escapement are sufficient to meet the reasonable requirements of predators.

The KYM has the general formula [1], where Y = yield, B_0 = median unexploited biomass, and γ = proportionality coefficient.

$$Y = \gamma B_0 \qquad [1]$$

Currently two values of γ are calculated. The first, γ_1, is chosen such that the probability of the spawning biomass dropping below 20% of its pre-exploitation median level over a 20-year harvesting period is 10%, and the second γ_2 is chosen so that the median krill escapement over a 20-year period is 75%. The lower of γ_1 and γ_2 is selected for the calculation of krill yield. These principles are applied so as to account for sustained consistency in catch with time while at the same time taking account of uncertainties in the estimators. This implies a conservative approach to the application of fishery models in pursuit of the precautionary principle; this approach is followed in the advice that comes forward from the Scientific Committee and is implemented by the Commission. Within this overall framework, due regard has to be taken of the requirements of dependent species – the ecosystem approach.

Early in the history of CCAMLR it became apparent that a wide range of interpretations can be put on the definition of an 'ecosystem approach.' At one extreme is an endeavor to understand all interactions in the food web in order to formulate management advice. Such an approach is favored by idealists on the one hand and those who did not wish to see any form of control on the other. It is impracticable because no advice would emerge in a timely manner. Recognizing this, CCAMLR has set up an Ecosystem Monitoring Program (CEMP) whereby certain features of a small suite of dependent species are monitored. Currently the CEMP species are Adelie, chinstrap, gentoo, and macaroni penguins; black-browed albatross; fur seal and crabeater seal. Key parameters associated with the ecology of each of these species are monitored according to a series of agreed CEMP protocols. The aim of the program is to determine how the dependent species perform in response to krill availability and how this is affected by fishing activities. CEMP parameters integrate krill availability over different time and space scales, varying from months and hundreds of kilometers in the case of the total mass of individual penguins on arrival at a colony at the start of breeding, to days and tens of kilometers in the case of foraging trip duration while feeding chicks. These provide indicators of overlap with commercial fishing. The other components monitored relate to the vital rates of dependent species – changes in population size, mortality, and recruitment – and progress is currently in hand to integrate these into the overall management process.

Recent papers have linked variations in krill distribution and standing stock to climate change and El Nĩno–Southern Oscillations through their effects on the ACC and sea ice regime. The CCAMLR management scheme implicitly takes account of such long-term environmental change because its regime can be adjusted to compensate as the monitored species and variables change with time.

See also

Crustacean Fisheries.

Further Reading

Information on the status of Southern Ocean resources and management decisions is provided in the Reports and Statistical Bulletins of the Commission and Scientific Committee published by CCAMLR, Hobart, Tasmania 7000, Australia.

Anon (1997) Estimates of catches of *Dissostichus eleginoides* inside and outside the CCAMLR Area. SC-CAMLR-XVI, Annex 5, Appendix D.

Constable AJ, de la Mare WK, Agnew DJ, Everson I, and Miller D (2000) Managing fisheries to conserve the Antarctic marine ecosystem: practical implementation of the Convention on the Conservation of Antarctic Marine Living Resources (CCAMLR). *ICES Journal of Marine Science* 57: 778–791.

El-Sayed SZ (ed.) (1994) *Southern Ocean Ecology: The BIOMASS Perspective*. Cambridge: Cambridge University Press.

Everson I (1978) Antarctic fisheries. *Polar Record* 19(120): 233–251.

Everson I (ed.) (2000) *Krill: Biology, Ecology and Fisheries*. Oxford: Blackwell Science.

Kock K-H (2000) *Antarctic Fish and Fisheries*. Cambridge: Cambridge University Press.

MARICULTURE

MARICULTURE DISEASES AND HEALTH

A. E. Ellis, Marine Laboratory, Aberdeen, Scotland, UK

Introduction

As with all forms of intensive culture where a single species is reared at high population densities, infectious disease agents are able to transmit easily between host individuals and large economic losses can result from disease outbreaks. Husbandry methods are designed to minimize these losses by employing a variety of strategies, but central to all of these is providing the cultured animal with an optimal environment that does not jeopardize the animal's health and well-being. All animals have innate and acquired defenses against infectious agents and when environmental conditions are good for the host, these defense mechanisms will provide protection against most infections. However, animals under stress have less energy available to combat infections and are therefore more prone to disease. Although some facilities on a farm may be able to exclude the entry of pathogens, for example hatcheries with disinfected water supplies, it is impossible to exclude pathogens in an open marine situation. Under these conditions, stress management is paramount in maintaining the health of cultured animals. Even then, because of the close proximity of individuals in a farm, if certain pathogens do gain entry they are able to spread and multiply extremely rapidly and such massive infectious burdens can overcome the defenses of even healthy animals. In such cases some form of treatment, or even better, prophylaxis, is required to prevent crippling losses. This article describes some of the management strategies available to fish and shellfish farmers in avoiding or reducing the losses from infectious diseases and some of the prophylactic measures and treatments. The most important diseases encountered in mariculture are summarized in **Table 1**.

Health Management

Facility Design

Farms and husbandry practices can be designed in such a way as to avoid the introduction of pathogens and to restrict their spread within a farm in a variety of ways.

Isolate the hatchery Infectious agents can be excluded from hatcheries by disinfecting the incoming water using filters, ultraviolet lamps or ozone treatments. It is also important not to introduce infections from other parts of the farm that may be contaminated. The hatchery then should stand apart and strict hygiene standards applied to equipment and personnel entering the hatchery. Some diseases cause major mortalities in young fry while older fish are more resistant. For example, infectious pancreatic necrosis virus (IPNV) causes mass mortality in halibut fry, but juveniles are much more resistant. It is vitally important therefore, to exclude the entry of IPNV into the halibut hatchery and as this virus has a widespread distribution in the marine environment, disinfection of the water supply may be necessary.

Hygiene practice Limiting the spread of disease agents on a farm include having hand nets for each tank and disinfecting the net after each use, disinfectant foot-baths at the farm entrance and between buildings, and restricted movement of staff, their protective clothing and equipment. Prompt removal of dead and moribund stock is essential as large numbers of pathogens are shed into the water from such animals. In small tanks this can easily be done using a hand net. In large sea cages, lifting the net can be stressful to fish and divers are expensive. Special equipment such as air-lift pumps, or specially designed cages with socks fitted in the bottom in which dead fish collect and which can be hoisted out on a pulley are more practical. Proper disposal of dead animals is essential. Methods such as incineration, rendering, ensiling and, on a small scale, in lime pits are recommended.

Husbandry and Minimizing Stress

Animals under stress are more prone to infectious diseases. However, it is not possible to eliminate all the procedures that are known to induce stress in aquaculture animals, as many are integral parts of aquaculture, e.g., netting, grading, and transport. Nevertheless, it is possible for farming practices to minimize the effects of these stressors and others, e.g., overcrowding and poor water quality, can be avoided by farmers adhering to the recommended

Table 1 Principal diseases of fish and shellfish in mariculture

Disease agent	*Host*	*Prevention/treatment*
Shellfish pathogens		
Protozoa		
Bonamia ostreae	European flat oyster	Exclusion
Marteilia refringens	Oyster, mussel	Exclusion
Bacteria:		
Aerococcus viridans (Gaffkaemia).	Lobsters	Improve husbandry
Fin-fish pathogens		
Viruses		
Infectious pancreatic necrosis	Salmon, turbot, halibut, cod, sea bass, yellowtail	Sanitary precautions, vaccinate
Infectious salmon anemia	Atlantic salmon	Sanitary precautions, eradicate
Pancreas disease	Atlantic salmon	Avoid stressors, vaccinate
Viral hemorrhagic septicemia	Turbot	Sanitary precautions
Viral nervous necrosis	Stripped jack, Japanese flounder, barramundi, sea bass, sea bream, turbot, halibut	Sanitary precautions
Bacteria		
Vibriosis (*Vibrio anguillarum*)	All marine species	Vaccinate, antibiotics
Cold water vibriosis (*Vibrio salmonicida*)	Salmon, trout, cod	Vaccinate, antibiotics
Winter ulcers (*Vibrio viscosus*)	Salmon	Avoid stress, vaccinate
Typical furunculosis (*Aeromonas salmonicida*)	Salmon	Vaccinate, antibiotics
Atypical furunculosis (*Aeromonas salmonicida achromogenes*)	Turbot, halibut, flounder, salmon	Vaccinate, antibiotics, avoid stress
Flexibacter maritimus	Sole, flounder, turbot, salmon, sea bass	Avoid stress, antibiotics
Enteric redmouth (*Yersinia ruckeri*)	Atlantic salmon	Vaccinate
Bacterial kidney disease (*Renibacterium salmoninarum*)	Pacific salmon	Exclude
Mycobacteriosis (*Mycobacterium marinum*)	Salmonids, sea bass, sea bream, stripped bass, cod, red drum, tilapia	Sanitary
Pseudotuberculosis (*Photobacterium damselae piscicida*)	Yellowtail, sea bass, sea bream	Vaccinate, sanitary
Piscirickettsiosis (*Piscirickettsia salmonis*)	Salmonids	Sanitary
Parasites		
Protozoan: many species; *Amyloodinium, Cryptobia, Ichthyobodo, Cryptocaryon, Tricodina*	Most species	Avoid stress, sanitary, chemical baths, e.g., formalin
Paramebic gill disease	Salmonids	Low salinity bath, H_2O_2
Crustacea: Sea lice	Salmonids, sea bass, sea bream	In-feed insecticides, H_2O_2, cleaner fish

limits for stocking densities, water flow rates and feeding regimes. In cases where stressors are unavoidable, farmers can adopt certain strategies to minimize the stress.

Withdrawal of food prior to handling Following feeding the oxygen requirement of fish is increased. Withdrawal of food two or three days prior to handling the fish will therefore minimize respiratory stress. It also avoids fouling of the water during transportation with fecal material and regurgitated food.

Use of anesthesia Although anesthetics can disturb the physiology of fish, light anesthesia can have a calming effect on fish during handling and transport and so reduce the stress resulting from these procedures.

Avoidance of stressors at high temperatures High temperatures increase the oxygen demand of animals and stress-induced mortality can result from respiratory failure at high temperatures. It is therefore safer to carry out netting, grading and transport at low water temperatures.

Avoidance of multiple stressors and permitting recovery The effects of multiple stressors can be additive or even synergistic so, for instance, sudden changes of temperature should be avoided during or after transport. Where possible, recovery from stress should be facilitated. Generally, the duration of the recovery period is proportional to the duration of the stressor. Thus, reducing the time of netting, grading or transport will result in recovery in a shorter time. The duration required for recovery to occur may be from a few days to two weeks.

Selective breeding In salmonids, it is now established that the magnitude of the stress response is a heritable characteristic and programs now exist for selecting broodstock which have a low stress response to handling stressors. This accelerates the process of domestication to produce stocks, which are more tolerant of aquaculture procedures with resultant benefits in increased health, survival and productivity. Such breeding programs have been conducted in Norway for some years and have achieved improvements in resistance to furunculosis and infectious salmon anemia virus (ISAV) in Atlantic salmon.

Management of the Pathogen

Breaking the pathogen's life cycle If a disease is introduced on to a farm, it is important to restrict the horizontal spread especially to different year classes of fish/shellfish. Hence, before a new year class of animals is introduced to a part of the farm, all tanks, equipment etc. should be thoroughly cleaned and disinfected. In sea-cage sites it is a useful technique to physically separate year classes to break the infection cycle.

Many pathogens do not survive for long periods of time away from their host and allowing a site to be fallow for a period of time may eliminate or drastically reduce the pathogen load. This practice has been very effective in controlling losses from furunculosis in marine salmon farms and also significantly reduces the salmon lice populations particularly in the first year after fallowing.

Eliminate vertical transmission Several pathogens may persist as an asymptomatic carrier state and be present in the gonadal fluids of infected broodstock and can infect the next generation of fry. In salmon farming IPNV may persist in or on the ova and disinfection of the eggs with iodine-based disinfectants (Iodophores) is recommended immediately after fertilization.

The testing of gonadal fluids for the presence of IPNV can also be carried out and batches of eggs from infected parents destroyed. IPNV has been associated with mass mortalities in salmon and halibut fry and has been isolated from a wide range of marine fish and shellfish, including sea bass, turbot, striped bass, cod, and yellowtail.

Avoid infected stock Many countries employ regulations to prevent the movement of eggs or fish that are infected with certain 'notifiable' diseases from an infected to a noninfected site. These policies are designed to limit the spread of the disease but require specialized sampling procedures and laboratory facilities to perform the diagnostic techniques. By testing the stock frequently they can be certified to be free from these diseases. Such certification is required for international trade in live fish and eggs but within a country it is widely practiced voluntarily because stock certified to be 'disease-free' command premium prices.

Eradication of infected stock Commercially this is a drastic step to take especially when state compensation is not usually available even when state regulations might require eradication of stock. This policy is usually only practiced rarely and when potential calamitous circumstances may result, for instance, the introduction of an important exotic pathogen into an area previously free of that disease. This has occurred in Scotland where the European Commission directives have required compulsory slaughter of turbot infected with viral hemorrhagic septicemia virus (VHSV) and Atlantic salmon infected with ISAV.

Treatments

Viral Diseases

There are no treatments available for viral diseases in aquaculture. These diseases must be controlled by husbandry and management strategies as described above, or by vaccination (see below).

Bacterial Diseases

Antibiotics can be used to treat many bacterial infections in aquaculture. They are usually mixed into the feed. Before the advent of vaccines against many bacterial diseases of fin fish, antibiotic treatments were commonly used. However, after a few years, the bacterial pathogens developed resistance to the antibiotics. Furthermore, there was a growing concern that the large amounts of antibiotics being used in aquaculture would have damaging effects on the environment and that antibiotic residues in fish flesh may have dangerous consequences for consumers by promoting the development of antibiotic resistant strains of human bacterial pathogens. These concerns have led to many restrictions on the use of antibiotics in aquaculture, especially in defining long withdrawal periods to ensure that carcasses for consumption are free of residues. These regulations have made the use of antibiotic treatments impractical for fish that are soon to be harvested for consumption but their use in the hatchery is still an important method of controlling losses from bacterial pathogens.

For most bacterial diseases of fish, vaccines have become the most important means of control (see below) and this has led to drastic reductions in the use of antibiotics in mariculture.

Parasite Diseases

Sea lice The most economically important parasitic disease in mariculture of fin fish is caused by sea lice infestation of salmon. These crustacean parasites normally infest wild fish and when they enter the salmon cages they rapidly multiply. The lice larvae and adults feed on the mucus and skin of the salmon and heavy infestations result in large haemorrhagic ulcers especially on the head and around the dorsal fin. These compromise the fish's osmoregulation and allow opportunistic bacterial pathogens to enter the tissues. Without treatment the fish will die.

A range of treatments are available and recently very effective and environmental friendly in-feed treatments such as 'Slice' and 'Calicide' have replaced the highly toxic organophosphate bath treatments. Hydrogen peroxide is also used. As a biological control method, cleaner fish are used but they are not a complete method of control.

Vaccination

Use of Vaccines

Vertebrates can be distinguished from invertebrates in their ability to respond immunologically in a specific manner to a pathogen or vaccine. Invertebrates, such as shellfish, only possess nonspecific defense mechanisms. In the strict definition, vaccines are used only as a prophylactic measure in vertebrates because a particular vaccine against a particular disease induces protection that is specific for that particular disease and does not protect against other diseases. Vaccines induce long-term protection and in aquaculture a single administration is usually sufficient to induce protection until the fish are harvested. However, vaccines also have nonspecific immunostimulatory properties that can also activate many nonspecific defense mechanisms. These can increase disease resistance levels but only for a short period of time. Thus, in their capacity to induce such responses they are also used in shellfish culture, especially of shrimps.

Current Status of Vaccination

In Atlantic salmon mariculture, vaccination has been very successful in controlling many bacterial diseases and has almost replaced the need for antibiotic treatments. In recent years vaccination of sea bass and sea bream has become common practice. Most of the commercial vaccines are against bacterial diseases because these are relatively cheap to produce. Obviously the cost per dose of vaccine for use in aquaculture must be very low and it is inexpensive to culture most bacteria in large fermenters and to inactivate the bacteria and their toxins chemically (usually with formalin). It is much more expensive to culture viruses in tissue culture and this has been a major obstacle in commercializing vaccines against virus diseases of fish. However, modern molecular biology techniques have made it possible to transfer viral genes to bacteria and yeasts, which are inexpensive to culture and produce large amounts of viral vaccine cheaply. A number of vaccines against viral diseases of Atlantic salmon are now becoming available. Currently available commercial vaccines for use in mariculture are summarized in **Table 2**.

Methods of Vaccination

There are two methods of administering vaccines to fish: immersion in a dilute suspension of the vaccine or injection into the body cavity. For practical

Table 2 Vaccines available for fish species in mariculture

Vaccines against	*Maricultured species*
Bacterial diseases	
Vibriosis	Salmon, sea bass, sea bream, turbot, halibut
Winter ulcers	Salmon
Coldwater Vibriosis	Salmon
Enteric redmouth	Salmon
Furunculosis	Salmon
Pseudotuberculosis	Sea bass
Viral diseases	
Infectious pancreatic necrosis (IPNV)	Salmon
Pancreas disease	Salmon
Infectious salmon anemia (ISAV)	Salmon

reasons the latter method requires the fish to be over about 15 g in weight.

Immersion vaccination is effective for some, but not all vaccines. The vaccine against the bacterial disease vibriosis is effective when administered by immersion. It is used widely in salmon and sea bass farming and probably could be administered by this route to most marine fish species. The vaccine against pseudotuberculosis can also be administered by immersion to sea bass. With the exception of the vaccine against enteric redmouth, which is delivered by immersion to fish in freshwater hatcheries, all the other vaccines must be delivered by injection in order to achieve effective protection.

Injection vaccination induces long-term protection and the cost per dose is very small. However, it is obviously very labor intensive. Atlantic salmon are usually vaccinated several months before transfer to sea water so that the protective immunity has time to develop before the stress of transportation to sea and exposure to the pathogens encountered in the marine environment.

Conclusions

It is axiomatic that intensive farming of animals goes hand in hand with culture of their pathogens. The mariculture of fish and shellfish has had severe problems from time to time as a consequence of infectious diseases. During the 1970s, *Bonamia* and *Marteilia* virtually eliminated the culture of the European flat oyster in France and growers turned to production of the more resistant Pacific oyster. In Atlantic salmon farming, Norway was initially plagued with vibriosis diseases and Scotland suffered badly from furunculosis in the late 1980s. These bacterial diseases have been very successfully brought under control by vaccines. However, there are still many diseases for which vaccines are not available and the susceptibility of Pacific salmon to bacterial kidney disease has markedly restricted the development of the culture of these fish species on the Pacific coast of North America. As new industries grow, new diseases come to the foreground, for instance piscirikettsia in Chilean salmon culture, paramoebic gill disease in Tasmanian salmon culture, pseudotuberculosis in Mediterranean sea bass and sea bream and Japanese yellowtail culture. Old diseases find new hosts, for example IPNV long known to affect salmon hatcheries, has in recent years caused high mortality in salmon postsmolts and has devastated several halibut hatcheries.

To combat these diseases and to ensure the sustainability of aquaculture great attention must be paid to sanitation and good husbandry (including nutrition). In some cases these are insufficient in themselves and the presence of certain enzootic diseases, or following their introduction, have made it impossible for certain species to be cultured, for example, the European flat oyster in France. The treatment of disease by chemotherapy, which was performed widely in the 1970s and 1980s, resulted in the induction of antibiotic-resistant strains of bacteria and chemoresistant lice. Furthermore, the growing concern for the environment and the consumer about the increasing usage of chemicals and antibiotics in aquaculture, led to increasing control and restrictions on their usage. This stimulated much research in the 1980s and 1990s into development of more environmentally and consumer friendly methods of control such as vaccines and immunostimulants. These have achieved remarkable success and the pace of current research in this area using biotechnology to produce vaccines more cheaply, suggests that this approach will allow continued growth and sustainability of fin-fish mariculture into the future.

See also

Crustacean Fisheries. Mariculture Overview. Molluskan Fisheries. Salmon Fisheries, Atlantic.

Further Reading

Bruno DW and Poppe TT (1996) *A Colour Atlas of Salmonid Diseases*. London: Academic Press.

Ellis AE (ed.) (1988) *Fish Vaccination*. London: Academic Press.

Lightner DV (1996) *A Handbook of Pathology and Diagnostic Procedures for Diseases of Cultured Penaeid Shrimps*. The World Acquaculture Society.

Roberts RJ (ed.) (2000) *Fish Pathology*, 3rd edn. London: W.B. Saunders.

MARICULTURE OF AQUARIUM FISHES

N. Forteath, Inspection Head Wharf, TAS, Australia

Introduction

Marine fishes and invertebrates have been kept in aquaria for decades. However, attempts to maintain marine species in a captive environment have been dependent on trial and error for the most part but it has been mainly through the attention and care of aquarists that our knowledge about many marine species has been obtained. This is particularly true of the charismatic syngnathids, which includes at least 40 species of sea horses.

During the past 30 years, technological advances in corrosion resistant materials together with advances in aquaculture systems have brought about a rapid increase in demand for large public marine aquarium displays, oceanariums and hobby aquaria suitable for colorful and exotic ornamental species. These developments have led to the establishment of important export industries for live fishes, invertebrates and so-called 'living rocks'. Attempts to reduce dependence on wild harvesting through the development of marine fish and invertebrate hatcheries met with limited success.

The availability of equipment, which greatly assists in meeting the water quality requirements for popular marine organisms, has turned the attention of aquarists towards maintaining increasingly complex living marine ecosystems and more exotic species. A fundamental requirement for the success of such endeavors is the need to understand species biology and interspecific relationships within the tank community. Modern marine aquarists must draw increasingly on scientific knowledge and this is illustrated below with reference to sea horse and coral reef aquaria, respectively.

History

The first scientific and public aquarium was built in the London Zoological Gardens in 1853. This facility was closed within a few years and another attempt was not undertaken until 1924. By the 1930s several public aquaria were built in other European capitals but by the end of World War II only that of Berlin remained. During the 1970s the scene was set for a new generation of public aquaria, several specializing in marine displays, and others becoming more popularly known as oceanariums due to the presence of marine mammals, displays of large marine fish and interactive educational activities.

Themes have added to the public interest. For example, Monterey Bay Aquarium exhibits a spectacular kelp forest, a theme repeated by several world class aquaria and Osaka Aquarium sets out to recreate the diverse environments found around the Pacific Ocean. Some of these aquaria have found that exhibits of species native to their location alone are not successful in attracting visitor numbers. The New Jersey Aquarium, for example, has been forced to build new facilities and tanks housing over 1000 brightly colored marine tropical fish with other ventures having to rely on the lure of sharks and touch pools.

The history of public aquaria has evolved from stand alone tank exhibits to massive 2–3 million liter tanks through which pass viewing tunnels. Once, visitors were content to be mere observers of the fishes and invertebrates but by the end of the millennium the emphasis changed to ensuring the public became actual participants in the aquarium experience. The modern day visitors seek as near an interactive experience as possible and hope to be transformed into the marine environment and witness for themselves the marine underwater world.

The concept of modern marine aquarium-keeping in the home has its origins in the United Kingdom and Germany. The United States is now the world's most developed market in terms of households maintaining aquaria, especially those holding exotic marine species: there are about 2.5 million marine hobby aquariums in the USA. In Holland and Germany, the emphasis has been on reef culture, a hobby which is becoming more widespread. The manufacture of products designed specifically for the ornamental fish trade first began in 1954 and scientific and technical advances during the 1960s brought aquarium keeping a very long way from the goldfish and goldfish bowl. The development of suitable materials for marine aquaria has been hampered by the corrosive nature and the toxicity of materials when immersed in sea water. One of the first authoritative texts on materials and methods for marine aquaria was written by Spotte in 1970, followed by Hawkins in 1981.

The volume of marine ornamental fish involved in the international trade is difficult to calculate accurately since records are poor. Current estimates are between 100 and 200 tonnes per annum which probably corresponds to more than 20 million individual specimens. The trade is highly dependent on harvesting from the wild. In Sri Lanka, Indonesia, the Philippines, Fiji and Cook Islands, the export of tropical reef species is now one of the most important export industries employing significant numbers of village people.

Members of the family Pomacentridae, in particular clown fishes, *Amphiprion* spp., and blue–green chromis, *Chromis viridis*, are central to the industry and cleaner wrasse, *Labroides dimidiatus*, flame angels, *Centropyge loriculus*, red hawks, *Neocirrhites armatus*, tangs, *Acanthus* spp., and seahorses, *Hippocampus* spp. are important. The fire shrimp, *Lysmata debelius*, is a major species with respect to invertebrates.

Historically, culture of marine ornamentals has not been in competition with the wild harvest industry. However, recent advances in aquaculture technology will undoubtedly enable more marine ornamentals to be farmed. One European hatchery already has an annual production of *Amphiprion* spp. which is 15 times that exported from Sri Lanka, whereas in Australia seahorse farming is gaining momentum. In both Europe and USA, commercial production of fire shrimp, *L. debelius*, is being attempted. To date the greatest impediment to culture lies in the fact that many popular marine ornamentals produce tiny, free-floating eggs and the newly hatched larvae either do not accept traditional prey used for rearing such as rotifers and brine shrimp, or the prey has proved nutritionally inadequate.

During the 1980s, the Dutch and German aquarists pioneered the development of miniature reef aquaria. Their success among other things, depended on efficient means of purifying the water. The Dutch developed a 'wet and dry' or trickle filter. These filters acted as both mechanical and biological systems. More compact and efficient trickle filter systems have been developed with the advent of Dupla bioballs during the 1990s. These aquaria are filled with so-called 'living rock' which is initially removed from coral reefs.

Coral culture *per se* has recently been established in the Philippines, Solomon Islands, Palau, Guam, and the United States and is aimed at reducing the need to remove living coral from natural reefs.

The upsurge in popularity in marine ornamentals over the past 30 years for both public and hobby aquaria has raised serious conservation concerns. There are calls for sustainable management of coral reefs worldwide and even bans on harvesting of all organisms including 'living rock'.

The sea horses in particular have received attention from aquarists and conservationists. Sea horses are one of the most popular of all marine species and are probably responsible for converting more aquarists to marine aquarium-keeping than any other fish. Unlike most other marine ornamentals sea horses have been bred in captivity for many years.

In 1996, the international conservation group TRAFFIC (the monitoring arm of the World Wide Trust for Nature) claimed sea horses were under threat from overfishing for use in traditional medicines, aquaria, and as curios. According to TRAFFIC at least 22 countries export sea horses, the largest known exporters being India, the Philippines, Thailand and Vietnam. Importers for aquaria include Australia, Canada, Germany, Japan, The Netherlands, United Kingdom, and the United States. The species commonly in demand are shown in **Table 1**.

The greatest problem for both sides in the debate over the trade in wild-caught marine ornamental fishes in general is a historical lack of scientific data on the biology of these animals both in their natural and captive environments. It is true to say that most of the information about marine ornamentals is derived from intelligence gathering by aquarists and scientific rigor has been applied in the case of only a few species. One such species is the pot-bellied sea horse, *Hippocampus abdominalis*, which has been studied both in the wild and captive environment for several years. Data on this species serve as useful

Table 1 Sea horse species commonly kept in aquaria

Aquarium common name	*Species name*	*Geographic reference*	*Size (cm)*	*Color*
Dwarf sea horse	*Hippocampus zosterae*	Florida coast	5	Green, gold, black, white
Northern giant	*H. erectus*	New Jersey coast	20	Mottled, yellow, red, black, white
Spotted sea horse	*H. reidi*	Florida coast	18	Mottled, white, black
Short-snouted	*H. hippocampus*	Mediterranean	15	Red/black
Mediterranean	*H. ramulosus*	Mediterranean	15	Yellow/green
Golden or Hawaiian	*H. kuda*	Western Pacific	30	Golden
Pot-bellied sea horse	*H. abdominalis*	Southern Australian coast	30	Mottled, gold, white, black

comparative tools for knowledge about other ornamental fishes.

The Pot-bellied Sea Horse and Other Aquarium Species

Species Suitability for Aquaria

Table 2 sets out major parameters affecting life support in marine aquaria. Factors about a species that it would be advantageous to know prior to selection for the aquarium are:

- water temperature range,
- water quality requirements,
- behavior and habitat (territorial, aggressive, cannibalistic, pelagic, benthic),
- diet,
- breeding biology,
- size and age,
- ability to withstand stress,
- health (resistance and susceptibility).

Water Temperature Range

The pot-bellied seahorse *H. abdominalis* has a broad distribution along the coastal shores of Australia, being found from Fremantle in Western Australia eastwards as far as Central New South Wales, all around Tasmania and also much of New Zealand. Within its range water temperatures may reach 28°C for several months and fall to 9°C for a few weeks. Acclimation trials in the laboratory and in home aquaria have shown that this species will live at water temperatures as high as 30°C and as low as 8°C. Unlike many marine ornamental fishes, *H. abdominalis* is eurythermal.

Many marine aquaria are kept between 24 and 26°C which is considered satisfactory for a number of marine ornamentals, however some species require higher temperatures, for example some butterfly fishes (Chaetodontidae) survive best at 29°C. Many tropical coral reef species are stenothermous and are difficult to maintain in temperate climates without accurate thermal control. The more eurythermous a species, the easier it will be to acclimate to a range of water temperature fluctuations.

Water Quality Requirements

Salinity The salinity of sea water may alter due to freshwater run-off or evaporation. Some marine species are less tolerant than others to salinity changes, and it is important to determine whether or not a species will survive even relatively minor changes. *H. abdominalis* is euryhaline, growing and breeding at salinities between 15–37 parts per thousand (‰). It is a coastal dwelling species being recorded at depths of 1–15 m and may be present in a range of habitats from estuaries, open rocky substrates and artificial harbors. Often, pot-bellied sea horses can be found attached to nets and cages used to farm other fish species. The somewhat euryecious behavior of this sea horse has probably resulted in its broad tolerance of salinities.

Many marine aquaria depend on artificial sea water which is purchased as a salt mixture and added to dechlorinated fresh water. Natural sea water is a complex chemical mixture of salts and trace elements. It has been shown that some artificial seawater mixtures are unsuitable for marine plants and even particular life stages of some fishes. Particular attention must be paid to trace elements in coral reef aquaria when using either natural or artificial salt water. Coral reef aquaria are also more sensitive to salinity changes than general fish aquaria. **Table 3** gives a useful saltwater recipe.

Table 2 Parameters affecting life support in marine aquaria

Parameter	*Factor*
Physical	
Temperature	
Salinity	
Particular matter	Composition
	Size
	Concentration
Light	Artificial/natural
	Photoperiod
	Spectrum
	Intensity
Water motion	Surge
	Laminar
	Turbulence
Chemical	
pH and alkalinity	
Gases	Total gas pressure
	Dissolved oxygen
	Un-ionized ammonia
	Hydrogen sulfide
	Carbon dioxide
Nutrients	Nitrogen compounds
	Phosphorus compounds
	Trace metals
Organic compounds	Biodegradable
	Nonbiodegradable
Toxic compounds	Heavy metals
	Biocides
Biological	
Bacteria	
Virus	
Fungi	
Others	

Table 3 The Wiedermann–Kramer saltwater formula

In 100 liters of distilled water:	
Sodium chloride (NaCl)	2765 g
Magnesium sulfate crystals ($MgSO_4 \cdot 7H_2O$)	706 g
Magnesium chloride crystals ($MgCl_2 \cdot 6H_2O$)	558 g
Calcium chloride crystals[a] ($Cacl_2 \cdot 6H_2O$)	154 g
Potassium chloride[a] (KCl)	69.7 g
Sodium bicarbonate ($NaHCO_3$)	25 g
Sodium bromide (NaBr)	10 g
Sodium bicarbonate ($NaCO_3$)	3.5 g
Boric acid (H_3BO_3)	2.6 g
Strontium chloride ($SrCl_2$)	1.5 g
Potassium iodate (KIO_3)	0.01 g

[a]The potassium chloride should be dissolved separately with some of the 100 liters of distilled water as should the calcium chloride. Add these after the other substances have been dissolved.

Other Water Quality Guidelines

Various tables have been provided setting out water quality guidelines for mariculture. **Table 4** is useful for well-stocked marine fish and coral reef aquaria but is possibly too rigid for lightly stocked fish tanks.

The toxicity of several parameters given in **Table 3** may be reduced by ensuring the water is always close to saturation with respect to dissolved oxygen concentration (DOC). Studies on *H. abdominalis* have indicated that this species becomes stressed when DOC falls below 85% saturation. Furthermore, the species is tolerant of much higher un-ionized ammonia (NH_3-N) levels when the water is close to DOC saturation. A combination of low dissolved oxygen (<80%) and NH_3N greater than 0.02 mg l^{-1} may result in high mortalities.

Ammonia is the major end product of protein metabolism in sea horses and most aquatic animals. It is toxic in the un-ionized form (NH_3). Ammonia concentration expressed as the NH_3 compound is converted into a nitrogen basis by multiplying by 0.822. The concentration of un-ionized ammonia depends on total ammonia, pH, temperature, and salinity. The concentration of un-ionized ammonia is equal to:

$$\text{Un-ionized ammonia (mgl}^{-1}\text{ as NH}_3\text{-N)} = (a)(\text{TAN})$$

where a = mole fraction of un-ionized ammonia and TAN = total ammonia nitrogen (mg l^{-1} as N).

Table 5 gives the mole fraction for given temperatures and pH in sea water.

The concentration of un-ionized ammonia is about 40% less in sea water than fresh water, but its toxicity is increased by the generally higher pH in the former. **Figure 1(A)** shows the operation of a simple subgravel filter which removes ammonia and nitrite

Table 4 Water quality levels for the aquarium

Parameter	*Level*
Dissolved oxygen	90–100% saturation (>6 mg l^{-1})
Ammonia	<0.02 mg l^{-1} NH_3-N
Nitrite	<0.1 mg l^{-1} NO_2-N
Hydrogen sulfide	<0.001 mg l^{-1} as H_2S
Chlorine residual	<0.001 mg l^{-1}
pH	7.8–8.2
Copper	<0.003 mg l^{-1}
Zinc	<0.0025 mg l^{-1}

Table 5 Mole fraction of un-ionized ammonia in sea water[a]

Temp (°C)	*pH*				
	7.8	*7.9*	*8.0*	*8.1*	*8.2*
20	0.0136	0.0171	0.0215	0.0269	0.0336
25	0.0195	0.0244	0.0305	0.0381	0.0475
30	0.0274	0.0343	0.0428	0.0532	0.0661

[a]Modified from Huguenin and Colt (1989).

by nitrification and **Figure 1(B)** shows the configuration of a power filter using both nitrification and absorption to remove ammonia. Both methods have been used to maintain water quality in marine ornamental aquaria.

Unfortunately, information on toxicity levels of ammonia for marine ornamentals is poorly documented but **Table 4** is probably a useful guide given the data on cultured species.

The pot-bellied sea horse is intolerant to even low levels of hydrogen sulfide (H_2S) (**Table 4**). This gas is difficult to measure at low levels thus care is required to avoid anoxic areas in aquaria. H_2S is almost certainly toxic to other marine ornamentals also, particularly reef dwellers.

Chlorine, copper and zinc have all proved toxic to *H. abdominalis* at levels exceeding those shown in **Table 4**. Chorine and copper are often used in aquaria: the former to sterilize equipment and the latter in treatment of various diseases. Furthermore, chlorine may be present in tap water when mixed with artificial seawater mixtures. Great care is required in the use of these chemicals. Available aquarium test kits are seldom sensitive enough to detect chronic chlorine concentrations. Often 1–5 mg l^{-1} sodium thiosulfate or sodium sulfite are used to remove chlorine but for some marine species these too may prove toxic. Bioassays for chlorine toxicity using marine ornamentals have not been carried out.

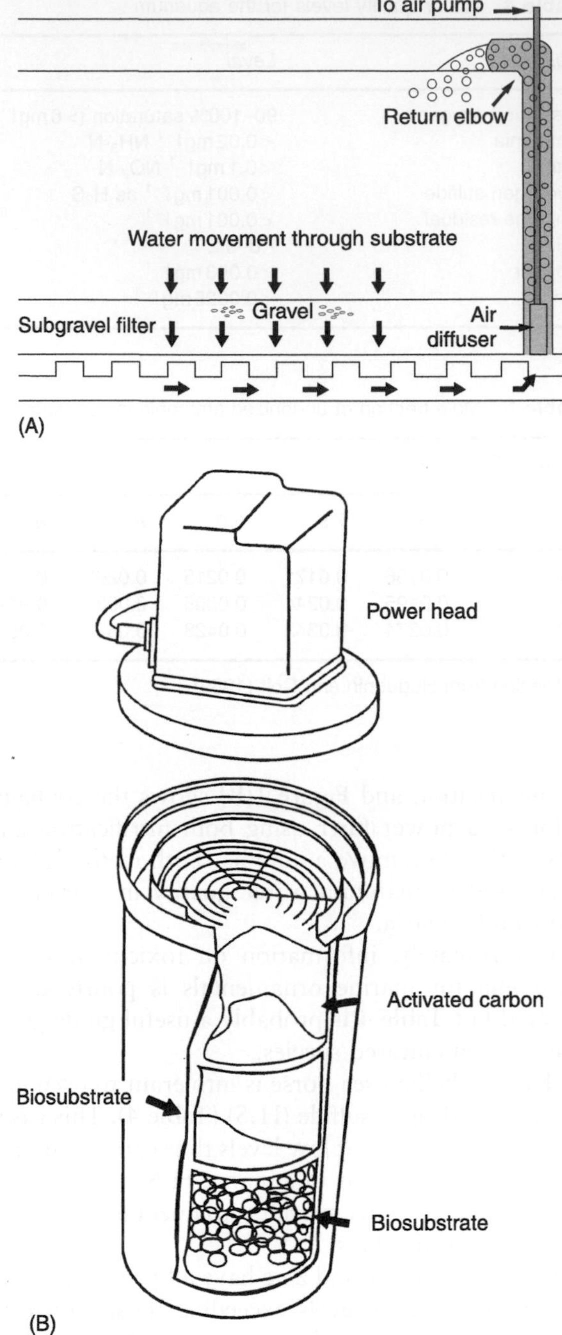

Figure 1 (A) Undergravel filtration within an aquarium tank. (B) Canister filter with power head and filter media chambers.

Copper toxicity can be significantly reduced with the addition of 1–10 mg l^{-1} of EDTA. EDTA is also a good chelating agent for zinc.

Behavior and Habitat

It is widely believed that sea horses spend their time anchored by their prehensile tails to suitable objects. This is not necessarily true. *H. abdominalis* is an active species feeding both in the water column and over the substratum. However, at night the fish 'roosts' often in association with other specimens. Furthermore, this species is remarkably gregarious and stocking levels as high as 10 fish, 6 cm in length per liter, have been regularly maintained in hatchery trials.

Although sea horses may tolerate the presence of several of their own species, their slow feeding behavior puts them at a competitive disadvantage in a mixed species aquarium, where faster feeders will ingest the sea horses' food.

Predation is a serious problem in marine aquaria. Sea horses are known to be prey for other fishes both in the wild and in aquaria. Members of the antennariids, particularly the sargassum fish, *Histrio histrio*, are known to feed on sea horses as are groupers (Serranidae) and trigger Fishes (Balistinae), flatheads (Platycephalidae) and cod (Moridae).

Territorial species are common among the coral-dwelling fishes and such behavior makes them difficult to keep in mixed-species aquaria. The blue damsel, *Pomacentrus coelestris*, shoals in its natural habitat but becomes pugnacious in the confines of the aquarium.

Several marine ornamentals seek protection or are cryptic. The majority of clown or anemone fish (*Amphiprion*) retreat into sea anemones if threatened, in particular, the anemones *Stoichaetis* spp., *Radianthus* spp., and *Tealia* spp., whereas some wrasse dive beneath sand when frightened. Cryptic coloration is seen in the sea horses and color changes have often been reported.

The cleaner wrasse, *Labroides dimidiatus*, lives in shoals over reefs but is successfully maintained singly in the aquarium, where even the most aggressive fish species welcome its attention. Other wrasses mimic the coloration and shape of *L. dimidiatus* simply to lure potential prey towards them.

Behavior and habitat of many marine ornamentals has been gleaned from observation only, but failure to understand these factors make fish-keeping difficult.

Diet

The dietary requirements are poorly known for marine ornamentals and many artificial feeds may do little more than prevent starvation without live or frozen feed supplements. Furthermore, a given species may require different foods at various life stages. Several stenophagic species are known in the aquarium mainly consisting of algal and live coral feeders, for example the melon butterfish *Chaetodon trifasciatus*. The diet of others may not even be known in spite of such fish being sold for the aquarium. The regal angelfish, *Pygoplites diacanthus*, seldom lives for long in the aquarium and dies from starvation.

Sea horses are easy to feed but require either live or frozen crustacea or small fish. The pot-bellied seahorse has been reared through all growth stages using diets of enriched brine shrimp, *Artemia salina*, live or frozen amphipods, small krill species, and fish fry. The ready acceptance and good growth rates recorded in hatchery-produced pot-bellied sea horses using 48-h-old enriched brine shrimp have resulted in significant numbers of sea horses being raised in at least one commercial farm. Apart from hatcheries for clown fishes and pot-bellied sea horses, the intensive culture of marine ornamentals has proved difficult due to a lack of suitable prey species for fry.

Breeding Behavior

Breeding behavior can be induced in several ornamental fishes with appropriate stimuli and environments. The easier species are sequential hermaphrodites such as serranids. The pomacentrids of the genus *Amphiprion* are a further good example. However, sea horses have been extensively studied.

The sea horses are unique in that the male receives, fertilizes and broods the eggs in an abdominal pouch following a ritualized dance. Much has been made of monogamy but as further studies are undertaken scientific support for such breeding behavior is being questioned. *H. abdominalis* is polygamous both in the wild and captivity.

The pot-bellied sea horse in captivity, at least, shows breeding behavior as early as four months of age and males may give birth to a single offspring; by one year of age males may give birth to as many as 80 fry and at two years of age 500 fry. Precocity in other marine ornamentals is not recorded but may exist.

Size and Age

The size and age of aquarium fish have been seldom studied scientifically but has been observed. Groupers (Serranidae) and triggerfishes (Balistinae) quickly outgrow aquaria, and some angelfishes and emperors may show dramatic color changes with age, becoming more or less pleasing to aquarists.

Longevity likewise is unknown for most marine ornamentals but the pygmy sea horse *H. zosterae* lives for no more than two years, whereas *H. abdominalis* may live for up to nine years.

Stress

Stress probably plays a pivotal role in the health of marine ornamentals but scientific studies have not been undertaken. The aquarist would do well to remember that stress suppresses aspects of the immune response of fishes and that studies on cultured species demonstrate that capture, water changes, crowding, transport, temperature changes, and poor water quality induce stress responses. Furthermore, stress can be cumulative and some species may be more responsive than others. Farmed species tend to show a higher stress threshold than wild ones. The potential advantage of purchasing hatchery-reared ornamentals (if available) are obvious, since survival in farmed stock should be greater than in wild fish held in aquaria.

Health Management

Good health management results from an understanding of the biological needs of a species. Treatment with chemicals is a short-term remedy only and the use of antibiotics may exacerbate problems through bacterial resistance.

A considerable number of pathogens have been recorded in marine aquarium fishes and include viruses, bacteria, protozoa, and metazoa. Most diseases have been shown not to be peculiar to a given species but epizootic. For example, several of the disease organisms recorded in sea horses, in particular *Vibrio* spp., protozoa and microsporidea, are known to infect other fish species also.

In coral reef aquaria, nonpathogenic diseases due to poor water quality may be common and in-depth knowledge pertaining to the husbandry of such systems is essential for their well-being.

Coral Reef Aquaria

The Challenge

The coral reef is one of the best-adapted ecosystems to be found in the world. Such reefs are biologically derived and the organisms which contribute substantially to their construction are hermatypic corals although ahermatypic species are present. Coral reefs support communities with a species diversity that far exceeds those of neighboring habitats and the symbiotic relationship between zooxanthellae and the scleractinian corals are central to the reef's well-being. Zooxanthellae are also present in many octocorallians, zoanthids, sea anemones, hydrozoans and even giant clams. As zooxanthellae require light for photosynthesis, the reef is dependent on clear water.

Coral reefs are further restricted by their requirement for warm water at 20–28°C, and the great diversity of life demands a plentiful supply of oxygen. The challenge for the aquarist lies in the need to match the physical parameters of the water in the aquarium as closely as possible with sea water of the reef itself.

Physical Considerations

Temperature The recommended water temperature for coral reef aquaria is a stable 24°C. The greater the temperature fluctuations the less the diversity of life the aquarium will support. At temperatures less than 18°C the reef will die and above 30°C increasing mortalities among zooxanthellae will occur leading to the death of hermatypic corals.

Light Water bathing coral reefs has a blue appearance which has been called the color of ocean deserts. Most of the primary production is the result of photosynthesis by benthic autotrophs (zooxanthellae) rather than drifting plankton. Photosynthetic pigments of zooxanthellae absorb maximally within the light wavelength bands that penetrate furthest into sea (400–750 μm) and therefore clear oceanic water is essential.

In the aquarium, both fluorescent and metal halides are available which will supply light at the correct wavelength. Actimic-03 fluorescents combined with white fluorescents are suitable. Typically three tubes, two actinic and one white, will be needed for a 200 liter tank. Metal halide lamps cannot be placed close to the tank because of heat and such lamps produce UV light which will destroy some organisms. A glass sheet placed over the aquarium will prevent UV penetration. One 175 W lamp is recommended per 60 cm of aquarium length.

Water movement Coral reefs are subjected to various types of water movements, namely surges, laminar currents, and turbulence. These water motions play an essential role by bringing oxygen to the corals and plants, removing detritus and, in the case of ahermatypic corals, transporting their food.

In the aquarium these necessary water movements must be present. Power head filters are available which produce acceptable surges and currents.

Biological Considerations

Nutrients Coral reefs are limited energy and nutrient traps: rather than being lost to deep water sediments, some organic compounds and nutrients are retained and recycled. However, water movements rid the reef of dangerous excess nutrients which might promote major macrophyte growth.

The coral reef aquarium soon becomes a nutritional soup if both organic compounds and nutrients are not recycled or removed. Although skilled and knowledgeable aquarists are able to use the biological components of the reef itself to produce an autotrophic system, most employ protein skimmers, mechanical filters and biological filters to prevent poisoning of the system. **Figure 2** represents nutrient cycling over a coral reef and **Figure 1** shows a suitable filter for reef aquaria. Removal of nitrate can be achieved through the use of specialized filters which grow denitrifying bacteria.

Living rock Living rock is dead compacted coral which has been colonized by various invertebrates. In addition, there will be algae and bacteria. Different sources of living rock will provide different populations of organisms. Over time the organisms which survived the transfer from reef to aquarium become established

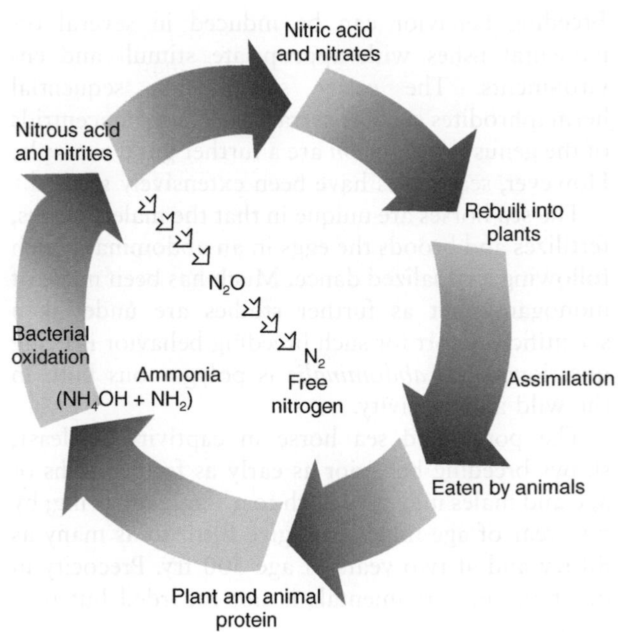

Figure 2 Nutrient cycling over a coral reef.

Table 6 Some additional faunal components for coral reef aquaria

Niche	*Common name*	*Scientific name*
Detritus feeder	Anemone shrimp	*Periclimenea brevicarpalis*
	Shrimp	*P. pedersoni*
	Fiddler crab	*Uca* spp.
Algal feeder	Tiger cowrie	*Cypraea tigris*
	Money cowrie	*C. moneta*
	Starfish	*Patiria* spp.
Plankton feeders	Mandarin fish	*Synchiropus splendidus*
	Psychedelic fish	*S. picturatus*
	Midas blenny	*Escenius midas*

and the aquarium is ready for corals. By the time the corals are placed in the aquarium all filter systems must be operating efficiently. Ahermatypic corals should be introduced first and placed in the darker regions of the tank since they feed on plankton. Once hermatypic corals are introduced appropriate blue light for at least 12 hours each day must be available, and strict water quality maintained (**Table 4**). Nitrate must not rise above $15\,\mathrm{mg\,l^{-1}}$ and pH fall below 8.

Additional faunal components The living rock will introduce various invertebrates. Additional species must be selected carefully and on a scientific rather than an esthetic basis. The introduction of detritus and algal feeders will probably be essential and coral eaters must be avoided. Fish species require high protein diets which will necessitate further reductions of ammonia, nitrite, and nitrate from the aquarium. The species given in **Table 6** might be considered but there are many others. Selection will depend on the inhabitants.

See also

Coral Reefs. Corals and Human Disturbance. Mariculture Diseases and Health. Mariculture Overview.

Further Reading

Barnes RSK and Mann KH (1995) *Fundamentals of Aquatic Ecology*. Oxford: Blackwell Scientific.

Emmens CW (1995) *Marine Aquaria and Miniature Reefs*. Neptune City: T.F.H. Publications.

Hawkins AD (1981) *Aquarium Systems*. New York: Academic Press.

Huguenin JE and Colt J (1989) *Design and Operating Guide for Aquaculture Seawater Systems*. Amsterdam: Elsevier.

Lawson TB (1995) *Fundamentals of Aquacultural Engineering*. New York: Chapman & Hall.

Spottle S (1979) *Seawater Aquariums – the Captive Environment*. New York: John Wiley.

Timmons MB and Losordo TM (1994) *Aquaculture Water Reuse Systems. Engineering Design and Management*. Amsterdam: Elsevier.

Untergasser D (1989) *Handbook of Fish Diseases*. Neptune City: T.F.H. Publications.

MARICULTURE OF MEDITERRANEAN SPECIES

G. Barnabé, Université de Montpellier II, France
F. Doumenge, Musée Océanographique de Monaco, Monaco

Basic Requirements

Obtaining Stock for Ongrowing

The starting point of any farming operation is the acquisition of stock for rearing; these may be spat for mollusks or alevins, fry or juveniles for fish.

From the wild For the Mediterranean mussel (*Mytilus galloprovincialis*), spat is always collected from the wild, from rocky shores or shallow harbors where they are abundant. Conditions are less favorable for oyster culture; the native (flat) oyster (*Ostrea edulis*) is captured only in the Adriatic and the other remnants of natural stocks are unable to support intensive culture. Spat from Japanese (cupped) oysters (*Crassostrea gigas*) has to be imported from the Atlantic coast. Clam culture utilizes both spat from Mediterranean species (*Tapes decussatus*) and a species originating in Japan (*Tapes philippinarum*) which has spread very rapidly, especially in the Adriatic.

Juvenile marine fish such as eels (*Anguilla anguilla*), mullets (*Mugil* sp.), gilthead sea bream (*Sparus aurata*) and sea bass (*Dicentrarchus labrax*) are traditionally captured in spring in the mouths of rivers or in traps or other places in protected lagoons in Italy; these form the basis of the valliculture of the northern Adriatic, a type of extensive fish culture. In practice, elvers and yellow eels are supplied mainly from fisheries in other Mediterranean lagoons, especially in France.

Bluefin tuna (*Thunnus thynnus*) for ongrowing are taken from the spring and summer fishery for juveniles weighing a few tens of kilograms by Spanish seine netters in the western Mediterranean and Croatians in the Adriatic.

Controlled reproduction in hatcheries The transition to intensive fish culture has been made possible by the control of reproduction in sea bass and sea bream (*see* Mariculture Overview). In 1999 around 100 hatcheries produced 209 million sea bass fry and 242 million gilthead sea bream fry; this represents respectively a doubling and a 50% increase on production in 1997. Hatcheries are very different from the structures used for ongrowing; France exports tens of millions of juveniles to all the Mediterranean countries.

Access to Technology

As an activity, aquaculture is becoming increasingly complex with regard to the technology associated with rearing as well as the interactions with the physical and socio-economic environment. It requires more specialized training than can be obtained in the traditional workplace. Management of a hatchery, monitoring of water quality, and genetic research are examples of the new requirements for training.

The transition to cage culture took place using cages designed and manufactured for salmonid culture. Manufacturing feed granules has developed only recently in countries such as Greece.

Feeding

Mollusk culture exploits the natural production of plankton and thus follows natural changes in productivity.

The extensive valliculture systems used in northern Italy in managed protected lagoons are based on natural production improved through control of the water (fish trapped in the channels communicating with the sea, overwintering in deep areas) and input of juveniles. Production remains low (20 kg ha^{-1} y^{-1}).

Intensive fish culture in cages uses pelleted protein-rich diets where fishmeal comprises 60% of the dry weight. However, bluefin tuna only consume whole fish or fresh or frozen cephalopods during ongrowing. Their conversion rate is exceptionally poor, of the order of 15–20% depending on water temperature.

Transport

The aquaculture cycle requires dependable and rapid transport to take juveniles to ongrowing facilities and, especially, to move the final production output which is generally valuable and perishable to the consumer market, and must remain chilled at all times.

For isolated sites, particularly small islands, there is always the problem that the transport of feed and various items of equipment is expensive.

Manpower Requirements

Hatcheries require a specialized and motivated workforce, as the production cycle for fry must not be interrupted. Sea sites are becoming ever more highly mechanized, even in countries such as Turkey where labor is cheap, because the volumes to be handled are increasing continuously.

Capital

Mediterranean aquaculture requires heavy investment both for setting up and for operating. It is dependent on international capital. Scandinavian, British, and Japanese interests play a considerable part, alongside local entrepreneurs.

Distinctive aspects

Favorable

Sheltered waters There are numerous large areas of sheltered water around the Mediterranean, the shoreline having been submerged and straightened out by the rapid rise in sea level following the last glaciation, 12 000 years ago. Nowadays, many bays provide big expanses of water which are both deep and sheltered (Carthagena, Toulon, La Spezia, Gaete, Naples, Trieste, Thessaloniki, etc.). These waters are productive, but are threatened by pollution.

Channels between the islands of the Dalmatian archipelago or in the Aegean Sea or the large gulfs surrounding the Hellenic peninsular (Patras, Corinth, Thessaloniki, Argolis, Evvoia, etc.), and to the north of Entail are perfectly suited to the types of aquaculture systems that benefit from a rapid turnover of water.

Above all, the input of sediment from deltas or the effects of littoral erosion have led to the build-up of sand bars forming lido-type complexes of impounded lagoons where the waters experience strong variations in salinity and high productivity because of the movements through gaps in the barriers.

High demand from local markets The Latin countries of the north-west Mediterranean and Adriatic have a culinary tradition based on the consumption of large quantities of seafood. Markets in Spain, Mediterranean France, and Italy pay high prices for the fresh products of mollusk culture and marine fish farming.

There is also the strong seasonal demand from more than 100 million tourists who travel around all the coasts and particularly the islands of the Mediterranean, except where there are political problems.

Unfavorable

Eutrophication, toxic blooms, bacterial pollution As elsewhere in the world, Mediterranean aquaculture cannot escape problems of disease. Nodovirus has hit fish in cages and two protozoans have decimated cultured stocks of the native oyster, *Ostrea edulis*.

Harmful blooms frequently affect lagoons where shellfish are cultured and the presence of serious pathogens (cholera, salmonella) has prohibited the use of certain zones for production.

Extreme temperatures In sheltered lagoons or enclosed waters (e.g., the Bougara Sea, Tunisia), temperatures may exceed 30°C in summer, causing harmful blooms which are prejudicial to the success of aquaculture. In the north-west Mediterranean and the northern Adriatic, the water temperature in the lagoons may drop to below + 5°C and ice may develop on the surface of the lagoons. In general, thermal conditions favor the eastern Mediterranean.

Competition for space (urbanization, tourism) Littoral space is under pressure from several users; tourism, navigation, and especially the extension of industrial and urban developments. Land bases are essential for mariculture. Lack of sites is further aggravated by complex regulations which are applied rigorously. Finally, the tendency to designate large areas of wetland and similar areas of sea for nature conservation removes significant areas from the expansion of aquaculture.

Dependence on fisheries When juvenile mollusks or fish for mariculture rearing are taken from the natural environment they are supplied by fisheries. This dependence is a major risk to the effective operation of the production process. In the absence of industrial pelagic fisheries there is no fishmeal production in the Mediterranean. In order to satisfy the demand linked to the expansion of fish culture, almost all components of the feed must be imported. This places production further under external control. Substitution of plant for animal protein in the diet is becoming necessary.

Cultural limits of the markets While the culture of Roman Catholic Christianity provides for a boosted local market, the Orthodox Christian culture is far less demanding in terms of seafood products. The Muslim countries of the southern coast and the Levantine Basin do not have the monetary resources nor, significantly, a dietary tradition adapted to the products of marine aquaculture. In contrast, the Mediterranean bluefin tuna achieves high prices on the Japanese sashimi market.

Production Systems

Mollusk Culture

The oyster reared everywhere (although sometimes in small quantities) is the cupped or Japanese oyster (*Crassostrea gigas*) introduced to Europe from Japan in the 1960s to replace the Portuguese oyster (*Crassostrea angulata*), stocks of which had been drastically reduced by disease. One single major center, the Etang de Thau, produces around 12 000 tonnes each year using spat taken from the Atlantic coast.

The native mussel (*Mytilus galloprovincialis*) is reared in small quantities around the Spanish coast (Ebro Delta, Mar Menor). In contrast, in Italy annual production is between 100 000 and 130 000 tonnes; the main sites are the Gulf of Taranto and the northern and mid-Adriatic, as well as several bays in the Tyrrhenean Sea (La Spezia, Gaete, and Naples). Along the French coast, only the Etang de Thau has a production of a few thousand tonnes. In Greece, the Gulf of Thessaloniki has a production of the same order.

For around 20 years, Japanese-type long lines have been installed in the open sea off Languedoc Rousillon in waters between 15 and 35 m deep. These structures resist the forces of the sea and production can reach tens of thousands of tonnes of mussels, but is strongly limited by storms and predation by sea bream.

Oysters and mussels are always cultured in suspension on ropes either in lagoons or in the sea. The spat of oyster is captured on the Atlantic coast on various substrates and is often attached with the help of quick-setting cement on the rearing bars, as this technique yields the best results. Other farmers leave the spat to develop on the mollusc shells where they have attached; the shells are placed in nets, spaced regularly, on the rearing ropes.

Mussel spat, collected from the wild, is placed in a tubular net, which is then attached to the rearing ropes. Mussels attach themselves to the artificial substrate with the aid of their byssus. It is therefore necessary to detach and clean the mussels once or twice during their growth.

Clams are especially abundant around the Italian Adriatic coast. The Japanese species, introduced as hatchery-reared spat at the beginning of the 1980s to seed protected areas, has spread very rapidly and invaded neighboring sectors. The density of these mollusks can reach 4000 individuals per square meter and production is around 40 000 tonnes per annum. A large part of the production is exported to Spain.

Culture of Sea Bass and Gilthead Sea Bream

Control and management of the spawning of sea bass and sea bream has made it possible to respond to the demand for high quality aquatic products that cannot be supplied on a regular basis from fisheries based on wild stocks. This production is based on hatcheries (*see* Mariculture Overview). Broodstock can be held in cages or in earth ponds. In such ponds, control of the photoperiod and temperature allows fertilized eggs to be obtained through almost the whole year. Larvae are reared in the same way as those of other marine fish (*see* Marine Fish Larvae) although sea bass can be fed on *Artemia nauplii* from first feeding; this does away with the burden of rearing rotifers in these hatcheries. The trend is towards enlargement to bring about economies of scale; some hatcheries produce 20 million juveniles each year; those producing fewer than 1–2 million juveniles per year are unlikely to be profitable.

Water is generally recycled within the hatchery to save energy and to maintain the stability of physicochemical characteristics. The quality of the water available in the natural environment determines the suitability of a site for a hatchery.

Almost all ongrowing is carried out in sea cages. All Mediterranean countries have contributed to the development and there is a tendency to move further and further east. Cages have been installed in sheltered coastal waters, but the scarcity of such sites and progress in cage technology are encouraging the development of exposed sites in the open sea. Techniques used in this type of farming resemble those used in the cage farming of salmonids. (*see* Salmonid Farming) and identical cages are used. These have a diameter of up to 20 m and are up to 10 m deep. Dry granular feed is used and the conversion rate is continuously improving (between 1.3 and 2).

Growth rate is determined by water temperature; in Greece a sea bass reaches a weight of 300 g in 11 months, but would take twice as long to reach the same weight on the French coast. This demonstrates the advantage of the eastern Mediterranean.

Other species, for which larval rearing is more difficult, are reared on a small scale using different techniques. These include the dentex (*Dentex dentex*), common (Couch's) sea bream (*Pagrus pagrus*) produced in mesocosms in Greece, and the greater amberjack (*Seriola dumerilii*), ongrown in Spain. Diversification of species offers the potential for increasing markets; this is becoming true for the meagre (*Argyrosomos regius*) and the red drum (*Scianops ocellatus*), and many other trials are taking place.

Ongrowing Bluefin Tuna

Japanese attempts at rearing larval bluefin tuna have not yet produced economically viable results. Mediterranean aquaculture depends on juveniles or sub-adults captured in the spring fishery, which are transported in nets supported by rafts to large cages where they will be kept for a few months. The towage, which must be at speeds of between 1 and 2 knots, may take several weeks.

The fish are harvested at the end of autumn and beginning of winter, before periods of low temperature and bad weather. In general, the tuna double their weight after 6–7 months of ongrowing and their flesh acquires the color and quality which puts them in demand for the Japanese sashimi market.

Regionalization

During the last 20 years, regional specialization has developed progressively, based on a balance of favorable and unfavorable factors for each of the types of mariculture.

The development of Mediterranean aquaculture production has thus been subject to an evolutionary process which has taken account not only of the major factors previously described but also conditions peculiar to each nation. This has produced contrasting situations, as decribed below.

France and Spain – Relatively Low Level of Aquaculture Development

This is due to a coincidence of relatively high levels of aquaculture development on the Atlantic coasts of both countries (mussels and fish in Galicia, mussels and oysters in the Bay of Biscay and the English Channel) and large quantities of imports supplementing regional production. Aquaculture developments remain small in size and are very spread out. One exception, demonstrating the possibilities that exist, is the success of the ongrowing of bluefin tuna in Spain in the Bay of Carthagena, from which 6000–7000 tonnes were sold in 1999–2000, for a revenue of over US $100 000 000.

The Size of the Italian Market

In the year 2000 only one third of the market was supplied by national production, in spite of rapid increases; production of sea bass (8800 tonnes) and gilthead sea bream (6200 tonnes) have both doubled from 1997. Italy absorbs almost all of the Maltese production (600 tonnes of sea bass and 1600 tonnes of sea bream in 2000); most of this comes from stock from eastern Mediterranean hatcheries.

The 40 000 tonnes of clams produced in the Adriatic saturate the market and the excess is exported to Spain. In spite of health problems, Italian mussel culture supports an annual market of 100 000–130 000 tonnes. In addition, the upper Adriatic has preserved the tradition of exploiting the valli and the lagoons.

In 2000 the Italian aquaculture market (by value) was supplied 40% by sea bass and sea bream (200 million lire), 33% by clams (165 million lire), and 27% by mussels (135 million lire).

The Dalmatian archipelago belonging to Croatia retains a sector of high quality traditional oyster culture. However, new possibilities have opened up with the transfer of technology and finance from Croatian emigrants to south Australia who have developed the ongrowing of bluefin tuna in the Ile de Kali, using the techniques they practiced in Port Lincoln. However, biological and logistic constraints are currently holding back development. Production from the Croatian businesses in 1999 and 2000 was limited to 1000 tonnes of small tuna (20–25 kg each); these receive relatively poor prices in Japan.

The Pioneering Front for Culture of Sea Bass and Sea Bream in the Eastern Mediterranean

This has reached Greece where production has gone from 1600 tonnes in 1990 to 6000 tonnes in 1992, 18 000 tonnes in 1996, 36 000 tonnes in 1998 and 56 000 tonnes in 2000. The movement then reached neighbouring Turkey, going from 1200 tonnes in 1992 to 3500 tonnes in 1994, reaching 12 500 tonnes in 1998 and 14 000 tonnes in 2000. Cyprus has also recently developed production which reached 1500 tonnes in 2000.

As part of this conquest of space, the first Greek seafish farms were installed between 1985 and 1995 in the west, in the Ionian islands and in Arcadia (center of the Peloponese). Then, via the Gulf of Corinth, they were joined from 1990 to 1995 by an active center in Argolis (east of the Peloponese). In addition, suitable sites for cages were found in the bays behind the barrier of the Isle of Evvoia. Finally, since 1990, developments have progressed to the Aegean around the archipelagos fringing the Peloponese. Between 1995 and 2000 Greek sites appear to have become saturated and mariculture has moved to the Turkish coast and the Anatolian bays of the Aegean coast.

Perspectives and Problems

Lack of sites

The area dedicated to intensive cage mariculture remains modest: the whole of the French marine fish culture, around 6000 tonnes annual production, occupies no more than 10 ha of sea and 5 ha of land.

Pollution Ascribed to Aquaculture

It is politically correct to speak of the pollution derived from intensive cage rearing. When this alleged pollution (from fish excreting mainly nitrogen and phosphorus) is ejected into oligotrophic open waters such as the Mediterranean it can be seen as a benefit rather than a nuisance. The FAO, elsewhere, has suggested that significant increases in fish catches occur in areas where human-derived wastes have increased in the Mediterranean.

Health Limits and Shellfish

Production of mussels fluctuates widely: pollution of coastal and lagoon waters (toxic plankton and bacterial pollution) regularly prevents their sale. Regular consumption of mussels could be dangerous as the species concentrates okadaic acid (a strong carcinogen) produced by toxic phytoplankton.

Marketing Problems

Overproduction of sea bass and gilthead sea bream has led to a periodic collapse in prices. This phenomenon appears to be due to lack of planning, organization, and commercial astuteness by the producers, as well as competition between countries operating within different economic frameworks (cost of manpower). The salmon market is characterized by identical examples. For bluefin tuna, dependence on a single, distant market (the sashimi market in Japan) makes ongrowing a risky activity but very profitable in economic terms.

Conclusions

Traditional mollusk culture maintains its position but is encountering problems of limited availability of water. Transfer out to the open sea which is less polluted has been piloted in Languedoc (France) for two decades, but has demonstrated neither the suitability of the techniques nor their profitability, and production has stagnated.

The explosive growth of the production of marine fish in cages can be said to demonstrate the true revolution in Mediterranean mariculture. This is based entirely on species with a high commercial value. This type of rearing has expanded eastwards from the European Mediterranean countries but has not yet reached the southern shore.

Markets, particularly the huge European market of 360 million inhabitants, are not yet saturated. Diversification of the species produced may open up new markets. The expansion of cage-based mariculture has not yet finished, while progress in technology is unpredictable.

The major missing element in Mediterranean aquaculture is the rearing of penaeids, in spite of several sporadic but insignificant attempts at production in Southern Italy and Morocco.

See also

Mariculture Overview. Salmonid Farming.

Further Reading

Barnabe G (1974) Mass rearing of the bass *Dicentrarchus labrax* L. In: Blaxter JHS (ed.) *The Early Life History of Fish*, pp. 749–753. Berlin: Springer-Verlag.

Barnabe G (1976) Ponte induite ET élevage des larves du Loup *Dicentrarchus labrax* (L.) et de la Dorade *Sparus aurata* (L.). *Stud. Rev. C.G.P.M. (FAO)* 55: 63–116.

Barnabe G (1990) Open sea aquaculture in the Mediterranean. In: Barnabe G (ed.) *Aquaculture*, pp. 429–442. New York: Ellis Horwood.

Barnabe G (1990) Rearing bass and gilthead bream. In: Barnabe G (ed.) *Aquaculture*, pp. 647–683. New York: Ellis Horwood.

Barnabe G and Barnabe-Quet R (1985) Avancement et amélioration de la ponte induite chez le Loup Dicentrarchus labrax (L) à l'aide d'un analogue de LHRH injecté. *Aquaculture* 49: 125–132.

Doumenge F (1991) Meditérranée. In: *Encyclopaedia Universalis* pp. 871–873.

Doumenge F (1999) L'aquaculture des thons rouges et son développement économique. *Biol Mar Medi* 6: 107–148.

Ferlin P and Lacroix D (2000) Current state and future development of aquaculture in the Mediterranean region. *World Aquaculture* 31: 20–23.

Heral M (1990) Traditional oyster culture in France. In: Barnabé G (ed.) *Aquaculture*, pp. 342–387. New York: Ellis Horwood.

MARICULTURE OVERVIEW

M. Phillips, Network of Aquaculture Centres in Asia-Pacific (NACA), Bangkok, Thailand

Introduction: Mariculture – A Growing Ocean Industry of Global Importance

Global production from aquaculture has grown substantially in recent years, contributing in evermore significant quantities to the world's supply of fish for human consumption. According to FAO statistics, in 2004, aquaculture production from mariculture was 30.2 million tonnes, representing 50.9% of the global total of farmed aquatic products. Freshwater aquaculture contributed 25.8 million tonnes, or 43.4%. The remaining 3.4 million tonnes, or 5.7%, came from production in brackish environments (**Figure 1**). Mollusks and aquatic plants (seaweeds), on the other hand, almost evenly make up most of mariculture at 42.9% and 45.9%, respectively. These statistics, while accurately reflecting overall trends, should be viewed with some caution as the definition of mariculture is not adopted consistently across the world. For example, when reporting to FAO, it is known that some countries report penaeid shrimp in brackish water, and some in mariculture categories. In this article, we focus on the culture of aquatic animals in the marine environment.

Nevertheless, the amount of aquatic products from farming of marine animals and plants is substantial, and expected to continue to grow. The overall production of mariculture is dominated by Asia, with seven of the top 10 producing countries within Asia. Other regions of Latin America and Europe however also produce significant and growing quantities of farmed marine product. The largest producer of mariculture products by far is China, with nearly 22 million tonnes of farmed marine species. A breakdown of production among the top 15 countries, and major commodities produced, is given in **Table 1**.

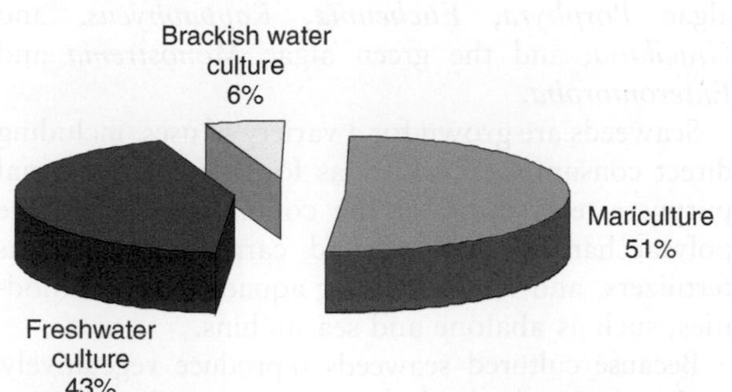

Figure 1 Aquaculture production by environment in 2004. From FAO statistics for 2004.

Commodity and System Descriptions

A wide array of species and farming systems are used around the world for farming aquatic animals and plants in the marine environment. The range of marine organisms produced through mariculture currently include seaweeds, mollusks, crustaceans, marine fish, and a wide range of other minor species such as sea cucumbers and sea horses. Various containment or holding facilities are common to marine ecosystems, including sea-based pens, cages, stakes, vertical or horizontal lines, afloat or bottom set, and racks, as well as the seabed for the direct broadcast of clams, cockles, and similar species.

Mollusks

Many species of bivalve and gastropod mollusks are farmed around the world. Bivalve mollusks are the major component of mariculture production, with most, such as the commonly farmed mussel, being high-volume low-value commodities. At the other end of the value spectrum, there is substantial production of pearls from farming, an extremely low-volume but high-value product.

Despite the fact that hatchery production technologies have been developed for many bivalves and gastropods, much bivalve culture still relies on collection of seedstock from the wild. Artificial settlement substrates, such as bamboo poles, wooden stakes, coconut husks, or lengths of frayed rope, are used to collect young bivalves, or spat, at settlements. The spat are then transferred to other grow-out substrates ('relayed'), or cultured on the settlement substrate.

Some high-value species (such as the abalone) are farmed in land-based tanks and raceways, but most mollusk farming takes place in the sea, where three major systems are commonly used:

- Within-particulate substrates – this system is used to culture substrate-inhabiting cockles, clams, and other species. Mesh covers or fences may be used to exclude predators.

Table 1 Top 15 mariculture producers

Country	*Production (tonnes)*	*Major species/commodities farmed in mariculture*
China	21 980 595	Seaweeds (kelp, wakame), mollusks (oysters, mussels, scallops, cockles, etc.) dominate, but very high diversity of species cultured, and larger volumes of marine fish and crustaceans
Philippines	1 273 598	Seaweeds (*Kappaphycus*), with small quantities of fish (milkfish and groupers) and mollusks
Japan	1 214 958	Seaweeds, mollusks, marine fish
Korea, Republic of	927 557	Seaweeds, mollusks, marine fish
Chile	688 798	Atlantic salmon, other salmonid species, smaller quantities of mollusks, and seaweed
Norway	637 993	Atlantic salmon, other salmonid species
Korea, Dem. People's Rep.	504 295	Seaweeds, mollusks
Indonesia	420 919	Seaweeds, smaller quantities of marine fish, and pearls
Thailand	400 400	Mollusks (cockle, oyster, green mussel), small quantities of marine fish (groupers)
Spain	332 062	Blue mussel dominates, but high diversity of fish and mollusks
United States of America	242 937	Mollusks (oysters) with small quantities of other mollusks and fish (salmon)
France	198 625	Mollusks (mussel and oyster) with small quantities of other mollusks and fish (salmon)
United Kingdom	192 819	Atlantic salmon and mussels
Vietnam	185 235	Seaweeds (*Gracilaria*) and mollusks
Canada	134 699	Atlantic salmon and other salmonid species, and mollusks
Total (all countries)	30 219 472	

From FAO statistics for 2004.

- On or just above the bottom – this culture system is commonly used for culture of bivalves that tolerate intertidal exposure, such as oysters and mussels. Rows of wooden or bamboo stakes are arranged horizontally or vertically. Bivalves may also be cultured on racks above the bottom in mesh boxes, mesh baskets, trays, and horizontal wooden and asbestos-cement battens.
- Surface or suspended culture – bivalves are often cultured on ropes or in containers, suspended from floating rafts or buoyant long-lines.

Management of the mollusk cultures involves thinning the bivalves where culture density is too high to support optimal growth and development, checking for and controlling predators, and controlling biofouling. Mollusk production can be very high, reaching 1800 tonnes per hectare annually. With a cooked meat yield of around 20%, this is equivalent to 360 tonnes of cooked meat per hectare per year, an enormous yield from a limited water area. Farmed bivalves are commonly sold as whole fresh product, although some product is simply processed, for example, shucked and sold as fresh or frozen meat. There has been some development of longer-life products, including canned and pickled mussels.

Because of their filter-feeding nature, and the environments in which they are grown, edible bivalves are subject to a range of human health concerns, including accumulation of heavy metals, retention of human-health bacterial and viral pathogens, and accumulation of toxins responsible for a range of shellfish poisoning syndromes. One option to improve the product quality of bivalves is depuration, which is commonly practiced with temperate mussels, but less so in tropical areas.

Seaweeds

Aquatic plants are a major production component of mariculture, particularly in the Asia-Pacific region. About 13.6 million tonnes of aquatic plants were produced in 2004. China is the largest producer, producing just less than 10 million tonnes. The dominant cultured species is Japanese kelp *Laminaria japonica*. There are around 200 species of seaweed used worldwide, or which about 10 species are intensively cultivated – including the brown algae *L. japonica* and *Undaria pinnatifida*; the red algae *Porphyra*, *Eucheuma*, *Kappaphycus*, and *Gracilaria*; and the green algae *Monostrema* and *Enteromorpha*.

Seaweeds are grown for a variety of uses, including direct consumption, either as food or for medicinal purposes, extraction of the commercially valuable polysaccharides alginate and carrageenan, use as fertilizers, and feed for other aquaculture commodities, such as abalone and sea urchins.

Because cultured seaweeds reproduce vegetatively, seedstock is obtained from cuttings. Grow-out is undertaken using natural substrates, such as long-lines, rafts, nets, ponds, or tanks.

Production technology for seaweeds is inexpensive and requires only simple equipment. For this reason, seaweed culture is often undertaken in relatively undeveloped areas where infrastructure may limit the development of other aquaculture commodities, for example, in the Pacific Island atolls.

Seaweeds can be grown using simple techniques, but are also subject to a range of physiological and pathological problems, such as 'green rot' and 'white rot' caused by environmental conditions, 'ice-ice' disease, and epiphyte growth. In addition, cultured seaweeds are often consumed by herbivores, particularly sea urchins and herbivorous fish species, such as rabbitfish.

Selective breeding for specific traits has been undertaken in China to improve productivity, increase iodine content, and increase thermal tolerance to better meet market demands. More recently, modern genetic manipulation techniques are being used to improve temperature tolerance, increase agar or carrageenan content, and increase growth rates. Improved growth and environmental tolerance of cultured strains is generally regarded as a priority for improving production and value of cultured seaweeds in the future.

Seaweed aquaculture is well suited for small-scale village operations. Seaweed fisheries are traditionally the domain of women in many Pacific island countries, so it is a natural progression for women to be involved in seaweed farming. In the Philippines and Indonesia, seaweed provides much-needed employment and income for many thousands of farmers in remote coastal areas.

Marine Finfish

Marine finfish aquaculture is well established globally, and is growing rapidly. A wide range of species is cultivated, and the diversity of culture is also steadily increasing.

In the Americas and northern Europe, the main species is the Atlantic salmon (*Salmo salar*), with smaller quantities of salmonids and species. Chile, in particular, has seen the most explosive growth of salmon farming in recent years, and is poised to become the number-one producer of Atlantic salmon. In the Mediterranean, a range of warmer water species are cultured, such as seabass and seabream.

Asia is again the major producer of farmed marine fish. The Japanese amberjack *Seriola quinqueradiata* is at the top of the production tables, with around 160 000 tonnes produced in 2004, but the region is characterized by the extreme diversity of species farmed, in line with the diverse fish-eating habits of the people living in the region. Seabreams are also common, with barramundi or Asian seabass (*Lates calcarifer*) cultured in both brackish water and mariculture environments. Grouper culture is expanding rapidly in Asia, driven by high prices in the live-fish markets of Hong Kong SAR and China, and the decreasing availability of wild-caught product due to overfishing. Southern bluefin tuna (*Thunnus mccoyii*) is cultured in Australia using wild-caught juveniles. Although production of this species is relatively small (3500–4000 tonnes per annum in 2001–03), it brings very high prices in the Japanese market and thus supports a highly lucrative local industry in South Australia. The 2003 production of 3500 tonnes was valued at US$65 million.

Hatchery technologies are well developed for most temperate species (such as salmon and seabream) but less well developed for tropical species such as groupers where the industry is still reliant on collection of wild fingerlings, a concern for future sustainability of the sector.

The bulk of marine fish are presently farmed in net cages located in coastal waters. Most cultured species are carnivores, leading to environmental concerns over the source of feed for marine fish farms, with most still heavily reliant on wild-caught so-called 'trash' fish. Excessive stocking of cages in coastal waters also leads to concern over water and sediment pollution, as well as impacts from escapes and disease transfer on wild fish populations.

Crustaceans

Although there is substantial production of marine shrimps globally, this production is undertaken in coastal brackish water ponds and thus does not meet the definition of mariculture. There has been some experimental culture of shrimp in cages in the Pacific, but this has not yet been commercially implemented. Tropical spiny rock lobsters and particularly the ornate lobster *Panulirus ornatus* are cultured in Southeast Asia, with the bulk of production in Vietnam and the Philippines. Lobster aquaculture in Vietnam produces about 1500 tonnes valued at around US$40 million per annum. Tropical spiny rock lobsters are cultured in cages and fed exclusively on fresh fish and shellfish.

In the medium to long term, it is necessary to develop hatchery production technology for seedstock for tropical spiny rock lobsters. There is currently considerable research effort on developing larval-rearing technologies for tropical spiny rock lobsters in Southeast Asia and in Australia. As in the case of tropical marine fish farming, there is also a need to develop less-wasteful and less-polluting diets to replace the use of wild-caught fish and shellfish as diets.

Other Miscellaneous Invertebrates

There are a range of other invertebrates being farmed in the sea, such as sea cucumbers, sponges, corals, sea horses, and others. Farming of some species has been ongoing for some time, such as the well-developed sea cucumber farming in northern China, but others are more recent innovations or still at the research stage. Sponge farming, for example, is generating considerable interest in the research community, but commercial production of farmed sponges is low, mainly in the Pacific islands. This farming is similar to seaweed culture as sponges can be propagated vegetatively, with little infrastructure necessary to establish farms. The harvested product, bath sponges, can be dried and stored and, like seaweed culture, may be ideal for remote communities, such as those found among the Pacific islands.

Environmental Challenges

Environmental Impacts

Mariculture is an important economic activity in many coastal areas but is facing a number of environmental challenges because of the various environmental 'goods' and 'services' required for its development. The many interactions between mariculture and the environment include impacts of: (1) the environment on mariculture; (2) mariculture on the environment; and (3) mariculture on mariculture.

The environment impacts on mariculture through its effects on water, land, and other resources necessary for successful mariculture. These impacts may be negative or positive, for example, water pollution may provide nutrients which are beneficial to mariculture production in some extensive culture systems, but, on the other hand, toxic pollutants and pathogens can be extremely damaging. An example is the farming of oysters and other filter-feeding mollusks which generally grow faster in areas where nutrient levels are elevated by discharge of wastewater from nearby centers of human population. However, excessive levels of human and industrial waste cause serious problems for mollusk culture, such as contamination with pathogens and toxins from dinoflagellates. Aquaculture is highly sensitive to adverse environmental changes (e.g., water quality and seed quality) and it is therefore in the long-term interests of mariculture farmers and governments to work toward protection and enhancement of environmental quality. The effects of global climate change, although poorly understood in the fishery sector, are likely to have further significant influences on future mariculture development.

The impacts of mariculture on the environment include the positive and negative effects farming operations may have on water, land, and other resources required by other aquaculturists or other user groups. Impacts may include loss or degradation of natural habitats, water quality, and sediment changes; overharvesting of wild seed; and introduction of disease and exotic species and competition with other sectors for resources. In increasingly crowded coastal areas, mariculture is running into more conflicts with tourism, navigation, and other coastal developments.

Mariculture can have significant positive environmental impacts. The nutrient-absorbing properties of seaweeds and mollusks can help improve coastal water quality. There are also environmental benefits from restocking of overfished populations or degraded habitats, such as coral reefs. For example, farming of high-value coral reef species is being seen as one means of reducing threats associated with overexploitation of threatened coral reef fishes traditionally collected for food and the ornamental trade.

Finally, mariculture development may also have an impact on itself. The rapid expansion in some areas with limited resources (e.g., water and seed) has led to overexploitation of these resources beyond the capacity of the environment to sustain growth, followed by an eventual collapse. In mariculture systems, such problems have been particularly acute in intensive cage culture, where self-pollution has led to disease and water-quality problems which have undermined the sustainability of farming, from economic and environmental viewpoints. Such problems emphasize the importance of environmental sustainability in mariculture management, and the need to minimize overharvesting of resources and hold discharge rates within the assimilative capacity of the surrounding environment.

The nature and the scale of the environmental interactions of mariculture, and people's perception of their significance, are also influenced by a complex interaction of different factors, such as follows:

- The technology, farming and management systems, and the capacity of farmers to manage technology. Most mariculture technology, particularly in extensive and semi-intensive farming systems, such as mollusk and seaweed farming, and well-managed intensive systems, is environmentally neutral or low in impact compared to other food production sectors.

- The environment where mariculture farms are located (i.e., climatic, water, sediment, and biological features), the suitability of the environment for the cultured animals and the environmental conditions under which animals and plants are cultured.
- The financial and economic feasibility and investment, such as the amount invested in proper farm infrastructure, short- versus long-term economic viability of farming operations, and investment and market incentives or disincentives, and the marketability of products.
- The sociocultural aspects, such as the intensity of resource use, population pressures, social and cultural values, and aptitudes in relation to aquaculture. Social conflicts and increasing consumer perceptions all play an important role.
- The institutional and political environment, such as government policy and the legal framework, political interventions, plus the scale and quality of technical extension support and other institutional and noninstitutional factors.

These many interacting factors make both understanding environmental interactions and their management (as in most sectors – not just mariculture) both complex and challenging.

Environmental Management of Mariculture

The sustainable development of mariculture requires adoption of management strategies which enhance positive impacts (social, economic, and environmental impacts) and mitigate against environmental impacts associated with farm siting and operation. Such management requires consideration of: (1) the farming activity, for example, in terms of the location, design, farming system, investment, and operational management; (2) the 'integration' of mariculture into the surrounding coastal environment; and (3) supporting policies and legislation that are favorable toward sustainable development.

Technology and Farming Systems Management

The following factors are of crucial importance in environmental management at the farm level:

- *Farm siting.* The sites selected for aquaculture and the habitat at the farm location play one of the most important roles in the environmental and social interactions of aquaculture. Farm siting is also crucial to the sustainability of an investment; incorrect siting (e.g., cages located in areas with unsuitable water quality) often lead to increased investment costs associated with operation and amelioration of environmental problems. Farms are better sited away from sensitive habitats (e.g., coral reefs) and in areas with sufficient water exchange to maintain environmental conditions. Problems of overstocking of mollusk culture beds are recognized in the Republic of Korea, for example, where regulations have been developed to restrict the areas covered by mollusk culture. For marine cage culture, one particularly interesting aspect of siting is the use of offshore cages, and new technologies developed in European countries are now attracting increasing interest in Asia.
- *Farm construction and design features.* Farm construction and system design has a significant influence on the impact of mariculture operations on the environment. Suitable design and construction techniques should be used when establishing new farms, and as far as possible seek to cause minimum disturbance to the surrounding ecosystems. The design and operation of aquaculture farms should also seek to make efficient use of natural resources used, such as energy and fuel. This approach is not just environmentally sound, but also economic because of increasing energy costs.
- *Water and sediment management.* Development of aquaculture should minimize impacts on water resources, avoiding impacts on water quality caused by discharge of farm nutrients and organic material. For sea-based aquaculture, where waste materials are discharged directly into the surrounding environment, careful control of feed levels and feed quality is the main method of reducing waste discharge, along with good farm siting. In temperate aquaculture, recent research has been responsible for a range of technological and management innovations – low-pollution feeds and novel self-feeding systems, lower stocking densities, vaccines, waste-treatment facilities – that have helped reduce environmental impacts. Complex models have also been developed to predict environmental impacts, and keep stocking levels within the assimilative capacity of the surrounding marine environment. In mariculture, there are also examples of integrated, polyculture, and alternate cropping farming systems that help to reduce impacts. For example in China and Korea, polyculture on sea-based mollusk and seaweed farms is practiced and for more intensive aquaculture operations, effluent rich in nutrients and microorganisms, is potentially suitable for culturing fin fish, mollusks, and seaweed.
- *Suitable species and seed.* A supply of healthy and quality fish, crustacean, and mollusk seed is

essential for the development of mariculture. Emphasis should be given to healthy and quality hatchery-reared stock, rather than collection from the wild. Imports of alien species require import risk assessment and management, to reduce risks to local aquaculture industries and native biodiversity.

- *Feeds and feed management.* Access to feeds, and efficient use of feeds is of critical importance for a cost-effective and environmentally sound mariculture industry. This is due to many factors, including the fact that feeds account for 50% or more of intensive farming costs. Waste and uneaten feed can also lead to undesirable water pollution. Increasing concern is also being expressed about the use of marine resources (fish meal as ingredients) for aquaculture feeds. One of the biggest constraints to farming of carnivorous marine fish such as groupers is feed. The development of sustainable supplies of feed needs serious consideration for future development of mariculture at a global level.
- *Aquatic animal health management.* Aquatic animal and plant diseases are a major cause of unsustainability, particularly in more intensive forms of mariculture. Health management practices are necessary to reduce disease risks, to control the entry of pathogens to farming systems, maintain healthy conditions for cultured animals and plants, and avoid use of harmful disease control chemicals.
- *Food safety.* Improving the quality and safety of aquaculture products and reducing risks to ecosystems and human health from chemical use and microbiological contamination is essential for modern aquaculture development, and marketing of products on domestic and international markets. Normally, seafood is considered healthy food but there are some risks associated with production and processing that should be minimized. The two food-safety issues, that can also be considered environmental issues, are chemical and biological. The chemical risk is associated with chemicals applied in aquaculture production and the biological is associated with bacteria or parasites that can be transferred to humans from the seafood products. Increasing calls for total traceability of food products are also affecting the food production industry such that consumers can be assured that the product has been produced without addition of undesirable or harmful chemicals or additives, and that the environments and ecosystems affected by the production facilities have not been compromised in any way.
- *Economic and social/community aspects.* The employment generated by mariculture can be highly significant, and globally aquaculture has become an important employer in remote and poor coastal communities. Poorly planned mariculture can also lead to social conflicts, and the future development and operation of mariculture farms must also be done in a socially responsible manner, as far as possible without compromising local traditions and communities, and ideally contributing positively. The special traditions of many coastal people and their relation with the sea in many places deserve particularly careful attention in planning and implementation of mariculture.

Planning, Policy, and Legal Aspects

Integrated Coastal Area Management

Effective planning processes are essential for sustainable development of mariculture in coastal areas. Integrated coastal area management (ICAM) is a concept that is being given increasing attention as a result of pressures on common resources in coastal areas arising from increasing populations combined with urbanization, pollution, tourism, and other changes. The integration of mariculture into the coastal area has been the subject of considerable recent interest, although practical experience in implementation is still limited in large measure because of the absence of adequate policies and legislation and institutional problems, such as the lack of unitary authorities with sufficiently broad powers and responsibilities.

Zoning of aquaculture areas within the coastal area is showing some success. In China, Korea, Japan, Hong Kong, and Singapore, there are now well-developed zoning regulations for water-based coastal aquaculture operations (marine cages, mollusks, and seaweeds). For example, Hong Kong has 26 designated 'marine fish culture zones' within which all marine fish-culture activities are carried out. In the State of Hawaii, 'best areas' for aquaculture have been identified, and in Europe zoning laws are being strictly applied to many coastal areas where aquaculture is being developed. Such an approach allows for mariculture to be developed in designated areas, reducing risks of conflicts with other coastal zone users and uses.

Policy and Legal Issues

While much can be done at farm levels and by integrated coastal management, government involvement through appropriate policy and legal instruments is important in any strategy for mariculture sustainability. Some of the important issues include legislation, economic incentives/disincentives, private sector/community participation in policy formulation, planning processes, research and knowledge transfer, balance between food and export earnings, and others.

While policy development and most matters of mariculture practice have been regarded as purely national concerns, they are coming to acquire an increasingly international significance. The implication of this is that, while previously states would look merely to national priorities in setting mariculture policy, particularly legislation/standards, for the future it will be necessary for such activities to take account of international requirements, including various bilateral and multilateral trade policies. International standards of public health for aquaculture products and the harmonization of trade controls are examples of this trend.

Government regulations are an important management component in maintaining environmental quality, reducing negative environmental impacts, and allocating natural resources between competing users and integration of aquaculture into coastal area management. Mariculture is a relative newcomer among many traditional uses of natural resources and has commonly been conducted within an amalgam of fisheries, water resources, and agricultural and industrial regulations. It is becoming increasingly clear that specific regulations governing aquaculture are necessary, not least to protect aquaculture development itself. Key issues to be considered in mariculture legislation are farm siting, use of water area and bottom in coastal and offshore waters; waste discharge, protection of wild species, introduction of exotic or nonindigenous species, aquatic animal health; and use of drugs and chemicals.

Environmental impact assessment (EIA) can also be an important legal tool which is being more widely applied to mariculture. The timely application of EIA (covering social, economic, and ecological issues) to larger-scale coastal mariculture projects can be one way to properly identify environmental problems at an early phase of projects, thus enabling proper environmental management measures to be incorporated in project design and management. Such measures will ultimately make the project more sustainable. A major difficulty with EIAs is that they are difficult (and generally impractical) to apply to smaller-scale mariculture developments, common throughout many parts of Asia, and do not easily take account of the potential cumulative effects of many small-scale farms. Strategic environmental assessment (SEA) can provide a broader means of assessing impacts.

Conclusion

Mariculture is and will increasingly become an important producer of aquatic food in coastal areas, as well as a source of employment and income for many coastal communities. Well-planned and -managed mariculture can also contribute positively to coastal environmental integrity. However, mariculture's future development will occur, in many areas, with increasing pressure on coastal resources caused by rising populations, and increasing competition for resources. Thus, considerable attention will be necessary to improve the environmental management of aquaculture through environmentally sound technology and better management, supported by effective policy and planning strategies and legislation.

See also

Mariculture, Economic and Social Impacts.

Further Reading

Clay J (2004) *World Aquaculture and the Environment. A Commodity by Commodity Guide to Impacts and Practices*. Washington, DC: Island Press.

FAO/NACA/UNEP/WB/WWF (2006) *International Principles for Responsible Shrimp Farming*, 20pp. Bangkok, Thailand: Network of Aquaculture Centres in Asia-Pacific (NACA). http://www.enaca.org/uploads/international-shrimp-principles-06.pdf (accessed Apr. 2008).

Hansen PK, Ervik A, Schaanning M, *et al.* (2001) Regulating the local environmental impact of intensive, marine fish farming-II. The monitoring programme of the MOM system (Modelling-Ongrowing fish farms-Monitoring). *Aquaculture* 194: 75–92.

Hites RA, Foran JA, Carpenter DO, Hamilton MC, Knuth BA, and Schwager SJ (2004) Global assessment of organic contaminants in farmed salmon. *Science* 303: 226–229.

Joint FAO/NACA/WHO Study Group (1999) Food safety issues associated with products from aquaculture. *WHO Technical Report Series 883*. http://www.who.int/foodsafety/publications/fs_management/en/aquaculture.pdf (accessed Apr. 2008).

Karakassis I, Pitta P, and Krom MD (2005) Contribution of fish farming to the nutrient loading of the Mediterranean. *Scientia Marina* 69: 313–321.

NACA/FAO (2001) Aquaculture in the third millennium. In: Subasinghe RP, Bueno PB, Phillips MJ, Hough C, McGladdery SE, and Arthur JR (eds.) *Technical*

Proceedings of the Conference on Aquaculture in the Third Millennium. Bangkok, Thailand, 20–25 February 2000, 471pp. Bangkok, NACA and Rome: FAO.

Naylor R, Hindar K, Flaming IA, *et al*. (2005) Fugitive salmon: Assessing the risks of escaped fish from net-pen aquaculture. *BioScience* 55: 427–473.

Neori A, Chopin T, Troell M, *et al*. (2004) Integrated aquaculture: Rationale, evolution and state of the art emphasizing sea-weed biofiltration in modern mariculture. *Aquaculture* 231: 361–391.

Network of Aquaculture Centres in Asia-Pacific (2006) Regional review on aquaculture development. 3. Asia and the Pacific – 2005. *FAO Fisheries Circular No. 1017/3*, 97pp. Rome: FAO.

Phillips MJ (1998) Tropical mariculture and coastal environmental integrity. In: De Silva S (ed.) *Tropical Mariculture*, pp. 17–69. London: Academic Press.

Pillay TVR (1992) *Aquaculture and the Environment*, 158pp. London: Blackwell.

Secretariat of the Convention on Biological Diversity (2004) Solutions for sustainable mariculture – avoiding the adverse effects of mariculture on biological diversity, *CBD Technical Series No. 12*. http://www.biodiv.org/doc/publications/cbd-ts-12.pdf (accessed Apr. 2008).

Tacon AJC, Hasan MR, and Subasinghe RP (2006) Use of fishery resources as feed inputs for aquaculture development: Trends and policy implications. *FAO Fisheries Circular No. 1018*. Rome: FAO.

World Bank (2006) *Aquaculture: Changing the Face of the Waters. Meeting the Promise and Challenge of Sustainable Aquaculture*. Report no. 36622. Agriculture and Rural Development Department, the World Bank. http://siteresources.worldbank.org/INTARD/Resources/Aquaculture_ESW_vGDP.pdf (accessed Apr. 2008).

Relevant Websites

http://www.pbs.org
– Farming the Seas, Marine Fish and Aquaculture Series, PBS.

http://www.fao.org/fi
– Food and Agriculture Organisation of the United Nations.

http://www.cbd.int
Jakarta Mandate, Marine and Coastal Biodiversity: Mariculture, Convention on Biological Diversity.

http://www.enaca.org
– Network of Aquaculture Centres in Asia-Pacific.

http://www.seaplant.net
– The Southeast Asia Seaplant Network.

http://www.oceansatlas.org
– UN Atlas of the Oceans.

MARICULTURE, ECONOMIC AND SOCIAL IMPACTS

C. R. Engle, University of Arkansas at Pine Bluff, Pine Bluff, AR, USA

Introduction

Mariculture is a broad term that encompasses the cultivation of a wide variety of species of aquatic organisms, including both plants and animals. These different products are produced across the world with a wide array of technologies. Each technology in each situation will entail various price and cost structures within different social contexts. Thus, each aquaculture enterprise will have distinct types and levels of economic and social impacts. Moreover, there are as many management philosophies, strategies, and business plans as there are aquaculture entrepreneurs. Each of these will result in different economic and social impacts. As an example, consider two shrimp farms located in a developing country that utilize the same production technology. One farm hires and trains local people as both workers and managers, invests in local schools and health centers, while paying local and national taxes. This farm will likely have a large positive social and economic impact. On the other hand, another farm that imports managers, pays the lowest wages possible, displaces local families through land acquisitions, pays few taxes, and exports earnings to developed countries may have negative social and perhaps even economic impacts.

Economic and social structure, interactions, and impacts are complex and interconnected, even in rural areas with seemingly simple economies. This article discusses a variety of types of impacts that can occur from mariculture enterprises.

Economic Impacts

Economic impacts begin with the direct effects from the sale of product produced by the mariculture operation. However, the impacts extend well beyond the effect of sales revenue to the farm. As the direct output of marine fish, shellfish, and seaweed production increases, the demand for supply inputs such as feed, fingerlings, equipment, repairs, transportation services, and processing services also increases. These activities represent indirect effects. Subsequent increased household spending will follow. As the industry grows, employment in all segments of the industry also grows and these new jobs create more income that generates additional economic activity. Thus, growth of the mariculture industry results in greater spending that multiplies throughout the economy.

Mariculture is an important economic activity in many parts of the world. **Table 1** lists the top 15 mariculture-producing countries in the world, both in terms of the volume of metric tons produced and the value in 2005. Its economic importance can be measured in total sales volume, total employment, or total export volume for large aquaculture industry sectors. Macroeconomic effects include growth that promotes trade and domestic resource utilization. Fish production in the Philippines, for example, accounted for 3.9% of gross domestic product (GDP) in 2001. In India, it contributed 1.4% of national GDP and 5.4% of agricultural GDP.

On the microlevel, incomes and livelihoods of the poor are enhanced through mariculture production. Small-scale and subsistence mariculture provides high-quality protein for household consumption,

Table 1 Top 15 mariculture-producing countries, with volumes (metric tons) produced in 2005 and value (in US)

Country	*Volume of production (metric ton)*	*Value of production ($1000 US)*
China	22 677 724	14 981 801
The Philippines	1 419 727	165 335
Japan	1 211 959	3 848 906
Korea, Rep.	1 042 142	1 317 250
Indonesia	950 819	338 093
Chile	889 867	3 069 169
Korea, Dem.	703 292	283 362
Norway	504 295	2 072 562
Thailand	656 636	58 587
France	347 750	571 543
Canada	216 103	503 974
Spain	136 724	262 394
United Kingdom	190 426	584 152
United States	161 339	235 912
Vietnam	134 937	158 800
Others	173 800	3 268 283
Total	31 417 540	31 720 123

Source: FishStat Plus (http://www.fao.org).

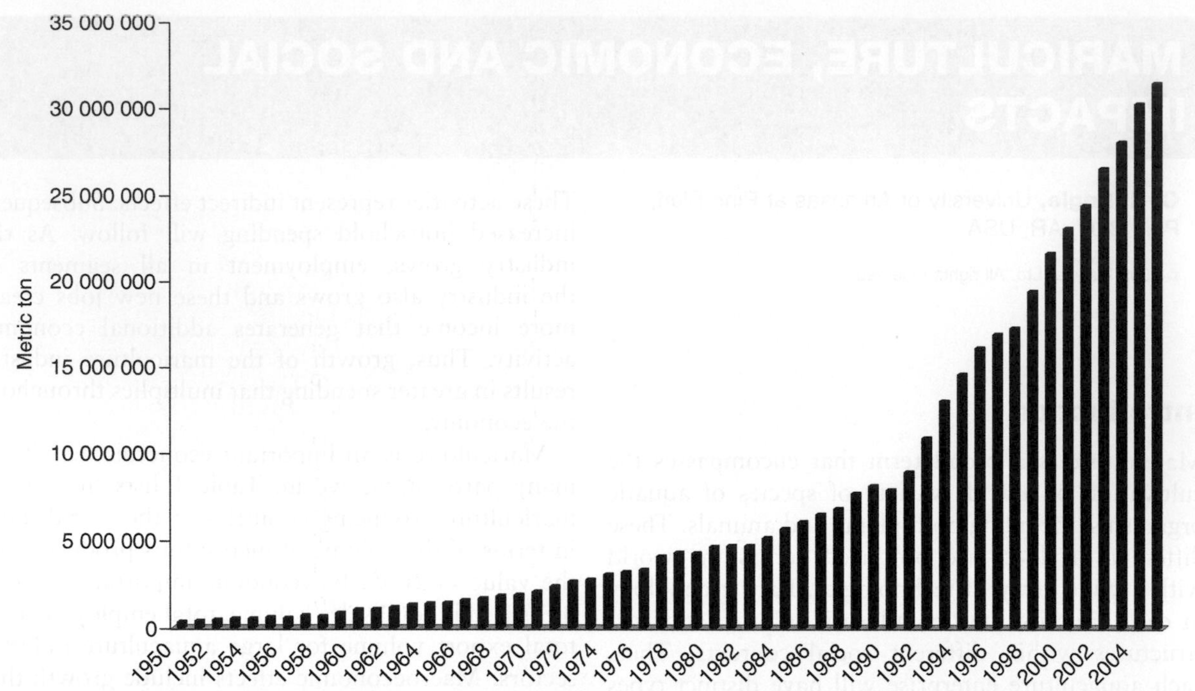

Figure 1 Growth of mariculture production worldwide, 1950–2005. Source: Food and Agriculture Organization of the United Nations.

generates supplemental income from sales to local markets, and can serve as a savings account to meet needs for cash during difficult financial times.

Income from Sales Revenue

According to the Food and Agriculture Organization of the United Nations, mariculture production grew from just over 300 000 metric ton in 1950 to 31.4 million metric ton valued at $31.7 billion in 2005 (**Figure 1**). **Figure 1** also shows that the total volume of mariculture production has tripled over the past decade alone. Sales revenue received by operators of mariculture businesses has increased rapidly over this same time period. The top mariculture product category in 2005, based on quantity produced, was Japanese kelp, followed in descending order of quantities produced by oysters, Japanese carpetshell, wakame, Yesso scallop, and salmon (**Table 2**). In terms of value, salmon was the top mariculture product produced in 2005, followed by oysters, kelp, and Japanese carpetshell. The top five marine finfish raised in 2005 (in terms of value) were Atlantic salmon, Japanese amberjack, rainbow trout, seabreams, and halibut.

Table 2 Top 15 mariculture species cultured, volume (metric tons), and value ($1000 US), 2005

Species	*Volume (metric ton)*	*Value ($1000 US)*
Atlantic salmon	1 216 791	4 659 841
Blood cockle	436 924	436 772
Blue mussel	391 210	385 131
Constricted tagelus	713 846	589 836
Green mussel	280 267	26 679
Japanese carpetshell	2 880 687	2 358 586
Japanese kelp	4 911 256	2 941 148
Laver (nori)	1 387 990	1 419 130
Pacific cupped oyster	4 496 196	3 003 831
Rainbow trout	183 575	762 707
Red seaweeds	866 383	121 294
Sea snails	238 331	131 188
Wakame	2 739 753	1 101 507
Warty gracilaria	985 667	394 257
Yesso scallop	1 239 811	1 677 870
Others	8 448 853	11 710 346
Total	31 417 540	31 720 123

Source: FishStat Plus (http://www.fao.org).

The increase in production and sales of mariculture products has resulted not only from the expansion of existing products and technologies (i.e., salmon, Japanese amberjack) but also from the rapid development of technologies to culture new species (i.e., cobia, grouper, and tuna). New mariculture farms have created new markets and opportunities for local populations, such as the development of backyard hatcheries in countries such as Indonesia. In Bali, for example, the development of small-scale hatcheries has resulted in substantial increases in income as compared to the more traditional coconut crops.

The immediate benefit to a mariculture entrepreneur is the cash income received from sale of the aquatic products produced. This income then is spent to pay for cash expenditures related to feed, fingerlings, repairs to facilities and equipment, fuel, labor, and other operating costs. Payments on loans will be made to financial lenders involved in providing capital for the facilities, equipment, and operation of the business. Profit remaining after expenses can be spent by the owner on other goods and services, invested back into the business, or invested for the future benefit of the owner.

Mariculture of certain species has begun to exceed the value of the same species in capture fisheries. Salmon is a prime example, in which the value of farmed salmon has exceeded that of wild-caught salmon for a number of years. A similar trend can be observed for the rapidly developing capture-based mariculture technologies. In the Murcia region of Spain, for example, the economic value of tuna farming now represents 8 times the value of regional fisheries.

Employment

Mariculture businesses generate employment for the owner of the business, for those who serve as managers and foremen, and those who constitute the principal workforce for the business. Studies in China, Vietnam, Philippines, Indonesia, and Thailand suggest that shrimp farming uses more labor per hectare than does rice farming.

Employment is generated throughout the market supply chain. Development of mariculture businesses increases demand for fry and fingerlings, feeds, construction, equipment used on the farm (trucks, tractors, aerators, nets, boats, etc.), and repairs to facilities and equipment. In Bangladesh, it is estimated that about 300 000 people derive a significant part of their annual income from shrimp seed collection.

Mariculture businesses also create jobs for fish collectors, brokers, and vendors. These intermediaries provide important marketing functions that are difficult for smaller-scale producers to handle. With declines in catch of some previously important fish products, fish vendors need new fish products to sell. Increasing volumes from mariculture can keep these marketers in business.

The employment stimulated by development of mariculture businesses is especially significant in economically depressed areas with high rates of unemployment and underemployment. Many mariculture activities develop in rural coastal communities where jobs are limited and mariculture can constitute a critical source of employment. Many jobs in fishing communities have been reduced as fishing opportunities have become more scarce. Given its relationship to the fishing industry and some degree of similarity in the skills required, mariculture businesses are often a welcome alternative for the existing skilled workforce.

Capture-based mariculture operations in particular provide job opportunities that require skill sets that are easily met by those who have worked in the fishing industry. The rapid development of capture-based tuna farms in the Mediterranean Sea provides a good example. In Spain, fishermen have become active partners in tuna farms. As a result, the number of specialized bluefin tuna boats has increased to supply the capture-based tuna farms in the area. These boats capture the small pelagic fish that are used as feed. In Croatia, trawlers transport live fish or feed to tuna farms off shore. This has generated important sources of employment in heavily depopulated Croatian islands. Tuna farming is labor-intensive, but offers opportunities for younger workers to develop new skills. Many of the employees on the tuna farms are young, 25–35 years old, who work as divers. The divers are used to crowd the tuna for hand-harvesting without stressing the fish, inspect for mortalities, and check the integrity of moorings. Working conditions on tuna farms are preferable to those on tuna boats because the hours and salaries are more regular. Workers can spend weekends on shore as compared to spending long periods of time at sea on tuna boats. The improved working conditions have improved social stability.

Formal studies of economic impacts have measured the effects on employment for fish farm owners and from the secondary businesses such as supply companies, feed mills, processing plants, transportation services, etc. In the Philippines, fish production accounted for 4.4% of total national employment in 2001. In the United States, catfish production accounted for nearly half of all employment in one particular county.

Tax Revenue

Tax policies vary considerably from country to country. Nevertheless, there is some form of tax structure in most countries that is used to finance local, regional, and national governments. Tax structures typically include some form of business tax in addition to combinations of property, income, value-added, or sales taxes. Tax revenue is the major source of revenue used for investments in roads and bridges, schools, health, and public sanitation facilities.

Mariculture businesses contribute to the tax base in that particular region or market. Moreover, the wages paid to managers and workers in the business are subject to income, sales, and property taxes. As the business grows, it pays more taxes. Increasing payrolls from increased employment by the company results in greater tax revenue from increased incomes, increased spending (sales taxes), and increased property taxes paid (as people purchase larger homes and/or more land).

Export Revenue and Foreign Exchange

The largest volume of mariculture production is in developing countries whereas the largest markets for mariculture products are in the more developed nations of the world. Thus, much of the flow of international trade in mariculture products is from the developing to the developed world. This trade results in export revenue for the companies involved. Exporters in the live grouper trade in Asia have been shown to earn returns on total costs as high as 94%.

Export volumes and revenue further generate foreign exchange for the exporting country. This is particularly important for countries that are dependent on other countries for particular goods and services. In order to import products not produced domestically, the country needs sufficient foreign exchange. Moreover, if the country exports products to countries whose currency is strong (i.e., has a high value relative to other forms of currency), the country can then import a relatively greater amount of product from countries whose currency does not have as high a value. Many low-income food-deficient countries use fish exports to pay for imports of low-value food commodities. For example, in 2000, Indonesia, China, the Philippines, India, and Bangladesh paid off 63%, 62%, 22%, 54%, and 32%, respectively, of their food import bills by exporting fish and fish products.

Economic Growth

Economic growth is generated from increased savings, investment, and money supply. The process starts with profitable businesses. Profits can either be saved by the owner or invested back into the business. Savings deposited in a bank or invested in stocks or bonds increase the amount of capital available to be loaned out to other potential investors. With increasing amounts of capital available, the cost of capital (the interest rate) is reduced, making it easier for both individuals and businesses to borrow capital. As a result, spending on housing, vehicles, property, and new and expanded businesses increases. This in turn increases demand for housing, vehicles, and other goods which increases demand for employment. As the economy grows (as measured by the GDP), the standard of living rises as income levels rise and citizens can afford to purchase greater varieties of goods and services. The availability of goods and services also increases with economic growth. Economic growth increases demand which also tends to increase price levels. As prices increase, businesses become more profitable and wages rise.

Economic growth is necessary for standards of living to increase. However, continuous economic growth means that prices also rise continuously. Continuous increases in prices constitute inflation that, if excessive, can create economic difficulties. It is beyond the scope of this article to discuss the mechanisms used by different countries to manage the economy on a national scale and how to manage both economic growth and inflation.

Economic Development

Economic development occurs as economies diversify, provide a greater variety of goods and services, and as the purchasing power of consumers increases. This combination results in a higher standard of living for more people in the country. Economic development results in higher levels of education, greater employment opportunities, and higher income levels. Communities are strengthened with economic development because increasing numbers of jobs result in higher income levels. Higher standards of living provide greater incentives for young people to stay in the area rather than outmigrate in search of better employment and income opportunities.

Studies of the economic impacts of aquaculture have highlighted the importance of backwardly linked businesses. These are the businesses that produce fingerlings, feed, and sell other supplies to the aquaculture farms. Studies of the linkages in the US catfish industry show that economic output of and the value added by the secondary businesses are greater than those of the farms themselves.

Economic Multiplier Effects

Each time a dollar exchanges hands in an economy, its impact 'multiplies'. This occurs because that same dollar can then be used to purchase some other item. Each purchase represents demand for the product being purchased. Each time that same dollar is used to purchase another good or service, it generates additional demand in the economy. Those businesses that have higher expenditures, particularly those with high initial capital expenditures, tend to have higher multiplier effects in the economy.

Aquaculture businesses tend to have high capital expenditures and, thus, tend to have relatively higher multipliers than other types of businesses, particularly other types of agricultural businesses. These high capital expenditures come from the expense of building ponds, constructing net pens or submerged cages, and the equipment costs associated with aerators, boats, trucks, tractors, and other types of equipment. As an example of how economic multipliers occur, consider a new shrimp farm enterprise. Construction of the ponds will require either purchasing bulldozers and bulldozer operators or contracting with a company that has the equipment and expertise to build ponds. The shrimp farm owner pays the pond construction company for the ponds that were built. The dollars paid to the construction company are used to pay equipment operators, fuel companies, for water supply and control pipes and valves, and repairs to equipment. The pond construction workers use the dollars received as wages to purchase more food, clothing for the family, school costs, healthcare costs, and other items. The dollars spent for food in the local market become income for the vendors who pay the farmers who raised the food. Those farmers can now pay for their seed and fertilizer, buy more clothing, and pay school and healthcare expenses, etc. In this expenditure chain, the dollar 'turned over' five times, creating additional economic demand each time.

Several formal studies of economic impacts from aquaculture have been conducted. These studies have shown large amounts of total economic output, employment, and value-added effects from aquaculture businesses. On the local area, aquaculture can generate the majority of economic output in a given district. Multipliers as high as 6.1 have been calculated for aquaculture activities on the local level.

Potential Negative Economic Impacts

Mariculture has been criticized by some for what economists call 'externalities'. Externalities refer to an effect on an individual or community other than the individual or community that created the problem. Pollution is often referred to as an externality. For example, if effluents discharged from a mariculture business create water-quality problems for another farm or community downstream, that farm or community will have to spend more to clean up that pollution. Yet, those costs are 'external' to the business that created the problem.

To the extent that a mariculture business creates an externality that increases costs for another business or community, there will be negative economic impacts. The primary potential sources of negative economic impacts from mariculture include: (1) discharge of effluents that may contain problematic levels of nutrients, antibiotics, or nonnative species; (2) spread of diseases; (3) use of high levels of fish meal in the diets fed to the species cultured; (4) clearing mangroves from coastal areas; and (5) consequences of predator control measures. The history of the development of mariculture includes cases in which unregulated discharge of untreated effluents from shrimp farms resulted in poor water-quality conditions in bays and estuaries. Since this same water also served as influent for other shrimp farms, its poorer quality stressed shrimp and facilitated spread of viral diseases. Economic losses resulted.

Concerns have been noted over the increasing demand for fish meal as mariculture of carnivorous species grows. The fear is that this increasing use will result in a decline in the stocks of the marine species that serve as forage for wild stocks and that are used in the production of fish meal. If this were to occur, economic losses could occur from declines in capture fisheries.

Loss of mangrove forests in coastal areas results in loss of important nursery grounds for a number of species and less protection during storms. The loss of mangrove areas is due to many uses, including for construction, firewood, for salt production, and others. If the construction of ponds for mariculture reduces mangrove areas, additional negative impacts will occur.

Control of predators on net pens and other types of mariculture operations often involves lethal methods. The use of lethal methods raises concerns related to biodiversity and viability of wild populations. While there is no direct link to other economic activities in many of these cases, the loss of ecosystem services from reductions in biodiversity may at some point result in other economic problems.

Social Impacts

Food Security

Aquaculture is an important food security strategy for many individuals and countries. Domestic production of food provides a buffer against interruptions in the food supply that can result from international disputes, trade embargoes, war, transportation accidents, or natural disasters. There are many examples of fish ponds being used to feed households during times of strife. Examples include fish pond owners in Vietnam during the war with the United States and during the Contra War in Nicaragua, among others.

Aquaculture has been shown to function as a food reserve for many subsistence farming households. Many food crops are seasonal in nature while fish in ponds or cages can be available year-round and used particularly during periods of prolonged drought or in between different crops. Pond levees also can provide additional land area to raise vegetable crops for household consumption and sale.

Nutritional Benefits

Fish are widely recognized to be a high-quality source of animal protein. Fish have long been viewed as 'poor people's food'. Subsistence farmers who raise fish, shellfish, or seaweed have a ready source of nutritious food for home consumption throughout the year. Particularly in Asia, fish play a vital role in supplying inexpensive animal protein to poorer households. This is important because the proportion of the food budget allocated to fish is higher among low-income groups. Moreover, it has been shown that rural people consume more fish than do urban dwellers and that fish producers consume more fish than do nonproducers. Those farmers who also supply the local market with aquatic products provide a supply of high-quality protein to other households in the area. Increasing supplies of farmed products result in lower prices in seafood markets. Tuna prices in Japan, for example, have been falling as a result of the increased farmed supply. This has resulted in making tuna and its nutritional benefits more readily available to middle-income buyers in Japan.

Health Benefits

In addition to the protein content, a number of aquatic products provide additional health benefits. Farmed fish such as salmon have high levels of omega-3 fatty acids that reduce the risk of heart disease. Products such as seaweeds are rich in vitamins. Mariculture can enable the poorest of the poor and even the landless to benefit from public resources. For example, cage culture of fish in public waters, culture of mollusks and seaweeds along the coast, and culture-based fisheries in public water bodies provide a source of healthy food for subsistence families.

Poverty Alleviation

As aquaculture has grown, its development has frequently occurred in economically depressed areas. Poverty alleviation is accompanied by a number of positive social impacts. These include improved access to food (that results in higher nutritional and health levels), improved access to education (due to higher income levels and ability to pay for fees and supplies), and improved employment opportunities. In Sumatra, Indonesia, profits from grouper farming enable members of Moslem communities to make pilgrimages to Mecca, enriching both the individuals and the community. In Vietnam, income from grouper hatcheries contributes 10–50% of the annual income of fishermen.

Enhancement of Capture Fisheries

Marine aquaculture can improve the condition of fisheries through supplemental stockings from aquaculture. This will help to meet the global demand for fish products. Increased mariculture supply relieves pressure on traditional protein sources for species such as live cod and haddock. Given that fish supplies from capture fisheries are believed to have reached or be close to maximum sustainable yield, mariculture can reduce the expected shortage of fish.

Potential Negative Social Impacts

Mariculture has been criticized for creating negative social impacts. These criticisms have tended to focus on displacement of people if large businesses buy land and move people off that land. Economic development is accompanied by social change and individuals and households frequently are affected in various ways by construction of new infrastructure and production and processing facilities. In developed countries, federal laws typically provide for some degree of compensation for land taken through eminent domain for construction of highways or power lines. However, developing countries rarely have similar mechanisms for compensating those who lose land. In situations in which the resources involved are in the public domain (such as coastal areas), granting a concession for a mariculture operation along the coast may result in reduced access for poor and landless individuals to collect shellfish and other foods for their households.

Resource ownership or use rights in coastal zones frequently are ambiguous. Many of these areas traditionally have been common access areas, but are now under pressure for development for mariculture activities. Displaced and poor migrant people are frequently marginalized in coastal lands and often depend to some degree on common resources. However, the extent of conflict over the use of resources is variable. Surveys in Asia reported less than 10% of farms experiencing conflicts in most countries, but higher incidents of conflict were reported in India (29%) and China (94%).

In the Mediterranean, capture-based mariculture has resulted in conflicts between local tuna fishermen who fish with long lines, and cage towing operations. Other conflicts have occurred in Croatia due to the smell and pollution during the summer from bluefin tuna farms. Uncollected fat skims on the sea surface that results from feeding oily trash fish spread outside the licensed zones onto beaches frequented by tourists. On the positive side, tourism in Spain was enhanced by offering guided tours to offshore cages.

Additional negative impacts could potentially include: (1) marine mammal entanglements; (2) threaten genetic makeup of wild fish stocks; (3) privatize what some think of as free open-access resource; (4) esthetically undesirable; and (5) negative impacts on commercial fishermen. For example, excessive collection of postlarvae and egg-laden female shrimp can result in loss of income for fishermen and reduction of natural shrimp and fish stocks.

Conclusions

The economic and social impacts of mariculture are variable throughout the world and are based on the range of technologies, cultures, habitats, land types, and social, economic, and political differences. Positive economic impacts include increased revenue, employment, tax revenue, foreign exchange, economic growth and development, and multiplicative effects through the economy. Negative economic effects could occur through excessive discharge of effluents, spread of disease, and through excessive use of fish meal. Mariculture enhances food security of households and countries and provides nutritional and health benefits while alleviating poverty. However, if earnings are exported and access to public domain resources restricted to poorer classes, negative social impacts can occur.

See also

Diversity of Marine Species. Mariculture Diseases and Health. Mariculture of Aquarium Fishes. Mariculture of Mediterranean Species. Mariculture Overview. Marine Fishery Resources, Global State of. Network Analysis of Food Webs.

Further Reading

Aguero M and Gonzalez E (1997) Aquaculture economics in Latin America and the Caribbean: A regional assessment. In: Charles AT, Agbayani RF, Agbayani EC, *et al.* (eds.) Aquaculture Economics in Developing Countries: Regional Assessments and an Annotated Bibliography. *FAO Fisheries Circular No. 932*. Rome: FAO.

Dey MM and Ahmed M (2005) *Special Issue: Sustainable Aquaculture Development in Asia. Aquaculture Economics and Management* 9(1/2): 286pp.

Edwards P (2000) Aquaculture, poverty impacts and livelihoods. *Natural Resource Perspectives* 56 (Jun. 2000).

Engle CR and Kaliba AR (2004) *Special Issue: The Economic Impacts of Aquaculture in the United States. Journal of Applied Aquaculture* 15(1/2): 172pp.

Ottolenghi F, Silvestri C, Giordano P, Lovatelli A, and New MB (2004) *Capture-Based Aquaculture. The Fattening of Eels, Groupers, Tunas and Yellowtails*, 308pp. Rome: FAO.

Tacon GJ (2001) Increasing the contribution of aquaculture for food security and poverty alleviation. In: Subasinghe RP, Bueno P, Phillips MJ, Hough C, McGladdery SE, and Arthur JR (eds.) Aquaculture in the Third Millennium. *Technical Proceedings of the Conference on Aquaculture in the Third Millennium*, pp. 63–72, Bangkok, Thailand, 20–25 February 2000. Bangkok and Rome: NACA and FAO.

Relevant Websites

http://www.fao.org
 – Food and Agriculture Organization of the United Nations.

http://www.worldfishcenter.org
 – WorldFish Center.

OCEAN RANCHING

A. G. V. Salvanes, University of Bergen, Bergen, Norway

Introduction

Ocean ranching is most often referred to as stock enhancement. It involves mass releases of juveniles which feed and grow on natural prey in the marine environment and which subsequently become recaptured and add biomass to the fishery. Releases of captive-bred individuals are common actions when critically low levels of fish species or populations occur either due to abrupt habitat changes, overfishing or recruitment failure from other causes. Captive-bred individuals are also introduced inside or outside their natural geographic range of the species to build up new fishing stocks.

At present 27 countries (excluding Japan) have been involved with ranching of over 65 marine or brackish-water species. Japan leads the world with approximately 80 species being ranched or researched for eventual stocking. This includes 20 shared with other nations and 60 additional species. **Table 1** shows an overview of the most important species worldwide. Many marine ranching projects are in the experimental or pilot stage. Around 60% of the release programs are experimental or pilot, 25% are strictly commercial, and 12% have commercial and recreational purposes. Only a few are dedicated solely to sport fish enhancement.

The success of ocean ranching relies on a knowledge of basic biology of the species that are captive bred, but also on how environmental factors and wild conspecifics and other species interact with the released. This article provides general information on life histories of the three major groups of animals that are being stocked, salmon, marine fish and invertebrates, and an overview of the status and success of ocean ranching programs. It also devotes a short section to the history of ocean ranching and a larger section on how success of stock enhancement is measured.

History

Ocean ranching has a long history going back to 1860–1880 that commenced with the anadromous salmonids in the Pacific. In order to restore populations that had been reduced or eliminated due to factors such as hydroelectric development or pollution, large enhancement programs were initiated on various Pacific salmon species mainly within the USA, Canada, USSR and Japan. In addition, transplantation of both Atlantic and Pacific salmonids to other parts of the world (e.g. Australia, New Zealand and Tasmania) with no native salmon populations was attempted.

Around 1900, ocean ranching was extended to coastal populations of marine fish. Because of large fluctuations in the landings of these, release programs of yolk-sac larvae of cod, haddock, pollack, place, and flounder were initiated in the USA, Great Britain, and Norway. It was intended that such releases should stabilize the recruitment to the populations and, thus, stabilize the catches in the coastal fisheries. There was, however, a scientific controversy of whether releases of yolk-sac larvae could have positive effects on the recruitment to these populations. In the USA the releases ceased by World War II without evaluation. In Norway evaluation was conducted in 1970 when it was shown to be impossible to separate the effect of releases of yolk-sac larvae from natural fluctuations in cod recruitment, a conclusion that led to termination of the program. Recent field estimates of the mortality of early life stages of cod suggest that only a handful of the 33 million larvae, an amount normally released, survive the three first months.

Larvae and juveniles of European lobster (*Homarus gammarus*) have been cultured and released along the coast of Norway for over 100 years. In 1889 newly hatched lobster eggs and newly settled juveniles were released in Southern Norway on an island which has its own continental shelf separated from the mainland. Because the larvae were too small to be tagged, no recapture could be registered. The first attempts at scallop enhancement occurred in 1900, but the activity ceased after a short time. From 1970, stock enhancement started on scallops on a commercial scale in Europe and Japan and from 1980 on giant clams in Indo-Pacific countries. Other invertebrate enhancement programs started on a commercial basis in 1963 when the government of Japan instituted such actions as a national policy to augment both marine fish populations and commercially interesting invertebrate populations (e.g. abalone, clams, sea urchins, shrimps and prawns).

Table 1 Main species which are captive bred and released in ocean ranching programs

English name	Scientific name	Country
Salmonids		
Atlantic salmon	*Salmo salar*	Norway, Iceland, UK
Pink salmon	*Oncorhynchus gorbuscha*	USA, Canada, Japan
Chinook	*O. tschawytcha*	USA, Canada
Chum salmon	*O. keta*	USA, Canada, Japan
Coho salmon	*O. kisutch*	USA, Canada, Japan
Sockeye salmon	*O. nerka*	USA, Canada, Japan
Masu salmon	*O. masou*	Japan
Marine fish		
Pacific herring	*Clupea pallasi*	Japan
Black sea bream	*Acanthopagrus schlegeli*	Japan
Red sea bream	*Pagrus major*	Japan
Sandfish	*Arctoscopus japonicus*	Japan
Jacopever	*Sebastes schlegeli*	Japan
Japanese flounder	*Paralichthys olivaceus*	Japan
Mud dab	*Limanda yokohamae*	Japan
Ocellate puffer	*Takifugu rubripes*	Japan
Striped jack	*Pseudocaranx dentex*	Japan
Yellow tail	*Seriola quinqueradiata*	Japan
Sea bass	*Lateolabrax japonicus*	Japan
Red drum	*Sciaenops ocellatus*	USA
Spotted sea trout	*Cynoscion nebulosus*	USA
Striped bass	*Morone saxatilis*	USA
Mullet	*Mugil cephalus*	Hawaii
Threadfin	*Polydactylus sexfilis*	Hawaii
Turbot	*Scophthalmus maximus*	Denmark, Spain
White sea bass	*Atractoscion nobilis*	USA
Whitefish	*Coregonus lavaretus*	Baltic
Cod	*Gadus morhua* L.	Norway, Sweden, Denmark, Faeroe Islands, USA
Invertebrates		
Kuruma shrimps	*Penaeus japonicus*	Japan
Chinese shrimps	*Penaeus chinensis*	Japan
Speckled shrimp	*Metapenaeus monoceros*	Japan
Mangrove crab	*Scylla serrata*	Japan
Swimming crab	*Portunus trituberculatus*	Japan
Blue crab	*Portunus pelagricus*	Japan
Japanese abalone	*Sulculus diversicolor*	Japan
Disk abalone	*Haliotis discus*	Japan
Yezo abalone	*Haliotis discus hannai*	Japan
Giant abalone	*Haliotis gigantea*	Japan
Spiny top shell	*Batillus cornutus*	Japan
Ark shell	*Scapharca broughtonii*	Japan
Scallop	*Patinopecten yessoensis*	Japan
Scallop	*Pecten maximus*	France, Ireland, Norway, UK
Hard clam	*Meretrix lusoria*	Japan
Hard clam	*Meretrix lamarckii*	Japan
Giant clam	*Tridacna maxima*	Indo-Pacific
Giant clam	*Tridacna derasa*	Indo-Pacific
Surf clam	*Spisula sachalinensis*	Japan
European lobster	*Homarus gammarus*	Norway, France, Ireland
Sea urchin	*Tripneustes gratilla*	Japan
Red sea urchin	*Pseudocentrotus depressus*	Japan
Sea urchin	*Strongylocentrotus intermedius*	Japan
Sea urchin	*Strongylocentrotus nudus*	Japan
Sea cucumber	*Stichopus japonicus*	Japan

Salmonids

Ocean ranching programs of salmon have the longest history and have been most comprehensive. Large programs occur in Japan, along the west coast of North America, and also in the Northern Atlantic. Seven species of salmon, which are all anadromous, have been used for releases of captive-bred individuals: Atlantic salmon (*Salmo salar*) which inhabit the northern Atlantic area and the northern Pacific species: pink (*Oncorhynchus gorbuscha*); Chinook (*O. tschawytcha*); chum (*O. keta*); coho (*O. kisutch*); sockeye (*O. nerka*); and the masu salmon (*O. masou*). Atlantic salmon juveniles remain in fresh water for 1–5 years before they smolt and migrate to the marine environment, whereas the pink and chum leave the rivers soon after they have resorbed the yolk-sac.

The ocean ranching program for chum salmon (*Oncorhynchus keta*) in Japan is typical of the way salmon programs are conducted and is the only program that is considered as economically successful to the operators. Chum fry are easy to produce, their survival is good and the program produces 40–50 million fish annually for the fishermen. The number of chum salmon released during the 15-year period 1980–1995 was stable and was about 2 billion per year, whereas the catches tripled during the same period. More than 90% of the chum salmon catches originated from released juveniles, suggesting an increase in the recapture rate from 1% to 3%. The success of ocean ranching of salmon in Japan was due to favorable environmental conditions over the Northern Pacific, but also due to improvements in marine ranching techniques such as egg collection, artificial diets and timing of release, improvements that also were supported by dietary, physiological and behavioral research.

Salmonids show a wide range of lifestyles and a high phenotypic plasticity. The degree of variation depends on the species. The pink salmon has few life history variants and spawns generally in the gravel of estuaries or the lower parts of North Pacific rivers and then dies. After emergence from the gravel the fry leave immediately for the ocean where they grow to maturity, then return to spawn in their natal gravel at 2 years old. The most complex lifestyles are those of steelhead trout (*Oncorhynchus mykiss*) ranging from anadromous to landlocked. The species spawn well upriver in the gravel of a stream. On emergence from the gravel, their fry may spend 1–6 years in the river before emigrating to the sea, or they may not emigrate at all, but mature in the juvenile freshwater habitat and reproduce there (in this case they are referred to as rainbow trout). The steelhead is generally iteroparous. The individuals that emigrate to the sea spend 1–6 years there feeding on prey organisms that are not harvested such as krill, copepods, amphipods, mesopelagic fish, but also on commercial species such as capelin, herring and squid before maturing and returning to breed. After spawning they may return to the sea again, mature, and return the next or later years. Through a combination of all these different developmental possibilities, a single cohort of steelhead may give rise to over 30 different life history types.

The length of the spawning period in anadromous salmonids varies and may exceed 3 months in wild populations. The individuals that return early in the spawning season generally produce offspring that hatch early, and grow faster than the fry of those that arrive later. The timing of the returns reflects individual differences in physiological and behavioral strategies for energy storage and usage, and differences in genetic control between individuals with different seasonal runs. The variation in seasonal runs in a population will have implications for survival probabilities of new recruits in fluctuating natural environments and thus for the stability of population sizes over generations. Populations that have long spawning periods exhibit a high degree of phenotypic variation whereas those that have a short spawning period have lower. The more variation there is, the more flexible will the population be in its adaptations to environmental fluctuations. The smaller the variation, the higher the possibility will be for a recruitment failure and for extinction if individuals are semelparous.

When entering the marine environment wild salmonids migrate to highly productive high sea areas where they grow fast. At the onset of sexual maturation most individuals return to the stream where they were spawned after migrating for several years and thousands of kilometers. Exceptions are jacks (males maturing at age 1 year) that remain at sea for a few months; some species remain in coastal waters rather than undergoing long oceanic migrations. Atlantic salmon can return after one sea winter (grilse), two sea-winters and to a lesser extent after three sea-winters. It has been shown that salmon have the ability to recognize distinctive odors of their home stream and that they become imprinted to this odor as they transform into smolts, just prior to their outward migration. Furthermore, there is evidence that salmon may respond to chemicals (pheromones) given off by conspecifics, such as juveniles inhabiting a stream, and that they are able to discriminate between water that contains fish from their own populations and those from other populations. These abilities allow a great deal of precision in homing. Most commercial harvesting occurs on the coast and recreational and native fishery in the rivers and streams.

Under the salmon enhancement programs it is assumed that the captive-bred released individuals have the same biological and genetic characteristics as wild individuals and that they undertake the same migration routes as their wild conspecifics, and return to their native rivers to spawn. However, it has not been possible to study whether released salmon migrate to the same high seas areas as wild individuals because the main fishery is on the coast and not offshore. The return rates of tagged wild and captive-bred individuals to the various rivers and streams have been compared and show that the homing instinct is not always 100% in any of the groups; the straying rate is three times as high for captive-bred fish with up to 13% recaptured in a different location than where the individuals were released. This suggests different homing instinct abilities in captive-bred and wild salmonids.

The nature of genetic variation within species that are part of ocean ranching programs was not appreciated when they started. The goal of ocean ranching was to restore wild stocks to their former abundance and so to restore the fishery to its former levels. No attention was paid to the number of spawners that were required in order to obtain sufficient genetic variation among captive-bred fish. Often the effective number of spawners was low, resulting in reduced genetic diversity of the population. Moreover, individuals from the broodstock were often taken from the first adults to return, and fry from early-spawning adults were hatched in captivity and released on the spawning grounds together with offspring from fish that had spawned in the wild. Such actions may have had a negative influence on the genetic diversity of the mixed population of wild and captive-bred fish because the fry that hatched early, grow faster, and can displace fry of later-spawning fish. For example, in the USA a decrease in the spawning season from 13 to 3 weeks over just 13 years has been demonstrated for coho salmon. It has also been reported that offspring of captive-bred steelhead that spawned naturally had much lower survival than those from wild fish. This could be a contributing factor to the continuing decline in enhanced salmon stocks despite large-scale releases.

Marine Fish

Ocean ranching is practiced on many marine fish in several countries; the Japanese programs are the largest. Captive-bred individuals of eleven marine fish species are released on a commercial basis in Japan (**Table 1**) with a total yearly release of 68 million fry or juveniles. Red sea bream (*Pagrus major*) and Japanese flounder (*Paralichthys olivaceus*) are released in highest quantities (22 million of each per year).

Juvenile Japanese flounder are 4–12 cm long when released. Those that are larger than 9 cm at release survive to commercial size. Captive-bred individuals show a high degree of partial or complete albinism on the ocular side and characteristic melanin deposits on the other side. These juveniles suffer from high mortality immediately after release (**Figure 1**). The proportion of the catches that have a captive origin is used to quantify impact on catches. Between 10 and 40% of the catches consist of released fish; but this program cannot be considered as a success. Despite releases having increased from 5 million to 22 million juveniles from 1985 to 1995, the total catch of flounder is decreasing (**Figure 2**).

There is also a size-dependent survival of red sea bream; up to 50% are reported in catches if fish were 4 cm at release. Enhancement of red sea bream has been conducted since 1974 and mostly 4 cm-long individuals are now released. However, the population continues to decline despite the releases (**Figure 3**).

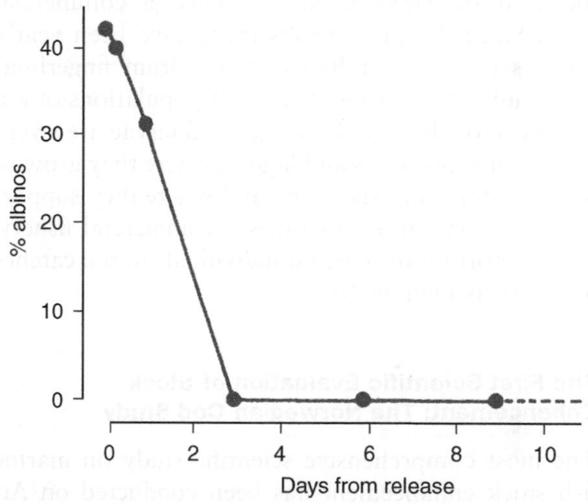

Figure 1 Change in the proportion of albino Japanese flounder juveniles in the catch for successive days after release. (After Blaxter, 2000.)

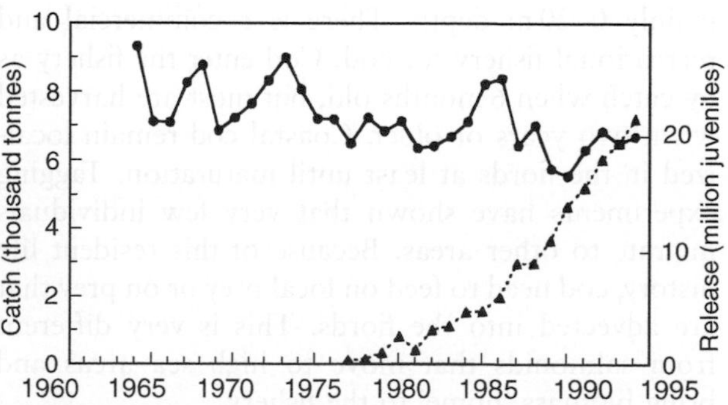

Figure 2 The catch (●) and number of released juveniles (▲) of flounder in Japan. (After Masuda and Tsukamoto, 1998; in Coleman *et al.*, 1998.)

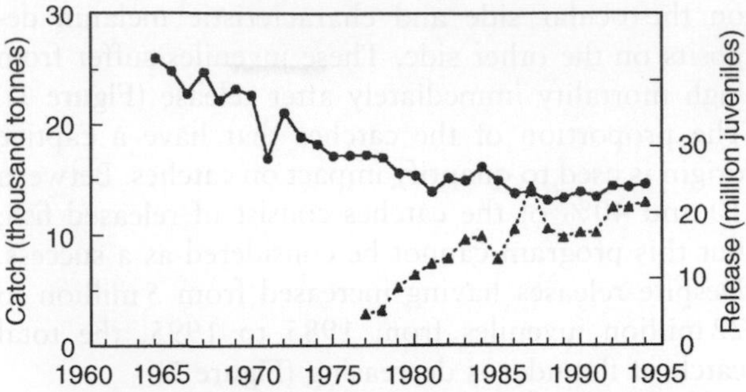

Figure 3 The catch (●) and number of released juveniles (▲) of red sea bream in Japan. (After Masuda and Tsukamoto, 1998; in Coleman *et al.*, 1998.)

Ten other marine fish species (**Table 1**) are used for ocean ranching, but most can be considered to be at an experimental stage. Only the red drum (*Sciaenops ocellatus*) and spotted sea trout (*Cynoscion nebulosus*), which are being released in Texas lagoons of the Gulf of Mexico, have reached a commercial scale. Since the early 1980s there have been yearly releases of 16–30 millions of red drum fingerlings and 5 million sea trout. The adult populations of red drum live offshore and the eggs and larvae are swept into the inshore areas and lagoon where they grow to juveniles for up to six years, and where they support a recreational, and sometimes a commercial fishery. The proportion of released individuals in the catches has been as high as 20%.

The First Scientific Evaluation of Stock Enhancement; The Norwegian Cod Study

The most comprehensive scientific study on marine fish stock enhancement has been conducted on Atlantic cod. Average age at maturity in coastal cod is 3 years. Spawning occurs at grounds located at *c.* 50 m depth and the spawning period is February–April. In western Norway juveniles settle in the shallow nearshore areas during summer and early fall and inhabit mainly 0–20 m depth. There is a commercial and recreational fishery for cod. Cod enter the fishery as by-catch when 6 months old, but most are harvested when two years or older. Coastal cod remain localized in the fiords at least until maturation. Tagging experiments have shown that very few individuals migrate to other areas. Because of this resident life history, cod need to feed on local prey or on prey that are advected into the fiords. This is very different from salmonids that move to high sea areas and bring biomass 'home' to the fishery.

Despite the failure of releases of yolk-sac larvae, new attempts with ocean ranching of cod commenced in Norway in 1983 first through large-scale experiments, but later as small-scale experiments also in Sweden, Denmark, the Faroe Islands and USA. The new approach was to release 6–9-month-old juveniles, 10–20 cm long. This was decided because a positive correlation had been documented between abundance at 6–9 months old and subsequent recruitment to the fishable population, and because large-scale production techniques for juvenile cod had been developed by the Institute of Marine Research in Norway.

The very extensive work on cod by the Norwegians is thought to represent the most thorough investigation of the possibilities and limitations of stock enhancement. A large governmental interdisciplinary research program was initiated in 1983 with experiments being carried out in a number of fiords along the coast. The aim was to develop full-scale production of juveniles, to develop techniques for mass marking and to design and carry out large-scale release experiments on the Norwegian coast. These were to be in association with wide-ranging field studies which were intended to evaluate the potential and limitations of sea ranching of cod from an ecological perspective, before ocean ranching was initiated on a commercial scale. Masfjorden was selected for the study of the whole ecosystem and its dynamic fluctuation in carrying capacity. Research here involved field studies before and after large-scale releases. In addition dynamic ecosystem simulation models were developed. The models integrated all relevant field data and were used to study the effect on fish production and carrying capacity of environmental and biological factors. It was shown that although juvenile wild cod populations were augmented after release of captive-bred cod, there was no sign of stock enhancement of 2-year-old cod when these should have recruited to the fishery. A recent time-series analysis of the survival of captive-bred cod from this fiord shows a high mortality rate in the spring at one year old. It is uncertain what causes this. One possibility is a higher mortality risk in captive-bred than wild cod in spring when there is a massive immigration of spurdog and also higher abundance of other predators. The modeling predicted that density-dependent growth, predation, and cannibalism restricted cod productivity, and that the most important environmental factor was non-local wind-driven fiord-coast advection of organisms at lower trophic levels. This limited the carrying capacity for fish at higher trophic levels in the fiord. The zooplankton *Calanus finmarchicus*, which in spring and summer occur in high abundance in the coastal waters of Norway, were exchanged between coast and fiord via advection of intermediate water-masses. These are the main prey of gobies, and gobies are the main prey of juvenile cod. If strong southerly winds occur frequently, they transport planktonic organisms into the fiord, whereas if strong northerly winds occur most frequently, they may reduce the abundance of planktonic organisms within the fiord and thus limit the carrying capacity for fish at higher trophic levels. This means that the carrying capacity of resident fish in local habitats is highly dependent on environmental processes that occur on a larger scale. The modeling also predicted that the distance from the coast affected the carrying capacity and individual growth of fish in the fiords with lower production and growth within the inner fiords than on the coast (**Figure 4**) because the advection rate and zooplankton density become damped with increasing distance from the coast (**Figure 5**). This was confirmed by empirical estimates of growth curves for cod (**Figure 6**).

As a consequence of the negative conclusion of the Norwegian experimental ocean ranching on cod, the releases were terminated and it was decided not to scale up to commercial level. The research has, however, not been wasted. New insights into fiord ecology and particularly the influence of environmental factors on ecosystem dynamics have been achieved.

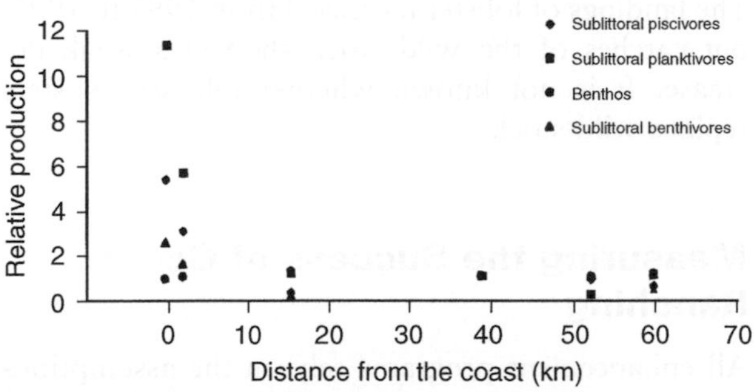

Figure 4 Simulated yearly production for five west Norwegian fiords located at different distances from the outer coast. Sublittoral planktivores refers to gobies, sublittoral piscivores refers to cod and other fish at the same trophic level. (After Salvanes *et al.*, 1995.)

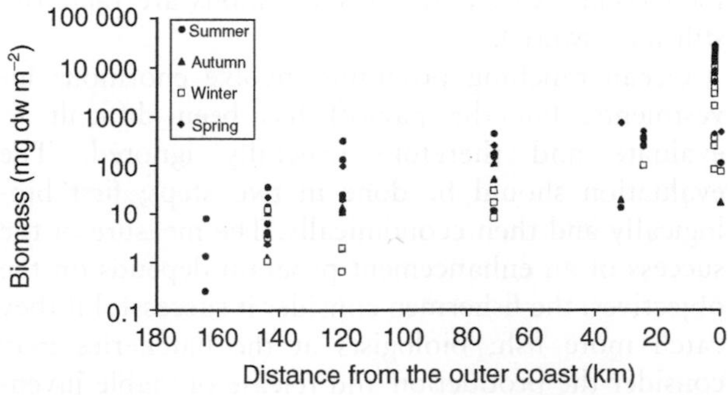

Figure 5 The density of *Calanus finmarchicus* as a function of the distance from the coast by season. Note the logarithmic scale on the biomass axis. (After Salvanes *et al.*, 1995.)

Marine Invertebrates

Enhancement programs on invertebrates are most comprehensive in Japan, including shrimps, crabs, abalone, clams, scallops, sea urchins and sea cucumber (**Table 1**). The present program started on a commercial basis around 1970. Annual releases of seedlings are 3.47×10^9, mostly of scallop (*Patinopecten yessoensis*) (3×10^9).

Scallop seabed culture of *Pecten maximus* started in Europe in France in 1980 and extended to Ireland, Scotland, and Norway. Each country produces 25 million juveniles per year for release. Juveniles are

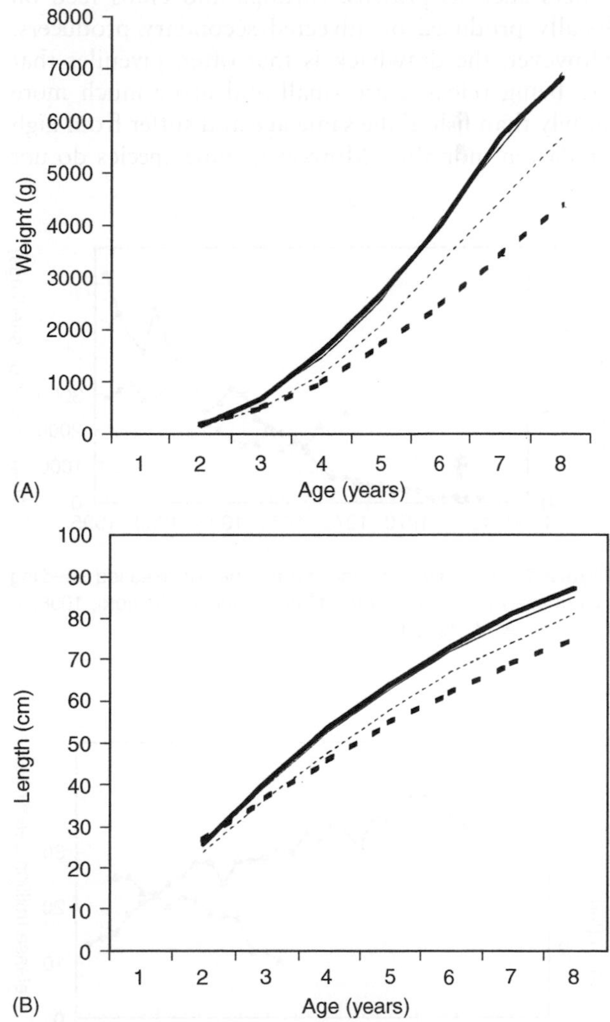

Figure 6 Growth curves ((A) weight; (B) length) for released and wild cod. ——, wild cod, outer coast; - - -, wild cod, inner fiord: ——, released cod, outer coast; - - -, released cod, inner fiord. (Modified from Svsånd *et al.*, 2000.)

first kept in intermediate culture for 1–2 years before release. In France 3 cm juveniles are released on seabeds, whereas in Norway they are 5–6 cm before being released. Ocean ranching with scallops in Japan is considered successful because the landings have increased from 5000 tonnes around 1970 to 200 000 tonnes in the mid 1990s (**Figure 7**). Other invertebrate populations in Japan are, however, still declining despite the high numbers of released seedlings. For example, the populations of abalone have more than halved over the last 25 years (**Figure 8**).

The benefits of releasing invertebrates is that they are either stationary or do not move far from the release point and that they feed lower in the food web than the fish and therefore utilize a larger proportion of the primary production. For example bivalves benefit from transferring locally produced algae to highly valuable biomass by filter feeding; others such as prawns, shrimps and crabs feed on locally produced or advected secondary producers. However, the drawback is that often juveniles that are being released are small and grow much more slowly than fish of the same age and suffer from high predation mortality. Moreover, some species do not move very much, and if juveniles are released in high densities, this could result in unwanted density-dependent mortality (predation, food competition). In addition, the production costs of viable seedlings are often high.

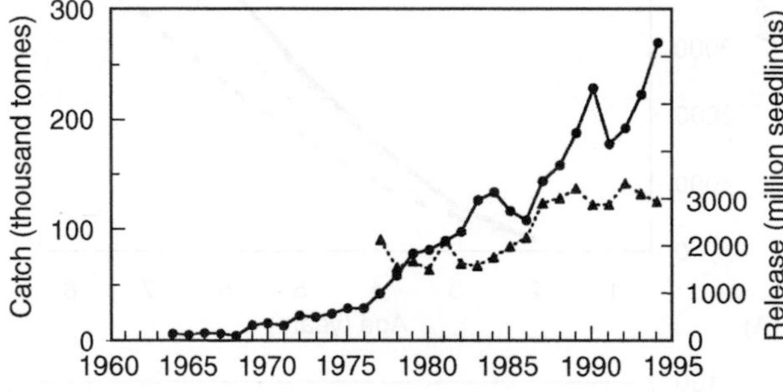

Figure 7 The catch (•) and the number of released seedling (▲) of scallop in Japan. (After Masuda and Tsukamoto, 1998; in Coleman *et al.*, 1998.)

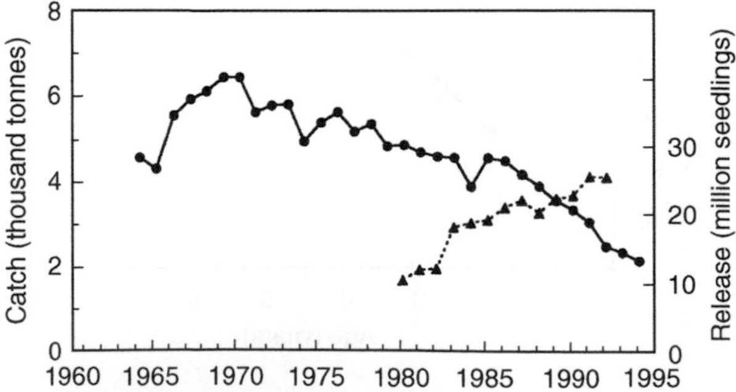

Figure 8 The catch (•) and the number of released seedlings (▲) of abalone in Japan. (After Masuda and Tsukamoto, 1998; in Coleman *et al.*, 1998.)

One example of a slow grower is the European lobster. In the 1970s, Norway succeeded in producing one-year-old juveniles for release, and soon thereafter a commercial lobster hatchery with a production capacity of 120 000 juveniles per year was built. These have been released along the Norwegian coast in attempts to rebuild depleted populations. A large-scale experiment has been conducted at Kvitsøy, a small island with its own continental shelf, located on the south-western coast of Norway. It took 5–8 years after the first large releases in 1985 and 1986 before the lobster recruited to the regulated commercial fishery. Although captive-bred individuals were not tagged, it was possible to distinguish them from the wild because most had developed two scissors claws instead of the normal one scissors and one crusher claw. From 1990 to 1994 the released juveniles were microtagged. In 1997 43% of the landings were of released lobster. The landings of lobster increased from 1995 to 1997, but catches of the wild stock showed a weak decrease. It is not known whether released lobsters replace wild stock.

Measuring the Success of Ocean Ranching

All enhancement programs rely on the assumptions that human operations have reduced populations to sizes below the carrying capacities, or that there are possibilities for increased production of the target species, and that releases of captive-bred juveniles will increase the number of adults and thus subsequent harvests. However, few attempts have been made to test these assumptions scientifically due to their complexity, and many questions are therefore still unanswered.

Ocean ranching programs involve enormous investments, but the pay-off has been difficult to evaluate and therefore generally ignored. The evaluation should be done in two steps; first biologically and then economically. The measure of the success of an enhancement program depends on the objectives: the fishermen consider it successful if they catch more fish; biologists at the hatcheries may consider the production and release of viable juveniles as a success; a conservation biologist will be happy if previously nearly extinct populations increase again; an ecologist will also demand that increased biological production on the target species does not have negative influence on wild conspecifics or on other parts of the ecosystem; and an economist will only consider it a success if there is a positive net pay-off of investments.

The first stage of an evaluation is to determine if there is a net biological benefit by: (1) estimating survival of released fish; (2) verifying survival over many generations; (3) measuring eventual negative effects on wild fish or on the ecosystem as a whole by covering questions such as: does individual variation in growth and abundance change? The final stage would then be to evaluate eventual economic benefit. The programs that occur today usually have immediacy and an applied aspect, and investments into experimental ecology to test major underlying assumptions for increased biological production have therefore been ignored. From a biological point of view this is very shortsighted as the absence of proper evaluation means that ocean ranching programs are conducted in an *ad hoc* way and this will result in a high chance of investment for nothing.

Fish populations have fluctuated in the periods before fishing technology allowed high fishing pressure. It has been reported that some of the marked shifts in sockeye salmon abundance over the last 300 years were associated with climatic changes long before humans were able to overfish. The way climate affects cycles in population sizes is, however, difficult to separate from human made effects. If fish are released because a population declines, and if such releases are conducted during a time period when there are unfavorable climatic conditions, the negative climatic factors may counteract positive enhancement effects. For example, the simulation modeling of the Masfjorden ecosystem showed that nonlocal wind-driven coast–fiord advection of organisms at lower trophic levels had a large impact on the carrying capacity of the cod populations. This means that carrying capacities for fish in coastal ecosystems fluctuate in an unpredictable manner. Simulations also suggest that with imperfect knowledge of the carrying capacity for juvenile cod – which represents more or less the current situation – it is impossible to conduct 'perfect releases' that match the carrying capacity. This means that there may be no payoff from releasing cod to increase a stock.

Other major questions concern the ecology of wild stocks, and also whether the captive-bred individual's genetic, physiological, morphological and behavioral traits are similar to those of wild conspecifics. If not, are these factors deviating so much that released individuals have poorer or better survival than wild conspecifics? Hence, will enhancement effects disappear soon after release or will released animals just replace wild and therefore not enhance total population sizes (released and wild) and harvests? For example in salmon groups size hierarchies among juveniles soon develop in their nursery streams; this also happens among adults in the spawning habitats. Large juveniles tend to monopolize the best feeding spots, and on the spawning beds the large adults get access to the best gravel beds and large males have higher mating success with females. Mature escaped farmed salmon enter Norwegian rivers. Here they often replace wild fish on the spawning grounds. This can be a long-term problem due both to the way the farmed salmon has been domesticated and selectively bred for rapid growth and late maturation, and to a life history that would not necessarily have any benefit under all environmental situations if they escape to the wild. Moreover, offspring of captive-bred steelhead that spawned naturally had much lower survival than those from the wild. It is possibile that instead of enhancing a population, large-scale releases of captive-bred individuals can first replace wild conspecifics, and thereafter become extinct over a few generations in extreme environmental conditions to which the released fish are not adapted. Hence, differences in the life history of captive-bred and wild fish can be a contributing factor to the continuing decline in enhanced salmon stocks despite large-scale releases. If enhancement programs are initiated without taking into consideration the genetic diversity the wild populations have evolved through generations and that is required for survival under extreme environmental fluctuations, the program may fail.

Another documented difference between captive-bred and wild fish is the capability of the latter to show flexible behavior under changing environmental conditions. Captive-bred cod and salmonids adapt more slowly than wild fish to novel food or to food encountered in a different way than previously. They also tend to take more risks than do wild fish. When captive-bred cod were released for sea ranching in Masfjorden off western Norway, it was evident from field samples that these fish had a different diet and higher mortality rate than wild cod at least for the first three days after release. Moreover, these captive-bred cod were released into the natural habitat in numbers that greatly exceeded that of wild cod of the same cohort. Despite this, released cod abundance declined sharply during the spring when the fish were one-year-old, six months after release. At release these fish were naive to the heterogeneous marine environment, to predators, and also to encounters with natural prey. It is possible that feeding behavior, and perhaps also other behavior, of captive-bred animals differed from that of wild cod over a much longer time than the first three days after release and that this could have made them more vulnerable toward piscivorous predators. Released salmonids also do less well than wild. Although they were able to adapt to novel and/or live prey by learning for some days or weeks before they were released, they took less prey, grew more slowly, had a higher mortality rate than wild, had a narrower range of dietary items and frequently lagged behind wild fish in switching to new prey items as these became abundant.

In the literature there is hardly any ocean ranching program that can be considered successful on both ecological and economic grounds. Only the chum salmon ranching program in Japan is considered economically successful, at least for the operators, because the cost of fry production is low and the fishery catches more fish (**Figure 9A**). There are, however, two possible drawbacks even for this program: the average size of individual chum has been decreasing (**Figure 9B**), and the change is correlated with the Japanese chum releases, and with the abundance of chum in the North Pacific. The

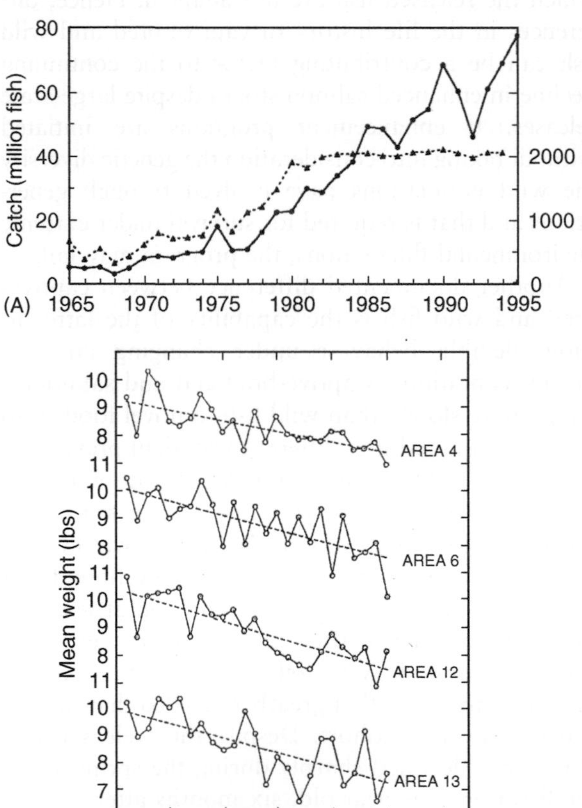

Figure 9 (A) The catch (•) and the number released of juveniles (▲) of chum salmon in Japan. (B) Changes in the mean weights of coho salmon harvested from the Pacific ocean between 1950 and 1981. (After Masuda and Tsukamoto, 1998; in Coleman *et al.*, 1998.)

economic value of individual fish has therefore decreased. Moreover, total Pacific chum production has only increased by 50% since 1970, and the North American production (largely natural) increased by about the same amount. It might thus be that the apparent increase in Japanese chum production could be due to favorable climatic conditions in combination with replacement of wild fish. It is, however, very difficult to separate enhancement and environmental effects. Nevertheless, cyclical changes in Pacific salmon populations have occurred earlier and long before humans were able to overfish or conduct captive breeding for release. This means that environmental factors would mask any cause of change in populations that are seen in populations chosen for ocean ranching. Another study, on chinook salmon of the Columbian river basin showed that the population continues to decline towards extinction despite the release of captive-bred individuals. It was concluded through simulation studies that the only way to reverse the decline would be to increase the survival of first-year fish, and the only way of doing this might be to remove the dams that had been constructed for hydropower production.

The prospects of marine ranching should be critically evaluated biologically and economically before commercial large-scale programs are initiated on a target species. The optimal species for stock enhancement should be easy and cheap to rear to release size, it should grow fast, have a low mortality rate and it should feed low in the food chain. It should also have a behavior that does not deviate from wild conspecifics or lead to negative effects on other species. However, the case studies reported here illustrate clear difficulties in increasing population sizes by releasing captive-bred individuals and show that hardly any commercial enhancement program can be regarded as clearly successful. Model simulations suggest, however, that stock-enhancement may be possible if releases can be made that match closely the current ecological and environmental conditions. However, this requires improvements of assessment methods of these factors beyond present knowledge. Marine systems tend to have strong nonlinear dynamics, and unless one is able to predict these dynamics over a relevant time horizon, release efforts are not likely to increase the abundance of the target population.

See also

Mariculture, Economic and Social Impacts. Salmonid Farming. Salmon Fisheries, Atlantic. Salmon Fisheries, Pacific.

Further Reading

Asplin L, Salvanes AGV, and Kristoffersen JB (1999) Non-local wind-driven fjord–coast advection and its potential effect on pelagic organisms and fish recruitment. *Fisheries Oceanography* 8: 255–263.

Blaxter JHS (2000) The enhancement of marine fish stocks. *Advances in Marine Biology* 38: 2–54.

Coleman F, Travis J and Thistle AB (1998) *Marine Stock Enhancement: A New Perspective*. Bulletin of Marine Science 62.

Danielssen DS, Howell BR, and Moksness E (1994) An International Symposium on Sea Ranching of Cod and other Marine Fish Species. *Aquaculture and Fisheries Management* 25 (Suppl. 1).

Finney B, Gregory-Eaves I, Sweetman J, Douglas MSV, and Smol JP (2000) Impacts of climatic change and fishing on Pacific salmon over the past 300 years. *Science* 290: 795–799.

Giske J and Salvanes AGV (1999) A model for enhancement potentials in open ecosystems. In: Howell BR, Moksness E, and Svsånd T (eds.) *Stock Enhancement and Sea Ranching*. Blackwell: Fishing, News Books.

Howell BR, Moksness E, and Svsånd T (1999) *Stock Enhancement and Sea Ranching*.

Kareiva P, Marvier M, and McClure M (2000) Recovery and management options for spring/summer chinnook salmon in the Columbia River basin. *Science* 290: 977–979.

Mills D (1989) *Ecology and Management of Atlantic Salmon*. London: Chapman Hall.

Ricker WE (1981) Changes in the average size and average age of Pacific salmon. *Canadian Journal of Fisheries and Aquatic Science* 38: 1636–1656.

Salvanes AGV, Aksnes DL, Foss JH, and Giske J (1995) Simulated carrying capacities of fish in Norwegian fjords. *Fisheries Oceanography* 4(1): 17–32.

Salvanes AGV, Aksnes DL, and Giske J (1992) Ecosystem model for evaluating potential cod production in a west Norwegian fjord. *Marine Ecology Progress Series* 90: 9–22.

Salvanes AGV and Baliño B (1998) Production and fitness in a fjord cod population: an ecological and evolutionary approach. *Fisheries Research* 37: 143–161.

Shelbourne JE (1964) The artificial propagation of marine fish. *Advances in Marine Biology* 2: 1–76.

Svsånd T, Kristiansen TS, Pedersen T, *et al.* (2000) The biological and economical basis of cod stock enhancement. *Fish and Fisheries* 1: 173–205.

Thorpe JE (1980) *Salmon Ranching*. London: Academic Press.

OYSTERS – SHELLFISH FARMING

I. Laing, Centre for Environment Fisheries and Aquaculture Science, Weymouth, UK

As I ate the oysters with their strong taste of the sea and their faint metallic taste that the cold white wine washed away, leaving only the sea taste and the succulent texture, and as I drank their cold liquid from each shell and washed it down with the crisp taste of the wine, I lost the empty feeling and began to be happy and to make plans. (Ernest Hemingway in *A Moveable Feast*).

A Long History

Oysters have been prized as a food for millennia. Carbon dating of shell deposits in middens in Australia show that the aborigines took the Sydney rock oyster for consumption in around 6000 BC. Oyster farming, nowadays carried out all over the world, has a very long history, although there are various thoughts as to when it first started. It is generally believed that the ancient Romans and the Greeks were the first to cultivate oysters, but some maintain that artificial oyster beds existed in China long before this.

It is said that the ancient Chinese raised oysters in specially constructed ponds. The Romans harvested immature oysters and transferred them to an environment more favorable to their growth. Greek fishermen would toss broken pottery dishes onto natural oyster beds to encourage the spat to settle.

All of these different cultivation methods are still practiced in a similar form today.

A Healthy Food

Oysters (**Figure 1**) have been an important food source since Neolithic times and are one of the most nutritionally well balanced of foods. They are ideal for inclusion in low-cholesterol diets. They are high in omega-3 fatty acids and are an excellent source of vitamins. Four or five medium-size oysters supply the recommended daily allowance of a whole range of minerals.

The Oyster

Oysters are one among a number of bivalve mollusks that are cultivated. Others include clams, cockles, mussels, and scallops.

Current Status

Oysters form the second most important group, after cyprinid fishes, in world aquaculture production.

Figure 1 A dish of Pacific oysters.

In 2004, 4.6 million tonnes were produced (Food and Agriculture Organization (FAO) data). Asia and the Pacific region produce 93.4% of this total production.

The FAO of the United Nations lists 15 categories of cultivated oyster species. The Pacific oyster (*Crassostrea gigas*) is by far the most important. Annual production of this species now exceeds 4.4 million metric tonnes globally, worth US$2.7 billion, and accounts for 96% of the total oyster production (see **Table 1**). China is the major producer.

The yield from wild oyster fisheries is small compared with that from farming and has declined steadily in recent years. It has fallen from over 300 000 tonnes in 1980 to 152 000 tonnes in 2004. In contrast to this, Pacific oyster production from aquaculture has increased by 51% in the last 10 years (see **Figure 2**).

General Biology

As might be expected, given a long history of cultivation, there is a considerable amount known about the biology of oysters.

The shell of bivalves is in two halves, or valves. Two muscles, called adductors, run between the inner surfaces of the two valves and can contract rapidly to close the shell tightly. When exposed to the air, during the tidal cycle, oysters close tightly to prevent desiccation of the internal tissues. They can respire anaerobically (i.e., without oxygen) when out of water but have to expel toxic metabolites when reimmersed as the tide comes in. They are known to be able to survive for long periods out of water at low temperatures such as those used for storage after collection.

Within the shell is a fleshy layer of tissue called the mantle; there is a cavity (the mantle cavity) between the mantle and the body wall proper. The mantle secretes the layers of the shell, including the inner nacreous, or pearly, layer. Oysters respire by using both gills and mantle. The gills, suspended within the mantle cavity, are large and function in food gathering (filter feeding) as well as in respiration. As water passes over the gills, organic particulate material, especially phytoplankton, is strained out and is carried to the mouth.

Table 1 The six most important oyster species produced worldwide

Oyster species	*Number of producing countries*	*Major producing countries (% of total)*	*Production (metric tonnes)*
Pacific oyster	30	China (85), Korea (5), Japan (5), France (2.5)	4 429 337
American oyster	4	USA (95), Canada (4.5)	110 770
Slipper oyster	1	Philippines (100)	15 915
Sydney oyster	1	Australia (100)	5600
European oyster	17	Spain (50), France (29), Ireland (8)	5071
Mangrove oyster	3	Cuba (99)	1184

Tonnages are 2004 figures (FAO data). All are cupped oyster species, apart from the European oyster, which is a flat oyster species.

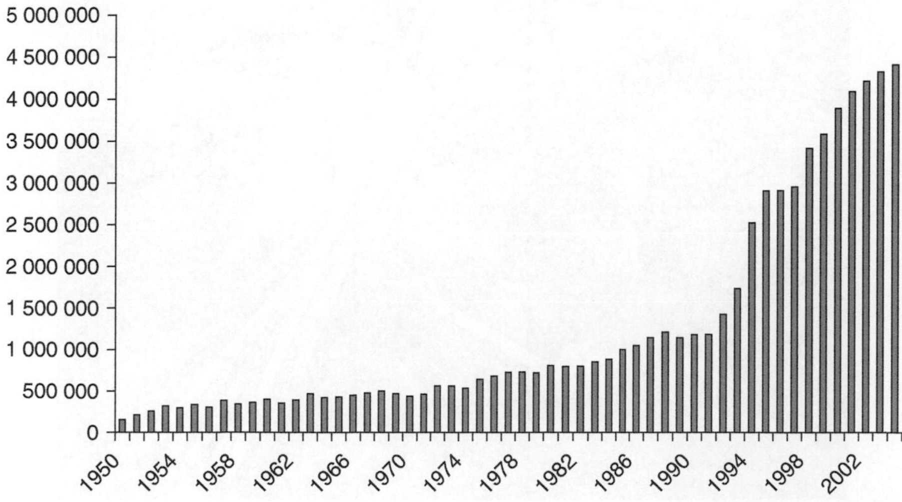

Figure 2 The increase in world production, in metric tonnes, of the Pacific oysters (1950–2004).

Oysters have separate sexes, but they may change sex one or more times during their life span, being true hermaphrodites. In most species, the eggs and sperm are shed directly into the water where fertilization occurs. Larvae are thus formed and these swim and drift in the water, feeding on natural phytoplankton. After 2–3 weeks, depending on local environmental conditions, the larvae are mature and they develop a foot. At this stage they sink to the seabed and explore the sediment surface until they find a suitable surface on which to settle and attach permanently, by cementation. Next, they go through a series of morphological and physiological changes, a process known as metamorphosis, to become immature adults. These are called juveniles, spat, or seed.

Cultivated species of flat oysters brood the young larvae within the mantle cavity, releasing them when they are almost ready to settle. Fecundity is usually related to age, with older and larger females producing many more larvae.

Growth of juveniles is usually quite rapid initially, before slowing down in later years. The length of time that oysters take to reach a marketable size varies considerably, depending on local environmental conditions, particularly temperature and food availability. Pacific oysters may reach a market size of 70–100 g live weight (shell-on) in 18–30 months.

Methods of Cultivation – Seed Supply

Oyster farming is dependent on a regular supply of small juvenile animals for growing on to market size. These can be obtained primarily in one of two ways, either from naturally occurring larvae in the plankton or artificially, in hatcheries.

Wild Larvae Collection

Most oyster farmers obtain their seed by collecting wild set larvae. Special collection materials, generically known as 'cultch', are placed out when large numbers of larvae appear in the plankton. Monitoring of larval activity is helpful to determine where and when to put out the cultch. It can be difficult to discriminate between different types of bivalve larvae in the plankton to ensure that it is the required species that is present but recently, modern highly sensitive molecular methods have been developed for this.

Various materials can be used for collection. Spat collected in this way are often then thinned prior to growing on (**Figure 3**). In China, coir rope is widely used, as well as straw rope, flax rope, and ropes woven by thin bamboo strips. Shells, broken tiles, bamboo, hardwood sticks, plastics, and even old tires can also be used. Coatings are sometimes applied.

Hatcheries

Restricted by natural conditions, the amount of wild spat collected may vary from year to year. Also, where nonnative oyster species are cultivated, there may be no wild larvae available. In these circumstances, hatchery cultivation is necessary. This also allows for genetic manipulation of stocks, to rear and maintain lines specifically adapted for certain

Figure 3 A demonstration of the range of oyster spat collectors used in France.

traits. Important traits for genetic improvement include growth rate, environmental tolerances, disease resistance, and shell shape.

Methods of cryopreservation of larvae are being developed and these will contribute to maintaining genetic lines. Furthermore, triploid oysters, with an extra set of chromosomes, can only be produced in hatcheries. These oysters often have the advantage of better growth and condition, and therefore marketability, during the summer months, as gonad development is inhibited.

Techniques for hatchery rearing were first developed in the 1950s and today follow well-established procedures.

Adult broodstock oysters are obtained from the wild or from held stocks. Depending on the time of year, these may need to be bought into fertile condition by providing a combination of elevated temperature and food (cultivated phytoplankton diets) over several weeks (see **Figure** 4). Selection of the appropriate broodstock conditioning diet is very important. An advantage of hatchery production is that it allows for early season production in colder climates and this ensures that seeds have a maximum growing period prior to their first overwintering.

For cultivated oyster species, mature gametes are usually physically removed from the gonads. This involves sacrificing some ripe adults. Either the gonad can be cut repeatedly with a scalpel and the gametes washed out with filtered seawater into a part-filled container, or a clean Pasteur pipette can be inserted into the gonad and the gametes removed by exerting gentle suction.

Broodstock can also be induced to spawn. Various stimuli can be applied. The most successful methods are those that are natural and minimize stress. These include temporary exposure to air and thermal cycling (alternative elevated and lowered water temperatures). Serotonin and other chemical triggers can also be used to initiate spawning but these methods are not generally recommended as eggs liberated using such methods are often less viable.

Flat oysters, of the genera *Ostrea* and *Tiostrea*, do not need to be stimulated to spawn. They will spawn of their own accord during the conditioning process as they brood larvae within their mantle cavities for varying periods of time depending on species and temperature.

The fertilized eggs are then allowed to develop to the fully shelled D-larva veliger stage, so called because of the characteristic 'D' shape of the shell valves (**Figure** 5).

These larvae are then maintained in bins with gentle aeration. Static water is generally used and this is exchanged daily or once every 2 days. Throughflow systems, with meshes to prevent loss of larvae, are also employed. Cultured microalgae are added into the tanks several times per day, at an appropriate

Figure 4 The SeaCAPS continuous culture system for algae. An essential element for a successful oyster hatchery is the means to cultivate large quantities of marine micro algae (phytoplankton) food species.

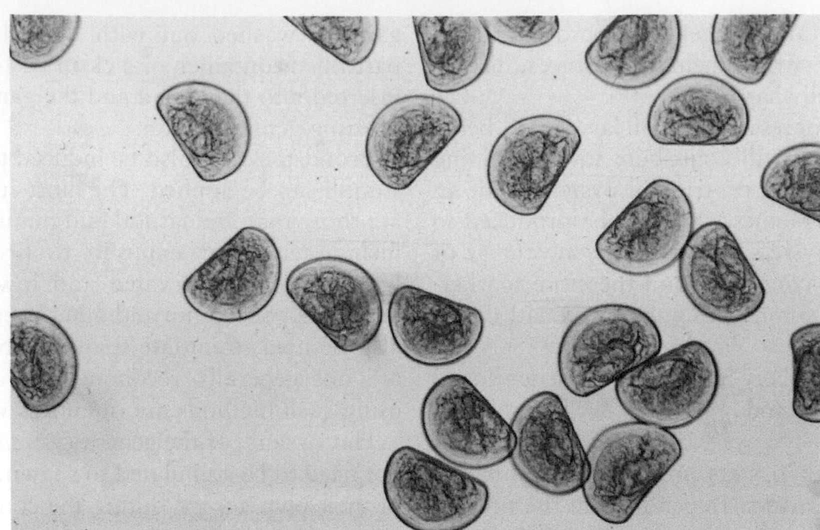

Figure 5 Pacific oyster D larvae. At this stage shell length is about 70 microns.

daily ration according to the number and size of the larvae.

The larvae eventually become competent to settle and for this, surfaces must be provided. The area of settlement surface is important. Types of materials in common usage to provide large surface areas for settlement include sheets of slightly roughened polyvinyl chloride (PVC), layers of shell chips and particles prepared by grinding aged, clean oyster shell spread over the base of settlement trays or tanks, bundles, bags, or strings of aged clean oyster shells dispersed throughout the water column, usually in settlement tanks or various plastic or ceramic materials coated with cement (lime/mortar mix).

For some of these methods, the oysters are subsequently removed from the settlement surface to produce 'cultchless' spat. These can then be grown as separate individuals through to marketable size and sold as whole or half shell, usually live.

Provision of competence to settle Pacific oyster larvae for remote setting at oyster farms is common practice on the Pacific coast of North America. Hatcheries provide the mature larvae and the farmers themselves set them and grow the spat for seeding oyster beds or in suspended culture.

In other parts of the world, hatcheries set the larvae, as described above, and grow the spat to a size that growers are able to handle and grow.

Oyster juveniles from hatchery-reared larvae perform well in standard pumped upwelling systems and survival is usually good, although some early losses may occur immediately following metamorphosis. Diet, ration, stocking density, and water flow rate are all important in these systems (**Figure 6**). They are only suitable for initial rearing of small seed. As the spat grow, food is increasingly likely to become limiting in these systems and they must be transferred to the sea for on-growing.

The size at which spat is supplied is largely dictated by the requirements and maturity of the grow-out industry. Seed native oysters are made available from commercial hatcheries at a range of sizes up to 25–30 mm. The larger the seed, the more expensive they are but this is offset by the higher survival rate of larger seed. Larger seed should also be more tolerant to handling.

Ponds

Pond culture offers a third method to provide seed. This was the method that was originally developed in Europe in early attempts to stimulate production following the decline of native oyster stocks in the late nineteenth century. Ponds of 1–10 ha in area and 1–3 m deep were built near to high water spring tides, filled with seawater, and then isolated for the period of time during which the oysters are breeding naturally, usually May to July. Collectors put into the ponds encourage and collect the settlement of juvenile oysters. There is an inherent limited amount of control over the process and success is very variable. In France, where spat collectors were also deployed in the natural environment, it was relatively successful and became an established method for a time. In the UK, spat production from ponds built in the early 1900s was insufficiently regular to provide a reliable supply of seed to the industry and the method was largely abandoned in the middle of the twentieth century in favor of the more controlled conditions available in hatcheries.

(a)

(b)

Figure 6 Indoor (a) and outdoor (b) oyster nursery system, in which seawater and algae food are pumped up through cylinders fitted with mesh bases and containing the seed.

Methods of Cultivation – On-growing

Various methods are available for on-growing oyster seed once this has been obtained, from either wild set larvae, hatcheries, or ponds.

Oysters smaller than 10 g need to be held in trays or bags attached to a superstructure on the foreshore until they are large enough to be put directly on to the substrate and be safe from predators, strong tidal and wave action, or siltation (**Figure 7**).

Tray cultivation of oysters can be successful in water of minimal flow, where water exchange is driven only by the rise and fall of the tide and gentle wave action. In such circumstances it may even be possible in open baskets.

Figure 7 The traditional bag and trestle method for on-growing Pacific oysters.

Figure 8 Open baskets for on-growing, as developed in Tasmania for Pacific oysters.

In more exposed sites, systems developed in Australia and employing plastic mesh baskets attached to or suspended from wires (see **Figure** 8) are becoming increasingly popular. It is claimed that an oyster with a better shape, free from worm (*Polydora*) infestation, will result from using these systems. *Polydora* can cause unsightly brown blemishes on the inner surface of the shell and decrease marketability of the stock. Less labor is required with these systems but they are more expensive to purchase initially.

In all of the above systems, the mesh size can be increased as the oysters grow, to improve the flow of water and food through the animals.

Holding seed oysters in trays or attached to the cultch material onto which they were settled, and suspended from rafts, pontoons, or long lines is an alternative method in locations where current speed will allow. The oysters should grow more quickly because they are permanently submerged but the shell may be thinner and therefore more susceptible to damage. Early stages of predator species such as crabs and starfish can settle inside containers, where they can cause significant damage unless containers are opened and checked on a regular basis. In the longer term the oysters will perform better on the seabed, although they can be reared to market size in the containers.

Protective fences can be put up around ground plots to give some degree of protection to smaller oysters from shore crabs. Potting crabs in the area of the lays is another method of control. The steps in the oyster cultivation process are shown as a diagram in **Figure 9**.

The correct stocking density is important so as not to exceed the carrying capacity of a body of water. When carrying capacity is exceeded, the algal population in the water is insufficient and growth declines. Mortalities also sometimes occur. Yields in extensively cultivated areas are usually about 25 tonnes per hectare per year but this can increase to 70 tonnes where individual plots are well separated.

In France, premium quality oysters are sometimes finished by holding them in fattening ponds known as claires (see **Figure 10**), in which a certain type of algae is encouraged to bloom, giving a distinctive green color to the flesh.

In the UK, part-grown native flat oysters from a wild fishery are relayed in spring into areas in which conditions are favorable to give an increase in meat yield over a period of a few months, prior to marketing in the winter.

There are proposals to establish standards for organic certification of mollusk cultivation but many oyster growers consider that the process is intrinsically organic.

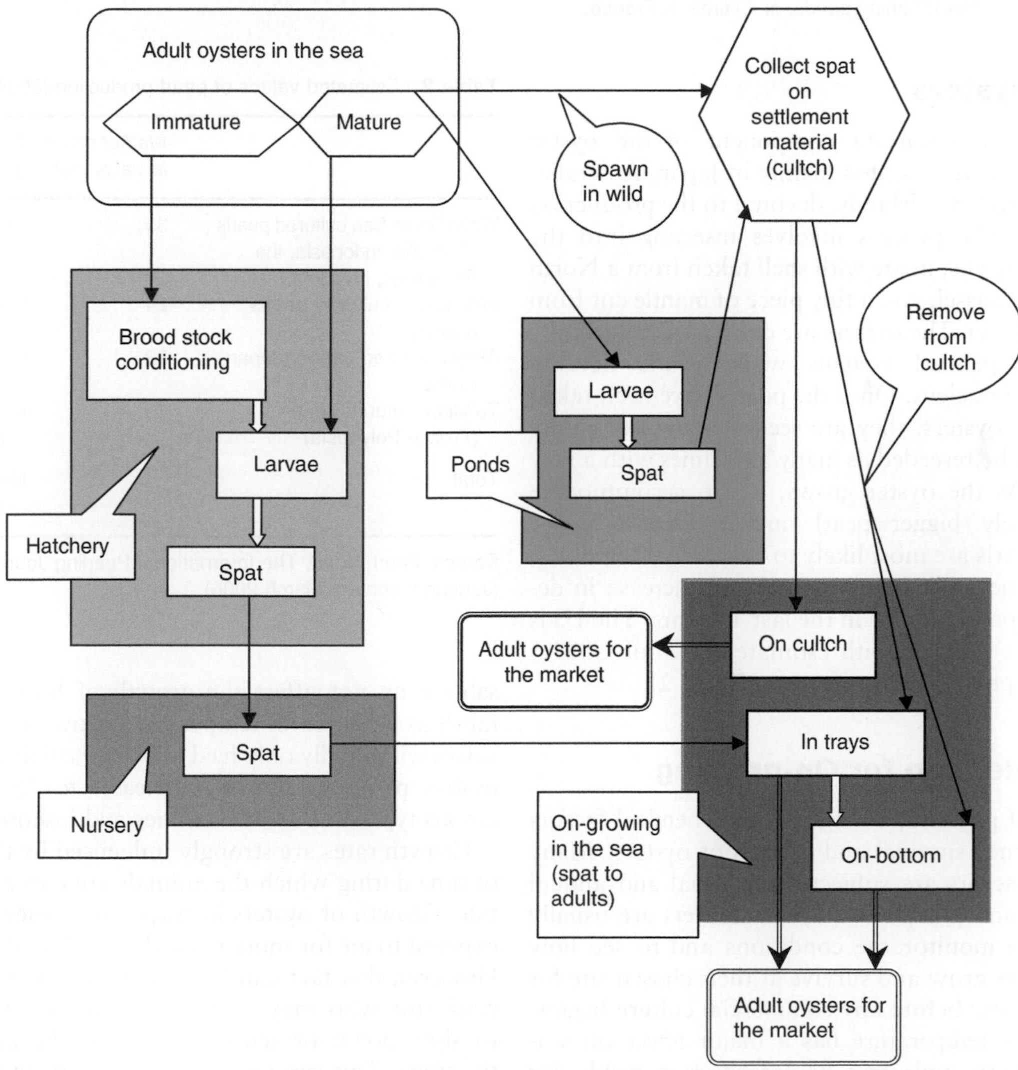

Figure 9 The steps and processes of oyster farming.

Figure 10 Oyster fattening ponds, or Claires, in France.

Pearl Oysters

There is an important component of the oyster farming industry, located mainly in Japan, Australia, and the South Sea Islands, devoted to the production of pearls. The process involves inserting into the oyster a nucleus, made with shell taken from a North American mussel, and a tiny piece of mantle cut from another oyster. The oysters are cultivated in carefully tended suspended systems while pearls develop around the nucleus. Once the pearls have been taken out of the oysters, they are seeded anew. A healthy oyster can be reseeded as many as 4 times with a new nucleus. As the oyster grows, it can accommodate progressively bigger pearl nuclei. Therefore, the biggest pearls are most likely to come from the oldest oysters. There has been a significant increase in demand for pearl jewelry in the last 10 years. The US is the biggest market, with estimated sales of US$1.5 billion of pearl jewelry per year (**Table 2**).

Table 2 Estimated values of pearl production (2004)

	Market share in value (%)	*Production value (2004)*
White South Sea cultured pearls (Australia, Indonesia, the Philippines, Myanmar)	35	US$220 million
Freshwater cultured pearls (China)	24	US$150 million
Akoya cultured pearls (Japan, China)	22	US$135 million
Tahitian cultured pearls (French Polynesia)	19	US$120 million
Total		US$625 million

Source: Pearl World, The International Pearling Journal (January/February/March 2006).

Site Selection for On-growing

A range of physical, biological, and chemical factors will influence survival and growth of oysters. Many of these factors are subject to seasonal and annual variation and prospective oyster farmers are usually advised to monitor the conditions and to see how well oysters grow and survive at their chosen site for at least a year before any commercial culture begins.

Seawater temperature has a major effect on seasonal growth and may be largely responsible for any differences in growth between sites. Changes in salinity do not affect the growth of bivalves by as much as variation in temperature. However, most bivalves will usually only feed at higher salinities. Pacific oysters prefer salinity levels nearer to 25 psu, conditions typical of many estuaries and inshore waters.

Growth rates are strongly influenced by the length of time during which the animals are covered by the tide. Growth of oysters in trays stops when they are exposed to air for more than about 35% of the time. However, this fact can be used to advantage by the cultivator who may wish, for commercial reasons, to slow down or temporarily stop the growth of the stock. This can be achieved by moving the stock higher up the beach. It is a practice that is routinely adopted in Korea and Japan for 'hardening off' wild-caught spat prior to sale.

Other considerations are related to access and harvesting. It is important to consider the type of equipment likely to be used for planting, maintenance, and harvesting, particularly at intertidal sites. Some beaches will support wheeled or tracked vehicles, while others are too soft and will require the use of a boat to transport equipment.

Risks

Oyster farming can be at risk from pollutants, predators, competitors, diseases, fouling organisms, and toxic algae species.

Pollutants

Some pollutants can be harmful to oysters at very low concentrations. During the 1980s, it was found that tributyl tin (TBT), a component of marine antifouling paints, was highly toxic to bivalve mollusks at extremely low concentrations in the seawater. Pacific oysters cultivated in areas in which large numbers of small vessels were moored showed stunted growth and thickening of the shell, and natural populations of flat oysters failed to breed (see **Figure 11**). The use of this compound on small vessels was widely banned in the late 1980s, and since then the oyster industry has recovered. The International Maritime Organization has since announced a ban for larger vessels as well. When marketed for consumption, oysters must meet a number of 'end product' standards. These include a requirement that the shellfish should not contain toxic or objectionable compounds such as trace metals, organochlorine compounds, hydrocarbons, and polycyclic aromatic hydrocarbons (PAHs) in such quantities that the calculated dietary intake exceeds the permissible daily amount.

Predators

In some areas oyster drills or tingles, which are marine snails that eat bivalves by rasping a hole through the shell to gain access to the flesh, are a major problem.

Competitors

Competitors include organisms such as slipper limpets (*Crepidula fornicata*), which compete for food and space. In silt laden waters, they also produce a muddy substrate, which is unsuitable for cultivation. They can be a significant problem. An example is around the coast of Brittany, where slipper limpets have proliferated in the native flat oyster areas during the last 30 years. In order to try to control this problem, approximately 40 000–50 000 tonnes of slipper limpets are harvested per year. These are taken to a factory where they are converted to a calcareous fertilizer.

Diseases

The World Organisation for Animal Health (OIE) lists seven notifiable diseases of mollusks and six of

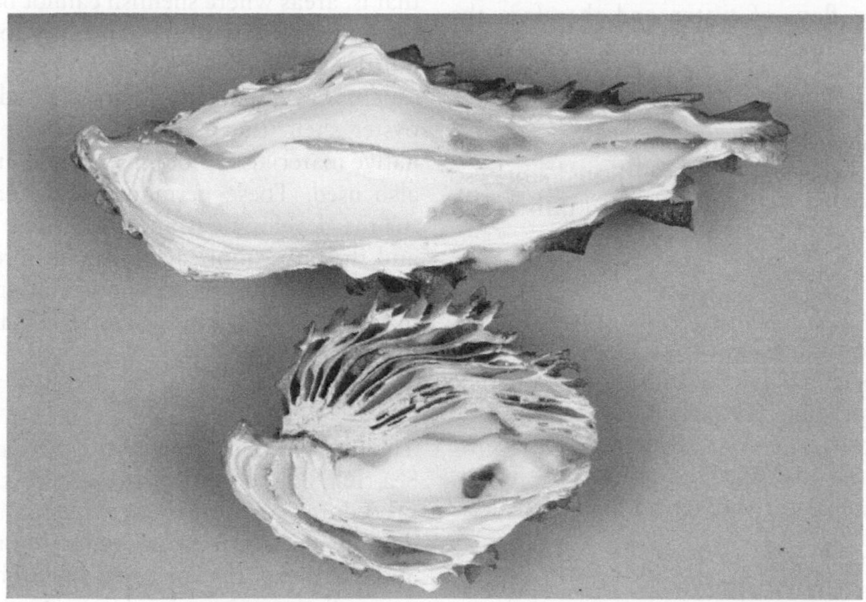

Figure 11 Compared with a normal shell (upper shell section), TBT from marine antifouling paints causes considerable thickening of Pacific oyster shells (lower section). These compounds are now banned.

these can infect at least one of the cultivated oyster species. Some of these diseases, often spread through national and international trade, have had a devastating effect on oyster stocks worldwide. For example, Bonamia, a disease caused by the protist parasitic organism *Bonamia ostreae*, has had a significant negative impact on *Ostrea edulis* production throughout its distribution range in Europe. Mortality rates in excess of 80% have been noted. The effect this can have on yields can be seen in the drastic (93%) drop in recorded production in France, from 20 000 tonnes per year in the early 1970s to 1400 tonnes in 1982. A major factor in the introduction and success of the Pacific oyster as a cultivated species has been that it is resistant to major diseases, although it is not completely immune to all problems.

There are reports of summer mortality episodes, especially in Europe and the USA, and infections by a herpes-like virus have been implicated in some of these. *Vibrio* bacteria are also associated with larval mortality in hatcheries.

The Fisheries and Oceans Canada website lists 53 diseases and pathogens of oysters. Juvenile oyster disease is a significant problem in cultivated *Crassostrea virginica* in the Northeastern United States. It is thought to be caused by a bacterium (*Roseovarius crassostreae*).

Fouling

Typical fouling organisms include various seaweeds, sea squirts, tubeworms, and barnacles. The type and degree of fouling varies with locality. The main effect is to reduce the flow of water and therefore the supply of food to bivalves cultivated in trays or on ropes and to increase the weight and drag on floating installations. Fouling organisms grow in response to the same environmental factors as are desirable for good growth and survival of the cultivated stock, so this is a problem that must be controlled rather than avoided.

Toxic Algae

Mortalities of some marine invertebrates, including bivalves, have been associated with blooms of some alga species, including *Gyrodinium aureolum* and *Karenia mikimoto*. These so-called red tides cause seawater discoloration and mortalities of marine organisms. They have no impact on human health.

Stock Enhancement

If we accept the FAO definition of aquaculture as "The farming of aquatic organisms in inland and coastal areas, involving intervention in the rearing process to enhance production and the individual or corporate ownership of the stock being cultivated," then in many cases stock enhancement can be included as a type of oyster farming.

Natural beds can be managed to encourage the settlement of juvenile oysters and sustain a fishery. Beds can be raked and tilled on a regular basis to remove silt and ensure that suitable substrates are available for the attachment of the juvenile stages. Adding settlement material (cultch) is also beneficial.

In some areas of the world, there has been a dramatic reduction in stocks of the native oyster species. This is attributed mainly to overexploitation, although disease is also implicated. Native oyster beds form a biotope, with many associated epifaunal and infaunal species. Loss of this habitat has resulted in a major decline in species richness in the coastal environment.

Considerable effort has been put into restoring these beds, using techniques that might fall under the definition of oyster farming. A good example is the Chesapeake Bay Program for the native American oyster *Crassostrea virginica*. Overfishing followed by the introduction of two protozoan diseases, believed to be inadvertently introduced to the Chesapeake through the importation of a nonnative oyster, *Crassostrea gigas*, in the 1930s, has combined to reduce oyster populations throughout Chesapeake Bay to about 1% of historical levels. The Chesapeake Bay Program is committed to the restoration and creation of aquatic reefs. A key component of the strategy to restore oysters is to designate sanctuaries, that is, areas where shellfish cannot be harvested. It is often necessary within a sanctuary to rehabilitate the bottom to make it a suitable oyster habitat. Within sanctuaries aquatic reefs are created primarily with oyster shell, the preferred substrate of spat. Alternative materials including concrete and porcelain are also used. These permanent sanctuaries will allow for the development and protection of large oysters and therefore more fecund and potentially disease-resistant oysters. Furthermore, attempts have been made at seeding with disease-free hatchery oysters.

Nonnative Species

The International Council for the Exploration of the Sea (ICES) Code of Practice sets forth recommended procedures and practices to diminish the risks of detrimental effects from the intentional introduction and transfer of marine organisms. The Pacific oyster, originally from Asia, has been introduced around the world and has become invasive in parts of Australia and New Zealand, displacing the native rock oyster in some areas. It is also increasingly becoming a problem in northern European coastal regions. In parts of France naturally recruited stock is competing with cultivated stock and has to be controlled.

The possibility of farming the nonnative Suminoe oyster (*Crassostrea ariakensis*) to restore oyster stocks in Chesapeake Bay is being examined and there are differing opinions about the environmental risks involved, with concerns of potential harmful effects on the local ecology.

It should also be noted that aquaculture has been responsible for the introduction of a whole range of passenger species, including pest species such as the slipper limpet and oyster drills, throughout the world.

Food Safety

Bivalve mollusks filter phytoplankton from the seawater during feeding; they also take in other small particles, such as organic detritus, bacteria, and viruses. Some of these bacteria and viruses, especially those originating from sewage outfalls, can cause serious illnesses in human consumers if they remain in the bivalve when it is eaten. The stock must be purified of any fecal bacterial content in cleansing (depuration) tanks (**Figure 12**) before sale for consumption. There are regulations governing this, which are based on the level of contamination of the mollusks.

Viruses are not all removed by normal depuration processes and they can cause illness if the bivalves are eaten raw or only lightly cooked, as oysters often are. These viruses can only be detected by using sophisticated equipment and techniques, although research is being carried out to develop simpler methods.

Finally, certain types of naturally occurring algae produce toxins, which can accumulate in the flesh of oysters. People eating shellfish containing these toxins can become ill and in exceptional cases death can result. Cooking does not denature the toxins responsible nor does cleansing the shellfish in depuration tanks eliminate them. The risks to consumers are usually minimized by a requirement for samples to be tested regularly. If the amount of toxin exceeds a certain threshold, the marketing of shellfish for consumption is prohibited until the amount falls to a safe level.

Figure 12 A stacked depuration system, suitable for cleansing oysters of microbiological contaminants. The oysters are held in trays in a cascade of UV-sterlized water.

See also

Crustacean Fisheries. Molluskan Fisheries.

Further Reading

Andersen RA (ed.) (2005) *Algal Culturing Techniques*. New York: Academic Press.

Bueno P, Lovatelli A, and Shetty HPC (1991) Pearl oyster farming and pearl culture, Regional Seafarming Development and Demonstration Project. *FAO Project Report No. 8*. Rome: FAO.

Dore I (1991) *Shellfish – A Guide to Oysters, Mussels, Scallops, Clams and Similar Products for the Commercial User*. London: Springer.

Gosling E (2003) *Bivalve Mollusks; Biology, Ecology and Culture*. London: Blackwell.

Helm MM, Bourne N, and Lovatelli A (2004) Hatchery culture of bivalves; a practical manual. *FAO Fisheries Technical Paper 471*. Rome: FAO.

Huguenin JE and Colt J (eds.) (2002) *Design and Operating Guide for Aquaculture Seawater Systems*, 2nd edn. Amsterdam: Elsevier.

Lavens P and Sorgeloos P (1996) Manual on the production and use of live food for aquaculture. *FAO Fisheries Technical Paper 361*. Rome: FAO.

Mann R (1979) Exotic species in mariculture. Proceedings of Symposium on Exotic Species in Mariculture: Case Histories of the Japanese Oyster, *Crassostrea gigas* (Thunberg), with implications for other fisheries. Woods Hole Oceanographic Institution, Cambridge Press.

Matthiessen G (2001) *Fishing News Books Series: Oyster Culture*. New York: Blackwell.

Spencer BE (2002) *Molluscan Shellfish Farming*. New York: Blackwell.

Relevant Websites

http://www.crlcefas.org

– Cefas is the EU community reference laboratory for bacteriological and viral contamination of oysters.

http://www.oie.int

– Definitive information on oyster diseases can be found on the website of The World Organisation for Animal Health (OIE).

http://www.pac.dfo-mpo.gc.ca

– Further information on diseases can be found on the Fisheries and Oceans Canada website.

http://www.cefas.co.uk

– The online magazine 'Shellfish News', although produced primarily for the UK industry, has articles of general interest on oyster farming. It can be found on the News/Newsletters area of the Cefas web site. A booklet on site selection for bivalve cultivation is also available at this site.

http://www.vims.edu

– The Virginia Institute of Marine Science has information on oyster genetics and breeding programs and the proposals to introduce the non-native *Crassostrea ariakensis* to Chesapeake Bay.

http://www.was.org

– The World Aquaculture Society website has a comprehensive set of links to other relevant web sites, including many national shellfish associations.

http://www.fao.org

– There is a great deal of information on oyster farming throughout the world on the website of The Food and Agriculture Organization of the United Nations (FAO). This website has statistics on oyster production, datasheets for various cultivated species, including Pacific oysters, and online manuals on the production and use of live food for aquaculture (*FAO Fisheries Technical Paper 361*), Hatchery culture of bivalves (*FAO Fisheries Technical Paper. No. 471*) and Pearl oyster farming and pearl culture (*FAO Project Report No. 8*).

SALMONID FARMING

L. M. Laird, Aberdeen University, Aberdeen, UK

Introduction

All salmonids spawn in fresh water. Some of them complete their lives in streams, rivers, or lakes but the majority of species are anadromous, migrating to sea as juveniles and returning to spawn as large adults after one or more years feeding. The farmed process follows the life cycle of the wild fish; juveniles are produced in freshwater hatcheries and smolt units and transferred to sea for ongrowing in floating sea cages. An alternative form of salmonid mariculture, ocean ranching, takes advantage of their accuracy of homing. Juveniles are released into rivers or estuaries, complete their growth in sea water and return to the release point where they are harvested.

The salmonids cultured in seawater cages belong to the genera *Salmo*, *Oncorhynchus*, and *Salvelinus*. The last of these, the charrs are currently farmed on a very small scale in Scandinavia; this article concentrates on the former two genera. The Atlantic salmon, *Salmo salar* is the subject of almost all production of fish of the genus *Salmo* (1997 worldwide production 640 000 tonnes) although a small but increasing quantity of sea trout (*Salmo trutta*) is produced (1997 production 7000 tonnes). Three species of *Oncorhynchus*, the Pacific salmon are farmed in significant quantities in cages, the chinook salmon (also known as the king, spring or quinnat salmon), *O. tshawytscha* (1997, 10 000 tonnes), the coho (silver) salmon, *O. kisutch* (1997, 90 000 tonnes) and the rainbow trout, *O. mykiss*. The rainbow trout (steelhead) was formerly given the scientific name *Salmo gairdneri* but following studies on its genetics and native distribution was reclassified as a Pacific salmon species. Much of the world rainbow trout production (1997, 430 000 tonnes) takes place entirely in fresh water although in some countries such as Chile part-grown fish are transferred to sea water in the same way as the salmon species.

Here, the history of salmonid culture leading to the commercial mariculture operations of today is reviewed. This is followed by an overview of the requirements for successful operation of marine salmon farms, constraints limiting developments and prospects for the future.

History

Salmonids were first spawned under captive conditions as long ago as the fourteenth century when Dom Pinchon, a French monk from the Abbey of Reome stripped ova from females, fertilized them with milt from males, and placed the fertilized eggs in wooden boxes buried in gravel in a stream. At that time, all other forms of fish culture were based on the fattening of juveniles captured from the wild. However, the large (4–7 mm diameter) salmonid eggs were much easier to handle than the tiny, fragile eggs of most freshwater or marine fish. By the nineteenth century the captive breeding of salmonids was well established; the main aim was to provide fish to enhance river stocks or to transport around the world to provide sport in countries where there were no native salmonids. In this way, brown trout populations have become established in every continent except Antarctica, sustaining game fishing in places such as New Zealand and Patagonia.

A logical development of the production of eggs and juveniles for release was to retain the young fish in captivity, growing them until a suitable size for harvest. The large eggs hatch to produce large juveniles that readily accept appropriate food offered by the farmer. In the early days, the fish were first fed on finely chopped liver and progressed to a diet based on marine fish waste. The freshwater rainbow trout farming industry flourished in Denmark at the start of the twentieth century only to be curtailed by the onset of World War I when German markets disappeared. The success of the Danish trout industry encouraged a similar venture in Norway. However, when winter temperatures in fresh water proved too low, fish were transferred to pens in coastal sea water. Although these pens broke up in bad weather, the practice of seawater salmonid culture had been successfully demonstrated. The next major steps in salmonid mariculture came in the 1950s and 1960s when the Norwegians developed the commercial rearing of rainbow trout and then Atlantic salmon in seawater enclosures and cages. Together with the development of dry, manufactured fishmeal-based diets this led to the industry in its present form.

Salmonid Culture Worldwide

Fish reared in seawater pens are subject to natural conditions of water quality and temperature. Optimum water temperatures for growth of most

Table 1 Production of four species of salmonids reared in seawater cages

	1988 production (t)	*1997 production (t)*
Atlantic salmon	112 377	638 951
Rainbow trout[a]	248 010	428 963
Coho salmon	25 780	88 431
Chinook salmon	4 698	9774

[a] Includes freshwater production

salmonid species are in the range 8–16°C. Such temperatures, together with unpolluted waters are found not only around North Atlantic and North Pacific coasts within their native range but also in the southern hemisphere along the coastlines of Chile, Tasmania, and New Zealand. Salmon and trout are thus farmed in seawater cages where conditions are suitable both within and outwith their native ranges.

Seawater cages are used for almost the entire sea water production of salmonids. A very small number of farms rear fish in large shore-based silo-type structures into which sea water is pumped. Such structures have the advantage of better protection against storms and predators and the possibility of control of environmental conditions and parasites such as sea lice. However, the high costs of pumping outweigh these advantages and such systems are generally now used only for broodfish that are high in value and benefit from controlled conditions.

The production figures for the four species of salmonid reared in seawater cages are shown in **Table 1**

Norway, the pioneering country of seawater salmonid mariculture, remains the biggest producer of Atlantic salmon. The output figures for 1997 show the major producing countries to be Norway (331 367 t), Scotland, UK (99 422 t), Chile (96 675 t), Canada (51 103 t), USA (18 005 t). Almost the entire farmed production of chinook salmon comes from Canada and New Zealand with Chile producing over 70 000 tonnes of coho salmon (1997 figures). Most of the Atlantic salmon produced in Europe is sold domestically or exported to other European countries such as France and Spain. Production in Chile is exported to North America and to Japan.

Seawater Salmonid Rearing

Smolts

Anadromous salmonids undergo physiological, anatomical and behavioral changes that preadapt them for the transition from fresh water to sea water. At this stage one of the most visible changes in the young fish is a change in appearance from mottled brownish to silver and herring-like. The culture of farmed salmonids in sea water was made possible by the availability of healthy smolts, produced as a result of the technological progress in freshwater units. Hatcheries and tank farms were originally operated to produce juveniles for release into the wild for enhancement of wild stocks, often where there had been losses of spawning grounds or blockage of migration routes by the construction of dams and reservoirs. One of the most significant aspects of the development of freshwater salmon rearing was the progress in the understanding of dietary requirements and the production of manufactured pelleted feed. The replacement of a diet based on wet trash fish with a dry diet also benefitted the freshwater rainbow trout farming industry, by improving growth and survival and reducing disease and the pollution of the watercourses receiving the outflow water from earth ponds.

It was found possible to transfer smolts directly to cages moored in full strength sea water. If the smolts are healthy and the transfer stress-free, survival after transfer is high and feeding begins within 1–3 days (Atlantic salmon).

The smolting process in salmonids is controlled by day length; natural seasonal changes regulate the physiological processes, resulting in the completion of smolting and seaward migration of wild fish in spring. For the first two decades of seawater Atlantic salmon farming, producers were constrained by the annual seasonal availability of smolts. These fish are referred to as 'S1s', being approximately one year post-hatching. This had consequences for the use of equipment (nonoptimal use of cages) and for the timing of harvest. Most salmon reached their optimum harvest size or began to show signs of sexual maturation at the same time of year; thus large quantities of fish arrived on the market together for biological rather than economic reasons. These fish competed with wild salmonids and missed optimum market periods such as Christmas and Easter.

Research on conditions controlling the smolting process enabled smolt producers to alter the timing of smolting by manipulating photoperiod. Compressing the natural year by shortening day length and giving the parr an early 'winter' results in S1/2s or S3/4s, smolting as early as six months after hatch. Similarly, by delaying winter, smolting can be postponed. Thus it is now possible to have Atlantic salmon smolts ready for transfer to sea water throughout the year. Although this benefits marketing it makes site fallowing (see below) more difficult than when smolt input is annual.

The choice of smolts for seawater rearing is becoming increasingly important with the establishment of controlled breeding programs. Few species of fish can be said to be truly domesticated. The only examples approaching domestication are carp species and, to a lesser degree, rainbow trout. Other salmon species have been captive bred for no more (and usually far less than) ten generations; the time between successive generations of Atlantic salmon is usually a minimum of three years which prevents rapid progress in selection for preferred characters although this is countered by the fact that many thousand eggs are produced by each female. Trials carried out mainly in Norway have demonstrated that several commercially important traits can be improved by selective breeding. These include growth rate, age at sexual maturity, food conversion efficiency, fecundity, egg size, disease resistance and survival, adaptation to conditions in captivity and harvest quality, including texture, fat, and color. All of these factors can also be strongly influenced by environmental factors and husbandry.

Sexual maturation before salmonids have reached the desired size for harvest has been a problem for salmonid farmers. Pacific salmon species (except rainbow trout) die after spawning; Atlantic salmon and rainbow trout show increased susceptibility to disease, reduced growth rate, deterioration in flesh quality and changes in appearance including coloration. Male salmonids generally mature at a smaller size and younger age than the females. One solution to this problem, routinely used in rainbow trout culture, is to rear all-female stocks, produced as a result of treating eggs and fry of potential broodstock with methyl testosterone to give functional males which are in fact genetically female. When crossed with normal females, all-female offspring are produced as the Y, male, sex chromosome has been eliminated. Sexual maturation can be eliminated totally by subjecting all-female eggs to pressure or heat shock to produce triploid fish. This is common practice for rainbow trout but used little for other salmonids, partly because improvements in stock selection and husbandry are overcoming the problem but also because of adverse press comment on supposedly genetically modified fish. This same reaction has limited the commercial exploitation of fast-growing genetically modified salmon, produced by the incorporation into eggs of a gene from ocean pout.

Site Selection

The criteria for the ideal site for salmonid cage mariculture have changed with the development of stronger cage systems, use of automatic feeders with a few days storage capacity and generally bigger and stronger boats, cranes and other equipment on the farm. The small, wooden-framed cages (typically 6 m × 6 m frame, 300 m^3 capacity) with polystyrene flotation required sheltered sites with protection from wind and waves greater than 1–2 m high. Recommended water depth was around three times the depth of the cage to ensure dispersal of wastes. This led to the siting of cages in inshore sites such as inner sea lochs and fiords. These sheltered sites had several disadvantages, notably variable water quality caused by runoff of fresh water, silt, and wastes from the land and susceptibility to the accumulation of feces and waste feed on the seabed because of poor water exchange. In addition, cage groups were often sited near public roads in places valued for their scenic beauty, attracting adverse public reaction to salmon farming.

Cages in use today are far larger (several thousand m^3 volume) and stronger. Frames are made from either galvanized steel or flexible plastic or rubber and can be designed to withstand waves of 5 m or more. Flotation collars are stronger and mooring systems designed to match cages to sites. Such sites are likely to provide more constant water quality than inshore sites; an ideal salmonid rearing site has temperatures of 6–16°C and salinities of 32–35‰ (parts per thousand). Rearing is thus moving into deeper water away from sheltered lochs and bays. However, there are still advantages to proximity to the coast; these include ease of access from shore bases, proximity to staff accommodation, reduction in costs of transport of feed and stock and ability to keep sites under regular surveillance. Other factors to be taken into account in siting cage groups are the avoidance of navigation routes and the presence of other fish or shellfish farms. Maintaining a minimum separation distance from other fish farms is preferred to minimize the risk of disease transfer; if this is not possible, farms should enter into agreements to manage stock in the same way to reduce risk. Models have been developed in Norway and Scotland to determine the carrying capacity of cage farm sites.

Current speed is an important factor in site selection. Water exchange through the cage net ensures the supply of oxygen to the stock and removal of dissolved wastes such as ammonia as well as feces and waste feed. Salmon have been shown to grow and feed most efficiently in currents with speeds equivalent to 1–2 body lengths per second. In an ideal site this current regime should be maintained for as much of the tidal cycle as possible. At faster current speeds the salmon will use more energy in swimming and cage nets will tend to twist, sometimes forming pockets and trapping fish, causing scale removal.

Some ideal sites may be situated near offshore islands; access from the mainland may require crossing open water with strong tides and currents making access difficult on stormy days. However, modern workboats and feeding barges with the capacity to store several days supply of feed make the operation of such sites possible.

The presence of predators in the vicinity is often taken as a criterion for site selection. Unprotected salmon cage farms are likely to be subject to predation from seals or, in Chile, sea lions, and birds, such as herons and cormorants. Such predators not only remove fish but also damage others, tear holes in nets leading to escapes and stress stock making it more susceptible to disease. Protection systems to guard against predators include large mesh nets surrounding cages or cage groups, overhead nets and acoustic scaring devices. When used correctly these can all be effective in preventing attacks. Attacks from predators are frequently reported to involve nonlocal animals, attracted to a food source. Because of this and the possibility of excluding and deterring predators, it seems that proximity to colonies is not necessarily one of the most important factors in determining site selection.

A further factor, which must be taken into account in the siting of cage salmonid farms, is the occurrence of phytoplankton blooms. Phytoplankton can enter surface-moored cages and can physically damage gills, cause oxygen depletion or produce lethal toxins that kill fish. Historic records may indicate prevalence of such blooms and therefore sites to be avoided although some cages are now designed to be lowered beneath the surface and operated as semi-submersibles, keeping the fish below the level of the bloom until it passes.

Farm Operation

Operation of the marine salmon farm begins with transfer of stock from freshwater farms. Where possible, transfer in disinfected bins suspended under helicopters is the method of choice as it is quick and relatively stress-free. For longer journeys, tanks on lorries or wellboats are used. The latter require particular vigilance as they may visit more than one farm and have the potential to transfer disease. Conditions in tanks and wellboats should be closely monitored to ensure that the supply of oxygen is adequate (minimum $6\,\mathrm{mg\,l^{-1}}$).

The numbers of smolts stocked into each cage is a matter for the farmer; some will introduce a relatively small number, allowing for growth to achieve a final stocking density of $10–15\,\mathrm{kg^{-3}}$ whereas others stock a greater number and split populations between cages during growth. This latter method makes better use of cage space but increases handling and therefore stress. Differential growth may make grading into two or three size groups necessary.

Stocking density is the subject of debate. It is essential that oxygen concentrations are maintained and that all fish have access to feed when it is being distributed. Fish may not distribute themselves evenly within the water column; because of crowding together the effective stocking density may therefore be a great deal higher than the theoretical one.

As with all farmed animals the importance of vigilance of behavior and health and the maintenance of accurate, useful records cannot be overemphasized. When most salmon farms were small, producing one or two hundred tonnes of salmon a year rather than thousands, hand feeding was normal; observation of stock during feeding provided a good indication of health. Today, fish are often fed automatically using blowers attached to feed storage systems. The best of these systems incorporate detectors to monitor consumption of feed and underwater cameras to observe the stock.

All of the nutrients ingested by cage-reared salmonids are supplied in the feed distributed. Typically, manufactured diets for salmonids will contain 40% protein (mainly obtained from fishmeal) and up to 30% oil, providing the source of energy, sparing protein for growth. Although very poorly digested by salmonids, carbohydrate is necessary to bind other components of the diet. Vitamins and minerals are also added, as are carotenoid pigments such as astaxanthin, necessary to produce the characteristic pink coloration of the flesh of anadromous salmonids. The feed used on marine salmon farms is nowadays almost exclusively a pelleted or extruded fishmeal-based diet manufactured by specialist companies. Feed costs make up the biggest component of farm operating costs, sometimes reaching 50%. It is therefore important to make optimum use of this valuable input by minimizing wastes. This is accomplished by ensuring that feed is delivered to the farm in good condition and handled with care to prevent dust formation, increasing the size of pellets as the fish grow and distributing feed to satisfy the appetites of the fish. Improvements in feed manufacture and in feeding practices have reduced feed conversion efficiency (feed input : increase in weight of fish) from 2 : 1 to close to 1 : 1. Such figures may seem improbable but it must be remembered that they represent the conversion of a nearly dry feed to wet fish flesh and other tissues.

The importance of maintaining a flow of water through the net mesh of the cages has been emphasized. Mesh size is generally selected to be the maximum capable of retaining all fish and preventing escapes. Any structure immersed in the upper few meters of coastal or marine waters will quickly be subjected to colonization by fouling organisms including bacteria, seaweeds, mollusks and sea squirts. Left unchecked, such fouling occludes the mesh, reducing water exchange and may place a burden on the cage reducing its resistance to storm damage. One of the most effective methods of preventing fouling of nets and moorings is to treat them with antifouling paints and chemicals prior to installation. However, one particularly effective treatment used in the early 1980s, tributyl tin, has been shown to have harmful effects on marine invertebrates and to accumulate in the flesh of the farmed fish; its use in aquaculture is now banned. Other antifoulants are copper or oil based; alternative, preferred methods of removing fouling organisms include lifting up sections of netting to dry in air on a regular basis or washing with high pressure hoses or suction devices to remove light fouling.

The aim of the salmonid farmer is to produce maximum output of salable product for minimum financial input. To do this, fish must grow efficiently and a high survival rate from smolt input to harvest must be achieved. Minimizing stress to the fish by reducing handling, maintaining stable environmental conditions and optimizing feeding practices will reduce mortalities. Causes of mortality in salmonid and other farms are reviewed elsewhere (*see* Mariculture Diseases and Health). It is vital to keep accurate records of mortalities; any increase may indicate the onset of an outbreak of disease. It is also important that dead fish are removed; collection devices installed in the base of cages are often used to facilitate this. Treatment of diseases or parasitic infestations such as sea lice (*Lepeophtheirus salmonis*, *Caligus elongatus*) is difficult in fish reared in sea cages because of their large volumes and the high numbers of fish involved. Some treatments for sea lice involve reducing the cage volume and surrounding with a tarpaulin so that the fish can be bathed in chemical. After the specified time the tarpaulin is removed and the chemical disperses into the water surrounding the cage. Newer treatments incorporate the chemicals in feed and are therefore simpler to apply. In the future, vaccines are increasingly likely to replace chemicals.

The health of cage-reared salmonids can be maintained by a site management system incorporating a period of fallowing when groups of cages are left empty for a period of at least three months and preferably longer. This breaks the life cycle of parasites such as sea lice and allows the seabed to recover from the nutrient load falling from the cages. Ideally a farmer will have access to at least three sites; at any given time one will be empty and the other two will contain different year classes, separated to prevent cross-infection.

Harvesting

Most of the farmed salmonids reared in sea water reach the preferred harvest size (3–5 kg) 10 months or more after transfer to sea water. Poor harvesting and handling methods can have a devastating effect on flesh quality, causing gaping in muscle blocks and blood spotting. After a period of starvation to ensure that guts are emptied of feed residues the fish are generally killed by one of two methods. One of these involves immersion in a tank of sea water saturated with carbon dioxide, the other an accurate sharp blow to the cranium. Both methods are followed by excision of the gill arches; the loss of blood is thought to improve flesh quality. It is important that water contaminated with blood is treated to kill any pathogens which might infect live fish.

Ocean Ranching

The anadromous behavior of salmonids and their ability to home to the point of release has been exploited in ocean ranching programs which have been operated successfully with Pacific salmon. Some of these programs are aimed at enhancing wild stocks and others are operated commercially. The low cost of rearing Pacific salmon juveniles, which are released into estuaries within weeks of hatching, makes possible the release of large numbers. In Japan over two billion juveniles are released annually; overall return rates have increased to 2%, 90% of which are chum (*Oncorhynchus keta*) and 8% pink (*Oncorhynchus gorbuscha*) salmon. The success of the operation depends on cooperation between those operating and financing the hatcheries and those harvesting the adult fish. The relatively high cost of producing Atlantic salmon smolts and the lack of control over harvest has restricted ranching operations.

See also

Mariculture Diseases and Health. Ocean Ranching. Salmon Fisheries, Atlantic. Salmon Fisheries, Pacific.

Further Reading

Anon (ed.) (1999) *Aquaculture Production Statistics 1988–1997*. Rome: Food and Agriculture Organization.

Black KD and Pickering AD (eds.) (1998) *Biology of Farmed Fish*. Sheffield Academic Press.

Heen K, Monahan RL, and Utter F (eds.) (1993) *Salmon Aquaculture*. Oxford: Fishing News Books.

Pennell W and Barton BA (eds.) (1996) *Principles of Salmonid Culture*. Amsterdam: Elsevier.

Stead S and Laird LM (In press) *Handbook of Salmon Farming Praxis*. Chichester: Springer-Praxis.

Willoughby S (1999) *Manual of Salmonid Farming*. Oxford: Blackwell Science.

SEAWEEDS AND THEIR MARICULTURE

T. Chopin and M. Sawhney, University of New Brunswick, Saint John, NB, Canada

Introduction: What are Seaweeds and their Significance in Coastal Systems

Before explaining how they are cultivated, it is essential to try to define this group of organisms commonly referred to as 'seaweeds'. Unfortunately, it is impossible to give a short definition because this heterogeneous group is only a fraction of an even less natural assemblage, the 'algae'. In fact, algae are not a closely related phylogenetic group but a diverse group of photosynthetic organisms (with a few exceptions) that is difficult to define, by either a lay person or a professional botanist, because they share only a few characteristics: their photosynthetic system is based on chlorophyll *a*, they do not form embryos, they do not have roots, stems, leaves, nor vascular tissues, and their reproductive structures consist of cells that are all potentially fertile and lack sterile cells covering or protecting them. Throughout history, algae have been lumped together in an unnatural group, which now, especially with the progress in molecular techniques, is emerging as having no real cohesion with representatives in four of the five kingdoms of organisms. During their evolution, algae have become a very diverse group of photosynthetic organisms, whose varied origins are reflected in the profound diversity of their size, cellular structure, levels of organization and morphology, type of life history, pigments for photosynthesis, reserve and structural polysaccharides, ecology, and habitats they colonize. Blue-green algae (also known as Cyanobacteria) are prokaryotes closely related to bacteria, and are also considered to be the ancestors of the chloroplasts of some eukaryotic algae and plants (endosymbiotic theory of evolution). The heterokont algae are related to oomycete fungi. At the other end of the spectrum (one cannot presently refer to a typical family tree), green algae (Chlorophyta) are closely related to vascular plants (Tracheophyta). Needless to say, the systematics of algae is still the source of constant changes and controversies, especially recently with new information provided by molecular techniques. Moreover, the fact that the roughly 36 000 known species of algae represent only about 17% of the existing algal species is a measure of our still rudimentary knowledge of this group of organisms despite their key role on this planet: indeed, approximately 50% of the global photosynthesis is algal derived. Thus, every second molecule of oxygen we inhale was produced by an alga, and every second molecule of carbon dioxide we exhale will be re-used by an alga.

Despite this fundamental role played by algae, these organisms are routinely paid less attention than the other inhabitants of the oceans. There are, however, multiple reasons why algae should be fully considered in the understanding of oceanic ecosystems: (1) The fossil record, while limited except in a few phyla with calcified or silicified cell walls, indicates that the most ancient organisms containing chlorophyll *a* were probably blue-green algae 3.5 billion years ago, followed later (900 Ma) by several groups of eukaryotic algae, and hence the primacy of algae in the former plant kingdom. (2) The organization of algae is relatively simple, thus helping to understand the more complex groups of plants. (3) The incredible diversity of types of sexual reproduction, life histories, and photosynthetic pigment apparatuses developed by algae, which seem to have experimented with 'everything' during their evolution. (4) The ever-increasing use of algae as 'systems' or 'models' in biological or biotechnological research. (5) The unique position occupied by algae among the primary producers, as they are an important link in the food web and are essential to the economy of marine and freshwater environments as food organisms. (6) The driving role of algae in the Earth's planetary system, as they initiated an irreversible global change leading to the current oxygen-rich atmosphere; by transfer of atmospheric carbon dioxide into organic biomass and sedimentary deposits, algae contribute to slowing down the accumulation of greenhouse gases leading to global warming; through their role in the production of atmospheric dimethyl sulfide (DMS), algae are believed to be connected with acidic precipitation and cloud formation which leads to global cooling; and their production of halocarbons could be related to global ozone depletion. (7) The incidence of algal blooms, some of which are toxic, seems to be on the increase in both freshwater and marine habitats. (8) The ever-increasing use of algae in pollution control, waste treatment, and biomitigation by developing balanced management practices such as integrated multi-trophic aquaculture (IMTA).

This chapter restricts itself to seaweeds (approximately 10 500 species), which can be defined as marine benthic macroscopic algae. Most of them are members of the phyla Chlorophyta (the green seaweeds of the class Ulvophyceae (893 species)), Ochrophyta (the brown seaweeds of the class Phaeophyceae (1749 species)), and Rhodophyta (the red seaweeds of the classes Bangiophyceae (129 species) and Florideophyceae (5732 species)). To a lot of people, seaweeds are rather unpleasant organisms: these plants are very slimy and slippery and can make swimming or walking along the shore an unpleasant experience to remember! To put it humorously, seaweeds do not have the popular appeal of 'emotional species': only a few have common names, they do not produce flowers, they do not sing like birds, and they are not as cute as furry mammals! One of the key reasons for regularly ignoring seaweeds, even in coastal projects is, in fact, this very problem of identification, as very few people, even among botanists, can identify them correctly. Reasons for this include: a very high morphological plasticity; taxonomic criteria that are not always observable with the naked eye but instead are based on reproductive structures, cross sections, and, increasingly, ultrastructural and molecular arguments; an existing classification of seaweeds that is in a permanent state of revisions; and algal communities with very large numbers of species from different algal taxa that are not always well defined. The production of benthic seaweeds has probably been underestimated, since it may approach 10% of that of all the plankton while only occupying 0.1% of the area used by plankton; this area is, however, crucial, as it is the coastal zone.

The academic, biological, environmental, and economic significance of seaweeds is not always widely appreciated. The following series of arguments emphasizes the importance of seaweeds, and why they should be an unavoidable component of any study that wants to understand coastal biodiversity and processes: (1) Current investigations about the origin of the eukaryotic cell must include features of present-day algae/seaweeds to understand the diversity and the phylogeny of the plant world, and even the animal world. (2) Seaweeds are important primary producers of oxygen and organic matter in coastal environments through their photosynthetic activities. (3) Seaweeds dominate the rocky intertidal zone in most oceans; in temperate and polar regions they also dominate rocky surfaces in the shallow subtidal zone; the deepest seaweeds can be found to depths of 250 m or more, particularly in clear waters. (4) Seaweeds are food for herbivores and, indirectly, carnivores, and hence are part of the foundation of the food web. (5) Seaweeds participate naturally in nutrient recycling and waste treatment (these properties are also used 'artificially' by humans, for example, in IMTA systems). (6) Seaweeds react to changes in water quality and can therefore be used as biomonitors of eutrophication. Seaweeds do not react as rapidly to environmental changes as phytoplankton but can be good indicators over a longer time span (days vs. weeks/months/years) because of the perennial and benthic nature of a lot of them. If seaweeds are 'finally' attracting some media coverage, it is, unfortunately, because of the increasing report of outbreaks of 'green tides' (as well as 'brown and red tides') and fouling species, which are considered a nuisance by tourists and responsible for financial losses by resort operators. (7) Seaweeds can be excellent indicators of natural and/or artificial changes in biodiversity (both in terms of abundance and composition) due to changes in abiotic, biotic, and anthropogenic factors, and hence are excellent monitors of environmental changes. (8) Around 500 species of marine algae (mostly seaweeds) have been used for centuries for human food and medicinal purposes, directly (mostly in Asia) or indirectly, mainly by the phycocolloid industry (agars, carrageenans, and alginates). Seaweeds are the basis of a multibillion-dollar enterprise that is very diversified, including food, brewing, textile, pharmaceutical, nutraceutical, cosmetic, botanical, agrichemical, animal feed, bioactive and antiviral compounds, and biotechnological sectors. Nevertheless, this industry is not very well known to Western consumers, despite the fact that they use seaweed products almost daily (from their orange juice in the morning to their toothpaste in the evening!). This is due partly to the complexity at the biological and chemical level of the raw material, the technical level of the extraction processes, and the commercial level with markets and distribution systems that are difficult to understand and penetrate. (9) The vast majority of seaweed species has yet to be screened for various applications, and their extensive diversity ensures that many new algal products and processes, beneficial to mankind, will be discovered.

Seaweed Mariculture

Seaweed mariculture is believed to have started during the Tokugawa (or Edo) Era (AD 1600–1868) in Japan. Mariculture of any species develops when society's demands exceed what natural resources can supply. As demand increases, natural populations frequently become overexploited and the need for the cultivation of the appropriate species emerges. At present, 92% of the world seaweed supply comes from cultivated species. Depending on the selected species, their biology, life history, level of tissue specialization, and the socioeconomic situation of the region where it is developed, cultivation technology (**Figures 1–6**) can be low-tech (and still extremely successful with highly efficient and simple culture techniques, coupled with intensive labour at low costs) or can become highly advanced and mechanized, requiring on-land cultivation systems for seeding some phases of the life history before growth-out at open-sea aquaculture sites. Cultivation and seed-stock improvement techniques have been refined over centuries, mostly in Asia, and can now be highly sophisticated. High-tech on-land cultivation systems (**Figure** 7) have been developed in a few rare cases, mostly in the Western World; commercial viability has only been reached when high value-added products have been obtained, their markets secured (not necessarily in response to a local demand, but often for export to Asia), and labor costs reduced to balance the significant technological investments and operational costs.

Because the mariculture of aquatic plants (11.3 million tonnes of seaweeds and 2.6 million tonnes of unspecified 'aquatic plants' reported by the Food and Agriculture Organization of the United Nations) has developed essentially in Asia, it remains mostly unknown in the Western World, and is often neglected or ignored in world statistics ... a situation we can only explain as being due to a deeply rooted zoological bias in marine academics, resource managers, bureaucrats, and policy advisors! However, the seaweed aquaculture sector represents 45.9% of the biomass and 24.2% of the value of the world mariculture production, estimated in 2004 at 30.2 million tonnes, and worth US$28.1 billion (**Table 1**). Mollusk aquaculture comes second at 43.0%, and the finfish aquaculture, the subject of many debates, actually only represents 8.9% of the world mariculture production.

The seaweed aquaculture production, which almost doubled between 1996 and 2004, is estimated at 11.3 million tonnes, with 99.7% of the biomass being

Figure 1 Long-line aquaculture of the brown seaweed, *Laminaria japonica* (kombu), in China. Reproduced by permission of Max Troell.

Figure 2 Long-line aquaculture of the brown seaweed, *Undaria pinnatifida* (wakame), in Japan. Photo by Thierry Chopin.

Figure 3 Net aquaculture of the red seaweed, *Porphyra yezoensis* (nori), in Japan. Photo by Thierry Chopin.

cultivated in Asia (**Table 2**). Brown seaweeds represent 63.8% of the production, while red seaweeds represent 36.0%, and the green seaweeds 0.2%. The seaweed aquaculture production is valued at US$5.7 billion (again with 99.7% of the value being provided by Asian countries; **Table 3**). Brown seaweeds dominate with 66.8% of the value, while red seaweeds contribute 33.0%, and the green seaweeds 0.2%. Approximately 220 species of seaweeds are cultivated worldwide; however, six genera (*Laminaria* (kombu; 40.1%), *Undaria* (wakame; 22.3%), *Porphyra* (nori; 12.4%), *Eucheuma*/*Kappaphycus* (11.6%), and *Gracilaria* (8.4%)) provide 94.8% of the seaweed aquaculture production (**Table 4**), and four genera (*Laminaria* (47.9%), *Porphyra* (23.3%), *Undaria* (17.7%), and *Gracilaria* (6.7%)) provide 95.6% of its value (**Table 5**). Published world statistics, which regularly mention 'data exclude aquatic plants' in their tables, indicate that in 2004 the top ten individual species produced by the global aquaculture (50.9% mariculture, 43.4% freshwater aquaculture, and 5.7% brackishwater aquaculture) were Pacific cupped oyster (*Crassostrea gigas* – 4.4 million tonnes), followed by three species of carp – the silver

Figure 4 Off bottom-line aquaculture of the red seaweed, *Eucheuma denticulatum*, in Zanzibar. Photo by Thierry Chopin.

Figure 5 Bottom-stocking aquaculture of the red alga, *Gracilaria chilensis*, in Chile. Photo by Thierry Chopin.

carp (*Hypophthalmichthys molitrix* – 4.0 million tonnes), the grass carp (*Ctenopharyngodon idellus* – 3.9 million tonnes), and the common carp (*Cyprinus carpio* – 3.4 million tonnes). However, in fact, the kelp, *Laminaria japonica*, was the first top species, with a production of 4.5 million tonnes.

Surprisingly, the best-known component of the seaweed-derived industry is that of the phycocolloids, the gelling, thickening, emulsifying, binding, stabilizing, clarifying, and protecting agents known as carrageenans, alginates, and agars. However, this component represents only a minor volume (1.26 million tonnes or 11.2%) and value (US$650 million or 10.8%) of the entire seaweed-derived industry (**Table 6**). The use of seaweeds as sea vegetable for direct human consumption is much more significant in tonnage (8.59 million tonnes or 76.1%) and value (US$5.29 billion or 88.3%). Three genera – *Laminaria* (or kombu), *Porphyra* (or nori), and *Undaria* (or wakame) – dominate the edible seaweed market. The phycosupplement industry is an emerging component. Most of the tonnage is used for the manufacturing of soil additives; however, the agrichemical and animal feed markets are comparatively much more lucrative if one considers the much smaller volume of seaweeds they require. The use of seaweeds in the development of pharmaceuticals and nutraceuticals, and as a source of pigments and

Figure 6 Net aquaculture of the green alga, *Monostroma nitidum*, in Japan. Photo by Thierry Chopin.

Figure 7 Land-based tank aquaculture of the red alga, *Chondrus crispus* (Irish moss), in Canada for high value-added sea-vegetable (Hana-nori) production. Photo by Acadian Seaplants Limited.

bioactive compounds is in full expansion. Presently, that component is difficult to evaluate accurately; the use of 3000 tonnes of raw material to obtain 600 tonnes of products valued at US$3 million could be an underestimation.

The Role of Seaweeds in the Biomitigation of Other Types of Aquaculture

One may be inclined to think that, on the world scale, the two types of aquaculture, fed and extractive, environmentally balance each other out, as 45.9% of the mariculture production is provided by aquatic plants, 43.0% by mollusks, 8.9% by finfish, 1.8% by crustaceans, and 0.4% by other aquatic animals. However, because of predominantly monoculture practices, economics, and social habits, these different types of aquaculture production are often geographically separate, and, consequently, rarely balance each other out environmentally, on either the local or regional scale (**Figure 8**). For example, salmon aquaculture in Canada represents 68.2% of the tonnage of the aquaculture industry and 87.2% of its farmgate value. In Norway, Scotland, and Chile,

Table 1 World mariculture production and value from 1996 to 2004 according to the main groups of cultivated organisms

	Production (%)			*% of value in*
	1996	*2000*	*2004*	*2004*
Mollusks	48	46.2	43.0	34.0
Aquatic plants	44	44.0	45.9	24.2
Finfish	7	8.7	8.9	34.0
Crustaceans	1	1.0	1.8	6.8
Other aquatic animals		0.1	0.4	1.0

Source: FAO (1998) *The State of World Fisheries and Aquaculture 1998.* Italy: Food and Agriculture Organization of United Nations. http://www.fao.org/docrep/w9900e/w9900e00.htm; FAO (2002) *The State of World Fisheries and Aquaculture 2002.* Italy: Food and Agriculture Organization of the United Nations. http://www.fao.org/docrep/005/y7300e/y7300e00.htm; and FAO (2006) *The State of World Fisheries and Aquaculture 2006.* Italy: Food and Agriculture Organization of the United Nations. http://www.fao.org/docrep/009/a0699e/A0699E00.htm (accessed Mar. 2008).

Table 2 Seaweed aquaculture production (tonnage) from 1996 to 2004 according to the main groups of seaweeds (the brown, red, and green seaweeds) and contribution from Asian countries

	Production (tonnes)		
	1996	*2000*	*2004*
Brown seaweeds			
World	4 909 269	4 906 280	7 194 316
Asia (%)	4 908 805	4 903 252	7 194 075
	(99.9)	(99.9)	(99.9)
Red seaweeds			
World	1 801 494	1 980 747	4 067 028
Asia (%)	1 678 485	1 924 258	4 035 783
	(93.2)	(97.2)	(99.2)
Green seaweeds			
World	13 418	33 584	19 046
Asia (%)	13 418	33 584	19 046
	(100)	(100)	(100)
Total			
World	6 724 181	6 920 611	11 280 390
Asia (%)	6 600 708	6 861 094	11 248 904
	(98.2)	(99.1)	(99.7)

Source: FAO (1998) *The State of World Fisheries and Aquaculture 1998.* Italy: Food and Agriculture Organization of United Nations. http://www.fao.org/docrep/w9900e/w9900e00.htm; FAO (2002) *The State of World Fisheries and Aquaculture 2002.* Italy: Food and Agriculture Organization of the United Nations. http://www.fao.org/docrep/005/y7300e/y7300e00.htm; and FAO (2006) *The State of World Fisheries and Aquaculture 2006.* Italy: Food and Agriculture Organization of the United Nations. http://www.fao.org/docrep/009/a0699e/A0699E00.htm (accessed Mar. 2008).

salmon aquaculture represents 88.8%, 93.3%, and 81.9% of the tonnage of the aquaculture industry, and 87.3%, 90.9%, and 95.5% of its farmgate value, respectively. Conversely, while Spain (Galicia) produces only 8% of salmon in tonnage (16% in farmgate value), it produces 81% of its tonnage in mussels (28% in farmgate value). Why should one think that the common old saying "Do not put all your eggs in one basket", which applies to agriculture and many other businesses, would not also apply to aquaculture? Having too much production in a single species leaves a business vulnerable to issues of sustainability because of low prices due to oversupply, and the possibility of catastrophic destruction of one's only crop (diseases, damaging weather conditions). Consequently, diversification of the aquaculture industry is imperative to reducing the economic risk and maintaining its sustainability and competitiveness.

Table 3 Seaweed aquaculture production (value) from 1996 to 2004 according to the main groups of seaweeds (the brown, red, and green seaweeds) and contribution from Asian countries

	Value (× US$1000)		
	1996	*2000*	*2004*
Brown seaweeds			
World	3 073 255	2 971 990	3 831 445
Asia (%)	3 072 227	2 965 372	3 831 170
	(99.9)	(99.8)	(99.9)
Red seaweeds			
World	1 420 941	1 303 751	1 891 420
Asia (%)	1 367 625	1 275 090	1 875 759
	(96.3)	(97.8)	(99.2)
Green seaweeds			
World	7263	5216	12 751
Asia (%)	7263	5216	12 751
	(100)	(100)	(100)
Total			
World	4 501 459	4 280 957	5 735 615
Asia (%)	4 447 115	4 245 678	5 719 680
	(98.8)	(99.2)	(99.7)

Source: FAO (1998) *The State of World Fisheries and Aquaculture 1998.* Italy: Food and Agriculture Organization of United Nations. http://www.fao.org/docrep/w9900e/w9900e00.htm; FAO (2002) *The State of World Fisheries and Aquaculture 2002.* Italy: Food and Agriculture Organization of the United Nations. http://www.fao.org/docrep/005/y7300e/y7300e00.htm; and FAO (2006) *The State of World Fisheries and Aquaculture 2006.* Italy: Food and Agriculture Organization of the United Nations. http://www.fao.org/docrep/009/a0699e/A0699E00.htm (accessed Mar. 2008).

Phycomitigation (the treatment of wastes by seaweeds), through the development of IMTA systems, has existed for centuries, especially in Asian countries, through trial and error and experimentation. Other terms have been used to describe similar systems (integrated agriculture-aquaculture systems (IAAS), integrated peri-urban aquaculture systems (IPUAS), integrated fisheries-aquaculture systems (IFAS), fractionated aquaculture, aquaponics); they can, however, be considered to be variations on the IMTA concept. 'Multi-trophic' refers to the incorporation of species from different trophic or nutritional levels in

Table 4 Production, from 1996 to 2004, of the eight genera of seaweeds that provide 96.1% of the biomass for the seaweed aquaculture in 2004

Genus	*Production (tonnes)*			*% of production in 2004*
	1996	*2000*	*2004*	
Laminaria (kombu)[a]	4 451 570	4 580 056	4 519 701	40.1
Undaria (wakame)[a]	434 235	311 125	2 519 905	22.3
Porphyra (nori)[b]	856 588	1 010 778	1 397 660	12.4
Kappaphycus/ Eucheuma[b]	665 485	698 706	1 309 344	11.6
Gracilaria[b]	130 413	65 024	948 292	8.4
Sargassum[a]	0	0	131 680	1.2
Monostroma[c]	8277	5288	11 514	0.1

[a]Brown seaweeds.
[b]Red seaweeds.
[c]Green seaweeds.
Source: FAO (1998) *The State of World Fisheries and Aquaculture 1998.* Italy: Food and Agriculture Organization of United Nations. http://www.fao.org/docrep/w9900e/w9900e00.htm; FAO (2002) *The State of World Fisheries and Aquaculture 2002.* Italy: Food and Agriculture Organization of the United Nations. http://www.fao.org/docrep/005/y7300e/y7300e00.htm; and FAO (2006) *The State of World Fisheries and Aquaculture 2006.* Italy: Food and Agriculture Organization of the United Nations. http://www.fao.org/docrep/009/a0699e/A0699E00.htm (accessed Mar. 2008).

Table 5 Value, from 1996 to 2004, of the eight genera of seaweeds that provide 99.0% of the value of the seaweed aquaculture in 2004

Genus	*Value (× US$1000)*			*% of value in 2004*
	1996	*2000*	*2004*	
Laminaria (kombu)[a]	2 875 497	2 811 440	2 749 837	47.9
Porphyra (nori)[b]	1 276 823	1 183 148	1 338 995	23.3
Undaria (wakame)[a]	178 290	148 860	1 015 040	17.7
Gracilaria[b]	60 983	45 801	385 794	6.7
Kappaphycus/ Eucheuma[b]	67 883	51 725	133 324	2.3
Sargassum[a]	0	0	52 672	0.9
Monostroma[c]	6622	1849	9937	0.2

[a]Brown seaweeds.
[b]Red seaweeds.
[c]Green seaweeds.
Source: FAO (1998) *The State of World Fisheries and Aquaculture 1998.* Italy: Food and Agriculture Organization of United Nations. http://www.fao.org/docrep/w9900e/w9900e00.htm; FAO (2002) *The State of World Fisheries and Aquaculture 2002.* Italy: Food and Agriculture Organization of the United Nations. http://www.fao.org/docrep/005/y7300e/y7300e00.htm; and FAO (2006) *The State of World Fisheries and Aquaculture 2006.* Italy: Food and Agriculture Organization of the United Nations. http://www.fao.org/docrep/009/a0699e/A0699E00.htm (accessed Mar. 2008).

the same system. This is one potential distinction from the age-old practice of aquatic polyculture, which could simply be the co-culture of different fish species from the same trophic level. In this case, these organisms may all share the same biological and chemical processes, with few synergistic benefits, which could potentially lead to significant shifts in the ecosystem. Some traditional polyculture systems may, in fact, incorporate a greater diversity of species, occupying several niches, as extensive cultures (low intensity, low management) within the same pond. The 'integrated' in IMTA refers to the more intensive cultivation of the different species in proximity to each other (but not necessarily right at the same location), connected by nutrient and energy transfer through water.

The IMTA concept is very flexible. IMTA can be land-based or open-water systems, marine or freshwater systems, and may comprise several species combinations. Ideally, the biological and chemical processes in an IMTA system should balance. This is achieved through the appropriate selection and proportioning of different species providing different ecosystem functions. The co-cultured species should be more than just biofilters; they should also be organisms which, while converting solid and soluble nutrients from the fed organisms and their feed into biomass, become harvestable crops of commercial value and acceptable to consumers. A working IMTA system should result in greater production for the overall system, based on mutual benefits to the cocultured species and improved ecosystem health, even if the individual production of some of the species is lower compared to what could be reached in monoculture practices over a short-term period.

While IMTA likely occurs due to traditional or incidental, adjacent culture of dissimilar species in some coastal areas (mostly in Asia), deliberately designed IMTA sites are, at present, less common. There has been a renewed interest in IMTA in Western countries over the last 30 years, based on the age-old, common sense, recycling and farming practice in which the byproducts from one species become inputs for another: fed aquaculture (fish or shrimp) is combined with inorganic extractive (seaweed) and organic extractive (shellfish) aquaculture to create balanced systems for environmental sustainability (biomitigation), economic stability (product diversification and risk reduction), and social acceptability (better management practices). They are presently simplified systems, like fish/seaweed/shellfish. Efforts to develop such IMTA systems are currently taking place in Canada, Chile, China, Israel, South Africa, the USA, and several European countries. In the future, more advanced

Table 6 Biomass, products, and value of the main components of the world's seaweed-derived industry in 2006

Industry component	*Raw material (wet tonnes)*	*Product (tonnes)*	*Value (US$)*
Sea vegetables	8.59 million	1.42 million	5.29 billion
Kombu (*Laminaria*)[a]	4.52 million	1.08 million	2.75 billion
Nori (*Porphyra*)[b]	1.40 million	141 556	1.34 billion
Wakame (*Undaria*)[a]	2.52 million	166 320	1.02 billion
Phycocolloids	1.26 million	70 630	650 million
Carrageenans[b]	528 000	33 000	300 million
Alginates[a]	600 000	30 000	213 million
Agars[b]	127 167	7630	137 million
Phycosupplements	1.22 million	242 600	53 million
Soil additives	1.10 million	220 000	30 million
Agrichemicals (fertilizers, biostimulants)	20 000	2000	10 million
Animal feeds (supplements, ingredients)	100 000	20 000	10 million
Pharmaceuticals nutraceuticals, botanicals, cosmeceuticals, pigments, bioactive compounds, antiviral agents, brewing, etc.	3000	600	3 million

[a]Brown seaweeds.
[b]Red seaweeds.
Source: McHugh (2003); Chopin and Bastarache (2004); and FAO (2004) *The State of World Fisheries and Aquaculture 2004*. Italy: Food and Agriculture Organization of the United Nations. http://www.fao.org/docrep/007/y5600e/y5600e00.htm (accessed Mar. 2008); FAO (2006) *The State of World Fisheries and Aquaculture 2006*. Italy: Food and Agriculture Organization of the United Nations. http://www.fao.org/docrep/009/a0699e/A0699E00.htm (accessed Mar. 2008).

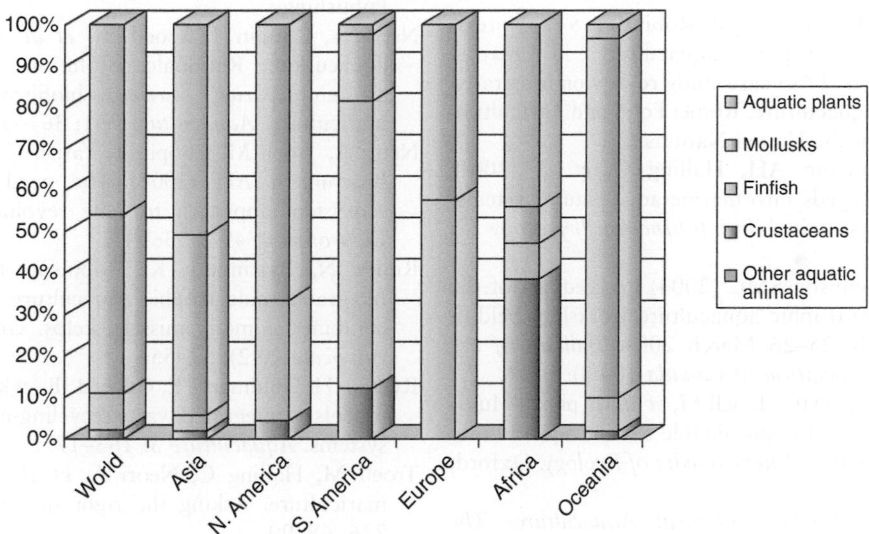

Figure 8 Distribution (%) of mariculture production among the main farmed groups of organisms, worldwide and on the different continents. The world distribution is governed by the distribution in Asia; however, large imbalances between fed and extractive aquacultures exist on the other continents. Source: FAO (2006) *FAO Fisheries Technical Paper 500: State of World Aquaculture 2006*. Rome: Food and Agriculture Organization of the United Nations. http://www.fao.org/docrep/009/a0874e/a0874e00.htm (accessed Mar. 2008).

systems with several other components for different functions, or similar functions but different size brackets of particles, will have to be designed.

Recently, there has been a significant opportunity for repositioning the value and roles seaweeds have in coastal ecosystems through the development of IMTA systems. Scientists working on seaweeds, and industrial companies producing and processing them, have an important role to play in educating the animal-dominated aquaculture world, especially in the Western World, about how to understand and take advantage of the benefits of such extractive organisms, which will help bring a balanced ecosystem approach to aquaculture development.

It is difficult to place a value on the phycomitigation industry, inasmuch as no country has yet implemented guidelines and regulations regarding nutrient discharge into coastal waters. Because the 'user pays' concept is expected to gain momentum as a tool in integrated coastal management, one should soon be able to put a value on the phycomitigation services of IMTA systems for improving water quality and coastal health. Moreover, the conversion of fed aquaculture by-products into the production of salable biomass and biochemicals used in the sea-vegetable, phycocolloid, and phycosupplement sectors should increase the revenues generated by the phycomitigation component.

See also

Mariculture Overview.

Further Reading

Barrington K, Chopin T, and Robinson S (in press) Integrated multi-trophic aquaculture in marine temperate waters. FAO case study review on integrated multi-trophic aquaculture. Rome: Food and Agriculture Organization of the United Nations.

Chopin T, Buschmann AH, Halling C, *et al.* (2001) Integrating seaweeds into marine aquaculture systems: A key towards sustainability. *Journal of Phycology* 37: 975–986.

Chopin T and Robinson SMC (2004) Proceedings of the integrated multi-trophic aquaculture workshop held in Saint John, NB, 25–26 March 2004. *Bulletin of the Aquaculture Association of Canada* 104(3): 1–84.

Chopin T, Robinson SMC, Troell M, *et al.* (in press) Multi-trophic integration for sustainable marine aquaculture. In: Jorgensen SE (ed.) *Encyclopedia of Ecology*. Oxford, UK: Elsevier.

Costa-Pierce BA (2002) *Ecological Aquaculture: The Evolution of the Blue Revolution*, 382pp. Oxford, UK: Blackwell Science.

Critchley AT, Ohno M, and Largo DB (2006) *World Seaweed Resources An Authoritative Reference System*. Amsterdam: ETI BioInformatics Publishers (DVD ROM).

FAO (1998) *The State of World Fisheries and Aquaculture 1998*. Italy: Food and Agriculture Organization of the United Nations. http://www.fao.org/docrep/w9900e/w9900e00.htm (accessed Jul. 2008).

FAO (2002) *The State of World Fisheries and Aquaculture 2002*. Italy: Food and Agriculture Organization of the United Nations. http://www.fao.org/docrep/005/y7300e/y7300e00.htm (accessed Jul. 2008)

FAO (2004) *The State of World Fisheries and Aquaculture 2004*. Italy: Food and Agriculture Organization of the United Nations. http://www.fao.org/docrep/007/y5600e/y5600e00.htm (accessed Mar. 2008).

FAO (2006) *FAO Fisheries Technical Paper 500: State of World Aquaculture 2006*. Rome: Food and Agriculture Organization of the United Nations. http://www.fao.org/docrep/009/a0874e/a0874e00.htm (accessed Mar. 2008).

FAO (2006) *The State of World Fisheries and Aquaculture 2006*. Italy: Food and Agriculture Organization of the United Nations. http://www.fao.org/docrep/009/a0699e/A0699E00.htm (accessed Mar. 2008).

Graham LE and Wilcox LW (2000) *Algae*, 699pp. Upper Saddle River, NJ: Prentice-Hall.

Lobban CS and Harrison PJ (1994) *Seaweed Ecology and Physiology*, 366pp. Cambridge, UK: Cambridge University Press.

McHugh DJ (2001) *Food and Agriculture Organization Fisheries Circular No. 968 FIIU/C968 (En): Prospects for Seaweed Production in Developing Countries*. Rome: FAO. http://www.fao.org/DOCREP/004/Y3550E/Y3550E00.HTM (accessed Mar. 2008).

McVey JP, Stickney RR, Yarish C, and Chopin T (2002) Aquatic polyculture and balanced ecosystem management: New paradigms for seafood production. In: Stickney RR and McVey JP (eds.) *Responsible Marine Aquaculture*, pp. 91–104. Oxon, UK: CABI Publishing.

Neori A, Chopin T, Troell M, *et al.* (2004) Integrated aquaculture: Rationale, evolution and state of the art emphasizing seaweed biofiltration in modern mariculture. *Aquaculture* 231: 361–391.

Neori A, Troell M, Chopin T, Yarish C, Critchley A, and Buschmann AH (2007) The need for a balanced ecosystem approach to blue revolution aquaculture. *Environment* 49(3): 36–43.

Ridler N, Barrington K, Robinson B, *et al.* (2007) Integrated multi-trophic aquaculture. Canadian project combines salmon, mussels, kelps. *Global Aquaculture Advocate* 10(2): 52–55.

Ryther JH, Goldman CE, Gifford JE, *et al.* (1975) Physical models of integrated waste recycling-marine polyculture systems. *Aquaculture* 5: 163–177.

Troell M, Halling C, Neori A, *et al.* (2003) Integrated mariculture: Asking the right questions. *Aquaculture* 226: 69–90.

Troell M, Rönnbäck P, Kautsky N, Halling C, and Buschmann A (1999) Ecological engineering in aquaculture: Use of seaweeds for removing nutrients from intensive mariculture. *Journal of Applied Phycology* 11: 89–97.

Relevant Website

http://en.wikipedia.org
– Integrated Multi-Trophic Aquaculture, Wikipedia.

METHODS OF ANALYSIS

FLUOROMETRY FOR BIOLOGICAL SENSING

D. J. Suggett and C. M. Moore, University of Essex, Colchester, UK

Introduction

Oceanography has been transformed through the use of fluorescence to assay biology. Light-absorbing pigments cause many organisms to naturally fluoresce. Primarily, but not exclusively, these pigments are associated with photosynthesizing organisms, such as algae, aquatic vascular plants, and aerobic anoxygenic photosynthetic (AAP) bacteria. Fluorescence is most commonly detected as the emission that follows 'active' excitation using an actinic light source. One of the major breakthroughs for oceanography occurred in the 1960s with the detection of actively induced chlorophyll *a* fluorescence *in situ*. Chlorophyll *a* is contained by all algae and cyanobacteria and thus provides a measure of abundance. However, at typical environmental temperatures the chlorophyll *a* fluorescence emission signature largely originates from oxygen evolving photosystem II (PSII). Consequently, the chlorophyll *a* fluorescence signal contains information that can be used to characterize the photosynthetic activity of this complex. Recent technological advances have provided twenty-first-century oceanography with an array of active chlorophyll *a* induction fluorometers that are used routinely to assess photosynthetic physiology. In addition, enhanced capacities within remote sensing platforms has enabled researchers to make major steps in using 'passive' fluorescence from chlorophyll *a*, the fluorescence that is stimulated as a result of natural excitation by the sun (solar-stimulated), to assess global photosynthetic activity.

In addition to the natural (auto-)fluorescent molecules, the process of introducing compounds into cells for binding to, and thus labeling, specific molecules has further expanded the tools available to oceanographic research. Such compounds are either themselves fluorescent or become fluorescent following a specific biological reaction and have opened up many new avenues for exploring physiology and taxonomy. Here we provide a brief synthesis of fluorescence techniques used to examine the biology of marine systems.

What Is Fluorescence?

Light absorption by chromophoric molecules (pigments) raises electrons within those molecules to an excited state or higher energy level. As the molecules de-excite, most of the energy is released as heat; however, a proportion of the energy is also released as light. Fluorescence is the emission of light at a longer wavelength following light absorption at a shorter wavelength. The fluorescence yield that arises from an excitation light source is typically referred to as *F*. Many factors including the optical geometry of the measurement system comprising the excitation light source, sample compartment, and detector will affect *F*. Typically, *F* is measured in relative units, although, for a given optical geometry (C), *F* will also vary according to the absorptivity (*A*) of the sample and the photon flux density (PFD) of the excitation light source:

$$F = \phi_F \cdot C \cdot A \cdot \mathrm{PFD}$$

The value of C depends on the proportion of fluoresced light that is intercepted by the detector and the efficiency of conversion of photons into an electric signal. Importantly, ϕ_F is the quantum yield of fluorescence, or the molar ratio of light emitted to light absorbed, a value that is unique to the inherent physical properties of the fluorescent molecule.

Estimating Abundance

Fluorescence is most commonly used to estimate the abundance of chlorophyll *a*. In the case of chlorophyll *a*, relaxation from excited state 1 to ground state results in dissipation of a small proportion (3–5%) of the excitation energy at wavelengths greater than *c.* 650 nm, that is, as red fluorescence. Most fluorometers deliver narrow-band blue light to correspond with wavelengths at which algae and plants exhibit maximum rates of light absorption (**Figure 1**). Subsequent detection of fluorescence is centered toward 680–700 nm to coincide with the peak emission by chlorophyll *a*. Almost all (>99%) of fluorescence at these wavelengths arises from PSII for the majority of algae and aquatic plants in nature.

When algal and plant pigments are extracted into solvents, the fluorescence that arises (*F*) is expected to exhibit a one-to-one relationship with the concentration of chlorophyll *a* (and their breakdown products). However, such a relationship is not

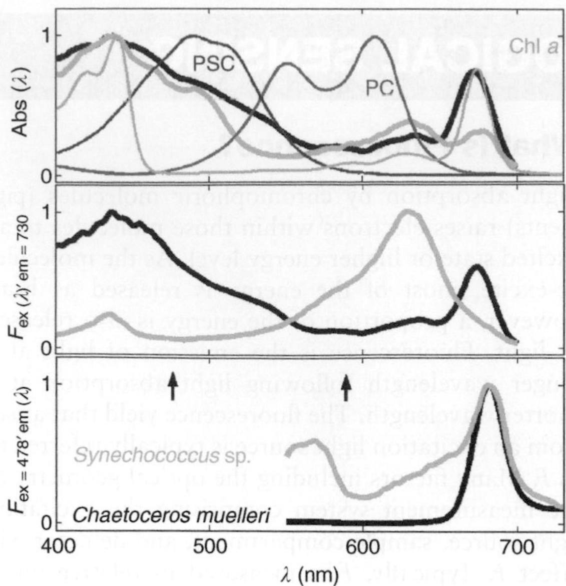

Figure 1 Absorption and fluorescence spectra for a diatom (*Chaetoceros muelleri*) and a phycocyanin-containing cyanobacteria (*Synechococcus* spp.): top panel – optical absorption spectra on intact cellular suspensions. Overlaid are absorption spectra for extracts of the predominant pigments, chlorophyll *a* (chl *a*) and photosynthetic carotenoids (PSC) for *Chaetoceros* and chl *a* and phycocyanin (PC) for *Synechococcus*; middle panel – the fluorescence emission at 730 nm following hyperspectral excitation. The two arrows are to demonstrate that excitation in the blue yields a higher F_{730} for *Chaetoceros* since excitation of chl *a* and PSC is favored while excitation in the orange yields a higher F_{730} for *Synechococcus* since excitation of chl *a* and PC is favored. In this way, fluorometers with fluorescence yields recorded for excitation at various wavelengths can provide some taxonomic discrimination; bottom panel – hyperspectral fluorescence emission following excitation by blue light, a wavelength commonly used for excitation sources of many commercial fluorometers. Most fluorometers are designed to measure fluorescence emission at *c.* 680 nm (chl *a*, e.g., *Chaetoceros*); however, fluorescence from PC can contaminate the chl *a* signal as is the case for *Synechococcus*. Maximum values for all spectra are scaled to a value of 1 for unity.

expected for natural intact samples. Viable cells modify the relationship between *F* and the chlorophyll *a* concentration as a result of variability of the amount of light absorbed, the proportion of absorbed light that is transferred to PSII, referred to as the transfer efficiency (Φ_t), as well as of ϕ_F. As a result, the ratio of *F* to chlorophyll *a* concentration can alter according to taxonomic (genetic) as well as physiological (acclimation and stress) variability. Consequently, chlorophyll *a* fluorescence in nature does not provide an absolute measurement of the amount of chlorophyll *a* in a water sample but can be used to infer changes of abundance.

Similar arguments hold for pigmented organisms other than aquatic plants and algae that fluoresce, most notably, corals and anemones (anthazoa). These organisms contain green fluorescence-like proteins (GFPs), a family of chromophoric molecules that contribute to the colorfulness of corals and fluoresce at wavelengths between 450 and 600 nm. Although the function of GFPs is widely debated, researchers have recently employed GFP fluorescence *in situ* to determine the abundance of polyps recruited onto coral reefs.

Phylogenetic Discrimination

In addition to chlorophyll *a*, algae and aquatic plants contain many other chromophoric molecules termed accessory pigments that act to supplement light capture. Light absorption by chlorophyll *a* is restricted to narrow wavebands centered on *c.* 440 and 670 nm; however, light spectra within aquatic systems are relatively broad (*c.* 300–750 nm). Accessory pigments absorb at wavelengths not targeted by chlorophyll *a* (**Figure 1**). Some accessory pigments have a high Φ_t with chlorophyll *a* and actively supplement the light that is harvested for photosynthesis, whereas others have a relatively low Φ_t and are termed photoprotective pigments. Importantly, a specific accessory pigment array is unique to each algal family providing a first-order taxonomic discrimination of complex algal communities. An accessory pigment that is excited with a light source tuned to the wavelength of peak absorption will induce a higher chlorophyll *a* fluorescence yield. Therefore, the chlorophyll *a* emission signature from sequential excitation of multiple wavelengths provides information on the relative abundance of specific accessory pigments and thus on certain algal families.

In a similar manner, fluorometers can be tuned to target-specific accessory pigments to assess changes in abundance of a specific algal group. Cyanophytes are a group of prokaryotic organisms that are also known as blue-green algae but are in fact bacteria that possess photosynthetic machinery, including chlorophyll *a* and accessory pigments. Blooms of cyanophytes occur frequently in coastal waters and lakes with many species producing substances that can prove highly toxic to other organisms, including humans if ingested in large quantities. The most abundant pigments in these organisms, collectively termed phycobilins, have a unique auto-fluorescence signature that can be easily detected in addition to that of chlorophyll *a* (**Figure 1**).

Prochlorophytes are oxygenic phototrophic prokaryotes and, along with the cyanophyte *Synechococcus*, represent the most abundant phytoplankton cells throughout the world's oceans. Cells of both phytoplankton groups are extremely small (0.2–2 μm) and require use of their fluorescence signatures for enumeration. Traditional microscope-based epifluorescence techniques rely on the fluorescence naturally generated by cells under actinic light; however, the fluorescence yielded by these phytoplankton groups is often too dim to make this approach viable for oceanography. Instead, flow cytometry (FCM), a technique of biomedical origin, is now commonly used. FCM employs a combination of light scattering by cells and auto-fluorescence by natural photosynthetic pigments (chlorophyll *a* and the orange fluorescing phycobilin, phycoerythrin) to enable both identification and enumeration of the different phytoplankton groups.

Modification of FCM via introduction of fluorescence tags into water samples can be used to discriminate other functional groups of aquatic organisms. For example, highly sensitive nucleic acid-specific fluorescent stains that target DNA or rRNA have also made possible the detection and enumeration of heterotrophic bacteria and most recently of viruses. Similarly, a more advanced technique, referred to as fluorescence *in situ* hybridization (FISH), can be used to selectively target regions of DNA or rRNA that consist of evolutionarily conserved and variable nucleotide sequences, thus enabling discrimination of cells at any taxonomic level ranging from kingdom to species. For example, the FISH technique has been modified to enable identification of the taxonomic and life cycle status of single coccolithophore cells collected from the ocean.

Physiological Applications

Many of the fluorescence-based approaches used to examine abundance and taxonomy have been further modified to provide insights into physiology. Applications of fluorescent tags, molecules, and dyes that can be covalently bound to sensing biomolecules are now widespread within online fiber-optic biosensors, DNA sequencing, DNA chips, protein detection, and immunoassays. Near-infrared (NIR) fluorescent dyes are often most desirable since nonspecific background fluorescence is considerably reduced in the NIR spectral range and hence sensitivity can be significantly improved. For example, NIR fluorescence dyes attached to esters can be used to examine the nature of intracellular signaling mechanisms. Recent advances in understanding the biological production of reactive oxygen species, such as singlet oxygen ($^1O_2$), superoxide (O_2^-), and hydrogen peroxide (H_2O_2), which can be destructive to proteins and lipids, has been achieved almost exclusively through use of fluorescent tags. Consequently, fluorescence-based approaches can be considered to be 'everyday' tools for examining physiology and molecular-based ecology. However, these approaches require removal of samples from nature. Perhaps the greatest advance for examining physiology *in situ* again takes advantage of the unique fluorescence characteristics of chlorophyll *a*.

Chlorophyll *a* fluorescence contains a measure of ϕ_F but represents just one of several pathways that can be used to dissipate absorbed excitation energy, the others being photochemistry (ϕ_P) and heat (ϕ_H). Conservation of energy requires that preferential use of one pathway decreases the use of the others such that $\phi_F + \phi_P + \phi_H = 1$. Therefore, an excitation protocol that changes the quantum (photon) yields of competing processes that affect ϕ_F can thus provide information on photosynthetic physiology. Such an approach is termed 'variable' fluorescence induction. Absolute quantum yields are not typically measured, nor do they need to be, provided that the fluorescence signals that are measured accurately report changes in the relative quantum yield of fluorescence.

One variable fluorescence induction approach that has become popular in oceanography is fast repetition rate (FRR) fluorometry and its derivatives. Here, the excitation that is delivered cumulatively closes the PSII reaction centers that dissipate excitons via photochemistry to raise *F* from an initial background (F_0) to a maximum (F_m) value (**Figure 2**). This process occurs within a single trapping event of the reaction center pool, termed a 'single turnover' (ST). A biophysical model describing the process of sequential reaction center closure is fit to this fluorescence transient to yield various parameters that describe the photochemical process (**Figure 2**). Another protocol commonly favored to assay PSII physiology is multiple turnover (MT) induction using an excitation pulse of longer duration to turn over the pool of reaction centers more than once (**Figure 2**). ST and MT protocols provide very different physiological information.

Use of variable fluorescence techniques to provide measures of photochemical efficiency, termed $F_v/F_m = (F_m - F_0)/F_m$, for algae and aquatic plants has become commonplace in oceanography (see **Figure 3**). Some of the earliest *in situ* measurements were made in the mid-1990s to demonstrate lower photosynthetic viability under limitation of nutrients, such as iron and nitrogen. Chlorophyll *a* fluorescence yields the same action spectrum as that of O_2 evolution confirming that it can provide a representative description of PSII photochemical efficiency. Several investigations have also demonstrated good agreement between F_v/F_m and

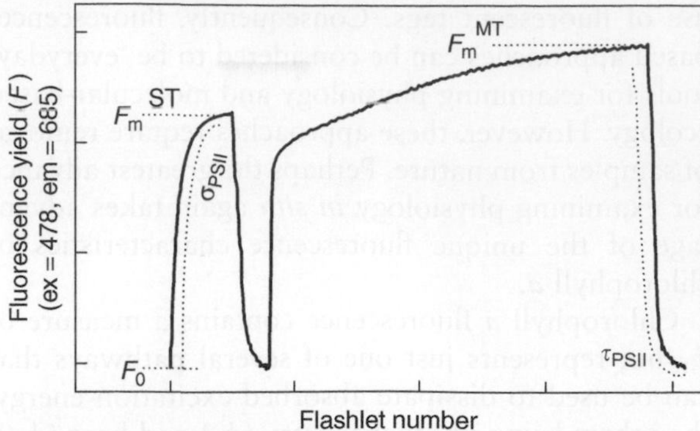

Figure 2 Fast repetition rate (FRR) style chl *a*-fluorescence induction for the microalga *Symbiodinium* spp. Each induction consists of a four-step sequence: (1) single turnover (ST) excitation from a short (*c.* 100 μs) pulse; (2) ST relaxation from a weak modulated light (*c.* 500 ms); (3) MT excitation from a longer (*c.* 600 ms) pulse; (4) MT relaxation from a weak modulated light (over 1 s). In this way the fluorescence is modulated in a controlled manner between the initial background and maximum yield, termed F_0 and F_m, respectively. A biophysical model describing the oxidation–reduction of PSII is fit to the induction profile to yield measures of various physiological parameters, notably, the effective absorption by PSII (σ_{PSII}) and the turnover time of electrons by PSII (τ). σ_{PSII} describes the target area available for light harvesting by the PSII reaction centers τ describing the rate at which acceptor molecules within PSII re-oxidize following excitation and thus the rate at which electrons can be funneled out of PSII. Both σ and τ characterize a wide range of acclimation and adaptation properties inherent to microalgae and AAPs.

the quantum yield of photosynthesis based on 'conventional' O_2 measurements. However, as has been identified from phytoplankton cultures in the laboratory, natural variability of F_v/F_m is notoriously difficult to interpret since ϕ_F can vary between taxa largely as a result of differences in photosynthetic architecture (**Figure 3**). Measurements of F_v/F_m also become problematic where a proportion of fluorescence that is induced does not originate from (active) chlorophyll *a*. One extreme example comes from cyanobacteria species containing a phycobilin called phycocyanin that often bloom in the Baltic Sea. Fluorescence emitted by phycocyanin is centered between *c.* 615 and 645 nm and has a waveband that overlaps with the fluorescence emitted by chlorophyll *a* (see **Figure 1**). Consequently, phycocyanin can lead to an elevated measure of the minimum fluorescence yield (F_0) that is detected and consequently to a lower F_v/F_m. Recently, fluorescence lifetime imaging (FLIM) has been introduced to avoid some of these issues associated with (what are essentially empirical) chlorophyll *a* induction kinetic measurments of F_v/F_m. FLIM is a technique for producing an image based on differences in the exponential decay rate of fluorescence and thus provides mechanistic information inherent to ϕ_F.

Measurements of PSII physiology as photochemical efficiency, effective absorption and electron turnover F_v/F_m, τ, and σ_{PSII} (**Figures 2 and 3**) are strictly obtained (1) in darkness either *in situ* at night and (2) in the laboratory upon discrete samples removed from the environment. Consequently, all parameters yield the characteristics inherent to the potential to process excitation energy. However, these parameters are all modified in the light as a result of changes to ϕ_P, ϕ_H, and ϕ_S, and hence the minimum and maximum fluorescence. Such measurements are important since they inform the 'functional' capacity of PSII to process electrons. Here, these parameters are denoted as F'_q/F'_v, τ', and σ'_{PSII}, respectively. Measurements of F'_q/F'_v represent the product of two other parameters of physiological interest, F'_q/F'_m (the PSII operating efficiency) and F'_v/F'_m (the maximum PSII yield under actinic light), which describe the contribution of photochemistry and heat to changes of F_v/F_m, respectively. Increasingly, these fluorescence parameters are being employed to address complex aspects of photosynthetic physiology that are observed in nature.

Algal and Aquatic Plant Productivity

Changes of abundance over time provide a simple index of productivity. Numerous laboratory studies have employed fluorescence to monitor phytoplankton abundance within cultures; however, such measurements rarely equate to the actual cellular growth rate since ϕ_F can vary according to the cellular growth phase. An alternative measurement of productivity can be obtained from transformation of the physiological parameters afforded from variable chlorophyll *a* fluorescence induction of PSII into photosynthesis rates.

Simple algorithms have been introduced that use FRR-style induction techniques to generate the gross photosynthetic electron transfer by PSII (PET, mol electrons (mol PSII reaction center)$^{-1}$ s^{-1}):

$$\mathrm{PET} = \sigma'_{\mathrm{PSII}} \cdot (F'_q/F'_v) \cdot \mathrm{PFD}$$

Measurements of PET *in situ* are relatively easy to make but surprisingly rarely reported, primarily since PET measurements are expressed in units that are difficult to interpret ecologically. Conversion of PET to a measure of productivity expressed in conventional units, such as O_2 evolution or CO_2 fixation, is not trivial for natural samples. Within PSII, four electron steps are required to evolve 1 mol of O_2; hence, PET should be proportional to 4 times the molar O_2 evolution by all functional PSII reaction centers (RCIIs). In the laboratory, the amount of O_2

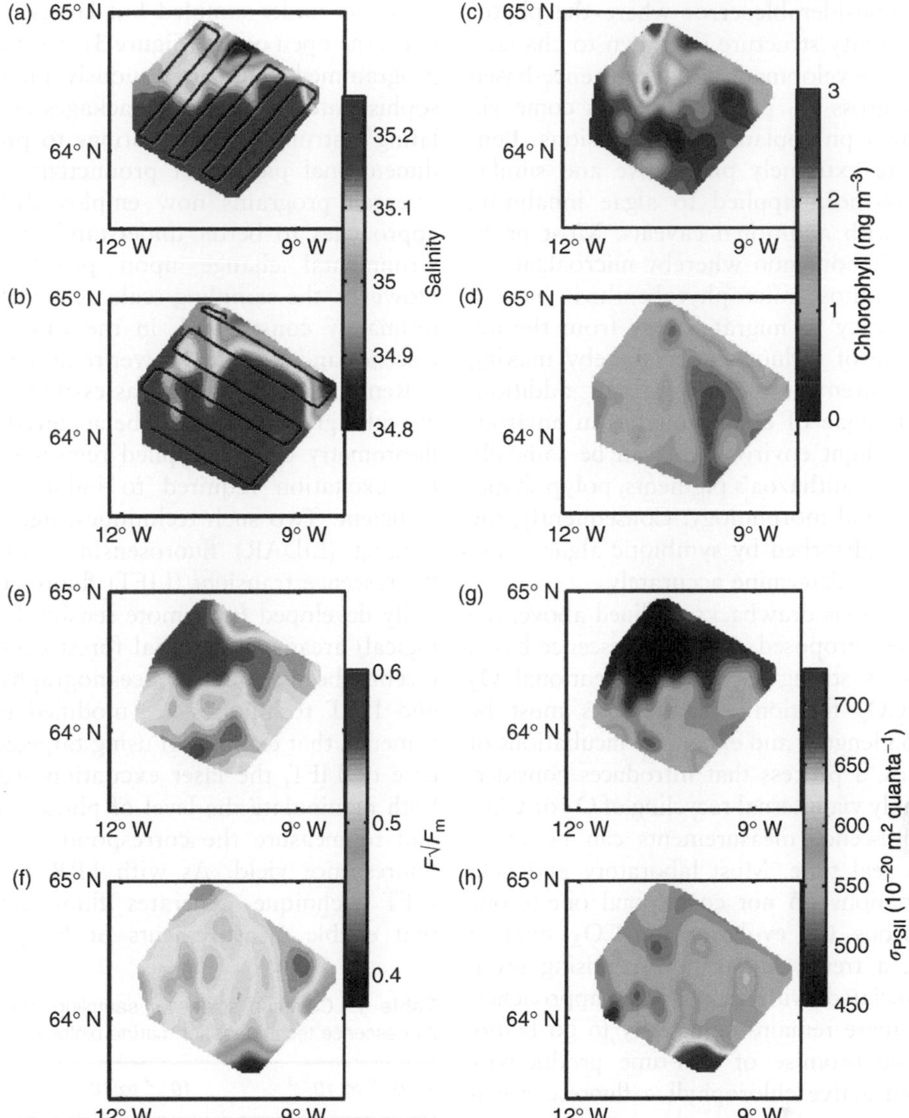

Figure 3 An example of small spatial and temporal resolution of phytoplankton physiology using active fluorescence: maps of salinity (a, b), chlorophyll (c, d), F_v/F_m (e, f), and σ_{PSII} (g, h), across a section of the Iceland-Faroes front during two surveys separated by around 2 weeks during June 2001 (Moore, unpublished). A bloom is observed along the frontal boundary and is associated with marked changes in physiological parameters (e–h) measured by active fast repetition rate fluorometry. These different 'physiological acclimation states' were associated with communities containing different phytoplankton species. Therefore, fluorescence-based physiological parameters contain information on both taxonomy (adaptation) and physiology (acclimation and/or stress).

evolved per unit chlorophyll *a* per ST saturating flash, termed the photosynthetic unit size of PSII (PSU_{PSII}), can be determined relatively easily. This measurement accounts both the yield of O_2 evolved per electron step and the ratio of RCIIs per unit chlorophyll *a*. Therefore, gross O_2 production by PSII per unit chlorophyll *a* ($P^{O_2}_{\mathrm{chl}\ a}$) can be obtained as

$$P^{O_2}_{\mathrm{chl}\ a} = \mathrm{PET} \cdot \mathrm{PSU}_{\mathrm{PSII}}$$

Of major concern for $P^{O_2}_{\mathrm{chl}\ a}$ determinations in oceanography is thus an accurate understanding of the variability of PSU_{PSII} in nature. Direct determination of PSU_{PSII} for natural phytoplankton samples has been achieved but only when the biomass was considerably high, such as within shelf seas or during blooms, to provide an adequate O_2 signal. Alternative (indirect) methods have been proposed to estimate PSU_{PSII} from knowledge of the spectral absorption and fluorescence excitation characteristics, σ_{PSII} and the pigment array. However, such alternative methods have only received limited validation. Most oceanographic investigations to date have simply assumed values for PSU_{PSII}, an approach that can introduce considerable error where the phytoplankton community structure is known to change.

Much of the development of fluorescence-based algorithms for gross O_2 productivity has come via laboratory-grown phytoplankton suspensions. Benthic habitats are extremely productive and similar algorithms have been applied to algae inhabiting substrates but with additional caveats. Most problematic is the phenomenon whereby microalgae inhabiting the benthos (microphytobenthos) have a frustrating tendency to migrate away from the actinic light source of a fluorometer, thereby making prolonged measurements inaccurate. In addition, symbiotic microalgae of corals inhabit an environment where the light environment can be radically altered by the host anthazoa's pigments, polyp shape, and $CaCO_3$ skeletal morphology. Consequently, the amount of light absorbed by symbiotic algae is extremely difficult to determine accurately.

Despite the various drawbacks outlined above, the major promise proposed for fluorescence-based productivity is as strong as ever. Conventional O_2 evolution or CO_2 fixation measurements must be performed upon lengthy and expensive incubations of discrete samples, a process that introduces considerable error notably via internal recycling of O_2 or CO_2. However, fluorescence measurements can be made *in situ* and in real time. Most laboratory and field $P^{O_2}_{\text{chl } a}$ determinations do not correspond one-to-one with simultaneous O_2 evolution or CO_2 fixation measurements, a trend that is not surprising given difficulties associated with the various approaches. Consequently, there remains some way to go before the considerable promise of real-time productivity estimation from active chlorophyll *a* fluorescence is achieved.

New *Optode* methods for measuring the net community O_2 evolution directly *in situ* also employ fluorescence. *Optode* technology is based on the ability of selected substances to act as dynamic fluorescence quenchers in the presence of oxygen. Not only do such methods provide highly accurate and sensitive O_2 measurements but also, upon coupling to *in situ* fluorescence-based $P^{O_2}_{\text{chl } a}$ determinations, may potentially provide important new insights into ecosystem function, metabolism, and trophic coupling.

Scales of Productivity Measurements

A second advantage afforded from fluorescence over conventional O_2 evolution or CO_2 fixation productivity determinations is an increase of both spatial and temporal scale that can be investigated. Such an advantage is particularly important for the relatively under-sampled but largest of aquatic systems, the open ocean (**Figure 3**). Fluorometers can be programmed to log continuously and integrated into sophisticated bio-optical packages as part of undulating instruments or moorings to provide a multidimensional picture of productivity. Indeed, many research programs now employ these larger-scale approaches to better understand the effects of environmental change upon primary productivity. However, the sampling scale achieved *in situ* is still ultimately constrained, in the case of active fluorescence and FCM, by power requirements (**Table 1**).

Remote sensing enables assessment of productivity at scales greater than can be achieved *in situ*. Active fluorometry can be applied remotely provided that the excitation required to induce fluorescence is sufficient. Two such techniques, light detection and ranging (LIDAR) fluorosensing and laser-induced fluorescence transient (LIFT) fluorescence, were initially developed for remote sensing large-scale (ecological) areas of terrestrial forest canopies but have recently been applied to oceanography. Both LIDAR and LIFT techniques are modified induction fluorometers that excite PSII using targeted lasers. In the case of LIFT, the laser excitation signal is used to both manipulate the level of photosynthetic activity and to measure the corresponding changes in the fluorescence yield. As with FRR fluorescence, the LIFT technique generates fluorescence transients that enable measurements of F_v/F_m, τ, and σ_{PSII}.

Table 1 Common scales of sampling (meters) with current fluorescence techniques for marine biological applications

$< 10^{-6}$ to 10^{-3}	10^{-3} to 10^{3}	$> 10^{3}$
Analysis of discrete samples:	*Small-scale in situ campaigns:*	*Large-scale in situ campaigns**
Fluorescence imaging microscopes (epifluorescence and active chl *a* fluorescence induction)	Induction fluorometers	*Remote sensing:*
Flow cytometers	Optodes	LIDAR and LIFT
Intracellular molecular tags	Flow cytometers	Satellite retrieval

The smallest scales are dominated by techniques that perform measurements at the single-cell (or intracellular) scale upon discrete water samples removed from nature. Mesoscales are dominated by instruments that can be deployed *in situ*, for example, variable chlorophyll *a* fluorometers and modified flow cytometers (*Cytobuoy* and *FlowCytobot* technologies), while the largest scales are dominated by remote sensing techniques. The asterisk indicates that any of the techniques could be applied but that the scale afforded is limited by sampling effort and power requirements.

Ultimately, power constraints and safety considerations also constrain measurement scales for these laser-based techniques.

Remote sensing of ocean productivity largely relies on satellite retrieval of ocean color reflectance by algal pigments. Productivity algorithms most simplistically combine chlorophyll *a* concentration and PFD; however, such an approach has proven difficult to retrieve quantitative information on algal physiology.

Fluorescence can be observed passively via remote sensing since algae and aquatic plants naturally fluoresce when they are excited by sunlight (solar stimulation). Chlorophyll *a* fluorescence is inherent to the water-leaving radiance with peak emission at 683–685 nm. By analogy to the fluorescence yield induced actively, the amount with which chlorophyll *a* fluorescence increases the water-leaving radiance (F) depends on several factors, including the specific absorption of chlorophyll (A^{chl}), fluorescence quantum efficiency (ϕ_F), amount of incident sunlight (PFD), in addition to various atmospheric effects (C^*):

$$F = \phi_F \cdot C^* \cdot A^{chl} \cdot \mathrm{PFD}$$

Chlorophyll *a* fluorescence (F) is quantified according to the fluorescence line height (FLH) above background levels at 678 nm, several nanometers below the wavelength of peak emission to avoid atmospheric oxygen absorption at 687 nm. Estimation of ϕ_F is achieved as $F/(C^* \cdot A^{chl} \cdot \mathrm{PFD})$, where $(C^* \cdot A^{chl} \cdot \mathrm{PFD})$ is the instantaneous absorbed radiation by phytoplankton, termed ARP. A semianalytical model that also retrieves chlorophyll *a* concentration from measurements of water-leaving radiance in all bands from 412 to 551 nm is used to determine the ARP. Primary productivity (PP) can then be determined as

$$\mathrm{PP} = \phi_P \cdot \mathrm{ARP}$$

Therefore, the relationship between ϕ_F and the quantum yield of PSII photochemistry (ϕ_P) must be known. As shown above, this relationship will be modified by ϕ_H and ϕ_S, parameters that are not easy to constrain for natural samples as a result of the huge taxonomic and physiological (both acclimation and inhibition) variability. Consequently, considerable effort is currently invested in trying to understand the relationship between ϕ_H and environment for key algal taxa. Despite these potential drawbacks, passive measurements of global chlorophyll *a* fluorescence show great potential for future observations of global productivity and may offer one of our most powerful monitoring tools as the Earth is subjected to a time of extreme climate change.

Nomenclature

A	absorptivity
A^{chl}	absorption specific to chlorophyll
C	geometry of fluorometer optics
C^*	atmospheric effects
F	fluorescence yield
F_m	maximum fluorescence yield
F_0	initial fluorescence yield
F'_q/F'_m	PSII operating efficiency
F'_q/F'_m	PSII photochemical efficiency (under actinic light)
F_v/F_m	PSII photochemical efficiency (dark adapted)
F'_v/F'_m	PSII maximum yield under actinic light
$P^{O_2}_{chl\ a}$	gross O_2 production by PSII per unit chlorophyll *a*
σ_{PSII}	PSII effective absorption cross section (dark adapted)
σ'_{PSII}	PSII effective absorption cross section (under actinic light)
τ	turnover time of electrons (dark adapted)
τ'	turnover time of electrons (under actinic light)
ϕ_F	quantum yield of fluorescence
ϕ_H	quantum yield of heat
ϕ_P	quantum yield of photochemistry
Φ_t	energy transfer efficiency between pigment molecules

See also

Coral Reefs. Phytoplankton Blooms.

Further Reading

Andersen RA (ed.) (2005) *Algal Culturing Techniques*. New York: Elsevier Academic Press.

Barbini R, Colao F, Fantoni R, Palucci A, and Ribezzo S (2001) Differential LIDAR fluorosensor system used for phytoplankton bloom and seawater quality monitoring in Antarctica. *International Journal of Remote Sensing* 22: 369–384.

Falkowski PG and Raven JA (1997) *Aquatic Photosynthesis*. New York: Blackwell.

Geider RJ and Osborne BA (1992) *Algal Photosynthesis: The Measurement of Algal Gas Exchange*. Boca Raton, FL: Chapman and Hall.

Groben R and Medlin L (2005) *In situ* hybridization of phytoplankton using fluorescently labeled rRNA probes. *Methods in Enzymology* 395: 299–310.

Hoge FE, Lyon PE, Swift RN, *et al.* (2003) Validation of Terra-MODIS phytoplankton chlorophyll fluorescence line height. I. Initial airborne LIDAR results. *Applied Optics* 42: 2767–2771.

Huot Y, Brown CA, and Cullen JJ (2005) New algorithms for MODIS sun-induced chlorophyll fluorescence and a comparison with present data products. *Limnology and Oceanography Methods* 3: 108–130.

Jeffrey SW, Mantoura RFC, and Wright SW (1997) *Phytoplankton Pigments in Oceanography.* Paris: UNESCO Publishing.

Kolber ZS, Klimov D, Ananyev G, Rascher U, Berry J, and Osmond B (2005) Measuring photosynthetic parameters at a distance: Laser induced fluorescence transient (LIFT) method for remote measurements of photosynthesis in terrestrial vegetation. *Photosynthesis Research* 84: 121–129.

Kolber ZS, Prášil O, and Falkowski PG (1998) Measurements of variable chlorophyll fluorescence using fast repetition rate techniques: Defining methodology and experimental protocols. *Biochimica Biophysica Acta* 1367: 88–106.

Labas YA, Gurskaya NG, Yanushevich YG, *et al.* (2002) Diversity and evolution of the green fluorescent protein family. *Proceedings of the National Academy of Sciences of the United States of America* 99: 4256–4261.

Moore CM, Suggett DJ, Hickman AE, *et al.* (2006) Phytoplankton photo-acclimation and photo-adaptation in response to environmental gradients in a shelf sea. *Limnology and Oceanography* 51: 936–949.

Papageorgiou GC and Govindjee (eds.) (2004) *Chlorophylla Fluorescence: A Signature of Photosynthesis.* Amsterdam: Springer.

Suggett DJ, Maberly SC, and Geider RJ (2006) Gross photosynthesis and lake community metabolism during the spring phytoplankton bloom. *Limnology and Oceanography* 51: 2064–2076.

Veldhuis MJW and Kraay GW (2000) Application of flow cytometry in marine phytoplankton research: Current applications and future perspectives. *Scientia Marina* 64: 121–134.

Relevant Websites

http://www.cytobuoy.com
– Environmental applications of flow cytometry, CytoBuoy.

http://www.esa.int
– European Space Agency Living Planet.

http://www.chelsea.co.uk
– Fast Repetition Rate Methods Manual, Chelsea Technologies Group.

http://modis.gsfc.nasa.gov
– MODIS web, NASA.

ZOOPLANKTON SAMPLING WITH NETS AND TRAWLS

P. H. Wiebe, Woods Hole Oceanographic Institution, Woods Hole, MA, USA
M. C. Benfield, Louisiana State University, Baton Rouge, LA, USA

Introduction

In the late 1800s and early 1900s, quantitative ocean plankton sampling began with non-opening/closing nets, opening/closing nets (mostly messenger-based), high-speed samplers, and planktobenthos net systems. Technology gains inelectrical/electronic systems enabled investigators to advance beyondsimple vertically or obliquely towed nets to multiple cod-end systems and multiple net systems in the 1950s and 1960s. Recent technological innovation has enabled net systems to be complemented or replaced by optical and acoustics-based systems. Multi-sensor zooplankton collection systems are now the norm and in the future, we can anticipate seeing the development of real-time four-dimensional plankton sampling and concurrent environmental measurements systems, and ocean-basin scale sampling with autonomous vehicles.

From the beginning of modern biological oceanography in the late 1800s, remotely operated instruments have been fundamental to observing and collecting zooplankton. For most of the past century, biological sampling of the deep ocean has depended upon winches and steel cables to deploy a variety of instruments. The development of quantitative zooplankton collecting systems began with Victor Hensen in the 1880s (**Figure 1A**). His methods covered the whole scope of plankton sampling from the building and handling of nets to the final counting of organisms in the laboratory.

Three kinds of samplers developed in parallel: waterbottle samplers that take discrete samples of a small volume of water (a few liters), pumping systems that sample intermediate volumes of water (tens of liters to tens of cubic meters), and nets of many different shapes and sizes that are towed vertically, horizontally, or obliquely and sample much larger volumes of water (tens to thousands of cubic meters) (**Table 1**). Net systems dominated the equipment normally used to sample zooplankton until recent technological developments enabled the use of high-frequency acoustics and optical systems as well.

Net Systems

A variety of net systems have been developed over the past 100 + years and versions of all of these devices are still in use today. They can be categorized into eight groups: non-opening/closing nets, simple opening/closing nets, high-speed samplers, neuston samplers, planktobenthos plankton nets, closing cod-end samplers, multiple-net systems, and moored plankton collection systems.

Non-opening/Closing Nets

Numerous variants of the simple non-opening/closing plankton net have been developed, which are principally hauled vertically. Most are simple ring-nets with mouth openings ranging from 25 to 113 cm in diameter and conical or cylinder-cone nets 300–500 cm in length. Among the ring-nets that have been widely used are the Juday net (**Figure 1B**), International Standard Net, the British N-series nets, the Norpac net, the Indian Ocean Standard net (**Figure 1C**), the ICITA net, the WP2 net, the CalCOFI net, and the MARMAP Bongo net (**Figure 1D**). Early nets were made from silk, but today nets are made from a square mesh nylon netting. Typical meshes used on zooplankton nets range from 150 μm to 505 μm, although larger and smaller mesh sizes are available. Most of these nets are designed to be hauled vertically. They are lowered to depth cod-end first and then pulled back to the surface with animals being caught on the way up. Others, such as the CalCOFI net and the Bongo net are designed to be towed obliquely from the surface down to a maximum depth of tow and then back to the surface. The Reeve net was a simple ring-net with a very large cod-end bucket designed to capture zooplankton alive. The Isaacs-Kidd midwater trawl (IKMT) has been used to collect samples of the larger macrozooplankton and micronekton. It has a pentagonal mouth opening and a dihedral depressor vane as part of the mouth opening. Four sizes of IKMTs, 3 foot (91 cm), 6 foot (183 cm), 10 foot (304 cm), and 15 foot (457 cm) are often cited.

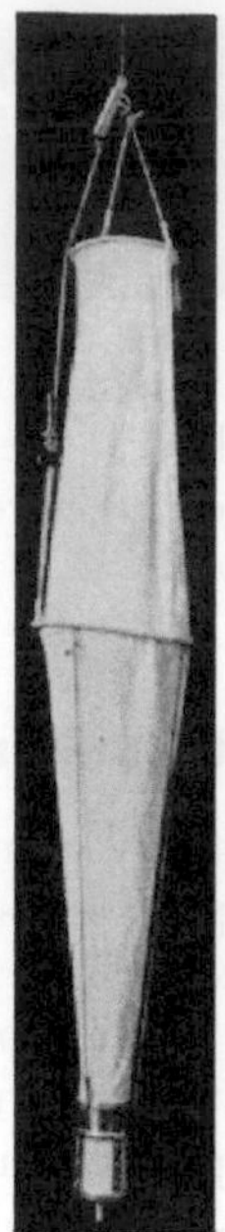

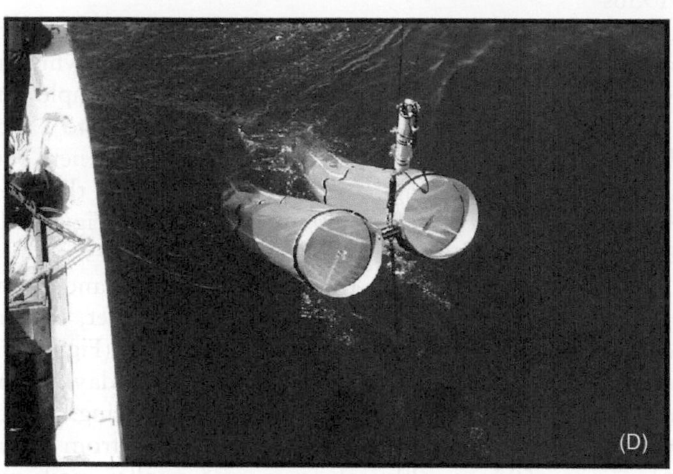

Figure 1 Some commonly used non-opening/closed nets. (A) The Hensen net. (Reproduced with permission from Winpenny, 1937.) (B) The Juday net; note the use of messenger release on this version of the net. (Reproduced with permission from Juday, 1916.) (C) The Indian Ocean Standard net. (Reproduced with permission from Currie, 1963.) (D) The Bongo net with CTD (*c.* 1999). (Photograph courtesy of P. Wiebe.) (E) The Tucker trawl. (Reproduced with permission from Tucker, 1951.)

Non-opening/closing nets with rectangular mouth openings were not widely used until the Tucker trawl was first described in 1951 (**Figure 1E**). This simple trawl design with a 180 cm × 180 cm mouth opening gave rise to a substantial number of opening/closing net systems described below.

Simple Opening/Closing Nets

The development of nets that could obtain depth-specific samples evolved from those of very simple design (a simple ring net) at an early stage. In the late 1800s and early 1900s, there was considerable effort

Table 1 Summary of zooplankton sampling gear types

Sampling gear	*Type of sampling*	*Size fraction*	*Resolving scale*		*Typical operating range*	
			Vertical	*Horizontal*	*Vertical*	*Horizontal*
Conventional methods						
Waterbottles	Discrete samples	Micro/meso	0.1–1 m	—	4000 m	—
Small nets	Vertically integrating	Micro/meso	5–100 m	—	500 m	—
Large nets	Vertical, obliquely Horizontally integrating	Meso/macro	5–1000 m	50–5000 m	1000 m	10 km
High-speed samplers	Obliquely, horizontally integrating	Meso/macro	5–200 m	500–5000 m	200 m	10 km
Pumps	Discrete samples	Micro/meso	0.1–100 m	—	200 m	—
Multiple net systems						
Continuous plankton recorder	Horizontally integrating	Meso	10–100 m	10–100 m	100 m	1000 km
Longhurst-Hardy plankton recorder	Obliquely, horizontally integrating	Meso	5–20 m	15–100 m	1000 m	10 km
MOCNESS	Obliquely, horizontally integrating	Meso/macro	1–200 m	100–2000 m	5000 m	20 km
BIONESS	Obliquely, horizontally integrating	Meso/macro	1–200 m	100–2000 m	5000 m	20 km
RMT	Obliquely, horizontally integrating	Meso/macro	1–200 m	100–2000 m	5000 m	20 km
Multinet	Vertically Obliquely, horizontally	Meso/macro	2–1000 m	100–2000 m	5000 m	5 km
Electronic optical or acoustical systems						
Electronic plankton-counter	High resolution in the horizontal/vertical plane	Meso	0.5–1 m	5–1000 m	300 m	100s of km
In situ silhouette camera net system	High resolution in the horizontal/vertical plane	Meso	0.5–1 m	5–1000 m	1000 m	10 km
Optical plankton counter	High resolution in the horizontal/vertical plane	Meso	0.5–1 m	5–1000 m	300 m	100s of km
Video plankton recorder	High resolution in the horizontal/vertical plane	Meso	0.01–1 m	5–1000 m	200 m	100s of km
Ichthyoplankton recorder	High resolution in the horizontal/vertical plane	Meso	0.1–1 m	5–1000 m	200 m	10 km
Multifrequency acoustic profiler system	High resolution in the horizontal/vertical plane	Meso/macro	0.5–1 m	5–1000 m	100 m	10 km
Dual-beam acoustic profiler	High resolution in the horizontal/vertical plane	Meso/macro	0.5–1 m	1–1000 m	800 m	100s of km
Split-beam acoustic profiler	High resolution in the horizontal/vertical plane	Meso/macro	0.5–1 m	1–1000 m	1000 m	100s of km
ADCP	High resolution in the horizontal/vertical plane	Meso/macro	10 m	5–500 m	500 m	100s of km

Most vertical nets are hauled at a speed of 0.5–1 m s^{-1}. Normal speed for horizontal tows are ~2 knots (1 m s^{-1}) and for high-speed samplers ~5 knots (2.6 m s^{-1}). For further categorization of pumping systems which are used by a number of investigators, reference is made to the review paper by Miller and Judkins (1981). (Reproduced with permission from Sameoto D, Wiebe P, Runge S, *et al.* (2000) Collecting zooplankton. In: Harris R, Wiebe P, Lenz J, Skjoldal HR, and Huntley M (eds.) *ICES Zooplankton Methodology Manual*, pp. 55–81. New York: Academic Press.)

to develop devices that closed or opened and closed nets at depth. Most employed mechanical release devices which were attached to the towing wire and activated by messengers traveling down the towing wire. The single-messenger Nansen closing mechanism and its variants were very popular during most of early to mid-twentieth century (**Figure 2**). Double-messenger systems that opened and then closed a net quickly followed. In the mid-1930s, the Leavitt net system became popular and variants of this system are still being used today (**Figure 2B**). Another popular system still in use today is the Clarke and Bumpus sampler, a two-messenger zooplankton collection system that can be deployed as multiple units on the wire and has a positive means of opening and closing the mouth of the net (**Figure 2C**).

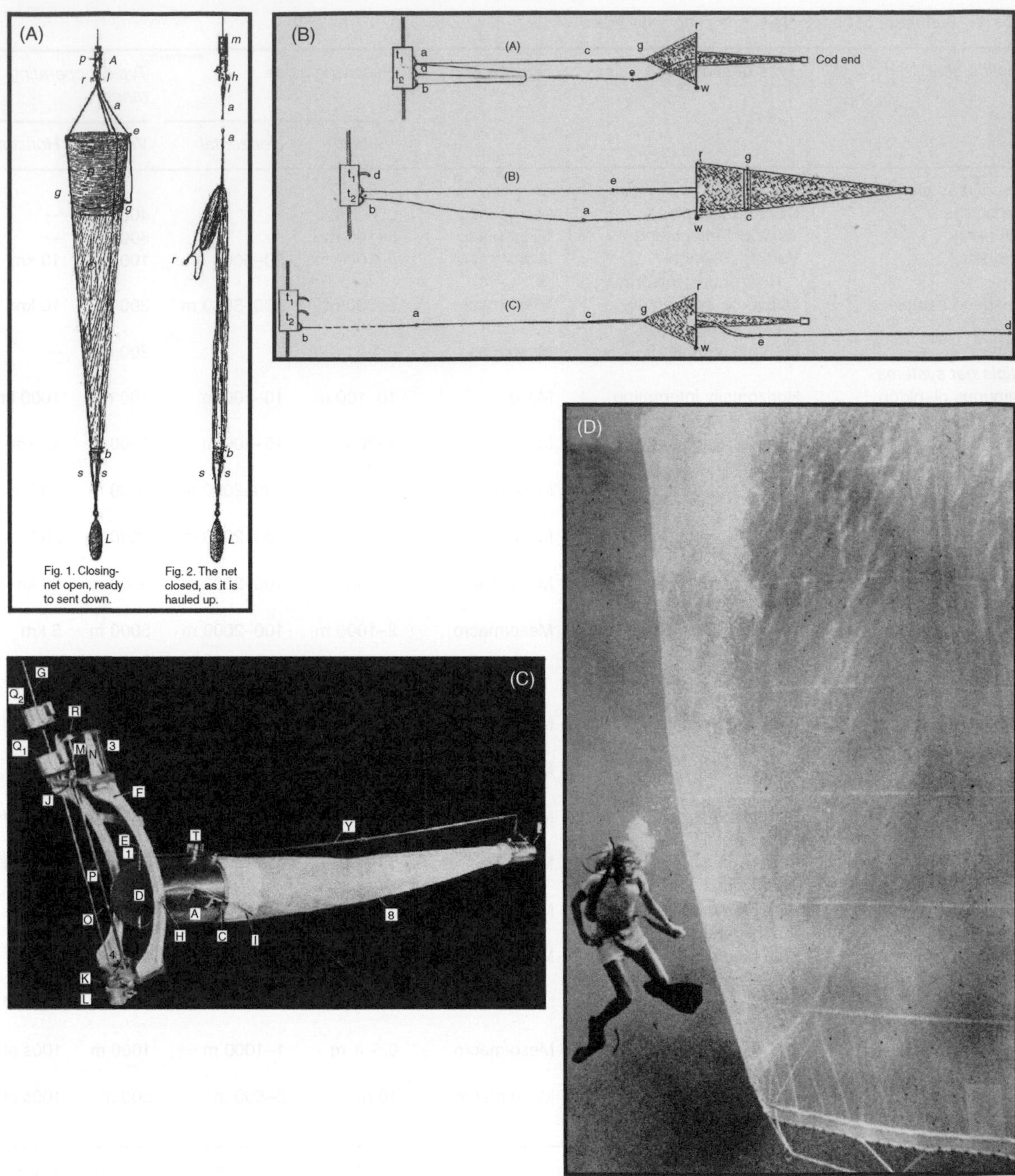

Figure 2 Some commonly used simple opening/closing nets. (A) The single-messenger Nansen closing net. (Reproduced with permission from Nansen, 1915.) (B) The two-messenger Leavitt net. (Reproduced with permission from Leavitt, 1935.) (C) The two-messenger Clarke-Bumpus net. (Reproduced with permission from Clarke and Bumpus, 1939.) The plankton purse seine (D) represents an unusual way to collect plankton from a specific region. (Reproduced with permission from Murphy and Clutter, 1972.)

Mechanical tripping mechanisms activated by pressure, by combinations of messengers and flowmeter revolutions, or clocks have also been devised.

Nontraditional approaches to collecting plankton include designs to catch plankton on the downward fall of the net rather than the reverse – so-called pop-down nets; to sample under sea ice using the English umbrella net; to sample plankton from several depths simultaneously, using a combination of nets and a pumping system; to sample plankton from the nuclear submarine, *SSN Seadragon*; to open and close a Tucker-style trawl using two towing cables, one for the top spreader bar and one for the bottom, with each cable going to a separate winch; and to capture plankton and fish larvae with a plankton purse seine (**Figure 2D**).

High-speed Samplers

Most of the net systems described above were towed at speeds <3 knots ($150\,cm\,s^{-1}$). High-speed samplers typically towed at speeds of 3–8 knots (150–$400\,cm\,s^{-1}$) were also developed in the late 1800s and early 1900s to sample in bad weather, for underway sampling between stations, or to reduce the effects of net avoidance by the larger zooplankton. The Hardy plankton indicator, developed in the 1920s, was the first widely used device. The original version was 17.8 cm in diameter and 91.4 cm in length with a circular filtering disk on which plankton were collected. It was subsequently modified (and renamed the standard plankton indicator) to make it smaller, more streamlined, and equipped with a depressor and stabilizing fins (**Figure 3**). An even smaller version, the Small Plankton Sampler, was developed. In the 1950s, it was further modified and named the Small Plankton Indicator, and in the 1960s, it was modified again so that multiple units could be used on the towing wire at speeds of 7–8 knots with a multiplane kit otter depressor at the end of the wire. Until the 1950s, only one high-speed collector was designed with a double-messenger system that enabled the mouth to be opened and closed; most could not make depth-specific collections.

The 'Gulf' series of high-speed samplers developed in the 1950s and early 1960s gave rise to a number of high-speed samplers still in use today. The first was the Gulf I-A which looked similar to earlier high-speed samplers. The Gulf III was a much larger high-speed sampler that was enclosed in a metal case. The Gulf V was an unencased and scaled-down version of the Gulf III (**Figure 3B**). The Gulf III and Gulf V samplers have been very popular, and have been modified numerous times. In the early 1960s, a five-bucket cod-end sampling device was added to the Gulf III that was electrically activated from a deck unit through two-conductor cable. HAI (shark) was the German version of the Gulf III built in the mid-1960s. A hemispherical nose cone and an opening/closing lid were added to the HAI. This German system evolved further when 'Nackthai' (naked shark), a modified Gulf V sampler, was developed in the late 1960s. Also in the 1960s, the British modified the Gulf III sampler, which was subsequently called the Lowestoft sampler (**Figure 3C**). Subsequently, the Lowestoft sampler was scaled down and made opened bodied; hence it became a modified Gulf V. The Ministry of Agriculture, Fisheries and Food MAFF/Guildline high-speed samplers, developed in the 1980s, were also modified Lowestoft samplers. These systems have a Guildline CTD sensor unit with oxygen, pH, and digital flowmeter as additional probes with telemetry through a conducting cable. Recently in the 1990s, the Gulf VII/Pro net and MAFF/Guildline high-speed samplers were developed that are routinely towed at 5–7 knots.

Other high-speed samplers were developed during the 1950s and 1960s, including a high-speed plankton sampler which could collect a series of samples during a tow; the 'Bary Catcher' that had an opening/closing mechanism in the mouth of the sampler (**Figure 3D**); a vertical high-speed sampler with a rectangular mouth opening that could be closed using the Juday method; an automatic high-speed plankton sampler with 21 small nets that were sequentially closed by means of a cam/screw assembly driven by a ships log (propellor); and the Clarke Jet net that was an encased high-speed sampler with an elaborate internal passageway designed to reduce the flow speed of water within the sampler to that normally experienced by a slowly towed net.

The continuous plankton recorder (CPR) is in a class by itself when it comes to high-speed plankton samplers, because it can take many samples and can be towed from commercial ships (**Figure 3E**). Originally built in the 1920s, it has evolved over the years to become the mainstay in a plankton survey program in the North Atlantic. This encased sampler weight 87 kg and is about 50 cm wide by 50 cm tall by 100 cm long. The $1.27\,cm \times 1.27\,cm$ rectangular aperture expands into a larger tunnel opening. The tunnel passes through the lower portion of the sampler and out of the back. Below the tunnel is one spool of silk gauze which threads across the tunnel and captures the plankton. A second spool of silk gauze lies above the tunnel and is threaded to meet the first gauze strip as it leaves the tunnel, sandwiching the plankton between the two strips. The gauze strips are wound up on a take-up spool which resides in a formalin-filled tank above the flow-through tunnel, preserving the plankton. The take-up spool is driven by a propellor on the back of the sampler behind the tail fins. This sampler is usually towed at 20 knots from commercial transport vessels at a fixed depth of about 10 m below the surface, thus it only samples the surface layer of the ocean. The undulating oceanographic recorder (UOR) was developed in the 1970s to extend the vertical sampling capability of high-speed plankton collection systems. The UOR carries sensors to measure

Figure 3 Some examples of high-speed plankton samplers. (A) The standard plankton indicator. (Reproduced with permission from Hardy, 1936.) (B) The encased Gulf III sampler. (Reproduced with permission from Gehringer, 1952.) (C) The open-bodied Lowestoft sampler (Gulf V type). (Reproduced with permission from Lockwood, 1974.) (D) The Bary catcher. (Reproduced with permission from Bary, 1958.) (E) The continuous plankton recorder (CPR). (Reproduced with permission from Hardy, 1936.)

temperature, salinity, and pressure; data are logged internally at 30 observations per minute. A propellor drives the rollers winding up the gauze and provides the power for the electronics.

Neuston Samplers

Nets to collect neuston, the zooplankton that live within a few centimeters of the sea surface, by-and-large are non-opening/closing. The first net specifically designed to sample zooplankton neuston was built in about 1960. A rectangular mouth opening design is typical of most of the systems. Neuston nets come either with a single net which collects animals right at the water surface or vertically stacked sets of two to six nets extending from the surface to about 100 cm depth (**Figure 4**). Normally they are towed from a vessel, but a 'push-net' was developed in the 1970s with a pair of rectangular nets positioned side-by-side in a framework and mounted in front of a small catamaran boat that pushed the frame through the water at ~2.6 knots.

Planktobenthos Plankton Nets

The ocean bottom is also special habitat structure for zooplankton, and gear to sample zooplankton living here ('planktobenthos') was developed early. The first nets were designed in the 1890s specifically to sample plankton living very near the bottom. Non-opening/closing systems were succeeded by samplers with mechanically operated opening/closing doors or with a self-closing device (**Figure 5A**).

An entirely different strategy has been to employ manned submersibles or deep-towed vehicles to collect deep-sea planktobenthos. A pair of nets mounted on the front of DSRV *Alvin* was used for making net collections at depths >1000 m in the 1970s; the pilot opened and closed the net (**Figure 5B**). A multiple net system was used on the Deep-Tow towed body. This system was attached to the bottom of the Deep-Tow and used for sampling within a few tens of meters above the deep-sea floor in the 1980s (**Figure 5C**). This net system was later adapted for use on DSRV *Alvin* for near-bottom studies of plankton in the vicinity of hydrothermal vent sites in the 1990s.

On other benthic habitats, such as coral reefs, fixed or stationary net systems which orient to the current's flow and filter out zooplankton drifting by, nets pushed by divers, and traps have been used to capture plankton close to the bottom. The Horizontal Plankton Sampler (HOPLASA) creates its own current to collect zooplankton on or near the bottom in coral reef areas with variable or little current flow (**Figure 5D**).

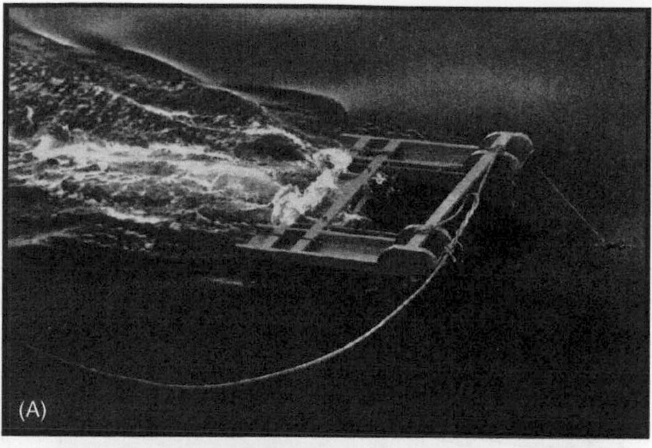

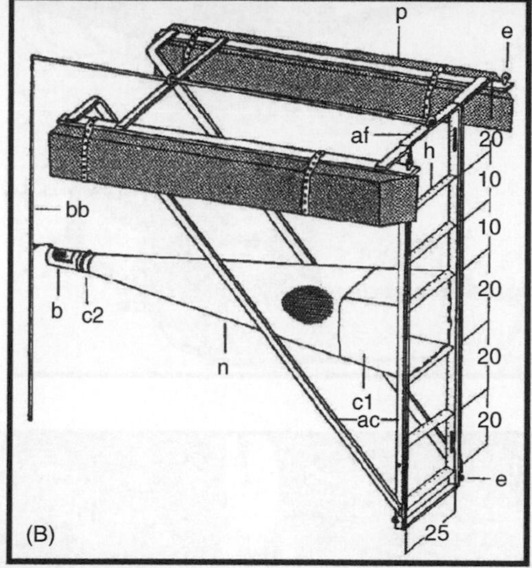

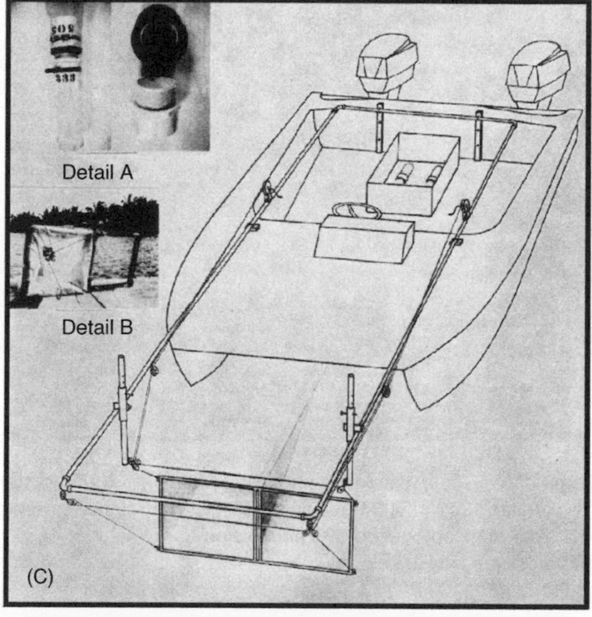

Figure 4 Neuston net samplers collect plankton living at the sea surface. (A) A single net system. (Reproduced with permission from David, 1965.) (B) A multinet system. (Reproduced with permission from Ellertsen, 1977.) (C) A push net. (Reproduced with permission from Miller, 1973.)

Figure 5 Some planktobenthos samplers. (A) Early system with opening/closing doors. (Reproduced with permission from Wickstead, 1953.) (B) DSR Alvin opening/closing system. (Reproduced with permission from Grice, 1972.) (C) The Deep-Tow multiple net system. (Reproduced with permission from Wishner, 1980.) (D) A system for coral reef sampling (HOPLASA). (Reproduced with permission from Rutzler, 1980.)

Closing Cod-end Systems

In the late 1950s and 1960s, conducting cables and transistorized electronics were beginning to be adapted for oceanographic use and sophisticated net systems began to do more than collect animals at specific depth intervals. Single nets equipped with closing cod-end devices preceded multiple net systems by only a few years. One of the first systems used a 1950s version of a serial device in the high-speed sampler that was mechanically driven by a propellor. Another had a pressure-actuated catch-dividing bucket (CDB) attached to the back of an IKMT (**Figure 6A**). The Mark III Discrete Depth Plankton Sampler (DDPS) also developed for use with an IKMT or a 1 m diameter net, had four catch chambers separated by solenoid-activated damper doors (**Figure 6B**). This latter system was one of the first to carry underwater electronics to sample depth and temperature, and to telemeter the data up a single conductor cable for display at the surface. The multiple plankton sampler (MPS, described below) was turned into a cod-end sampler for an IKMT and later modified by adding environmental sensors and an electronically controlled opening/closing mechanism.

The Longhurst-Hardy plankton recorder (LHPR), a modification of the CPR, was developed in the 1960s (**Figure 6D**). The recorder box was attached to the back end of a net and gauze strips in the box were advanced in discrete steps (15 s to 60 s) by an electronics package on the tow frame; data on pressure, temperature, and flow were logged on an internal recorder; power was supplied by a NICAD battery pack. The LHPR was redesigned in the 1970s to reduce problems with hang-ups and stalling of animals in the net which caused smearing of the

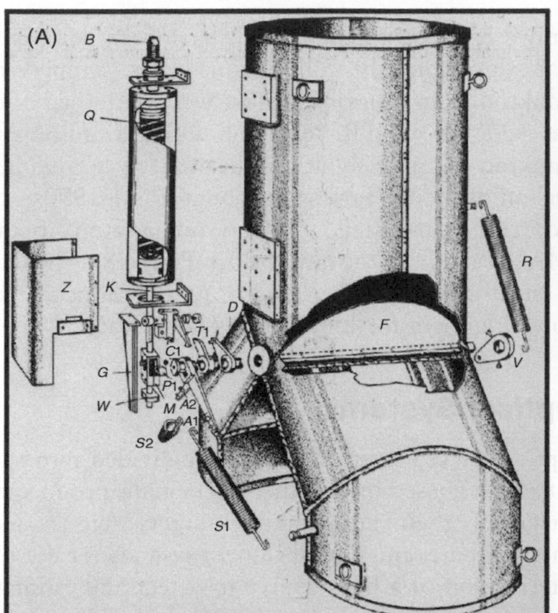

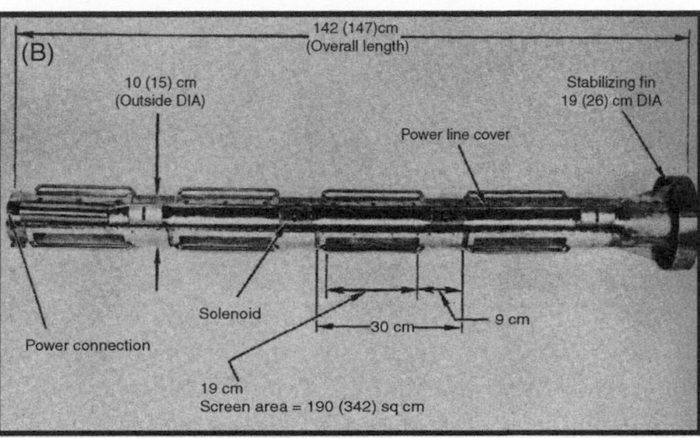

Figure 6 Some discrete-depth samplers using a closing cod-end. (A) The catch dividing cod-end. (Reproduced with permission from Foxton, 1963.) (B) The Mark III multiple cod-end bucket. (Reproduced with permission from Aron *et al.* 1964.) (C) ARIES. (Reproduced with permission from Dunn *et al.* 1993.) (D) A version of the LHPR. (Photograph courtesy of J. Smith, 1966.)

distributions of animals and losses of animals from the recorder box. The modified LHPR was used without a net on the conning-tower of the US Navy research submarine *Dolphin* in the 1980s. Another modification of the LHPR was made by the British in 1980s. They used an unenclosed Lowestoft sampler to mount a pair of recorder boxes to collect meso- and micro-zooplankton. The system acoustically telemetered depth, flow, and temperature. It also carried a chlorophyll sensor with a recorder system. The LHPR was further modified for use in catching Antarctic krill. A descendant of the LHPR developed in the 1990s is the Autosampling and Recording Instrumental Environmental Sampler (ARIES) (**Figure 6C**). This cod-end plankton sampling device is a stretched version of the Lowestoft-modified Gull III frame. It has a multiple cod-end system, water sampler, data logger, and an acoustical telemetry system.

Multiple Net Systems

The development of multiple net systems began with the simple non-opening/closing Tucker trawl system. In the mid-1960s, timing clocks were used to open and close the Tucker trawl mouth. Then late in the 1960s, the British rectangular mouth opening trawl (RMT), which was opened and closed acoustically, was developed. The RMT was expanded into the NIO Combination Net (RMT 1 + 8), which carries nets with 1 m^2 and 8 m^2 mouth openings (**Figure 7A**). This was expanded into a multiple net system with three sets of 1 m and 8 m nets controlled acoustically. The acoustic command and telemetry system for the RMT 1 + 8 was replaced in the 1990s by a micro-computer-controlled unit connected by conducting cable to an underwater electronics unit.

In a parallel development in the 1970s, a five-net and a nine-net Tucker Multiple Net Trawl was developed on the West Coast of the USA. The system was powered electrically through conducting wire and controlled from the surface. A modified Tucker trawl system, the Multiple Opening/Closing Net and Environmental Sensing System (MOCNESS), with nine nets and a rigid mouth opening was built soon after on the US east coast (**Figure 7E**). The current versions of the MOCNESS are computer-controlled (**Table 2**). Sensors include pressure, temperature, conductivity, fluorometer, transmissometer, oxygen, and light.

The design of the Bé multiple plankton sampler (MPS) (**Figure 7B**), initially messenger operated in the late 1950s and then pressure-actuated in the 1960s, was the basis for the Bedford Institute of Oceanography Net and Environmental Sensing System (BIONESS), with 10 nets, developed in the 1980s (**Figure 7D**). A modified version of the MPS was developed in Germany at about the same time and named the Multinet; it carried five nets, which were opened and closed electronically via conducting cable (**Figure 7C**). A scaled-up version of BIONESS built in the 1990s was the Large Opening Closing High Speed Net and Environmental Sampling System (LOCHNESS). Another variant of the MPS was the Ocean Research Institute's (Japan) vertical multiple plankton sampler developed in the 1990s in which the nets are opened/closed by surface commands transmitted via conducting cable to an underwater unit.

Moored Plankton Collection Systems

Only a few instrument systems have been developed that autonomously collect time-series samples of plankton from moorings. Most were patterned after the CPR or LHPR (e.g. the O'Hara automatic plankton sampler built in the 1980s; a modified version of the O'Hara system built in the 1990s; the moored, automated, serial zooplankton pump (MASZP) built in the late 1980s) (**Figure 8**). The lack of such systems may be due to the difficulty of powering them for long periods underwater.

Optical Systems

Optical survey instruments can be divided into two categories, based on whether the systems produce an image of their zooplankton targets (e.g. video, photographic, and digital camera systems) or use the interruption of a light source to detect and estimate the size of particles (e.g. the optical plankton counter). The first attempts to quantify plankton optically appear to have been made in the 1950s using a beam of light projected into the chamber from a 300 W mercury vapor lamp and a Focabell camera (Orion Camera, Tokyo).

Image-forming Systems Mounted on Non-opening/Closing Nets

In the 1980s, a 35 mm still camera with a high-capacity film magazine in front of the cod-end of a plankton net attached to a rigid frame was used to take *in situ* silhouette photographs of zooplankton as they passed into the cod-end. This was a field application of the laboratory-based silhouette photography system developed in the late 1970s. The camera provided a series of photographic images at points along the trajectory of the net separated by <1 m. In the development of the ichthyoplankton recorder, the still camera was replaced with a video

Figure 7 Some examples of multiple net plankton sampling systems. (A) The RMT 1 + 8. (Reproduced with permission from Baker, 1973.) (B) The Bé net. (Reproduced with permission from Bé, 1959.) (C) The Multinet. (Photograph courtesy of B. Niehof.) (D) The BIONESS. (Photograph courtesy of P. Wiebe, 1993.) (E) The 1 m^2 MOCNESS. (Photograph courtesy of Wiebe, 1998.)

camera, which was located in front of the cod-end of a high-speed Gulf V-type net (*Nackthai*). It had an estimated horizontal spatial resolution of 3 cm. One consequence of going from camera film to video tape was a loss of image resolution.

Stand-alone Image-forming Systems

The video plankton recorder (VPR) was developed in the early 1990s as a towed instrument capable of imaging zooplankton within a defined volume of water (**Figure 9A**). The original VPR had four video cameras; each camera imaged concentrically located volumes of water ranging from 1 ml to 1000 ml, but it has been modified to a one- or two-camera system. It has been possible to image undisturbed animals in their natural orientations. The current VPR image processing system is capable of digitizing each video field in real time and scanning the fields for targets using user-defined search criteria for

Table 2 MOCNESS system dimensions and weights

System	*Number of nets*	*Width of frame (m)*	*Height of frame (m)*	*Net width (m)*	*Mouth area at 45° towing angle (m)*	*Length of net (m)*	*Approx. weight in air (kg)*	*Rec. wire diameter (mm)*
MOCNESS-1/4	9	0.838	1.430	0.50	0.5	6.00	70	6.4
MOCNESS-1/4-Double	18/20	1.430	1.430	0.50	0.5	6.00	155	7.4
MOCNESS-1	9	1.240	2.870	1.00	1.0	6.00	150	7.4
MOCNESS-1-Double	18/20	2.560	2.870	1.00	1.0	6.00	320	12.1
MOCNESS-2	9	1.650	3.150	1.41	2.0	6.00	210	11.8
MOCNESS-4	6	2.140	4.080	2.00	4.0	8.44	460	11.8
MOCNESS-10	6	3.410	4.690	3.17	10.0	18.25	640	11.8
MOCNESS-20	6	5.500	7.300	4.47	20.0	14.50	940	17.3

The MOCNESS systems are denoted by the mouth area when being towed. Thus a MOCNESS-1/4 has a 0.25 m^2 mouth opening. The 'Double' systems have two sets of nets side-by-side in a single rigid framework. Nets can be opened and closed on one side and then opened and closed on the other.

brightness, focus, and size. The targets are identified using a zooplankton identification program to provide near-real-time maps of the zooplankton distributions.

A number of VPR-based systems are currently in operation or under development: a single-camera system is mounted on the BIOMAPER II vehicle (described below); an internally recording VPR has been constructed and used to quantify radiolarians and foraminiferans; and one has been mounted on a 1 m^2 MOCNESS net system to map the fine-scale distributions of the larval cod prey items. A moored system called the Autonomous Vertically Profiling Plankton Observatory (AVPPO) utilizes an internally recording, two-camera VPR, and has been deployed in coastal waters off New England.

Image resolution constraints inherent in the use of standard video formats have driven the development

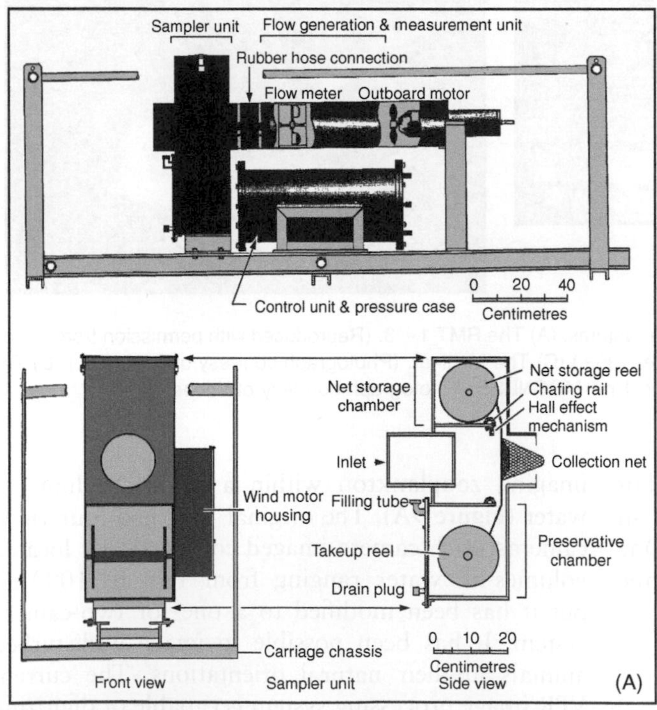

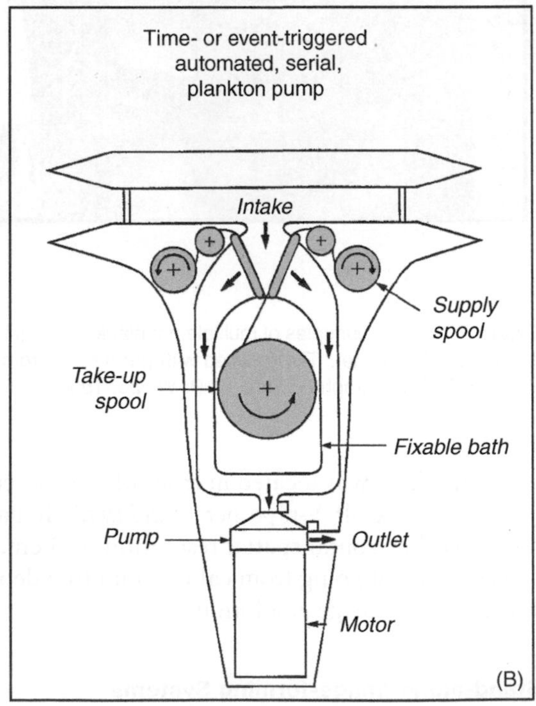

Figure 8 Two examples of moored plankton collecting systems. (A) A modified version of the O'Hara sampler. (Reproduced with permission from Lewis and Heckl, 1991.) (B) MASZP. (Reproduced with permission from Doherty *et al.* 1993.)

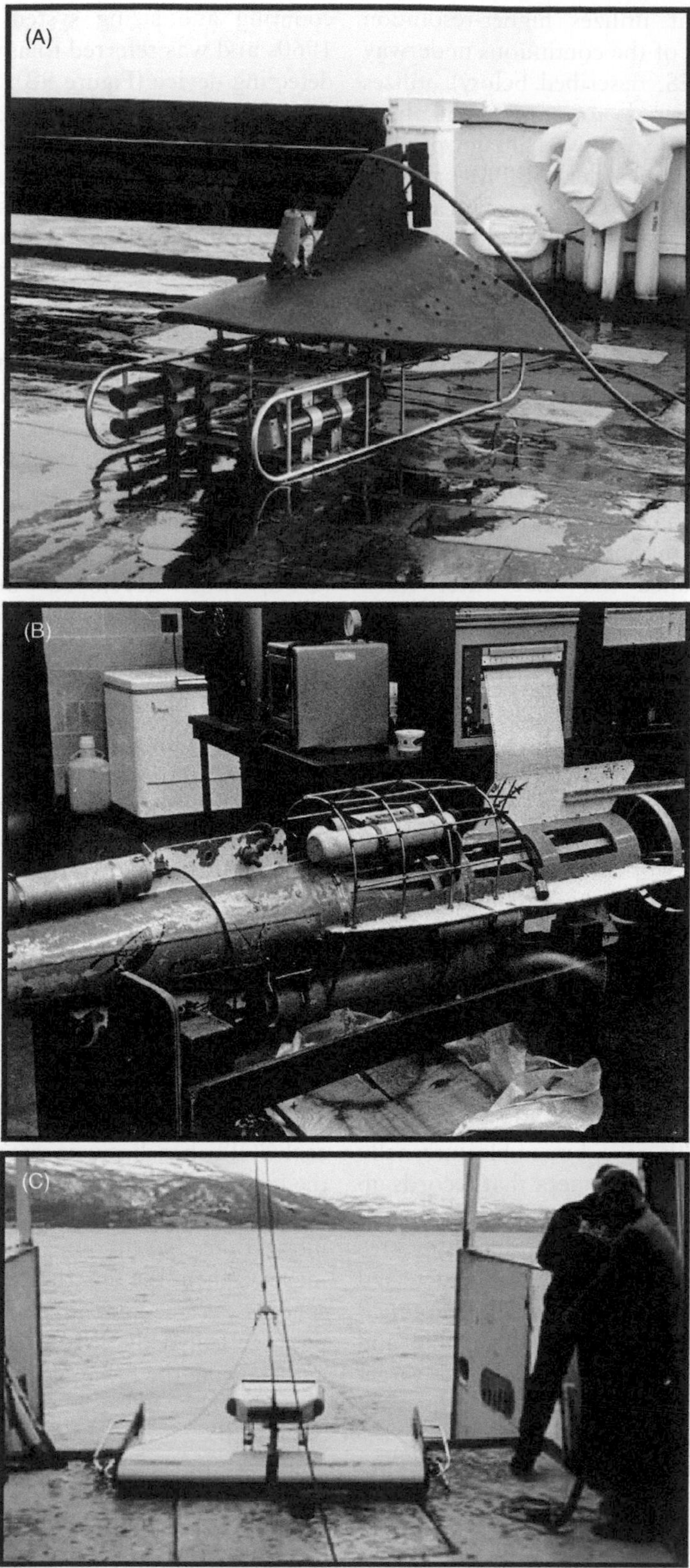

Figure 9 Examples of optical or electrical systems for collecting zooplankton data. (A) The VPR. (Photograph courtesy of P. Alatalo, 1999.) (B) The *in-situ* zooplankton detecting device. (Photograph courtesy of P. Wiebe, *c.* 1972.) (C) The optical plankton counter (OPC). (Photograph courtesy of M. Zhou, 2000.)

of optical systems that utilizes higher-resolution formats. A modification of the continuous underway fish egg sampler (CUFES, described below) utilizes a line-scanning digital camera to quantify the abundances of fish eggs. The shadowed image particle profiling and evaluation recorder (SIPPER) utilizes high-resolution digital line-scanning cameras to quantify zooplankton passing through a laser light sheet. The SIPPER has been mounted either on a towed vehicle called the high-resolution sampler (HRS) or an AUV.

The need for systems to quantify the abundance of 'marine snow' prompted development of profiling systems based on both still and video cameras. In the 1980s, a profiling system called the large amorphous aggregates (LAA) camera was constructed which employed a photographic camera and a pair of strobes to photograph marine aggregates. A video profiling instrument called the underwater video profiler (UVP) has been used to quantify the vertical distribution and size frequency of marine snow, and to examine the distributions of macrozooplankton. The UVP consists of a Hi-8 video camera imaging a collimated light sheet coupled with a CTD, data logger, and batteries. A profiling system called ZOOVIS recently has been developed around a high resolution (2048×2048 pixel) digital camera and CTD linked to a surface workstation via a fiber-optic cable. A color video camera has been mounted on the front of a Sea Owl II remotely operated vehicle (ROV) and used to quantify the vertical distribution of gelatinous zooplankton off the west coast of Sweden.

Still holographic imaging of plankton in a laboratory was first reported in 1966. It was refined in the 1970s to record movies of live plankton in the laboratory. In the 1990s, a submersible internally recording in-line holographic camera that records up to 300 holograms on a film emulsion was developed.

Many zooplankton produce or induce the production of bioluminescent light that can be detected with sensitive CCD cameras. One system is mounted on the Johnson SeaLink manned submersible and consists of an intensified silicon-intensified target (ISIT) video camera mounted on and aimed forward at a 1 m diameter transect screen to quantify the distribution, abundance, and identities of bioluminescent zooplankton.

Particle Detection Systems

Particle detection systems refer to non-image-forming devices that utilize interruption of an electrical current or a light beam to detect and estimate the size of a passing particle. The first *in situ* particle counting and sizing system appeared in the late 1960s and was referred to as the *in situ* zooplankton detecting device (**Figure 9B**). A shipboard version of the device was connected to a continuously pumped stream of water and employed to analyze spatial heterogeneity of zooplankton in surface waters in relation to chlorophyll fluorescence and temperature. A version of this conductive zooplankton counter was deployed aboard a Batfish towed vehicle in the 1980s.

A second group of particle detectors utilized photodetectors rather than changes in voltage. The Opto-Electronic Plankton Sizer was a laboratory-based system designed in the 1970s to automate the measurement of preserved plankton samples. The HIAC particle size analyzer was modified at the Lowestoft Laboratory during the late 1970s for plankton counting. The optical plankton counter (OPC) was developed during the mid-1980s (**Figure 9C**). This instrument measures changes in the intensity of a light beam that occur when a particle crosses the beam. The OPC has been mounted on a variety of towed platforms or in shore-based or shipboard applications. The OPC has also been incorporated into a shipboard device called the continuous underway fish egg sampling system (CUFES) which enumerates the distribution and abundance of fish eggs in surface waters. In spite of the prevalence of OPC systems in current use, interpretation of OPC data remains a subject of some controversy.

Optical Instruments for Nonquantitative Studies

The ecoSCOPE is an optical video-endoscope that enables direct observation of predator–prey interactions between juvenile fish and zooplankton. The ecoSCOPE has been operated from an ROV, from the keel of a sailing vessel, and in towed and moored modes, but the best recordings of predator/prey interactions have come from free-drifting deployments, when the instrument was hovering within schools of feeding juvenile herring. A software package called dynIMAGE animates sequential images keeping the fish and its prey in the middle of the viewing field.

Optical sensors can provide valuable ground-truthing for acoustical sensors. In the 1990s, a megapixel digital still camera was mounted on a FishTV sonar array and the resulting system was named the Optical-Acoustical Submersible Imaging System (OASIS). In this system, high acoustic returns are used to trigger the camera taking a picture of the acoustical target. An analog video camera aimed at the focal point of an acoustic array mounted on the front of a MAXRover ROV has been used to take pictures of individual zooplankton passing through the acoustic beam.

High-frequency Acoustics

High-frequency acoustics (≥38–1000 kHz) provide the foundation for another class of tools to study zooplankton. The utility of the acoustic systems derives from their ability to operate with high ping rates and precision range-gating. Mapping planktonic distributions on a wide range of space and timescales is becoming possible because of the continued development of acoustics systems and appropriate ground-truthing methods. There are two fundamental measurements: volume backscattering (integration of the energy return from all individuals in a given ensonified volume, i.e. echo integration) and target strength (echo strength from an individual). Statistical procedures have been developed to estimate animal assemblage size distribution using the data from single-beam transducers. In some cases, it is possible to extract estimates of animal target strength distribution in addition to volume backscattering from a series of single-beam transducers operating at different frequencies. Multibeam acoustical systems provide a direct means of determining individual target strength (TS). The two current designs, dual-beam and split-beam, both provide a hardware solution to the problem of TS determination.

The Current State of Plankton Sampling Systems

The diversity of zooplankton samplers in use today reflects the fact that no single collection system adequately samples all zooplankton. Non-opening/closing nets, such as the WP2, the modified Juday net, and the Bongo net, are used in large ocean surveys. Simple, double-messenger opening/closing nets similar to those developed in the first half of the last century are still manufactured and used. The Multinet, RMT 1+8, BIONESS, and MOCNESS are widely used multiple-net systems that also carry additional sensors to measure other water properties. Plankton pumps are also being used, especially to collect micro-zooplankton.

The advent of high-speed computers and towing cables with optical fibers and electrical conductors have enabled development of multi-sensor towed systems which provide real-time data while the instrument package is deployed. The MOCNESS has been equipped with a high-frequency acoustic system for forward or sideways range-gated viewing (**Figure 10A**). An EG&G Edgerton model 205 camera and aflash light were mounted on the top of a modified MOCNESS and on the top of BIONESS to take black and white photographs about 2 m in front of the net mouth. The BIONESS has also been equipped with an OPC and video lighting system, and used in conjunction with an echosounder.

The BIo-Optical Multi-frequency Acoustical and Physical Environmental Recorder – BIOMAPER II – was developed to conduct high-speed, large-area surveys of zooplankton and environmental property distributions to depths of 500 m (**Figure 10B**). Mounted inside are a multi-frequency sonar (upwards-looking and downwards-looking pairs of transducers operating at five frequencies: 43, 120, 200, 420, and 1000 kHz), an environmental sensor package (CTD, fluorometer, transmissometer), and several other bio-optical sensors (down- and upwelling spectral radiometers, spectrally matched attenuation, and absorption meters). A single-camera video plankton recorder (VPR) system is mounted above and just forward of the nose piece. The lower four acoustical frequencies involve split-beam technology and are able to make target strength and echo integration measurements.

A variety of vehicles have been built that actively change their vertical position without changing the towing wire length. Examples for surveying zooplankton include the undulating oceanographic recorder and SeaSoar equipped with optical (VPR and OPC) and/or acoustical (the Tracor Acoustical Profiling System, TAPS). Remotely operated vehicles (ROVs) have also been equipped with acoustical and video systems to study zooplankton. A SeaRover ROV was equipped with the same dual-beam acoustic system and environmental sensors. A VPR rigged to provide 3-D images of plankton and an environmental sensor package (temperature, conductivity, pressure, fluorescence) were mounted on the front of the ROV JASON and on the SeaRover ROV (**Figure 10C**). FishTV (FTV) has been used on a Phantom IV ROV and a combination of acoustics and video has been used on the front of a MAXRover ROV. Dual-beam acoustics (420 and 1000 kHz) have also been deployed on the DSRV Johnson SeaLink.

Future Developments

The future promises vastly increased application of remote sensing techniques and sensor development, and real-time data telemetry, processing, and display. Three-dimensional (space) and four-dimensional visualization (space and time) of biological and

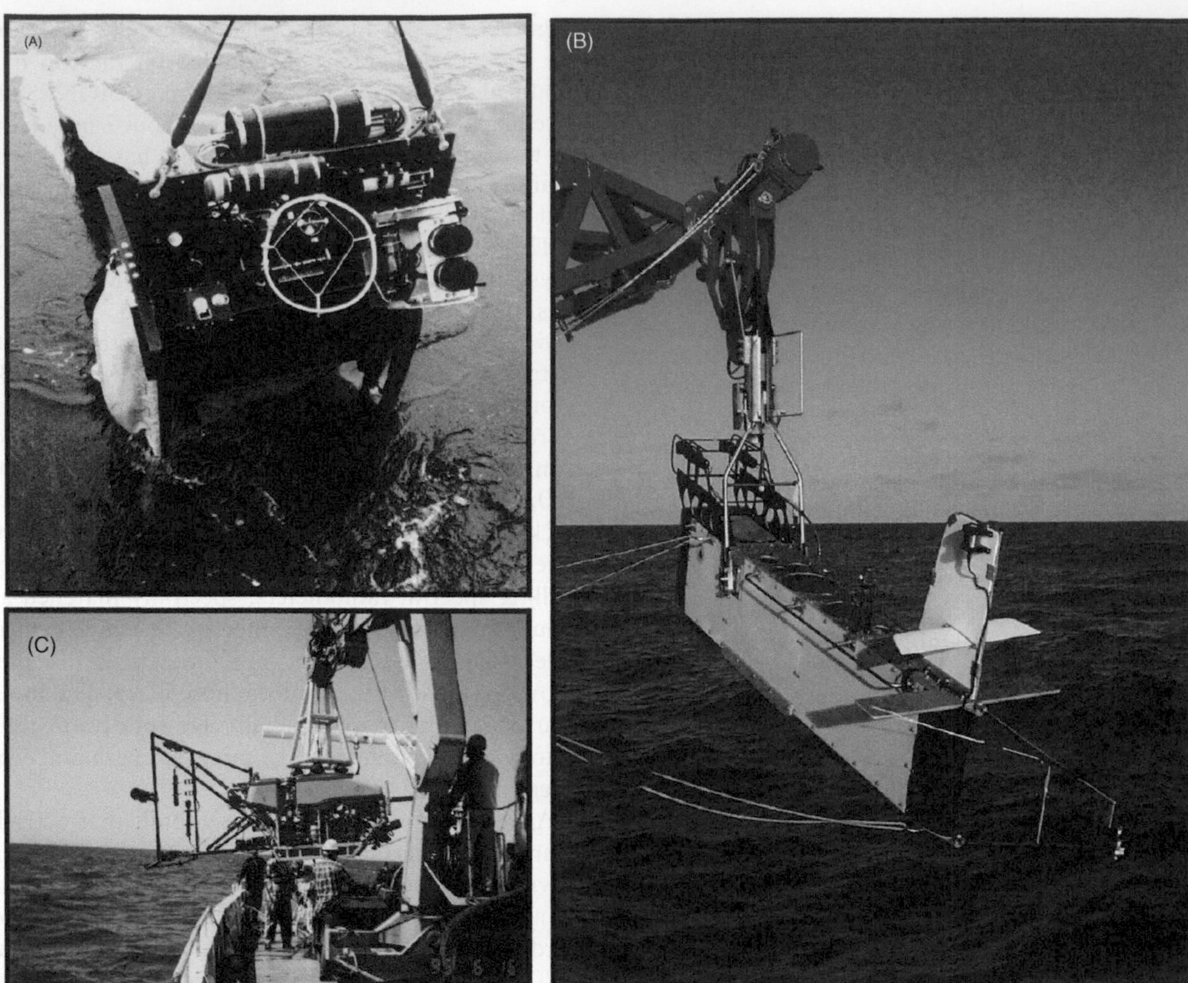

Figure 10 Examples of multi-sensor plankton sampling systems. (A) MOCNESS with a dual-beam acoustic system. (Photograph courtesy of P. Wiebe, 1994.) (B) BIOMAPER-II. (Photograph courtesy of P. Wiebe, 1999.) (C) The JASON-ROV with 3-D VPR system. (Photograph courtesy of P.Alatalo, 1995.)

acoustic data are also an increasingly important aspect of data processing. For a number of research programs today, the development of an image of the spatial arrangement of organisms is but the first step in efforts to study and understand their relationships to each other and to their environment. Thus, there is need for real-time 3-D and 4-D images.

Autonomous self-propelled vehicles (AUVs) have only recently begun to be used widely to gather-oceanographic data. The remote environmental measuring units (REMUS) are a new class of small AUVs which can carry an impressive array of environmental sensors including a VPR. Another class of autonomous vehicles is epitomized by the autonomous benthic explorer (ABE), which is equipped with precise navigation and control systems that enable it to descend to a worksite, navigate preset tracklines or terrain-follow, and find a docking station. A much larger AUV which has been employed for biological studies is the Autosub-1 that carries a gyrocompass, ADCP, an echosounder, and acoustic telemetry and surface radio electronics. It can be programmed to run a geographically based course using GPS surface positions and dead reckoning.

The autonomous Lagrangian circulation explorer (ALACE) and the more recently developed profiling version (PALACE) floats that carry temperature and conductivity probes are vertically migrating neutrally buoyant drifters. They track the movements of water at depths between the surface and 1000–2000 m depth. Hundreds to thousands of the PALACE floats will be deployed over the next few years and it is expected that they will become a mainstay in the Global Ocean Observing System (GOOS). The next generation of neutrally buoyant floats is an autonomous glider named SPRAY. SPRAY will be able to sail along specific preprogrammed tracklines. A further step in their development is to provide biological instrumentation to complement the physical sensors.

High-resolution optical systems, suchas the VPR, combined with computer-based identification programs can now provide higher level taxa identifications in near-real time. Classification of species using acoustic signatures is less well developed and it now seems unlikely that the technology to develop species-specific acoustic signatures will be developed soon. Molecularly based species identification is likely to make significant strides in the next decade. It is now conceivable that this information will enable simultaneous analysis, identification, and quantification of all species occurring in a zooplankton sample.

See also

Continuous Plankton Recorders. Grabs for Shelf Benthic Sampling.

Further Reading

Harris RP, Wiebe PH, Lenz J, Skjoldal HR, and Huntley M (eds.) (2000) *ICES Zooplankton Methodology Manual.* New York: Academic Press.

Kofoid CA (1991) On a self-closing plankton net for horizontal towing. *University of California Publications in Zoology* 8: 312–340.

Miller CB and Judkins DC (1981) Design of pumping systems for sampling zooplankton with descriptions of two high-capacity samplers for coastal studies. *Biol Oceanogr* 1: 29–56.

Omori M and Ikeda Y (1976) *Methods in Marine Zooplankton Ecology.* New York: John Wiley & Sons.

Schulze PC, Strickler JR, Bergström BI, *et al.* (1992) Video systems for in situ studies of zooplankton. *Arch Hydrobiol Beih Ergebn Limnol* 36: 1–21.

Sprules WG, Bergström B, Cyr H, *et al.* (1992) Non-video optical instruments for studying zooplankton distribution and abundance. *Arch Hydrobiol Beih Ergebn Limnol* 36: 45–58.

Tranter DJ (ed.) (1968) *Part I. Reviews on Zooplankton Sampling Methods. Monographs on Oceanographic Methodology, Zooplankton Sampling.* UNESCO

CONTINUOUS PLANKTON RECORDERS

A. John, Sir Alister Hardy Foundation for Ocean Science, Plymouth, UK
P. C. Reid, SAHFOS, Plymouth, UK

Introduction

The Continuous Plankton Recorder (CPR) survey is a synoptic survey of upper-layer plankton covering much of the northern North Atlantic and North Sea. It is the longest running and the most geographically extensive of any routine biological survey of the oceans. Over 4 000 000 miles of towing have resulted in the analysis of nearly 200 000 samples and the routine identification of over 400 species/groups of plankton. Data from the survey have been used to study biogeography, biodiversity, seasonal and interannual variation, long-term trends, and exceptional events. The value of such an extensive time-series increases as each year's data are accumulated. Some recognition of the importance of the CPR survey was achieved in 1999 when it was adopted as an integral part of the Initial Observing System of the Global Ocean Observing System (GOOS).

History

The CPR prototype was designed by Alister Hardy for operation on the 1925–27 Discovery Expedition to the Antarctic, as a means of overcoming the problem of patchiness in plankton. It consisted of a hollow cylindrical body tapered at each end, weighted at the front and with a diving plane, horizontal tail fins, and a vertical tail fin with a buoyancy chamber on top (**Figure 1A**). Hardy designed a more compact version with a smaller sampling aperture for use on merchant ships and this was first deployed on a commercial ship in the North Sea in September 1931 (**Figure 1B**). During the 1980s the design was modified further to include a box-shaped double tail-fin that provides better stability when deployed on the faster merchant ships of today (**Figure 1C**). The space within this tail-fin is used in some machines to accommodate physical sensors and flowmeters. The normal maximum tow distance for a CPR is approximately 450 nautical miles (834 km).

By the late 1930s there were seven CPR routes in the North Sea and one in the north-east Atlantic; in 1938 CPRs were towed for over 30 000 miles. After a break for the Second World War, the survey restarted in 1946 and expanded into the eastern North Atlantic. Extension of sampling into the western North Atlantic took place in 1958. The survey reached its greatest extent from 1962 to 1972 when CPRs were towed for at least 120 000 nautical miles annually. Sampling in the western Atlantic, which had been suspended due to funding problems in 1986, recommenced in 1991 and is still ongoing. **Figure 2A** shows the extent of the survey in 1999.

Initially based at the University College of Hull, the survey moved to Leith, Edinburgh in 1950 under the management of the Scottish Marine Biological Association (now the Scottish Association for Marine Science). In 1977 it finally moved to Plymouth as part of the Institute for Marine Environmental Research (now Plymouth Marine Laboratory). After a short period of uncertainty in the late 1980s, when the continuation of the survey was threatened, the Sir Alister Hardy Foundation for Ocean Science (SAHFOS) was formed in November 1990 to operate the survey. Since 1931 more than 200 merchant ships, ocean weather ships, and coastguard cutters – known as 'ships of opportunity' – from many nations have towed CPRs in a voluntary capacity to maintain the survey. The Foundation is greatly indebted to the captains and crews of all these towing ships and their shipping and management companies, without whom the survey could not continue.

During the 1990s CPRs were towed by SAHFOS in several other areas, including the Mediterranean (1998–99), the Gulf of Guinea (1995–99), the Baltic (1998–99), and the Indian Ocean (1999). A separate survey by the National Oceanic and Atmospheric Administration/National Marine Fisheries Service using CPRs along the east coast of the USA off Narragansett has been running since 1974; CPRs are currently towed on two routes in the Middle Atlantic Bight. Following a successful 2000 mile trial tow in the north-east Pacific from Alaska to California in July–August 1997, a 2-year survey by SAHFOS using CPRs in the north-east Pacific started in March 2000. In addition to five tows per year on the Alaska–California route, there is one 3000 mile tow annually east–west from Vancouver to the north-west Pacific (**Figure 2B**). A 'sister' survey, situated in the Southern Ocean south of Australia between 60°E and 160°E, is operated by the Australian Antarctic Division. In this survey CPRs have been deployed since the early 1990s on voyages between Tasmania and stations in the Antarctic.

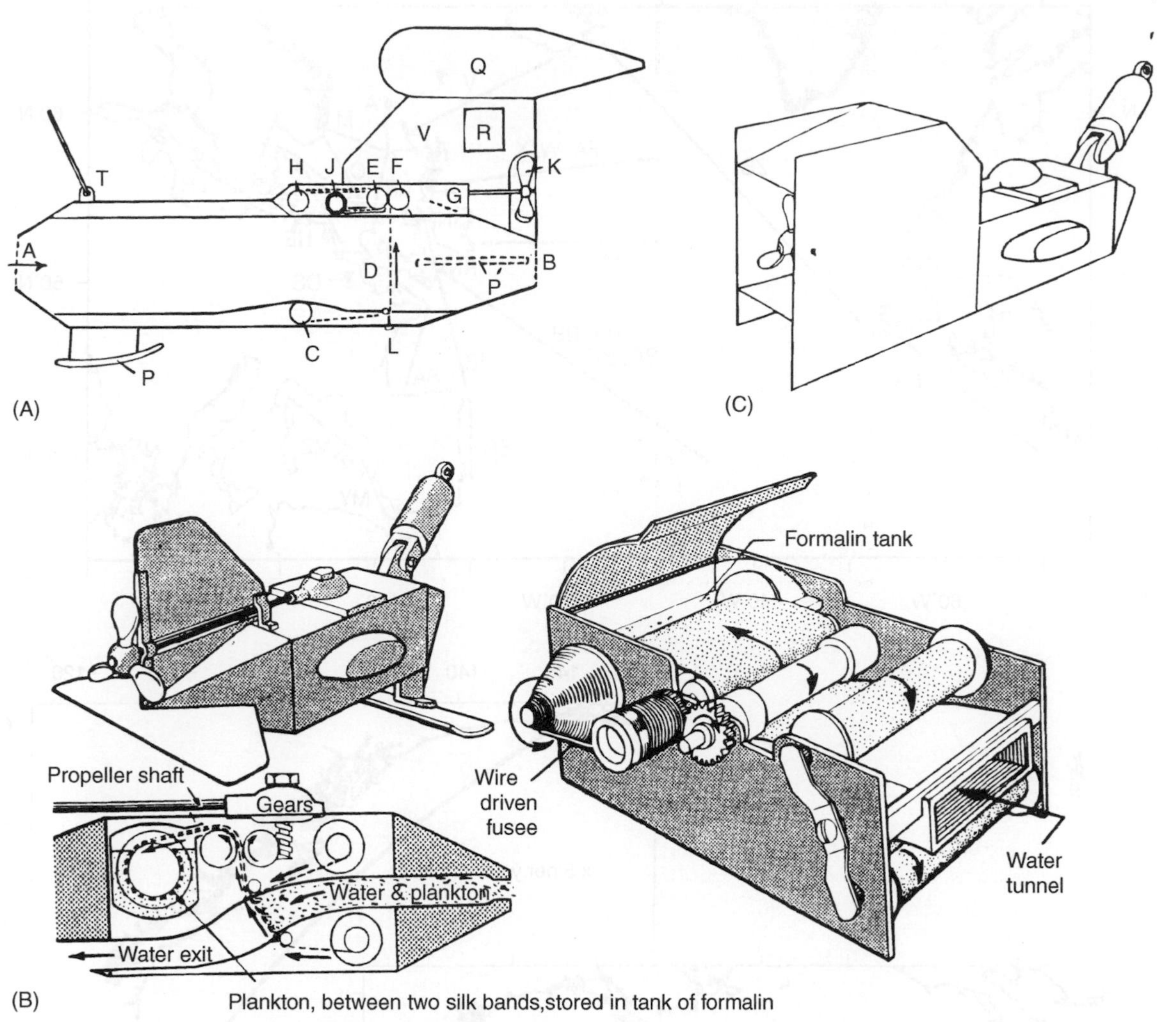

Figure 1 (A) Diagram of the first Continuous Plankton Recorder used on 'Discovery'. (Reproduced with permission from Hardy, 1967). (B) The 'old' CPR, used up to around 1983, showing the internal filtering mechanism. (C) The CPR in current use, with the 'box' tail-fin.

As the operator of a long-term international survey, which has sampled in most of the world's oceans, SAHFOS regularly trains its own staff in plankton identification. In recent years SAHFOS has also trained scientists from the following 10 countries: Benin, Cameroon, Côte d'Ivoire, France, Finland, Ghana, Italy, Nigeria, Thailand, and the USA.

The Database and Open Access Data Policy

The CPR database is housed on an IBM-compatible PC and stored in a relational Microsoft Access DATABASE system. Spatial and temporal data are stored for every sample analysed by the CPR survey since 1948. This amounts to >175 000 samples, with around 400 more samples added per month. There are more than two million plankton data points in the database, which also contains supporting information, including sample locations, dates and times of samples, a taxon catalog, and analyst details. In the near future it will also hold additional conductivity, temperature, and depth (CTD) data. Routine processing procedures ensure that, despite various operational difficulties, the previous year's data are usually available in the database within 9 months.

In 1999 SAHFOS adopted a new open access data policy, i.e. data are freely available to all users worldwide, although a reasonable payment may be incurred for time taken to extract a large amount of data. The only stipulation is that users have to sign a SAHFOS Data Licence Agreement. Details of the database can be found on the web site: http://www.npm.ac.uk/sahfos/. This site advertises the availability of data and allows requests for data to be made easily.

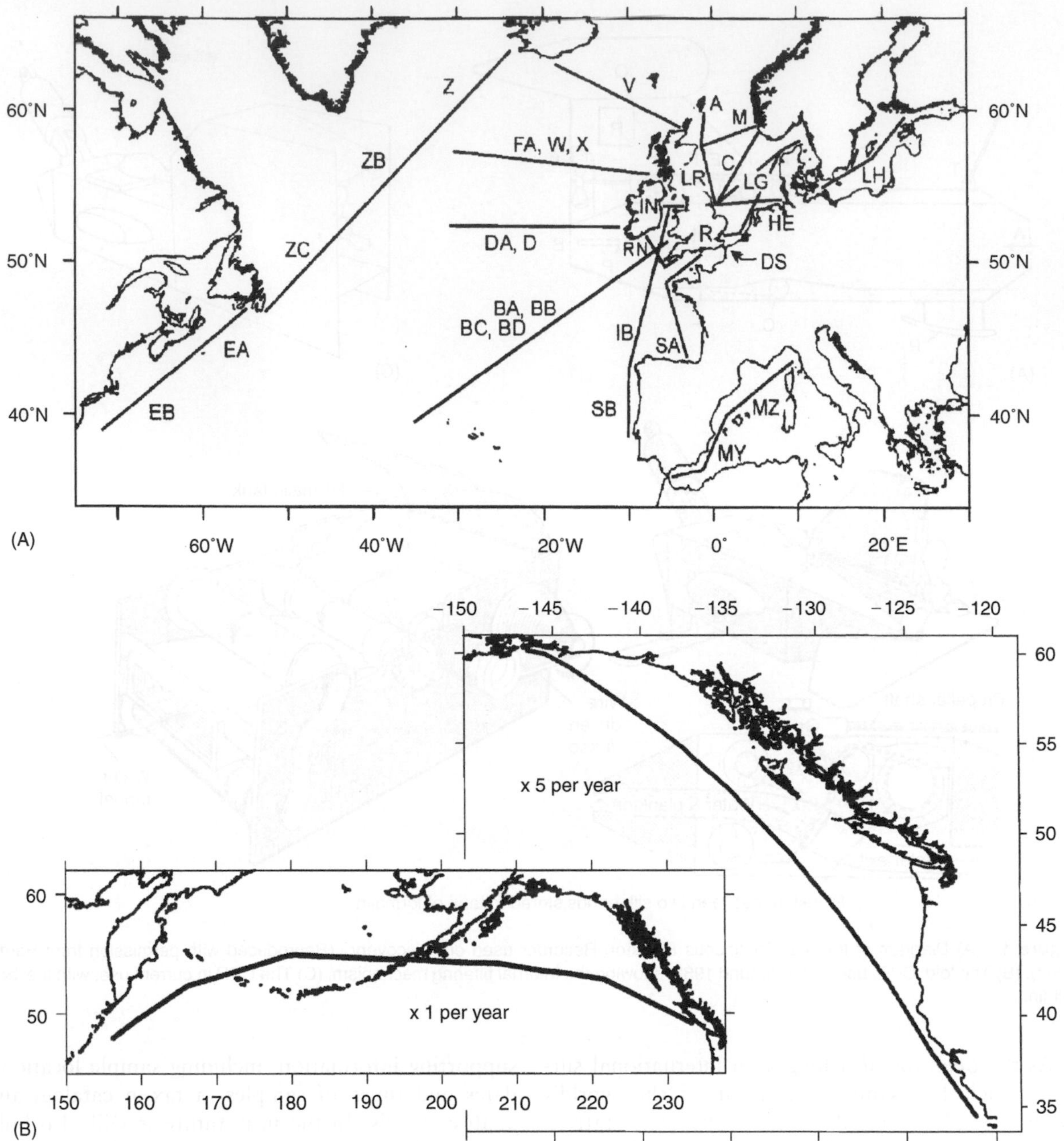

Figure 2 (A) CPR routes in 1999/2000. (B) CPR routes in the Pacific Ocean towed in 2000 and 2001.

The CPR Bibliography, which is available on the SAHFOS web site, lists over 500 references using results from the survey. During the early years many of the papers based on CPR data were published in the 'in-house' journal *Hull Bulletins of Marine Ecology*, which from 1953 onwards became the *Bulletins of Marine Ecology*; this was last published in 1980.

Methods

Merchant ships of many nations tow CPRs each month along 20–25 standard routes (**Figure 2A**) at a depth of 6–10 m. Water enters the CPR through a 12.7 mm square aperture and travels down a tunnel that expands to a cross-section of 50 × 100 mm, where it passes through a silk filtering mesh with a mesh size of approximately 280 μm. The movement of the CPR through the water turns a propeller that drives a set of rollers and moves the silk across the tunnel. At the top of the tunnel the filtering silk is joined by a covering silk and both are wound onto a spool located in a storage chamber containing formaldehyde solution. The CPRs are then returned to SAHFOS in Plymouth for examination. The green coloration of each silk is visually assessed by reference to a standard color scale; this is known as 'Phytoplankton Color' and gives a crude measure of total phytoplankton biomass. The silks are then cut into sections corresponding to 10 nautical miles (18.5 km) of tow and are distributed randomly to a team of 10–12 analysts. The volume of water filtered per 10-nautical-mile sample is approximately $3\,m^3$. Phytoplankton, small zooplankton (<2 mm in size) and larger zooplankton (>2 mm) are then identified and counted in a three-stage process. Over 400 different taxa are routinely identified during the analysis of samples and the recent expansion of the survey into tropical waters and the Pacific Ocean will certainly increase this figure.

A detailed and thorough quality control examination is carried out by the most experienced analyst on the completed analysis data. Apparently anomalous results are rechecked by the original analyst and the data are altered accordingly where necessary. This system ensures consistency of the data and acts as 'in-service' training for the less experienced analysts.

Instrumentation

On certain routes CPRs carry additional equipment to obtain physical data. In the past temperature has been recorded on certain routes in the North Sea using Braincon^{TM} recording thermographs, prototype electronic packages, and Aquapacks^{TM}. Aquapacks record temperature, conductivity, depth, and chlorophyll fluorescence. These are now deployed on CPR routes off the eastern coast of the USA, in the southern Bay of Biscay and, until November 1999, in the Gulf of Guinea. Vemco^{TM} minilogger temperature sensors are used on routes from the UK to Iceland, and from Iceland to Newfoundland. In order to measure flow rate through the CPR, electromagnetic flowmeters are used on some routes. Such recording of key physical and chemical variables simultaneously with abundance of plankton enhances our ability to interpret observed changes in the plankton.

Results and Applications of the Data

The long-term time-series of CPR data acts as a baseline against which to measure natural and anthropogenic changes in biogeography, biodiversity, seasonal variation, inter-annual variation, long-term trends, and exceptional events. The results have applications to studies of eutrophication and are increasingly being applied in statistical analysis of plankton populations and modeling. Some examples are given below.

Another possible application, in the context of the new Pacific CPR programme, is an inter-comparison with data from the CalCOFI Program, the only other existing decadal-scale survey in the world sampling marine plankton. This survey has taken monthly or quarterly net samples from 1949 to the present over an extensive grid of stations off the west coast of California. In the majority of samples the zooplankton has been measured only as displacement volume, rather than being identified to species, but concurrently measured physical and chemical data are more extensive.

Biogeography of Marine Plankton

Much of the early work of the survey focused on biogeography. Using Principal Component Analysis, Colebrook was able to distinguish five main geographical distribution patterns in the plankton – northern oceanic, southern oceanic, northern intermediate, southern intermediate, and neritic. Two closely related species of calanoid copepod – *Calanus finmarchicus* and *C. helgolandicus* – which co-occur in the North Atlantic and are morphologically very similar, show very different distributions (**Figure 3**). *C. finmarchicus* is a cold-water species whose center of distribution lies in the north-west Atlantic gyre and the Norwegian Sea ('northern oceanic'). In contrast, *C. helgolandicus* is a warm–temperate water species occurring in the Gulf Stream, the Bay of Biscay and the North Sea ('southern intermediate'). These different distribution patterns are reflected in their life histories; *C. finmarchicus* overwinters in deep waters off the shelf edge, whereas *C. helgolandicus* overwinters in shelf waters.

A new species of marine diatom, *Navicula planamembranacea* Hendey, was first described from CPR samples taken in 1962. The species was found to have a wide distribution in the western North Atlantic from Newfoundland to Iceland.

An atlas of distribution of 255 species or groups (taxa) of plankton recorded by the CPR survey between 1958 and 1968 was published by the Edinburgh Oceanographic Laboratory in 1973. An updated version of this atlas, covering more than 40 years of CPR data and over 400 taxa, is in preparation.

Phytoplankton, Zooplankton, Herring, Kittiwake Breeding Data, and Weather

A study in the north-eastern North Sea found that patterns of four time-series of marine data and weather showed similar long-term trends. Covering

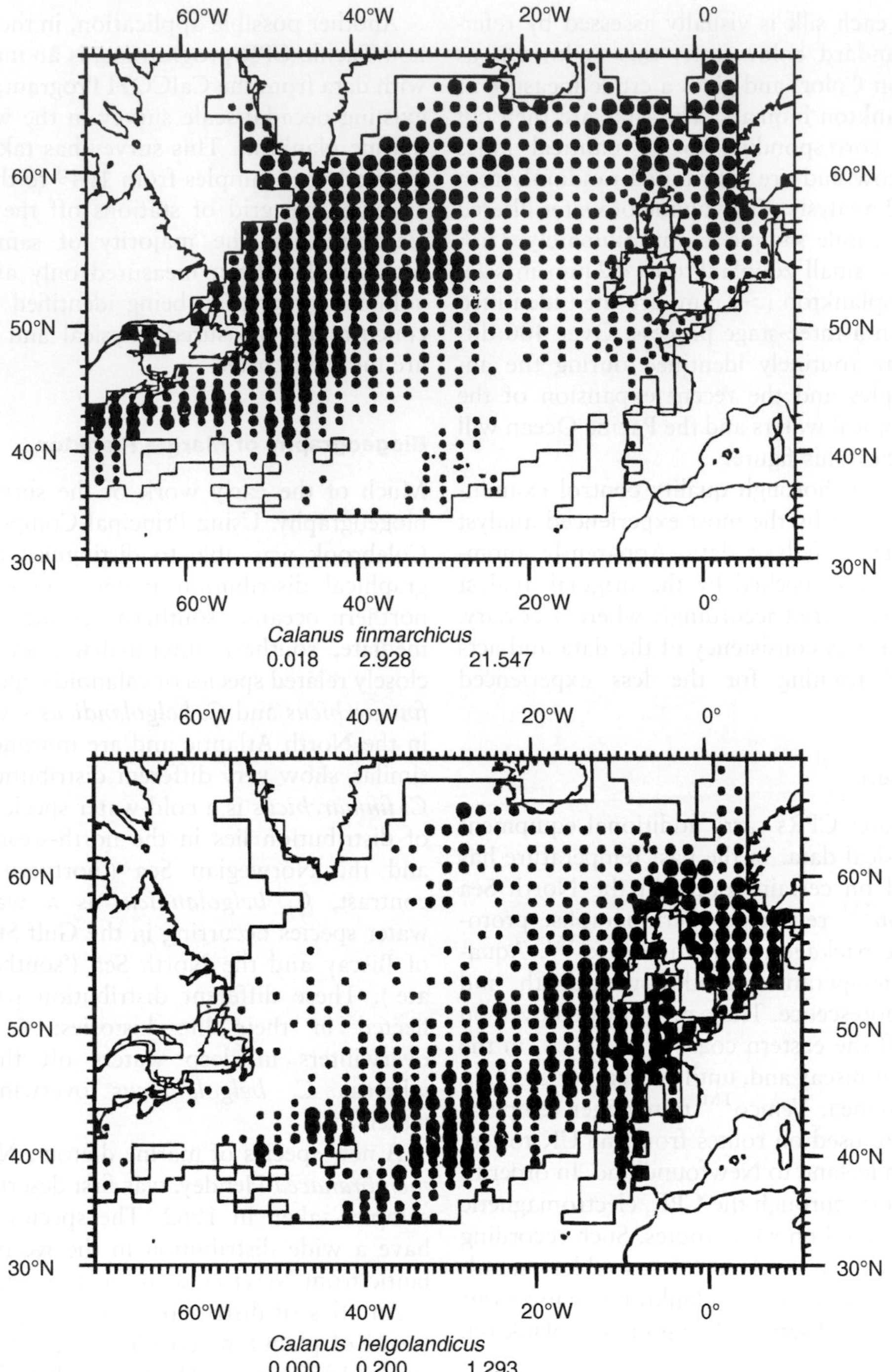

Figure 3 Distribution of *Calanus finmarchicus* and *C. helgolandicus* recorded in CPR samples from 1958 to 1994.

the period 1955–87, these trends were found in the abundance of phytoplankton and zooplankton (as measured by the CPR), herring in the northern North Sea, kittiwake breeding success (laying date, clutch size, and number of chicks fledged per pair) at a colony on the north-east coast of England, and the frequency of westerly weather (Figure 4). The mechanisms behind the parallelism in these data over the 33-year period are still not fully understood.

Calanus and the North Atlantic Oscillation

The North Atlantic Oscillation (NAO) is a large-scale alternation of atmospheric mass between

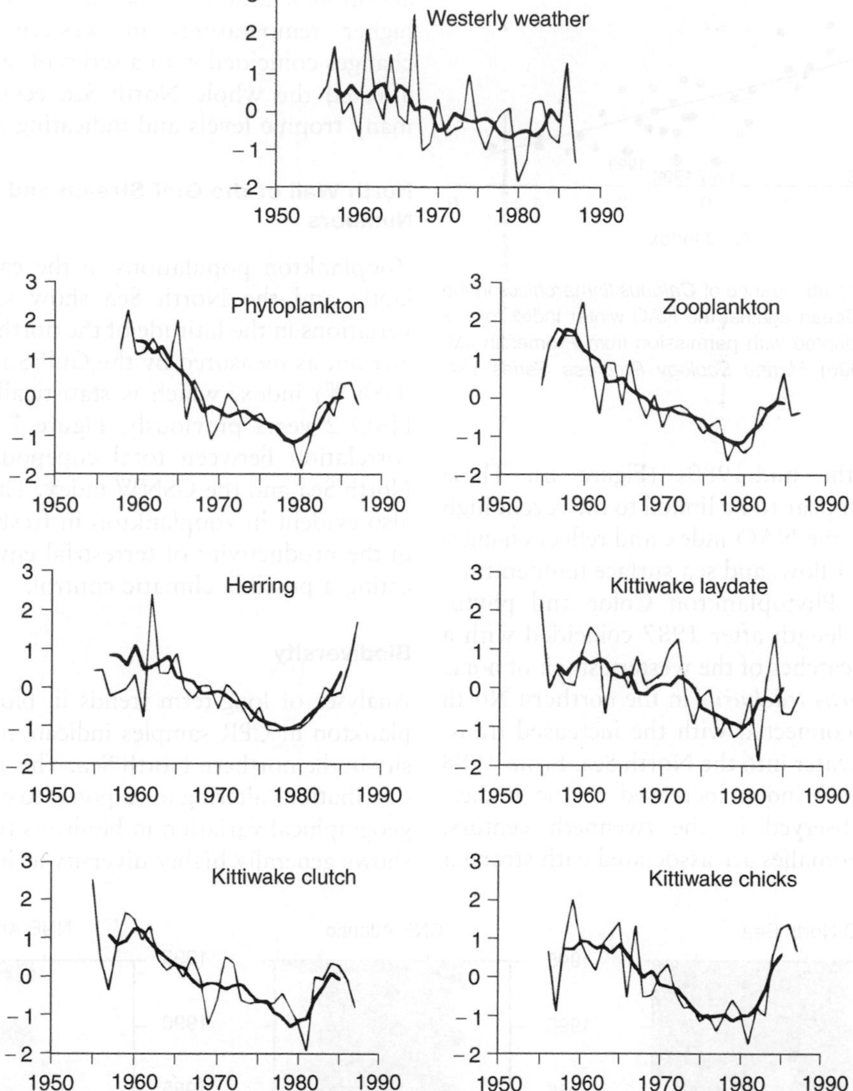

Figure 4 Standardized time-series and 5-year running means for frequency of westerly weather, and for abundances of phytoplankton, zooplankton, herring, and three parameters of kittiwake breeding (laying date, clutch size, and number of chicks fledged per pair), from 1955 to 1987. (Reproduced with permission from Aebischer NJ *et al.* (1990) *Nature* 347: 753–755.)

subtropical high surface pressure, centred on the Azores, and subpolar low surface pressures, centred on Iceland. The NAO determines the speed and direction of the westerly winds across the North Atlantic, as well as winter sea surface temperature. The NAO index is the difference in normalized sea level pressures between Ponta Delgadas (Azores) and Akureyri (Iceland). There is a close association between the abundance of *Calanus finmarchicus* and *C. helgolandicus* in the north-east Atlantic and this index (**Figure** 5). At times of heightened pressure difference between the Azores and Iceland, i.e. a high, positive NAO index, there is low abundance of *C. finmarchicus* and high abundance of *C. helgolandicus*; during a low, negative NAO index the reverse is true. However, since 1995 this strong *Calanus*/NAO relationship has broken down and the causes of this are presently unknown. It suggests a change in the nature of the link between climate and plankton in the north-east Atlantic.

North Sea Ecosystem Regime Shift

Recent studies have shown changes in CPR Phytoplankton Color, a visual assessment of chlorophyll, for the north-east Atlantic and the North Sea. In the central North Sea and the central north-east Atlantic an increased season length was strikingly evident after the mid-1980s. In contrast, in the north-east Atlantic north of 59°N Phytoplankton Color

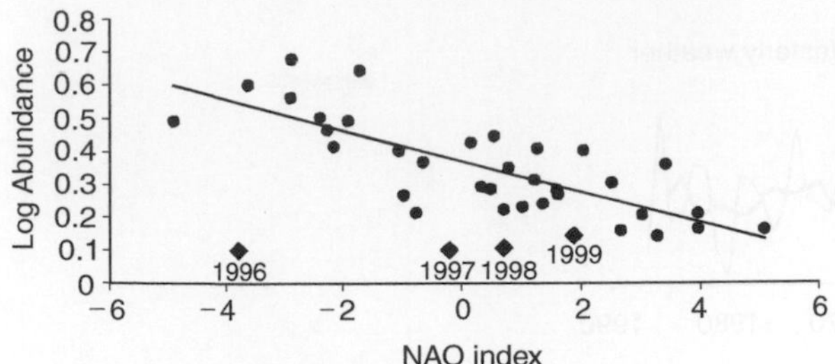

Figure 5 Annual log abundance of *Calanus finmarchicus* in the north-east Atlantic Ocean against the NAO winter index for the period 1962–99. (Adapted with permission from Fromentin JM, and Planque B (1996) *Marine Ecology Progress Series* 134: 111–118.)

declined after the mid-1980s (**Figure 6**). These changes in part appear to be linked to the recent high positive phase of the NAO index and reflect changes in mixing, current flow, and sea surface temperature. The increase in Phytoplankton Color and phytoplankton season length after 1987 coincided with a large increase in catches of the western stock of horse mackerel *Trachurus trachurus* in the northern North Sea, apparently connected with the increased transport of Atlantic water into the North Sea. From 1988 onwards the NAO index increased to the highest positive level observed in the twentieth century. Positive NAO anomalies are associated with stronger and more southerly tracks of the westerly winds and higher temperatures in western Europe. These changes coincided with a series of other changes that affected the whole North Sea ecosystem, affecting many trophic levels and indicating a regime shift.

North Wall of the Gulf Stream and Copepod Numbers

Zooplankton populations in the eastern North Atlantic and the North Sea show similar trends to variations in the latitude of the north wall of the Gulf Stream, as measured by the Gulf Stream North Wall (GSNW) index, which is statistically related to the NAO 2 years previously. **Figure** 7 shows the close correlation between total copepods in the central North Sea and the GSNW index. This relationship is also evident in zooplankton in freshwater lakes and in the productivity of terrestrial environments, indicating a possible climatic control.

Biodiversity

Analyses of long-term trends in biodiversity of zooplankton in CPR samples indicate increases in diversity in the northern North Sea. This may be related to distributions altering in response to climatic change as geographical variation in biodiversity of the plankton shows generally higher diversity at low latitudes than

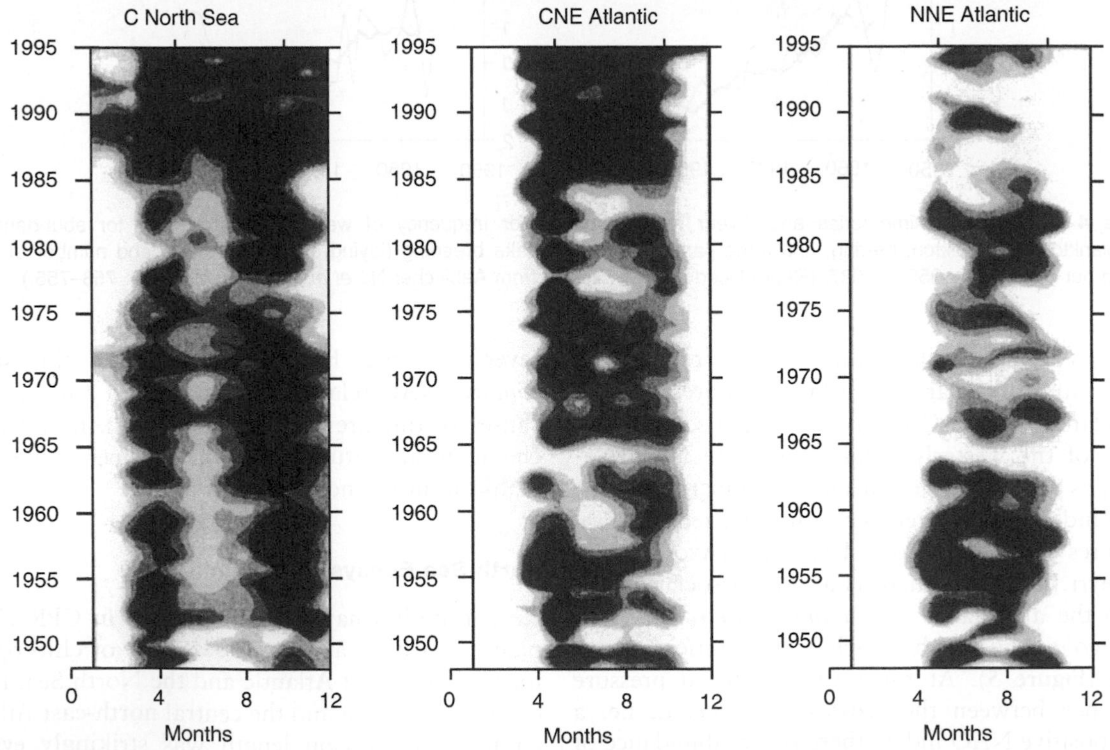

Figure 6 Contour plots of mean monthly Phytoplankton Color during 1948–95 for the central North Sea, and for the central and northern north-east Atlantic. (Reproduced with permission from Reid PC *et al.* (1998) *Nature* 391: 546.)

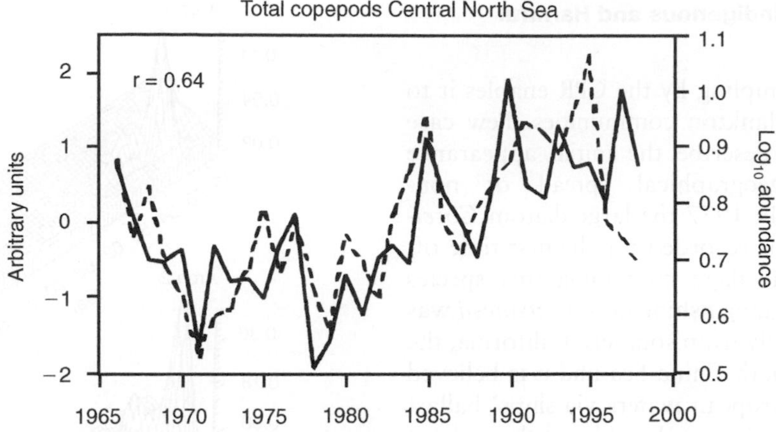

Figure 7 The latitude of the Gulf Stream (the GSNW index 'arbitrary units', broken line) compared with the abundance of total copepods in the central North Sea (solid line). Adapted with permission from Taylor AH *et al.* (1992) *Journal of Mar. Biol. Ass.* UK 72: 919–921.

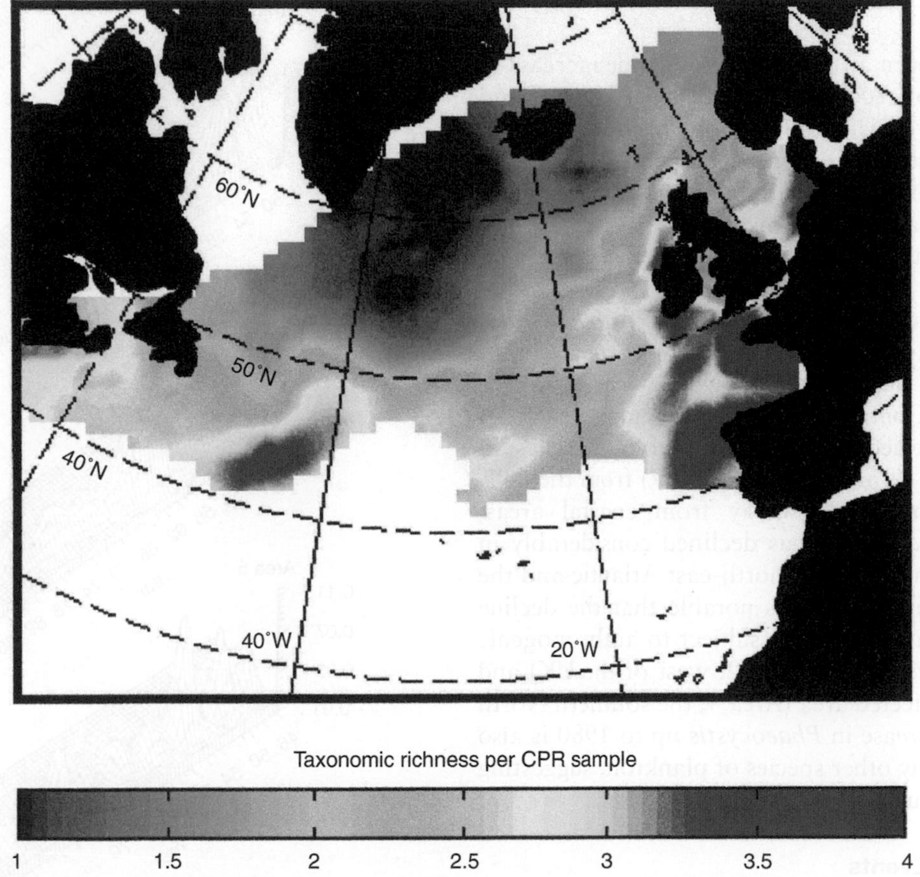

Figure 8 The biodiversity (taxonomic richness) of calanoid copepods in the CPR sampling area. (Adapted with permission from Beaugrand *et al.* (2000) *Marine Ecology Progress Series* 204: 299–303.)

at high latitudes. Calanoid copepods are the dominant zooplankton group in the North Atlantic and the large data set from the CPR survey has been used to map their diversity. This has demonstrated a pronounced local spatial variability in biodiversity. Higher diversity was found in the Gulf Stream extension, the Bay of Biscay, and along the southern part of the European shelf. Cold water south of Greenland, east of Canada, and west of Norway was found to have the lowest diversity (**Figure 8**).

Monitoring for Nonindigenous and Harmful Algal Blooms

The regularity of sampling by the CPR enables it to detect changes in plankton communities. Few case histories exist that describe the initial appearance and subsequent geographical spread of nonindigenous species. In 1977 the large diatom *Coscinodiscus wailesii* was recorded for the first time off Plymouth, when mucilage containing this species was found to be clogging fishing nets. *C. wailesii* was previously known only from southern California, the Red Sea, and the South China Sea and it is believed that it arrived in European waters via ships' ballast water. Since then the species has spread throughout north-west European waters and has become an important contributor to North Sea phytoplankton biomass, particularly in autumn and winter. Such introduced species can, on occasions, have considerable ecological and economic effects on regional ecosystems.

There has been an apparent worldwide increase in the number of recorded harmful algal blooms and the CPR survey is ideally placed to monitor such events. The serious outbreak of paralytic shellfish poisoning that occurred in 1968 on the north-east coast of England was shown by CPR sampling to have been caused by the dinoflagellate *Alexandrium tamarense*.

Increased nutrient inputs into the North Sea since the 1950s have been linked with an apparent increase in the haptophycean alga *Phaeocystis*, particularly in Continental coastal regions where it produces large accumulations of foam on beaches. In contrast, long-term records (1946–87) from the CPR survey, which samples away from coastal areas, show that *Phaeocystis* has declined considerably in the open-sea areas of the north-east Atlantic and the North Sea (**Figure 9**). It is notable that the decline occurred both in areas not subject to anthropogenic nutrient inputs (Areas 1 and 2, west of the UK) and in the most affected area (Area 4, the southern North Sea). This decrease in *Phaeocystis* up to 1980 is also shown by many other species of plankton, suggesting a common causal relationship.

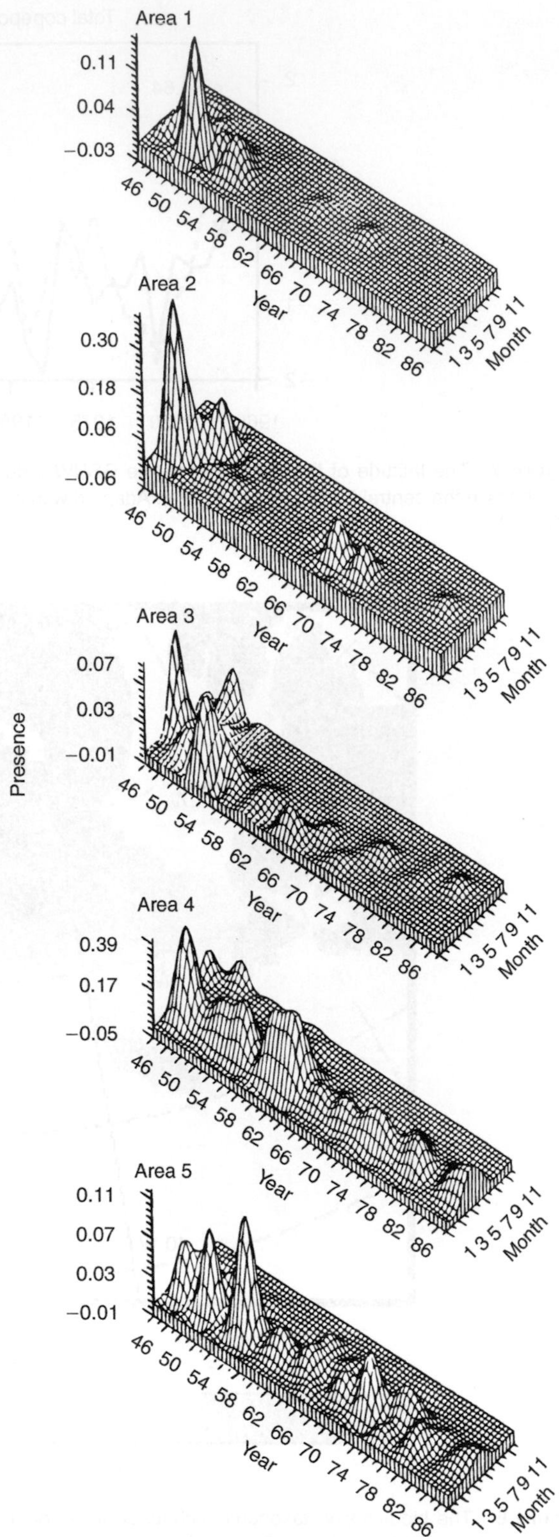

Figure 9 Presence of *Phaeocystis* in five areas of the north-east Atlantic Ocean and the North Sea. Data are plotted for each month for 1946–87 inclusive. (Reproduced with permission from Owens NJP *et al.* (1989) *Journal of Mar. Biol. Ass. UK* 69: 813–821.)

Exceptional Events

Doliolids are indicators of oceanic water and in CPR samples are normally found to the west and south-west of the British Isles; they occur only sporadically in the North Sea and are rarely recorded in the central or southern North Sea. On two occasions in recent years, in October–December 1989 and September–October 1997, the doliolid *Doliolum nationalis* was recorded in CPR samples taken in the

German Bight, accompanied by other oceanic indicator species, suggesting a strong influx of north-east Atlantic water into the North Sea. Both these occasions coincided with higher than average sea surface temperature and salinities.

Summary and the Future

The long-term time-series of CPR data have been used in many different ways:

- mapping the geographical distribution of plankton
- a baseline against which to measure natural and anthropogenically forced change, including eutrophication and climate change
- linking of plankton and environmental forcing
- detecting exceptional events in the sea
- monitoring for newly introduced and potentially harmful species.

In the future new applications of CPR data may include:

- use as 'sea-truthing' for satellites
- regional assessment of plankton biodiversity
- regional studies of responses to climate change
- as input variables to predictive modeling for fish stock and ecosystem management
- for construction and validation of new models comparing ecosystems of different regional seas.

The CPR survey has gathered nearly 70 years of data on marine plankton throughout the North Atlantic Ocean, and has recently extended into the North Pacific Ocean. Alister Hardy's simple concept in the 1920s has succeeded in providing us with a unique and valuable long-term data set. There is increasing worldwide concern about anthropogenic effects on the marine ecosystem, including eutrophication, overfishing, pollution, and global warming. The data in the CPR time-series is being used more and more widely to investigate these problems and now plays a significant role in our understanding of global ocean and climate change.

See also

Diversity of Marine Species. Ecosystem Effects of Fishing. Large Marine Ecosystems. Phytoplankton Blooms.

Further Reading

Colebrook JM (1960) Continuous Plankton Records: methods of analysis, 1950–59. *Bulletins of Marine Ecology* 5: 51–64.

Gamble JC (1994) Long-term planktonic time series as monitors of marine environmental change. In: Leigh RA and Johnston AE (eds.) *Long-term Experiments in Agricultural and Ecological Sciences*, pp. 365–386. Wallingford: CAB International.

Glover RS (1967) The continuous plankton recorder survey of the North Atlantic. *Symp. Zoological Society of London* 19: 189–210.

Hardy AC (1939) Ecological investigations with the Continuous Plankton Recorder: object, plan and methods. *Hull Bulletins of Marine Ecology* 1: 1–57.

Hardy AC (1956) *The Open Sea: Its Natural History. Part 1: The World of Plankton.* London: Collins.

Hardy AC (1967) *Great Waters*. London: Collins.

IOC and SAHFOS (1991) *Monitoring the Health of the Ocean: Defining the Role of the Continuous Plankton Recorder in Global Ecosystem Studies*. Paris: UNESCO.

Oceanographic Laboratory, Edinburgh (1973) Continuous plankton records: a plankton atlas of the North Atlantic and the North Sea. *Bulletins of Marine Ecology* 7: 1–174.

Reid PC, Planque B, and Edwards M (1998) Is observed variability in the observed long-term results of the Continuous Plankton Recorder survey a response to climate change? *Fisheries Oceanography* 7: 282–288.

Warner AJ and Hays GC (1994) Sampling by the Continuous Plankton Recorder survey. *Progress in Oceanography* 34: 237–256.

GRABS FOR SHELF BENTHIC SAMPLING

P. F. Kingston, Heriot-Watt University, Edinburgh, UK

Introduction

The sedimentary environment is theoretically one of the easiest to sample quantitatively and one of the most convenient ways to secure such samples is by means of grabs. Grab samplers are used for both faunal samples, when the grab contents are retained in their entirety and then sieved to remove the biota from the sediment, and for chemical/physical samples when a subsample is usually taken from the surface of the sediment obtained. In both cases, the sampling program is reliant on the grab sampler taking consistent and relatively undisturbed sediment samples.

Conventional Grab Samplers

The forerunner of the grab samplers used today is the Petersen grab, designed by C.G.J. Petersen to conduct benthic faunal investigations in Danish fiords in the early part of the twentieth century. It consisted of two quadrant buckets that were held in an open position and lowered to the seabed (**Figure 1**). On the bottom, the relaxing of the tension on the lowering warp released the buckets and subsequent hauling caused them to close before they left the bottom. The instrument is still used today but is seriously limited in its range of usefulness, working efficiently only in very soft mud.

Petersen's grab formed the basis for the design of many that came after. One enduring example is the van Veen grab, a sampler that is in common use today (**Figure 2**). The main improvement over Petersen's design is the provision of long arms attached to the buckets to provide additional leverage to the closing action. The arms also provided a means by which the complex closing mechanism of the Petersen grab could be simplified with the hauling warp being attached to chains on the ends of the arms. The mechanical advantage of the long arms can be improved further by using an endless warp rig; this has the added advantage of helping to prevent the grab being jerked off the bottom if the ship rolls as the grab is closing. The van Veen grab was designed in 1933 and is still widely used in benthic infaunal studies owing to its simple design, robustness, and digging efficiency. The van Veen grab

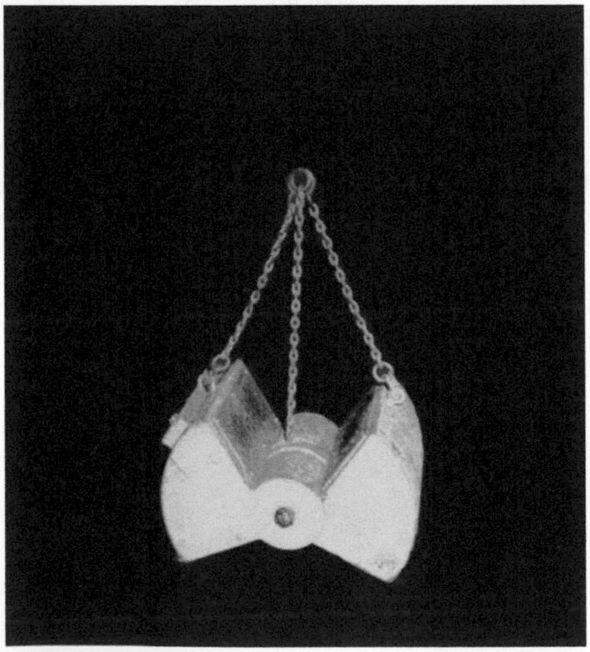

Figure 1 Petersen grab.

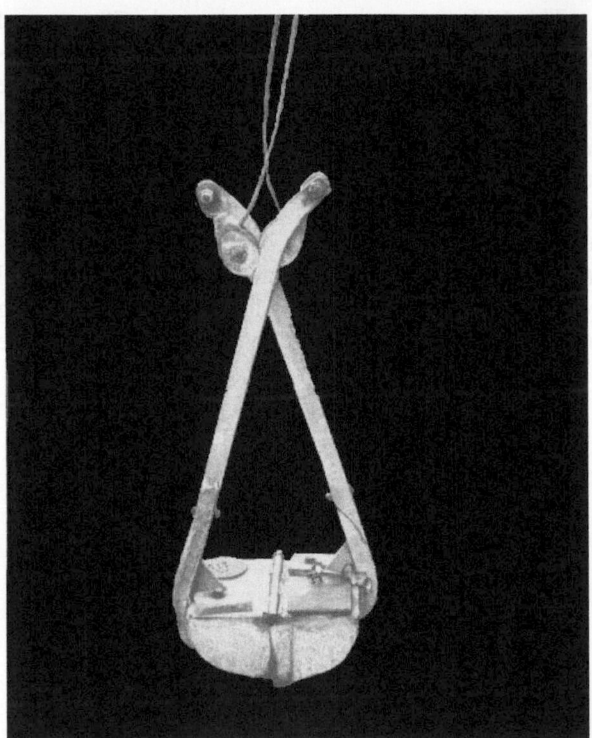

Figure 2 van Veen grab.

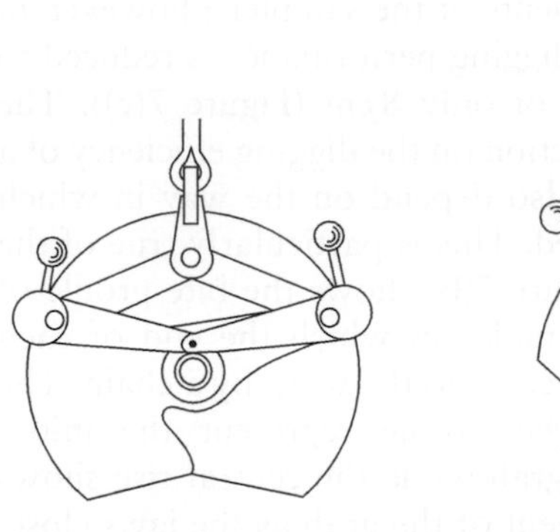
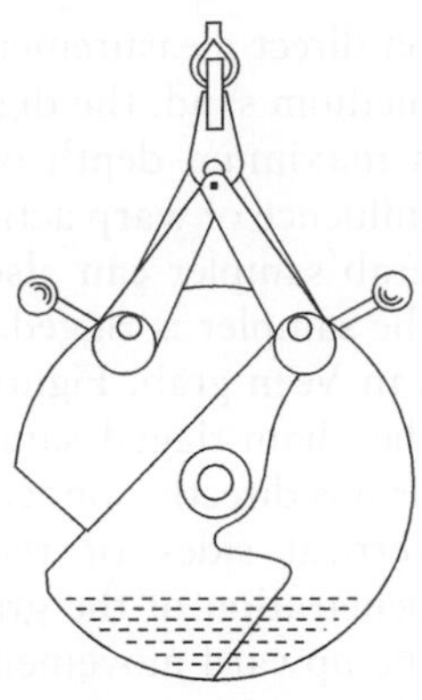

Figure 3 Diagram of a Hunter grab. Reproduced from Hunter B and Simpson AE (1976) A benthic grab designed for easy operation and durability. *Journal of the Marine Biological Association* 56: 951–957.

typically covers a surface area of $0.1\,m^2$, although instruments of twice this size are sometimes used.

A more recent design of frameless grab is the Hunter grab (**Figure 3**). This is of a more compact design than the van Veen. The jaws are closed by levers attached to the buckets in a parallelogram configuration, giving the mechanism a good overall mechanical advantage. The closing action requires no chains or pulleys and the instrument can be operated by one person. Its disadvantage is that the bucket design does not encourage good initial penetration of the sediment, which is important in hard-packed sediments.

A disadvantage of the grab samplers discussed so far is that there is little latitude for horizontal movement of the ship while the sample is being secured: the smallest amount of drift and the sampler is likely to be pulled over. The Smith–McIntyre grab was designed to reduce this problem by mounting the grab buckets in a stabilizing frame (**Figure 4**). Initial penetration of the leading edge of the buckets is assisted by the use of powerful springs and the buckets are closed by cables pulling on attached short arms in a similar way to that on the van Veen grab. The driving springs are released by two trigger plates, one on either side of the supporting frame to ensure that the sampler is resting flat on the seabed before the sample is taken. In firm sand the Smith–McIntyre grab penetrates to about the same depth of sediment as the van Veen grab. Its main disadvantage is the need to cock the spring mechanism on deck before deploying the sampler, a process that can be quite hazardous in rough weather.

The Day grab is a simplified form of the Smith–McIntyre instrument in which the trigger and closing mechanism remains the same, but without spring assistance for initial penetration of the buckets

Figure 4 Smith–McIntyre grab.

Figure 5 Day grab.

(**Figure 5**). The Day grab is widely used, particularly for monitoring work, despite its poor performance in hard-packed sandy sediments.

Most of the grabs thus far discussed have been designed to take samples with a surface area of 0.1 or $0.2\,m^2$. The Baird grab, however, takes samples of $0.5\,m^2$ by means of two inclined digging plates that are pulled together by tension on the warp (**Figure 6**). The grab is useful where a relatively large surface area needs to be covered, but has the disadvantage of taking a shallow bite and having the surface of the sample exposed while it is being hauled in.

Warp Activation

All the grabs described above use the warp acting against the weight of the sampler to close the jaws. However, direct contact with the vessel on the surface during the closure of the grab mechanism poses several problems.

Warp Heave

As tension is taken up by the warp to close the jaws, there is a tendency for the grab to be pulled up off the bottom, resulting in a shallower bite than might be expected from the geometry of the sampler. This tendency is related to the total weight of the sampler and the speed of hauling and is exacerbated by firm sediments. For example, the theoretical maximum depth of bite of a 120-kg Day grab is 13 cm (based on direct measurements of the sampler); however, in medium sand, the digging performance is reduced to a maximum depth of only 8 cm (**Figure 7(e)**). The influence of warp action on the digging efficiency of a grab sampler can also depend on the way in which the sampler is rigged. This is particularly true of the van Veen grab. **Figure 7(b)** shows the bite profile of the chain-rigged sampler in which the end of each arm is directly connected to the warp by a chain. The vertical sides of the profile represent the initial penetration of the grab while the central rise shows the upward movement of the grab as the jaws close. **Figure 7(c)** shows the bite profile of a van Veen of similar size and weight (30 kg) rigged with an endless warp in which the arms are closed by a loop of wire passing through a block at the end of each arm (as in **Figure 2**). The vertical profile of the initial penetration is again apparent; however, in this case, the overall depth of the sampler in the sediment is maintained as the jaws close. The endless warp rig increases the mechanical advantage of the pull of the warp while decreasing the speed at which the jaws are closed. The result is that the sampler is 'insulated' from surface conditions to a greater extent than when chain-rigged, giving a better digging efficiency.

Figure 6 Baird grab.

Grab 'Bounce'

In calm sea conditions it is relatively easy to control the rate of warp heave and obtain at least some consistency in the volume of sediment secured. However, such conditions are seldom experienced in the open sea where it is more usual to encounter wave action. Few ships used in offshore benthic

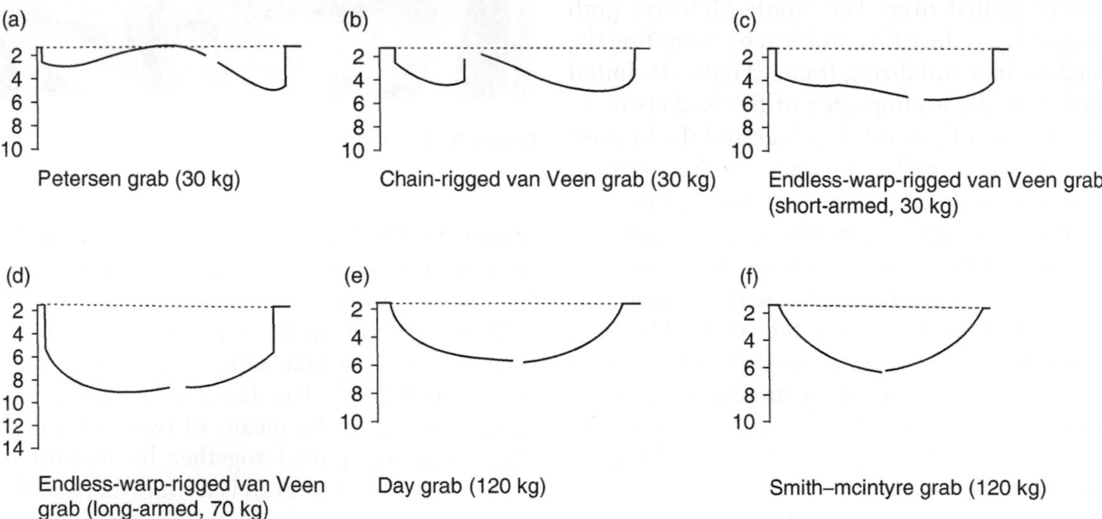

Figure 7 Digging profiles of a range of commonly used benthic grab samplers obtained in a test tank using a fine sand substratum. (a) Peterson grab (30 kg); (b) chain-rigged van Veen grab (30 kg); (c) endless-warp-rigged van Veen grab (short-armed, 30 kg); (d) endless-warp-rigged van Veen grab (long-armed, 70 kg); (e) Day grab (120 kg); (f) Smith–McIntyre grab (120 kg).

studies are fitted with winches with heave compensators so that the effect of ship's roll is to introduce an erratic motion to the warp. This may result in the grab 'bouncing' off the bottom where the ship rises just as bottom contact is made, or in the grab being snatched off the bottom where the ship rises just as hauling commences. In the former instance, it is unlikely that any sediment is secured; in the latter, the amount of material and its integrity as a sample will vary considerably, depending on the exact circumstances of its retrieval.

The intensity of this effect will depend on the severity of the weather conditions. **Figure 8** shows the relationship between wind speed and grab failure rate, which is over 60% of hauls at wind force 8. What is of more concern to the scientist attempting to obtain quantitative samples is the dramatic increase in variability with increase in wind speed with a coefficient of variation between 20 and 30 at force 7. The high cost of ship-time places considerable pressure on operators to work in as severe weather conditions as possible and it is not unusual for sampling to continue in wind force 7 conditions with all its disadvantages.

Drift

For a warp-activated grab sampler to operate efficiently it should be hauled with the warp positioned vertically above. Where there is a strong wind or current, these conditions may be difficult to achieve. The result is that the grab samplers are pulled on to their sides. This is a particular problem with samplers, such as the van Veen grab, that do not have stabilizing frames. Diver observations have shown, however, that at least in shallow water, where the drift effect is at its greatest on the bottom, even the framed heavily weighted Day and Smith–McIntyre grabs can be toppled.

Initial Penetration

It is clear that the weight of the sampler is an important element in determining the volume of the sample secured. Much of the improved digging efficiency of the van Veen grab shown in **Figure 7(d)** can be attributed to the addition of an extra 40 kg of weight which increased the initial penetration of the sampler on contact with the sediment surface.

Initial penetration is one of the most important factors in the sequence of events in grab operation, determining the final volume of sediment secured. **Figure 9** shows the relationship between initial penetration and final sample volume obtained for a van Veen grab. Over 70% of the final volume is determined by the initial penetration. Subsequent digging of the sampler is hampered, as already shown, by the pull of the warp.

For most benthic faunal studies it is important for the sampler to penetrate at least 5 cm into the sediment (for a 0.1 m^2 surface area sample this gives 5 l of sample). In terms of number of species and individuals, over 90% of benthic macrofauna are found in the top 4–5 cm of sediment. **Figure 10** shows how the number of individuals relates to average sample

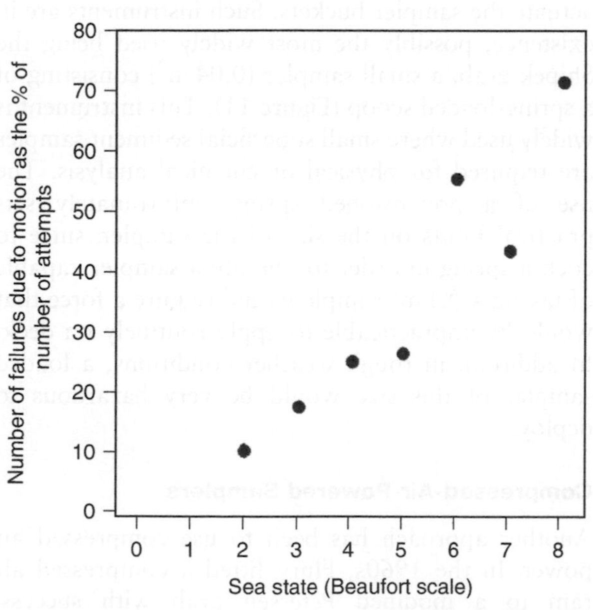

Figure 8 Relationship between wind speed and grab failure rate.

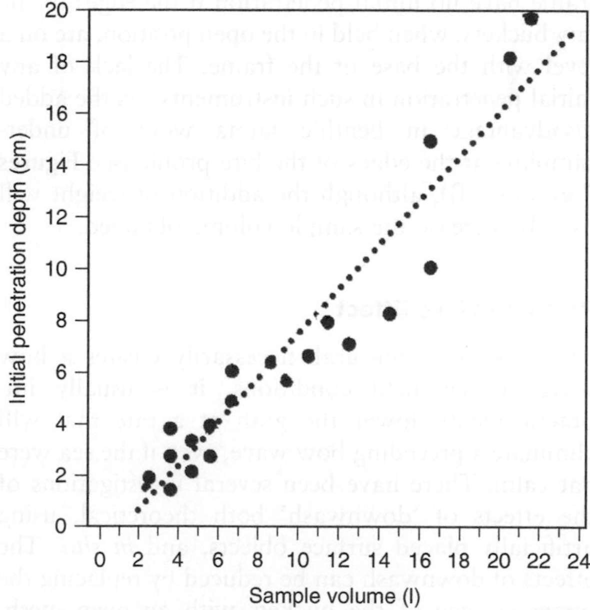

Figure 9 Relationship between initial penetration of a van Veen grab sampler and volume of sediment secured.

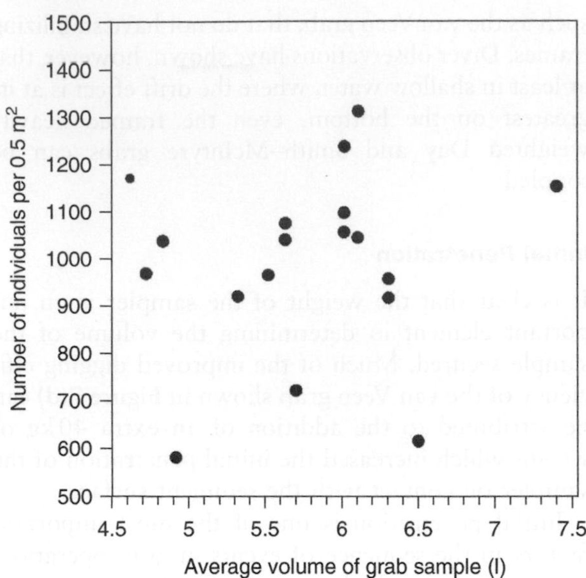

Figure 10 Relationship between number of benthic fauna individuals captured and sample volume for a boreal offshore sand substratum.

volume for 18 stations in the southern North Sea (each liter recorded represents 1 cm of penetration). Although there is a considerable variation in the numbers of individuals between stations, there is no significant trend linking increased abundance with increased sample volume penetration. No sample volumes of less than 4.5 l were taken, indicating that at that level of penetration most of the fauna were being captured.

Samplers in which the jaws are held rigidly in a frame have no initial penetration if the edges of the jaw buckets, when held in the open position, are on a level with the base of the frame. The lack of any initial penetration in such instruments has the added disadvantage in benthic fauna work of under-sampling at the edges of the bite profile (see **Figures** 7(e) and 7(f)), although the addition of weight will usually increase the sample volume obtained.

Pressure Wave Effect

The descent of the grab necessarily creates a bow wave. Under field conditions, it is usually impracticable to lower the grab at a rate that will eliminate a preceding bow wave, even if the sea were flat calm. There have been several investigations of the effects of 'downwash' both theoretical, using artificially placed surface objects, and *in situ*. The effects of downwash can be reduced by replacing the upper surface of the buckets with an open mesh. Although there is still a considerable effect on the surface flock layer (rendering the samples of dubious

Figure 11 Shipek grab.

value for chemical contamination studies), the effect on the numbers of benthic fauna is generally very small.

Self-Activated Bottom Samplers

There can be little doubt that one of the most important factors responsible for sampler failure or sample variability in heavy seas is the reliance of most presently used instruments on warp-activated closure. The most immediate and obvious answer to this problem is to make the closing action independent of the warp by incorporating a self-powering mechanism.

Spring-Powered Samplers

One solution to the problem is to use a spring to actuate the sampler buckets. Such instruments are in existence, possibly the most widely used being the Shipek grab, a small sampler ($0.04\,m^2$) consisting of a spring-loaded scoop (**Figure 11**). This instrument is widely used where small superficial sediment samples are required for physical or chemical analysis. The use of a pretensioned spring unfortunately sets practical limits on the size of the sampler, since to cock a spring in order to operate a sampler capable of taking a $0.1\,m^2$ sample would require a force that would be impracticable to apply routinely on deck. In addition, in rough weather conditions, a loaded sampler of this size would be very hazardous to deploy.

Compressed-Air-Powered Samplers

Another approach has been to use compressed air power. In the 1960s, Flury fitted a compressed air ram to a modified Petersen grab with success. However, the restricted depth range of the instrument and the inconvenience of having to recharge the air reservoir for each haul limited its potential for routine offshore work.

Hydraulically Powered Samplers

Hydraulically powered grabs are commonly used for large-scale sediment shifting operations such as seabed dredging. The Bedford Institute of Oceanography, Nova Scotia, successfully scaled down this technology to that of a practical benthic sampler. Their instrument is relatively large, standing 2.5 m high and weighing some 1136 kg. It covers a surface area of 0.5 m^2 and samples to a maximum sediment depth of 25 cm. At full penetration, the sediment volume taken is about 100 l. The buckets are driven closed by hydraulic rams powered from the surface. The grab is also fitted with an underwater television camera which allows the operator to visually select the precise sampling area on the seabed, close and open the bucket remotely, and verify that the bucket closed properly prior to recovery. The top of the buckets remain open during descent to minimize the effect of downwash and close on retrieval to reduce washout of the sample on ascent. The current operating depth of the instrument is 500 m. The instrument has been successfully used on several major offshore studies, but does require the use of a substantial vessel for its deployment.

Hydrostatically Powered Samplers

Hydrostatically powered samplers use the potential energy of the difference in hydrostatic pressure at the sea surface and the seabed. The idea of using this power source is not new. In the early part of the twentieth century, a 'hydraulic engine' was in use by marine geologists that harnessed hydrostatic pressure to drive a rock drill. Hydrostatic power has also been used to drive corers largely for geological studies. However, these instruments were principally concerned with deep sediment corers and were not designed to collect macrofauna or material at the sediment–water interface.

A more recent development has been that of a grab built by Heriot-Watt University, Edinburgh. The sampler uses water pressure difference to operate a hydraulic ram that is activated when the grab reaches the seabed. **Figure 12** shows the general layout of the instrument. Water enters the upper chamber of the cylinder when the sampler is on the seabed, forcing down a piston that is connected to a system of levers that close the jaws. The actuating valve is held shut by the weight of the sampler and there is a delay mechanism to prevent premature closure of the jaws resulting from 'bounce'. Back on the ship, the jaws are held shut by an overcenter locking mechanism

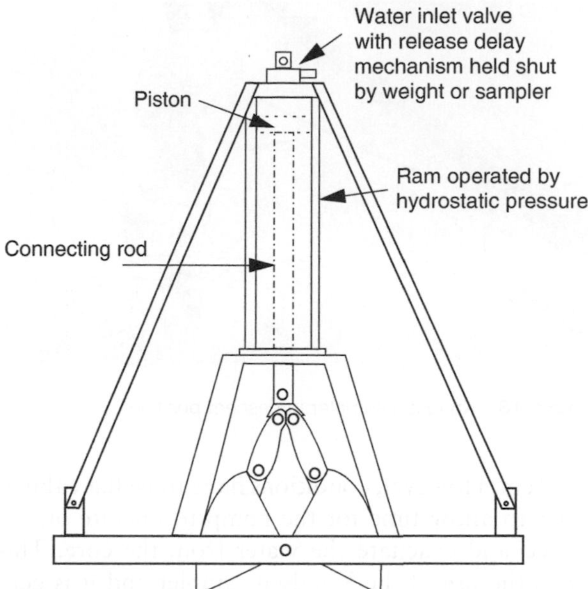

Figure 12 Diagram of a grab sampler using hydrostatic pressure to close the jaws.

and, on release, are drawn open by reversal of the piston motion from air pressure built up on the underside of the piston during its initial power stroke. Since the powering of the grab jaws is independent of the warp, the sampler may be used successfully in a much wider range of surface weather conditions than conventional grabs.

Alternatives to Grab Samplers

Ideally a benthic sediment sample for faunal studies should be straight-sided to the maximum depth of its excavation and should retain the original stratification of the sediment. Grab samplers by the very nature of their action will never achieve this end.

Suction Samplers

One answer to this problem is to employ some sort of corer designed to take samples of sufficient surface area to satisfy the present approaches to benthic studies. The Knudsen sampler is such a device and is theoretically capable of taking the perfect benthic sample. It uses a suction technique to drive a core tube of 0.1 m^2 cross-sectional area 30 cm into the sediment. Water is pumped out of the core tube on the seabed by a pump that is powered by unwinding a cable from a drum. The sample is retrieved by pulling the core out sideways using a wishbone arrangement and returning it to the surface bottom-side up (**Figures 13** and **14**). Under ideal conditions, the device will take a straight-sided sample to a depth

Figure 13 Knudsen sampler in descent position.

Figure 14 Knudsen sampler in ascent position.

of 30 cm. However, conditions have to be flat calm in order to allow time for the pump to operate on the seabed and evacuate the water from the core. This limits the use of the Knudsen sampler and it is generally not suitable for sampling in unsheltered conditions offshore. Mounting the sampler in a stabilizing frame can improve its success rate and it is used regularly for inshore monitoring work where it is necessary to capture deep burrowing species.

Spade Box Samplers

Another approach to the problem is to drive an open-ended box into the sediment, using the weight of the sampler, and arrange for a shutter to close off the bottom end. The most widespread design of such an instrument is that of the spade box sampler, first described by Reineck in the 1950s and later subjected to various modifications. The sampler consists of a removable steel box open at both ends and driven into the sediment by its own weight. The lower end of the box is closed by a shutter supported on an arm pivoted in such a way as to cause it to slide through the sediment and across the mouth of the box (**Figure 15**). As with the grab samplers previously described, the shutter is driven by the act of hauling on the warp with all the attendant disadvantages. Nevertheless, box corers are very successful and are used widely for obtaining relatively undisturbed samples of up to $0.25\,m^2$ surface area (**Figure 16**). One big advantage of the box sampler is that the box can usually be removed with the sample and its overlying water left intact, allowing detailed studies of the sediment surface. Furthermore, it is possible to subsample using small-diameter corers for studies of chemical and physical characteristics. Despite their potential of securing the 'ideal' sediment sample, box corers are rarely used for routine benthic monitoring work. This is largely because of their size (a box corer capable of taking a $0.1\,m^2$

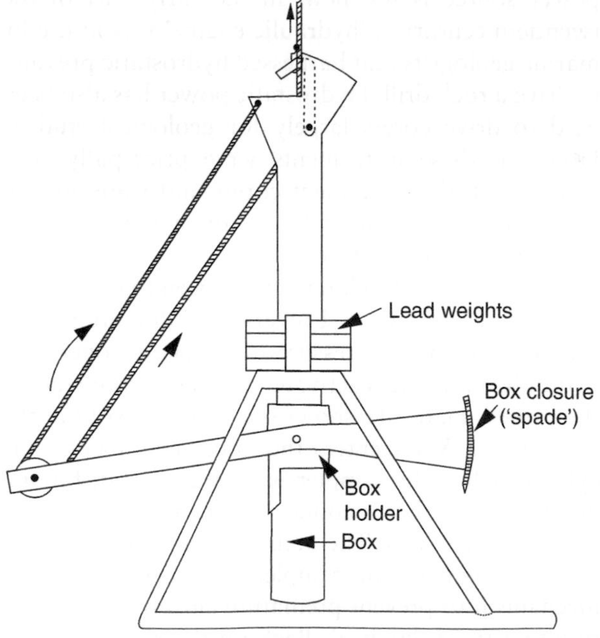

Figure 15 Diagram of a Reineck spade box sampler.

Figure 16 A 0.25 m^2 spade box sampler.

sample weighs over 750 kg and stands 2 m high) and the difficulty in deployment and recovery in heavy seas.

Precision Corers

For chemical monitoring, it is important that the sediment–water interface is maintained intact, for it is the surface flock layer that will contain the most recently deposited material. Unfortunately, such undisturbed samples are rarely obtained using grab samplers or box corers. Precision corers are capable of securing undisturbed surface sediment cores; however, they are unsuitable for routine offshore work because of the time taken to secure a sample on the seabed and dependence on warp-activated closure. Additionally, the cross-sectional area of the core (0.002–0.004 m^2) would necessitate the taking of large numbers of replicate samples in order to capture sufficient numbers of benthic macrofauna to be useful. This would be impracticable given the time taken to take a single sample. Large multiple precision corers have been constructed; these are usually too large and difficult to deploy for routine macrobenthos sampling. At present, there is no instrument that fulfils the requirements for a quick turnaround precision multiple corer for offshore sampling.

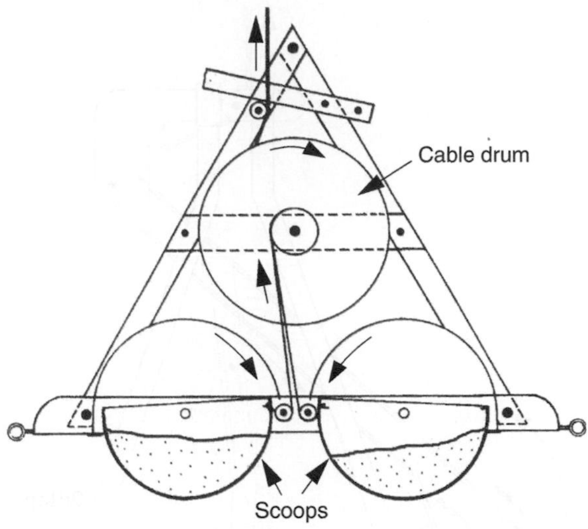

Figure 17 Diagram of a Holme scoop. Reproduced from Holme NA (1953) The biomass of the bottom fauna in the English Channel off Plymouth. *Journal of the Marine Biological Association* 32: 1–49.

Sampling Difficult Sediments

Most of the samplers so far discussed operate reasonably well in mud or sand substrata. Few operate satisfactorily in gravel or stony mixed ground either because the bottom is too hard for the sampler to penetrate the substratum or because of the increased likelihood of a stone holding the jaws open when they are drawn together. To get around this problem various types of scoops have been devised. The Holme grab has a double scoop action with two buckets rotating in opposite directions to minimize any lateral movement during digging. The scoops are closed by means of a cable and pulley arrangement (**Figure 17**) and simultaneously take two samples of 0.05 m^2 surface area.

The Hamon grab, which has proved to be very effective in coarse, loose sediments, takes a single rectangular scoop of the substratum covering a surface area of about 0.29 m^2. The scoop is forced into the sediment by a long lever driven by pulleys that are powered by the pull of the warp (**Figure 18**). Although the samples may not always be as consistent as those from a more conventional grab sampler, the Hamon grab has found widespread use where regular sampling on rough ground is impossible by any other means.

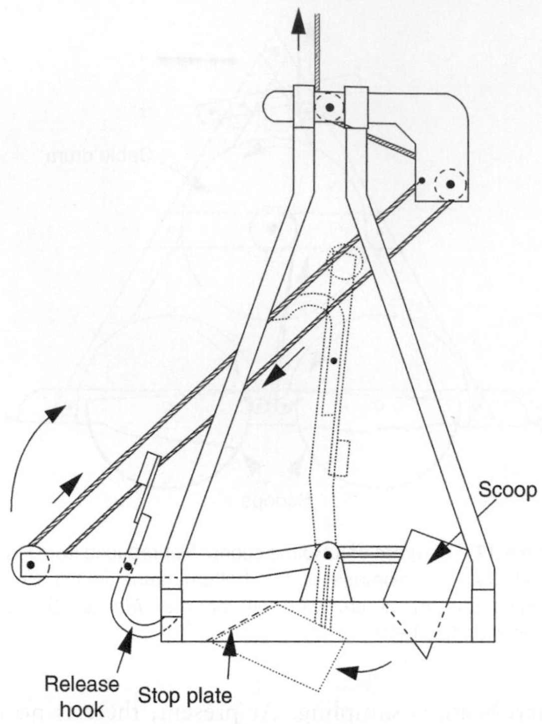

Figure 18 Diagram of a Hamon scoop.

Present State of Technology

It is perhaps surprising that given the high state of technology of survey vessels, position fixing, and analytical equipment, the most commonly used samplers are relatively primitive (being designed some 40 or more years ago). Yet the quality of the sample is of fundamental importance to any research or monitoring work. Currently the most popular instruments are grab samplers, probably because of their wide operational weather window and apparent reliability. Samplers such as the Day, van Veen, and Smith–McIntyre grab samplers are still routinely used for sampling sediment for chemical and biological analysis despite their well-documented shortcomings. As discussed earlier, the most important of these are the substantial downwash that precedes the sampler as it descends and the disturbance of the trapped sediment layers by the closing action of the jaws. Both chemical and meiofaunal studies are particularly vulnerable to these.

Although box corers go some way to reducing disturbance of the sediment strata, the all-important surface flocculent layer is invariably washed away. A big disadvantage of the box corer for routine offshore work is that it is sensitive to weather conditions; in addition, the larger instruments do not perform well on sand substrata.

A generic disadvantage of most samplers presently in use is that they rely on slackening of the warp to trigger the action and the heave of the warp to drive the closing mechanism. In calm conditions, this presents no great problem, but with increasing sea state the vertical movement of the warp decreases reliability dramatically until the variability and finally failure rate of hauls make further sampling effort fruitless.

Specific Problems, Requirements, and Future Developments

Chemical Studies

Studies involving sediment chemistry require precision sampling if undisturbed samples at the sediment/water interface are to be obtained. This can be critical, particularly when the results of recent sedimentation are of interest. The impracticality of using existing hydraulically damped corers such as the Craib corer for offshore work has led to the widespread use of less 'weather sensitive' devices such as spade box corers and grabs for routine monitoring purposes. However, studies carried out have shown that these samplers produce a considerable 'downwash effect', blowing the surface flocculent layer away before the sample is secured. This can have serious consequences if any meaningful estimation of the surface chemistry of the sediment is desired. Repetitive and accurate sampling is also a prerequisite for determining spatial and temporal change in sediment chemistry.

Meiofauna Studies

Meiofauna has increasingly been shown to have potential as an important tool in benthic monitoring. One of the major factors limiting its wider adoption is the lack of a suitable sampler. Although instruments such as the Craib corer and its multicorer derivatives are capable of sampling the critically important superficial sediment layer, these designs provide a poor level of success on harder sediments and in anything but near-perfect weather conditions. They also have a slow turnaround time. Box corers are widely used as an alternative; however, they are known to be unreliable in their sampling of meiofauna. As with the macrobenthos, meiobenthic patchiness results in low levels of precision of abundance estimates unless large numbers of samples are taken.

Macrobenthic Studies

The measurement and prediction of spatial and temporal variation in natural populations are of great importance to population biologists, both for fundamental research into population dynamics and productivity and in the characterization of benthic communities for determining change induced by environmental impact. Though always an important consideration, cost-effectiveness of sampling and sample processing is not so crucial in fundamental research, since time and funding may be tailored to fit objectives. This is rarely the case in routine monitoring work where often resolution and timescales have to fit the resources available.

Benthic fauna are contagiously distributed and to sample such communities with a precision that will enable distinction between temporal variation and incipient change resulting from pollution effects, it is generally accepted that five replicate $0.1\,m^2$ hauls from each station are necessary (giving a precision at which the standard error is no more than 20% of the mean). This frequency of sampling requires approximately 10–15 man-days of sediment faunal analysis per sample station. While this may be acceptable in community structure studies in which time and manpower (and thus cost) are not a primary consideration, this high cost of analyzing samples is of importance in routine monitoring studies and has led monitoring agencies and offshore operators to reduce sampling frequency on cost grounds to as few as two replicates per station. This reduces the precision with which faunal abundance can be estimated to a level at which only gross change can be demonstrated. However, sampling to an acceptable precision may be achieved from an area equivalent to $0.1\,m^2$ if a smaller sampling unit is used. For example, 50 5-cm core samples (with a similar total surface area) have been shown to give a similar precision to that of 5 to 12 $0.1\text{-}m^2$ grab samples. Thus a similar degree of precision may be obtained for around one-fifth to one-twelfth the analytical costs using a conventional approach.

The problem is to be able to secure the 50 core samples per site that would be needed for the macrofaunal monitoring in a timescale that would be realistic offshore. At present, there is no instrument capable of supporting such a sampling demand and operating in the range of sediment types that wide-scale monitoring studies demand. Clearly a single core sampler would be impracticable, and one must look to the future development of a multiple corer that is capable of flexibility in its operation, which allows a quick on-deck turnaround between hauls.

Further Reading

Ankar S (1977) Digging profile and penetration of the van Veen grab in different sediment types. *Contributions from the Askö Laboratory, University of Stockholm, Sweden* 16: 1–12.

Beukema JJ (1974) The efficiency of the van Veen grab compared with the Reineck box sampler. *Journal du Conseil Permanent International pour l'Exploration de la Mer* 35: 319–327.

Eleftheriou A and McIntyre AD (eds.) (2005) *Methods for the Study of the Marine Benthos*. Oxford, UK: Blackwell.

Flury JA (1967) Modified Petersen grab. *Journal of the Fisheries Research Board of Canada* 20: 1549–1550.

Holme NA (1953) The biomass of the bottom fauna in the English Channel off Plymouth. *Journal of the Marine Biological Association* 32: 1–49.

Hunter B and Simpson AE (1976) A benthic grab designed for easy operation and durability. *Journal of the Marine Biological Association* 56: 951–957.

Riddle MJ (1988) Bite profiles of some benthic grab samplers. *Estuarine, Coastal and Shelf Science* 29(3): 285–292.

Thorsen G (1957) Sampling the benthos. In: Hedgepeth JW (ed.) *Treatise on Marine Ecology and Paleoecology, Vol. 1: Ecology*. Washington, DC: The Geological Society of America.

BIOACOUSTICS

P. L. Tyack, Woods Hole Oceanographic Institution, Woods Hole, USA

Introduction

The term 'bioacoustics' has two different usages in ocean sciences. Biological oceanographers use active sonars to map organisms in the sea. Since they use sound to detect marine life, they often call this approach 'bioacoustics'. The other sense of bioacoustics involves studying how animals use sound themselves in the ocean. This is the kind of bioacoustics covered in this article.

Humans are visual animals, and we think of vision as a primary distance sense because light carries so well in terrestrial environments. However, light is useful for vision under the sea only over ranges of tens of meters at best. Sound, on the other hand, propagates extremely well in water – that is why oceanographers so often select sound as a medium for exploring the sea or for communicating under the sea. Sound propagates so well under water that a depth charge exploded off Australia can be heard in Bermuda. Just as we can hear well but emphasize vision, so many marine mammal species see well but emphasize hearing. It is possible to gauge the relative importance of audition versus vision in animals by comparing the number of nerve fibers in the auditory versus the optic nerves. Of all marine mammals, the cetaceans are the most specialized to use sound. Most cetaceans have auditory:optic ratios of fiber counts that are 2–3 times those of land mammals, suggesting that audition is more important than vision. Some cetaceans also use sound to echolocate. Dolphins have a large repertoire of vocalizations spanning frequencies from below 100 Hz to over 100 kHz, and dolphins have evolved high-frequency echolocation similar to some human-made sonars and to the biosonar used by bats.

Marine mammals not only hear well, they are also very vocal animals. The sounds of marine mammals are now well known, but the first recordings identified from a marine mammal species were only made in the late 1940s. In the 1950s and 1960s, there was rapid growth in studies of how dolphins echolocate using high-frequency click sounds and of field studies associating different sounds with different species of marine mammal. Marine mammal bioacoustics during this period was concerned primarily with identifying which species produced which sounds heard under water. Much of this research was funded by naval research organizations because biological sources of noise can interfere with military use of sound in the sea.

Elementary Acoustics

Sound consists of mechanical vibrations that propagate through a medium. Sound induces movements or displacements of the particles in the medium. Imagine a small sphere that expands to create a denser area. This compression will propagate as particles are displaced in the direction of propagation. If the sphere then contracts, it can create an area of rarefaction, or lower density, and this also can propagate outward. These compressions or rarefactions can be expressed in terms of particle displacement or as a pressure differential.

Now imagine a sound source that creates a series of compressions and rarefactions that propagate through the medium. A source with a purely sinusoidal pattern of compression and rarefaction would produce energy at only one frequency. The frequency of this sound is measured in cycles per second. A sound that takes t seconds to make a full cycle has a frequency $f = t^{-1}$. Older references may refer to frequency in cycles per second, but the modern unit of frequency is the Hertz (Hz) and a frequency of 1000 Hz is expressed as one kiloHertz (1 kHz). If a sound took 1 s for a full cycle, it would have a frequency of 1 Hz. The wavelength of a tonal sound is the distance from one measurement of the maximum pressure to the next maximum. The speed of sound is approximately $1500\,\mathrm{m\,s^{-1}}$ in water, roughly five times the value in air, $340\,\mathrm{m\,s^{-1}}$. The speed of a sound c is related in a simple way to the frequency f and the wavelength λ by $c = \lambda f$. An under-water sound with $f = 1\,\mathrm{Hz}$ would have $\lambda = 1500\,\mathrm{m}$; for $f = 1500$ Hz, $\lambda = 1\,\mathrm{m}$. Not all sounds have energy limited to one frequency. Sounds that have energy in a range of frequencies, say in the frequency range between 2000 and 3000 Hz (2 and 3 kHz), would be described as having a bandwidth of 1 kHz.

One can imagine a sound wave as a growing sphere propagating outward from a compression or rarefaction generated by a point source. The initial movement of the source will have transmitted a certain amount of energy to the medium. If none of this energy is lost as the sound propagates, then it will be evenly diluted over the growing sphere. The acoustic intensity is defined as the amount of energy flowing through an area over a unit of time. As the sphere increases in radius from 1 to r, the surface area increases to $4\pi r^2$. The intensity of a sound thus declines as the inverse of the square of the range from the source (r^{-2}). A sound in the middle of the ocean can be thought of as spreading in this way until it encounters a boundary such as the surface or seafloor that might cause reflection, or an inhomogeneity in the medium that might cause refraction. One fascinating acoustic feature of the deep ocean is that sound rays propagating upward may refract downward as they encounter warmer water near the surface, and downward-propagating rays will refract upward as they encounter denser water at depth. When one is far from a sound source compared to the ocean depth, the sound energy may be concentrated by refraction in the deep ocean sound channel. This sound can be thought of as spreading in a plane, to a first approximation. In this case, sound intensity would decline as the inverse of the first power of the range, or r^{-1}. This involves much lower loss than the inverse square spreading loss in an unbounded medium.

Sound spreading is a 'dilution' factor and is not a true loss of sound energy. Absorption, on the other hand, is conversion of acoustic energy to heat. The attenuation of sound due to absorption is a constant per unit distance, but this constant is dependent upon signal frequency. While absorption yields trivial effects at frequencies below 100 Hz, it can significantly limit the range of higher frequencies, particularly above 40 kHz or so. A 100 Hz sound can travel over a whole ocean basin with little absorption loss, while a 100 kHz sound would lose half its energy just traveling about 100 m.

Relating Acoustic Structure to Biological Function of Marine Mammal Calls

Understanding the physics of sound in the sea can help us understand why animals make the kinds of sound they do. For example, the calls of baleen whales are low-frequency because they are adapted for long-range propagation in the deep sea. Large baleen whales have evolved abilities to produce and to hear low-frequency calls well-suited for long-range communication. Blue whales and fin whales produce the lowest-frequency signals of all marine mammals, so low that humans can barely hear them. The long moans of blue whales, *Balaenoptera musculus*, have fundamental frequencies in the 14–36 Hz band and they last several tens of seconds. The pulses of finback whales, *Balaenoptera physalus*, range roughly between 15 and 30 Hz and last on the order of 1 s. Particularly during the breeding season in mid-latitudes, finbacks produce series of pulses in a regularly repeating pattern in bouts that may last many days.

These loud low-frequency sounds appear to be specialized for long-range propagation in the sea. Absorption is negligible at the frequencies of these sounds. While acoustic models predicted that these sounds could be detected at ranges of hundreds of kilometers, it is only recently that this has been confirmed empirically. During the Cold War, the US Navy developed bottom-mounted hydrophones to locate ships and to track them. After the end of the Cold War, these sophisticated systems were made available to biologists, who have worked with Navy personnel to locate and track whales over long ranges, including one whale tracked for more than 1700 km over 43 days (**Figure 1**). These arrays have proven capable of detecting whales at ranges of hundreds to thousands of kilometers, as was predicted by the earlier acoustic models.

The physics of sound can also help explain why dolphins specialize in high-frequency sounds. Dolphins can detect distant objects acoustically by producing loud clicks and then listening for echoes. The clicks used by dolphins for echolocation have been well described. The echolocation clicks of bottlenose dolphins are very short (< 100 μs), with a rapid rise-time and a relatively broad bandwidth from several tens of kilohertz up to near 150 kHz (**Figure 2A**).

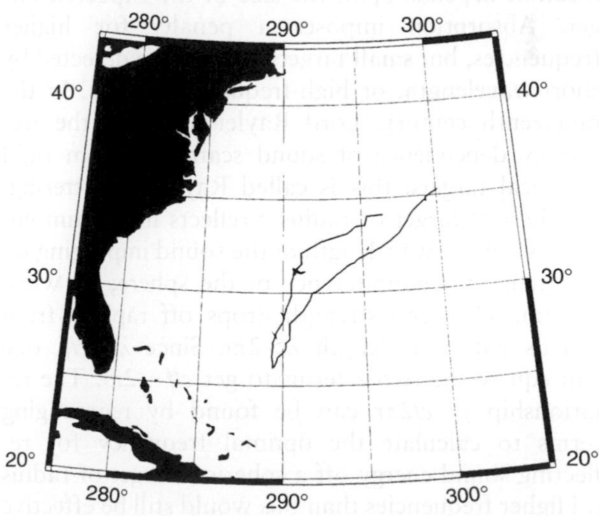

Figure 1 Track of a calling blue whale, *Balaenoptera musculus*, as it swam 1700 km over 43 days. The whale was tracked using the Integrated Underwater Sound Surveillance System (IUSS) of the US Navy. (From Figure 4.17 of Au *et al.* (2000).)

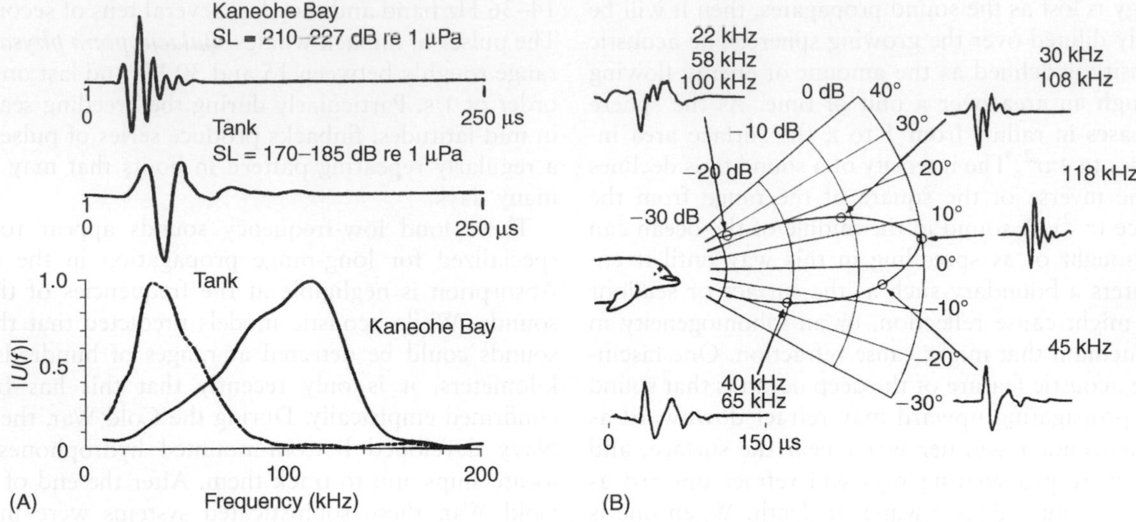

Figure 2 (A) Waveform and spectrum of echolocation clicks of bottlenose dolphins, *Tursiops truncatus*, in open ocean (Kaneohe Bay) and in a tank. The spectrum of the click from the tank (indicated with a dashed line) has a lower frequency peak at 40 kHz. (B) Beam pattern of *Tursiops* echolocation clicks. (µPa = micropascal, reference for sound pressure measurements. SL = source level.) ((A) from Figure 9.1, (B) from Figure 9.5 of Au *et al.* (2000).)

Captive dolphins in a reverberant pool make clicks that are less loud and lower in frequency than dolphins working on long-range echolocation in an open bay. The high-frequency components of these clicks are highly directional. If one moves 10 degrees off the axis of the beam, the click energy is halved and the click contains energy at lower frequencies (**Figure 2B**). The detection abilities of echolocating dolphins are truly remarkable. For example, trained bottlenose dolphins can detect a 2.54 cm solid steel sphere at 72 m, nearly a football field away.

The optimal frequency of a sound used for echolocation depends upon the size of the expected target. Absorption imposes a penalty for higher frequencies, but small targets can best be detected by short-wavelength, or high-frequency, signals. In the nineteenth century, Lord Rayleigh solved the frequency dependence of sound scattering from rigid spherical targets; this is called Rayleigh scattering. A spherical target of radius r reflects maximum energy when the wavelength of the sound impinging on it equals the circumference of the sphere, or when $\lambda = 2\pi r$. The echo strength drops off rapidly from signals with wavelength $\lambda > 2\pi r$. Since $\lambda = c/f$, one can equate the two λ terms to get $c/f = 2\pi r$. The relationship $f = c/2\pi r$ can be found by rearranging terms to calculate the optimal frequency for reflecting sound energy off a spherical target of radius r. Higher frequencies than this would still be effective sonar signals, but frequencies below f would show a strong decrease in effectiveness with decreasing frequency. A dolphin echolocating on rigid targets with a 'radius' of 0.5 cm should use a frequency $f \geq c/2\pi r = 1500/(2\pi \times 0.005) \sim 50\text{kHz}$. This is within the frequency range of dolphin echolocation clicks, which include energy up to about 150 kHz. This upper frequency is appropriate for detecting spherical targets with radii as small as 1.5 mm. The hearing of dolphins is also most sensitive at frequencies of roughly 50–100 kHz. If dolphins have a need to echolocate on rigid targets with sizes in the 1 cm range, that helps explain why their echolocation system emphasizes these high frequencies.

Marine Mammal Hearing

In order to detect sound, animals require a receptor that can transduce the forces of particle motion or pressure changes into neural signals. Most mechanoreceptors in animals involve cells with hairlike cilia on their surfaces. As these cilia move, the electric potential between the inside and the outside of the receptor cells changes, and this potential difference modifies the rate of nerve impulses that signal other parts of the nervous system.

Terrestrial mammals evolved an ear that is divided into three sections: the outer, middle, and inner ear. The outer ear and middle ear function in terrestrial mammals to transduce airborne sound into vibrations of a fluid the inner ear of mammals which contains the cochlea, the organ in which sound energy is converted into neural signals. Sound enters the cochlea via the oval window and causes a membrane, called the basilar membrane, to vibrate. This membrane is mechanically tuned to vibrate at different frequencies. Near the oval window, the basilar membrane is stiff and narrow, causing it to vibrate when excited with high frequencies. Farther into the cochlea, the basilar membrane becomes wider and 'floppier', making it more sensitive to lower frequencies. Sensory cells at different positions along the basilar membrane are excited by different frequencies, and their rate of firing is proportional to the amount of sound energy in the frequency band to which they are sensitive.

Marine mammals share basic patterns of mammalian hearing but also have varying adaptations for listening under water as opposed to in air. All marine mammals other than sirenians, the sea otter, and cetaceans spend critical parts of their lives on land or ice and some phocid seals communicate both in air and under water. The relative importance of hearing in air and under water has been compared for three pinniped species whose hearing has been tested in both environments. The California sea lion (*Zalophus californianus*) is adapted to hear best in air; the harbor seal (*Phoca vitulina*) can hear equally well in air and under water; and the northern elephant seal (*Mirounga angustirostris*) has an auditory system adapted for under water sensitivity at the expense of aerial hearing.

The eardrum and middle ear in terrestrial mammals functions to efficiently transmit airborne sound to the inner ear where the sound is detected in a fluid. No such matching is required for an animal living in the water, and cetaceans, which are adapted exclusively for listening under water, do not have an air-filled external ear canal. The problem for cetaceans is isolating the ears acoustically, and the inner ear is surrounded by an extremely dense bone that is isolated from the skull. High-frequency sound is thought to enter the dolphin head through a thin section of bone in the lower jaw and is conducted to the inner ear via fatty tissue that acts as a waveguide.

Hearing abilities have been tested for those species of marine mammals that can be held in captivity. **Figure 3** shows audiograms from a dolphin, porpoise, and several pinnipeds. As discussed above, dolphins have hearing specialized to hear very high frequencies up to ten times the upper limit of human hearing. Seals have less acute hearing than do dolphins and they are less able to hear the highest frequencies. The frequency range of hearing has never been tested in baleen whales. Hearing is usually tested by training an animal, and baleen whales are so big that only a few have been kept for short periods in captivity. However, both their low-frequency vocalizations and the frequency tuning of their cochlea suggest they are specialized for low-frequency hearing.

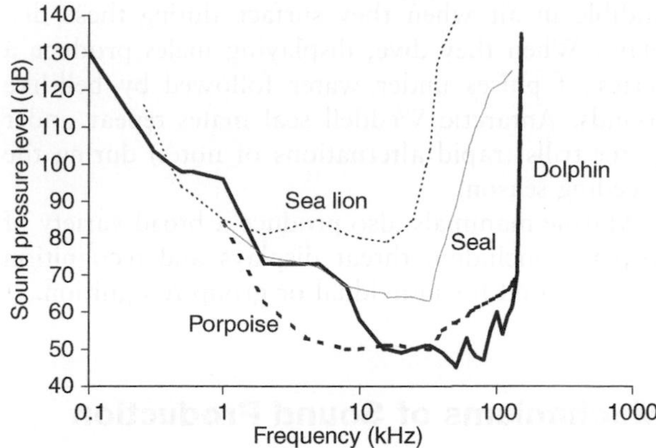

Figure 3 Audiograms from a variety of marine mammals: dolphin *Tursiops truncatus*; porpoise *Phocoena phocoena*; sea lion *Zalophus californianus*; seal *Phoca vitulina*.

Mammalian hearing is designed to analyze the frequency content of sound. Among mammals, dolphins have extraordinarily good abilities of discriminating different frequencies. They can detect a change of as little as 0.2% in frequency, which is close to the resolution of human hearing.

Vocalizations of Marine Mammals

When terrestrial carnivores and ungulates invaded the sea, they encountered new constraints and opportunities for sensing signals. The sirenians, cetaceans, phocid seals, and the walrus (*Odobenus rosmarus*) evolved specializations for using sound to communicate under water and to explore the marine environment; other taxa, including the otariid pinnipeds, sea otter (*Enhydra lutra*) and polar bear (*Ursus maritimus*), vocalize mainly in air. As with hearing, cetaceans show the most elaborate and extreme specializations for acoustic communication under water.

The best-known acoustic displays of marine mammals are the reproductive advertisement displays called songs. The songs of humpback whales are the best known advertisement display in the cetaceans, but bowhead whales also sing. Male seals of some species repeat acoustically complex songs during the breeding season. Songs are particularly common among seals that inhabit polar waters and that haul out on ice. The songs of bearded seals, *Erignathus barbatus*, are produced by sexually mature adult males and are heard most frequently during the peak of the breeding season. Male walruses, *Odobenus rosmarus*, also produce complex visual and acoustic displays near herds of females during their breeding season. They use their lips to whistle, and also produce loud sounds of breathing that are audible in air when they surface during these displays. When they dive, displaying males produce a series of pulses under water followed by bell-like sounds. Antarctic Weddell seal males repeat under water trills (rapid alternations of notes) during the breeding season.

Marine mammals also produce a broad variety of displays, including threat displays and recognition displays used for individual or group recognition.

Mechanisms of Sound Production

Most terrestrial mammals produce vocal sounds by vibrating vocal cords in the larynx. It is thought that the polar bear and most pinnipeds make sounds using similar mechanisms. Some adaptations for diving may affect vocalization mechanisms in pinnipeds. Pinnipeds have a more flexible trachea than do terrestrial mammals, so that air inside can compress during a dive, and they have a wider trachea to allow higher rates of air flow. Most pinnipeds can vocalize under water without emitting bubbles; some species have sacs attached to the trachea or upper respiratory sac, but the role of these in vocalization has not been determined. Walruses have many ways of producing sounds. They produce gonglike impulse sounds using specialized pharyngeal sacs, and can even use their lips to whistle in air.

Odontocetes have well developed vocal folds in the larynx, but most biologists argue that odontocetes produce sounds as air flows past the nasal plugs or phonic lips in the upper nasal passages (**Figure 4A**). Mechanisms for sound production must also match the acoustic impedance to the medium of air or sea water, and they may function to direct some sounds in a beam. The beam pattern of dolphin clicks (shown in **Figure 2B**) stems from a complex interaction of reflection from the skull and air sacs, coupled with refraction in soft tissues (**Figure 4A**).

There is a more detailed model of sound production for sperm whales (*Physeter macrocephalus*) than for other cetacean species. Sperm whales have a large organ called the spermaceti organ, which lies dorsal and anterior to the skull (**Figure 4B**). Below the spermaceti organ is the 'junk', which is composed of a series of fatty structures separated by dense connective tissue. The primary vocalizations of sperm whales are distinctive clicks comprising a burst of pulses with equally spaced interpulse intervals (IPIs). Bioacousticians suggest that these regular IPIs may result from reverberation within the

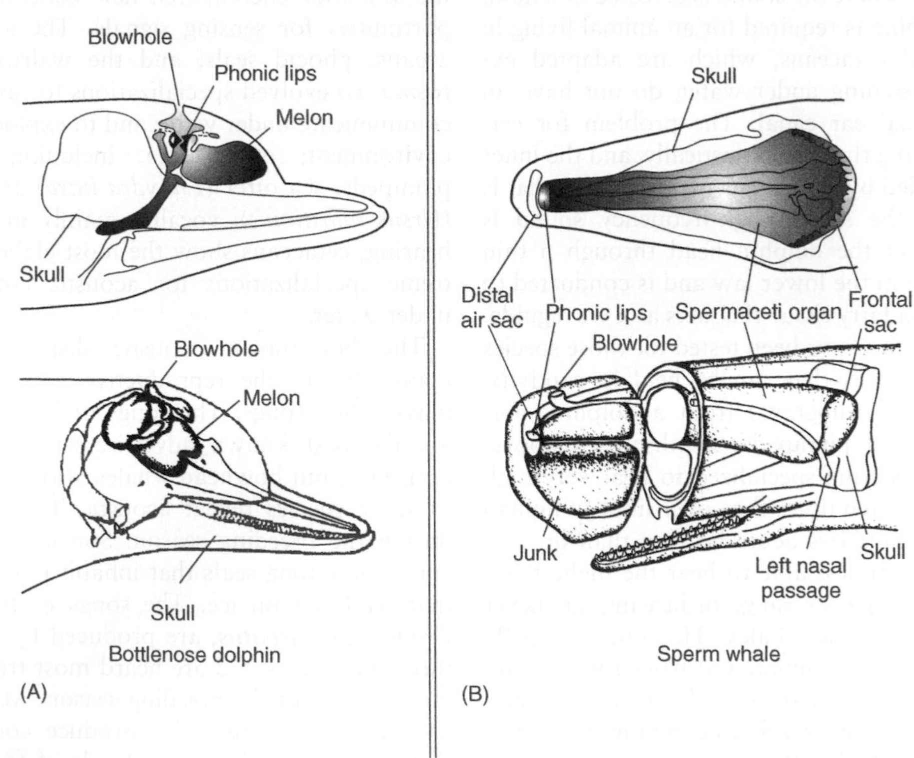

Figure 4 Functional anatomy of sound production in two odontocete cetaceans: (A) bottlenose dolphin *Tursiops truncatus*; (B) sperm whale *Physeter macrocephalus*. (Adapted from Figures 1.4 and 3.1 of Au *et al.* (2000).)

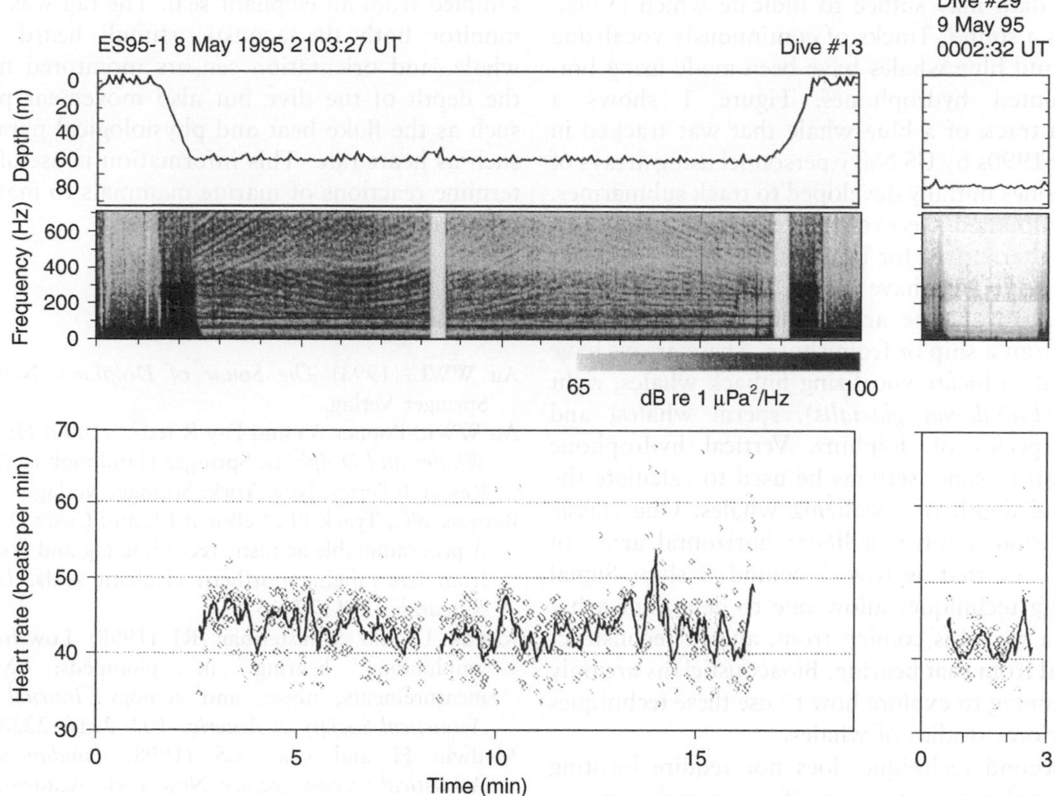

Figure 5 Data on dive profile, acoustic record, and acoustically determined heart rate from a tag on an elephant seal. The acoustic record shows a vessel passing. The closest point of approach occurs at the minimum frequency of the 'U' shaped pattern in the spectrogram at about 6 min. The heart rate differs little from the surrounding times or from a quiet period in a later dive from the same seal. (μPa = micropascal, reference for sound pressure measurements.) (Adapted from Figure 8 of Burgess *et al.* (1998).)

spermaceti organ. The frontal sac at the posterior end of the spermaceti organ has been suggested as a potential reflector of sound and the distal sac as a partial reflector of sound at the anterior end (**Figure 4B**). The source of the sound energy in the click is thought to come from a strong valve (phonic lips) in the right nasal passage at the anterior end of the spermaceti organ (**Figure 4B**). This sound production model suggests that some of the energy from the first pulse within the click is transmitted directly into the water. The remaining pulses are hypothesized to occur as some of the sound energy passes through the anterior reflector into the ocean at each reflection there.

Methods for Bioacoustic Research

It has been difficult to integrate visual observation of social behavior with patterns of vocalization in submerged mammals because it is difficult to identify which animal within an interacting group produces a sound under water. Biologists studying terrestrial animals take it for granted that they can identify which animal is vocalizing by using their own ears to locate the source of a sound and then looking for movements associated with sound production. Humans cannot locate sounds under water in the same way that they locate airborne sounds. Furthermore, marine mammals seldom produce visible motions coordinated with sound production under water. It is even more difficult to attempt behavioral observations on marine mammals during a dive when they are out of sight. The need for some technique to track behavior during a dive and to identify which cetacean produces which sound during normal social interaction has been discussed for over three decades. Two different approaches have emerged: (1) passive acoustic location of sound sources using an array of hydrophones; (2) recording information about behavior and sound production by attaching a tag onto the animal.

Acoustic location of vocalizing animals is a useful method for identifying which animal is producing a sound. It involves no manipulation of the animals, merely placement of hydrophones near them. In some applications, animals may vocalize frequently enough and be sufficiently separated that source location data may suffice to indicate which animal produces a sound. Tracks of continuously vocalizing finback and blue whales have been made using bottom-mounted hydrophones. **Figure 1** shows a 1700 km track of a blue whale that was tracked in the early 1990s by US Navy personnel using arrays of hydrophones initially developed to track submarines. Bottom-mounted recording devices are proving cheaper alternatives for biologists today.

Bioacousticians have also developed smaller, portable hydrophone arrays that can be deployed rapidly from a ship or from shore. These arrays have been used to locate vocalizing finback whales, right whales (*Eubalaena glacialis*), sperm whales, and several species of dolphins. Vertical hydrophone arrays can in some settings be used to calculate the range and depth of vocalizing whales. One classic configuration involves a linear horizontal array of hydrophones that is towed behind a ship. Signal processing techniques allow one to determine what bearing a sound is coming from, and to reconstruct the signal from that bearing. Bioacousticians are only just beginning to explore how to use these techniques in behavioral studies of whales.

The second technique does not require locating each animal within a group. If an animal carries a telemetry device that transmits acoustic data recorded at the animal, then the device can record all vocalizations of the animals along with most everything else it hears. This kind of tag can also record depth of dive, movement and orientation of the tagged animal. However, it is difficult to telemeter information through sea water, and marine mammals might sense many of the signals one might want to use for telemetry. These problems with telemetry have led biologists to develop recoverable tags that record data while on an animal, but that need to be recovered from the animal in order for the data to be downloaded. Recently, biologists have had successful programs recovering such tags from many different kinds of marine mammal. Recoverable acoustic tags may have scientific uses well beyond identifying vocalizations. **Figure 5** shows acoustic and dive data sampled from an elephant seal. The tag was able to monitor both the acoustic stimuli heard by the whale, and orientation sensors monitored not just the depth of the dive but also movement patterns such as the fluke beat and physiological parameters such as heart rate. This information is useful to determine reactions of marine mammals to man-made noise, an issue of growing concern.

Further Reading

Au WWL (1993) *The Sonar of Dolphins*. New York: Springer Verlag.

Au WWL, Popper AS and Fay R (eds.) (2000) *Hearing by Whales and Dolphins*. Springer Handbook of Auditory Research Series. New York: Springer Verlag.

Burgess WC, Tyack PL, LeBoeuf BJ, and Costa DP (1998) A programmable acoustic recording tag and first results from free-ranging northern elephant seals. *Deep-Sea Research* 45: 1327–1351.

Kastak D and Schusterman RJ (1998) Low-frequency amphibious hearing in pinnipeds: Methods, measurements, noise, and ecology. *Journal of the Acoustical Society of America* 103: 2216–2228.

Medwin H and Clay CS (1998) *Fundamentals of Acoustical Oceanography*. New York: Academic Press.

Miller P and Tyack PL (1998) A small towed beamforming array to identify vocalizing resident killer whales (*Orcinus orca*) concurrent with focal behavioral observations. *Deep-Sea Research* 45: 1389–1405.

Rayleigh, Lord (1945) *The Theory of Sound*. New York: Dover.

Tyack P (1998) Acoustic communication under the sea. In: Hopp SL, Owren MJ, and Evans CS (eds.) *Animal Acoustic Communication: Recent Technical Advances*, pp. 163–220. Heidelberg: Springer Verlag.

Tyack PL (2000) Functional aspects of cetacean communication. In: Mann J, Connor R, Tyack PL, and Whitehead H (eds.) *Cetacean Societies: Field Studies of Dolphins and Whales*, pp. 70–307. Chicago: University of Chicago Press.

Wartzok D and Ketten DR (1999) Marine mammal sensory systems. In: Reynolds JE III and Rommel SA (eds.) *Biology of Marine Mammals*, vol. 1, pp. 117–175. Washington, DC: Smithsonian Press.

MARINE MESOCOSMS

J. H. Steele, Woods Hole Oceanographic Institution, MA, USA

Controlled experiments are the basis of the scientific method. There are obvious difficulties in using this technique when dealing with natural communities or ecosystems, given the great spatial and temporal variability of their environment. On land the standard method is to divide an area of ground, say a field, into a large number of equal plots. Then with a randomized treatment, such as nutrient addition, it is possible to replicate growth of plants and animals over a season.

It is apparent that this approach is not possible in the open sea because of continuous advection and dispersion of water and the organisms in it. Bottom-living organisms are an exception, especially these living near shore, so there have been a wide range of experiments on rocky shores, salt marshes, and sea grasses. But even there, the critical reproductive period for most animals involves dispersion of the larvae in a pelagic phase. Also these experiments require continuous exchange of sea water.

For the completely pelagic plants and animals, short-term experiments – usually a few days – on single species are used to study physiological responses. There can be 24-hour experimental measurements of the rates of grazing of copepods on phytoplankton in liter bottles. But for studies of longer-term interactions, much larger volumes of

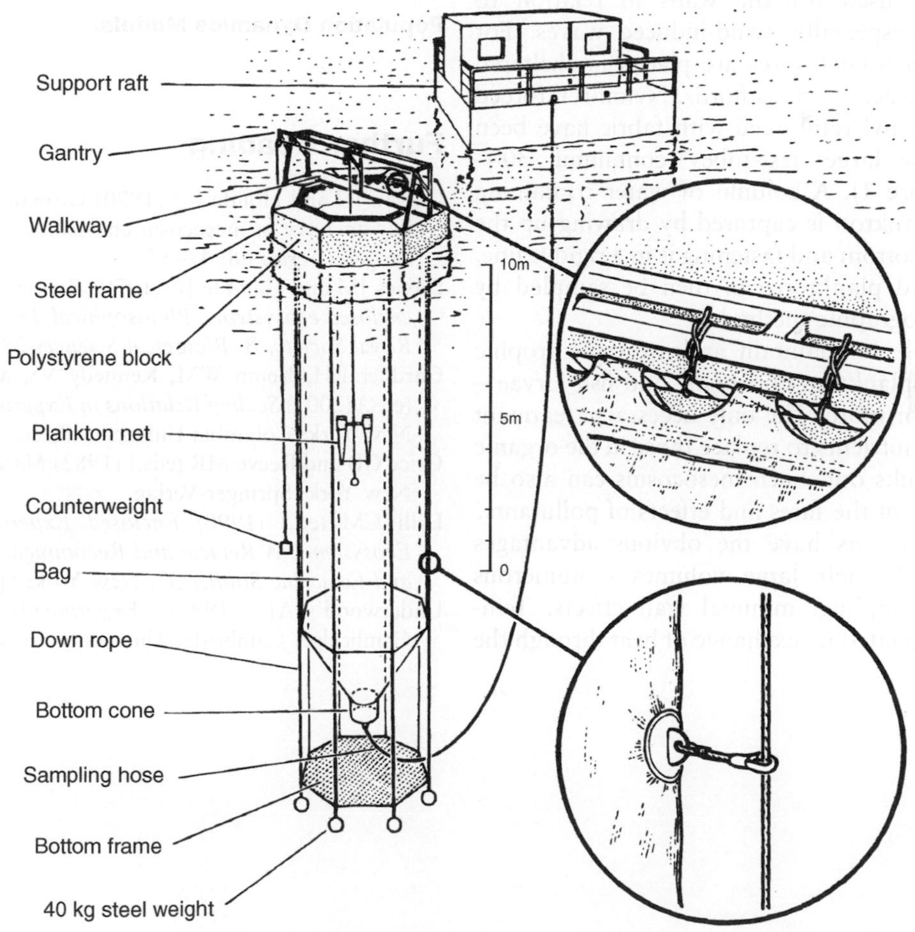

Figure 1 The design of a mesocosm used in Loch Ewe, Scotland for studies of the dynamics of plankton communities and of fish larval growth and mortality (adapted from Davies and Gamble, 1979).

water are necessary, to contain whole communities and to minimize wall effects of the containers.

To this end 'mesocosms' – containers much larger than can fit into the normal laboratory – have been used in a variety of designs and for a diversity of purposes. The first choice is whether to construct these on land, at the sea's edge, or to immerse them in the sea. The former has advantages in durability, ease of access, and re-use. There are constraints on the volumes that can be contained, difficulties in temperature control, and, especially, problems in transferring representative marine communities from the sea to the tanks. This approach was used originally in tall relatively narrow tanks to study populations of copepods and fish larvae; in particular to experiment on factors such as light that control vertical migration. Another use of such large tanks is to study the effect of pollutants on communities of pelagic and benthic organisms.

These shore-based tanks are limited by the weight of water, usually to volumes of 10–30 m^3. Enclosures immersed in the sea do not have this constraint. Instead the problems concern the strength of the flexible materials used for the walls in relation to currents and, especially, wind-induced waves. For this reason, such enclosures are placed in sheltered semienclosed places such as fiords. Nylon-reinforced polythene or vinyl reinforced with fabric have been used for these large 'test-tubes' containing 300–3000 m^3 (**Figure 1**). A column of water containing the natural plankton is captured by drawing up the bag from the bottom and fastening it in a rigid frame. The water and plankton can then be sampled by normal oceanographic methods.

It is possible to maintain at least three trophic levels – phytoplankton, copepods, and fish larvae – for 100 days or more. The only necessary treatment is addition of nutrients to replace those in the organic matter that sinks out. Such mesocosms can also be used for study of the fates and effects of pollutants.

These mesocosms have the obvious advantages associated with their large volumes – numerous animals for sampling, minimal wall effects. Temperature is regulated by exchange of heat through the walls. But they have various drawbacks. Not only is advection suppressed but vertical mixing decreases so that the outside physical conditions are not reproduced. The greatest disadvantage, however, is lack of adequate replication. There have been only three to six of these mesocosms available for any experiment and pairs did not often agree closely. Thus each tube represents an ecosystem on its own rather than a replicate of a larger community.

The need for experimental results at the community level represents an unresolved problem in biological oceanography. There are smaller-scale experiments continuing. Open mesh containers through which water and plankton pass can be a compromise for the study of small fish and fish larvae. It is now possible to mark a body of water with very sensitive tracers and follow the effects on plankton of the addition of nutrients, specifically iron, for several weeks. The concatenation of these results may have to depend on computer simulations.

See also

Population Dynamics Models.

Further Reading

Cowan JH and Houde ED (1990) Growth and survival of bay anchovy in mesocosm enclosures. *Marine Ecology Progress Series* 68: 47–57.

Davies JM and Gamble JC (1979) *Experiments with large enclosed ecosystems. Philosophical Transcations of the Royal Society, B. Biological Sciences* 286: 523–544.

Gardner RH, Kemp WM, Kennedy VS, and Peterson JS (eds.) (2001) *Scaling Relations in Experimental Ecology.* New York: Columbia University Press.

Grice GD and Reeve MR (eds.) (1982) *Marine Mesocosms.* New York: Springer-Verlag.

Lalli CM (ed.) (1990) *Enclosed Experimental Marine Ecosystems: A Review and Recommendations. Coastal and Estuarine Studies* 37. New York: Springer-Verlag.

Underwood AJ (1997) *Experiments in Ecology.* Cambridge: Cambridge University Press.

MESOCOSMS: ENCLOSED EXPERIMENTAL ECOSYSTEMS IN OCEAN SCIENCE

J. E. Petersen, Oberlin College, Oberlin, OH, USA
W. M. Kemp, University of Maryland Center for Environmental Science, Cambridge, MD, USA

Introduction: Experimental Ecosystems as Tools for Aquatic Research

Within the last few decades there has been a clear trend within ecological science of growing reliance on manipulative experiments as a means of testing ecological theory. Many approaches are available for experimentation. An important distinction can be drawn between field and laboratory-based experiments. In field experiments, either parts of nature or whole, naturally bounded ecosystems are manipulated in place while similar areas are left as controls. In laboratory experiments, organisms, communities, and the physical substrate are transported to controlled facilities. A second distinction can be drawn between experiments in which organisms and materials freely exchange between the experiment and surrounding environment and those in which organisms and materials are enclosed and isolated either in a laboratory setting or with physical boundaries imposed in the field. The term 'enclosed experimental ecosystem' is used when the goal of an enclosure experiment, conducted in either laboratory or field conditions, is to explore interactions among organisms or between organisms and their chemical and physical environment. Because enclosed experimental ecosystems are intended to serve as miniaturized worlds for studying ecological processes, they are often called 'microcosms' or 'mesocosms'.

Enclosed experimental ecosystems have become widely used research tools in oceanographic and freshwater sciences because they allow for a relatively high degree of experimental control and replication necessary for hypothesis testing while still capturing dynamics that emerge from ecosystem-level interactions between organisms and their physical and chemical environments. They provide a bridge between observational field studies and process-oriented lab research. Mesocosms have been used to conduct experiments on a broad range of aquatic habitats. Over the last 30 years, enclosed experimental ecosystems have become important tools in both coastal and open ocean contexts to address critical research questions in the fields of chemical and physical oceanography, ecotoxicology, fisheries science, and basic and applied ecology (**Figure 1**).

Two fundamental objectives of ecological experiments are to achieve high levels of control and realism. Control refers to the ability to relate cause and effect, to manipulate, to replicate, and to repeat experiments; realism is a measure of the degree to which results accurately mimic the dynamics of particular natural ecosystems. Trade-offs between control and realism are inherent in different experimental approaches; experiments conducted within nature tend to maximize realism, whereas physiological experiments in the laboratory allow for the highest degree of experimental control. In theory, mesocosms provide intermediate levels of both control and realism (**Figure 2**).

Scale Is a Crucial Issue in Mesocosm Research

Scale is a crucial issue for all ocean scientists and has particular implications for researchers using enclosed experimental ecosystems. How can large-scale processes be simulated and incorporated into enclosed experimental ecosystems so as to maximize realism? How can research findings be quantitatively extrapolated from small, often simplified experimental ecosystems up to whole natural ecosystems? For that matter, how can information gleaned from research in one type of ecosystem be extrapolated to other natural ecosystems that differ in scale? Recent research indicates that scale effects can be parsed into 'fundamental effects', that are evident in both natural and experimental ecosystems, and 'artifacts of enclosure', that are solely attributable to the artificial environment in mesocosms. A key objective of this contribution to the encyclopedia is to review the ways in which mesocosm experiments have been used to study the marine environment and to suggest ways in which scaling considerations can be used to improve the use of mesocosm's research tools.

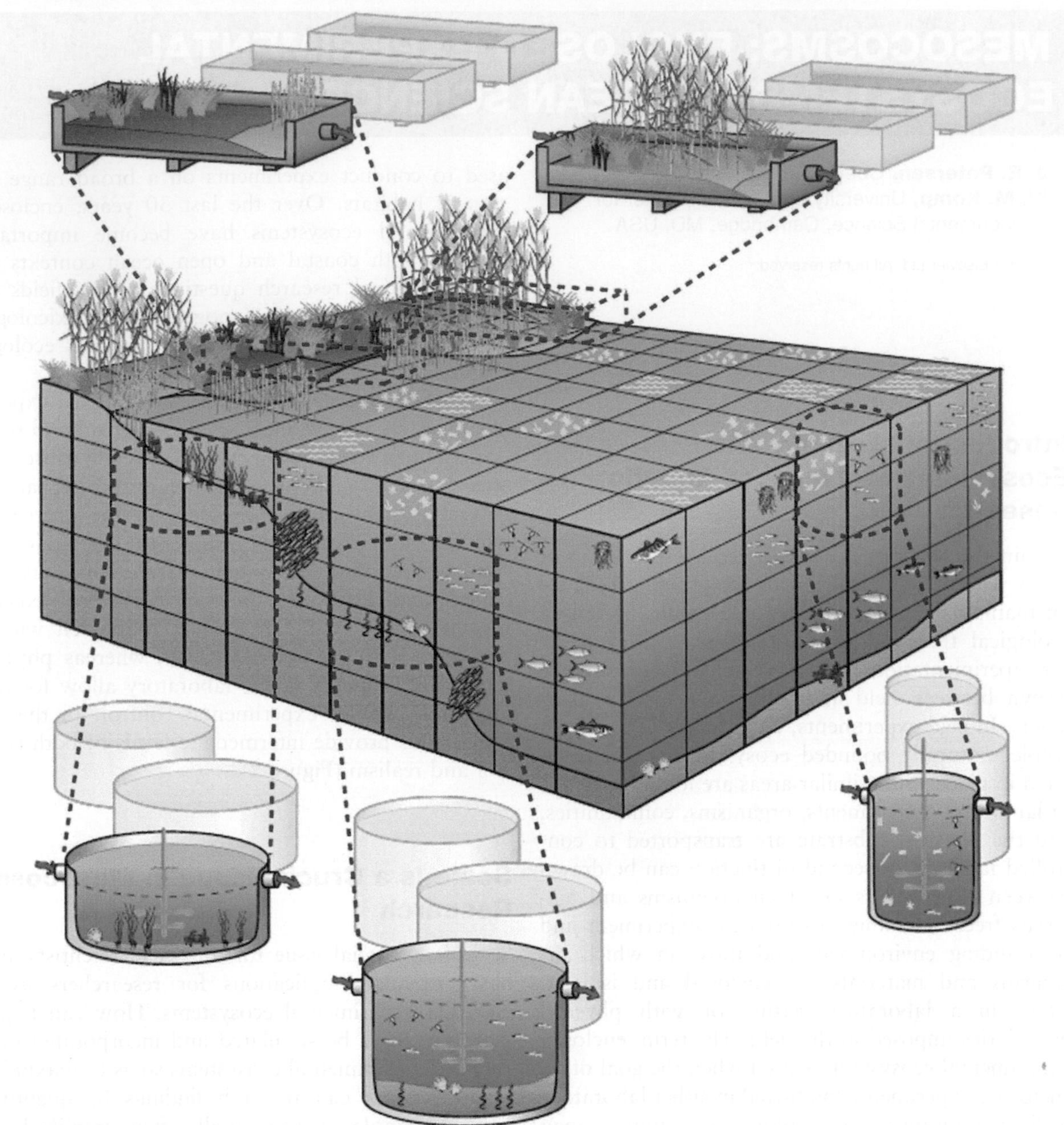

Figure 1 Enclosed experimental ecosystems provide a means of conducting controlled, replicated experiments to reveal processes and interactions that occur within different marine habitats. Adapted from Petersen JE, Kennedy VS, Dennison WC, and Kemp WM (eds.) (2008) *Enclosed Experimental Ecosystems and Scale: Tools for Understanding and Managing Coastal Ecosystems.* New York: Springer.

History and Applications

There is a rich history in the use of enclosed experimental ecosystems. The initial concept of microcosms, as hierarchically nested miniature worlds contained successively within larger worlds, has been credited to early Greek philosophers including Aristotle. Although it is difficult to date the initial scientific uses of enclosed experimental ecosystems, small glass jars and other containers were routinely used as experimental ecosystems by the middle of the twentieth century. H.T. Odum and his colleagues were pioneers and proponents of the use of mesocosms to study aquatic ecosystems. They constructed a wide variety of experimental ecosystems including laboratory streams, containers with planktonic and vascular plant communities, and shallow outdoor ponds containing oysters and/or seagrasses. Although the word 'microcosm' was used initially to describe virtually all experimental ecosystems, the term 'mesocosm' was later adopted to

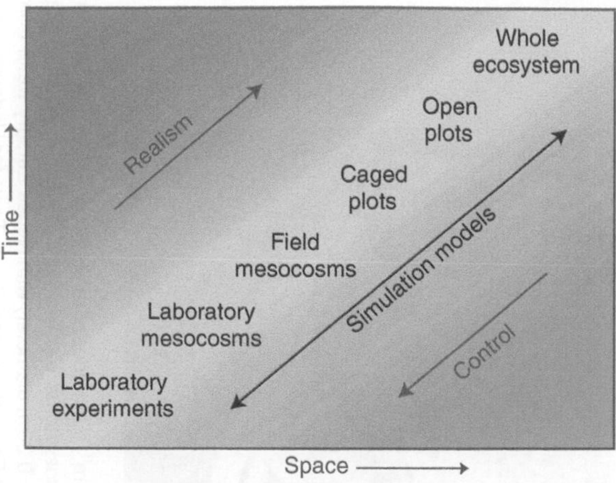

Figure 2 As the scale of experiments increases from simple laboratory experiments to complex whole-ecosystem manipulations, greater realism is possible, but control over experimental conditions declines. Simulation models can be used to synthesize and integrate results from all types of studies. Adapted from Petersen JE, Kennedy VS, Dennison WC, and Kemp WM (eds.) (2008) *Enclosed Experimental Ecosystems and Scale: Tools for Understanding and Managing Coastal Ecosystems*. New York: Springer.

distinguish larger experimental units from smaller bench-top laboratory systems.

Some have suggested that experimental manipulations of whole aquatic ecosystems in nature are always preferable to mesocosm studies. However, the characteristically steep spatial gradients, three-dimensional water exchanges, lack of boundaries, and natural variability make such whole ecosystem manipulations extremely difficult to accomplish in coastal and open ocean environments, leaving mesocosms as critical tools for controlled experimentation. A series of books devoted to aspects of experimental aquatic ecosystems mark recent progress with this research approach.

There are diverse styles and applications of enclosed experimental ecosystems (**Figure 3**). During the last four decades, experimental microcosms and mesocosms have been developed in a diversity of sizes, shapes, and habitats to address a broad range of research questions. Small (~0.5 l) laboratory chemostat flasks have been widely used by R. Margalef and others to study plankton community dynamics, while large (30–1300 m^3) plastic bag enclosures have been deployed *in situ* by J. Gamble, G. Grice, D. Menzel, T. Parsons, M. Reeve, J. Steele, J. Strickland, and others to study pelagic (in some cases including benthic) coastal ecosystems in Europe and North America. Similarly, mesocosm shapes vary from the tall and relatively narrow (23 m high × 9.5 m deep) *in situ* plankton bags used by Gamble, Steele, and their colleagues to broad (350 m^2 surface), shallow (1 m deep) estuarine ponds used by R. Twilley and others. Mesocosms have been constructed to study diverse marine habitats, including planktonic regions of oceans and estuaries, deep benthos, shallow tidal ponds, coral reefs, salt marshes, and seagrasses.

Composition and organization of experimental ecological communities range broadly and include: simple 'gnotobiotic' ecosystems where all species are selected and identified; interconnected microcosms, each containing a different trophic level; intact 'undisturbed' columns of sediment and overlying water extracted and contained; and tidal ponds with 'self-organizing' communities developed by seeding with diverse inoculant communities taken from different natural ecosystems.

Marine mesocosms have been used effectively to address a range of theoretical and applied scientific questions. Early studies using *in situ* bag enclosures (e.g., Controlled Ecosystem Pollution Experiment (CEPEX), Loch Ewe Enclosures, Kiel Plankton Towers) examined planktonic food web responses to nutrient enrichment and introduction of toxic contaminants (e.g., copper, mercury). These experiments were designed to assess the effects of both 'bottom-up' (resource-limited) and 'top-down' (herbivore- and predator-determined) controls (e.g., **Figures 3(a)** and **4**). Although these studies were very instructive, difficulties in controlling mixing regimes and lack of replication of treatments tended to limit interpretation of results. Later studies, notably the land-based Marine Ecosystem Research Laboratory (MERL), employed mechanical mixing, added intact sediments, and increased replication (**Figure 3(b)**). MERL systems were used by S. Nixon, C. Oviatt, and their colleagues to investigate trophic and biogeochemical responses to similar treatments including N, P, and Si enrichment, crude oil contamination, filter-feeding, and storm mixing events. The versatile and permanent MERL facility allowed investigators to explore interactions between pelagic and benthic communities that are critical in the dynamics of shallow coastal ecosystems (e.g., **Figure 5**).

The Challenges and Opportunities of Scale in Mesocosm Research

Two parallel trends in ecology during the last 20 years have been an increased use of mesocosms as research tools (**Figure 6(a)**) and an increased recognition of the importance of scale as a determinant of the patterns and processes observed in natural ecosystems (**Figure 6(b)**). As we have discussed, mesocosms have become widely used and accepted tools in ocean science because they provide a means of conducting

Figure 3 Marine mesocosm facilities have taken diverse forms including (a) Controlled Ecosystem Pollution Experiment (CEPEX, 1300 m^3, 17 m deep, 3 m diameter) system in Saanich Inlet, British Columbia, 1978; (b) Marine Ecosystem Research Laboratory (MERL, 13 m^3, 5 m deep, 1.8 m diameter) experimental ecosystems established in 1980; (c) rocky littoral mesocosms (23 m^3, 4.7 m long, 3.6 m wide, 1.3 m deep) at Solbergstrand, Norway; (d) Plankton community mesocosms (55 l, 0.77 m deep, 0.30 m diameter) with Neuse Estuary water from University of North Carolina. Adapted from Petersen JE, Kennedy VS, Dennison WC, and Kemp WM (eds.) (2008) *Enclosed Experimental Ecosystems and Scale: Tools for Understanding and Managing Coastal Ecosystems*. New York: Springer.

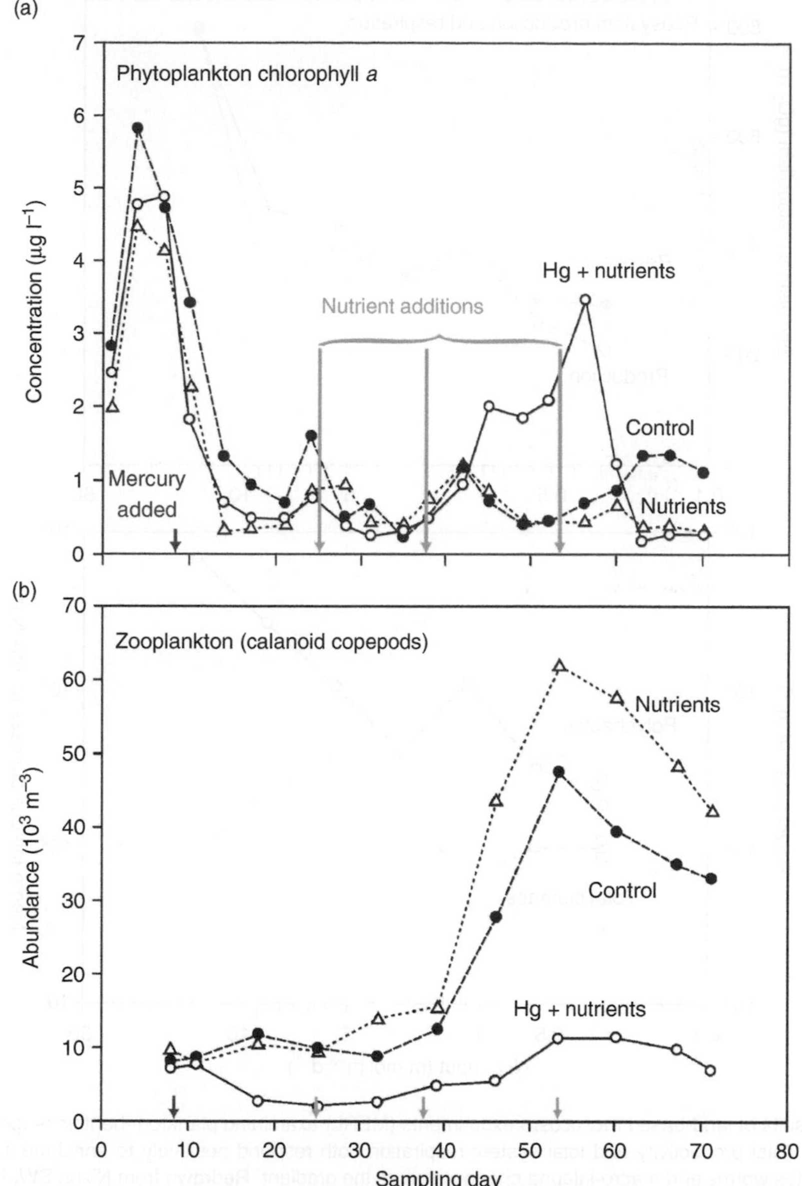

Figure 4 Example results of *in situ* mesocosm experiments (CEPEX) designed to investigate 'top-down' (predator) and 'bottom-up' (nutrient) controls on phytoplankton. Inorganic nutrients were added (on days 25, 37, and 53) to two of three mesocosms to stimulate primary productivity (a). Mercury was added to one of these mesocosms (on day 9) to reduce zooplankton abundance (b). Although the experiments incorporated no replication, the findings contributed to our understanding of the importance of top-down control. Redrawn from Grice GD, Reeve MR, Koeller P, and Menzel DW (1977) The use of large volume, transparent, enclosed sea-surface water columns in the study of stress on plankton ecosystems. *Helgolander Wissenschaftliche Meeresuntersuchungen* 30: 118–133.

ecosystem-level experiments under replicated, controlled, and repeatable conditions. The focus on scale can be traced to a number of factors including: theoretical and technological advances that increase our understanding of causal linkages between local, regional, and global phenomena; a growing awareness of human impact at all scales; and the formalization of scale as a legitimate topic of inquiry within the emerging field of landscape ecology. This emphasis on scale is evidenced by the steady increase in the number of journal articles listing 'scale' as a keyword (**Figure 6(b)**) and in the publication of a number of new books devoted to scaling theory.

It has long been recognized that scale is an inherent design problem that may confound the interpretation of results from experimental ecosystem studies. Since their use first became prevalent in the 1970s, researchers have expressed concerns regarding scaling problems associated with mesocosms including the effects of: reduced system size and short timescale of

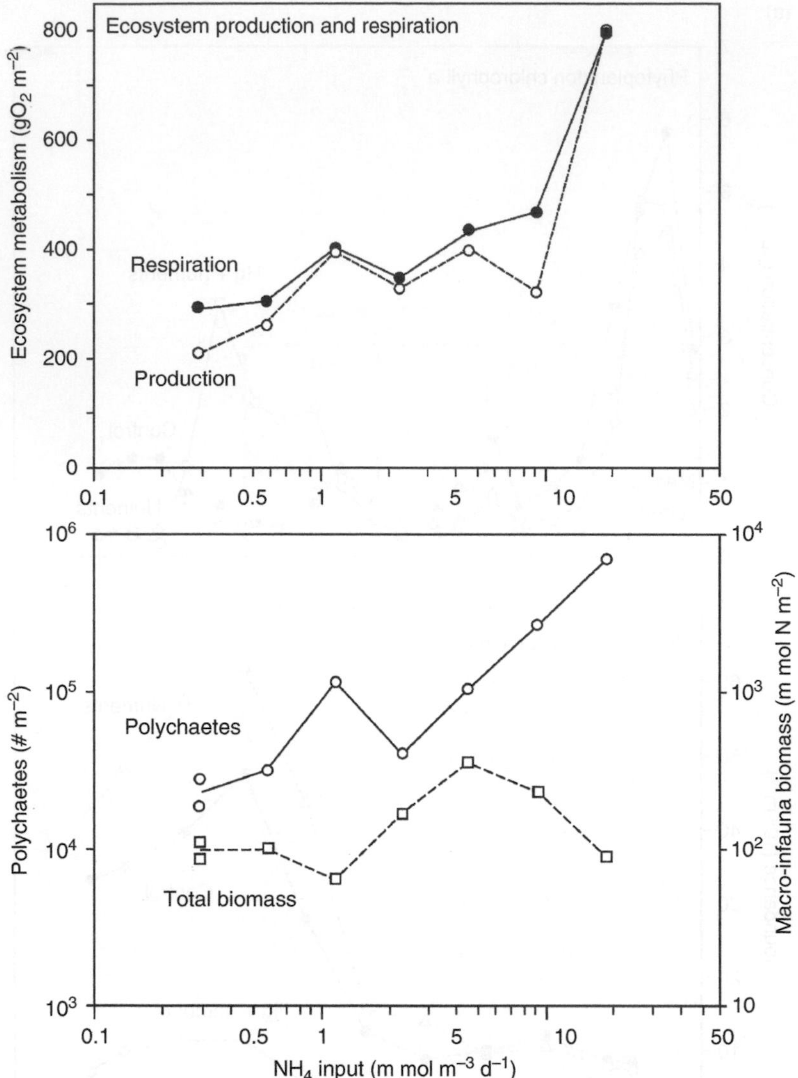

Figure 5 Example results of land-based mesocosm experiments (MERL) examining plankton–benthic responses to different levels of nutrient enrichment. Total productivity and total system respiration both respond positively to enrichment. However, the relative importance of polychaetes worms and macro-infauna changes across the gradient. Redrawn from Nixon SW, Pilson MEQ, and Oviatt CA (1984) Eutrophication of a coastal marine ecosystem – an experimental study using the MERL microcosms. In: Fasham MJR (eds.) *Flows of Energy and Materials in Marine Ecosystems: Theory and Practice*, pp. 105–135. New York: Plenum.

experiments, reduced ecological complexity, wall growth, limitations on animal movements, distorted mixing regimes, and unrealistic water exchange rates. A few investigators have used a simple idea of mesocosm calibration, where key properties are adjusted in experimental systems to mimic conditions in the natural environment. However, the majority of early mesocosm studies skirted the question of scaling and the problem of extrapolation altogether. By the end of the 1980s, it was clear that further progress in the application of experimental ecosystem methods to aquatic science would require focused quantitative study of how scale affects behavior in natural and experimental ecosystems and how experimental ecosystems might be better designed to account for scale. The development of systematic techniques for extrapolating results from small experimental ecosystem studies to conditions in nature at large remains an active area of research. Recent research (e.g., at the Multiscale Experimental Ecosystem Research Center (MEERC)) has focused on developing quantitative and systematic approaches for the design and interpretation of experimental ecosystem research with a particular focus on the problem of scale.

Several scaling concerns must be addressed when using mesocosm results to predict effects in natural aquatic ecosystems. The first and most obvious is

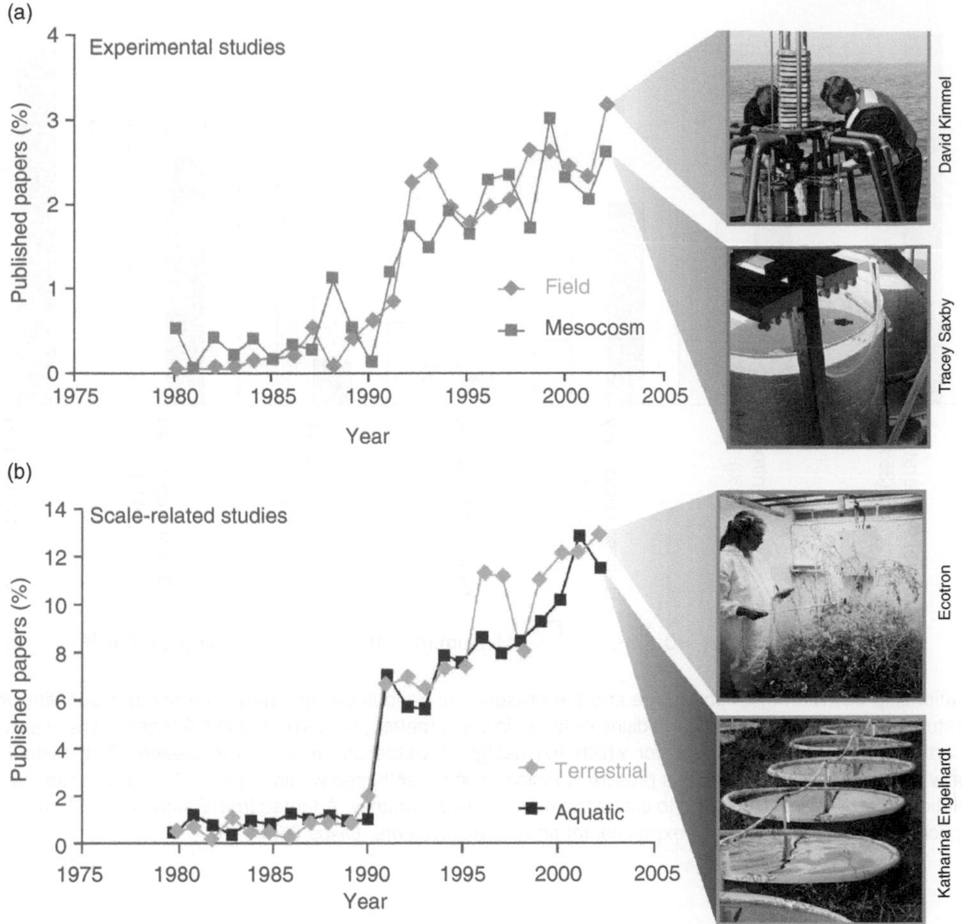

Figure 6 (a) Trends in use of field experiments and mesocosms in ecological studies as revealed from keyword searches in ecological journals. Note that field experiments and mesocosm studies are not mutually exclusive categories because the latter can be used in the field. The patterns suggest an increasing reliance on both categories of experimentation. (b) Trends in scale studies in ecology based on separate searches conducted by year for the term 'scale' in keywords and abstracts of journals emphasizing terrestrial research (*Ecology*, *Oikos*, *Oecologia*) and journals publishing only aquatic research (*Limnology and Oceanography*, *Marine Ecology Progress Series*). The number of papers identified in each year was then standardized to the total number of papers published for that year in those journals and expressed in the graph as a percent. Adapted from Petersen JE, Cornwell, JC, and Kemp WM (1999) Implicit scaling in the design of experimental aquatic ecosystems. *Oikos* 85: 3–18.

that experimental systems are constrained in size and duration. An extensive literature review revealed a median experimental duration of 49 days and median volume of $1.7\,m^3$; aquatic mesocosm experiments are brief and small relative to the natural scales that characterize many important ecological processes of interest. A second problem is the presence of walls, which restrict biological, material, and energy exchange with the outside world and provide a substrate for growth of undesirable but potentially influential organisms on this artificial edge habitat. A third problem is that a host of experimental design decisions – such as how many replicates to include per treatment and whether to control light, mixing, and other properties – tend to vary together with choices of size, duration, and ecological complexity (**Figures** 7 and 8). Finally, the relative importance of the air–water area, sediment–water area, and wall area, in relation to each other and to water and sediment volume, changes with physical dimensions. Unfortunately, parallel scaling problems also exist for field experiments. For example, replication tends to decrease with increasing plot size and experimental lakes and field plots tend to be orders of magnitude smaller than the natural systems for which inferences are drawn.

An analysis of aquatic studies conducted in cylindrical planktonic–benthic mesocosms reveals that in designing experimental ecosystems, researchers gravitate toward a depth/radius ratio of approximately 4.5 (**Figure** 9). As a consequence of this bias, in general, larger mesocosms are simultaneously less

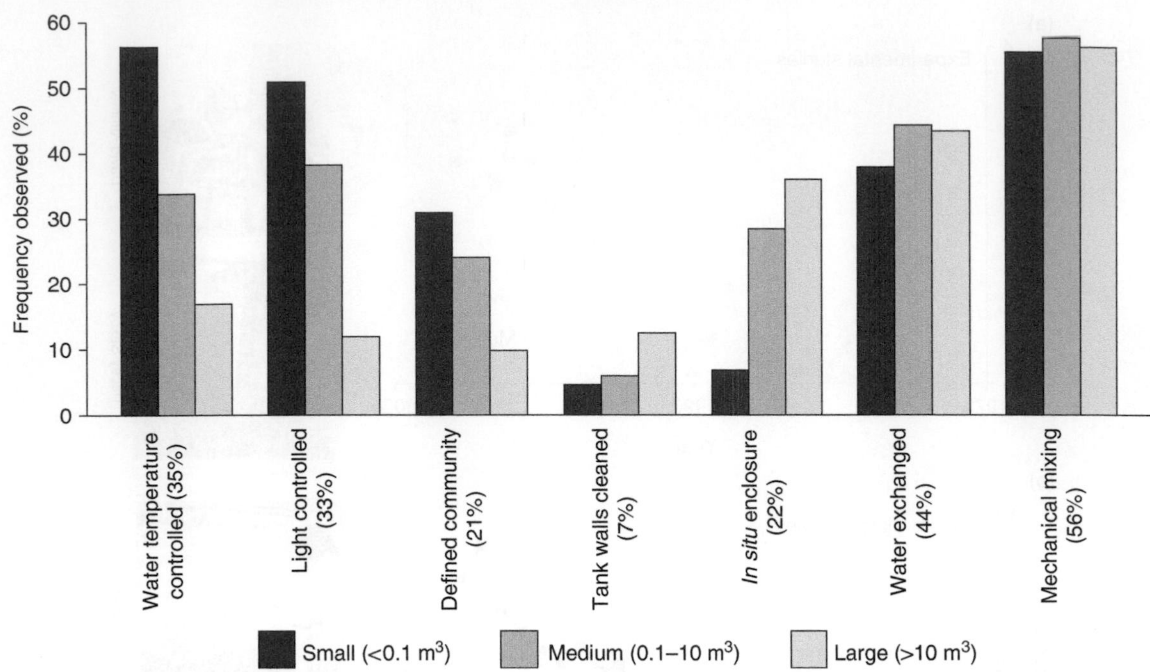

Figure 7 Relationship between mesocosm size and the presence of various design characteristics in a quantitative review of the mesocosm literature. Size categories (small, medium, or large, in cubic meters) are indicated in the legend. The *y*-axis represents the percentage of articles in a given size class for which the design characteristic indicated is present. The overall percentage of experiments for which a given characteristic is present is indicated in parentheses within the key. 'Defined community' indicates that individual populations were selectively added to create the mesocosm community. Adapted from Petersen JE, Cornwell JC, and Kemp WM (1999) Implicit scaling in the design of experimental aquatic ecosystems. *Oikos* 85: 3–18.

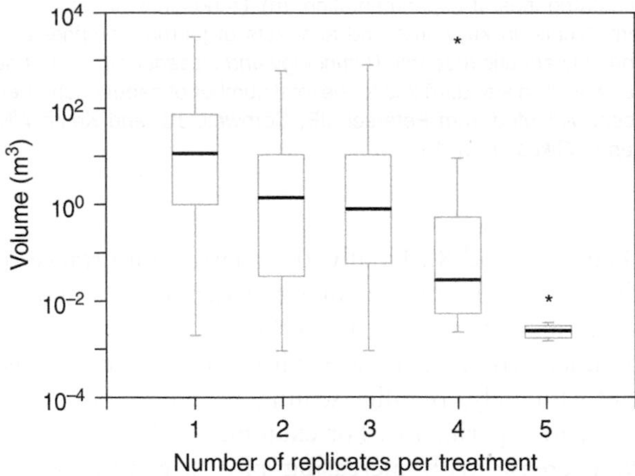

Figure 8 Plot of mesocosm volume vs. number of replicates per treatment. Median values are represented by the bar within a box, and the 75th and 25th percentiles (i.e., the interquartile range) by the top and bottom of the box. The ends of the 'whiskers' represent the farthest data point within a span that extends 1.5 times the interquartile range from the 75th and 25th percentiles. Data outside this span are graphed with asterisks. Adapted from Petersen JE, Cornwell JC, and Kemp WM (1999) Implicit scaling in the design of experimental aquatic ecosystems. *Oikos* 85: 3–18.

influenced by wall artifacts, have less sediment area per unit volume, and have less surface area available for gas and light exchange per unit volume than do smaller systems (**Figures 9(b)** and **9(c)**). Collectively, these scaling attributes can potentially confound interpretation, comparison, and extrapolation of findings from mesocosm experiments.

One might conclude from the preceding figures and discussion that reductions, artifacts, co-variation, and distortions in scale pose an almost insurmountable obstacle to designing mesocosm studies to examine oceanic processes. Alternatively, these problems can be viewed as interesting research opportunities to advance our theoretical and practical understanding of the 'science of scale'. A variety of mesocosm scaling experiments have been designed to shed light on two classes of effects: 'fundamental effects of scale' evident in both natural and experimental ecosystems (e.g., the effects of water depth), and 'artifacts of enclosure' attributable to the artificial environment in experimental ecosystems (e.g., the effects of wall growth). In these experiments, ecological responses are measured in relation to manipulations in experimental scales

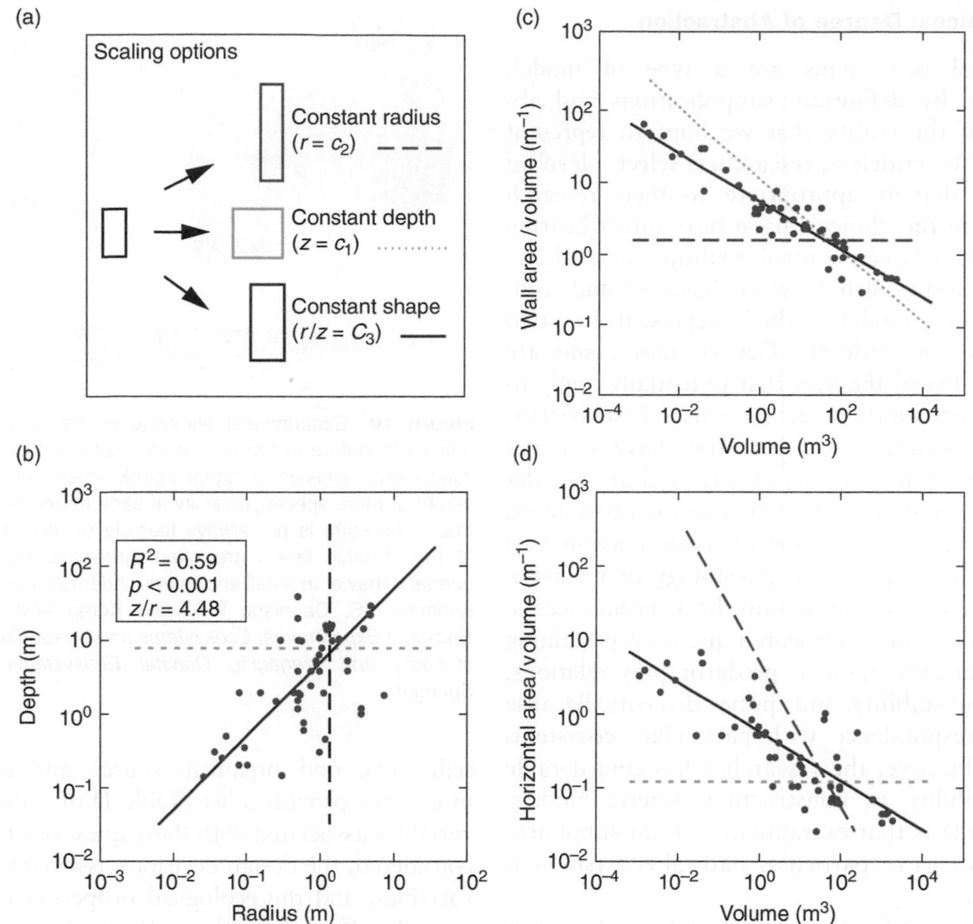

Figure 9 (a) Available options for conserving characteristic length relationships as the size of a cylindrical mesocosm is increased. (b) Relations between depth and radius for the cylindrical mesocosms in the ecological literature. Dots are physical dimension data from a comprehensive literature review of experiments conducted in mesocosms. (c) Surface areas of the vertical walls vs. volume. (d) Surface area of bottom and top vs. mesocosm volume. Dotted (green) lines represent scaling for constant depth and are placed at values corresponding with median depth. Dashed (red) lines that represent scaling for constant radius are placed at median radius. The solid (blue) lines represent scaling for constant shape and are derived from linear regression of radius (r) vs. depth (z), with statistics provided in (b). A clear implicit bias is evident toward scaling for constant shape. Adapted from Petersen JE, Cornwell JC, and Kemp WM (1999) Implicit scaling in the design of experimental aquatic ecosystems. *Oikos* 85: 3–18.

(i.e., time, space, and complexity) for a variety of coastal habitat types. Such studies suggest that it is possible to improve substantially the design of experimental ecosystems. Selected examples are discussed in the sections that follow.

Effective Design of Enclosed Experimental Ecosystems

There are many issues and questions that must be considered in the design of enclosed experimental ecosystems. Design decisions are important because they affect how results can be interpreted and extrapolated to nature. Optimal design is determined by the research question under consideration. The processes, organisms, and habitats associated with this question determine the appropriate size, shape, duration, and complexity for the experimental ecosystem. Even within a given ecosystem type, there is no single best design that will suit all research goals. Typically, the choices made will reflect a balance between three competing objectives: (1) control (the ability to relate cause and effect, to manipulate, to replicate, and to repeat experiments); (2) realism (the degree to which results accurately mimic nature); and (3) generality (the breadth of different systems to which results are applicable). There are, however, specific tools and guidelines available to aid in the experiment design process for enhancing the probability of research success. The sections below provide guidance on critical issues that must be considered.

Design Choices: Degree of Abstraction

Experimental ecosystems are a type of model. Models are, by definition, simplifications and abstractions of the reality that we hope to represent with them. As modelers, researchers select a level of abstraction that is appropriate to their research question, and the choices made have direct bearing on trade-offs between control, realism, and scale.

One can distinguish between 'generic' and 'ecosystem-specific' models, which represent the two extremes in this trade-off. Generic mesocosms are used to test broad theories that potentially apply to many different kinds of ecosystems. These systems tend to be small, highly artificial, have minimal physical and biological complexity, and are not designed to represent particular natural ecosystems. This is the ecological analog of using a worm as a model for studying human physiology or behavior. In ecology, generic mesocosms have been successfully applied to address research questions pertaining to ecosystem development, predator–prey relations, stress, system stability, and species diversity. Because precise correspondence with particular ecosystems is not an objective, the researcher has considerable design flexibility in constructing generic models. The downside is that extrapolation from simplified, abstract systems to particular natural ecosystems is challenging.

Ecosystem-specific mesocosms are used to test hypotheses linked to particular types of ecosystems. This is the ecological analog of using chimpanzees to study human physiology and behavior. To achieve the higher degree of realism required, these systems must incorporate the essential physical and biological features that control the dynamics in the systems that they represent. Various ecosystem-specific models have been constructed, ranging from coral reefs to coastal plain estuaries. As the desired degree of specificity and desired level of realism increase, so does the complexity of engineering necessary to achieve realistic ecological conditions (e.g., **Figure 10**).

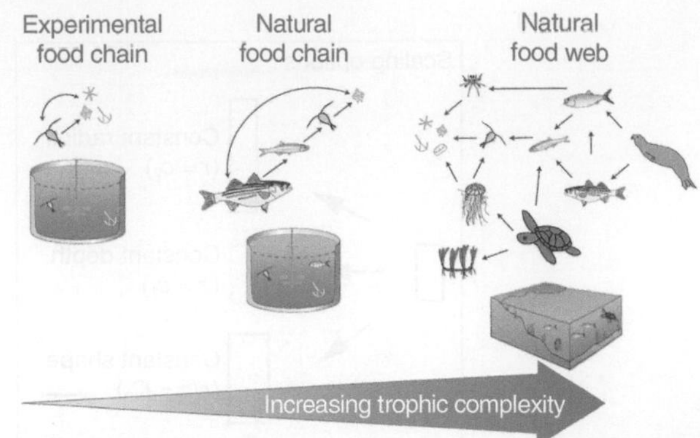

Figure 10 Experimental ecosystems are typically simplified relative to nature in terms of biodiversity and trophic (feeding) complexity. Inclusion of higher trophic levels (increased trophic depth) or more species diversity at each trophic level (increased trophic breadth) is not always feasible or desirable. Predators at high trophic levels are often large and may not exhibit normal behavior in small enclosures. Adapted from Petersen JE, Kennedy VS, Dennison WC, and Kemp WM (eds.) (2008) *Enclosed Experimental Ecosystems and Scale: Tools for Understanding and Managing Coastal Ecosystems.* New York: Springer.

Design Choices: Physical Characteristics

In addition to questions related to appropriate degree of abstraction and ecosystem type, researchers face crucial design questions regarding the physical characteristics of the experimental ecosystem. For example, what are the minimum system size, experimental duration, and ecological complexity necessary to answer the research question? How will the experimental system address each of the following design decisions: light source, mixing, temperature, exchange of water and constituents, inclusion of sediments, and organism source and introduction mode? We provide a list (**Table 1**) of some of the key variables associated with these questions that must be considered, the design decisions associated with these variables, and the ecological properties that are potentially affected by these design decisions. Choices related to physical characteristics are obviously also dependent on resources available in terms of funds, time, equipment, and support personnel.

Design Choices: Mixing and Exchange

Mixing and exchange of water and associated constituents are particularly important factors to consider in the design of enclosed experimental ecosystems. A core objective of mesocosm experiments is to isolate biological, chemical, and physical conditions to facilitate controlled manipulative experiments. This act of isolation can, however, create conditions within the mesocosm that are very different from those in nature, thereby distorting the dynamics observed in these experiments. Exchange can be defined as the net transport of water and its constituents through a system. Mixing can be defined as the physical movement of the water and its constituents within the system, generating turbulence within the fluid and homogenization of the constituents. Mixing and exchange are important aspects of natural marine ecosystems from the largest to the smallest of scales. Depending on how the system of interest is defined, mixing at one scale can sometimes be considered exchange at another scale.

Table 1 Key variables to consider in the design of experimental ecosystems

Variable	*Design decisions*	*Properties affected*
Size	Volume, depth, radius, surface area	Relative dominance of pelagic, benthic, and emergent producer communities, wall growth, temperature oscillations
Time	Experimental duration, timing of perturbation, sampling frequency	Ecological dynamics and life cycle of organism included in experiment, ability to detect seasonal and long-term effects, influence of experimental artifacts
Mixing	Vertical and horizontal mixing environment, mechanical mixing apparatus employed	Pelagic–benthic interactions, feeding rates and behavior, access to nutrients, artifacts, and potential mortality associated with mechanical devices
Materials exchange	Frequency, magnitude, variability chemical composition, biological composition	Recolonization rates, flushing of planktonic organisms, selection for particular organisms and communities
Light	Natural or artificial, intensity, spectral properties	Primary productivity, producer community composition, water temperature
Walls	Construction materials, whether to clean, cleaning frequency	Relative dominance of wall growth, light environment
Temperature	Whether to control, how to control	Rate of biogeochemical activity, selection for particular organisms
Ecological complexity	Species and functional group diversity, number of habitats and biogeochemical environments included	Primary productivity, trophic dynamics, biogeochemical pathways
Sediments	From nature or synthesized, intact or homogenized, particle size, organic matter content, organisms included	Pelagic–benthic interactions, vascular plant growth, primary productivity

Adapted from Petersen JE, Kennedy VS, Dennison WC, and Kemp WM (eds.) (2008) *Enclosed Experimental Ecosystems and Scale: Tools for Understanding and Managing Coastal Ecosystems*. New York: Springer.

For example, mixing of surface and bottom waters can be thought of as an exchange that delivers nutrient-rich bottom water to the surface. Mesocosms need to be designed to either include or simulate the variety and magnitude of exchange and mixing that occur in the natural ecosystems that they are designed to represent.

At intermediate (meso-)scales, mixing and exchange are crucial in estuaries and coastal waters where fresh and saltwater interact. Exchange and mixing of water are intricately linked processes that determine the estuary's flushing rate, and in so doing they play a major role in its biological productivity and its susceptibility to pollution effects.

At very small scales, microscopic organisms are influenced by relative motion of the fluid (shear) that is directly related to mixing intensity. Small-scale mixing renews nutrient and food supplies, affects contact between predators and prey, and may be a source of physical stress at high levels (**Table 2**). Mesocosm experiments indicate that mixing intensity can have a negative effect on copepod abundance and a highly negative effect on gelatinous zooplankton (**Figure 11**).

At the interfaces between water and fixed solid surfaces, boundary layers (regions of reduced mixing) are formed due to effects of friction. Experimental ecosystems will generally require special mixing mechanisms to minimize boundary layers at their walls and mimic natural boundary layers near the sediment surface (benthic boundary layers).

A variety of engineering approaches can be taken to mix water in mesocosms. Spinning paddles and discs, mechanical plungers, bubbling, and water pumping have all been used as approaches to generating mixing in the water column (**Figure 12**). A range of techniques can be used to characterize the mesocosm mixing environment, including current meters and acoustic Doppler current profilers, as well as measurements of dye dispersion and gypsum dissolution. Scale models can be developed and used to explore mixing characteristics before full-scale experimental ecosystems are built. A range of investigations in various mesocosm systems (e.g., CEPEX, Loch Ewe, MERL, MEERC) have demonstrated the physical and ecological effects of alternative mixing regimes. The goal of these studies is to characterize the mixing environment within the water column and the mixing and flow environment across the bottom so that key mixing parameters (e.g., turbulent energy dissipation, vertical mixing rate) can be matched to natural conditions. Mesocosm researchers should familiarize themselves with the mixing literature as it relates to the design of these systems.

Table 2 Empirically determined effects of mixing on phytoplankton, zooplankton, and ecosystem processes

Variable	*Relationship*[a]
Phytoplankton	
Settling rate	(−)
Cell size	(+)
Cell abundance	(+) or (0)
Chlorophyll *a*	(+)
Cell growth	(+) or (−)
Diatom/flagellate	(+)
Species composition	(√)
Nutrient uptake	(+) or (−)
Timing of bloom	(√)
Microzooplankton (protozoa)	
Predation/grazing rate	(+), (−), or (0)
Growth rate (numbers)	(+)
Cell size	(−)
Macrozooplankton (copepods)	
Abundance/biomass	(−) or (+)
Metabolic rate	(+)
Excretion rate	(+)
Predation/grazing rate	(+) or (−)
Growth rate	(+)
Development rate	(+)
Age structure	(√)
Sex ratio	(√)
Ecosystem	
Community productivity	(+), (−), or (0)
Ecosystem productivity	(+)
Ecosystem *R*	(+)
Nutrient dynamics	(√)

[a](+) symbol indicates a positive relationship between the variable and turbulence, (−) indicates a negative relationship, (√) indicates the presence of a relationship, (0) indicates no relationship. Because mixing levels used in individual experiments included in this analysis ranged from no mixing to unrealistically high levels atypical of nature, this table can only be considered a rough summary of findings.
Citations to studies in this analysis are included in Petersen JE, Sanford LP, and Kemp WM (1998) Coastal plankton responses to turbulent mixing in experimental ecosystems. *Marine Ecology Progress Series* 171: 23–41.

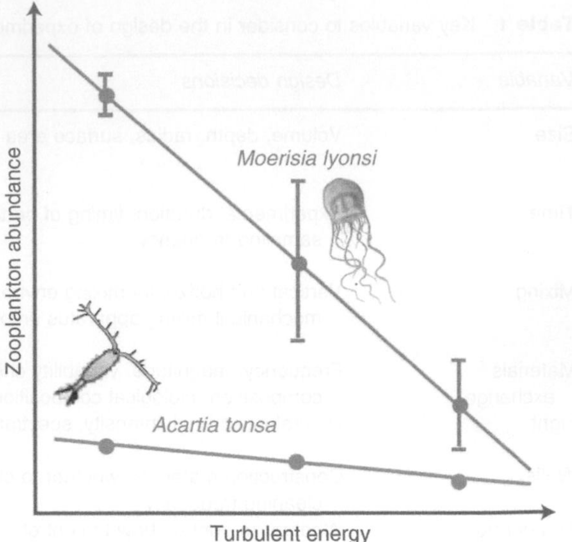

Figure 11 Relationships between the abundance of *Moerisia lyonsi* and *Acartia tonsa*, and the turbulent energy dissipation rate (ϵ) in the three mixing treatments. Turbulent energy dissipation is one of a number of important parameters that can be used to match conditions in nature and mesocosms. Data from Petersen JE, Sanford LP, and Kemp WM (1998) Coastal plankton responses to turbulent mixing in experimental ecosystems. *Marine Ecology Progress Series* 171: 23–41.

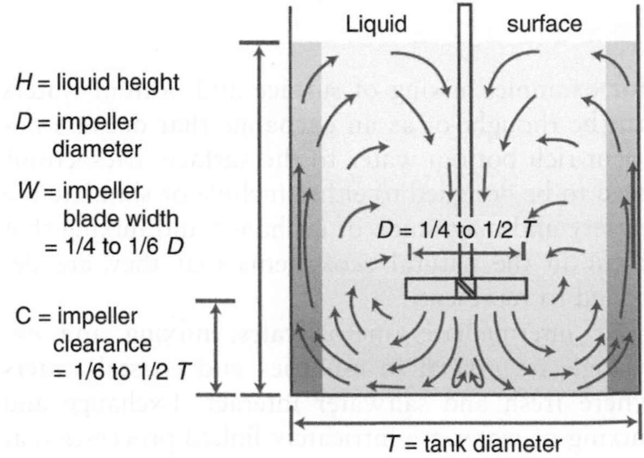

Figure 12 Typical water flow patterns generated in a mesocosm provided with a single rotating axial impeller. Adapted from Petersen JE, Kennedy VS, Dennison WC, and Kemp WM (eds.) (2008) *Enclosed Experimental Ecosystems and Scale: Tools for Understanding and Managing Coastal Ecosystems*. New York: Springer.

The rate at which water is exchanged with surrounding ecosystems is a physical feature that controls many important processes in marine systems. Indeed, the relatively high rate of primary and secondary productivity typical of coastal ecosystems is often attributed to large material exchange resulting from their position at the interface between the watershed and open ocean. Although exchange incorporates both temporal and spatial scale, it is often convenient to express water exchange in terms of 'residence time' (i.e., time required for incoming water to replace the entire volume of the basin or container), or alternatively as 'exchange rate' (i.e., (residence time)$^{-1}$).

Residence time is an important scaling factor to consider in natural and experimental ecosystems because it determines whether a system is dominated by internal or external processes. The actual time a substance or organism resides in the system depends on the combination of flow rate and the rate of reaction, growth, or death inside the system. Flow-through 'chemostat' experiments are commonly used to study phytoplankton growth; however, few ecosystem-level studies have attempted to simulate exchange rates that characterize specific natural

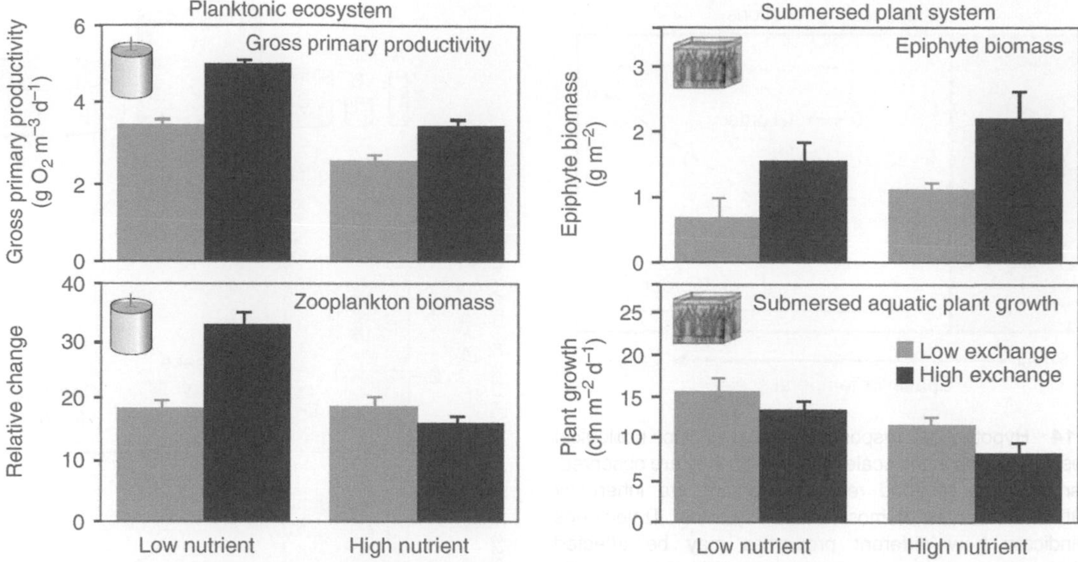

Figure 13 Effects of water exchange rate and nutrient concentration of inflowing waters on gross primary productivity and zooplankton biomass in planktonic experimental ecosystems (left panels) and on competition between aquatic grasses and epiphytes growing on plant leaves (right panels). Values presented are experimental means ±SE. Adapted from Petersen JE, Kennedy VS, Dennison WC, and Kemp WM (eds.) (2008) *Enclosed Experimental Ecosystems and Scale*: *Tools for Understanding and Managing Coastal Ecosystems*. New York: Springer.

ecosystems, and fewer still have explicitly assessed the effects of different exchange rates on ecological dynamics.

The studies that have been conducted indicate that variations in water exchange rate can have substantial effects on ecological dynamics observed in both planktonic and seagrass mesocosms (**Figure 13**). The specific impacts of exchange rates are regulated by the nature of the constituents being exchanged with the water, and the organisms present within the system. Depending on the actual conditions and the organisms involved, variations in water exchange sometimes have counteracting effects. For example, exchange can deliver nutrients or other resources to a system and at the same time flush out mobile organisms that might utilize those resources. The effects of exchange are distinct for systems dominated by planktonic primary producers from those that are dominated by stationary producers. It is also important to recognize that variability in exchange rates can be as important in controlling ecological dynamics as the mean rates of exchange. The various effects of exchange must be taken into careful consideration in the design of experimental ecosystems.

Scaling Considerations in Design and Extrapolation

Even in the case of 'ecosystem-specific' mesocosms that are designed to match precisely certain natural habitats (see section above on abstraction), experimental systems will generally be far smaller than the natural ecosystem that they are intended to represent. Scaling theory suggests that certain patterns and processes only become evident as system size and duration are increased beyond thresholds. Furthermore, scaling responses are often nonlinear and unique for specific variables. Thus, for example, patterns determined to be scale-dependent in mesocosm experiments may become scale-independent at the larger scales of natural systems (solid line in **Figure 14**). Likewise, relationships seen as scale-independent in mesocosms may change with scale in larger natural ecosystems (dashed line in **Figure 14**). Finally, it is possible that thresholds exist over which small changes in scale result in dramatic and possibly discontinuous changes in ecological dynamics.

Given these possibilities, special attention is necessary to account for the potential scale dependence of observations made in mesocosms. Spatial scaling relationships such as those established between water depth and both phytoplankton primary productivity and zooplankton biomass (**Figure 15**) provide a basis for quantitative extrapolation. Although less information is available, it is clear that temporal as well as spatial dynamics can also profoundly affect experimental outcomes (**Figure 16**). In most cases, experimental interpretations and conclusions must be qualified with the acknowledgement that precise effects of scale are yet to be known.

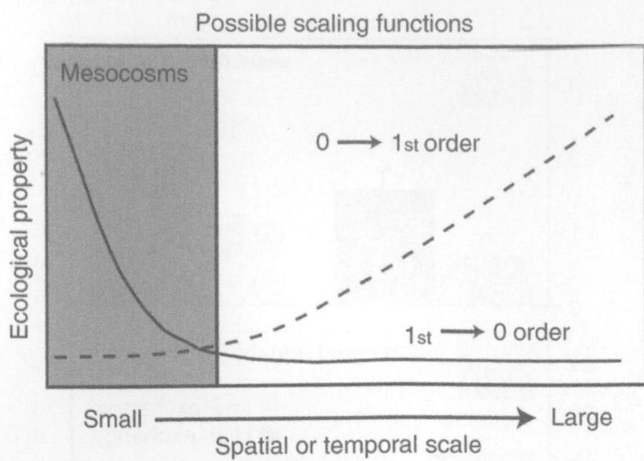

Figure 14 Hypothetical responses of two distinct ecological properties to changes in the scales over which they are observed. Mesocosms scales (shaded region of graph) are inherently smaller than the scales of most natural systems. Trajectories shown indicate how different properties may be affected differently by changes in scale. From Kemp WM, Petersen JE, and Gardner RH (2001) Scale-dependence and the problem of extrapolation: Implications for experimental and natural coastal ecosystems. In: Gardner RH, Kemp WM, Kennedy VS, and Petersen JE (eds.) *Scaling Relations in Experimental Ecology*, pp. 3–57. New York: Columbia University Press.

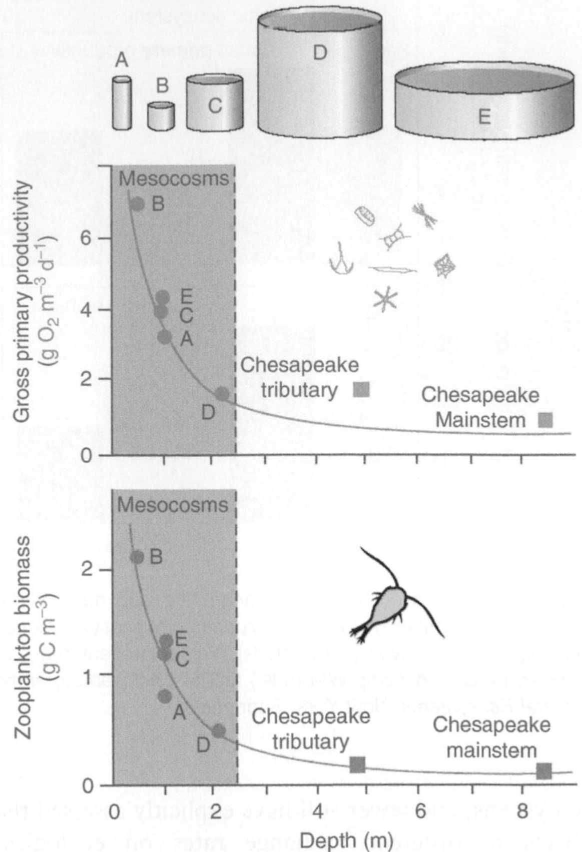

Figure 15 Variations in primary productivity and depth with changes in water column depth for five experimental and two natural estuarine ecosystems with similar salinity. Experimental ecosystems have five different sizes or shapes and the estuarine sites are in the mainstream and a tributary of Chesapeake Bay. Data for gross primary productivity (GPP per unit water volume) are mean values measured from changes in dissolved oxygen concentration. Data are from Petersen J, Chen C-C, and Kemp WM (1997) Scaling aquatic primary productivity: Experiments under nutrient- and light-limited conditions. *Ecology* 78: 2326–2338. From Kemp WM, Petersen JE, and Gardner RH (2001) Scale-dependence and the problem of extrapolation: Implications for experimental and natural coastal ecosystems. In: Gardner RH, Kemp WM, Kennedy VS, and Petersen JE (eds.) *Scaling Relations in Experimental Ecology*, pp. 3–57. New York: Columbia University Press.

The evidence that we have presented thus far implies that mesocosms are inherently distorted representations of nature. A key question then is, can we somehow compensate for these distortions in the design and interpretation of experiments? The term 'dimensional analysis' encompasses a variety of techniques that are based on the proposition that universal relationships should apply regardless of the dimensions of a particular system under investigation. In general, the technique involves developing dimensionless relationships that capture the balance between processes or forces governing the dynamics of a particular system. Dimensional analysis provides a potentially valuable tool for designing experimental ecosystems so that they retain key features of nature. For example, spatially patchy distributions of resources and predators in natural ecosystems may be simulated in mesocosms by creating an exchange regime that is pulsed over time. Similarly, the effects of patchy schools of plankton-eating fish on plankton community dynamics can be simulated experimentally with periodic additions and then removal of fish from the tank. In these cases, temporal variability is substituted for spatial heterogeneity, and the dimensional properties conserved in the mesocosm study are both the duration and frequency of contact between organisms, resources, and predators.

Simulation models provide an additional tool that can be used to improve both the design and interpretation of mesocosm research. Given the importance of spatial heterogeneity in controlling ecological dynamics, coupling mesocosms with spatially explicit dynamic simulation models may become an increasingly valuable approach to ecological research. In this approach, mesocosms can be thought of as individual cells (grain) within a heterogeneous matrix of different habitats that cover broad spatial extent (**Figure 1**). Likewise, models can be used to explore effects of temporal variability that are difficult to incorporate in the design of mesocosm studies. Numerical models offer an excellent tool for exploring nonlinear feedback effects at scales that are larger than individual mesocosms.

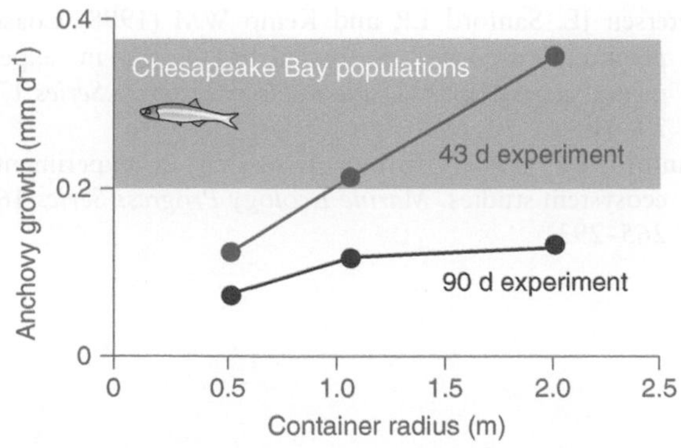

Figure 16 Variations in mean growth of bay anchovy, *Anchoa mitchelli*, with size (radius) of cylindrical mesocosms and with duration of experiment. In smaller containers and in longer experiments, fish exhibit lower growth rate. Shaded area indicates the range of growth rates measured in natural coastal waters. Only in the larger containers and shorter experiments were bay anchovy growth rates comparable to those reported for the estuarine waters that serve as natural habitat for these fish. Adapted from Mowitt WP, Houde ED, Hinkle D, and Sanford A (2006) Growth of planktivorouos bay anchovy *Anchoa mitchelli*, top-down control, and scale-dependence in estuarine mesocosms. *Marine Ecology Progress Series* 308: 255–269.

Conclusions

Enclosed experimental ecosystems have become crucial tools for conducting controlled and repeatable studies of the ocean environment. Those who use mesocosms as research tools and those who use the results of mesocosm experiments need to understand that experimental design choices have important implications for interpretation. Mesocosms are model ecosystems and as such they represent imperfect representations of nature. A great deal is now known about how to design these experimental ecosystems, so that they capture the essential features of nature. Much remains to be learned. The information presented in this article is intended to provide the reader with an introduction to some of the key issues in mesocosm research. The interested reader is encouraged to explore the more detailed information provided in the 'Further reading section'.

See also

Ocean Biogeochemistry and Ecology, Modeling of. Patch Dynamics.

Further Reading

Adey WH and Loveland K (1991) *Dynamic Aquaria: Building Living Ecosystems*. San Diego, CA: Academic Press.

Beyers RJ and Odum HT (1993) *Ecological Microcosms*. New York: Springer.

Gardner RH, Kemp WM, Kennedy VS, and Petersen JE (eds.) (2001) *Scaling Relations in Experimental Ecology*. New York: Columbia University Press.

Giesy JPJ (ed.) (1980) *Microcosms in Ecological Research*. Springfield, VA: National Technical Information Service.

Graney RL, Kennedy JH, and Rodgers JH, Jr. (eds.) (1994) *Aquatic Mesocosm Studies in Ecological Risk Assessment*. Boca Raton, FL: CRC Press.

Grice GD and Reeve MR (eds.) (1982) *Marine Mesocosms: Biological and Chemical Research in Experimental Ecosystems*. New York: Springer.

Grice GD, Reeve MR, Koeller P, and Menzel DW (1977) The use of large volume, transparent, enclosed sea-surface water columns in the study of stress on plankton ecosystems. *Helgolander Wissenschaftliche Meeresuntersuchungen* 30: 118–133.

Kemp WM, Lewis MR, Cunningham FF, Stevenson JC, and Boynton W (1980) Microcosms, macrophytes, and hierarchies: Environmental research in the Chesapeake Bay. In: Giesy JPJ (ed.) *Microcosms in Ecological Research*, pp. 911–936. Springfield, VA: National Technical Information Service.

Kemp WM, Petersen JE, and Gardner RH (2001) Scale-dependence and the problem of extrapolation: Implications for experimental and natural coastal ecosystems. In: Gardner RH, Kemp WM, Kennedy VS, and Petersen JE (eds.) *Scaling Relations in Experimental Ecology*, pp. 3–57. New York: Columbia University Press.

Lalli CM (ed.) (1990) *Enclosed Experimental Marine Ecosystems: A Review and Recommendations*. New York: Springer.

Mowitt WP, Houde ED, Hinkle D, and Sanford A (2006) Growth of planktivorouos bay anchovy *Anchoa mitchelli*, top-down control, and scale-dependence in estuarine mesocosms. *Marine Ecology Progress Series* 308: 255–269.

Nixon SW, Pilson MEQ, and Oviatt CA (1984) Eutrophication of a coastal marine ecosystem – an experimental study using the MERL microcosms. In: Fasham MJR (ed.) *Flows of Energy and Materials in Marine Ecosystems: Theory and Practice*, pp. 105–135. New York: Plenum.

Odum EP (1984) The mesocosm. *BioScience* 34: 558–562.

Oviatt C (1994) Biological considerations in marine enclosure experiments: Challenges and revelations. *Oceanography* 7: 45–51.

Petersen J, Chen C-C, and Kemp WM (1997) Scaling aquatic primary productivity: Experiments under nutrient- and light-limited conditions. *Ecology* 78: 2326–2338.

Petersen JE, Cornwell JC, and Kemp WM (1999) Implicit scaling in the design of experimental aquatic ecosystems. *Oikos* 85: 3–18.

Petersen JE and Hastings A (2001) Dimensional approaches to scaling experimental ecosystems: Designing mousetraps to catch elephants. *American Naturalist* 157: 324–333.

Petersen JE, Kemp WM, Bartleson R, *et al.* (2003) Multiscale experiments in coastal ecology: Improving realism and advancing theory. *BioScience* 53: 1181–1197.

Petersen JE, Kennedy VS, Dennison WC, and Kemp WM (eds.) (2008) *Enclosed Experimental Ecosystems and Scale: Tools for Understanding and Managing Coastal Ecosystems*. New York: Springer.

Petersen JE, Sanford LP, and Kemp WM (1998) Coastal plankton responses to turbulent mixing in experimental ecosystems. *Marine Ecology Progress Series* 171: 23–41.

Sanford LP (1997) Turbulent mixing in experimental ecosystem studies. *Marine Ecology Progress Series* 161: 265–293.

ECOLOGICAL MODELS

SMALL-SCALE PATCHINESS, MODELS OF

D. J. McGillicuddy Jr., Woods Hole Oceanographic Institution, Woods Hole, MA, USA

Introduction

Patchiness is perhaps the most salient characteristic of plankton populations in the ocean. The scale of this heterogeneity spans many orders of magnitude in its spatial extent, ranging from planetary down to microscale (**Figure 1**). It has been argued that patchiness plays a fundamental role in the functioning of marine ecosystems, insofar as the mean conditions may not reflect the environment to which organisms are adapted. For example, the fact that some abundant predators cannot thrive on the mean concentration of their prey in the ocean implies that they are somehow capable of exploiting small-scale patches of prey whose concentrations are much larger than the mean. Understanding the nature of this patchiness is thus one of the major challenges of oceanographic ecology.

The patchiness problem is fundamentally one of physical–biological–chemical interactions. This interconnection arises from three basic sources: (1) ocean currents continually redistribute dissolved and suspended constituents by advection; (2) space–time fluctuations in the flows themselves impact biological and chemical processes; and (3) organisms are capable of directed motion through the water. This tripartite linkage poses a difficult challenge to understanding oceanic ecosystems: differentiation between the three sources of variability requires accurate assessment of property distributions in space and time, in addition to detailed knowledge of organismal repertoires and the processes by which ambient conditions control the rates of biological and chemical reactions.

Various methods of observing the ocean tend to lie parallel to the axes of the space/time domain in which these physical–biological–chemical interactions take place (**Figure 2**). Given that a purely observational approach to the patchiness problem is not tractable with finite resources, the coupling of models with observations offers an alternative which provides a context for synthesis of sparse data with articulations of fundamental principles assumed to govern functionality of the system. In a sense, models can be used to fill the gaps in the space/time domain shown in **Figure 2**, yielding a framework for exploring the controls on spatially and temporally intermittent processes.

The following discussion highlights only a few of the multitude of models which have yielded insight into the dynamics of plankton patchiness. Examples have been chosen to provide a sampling of scales which can be referred to as 'small' – that is, smaller than the planetary scale shown in **Figure 1A**. In addition, this particular collection of examples is intended to furnish some exposure to the diversity of modeling approaches which can be brought to bear on the problem. These approaches range from abstract theoretical models intended to elucidate specific processes, to complex numerical formulations which can be used to actually simulate observed distributions in detail.

Formulation of the Coupled Problem

A general form of the coupled problem can be written as a three-dimensional advection-diffusion-reaction equation for the concentration c_i of any particular organism of interest:

$$\underbrace{\frac{\partial C_i}{\partial t}}_{\text{localrateofchange}} + \underbrace{\nabla \cdot (\mathrm{v}C_i)}_{\text{advention}} - \underbrace{\nabla \cdot (K\nabla C_i)}_{\text{diffusion}} = \underbrace{R_i}_{\text{biologicalsources/sinks}} \quad [1]$$

where the vector v represents the fluid velocity plus any biologically induced transport through the water (e.g., sinking, swimming), and K the turbulent diffusivity. The advection term is often written simply as $\mathrm{v} \cdot \nabla C_i$ because the ocean is an essentially incompressible fluid (i.e., $\nabla \cdot \mathrm{v} = 0$). The 'reaction term' R_i on the right-hand side represents the sources and sinks due to biological activity.

In essence, this model is a quantitative statement of the conservation of mass for a scalar variable in a fluid medium. The advective and diffusive terms simply represent the redistribution of material caused by motion. In the absence of any motion, eqn [1] reduces to an ordinary differential equation describing the biological and/or chemical dynamics. The reader is referred to the review by Donaghay and Osborn for a detailed derivation of the advection-diffusion-reaction equation, including explicit treatment of the Reynolds decomposition for biological and chemical scalars (see Further Reading).

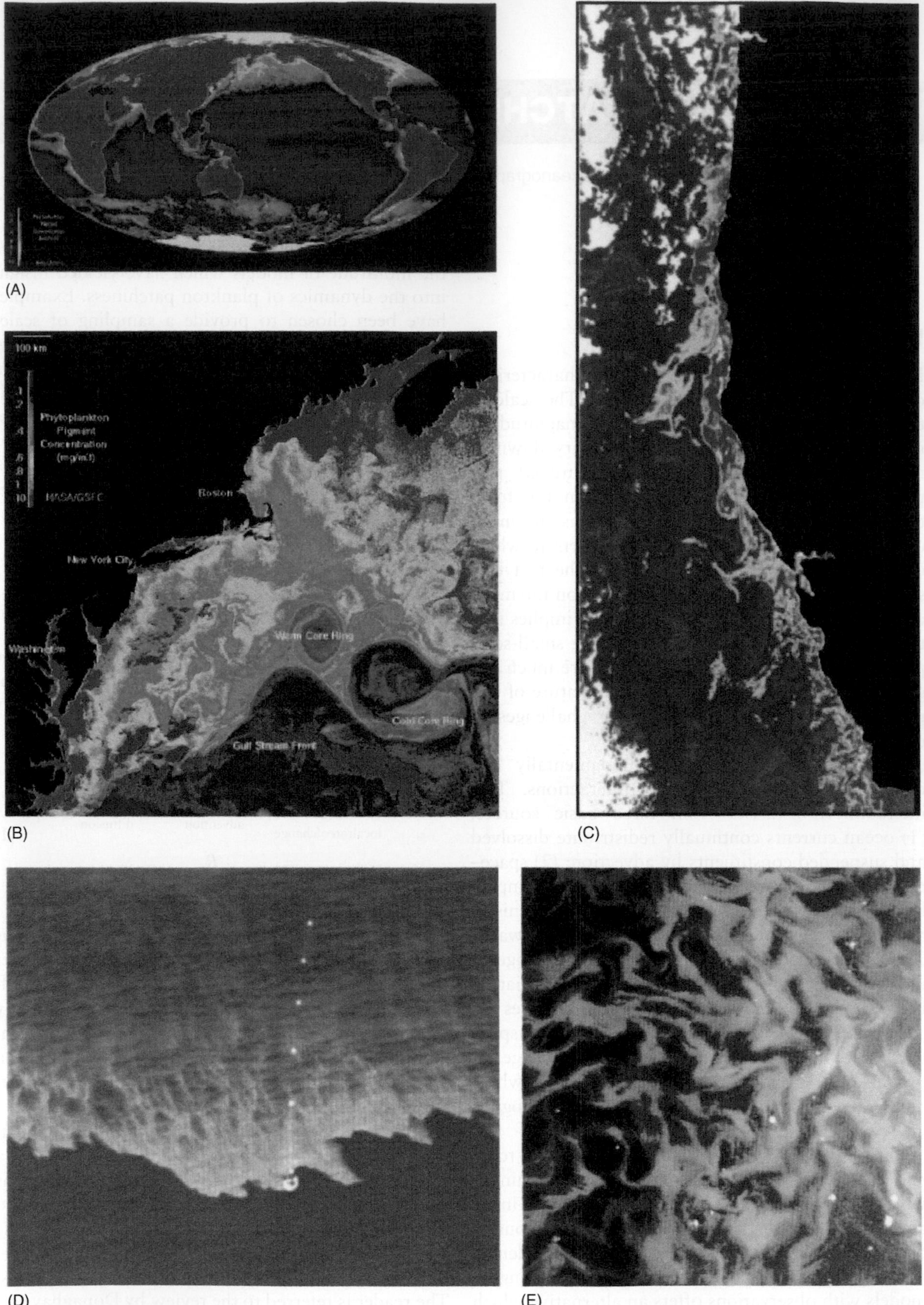

Figure 1 Scales of plankton patchiness, ranging from global down to 1 cm. (A–C) Satellite-based estimates of surface-layer chlorophyll computed from ocean color measurements. Images courtesy of the Seawifs Project and Distributed Active Archive Center at the Goddard Space Flight Center, sponsored by NASA. (D) A dense stripe of *Noctiluca scintillans*, 3 km off the coast of La Jolla. The boat in the photograph is trailing a line with floats spaced every 20 m. The stripe stretched for at least 20 km parallel to the shore (photograph courtesy of P.J.S. Franks). (E) Surface view of a bloom of *Anabaena flos-aquae* in Malham Tarn, England. The area shown is approximately 1 m^2 (photograph courtesy of G.E. Fogg).

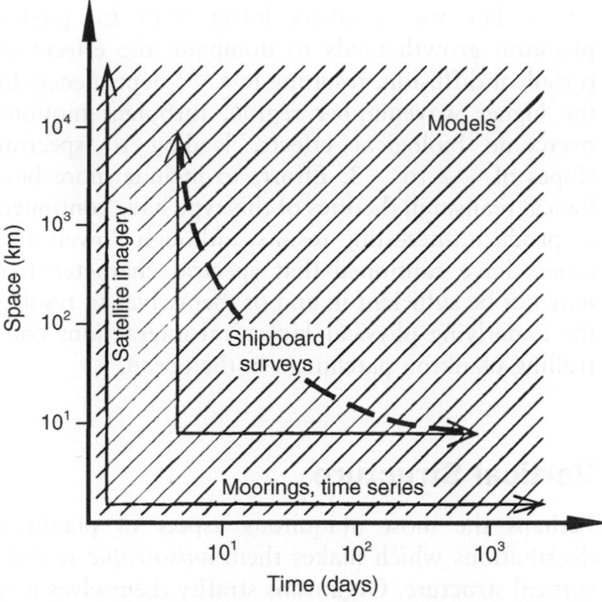

Figure 2 Space–time diagram of the scales resolvable with current observational capabilities. Measurements tend to fall along the axes; the dashed line running between the 'shipboard survey' axes reflects the trade-off between spatial coverage and temporal resolution inherent in seagoing operations of that type. Models can be used to examine portions of the space–time continuum (shaded area).

Any number of advection–diffusion–reaction equations can be posed simultaneously to represent a set of interacting state variables C_i in a coupled model. For example, an ecosystem model including nutrients, phytoplankton, and zooplankton (an 'NPZ' model) could be formulated with $C_1 = dN, C_2 = P$ and $C_3 = Z$. The biological dynamics linking these three together could include nutrient uptake, primary production, grazing, and remineralization. R_i would then represent not only growth and mortality, but also terms which depend on interactions between the several model components.

Growth and Diffusion – the 'KISS' Model

Some of the earliest models used to investigate plankton patchiness dealt with the competing effects of growth and diffusion. In the early 1950s, models developed independently by Kierstead (KI) and Slobodkin (S) and Skellam (S) – the so-called 'KISS' model – were formulated as a one-dimensional diffusion equation with exponential population growth and constant diffusivity:

$$\frac{\partial C}{\partial t} - K\frac{\partial^2 C}{\partial x^2} = \alpha C \qquad [2]$$

Note that this model is a reduced form of eqn [1]. It is a mathematical statement that the tendency for organisms to accumulate through reproduction is counterbalanced by the tendency of the environment to disperse them through turbulent diffusion. Seeking solutions which vanish at $x = 0$ and $x = L$ (thereby defining a characteristic patch size of dimension L), with initial concentration $C(x, 0) = f(x)$, one can solve for a critical patch size $L = \pi(K/\alpha)^{\frac{1}{2}}$ in which growth and dispersal are in perfect balance. For a specified growth rate α and diffusivity K, patches smaller than L will be eliminated by diffusion, while those that are larger will result in blooms. Although highly idealized in its treatment of both physical transport and biological dynamics, this model illuminates a very important aspect of the role of diffusion in plankton patchiness. In addition, it led to a very specific theoretical prediction of the initial conditions required to start a plankton bloom, which Slobodkin subsequently applied to the problem of harmful algal blooms on the west Florida shelf.

Homogeneous Isotropic Turbulence

The physical regime to which the preceding model best applies is one in which the statistics of the turbulence responsible for diffusive transport is spatially uniform (homogeneous) and has no preferred direction (isotropic). Turbulence of this type may occur locally in parts of the ocean in circumstances where active mixing is taking place, such as in a wind-driven surface mixing layer. Such motions might produce plankton distributions such as those shown in **Figure 1E**.

The nature of homogeneous isotropic turbulence was characterized by Kolmogoroff in the early 1940s. He suggested that the scale of the largest eddies in the flow was set by the nature of the external forcing. These large eddies transfer energy to smaller eddies down through the inertial subrange in what is known as the turbulent cascade. This cascade continues to the Kolmogoroff microscale, at which viscous forces dissipate the energy into heat. This elegant physical model inspired the following poem attributed to L. F. Richardson:

> Big whorls make little whorls
> which feed on their velocity;
> little whorls make smaller whorls,
> and so on to viscosity.

Based on dimensional considerations, Kolmogoroff proposed an energy spectrum E of the form

$$E(k) = A\varepsilon^{\frac{2}{3}}k^{\frac{-5}{3}}$$

where k is the wavenumber, ε is the dissipation rate of turbulent kinetic energy, and A is a dimensionless constant. This theoretical prediction was later borne out by measurements, which confirmed the 'minus five-thirds' dependence of energy content on wavenumber.

In the early 1970s, Platt published a startling set of measurements which suggested that for scales between 10 and 10^3 m the variance spectrum of chlorophyll in the Gulf of St Lawrence showed the same $-5/3$ slope. On the basis of this similarity to the Kolmogoroff spectrum, he argued that on these scales, phytoplankton were simply passive tracers of the turbulent motions. These findings led to a burgeoning field of spectral modeling and analysis of plankton patchiness. Studies by Denman, Powell, Fasham, and others sought to formulate more unified theories of physical–biological interactions using this general approach. For example, Denman and Platt extended a model for the scalar variance spectrum to include a uniform growth rate. Their theoretical analysis suggested a breakpoint in the spectrum at a critical wavenumber k_c (**Figure 3**), which they estimated to be in the order of $1\,\text{km}^{-1}$ in the upper ocean. For wavenumbers lower than k_c, phytoplankton growth tends to dominate the effects of turbulent diffusion, resulting in a k^{-1} dependence. In the higher wavenumber region, turbulent motions overcome biological effects, leading to spectral slopes of -2 to -3. Efforts to include more biological realism in theories of this type have continued to produce interesting results, although Powell and others have cautioned that spectral characteristics may not be sufficient in and of themselves to resolve the underlying physical–biological interactions controlling plankton patchiness in the ocean.

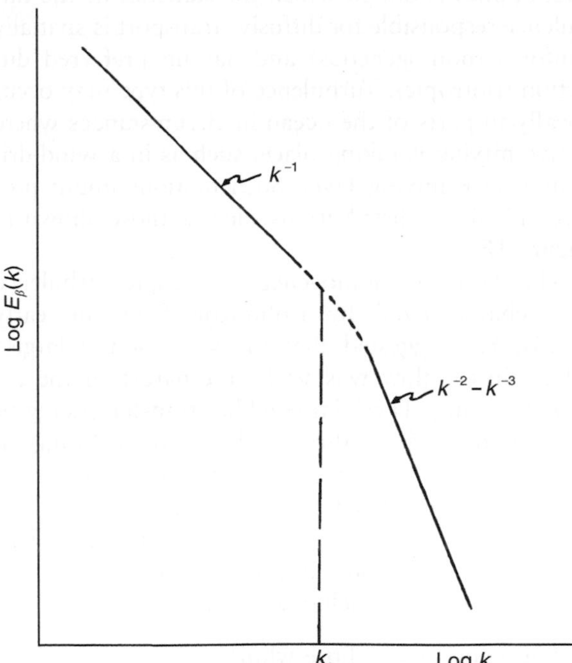

Figure 3 A theoretical spectrum for the spatial variability of phytoplankton, $E_\beta(k)$, as a function of wavenumber, k, displayed on a log-log plot. To the left of the critical wavenumber k_c, biological processes dominate, resulting in a k^{-1} dependence. The high wavenumber region to the right of k_c where turbulent motions dominate, has a dependence between k^{-2} and k^{-3}. (Reproduced with permission from Denman KL and Platt T (1976). The variance spectrum of phytoplankton in a turbulent ocean. *Journal of Marine Research* 34: 593–601.)

Vertical Structure

Perhaps the most ubiquitous aspect of plankton distributions which makes them *anisotropic* is their vertical structure. Organisms stratify themselves in a multitude of ways, for any number of different purposes (e.g., to exploit a limiting resource, to avoid predation, to facilitate reproduction). For example, consider the subsurface maximum which is characteristic of the chlorophyll distribution in many parts of the world ocean (**Figure 4**). The deep chlorophyll maximum (DCM) is typically situated below the nutrient-depleted surface layer, where nutrient concentrations begin to increase with depth. Generally this is interpreted to be the result of joint resource limitation: the DCM resides where nutrients are abundant and there is sufficient light for photosynthesis. However, this maximum in chlorophyll does not necessarily imply a maximum in phytoplankton biomass. For example, in the nutrient-impoverished surface waters of the open ocean, much of the phytoplankton standing stock is sustained by nutrients which are rapidly recycled; thus relatively high biomass is maintained by low ambient nutrient concentrations. In such situations, the DCM often turns out to be a pigment maximum, but not a biomass maximum. The mechanism responsible for the DCM in this case is photoadaptation, the process by which phytoplankton alter their pigment content according to the ambient light environment. By manufacturing more chlorophyll per cell, phytoplankton populations in this type of DCM are able to capture photons more effectively in a low-light environment.

Models have been developed which can produce both aspects of the DCM. For example, consider the nutrient, phytoplankton, zooplankton, detritus (NPZD) type of model (**Figure 5**) which simulates the flows of nitrogen in a planktonic ecosystem. The various biological transformations (such as nutrient uptake, primary production, grazing, excretion, etc.)

are represented mathematically by functional relationships which depend on the model state variables and parameters which must be determined empirically. Doney *et al.* coupled such a system to a one-dimensional physical model of the upper ocean (**Figure 6**). Essentially, the vertical velocity (w) and diffusivity fields from the physical model are used to drive a set of four coupled advection-diffusion-reaction equations (one for each ecosystem state variable) which represent a subset of the full three-dimensional eqn [1]:

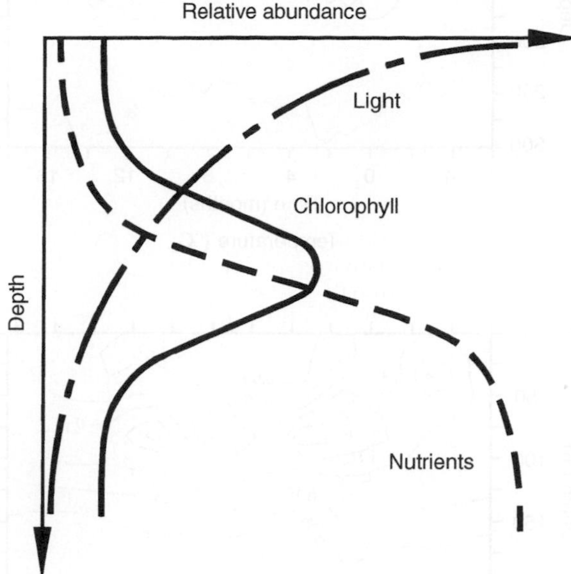

Figure 4 Schematic representation of the deep chlorophyll maximum in relation to ambient light and nutrient profiles in the euphotic zone (typically 10s to 100s of meters in vertical extent).

$$\frac{\partial C_i}{\partial t} + w\frac{\partial C_i}{\partial z} - \frac{\partial}{\partial z}\left(K\frac{\partial C_i}{\partial z}\right) = R_i \qquad [3]$$

The R_i terms represent the ecosystem interaction terms schematized in **Figure 5**. Using a diagnostic photoadaptive relationship to predict chlorophyll from phytoplankton nitrogen and the ambient light and nutrient fields, such a model captures the overall character of the DCM observed at the Bermuda Atlantic Time-series Study (BATS) site (**Figure 6**).

Broad-scale vertical patchiness (on the scale of the seasonal thermocline) such as the DCM is accompanied by much finer structure. The special volume of *Oceanography* on 'Thin layers' provides an excellent overview of this subject, documenting small-scale vertical structure in planktonic populations of many different types. One particularly striking example comes from high-resolution fluorescence measurements (**Figure 7A**). Such profiles often show strong peaks in very narrow depth intervals, which

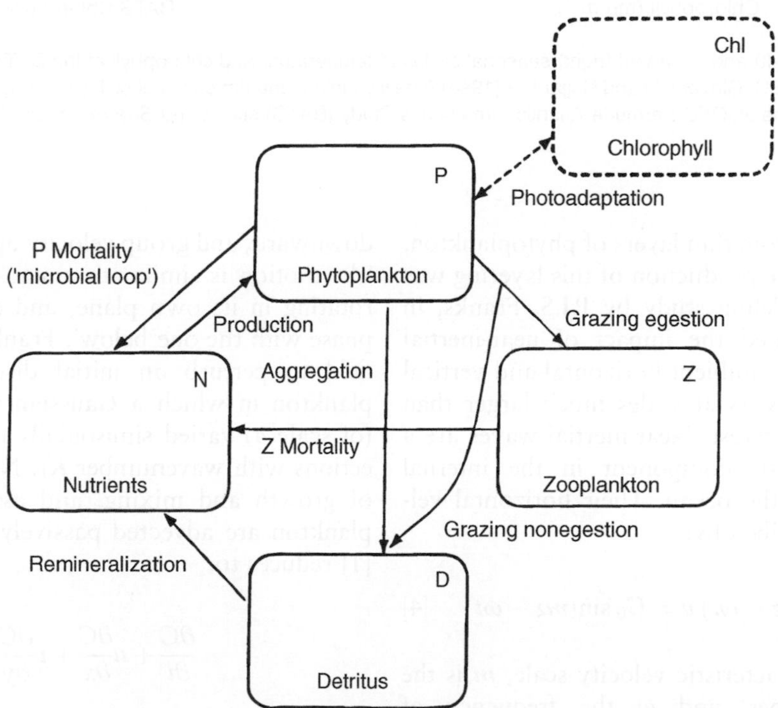

Figure 5 A four-compartment planktonic ecosystem model showing the pathways for nitrogen flow. (Reproduced with permission from Doney SC, Glover DM and Najjar RG (1996) A new coupled, one-dimensional biological–physical model for the upper ocean: applications to the JGOFS Bermuda Atlantic Time-series Study (BATS) site. *Deep-Sea Research II* 43: 591–624.)

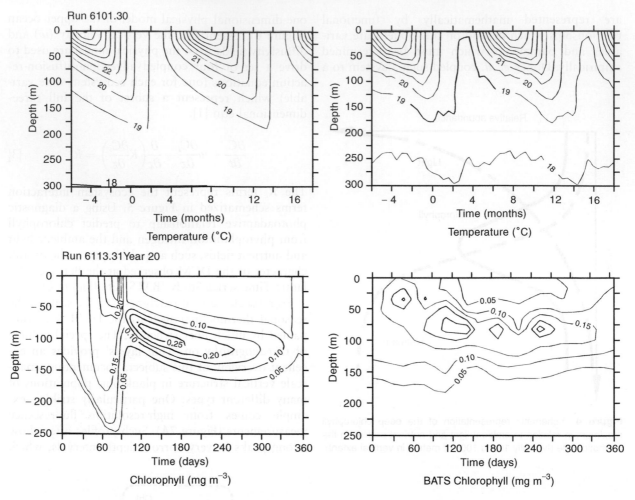

Figure 6 Simulated (left) and observed (right) seasonal cycles of temperature and chlorophyll at the BATS site. (Reproduced with permission from Doney SC, Glover DM and Najjar RG (1996) A new coupled, one-dimensional biological–physical model for the upper ocean: applications to the JGOFS Bermuda Atlantic Time-series Study (BATS) site. *Deep-Sea Research II* 43: 591–624.)

presumably result from thin layers of phytoplankton. A mechanism for the production of this layering was identified in a modeling study by P.J.S. Franks, in which he investigated the impact of near-inertial wave motion on the ambient horizontal and vertical patchiness which exists at scales much larger than the thin layers of interest. Near-inertial waves are a particularly energetic component in the internal wave spectrum of the ocean. Their horizontal velocities can be described by:

$$u = U_0 \cos(mz - \omega t) \quad v = U_0 \sin(mz - \omega t) \qquad [4]$$

where U_0 is a characteristic velocity scale, m is the vertical wavenumber, and ω the frequency of the wave. This kinematic model prescribes that the velocity vector rotates clockwise in time and counterclockwise with depth; its phase velocity is downward, and group velocity upward. In his words, 'the motion is similar to a stack of pancakes, each rotating in its own plane, and each slightly out of phase with the one below'. Franks used this velocity field to perturb an initial distribution of phytoplankton in which a Gaussian vertical distribution (of scale σ) varied sinusoidally in both x and y directions with wavenumber K_P. Neglecting the effects of growth and mixing, and assuming that phytoplankton are advected passively with the flow, eqn [1] reduces to:

$$\frac{\partial C}{\partial t} + u\frac{\partial C}{\partial x} + v\frac{\partial C}{\partial y} = 0 \qquad [5]$$

Plugging the velocity fields [4] into this equation, the initial phytoplankton distribution can be integrated forward in time. This model demonstrates the

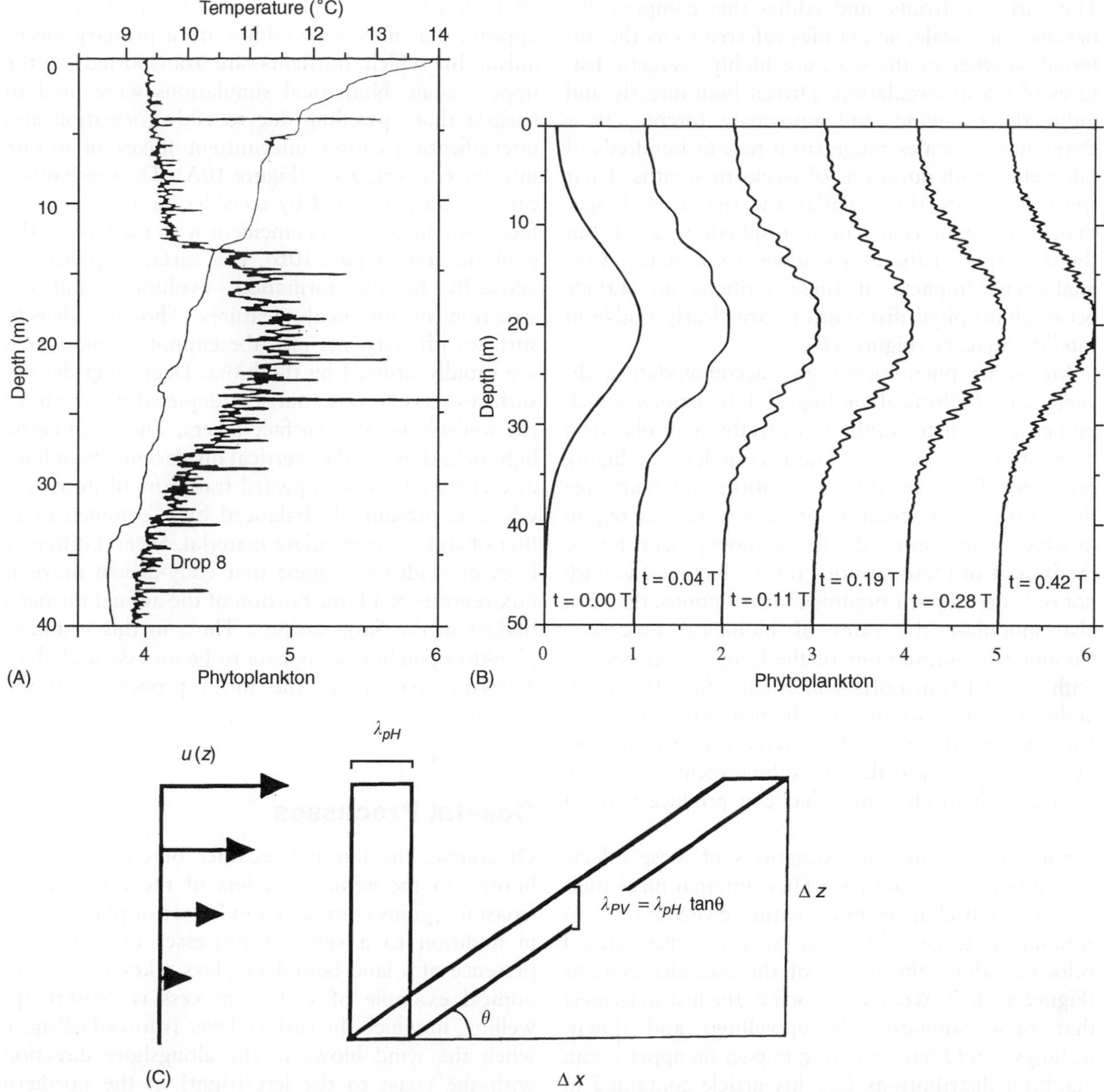

Figure 7 (A) Observations of thin layer structure in a high-resolution profile of fluorescence (thick line, arbitrary units). The thin line shows the corresponding temperature structure. (Data courtesy of Dr T. Cowles.) (B) Simulated vertical profiles of phytoplankton concentration (arbitrary units) at six sequential times. Each profile is offset from the previous by 1 phytoplankton unit. The times are given as fractions of the period of the near-inertial wave used to drive the model. (C) A schematic diagram of the layering process. Vertical shear stretches a vertical column of a property horizontally through an angle θ, creating a layer in the vertical profile. (Reproduced with permission from Franks PJS (1995) Thin layers of phytoplankton: a model of formation by near-inertial wave shear. *Deep-Sea Research I* 42: 75–91.)

striking result that such motions can generate vertical structure which is much finer than that present in the initial condition (**Figure 7B**). Analysis of the simulations revealed that the mechanism at work here is simple and elegant: vertical shear can translate horizontal patchiness into thin layers by stretching and tilting the initial patch onto its side (**Figure 7C**).

Mesoscale Processes: The Internal Weather of the Sea

Just as the atmosphere has weather patterns that profoundly affect the plants and animals that live on the surface of the earth, the ocean also has its own set of environmental fluctuations which exert fundamental control over the organisms living within it.

The currents, fronts, and eddies that comprise the oceanic mesoscale, sometimes referred to as the 'internal weather of the sea', are highly energetic features of ocean circulation. Driven both directly and indirectly by wind and buoyancy forcing, their characteristic scales range from tens to hundreds of kilometers with durations of weeks to months. Their space scales are thus smaller and timescales longer than their counterparts in atmospheric weather, but the dynamics of the two systems are in many ways analogous. Impacts of these motions on surface ocean chlorophyll distributions are clearly visible in satellite imagery (**Figure 1B**).

Mesoscale phenomenologies accommodate a diverse set of physical–biological interactions which influence the distribution and variability of plankton populations in the sea. These complex yet highly organized flows continually deform and rearrange the hydrographic structure of the near-surface region in which plankton reside. In the most general terms, the impact of these motions on the biota is twofold: not only do they stir organism distributions, they can also modulate the rates of biological processes. Common manifestations of the latter are associated with vertical transports which can affect the availability of both nutrients and light to phytoplankton, and thereby the rate of primary production. The dynamics of mesoscale and submesoscale flows are replete with mechanisms that can produce vertical motions.

Some of the first investigations of these effects focused on mesoscale jets. Their internal mechanics are such that changes in curvature give rise to horizontal divergences which lead to very intense vertical velocities along the flanks of the meander systems (**Figure 8**). J. D. Woods was one of the first to suggest that these submesoscale upwellings and downwellings would have a strong impact on upper ocean plankton distributions (see his article contained in the volume edited by Rothschild; see Further Reading). Subsequent modeling studies have investigated these effects by incorporating planktonic ecosystems of the type shown in **Figure 5** into three-dimensional dynamical models of meandering jets. Results suggest that upwelling in the flank of a meander can stimulate the growth of phytoplankton (**Figure 9**). Simulated plankton fields are quite complex owing to the fact that fluid parcels are rapidly advected in between regions of upwelling and downwelling. Clearly, this complicated convolution of physical transport and biological response can generate strong heterogeneity in plankton distributions.

What are the implications of mesoscale patchiness? Do these fluctuations average out to zero, or are they important in determining the mean characteristics of the system? In the Sargasso Sea, it appears that mesoscale eddies are a primary mechanism by which nutrients are transported to the upper ocean. Numerical simulations were used to suggest that upwelling due to eddy formation and intensification causes intermittent fluxes of nitrate into the euphotic zone (**Figure 10A**). The mechanism can be conceptualized by considering a density surface with mean depth coincident with the base of the euphotic zone (**Figure 10B**). This surface is perturbed vertically by the formation, evolution, and destruction of mesoscale features. Shoaling density surfaces lift nutrients into the euphotic zone which are rapidly utilized by the biota. Deepening density surfaces serve to push nutrient-depleted water out of the well-illuminated surface layers. The asymmetric light field thus rectifies vertical displacements of both directions into a net upward transport of nutrients, which is presumably balanced by a commensurate flux of sinking particulate material. Several different lines of evidence suggest that eddy-driven nutrient flux represents a large portion of the annual nitrogen budget in the Sargasso Sea. Thus, in this instance, plankton patchiness appears to be an essential characteristic that drives the mean properties of the system.

Coastal Processes

Of course, the internal weather of the sea is not limited to the eddies and jets of the open ocean. Coastal regions contain a similar set of phenomena, in addition to a suite of processes in which the presence of a land boundary plays a key role. A canonical example of such a process is coastal upwelling, in which the surface layer is forced offshore when the wind blows in the alongshore direction with the coast to the left (right) in the northern (southern) hemisphere. This event triggers upwelling of deep water to replace the displaced surface water. The biological ramifications of this were explored in the mid-1970s by Wroblewski with one of the first coupled physical–biological models to include spatial variability explicitly. Configuring a two-dimensional advection-diffusion-reaction model in vertical plane cutting across the Oregon shelf, he studied the response of an NPZD-type ecosystem model to transient wind forcing. His 'strong upwelling' case provided a dramatic demonstration of mesoscale patch formation (**Figure 11**). Deep, nutrient-rich waters from the bottom boundary layer drawn up toward the surface stimulate a large increase in primary production which is restricted to within 10 km of the coast. The phytoplankton distribution reflects

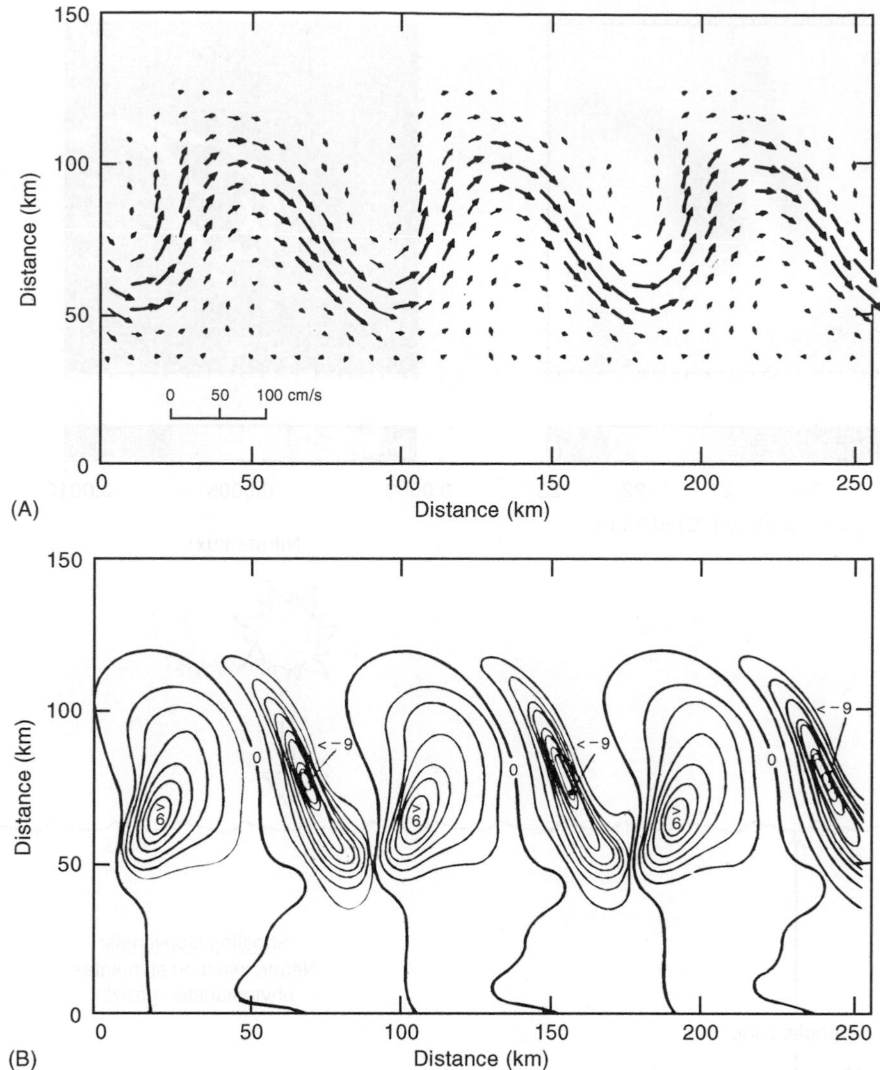

Figure 8 Simulation of a meandering mesoscale jet: (A) velocity on an isopycnal surface with a mean depth of 20 m; (B) vertical velocity (m d^{-1}) on the same isopycnal surface as in (A). Note the consistent pattern of the vertical motion with respect to the structure of the jet. (Reproduced with permission from Woods JD (1988) Mesoscale upwelling and primary production. In: Rothschild BJ (ed) *Toward a Theory on Physical–Biological Interactions in the World Ocean*. London: Kluwer Academic.)

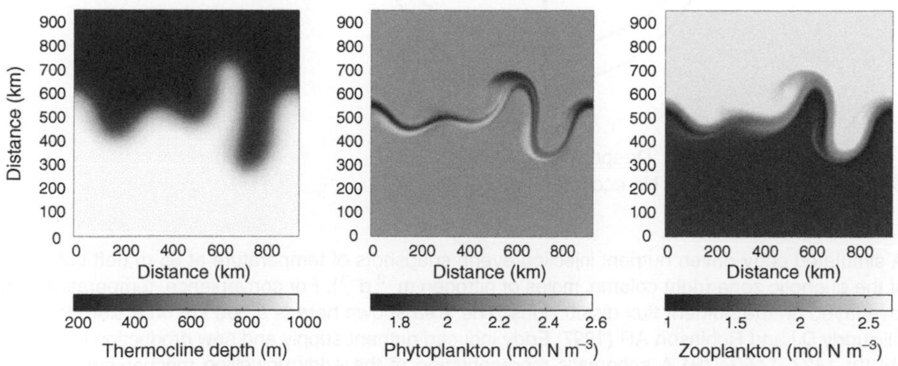

Figure 9 Results from a coupled model of the Gulf Stream: thermocline depth (left), phytoplankton concentration (middle), and zooplankton concentration (right). (Courtesy of GR Flierl, Massachusetts Institute of Technology).

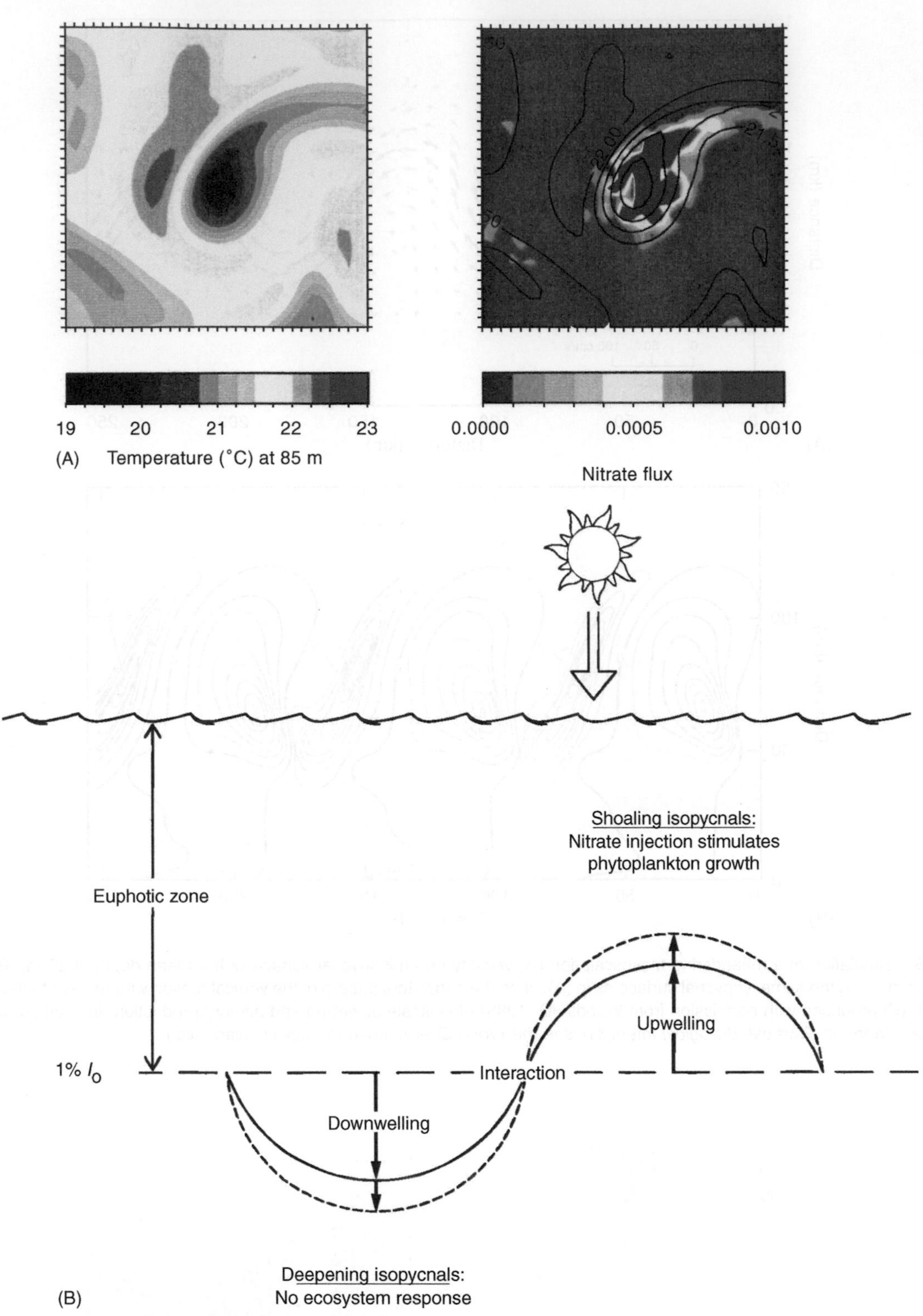

Figure 10 (A) A simulated eddy-driven nutrient injection event: snapshots of temperature at 85 m (left column, °C) and nitrate flux across the base of the euphotic zone (right column, moles of nitrogen m^{-2} d^{-1}). For convenience, temperature contours from the left-hand panels are overlayed on the nutrient flux distributions. The area shown here is a 500 km on a side domain. (The simulation is described in McGillicuddy DJ and Robinson AR (1997) Eddy induced nutrient supply and new production in the Sargasso Sea. *Deep-Sea Research I* 44(8): 1427–1450.) (B) A schematic representation of the eddy upwelling mechanism. The solid line depicts the vertical deflection of an individual isopycnal caused by the presence of two adjacent eddies of opposite sign. The dashed line indicates how the isopycnal might be subsequently perturbed by interaction of the two eddies. (Reproduced with permission from McGillicuddy DJ *et al.* (1998) Influence of mesoscale eddies on new production in the Sargasso Sea. *Nature* 394: 263–265.)

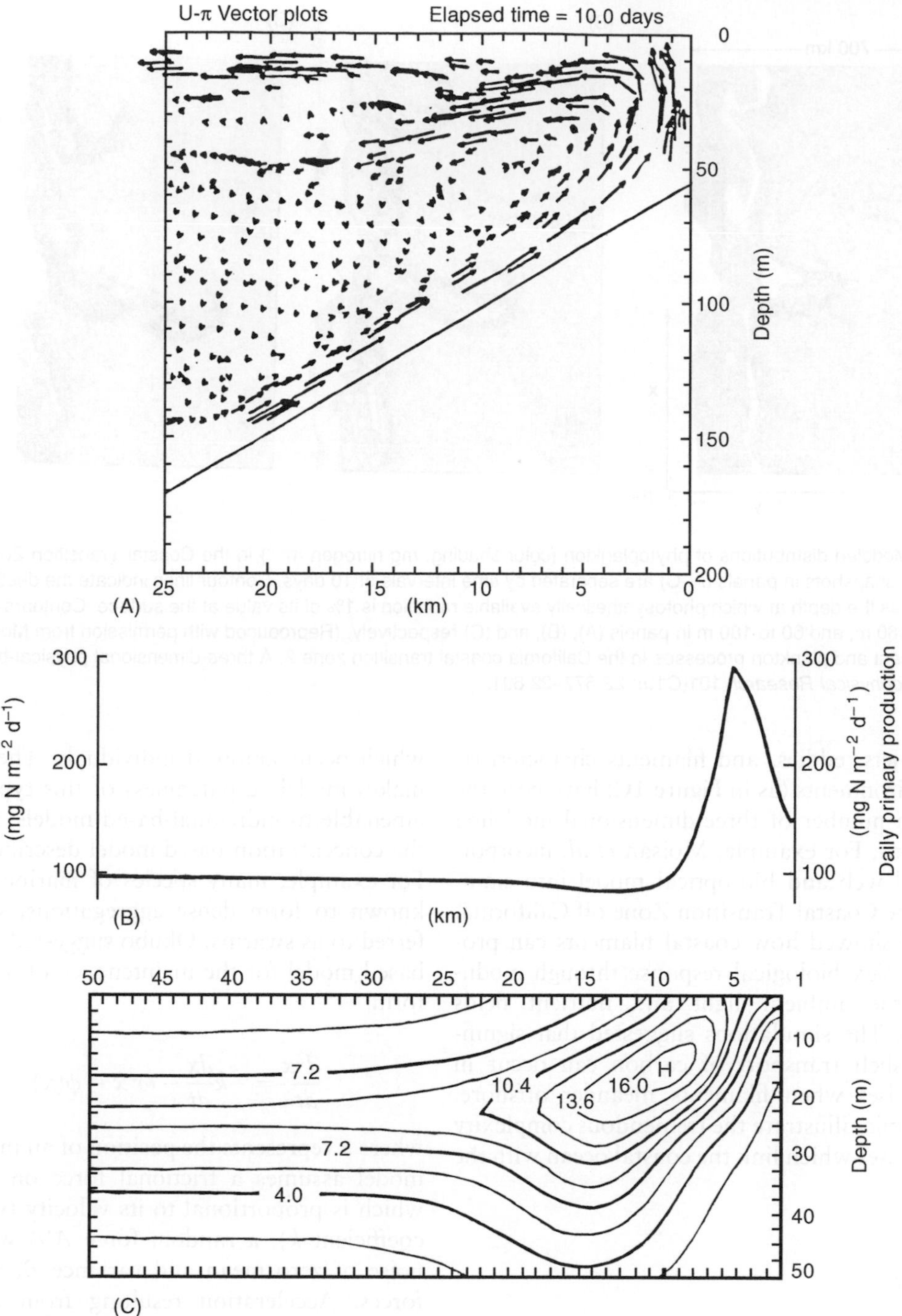

Figure 11 Snapshot from a two-dimensional coupled model of coastal upwelling. (A) Circulation in the transverse plane normal to the coast (maximum horizontal and vertical velocities are −6.1 and 0.05 cm s^{-1}, respectively); (B) daily gross primary production; (C) phytoplankton distribution (contour interval is 1.6 μg) at N l^{-1}. (Adapted with permission from Wroblewski JS (1977) A model of plume formation during variable Oregon upwelling. *Journal of Marine Research* 35(2): 357–394.)

the localized enhancement of production, in addition to advective transport of the resultant biogenic material. Note that the highest concentrations of phytoplankton are displaced from the peak in primary production, owing to the offshore transport in the near-surface layers.

Although Wroblewski's model was able to capture some of the most basic elements of the biological response to coastal upwelling, its two-dimensional formulation precluded representation of alongshore variations which can sometimes be as dramatic as those in the cross-shore direction. The complex set of

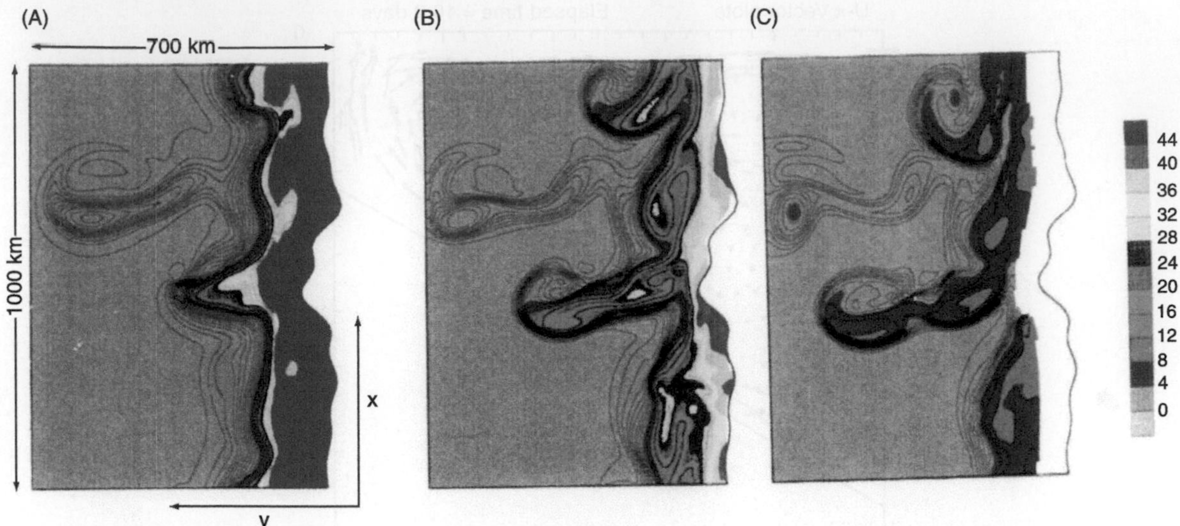

Figure 12 Modeled distributions of phytoplankton (color shading, mg nitrogen m^{-3}) in the Coastal Transition Zone off California. Instantaneous snapshots in panels (A–C) are separated by time intervals of 10 days. Contour lines indicate the depth of the euphotic zone, defined as the depth at which photosynthetically available radiation is 1% of its value at the surface. Contours range from 30 to 180 m, 40 to 180 m, and 60 to 180 m in panels (A), (B), and (C) respectively. (Reproduced with permission from Moisan *et al.* (1996) Modeling nutrient and plankton processes in the California coastal transition zone 2. A three-dimensional physical-bio-optical model. *Journal of Geophysical Research* 101(C10): 22 677–22 691.

interacting jets, eddies, and filaments characteristic of such environments (as in **Figure 1C**) have been the subject of a number of three-dimensional modeling investigations. For example, Moisan *et al.* incorporated a food web and bio-optical model into simulations of the Coastal Transition Zone off California. This model showed how coastal filaments can produce a complex biological response through modulation of the ambient light and nutrient fields (**Figure 12**). The simulations suggested that significant cross-shelf transport of carbon can occur in episodic pulses when filaments meander offshore. These dynamics illustrate the tremendous complexity of the processes which link the coastal ocean with the deep sea.

Behavior

The mechanisms for generating plankton patchiness described thus far consist of some combination of fluid transport and physiological response to the physical, biological, and chemical environment. The fact that many planktonic organisms have behavior (interpreted narrowly here as the capability for directed motion through the water) facilitates a diverse array of processes for creating heterogeneity in their distributions. Such processes pose particularly difficult challenges for modeling, in that their effects are most observable at the level of the population, whereas their dynamics are governed by interactions which occur amongst individuals. The latter aspect makes modeling patchiness of this type particularly amenable to individual-based models, in contrast to the concentration-based model described by eqn [1]. For example, many species of marine plankton are known to form dense aggregations, sometimes referred to as swarms. Okubo suggested an individual-based model for the maintenance of a swarm of the form:

$$\frac{d^2x}{dt^2} = -k\frac{dx}{dt} - \omega^2 x - \phi(x) + A(t) \qquad [6]$$

where x represents the position of an individual. This model assumes a frictional force on the organism which is proportional to its velocity (with frictional coefficient k), a random force $A(t)$ which is white noise of zero mean and variance B, and attractive forces. Acceleration resulting from the attractive forces is split between periodic (frequency ω) and static ($\phi(x)$) components. The key aspect of the attractive forces is that they depend on the distance from the center of the patch. A Fokker-Planck equation can be used to derive a probability density function:

$$p(x) = p_0 \exp\left(-\frac{\omega^2}{2B}x^2 - \int \frac{\phi(x)}{b}\mathrm{d}x\right) \qquad [7]$$

where p_0 is the density at the center of the swarm. Thus, the macroscopic properties of the system can

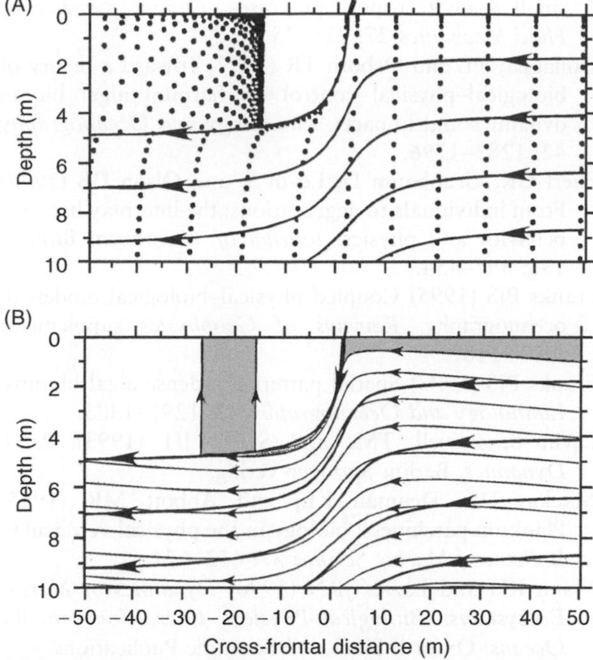

Figure 13 Surface-seeking organisms aggregating at a propagating front. Modeled particle locations (dots, panel (A)) and particle streamlines (thin lines, panel (B)) in the cross-frontal flow. The front is centered at $x = 0$, and the coordinate system translates to the right with the motion of the front. Flow streamlines are represented in both panels as bold lines; they differ from particle streamlines due to propagation of the front. The shaded area in (B) indicates the region in which cells are focused into the frontal zone, forming a dense band at $x = -20$ m. (Reproduced with permission from Franks, 1997.)

be related to the specific set of rules governing individual behavior. Okubo has shown that observed characteristics of insect swarms compare well with theoretical predictions from this model, both in terms of the organism velocity autocorrelation and the frequency distribution of their speeds. Analogous comparisons with plankton have proven elusive owing to the extreme difficulty in making such measurements in marine systems.

The foregoing example illustrates how swarms can arise out of purely behavioral motion. Yet another class of patchiness stems from the joint effects of behavior and fluid transport. The paper by Flierl *et al.* is an excellent reference on this general topic (see Further Reading). One of the simplest examples of this kind of process arises in a population which is capable of maintaining its depth (either through swimming or buoyancy effects) in the presence of convergent flow. With no biological sources or sinks, eqn [1] becomes:

$$\frac{\partial C}{\partial t} + \mathrm{v} \cdot \nabla_H C + C \nabla_H \cdot \mathrm{v} - \nabla \cdot (K \nabla C) = 0 \quad [8]$$

where ∇_H is the vector derivative in the horizontal direction only. Because vertical fluid motion is exactly compensated by organism behavior (recall that the vector v represents the sum of physical and biological velocities), two advective contributions arise from the term $\nabla \cdot (\mathrm{v}C)$ in eqn [1]: the common form with the horizontal velocity operating on spatial gradients in concentration, *plus* a source/sink term created by the divergence in total velocity (fluid + organism). The latter term provides a mechanism for accumulation of depth-keeping organisms in areas of fluid convergence. It has been suggested that this process is important in a variety of different oceanic contexts. In the mid-1980s, Olson and Backus argued it could result in a 100-fold increase in the local abundance of a mesopelagic fish *Benthosema glaciale* in a warm core ring. Franks modeled a conceptually similar process with a surface-seeking organism in the vicinity of a propagating front (**Figure 13**). Simply stated, upward swimming organisms tend to accumulate in areas of downwelling. This mechanism has been suggested to explain spectacular accumulations of motile dinoflagellates at fronts (**Figure 1D**).

Conclusions

The interaction of planktonic population dynamics with oceanic circulation can create tremendously complex patterns in the distribution of organisms. Even an ocean at rest could accommodate significant inhomogeneity through geographic variations in environmental variables, time-dependent forcing, and organism behavior. Fluid motions tend to amalgamate all of these effects in addition to introducing yet another source of variability: space–time fluctuations in the flows themselves which impact biological processes. Understanding the mechanisms responsible for observed variations in plankton distributions is thus an extremely difficult task.

Coupled physical–biological models offer a framework for dissection of these manifold contributions to structure in planktonic populations. Such models take many forms in the variety of approaches which have been used to study plankton patchiness. In theoretical investigations, the basic dynamics of idealized systems are worked out using techniques from applied mathematics and mathematical physics. Process-oriented numerical models offer a conceptually similar way to study systems that are too complex to be solved analytically. Simulation-oriented models are aimed at reconstructing particular data sets using realistic hydrodynamic forcing pertaining to the space/time domain of interest. Generally speaking, such models tend to be quite complex because of the multitude of processes which must be included to simulate observations made in the natural environment. Of course, this complexity makes diagnosis of the coupled system more challenging. Nevertheless, the combination of models and observations provides a unique context for the synthesis of necessarily sparse data: space–time continuous representations of the real ocean which can be diagnosed term-by-term to reveal the underlying processes. Formal union between models and observations is beginning to occur through the emergence of inverse methods and data assimilation in the field of biological oceanography. 410 provides an up-to-date review of this very exciting and rapidly evolving aspect of coupled physical–biological modeling.

Although the field is more than a half-century old, modeling of plankton patchiness is still in its infancy. The oceanic environment is replete with phenomena of this type which are not yet understood. Fortunately, the field is perhaps better poised than ever to address such problems. Recent advances in measurement technologies (e.g., high-resolution acoustical and optical methods, miniaturized biological and chemical sensors) are beginning to provide direct observations of plankton on the scales at which the coupled processes operate. Linkage of such measurements with models is likely to yield important new insights into the mechanisms controlling plankton patchiness in the ocean.

See also

Biogeochemical Data Assimilation. Continuous Plankton Recorders. Fluorometry for Biological Sensing. Patch Dynamics. Phytoplankton Blooms. Small-Scale Physical Processes and Plankton Biology.

Further Reading

Denman KL and Gargett AE (1995) Biological–physical interactions in the upper ocean: the role of vertical and small scale transport processes. *Annual Reviews of Fluid Mechanics* 27: 225–255.

Donaghay PL and Osborn TR (1997) Toward a theory of biological–physical control on harmful algal bloom dynamics and impacts. *Limnology and Oceanography* 42: 1283–1296.

Flierl GR, Grunbaum D, Levin S, and Olson DB (1999) From individuals to aggregations: the interplay between behavior and physics. *Journal of Theoretical Biology* 196: 397–454.

Franks PJS (1995) Coupled physical–biological models in oceanography. *Reviews of Geophysics* supplement: 1177–1187.

Franks PJS (1997) Spatial patterns in dense algal blooms. *Limnology and Oceanography* 42: 1297–1305.

Levin S, Powell TM, and Steele JH (1993) *Patch Dynamics*. Berlin: Springer-Verlag.

Mackas DL, Denman KL, and Abbott MR (1985) Plankton patchiness: biology in the physical vernacular. *Bulletin of Marine Science* 37: 652–674.

Mann KH and Lazier JRN (1996) *Dynamics of Marine Ecosystems: Biological–Physical Interactions in the Oceans*. Oxford: Blackwell Scientific Publications.

Okubo A (1980) *Diffusion and Ecological Problems: Mathematical Models*. Berlin: Springer-Verlag.

Okubo A (1986) Dynamical aspects of animal grouping: swarms, schools, flocks and herds. *Advances in Biophysics* 22: 1–94.

Robinson AR, McCarthy JJ, and Rothschild BJ (2001) *The Sea: Biological–Physical Interactions in the Ocean*. New York: John Wiley and Sons.

Rothschild BJ (1988) *Toward a Theory on Biological–Physical Interactions in the World Ocean*. Dordrecht: D. Reidel.

Oceanography Society (1998) *Oceanography 11(1): Special Issue on Thin Layers*. Virginia Beach, VA: Oceanography Society.

Steele JH (1978) *Spatial Pattern in Plankton Communities*. New York: Plenum Press.

Wroblewski JS and Hofmann EE (1989) U.S. interdisciplinary modeling studies of coastal–offshore exchange processes: past and future. *Progress in Oceanography* 23: 65–99.

POPULATION DYNAMICS MODELS

Francois Carlotti, C.N.R.S./Université Bordeaux 1, Arachon, France

Introduction

The general purpose of population models of plankton species is to describe and eventually to predict the changes in abundance, distribution, and production of targeted populations under forcing of the abiotic environment, food conditions, and predation. Computer-based approaches in plankton ecology were introduced during the 1970s with the application of population models to investigate large-scale population phenomena by the use of mathematical models.

Today, virtually every major scientific research project of population ecology has a modeling component. Population models are built for three main objectives: (1) to estimate the survival of individuals and the persistence of populations in their physical and biological environments, and to look at the factors and processes that regulate their variability; (2) to estimate the flow of energy and matter through a given population; and (3) to study different aspects of behavioral ecology. The study of internal properties of a population, like the various effects of individual variability, and the study of interactions between populations and successions of population are also topics related to population models. The field of biological modeling has diversified and, at present, complex mathematical approaches such as neural networks, genetic algorithms, and dynamical optimization are coming into use, along with the application of supercomputers. However, the use of models in marine research should always be accompanied by extensive field data and laboratory experiments, for initialization, verification or falsification, or continuous updating.

Approach for Modelling Plankton Populations

Population Structure and Units

A population is defined as a group of living organisms all of one species restricted to a given area and with limited exchanges of individuals from other populations. The first step in building a population model is to identify state variables (components of the population) and to describe the interactions between these state variables and external variables of the system and among the components themselves. The components of a population can be (1) the entire population (one component); (2) groups of individuals identified by a certain states: developmental stages, weight or size classes, age classes (fixed numbers of components); or (3) all individuals (varying numbers of components).

The usual unit in population dynamics models is the number of individuals per volume of water, but the population biomass can also be used (in g biomass or carbon (C) or nitrogen (N)). When all individuals or groups of individuals are represented, the individual weight can be considered as a state variable. The forcing variables influencing the population dynamics are biological factors, mainly nutrients and predators, and physical factors, mainly temperature, light, advection, and diffusion.

Individual and Demographic Processes

Population models usually work with four major processes: individual growth, development, reproduction, and mortality. Growth is computed by the rate of individual weight change. Development is represented by the change of states (phases in phytoplankton and protozoan cell division, developmental stages in zooplankton) through which each individual progresses to reach maturity. Reproduction is represented by the production of new individuals. Mortality induces loss of individuals, and can be divided in two components: natural physiological mortality (due, for instance, to starvation) and mortality due to predation. Combination of the four processes permits one to stimulate (1) increase in terms of number of individuals in the population, (2) the body growth of these individuals, and (3) by combination of the two previously simulated values, the increase in total biomass of the population (which is usually termed 'population growth').

Plankton Characteristics

Essential information to be built into models of plankton organisms are (1) the individual life duration (a few hours for bacteria; one to a few days for phytoplankton and unicellular animals; several weeks to years for zooplankton and ichthyoplankton organisms); (2) the range of change in size or weight between the beginning and the end of a life cycle; and

(3) the number of individuals produced by a mother individual (from two individuals up to thousands of individuals). When developmental stages in the life cycle are identified, the stage durations are needed.

The observation time step has to be defined to adequately follow the timescale of the chosen variables, and thus should be smaller than the duration of the shortest phases.

Plankton Population Models

The most modeled component in marine planktonic ecosystems is phytoplankton production. Most of the phytoplankton models simulate the growth of phytoplankton as a whole, using only the process of photosynthesis. Few models deal with phytoplankton population growth dynamics at the species level. Existing models of other unicellular plankton organisms (bacterioplankton, species of microzooplankton) usually treat them as a single unit, except for a few models simulating phytoplankton and microbial cell cycles. In contrast, mesozooplanktonic organisms, including the planktonic larval stages of benthic species (meroplankton), and fish that have complex life cycles are extensively modeled at the population level.

Dynamics of Single Species

Population Models Described by the Total Density

When a population is observed at timescales much larger than the individual life span, and on a large number of generations, models with one variable (the total number of individuals or the total biomass in that population) are the simplest. These models postulate that the rate of change of the population number, N, is proportional to N (eqn [1], where r is the difference between birth and death rates).

$$\frac{dN}{dt} = rN \quad [1]$$

The logistic equation ([2]) represents limitation due to the resources or space (see **Figure 1**).

$$\frac{dN}{dt} = rN\left(1 - \frac{N}{K}\right) \quad [2]$$

where K is the carrying capacity.

Population growth of bacteria, phytoplankton, or microzooplankton can be simulated adequately by the logistic equation. With addition of a time delay term into the logistic equation, oscillations of the population can be represented.

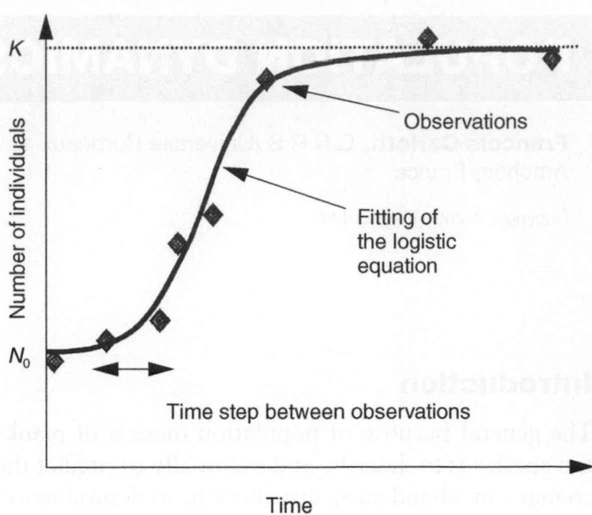

Figure 1 The growth of a plankton population with density regulation and its fitting by the logistic equation.

Population Models of Organisms with Description of the Life Cycle

If observations of a population are made with a time step shorter than the life cycle duration (**Figure 2**), the population development pattern is a succession of periods with decreasing abundance of individuals due to mortality, and increasing abundance due to recruitment of new individuals in periods of reproduction (cell division or egg production). Recruitment is defined as the input flux of individuals in a given state (stage, size class, etc.).

To represent such patterns, it is necessary to identify different phases in the life cycle, based on age, size, developmental stages, and so on. Two types of models can be developed:

- Structured population models, which consider the flux of individuals through different classes
- Individual-based models, which simulate birth, growth and development through stages and death of each individual

The major distinction between physiologically structured population models and individual-based models in a stricter sense is that individual-based models track the fate of all individuals separately over time, while physiologically structured population models follow the density of individuals of a specific type (age or size classes, stages). These models are particularly used for representing the complex life cycles of zooplankton and ichthyoplankton, but have also been useful for studying the population growth of bacteria, phytoplankton, and microzooplankton, particularly division synchrony in controlled conditions (chemostat).

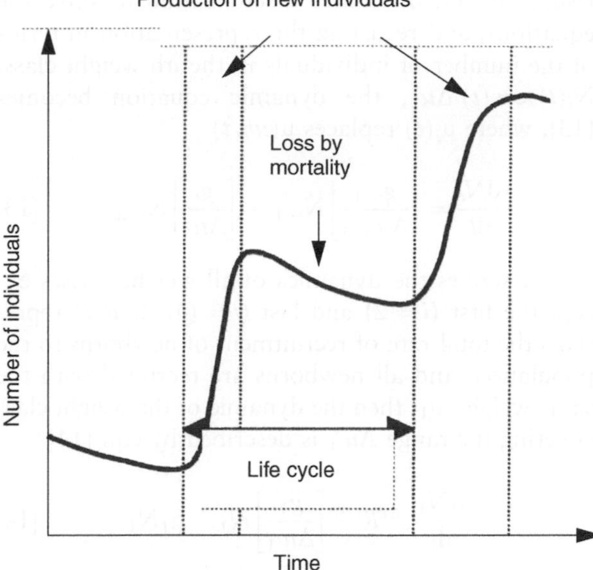

Figure 2 Total population abundance in controlled conditions. The population is initiated with newborn individuals, and decreases, first owing to mortality, until the maturation period, when a new generation is produced. At the end of the recruitment period of new individuals of second generation, the total abundance decreases, again owing to mortality. Owing to individual variability in development, loss of synchronism in population induces a broader recruitment period for the third generation. In nonlimiting food conditions, recruitment of new individuals is continued after few generations. The three generations correspond to the exponential phase presented in **Figure 1** (i.e., without density regulation).

Structured Population Models

A population can be structured with respect to age (age-structured population models), stage (stage-structured population models), or size or weight (size or weight-structured population models). Two types of equations systems are usually used: matrix models, which are discrete-time difference equation models, and continuous-time structured population models.

Matrix models constitute a class of population models that incorporate some degree of individual variability. Matrix models are powerful tools for analyzing, for example, the impact of life history characteristics on population dynamics, the influence of current population state on its growth potential, and the sensitivity of the population dynamics to quantitative changes in vital rates. Matrix models are convenient for cases where there are discrete pulses of reproduction, but not for populations with continuous reproduction. They are not suitable for studying the dynamics of populations that live in fluctuating environments.

The Leslie matrix is the simplest type of age-structured dynamic considering discrete classes. Suppose there are m age classes numbered 1, 2, …, m, each covering an interval τ. If $N_{j,t}$ denotes the number of individuals in age class j at time t and G_j denotes the fraction of the population in this age class that survive to enter age class $j+1$, then eqn [3] applies.

$$N_{j+1,\,t+1} = G_j N_{j,t} \qquad [3]$$

Individuals of the first age class are produced by mature individuals from older age classes and eqn [4] applies, where F_j is the number of age class 1 individuals produced per age class i individual during the time step τ.

$$N_{1,t+1} = \sum_{j=1}^{m} F_j N_{j,t} \qquad [4]$$

The system of eqns [3] and [4] can be written in matrix form (eqn [5]).

$$\begin{bmatrix} N_1 \\ N_2 \\ N_3 \\ \vdots \\ N_m \end{bmatrix}(t+1) = \begin{bmatrix} 0 & F_2 & F_3 & \cdots & F_m \\ G_1 & 0 & 0 & \cdots & 0 \\ 0 & G_2 & 0 & \cdots & 0 \\ \vdots & \ddots & \ddots & \cdots & \vdots \\ 0 & 0 & & G_{m-1} & 0 \end{bmatrix} \begin{bmatrix} N_1 \\ N_2 \\ N_3 \\ \vdots \\ N_m \end{bmatrix}(t) \qquad [5]$$

The Leslie matrix can easily be modified to deal with size classes, weight classes, and developmental stages as the key individual characteristics of the population. Organisms grow through a given stage or size/weight class for a given duration.

There are several variations of matrix models, differing mainly in the expression of vital rates, which can vary with time depending on external factors (e.g., temperature, food concentration, competitors, predators) or internal (e.g., density-dependent) factors.

The earlier type of continuous-time structured model is usually referred to as the McKendrick–von Foerster equation, and uses the age distribution on a continuous-time basis in partial differential equations. This type of model has been developed to the extent that it can be used to describe population dynamics in fluctuating environments. In addition, it also applies to situations in which more than one physiological trait of the individuals (e.g., age, size, weight, and energy reserves) have strong influences on individual reproduction and mortality. The movement of individuals through the different structural classes is followed over time. Age and weight are continuous variables, whereas stage is a discrete variable.

The general equation is eqn [6], where n is abundance of individuals of age a and mass m at time t.

$$\frac{\partial n(a,w,t)}{\partial t}+\frac{\partial n(a,w,t)}{\partial a}+\frac{\partial g(a,w,t)n(a,w,t)}{\partial w} = -\mu(a,w,t)n(a,w,t) \quad [6]$$

where $\mu(a,w,t)$ is the death rate of the population of age a, weight w at time t.

The von Foerster equation describes population processes in terms of continuous age and time (age-structured models) according to eqn [7].

$$\frac{\partial n(a,t)}{\partial t}+\frac{\partial n(a,t)}{\partial a} = -\mu(a,t)n(a,t) \quad [7]$$

The equation has both an initial age structure φ at $t=0$ (eqn [8]) and a boundary condition of egg production at $a=0$ (eqn [9]).

$$n(a,0)=\varphi_0(a) \quad [8]$$

$$n(0,t)=\int_0^\infty F(a,S_R)n(a,t)\mathrm{d}a \quad [9]$$

F is a fecundity function that depends on age (a) and the sex ratio of the population S_R. These kinds of equations are mathematically and computationally difficult to analyze, especially if the environment is not constant.

The same type of equation as [7] can be used where the age is replaced by the weight (weight-structured models (eqn [10]).

$$\frac{\partial n(w,t)}{\partial t}+\frac{\partial g(w,T,P)n(w,t)}{\partial w} = -\mu(w,t)n(w,t) \quad [10]$$

The weight of the individual w and the growth g are influenced by the temperature T, the food P, and by the weight itself through allometric metabolic relationships.

The equation has both an initial age structure φ at $t=0$ (eqn [11]) and a boundary condition of egg production at $w=w_0$ (eqn [12]).

$$n(w,0)=\varphi_0(w) \quad [11]$$

$$N(0,t)=\int_0^\infty F(w,S_R)n(w,t)\mathrm{d}w \quad [12]$$

F is the fecundity function, which depends on weight (w) and the sex ratio of the population S_R.

The numerical realization of this equation requires a representation of the continuous distribution $n(w,t)$ by a set of discrete values $n_i(t)$ that are spaced along the weight axis at intervals $\Delta w_i = w_{i+1} - w_i$. Using upwind difference discretization to solve the equations, and recasting the representation in terms of the number of individuals in the ith weight class, $N_i(t) \approx n_i(t)\,\Delta w_i$, the dynamic equation becomes [13], where $\mu_i(t)$ replaces $\mu(w_i,t)$

$$\frac{\mathrm{d}N_i}{\mathrm{d}t}=\left[\frac{g_{i-1}}{\Delta w_{i-1}}\right]N_{i-1}-\left[\frac{g_i}{\Delta w_i}\right]N_{i-\mu_i N_i} \quad [13]$$

This describes the dynamics of all weight classes except the first ($i=2$) and last ($i=Q$). If $R(t)$ represents the total rate of recruitment of newborns to the population, and all newborns are recruited with the same weight w_1, then the dynamic of the weight class covering the range Δw_1 is described by eqn [14].

$$\frac{\mathrm{d}N_1}{\mathrm{d}t}=R-\left[\frac{g_1}{\Delta w_1}\right]N_1-\mu_1 N_1 \quad [14]$$

If we assume that individuals in only the Qth weight class are adult, and that adult individuals expend all assimilated energy on reproduction rather than growth, the population dynamics of the adult population is given by eqn [15].

$$\frac{\mathrm{d}N_Q}{\mathrm{d}t}=\left[\frac{g_{Q-1}}{\Delta w_{Q-1}}\right]N_{Q-1}-\mu_Q N_Q \quad [15]$$

The rate of recruitment of newborns to the population is given by [16] where $\beta(t)$ represents the per capita fecundity of an average adult at time t.

$$R(t)=\beta(t)N_Q(t) \quad [16]$$

The weight intervals Δw_i increase with class number i as an allometric function. The growth rate $g(w,t)$ can be calculated by a physiological model.

Stage-structured Population Models

Plankton populations often have continuous recruitment and are followed in the field by observing stage abundances over time. A large number of zooplankton population models deal with population structures in term of developmental stage, using ordinary differential equations (ODEs).

A single ODE can be used to model each development stage or group of stages: for instance, a copepod population can be subdivided into four groups: eggs, nauplii, copepodites, and adults. The equation system is eqn [17]–[20], where R is recruitment, α is the transfer rate to next stage, and μ is the mortality rate.

$$\text{Eggs} \quad \frac{\mathrm{d}N_1}{\mathrm{d}t}=R-\alpha_1 N_1-\mu_1 N_1 \quad [17]$$

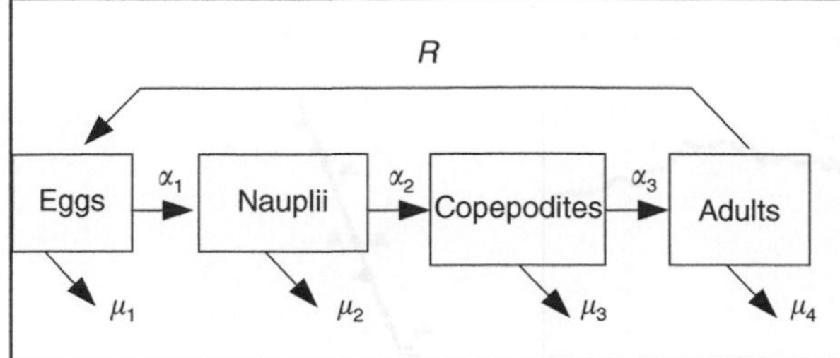

Figure 3 Schematic representation of the population dynamics mathematically represented by eqn [17]–[20]. α_i = transfer rate of stage i to stage $i+1$; μ_i = mortality rate in stage i; R = recruitment: number of eggs produced by females per day.

$$\text{Nauplii} \quad \frac{dN_2}{dt} = \alpha_1 N_1 - \alpha_2 N_2 - \mu_2 N_2 \qquad [18]$$

$$\text{Copepodids} \quad \frac{dN_3}{dt} = \alpha_2 N_2 - \alpha_3 N_3 - \mu_3 N_3 \qquad [19]$$

$$\text{adults} \quad \frac{dN_4}{dt} = \alpha_3 N_3 - \mu_4 N_4 \qquad [20]$$

The system of ODEs is solved by Euler or Runge–Kutta numerical integration methods, usually with a short time step (approximately 1 hour).

In the model presented in **Figure 3**, the transfer rate of animals from stage to stage and the mortality at each stage are expressed as simple linear functions, which induce a rapidly stable stage distribution. To represent the delay of growth within a stage, more refined models consider age-classes within each stage, or systems of delay differential equations. They have a high degree of similarity with observed cohort development in mesocosms or closed areas (**Figure 4**).

Individual-based Models of a Population

Individual-based models (IBMs) describe population dynamics by simulating the birth, development, and eventual death of a large number of individuals in the population. IBMs have been developed for phytoplankton, zooplankton, meroplanktonic larvae, and early life history of fish populations. Object-oriented programming (OOP) and cellular automata techniques have been applied to IBMs.

As powerful computers become more accessible, numerous IBMs of plankton populations have been developed, mainly to couple them with 1D-mixed layer models (phyto- and zooplankton) and circulation models (zooplankton).

IBMs treat populations as collections of individuals, with explicit rules governing individual biology and interactions with the environment. Each biological component can change as a function of the others. Each individual is represented by a set of variables that store its *i*-state (age, size, weight, nutrient or reserve pool, etc.). These variables may be grouped together in some data structure that represents a single individual, or they may be collected into arrays (an array of all the ages of the individuals, an array of all the sizes of the individuals, etc.), in which case an individual is an index number in the set of arrays. The *i*-state of an individual changes as a function of the current *i*-state, the interactions with other individuals, and the state of the local environment. The local environment can include prey and predator organisms that do not warrant explicit representation as individuals in the model. Population-level phenomena (e.g., temporal or spatial dynamics) or vital rates can then be inferred directly from the contributions of individuals in the ensemble.

The model starts with an initial population and the basic environment, then monitors the changes of each individual. At any time t, the *i*-state of individual j changes as eqn [21].

$$X_{i,j}(t) = X_{i,j}(t - \mathrm{d}t) + f\left(X_{1,j}(t - \mathrm{d}t), \ldots, X_{i,j}(t - \mathrm{d}t)\right) \qquad [21]$$

$X_{i,j}(t)$ is the value of the *i*-state of individual j, and f is the process modifying $X_{i,j}$, as a function of the values of different *i*-states of the organism and external parameters such as the temperature. When the fate of all individuals during the time-step dt has been calculated, the changes to the environment under the effects of individuals can be updated. Any stochastic process can be added to eqn [21].

This type of model can add a lot of detail in the representation of physiological functions. Individual growth can be calculated as assimilation less metabolic loss, and the interindividual variation in physiology can be represented by adding stochastic processes or parameters describing the characteristics of each individual (growth and development parameters, mortality coefficient, and parameters connected with reproduction). The end results are unique life histories, which when considered as a whole give rise to growth/size distributions that provide a measure of the state of the population.

Calibration of Parameters

Parametrization of a model can range from very simplistic to extremely complex depending upon the amount of information known about the population under consideration. Bioenergetic processes (ingestion, egestion, excretion, respiration, and egg

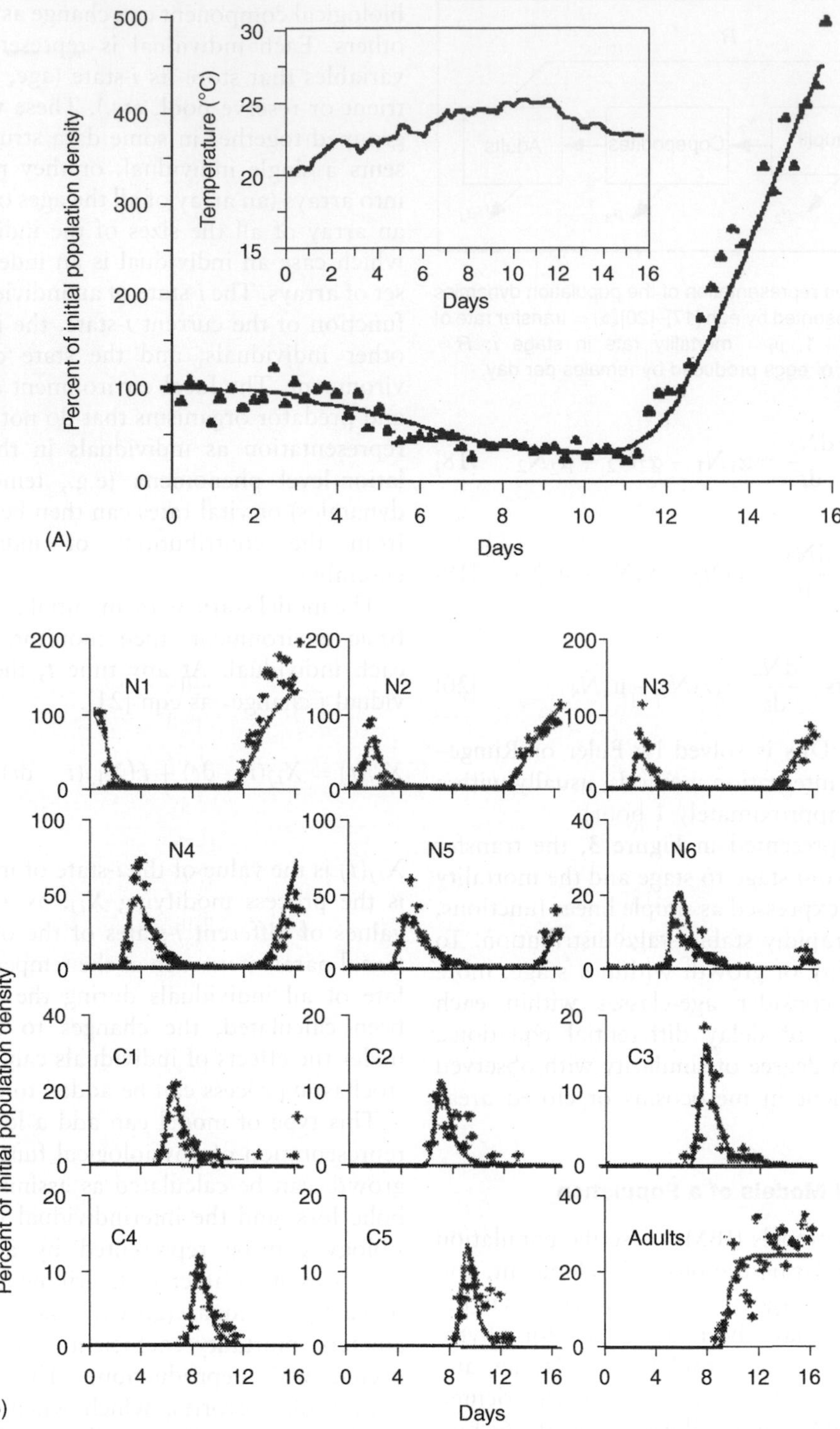

Figure 4 Simulation of the cohort development of the copepod *Euterpina acutifrons* in mesocosms with a structured stage and age-within-stage model. (A) Total population during development, with variable temperature and constant food supply (points = experimental data; line = simulation). The initial density decreases owing to mortality and then increases to newborn individuals (as the first part in **Figure 2**). (B) Naupliar stages (N1 to N6), copepodite stages (C1 to C5), and adults during development (points = experimental data; line = simulation). The simulation start with similar N1 of same age. Reproduced from Carlotti F and Sciandra A, 1989. Population dynamics model of *Euterpina acutifrons* (Copepoda: Harpacticoida) coupling individual growth and larval development. *Mar. Ecol. Prog. Ser.*, 56, 3, 225–242.

production) are usually modeled from experimental results, whereas biometrics (size, weight,...) and demographic (development rate, mortality rate, ...) parameters are estimated by combining data from life tables collected in the field or from laboratory studies.

To solve for the unknown parameters, new techniques have been developed such as inverse methods and data assimilation by fitting simulations to data.

Spatial Distribution of Single Plankton Populations

An important development in plankton population modeling is to make full use of the increased power of computers to simulate the dynamics of plankton (communities or populations) in site-specific situations by coupling biological and transport models, giving high degrees of realism for interpreting plankton population growth, transport, spatial distribution, dispersion, and patchiness. Structured population models and individual-based models allow detailed simulations of zooplankton populations in different environmental conditions. Physical–biological models of various levels of sophistication have been developed for different regions of the ocean.

Spatial Plankton Dynamics with Advection–Diffusion–Reaction Equations

Equation [22] is the general physical–biological model equation used to describe the interaction between physical mixing and biology.

$$\frac{\partial C}{\partial t} + \nabla \cdot (v_{\mathrm{a}} C) - \nabla \cdot (K \nabla C) = \text{‘biological teams’} \qquad [22]$$

$C(x, y, z, t)$ is the concentration of the biological variable, which is a functional group (phytoplankton, microzooplankton, or zooplankton), a species or a developmental stage, or a size class (in which case the number of equations would equal the number of stages or size classes) at position x, y, z at time t. The concentration can be expressed as numbers of organisms or biomass of organisms per unit volume. v_a (u_a, v_a, w_a) represents the advective fluid velocities in x, y, z directions. K_x, K_y, K_z are diffusivities in x, y, z directions. $\Delta = (\partial/\partial x,\ \partial/\partial y,\ \partial/\partial z)$ is the Laplacian operator.

On the left-hand side of eqn [22], the first term is the local change of C, the second term is advection caused by water currents, and the third term is the diffusion or redistribution term. The right-hand side of eqn [22] has the biological terms that represent the sources and sinks of the biological variable at position x, y, z as a function of time.

The biological terms may or may not include a velocity component (swimming of organisms, migrations, sinking, ...), and the complexity of the biological representation can vary from the dispersion of one (the concentration of a cohort) to detailed population dynamics. Physical–biological models of various levels of sophistication have been developed recently for different regions of the ocean.

Biological models can be configured as compartmental ecosystem models in an upper-ocean mixed layer, where phyto-, microzoo-, and mesozooplankton are represented by one variable. In extended cases, the model takes into account several size classes of phyto-, microzoo-, and zooplankton. Such types of ecosystem model have been coupled to one-dimensional physical, and embedded into two-dimensional and three-dimensional circulation.

Studies of plankton population distribution in regions where plankton may be aggregated (e.g., upwelling and downwelling regions, Langmuir circulations, eddies) can be undertaken with populations described by equations of the McKendrick–von Foester type coupled with 2D or 3D hydrodynamical models.

In 1982, Wroblewski presented a clear example with a stage-structured population model of *Calanus marshallae*, a copepod species, embedded in a circulation system simulating the upwelling off the Oregon coast. Simulations of the dynamics focused on the interaction between diel vertical migration and offshore surface transport.

The zonal distribution of the life stage categories C_i of *C. marshallae* over the Oregon continental shelf was modeled by the two-dimensional (x, z, t) equation [23], where w_{bi} is the vertical swimming speed of the ith stage, assumed to be a sinusoidal function of time: $w_{bi} = w_{si} \sin(2\pi t)$, with w_{si} the maximum vertical migration speed of the ith stage.

$$\begin{aligned} &\frac{\partial C_i(x,z,t)}{\partial t} + \frac{\partial [u_a(x,z,t) C_i(x,z,t)]}{\partial x} \\ &\quad + \frac{\partial [w_a(x,z,t) C_i(x,z,t)]}{\partial z} \\ &\quad - \frac{\partial}{\partial x}\left[K(x,t)\frac{\partial C_i(x,z,t)}{\partial x}\right] - \frac{\partial}{\partial z}\left[K(z,t)\frac{\partial C_i(x,z,t)}{\partial z}\right] \\ &= \text{population dynamics} + \frac{\partial [w_{\mathrm{b}i}(x,z,t) C_i(x,z,t)]}{\partial z} \end{aligned} \qquad [23]$$

The population dynamics model was presented in eqns [17]–[20].

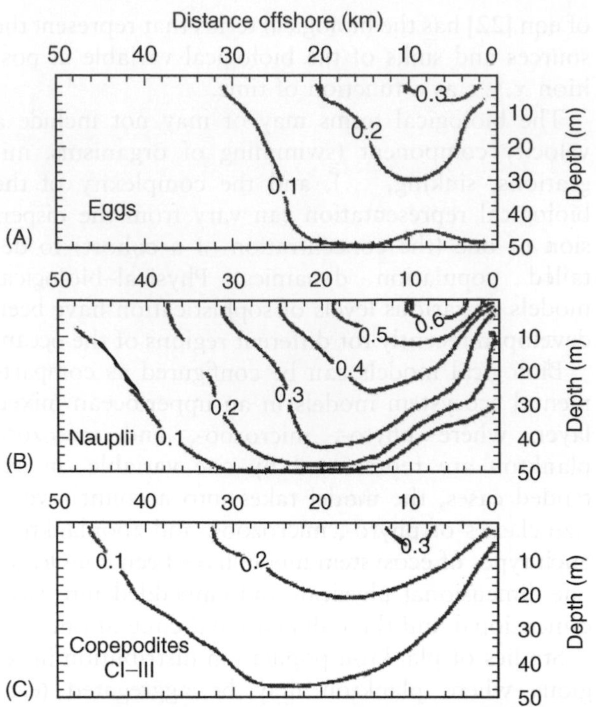

Figure 5 Model of *Calanus marshallae* in the Oregon upwelling zone. The figure shows the simulated zonal distribution of (A) eggs, (B) nauplii, and (C) early copepodites at noon on 15 August. Concentrations of each stage are expressed as a fraction of the total population (all stages m^{-3}). (Reproduced with permission from Wroblewski JS, 1982. Interaction of currents and vertical migration in maintaining *Calanus marshallae* in the Oregon upwelling zone—a simulation. *Deep Sea Research* 29: 665–686.

The upwelling zone extended 50 km from the coast down to a depth of 50 m, and was divided into a grid with spacing 2.5 m in depth and 1 km in the horizontal. The author used a finite-difference scheme with a time step of 1 h, which fell within the bounds for computational stability (**Figure 5**).

Coupling IBMs and Spatially Explicit Models

Individual-based models are more and more frequently used to assess the influence of space on the population dynamics, namely on the time course of population abundances and the pattern formation of populations in their habitats. This approach uses simulated currents from sophisticated 3D hydrodynamic models driving Lagrangian models of particle trajectories to examine dispersion processes.

The approach is relatively straightforward and is a first step in formulating spatially explicit individual based models. Given a 'properly resolved' flow field, particle (larval fish/zooplankton/meroplanktonic larvae) trajectories are computed (generally with standard Runge–Kutta integration methods of the velocity field). Specifically, hydrodynamic models provide the velocity vector $\mathrm{v} = (u, v, w)$ as a function of location $x = (x, y, z)$ and time t, and the particle trajectories are obtained from the integration of eqn [24].

$$\frac{\mathrm{d}x}{\mathrm{d}t} = \mathrm{v}(x, y, z, t) \qquad [24]$$

The simplest model of dispersion is a random walk model in which individuals move along a line from the same starting position. These trajectories could be modified by turbulent dispersion as described in the section below. Once the larval/particle position is known, additional local physical variables can be estimated along the particle's path: e.g., temperature, turbulence, light, etc., and input to the IBM. The physical quantities are then included in biological (physiological or behavioural) formulations of IBMs (see below).

Simulations considering trajectories of plankton as passive particles are a necessary step before considering any active swimming capability of planktonic organisms. They show the importance of physical features in the aggregation or dispersion of the particles.

Plankton transport models that include biological components typically use a prescribed vertical migration strategy for all or part of an animal's life history or a vertical motion (sinking or swimming) that is determined by animal's development and growth. The simulated plankton distributions from these models tend to compare better with observed distributions than do models that use passive particles. Sensitivity studies show that behavior is an important factor in determining larval transport and/or retention.

The coupling of IBMs of zooplankton and fish populations and 3D circulation models is a recent field of study, even for fish models. Generally, models that describe the spatial heterogeneity of the habitat have been designed to answer questions about the spatial and temporal distribution of a population rather than questions about the numbers and characteristics of surviving individuals. They allow us to explore the potential effects of habitat alteration on these populations. Using this approach, biological mechanisms that are strongly dependent on habitat and that are not fully understood could be studied by examining different scenarios.

Modeling Behavioral Mechanisms, Aggregation and Schooling, and Patches

Different types of models have been built, some of them focusing on the structure and shape of aggregations depending on internal and external physical forces, others dealing with the benefits for individuals of living in groups with regard to feeding (foraging models) and to predation. The Lagrangian approach can take into account the behavior of individual organisms and the effects of the physical environment upon them. Although Eulerian approaches are mathematically tractable, the methods do not explicitly address the density dependence of aggregating individual behavior within a patch.

Dynamic optimization allows descriptions of the internal state of individuals, which may lead to both variable and fluctuating motivations among individuals over short time periods.

Interactions between Populations

Models with Plankton Populations in Interaction

Simple models of two species interactions take the form of eqns [25] and [26].

$$\frac{dN_1}{dt} = r_1 N_1 - k_1 N_1 N_2 \qquad [25]$$

$$\frac{dN_2}{dt} = r_2 N_2 - k_2 N_1 N_2 \qquad [26]$$

These population models represent some special experimental situations or typical field situations. Interactions between two species have been rarely treated by population models with description of the life cycle, although structured population models as well as IBM models can represent interactions between species such as predation, parasitism, or even cannibalism.

As an example, Gaedke and Ebenhöh presented in 1991 a study on the interaction between two estuarine species of copepods, *Acartia tonsa* and *Eurytemora affinis*. They first used a simple model based on eqns [25] and [26] including (a) predation (including self-predation of immature stages) by *Acartia* on the two, (b) a term of biomass gain of *Acartia* by this predation, and (c) a density-dependent loss term caused by predation by invertebrates or by starvation of the two species. This simple model did not result in stable coexistence between the two species with a reasonable parameter range under steady-state conditions.

These authors then used two-stage-structured population models with stage-specific interactions (with similar equations to [17]–[20]) allowing the predation of large individuals of *A. tonsa* (copepodites 4 to adults) on nauplii of both species to be represented. The results of this detailed numerical model were compared with results obtained using the simpler model with two variables. The predation on nauplii by *Acartia tonsa* appears to be key factor in the interaction of the two copepod populations.

Food Webs with Population Models

Structured models should be chosen to stimulate the dynamics of several interacting species. The stage-based approach will be acceptable with few species, but quickly become intractable with increasing numbers of species. In this case, a community model based on size structure and using prey–predator size ratio is the alternative approach. There is a continuum of models from detailed size spectrum structure up to large size classes representing functional (trophic) groups in food web models. The detailed size spectrum approach is particularly useful when simulating the predation of a fish cohort on its prey, whereas large functional groups are required for large-scale ecosystem models. Numerous examples include models with size structure of herbivorous zooplankton populations and their prey, and their interactions, in a nutrient–phytoplankton–herbivore–carnivore dynamics model. Size-based plankton model with large entities consider the size range 0.2–2000 μm, picophytoplankton, bacterioplankton, nanophytoplankton, heterotrophic flagellates, phytoplankton, microzooplankton, and mesozooplankton.

See also

Biogeochemical Data Assimilation. Lagrangian Biological Models. Large Marine Ecosystems. Marine Mesocosms. Microbial Loops. Network Analysis of Food Webs. Ocean Gyre Ecosystems. Patch Dynamics. Polar Ecosystems. Small-scale Patchiness, Models of. Small-Scale Physical Processes and Plankton Biology. Upwelling Ecosystems.

Further Reading

Carlotti F, Giske J, and Werner F (2000) Modelling zooplankton dynamics. In: Harris RP, Wiebe P, Lenz J, Skjoldal HR, and Huntley M (eds.) *Zooplankton Methodology Manual*, pp. 571–667. New York: Academic Press.

Caswell H (1989) *Matrix Population Models: Construction, Analysis and Interpretation*. Sunderland, MA: Sinauer Associates.

Coombs S, Harris R, Perry I and Alheit J (1998) *GLOBEC* special issue, vol. 7 (3/4).

DeAngelis DL and Gross LJ (1992) *Individual-based Models and Approaches in Ecology: Populations,*

Communities and Ecosystems. New York: Chapman and Hall.

Hofmann EE and Lascara C (1998) Overview of interdisciplinary modeling for marine ecosystems. In: Brink KH and Robinson AR (eds.) *The Sea*, vol. 10, Ch. 19, pp. 507–540. New York: Wiley.

Levin SA, Powell TM, and Steele JH (1993) *Patch Dynamics*. Berlin: Springer-Verlag. Lecture Notes in Biomathematics 96..

Mangel M and Clark CW (1988) *Dynamic Modeling in Behavioral Ecology*. Princeton, NJ: Princeton University Press.

Nisbet RM and Gurney WSC (1982) *Modelling Fluctuating Populations*. Chichester: Wiley.

Renshaw E (1991) *Modelling Biological Populations in Space and Time*. Cambridge: Cambridge University Press. Cambridge Studies in Mathematical Biology.

Tuljapurkar S and Caswell H (1997) *Structured-population Models in Marine, Terrestrial, and Freshwater Systems*. New York: Chapman and Hall. Population and Community Biology Series 18.

Wood SN and Nisbet RM (1991) *Estimation of Mortality Rates in Stage-structured Populations*. Berlin: Springer-Verlag. Lecture Notes in Biomathematics 90.

Wroblewski JS (1982) Interaction of currents and vertical migration in maintaining *Calanus marshallae* in the Oregon upwelling zone – a simulation. *Deep Sea Research* 29: 665–686.

LAGRANGIAN BIOLOGICAL MODELS

D. B. Olson, C. Paris, and R. Cowen, University of Miami, Miami, FL, USA

Introduction

The Swiss mathematician Leonhard Euler (1707–1783) derived the formulations for describing fluid motion by either measuring the properties of the fluid at a fixed point overtime or alternatively following the trajectory of a parcel of fluid as it is carried with the flow. The first of these is known as the Eulerian description of the flow, while the method following a material parcel or particle is known as the Lagrangian description after the French mathematician Joseph Lagrange (1736–1813). Most of the theory used to model ocean currents is posed in an Eulerian frame because of the difficulties in solving the momentum equations in the complicated matrices that arise in the Lagrangian form of the equations. However, it is often useful to use the Lagrangian frame of reference when considering the manner in which mixing occurs in turbulent flows such as those found in the oceans. It is also common to measure these flows by using drifters or floats that trace out oceanic currents. As shown below, the Lagrangian description is also conducive to handling models of marine populations in many cases. This is especially true when the models include quantities that structure the population, such as age, genetics, or physiological traits that depend upon the history of individual organisms that are carried in or swim through oceanic flows. Organisms that drift freely with the currents are termed planktonic, while those that can swim effectively are termed nektonic fauna. Here Lagrangian methods for considering populations of both plankton and nekton are given. Much of the detailed formalism can be found in Okubo (1980) (*see* Further Reading). The present discussion highlights the application of these methods to marine population dynamics.

Comparing the Eularian and Lagrangian Formulations

To understand the difference between the Lagrangian and Eulerian formulations, consider the population dynamic equations for marine organisms. If the ith population is made up of N_i individuals, one can write an equation for each individual. This will include each organism's position, $X_m(t)$, as a function of time t. The total, or Lagrangian, derivative of $X_m(t)$ with respect to time, dX_m/dt, gives the individual's velocity, $V_m(t)$. This can be separated into the influence of the advection of the organism by ocean currents, U (X_m, t), and a swimming contribution, $U_s(X_m, B, t)$, where B (X_m, t) is a vector of behavioral clues. These clues involve both physical and biotic components of the environment. The acceleration of the individual organism is then derived by carrying out another differentiation in time (eqn [1]).

$$\frac{dV_m}{dt} = \frac{\partial U}{\partial t} + \frac{\partial U_s}{\partial t} + U \bullet \nabla V_m + U_s \bullet \nabla V_m = F(B) \qquad [1]$$

∇ is the spatial gradient operator and F is the gravitational and viscous forces imposed on the organism as well as behavioral responses, i.e., swimming. Notice that the Lagrangian derivative on the left leads to a set of Eulerian terms that involve spatial gradients in the fluid velocity and the behavioral clues on the right side of the equation. This equation fully expresses the motion of an individual. To the equation describing the organism's motion, a set of state relations must be added expressing changes in its physiological state, its age or stage, and the probabilities of its death and reproduction to explain population dynamics. Such a model considering the conditions of each individual explicitly in a population is called an individual-based model (IBM).

Individual-based models provide a method for understanding behavior and small groups of organisms as discussed below. For large populations, however, the number of equations involved becomes impossible to handle. It is therefore common to introduce the concept of organism density, $n_i = N_i/A$, where N_i is the number of individuals and A is a given a real measure. The density of the ith taxon is then measured in numbers per square kilometer of ocean surface area. This leads to a continuous spatial field equation. It is typical to consider the mean field of population density and perturbations about it so $n_i = \langle n_i \rangle + n_i'$. Here the first term is the mean population density and the second the perturbations (or variance) about the mean; the mean is over the population. The same separation can be done for the velocity components such that $U = \langle U \rangle + U'$ and $U_s = \langle U_s \rangle + U_s'$. Equation [1] above involves

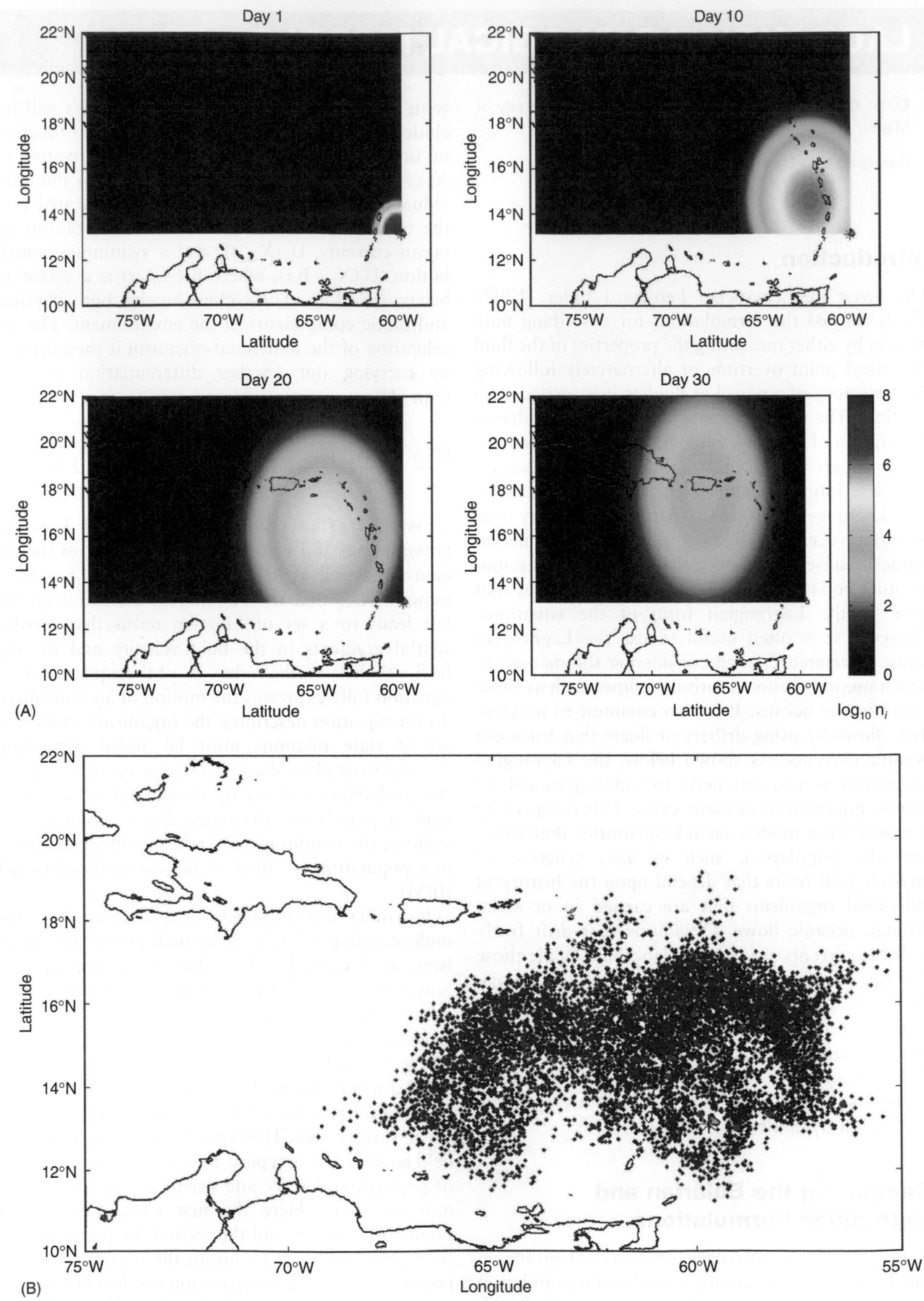
Day 1
Day 10
Day 20
Day 30
22°N
20°N
18°N
16°N
14°N
12°N
10°N
75°W
70°W
65°W
60°W
55°W
Longitude
Latitude
8
6
4
2
0
$\log_{10} n_i$
(A)
(B)

products of velocity components with each other such that the contributions to the mean motion of the population and therefore its average spread will involve both the mean velocities and correlations between velocity perturbations. In the field equations these correlation's between fluctuations in the turbulent fluid velocity or swimming behavior will lead to turbulent and behavioral diffusion. It is typical to introduce diffusivities, K for the turbulence and K_s for the behavioral related dispersion. There is also a correlation between the variations in the environmental factors, including the distance to members of the same species that come into effect given $B = \langle B \rangle + B'$. The resulting field equation for the expected mean density of an organism is then given by eqn [2].

$$\frac{dn_i}{dt} = \frac{\partial n_i}{\partial t} + V_m \bullet \nabla n_i = F(B) + \nabla[(\kappa + \kappa_s)\nabla n_i] \quad [2]$$

The κ are inside the spatial gradient operators to denote that they are functions of space. Taking the κ_s term all the way inside the ∇ operators allows density or schooling effects on population density through behavioral preferences for nearest neighbor distance. The results of the field model versus the Lagrangian model following individuals can lead to impressive differences (**Figure 1**).

The diffusion model in **Figure 1A** using the equations above lead to a finite probability of finding organisms everywhere in a domain immediately. The Lagrangian treatment limits an organism's spread to the fastest velocities present, so that it takes a finite time for spread. It is important to note, however, that there must also be population losses that are more abrupt than simple exponential decrease of the population across habitat boundaries. This occurs in most population parametrizations such as a logistic growth with linear mortality to achieve realistic population distributions. In the real ocean, while it takes much longer than in the analytical diffusion model, the Lagrangian motions will still lead to a finite possibility of finding organisms everywhere in the domain at large timescales, unless mortality is properly treated.

Simple Models in the Lagrangian Frame

The most important issue in modeling marine populations is providing an accurate depiction of the physics, the biology, and the intricate biological/physical interactions that occur. These are represented in the equations above by mean quantities acting on means or perturbations and then by correlations between both physical and biological perturbations. One use of the Lagrangian description of motion is to simulate these interactions along a fluid trajectory in the case of the plankton (**Figures 2** and **3**). In this case a simple meandering current and its impact on populations is envisioned. The calculation involved is a simple integration that in **Figure 2** reveals the basic response without biological non linearities. In **Figure 3** the impact of a primary production response as seen in **Figure 2** on a density-dependent population conceals a set of more interesting patterns, including extinction. The situation in **Figures 2** and **3** involves dynamics that allow exact calculations, i.e, in this case simple integrations of the functions without any use of numerics. This sort of analysis is recommended for testing morecomplicated cases where numerical methods become a major issue. Essentially these simple applications use the Lagrangian frame of viewing advection as a means of allowing simple calculations of population dynamics. The Lagrangian frame becomes indispensable when structured populations are considered.

Simulations of Populations with Demographic Structure

The Lagrangian description of the path that biological entities follow through the ocean environment becomes the only feasible method for treating population dynamics where the detailed history of the populations' interaction with the physical environment and other populations are important. Early works in this area include the plankton models of Wolf and Woods following the details of mixed layer and thermocline development in the North Atlantic over many seasons and the work on zooplanktonin the coastal environment by Hoffman *et al.* The problem becomes that the Lagrangian

Figure 1 (Left) (A) Simulation of larval reef fish drift from Barbados using the mean flow into the eastern Caribbean and a κ of 5000 m^2 s^{-1} at 1, 10, 20, and 30 days after spawning. A typical larval mortality rate $\mu = 0.2$ (or ~18% per day) is applied to larval abundance (N_i). (B) Lagrangian simulation of the same case with trajectories computed from an oceanic general circulation model at day 30 (Cowen *et al.*, 2000). The survivors are indicated by red dots after applying the same $\mu = 0.2$. Note that none are on suitable island habitat after 30 days. For the diffusive case (A) there is a finite probability of finding larvae well beyond the range of any of the simulated trajectories. In this case the mortality is truncated by the 30-day duration of planktonic behavior in the fish's assumed development, i.e., after this time there is assumed recruitment to an island or death.

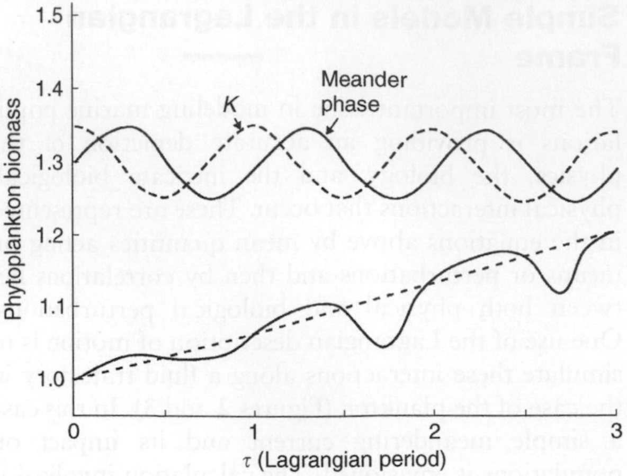

Figure 2 Response of phytoplankton to a simple meandering current. The meander is assumed to set up a simple sinusoidal series of upwelling (high) and down welling (low) at meander crests and troughs, respectively, that provide nutrients upon the upwelling phase. The model assumes logistic phytoplankton response to a sinusoidal carrying capacity (K) and is solved analytically by simple integration of the Lagrangian equation in time. The forcing function involves linear response to a cosine shown in the figure.

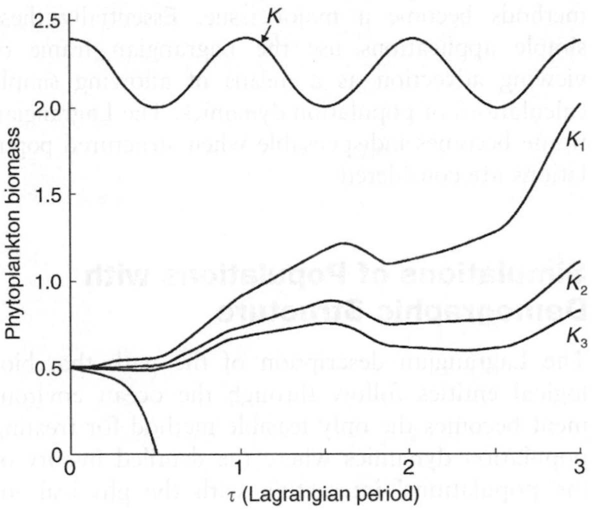

Figure 3 A calculation of the zooplankton response to the phytoplankton distribution in a meander like that treated in **Figure 2**. Here the integration in time along trajectories is slightly more complicated but still analytical. The zooplankton response is parametrized by a Hollings type 2 curve such that the time dependence of zooplankton (Z) is governed by the equation below.

$$\frac{dZ}{dt} = r\frac{ZP}{K_0 + P} - dZ$$

Here r is a growth rate, P is the sinusoidal phytoplankton field, K_0 is a half-saturation term, and d is the death rate for Z. The pattern of the carrying capacity, K is shown at the top of figure. Four different Ks are shown with different magnitudes K_1–K_3 and fourth that goes extinct. See Olson and Hood (1994) and the literature cited there for further discussion of meander impacts on marine ecosystems.

frame (i.e., following individual trajectories of individuals or individual subpopulations) is the only way to track the biological dynamics where the past history plays a major role in determining the present dynamics. These historical parameters may involve the past history of nutrient or forage availability, the temperature and salinity encountered over the development of the organisms, or the past history of selection on the genetic structure of populations when reproduction occurs.

As an example of a structured population simulation in a turbulent ocean gyre, a simulation of a population of physiologically and genetically structured pelagic copepods is described. The population

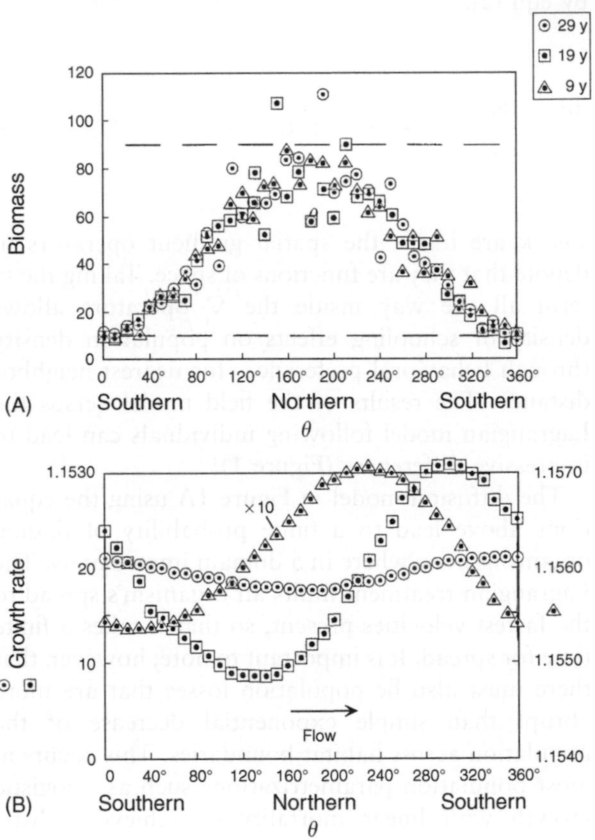

Figure 4 (A) Biomass of copepods per m^2 of surface area as a function of distance (θ) around the gyre, at 10-year intervals. Gyre circulation time is 3 years. The carrying capacity at the northern and southern ends of the gyre are indicated by dashed lines. Note that mixing can cause a region to exceed the carrying capacity. The variation in populations is largest at the highest carrying capacity. (B) The mean growth rate (biomass per day) in different segments around the gyre at the three times. The scale at 9 years is 10 times that for the later times (scale at right). Growth rate is going slowly to a constant or fixed state as expected from population genetics grounds. The distribution of growth rates matches the cosine nature of the carrying capacity distribution, but is fully advective in the sense that these patterns advect with the mean flow around the gyre. The direction of this drift is indicated by an arrow.

is based on the properties of *Nanocalanus minor*, a copepod found in the subtropical gyre in the North Atlantic. The model is designed to consider the dynamics behind the mitochondrial DNA patterns found in this population in the Gulf Stream. The Gulf Stream and its recirculation gyre are treated as a circular flow with superimposed turbulence. The copepods are simulated as subpopulations, each carried on Lagrangian particles advected in this flow. The populations are subject to a carrying capacity, $K = K_0 + K' sin(\theta/2)$, that is high in the northern Gulf Stream and low in the oligotrophic southern portions of the gyre. The population has variable growth rates that are controlled genetically. The growth potential is determined by the statistics of the local breeding subpopulations and by selection induced by competition at a given location and time for food. Selection is local since there is not an optimal growth rate in the sense that it pays to have a high reproductive potential in the northern gyre under low population densities. The offspring of such a population are inevitably at a disadvantage, however, when advected into the southern oligotrophic reaches of the gyre. Since the resulting genetic and physiological attributes at a location depend upon the past history of all of the subpopulations contributing to the interaction at a given time, this sort of simulation becomes computationally impossible in a Eulerian frame. The population distributions in **Figure 4**, done in a Lagrangian simulation, takes only an hour on a laptop computer. The details of a suite of such simulations are currently being compared to population density and gene sequences.

Conclusions

The use of Lagrangian particle-following simulations in modeling population dynamics allows several advantages over Eularian fixed-grid calculations. For simple models the advantage is that the population equations can be simply integrated in time. As new techniques for tracking fluid parcels and therefore planktonic trajectories or individual large pelagic fish or whales become more available, models using real trajectories will become possible. The other advantage that direct Lagrangian simulation of turbulent dispersal of organisms has is that it overcomes the problems that advection/diffusion schemes have with population densities at large distances from their source. Finally, the largest promise in Lagrangian simulations is their use in models that explicitly treat the demographic traits of populations. With the ever-increasing realism in physical models of the marine environment and Lagrangian population models, new insights into marine population dynamics are possible.

See also

Population Dynamics Models.

Further Reading

Bucklin A, LaJeunesse TC, Curry E, Wallinga J, and Garrison K (1996) Molecular genetic diversity of the copepod, *Nannocalanus minor*: genetic evidence of species and population structure in the N. Atlantic Ocean. *Journal of Marine Research* 54: 285–310.

Carlotti F (1996) A realistic physical-biological model for *Calanus finmarchicus* in the North Atlantic. A conceptual approach. *Ophelia* 44: 47–58.

Carlotti F and Wolf KU (1998) A Lagrangian ensemble model of *Calanus finmarchicus* coupled with a 1-D ecosystem model. *Fishers and Oceanography* 7(7): 191–204.

Cowen RK, Lwiza KMM, Sponaugle S, Paris CB, and Olson DB (2000) Connectivity of marine populations: Open or closed? *Science* 287: 857–859.

Flierl G, Grünbaum D, Levin S, and Olson DB (1999) From individuals to aggregation: the interplay between behavior and physics. *Journal of Theoretical Biology* 196: 397–454.

Hoffmann EE, Halstrom KS, Moisan JR, Haidvogel DB, and Mackas DL (1991) Use of simulated drifter tracks to investigate general transport patterns and residence times in the coastal transition zone. *Journal of Geophysical Research* 96: 15041–15052.

Metz JAJ and Diekmann O (eds.) (1986) *The Dynamics of Physiologically Structured Populations, 68*. Berlin: Springer-Verlag.

Okubo A (1980) *Diffusion and Ecological Problems: Mathematical Models, 10*. Berlin: Springer-Verlag.

Olson DB and Hood RR (1994) Modelling pelagic biogeography. *Progress in Oceanography* 34: 161–205.

Wolf KU and Woods JD (1998) Lagrangian simulation of primary production in the physical environment: The deep chlorophyll maximum and nutricline. In: Rothschild BJ (ed.) *Toward a Theory on Biological–Physical Interactions in the World Ocean*, pp. 51–70. Dordrecht: Kluwer Academic.

OCEAN BIOGEOCHEMISTRY AND ECOLOGY, MODELING OF

N. Gruber, Institute of Biogeochemistry and Pollutant Dynamics, ETH Zurich, Switzerland
S. C. Doney, Woods Hole Oceanographic Institution, Woods Hole, MA, USA

Introduction

Modeling has emerged in the last few decades as a central approach for the study of biogeochemical and ecological processes in the sea. While this development was facilitated by the fast development of computer power, the main driver is the need to analyze and synthesize the rapidly expanding observations, to formulate and test hypotheses, and to make predictions how ocean ecology and biogeochemistry respond to perturbations. The final aim, prediction, has gained in importance recently as scientists are increasingly asked by society to investigate and assess the impact of past, current, and future human actions on ocean ecology and biogeochemistry.

The impact of the carbon dioxide (CO_2) that humankind has emitted and will continue to emit into the atmosphere for the foreseeable future is currently, perhaps, the dominant question facing the marine biogeochemical/ecological research community. This impact is multifaceted, and includes both direct (such as ocean acidification) and indirect effects that are associated with the CO_2-induced climate change. Of particular concern is the possibility that global climate change will lead to a reduced capacity of the ocean to absorb CO_2 from the atmosphere, so that a larger fraction of the CO_2 emitted into the atmosphere remains there, further enhancing global warming. In such a positive feedback case, the expected climate change for a given CO_2 emission will be larger relative to a case without feedbacks. Marine biogeochemical/ecological models have played a crucial role in elucidating and evaluating these processes, and they are increasingly used for making quantitative predictions with direct implications for climate policy.

There are many other marine biogeochemical and/or ecological problems related to human activities, for which models play a crucial role assessing their importance and magnitude and devising possible solutions. These include, for example, coastal eutrophication, overfishing, and dispersion of invasive species. The use of marine biogeochemical/ecological models is now so pervasive that practically every field of oceanography is on this list.

The aim of this article is to provide an introduction and overview of marine biogeochemical and ecological modeling. Given the breadth of modeling approaches in use today, this overview can by design not be inclusive and authoritative. We rather focus on some basic concepts and provide a few illustrative applications. We start with a broader description of the marine biogeochemical/ecological challenge at hand, and then introduce basic concepts used for biogeochemical/ecological modeling. In the final section, we use a number of examples to illustrate some of the core modeling approaches.

The Marine Ecology and Biogeochemistry Challenge

The complexity of the ocean biogeochemical/ecological problem is daunting, as it involves a complex interplay among biology, physical variability of the oceanic environment, and the interconnected cycles of a large number of bioactive elements, particularly those of carbon, nitrogen, phosphorus, oxygen, silicon, and iron (**Figure 1**). Furthermore, the ocean is an open system that exchanges mass and many elements with the surrounding realms, such as the atmosphere, the land, and the sediments.

The engine that sets nearly all of these cycles into motion is the photosynthetic fixation of dissolved inorganic carbon and many other nutrient elements into organic matter by phytoplankton in the illuminated upper layers of the ocean (euphotic zone). The net rate of this process (i.e., net primary production) is distributed heterogeneously in the ocean, primarily as a result of the combined limitation of nutrients and light. This results in similar heterogeneity in surface chlorophyll, a direct indicator of the amount of phytoplankton biomass (**Figure 2(a)**). The large-scale surface nutrient distributions, in turn, reflect the balance between biological removal and the physical processes of upwelling and mixing that transport subsurface nutrient pools upward into the euphotic zone (**Figure 2(b)**). In addition to the traditional macronutrients (nitrate, phosphate, silicate etc.), growing evidence shows that iron limitation is

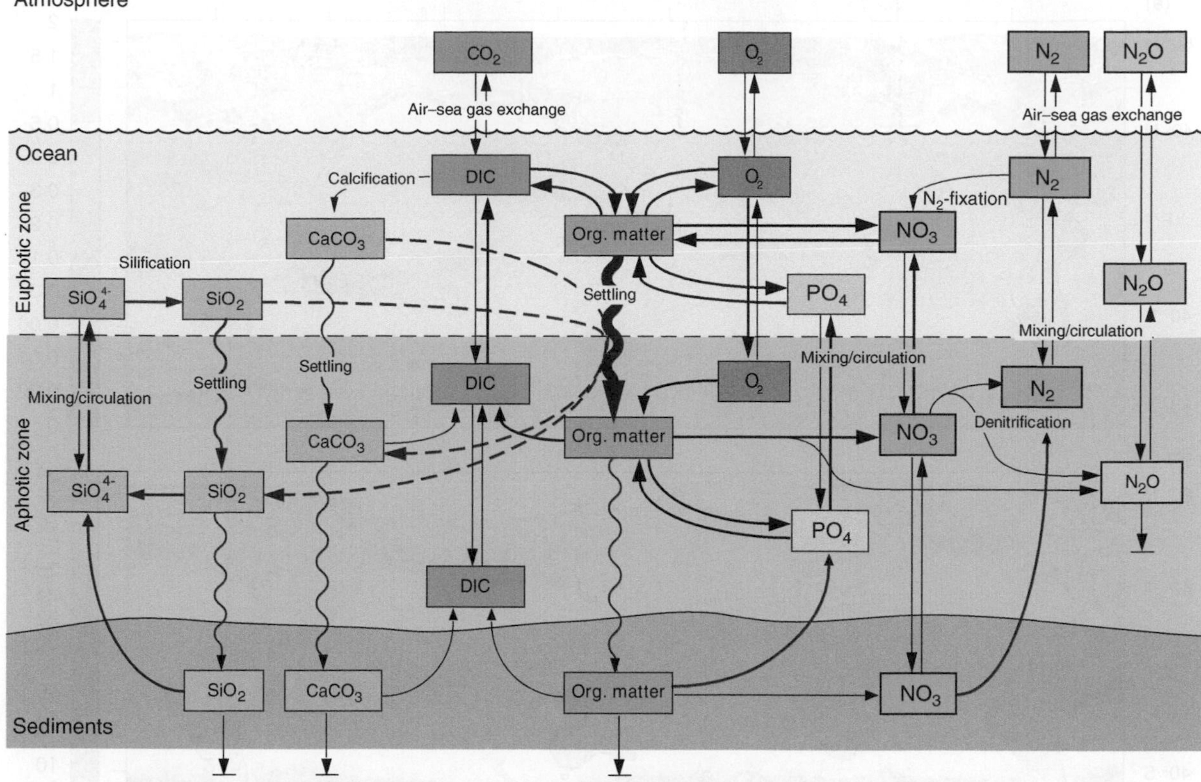

Figure 1 Schematic diagram of a number of key biogeochemical cycles in the ocean and their coupling. Shown are the cycles of carbon, oxygen, phosphorus, nitrogen, and silicon. The main engine of most biogeochemical cycles in the ocean is the biological production of organic matter in the illuminated upper ocean (euphotic zone), part of which sinks down into the ocean's interior and is then degraded back to inorganic constituents. This organic matter cycle involves not only carbon, but also nitrogen, phosphorus, and oxygen, causing a tight linkage between the cycles of these four elements. Since some phytoplankton, such as diatoms and coccolithophorids, produce shells made out of amorphous silicon and solid calcium carbonate, the silicon and $CaCO_3$ cycles also tend to be closely associated with the organic matter cycle. These ocean interior cycles are also connected to the atmosphere through the exchange of a couple of important gases, such as CO_2, oxygen, and nitrous oxide (N_2O). In fact, on timescales longer than a few decades, the ocean is the main controlling agent for the atmospheric CO_2 content.

a key factor governing net primary production in many parts of the ocean away from the main external sources for iron, such as atmospheric dust deposition and continental margin sediments. The photosynthesized organic matter is the basis for a complex food web that involves both a transfer of the organic matter toward higher trophic levels as well as a microbial loop that is responsible for most of the breakdown of this organic matter back to its inorganic constituents.

A fraction of the synthesized organic matter (about 10–20%) escapes degradation in the euphotic zone and sinks down into the dark aphotic zone, where it fuels the growth of microbes and zooplankton that eventually also remineralize most of this organic matter back to its inorganic constituents. The nutrient elements and the dissolved inorganic carbon are then eventually returned back to the surface ocean by ocean circulation and mixing, closing this 'great biogeochemical loop'. This loop is slightly leaky in that a small fraction of the sinking organic matter fails to get remineralized in the water column and is deposited onto the sediments. Very little organic matter escapes remineralization in the sediments though, so that only a tiny fraction of the organic matter produced in the surface is permanently removed from the ocean by sediment burial. In the long-term steady state, this loss of carbon and other nutrient elements from the ocean is replaced by the input from land by rivers and through the atmosphere.

Several marine phytoplankton and zooplankton groups produce hard shells consisting of either mineral calcium carbonate ($CaCO_3$) or amorphous silica, referred to as 'opal' (SiO_2). The most important producers of $CaCO_3$ are coccolithophorids, while most marine opal stems from diatoms. Both groups are photosynthetic phytoplankton, highlighting the extraordinary importance of this trophic group for marine ecology and biogeochemistry. Upon the death of the mineral-forming organisms, these minerals

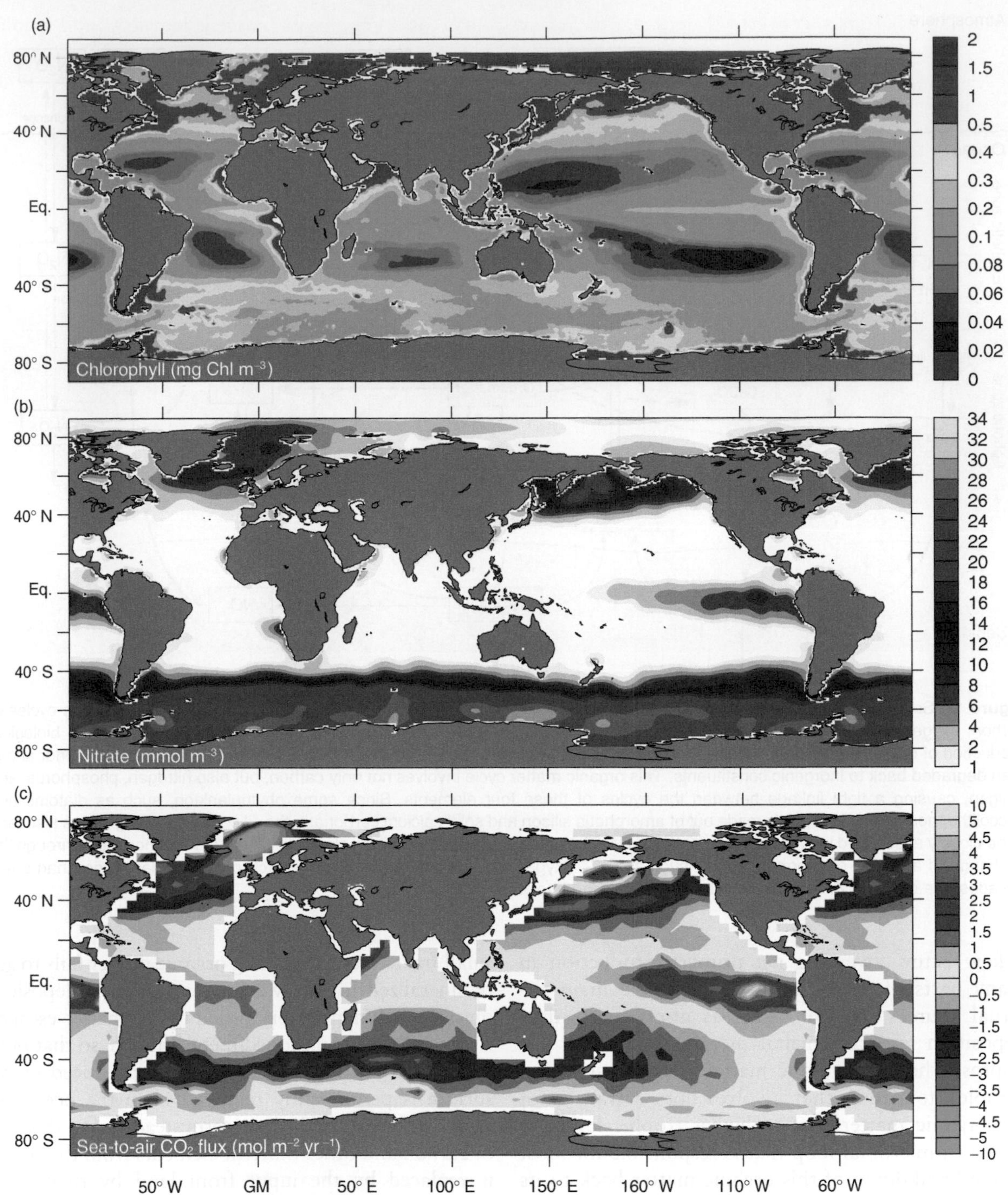

Figure 2 Global maps of the distribution of three key biogeochemical/ecological modeling targets. (a) Annual mean distribution of near-surface chlorophyll (mg Chl m^{-3}) as measured by the *SeaWiFS* satellite. (b) Annual mean surface distribution of nitrate (mmol^{-3}) compiled from *in situ* observations (from the World Ocean Atlas). (c) Annual mean air–sea CO_2 flux (mol m^{-2} yr^{-1}) for a nominal year of 2000 derived from a compilation of *in situ* measurements of the oceanic pCO_2 and a wind-speed gas exchange parametrization. Data provided by *T. Takahashi*, Lamont Doherty Earth Observatory of Columbia University.

also sink into the ocean's interior, thereby often acting as 'ballast' for organic matter.

This great biogeochemical loop also has a strong impact on several gases that are dissolved in seawater and exchange readily with the atmosphere, most importantly CO_2 and oxygen (O_2). Biological processes have opposite effects on these two gases: photosynthesis consumes CO_2 and liberates O_2, while respiration and bacterial degradation releases CO_2 and consumes O_2. As a result, one tends to find an inverse relationship in the oceanic distribution of these two gases. CO_2 also sets itself apart from O_2 in that it reacts readily with seawater. In fact, due to the high content of alkaline substances in the ocean, this reaction is nearly complete, so that only around 1% of the total dissolved inorganic carbon in the ocean exists in the form of CO_2.

The exchange of CO_2 across the air–sea interface is a prime question facing ocean biogeochemical/ ecological research, particularly with regard to the magnitude of the oceanic sink for anthropogenic CO_2. The anthropogenic CO_2 sink occurs on top of the natural air–sea CO_2 fluxes characterized by oceanic uptake in mid-latitudes and some high latitudes, and outgassing in the low latitudes and the Southern Ocean. This distribution of the natural CO_2 flux is the result of an interaction between the exchange of heat between the ocean and the atmosphere, which affects the solubility of CO_2, and the great biogeochemical loop, which causes an uptake of CO_2 from the atmosphere in regions where the downward flux of organic carbon exceeds the upward supply of dissolved inorganic carbon, and an outgassing where the balance is the opposite. This flux has changed considerably over the last two centuries in response to the anthropogenically driven increase in atmospheric CO_2 that pushes additional CO_2 from the atmosphere into the ocean. Nevertheless, the pattern of the resulting contemporary air–sea CO_2 flux (**Figure 2(c)**) primarily still reflects the flux pattern of the natural CO_2 fluxes, albeit with a global integral flux into the ocean reflecting the oceanic uptake of anthropogenic CO_2.

Given the central role of the great biogeochemical loop, any modeling of marine biogeochemical/ecological processes invariably revolves around the modeling of all the processes that make up this loop. As this loop starts with the photosynthetic production of organic matter by phytoplankton, biogeochemical modeling is always tightly interwoven with the modeling of marine ecology, especially that of the lower trophic levels.

Core questions that challenge marine biogeochemistry and ecology are as follows:

1. What controls the mean concentration and three-dimensional (3-D) distribution of bioreactive elements in the ocean?
2. What controls the air–sea balance of climatically important gases, that is, CO_2, N_2O, and O_2?
3. What controls ocean productivity, the downward export of organic matter, and the transfer of organic matter to higher trophic levels?
4. How do ocean biogeochemistry and ecology change in time in response to climate dynamics and human perturbations?

Modeling represents a powerful approach to studying and addressing these core questions. Modeling is by no means the sole approach. In fact, integrated approaches that combine observational, experimental, and modeling approaches are often necessary to tackle this set of complex problems.

What Is a Biogeochemical/Ecological Model?

At its most fundamental level, a model is an abstract description of how some aspect of nature functions, most often consisting of a set of mathematical expressions. In the biogeochemical/ecological modeling context, a model usually consists of a number of partial differential equations, which describe the time and space evolution of a (limited) number of ecological/biogeochemical state variables. As few of these equations can be solved analytically, they are often solved numerically using a computer, which requires the discretization of these equations, that is, they are converted into difference equations on a predefined spatial and temporal grid.

The Art of Biogeochemical/Ecological Modeling

A model can never fully represent reality. Rather, it aims to represent an aspect of reality in the context of a particular problem. The art of modeling is to find the right level of abstraction, while keeping enough complexity to resolve the problem at hand. That is, marine modelers often follow the strategy of Occam's razor, which states that given two competing explanations, the one that is simpler and makes fewer assumptions is the more likely to be correct. Therefore, a typical model can be used only for a limited set of applications, and great care must be used when a model is applied to a problem for which it was not designed. This is especially true for biogeochemical/ecological models, since their underlying mathematical descriptions are for the most part not based on first principles, but often derived from empirical relationships. In fact, marine biogeochemical/ecological modeling is at present a data-limited activity because we lack data to formulate and parametrize key processes and/or to evaluate the model predictions. Further, significant simplifications are often made to make the problem more tractable. For example, rather than treating individual organisms or even species, model variables often aggregate entire functional groups into single boxes (e.g., photosynthetic organisms, grazers, and detritus decomposers), which are then simulated as bulk concentrations (e.g., mol C m^{-3} of phytoplankton).

Despite these limitations, models allow us to ask questions about the ocean inaccessible from data or experiments alone. In particular, models help researchers quantify the interactions among multiple processes, synthesize diverse observations, test hypotheses, extrapolate across time – space scales, and predict past and future behavior. A well-posed model encapsulates our understanding of the ocean in a mathematically consistent form.

Biogeochemical/Ecological Modeling Equations and Approaches

In contrast with their terrestrial counterparts, models of marine biogeochemical/ecological processes must be coupled to a physical circulation model of some sort to take into consideration that nearly all relevant biological and biogeochemical processes occur either in the dissolved or suspended phase, and thus are subject to mixing and transport by ocean currents. Thus, a typical coupled physical–biogeochemical/ecological model consists of a set of time-dependent advection, diffusion, and reaction equations:

$$\frac{\partial C}{\partial t} + \text{Adv}(C) + \text{Diff}(C) = \text{SMS}(C) \qquad [1]$$

where C is the state variable to be modeled, such as the concentration of phytoplankton, nutrients, or dissolved inorganic carbon, often in units of mass per unit volume (e.g., mol m^{-3}). Adv(C) and Diff(C) are the contributions to the temporal change in C by advection and eddy-diffusion (mixing), respectively, derived from the physical model component. The term SMS(C) refers to the 'sources minus sinks' of C driven by ecological/biogeochemical processes. The SMS term often involves complex interactions among a number of state variables and is provided by the biogeochemical/ecological model component.

Marine biogeochemical/ecological models are diverse, covering a wide range of complexities and applications from simple box models to globally 4-D- (space and time) coupled physical–biogeochemical simulations, and from strict research tools to climate change projections with direct societal implications. Model development and usage are strongly shaped by the motivating scientific or policy problems as well as the dynamics and time–space scales considered. The complexity of marine biogeochemical/ecological models can be organized along two major axes (**Figure 3**): the physical complexity, which determines how the left-hand side of [1] is computed, and the biogeochemical/ecological complexity, which determines how the right-hand side of [1] is evaluated.

Due to computational and analytical limitations, there is often a trade-off between the physical and biogeochemical/ecological complexity, so that models of the highest physical complexity are often using relatively simple biogeochemical/ecological models and vice versa (**Figure 4**). Additional constraints arise from the temporal domain of the integrations. Applications of coupled physical–biogeochemical/ecological models to paleoceanographic questions require integrations of several thousand years. This can only be achieved by reducing both the physical and the biogeochemical/ecological complexity (**Figure 4**). At the same time, the continuously increasing computational power has permitted researchers to push forward along both complexity axes. Nevertheless, the fundamental tradeoff between physical and biogeochemical/ecological complexity remains.

In addition to the physical and biogeochemical/ecological complexity, models can also be categorized with regard to their interaction with observations. In the case of 'forward models', a set of equations in the form of [1] is integrated forward in time given initial and boundary conditions. A typical forward problem is the prediction of the future state of ocean biogeochemistry and ecology for a certain evolution of the Earth's climate. The solutions of such forward models are the time–space distribution of the state variables as well as the implied fluxes. The expression 'inverse models' refers to a broad palette of modeling approaches, but all of them share the goal of optimally combining observations with knowledge about the workings of a system as embodied in the model. Solutions to such inverse models can be improved estimates of the current state of the system (state estimation), improved estimates of the initial or boundary conditions, or an optimal set of parameters. A typical example of an inverse model is the optimal determination of ecological parameters, such as growth and grazing rates, given, for example, the observed distribution of phytoplankton, zooplankton, and nutrients.

Examples

Given the large diversity of marine biogeochemical/ecological models and approaches, no review can do

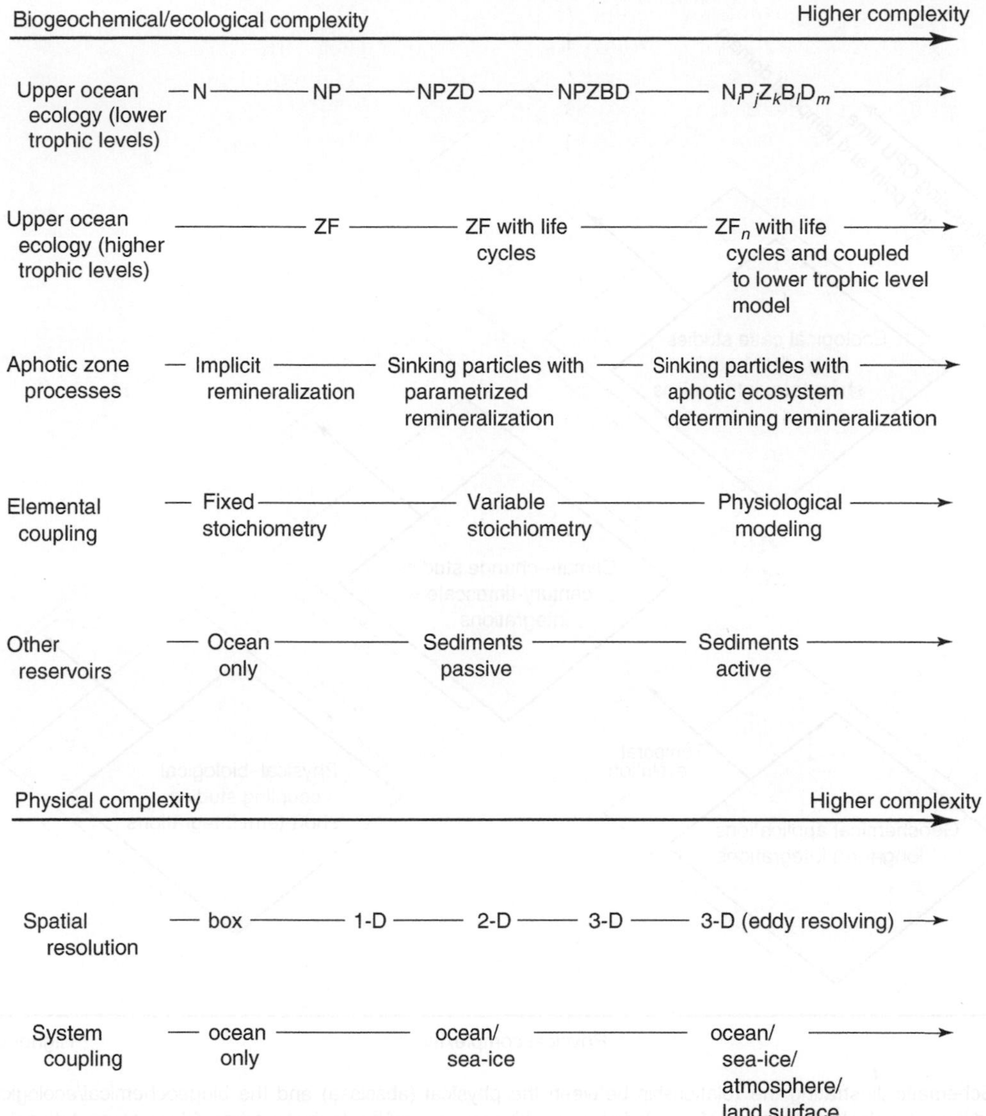

Figure 3 Schematic diagram summarizing the development of complexity in coupled physical–biogeochemical/ecological models. The two major axes of complexity are biogeochemistry/ecology and physics, but each major axis consists of many subaxes that describe the complexity of various subcomponents. Currently existing models fill a large portion of the multidimensional space opened by these axes. In addition, the evolution of models is not always necessarily straight along any given axis, but depends on the nature of the particular problem investigated. N, nutrient; P, phytoplankton; Z, zooplankton; B, bacteria; D, detritus; F, fish.

full justice. We restrict our article here to the discussion of four examples, which span the range of complexities as well as have been important milestones in the evolution of biogeochemical/ecological modeling.

Box Models or What Controls Atmospheric Carbon Dioxide?

Our aim here is to develop a model that explains how the great biogeochemical loop controls atmospheric CO_2. The key to answering this question is a quantitative prediction of the surface ocean concentration of CO_2, as it is this surface concentration that controls the atmosphere–ocean balance of CO_2. The simplest models used for such a purpose are box models, where the spatial dimension is reduced to a very limited number of discrete boxes. Such box models have played an important role in ocean biogeochemical/ecological modeling, mostly because their solutions can be readily explored and understood. However, due to the dramatic reduction of complexity, there are also important limitations, whose consequences one must keep in mind when interpreting the results.

Box models can be formally derived from the tracer conservation eq [1] by integrating over

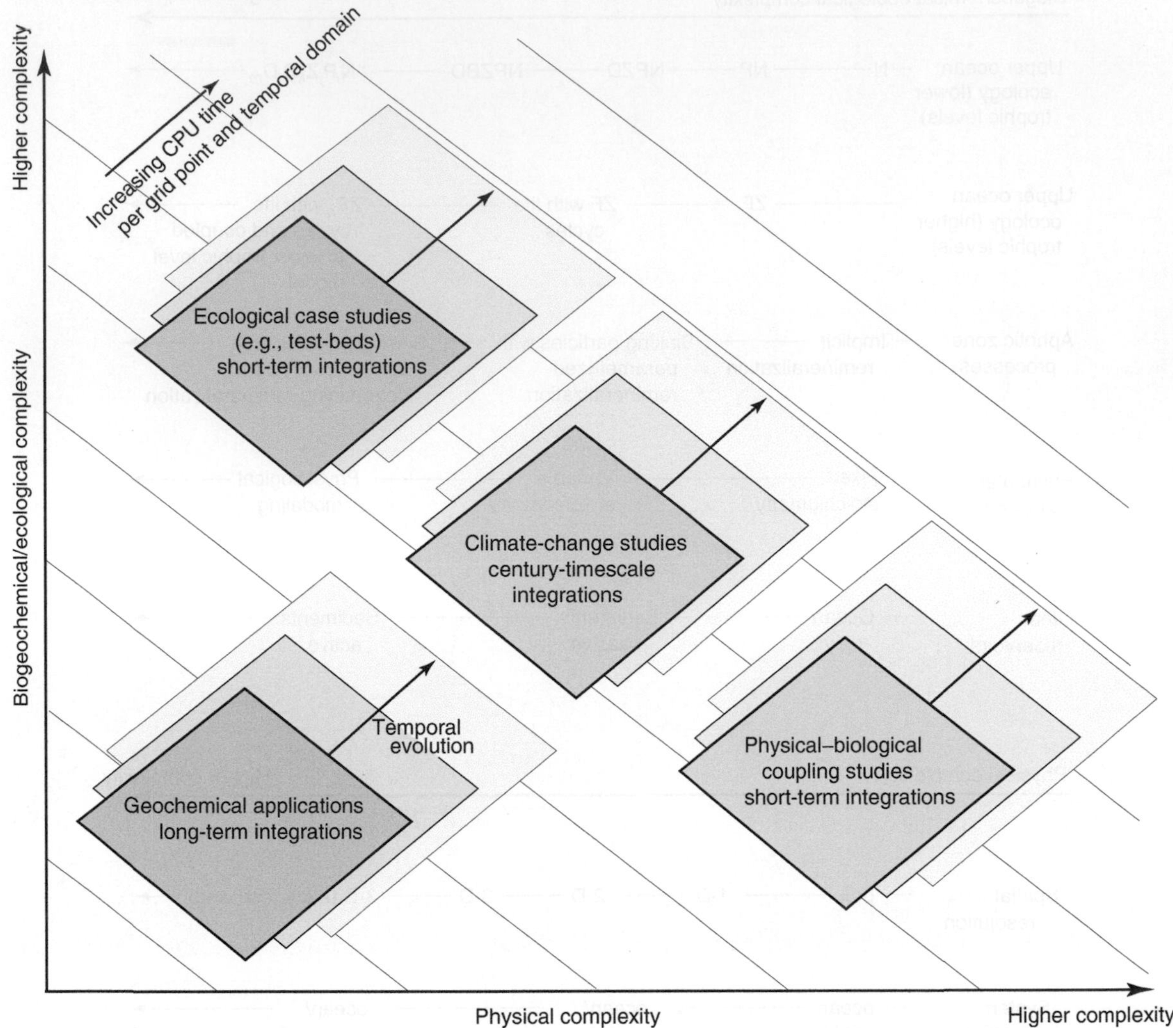

Figure 4 Schematic illustrating the relationship between the physical (abscissa) and the biogeochemical/ecological complexity (ordinate) of different typical applications of coupled physical–biogeochemical/ecological models. Given computational and analytical constraints, there is often a trade-off between the two main complexities. The thin lines in the background indicate the CPU time required for the computation of a problem over a given time period.

the volume of the box, V, and by applying Gauss' (divergence) theorem, which states that the volume integral of a flux is equal to the flux in normal direction across the boundary surfaces, S. The resulting integral form of [1] is:

$$\int \frac{\partial C}{\partial t}\,dV + \oint (\mathrm{Adv}_n(C) + \mathrm{Diff}_n(C))\,dS = \int \mathrm{SMS}(C)\,dV \qquad [2]$$

where $\mathrm{Adv}_n(C)$ and $\mathrm{Diff}_n(C)$ are the advective and diffusive transports in the normal direction across S. If we assume that the concentration of tracer C as well as the SMS term within the box are uniform, [2] can be rewritten as

$$V \cdot \frac{d\overline{C}}{dt} + \sum_i (T_{i,\mathrm{out}} \cdot \overline{C} - T_{i,\mathrm{in}} \cdot C_i + v_i(\overline{C} - C_i)) = V \cdot \mathrm{SMS}(\overline{C}) \qquad [3]$$

where $\overline{C}$ is the mean concentration within the box. We represented the advective contributions as products of a mass transport, T (dimensions volume time^{-1}), with the respective upstream concentration, separately considering the mass transport into the box and out of the box across each interface i, that is, $T_{i,\mathrm{in}}$ and $T_{i,\mathrm{out}}$. The eddy-diffusion contributions are parametrized as products of a mixing coefficient, v_i (dimension volume time^{-1}), with the gradient across each interface ($\overline{C} - C_i$). For simplicity, we subsequently drop the overbar, that is, use C instead of $\overline{C}$.

The simplest box-model representation of the great biogeochemical loop is a two-box model, wherein the ocean is divided into a surface and a deep box, representing the euphotic and aphotic zones, respectively (**Figure 5(a)**). Mixing and transport between the two boxes is modeled with a mixing term, v. The entire complexity of marine ecology and biogeochemistry in the euphotic zone is reduced to a single term, Φ, which represents the flux of organic matter out of the surface box and into the deep box, where the organic matter is degraded back to inorganic constituents.

Let us consider the deep-box balance for a dissolved inorganic bioreactive element, C_d (such as nitrate, phosphate, or dissolved inorganic carbon):

$$V_d \frac{dC_d}{dt} + v \cdot (C_d - C_s) = \Phi \qquad [4]$$

where V_d is the volume of the deep box and C_s is the dissolved inorganic concentration of the surface box. The steady-state solution of [4], that is, the solution when the time derivative of C_d vanishes ($dC_d/dt = 0$),

$$v \cdot (C_d - C_s) = \Phi \qquad [5]$$

represents the most fundamental balance of the great biogeochemical loop. It states that the net upward transport of an inorganic bioreactive element by physical mixing is balanced by the downward transport of this element in organic matter. This steady-state balance [5] has several important applications. For example, it permits us to estimate the downward export of organic matter by simply analyzing the vertical gradient and by estimating the vertical exchange. The balance also states that the magnitude of the vertical gradient in bioreactive elements is proportional to the strength of the downward flux of organic matter and inversely proportional to the mixing coefficient. Since nearly all oceanic inorganic carbon and nutrients reside in the deep box, that is, $V_d \cdot C_d \gg V_s \cdot C_s$, one can assume, to first order, that C_d is largely invariant, so that the transformed balance for the surface ocean concentration

$$C_s = C_d - \frac{\Phi}{v} \qquad [6]$$

reveals that the surface concentration depends primarily on the relative magnitude of the organic matter export to the magnitude of vertical mixing. Hence, [6] provides us with a first answer to our challenge: it states that, for a given amount of ocean mixing, the surface ocean concentration of inorganic carbon, and hence atmospheric CO_2, will decrease with increasing marine export production. Relationship [6] also states that for a given magnitude of marine export production, atmospheric CO_2 will increase with increased mixing.

While very powerful, the two-box model has severe limitations. The most important one is that this model does not consider the fact that the deep ocean exchanges readily only with the high latitudes, which represent only a very small part of the surface ocean. The exchange of the deep ocean with the low latitudes is severely limited because diapcyncal mixing in the ocean is small. Another limitation is that the one-box model does not take into account that the nutrient concentrations in the high latitudes tend to be much higher than in the low latitudes (**Figure 2(c)**).

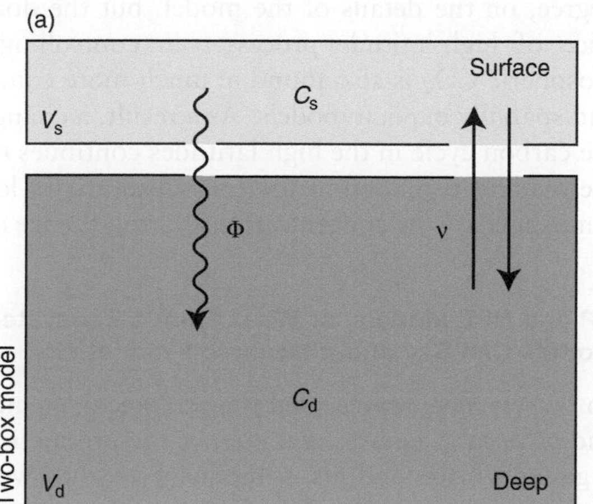

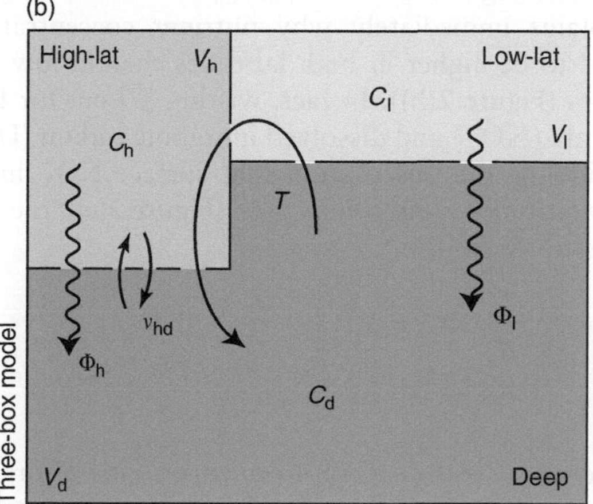

Figure 5 Schematic representation of the great biogeochemical loop in box models: (a) two-box model and (b) three-box model. C, concentrations; V, volume; v, exchange (mixing) coefficient; Φ, organic matter export fluxes; T, advective transport.

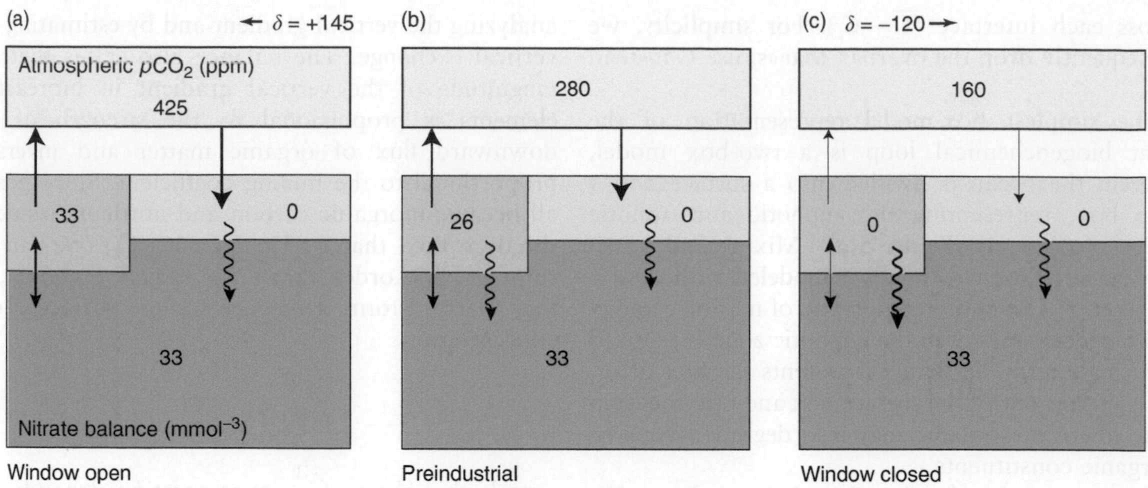

Figure 6 Schematic illustration of the impact of the high-latitude (Southern Ocean) window on atmospheric CO_2. In (a) the high-latitude window is open, high-latitude nitrate is high, and atmospheric CO_2 attains 425 ppm. Panel (b) represents the preindustrial ocean, where the window is half open, nitrate in the high latitudes is at intermediate levels, and atmospheric CO_2 is 280 ppm. In (c), the high-latitude window is closed, high-latitude nitrate is low, and atmospheric CO_2 decreases to about 160 ppm.

An elegant solution is the separation of the surface box into a high-latitude box, h, and a low-latitude box, l (**Figure 5(b)**). Intense mixing is assumed to occur only between the deep and the high-latitude boxes, v_{hd}. A large-scale transport, T, is added to mimic the ocean's global-scale overturning circulation. As before, the complex SMS terms are summarized by Φ_l and Φ_h. In steady state, the surface concentrations are given by

$$C_l = C_d - \frac{\Phi_l}{T}$$
$$C_h = C_d - \frac{\Phi_h + \Phi_l}{T + v_{hd}} \qquad [7]$$

which are structurally analogous to [6], but emphasize the different dynamics setting balances in the low and high latitudes. With $v_{hd} \gg T$ and $\Phi_l \approx \Phi_h$, [7] explains immediately why nutrient concentrations tend to be higher in high latitudes than in low latitudes (**Figure 2(b)**). In fact, writing [7] out for both nitrate (NO_3^-) and dissolved inorganic carbon (DIC), and using the observation that surface NO_3^- in the low latitudes is essentially zero (**Figure 2(c)**), the DIC equations are

$$DIC_l = DIC_d - r_{C:N} \cdot [NO_3^-]_d$$
$$DIC_h = DIC_d - \frac{\Phi_h + r_{C:N} \cdot T \cdot [NO_3^-]_d}{T + v_{hd}} \qquad [8]$$

where $r_{C:N}$ is the carbon-to-nitrogen ratio of organic matter. Assuming that the deep-ocean nitrate concentration is time invariant, the analysis of [8] reveals that the low-latitude DIC concentration, DIC_l, is more or less fixed, while the high-latitude DIC concentration, DIC_h, can be readily altered by changes in either the high-latitude export flux, Φ_h, or high-latitude mixing, v_{hd}. Furthermore, if one considers the fact that the rapid communication between the high-latitude ocean and the deep ocean makes the high latitudes the primary window into the oceanic reservoir of inorganic carbon, it becomes clear that it must be the high latitudes that control atmospheric CO_2 on the millenial timescales where the steady-state approximation is justified. This high-latitude dominance in controlling atmospheric CO_2 is illustrated in **Figure 6**, which demonstrates that atmospheric CO_2 can vary between about 160 and 425 ppm by simply opening or closing this high-latitude window. The exact magnitude of the atmospheric CO_2 change depends, to a substantial degree, on the details of the model, but the dominance of high-latitude processes in controlling atmospheric CO_2 is also found in much more complex and spatially explicit models. As a result, a change in the carbon cycle in the high latitudes continues to be the leading explanation for the substantially lower atmospheric CO_2 concentrations during the ice ages.

NP and NPZ Models, or What Simple Ecosystem Models Can Say about Oceanic Productivity

So far, we have represented the ecological, chemical, and physical processes that control the production of organic matter and its subsequent export with a single parameter, Φ. Clearly, in order to assess how marine biology responds to climate change and other perturbations, it is necessary to resolve these processes in much more detail. Two fundamentally different approaches have been developed to model such ecological processes: concentration-based models and individually based models (IBMs). In the latter, the model's equations represent the growth and losses of individual organisms, and the model then simulates the evolution of a large number of these individuals through space and time. This approach is most commonly used for the modeling of organisms at higher trophic levels, where life cycles play an important role (e.g., models of zooplankton, fish, and marine mammals). In contrast, concentration-based models are almost exclusively used for the modeling of the lower trophic levels, in particular to represent the interaction of light, nutrient, and grazing in controlling the growth of phytoplankton, that is, marine primary production.

The simplest such model considers just one limiting nutrient, N, and one single phytoplankton group, P (see also **Figure** 7(**a**)):

$$\begin{aligned} \mathrm{SMS}(P) &= V_{\max} \cdot \frac{N}{K_{\mathrm{N}}+N} \cdot P - \lambda_{\mathrm{P}} \cdot P \\ \mathrm{SMS}(N) &= -V_{\max} \cdot \frac{N}{K_{\mathrm{N}}+N} \cdot P + \mu_{\mathrm{P}} \cdot \lambda_{\mathrm{P}} \cdot P \end{aligned} \qquad [9]$$

where the first term on the right-hand side of the phytoplankton eqn [9] is net phytoplankton growth (equal to net primary production) modeled here as a function of a nutrient-saturated growth rate $V_{\max}$, and a hyperbolic dependence on the *in situ* nutrient concentration (Monod-type), with a single parameter, the half-saturation constant, K_{N}. The second term is the net loss due to senescence, viral infection, and grazing, modeled here as a linear process with a loss rate λ_{P}. A fraction μ_{P} of the phytoplankton loss is assumed to be regenerated inside the euphotic zone, while a fraction $(1-\mu_{\mathrm{P}})$ is exported to depth (**Figure** 7(**a**)). These two parametrizations for phytoplankton growth and losses reflect a typical situation for marine ecosystem models in that the growth terms are substantially more elaborate, reflecting the interacting influence of various controlling parameters, whereas the loss processes are highly simplified. This situation also reflects the fact that the processes controlling the growth of marine organisms are often more amenable to experimental studies, while the loss processes are much harder to investigate with careful experiments.

When the SMS terms of [10] are inserted into the full tracer conservation eqn [1], the resulting equations form a set of coupled partial differential equations with a number of interesting consequences in steady state as shown in **Figure** 8(**a**). Below a certain nutrient concentration threshold, phytoplankton growth is smaller than its loss term, so that the resulting steady-state solution is $P=0$, that is, no phytoplankton. Above this threshold, the abundance of phytoplankton (and hence primary production) increases with increasing nutrient supply in a linear manner. Phytoplankton is successful in reducing the dissolved inorganic nutrient concentration to low levels, so that the steady state is characterized by most nutrients residing in organic form in the phytoplankton pool. This NP model reflects a situation where marine productivity is limited by bottom-up processes, that is, the supply of the essential nutrients. Comparison with the nutrient and phytoplankton distribution shown in (**Figure** 2(**b**)) shows that this model could explain the observations in the subtropical open ocean, where nutrients are indeed drawn down to very low levels. However, the NP model would also predict relatively high P levels to go along with the low nutrient levels, which is clearly not observed (**Figure** 2(**a**)). In addition, this NP model also fails clearly to explain the high-nutrient regions of the high latitudes. However, we have so far neglected the limitation by light, as well as the impact of zooplankton grazing.

The addition of a zooplankton compartment to the NP model (**Figure** 7(**b**)) dramatically shifts the nutrient allocation behavior (**Figure** 8(**b**)). In this case, as the nutrient loading increases, the phytoplankton abundance gets capped at a certain level by zooplankton grazing. Due to the reduced levels of biomass, the phytoplankton is then no longer able to consume all nutrients at high-nutrient loads, so that an increasing fraction of the supplied nutrients remains unused. This top-down limitation situation could therefore, in part, explain why macronutrients in certain regions remain untapped. Most recent research suggests that micronutrient (iron) limitation, in conjunction with zooplankton grazing, plays a more important role in causing these high-nutrient/low chlorophyll regions. Another key limitation is light, which has important consequences for the seasonal and depth evolution of phytoplankton.

Global 3-D Modeling of Ocean Biogeochemistry/Ecology

The coupling of relatively simple NPZ-type models to global coarse-resolution 3-D circulation models has proven to be a challenging task. Perhaps the most important limitation of such NPZ models is the fact that all phytoplankton in the ocean are represented by a single phytoplankton group, which is grazed upon by a single zooplankton group. This means that the tiny phytoplankton that dominate the relatively nutrient-poor central gyres of the ocean, and the

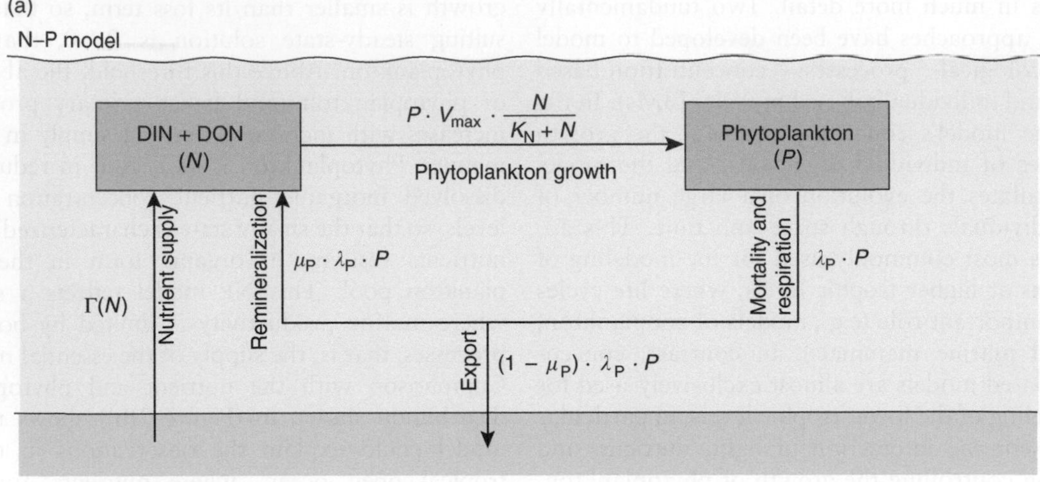

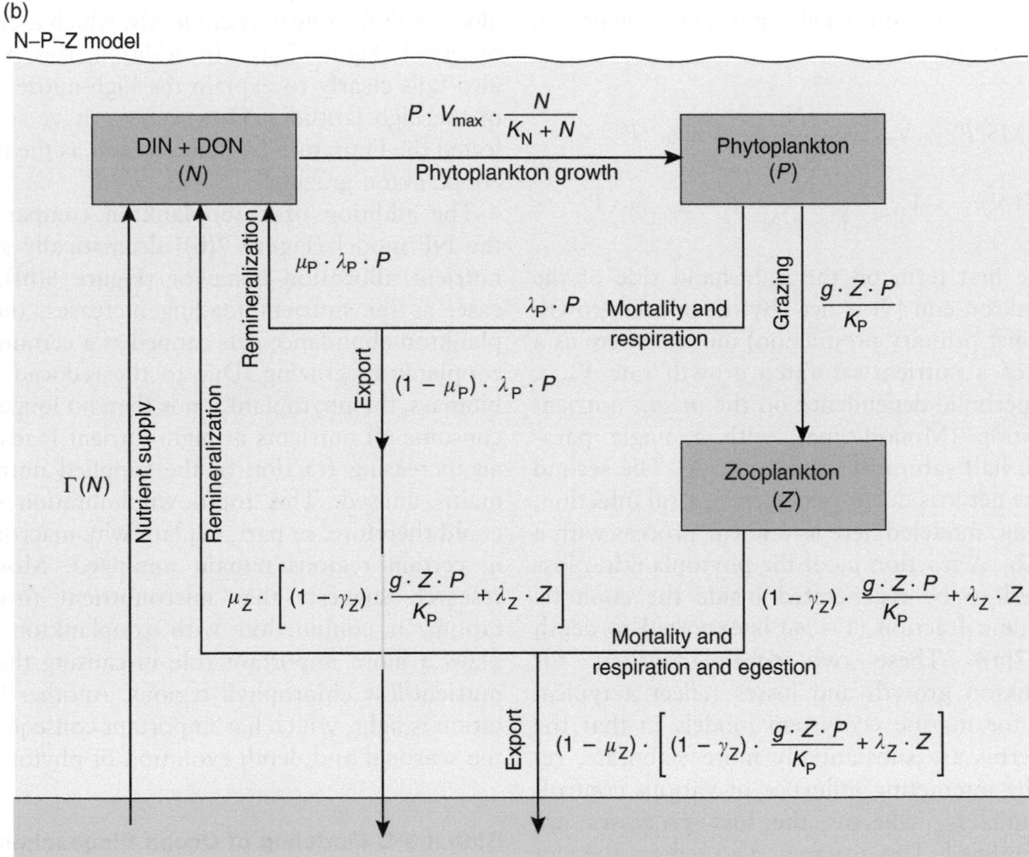

Figure 7 A schematic illustration of (a) a nitrate–phytoplankton model and (b) a nitrate–phytoplankton–zooplankton model. The mathematical expressions associated with each arrow represent typical representations for how the individual processes are parametrized in such models. V_{max}, maximum phytoplankton growth rate; K_N, nutrient half saturation concentration; g, maximum zooplankton growth rate; K_P, half saturation concentration for phytoplankton grazing; γ_Z, zooplankton assimilation efficiency; λ_P, phytoplankton mortality rate; λ_Z, zooplankton mortality rate; μ_P, fraction of dead phytoplankton nitrogen that is remineralized in the euphotic zone; μ_Z, as μ_P, but for zooplankton. Adapted from Sarmiento JL and Gruber N (2006) *Ocean Biogeochemical Dynamics*, 526pp. Princeton, NJ: Princeton University Press.

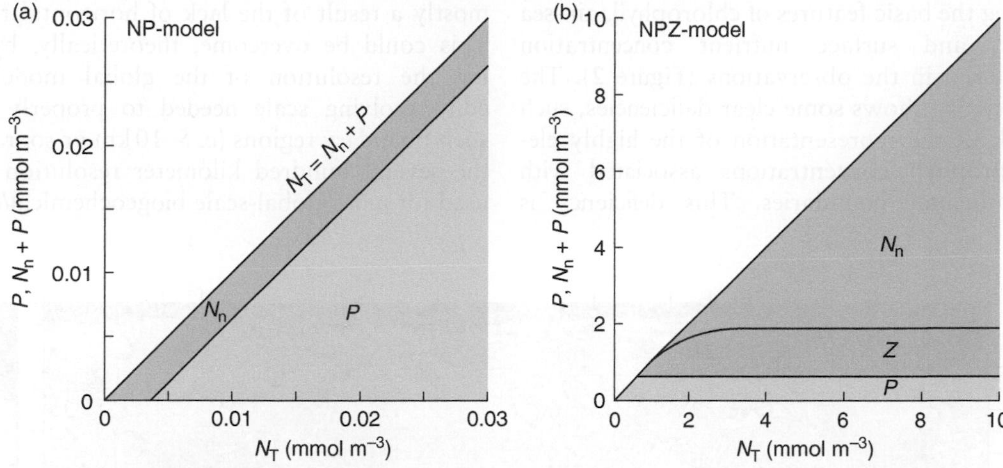

Figure 8 Nitrogen in phytoplankton (*P*), zooplankton (*Z*), and nitrate (N_n) as a function of total nitrogen. (a) Results from a two-component model that just includes N and P. (b) Results from a three-component model that is akin to that shown in **Figure 7(b)**. The vertical axis is concentration in each of the components plotted as the cumulative amount. Adapted from Sarmiento JL and Gruber N (2006) *Ocean Biogeochemical Dynamics*, 526pp. Princeton, NJ: Princeton University Press.

large phytoplankton that dominate the highly productive upwellling regions are represented with a single group, whose growth characteristics cannot simultaneously represent both. An additional challenge of the large-scale coupling problem is the interaction of oceanic circulation/mixing with the marine ecological and biogeochemical processes. Often, small deficiencies in the physical model get amplified by the ecological/biogeochemical model, so that the resulting fields of phytoplankton abundance may diverge strongly from observations.

The latter problem is addressed by improving critical aspects of the physical model, such as upper ocean mixing, atmospheric forcing, resolution, and numerical tracer transport algorithms, while the first problem is addressed by the consideration of distinct plankton functional groups. These functional groups distinguish themselves by their different size, nutrient requirement, biogeochemical role, and several other characteristics, but not necessarily by their taxa. Currently existing models typically consider the following phytoplankton functional groups: small phytoplankton (nano- and picoplankton), silicifying phytoplankton (diatoms), calcifying phytoplankton (coccolithophorids), N_2-fixing phytoplankton, large (nonsilicifying) phytoplankton (e.g., dinoflagellates and phaeocystis), with the choice of which group to include being driven by the particular question at hand. The partial differential equations for the different phytoplankton functional groups are essentially the same and follow a structure similar to [9], but are differentiated by varying growth parameters, nutrient/light requirements, and susceptibility to grazing. At present, most applications include between five and 10 phytoplankton functional groups. Some recent studies have been exploring the use of several dozen functional groups, whose parameters are generated stochastically with some simple set of rules that are then winnowed via competition for resources among the functional groups. Since the different phytoplankton functional groups are also grazed differentially, different types of zooplankton generally need to be considered as well, though fully developed models with diverse zooplankton functional groups are just emerging. An alternative is to use a single zooplankton group, but with changing grazing/growth characteristics depending on its main food source. A main advantage of models that build on the concept of plankton functional groups is their ability to switch between different phytoplankton community structures and to simulate their differential impact on marine biogeochemical processes. For example, the nutrient poor subtropical gyres tend to be dominated by nano- and picoplankton, whose biomass tends to be tightly capped by grazing by very small zooplankton. This prevents the group forming blooms, even under nutrient-rich conditions. By contrast, larger phytoplankton, such as diatoms, can often escape grazing control (at least temporarily) and form extensive blooms. Such phytoplankton community shifts are essential for controlling the export of organic matter toward the abyss, since the export ratio, that is, the fraction of net primary production that is exported, tends to increase substantially in blooms.

Results from the coupling of such a multiple functional phytoplankton ecosystem model and of a relatively simple biogeochemical model to a global 3-D circulation model show substantial success in representing the basic features of chlorophyll, air–sea CO_2 flux, and surface nutrient concentration (**Figure 9**) seen in the observations (**Figure 2**). The comparison also shows some clear deficiencies, such as the lack of the representation of the highly elevated chlorophyll concentrations associated with many continental boundaries. This deficiency is mostly a result of the lack of horizontal resolution. This could be overcome, theoretically, by increasing the resolution of the global models to the eddy-resolving scale needed to properly represent such boundary regions (*c.* 5–10 km or so, rather than the several hundred kilometer resolution currently used for most global-scale biogeochemical/ecological

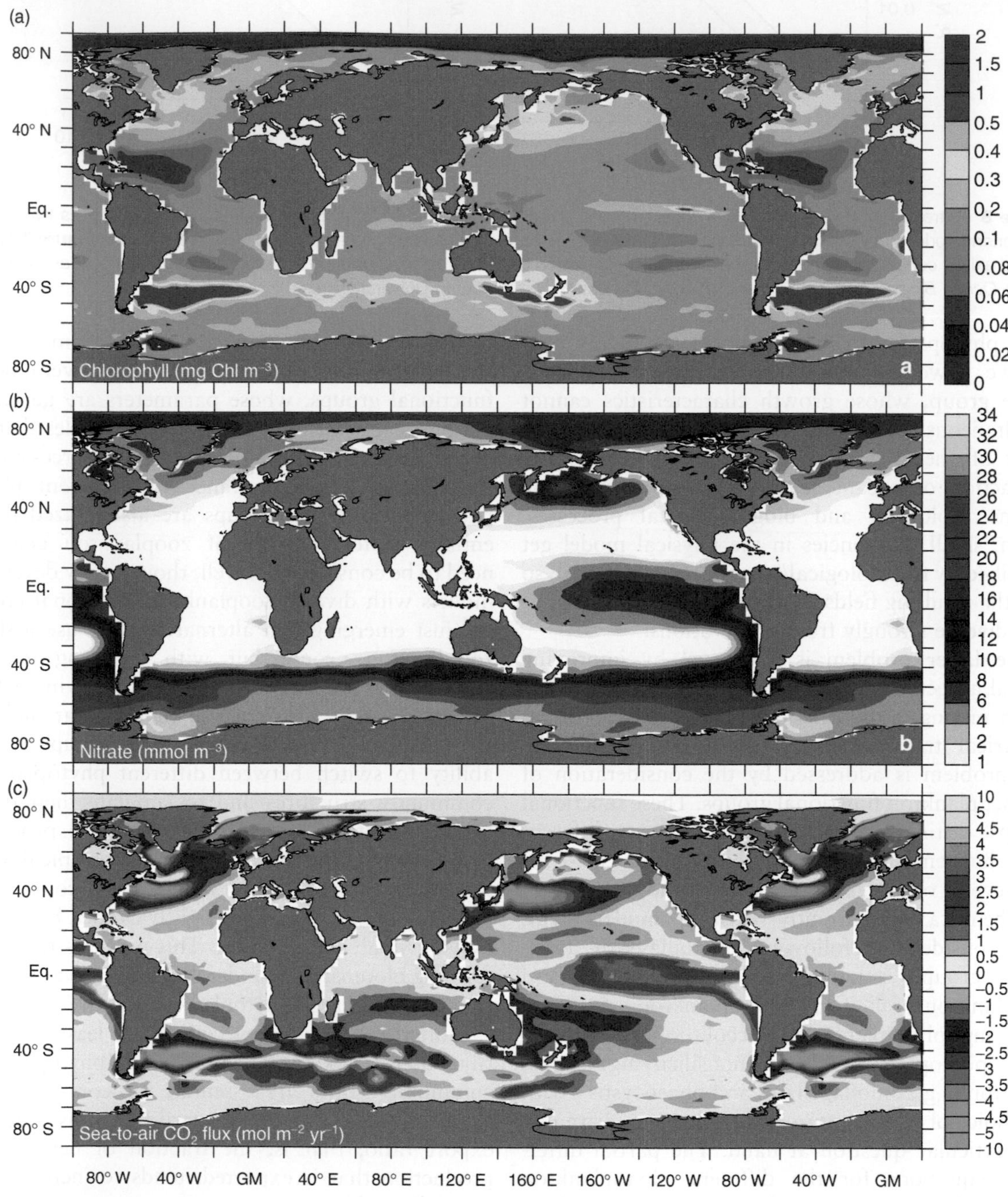

Figure 9 Global maps of model simulated key biogeochemical/ecological targets (cf. **Figure 2**) from the National Center for Atmospheric Research (NCAR) Community Climate System Model (CCSM): (a) annual mean distribution of near-surface chlorophyll (mg Chl m^{-3}); (b) annual mean surface distribution of nitrate (mmol^{-3}); and (c) annual mean air–sea CO_2 flux (mol m^{-2} yr^{-1}) for the period 1990 until 2000. All model results stem from a simulation, in which a multiple functional phytoplankton group model was coupled to a global 3-D ocean circulation model. Results provided by S.C. Doney and I. Lima, WHOI.

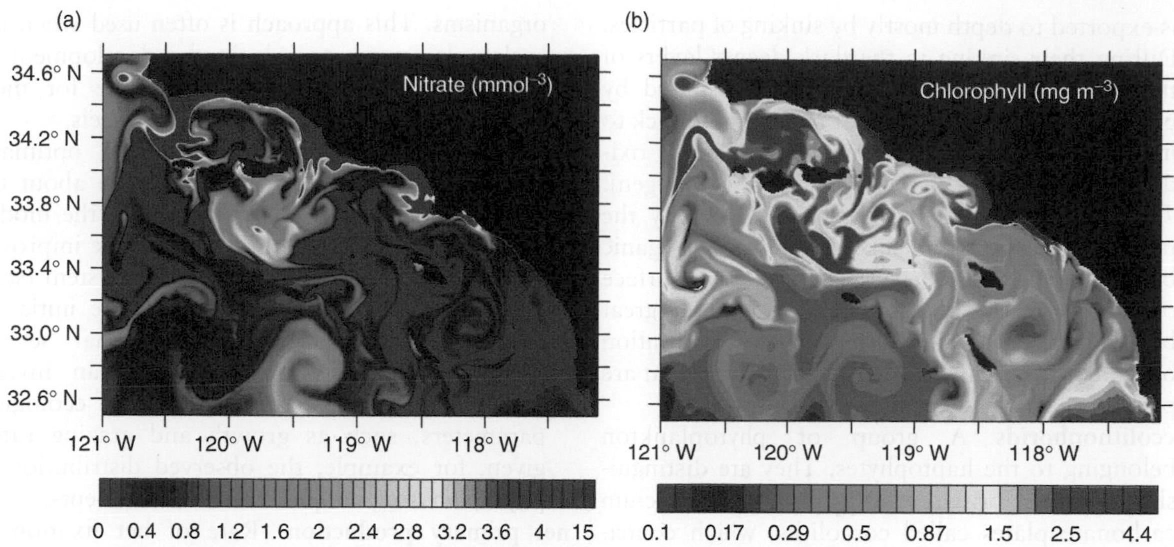

Figure 10 Snapshots of (a) surface nitrate and (b) surface chlorophyll as simulated by a 1-km, that is, eddy-resolving, regional model for the Southern California Bight. The snapshots represent typical conditions in response to a spring-time upwelling event in this region. The model consists of a multiple phytoplankton functional group model that has been coupled to the Regional Ocean Modeling System (ROMS). Results provided by H. Frenzel, UCLA.

simulations). This is currently feasible for only very short periods, so that limited domain models are often used instead.

Eddy-resolving Regional Models

Limited domain models, as shown in **Figure 10** for the Southern California Bight, that is, the southernmost region of the West Coast of the US, can be run at much higher resolutions over extended periods, permitting the investigation of the processes occurring at the kilometer scale. The most important such processes are meso- and submesoscale eddies and other manifestations of turbulence, which are ubiquitous features of the ocean. The coupling of the same ecosystem/biogeochemical model used for the global model shown in **Figure 9** to a 1-km resolution model of the Southern California Bight reveals the spatial richness and sharp gradients that are commonly observed (**Figure 10**) and that emerge from the intense interactions of the physical and biogeochemical/ecological processes at the eddy scale. Since such mesoscale processes are unlikely to be resolved at the global scale for a while, a current challenge is to find parametrizations that incorporate the net impact of these processes without actually resolving them explicitly.

Future Directions

The modeling of marine biogeochemical/ecological processes is a new and rapidly evolving field, so that predictions of future developments are uncertain. Nevertheless, it is reasonable to expect that models will continue to evolve along the major axes of complexity, including, for example, the consideration of higher trophic levels (**Figure 4**). One also envisions that marine biogeochemical/ecological models will be increasingly coupled to other systems. A good example are Earth System Models that attempt to represent the entire climate system of the Earth including the global carbon cycle. In those models, the marine biogeochemical/ecological models are just a small part, but interact with all other components, possibly leading to complex behavior including feedbacks and limit cycles. Another major anticipated development is the increasing use of inverse modeling approaches to optimally make use of the increasing flow of observations.

Glossary

advection The transport of a quantity in a vector field, such as ocean circulation.

anthropogenic carbon The additional carbon that has been released to the environment over the last several centuries by human activities including fossil-fuel combustion, agriculture, forestry, and biomass burning.

biogeochemical loop, great A set of key processes that control the distribution of bioreactive elements in the ocean. The loop starts with the photosynthetic production of organic matter in the light-illuminated upper ocean, a fraction of which is exported to depth mostly by sinking of particles. During their sinking in the dark deeper layers of the ocean, this organic matter is consumed by bacteria and zooplankton that transform it back to its inorganic constituents while consuming oxidized components in the water (mostly oxygen). The great biogeochemical loop is closed by the physical transport and mixing of these inorganic bioreactive elements back to the near-surface ocean, where the process starts again. This great biogeochemical loop also impacts the distribution of many other elements, particularly those that are particle reactive, such as thorium.

coccolithophorids A group of phytoplankton belonging to the haptophytes. They are distinguished by their formation of small mineral calcium carbonate plates called coccoliths, which contribute to the majority of marine calcium carbonate production. An example of a globally significant coccolithophore is *Emiliania huxleyi*.

diatoms A group of phytoplankton belonging to the class Bacillariophyceae. They are one of the most common types of phytoplankton and distinguish themselves through their production of a cell wall made of amorphous silica, called opal. These walls show a wide diversity in form, but usually consist of two symmetrical sides with a split between them.

diffusion The transport of a quantity by Brownian motion (molecular diffusion) or turbulence (eddy diffusion). The latter is actually some form of advection, but since its net effect is akin to diffusion, it is often represented as a diffusive process.

export production The part of the organic matter formed in the surface layer by photosynthesis that is transported out of the surface layer and into the interior of the ocean by particle sinking, mixing, and circulation, or active transport by organisms.

models, concentration based Models, in which the biological state variables are given as numbers per unit volume, that is, concentration. This approach is often used when physical dispersion plays an important role in determining the biological and biogeochemical impact of these state variables. Concentration-based models are most often used to represent lower trophic levels in the ocean.

models, forward Models, in which time is integrated forward to compute the temporal evolution of the distribution of state variables in response to a set of initial and boundary conditions.

models, individually based Models, in which the biological state variables represent individual organisms or a small group of individual organisms. This approach is often used when life cycles play a major role in the development of these organisms, such as is the case for most marine organisms at higher trophic levels.

models, inverse Models that aim to optimally combine observations with knowledge about the workings of a system as embodied in the model. Solutions to such inverse models can be improved estimates of the current state of the system (state estimation), improved estimates of the initial or boundary conditions, or an optimal set of parameters. A typical example of an inverse model is the optimal determination of ecological parameters, such as growth and grazing rates, given, for example, the observed distribution of phytoplankton, zooplankton, and nutrients.

net primary production Rate of net fixation of inorganic carbon into organic carbon by autotrophic phytoplankton. This net rate is the difference between the gross uptake of inorganic carbon during photosynthesis and autotrophic respiration.

organisms, autotrophic An autotroph (from the Greek *autos* = self and *trophe* = nutrition) is an organism that produces organic compounds from carbon dioxide as a carbon source, using either light or reactions of inorganic chemical compounds, as a source of energy. An autotroph is known as a producer in a food chain.

organisms, heterotrophic A heterotroph (Greek *heterone* = (an)other and *trophe* = nutrition) is an organism that requires organic substrates to get its carbon for growth and development. A heterotroph is known as a consumer in the food chain.

plankton Organisms whose swimming speed is smaller than the typical speed of ocean currents, so that they cannot resist currents and are hence unable to determine their horizontal position. This is in contrast to nekton organisms that can swim against the ambient flow of the water environment and control their position (e.g., squid, fish, krill, and marine mammals).

plankton, phytoplankton Phytoplankton are the (photo)autotrophic components of the plankton. Phytoplankton are pro- or eukaryotic algae that live near the water surface where there is sufficient light to support photosynthesis. Among the more important groups are the diatoms, cyanobacteria, dinoflagellates, and coccolithophorids.

plankton, zooplankton Zooplankton are heterotrophic plankton that feed on other plankton. Dominant groups include small protozoans or metazoans (e.g., crustaceans and other animals).

See also

Biogeochemical Data Assimilation.

Further Reading

DeYoung B, Heath M, Werner F, Chai F, Megrey B, and Monfray P (2004) Challenges of modeling ocean basin ecosystems. *Science* 304: 1463–1466.

Doney SC (1999) Major challenges confronting marine biogeochemical modeling. *Global Biogeochemical Cycles* 13(3): 705–714.

Fasham M (ed.) (2003) *Ocean Biogeochemistry.* New York: Springer.

Fasham MJR, Ducklow HW, and McKelvie SM (1990) A nitrogen-based model of plankton dynamics in the oceanic mixed layer. *Journal of Marine Systems* 48: 591–639.

Glover DM, Jenkins WJ, and Doney SC (2006) Course No. 12.747: Modeling, Data Analysis and Numerical Techniques for Geochemistry. http://w3eos.whoi.edu/12.747 (accessed in March 2008).

Rothstein L, Abbott M, Chassignet E, *et al.* (2006) Modeling ocean ecosystems: The PARADIGM Program. *Oceanography* 19: 16–45.

Sarmiento JL and Gruber N (2006) *Ocean Biogeochemical Dynamics*, 526pp. Princeton, NJ: Princeton University Press.

Relevant Websites

http://www.uta.edu
– Interactive models of ocean ecology/biogeochemistry.

BIOGEOCHEMICAL DATA ASSIMILATION

E. E. Hofmann and M. A. M. Friedrichs,
Old Dominion University, Norfolk, VA, USA

Introduction

Data assimilation is the systematic use of data to constrain a mathematical model. It is assumed that the dynamics that are responsible for a particular process or distribution are inherent in the data. By inputting data of various types into a mathematical model, the model, which is a truncated version of the real world, will more accurately stimulate a particular environment or situation. Through data assimilation, the hindcast, nowcast, and/or forecast of the model will be improved.

Data assimilation was first used in the 1960s in numerical weather forecasting models, with the goal of providing short-term predictions of meteorological conditions. The use of data assimilation techniques was made feasible by the development of a worldwide atmospheric data network that could provide the measurements needed. Data assimilation provided a methodology for using these observations to improve the forecasting skills of the operational models. Although weather forecasts are now taken for granted, to a large extent the accuracy of these forecasts results from assimilation of meteorological observations.

In the 1970s, numerical ocean general circulation models (OGCMs) became an important tool for understanding ocean circulation processes. Initial applications of these models focused on simulation of the large-scale structure of ocean currents. From these simulations, the limitations of the OGCMs were clear. Data assimilation was looked to as an approach for constraining these dynamical models with available data. For example, data assimilation could be used to quantitatively and systematically test and improve poorly known sub-grid-scale parametrizations and boundary conditions that are so abundant in OGCMs. With recent advances in data availability, it is also now feasible to use data-assimilative OGCMs for making forecasts of the ocean state, such as the El-Niño–La Niña cycle in the equatorial Pacific Ocean.

Implementing data-assimilative biogeochemical models has been problematic because of the paucity of adequate data. Historically, biological and chemical data were obtained almost exclusively by ship surveys, and thus were extremely limited in both space and time. However, advances in satellite and mooring instrumentation, as well as in the understanding of the structure and function of marine ecosystems, now makes it feasible to begin the development of data-assimilative biogeochemical models. As a result, since the mid 1990s there has been a dramatic increase in the types of data that are input into marine ecosystem models, and the development of robust and varied approaches for assimilating these data. This research provides a framework for future studies of biogeochemical data assimilation and predictive biogeochemical modeling that will inevitably play a major role in the next generation of large interdisciplinary oceanographic observational programs.

The following section provides a brief history of how the field of marine biogeochemical modeling has matured as more and more data have become available. This is followed by a description of some data assimilation methods and specific examples of how two of these methods can be used in conjunction with a simple marine ecosystem model. The final section provides a summary.

Biogeochemical Models and Data Availability

Mathematical models provide a quantitative framework for investigating processes that are responsible for the biological and chemical distributions that underlie the structure and function of marine ecosystems. Mathematical models were first used to study marine ecosystems in the late 1940s and these models had their basis in the predator–prey models developed in the early 1900s. These early modeling attempts were focused on understanding the processes that allow large blooms of phytoplankton and zooplankton to occur. The models were simple in nature, including only average population characteristics, basic biological processes resulting in plant and animal growth, and interactions at the lowest trophic levels, e.g., primary and secondary producers. Effects of environmental factors such as temperature and circulation, which are important in marine systems, were not explicitly included in these models.

The following generation of models included more complex biogeochemical processes, differentiation of species, and coupling of the marine biogeochemical system to circulation models. These more realistic models were made possible by the development of large multidisciplinary oceanographic programs in the 1970s, which, for the first time, provided concurrent physical, biological, and chemical measurements that could be input to coupled circulation–biogeochemical models. The resulting models clearly demonstrated the utility of modeling for integrating and synthesizing large multidisciplinary oceanographic datasets. However, more importantly, the realism of the simulated distributions obtained from the coupled modeling efforts helped establish this approach as an important research tool for understanding marine biogeochemical systems. Present-day multidisciplinary oceanographic programs now routinely include a mathematical modeling component.

At the time coupled circulation–biogeochemical models were being developed, significant advances were being made in the measurements of biological and chemical distributions in the ocean. In the 1980s, the Coastal Zone Color Scanner satellite was launched, which provided large-scale ocean color distributions, from which phytoplankton chlorophyll distributions, and their evolution over space and time, could be derived. The ocean color data also facilitated making inferences about the relative roles of circulation versus biogeochemical processes in controlling phytoplankton distributions. The availability of large-scale observations of chlorophyll has been enhanced with the subsequent launch of the Sea-viewing Wide Field-of-view Sensor in 1997. Instrumentation capable of providing biological measurements at fine space and time scales, e.g., moored optical and acoustic measurements, now provide *in situ* observations that can be combined with ocean color to reveal a more complete view of chlorophyll distributions. Also, moored buoy arrays provide concurrent physical data. Thus, high-resolution datasets that can be used to study marine systems are becoming increasingly available.

These new, high-quality datasets can now be used to validate coupled circulation–biogeochemical models. In many cases simulated distributions from the models reproduce many of the features seen in the ocean color observations. However, the simulated fields are often unable to reproduce the variability observed on short space (tens of kilometers) and time (days) scales. There are many potential explanations for these discrepancies, including mismatches in the space and time scales that the model resolves versus those resolved by the measurements. There may also be inconsistencies between the model structure and the observations. For instance, specific parameter values or choices for empirical formulations, forcing functions, or initial conditions may be in error.

Along with these new types of data comes the possibility of developing a new generation of data-assimilative biogeochemical models that will not only be better able to reproduce the observed variability in biogeochemical fields but may also have the potential for significantly improving the accuracy of model predictions. In the 1990, researchers began to investigate the use of data assimilation as an approach for improving these coupled models, and as an approach for making better use of the many types of environmental and biogeochemical data that are becoming available.

Data Assimilation Methods

Many techniques exist for systematically combining data with mathematical models. The development and refinement of many of these techniques have been through the use of meteorological and oceanic general circulation models. These data assimilation methods are just starting to be tested in marine biogeochemical models; however, it is not clear whether the same methods can be used with these multidisciplinary models, since biogeochemical ocean models differ substantially from their physical counterparts. For instance, biological systems have no analogue to the Navier–Stokes equations that form the basis for fluid dynamics. Thus, biogeochemical models are by necessity largely empirical and nonlinear, and abound with poorly known formulations. For example, such models typically include large numbers of parameters that are difficult (*in situ* growth rates), or even impossible (mortality rates) to measure with current oceanographic instrumentation. The timescales of these models are also typically short, since the model must resolve the rate at which populations double in number, which for most of the abundant phytoplankton species is one day or less.

Because of these innate differences between physical models and biogeochemical models, the application of data assimilation techniques to biogeochemical ocean models, and specifically to marine ecosystem models, presents many exciting new challenges. Although considerable effort will undoubtedly be put into developing new assimilation schemes specific to these types of models, the data-assimilative marine ecosystem models that have been developed using existing assimilation methods already show much promise.

One of the most straightforward methods that has been used to combine model dynamics with data entails simply replacing the model solution with data whenever such information is available. This technique, referred to as data insertion, integrates the model forward in time until additional observations become available, at which point the model is reinitialized and the process is repeated. A basic assumption underlying this method is that there is adequate knowledge of the governing model dynamics and parameter values.

This technique has been used to estimate velocity fields by inserting temperature and salinity data into relatively complex physical oceanographic models. In these analyses, however, model–data inconsistencies caused the resulting simulations to compare poorly with observations. This led to the development of a technique in which the model solution is 'nudged' toward observations whenever they become available, instead of being directly replaced by the observations. Although this more gentle method of nudging may provide a significant improvement in simulation skill, like data insertion, it still lacks a means by which information on data uncertainty can be incorporated, and does not provide an estimate of the errors in the resulting solution.

More advanced assimilation schemes, such as optimal interpolation and Kalman filtering, have been successfully applied by meteorologists, yet hold little hope for marine ecosystem models because of the inherent nonlinearities of biological systems. Instead, variational schemes, which have recently been applied to nonlinear physical oceanographic systems, may be more applicable to multidisciplinary problems containing biological components. These variational methods of data assimilation, such as the adjoint method and simulated annealing, have their basis in optimization theory and rely on minimizing the differences between observed and simulated quantities, pursuant to predetermined minimization criteria. At the most basic level, these methods can be thought of as nonlinear least-squares analyses, which determine the optimal solution (including parameter values and initial and/or boundary conditions) that maximizes agreement between the model simulation and observations.

The adjoint method is a variational scheme that has found considerable success in the field of physical oceanography. Although this method is now also being used in marine ecosystem modeling, the nonlinear nature of these types of models may result in the recovery of suboptimal parameter sets. Simulated annealing is another assimilation scheme that has been used with data-assimilative ecosystem models. Although this method is typically capable of recovering a single optimal parameter set, the stochastic, 'random-walk' nature of simulated annealing causes this technique to be considerably less efficient than the adjoint method. As a result, simulated annealing may be computationally too intensive to be of use in large-scale marine biogeochemical assimilation analyses.

Numerical Twin Experiments

Before the application of data assimilation techniques to marine biogeochemical models becomes routine, a number of methodological questions need to be addressed. For instance, how many data are needed for these studies? What types of sampling strategies are optimal? What level of uncertainty in the data can be tolerated? Are these models too complex, or too simple? What types of data assimilation schemes will work best for these highly nonlinear models?

One method for addressing such methodological questions is through the use of identical twin experiments. In an identical twin experiment, the model is initially run using best estimates for the model parameters in order to provide a 'true' simulated time-series. This time-series is subsampled to generate a synthetic data set (**Figure 1A**). The model is then run a second time, using an imperfect parameter set in order to generate a 'reference' (no assimilation) time-series. This same imperfect

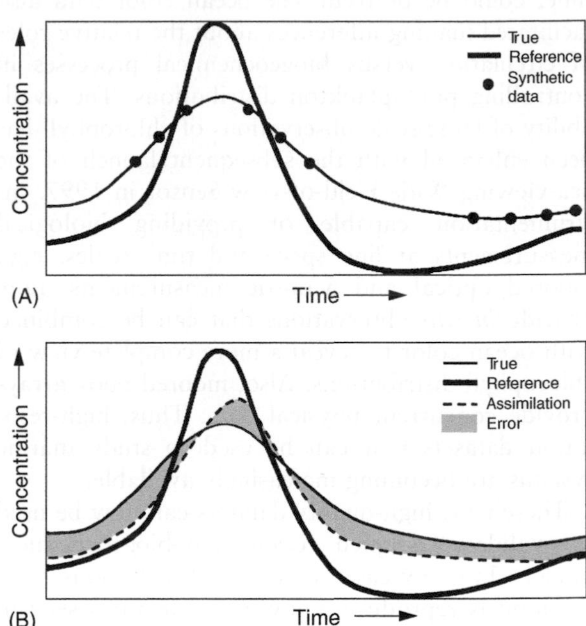

Figure 1 Schematic illustrating the implementation of an identical twin experiment. The true simulation (thin solid line) represents the solution for one component of the model (e.g., phytoplankton, zooplankton, or nutrient concentration) obtained using the best estimates of the model parameters and initial conditions. A second simulation, using a different parameter set or initial conditions, provides a reference simulation (thick solid line). (A) The true simulation is subsampled to create a synthetic data set. (B) The assimilation of the synthetic data into the reference simulation results in a third model solution (dashed line). The difference between this solution and the true solution (shaded region) is a measure of the error in the data-assimilative solution.

parameter set is used in the third and final model run, but this time the synthetic data are assimilated into the model (**Figure 1B**). The success of the assimilation process is judged by the difference between these results and the true simulation, and is typically normalized by the difference between the true and reference simulations.

Identical twin experiments are a necessary precursor to true data-assimilative model runs, and have the potential to provide considerable insight into a number of important issues regarding the assimilation process. For instance, they can be used to rigorously compare different assimilation schemes, to determine optimal sampling strategies, and to assess the effects of assimilating observations that are associated with known levels of noise. Furthermore, identical twin experiments can be invoked to determine whether a certain set of model parameters can be estimated independently, and thus whether or not a given model may need to be simplified. Although the utility of identical twin experiments is well accepted within the fields of meteorology and physical oceanography, this approach has only recently been applied to ecosystem modeling analyses. The two examples described below illustrate some of the strengths and weaknesses of this approach for understanding data-assimilative marine biogeochemical models.

Example 1: Data Insertion and Nudging

The pros and cons of using data insertion or 'nudging' to assimilate biogeochemical data can be illustrated using a three-component marine ecosystem box model (nitrogen, phytoplankton, zooplankton). Simulations with a non-data-assimilative version of this model (**Figure 2**) show the behavior that is expected in this type of marine system. Nitrogen concentrations decrease over time, as nitrogen is used to support a bloom of phytoplankton. Zooplankton, the primary grazer of the phytoplankton, blooms subsequently and results in a decrease of the phytoplankton.

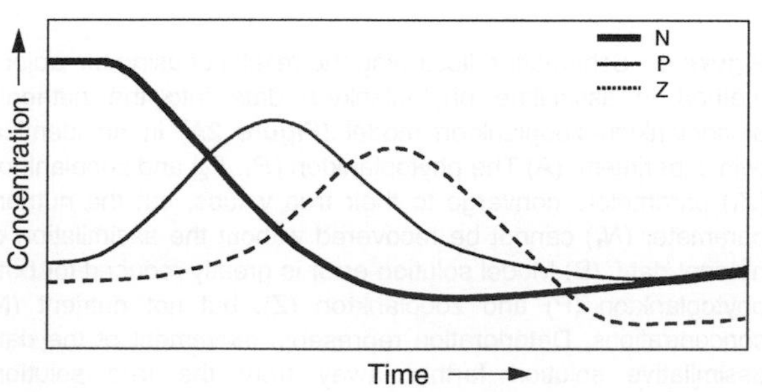

Figure 2 Schematic of the time evolution of nutrient (N), phytoplankton (P), and zooplankton (Z) concentrations, obtained from a marine ecosystem model simulation.

The accuracy of this basic simulation can be improved by inserting phytoplankton concentrations, such as those derived from ocean color measurements. Results of an identical twin experiment demonstrate that at times when average chlorophyll concentrations are inserted into the box model (e.g., every other day in **Figure 3**), the error in the simulation of phytoplankton decreases to zero. Nudging yields similar results, except that the error in the phytoplankton field would be reduced to a fixed nonzero value, dependent upon the strength of the nudging.

One requirement for data insertion or nudging is that data must be available on timescales coincident with those of the dominant biological processes. Because biological processes, such as phytoplankton growth, have timescales of 1–2 days in many regions of the ocean, data with this level of time resolution are required for data insertion methods to adequately represent the biological dynamics. As illustrated in **Figure 3A**, the improvement in simulation skill lasts only 1 to 2 days beyond the point at which phytoplankton data are no longer available for insertion. Hence, when using these methods, fully data-assimilative marine biogeochemical models can potentially create a huge demand on data resources.

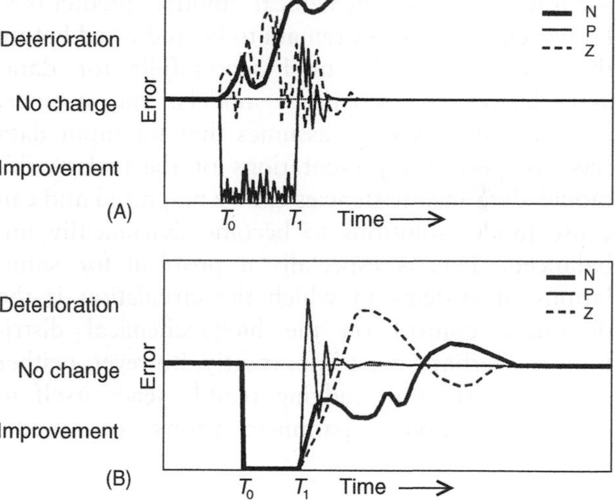

Figure 3 Schematics of the change in model solution error obtained in an identical twin experiment, when data are inserted into the nutrient–phytoplankton–zooplankton model (**Figure 2A**) between (insert eqn) and (insert eqn). (A) Only phytoplankton (P) observations are inserted every other day, and (B) nutrient (N), phytoplankton, and zooplankton (Z) observations are inserted daily. Deterioration represents movement of the data-assimilative solution farther from the true solution; improvement represents convergence of the data-assimilative and true solutions. No change occurs when the data-assimilative model solution remains the same as in the non-data-assimilative (reference) solution.

Another primary factor limiting the use of data insertion methods for marine biogeochemical models is that these models usually consist of many components, all of which must be updated to be in balance with the assimilated data. For instance, although estimates of phytoplankton biomass are improved as a result of the assimilation of phytoplankton data, the accuracy of the other model components is reduced (**Figure 3A**). The error produced in the other model state variables is variable, and depends upon the level of coupling between the various components of the marine biogeochemical system. This problem may be alleviated if data exist for the other model components, since the assimilation of these additional data (**Figure 3B**) can substantially reduce errors in the other model components. Since the adjustment timescales for the model components differ (**Figure 3B**), the insertion of data may be required at varying time intervals for the different model components. Unfortunately, data sufficient to update the other ecosystem model components often do not exist. In this case *ad hoc* approaches, perhaps based on maintaining ratios between different ecosystem components, can be invoked; however, such approaches also have the potential to introduce errors that may negate the gains made through data assimilation.

Thus, data insertion and nudging are easy to implement and in certain instances may improve the accuracy of biogeochemical model predictions. However, many issues remain to be addressed before this method can be used successfully for data-assimilative biogeochemical models. For instance, because data insertion assumes that the input data sets are perfect representations of the real world, model–data inconsistencies can be magnified and can cause model solutions to become dynamically unbalanced. This is especially a problem for simulations of systems in which the circulation is the dominant control on the biogeochemical distributions. Perhaps most importantly, however, neither data insertion nor nudging readily lends itself to improving model parametrizations or model structure.

Example 2: Adjoint Method

Even a relatively simple marine ecosystem model, such as the model shown in **Figure 2**, typically contains 10–20 model parameters that must be specified for a given simulation. This is a crucial aspect of ecosystem modeling, since even small changes in some of these parameters may result in large differences in simulation results. Unfortunately, values for these parameters are often poorly constrained in space and time, and some, such as *in situ* zooplankton mortality rates, are virtually unknown. Thus, the specification of an optimal parameter set for a given biogeochemical model is a challenge at best, and in most cases, nearly impossible. For these reasons, the adjoint method, which searches parameter space to find a parameter set that minimizes model–data misfits, holds considerable potential for use in data-assimilative marine biogeochemical models.

The utility of the adjoint method can be demonstrated by applying identical twin experiments to the three-component ecosystem model (**Figure 2**). Because marine biogeochemical models typically contain many parameters that are very highly correlated, it is often not possible to recover a parameter set in its entirety. Therefore, sensitivity or correlation analyses can be performed in order to choose a subset of relatively uncorrelated parameters that will be recovered.

If an identical twin experiment is carried out in which phytoplankton data are assimilated every other day using the adjoint method, phytoplankton and zooplankton parameters may be recovered precisely (**Figure 4**). As a result, the errors in the phytoplankton

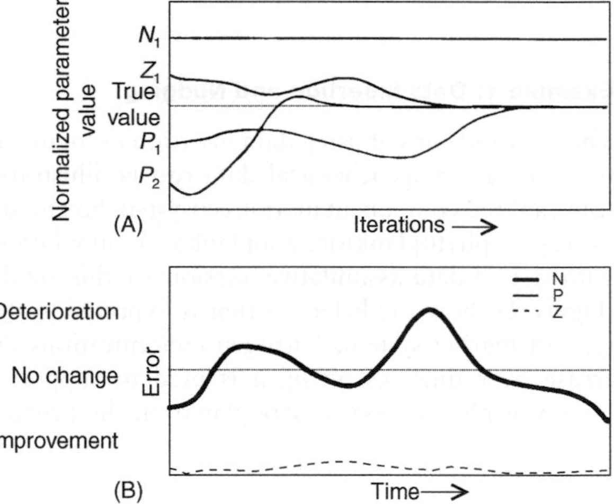

Figure 4 Schematics illustrating the results of using the adjoint method to assimilate phytoplankton data into the nutrient–phytoplankton–zooplankton model (**Figure 2A**) in an identical twin experiment. (A) The phytoplankton (P_1, P_2) and zooplankton (Z_1) parameters converge to their true values, but the nutrient parameter (N_1) cannot be recovered without the assimilation of nutrient data. (B) Model solution error is greatly reduced for both phytoplankton (P) and zooplankton (Z), but not nutrient (N) concentrations. Deterioration represents movement of the data assimilative solution farther away from the true solution; improvement represents convergence of the data assimilative and true solutions. No change occurs when the data-assimilative model solution remains the same as the non-data-assimilative (reference) solution.

and zooplankton simulations will be significantly reduced. (In **Figure 4B** the error in the phytoplankton is shown to be zero for the entire model run.) However, if no nutrient data are assimilated, the parameter(s) on which nutrient concentration is most highly dependent, e.g., N_1 (**Figure 4A**), will not be recovered, and therefore no significant improvement in the simulation skill of the nutrient component will result (**Figure 4B**). If synthetic phytoplankton, zooplankton, and nutrient data are all available for assimilation, all parameters may be recovered precisely, and the errors in all model components may be reduced to zero for the entire model run.

Research on data-assimilative marine ecosystem models has shown that under certain conditions the results described above are characteristic of those obtained when real data are assimilated using the adjoint method. However, in other instances it is possible that the assimilative model may fail to recover an optimal parameter set. This can occur even if the model has been well tested and calibrated, and implies that the model is in some way inconsistent with the assimilated data set. For instance, changes in plankton community dominance might result in inconsistencies in the model and data that cannot be resolved simply through data assimilation. If this is the case, it may be possible to isolate the specific model assumption(s) that have been violated, e.g., the assumption of a constant species composition, to reformulate the model in a more realistic fashion and to repeat the assimilation analysis in order to test this hypothesis.

Sometimes the adjoint method may recover multiple parameter sets, each dependent on the initial choices made for the model parameters. In these situations, rigorous approaches for choosing between the possible parameter sets are required. One approach is to establish a specific uncertainty range, either from experimental or from theoretical considerations, for each parameter that is allowed to vary in the adjoint analysis. Alternatively, the optimal parameter set could be selected on the basis of the ability of each parameter set in simulating an independent data set.

Summary

Data assimilation techniques for marine biogeochemical models are just beginning to be explored. Initial results are encouraging and data assimilation approaches, such as adjoint methods, hold great promise for improving the capability of these models. For instance, recent analyses of data-assimilative biogeochemical models demonstrate that the assimilation of biogeochemical data can reduce model–data misfit by recovering optimal parameter sets using multiple types of data. Perhaps even more importantly, these data assimilation analyses can demonstrate whether or not a given model structure is consistent with a specific set of observations. When model and data are shown to be consistent, the specific mechanisms underlying observed patterns in simulated distributions can be identified. If a model is determined to be inconsistent with observations, it may be possible to isolate the specific model assumption that has been violated, and to reformulate the model in a more realistic fashion. Thus, although the assimilation of data into a marine biogeochemical model cannot necessarily overcome inappropriate model dynamics and structure, it can serve to guide model reformulation.

During the 1990s, large interdisciplinary oceanographic programs included model prediction and forecasting as specific research objectives. However, new studies are revealing that much more work needs to be performed before this becomes a realistic and achievable goal. Until high-resolution biological and chemical data are available over large regions of the ocean, and until a much clearer understanding of the intricacies of marine ecosystems is attained, data assimilation in biogeochemical models will be more useful for model improvement and parameter estimation than for model prediction and forecasting. By providing a means for recovering the best-fit set of parameters for a given model, certain assimilation techniques may prove to be a crucial tool for marine biogeochemical modelers.

The importance of inclusion of data in all steps of model development and implementation cannot be emphasized enough. It is through model and data comparisons that models are advanced and better observation systems are developed. Therefore, an important aspect of furthering the development of predictive marine biogeochemical models is recognizing the need for interdisciplinary multiscale observational and experimental networks. The availability of such data will necessitate the development of techniques for input of these data into models, and facilitate the development of data-assimilative marine biogeochemical models.

See also

Population Dynamics Models.

Further Reading

Fasham MJR and Evans GT (1995) The use of optimization techniques to model marine ecosystem dynamics at the JGOFS station at 47°N 20°W.

Philosophical Transactions of the Royal Society of London, Series B 348: 203–209.

Ishizaka J (1993) Data assimilation for biogeochemical models. In: Evans GT and Fasham MJR (eds.) *Towards a Model of Ocean Biogeochemical Processes*, pp. 295–316. New York: Springer-Verlag.

Lawson LM, Spitz YH, Hofmann EE, and Long RB (1995) Data assimilation applied to a simple predator–prey model. *Bulletin of Mathematical Biology* 57: 593–617.

Matear RJ (1995) Parameter optimization and analysis of ecosystem models using simulated annealing: a case study at Station P. *Journal of Marine Research* 53: 571–607.

APPENDIX

APPENDIX 9. TAXONOMIC OUTLINE OF MARINE ORGANISMS

L. P. Madin, Woods Hole Oceanographic Institution, Woods Hole, MA 02543, USA

Introduction

This appendix is intended as a brief outline of the taxonomic categories of marine organisms, providing a classification and description of groups that are mentioned elsewhere in the Encyclopedia. It is presented at the level of Phylum (or equivalent) in most cases, but to the level of Class in large groups with many marine species. The few categories that contain no known marine species are not included in this list.

The outline is intended as a list rather than any kind of phylogenetic tree. The three Domains listed are now considered, on genetic evidence, to be the three primary categories of living organisms on Earth. The definitions and relationships of many higher taxonomic groups are currently in flux owing to the advent of new molecular genetic data. This is particularly true for the Archaea, Bacteria, and Protista (also called Protozoa or Protoctista), in which numerous phylogenetic lineages have been identified in recent years that are neither consistent with classical categories based on morphological characters nor arranged in a comparable hierarchy. The accepted systematics of the Protista will probably change in the near future, reflecting this new genetic information. However, as these new categories are still being defined, and have not yet come into common usage in ocean sciences, this outline retains a more classical and widely known system for protists (Margulis and Schwartz, 1988), with names that will more likely be familiar to marine scientists. Where possible, the approximate number of known (named) species is given, with an indication of how many of these are marine. For many groups the described species may be only a small fraction of the probable true number of species.

It should be apparent from perusal of the list that the ocean environment is home to the vast majority of the biological diversity on Earth. Notwithstanding the large numbers of insect species and flowering plants on land, the remarkable diversity of different body plans defining the higher taxonomic levels occurs almost entirely in the sea.

Readers interested in recent work on the phylogeny and classification of any of these organisms are directed to the Further Reading list. Often the most up-to-date information will be found on the Web sites listed.

DOMAIN ARCHAEA

Prokaryotic cells having particular cell lipids and genetic sequences that distinguish them from all other organisms. Many are extremophiles living at high temperatures or under unusual chemical conditions. Few have been cultured. "Species" are defined and classification is based mainly on molecular evidence. Increasing numbers of Archaea are being detected in marine environments.

Korarchaeota

A poorly-known group of hyperthermophilic organisms thought to be near the evolutionary base of the Archaea, and perhaps the most primitive organisms known.

Crenarchaeota

Primarily hyperthermophilic forms, including many sulfur-reducing species, but also species living at very low temperatures in the ocean.

Euryarachaeota

A group containing methane-producing forms and others that live in extremely saline conditions.

DOMAIN BACTERIA

A tremendously diverse domain of prokaryotic cells with particular genetic sequences and cell constituents. All photosynthetic and pathogenic forms are in this domain. A dozen major phylogenetic groups can be identified on genetic evidence, of which all but one include marine forms. Numbers of species are almost impossible to specify.

Aquificales

Primitive hyperthermophiles with a chemoautotrophic metabolism. Known from hydrothermal environments.

Green non-sulfur bacteria

A group including some photosynthetic and multicellular filamentous forms. Related forms may have produced ancient stromatolite structures in shallow seas.

Proteobacteria

Purple bacteria; a large, metabolically and morphologically diverse group including some photosynthetic, chemoautotrophic, and nitrogen-fixing forms that occur in the ocean. There are also many pathogenic species.

Gram-positive bacteria

Common soil-dwelling forms, but some also marine. Many form resting stages called endospores enabling survival under harsh conditions.

Cyanobacteria

An ancient and diverse lineage of photosynthetic bacteria that include some of the most abundant and important primary producers in the ocean.

Bacteroides, Flavobacterium and related forms

Gram-negative microbes occurring in freshwater and marine environments, as well as soils and deep-sea sediments.

Green sulfur bacteria

Anaerobic forms that oxidize sulfur compounds to elemental sulfur and occupy a variety of marine environments.

Deinococci

Highly radiation-resistant cells, which include forms found in terrestrial and marine thermal springs.

Planctomyces

One of only two groups having cell walls that lack the component peptoglycan; found in freshwater and marine environments.

Thermotoga

A group of anaerobic microbes found at shallow and deep-sea hydrothermal vents, and which includes some of the most thermophilic bacteria known.

Spirochaetes

Flagellated cells with a unique helical morphology; most are parasites and some inhabit marine animals.

DOMAIN EUKARYA

Single-celled or multicelled organisms with membrane-bounded nuclei containing the genetic material, and other specialized cellular organelles. Cell division is by some form of mitosis, and metabolism is usually aerobic. All protists, plants and animals are eukaryotes.

PROTISTA (also Protoctista, Protozoa)

Single-celled (and some multicellular) organisms that are neither animals nor true plants. It is a diverse and polyphyletic group that includes protozoans, algae, seaweeds and slime molds. These categories are defined mostly by morphology.

Phylum Dinoflagellata

Marine protozoans having unusual organization of their DNA, two locomotory flagella, and frequently encased in a rigid test. Dinoflagellates can be photosynthetic or heterotrophic; some live symbiotically in the tissues of other organisms. Approximately 3000 species, mostly marine.

Phylum Rhizopoda (also Amoebozoa)

Amoebas; single-celled organisms lacking flagella or cilia and moving by pseudopodia. Some are encased in a test. There are several thousand species in terrestrial, freshwater and marine environments, as well as parasitic and pathogenic forms.

Phylum Chrysophyta

A large group of unicellular algae that lack sexual stages and reproduce only by asexual division and formation of "swarmer" cells for dispersal. Most live in fresh water; the silicoflagellates are the only marine representatives.

Phylum Haptophyta

Marine photosynthetic cells that alternate between a flagellated free-swimming stage and a resting coccolithophorid stage covered with calcareous plates. These stages were previously thought to be distinct species. Several hundred species.

Phylum Euglenophyta

Flagellated protozoa that may be photosynthetic or heterotrophic, solitary, or colonial. They have flexible cell walls made of protein, and complex internal organelles. About 800 species including some marine.

Phylum Cryptophyta

Also called cryptomonads, these cells are widely distributed in fresh and salt water, and as internal parasites. They lack sexual reproduction and swim with two flagella.

Phylum Zoomastigina

Nonphotosynthetic cells with one to many flagella, including free-living, symbiotic, and parasitic species, with both sexual and asexual reproduction. The group is probably polyphyletic and includes hundreds of described species, with probably thousands more unknown; many are marine.

Phylum Bacillariophyta

The diatoms; photosynthetic cells enclosed in elaborate siliceous tests consisting of two halves or valves. Diatoms are important components of the food chain in marine and fresh water, with about 12 000 species.

Phylum Phaeophyta

The brown algae and seaweeds; macroscopic, photosynthetic, multicellular plantlike forms inhabiting intertidal, subtidal, or pelagic marine environments. About 1500 species, all marine.

Phylum Rhodophyta

Red algae; complex multicellular seaweeds inhabiting intertidal and subtidal environments worldwide. All contain particular photosynthetic pigments that give them a reddish color. About 4000 species, mainly marine.

Phylum Chlorophyta

Green algae, including single-celled forms and some multicellular seaweeds. They are important primary producers in all aquatic environments. There are about 7000 species, including many marine forms.

Phylum Actinopoda

Heterotrophic protozoa having long filamentous cytoplasmic extensions called axopods, supported by silica-based or strontium-based skeletal elements. Important marine forms are the radiolarians and acantharians. Some harbor algal symbionts or form macroscopic colonies. About 4000 species, mostly marine.

Phylum Foraminifera

Amoeboid protozoans with internally chambered, calcified, or agglutinated shells or tests. All are marine, living in benthic and pelagic habitats. About 4000, mainly benthic, extant species, but 30 000 fossil species.

Phylum Ciliophora

The ciliates; single-celled organisms covered with short cilia that are used for locomotion and/or food gathering. Ciliate cells have two nuclei and reproduce by fission. The 8000 species live in freshwater and marine environments.

Phylum Cnidosporidia

A diverse, polyphyletic group of heterotrophic microbes that are parasites and pathogens of animals, including many marine invertebrates and fish. About 850 species.

Phylum Labyrinthulomycota

Slime nets; colonial protozoans that construct networks of slime pathways on the surface of various substrates. The osmotrophic cells move along the slimeways toward food sources. Only about 10 described species, including marine forms.

PLANTAE

Plants are multicellular, photosynthetic organisms that develop from an embryo that is produced by sexual fusion. Plant cells have rigid cell walls and contain chloroplasts where photosynthesis occurs. Most of the 235 000 described species are terrestrial, with a few secondarily adapted to shallow marine environments.

Phylum Angiospermophyta

Flowering plants; virtually all the familiar grasses, flowers, vegetables, shrubs, and trees on Earth belong in this group, comprising about 230 000 species. A few grasses live in salt marsh and shallow subtidal marine environments.

FUNGI

Multicellular organisms that are neither motile nor photosynthetic, and form spores for reproduction. The basic structural elements of fungi are threadlike hyphae, which are partially divided into separate cells. Fungi range from microscopic yeasts and molds to large mushrooms and shelf fungi, and include some pathogenic forms. The vast majority of the 100 000 species are terrestrial.

Phylum Ascomycota

A diverse group including yeasts, molds, and truffles. In the marine environment filamentous ascomycotes grow and feed on decomposing plant material. A few of the 30 000 species are found in marine environments.

Phylum Basidiomycota

Complex, mainly terrestrial fungi including rusts, smuts, and mushrooms. Of some 25 000 species, only a handful are known from the marine environment, where they grow on marine grasses.

ANIMALIA

Animals are motile, heterotrophic, multicellular organisms, all of which develop from a ball of cells called a blastula, which originates by fusion of gametes. Most animals have complex tissues, organs, and organ systems, and higher animals have well-developed nervous and sensory capabilities.

Phylum Placozoa A simple, tiny multicellular marine organism resembling a large amoeba, lacking tissues or organs. Only one species known.

Phylum Porifera There are about 10 000 species of sponges, animals with skeletons composed of spicules, but which lack tissues, organs, or definite symmetry. Sponges have free-swimming larvae and sessile adults that filter-feed. All but a few hundred species are marine.

Class Calcarea. Sponges with skeletons made up of calcareous spicules. About 500 species, all marine.

Class Demospongia. Sponges with skeletons of spongy protein and/or silica, mainly marine. About 9500 species.

Class Hexactinellida. Glass sponges, with skeletons of six-rayed silica spicules. About 50 deep-sea species.

Phylum Cnidaria Radially symmetric animals with distinct tissues, including the jellyfishes, corals, anemones, and hydroids. All cnidarians are predators, using cnidocysts (nematocysts) to sting prey. Body forms include the polyp and medusa. Over 10 000 described species, nearly all marine. The group Myxozoa, previously considered protozoans or degenerate metazoans, are now thought to belong with the Cnidaria.

Class Anthozoa. Corals and sea anemones, having the polyp form only. About 6200 benthic marine species.

Class Hydrozoa. Most hydrozoans have a life cycle that alternates between an asexual polyp (hydroid) stage and a free-swimming, sexual medusa. Hydroid stages are usually colonial. Some coastal hydroid species lack the medusa and some oceanic species lack the hydroid. About 3000 species, nearly all marine.

Class Scyphozoa. More complex, larger jellyfish with simpler or absent polyp stages. About 200 marine species.

Class Cubozoa. Medusae with cuboidal body shape, well-developed nervous system and eyes, and highly toxic nematocysts. About 30 mainly tropical species.

Phylum Ctenophora Comb jellies; transparent gelatinous animals that use fused plates of cilia (comb plates) for locomotion and sticky tentacles to capture prey. They have biradial symmetry and a more complex digestive system than Cnidarians. All 100 species are marine, mainly planktonic.

Phylum Rhombozoa Simple, microscopic organisms that live as internal parasites in the kidneys of cephalopods. They have complex life cycles, and the group is sometimes considered a class of the phylum Mesozoa, along with the Orthonectida. About 65 species.

Phylum Orthonectida About 20 species of simple, small organisms that are internal parasites of various marine worms, mollusks, and echinoderms.

Phylum Platyhelminthes The flatworms, bilaterally symmetrical worms with three cell layers and distinct tissues, but no body cavity (coelom) and guts with only one opening. Most of the approximately 18 000 species are parasites in a wide range of hosts, but there are many free-living forms in all environments.

Class Turbellaria. Free-living flatworms that are mainly predators or scavengers of other small organisms. Most are hermaphroditic. About 4500 species, including many marine forms.

Class Monogenea. Ectoparasitic flatworms, mainly on skin or gills of marine fishes. Although previously included in the Trematoda, this group now appears to be evolutionarily distinct. About 1100 described species, but possibly many more.

Class Trematoda. Parasitic flatworms or flukes, having digestive systems and complex life cycles, often among alternating hosts. There are about 8000 species of flukes, which infect both invertebrate and vertebrate hosts and cause some human diseases.

Class Cestoda. Tapeworms; parasitic, segmented flatworms that lack digestive systems and live in the alimentary tracts of vertebrate hosts, including humans. About 5000 species, some in marine fishes or turtles.

Phylum Nemertea Ribbon worms; long unsegmented worms with a complete digestive tract and a large cavity containing a proboscis that can be extended to sample the environment or capture prey. There are about 900 species, mainly benthic marine forms, but some freshwater or terrestrial.

Phylum Gnathostomulida Minute, wormlike animals that live interstitially in marine sands and sediments. They feed on bacteria and protozoa using a specialized jaw, and are hermaphroditic. About 100 described species, but probably many more undiscovered.

Phylum Gastrotricha Small wormlike organisms in freshwater and marine environments, living in sediments or on plants or animals. They feed on bacteria, protozoa, and detritus, using cilia to collect particles. About 500 species, half of them marine.

Phylum Rotifera Small aquatic organisms with ciliated structures and complex jaws at the head. They have internal organs and complete guts, and feed either on particles or on small animals. Reproduction is sexual, but males are rare or unknown in many species. Of the 2000 species only about 50 are marine.

Phylum Kinorhyncha Small, segmented animals with external spines that live interstitially in marine sediments or on the surfaces of seaweeds or sponges. There are about 150 species known.

Phylum Loricifera Microscopic marine animals encased in a covering of spiny plates called a lorica, into which the head and neck can retract. Described only in 1983, the 10 known species of loriciferans live between and clinging to sand grains.

Phylum Acanthocephala Parasitic worms in the guts of vertebrates, where they anchor to the intestine wall by spines on their head. About 1100 species., some living in marine fishes, turtles, and mammals.

Phylum Cycliophora Described in 1995, this phylum comprises one known species, a microscopic animal that lives attached to the mouthparts of lobsters and collects particulate food. The life cycle is unusual and complex, with sexual and asexual stages.

Phylum Entoprocta Small filter-feeding animals on stalks that live attached to various substrates either as single organisms or as colonies. A ring of ciliated tentacles surrounds the mouth and anus and creates water currents to collect food. About 150 species, all but one marine.

Phylum Nematoda Roundworms; unsegmented worms with a layered cuticle, which molts during growth. Nematodes are among the most ubiquitous and numerous animals on Earth. They live in all environments and as parasites of most plants and animals. Of the 16 000 described species, a few thousand are marine. It is likely that many times more species exist.

Phylum Nematomorpha Long, wiry, unsegmented worms, sometimes called horsehair worms. The gut is reduced or absent. Larval stages are internal parasites in arthropods and adults do not feed at all. A few of the 325 species are marine.

Phylum Bryozoa Also called the Ectoprocta, a group of small colonial organisms that filter-feed using a tentaculate structure called the lophophore. Individual bryozoans are encased in tubular or boxlike housings and reproduce asexually to produce encrusting or plumose colonies attached to hard substrates. About 5000 species, all but 50 are marine.

Phylum Phoronida Phoronids are tube-dwelling marine worms that also use a lophophore to collect particulate food. They are common in mud or sand, or attached to rocks or pilings. About a dozen widely distributed species are known.

Phylum Brachiopoda Brachiopods are lophophorate, filter-feeding animals whose bodies are enclosed in bivalve shells. Most live secured by a stalk to hard substrates or in sediments, at depths from intertidal to 4000 m. Only about 335 living species, but over 30 000 fossil ones known. The living genus *Lingula* dates back over 400 million years.

Phylum Mollusca A large and diverse phylum containing the familiar clams, snails, squid, and octopus. Mollusks possess mantle tissue that secretes a carbonate shell around the body, a belt of teeth called the radula for feeding, and a muscular foot variously modified for digging, crawling, or swimming. A diverse, widespread, and economically important group, mollusks have a long and complex taxonomic history, with between 50 000 and 100 000 described species. Most mollusks are marine but there are many freshwater and terrestrial snails

Class Monoplacophora. Small, single-shelled animals living on hard surfaces, usually in the deep sea. Primitive in structure and thought to be similar to ancestral forms. Only 11 known species.

Class Aplacophora. Small wormlike animals with calcareous spicules but no true shell. They lack the typical molluscan foot, but creep with cilia. About 250 species are known from various benthic marine environments.

Class Caudofoveata. Shell-less wormlike animals that live in burrows in deep-sea sediments. Little is known of the ecology of the 70 known species.

Class Polyplacophora. The chitons; mollusks having a shell of eight overlapping, articulating plates. All are marine and most live on intertidal or subtidal rocks, where they feed by scraping algae with their radulas. About 600 species.

Class Gastropoda. The largest and most diverse class of mollusks, gastropods include aquatic and terrestrial snails, slugs, limpets, and nudibranchs. In most, the body sits on a muscular foot used for locomotion, and is enclosed in a conical or coiled shell. Gastropods may be filter feeders, grazers, or predators. By various counts there are 40 000 to 80 000 species, about half of them marine.

Class Bivalvia. Bivalves or Pelecypods, mollusks with the body enclosed between two valves or shells hinged together, and closed by an adductor muscle. Most are filter feeders, drawing water into the shell cavity and filtering particles from it. Some bivalves attach to surfaces; others burrow into sediments. Most of the 8000 species are marine.

Class Scaphopoda. Mollusks with a conical, tusk-shaped shell that is open at both ends. Scaphopods burrow into marine sediments and collect small food organisms with specialized tentacles. About 350 species.

Class Cephalopoda. Squid, octopus, and *Nautilus*; in cephalopods the molluscan foot is modified into tentacles surrounding the mouth. Cephalopods are actively swimming predators with highly developed nervous and sensory systems. *Nautilus* and most extinct cephalopods have external chambered shells, while squid have reduced internal skeletons and octopus have none. All 650 species are marine.

Phylum Priapulida Marine worms that burrow into sediments with only their mouths exposed at the surface. They are predatory on other small worms. The 10 known species live from estuarine to abyssal environments.

Phylum Sipuncula About 320 species of unsegmented marine worms, with a retractable proboscis called the introvert. They are benthic, often living in sediments or among other animals. Tentacles around the mouth collect detritus and other particulate food.

Phylum Echiura Unsegmented, benthic marine worms having an extensible proboscis that is used to collect detrital food. They live mainly within burrows in sediments. Considered by some to be a class of the Annelida. About 140 species.

Phylum Annelida Segmented worms, a large group of diverse species, most having bodies divided into segments by internal septa, and with chitinous setae on the exterior body. There are about 12 000 species of annelids, in all aquatic and terrestrial environments.

Class Polychaeta. Worms usually with distinct head region, numerous setae and paddle- or leg-like parapodia for locomotion. The group includes mobile, burrowing, attached, and symbiotic forms, feeding as predators, scavengers, filter or deposit feeders. About 8000 species, almost all marine.

Class Oligochaeta. Worms lacking parapodia and with few setae; terrestrial forms include earthworms. Most of the 3100 species are freshwater or terrestrial.

Class Hirudinea. Leeches; the body is not segmented internally and lacks setae on the exterior. Most are ectoparasites, feeding on blood of other animals, but some are predators. About 500 species, many marine.

Phylum Tardigrada Water bears; minute animals with eight short legs that live in aquatic or moist terrestrial environments and suck juices from plants or animals. They are able to remain in a dried state for long periods, returning to active metabolism on rehydration. About 550 species, a few of them marine.

Phylum Arthropoda The arthropods, or jointed-leg animals, are one of the most successful and widespread metazoan groups. All possess segmented bodies and articulated exoskeletons, which are molted during growth. Insects and arachnids are the dominant arthropods on land, but almost entirely absent from the sea, where crustaceans predominate. About 1 million described species, mostly insects, but many more probably exist.

Class Merostomata. An ancient group of chelicerates now containing only 4 species of horseshoe crabs. They live in subtidal environments, and feed as predators and scavengers.

Class Pycnogonida. Sea spiders; lacking the well-developed head, thorax, and abdomen of other arthropods. About 1000 species, entirely marine, which feed on body fluids of other animals and plants.

Class Crustacea. Largely aquatic arthropods including copepods, amphipods, shrimp, barnacles, crabs, and lobster. Often with a calcified carapace covering the segmented body. All forms have a nauplius larva as the first of many molt stages. About 35 000 species, almost entirely marine, in all habitats.

Phylum Pogonophora Thin, wormlike animals that live in tubes in sediment or attached to benthic surfaces. Pogonophorans lack digestive systems and obtain nutrition by absorption of dissolved organic nutrients. Pogonophorans are thought by some to be aberrant annelids. About 100 species, mostly in deep water.

Phylum Vestimentifera Closely related to pogonophorans, these larger marine worms rely on symbiotic bacteria in their tissues to generate nutrition from the metabolism of inorganic chemical compounds. They are best known from deep-sea hydrothermal vents, where they can be over 2 m long. About a dozen species have been described.

Phylum Echinodermata An entirely marine phylum including the sea stars, urchins, and brittle stars. All have a five-part radial symmetry, a water-vascular system, tube feet used for locomotion, respiration and feeding, and a skeleton made of minute calcareous ossicles or spicules. Over 6000 species.

Class Crinoidea. Most ancient of the living echinoderms, crinoids have multiple pinnate arms used for filtering food particles from the water. Some are attached to the bottom by a stalk, others swim by movement of the arms. About 600 subtidal and deep-sea species.

Class Asteroidea. The seastars, most with five radial arms, are slow-moving predators in intertidal and subtidal environments. About 1500 species.

Class Ophiuroidea. Brittle stars, having five slender and flexible arms radiating from a central disk. Some are deposit or filter feeders, others predatory. About 2000 species, including many deep-sea forms.

Class Concentricycloidea. Small discoidal organisms known only from submerged wood in the deep sea. They lack five-part symmetry and arms, and the body is covered by overlapping calcareous plates. Two known species.

Class Echinoidea. Sea urchins; with a rigid, globular or flattened test made of calcareous ossicles and a complex mouth structure for grazing and chewing. Most are free-living in subtidal environments, but some burrow in sediments or rock. Approximately 950 species.

Class Holothuroidea. Sea cucumbers; with an elongate, flexible body, bilateral symmetry and no arms. Most of the 1500 species are benthic deposit or filter feeders.

Phylum Chaetognatha Arrow worms; planktonic marine organisms that are predatory on small zooplankton, using chitinous spines around the mouth to catch prey. The 100 species are mainly planktonic with a few benthic forms.

Phylum Hemichordata Wormlike marine organisms that burrow in sediments or form colonies on hard substrates. Most are deposit or suspension feeders. About 100 species from shallow tropics to deep sea.

Phylum Chordata A large and diverse phylum including the familiar vertebrates. All have a dorsal nerve cord that can form a brain, a notochord that becomes the vertebral column in vertebrates, and gill slits in the throat at some stage of development. Chordates live in all environments and are one of the most successful and widespread groups. Perhaps 45 000 species, about half marine forms (mainly fish).

Subphylum Urochordata The tunicates; sessile or motile animals with the body enclosed in a tough, flexible tunic. Most are filter feeders. The notochord and dorsal nerve are seen only in larval stages, and sessile adults may be asymmetrical in form. About 3000 species, all marine.

Class Ascidiacea. Sea squirts; sessile, filter-feeding tunicates that live mainly on hard benthic substrates. About 2700 species.

Class Sorberacea. A small group of solitary, deep-sea tunicates that appear to prey on live organisms instead of filter-feeding.

Class Larvacea. Minute planktonic tunicates (also called appendicularians) with small bodies and long tails. They filter feed using an external mucous structure that concentrates small particulate material for ingestion by the larvacean. About 200 species.

Class Thaliacea. Pelagic tunicates with gelatinous, transparent bodies. They filter feed by creating a water current through their bodies. All have complex life-cycles with sexual and asexual, solitary, and colonial stages. About 100 species.

Subphylum Cephalochordata Lancelets or "Amphioxus"; small fish-shaped animals with notochords extending the length of the body. They burrow into substrates with the head end exposed and filter particulate food. About 20 species.

Subphylum Vertebrata Chordates with a backbone replacing the notochord, and a distinct head region with brain. Approximately 42 000 species in all environments.

Class Agnatha. Lampreys and hagfish; eel-like jawless fishes without scales, bones or fins. Most are scavengers or parasites on other fish. About 60 marine and freshwater species.

Class Chondrichthyes. Sharks and rays; fish with cartilaginous bones and small denticle scales embedded in the skin. The 850 species are virtually all marine.

Class Osteichthyes. The bony fishes; having bone skeletons, scales and often air bladders for buoyancy. Highly diverse and widely distributed in all marine and freshwater habitats, with about 25 000 species.

Class Reptilia. Turtles, snakes and lizards. Most of the 6000 species are terrestrial except for a few marine turtles, crocodiles, and snakes.

Class Aves. Birds; about 9000 species in all terrestrial habitats and many marine forms including penguins, albatrosses, gulls, etc.

Class Mammalia. Four legged, endothermic animals usually with fur or hair, which mainly give live birth and suckle the young. Most of the 4500 species are terrestrial; marine forms include whales, dolphins, seals, and otters.

Further Reading

Atlas RM (1997) *Principles of Microbiology.* Dubuque, IA: William C. Brown.

Brusca RC and Brusca GJ (1990) *Invertebrates.* Sunderland, MA: Sinauer Associates.

Cavalier-Smith T (1998) A revised six-kingdom system of life. *Biological Reviews of the Cambridge Philosophical Society* 73: 203–266.

Margulis L, Corliss JO, Melkonian M, and Chapman DJ (1990) *Handbook of Protoctista.* Boston: Jones and Bartlett.

Margulis L and Schwartz KV (1988) *Five Kingdoms: An Illustrated Guide to the Phyla of Life on Earth.* NewYork: WH Freeman.

Nielsen C (1995) *Animal Evolution: Interrelationships of the Living Phyla.* Oxford: Oxford University Press.

Patterson DJ (1999) The diversity of eukaryotes. *American Naturalist* 154(supplement): S96–S124.

Pechenik JA (2000) *Biology of the Invertebrates*. Boston: McGraw-Hill.
Williams DD (2000) *Invertebrate Phylogeny*. Scarborough: CITD Press, University of Toronto. CD ROM.

Websites

"Microscope": http://www.mbl.edu/baypaul/microscope/general/page_01.htm
"Tree of Life" http://phylogeny.arizona.edu/tree/phylogeny.html

INDEX

Notes

Cross-reference terms in italics are general cross-references, or refer to subentry terms within the main entry (the main entry is not repeated to save space). Readers are also advised to refer to the end of each article for additional cross-references - not all of these cross-references have been included in the index cross-references.

The index is arranged in set-out style with a maximum of three levels of heading. Major discussion of a subject is indicated by bold page numbers. Page numbers suffixed by T and F refer to Tables and Figures respectively. vs. indicates a comparison.

This index is in letter-by-letter order, whereby hyphens and spaces within index headings are ignored in the alphabetization. For example, 'oceanography' is alphabetized before 'ocean optics.' Prefixes and terms in parentheses are excluded from the initial alphabetization.

Where index subentries and sub-subentries pertaining to a subject have the same page number, they have been listed to indicate the comprehensiveness of the text.

Abbreviations used in subentries

AUV - autonomous underwater vehicle
CPR - continuous plankton recorder
DIC - dissolved inorganic carbon
ENSO - El Niño Southern Oscillation
ROV - remotely operated vehicle
SST - sea surface temperature

Additional abbreviations are to be found within the index.

A

C

G

H

I

J

K

L

M

N

O

P

Q

R

S

T

U

V

Printed and bound by CPI Group (UK) Ltd, Croydon, CR0 4YY

Printed and bound by CPI Group (UK) Ltd, Croydon, CR0 4YY
03/10/2024
01040311-0009